T0181729

Lecture Notes in Computer Science 11446

More information about this series at http://www.springer.com/series/7409

Guoliang Li · Jun Yang ·
Joao Gama · Juggapong Natwichai ·
Yongxin Tong (Eds.)

Database Systems for Advanced Applications

24th International Conference, DASFAA 2019 Chiang Mai, Thailand, April 22–25, 2019 Proceedings, Part I

Editors
Guoliang Li
Tsinghua University
Beijing, China

Jun Yang
Duke University
Durham, NC, USA

Joao Gama
University of Porto
Porto, Portugal

Juggapong Natwichai
Chiang Mai University
Chiang Mai, Thailand

Yongxin Tong
Beihang University
Beijing, China

ISSN 0302-9743 ISSN 1611-3349 (electronic)
Lecture Notes in Computer Science
ISBN 978-3-030-18575-6 ISBN 978-3-030-18576-3 (eBook)
https://doi.org/10.1007/978-3-030-18576-3

LNCS Sublibrary: SL3 – Information Systems and Applications, incl. Internet/Web, and HCI

This Springer imprint is published by the registered company Springer Nature Switzerland AG
The registered company address is: Gewerbestrasse 11, 6330 Cham, Switzerland

Preface

The International Conference on Database Systems for Advanced Applications DASFAA provides a leading international forum for discussing the latest research on database systems and advanced applications. DASFAA 2019 provided a forum for technical presentations and discussions among database researchers, developers, and users from academia, business, and industry, which showcases state-of-the-art R&D activities in database systems and their applications. The conference's long history has established the event as the premier research conference in the database area.

On behalf of the DASFAA 2019 program co-chairs, we are pleased to welcome you to the proceedings of the 24th International Conference on Database Systems for Advanced Applications DASFAA 2019, held during April 22–25, 2019, in Chiang Mai, Thailand. Chiang Mai is the largest city in northern Thailand. It is the capital of Chiang Mai Province and was a former capital of the kingdom of Lan Na 1296–1768, which later became the Kingdom of Chiang Mai, a tributary state of Siam from 1774 to 1899, and finally the seat of princely rulers until 1939. It is 700 km north of Bangkok near the highest mountains in the country. The city sits astride the Ping River, a major tributary of the Chao Phraya River.

We received 501 research paper submissions, each of which was assigned to at least three Program Committee (PC) members and one SPC member. The thoughtful discussion on each paper by the PC with facilitation and meta-review provided by the SPC resulted in the selection of 92 full research papers (acceptance ratio of 18%) and 64 short papers (acceptance ratio of 28%). In addition, we included 13 demo papers and six tutorials in the program. This year the dominant topics for the selected papers included big data, machine learning, graph and social network, recommendation, data integration and crowd sourcing, and spatial data management.

Three workshops are selected by the workshop co-chairs to be held in conjunction with DASFAA 2019, including BDMS: the 6th International Workshop on Big Data Management and Service; BDQM: the 4th Workshop on Big Data Quality Management; GDMA: the Third International Workshop on Graph Data Management and Analysis. We received 26 workshop paper submissions and accepted 14 papers.

The conference program included three keynote presentations by Prof. Anthony K. H. Tung National University of Singapore, Prof. Lei Chen The Hong Kong University of Science and Technology, and Prof. Ashraf Aboulnaga Qatar Computing Research Institute.

We wish to thank everyone who helped with the organization including the chairs of each workshop, demonstration chairs, and tutorial chairs and their respective PC members and reviewers. We thank all the authors who submitted their work, all of which contributed to making this part of the conference a success. We are grateful to the general chairs, Xue Li from The University of Queensland, Australia, and Nat Vorayos from Chiang Mai University, Thailand. We thank the local Organizing Committee chairs, Juggapong Natwichai and Krit Kwanngern from Chiang Mai

University, Thailand, for their tireless work before and during the conference. Special thanks go to the proceeding chairs, Yongxin Tong Beihang University, China and Juggapong Natwichai Chiang Mai University, Thailand, for producing the proceedings.

We hope that you will find the proceedings of DASFAA 2019 interesting and beneficial to your research.

March 2019

Guoliang Li
Joao Gama
Jun Yang

Organization

Honorary Co-chairs

Sampan Singharajwarapan	Chiang Mai University, Thailand
Bannakij Lojanapiwat	Chiang Mai University, Thailand

General Co-chairs

Xue Li	The University of Queensland, Australia
Nat Vorayos	Chiang Mai University, Thailand

Program Co-chairs

Guoliang Li	Tsinghua University, China
Joao Gama	University of Porto, Portugal
Jun Yang	Duke University, USA

Industrial Track Co-chairs

Cong Yu	Google Inc., USA
Fan Jiang	Alibaba, China

Workshop Co-chairs

Qun Chen	Northwestern Polytechnical University, China
Jun Miyazaki	Tokyo Institute of Technology, Japan

Demo Track Co-chairs

Nan Tang	Qatar Computing Research Institute, Qatar
Masato Oguchi	Ochanomizu University, Japan

Tutorial Chair

Sen Wang	The University of Queensland, Australia

Panel Chair

Lei Chen	Hong Kong University of Science and Technology, SAR China

PhD School Co-chairs

Sarana Nutanong	Vidyasirimedhi Institute of Science and Technology, Thailand
Xiaokui Xiao	National University of Singapore, Singapore

Proceedings Co-chairs

Yongxin Tong	Beihang University, China
Juggapong Natwichai	Chiang Mai University, Thailand

Publicity Co-chairs

Nah Yunmook	Dankook University, South Korea
Ju Fan	Renmin University of China, China
An Liu	Soochow University, China
Chiemi Watanabe	University of Tsukuba, Japan

Local Organization Co-chairs

Juggapong Natwichai	Chiang Mai University, Thailand
Krit Kwanngern	Chiang Mai University, Thailand

DASFAA Liaison

Xiaofang Zhou	The University of Queensland, Australia

Regional Coordinator

Pruet Boonma	Chiang Mai University, Thailand

Web Master

Titipat Sukhvibul	Chiang Mai University, Thailand

Senior Program Committee Members

Philippe Bonnet	IT University of Copenhagen, Denmark
K. Selçuk Candan	Arizona State University, USA
Bin Cui	Peking University, China
Dong Deng	Rutgers University, USA
Yunjun Gao	Zhejiang University, China
Tingjian Ge	University of Massachusetts, Lowell, USA
Zhiguo Gong	Macau University, SAR China
Wook-Shin Han	Pohang University of Science and Technology, South Korea

Zi Huang	The University of Queensland, Australia
Wen-Chih Peng	National Chiao Tung University, Taiwan
Florin Rusu	UC Merced, USA
Chaokun Wang	Tsinghua University, China
Hongzhi Wang	Harbin Institute of Technology, China
Xiaokui Xiao	National University of Singapore, Singapore
Nan Zhang	The George Washington University, USA

Program Committee

Alberto Abelló	Universitat Politècnica de Catalunya, Spain
Marco Aldinucci	University of Turin, Italy
Laurent Anne	University of Montpellier, LIRMM, CNRS, France
Akhil Arora	École polytechnique fédérale de Lausanne, Switzerland
Zhifeng Bao	Royal Melbourne Institute of Technology, Australia
Ladjel Bellatreche	ISAE-ENSMA, France
Yi Cai	South China University of Technology, China
Andrea Cali	Birkbeck University of London, UK
Yang Cao	Kyoto University, Japan
Yang Cao	University of Edinburgh, UK
Huiping Cao	New Mexico State University, USA
Lei Cao	Massachusetts Institute of Technology, USA
Barbara Catania	University of Genoa, Italy
Chengliang Chai	Tsinghua University, China
Lijun Chang	The University of Sydney, Australia
Cindy Chen	University of Massachusetts, Lowell, USA
Yueguo Chen	Renmin University of China, China
Reynold Cheng	The University of Hong Kong, SAR China
Fei Chiang	McMaster University, Canada
Lingyang Chu	Simon Fraser University, Australia
Antonio Corral	University of Almeria, Spain
Lizhen Cui	Shandong University, China
Hasan Davulcu	Arizona State University, USA
Sabrina De Capitani di Vimercati	Università degli Studi di Milano, Italy
Zhiming Ding	Institute of Software, Chinese Academy of Sciences, China
Eduard Dragut	Temple University, USA
Lei Duan	Sichuan University, China
Amr Ebaid	Purdue University, USA
Ahmed Eldawy	University of California, Riverside, USA
Damiani Ernesto	University of Milan, Italy
Liyue Fan	University at Albany SUNY, USA
Ju Fan	Renmin University of China, China
Elena Ferrari	University of Insubria, Italy
Yanjie Fu	Missouri University of Science and Technology, USA

Xiaofeng Gao	Shanghai Jiaotong University, China
Yunjun Gao	Zhejiang University, China
Boris Glavic	Illinois Institute of Technology, USA
Neil Gong	Iowa State University, USA
Yu Gu	Northeastern University, China
Donghai Guan	Nanjing University of Aeronautics and Astronautics, China
Shuang Hao	Beijing Jiaotong University, China
Yeye He	Microsoft Research, USA
Huiqi Hu	East China Normal University, China
Juhua Hu	University of Washington, USA
Chao Huang	University of Notre Dame, USA
Matteo Interlandi	Microsoft, USA
Saiful Islam	Griffith University, Australia
Peiquan Jin	University of Science and Technology of China, China
Cheqing Jin	East China Normal University, China
Latifur Khan	The University of Texas at Dallas, USA
Peer Kroger	Ludwig-Maximilians-Universität München, Germany
Jae-Gil Lee	Korea Advanced Institute of Science and Technology, South Korea
Sang Won Lee	Sungkyunkwan University, South Korea
Young-Koo Lee	Kyung Hee University, South Korea
Zhixu Li	Soochow University, China
Lingli Li	Heilongjiang University, China
Jianxin Li	Deakin University, Australia
Jian Li	Tsinghua University, China
Cuiping Li	Renmin University of China, China
Qingzhong Li	Shandong University, China
Zheng Li	Amazon, USA
Bohan Li	Nanjing University of Aeronautics and Astronautics, China
Guoliang Li	Tsinghua University, China
Xiang Lian	Kent State University, USA
Chunbin Lin	Amazon AWS, USA
An Liu	Soochow University, China
Qing Liu	Data61, CSIRO, Australia
Eric Lo	Chinese University of Hong Kong, SAR China
Guodong Long	University of Technology Sydney, Australia
Cheng Long	Nanyang Technological University, Singapore
Hua Lu	Aalborg University, Denmark
Shuai Ma	Beihang University, China
Sofian Maabout	University of Bordeaux, France
Ioana Manolescu	Inria Saclay, France
Xiangfu Meng	Liaoning Technical University, China
Jun-Ki Min	Korea University of Technology and Education, South Korea

Yang-Sae Moon	Kangwon National University, South Korea
Parth Nagarkar	New Mexico State University, USA
Yunmook Nah	Dankook University, South Korea
Svetlozar Nestorov	Loyola University Chicago, USA
Quoc Viet Hung Nguyen	Griffith University, Australia
Baoning Niu	Taiyuan University of Technology, China
Kjetil Norvag	Norwegian University of Science and Technology, Norway
Sarana Yi Nutanong	City University of Hong Kong, SAR China
Vincent Oria	New Jersey Institute of Technology, USA
Xiao Pan	Shijiazhuang Tiedao University, China
Dhaval Patel	IBM TJ Watson Research Center, USA
Dieter Pfoser	George Mason University, USA
Silvestro Roberto Poccia	University of Turin, Italy
Weining Qian	East China Normal University, China
Tieyun Qian	Wuhan University, China
Shaojie Qiao	Chengdu University of Information Technology, China
Lu Qin	University of Technology Sydney, Australia
Xiaolin Qin	Nanjing University of Aeronautics and Astronautics, China
Weixiong Rao	Tongji University, China
Oscar Romero	Universitat Politècnica de Catalunya, Spain
Sourav S. Bhowmick	Nanyang Technological University, Singapore
Babak Salimi	University of Washington, USA
Simonas Saltenis	Aalborg University, Denmark
Claudio Schifanella	University of Turin, Italy
Marco Serafini	University of Massachusetts Amherst, USA
Zechao Shang	University of Chicago, USA
Xuequn Shang	Northwestern Polytechnical University, China
Shuo Shang	King Abdullah University of Science and Technology, Saudi Arabia
Wei Shen	Nankai University, China
Yanyan Shen	Shanghai Jiao Tong University, China
Kyuseok Shim	Seoul National University, South Korea
Alkis Simitsis	Hewlett Packard Enterprise, USA
Rohit Singh	Uber AI Labs, USA
Chunyao Song	Nankai University, China
Shaoxu Song	Tsinghua University, China
Weiwei Sun	Fudan University, China
Hailong Sun	Beihang University, China
Na Ta	Renmin University of China, China
Jing Tang	National University of Singapore, Singapore
Chaogang Tang	China University of Mining and Technology, China
Nan Tang	Qatar Computing Research Institute, Qatar
Shu Tao	IBM Research, USA

Saravanan Thirumuruganathan	Qatar Computing Research Institute, Qatar
Yongxin Tong	Beihang University, China
Hanghang Tong	Arizona State University, USA
Ismail Toroslu	Middle East Technical University, Turkey
Vincent Tseng	National Chiao Tung University, Taiwan
Leong Hou U.	University of Macau, SAR China
Qian Wang	Yanshan University, China
Sibo Wang	The Chinese University of Hong Kong, SAR China
Xiaoling Wang	East China Normal University, China
Xin Wang	Tianjin University, China
Ning Wang	Beijing Jiaotong University, China
Sen Wang	The University of Queensland, Australia
Jiannan Wang	Simon Fraser University, Australia
Jianmin Wang	Tsinghua University, China
Wei Wang	National University of Singapore, Singapore
Shi-ting Wen	Zhejiang University, China
Yinghui Wu	Washington State University, USA
Kesheng Wu	Lawrence Berkeley National Lab, USA
Mingjun Xiao	University of Science and Technology of China, China
Xike Xie	University of Science and Technology of China, China
Jianliang Xu	Hong Kong Baptist University, SAR China
Guandong Xu	University of Technology Sydney, Australia
Jianqiu Xu	Nanjing University of Aeronautics and Astronautics, China
Yajun Yang	Tianjin University, China
Yu Yang	Simon Fraser University, Australia
Bin Yao	Shanghai Jiaotong University, China
Lina Yao	University of New South Wales, Australia
Minghao Yin	Northeast Normal University, China
Hongzhi Yin	The University of Queensland, Australia
Man Lung Yiu	Hong Kong Polytechnic University, SAR China
Zhiwen Yu	South China University of Technology, China
Minghe Yu	Northeast University, China
Xiangyao Yu	Massachusetts Institute of Technology, USA
Jeffrey Xu Yu	Chinese University of Hong Kong, SAR China
Yi Yu	National Institute of Informatics, Japan
Ge Yu	Northeast University, China
Ye Yuan	Beijing Institute of Technology, China
Xiaowang Zhang	Tianjin University, China
Wei Zhang	East China Normal University, China
Yan Zhang	Peking University, China
Dongxiang Zhang	University of Electronic Science and Technology of China, China
Richong Zhang	Beihang University, China
Detian Zhang	Soochow University, China

Rui Zhang	University of Melbourne, Australia
Ying Zhang	University of Technology Sydney, Australia
Wenjie Zhang	University of New South Wales, Australia
Yong Zhang	Tsinghua University, China
Xiang Zhao	National University of Defence Technology, China
Lei Zhao	Soochow University, China
Jun Zhao	Nanyang Technological University, Singapore
Kai Zheng	University of Electronic Science and Technology of China, China
Yudian Zheng	Twitter, USA
Bolong Zheng	Aalborg University, Denmark
Rui Zhou	Swinburne University of Technology, Australia
Xiangmin Zhou	Royal Melbourne Institute of Technology, Australia
Yongluan Zhou	University of Copenhagen, Denmark
Yuanyuan Zhu	Wuhan University, China
Yan Zhuang	Tsinghua University, China

Demo Program Committee

Lei Cao	Massachusetts Institute of Technology, USA
Yang Cao	University of Edinburgh, UK
Lijun Chang	The University of Sydney, Australia
Yueguo Chen	Renmin University of China, China
Fei Chiang	McMaster University, Canada
Bolin Ding	Alibaba, USA
Eduard Dragut	Temple University, USA
Amr Ebaid	Purdue University, USA
Ahmed Eldawy	University of California, Riverside, USA
Miki Enoki	IBM Research, Japan
Yunjun Gao	Zhejiang University, China
Tingjian Ge	University of Massachusetts, Lowell, USA
Yeye He	Microsoft Research, USA
Matteo Interlandi	Microsoft, USA
Zheng Li	Amazon, USA
Jianxin Li	Deakin University, Australia
Shuai Ma	Beihang University, China
Hidemoto Nakada	National Institute of Advanced Industrial Science and Technology, Japan
Abdulhakim Qahtan	Qatar Computing Research Institute, Qatar
Lu Qin	University of Technology Sydney, Australia
Jorge-Arnulfo Quiane-Ruiz	Qatar Computing Research Institute, Qatar
Elkindi Rezig	Massachusetts Institute of Technology, USA
Marco Serafini	University of Massachusetts Amherst, USA
Zechao Shang	University of Chicago, USA
Giovanni Simonini	Massachusetts Institute of Technology, USA
Rohit Singh	Uber AI Labs, USA

Additional Reviewers

Contents – Part I

Crowdsourcing

Data Integration

Embedding

Graphs

Knowledge Graph

Contents – Part II

Machine Learning

Privacy and Graph

Recommendation

Social Network

Spatial

Spatio-Temporal

Big Data

Accelerating Real-Time Tracking Applications over Big Data Stream with Constrained Space

Guangjun Wu[1], Xiaochun Yun[1], Shupeng Wang[1(✉)], Ge Fu[2(✉)], Chao Li[2], Yong Liu[1], Binbin Li[1], Yong Wang[1], and Zhihui Zhao[1]

[1] Institute of Information Engineering, Chinese Academy of Sciences, Beijing, China
{wuguangjun,wangshupeng}@iie.ac.cn

[2] National Computer Network Emergency Response Technical Team/Coordination Center of China, Beijing, China
fg@cert.org.cn

Abstract. Existing approaches are insufficient to provide real-time results for tracking applications against *big* and *fast* data streams. In this paper, we leverage freshness sensitive properties of tracking applications and propose an approximate query answering approach, called FS-Sketch, to accelerating real-time temporal queries over big data streams. FS-Sketch constructs its sketch over high-speed data streams via composed online sampling strategies, including sliding-window sampling and space-constrained sampling. Furthermore, FS-Sketch can compress its sketch into constrained space dynamically via utilizing time-decayed mechanism. We evaluate performance of FS-Sketch using real-world and synthetic datasets. FS-Sketch can respond temporal queries within 2 ms from 1.4 billion records with accurate estimates. Meanwhile, FS-Sketch can also outperform the state-of-the-art big data analytical system (Spark) by 5 orders of magnitude on response time when we query over TB-scale real-world datasets.

Keywords: Approximate answering · Big data query · Data streams · Distributed computing · Networks

1 Introduction

Recently, many network applications produce high-speed and continuous data streams. It is necessary to perform tracking applications to monitor, track and explore time-evolving anomalies over the big and fast data streams, and provide interactive queries for users. Typical examples include finding frequency of a term in all of the posted tweets in the last ten hours and returning the number of page-views in large-scale websites from 01/05/2013 to 08/05/2013. We call

Supported by the National Key Research and Development Program of China (2016YFB0801305). S. Wang and G. Fu are corresponding authors of this paper.

G. Li et al. (Eds.): DASFAA 2019, LNCS 11446, pp. 3–18, 2019.
https://doi.org/10.1007/978-3-030-18576-3_1

such operations *temporal aggregation queries* (TAQs). We have seen a flurry of activities in the area of building large-scale analytics for big data query and processing [10,12,25]. Examining current tracking and exploring applications in detail allows us to understand the challenging problems which we confront: Firstly, the key operations of network tracking applications are to apply flexible aggregation functions, such as count, sum, avg and quantiles etc., on streaming elements within a time-interval of interest. Secondly, the drivers of tracking applications are usually deployed over continuous data streams using small space, yet they require real-time (or near real-time) query response time for emergent events.

An exact solution for obtaining temporal statistics over data stream costs linear time and space [5,17]. Therefore, methods of *approximate query processing* (AQP) are essential for dealing with massive and high-speed data streams in current commercial product [16] and academic prototypes [5,11,14,22]. Currently, there are mainly two types of AQP techniques deployed over data streams: (1) fingerprint-based synopsis, and (2) sampling-based synopsis respectively. The first type of synopsis builds fingerprints of input elements by hash function(s), such as Count-Mini sketch [6], Bloom filters [8] and ECM-Sketch [14] etc. These techniques are space-efficient and can be optimized for inserting and searching operations, while they can not provide the capability of searching aggregates within any temporal interval of interest. Moreover, they are often hard to be resized for continuously inputs with error-defined estimates.

Methods of sampling-based synopsis select a small number of representative items from data streams with adjusted weights, and can support a wide variety of queries. However, current methods make different trade-offs between samples size and query accuracy they support when confronting continuous data streams. For details, let N be value of summarization of input items, $\hat{E}$ be an estimate for true value E, R be an additive error of estimation (i.e., $R = |\hat{E} - E|$), and ϵ be a relative error of estimation ($\epsilon = |\hat{E} - E|/E$, $\epsilon \leq 1$). On one end of the spectrum, existing sampling approaches over synchronous data streams [9,14] and asynchronous data streams [18,22] exhibit error-guaranteed accuracy estimates for current window, i.e., $R \leq \epsilon E$, while they usually discard (or weight zero) items which are out-side of the window, and they can not provide estimates for long-lifetime elements. On the other end of the spectrum, such as variance-optimal sampling and structure-aware sampling [3], these algorithms construct space constrained summaries to provide estimates with an upper-bounded error for long-lifetime elements, i.e., $R \to \epsilon N$, while they can not report emerging events precisely from fresher items, for ϵN is usually too large to accurately measure emerging events.

We notice that many tracking operations have freshness sensitive properties, i.e., they are sensitive to emergent events and can tolerate some error for exploring long-lifetime events [15,19,21]. In this paper, we leverage freshness sensitive properties of tracking applications and design an accuracy-decayed sketching approach to improve query quality and query efficiency over continuous data streams. We call it *Freshness Sensitive Sketch* (FS-Sketch). The contributions are as follows:

1. We present a novel composed online sampling technique to conquer the requirement of flexible temporal queries within any time-interval of interest. We first extract samples from data streams in a symmetrical sliding-window manner to support queries with temporal length larger than a window. Moreover, for discarded elements of a window, we compress them dynamically into constrained space according to a running bounded error. This technique makes our approach appropriate for obtaining error-bounded temporal statistics within any time interval of interest.
2. We design an accuracy-decayed sketch compression method to strive to achieve a better balance between sketch size and sketch accuracy it supports over large-scale datasets. FS-Sketch maintains samples in an accuracy-decayed manner: fresher windows maintain more samples for accurate query answers, while older windows keep fewer samples to improve spatial consumption with upper bounded error. Moreover, the decayed properties, such as query accuracy and space consumption, can be depicted and maintained by general time-decayed functions.
3. We present detailed theoretical and experimental analysis to evaluate usability of our approach. We also compare our prototype with big data analytics, such as MapReduce, Spark, Spark Streaming under production environments using real-world datasets. Our approach is more appropriate for solving big data streams and can provide real-time responses for temporal queries with accurate estimates.

We implement prototype of FS-Sketch on Linux platform, and compare it with big data analytics, such as MapReduce, Spark and Spark streaming, over real-world and synthetic datasets. The experimental results validate the efficiency and effectiveness of our approach. FS-Sketch only costs 2 ms for an ad-hoc temporal aggregation queries over 1.4 billion records. When compared to Spark, FS-Sketch can even achieve 4–5 orders of magnitude improvement on query response time for tracking applications using TB-scale real-world datasets.

2 Approach Overview

2.1 Problem Statement

As mentioned earlier, a tracking application is sensitive to emergent events and can tolerate some error for long-lifetime events. We first depict freshness sensitive properties of tracking applications as follows:

Problem 1. The input to the problem is a stream of elements (v_1, t_{s_1}), (v_2, t_{s_2}), (v_3, t_{s_3}), ... The goal is to compute an aggregate for a temporal query Q, which searches within a time-interval $[c-w, c]$ of interest using constrained space, where c is an arbitrary time point on time-scales and w is the length of a time-interval. We are interested in freshness sensitive error $R_{g(T)}$, such that $R_{g(T)}$ increases according to a time-decayed function $g(T)$ which progresses from current time

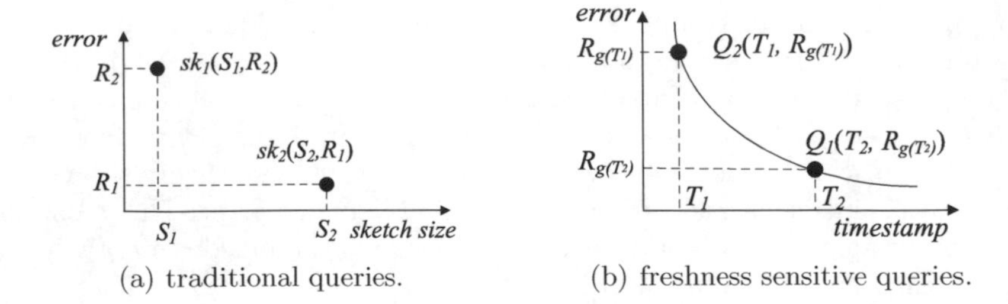

(a) traditional queries. (b) freshness sensitive queries.

Fig. 1. Terminologies of freshness sensitive queries.

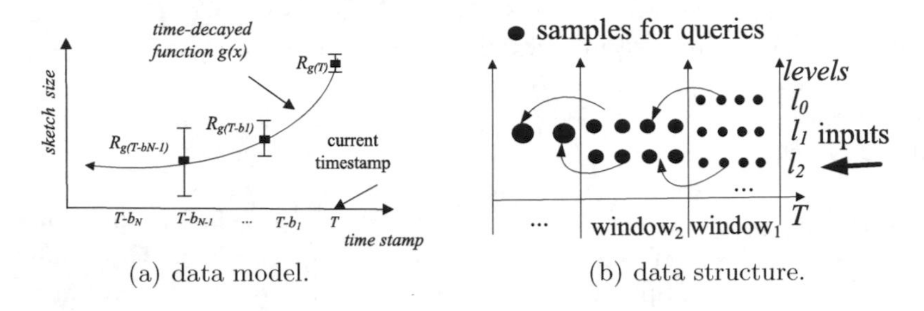

(a) data model. (b) data structure.

Fig. 2. Design principles of FS-sketch.

T. The time-decayed query accuracy $R_{g(T)}$ of an estimate $\hat{E_Q}$ of the query Q for true value E_Q is depicted as

$$E_Q \in [\hat{E_Q} - R_{g(T)}, \hat{E_Q} + R_{g(T)}]. \tag{1}$$

2.2 Design Principle

We now present principle of FS-Sketch for solving continuous data streams using constrained space and returning real-time results with time-decayed accuracy. Traditional sampling techniques provide uniform quality of query accuracy when they solve large scale datasets. An intuitive example is shown in Fig. 1(a), let sk_1 and sk_2 be two sketches built from the same sampling technique, and sk_2 can provide more accurate results than sk_1 ($R_1 < R_2$), iff. sk_2 consumes more space than sk_1 ($S_2 > S_1$). Previous techniques can not compress older samples dynamically to improve space consumption and provide more accurate estimates for queries of fresher elements.

As with freshness sensitive queries in tracking applications, i.e., $Q_1(T_2, R_{g(T_2)})$ and $Q_2(T_1, R_{g(T_1)})$ in Fig. 1(b), Q_1 searches more fresher elements than Q_2 ($T_2 > T_1$) and requires more accurate estimates than Q_2 ($R_{g(T_2)} < R_{g(T_1)}$). Note that Q_1 and Q_2 might be a same query which is conducted at different timestamps T_2, T_1 respectively, while the differences among query accuracy requirements are owning to the differences of query times.

We propose a novel sketching model whose additive error $R_{g(x)}$ changes with a time-decayed function $g(x)$ and it costs constrained space (as shown in Fig. 2(a), (b)). The $g(x)$ could be a general time-decayed function, such as a polynomial decayed function ($g(x) = \frac{1}{x^\alpha}, \alpha > 0$) or a polyexponential decayed function ($g(x) = x^k e^{-x}/k!$, $k = 1, 2, 3, ...$), and we can compute *prior* space consumption of a sketch by summarization of $g(x)$ at data domain $[0, N-1]$.

In general, polynomial decayed function is regarded as a common case. If we define $g(T_i) = (1 - 1/(T - T_i)^\alpha)$, $\alpha > 0$, as the time-decayed function at time point T_i, S_1 be the space of the first window, the space of FS-Sketch can be depicted as $S = \frac{S_1(1-\alpha^n)}{1-\alpha}$, where n is the number of windows in the sketch. Meanwhile, the additive error $R_{g(x)}$ can be depicted as $R_{g(T_i)} = U \times g(T_i) = U(1-1/(T - T_i)^\alpha)$, where U is an upper bound of estimation, T_i is the timestamp for a query, $T_i < T$. For fresher elements queries, $T - T_i \to 1$ and $R_{g(T_i)} \to 0$, while for older elements queries, $T - T_i \to T - 1$ and $R_{g(T_i)} \to U$.

The key idea of FS-sketch implementation is that we first define parameters (ϵ, S) for a driver, where ϵ is minimum relative error for tracking application, S is configured space for a sketch, and then we can select a time-decayed function $g(x)$ to fit the space limitation of S. For details, we utilize a composed sampling technique to meet the time-decayed mechanism as time progresses. For fresher elements, we maintain data streams in the form of sliding-window to sustain accurate estimation, i.e., $R \leq \epsilon E$, meanwhile for long-lifetime elements, we use space-constrained sampling techniques iteratively to sustain error-bounded estimation, i.e., $R \leq U = \epsilon N$. As time progresses, the estimation error of a sliding-window changes, and the decayed samples from a window are merged incrementally into the space-constrained samples set to keep the size of the sketch. Notice that other AQP techniques, e.g., Bernoulli sampling, exponential histogram, random waves [9] can also be tailored and incorporated into the model of FS-Sketch.

In order to boost the performance of FS-Sketch construction for high-speed data stream processing, we develop an ϵ-approximate solution to merge samples with the accuracy-decayed policy as time progresses. Let $g'(T_j)$ be the approximate function of $g(T_j)$, which ensures that $\forall\ T_j$, $T_j < T$ such that

$$\frac{\sum_{x=1}^{j} g'(T - T_j)}{\sum_{x=1}^{j} g(T - T_j)} \leq 1 + \epsilon. \tag{2}$$

We employ the approximate solution for accuracy-decayed sketch compression, and it would improve the space of computation from $O(\log^2(N))$ to $O(\log(N))$ and augment the usability of our approach in data stream processing significantly.

3 Approach Design

In this section, we present details of FS-Sketch implementation, including composed sampling strategies, sketch construction and query operations respectively. The proposed techniques all target at maintaining time-evolving statistics over continuous data streams and answering freshness sensitive queries using small space.

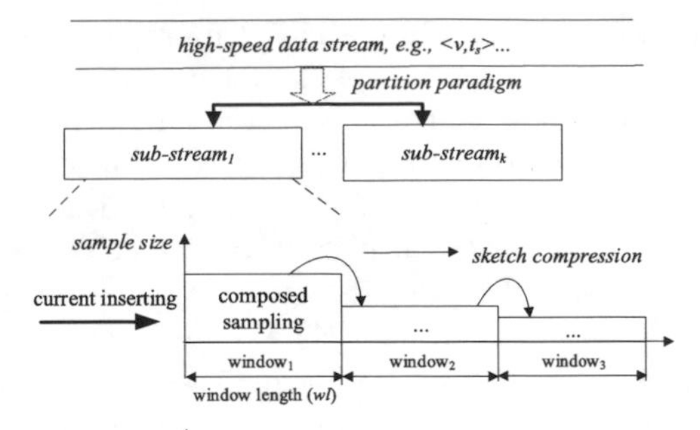

Fig. 3. The architecture of FS-Sketch in distributed scenarios.

3.1 Software Architecture

FS-Sketch is a composed sketch, which comprises a series of continuous temporal windows. A window is depicted by a regular length of time-interval. We term the interval as window length (wl) in the following discussion, and difference of samples in a window is no bigger than wl. Figure 3 presents the basic idea of FS-Sketch deployed over a high-speed data stream. A logical stream is partitioned into sub-streams. Usually, a hash-based function is used to boot the process of sub-streams assignment.

An input element of a sub-stream is inserted into the first window of a local sketch. When the length of the first window is bigger than wl, the following elements are inserted into a newly created window, and the former window is inactive for writing and becomes an older window. We design a freshness sensitive sketch compression method, such that the samples size and query accuracy of a window are depicted by data freshness simultaneously. The size of an older window is compressed dynamically according to its freshness.

To answer flexible predictions of a request as defined in Problem 1, we propose composed sampling methods to select elements from data stream and arrange them in a symmetric sliding-window manner. The design of symmetric sliding-window enables our approach to provide error-guaranteed estimates in time-interval $[StartTs; *]$ or $[*; EndTs]$, where * is an arbitrary time point in a window. For the discarded elements from sliding-window, we compress them into a set, which is tailored by a running bounded value R_k, using k samples. The space-constrained set can answer subset summarization queries with R_k as the upper-bounded error. Therefore, FS-Sketch can answer requests of TAQs within any time-interval of interest, whether the queries intervals are larger than wl or not, with time-decayed errors.

3.2 Composed Sampling Strategies

We now present details of FS-Sketch implementation. We first describe the online sampling techniques, including sliding-window sampling and space-constrained sampling, which are used to select elements from data streams and arrange them in a space-constrained window.

Sliding-Window Sampling. We extend methods of symmetrical sliding-windows sampling presented in [22] to extract elements from streams, and tailor them to high-speed data streams processing. We keep elements in L-'levels' for the temporal aggregation problem, $L = \lceil \log N_s \rceil$, where N_s is the value summarization of elements. Within a level, it maintains the first l elements and outdated elements are discarded directly. The previous literature has proved that when we keep l samples at each level, $l = O(\lceil \frac{1}{2\epsilon})\rceil$, we can obtain error-guaranteed estimates within time-interval [*, T] for streams with time-series model, where '*' is any time point in the window, and T is current time. Furthermore, we can also increase the size of a level l to $O(\frac{1}{\epsilon^2})$, and we can achieve probabilistic error guaranteed for streams with out-of-order series [9].

For an input element e_i, $e_i = (v_i, t_{si})$, it is arranged at 'level'-l within the first window. The 'level'-l keeps samples which make l be the largest number for $2^l < N_s(T_i)$ at time stamp T_i, where $N_s(T_i)$ is summarization of elements at T_i in data streams, i.e., $N_s(T_i) = \sum_{j=1}^{T_i} v_j$. The $N_s(T_i)$ is called *rank* of T_i in the following discussion. A window works with $\lceil log(N_s) \rceil$ levels, $N_s = max\{N_s(T_i)\}$. As shown in Fig. 2(b), the numbers of levels in a window are labeled by l_0, l_1, l_2..., respectively. A hash-based algorithm can insert an element into the sketch using $O(1)$ time. Within a level, we select and keep $\lceil \frac{1}{2\epsilon} \rceil$ exact elements which are close to current time point. When the level size is larger than $\lceil \frac{1}{2\epsilon} \rceil$, the outdated elements will be pushed into space-constrained synopsis. The size of samples kept in the sliding-window can be depicted as $\lceil \frac{\log N_s}{2\epsilon} \rceil$ totally.

Space-Constrained Sampling. As with long-lifetime elements, our aim is to provide weight-bounded estimates with accuracy-decayed policy using constrained space. We need to compute a running upper-bound R_k for the samples set using constrained space k on the fly. We first compute the upper-bound R_k via tracking the total weight of all discarded elements from sliding-window synopsis and maintain $O(k)$ samples in the summaries. When a discarded element from sliding-window is inserted into the summaries and its timestamp is fresher than the time when R_k is computed, we recompute R_k by samples in the summaries, and this can be computed in $O(1)$ time [3]. The new value of R_k is upwards on the go. After the new R_k is obtained, we adjust weights of k samples in the samples set, and keep the current timestamp for R_k until next insertion.

For details, when a new sample is inserted into sample set of the space-constrained summaries, we eject a sample from the summaries to keep the k samples on the fly. We use time-decayed range cost ρ_g with each possible pairs (e_{i1}, e_{i2}) and select the smallest range cost as a candidate. We accumulate weight of the smaller sample of the candidate into the bigger one, and eject the smaller one from the summaries directly. An optimal candidate pairs selection method can be found in literature [3], which costs $O(k^2)$ time for the summaries updating.

An more efficient and heuristics way to compute range cost $\rho_{g(.)}$ for a pair (e_{i1}, e_{i2}) is that we can use summarization and production along with the time-decayed function $g(.)$ as range cost (shown in Eq. 3). It can also achieve appropriate accuracy in experimental evaluation, moveover, it can improve the updating time from $O(k^2)$ to $O(k)$ [3].

$$\rho_{g(x)|x=i_1}(e_{i1}, e_{i2}) = \frac{1}{3}e_{i1}e_{i2}(e_{i1} + e_{i2}) \times g(i_1). \tag{3}$$

3.3 Compression

We incorporate the time-decayed mechanism into FS-Sketch organization to improve the space consumption for long-lifetime windows. Note that this method is interesting itself, since the sketch structure needs to be compressed iteratively and provide deterministic error for temporal queries in each round of compression. We formalize the time-decayed function $g(x)$, with the following property: for any time interval w, the ratio $g(x)/g(x+w)$ is no-increasing with x (x is an integer and $x > 0$).

The structure of FS-Sketch constitutes sketches sk_1, sk_2, ..., $sk_{T/w}$, and $[0, w]$, $[w+1, 2w]$, ..., $[mw, T]$ are m corresponding windows to the sketches respectively. Let sk_i be the sketch for the ith window, and $(i-1)w+1$ and iw are the start time and end time of the window, $i \geq 1$. The decayed accuracy of sketch sk_i can be described by a triple (ϵ_i, R_i, r_i), where ϵ_i and R_i are sampling parameters in sliding-window sampling and space-constrained sampling respectively, and r_i is the decay ratio computed by time-decayed function.

We consider an incremental method to maintain sketches in different time intervals. For current window (ϵ_T, $R_{k,T}$, r_T), we define the basic sampling parameters ϵ_T and $R_{k,T}$, such that we keep $O(\frac{1}{\epsilon}\log N_s)$ samples in the first sliding-window and k samples in constrained space set. Meanwhile, the decay ratio would be $r_T = g(1)$. For the ith window sk_i, $i > 1$, we first compute the decay ratio r_i of sk_i. A simple method is that we can compute the decay ratio r_i of sk_i by time-decayed function $g(T-x)|_{x=iw-1}$. So as the sampling parameters, such that $\epsilon_i = \epsilon_T \times r_i$, $R_i = R_{k,T} \times r_i$. When we obtain the new sampling parameters ϵ_i and R_i, we can compress the sketch accordingly.

For samples compression within sliding-windows, let ϵ_i^{new} be the new error parameter computed by time-decayed ratio in window w_i, and ϵ_i is the previous error parameter, $\epsilon_i^{new} \geq \epsilon_i$. The number of samples in a level of sliding-window is related to the error parameter. Thus, we can reset samples size of a level according to the new parameter (ϵ_i^{new}). We only keep $O(\frac{1}{\epsilon_i^{new}})$ samples in each level, i.e., we can delete the samples after the position of $O(\frac{1}{\epsilon_i^{new}})$ directly.

As shown in Fig. 4, the samples in a symmetrical sliding-window can be maintained at two dequeues fiq and liq. When we organize the sketch using the new parameter ϵ_i^{new}, we just keep $\frac{1}{2}(1+\frac{1}{\epsilon_i^{new}})$ samples at the head of each queue, and remove the samples after the position of $\frac{1}{2}(1+\frac{1}{\epsilon_i^{new}})$ directly. The window w_i can provide error-guaranteed estimates with relative error less than ϵ_i^{new}. For example, when $r = 2$, the window w_i can provide estimates with relative error less than $2\epsilon_i$, and the rate of space improvement will be $\frac{2(\epsilon_i+1)}{2\epsilon_i+1}$. When ϵ_i is small enough, $\epsilon_i \ll 1$, nearly a half of samples are removed from the window w_i. An initiative example is shown in Fig. 4, and the gray part of a level denotes the discarded samples in the process of sketch compression.

Algorithm 1. SketchCompression(sk_i, r_i).

input : (sk_i, r_i);
r_i: the compression ratio of sk_i decaying from the first window (ϵ_T, R_T).
output: (sk_i).

1 $\epsilon_i^{new} \leftarrow \epsilon_T \times r_i$;
2 $R_i \leftarrow R_T \times r_i$;
3 $n \leftarrow \lceil \frac{1+1/\epsilon_i^{new}}{2} \rceil$;
4 **foreach** *each level* l *in* sk_i **do**
5 $\quad n_l \leftarrow |fiq[l]|$;
6 $\quad n_2 \leftarrow |liq[l]|$;
7 $\quad$ **if** $n_1 + n_2 > 2n$ **then**
8 $\quad\quad$ Remove samples whose positions are larger than n from tail of $fiq[l]$ and $liq[l]$;
9 $\quad\quad$ Insert the discarded samples into space-constrained summaries;
10 $\quad\quad$ Compress space-constrained summaries using R_i;

11 **return** sk_i.

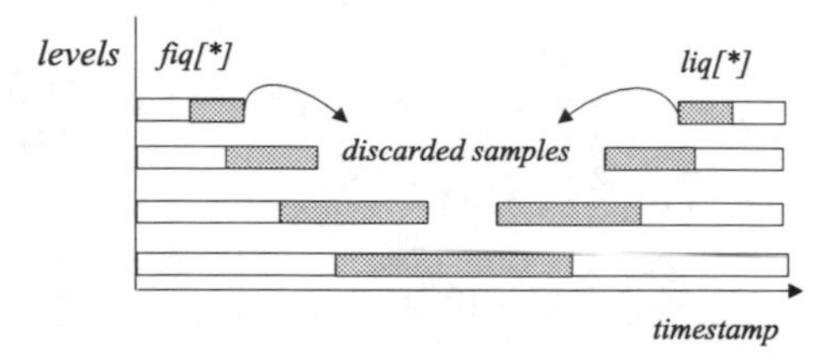

Fig. 4. Sketch compression in a window.

For the sketch compression within space-constrained summaries, we resample the summaries by the new parameter of R_i, $R_i = R_T \times r_i$. A more efficient process for the sub-sampling can be carried out by the streaming probability aggregation process. We set the inclusive probability p_i of a sample as $p_i = min\{a_i/R_i, 1\}$, and accumulate the p_i to the next sample when the sample is not included into the summaries, where a_i is adjusted weight of the sample. We introduce the process of sketch compression in Algorithm 1 for a sketch sk_i. This algorithm makes fresher windows keep more samples to improve query accuracy and older windows keep fewer samples to improve space consumption.

We design an ϵ-approximate function $g(x)'$ for $g(x)$, such that $\sum_{x=1}^{T} g(x)'/\sum_{x=1}^{T} g(x) = 1 + \epsilon$. When defined with the deterministic parameter ϵ, the $g(x)'$ is also a time-decayed function, such that $g(x)'/g(x+w)' \geq 1$, for any $w > 0$.

An efficient implementation of the approximate time-decayed function $g'(x)$ is that we can build a Merging-Based Exponential Histogram (WBEH) to achieve ϵ-approximation of a time-decayed function using $O(\log(n))$ time over data streams [4]. We briefly present the key points of the WBEH for implementing the function $g(x)'$. Within WBEH, the buckets are B_1, B_2, ..., B_i, ..., and if B_1 is the current

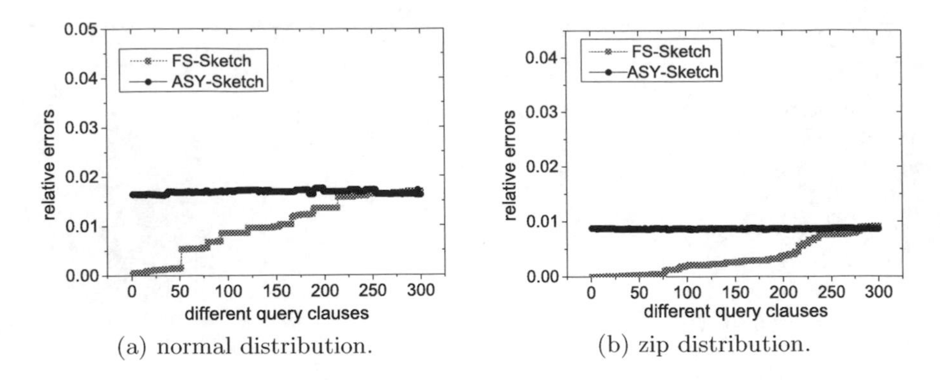

(a) normal distribution. (b) zip distribution.

Fig. 5. Compared with ASY-Sketch over synthetic datasets.

bucket, we have that $T \bmod b_1 = 0$, where T is current timestamp. Whenever, there is a b_i at the two consecutive buckets $[a_s, a_t]$ and $[a_t + 1, a_e]$, such that $b_i \leq T - a_t$ and $T - a_s \leq b_{i+1} - 1$, the two buckets are merged into one bucket $[a_s, a_e]$. And all the samples arranged in bucket $[a_s,\ a_e]$ will share the same decayed ratio, such that if the start time of sk_i belongs to a time interval divided by the anchor points $[b_{j-1},\ b_j - 1]$, $b_j - 1 \in [a_s,\ a_e]$, and the decay ratio r_i of sk_i will be approximately computed by the estimate in MBEH using buckets $[a_s,\ a_e]$. As proved in LEMMA 5.1 of previous work [4], the MBEH can solve the time-decayed counting and summarization problems using $O(\log\log(N)\log(D(g)))$ space, where $D(g) = g(1)/g(N)$.

4 Analysis

We now describe the query accuracy, as well as time and space complexity of our approach when answering freshness sensitive queries for tracking applications. Here, we just present brief description of query accuracy and detailed proof is presented in the Appendix. Let $[c - w, c]$ be the queried time-interval, wl be the window length, ϵ be the minimum error defined in a tracking application, and $\hat{E}$ be the estimate. We can depict query accuracy in the following cases:

1. For queries within fresher elements, i.e. $c = T$ and $w > 0$, FS-Sketch can obtain estimates with deterministic relative error ϵ, such that $E \in [(1 - \epsilon)\hat{E}, (1 + \epsilon)\hat{E}]$;
2. For queries within large-scale and long-lifetime elements, i.e. $w \geq wl$, FS-Sketch can also obtain deterministic relative error ϵ_i, $E \in [(1-\epsilon_i)\hat{E}, (1+\epsilon_i)\hat{E}]$, ϵ_i is the error parameter of the ith window of sketch sk_i such that $(c-w) \in sk_i$;
3. For queries within smaller time-interval elements, i.e. $w < wl$, we use the weight-bounded synopsis to achieve upper-bounded error estimate, such that $E \in [\hat{E} - R_{g(i)}, \hat{E} + R_{g(i)}]$, and $R_{g(i)}$ is an upper bounded value of the ith window.

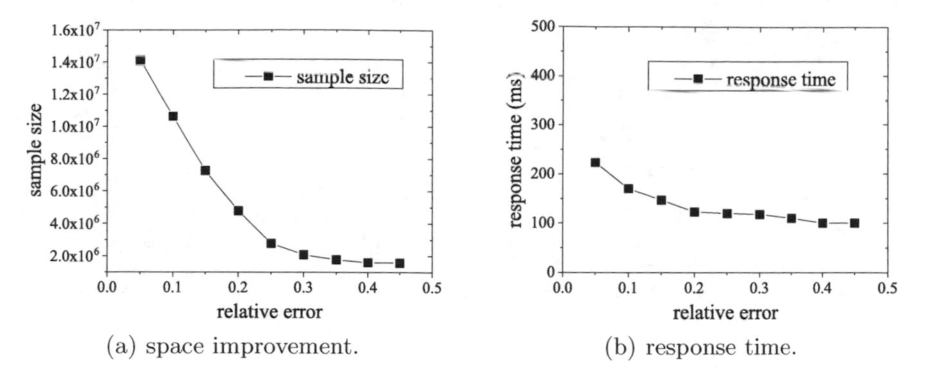

(a) space improvement. (b) response time.

Fig. 6. Sketch compression evaluation.

For complexity analysis, FS-sketch keeps $O(\lceil \frac{1}{2\epsilon} \log N \rceil + k)$ samples in a window, and it costs logarithmic space complexity with rank (N) of the data streams. When an element is inserted into the sketch, it is first inserted into deterministic synopsis and the outdated element is pushed into weight-bounded synopsis. It costs $O(1)$ time to insert an element into a level of the deterministic synopsis and discard an outdated element from the tail of the level. The updating operation in space-constrained synopsis is relative complex. The approximate time-decayed function can be computed using $O(\log \log(N) \log(D(g)))$ time when two consecutive buckets merging. We get an upper bound R_k of the synopsis using $O(1)$ time, and thus we need $O(k)$ time to update the weight of samples in space-constrained synopsis. Therefore, the worst case updating time is $O(\log \log(N) \log(D(g) + k))$ for an element inserting and the average time is $O(k)$ for each inserting.

5 Experimental Evaluation

We implement FS-Sketches on Linux platform using Java 1.8 packages with 64-bit addressing. The experiments are performed on a cluster of 11 machines, each server with 32 GB RAM and 8 × 2.0 GHz CPU. We conduct the evaluations on real-world and synthetic datasets, and compare our approach with sliding-window algorithms and the state-of-the-art big data analytics to demonstrate the effectiveness and efficiency of our approach.

We use real-world datasets, which are hourly page-views from Wikipedia [13]. We select 8 days page-view traffics (100 GB uncompressed data, 1.4 billion records), and 8 weeks of page-view traffics, nearly 1TB uncompressed data, to perform the examination. We also generate synthetic datasets to simulate the high-speed data streams to overcome limitations of the real-world datasets. We generate two types of synthetic datasets, which obey the Normal Form distribution $N(\mu = 1000, \delta = 50)$ and the Zipf distribution $Z(deg = 0.5)$. We use the number of records in a window to represent the velocity of data streams. The

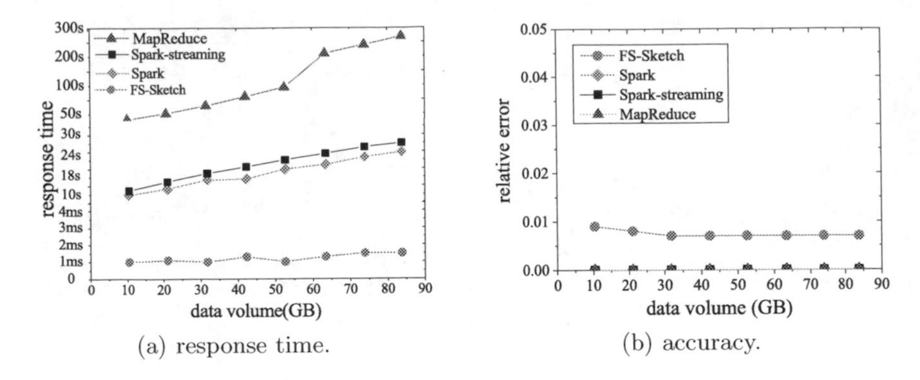

(a) response time. (b) accuracy.

Fig. 7. Compared with big data analytic systems over real-world datasets.

number of records changes from 2^4 to 2^{30} in a window (1 million records per second).

As explained in Sect. 3, FS-Sketch incorporates sliding-window model into our approach to support temporal summarization over data streams. ASY-Sketch [18] is also a useful method, which is an extension from randomized waves to support values summarization over data streams. We generate 300 random temporal queries and compute relative error using the two approaches under different freshness sensitive queries. Each query searches within a time-interval $[T-w, T]$, where T is the current timestamp and w is the length of a time-interval, $w > 0$. We increase w gradually to examine the query accuracy with data freshness changing. We conduct the testing under two types of synthetic data streams using same space overhead, and the results are shown in Fig. 5(a) and (b). Compared with ASY-Sketch, FS-Sketch keeps more samples for fresher elements and achieves more accurate estimates for fresher elements queries. As the length of the queried time-interval (w) increases, the relative error of estimates of FS-Sketch increases gradually. When w is close to the window length, the estimation error of FS-Sketch is slightly bigger than ASY-Sketch.

Finally, we evaluate the space improvement when we use the time-decayed mechanism to compress local sketch. We focus our evaluation on the aspects of the space improvement and time costed in the process of sketch compression. After we load the high-speed data stream into a local sketch, we increase error parameter and compress sketch gradually. We present the space improvement and time used in the process of sketch compressing. In Fig. 6(a), we present the sample size and the corresponding relative error in a window. In the evaluation, when the relative error ϵ changes from 0.05 to 0.5, the size of samples decreases from 1.4×10^7 to 1.5×10^6 accordingly. It achieves nearly linear improvement on space consumption. We also notice that when the error becomes large enough, the improvement is not so rapidly. Because the local sketch includes two summaries, i.e., leveled samples and space-constrained samples, and their size is changing inversely with the error-parameter respectively. To examine the efficiency of sketch compression, we reconstruct local sketch at different error parameters gradually. We just remove samples from the tail of levels in randomized waves, and it can be carried out efficiently in structure of FS-Sketch. As shown in Fig. 6(b), the sketch can be reconstructed in 100 ms–200 ms under the loaded data streams. Therefore, the process of sketch compression is a useful method to improve space consumption when the error ϵ is small. Users can achieve a better balance between query accuracy and space consumption when they confront large-scale datasets analytics with small space.

In order to evaluate the macro properties of FS-Sketch, we implement FS-Sketch on Linux platform and compare it with Spark-Streaming, Spark and MapReduce when processing the real-world datasets. These systems are the state-of-the-art systems for big data analytics. We load the real-world traffics of page-view into the FS-Sketch, Spark, Spark-Streaming, and HDFS respectively, and carry out a same request of TAQ in these systems. Spark and Spark-Streaming are both efficient distributed memory-computing frameworks. Spark transforms the fields of interest into RDDs. Once the RDDs are created in the memory, they can be reused for further queries. Spark-Streaming is based on the framework of Spark. It further transforms the RDDs into DStreams by parameters of window length and temporal intervals in the window.

We cache the DStreams in a queue for a request of TAQ. In our experiments, the production cluster is configured with 12 workers, and the total configured memory in the 11 servers is nearly 100 GB. It is a time-consuming operation to load data from the HDFS into memory in the framework of MapReduce. It costs nearly 300s to respond a request of TAQ in MapReduce. Spark and Spark-Streaming can improve query performance by one order of magnitude than MapReduce. They cost about 25 s to obtain an exact answer for a request of TAQ.

FS-Sketch can obtain nearly real-time responses of queries over different input size. As shown in Fig. 7(a), FS-Sketch costs about 2 ms to respond a request of TAQ. When the size of input increases, the cost time is nearly unchanged. Meanwhile, the relative error is less than 0.01 in the testing as shown in Fig. 7(b).

6 Related Work

The *temporal aggregation query* (TAQ) problem has been study explicitly and implicitly in previous literatures [9,11,17,22,23]. An exact solution for obtaining statistics of TAQ costs linear time and space [5,17]. Traditional approximate engines focus on evaluating ad-hoc queries over static datasets [23]. Serval emerging applications require answering on dynamic streaming data, which is highly distributed and constantly updated. The techniques of sketching over dynamic data streams have been proposed in recent years. The sliding-window is a well-known streams processing model which focuses on computing estimates for elements seen so far. Many sliding-window methods have been explored over the past decades, such as Exponential Histogram [7], random waves [9] and asynchronous streaming sketch [18] etc. However, traditional sliding-window models are designed as one sketch for one operator service schema. For example, Exponential Histogram [7] and random waves [9] are efficient on answering aggregation queries. The G-K and q-digits algorithms can provide an ϵ-approximation estimate for quantiles queries [20]. Composed sketches are also proposed by Arasu and Manku within the context of sliding-window to solve the approximate counts and quantiles in a same structure [2]. The preceding studies provide theories and baseline algorithms for Approximate Query Processing. We combine the core idea of these techniques to produce composed sampling strategies and improve query accuracy for practical data streams processing.

Big data analytic systems, based on Hadoop, have experienced tremendous growth over past few years. Many approximate answering engines have been built on top of the Hadoop software stacks, but few of them focused on low-latency query processing requirements of high-speed data streams. For example, the Hadoop-based approximate answering engines, such as BlinkDB [1] and G-OLA tools [24], extract offline samples from HDFS and then support OLAP queries over the samples with relative complex clauses. The latency of offline sampling techniques make these approximate answering engines do not meet the strict low-latency requirement of analytical queries over data streams.

7 Conclusion

Network tracking applications need to provide capability of real-time monitoring and tracking emergent events over continuous data streams using limited resources. In this paper, we propose FS-Sketch which utilizes time-decayed mechanism to improve query quality and query efficiency when confronting the big and fast data streams using constrained space. The theoretical analyses and experimental results validate the efficiency and effectiveness of our approach. As future work, we plan to consider more query optimization strategies into FS-Sketch and incorporate FS-Sketch into production stream-computing systems to deal with streams with different arrival rates.

Appendix

We now describe the query accuracy of FS-Sketch when solving time-decayed query problems. For a TAQ searching in a time-interval $[c-w, c]$, where w is a length of a time-interval, and we have:

Theorem 1. FS-Sketch can provide an estimate $\hat{E}$ of true value E for fresher TAQs with deterministic relative error ϵ, such that $E \in [(1-\epsilon)\hat{E}, (1+\epsilon)\hat{E}]$.

Proof. We arrange fresher $\lceil \frac{1}{2\epsilon} \rceil$ elements in the model of time-series sliding-window [9]. For a TAQ, we search elements in $[T-w, T]$, and the synopsis can provide estimates with relative error less than ϵ, and thus the additive error is less than $\epsilon\hat{E}$. □

Theorem 2. FS-Sketch can provide a time-decayed additive error R_g for long lifetime TAQs, such that additive error $E \in [\hat{E} - R_g, \hat{E} + R_g]$, meanwhile R_g is less than an upper bounded value U.

Proof. We select a sample pair with minimum range cost $\rho_g(e_{i1}, e_{i2})$ as a candidate to eject to keep synopsis size. We notice that our selectivity mechanism can be considered as an invariant of space-constrained sampling technique [3], which can support any key-range sum query with upper bound R_k using k samples. If we change the samples size according to the time-decayed function $g(x)$, we can provide estimates for a TAQ with additive error less than R_g. For exploring query over long lifetime elements, the additive error is less than an upper bound U, such that $U = Max\{\bigcup_{i=1}^{j} R_{g(i)}\}$, where $1, 2, ..., j$ are windows satisfy the query prediction. □

Notice that the space-constrained summaries can be considered an invariant of WB-summaries [3], which can support any key-range sum query with bounded with R using k samples.

References

1. Agarwal, S., Mozafari, B., Panda, A., Milner, H., Madden, S., Stoica, I.: Blinkdb: queries with bounded errors and bounded response times on very large data. In: Proceedings of the 8th ACM European Conference on Computer Systems, EuroSys 2013, pp. 29–42. ACM, New York (2013)
2. Arasu, A., Manku, G.S.: Approximate counts and quantiles over sliding windows. In: Proceedings of the Twenty-Third ACM SIGMOD-SIGACT-SIGART Symposium on Principles of Database Systems, PODS 2004, pp. 286–296 (2004)
3. Cohen, E., Cormode, G., Duffield, N.: Structure-aware sampling on data streams. In: Proceedings of the ACM SIGMETRICS Joint International Conference on Measurement and Modeling of Computer Systems, SIGMETRICS 2011, pp. 197–208. ACM, New York (2011). https://doi.org/10.1145/1993744.1993763
4. Cohen, E., Strauss, M.: Maintaining time-decaying stream aggregates. In: Proceedings of the Twenty-Second ACM SIGMOD-SIGACT-SIGART Symposium on Principles of Database Systems, PODS 2003, pp. 223–233. ACM, New York (2003). https://doi.org/10.1145/773153.773175
5. Cormode, G., Garofalakis, M., Haas, P.J., Jermaine, C.: Synopses for massive data: samples, histograms, wavelets, sketches. Found. Trends Databases **4**, 1–294 (2012)
6. Cormode, G., Muthukrishnan, S.: An improved data stream summary: the count-min sketch and its applications. In: Farach-Colton, M. (ed.) LATIN 2004. LNCS, vol. 2976, pp. 29–38. Springer, Heidelberg (2004). https://doi.org/10.1007/978-3-540-24698-5_7
7. Datar, M., Gionis, A., Indyk, P., Motwani, R.: Maintaining stream statistics over sliding windows: (extended abstract). In: Proceedings of the Thirteenth Annual ACM-SIAM Symposium on Discrete Algorithms, SODA 2002, pp. 635–644 (2002)
8. Fu, Y., Biersack, E.: Tree-structured bloom filters for joint optimization of false positive probability and transmission bandwidth. SIGMETRICS Perform. Eval. Rev. **43**(1), 437–438 (2015)
9. Gibbons, P., Tirthapura, S.: Distributed streams algorithms for sliding windows. Theory Comput. Syst. **37**(3), 457–478 (2004)
10. Goodstein, M.L., Chen, S., Gibbons, P.B., Kozuch, M.A., Mowry, T.C.: Chrysalis analysis: incorporating synchronization arcs in dataflow-analysis-based parallel monitoring. In: Proceedings of the 21st International Conference on Parallel Architectures and Compilation Techniques, PACT 2012, pp. 201–212. ACM, New York (2012)

11. Gordevičius, J., Gamper, J., Böhlen, M.: Parsimonious temporal aggregation. VLDB J. **21**(3), 309–332 (2012)
12. Katsipoulakis, N.R., Thoma, C., Gratta, E.A., Labrinidis, A., Lee, A.J., Chrysanthis, P.K.: Ce-storm: confidential elastic processing of data streams. In: Proceedings of the 2015 ACM SIGMOD International Conference on Management of Data, SIGMOD 2015, pp. 859–864. ACM, New York (2015)
13. Mituzas, D.: Page view statistics for wikimedia projects. http://dumps.wikimedia.org/other/pagecounts-raw/
14. Papapetrou, O., Garofalakis, M., Deligiannakis, A.: Sketching distributed sliding-window data streams. VLDB J. **24**(3), 345–368 (2015)
15. Preis, T., Moat, H.S., Stanley, E.H.: Quantifying trading behavior in financial markets using Google trends. Sci. Rep. **3**, 1684 (2013)
16. Ramnarayan, J., et al.: Snappydata: a hybrid transactional analytical store built on spark. In: Proceedings of the 2016 International Conference on Management of Data, SIGMOD 2016, pp. 2153–2156. ACM (2016)
17. Tao, Y., Xiao, X.: Efficient temporal counting with bounded error. VLDB J. **17**(5), 1271–1292 (2008)
18. Tirthapura, S., Xu, B., Busch, C.: Sketching asynchronous streams over a sliding window. In: Proceedings of the Twenty-Fifth Annual ACM Symposium on Principles of Distributed Computing, PODC 2006, pp. 82–91. ACM, New York (2006)
19. Wang, H., Can, D., Kazemzadeh, A., Bar, F., Narayanan, S.: A system for real-time Twitter sentiment analysis of 2012 u.s. presidential election cycle. In: Proceedings of the ACL 2012 System Demonstrations, ACL 2012, pp. 115–120. Association for Computational Linguistics, Stroudsburg (2012)
20. Wang, L., Luo, G., Yi, K., Cormode, G.: Quantiles over data streams: an experimental study. In: Proceedings of the 2013 ACM SIGMOD International Conference on Management of Data, SIGMOD 2013, pp. 737–748. ACM, New York (2013)
21. Wang, Z., Quercia, D., Séaghdha, D.O.: Reading tweeting minds: real-time analysis of short text for computational social science. In: Proceedings of the 24th ACM Conference on Hypertext and Social Media, HT 2013, pp. 169–173. ACM (2013)
22. Wu, G., et al.: Supporting real-time analytic queries in big and fast data environments. In: Candan, S., Chen, L., Pedersen, T.B., Chang, L., Hua, W. (eds.) DASFAA 2017. LNCS, vol. 10178, pp. 477–493. Springer, Cham (2017). https://doi.org/10.1007/978-3-319-55699-4_29
23. Yun, X., Wu, G., Zhang, G., Li, K., Wang, S.: Fastraq: a fast approach to range-aggregate queries in big data environments. IEEE Trans. Cloud Comput. (TCC) **3**(2), 206–218 (2015). https://doi.org/10.1109/TCC.2014.2338325
24. Zeng, K., Agarwal, S., Dave, A., Armbrust, M., Stoica, I.: G-ola: generalized on-line aggregation for interactive analysis on big data. In: Proceedings of the 2015 ACM SIGMOD International Conference on Management of Data, SIGMOD 2015, pp. 913–918. ACM, New York (2015)
25. Zhang, Y., Chen, S., Wang, Q., Yu, G.: i2MapReduce: incremental mapreduce for mining evolving big data. IEEE Trans. Knowl. Data Eng. **27**(7), 1906–1919 (2015)

A Frequency Scaling Based Performance Indicator Framework for Big Data Systems

Chen Yang[1,3], Zhihui Du[2], Xiaofeng Meng[1(✉)], Yongjie Du[1], and Zhiqiang Duan[1]

[1] School of Information, Renmin University, Beijing, China
xfmeng.ruc@gmail.com
[2] Department of Computer Science and Technology, Tsinghua University, Beijing, China
[3] School of Software, Zhengzhou University of Light Industry, Zhengzhou, China

Abstract. It is important for big data systems to identify their performance bottleneck. However, the popular indicators such as resource utilizations, are often misleading and incomparable with each other. In this paper, a novel indicator framework which can directly compare the impact of different indicators with each other is proposed to identify and analyze the performance bottleneck efficiently. A methodology which can construct the indicator from the performance change with the CPU frequency scaling is described. Spark is used as an example of a big data system and two typical SQL benchmarks are used as the workloads to evaluate the proposed method. Experimental results show that the proposed method is accurate compared with the resource utilization method and easy to implement compared with the white-box method. Meanwhile, the analysis with our indicators leads to some interesting findings and valuable performance optimization suggestions for big data systems.

1 Introduction

Big data systems for large-scale data processing are now in widespread use. To improve their performance, both academia and industry have expended a great deal of effort in identifying their performance bottleneck. The more time a specified resource is used to execute a workload, the larger impact it can change the total performance, and vice versa. The resource with the highest impact (the longest time consumption) is the bottleneck. How to evaluate the resource impact on performance is an essential work to design the big data systems.

Most big data systems use Mapreduce-like frameworks, such as Apache Hadoop and Apache Spark. They allow distributed computing [11] across clusters and always parallelize the use of four major system resources, including CPU, main memory (memory for short), disk and network. It is complex to directly measure the time consumed on different major resources for big data systems. Many researchers [2,9,18,21,24] use resource utilizations as indicators to evaluate the resource impact. Many measurement tools have also been developed to

G. Li et al. (Eds.): DASFAA 2019, LNCS 11446, pp. 19–35, 2019.
https://doi.org/10.1007/978-3-030-18576-3_2

monitor the different utilizations [10,16]. However, picking the resource having the greatest utilization as the bottleneck is often not correct and misleading. Different resource utilizations are incomparable with each other, due to different means. For example, the CPU utilization measures percentage of the CPU usage time and the disk bandwidth utilization measures the percentage of the used bandwidth. They are not based on the same metric.

Some researches [18] measure the time consumed on the specified resource directly by adding the fine-gained instrumentations into systems, currently only for the disk and network. However, such white-box approaches are too detailed and complicated to implement them easily. In addition, the results are also not accurate. Based on our experiments, the time consumed on I/O resources may be underestimated by 1.6× (see Sect. 5.5 for details), causing the bottleneck to be misidentified.

Some comparable metrics which can locate the performance bottleneck in an easy way are necessary for big data systems. Unfortunately, the existing approaches cannot work well. We propose a comparable analysis method to handle this problem instead of the utilization or white-box method. Employing the CPU frequency scaling performance results, our approach can separate the impact of different resources and construct corresponding indicators derived from the same metric. So the value of our indicators are comparable and it is easy to analyze the performance bottleneck based on the proposed indicator framework. The major contributions of this paper are as follows.

- A methodology is proposed to capture the degree of performance impact by measuring how the performance is close to linear speedup when improving the CPU frequency.
- Based on the proposed methodology, a comparable performance indicator framework as a black-box approach to quantify the impacts of four major resources on big data systems is built.
- The proposed framework has been employed on a typical Spark based big data system to evaluate its accuracy and efficiency. Furthermore, many interesting findings are gained and many valuable performance optimization suggestions are proposed to help users tune big data systems like Spark.

The rest of the paper is organized as follows. Section 2 describes the related work. Section 3 presents our approach. Section 4 describes the experimental method. Section 5 presents our experimental results along with a detailed analysis. Section 6 discusses how to use our indicator framework efficiently. Section 7 summaries our work.

2 Related Work

The existing researches have extensively studied on the resource impact on the system performance in various flavors, including (1) hardware event counters, (2) resource utilization and (3) resource score. However, they do not provide an easy and comparable approach to evaluate the impacts of four major resources of big data systems.

Hardware Event Counters. The current computer system provides lots of performance event counters from hardware layer. Hardware events are excellent at capturing how a given piece of hardware is used. Many tools can collect them to help users analyze the performance, such as Perf [3] in Linux core and Dtrace [8], etc. Although they can dynamically trace the system runtime with a low overhead, they never provide the analytical approaches which can quantify the resource bottleneck. In addition, many works focus on interpreting these event counts to analysis the performance bottleneck, including experimental approaches [20,22] and modeling approaches [13,25]. However, they currently focus on the low-level indicators. Our approach can generate high-level indicators.

Resource Utilization. For profiling a given program to find the bottleneck of major resources, a basic idea is to use resource utilizations. Lots of works [9,18,21] consider a bottleneck resource with a high resource usage and vice versa. Others [2,24] simply optimize resource allocation on the basis of resource utilization. However, an important misleading is that resource utilizations are incomparable with each other. The highest utilization might not mean the bottleneck. It might lead to incorrect conclusions of the above works. For example, the blocked time analysis method [18] considers that CPU is the bottleneck resource of Spark by a high CPU utilization in its experiments. Actually, we find that it ignores the memory impact because the classic CPU utilization contains the memory stall cycle (see Sect. 5.1 for details).

Resource Score. For keeping the comparability, many works give each resource a score and use the score as the resource impact. MIA [26] uses the stochastic gradient boosted regression tree to assign the existing indicators the new scores. The new scores are to measure the importance of existing indicators. However, many existing indicators, such as resource utilization, may be not strongly related to the resource impact, so that it cannot correctly find the bottleneck resource. The main goal of our new indicators is to find the bottleneck. Another approach is to run the elaborate benchmark and give a specified resource a unique score to represent the resource performance, such as Spec score for CPU [5]. [12] uses the similar scores to compare the performance of the same resource on different cloud instance types. However, those methods can only evaluate the physics performance of given resources, not the resource impact on a given system.

3 Methodology for Building a Comparable Performance Indicator Framework

In this section, we will propose our methodology on how to build a comparable performance indicator framework to identify the performance bottleneck of a big data system.

3.1 Problem Formulization

In this paper, we focus on analyze the resource impact on the end-to-end performance of big data systems using Mapreduce-like framework as the processing engine. We assume that the systems run on a homogeneous cluster with a given resource provisioning and data size. In addition, the system parameters about the resource allocation are also fixed. These requirements mean that we only concern the resource impact on the system under the given configuration. In addition, the load is equally divided among all the tasks, just most of the Spark systems done.

For a given cluster, four major resources are formalized as a vector $R_b = <c_b, m_b, d_b, n_b>$ as the base resource scheme, which represents the CPU including the on-chip cache hierarchy, main memory, disk and network, respectively. Noting that when the CPU is given, we specify c_b as the CPU frequency. With the improvement of CPU frequency, we assume that the memory performance is little or no change and the performance of on-chip cache hierarchy is linear correlation in our cluster. This hypothesis is widely accepted for most x86_64 computes [14]. Thus, for a full CPU-intensive workload (i.e., only using the CPU), the performance should change linearly with the CPU frequency scaling.

Cited above, our methodology is to observe the performance improvement with the CPU frequency scaling. We first define the performance improvement. For a given cluster, $CF = \{c_1, c_2, ..., c_l\}$ is the CPU frequency set from the same CPU where $c_j \geq c_i$ if $j \geq i$ and $c_1 \geq c_b$. We can easily scale the CPU frequency on modern CPUs. $DB = \{d_1, d_2, ..., d_m\}$ is the disk set, where the performance of $\forall d_j \in DB$ is better than d_b. $NB = \{n_1, n_2, ..., n_z\}$ is the network bandwidth set, where the performance of $\forall n_k \in NB$ is faster than n_b. The set about memory has not been defined due to not upgrade memory.

$RT(c, d, n)$ is the running time of one workload, where the resource scheme $<c, m_b, d, n>$ is configured to a cluster. When the CPU frequency goes up to $c_i \in CF$ from c_b and the other resources are fixed, we can define the CPU performance improvement degree CPI as

$$CPI(c_i, d, n) = 1 - \frac{RT(c_i, d, n)}{RT(c_b, d, n)}, \tag{1}$$

where $CPI \in [0, 1)$. If it is closer to 1, the performance improvement is higher.

3.2 Performance Indicator Definition

For a given big data system, the execution time can be formalized as follows.

$$RT = \theta_1 \frac{scale}{machine} + \theta_2 \log(machine) + \theta_3 machine + \theta_4, \tag{2}$$

where *scale* is the data size and *machine* is the cluster size [23]. The first item is the computation time, including CPU impact and memory impact, i.e., the time consumed on the CPU and the memory. The rest are communication time

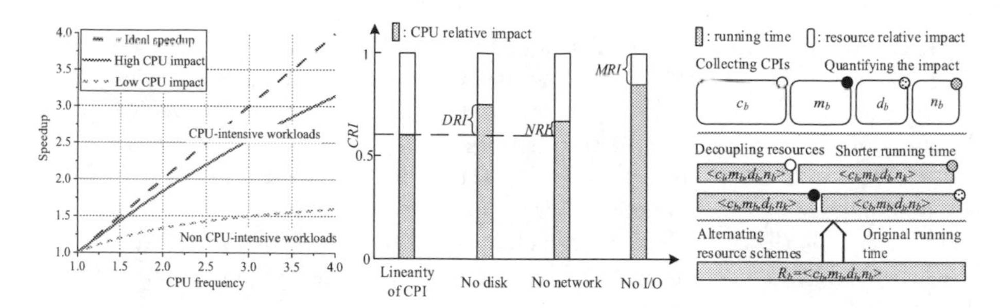

Fig. 1. Left: The speedup (i.e., $RT(c_b, d, n)/RT(c_i, d, n), c_b = 1.0$) on different kinds of workloads when improving the CPU frequency. Middle: The impacts of non-CPU resources are derived from the variation of CPU impact. Right: We decouple the resource impacts by alternating resource schemes and each resource scheme corresponds to an evaluation of resource impact.

and fixed cost, mainly including disk impact and network impact, i.e., the time consumed on the disk and the network.

Cited above, both data size and cluster size are fixed in our scenario. By improving CPU frequencies, we can only reduce the CPU impact of Eq. (2), causing the performance improvement. In this way, we can demonstrate the relation between the system performance and CPU frequency. For easy understanding, we use the speedup, not CPI in Fig. 1 but their features are similar. For a CPU-intensive system, it will be always on-CPU and rarely be blocked by I/O or memory stalls. The CPU is the only limiting resource. Therefore, the speedup is linearly proportional to the improvement of the CPU frequency in the ideal case. Obviously, if the CPU impact is low at Eq. (2), CPU frequency scaling will have little impact on the performance (i.e., low speedup), showing the high non-CPU impact. It motivates us to understand both CPU impact and non-CPU impact by observing the non-linear change in performance.

CPU Relative Impact. With the improvement of CPU frequency, we can define the linearity of performance improvement as CPU relative impact (CRI) to correlate the CPU impact. For assigning $CRI \in [0, 1]$, we define $1 - c_b/c_i$ ($c_i \in CF$) as the upper bound of the performance improvement. If $CPI(c_i, d, n)$, instead of speedup, is closer to $1 - c_b/c_i$, systems are more CPU-intensive. If $CPI(c_i, d, n)$ is closer to 0, systems are not CPU-intensive. To describe this relationship, we formalize CRI on R_b as

$$CRI(R_b) = \frac{1}{l} \sum_{c_i \in CF} \frac{CPI(c_i, d, n)}{1 - c_b/c_i}, \tag{3}$$

where $l = |CF|$ is the number of alternative CPU frequencies, and $CRI \in [0, 1]$. For $\forall c_i \in CF$, if $CPI(c_i, d, n) = 1 - c_b/c_i$, then $CRI = 1$, the workloads will be full CPU-intensive. On the other extreme, for $\forall c_i \in CF$, if $CPI(c_i, d, n) = 0$ then $CRI = 0$, the CPU has no impact on the system performance.

Disk Relative Impact. If the disk is upgraded to $d_j \in DB$, the upper bound of the performance improvement being similar to $1 - c_b/c_i$ is unknown, so we cannot use a method similar to Eq. (3) to evaluate the disk relative impact (DRI). Actually, if we could eliminate the disk blocked time, the system will tend to more CPU-intensive leading that CRI will be higher. It in essence correlates the disk impact to the change of CRI by the CPU frequency scaling. Thus, we can identify the disk impact from the change of CRI. We can eliminate the disk blocked time by upgrading the disk in Fig. 1 and use the increment of CRI to define DRI as

$$DRI(R_b) = \max_{d_j \in DB}(CRI\,(c_b, m_b, d_j, n_b) - CRI\,(R_b)), \tag{4}$$

where $m = |DB|$ is the number of alternative disks, and $DRI \in [0, 1]$. If $DRI \to 0$, the disk has no impact on the system performance. On the other extreme, if $DRI \to 1$, the system is full disk-intensive. In addition, the upgraded disks may introduce different performance improvements due to sequential and random access. However, the precision of DRI is dependent on the performance of upgraded disk, so that the equation suggests that the optional disk should maximize CRI, otherwise the evaluated DRI will be small.

Network Relative Impact. The same method for the disk can be used to evaluate the network relative impact (NRI) as

$$NRI(R_b) = \max_{n_k \in NB}(CRI\,(c_b, m_b, d_b, n_k) - CRI\,(R_b)), \tag{5}$$

where $z = |NB|$ is the number of alternative networks, and $NRI \in [0, 1]$. This is similar to DRI, where $NRI \to 1$ represents highly network-intensive systems and vice versa.

Memory Relative Impact. Because the performance of different consumer memories are so close to each other, we cannot identify the memory impact by upgrading the memory hardware. For example, our test finds that STREAM [6] (an intensive memory access benchmark) with DDR3-1600 RAM is only 4.2% faster than with DDR3-1333 RAM. Thus, the performance improvement is hard to be observed by using the faster memory. From another perspective, we can eliminate the I/O impact (disk and network) as much as possible, leading the system to be only impacted by the CPU and memory. Based on this observation, we define memory relative impact (MRI) as

$$MRI(R_b) = 1 - \max_{d_j \in DB, n_k \in NB}(CRI\,(c_b, m_b, d_j, n_k)), \tag{6}$$

where $MRI \in [0, 1]$. A high MRI means a memory-intensive big data system.

4 Experimental Method

In this section, we use our approach to analyze Spark's performance, including two running modes, which have different performance characteristics, so that they can be considered as two systems. The detailed cluster setup and running mode are as follows.

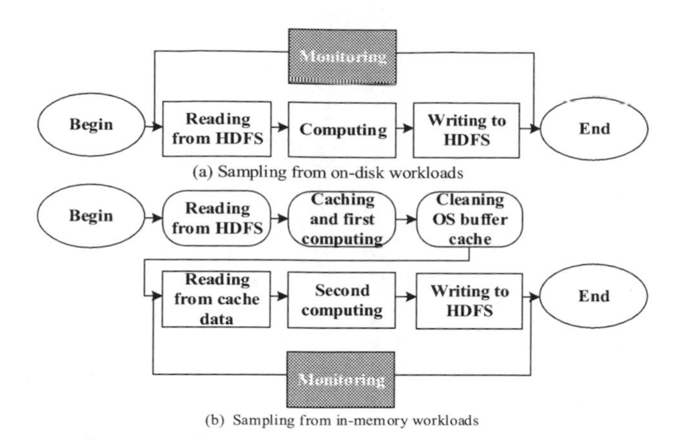

Fig. 2. The procedures of disk mode and memory mode

4.1 Cluster Setup

In our experiment, the baseline resource scheme is set as $R_b = \{1.2\,\text{GHz}, DDR3\text{-}1600, HDD, 1\,\text{Gbps}\}$ and other resource sets are $CF = \{2.4\,\text{GHz}, 3.6\,\text{GHz}\}$, $DB = \{SSD\}$ and $NB = \{5\,\text{Gbps}, 10\,\text{Gbps}\}$. The processor is Intel i7-4790, which has 8 logical CPU cores. We investigate the long-term scientific data services of our collaborators and find that the CPU frequency is usually set at a low level to save energy. Therefore, we set the base CPU frequency to 1.2 GHz to get close to the production environment. For disk, we upgrade the disk by replacing HDD with SSD that has the same capacity. The performance of SSD can eliminate most of disk blocked time for our experimental setup. In addition, the software environment, cluster configuration, and data distribution are identical between HDD and SSD. For network, we have a fiber-optic 10 Gbps network environment. Here, 1 Gbps and 5 Gbps can be obtained by the network speed limit using `tc`. Our cluster has 10 nodes with 1 master and 9 slaves, where every node has 1 processor, 32 GB RAM and two 500 GB SATA disks.

We build our system environment by using stable versions of the software (Apache Spark 1.6.3, Apache Hadoop 2.6.0, and 64-bit Ubuntu 14.04 Server), where Hadoop and Spark use the same machine as the master node. We run Spark on the standalone and create one worker configured with 8 threads and 28 GB RAM (i.e., one thread per CPU). We store the input data in HDFS (Hadoop Distributed File System), and the output data will also be written to HDFS.

We use two SQL benchmarks, i.e., BDBench [19] (Big Data Benchmark, 50 GB Gzip data) and TPC-DS [17] (40 GB Parquet data [4]). BDBench has 9 queries and in TPC-DS we choose 42 queries. The chosen benchmarks are comprehensive and can cover most of the basic operations for big data.

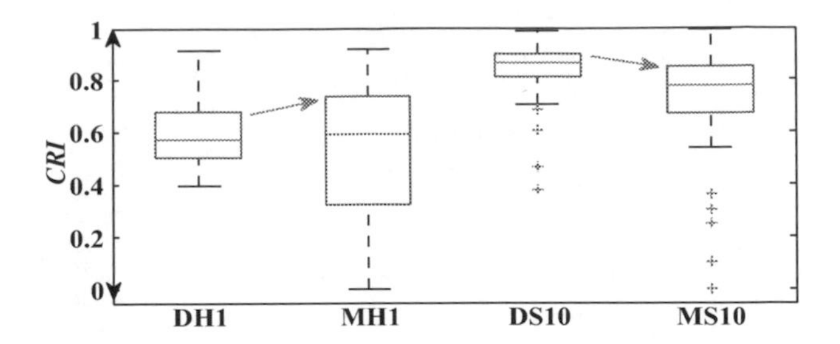

Fig. 3. CPU relative impact. On the x-axis, "DH1" and "MH1" represent disk mode and memory mode on HDD and 1 Gbps, respectively. "DS10" and "MS10" are disk mode and memory mode on SSD and 10 Gbps, respectively.

Table 1. Average resource relative impact on R_b

Resource relative impact	Running mode	BDBench	TPC-DS	Avg
CRI	Disk mode	0.73	0.58	0.61
	Memory mode	0.55	0.52	0.53
MRI	Disk mode	0.04	0.18	0.16
	Memory mode	0.18	0.31	0.3
DRI	Disk mode	0.17	0.25	0.24
	Memory mode	0.19	0.2	0.2
NRI	Disk mode	0.04	0.015	0.02
	Memory mode	0.06	0.06	0.06

4.2 Running Mode

Every query is finally parsed into on-disk and in-memory workloads (i.e., in-memory analytics) by Spark. For disk mode, on-disk workloads read the input data from HDFS and write the output data to HDFS, and we collect the running time over the whole process, as shown in Fig. 2(a). For memory mode, in-memory workloads read input data from the memory and write output data to HDFS. As shown in Fig. 2(b), we have to run every workload twice because the cache function is not a action operator [1]. The first running is to cache data and the second running is just called as the memory mode, which will be monitored.

5 Experimental Results and Analysis

In this section, we split the whole experiment into five parts to study the resource impact on Spark. We compare our performance indicator framework with the resource utilization (e.g., CPU utilization, disk bandwidth utilization and network bandwidth utilization) and the time blocked white box analysis method [18].

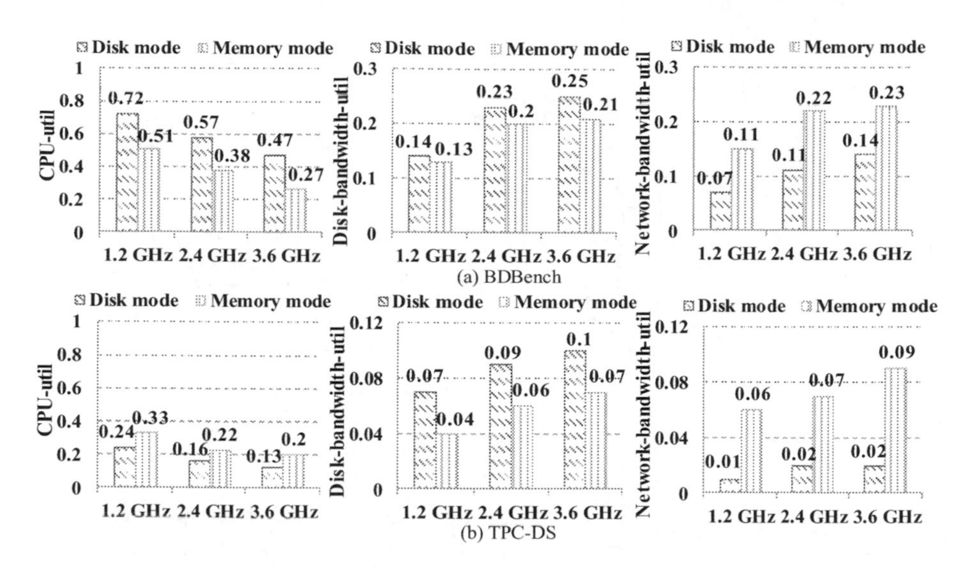

Fig. 4. Different resource utilizations on HDD and 1 Gbps

5.1 CPU Impact Analysis

Table 1 shows CRI solved by Eq. (3), where the label "Avg" represents the average of both benchmarks. Overall, CRI is, on average, 0.57 for both running modes, where 76% queries in disk mode are CPU-intensive and 64% queries in memory mode are CPU-intensive, i.e., $CRI \geq 0.5$. It suggests that the CPU is the bottleneck for Spark. Curiously, for both benchmarks, CRI in memory mode is always lower than it in disk mode. This implies that Spark is more CPU-intensive when reading input data from the disk. We put BDBench and TPC-DS together to find that CRI in memory mode tends to two extreme poles, as shown in "DH1" and "MH1" of Fig. 3.

When $CRI \geq 0.6$, the approximate median 51% of in-memory workloads are only slightly greater than the 45% of on-disk workloads. This phenomenon shows the memory mode has more CPU-intensive workloads. However, too many in-memory workloads are low CPU-intensive. When $CRI \leq 0.4$, 27% are in-memory workloads (approximately 87% of them in TPC-DS), far more than the 2% of on-disk workloads. This phenomenon that reading data from memory has a lower CPU impact can be be reasonably explained as follows.

- Reading the cache data causes a lower LLC (Last Level Cache) hit rate, so the memory stall time gets longer. Details are provided in Sect. 5.2.
- The high CRI is incurred when decompressing input data in disk mode, but it is not required for memory mode. Relatively speaking, memory mode will show the low CPU impact. Details are provided in Sect. 5.3.
- Spark is blocked by the network I/O more frequently in memory mode. This phenomenon can also reduce CRI. Details are provided in Sect. 5.4.

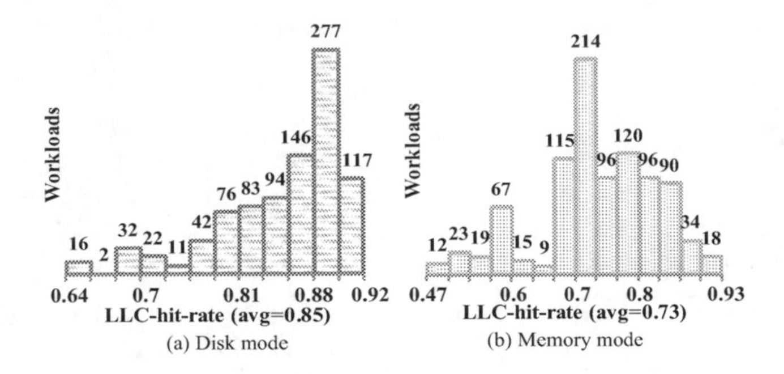

Fig. 5. Distributions of LLC hit rate on all of resource schemes

Why CPU Utilization Is Not a Good Representation of the CPU Impact on Performance? For most of performance analysis scenarios, the CPU utilization (CPU-util) is usually used to evaluate the CPU impact on performance. A high CPU utilization means that applications are more CPU-intensive and vice versa. Strictly speaking, it is not accurate enough.

In Fig. 4, for TPC-DS CPU-util in memory mode is greater than it in disk mode. This trend contradicts *CRI*. The reason is that LLC misses are more frequent causing the high memory stall time in memory mode of TPC-DS and CPU-util includes the memory stall time. Too many stall cycles cause a high CPU-util. However, the memory stall time should be the memory impact, not the CPU impact. This finding suggests that a high CPU-util cannot always represent a high CPU impact.

For both benchmarks, especially in TPC-DS, CPU-util is very low, but *CRI* suggests that the CPU is the bottleneck. This is also contradictory. The reason is that CPU-util only shows the system impact on the CPU (the CPU usage), not the CPU impact on the system (the percentage of the CPU usage time). For a multicore processor, in most cases, CPU-util is the average utilization of all cores. For Spark, because of scheduling delay and task difference, the scheduler cannot ensure that the task threads always run on all cores. Thus, in our experiments, it is normal that some CPU cores are idle but other CPU cores are always in use, causing the low CPU-util. However, it does not mean that the CPU consumes less time than other resources. This finding suggests that a low CPU-util cannot represent a low CPU impact.

Suggestion. CPU-util and *CRI* can be combined to help users give the optimized suggestions. For example, the high CPU-util and the low *CRI* may imply that the system has a weak memory management strategy. The low CPU-util and the high *CRI* may imply that the CPU cores are not fully used.

5.2 Memory Impact Analysis

As shown in Fig. 3, we focus on "DS10" and "MS10". There is an abnormal phenomenon. After we only upgrade the I/O resources, *CRI* in memory mode shows a downward trend relative to CRI in disk mode, compared with the upward trend on HDD and 1 Gbps. It reveals that the memory impact in memory mode is 2-4.5$\times$ greater than it in disk mode. Based on this, we determine MRI by Eq. (6), as shown in Table 1. Overall, the average MRI is 0.23. For both benchmarks, MRI in memory mode is always greater than it in disk mode. This shows that reading the cache data makes Spark more memory-intensive.

Why Is Memory Mode More Memory-Intensive? It is worth to note the LLC hit rate. In Fig. 5, we demonstrate the distribution of LLC hit rate, where the interval of every bar is nearly equal. For memory mode, the average and highest bars have 14–21% deteriorations compared with those in disk mode. This shows that the performance of LLC hit in memory mode is weak.

Especially when we run TPC-DS on SSD and 1 Gbps, memory mode (Running time is average 56.7 s) is unexpectedly slower than disk mode (average 55.7 s). The high overhead of moving data into the CPU even exceeds the advantage of caching data in memory, causing caching data to have no effect. Therefore, for memory mode reading cache data causes MRI to be increased.

The cache operation for Spark 1.6.3 is important because it is the basis for in-memory data analytics. In Spark SQL, the main idea of the cache strategy for structured data is as follows. When data are cached in memory, Spark stores them in a two-dimensional array into a columnar format. This data structure is conducive to in-memory compression. However, when the data is processed, Spark must transform them from columnar format into a row format. This transformation can break the data locality, leading to a reduction of the LLC hit rate.

Suggestion. The high MRI implies that the performance of Spark SQL in memory mode can be improved by optimizing the cache operation. Actually, we find that reading cache data in memory mode has to transform a columnar array into a row array, causing the performance reduction.

5.3 Disk Impact Analysis

We solve DRI using Eq. (4), and the average is 0.22 for both modes. Disk mode is more disk-intensive than memory mode, as shown in Table 1. This trend is also demonstrated in Fig. 6(a). However, for different benchmarks, the trend is the opposite. For BDBench, memory mode is more disk-intensive. For TPC-DS, disk mode is more disk-intensive. Especially for BDBench, the abnormal trend of DRI suggests that reading input data from disk does not necessarily represent that the system is more disk-intensive. We also show the traditional indicator (i.e., disk bandwidth utilization) in Fig. 4. It suggests that disk mode is more disk-intensive for BDBench, being different from the trend of DRI. Actually, we think the disk bandwidth utilization cannot accurately reflect the disk impact.

What Factors Cause the Difference of DRI? Combined with our experiments, we summarize two reasons causing different disk impacts for both benchmarks as follows.

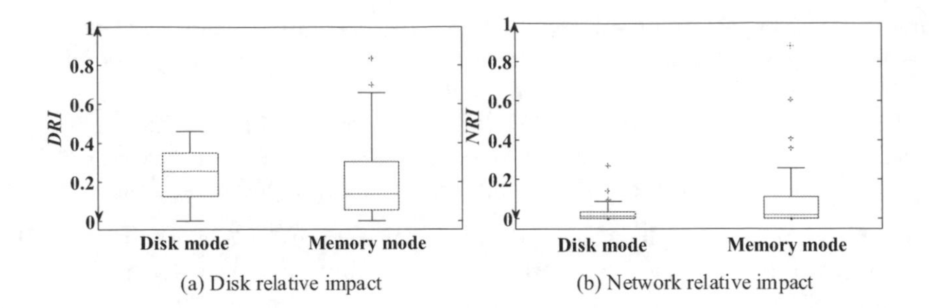

(a) Disk relative impact (b) Network relative impact

Fig. 6. Disk relative impact and network relative impact

Compression for BDBench. Memory mode is more disk-intensive in BDBench. The additional decompression in disk mode increases much computation, causing the relatively low disk impact. The only difference between disk mode and memory mode is reading compression data from the disk or not. CRI in disk mode is 0.73, far more than 0.55 in memory mode, showing that many CPU cycles are used for the decompression. Relatively speaking, the disk impact is low in disk mode, compared with memory mode. Therefore, the high disk bandwidth utilization does not mean the high disk impact.

Short Tasks for TPC-DS. Disk mode is more disk-intensive in TPC-DS. Unlike the long-calculation tasks in BDBench, the short tasks in TPC-DS easily are blocked by I/O due to the less overlap. The current big data systems leverage the asynchronous I/O mechanism to overlap the computation and I/O request to improve the performance. However, in disk mode of TPC-DS the CPU overhead is relatively low for each task, causing the overlap to be disabled. Thus, the disk impact increases. For example, the low CPU-util in TPC-DS is on average 17% in disk mode and 25% in memory mode, showing the lower opportunity for overlap. In addition, TPC-DS has many short tasks (e.g., 0.7 s per task in TPC-DS, compared with 9.1 seconds per task in BDBench) also showing the less overlap. It leads to an interesting phenomenon that DRI is excellent in TPC-DS, but the disk-bandwidth-util is very low, as shown in Fig. 4. Therefore, the low disk bandwidth utilization does not mean the low disk impact.

Suggestion. The high disk-bandwidth-util and the low DRI imply that the system has a good I/O performance with less disk blocked time when reading much data. In contrast, the low disk-bandwidth-util and the high DRI mean a weak I/O performance. The SQL's optimizer should build the long computing tasks or merge I/O requests as much as possible to maximize the overlap.

5.4 Network Impact Analysis

As shown in Table 1, we solve NRI using Eq. (5), and the average is 0.04 for both modes, which is minimal within the four major resources. NRI in memory mode is 1.5-4$\times$ greater than it in disk mode in Fig. 6(b). The network bandwidth utilization in Fig. 4 also has the same trend.

Why Does Memory Mode Have a Higher NRI? Spark uses the network in the following three stages. (1) The input data may be read from remote disks, but this rarely happens (ie.g., only 5% of the data from remote disks in our cluster), because HDFS preferentially reads input data from local disks. (2) The shuffle read stage needs to read data from both local and remote disks. (3) Writing output data to HDFS needs to backup two duplicates to remote disks. Because HDFS rarely reads input data from remote disks, the network I/O has nothing to do with the decompression. Moreover, shuffle and output stages are the same in both modes. Thus, the amount of data transferred over the network is nearly equal in both modes (The difference is, on average, 5.6% in our cluster). Due to the shorter running time in memory mode, more data are transferred over the network per second. This is manifested as a higher network impact.

Suggestion. Memory mode needs to transfer data over the network more frequently, causing the higher NRI. Combined with the result that BDBench's memory mode have a higher disk impact than disk mode in the previous section, it is actually necessary for users to pay more attention to the I/O impact, rather than the CPU impact for in-memory analytics in some cases.

5.5 Inaccuracy of Time Blocked White-Box Method

The blocked time analysis method [18] for Spark is used for analyzing the impacts of the disk and network. It collects the I/O blocked time by adding some instrumentations into the system and simplifies part of shuffle I/O into the upper bound of the disk I/O or network I/O. Finally, simulate the infinitely fast disk or network by ignoring I/O blocked time to evaluate Spark's maximum performance improvement. Actually, it mainly evaluates the I/O impact, i.e., both disk and network.

This method relies on adding some instrumentations into HDFS's core to get the blocked time when Spark accesses HDFS. However, the corresponding codes have not been opened, so that only shuffle I/O can be profiled [7]. Even so, we design several cases to illustrate the limitations of this approach.

Major Page Faults. Spark is usually impacted by the external factors, such as OS. The intra-system instrumentations cannot monitor them. For example, the system execution is not only blocked when reading data from disk, may also be blocked due to major page faults issued by OS. We design a simple experiment to demonstrate this problem. In BDBench, q3C is the most complex query. When we use 56 GB compressed data to run Spark with q3C in our cluster, Spark will be starved for memory. We run q3C without output in memory mode on R_b. For contrast, this query also runs when I/O resources are upgraded to SSD and 10 Gbps. Time blocked analysis method shows that q3C can be sped up by 48.6% (<50%), suggesting that q3C might not be the I/O bottleneck, but it is actually sped up by 77.7%, suggesting that q3C is definitely the I/O bottleneck. The I/O impact is underestimated by 1.6×. Actually, Spark on HDD is slowed down due to major page faults (6,394 per node) but it can be significantly sped up by SSD.

This phenomenon is ubiquitous. In 42 queries of TPC-DS, approximately 79% of them have major page faults. Overall, the I/O impact might be underestimated.

Compared with the blocked time analysis method, our indicator framework does not only focus on the impacts of four major resources, but it can also evaluate the latent I/O impact by upgrading I/O resources.

6 Discussion

In this section, we discuss the major characteristics of our indicator framework and how to use our method efficiently.

Comparability. Our indicator framework is built from CRI, so their values are comparable. Thus, the greatest one can be identified as the bottleneck. It is noted that the sum of them is not necessarily equal to 1. When we upgrade the disk and network simultaneously for calculating MRI, the improvement of CRI may be not equal to the sum of DRI and NRI by upgrading the disk or network separately.

Scalability. The resource replacement method may limit our indicators on large-scale clusters. Our indicators are only dependent on the end-to-end performance in essence. Thus, we can leverage the performance prediction technique to achieve the scalability. For example, Ernest [23] can predict the end-to-end performance of the large-scale MapReduce-like workloads by training a performance model with the performance data from different small-scale clusters. Thus, we can run the system on small-scale clusters with our indicator framework and train the performance model by Ernest. Further, we can predict the resource impact on large-scale clusters.

Cost. For our indicator framework, the major cost is upgrading the disk and network. However, it is not necessary to upgrade I/O resources for CPU-intensive applications, e.g., $CRI > 0.5$. This observation is very helpful to reduce the usage cost of our indicators.

Accuracy. The workload and the alternative resources can affect the accuracy of the value of our indicators. If the load can be equally divided among all the tasks, our indicator framework will identify the performance change as accurate as possible when upgrading resources. To evaluate the linearity of performance improvement with CRI, we run the system at different $c_i \in CF$ and use the average as the CRI to improve the accuracy. For three other indicators, it is easily to achieve the upgrade of I/O resources. For the disk, both SSD and main memory are used as d_j. The RamDisk technique [15] supported by Linux core can use the main memory as the disk. For the network, both fiber-optic 10 Gbps network or faster InfiniBand network architecture can be used as n_k.

7 Summary

In this paper, we propose a performance indicator framework to evaluate the relative impact of four major resources on big data systems. Values of different indicators can be built based on measuring the CPU frequency scaling performance results. Many experiments are done to verify our approach and Spark's performance is analyzed in depth. We summary the most important advantages of our framework. In addition, many interesting findings are found and some valuable suggestions are given to help users tune their Spark.

Advantages. First, our four indicators are comparable with each other because they are derived from the same metric. The feature ensures the bottleneck can be easily found through our approach. Second, our indicators are more accurate compared with the resource utilization and the existing white-box approach because our approach is strongly related to the time consumed on the specified resource. Therefore, our approach can easily find the underlying performance issues. Third, our approach is also easy to implement relying on the general CPU frequency scaling technology.

Findings. (1) The CPU impact may go down when Spark reads data from memory instead of disk, because lower LLC hit rate, no data decompress, and more frequent network blocking will happen. (2) Using CPU utilization as CPU performance impact indicator is often misleading because long memory stall time may lead to high CPU utilization and unbalanced tasks/threads scheduling on multicore systems will lead to low CPU utilization. (3) Reading data from memory will significantly increase the memory impact by 2-4.5× because lower LLC hit rate will happen. Sometimes the performance will be lower than reading data from disk because data locality is broken. (4) Disk bandwidth utilization is also often misleading to identify the disk impact because it cannot show how much disk time can be overlapped with CPU time. (5) The network impact is often the lowest for most Spark big data systems. But its value can be increased by 1.5-4× when Spark reads data from memory.

Suggestions. Even though resource utilizations are often misleading, with the help of our indicator framework, it is easy for us to not only find the cause, but also give the method to handle the problem. So we can combine the two methods together to identify some Spark's potential problems on the memory management strategy, the scheduler and the SQL's optimizer. Some specific tuning suggestions can also be given from our work. For example, users should pay more attention to the impact of I/O resources when executing in-memory analytics, rather than ignore them.

Acknowledgement. This research was partially supported by the grants from National Key Research and Development Program of China (No. 2016YFB1000602, 2016YFB1000603); Natural Science Foundation of China (No. 91646203, 61532016, 61532010, 61379050, 61762082); Fundamental Research Funds for the Central Universities, Research Funds of Renmin University (No. 11XNL010); and Science and Technology Opening up Cooperation project of Henan Province (172106000077).

References

1. Apache spark. http://spark.apache.org/
2. Google vm rightsizing service. https://cloud.google.com/compute/docs/instances/viewing-sizing-recommendations-for-instances
3. Linux perf subsystem. https://perf.wiki.kernel.org/index.php/Main_Page
4. Parquet. http://parquet.apache.org/
5. Spec. http://www.spec.org/
6. Stream. http://www.cs.virginia.edu/stream/
7. Trace-analysis. https://github.com/kayousterhout/trace-analysis
8. Cantrill, B., Shapiro, M.W., Leventhal, A.H., et al.: Dynamic instrumentation of production systems. In: USENIX Annual Technical Conference, General Track, pp. 15–28 (2004)
9. Conley, M., Vahdat, A., Porter, G.: Achieving cost-efficient, data-intensive computing in the cloud. In: Proceedings of the Sixth ACM Symposium on Cloud Computing, pp. 302–314 (2015)
10. Dai, J., Huang, J., Huang, S., Huang, B., Liu, Y.: Hitune: dataflow-based performance analysis for big data cloud, pp. 87–100 (2011)
11. Dean, J., Ghemawat, S.: Mapreduce: simplified data processing on large clusters. In: Proceedings of Operating Systems Design and Implementation, vol. 51, no. 1, pp. 107–113 (2004)
12. Dittrich, J.: Runtime measurements in the cloud: observing, analyzing, and reducing variance. VLDB Endow. **3**, 460–471 (2010)
13. Gao, F., Sair, S.: Long-term performance bottleneck analysis and prediction. In: International Conference on Computer Design, pp. 3–9 (2007)
14. Hackenberg, D., Molka, D.: Memory performance at reduced CPU clock speeds: an analysis of current x86_64 processors. In: Workshop on Power-Aware Computing Systems, HotPower, pp. 5–9 (2012)
15. Koutoupis, P.: The linux ram disk. Linux+ Magzine, pp. 36–39 (2009)
16. Massie, M.L., Chun, B.N., Culler, D.E.: The ganglia distributed monitoring system: design, implementation, and experience. Parallel Comput. **30**(7), 817–840 (2004)
17. Nambiar, R.O., Poess, M.: The making of TPC-DS. In: International Conference on Very Large Data Bases, pp. 1049–1058 (2006)
18. Ousterhout, K., Rasti, R., Ratnasamy, S., Shenker, S., Chun, B.G.: Making sense of performance in data analytics frameworks. In: 12nd USENIX Symposium on Networked Systems Design and Implementation, pp. 293–307 (2015)
19. Pavlo, A., et al.: A comparison of approaches to large-scale data analysis. In: ACM SIGMOD International Conference on Management of Data, pp. 165–178 (2009)
20. Sambasivan, R.R., et al.: Diagnosing performance changes by comparing request flows. In: USENIX Conference on Networked Systems Design and Implementation, pp. 43–56 (2011)
21. Shi, J., et al.: Clash of the titans: Mapreduce vs. spark for large scale data analytics. Proc. VLDB Endow. **8**(13), 2110–2121 (2015)
22. Sridharan, S., Patel, J.M.: Profiling R on a contemporary processor. Proc. VLDB Endow. **8**(2), 173–184 (2014)
23. Venkataraman, S., Yang, Z., Franklin, M.J., Recht, B., Stoica, I.: Ernest: efficient performance prediction for large-scale advanced analytics. In: Proceedings of USENIX Symposium on Networked System Design and Implementation, pp. 363–378 (2016)

24. Wang, C., Meng, X., Guo, Q., Weng, Z., Yang, C.: Automating characterization deployment in distributed data stream management systems. IEEE Trans. Knowl. Data Eng. **29**(12), 2669–2681 (2017)
25. Yoo, W., Larson, K., Baugh, L., Kim, S., Campbell, R.H.: ADP: automated diagnosis of performance pathologies using hardware events. In: ACM Sigmetrics/Performance Joint International Conference on Measurement and Modeling of Computer Systems, pp. 283–294 (2012)
26. Zhibin, Y., Xiong, W., Eeckhout, L., Bei, Z., Mendelson, A., Chengzhong, X.: Mia: metric importance analysis for big data workload characterization. IEEE Trans. Parallel Distrib. Syst. **29**(6), 1371–1384 (2018)

A Time-Series Sockpuppet Detection Method for Dynamic Social Relationships

Wei Zhou[1,2], Jingli Wang[1,2], Junyu Lin[1,2], Jiacheng Li[1,2], Jizhong Han[1], and Songlin Hu[1,2](✉)

[1] Institute of Information Engineering, Chinese Academy of Sciences, Beijing 100093, China
{zhouwei,wangjingli,linjunyu,lijiacheng,hanjizhong,husonglin}@iie.ac.cn
[2] School of Cyber Security, University of Chinese Academy of Sciences, Beijing 100049, China

Abstract. Multiple identity deception, called as sockpuppet, has been commonly used in online social media to spread rumours, publish hate speeches or evade censors. Current works are continually making efforts to detect sockpuppets based on verbal, non-verbal or network-structure features. Network structure has attracted much attention, while the time series dynamic characteristic of sockpuppet network has not been considered. With our observation, after being blocked, a puppetmaster tends to recover previous social relationships as soon as possible to maintain the propagation influence. The earlier the relationship is recovered, the more important it is. To take advantage of this dynamic nature, a time-series sockpuppet detection method is proposed. We first design a weight representation method to record the dynamic growth of sockpuppet's social relationships and then transfer sockpuppets detection to a similarity time-series analysis problem. The experiments on two real-world datasets of Sina Weibo demonstrate that our method obtains excellent detection performance, significantly outperforming previous methods.

Keywords: Identity deception · Time-series sockpuppet detection · Dynamic social network structure

1 Introduction

Online social media has become a daily part of people's lives due to its convenience on information spread. The existence of malicious accounts on the online social media leads to serious risks. Malicious accounts utilize online social media to spread unwelcome information such as spamming [1], fraud [10], cyberbullying [3], hate speech [9], discrimination [22], etc. When the malicious accounts are detected and blocked, they create some new accounts called sockpuppets [14] to continue spreading information. We broadly define a *sockpuppet* as a user account that is controlled by an individual (or *puppetmaster*) who controls at

W. Zhou and J. Wang—These authors contributed equally to the work.

G. Li et al. (Eds.): DASFAA 2019, LNCS 11446, pp. 36–51, 2019.
https://doi.org/10.1007/978-3-030-18576-3_3

least one other user account. A sockpuppet pair means two or more sockpuppets controlled by a puppetmaster. To cope with this serious risk, it is essential to provide an accurate and effective detection method to find sockpuppets and sockpuppet pairs among a large number of accounts in social media.

Some existing methods for sockpuppets detection exploit verbal features, including characters, sentences, words, tokens, etc. Their effectiveness is affected by computational efficiency, complexity of practical implementation and the availability of the appropriate data [15]. In addition, some non-verbal behaviors are introduced in sockpuppets detection, such as the number of revisions, the total number of bytes added and removed on Wikipedia [13]. Compared with verbal methods, non-verbal ones are more computationally efficient [8]. However, non-verbal features dug from one platform (e.g.Wikipedia) may not be fit for other platforms (e.g. Twitter) and these methods only focus on the sockpuppet account but not sockpuppet pairs [23]. Moreover, network-structure method based on community detection has been proposed to detect sockpuppets [2], while the time series dynamic characteristic of sockpuppet network has not been considered [20]. Dynamic characteristic of relationship can indicate the sockpuppets's nature features, who prefer to establish the relationship according to the different importance.

In this paper, by delving into the sockpuppets' temporal character of social networks, we observe that sockpuppets tend to keep similar propagation influence with previously blocked accounts by dynamically recovering similar social networks in temporal order. To take advantage of this dynamic nature, a time-series sockpuppet detection method is proposed. We first design a weight representation method to record the dynamic growth of sockpuppet's social relationships and then transfer sockpuppets detection to a similarity time-series analysis problem. To summarize, the main contributions of this study can be summarized as follows:

- We observe that after being blocked, a puppetmaster tends to recover previous social relationships as soon as possible to maintain the propagation influence. The earlier the relationship is recovered, the more important it is.
- Further, we exploit a time-series sockpuppet detection method for dynamic social networks based on the observation. We first design a weight representation method to record the dynamic growth of sockpuppet's social relationships and then transfer sockpuppets detection to a similarity time-series analysis problem. As far as we know, this is the first work to introduce the dynamic social structure to sockpuppet detection field.
- To validate the effectiveness of our method, we collect two real-world, publicly available data from Sina Weibo. The experimental results have achieved a precision of 0.901 and recall at 0.925, for a best F1 of 0.913 on datasets, demonstrating that our proposed method can be more widely used and is significantly better than the previous methods.

The rest of this paper is organized as follows: Sect. 2 is about the current techniques to detect sockpuppets. Section 3 explores social networks of accounts and social structure features. And, Sect. 4 introduces our structure-based online method and its efficiency. Then, Sect. 5 presents our dataset and experimental results. Finally, in Sect. 6 we conclude and show our future directions.

2 Related Work

Sockpuppet detection is to detect whether a given account is a sockpuppet or not [8,12,23], or whether a pair is controlled by the same puppetmaster [2,5,14,15, 18]. The existing sockpuppet detection methods are roughly classified into verbal communication, non-verbal behaviors, and network structure-based methods.

2.1 Verbal Features

For verbal features, in Wikipedia, according to alphabet count, number of tokens, emoticons count or the use of words without vowels, Solorio et al. [14] explored sockpuppet detection on 77 sockpuppets. Hosseinia et al. [5] proposed a transduction scheme, spy induction that leverages the diversity of authors in the unlabeled test set by sending a set of spies (positive samples) from training set to retrieve hidden samples in the unlabeled test set using nearest and farthest neighborsand. For the detection efficiency, better sockpuppets seeds get higher accuracy. Character n-grams, as a feature set for authorship attribution, also can identify sockpuppets on the basis of similarity scores [6]. As a matter of fact, because the verbal communication is static and assumes that sockpuppet has similar linguistic styles [18,21], the accuracy of detection will be affected by the behaviors of puppetmasters by changing their writing styles on purpose [8].

2.2 Non-verbal Behavior

Tsikerdekis et al. [8] studied the posting behaviors including the number of revisions, the total number of bytes added or removed on Wikipedia, etc. Yamak et al. [23] proposed an automatic detection approach for sockpuppets on Wikipedia, based on features of the number of user's contributions, the frequency of revert after each contribution in the articles, etc. Compared with verbal-based methods, the above methods are computationally efficient, and the reason is that these methods do not have to process large amounts of text data and ignore the link between the sockpuppets from the same puppetmaster and do not group the sockpuppets.

2.3 Network Structure

Existing network structure-based detection methods are subjectively based on user views or emotional similarities. Bu et al. [21] proposed a sockpuppet detection algorithm based on authorship-identification techniques and relationship analysis. The relationships between two accounts are built if they have a similar attitude to most topics and similar writing styles. Then, the relationship-based community detection is performed to identify sockpuppets. Besides, Kumar et al.

[11] created a reply network on The AV Club discussion community and observed that the nodes denoting socpuppets were more central and highly active. As a result, the above network structure-based methods focused on establishing social relationship and identify sockpuppets based on the relationships, where ignoring the time series dynamic characteristic of sockpuppet network.

3 Time-Series Social Relationship

We observe that, after being blocked, a puppetmaster tends to recover previous social relationships as soon as possible to maintain the propagation influence. The earlier the relationship is recovered, the more important it is. In this section, we introduce our observation in details and analyze the difference of dynamically social networks between sockpuppet pairs and sockpuppet-ordinary pairs. The sockpuppets we analyzed in this section are widely known or have been published.

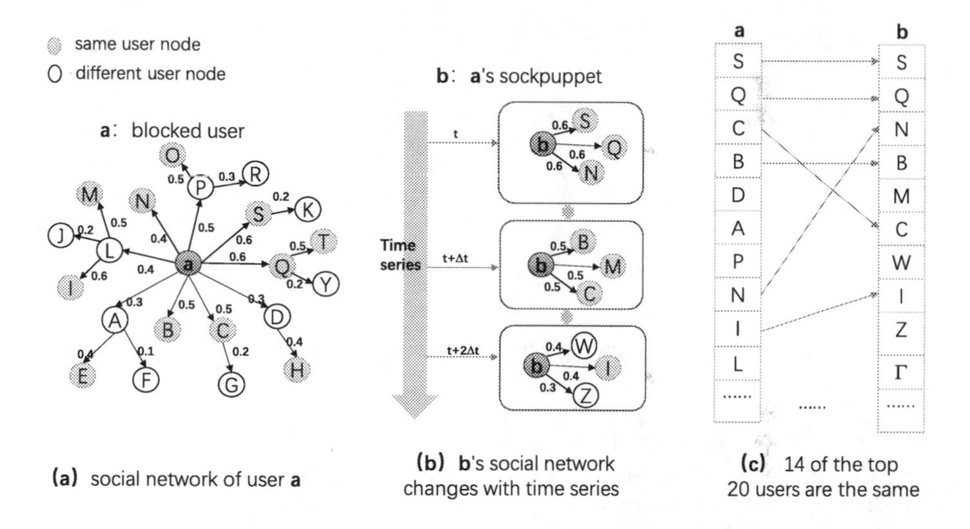

Fig. 1. An example of dynamic social relationship of a sockpuppet pair

In order to get an insight into the relationships of sockpuppets, we record the dynamic growth of sockpuppet's and ordinary account's social relationships as the time goes. For example, as Fig. 1(a) shown, we suppose *Account* $\boldsymbol{a}$ is a blocked one with many relationships to the other accounts and the weights on relationships mean their importance. If *Account* $\boldsymbol{b}$ is $\boldsymbol{a}$'s sockpuppet, we can observe that as time goes, there are many same relationships with account $\boldsymbol{a}$ established around $\boldsymbol{b}$ and the more important the relationship is, the earlier it is established, shown in Fig. 1(b). At $\boldsymbol{t}$ time, there are only $\boldsymbol{S}$, $\boldsymbol{Q}$, $\boldsymbol{N}$ in the same relationships. While $t+\Delta t$, accounts $\boldsymbol{B}$, $\boldsymbol{M}$, $\boldsymbol{C}$ are added and the number of same relationships increases to 6. As the time goes, $\boldsymbol{W}$, $\boldsymbol{I}$, $\boldsymbol{Z}$ become $\boldsymbol{b}$'s neighbors and so on. In sum, we can conclude that $\boldsymbol{a}$ and $\boldsymbol{b}$ have 14 identical relationships in the top-20 neighbors [17], as shown in Fig. 1(c).

The situation is different for sockpuppet-ordinary pair. In Fig. 2, the sockpuppet-ordinary pair (sockpuppet *Account* $\boldsymbol{a}$ and ordinary *Account* $\boldsymbol{c}$) interact with different neighbors. After a while, the number of their identical relationships is only one.

By comparison of this two situations, we can observe that the social relationship formed in the early days is more important because it represents the social relationship that the puppet master urgently needs to recover.

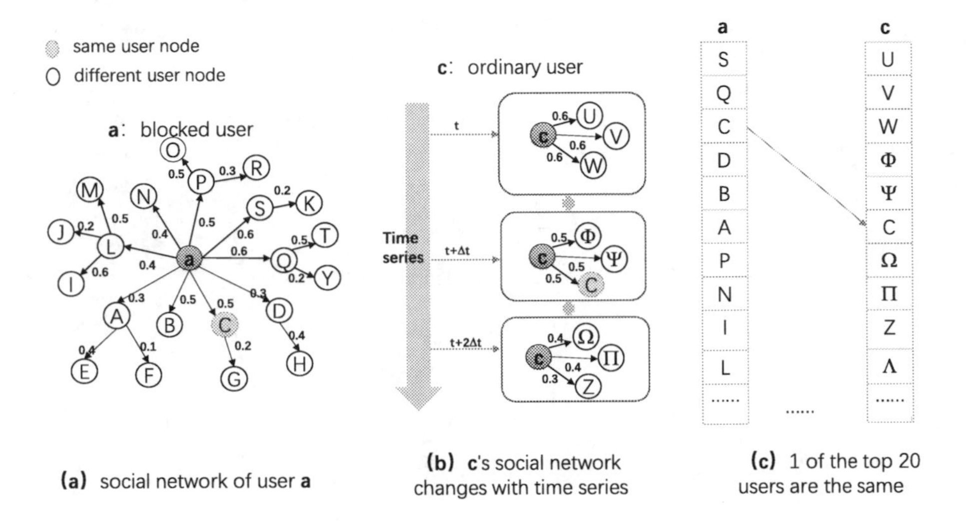

Fig. 2. An example of network structure between a sockpuppet-ordinary pair

Over all, we randomly collect a total of 284 accounts, consisting of 93 sockpuppet pairs and ordinary accounts from Sina Weibo. Then, we analyze the time series dynamic characteristic of sockpuppet pairs and sockpuppet-ordinary ones.

Based on statistics, some findings are as follows:

- The sockpuppet pairs are more likely to establish similar social networks. The average number of identical neighbors contained in the interaction network of the sockpuppet pair is much larger than that of the sockpuppet-ordinary pair (83 s. 7), shown in Fig. 3a. This suggests that after the new sockpuppet is registered, the puppetmaster would like to recover previous social relationships in order to continue spreading their malicious posts.
- The earlier the relationship is recovered, the more important it is. As the time goes, we can find that the number of establishing more important relationships firstly is larger (65 vs. 11 vs. 7), shown in Fig. 3b.

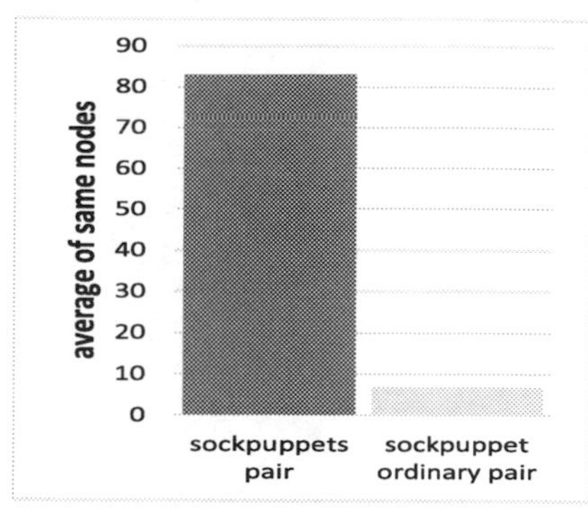

(a) The difference between sockpuppet pair and sockpuppet-ordinary pair

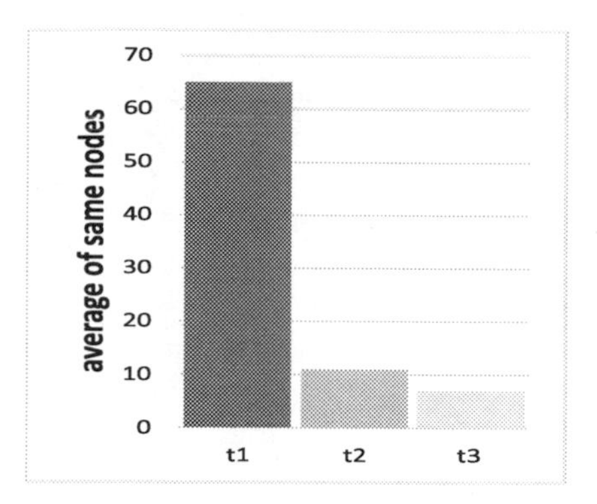

(b) The difference with time going by: t1 → t2 → t3.

Fig. 3. The statistical results of differences between sockpuppet pairs and sockpuppet-ordinary ones.

4 Sockpuppet Detection Method for Dynamic Social Relationships

To take advantage of this dynamic nature, a time-series sockpuppet detection method is proposed. We first construct the social graph and then design a weight representation method to record the dynamic growth of sockpuppet's social relationships. Furthermore, we formalize sockpuppets detection as a similarity time-series analysis problem. Finally, to prove the efficiency of our method, the time complexity is further calculated.

4.1 Graph Construction

All the symbols we used are listed in Table 1. In this subsection, we build the social relationships for each account, which is the interaction graph as defined below:

Interaction Graph. We represent the interactions as a link in the graph. A directed and weighted interaction graph of the center node u is defined as $G_u = (V_u, E_u, W_u)$, where (1) V_u is a set of ordinary nodes and the center one; (2) $E_u \subseteq V_u \times V_u$ is a set of edges between nodes in V_u; (3) W_u is a set of weight of edges. In our model, nodes are accounts and edges represent the interaction relationship between these nodes. Figure 4 represents an example of two interaction graphs with center nodes u, v, and there are four identical nodes (node B, C, E, H). The interaction relationship does not limit by one hop and the weight between two nodes with no direct connection can be computed by combining all the weights of their path. In order to represent the intimacy degree

Table 1. Symbol definition

Number	Symbol	Description
1	G_u	The interaction graph of the center node u
2	V_u	A set of nodes that have interactions with the center node u
3	W_u	A set of weight of edges with the center node u
4	u_i	The ith element of V_u
5	A	The set of interactions between two accounts
6	A_i	The ith element of A
7	$w_{u_i}^{u_j}$	The weight between u_j and u_i
8	p_u^a	A shortest path from node u to a, $a \in V_u$
9	$X_{u,v}$	The set $V_u \bigcap V_v$
10	x_i	The ith element of X
11	k	The account to be detected
12	B	The set of all blocked users
13	b_i	The ith element of the blocked users
14	S_k	The set of k's sockpuppets

of two users, the weight $w_{u_i}^{u_j}$ of $e_{u_i}^{u_j}$ $(u_j, u_i \in V)$ is calculated as $F(u_i, u_j)$ the logarithm of the value of each interaction type between u_i and u_j:

$$w_{u_i}^{u_j} = F(u_i, u_j) = \lg \sum_{i=1}^{|A|} Type_{A_i}(u_i, u_j) \tag{1}$$

where A represents the set of interactions between two accounts (u_i, u_j), and the more interactions means the larger weight. The value of an interaction type is obtained by the function $Type(\cdot)$. Inside, $Type_{A_i}(\cdot)$ represents the value of this interaction type (A_i). Particularly, the interaction in social media can be built passively or actively, e.g. Weibo Search, Weibo Recommending on Sina Weibo. And, the more active the interaction type is, the larger the value of the function $Type(\cdot)$ is, because the more active type indicates that the user is eager to build relationships with the other.

4.2 Interaction Graph with Time Series

In this subsection, we encode time series into the constructed social graph by adjusting the weights of relationships.

$$w_{u_i}^{u_j} = (1 - \lambda) * F(u_i, u_j) + \lambda * K(u_i, u_j) \tag{2}$$

Where λ represents the proportion of the time series in the weight and $K(\cdot)$ represents the time series of two nodes forming the edges in the social network, and the specific calculation method is as follows:

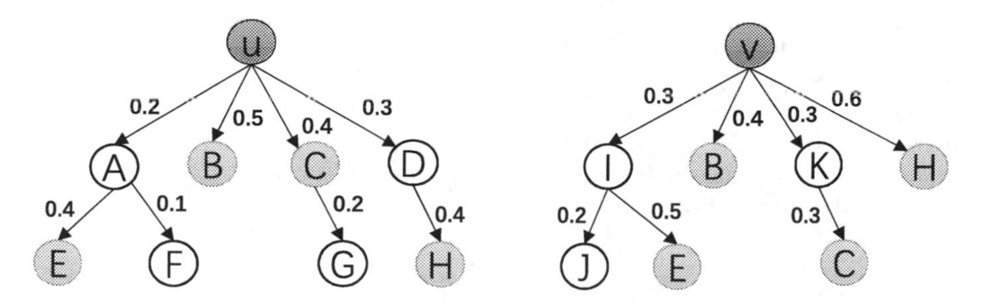

Fig. 4. The social network graph of center nodes u and v

$$K(u_i, u_j) = \frac{\sum_{i=1}^{|A|} Time_{A_i}(u_i, u_j)}{|A|} \tag{3}$$

where the value of an interaction time is obtained by the function $Time(\cdot)$. Inside, $Time_{A_i}(\cdot)$ represents the value of this time series (A_i). Particularly, the sooner an interaction builds on social media, the larger the value of the function $Time(\cdot)$ is, because the earlier interactions indicate that the user is eager to build relationships with others.

4.3 Graph Similarity with Dynamic Social Relationship

In this section, we formalize sockpuppets detection as a similarity time-series analysis problem. We first apply the shortest path hops and the weight sum of edges between two nodes to show the degree of relationships between accounts and the degree function $\Phi(u_i, u_j)$ can be expressed as:

$$\Phi(u_i, u_j) = \frac{w_{u_i}^{u_j}}{|p_{u_i}^{u_j}|^2} \tag{4}$$

Ruan et al. proposed to use Jaccard coefficient as defined below to calculate the graph similarity [19] as below:

$$jaccard(G_u, G_v) = \frac{|V_u \bigcap V_v|}{|V_u \bigcup V_v|} = \frac{|X_{u,v}|}{|V_u \bigcup V_v|} \tag{5}$$

Our method improves the Jaccard coefficient, which combines with the degree of relationship between each of the same nodes. Therefore, the method calculates the interaction graph similarity $H(u, v)$ as follows:

$$H(u, v) = \frac{\sum_{i=1}^{|X|} (\Phi(u, x_i) + \Phi(v, x_i))}{|V_u \bigcup V_v|} \tag{6}$$

Algorithm 1. Detect whether a newly registered account is a sockpuppet and find his/her sockpuppets

Input: newly registered account k, all blocked users B
Output: sockpuppet set S_k or ordinary users
1: Given account k and all blocked users B, construct their interaction networks G_k, G_{b_i}, and b_i is one of the blocked users
2: $S_k = NULL$
3: LOOP: b_i in B
4: LOOP: node j in G_k
5: compare node j with nodes in G_{b_i} to find same interaction nodes
6: END LOOP
7: calculate relationship degree, $\Phi(u_i, u_j)$
8: calculate interaction graph similarity, $H(k, b_i)$
9: classification algorithm combined with graph similarity to find the association between accounts
10: if graphs are similar
11: $S_k = S_k + b_i$
12: END LOOP
13: if $S_k == NULL$
14: return k is an ordinary account
15: else
16: return S_k

4.4 Efficiency

This subsection describes the efficiency of our method. In the online platform, accounts will be blocked when they spread rumors or post malicious contents. They will register new accounts to continue their purposes. Our method detects these accounts by social network structures, as shown in the Algorithm 1. Our method just needs to match the social network subgraphs between the account (k) with each blocked user (b_i) based on a classification algorithm by comparing each node in the G_k with G_{b_i}. The procedure requires to be continuously executed until all blocked users have been matched to find k's all sockpuppets. The time consumption depends on the number of the blocked accounts and the size of the social graph of account k. Therefore, the time complexity of the proposed algorithm is:

$$O(Algorithm \quad 1) = O(|B| * |G_k|) \tag{7}$$

5 Experiment and Results

5.1 Data Collection and Analysis

Data. To demonstrate our method's effectiveness, we collected public data from Sina Weibo as our experimental data, which is different from the analytical data we mentioned above. In this platform, users can establish social relationships with others to share information [4]. From the website, we collect each user's homepage which contains the user profile (e.g. username, gender, description, etc.), posts information, and interaction users.

Identifying Sockpuppets. For our method, we need an accurately classified sockpuppet dataset. In previous work, there are many ways to identify sockpuppets. We combine these methods. According to observation, some sockpuppets will announce their own sockpuppet information in their posts to help others finding them. Based on this information, some other user information is also used to help label, such as the username, avatar, registration time, unhidden login IP and posts. Once different user accounts have announced their sockpuppet information and have quite similar even identical profile information (e.g., the login within a short time interval many times, their login IPs are similar, and their avatar are very similar as well), they are considered to be controlled by the same puppetmaster [2].

Dataset Construction. In order to verify the effectiveness of our method, we choose users of different time periods, which can construct two datasets, summarized in Table 2. According to the above method to judge sockpuppets, our datasets contain 4035 sockpuppets and 48309 ordinary users, including their profiles, posts and social network information.

In this paper, for the convenience of calculation, we only regard the interaction network that the path hops from the center node to other nodes are no more than two.

Table 2. Statistics of dataset 1 and dataset 2

	Dataset 1	Dataset 2
# of sockpuppets	1905	2130
# of ordinary users	19763	28546
# of interaction nodes of sockpuppets	64873	105347
# of posts of sockpuppets	143065	150899
sockpuppets ratio	9.64%	7.46%

5.2 Experimental Design

In our experiment, the interaction types can be divided into Weibo Search Following, Weibo Recommending Following, Forwarding, etc., which are assigned different function values. For the convenience of the experiment, we only consider user nodes that have 2 hops to a center node in its social network.

In the previous work, Solorio et al. [14] evaluated their proposed model using only Support Vector Machine (SVM), while Tsikerdekis et al. [8] used Support

Vector Machine, Random Forest (RF), Logistic Regression (LR) and Adaptive Boosting (ADA) to evaluate their models, where ADA performs best. Besides, Yamak et al. [23] used six classification algorithms, and the RF performed best. Therefore, to further verify the effectiveness of our method with different algorithms, we employed the following four classification algorithms: LR, SVM, RF, ADA. To identify the different performances of our features, we have firstly designed two models: interaction graph with interaction type (*M1*); interaction graph with time series (*M2*).

We applied 10-fold cross-validation procedure to achieve our experimental method. The procedure splits the data into ten parts, taking nine of them as the training set whereas the other as the testing set. The procedure is sequentially executed until all possible ten combinations have been applied. Our experiment results are the average of ten executions.

To more comprehensively evaluate the efficiency of our models, we derived results to use following metrics: ***Recall*** (the fraction of sockpuppet pairs in the sample is predicted correctly), ***Precision*** (the fraction of positive predictions that are valid sockpuppet pairs), ***F1*** (the harmonic mean that combines recall and precision), ***Accuracy*** (the fraction of true positives and true negatives over the total number of data), ***False positive rate*** (FPR) (the fraction of falsely identified sockpuppet pairs).

5.3 Performance Evaluation Comparison

Interaction Graph Similarity with Interaction Type (M1). Let $\lambda = 0$, which means that we only consider building the interaction type through the similarity of the interaction graph (M1). Referring to *M1* in Table 3a, the performance on dataset 1 is surprisingly high compared with the precision at 0.960, the recall at 0.738 and the *F1* at 0.835 through *ADA*. Relative to the dataset 2, the performance of *M1* has decreased. But, on two datasets, the interaction graph similarity can get good performance with ADA.

In a word, the result further illustrates the fact that the feature we proposed is extremely effective, because the feature reflects the purpose of sockpuppets.

Table 3. The result of *M1*

(a) dataset 1

	P	R	F1	Acc	FPR
LR	0.978	0.692	0.811	0.838	0.015
SVM	0.958	0.708	0.814	0.838	0.03
RF	0.940	0.723	0.817	0.838	0.046
ADA	0.960	0.738	**0.835**	0.854	0.031

(b) dataset 2

	P	R	F1	Acc	FPR
LR	0.938	0.692	0.796	0.823	0.046
SVM	0.902	0.708	0.793	0.815	0.077
RF	0.873	0.738	0.800	0.815	0.108
ADA	0.860	0.754	**0.803**	0.815	0.123

Table 4. The result of *M2*

(a) dataset 1

	P	R	F1	Acc	FPR
LR	0.877	0.905	0.891	0.889	0.126
SVM	0.861	0.916	0.888	0.884	0.147
RF	0.923	0.884	0.903	0.905	0.073
ADA	0.896	0.923	**0.909**	0.908	0.108

(b) dataset 2

	P	R	F1	Acc	FPR
LR	0.901	0.846	0.873	0.877	0.092
SVM	0.889	0.864	0.875	0.877	0.108
RF	0.919	0.877	0.898	0.900	0.077
ADA	0.921	0.892	**0.906**	0.908	0.077

Interaction Graph Similarity with Time Series (M2). Let $\lambda = 1$, which means that we only consider building the time series through the similarity of the interest graph (M2), whose precision is 0.896, recall 0.923, *F1* 0.909 through ADA on dataset 1, shown in Table 4a. Comparing *M2* with *M1*, the best recall almost increases by 18% on dataset 1 and 14% on dataset 2, which reveals that the dynamic social network is booming in detecting more sockpuppet pairs. According to analysis, in the early days, the sockpuppet will quickly resume his social network. The earlier the time represents, the more important this person is. Even if the social network intersection between the two sockpuppets is relatively small, if the weight values are large, it proves that these nodes are more important and can make up for the smaller intersection defects. Therefore, adding timing features to our method can better find more sockpuppet pairs.

Discussion of Results. These results show that ADA algorithm appears to provide almost the best performance and our method provides the best balance between recall and precision whereas maintaining the highest achieved accuracy. The recall level is relatively high (0.923) on dataset 1, which means that most sockpuppets are picked up by our proposed model. On dataset 2, the experimental performance is also excellent and optimal. This indicates that our method is effective in detecting sockpuppets. Interaction graph similarity is the basis of our method and is important to ensure the validity of our method. In addition, our method has a relatively high precision with LR, thus if we tend to detect whether an account is a sockpuppet of blocked users, this method is precise and can get results quickly. Above all, our method can not only detect sockpuppets, but also can find out the puppetmaster's sockpuppets accurately from a large number of users. Besides, puppetmasters have a difficulty to skip from the detection and to change their social network structures if they wish to achieve their objectives. Therefore, our method can guarantee the detection effect and be disposed on different social platforms to meet different demands with different models.

5.4 Parameter Analysis

According to the above models, we find that the change of parameters has a great influence on the experiment. The time series can effectively improve the experimental results. Therefore, in this sub-section, we want to find the best parameters to balance the interaction type and time series. Specifically, we track how the sockpuppet detection quality changes when the value of λ is varied from 0.1 to 0.9 with a step length of 0.2. According to the above experimental results, the ADA algorithm performs best, so in this sub-section we use ADA as the classification algorithm.

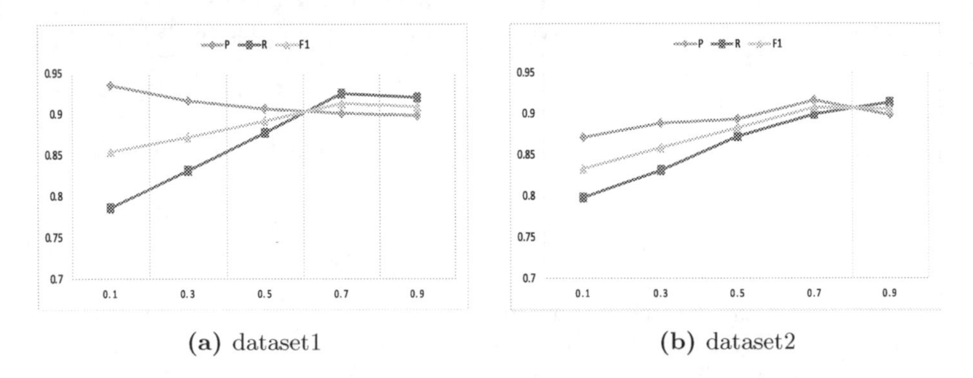

(a) dataset1 **(b)** dataset2

Fig. 5. Effect of varying λ on performance

From Fig. 5, performance on two datasets is constantly improving when λ increases (i.e. more weight assigned to time series). But when $\lambda = 0.7$, the performance is the best with the precision of 0.901, recall at 0.925, F1 of **0.913** on dataset 1 and the precision of 0.916, recall at 0.899, F1 of **0.907** on dataset 2. From the experimental results, we can see that the time series can effectively improve the recall rate, but the impact on accuracy is not very large.

This result indicates that time series helps us detecting more sockpuppets and is important to find more sockpuppets. Some newly created sockpuppets have not yet recovered to a large enough social network, but if existing nodes account for a large proportion in the network, they can also help us judge it as a sockpuppet. Therefore, the reasonable weight distribution can achieve the best experimental performance.

5.5 Literature Comparison

In order to investigate the superiority of our method, we assess the effectiveness of different types of information on sockpuppet detection. In particular, we compare our proposed method with those of four previous approaches on our two datasets.

- **Profile Similarity (PS):** This method applys the user profile (the username, gender and description, etc.) similarity to detecting the sockpuppet pairs because sockpuppets' profiles to a certain extent will reflect the puppetmasters' lexical preferences with character n-grams [16].

- **Verbal Features (VF):** Verbal features [14], such as total number of words, sentences, punctuations, are used to capture the similarity of linguistic styles between accounts.
- **Non-verbal Behavior (NVB):** Simple and more complex variables are used to represent users' posting behaviors [7], such as the time interval of posting blogs, the number of posts in one day and so on.
- **Similar-orientation Network (SON):** The approach first analyzes users' sentiment orientation to topics based on emotional phases [2], and then evaluates the similarity between sentiment orientations of account pairs to build a similar-orientation network. Finally, community detection is adopted to detect sockpuppets in the network.

Table 5. Methods comparison on two datasets

Model	Dataset 1			Dataset 2			Efficiency for a newly account
	P	R	F1	P	R	F1	
PS	0.727	0.615	0.667	0.726	0.602	0.658	$O(B * C)$
VF	0.732	0.745	0.739	0.753	0.625	0.683	$O(B * R * F)$
NVB	0.759	0.745	0.752	0.770	0.648	0.704	$O(B * R)$
SON	0.719	0.631	0.672	0.726	0.602	0.658	$O(B * (R + T))$
Our method	0.960	0.738	0.835	0.860	0.754	0.803	$O(B * \|G_k\|)$
Our method with $\lambda = 0.7$	0.901	0.925	**0.913**	0.916	0.899	**0.907**	$O(B * \|G'_k\|)$

In the above table, B is the number of all users, C means the number of n-grams, R is all posts made by each user and F is the number of features. T is the number of topics, $|G_k|$ is the number of all social nodes and $|G'_k|$ is the number of all social nodes where $|G'_k| \ll |G_k|$. We compare our method with these above methods and ADA is selected as classification model. The results on two datasets are shown in Table 5. We have the following observations:

- With different types of the features considered, our method with time series obtains the best *F1* by approaching to 0.913 and is suitable for the online platform with low time complexity.
- The *Profile Similarity* achieves the worst performance. Observing from social media platform, the profiles of many sockpuppets are missing or different, which leads to low recall. It also indicates that smart sockpuppets tend to empty or change their profiles irregularly to avoid being blocked. In our method, we just focus on the social network structure, and ignore verbal characters. It will help us diminish the misleading of primer deliberated disguise.
- *Verbal Features* and *Non-verbal Behaviors* achieve better performance compared to *Profile Similarity*, because they align with the purpose of puppet-masters and indicate the hidden information of sockpuppets. However, they ignore the important social network feature controlled by the same puppet-master, which is not easy to be disguised.

- The *Similar-orientation Network* method uses network structure built by similar opinions. The network assumes that two users have links if they have similar orientations to most topics. However, this assumption ignore several features of sockpuppets. In our method, we focus on the social network graph similarity of each sockpuppet. It reflects that sockpuppets will construct the similarity social network because sockpuppets want to maintain influence. And, our method analyzes the characteristics from the purpose of sockpuppet creation, which gets better performance.

6 Conclusion

In this paper, we deeply observe social structure of sockpuppets and propose an online method to detect sockpuppets by introducing structural time-series features. The experimental results have achieved a precision of 0.901 and recall at 0.925, for a best F1 of 0.913, which can certify the method we proposed is efficient, relatively to other methods. In addition, our work can be used to online detect sockpuppets of large-scale social media platform with low time complexity. In the future work, we plan to further study other features of social relationships to help improving the efficiency, including the recovering speed, interval, etc.

Acknowledgment. This research is supported in part by the National Key Research and Development Program of China (No. 2017YFB1010000).

References

1. Java, A., Song, X., Finin, T., Tseng, B.: Why we Twitter: an analysis of a microblogging community. In: Zhang, H., et al. (eds.) SNAKDD/WebKDD -2007. LNCS (LNAI), vol. 5439, pp. 118–138. Springer, Heidelberg (2009). https://doi.org/10.1007/978-3-642-00528-2_7
2. Liu, D., Wu, Q., Han, W., et al.: Sockpuppet gang detection on social media sites. Front. Comput. Sci. **10**(1), 124–135 (2016)
3. Dani, H., Li, J., Liu, H.: Sentiment informed cyberbullying detection in social media. In: Ceci, M., Hollmén, J., Todorovski, L., Vens, C., Džeroski, S. (eds.) ECML PKDD 2017. LNCS (LNAI), vol. 10534, pp. 52–67. Springer, Cham (2017). https://doi.org/10.1007/978-3-319-71249-9_4
4. Li, Y., Cai, Y., Leung, H., Li, Q.: Improving short text modeling by two-level attention networks for sentiment classification. In: Pei, J., Manolopoulos, Y., Sadiq, S., Li, J. (eds.) DASFAA 2018. LNCS, vol. 10827, pp. 878–890. Springer, Cham (2018). https://doi.org/10.1007/978-3-319-91452-7_56
5. Hosseinia, M., Mukherjee, A.: Detecting Sockpuppets in Deceptive Opinion Spam. arXiv preprint arXiv:1703.03149 (2017)
6. Koppel, M., Schler, J., Argamon, S.: Authorship attribution in the wild. Lang. Resour. Eval. **45**(1), 83–94 (2011)
7. Tsikerdekis, M., Zeadally, S.: Online deception in social media. Commun. ACM **57**(9), 72–80 (2014)

8. Tsikerdekis, M., Zeadally, S.: Multiple account identity deception detection in social media using non-verbal behavior. IEEE Trans. Inf. Forensics Secur. **9**(8), 1311–1321 (2014)
9. Djuric, N., Zhou, J., Morris, R., et al.: Hate speech detection with comment embeddings. In: Proceedings of the 24th International Conference on World Wide Web, pp. 29–30. ACM (2015)
10. Afroz, S., Brennan, M., Greenstadt, R.: Detecting hoaxes, frauds, and deception in writing style online. In: 2012 IEEE Symposium on Security and Privacy (SP), pp. 461–475. IEEE (2012)
11. Kumar, S., Cheng, J., Leskovec, J., et al.: An army of me: sockpuppets in online discussion communities. In: Proceedings of the 26th International Conference on World Wide Web. International World Wide Web Conferences Steering Committee, pp. 857–866 (2017)
12. Maity, S.K., Chakraborty, A., Goyal, P., et al.: Detection of sockpuppets in social media. In: Companion of the 2017 ACM Conference on Computer Supported Cooperative Work and Social Computing, pp. 243–246. ACM (2017)
13. Meservy, T.O., Jensen, M.L., Kruse, J., et al.: Deception detection through automatic, unobtrusive analysis of nonverbal behavior. IEEE Intell. Syst. **20**(5), 36–43 (2005)
14. Solorio, T., Hasan, R., Mizan, M.: A case study of sockpuppet detection in Wikipedia. In: Proceedings of the Workshop on Language Analysis in Social Media, pp. 59–68 (2013)
15. Solorio, T., Hasan, R., Mizan, M.: Sockpuppet detection in Wikipedia: a corpus of real-world deceptive writing for linking identities. arXiv preprint arXiv:1310.6772 (2013)
16. Keselj, V., Peng, F., Cercone, N., Thomas, C.: N-gram based author profiles for authorship attribution. In: Proceedings of the Pacific Association for Computational Linguistics, pp. 255–264 (2003)
17. Wang, X., Gao, H., Wang, J., Yue, T., Li, J.: Detecting top-k active inter-community jumpers in dynamic information networks. In: Pei, J., Manolopoulos, Y., Sadiq, S., Li, J. (eds.) DASFAA 2018. LNCS, vol. 10827, pp. 538–546. Springer, Cham (2018). https://doi.org/10.1007/978-3-319-91452-7_35
18. Zheng, X., Lai, Y.M., Chow, K.P., et al.: Sockpuppet detection in online discussion forums. In: Intelligent Information Hiding and Multimedia Signal Processing (IIH-MSP), pp. 374–377. IEEE (2011)
19. Ruan, Y., Fuhry, D., Parthasarathy, S.: Efficient community detection in large networks using content and links. In: Proceedings of the 22nd International Conference on World Wide Web, pp. 1089–1098. ACM (2013)
20. Yu, Y., Yao, H., Wang, H., Tang, X., Li, Z.: Representation learning for large-scale dynamic networks. In: Pei, J., Manolopoulos, Y., Sadiq, S., Li, J. (eds.) DASFAA 2018. LNCS, vol. 10828, pp. 526–541. Springer, Cham (2018). https://doi.org/10.1007/978-3-319-91458-9_32
21. Bu, Z., Xia, Z., Wang, J.: A sock puppet detection algorithm on virtual spaces. Knowl.-Based Syst. **37**, 366–377 (2013)
22. Waseem, Z., Hovy, D.: Hateful symbols or hateful people? Predictive features for hate speech detection on Twitter. In: Proceedings of the NAACL Student Research Workshop, pp. 88–93 (2016)
23. Yamak, Z., Saunier, J., Vercouter, L.: Detection of multiple identity manipulation in collaborative projects. In: Proceedings of the 25th International Conference Companion on World Wide Web. International World Wide Web Conferences Steering Committee, pp. 955–960 (2016)

Accelerating Hybrid Transactional/Analytical Processing Using Consistent Dual-Snapshot

Liang Li[1], Gang Wu[1,3], Guoren Wang[2](✉), and Ye Yuan[1]

[1] Computer Science and Engineering, Northeastern University, Shenyang, China
liliang@stumail.neu.edu.cn

[2] Computer Science and Technology, Beijing Institute of Technology, Beijing, China
wanggr@bit.edu.cn

[3] State Key Laboratory for Novel Software Technology, Nanjing University, Nanjing, China

Abstract. To efficiently deal with OLTP and OLAP workload simultaneously, the conventional method is to deploy two separate processing systems, *i.e.*, the transaction-friendly designed OLTP system and the analytic-friendly optimized OLAP system. To maintain the freshness of the data, extra ETL tools are required to propagate data from the OLTP to the OLAP system. However, low-speed ETL processing is the bottleneck in those business decision support systems. As a result, there has been tremendous interest in developing hybrid transactional/analytical processing (HTAP) systems. This paper proposes a wait-free HTAP (WHTAP) architecture, that can perform both OLTP and OLAP requests efficiently in a wait-free form. We develop and evaluate a prototype WHTAP system. Our experiments show that the system can obtain a similar OLTP performance as the TicToc system and a four to six times acceleration in analytical processing at the same time.

Keywords: HTAP · In-memory database systems · Transaction concurrency control

1 Introduction

From a database perspective, there are two types of workloads, *i.e.*, online transaction processing (OLTP) and online analytical processing (OLAP). It is well known that OLTP and OLAP are difficult to process uniformly for a given system. The conventional approach for dealing with such hybrid workloads is to maintain a data warehouse for OLAP that is independent of the OLTP system. The data generated by OLTP systems are periodically transferred into the OLAP systems for analytical processing in a batch-wise fashion (*i.e.*, ETL tools), where ETL tools provide both excellent performance isolation between

G. Li et al. (Eds.): DASFAA 2019, LNCS 11446, pp. 52–69, 2019.
https://doi.org/10.1007/978-3-030-18576-3_4

the two workloads and the ability to tune each system independently. However, the traditional methods are not always timely enough when making business decisions.

In the last few years, organizations have grown increasingly interests in conducting real-time analyses of, and hence making decisions based on, up-to-date sets of raw data. As a result, the freshness of the data used for analysis becomes increasingly important. Recently, there has been tremendous interest in developing hybrid transactional/analytical processing (HTAP) systems. Gartner coined the term "(HTAP)" to describe this new type of database [8]. A similar term used to describe this type of processing is **operational analytics** [1], which indicates that insight and decision-making occur instantaneously with a transaction. HTAP offers 3 advantages: **1. Lower implementation complexity.** The implementation of HTAP no longer requires the consideration of data movement between transactional and analytical databases. **2. Lower analytic latency.** Analytic queries can access the latest transactional data to any extent needed, guaranteeing freshness in decision-making activities. **3. Less data duplication.** Since data movement is avoided, it is possible to reduce data duplication.

Fortunately, the recently emerged database technologies, such as in-memory computing (IMC), have brought new opportunities for the support of hybrid workloads within one database instance. Currently, there are two main types of HTAP implementations: *true-HTAP* and *loose-form HTAP*.

1.1 true-HTAP

To handle the OLTP and OLAP workloads in one single system, multi-version concurrency control (MVCC) is the currently accepted approach [30]. MVCC guarantees a high degree of parallelism, since writes are never blocked by reads. If a tuple is updated, a new physical version of this tuple is created and stored alongside the old version by engaging a version chain, which allows the old version to remain available for the allowed readers.

Limitations. MVCC does not distinguish between transaction types. Both short-running OLTP transactions and long-running read-oriented OLAP queries are treated in the same way on the same database. Obviously, it is inefficient to treat both OLTP and OLAP workloads similarly. First, scan-heavy OLAP queries require a substantial amount of time to deal with long version chains [30]. Second, costly garbage collections are inevitable in the recycling of unattended data.

Therefore, uniform processing approaches, such as MVCC, cannot meet the requirements of HTAP workloads, which consist of transactions of inherently different natures.

1.2 Loose-Form HTAP

As mentioned above, *true-HTAP*, which treats OLTP and OLAP equally, is not a suitable solution for the unsatisfactory performance of this approach. To the best of our knowledge, recently published studies typically classify queries based on the query type and execute the queries in separate replicas within a single system (*loose-form*). Two methods are used to organize the replicas of OLTP and OLAP.

- **Copy on Write.** The system call of fork obtain a virtual snapshot of the OLTP data as an OLAP replica, as in Hyper [13]. However, OLTP and OLAP replica data own the same physical data layouts, which is not effective for optimizing these methods simultaneously. Furthermore, the granularity of the fork is too coarse, which significantly influences performance based on dataset size. In this sense, the fork is not a particularly flexible solution. Anker [37] proposed a fine-grained system-level virtual snapshot system call (similar to the fork), but this method must be applied at the column storage level.
- **Recording and merging delta snapshot.** The recently published AIM [2] and BatchDB [26] both employ the delta snapshot method for data propagation. These methods require additional memory space to record transactions, generate delta snapshots over certain intervals and then merge the snapshots into the OLAP datasets. Note that BatchDB exploits a special in-memory log to record delta updates as a kind of delta snapshot. The most typical industry system is SAP HANA [7,38]. HANA holds a write-optimized delta store to collect the insert and delete operations and then merges the stores into a read-optimized and immutable main store.

1.3 Motivation

Apparently, the delta snapshot scheme is more flexible. However, two challenges are encountered when using delta snapshots. (1) It is difficult to record delta snapshots without significantly affecting the OLTP's latency and throughput. (2) It is difficult to avoid merging snapshots without blocking OLAP queries. For interactive applications, it is better to run both OLTP and OLAP in a wait-free manner. To the best of our knowledge, the state-of-the-art systems both in academia and industry have not yet addressed the wait-free execution of OLTP and OLAP. Is it possible to run both transactions and analysis in a wait-free manner and to perform transactions and analysis simultaneously while ensuring good performance?

This paper presents a wait-free HTAP (WHTAP) architecture. The key feature of the WHTAP architecture is a primary-secondary replication plus a dual snapshot. The primary replica is dedicated for OLTP, whereas the secondary replica is dedicated to OLAP workloads. The OLTP replica and OLAP replica are then connected by a dual-snapshot. Hence, WHTAP can independently optimize OLTP and OLAP according to their workload characteristics while physically isolating their resources. WHTAP not only guarantees both short transactions and long-running analytical queries execute in wait-free manner, but also ensures data freshness besides the consistency and isolation.

In summary, this work results in the following contributions:

- **Dual snapshot-based engine.** We design a dual snapshot structure in the storage engine to ensure the data freshness and wait-free features of WHTAP (Sect. 3).
- **LSM-like query layer.** To ensure that the analytical queries can always access the latest data, we propose an LSM-like query algorithm with a state controller, which ensures that analysis transactions run in a wait-free mode and access the freshest data (Sect. 4).
- **High-performance wait-free HTAP (WHTAP) system.** We implement a prototype and open the source code on Github[1], which can ensure the serialization of OLTP and OLAP at the snapshot isolation level. Compared with the traditional single-engine (*true-HTAP*) method, WHTAP can obtain a similar performance as previously documented for OLTP while delivering an OLAP performance that is approximately 4 to 6 times better relative to values recorded when using single engine. (Sect. 5).

The remainder of the paper is organized as follows: Sect. 2 shows the architecture of the WHTAP system, and Sects. 3 and 4 give the details of the components of WHTAP. Section 5 evaluates several experiments with a modified YCSB benchmark. Related work is discussed in Sect. 6. Section 7 presents our conclusion.

2 Dual Snapshot-Based Architecture

In this paper, we propose a dual delta snapshot-based structure for managing mixed workloads. In this structure, both the transactions and analytical queries are handled in isolation mode. In other words, distinct data replicas are used by OLTP and OLAP.

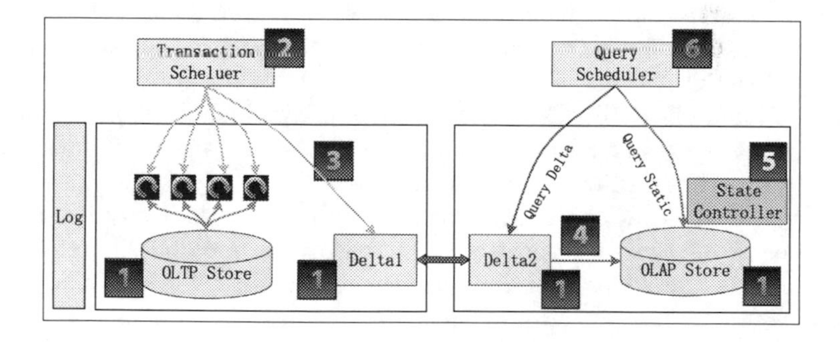

Fig. 1. Architecture of WHTAP

At the highest level, the system runs alternately between two periods. As shown in Fig. 1, we assume that in period one, the transaction data are recorded

[1] https://github.com/bombehub/WHTAP.git.

into delta snapshots $delta_1$ and $delta_2$, which will be merged into the OLAP data store. Then, in the next period, the roles of $delta_1$ and $delta_2$ are exchanged. That is, the delta snapshot remains $delta_2$, while $delta_1$ is merged into the OLAP data store. The data freshness of the WHTAP system can be guaranteed by minimizing the duration of each period.

As is known, taking a delta snapshot will lead to a throughput loss and latency spike, and merging a delta snapshot will break the OLAP's read-only feature. Therefore, the challenge in design is how to record a delta snapshot (Component 3) but not block the OLTP from running, followed by merging the delta snapshot (Component 4) but not blocking the OLAP running at the same time.

To overcome those challenge, we propose an architecture named WHTAP that possesses six total components: the logical components of OLTP are dual-snapshot storage engine and delta snapshot recording, and the OLAP part includes snapshot compaction, LSM-like query layer, and state controller.

- **Storage Engine.** OLTP and OLAP are stored in 2 replicas. Additionally, space to recode the delta data is necessary. This module is primarily responsible for the organization and storage of data.
- **Transaction Concurrency Control.** This control relates to scheduling of the running logic of the OLTP workload and guarantees that the structure is serializable and meets the desired performance.
- **Delta Snapshot.** To ensure the data freshness, the running data of the transaction must be transformed into an OLAP Store. To collect the snapshot, three difficulties must be overcome. First, the linear serializability between the OLTP and delta snapshot must be ensured; second, the litter performance must be disregarded as much as possible; and third, it is important not to cause a significant latency spike.
- **Compact Snapshot.** The dual snapshot structure must periodically merge the frozen snapshot data into the OLAP data. The difficulty that must be overcome here is merging of the snapshot without blocking the normal execution of the OLAP queries.
- **State Controller.** The WHTAP system maintains a state controller to ensure that the query runs correctly and is wait-free, and that the snapshot is merged within a suitable amount of time.
- **Query Execution.** The OLAP data are updated periodically by the delta snapshot data, thus ensuring the high performance of OLAP queries and guaranteeing the snapshot isolation level. Doing so while also maintaining the wait-free feature remains a challenge that must be addressed.

3 OLTP Components

3.1 Storage Engine

It is better for a system to expose one single schema while maintaining two data replicas for OLTP and OLAP. The storage engine should also employ two extra

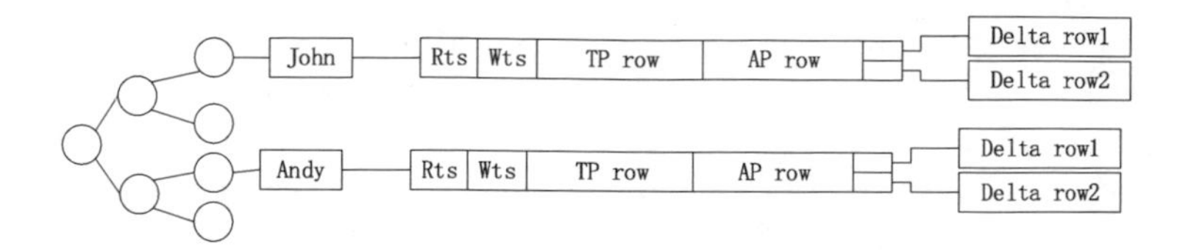

Fig. 2. Dual snapshot-based storage engine.

memory spaces to record the delta snapshots. For each tuple, we use two counters to represent the read and write timestamps for concurrency control, similar to the approach presented in TicToc [42]. The storage engine is organized in the form of a tuple level. Each logical tuple contains two static in-memory rows that store the OLTP and OLAP data, one set per row. Additionally, the logical tuple holds two copies of dynamic data, $delta_1$ and $delta_2$ (Fig. 2). The dynamic data are used to record increments in an alternating fashion. The use of dynamic memory can save memory footprint as desired, while static memory can help avoid memory allocation times.

Many indexes have been designed to organize in-memory row data, such as the Adaptive Radix Tree (ART) [18,19], BwTree [20], Masstree [27], and SkipList [34]. [40] gives an in-depth performance evaluation of the state-of-the-art indexes. How the appropriate index is chosen is beyond the scope of this paper. To access the table in a fast and easy manner, we can exploit a hash index or use a memory-optimized b+ tree for organizing the tuples. The OLTP, OLAP and dual delta snapshot components share the same index.

3.2 Transactions and Concurrency Control

Component 2 (Fig. 1) entails choosing an OCC-based protocol to schedule the OLTP workload. One of the largest advantages of using the optimistic concurrency control protocol for main memory DBMSs is that the contention period is short, because transactions write their updates to shared memory only at the commit time [15]. This outstanding feature could be combined with a recording snapshot, *i.e.*, each write operation occurring in the write phase will be accompanied by a **write_delta** (details are available in Sect. 3.3) operation synchronously. As a result, we chose OCC as WHTAP's transaction execution protocol. Our baseline OCC protocol is specified in TicToc [42], which details a new concurrency control algorithm that achieves higher concurrency than state-of-the-art T/O schemes. Another popular OCC protocol that can be used to replace TicToc is Silo [39]. Although our algorithms can also be adapted to the multiple-version concurrent control, we abandon MVCC in designing our WHTAP system due to its high overhead memory cost.

3.3 Delta Snapshot

Our concurrency control algorithm falls under the framework of TicToc, although we must record the delta snapshots in conjunction with transaction execution.

We integrate Pingpong [3] (or Hourglass [24]) with TicToc. Each OLTP's write instruction (only appearing in the write phase) should not only write to the OLTP data but also write to a particular delta snapshot according to the pointer, as introduced in Pingpong. Both Pingpong and Hourglass can be combined with a virtual snapshot approach [23,35], *i.e.*, when we take virtual snapshots into consideration, once the pointers have been exchanged, this approach only affects the behavior of the new transactions and has no effect on the current running transactions.

Algorithm 1. Transaction Execution Thread

```
   Input: Transaction T
1  Read Phase
2  if Validation Passed then
3      pointer = p_update
4      for each request in T.writeset do
5          write(index(request.key))
6          malloc_delta(pointer)
7          write_delta(index(request.key), pointer)
8          KeySet.insert(request.key)
9          index(request.key).wts = commit_ts
10         index(request.key).rts = commit_ts
11         unlock(index(request.key))
```

Since TicToc accesses OLTP data only during the write phases, we can improve the algorithms by making WHTAP only determine to which period (odd or even period) the time point (at the beginning of the write phase) belongs. That is, pointer swapping (see Algorithm Sect. 2 Line 5) only affects the transactions that extend into the write phase after the pointer swapping moment.

Algorithm 1 presents the details of the transactions' concurrency control and the delta recording process. The read and validation phase is the same with TicToc, so the details regarding line 1 and line 2 can be found in [42]. The primary difference between WHTAP and TicToc exists in the write phase. Once deep into the write phase, the transaction first decides which period it belongs to (line 3) and then determines whether the delta snapshot should be recorded in $delta_1$ or $delta_2$. The thread commits the writeset data to OLTP data (line 5) and the corresponding periodic delta snapshot (lines 6 and 7). Memory needs to be allocated dynamically (line 6) before each delta recording operation. Considering that malloc is very time-consuming, we set $delta_1$ and $delta_2$ to be static memory spaces. Since the storage engine shares the index structure, our approach requires an additional KeySet to record the modified keys (line 8) during the current period. If we maintain the index structure separately for the delta structure, then the KeySet here is unnecessary.

4 OLAP Components

4.1 Compact Snapshot and State Controller

Section 3.3 describes an approach for freezing delta snapshots interchangeably. With the help of the dual snapshot scheme, it is possible to always maintain the data freshness of the OLAP component by frequently compacting the frozen data in the snapshot into the OLAP store. However, the approach that must be used to compact the delta snapshot into the existing OLAP data without blocking the execution of the OLAP query and while maintaining the model's high performance is a challenge.

We propose to solve the problem and make OLAP wait-free by distinguishing and scheduling five states/phases, *i.e.*, *NORMAL*, *FROZEN*, *WAITING*, *COMPACTION*, and *GC* phase, between consecutive pointer swapping events.

***NORMAL* Phase.** In the *NORMAL* phase, every update transaction that reaches the write phase should write to both the OLTP store and the delta snapshot (lines 5 and 7 in the Algorithm 1). Long-running OLAP queries must directly consume data in the OLAP store in a consistent manner. Therefore, the time duration of this phase becomes a key factor with regard to the data freshness of OLAP.

***FROZEN* Phase.** The *FROZEN* phase occurs after the *NORMAL* phase. The delta snapshot must be frozen in this phase. Accordingly, The first step is to swap the *p_update* and *p_delta* pointers, in other words, to exchange the dual snapshot roles. As an exception, the transactions with write phases that begin in the *NORMAL* phase but have not yet been committed at this point (called *dirty* transaction) should ignore the role change operations. The completion of all the dirty transactions denote the end time point of the *FROZEN* phase. At the end boundary of the *FROZEN* phase, the delta snapshot is frozen, meaning that no OLTP transaction should write data to this dataset. We can see that at the beginning of the *FROZEN* phase, all of the new OLTP transaction write data should be allowed to write to another delta snapshot.

A global counter is needed to identify the end times of all of the dirty transactions. The use of a global counter will affect performance, this element behaves as a bottleneck in a multi-core system [39]. Because the OLTP transaction is usually short, we assume that the OLTP transaction is less than 1 μs, we simply wait for 1 μs because the transaction time under the memory database will be very short. It should be noted that this time can be directly adjusted. The specific code corresponds to line 6 in the Algorithm 2.

***WAITING* Phase.** The Waiting phase begins at the completion of the *FROZEN* phase and includes all transactions started when the *NORMAL* phase commenced. At the beginning of the *WAITING* phase, the delta frozen snapshot is generated. The snapshot cannot be compacted into the OLAP data because several active OLAP queries are still running; hence, we must wait for all active OLAP queries to finish (*i.e.*, *static_counter* = 0). Because those transactions are queries from OLAP data, the data should not be able to write. Once those queries have finished, we reach the end point of the *WAITING* phase.

We innovatively introduce a Log Structure Merge tree [12] approach for query execution. Here, the frozen delta snapshot can be regarded as a *MemTable* like that in the LSM tree, while OLAP data storage can be regarded as an *SSTable*. Once the delta snapshot is generated (*i.e.*, has been frozen), the system compacts the snapshot into the OLAP data. Any query transaction must first search the delta and then access the OLAP data if the result is not found in the delta. The queries that begin in the *WAITING* phase should query the delta first. For the wait-free consideration, since the "MemTable" has been generated, we can query more fresh data from the delta and then can merge the data into the OLAP (SSTable).

***COMPACTION* Phase.** The *WAITING* phase ends when all queries begun in the *GC* and *NORMAL* and *FROZEN* phases have been committed. Next, the system enters the *COMPACTION* phase. Once this phase begins, the delta data are traversed sequentially according to the keyset, then compacted into the OLAP data set. If the OLAP storage engine is column-major designed, this approach will be more complicated.

For querying transactions, we still require an LSM-like query approach. First, we query the frozen delta, and if the result is found, we return it directly; otherwise, we continue the search in the OLAP storage. We can even add a bloom filter to the frozen delta.

Note that in this process, because our OLAP query looks up the delta first, if the query finds the result and hence does not need to look up the OLAP data, then the compaction work can be conducted in a lock-free manner and will not cause any blocks.

***GC* Phase.** GC is the last phase of the cycle and immediately follows the *COMPACTION* phase. Hence, any query beginning in the *GC* phase can read the latest data directly from the OLAP. As the frozen delta snapshot has been successfully compacted, the snapshot can be released just like the process of "Garbage Collection". The process remains in the *GC* phase until the completion of all queries issued during the Waiting and Compaction phases. When the *GC* phase finished, the system has now entered the next cycle of the *NORMAL* phase.

Algorithm Code. As shown in the Algorithm 2, the system alternates between five phases. From the normal to the frozen phase (line 3), we can control the duration of this phase. If the time duration is too long, the data of the OLAP will be out of date, because the data freshness will not be good enough. Once the system enters the *FROZEN* phase, we must exchange the roles of the dual-snapshot and wait for the dirty transaction to commit. Note that in line 6 of the code, for performance reasons, we can directly wait for a period of time (such as 1 μs) to replace the function to ensure that the dirty transactions are all committed. For the duration of the waiting phase, we need to use the counter

Algorithm 2. State Controller and Compaction Work

```
1  while true do
2      State = NORMAL
3      Waiting for the delta snapshot frozen signal
4      State = FROZEN
5      Swap(p_update, p_delta)
6      Wait Until Dirty Transaction Finished
7      State = WAITING
8      wait until static_counter = 0
9      State = COMPACTION
10     Compact(p_delta, AP)
11     State = COMPLETE
12     Gargage_Colection(p_delta)
13     wait until delta_counter = 0
```

to determine whether the transactions (which were queried directly from the OLAP) are all committed. The compact function (line 10) must be used to merge the data by engaging KeySet. In the GC phase, the main goal is to free useless data and wait until the LSM-like queries have finished.

4.2 LSM-Like Query Layer

In Sect. 3.2, we discuss approaches to handling the OLTP workload; in this part, we describe the OLAP workload running process.

Once a query request is accepted, the system detects the state phase in which it was begun. For performance reasons, only in the *WAITING* and *COMPACTION* phases do we execute the LSM-like query strategies; in contrast, in the *NORMAL*, *FROZEN* and *GC* phases, we query the OLAP data directly. As shown in the Algorithm 3, line 5 **Query_Static()** and line 11 **Query_DeltaFirst()** represent two different query strategies. The 2 counters, static_counter and delta_counter, are essential for identifying the state in which the system is functioning.

Algorithm 3. Query Execution Thread

```
   Input: Query Q
1  start_state = State
2  if start_state = GC||NORMAL||FROZEN then
3      fetch_and_add(static_counter)
4      for each request in Q do
5          Query_Static(index(request.key))
6      fetch_and_sub(static_counter)
7  else
8      if start_state = WAITING||COMPACTION then
9          fetch_and_add(delta_counter)
10         for each request in Q do
11             Query_DeltaFirst(index(request.key)) //LSM-alike query
12         fetch_and_sub(delta_counter)
```

The performance associated with looking up a delta first is worse (slower) than the performance achieved by querying OLAP data directly. Fortunately, **Query_DeltaFirst()** is a relatively rare event, because the frozen and compact phases are short. The next section elaborates on this idea.

4.3 Running Example

To enhance clarity in our explanation of the system, the Fig. 3 is presented, detailing a running example of our WHTAP algorithm. This figure fully illustrates the five phases in a cycle and shows several types of transactions running.

As shown in the Fig. 3, green represents each transaction that successfully enters the write phase within the current cycle. Write transactions for the previous period are shown in yellow. After the system triggers the frozen signal (at time t_1), the new transaction entering the write phase (green, T_2, T_3, T_4, T_5, T_6, T_7) records the delta into another data set. Transactions that have not been completed (T_2) continue to execute; once they are finished, the system exits the frozen phase.

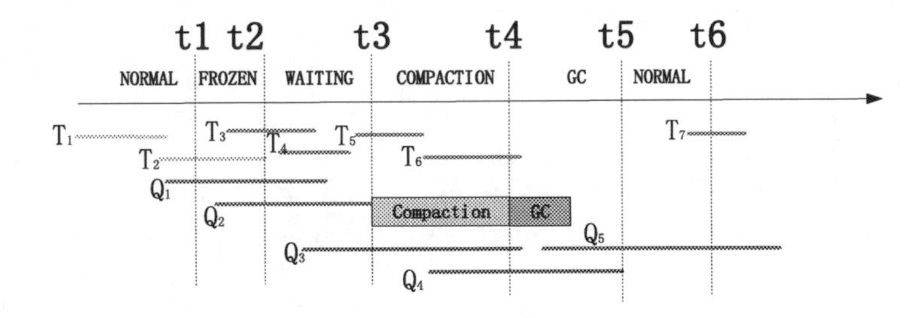

Fig. 3. Running example of WHTAP. (Color figure online)

At time t_2, the system has obtained a stable delta snapshot of the previous cycle. Next, we can compact the delta snapshot from the previous cycle into the OLTP storage. Unfortunately, several active OLAP queries (Q_1, Q_2) remain uncommitted. As a result, the delta snapshot may overwrite this part of the data during direct compact, causing a query error. Therefore, we must invoke a waiting phase to wait for the blue query transaction in the figure to complete. For queries that are started during the waiting phase (Q_3), we need to query the delta snapshot first to execute the OLAP queries, since this approach will ensure that the write phase of the *COMPACT* phase can merge in a lock-free manner. Once transactions Q_1 and Q_2 are committed, the system can start the compact operation. The yellow rectangle in the figure represents the *COMPACT* operation, which basically merges the incremental transaction data for the previous cycle into the OLAP store. Similarly, any query started in this phase is still queried in a similar way as those in LSM. Once this phase is over and the incremental data from the previous cycle have become useless, we can naturally delete these data (blue rectangle). At the same time, we must also wait for the end of the transaction for the LSM-style query (Q_4). Once that point has been reached, we can proceed to the next normal phase of the cycle and once more wait for the frozen signal to trigger.

5 Experimental Study

This section implements and evaluates the WHATP prototype. The prototype is based on DBx1000 OLTP DBMS [41]. To integrate our WHTAP algorithm into the DBx1000 system, we need to modify the storage engine and the concurrent processing of the OLTP transaction and then must separate the OLAP queries from the OLTP component, and add the state controller and LSM-like OLAP query methods. The prototype allows us to compare five approaches all within the same system: TicToc [42], SILO [39], HEKATON [5], Basic MVCC and WHTAP. All of the experiments are run on a high-end server, which is equipped with two E7-4820 CPU sockets each with 40 physical sockets, 512 GB of memory, and 1 TB of hard disk drive space. CentOS 7.3 X86_64 with Linux kernel 3.10 and g++ 4.8 is installed.

5.1 Benchmark Setup

The experiments focus on OLTP and OLAP mixed transactions. The performance of the HTAP workloads was evaluated with the YCSB [4] benchmark. Because YCSB is used to evaluate the OLTP workload, we must modify the origin benchmark to support the evaluation of both transaction and long-running read-only queries. First, we prepared a single table database for the experiments. The schema of the database table was configured to own 11 columns. The first column is the primary key, and the following 10 columns store randomly generated string values. Initially, 10M records were bulk-loaded into the table. The accesses (reads or writes) of records follow a Zipfian distribution that can be controlled by parameter θ (reflecting the level of contention).

For OLTP transactions, 16 accesses per transaction (50% reads and 50% writes) with a hotspot of 10% tuples accessed by 75% of all queries ($\theta = 0.9$). For OLAP queries, 48 queries (100% reads) per transaction and a uniform access distribution ($\theta = 0$).

The first group evaluation fixes the number of threads in OLAP and tests the performance impact of the number of OLTP thread. In the second group evaluation, the number of threads in the OLTP is fixed, and the performance impact of the number of OLAP threads is tested.

5.2 Fixing OLAP Threads

We fix the number of OLAP threads to 8 and then fix the length of each OLAP query to 48. Figure 4(a) shows the performance results of OLTP when the number of OLAP threads is fixed. The horizontal axis corresponds to the number

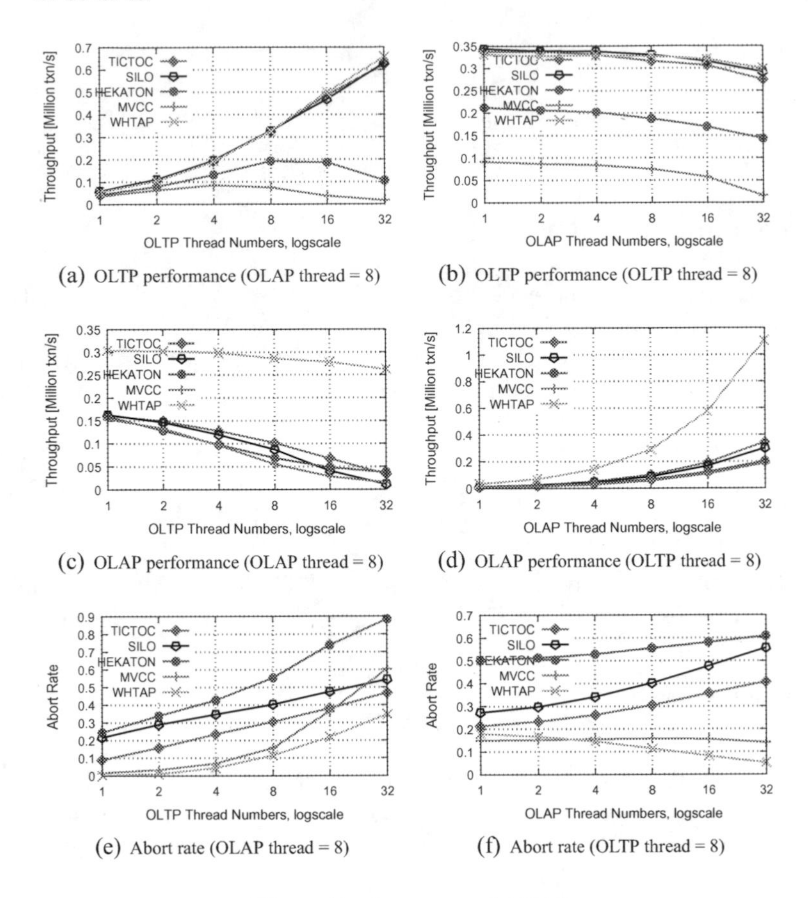

Fig. 4. Performance of HTAP

of threads in OLTP from 1 to 32, which shows that the OLAP workload has little impact on OLTP performance. Figure 4(c) shows the performance results of OLAP in which the number of OLAP threads is 8, and the number of OLTP threads is from 1 to 32. The performance of OLAP is 2–3 times that of other algorithms. Increasing the number of OLTP threads in the WHTAP algorithm has little impact on OLAP performance, which effectively does not decrease with the increase of OLTP threads. In the traditional concurrency control algorithm, as the number of OLAP threads is increased, the performance decreases of the algorithm decreases significantly. In particular, even MVCC and HEKATON are suitable for reading operations, but the performance of OLAP remains poor due to version chain scanning. Figure 4(e) gives the effect of the number of OLAP threads on the abort rate. As it can be seen, our algorithm has the lowest abort rate (substantially lower than that of TicToc), because our algorithm's OLAP thread is executed on the snapshot and does not require abort at all.

5.3 Fixing OLTP Threads

Figure 4(b) shows the performance results of OLTP when the number of OLTP threads is fixed. The horizontal axis corresponds to the number of threads of OLAP and ranges from 1 to 32. When the OLAP's workload is increased, the OLTP performance decreases accordingly. The performance of the OCC-based single-version concurrent control algorithm should be comparatively worse. As the number of OLAP threads increases, the performance of OLTP decreases. MVCC decreases significantly, while the OCC scheme decreases by a small amount. The OLTP workload performance of WHTAP is slightly worse than that of TicToc. Figure 4(d) shows the performance results of OLAP when the number of OLTP threads is fixed. We can see that the performance of WHTAP has increased significantly, mainly because the algorithm benefits from technologies that are processed separately from OLTP and OLAP. When the number of OLAP threads is 32, WHTAP has 3.2 times the performance of TicToc and 5.5 times that of HEKATON. Figure 4(f) shows the number of system rollback transactions when the number of OLTP threads is fixed. As the OLAP workload increases, the number of WHTAP rollback transactions is trending downward, because OLAP transactions are certain to execute and do not abort. The read-only queries of other algorithm can also result in abort events.

Combining the above results, we obtain the following findings:

When both OLTP and OLAP workloads exist at the same time in a given scenario, WHTAP outperforms other algorithms. The performance of the OLTP part is close to that of TicToc (Fig. 4(a) and (b)), although WHTAP has an absolute advantage over the performance of the OLAP part (Fig. 4(c) and (d)).

- In the WHTAP algorithm, OLTP and OLAP workloads have the least impact on each other's performance (Fig. 4(c) and (b)).

6 Related Work

HTAP. Hybrid transaction/analytical processing (HTAP) was first defined by Gartner Inc. [8,9] and referred to a novel architecture that could effectively combine both types of processing to fulfill the emerging requests in informative and real-time decision making activities. A recent works [1,10,31] survey on this topic was conducted. HTAP emphasizes two main points: data freshness and unified data representation. Since the 2000s, many systems have targeted the HTAP market. Typical systems include SAP HANA [7,38], Hyper [30] and HYRIES [11], Peleton [32]. The other systems (including MemSQL [28], Hekaton [5] and Apollo [16]) can support OLTP and OLAP workload separately.

The effective design of an HTAP system remains an open problem that can be broadly divided into three categories. (1) One straightforward method is to use a single system to process the mixed workload. Hyper [30] proposed a novel MVCC implementation that can update in place and store prior versions before image deltas, enabling both an efficient scan execution and the fine-grained serializability validation needed for the rapid processing of point access transactions. From the NoSQL side, Pilman et al. [33] demonstrated how scans can be efficiently implemented on a key value store (KV store) to enable more complex analytics with large and distributed KV stores. (2) The second method is to create a copy of the main database, which is then used as an OLAP dataset. Specifically, Hyper [13] and Swingdb [29] use such an approach. (3) The 3rd method is to generate a delta snapshot, which is then periodically merged into the second replica. Specifically, BatchDB and SAP HANA do this.

In-Memory Concurrency Control. The traditional OCC [15] algorithm is found to be effective in the control of concurrency in OLTP with the advent of new high-performance DBMSs. For example, OCC variants are commonly used in in-memory database scenarios and include examples such as TicToc [42] and SILO [39].

MVCC and its variants constitute another type of concurrency control algorithm that has been widely adopted in in-memory databases. Such algorithms can be found in Hekaton [5,17], HyPer [30], Bohm [6], Deuteronomy [21,22] and ERMIA [14] and Cicada [25]. For a record, MVCC creates multiple version copies to reduce conflict between transactions and allow access to earlier versions of a record.

Frequent Snapshot. To obtain a consistent snapshot, Salles *et al.* [36] provides a comparison of several state-of-the-art snapshot algorithms. In this work, the authors concluded that Naive Snapshot is good for high-throughput workloads with small datasets, whereas Copy On Update is more widely applicable. Swingdb [29] and Hyper [13] work by modifying the Linux kernel to support a fine-grained fork-like system that is then called to generate the snapshot. Zigzag [3] was developed for use in MMO game scenarios, although it is suitable for use only with small datasets and is not a more generally applicable algorithm. Pingpong, Hourglass and Piggyback [24] all use a kind of pointer swapping technique to generate snapshots. Moreover, both Pingpong and Hourglass can be used to generate a delta snapshot, and therefore are useful with HTAP systems. It is important to note that all of these snapshot algorithms depend on a physical consistent time-point. In contrast, for more widely applicable cases, such as OLTP, to establish a physical consistent state in a running system, system blocks must be introduced. In the most recent work on this subject, CALC [35], the authors invent a virtual snapshot idea to solve the problem; in this case, it would be possible to integrate both Pingpong and Hourglass with this idea.

7 Conclusion

In this paper, we proposed a wait-free HTAP control protocol and implemented a prototype (WHTAP). We used WHTAP to exploit a dual-snapshot structure to isolate OLTP and OLAP workloads. The OLTP transaction was able to run in a serializable context, whereas OLAP was able to yield isolated snapshots. Our prototype enabled both OLTP and OLAP to simultaneously function in a high throughput manner, furthermore, both OLTP and OLAP could be executed in a wait-free manner. We evaluated the performance of our prototype using the modified YCSB benchmark to validate our performance with regard to the HTAP workload. Compared with the TicToc concurrency control protocol, WHTAP can achieve an OLTP performance that is similar to that of TicToc, and in contrast, the performance of OLAP with our algorithm is approximately four times that of TicToc. We strongly recommend that developers in scenarios in which significant data quantities are processed use the WHTAP architecture for scenarios that require real-time analysis and processing, such as "Amazon's Prime Day".

Acknowledgements. Guoren Wang is the corresponding author of this paper. Guoren Wang is supported by the NSFC (Grant No. U1401256, 61732003, 61332006 and 61729201). Gang Wu is supported by the NSFC (Grant No. 61872072) and the State Key Laboratory of Computer Software New Technology Open Project Fund (Grant No. KFKT2018B05).

References

1. Bohm, A., Dittrich, J., Mukherjee, N., Pandis, I., Sen, R.: Operational analytics data management systems. PVLDB **9**(13), 1601–1604 (2016)
2. Braun, L., et al.: Analytics in motion: high performance event-processing and real-time analytics in the same database. In: SIGMOD, pp. 251–264 (2015)
3. Cao, T., et al.: Fast checkpoint recovery algorithms for frequently consistent applications. In: SIGMOD, pp. 265–276 (2011)
4. Cooper, B.F., Silberstein, A., Tam, E., Ramakrishnan, R., Sears, R.: Benchmarking cloud serving systems with YCSB. In: SoCC, pp. 143–154. ACM (2010)
5. Diaconu, C., et al.: Hekaton: SQL server's memory-optimized OLTP engine. In: SIGMOD, pp. 1243–1254. ACM (2013)
6. Faleiro, J.M., Abadi, D.J.: Rethinking serializable multiversion concurrency control. PVLDB **8**(11), 1190–1201 (2015)
7. Farber, F., Cha, S.K., Primsch, J., Bornhovd, C., Sigg, S., Lehner, W.: SAP HANA database: data management for modern business applications. Sigmod Rec. **40**(4), 45–51 (2012)
8. Gartner: hybrid transaction/analytical processing will foster opportunities for dramatic business innovation. https://www.gartner.com/doc/2657815/hybrid-transactionanalytical-processing-foster-opportunities
9. Gartner: market guide for HTAP-enabling in-memory computing technologies. https://www.gartner.com/doc/3599217/market-guide-htapenabling-inmemory-computing
10. Giceva, J., Sadoghi, M.: Hybrid OLTP and OLAP. In: Sakr, S., Zomaya, A. (eds.) EBDT. Springer, Cham (2018). https://doi.org/10.1007/978-3-319-63962-8
11. Grund, M., Krüger, J., Plattner, H., Zeier, A., Cudre-Mauroux, P., Madden, S.: HYRISE: a main memory hybrid storage engine. PVLDB **4**(2), 105–116 (2010)
12. Jagadish, H.V., Narayan, P.P.S., Seshadri, S., Sudarshan, S., Kanneganti, R.: Incremental organization for data recording and warehousing. In: VLDB 1997 (1997)

13. Kemper, A., Neumann, T.: HyPer: a hybrid OLTP&OLAP main memory database system based on virtual memory snapshots. In: ICDE, pp. 195–206. IEEE (2011)
14. Kim, K., Wang, T., Johnson, R., Pandis, I.: ERMIA: fast memory-optimized database system for heterogeneous workloads. In: SIGMOD, pp. 1675–1687 (2016)
15. Kung, H.T., Robinson, J.T.: On optimistic methods for concurrency control. ACM Trans. Database Syst. **6**(2), 213–226 (1981)
16. Larson, P.A., Birka, A., Hanson, E.N., Huang, W., Nowakiewicz, M., Papadimos, V.: Real-time analytical processing with SQL server. Proc. VLDB Endow. **8**(12), 1740–1751 (2015)
17. Larson, P., Blanas, S., Diaconu, C., Freedman, C.S., Patel, J.M., Zwilling, M.: High-performance concurrency control mechanisms for main-memory databases. PVLDB **5**(4), 298–309 (2011)
18. Leis, V., Kemper, A., Neumann, T.: The adaptive radix tree: ARTful indexing for main-memory databases. In: ICDE. pp. 38–49 (2013)
19. Leis, V., Scheibner, F., Kemper, A., Neumann, T.: The art of practical synchronization. In: DaMoN, pp. 1–8 (2016)
20. Levandoski, J.J., Lomet, D.B., Sengupta, S.: The Bw-Tree: a B-tree for new hardware platforms. In: ICDE, pp. 302–313 (2013)
21. Levandoski, J.J., Lomet, D.B., Sengupta, S., Stutsman, R., Wang, R.: High performance transactions in deuteronomy. In: CIDR (2015)
22. Levandoski, J.J., Lomet, D.B., Sengupta, S., Stutsman, R., Wang, R.: Multi-version range concurrency control in deuteronomy. PVLDB **8**(13), 2146–2157 (2015)
23. Li, L., Wang, G., Wu, G., Yuan, Y., Chen, L., Lian, X.: A comparative study of consistent snapshot algorithms for main-memory database systems. ArXiv e-prints, October 2018
24. Li, L., Wang, G., Wu, G., Yuan, Y.: Consistent snapshot algorithms for in-memory database systems: experiments and analysis. In: ICDE, pp. 1284–1287 (2018)
25. Lim, H., Kaminsky, M., Andersen, D.G.: Cicada: dependably fast multi-core in-memory transactions. In: SIGMOD, pp. 21–35 (2017)
26. Makreshanski, D., Giceva, J., Barthels, C., Alonso, G.: BatchDB: efficient isolated execution of hybrid OLTP+OLAP workloads for interactive applications. In: SIGMOD, pp. 37–50 (2017)
27. Mao, Y., Kohler, E., Morris, R.T.: Cache craftiness for fast multicore key-value storage. In: ACM European Conference on Computer Systems, pp. 183–196 (2012)
28. MemSQL: MemSQL. https://www.memsql.com/
29. Meng, Q., Zhou, X., Chen, S., Wang, S.: SwingDB: an embedded in-memory DBMS enabling instant snapshot sharing. In: Blanas, S., Bordawekar, R., Lahiri, T., Levandoski, J., Pavlo, A. (eds.) IMDM/ADMS -2016. LNCS, vol. 10195, pp. 134–149. Springer, Cham (2017). https://doi.org/10.1007/978-3-319-56111-0_8
30. Neumann, T., Muhlbauer, T., Kemper, A.: Fast serializable multi-version concurrency control for main-memory database systems. In: SIGMOD, pp. 677–689 (2015)
31. Ozcan, F., Tian, Y., Tozun, P.: Hybrid transactional/analytical processing: a survey. In: SIGMOD, pp. 1771–1775 (2017)
32. Pelotondb: Pelotondb. https://pelotondb.io/
33. Pilman, M., Bocksrocker, K., Braun, L., Marroquín, R., Kossmann, D.: Fast scans on key-value stores. PVLDB **10**(11), 1526–1537 (2017)
34. Pugh, W.: Skip lists: a probabilistic alternative to balanced trees. In: The Workshop on Algorithms & Data Structures, pp. 668–676 (1990)

35. Ren, K., Diamond, T., Abadi, D.J., Thomson, A.: Low-overhead asynchronous checkpointing in main-memory database systems. In: SIGMOD, pp. 1539–1551. SIGMOD (2016)
36. Salles, M.A.V., et al.: An evaluation of checkpoint recovery for massively multiplayer online games. PVLDB **2**(1), 1258–1269 (2009)
37. Sharma, A., Schuhknecht, F.M., Dittrich, J.: Accelerating analytical processing in MVCC using fine-granular high-frequency virtual snapshotting. In: SIGMOD (2017)
38. Sikka, V., Farber, F., Goel, A.K., Lehner, W.: SAP HANA: the evolution from a modern main-memory data platform to an enterprise application platform. PVLDB **6**(11), 1184–1185 (2013)
39. Tu, S., Zheng, W., Kohler, E., Liskov, B., Madden, S.: Speedy transactions in multicore in-memory databases. In: SOSP, pp. 18–32 (2013)
40. Xie, Z., Cai, Q., Chen, G., Mao, R., Zhang, M.: A comprehensive performance evaluation of modern in-memory indices. In: ICDE, pp. 641–652. IEEE (2018)
41. Yu, X., Bezerra, G., Pavlo, A., Devadas, S., Stonebraker, M.: Staring into the abyss: an evaluation of concurrency control with one thousand cores. PVLDB **8**(3), 209–220 (2014)
42. Yu, X., Pavlo, A., Sanchez, D., Devadas, S.: TicToc: time traveling optimistic concurrency control. In: SIGMOD, pp. 1629–1642 (2016)

HSDS: An Abstractive Model for Automatic Survey Generation

Xiao-Jian Jiang[1,2], Xian-Ling Mao[1(✉)], Bo-Si Feng[1], Xiaochi Wei[3], Bin-Bin Bian[1], and Heyan Huang[1]

[1] Department of Computer Science and Technology, Beijing Institute of Technology, Beijing 100081, China
{xjjiang,maoxl,2120160986,bianbinbin,hhy63}@bit.edu.cn
[2] CETC Big Data Research Institute, Guiyang 550008, China
[3] Baidu Inc., Beijing 100193, China
weixiaochi@baidu.com

Abstract. Automatic survey generation for a specific research area can quickly give researchers an overview, and help them recognize the technical developing trend of the specific area. As far as we know, the most relevant study with automatic survey generation is the task of automatic related work generation. Almost all existing methods of automatic related work generation extract the important sentences from multiple relevant papers to assemble a related work. However, the extractive methods are far from satisfactory because of poor coherence and readability. In this paper, we propose a novel abstractive method named **H**ierarchical **S**eq2seq model based on **D**ual **S**upervision (HSDS) to solve problems above. Given multiple scientific papers in the same research area as input, the model aims to generate a corresponding survey. Furthermore, we build a large dataset to train and evaluate the HSDS model. Extensive experiments demonstrate that our proposed model performs better than the state-of-the-art baselines.

Keywords: Abstractive · Survey · Dual Supervision

1 Introduction

It is very important to make a survey for a specific research area, because the researchers can quickly grasp the corresponding area and recognize the technical developing trend through reading the survey. However, current surveys are almost made by reading lots of papers and summarizing them manually, which is time-consuming and costly. Taking the Gartner Inc.[1] for example, it employs about 6,600 professionals, including more than 1,500 research analysts and consultants, to spend a lot of time reading and summarizing the latest papers into surveys. Therefore, it is highly desirable to automatically make surveys.

[1] https://www.gartner.com/en.

G. Li et al. (Eds.): DASFAA 2019, LNCS 11446, pp. 70–86, 2019.
https://doi.org/10.1007/978-3-030-18576-3_5

The related work can be regarded as a survey for a specific area, thus the related work is a good annotated survey for its corresponding cited papers. To study how to automatically obtain a related work, several methods [5,10,11] have been proposed. They extract important sentences from multiple original papers to assemble related work. Unfortunately, these works belong to extractive methods, which simply select and reassemble sentences from the original papers. They hence face the drawbacks of information redundancy and incoherence between sentences. For example, we compare the results of survey generated by state-of-the-art extractive method called SummaRuNNer [16] and man-made survey in Fig. 1. In the result of extractive survey, the first and the second sentence have the same sentence pattern "*We propose (introduce) ...*" without considering the real interaction between these two works. However, the concatenation between the descriptions of different works in the man-made survey is much smoother because of "*This trend*". What's more, extractive survey selects one sentence from one of the original papers each time, thus, each sentence in extractive survey has only a citation. While man-made survey can simultaneously summarize one or more works within a sentence. The above two shortcomings make the extractive survey far from satisfactory.

Extractive Survey（SummaRuNNer）：① *We propose* a strongly performing method that scales to such datasets and learns a low-dimensional joint embedding space for both images and annotations [1]. ② *We introduce* a max-margin structure prediction architecture based on recursive neural networks that can successfully recover such structure both in complex scene images as well as sentences [2].
Man-Made Survey: ① *In the last few years,* neural networks based on dense vector representations have been producing superior results on various NLP tasks [1, 2, 3]. ② *This trend* is sparked by the success of word embeddings [4, 5] and deep learning methods [6].

Fig. 1. Comparison of survey generated by the state-of-the-art extractive method (SummaRuNNer) and man-made survey with the same multiple original papers.

In this paper, we regard automatic survey generation as a task of multiple papers' summarization and propose an abstractive method for the task, named **H**ierarchical **S**eq2seq model based on **D**ual **S**upervision (HSDS). Compared with traditional abstractive models [14,19,22], the HSDS model use the hierarchical structure to learn the more precise vector representations for original papers and target survey. Moreover, we introduce the dual supervised information including citation supervision and text supervision, and design a multi-task learning framework to train the HSDS model. Concretely, we first use a hierarchical Recurrent Neural Network (RNN) to encode input words and papers respectively. Then, the attention based RNN is utilized as a sentence decoder and a word decoder respectively. Particularly, in the sentence decoder, we use citations as intermediate supervised information and introduce the residual mechanism [9] to decide which papers should be cited by current decoding sentence. And in word decoder, the texts are sequentially generated according to the content of the cited papers.

Extensive experimental results demonstrate that our HSDS model outperforms state-of-the-art baselines.

2 Related Work

Automatic survey generation can be regarded as a summarization of multiple relevant papers in a specific area. In this section, we introduce relevant studies on Multi-document Summarization and Automatic Survey Generation.

2.1 Multi-document Summarization

Multi-document summarization aims to summarize multiple original relevant documents to a summary. As a long text generation task, the existing multi-document summarization methods are almost extractive. The extractive methods can be divided into two folds, i.e., unsupervised and supervised methods. Some prior unsupervised works [2,8] used lexical and grammatical structure of documents to synthesize summaries. After that, combinatorial optimization and approximation methods have been widely used for multi-document summarization, e.g., ILP [7], TextRank [15], and sparse coding [13]. In contrast to unsupervised approaches, supervised methods treat document summarization as a sequence classification task. [21] used Conditional Random Fields to binary-classify sentences sequentially. Other representative methods such as R2N2 [3] and GCN [23] used deep neural networks to extract important sentences.

2.2 Automatic Survey Generation

The related work can be regard as a survey for a specific area. As far as we know, some past studies focused on generating a related work via extracting important sentences from multiple original papers. The earliest method ReWoS took in a set of keywords arranged in a hierarchical fashion that describes a target paper's topics to extract related works [10]. The later work [11] improves that by considering the content of the target paper and creating topic-biased related works using PLSA model. Recently, [5] used reference papers cited in related works as multiple original papers and proposed a graph based comparative summarization approach to generate related works. However, all those studies are based on extractive methods and the generated related works have a poor coherence and readability. Therefore, we explore an abstractive method for automatic survey generation in this paper.

3 Our Proposed Model

The input of HSDS is multiple relevant papers $d = \{d_1, d_2, \cdots, d_N\}$ in a specific area, where N is the number of relevant papers in a research area, $d_i = \{x_{i,1}, x_{i,2}, \cdots, x_{i,N_i}\}$, where N_i is the number of words in d_i. The target is

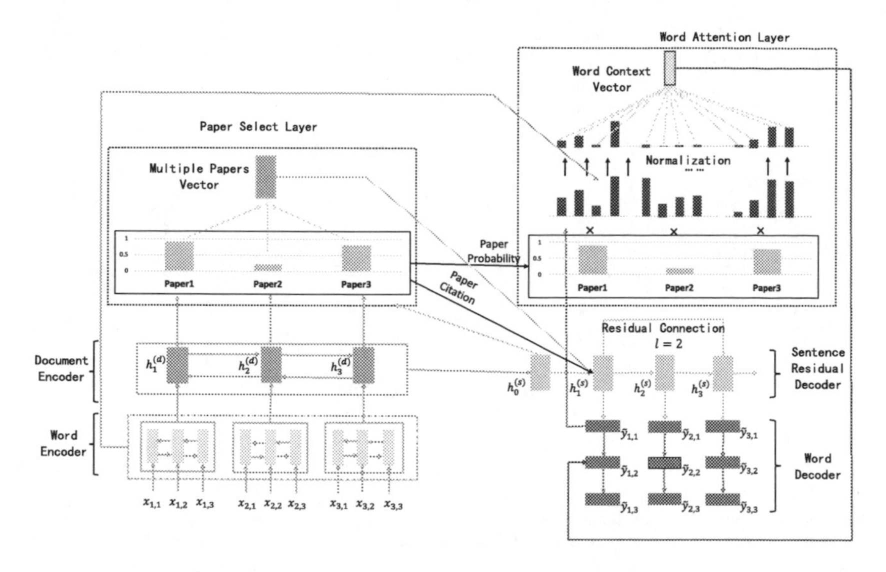

Fig. 2. The overall framework of HSDS model.

a corresponding survey $s = \{s_1, s_2, \cdots, s_L\}$, where L is the number of sentences in a survey, $s_i = \{y_{i,1}, y_{i,2}, \cdots, y_{i,L_i}\}$, where L_i is the number of words in s_i.

An overall framework of our HSDS model is shown in Fig. 2. For automatic survey generation, the HSDS model exceeds traditional abstractive models because of following three aspects: **(1)** The hierarchical structure in HSDS can capture hierarchical information among words, sentences and documents and generate more precise vector representations. **(2)** The residual connection in sentence decoder helps to accelerate convergence of citation generation. **(3)** The multi-task learning framework based on the dual supervised information including citation supervision and text supervision can promote mutually to generate a better survey. The three key modules will be introduced in Sects. 3.1, 3.2, 3.3 and 3.4 respectively.

3.1 Hierarchical Encoder

The HSDS model utilizes the hierarchical encoder to map input words and papers into the more precise vector representations, while traditional abstractive models regard multi-document summarization as a single document summarization and lose the hierarchical information between them. The hierarchical encoder contains word encoder and document encoder.

Word Encoder. Word encoder is based on the bidirectional Gated Recurrent Unit (BiGRU). The BiGRU contains a forward GRU and a backward GRU. For each original paper words sequence $\{\mathbf{x}_{i,1}, \mathbf{x}_{i,2}, \cdots, \mathbf{x}_{i,N_i}\}$, the BiGRU get forward hidden states sequence $(\overrightarrow{\mathbf{h}}_{i,1}, \overrightarrow{\mathbf{h}}_{i,2}, \cdots, \overrightarrow{\mathbf{h}}_{i,N_i})$ and backward hidden states sequence $(\overleftarrow{\mathbf{h}}_{i,1}, \overleftarrow{\mathbf{h}}_{i,2}, \cdots, \overleftarrow{\mathbf{h}}_{i,N_i})$. Then each forward and backward GRU state are combined to a single state using a simple feed-forward network:

$$\mathbf{h}_{i,j}^{(x)} = \mathrm{relu}(\mathbf{W}_e[\overrightarrow{\mathbf{h}}_{i,j}; \overleftarrow{\mathbf{h}}_{i,j}] + \mathbf{b}_e) \tag{1}$$

where $\mathbf{W}_e$ is the weight matrix and $\mathbf{b}_e$ is the bias vector, $[;]$ is the concatenation operation and $\mathbf{h}_{i,j}^{(x)}$ is the hidden state of input word $x_{i,j}$. The $\overrightarrow{\mathbf{h}}_{i,N_i}$ and $\overleftarrow{\mathbf{h}}_{i,1}$ are used to combine a embedding representation $\mathbf{z}_i$ of the document d_i, denoted as:

$$\mathbf{z}_i = \mathrm{relu}(\mathbf{W}_x[\overrightarrow{\mathbf{h}}_{i,N_i}; \overleftarrow{\mathbf{h}}_{i,1}] + \mathbf{b}_x) \tag{2}$$

where $\mathbf{W}_x$ is the weight matrix and $\mathbf{b}_x$ is the bias vector.

Document Encoder. For original papers sequence representation $(\mathbf{z}_1, \mathbf{z}_2, \cdots, \mathbf{z}_N)$ which is obtained by word encoder, we use BiGRU to get the forward GRU state $(\overrightarrow{\mathbf{h}}_1, \overrightarrow{\mathbf{h}}_2, \cdots, \overrightarrow{\mathbf{h}}_N)$ and backward GRU state $(\overleftarrow{\mathbf{h}}_1, \overleftarrow{\mathbf{h}}_2, \cdots, \overleftarrow{\mathbf{h}}_N)$. Then they are combined to get the hidden state sequence $(\mathbf{h}_1^{(d)}, \mathbf{h}_2^{(d)}, \cdots, \mathbf{h}_N^{(d)})$:

$$\mathbf{h}_i^{(d)} = \mathrm{relu}(\mathbf{W}_d[\overrightarrow{\mathbf{h}}_i; \overleftarrow{\mathbf{h}}_i] + \mathbf{b}_d) \tag{3}$$

where $\mathbf{W}_d$ is the weight matrix and $\mathbf{b}_d$ is the bias vector. Furthermore, the $\overrightarrow{\mathbf{h}}_N$ and $\overleftarrow{\mathbf{h}}_1$ are used to get the embedding representation $\mathbf{I}$ of multiple papers:

$$\mathbf{I} = \mathrm{relu}(\mathbf{W}_m[\overrightarrow{\mathbf{h}}_N; \overleftarrow{\mathbf{h}}_1] + \mathbf{b}_m) \tag{4}$$

where $\mathbf{W}_m$ is the weight matrix and $\mathbf{b}_m$ is the bias vector.

3.2 Hierarchical Decoder

Similar to hierarchical encoder, the decoder is also divided into two steps: sentence residual decoder and word decoder.

Sentence Residual Decoder. Sentence residual decoder is used to generate representation of each sentence in a survey. Compared with traditional decoder, the sentence residual decoder has innovations in two aspects: **(I)** The attention mechanism in sentence residual decoder is taken as a selective gate to decide which original papers should be cited by current sentence, thus it can be trained according to the citations in a target survey. **(II)** Considering that sentences with different citations may be not adjacent, we introduce the residual connection to depict the phenomenon.

Following above **(I)**, at each sentence decoding time step t, the unidirectional GRU reads the previous sentence hidden state $\mathbf{h}_{t-1}^{(s)}$, previous sentence vector $\mathbf{v}_{t-1}$ (while training, this is the mean vector of all words of the previous sentence in target survey; at test time it is the previous sentence representation emitted by the sentence decoder) and previous context vector $\mathbf{c}_{t-1}$ to compute current hidden state $\mathbf{h}_t^{(s)}$. We initialize the $\mathbf{h}_0^{(s)}$ according to multiple original papers representation $\mathbf{I}$ obtained by document encoder:

$$\mathbf{h}_t^{(s)} = \mathrm{GRU}(\mathbf{v}_{t-1}, \mathbf{h}_{t-1}^{(s)}, \mathbf{c}_{t-1}) \tag{5}$$

$$\mathbf{h}_0^{(s)} = \mathrm{relu}(\mathbf{W}_s\mathbf{I} + \mathbf{b}_s) \tag{6}$$

where $\mathbf{W}_s$ is the weight matrix and $\mathbf{b}_s$ is the bias vector.

For current time step t, the context vector $\mathbf{c}_t$ is calculated through a soft selective mechanism which is achieved by a special attention function:

$$a_i^t = \mathrm{sigmoid}(\mathbf{v}_{a1}^T \tanh(\mathbf{W}_{a1}\mathbf{h}_t^{(s)} + \mathbf{U}_{a1}\mathbf{h}_i^{(d)} + \mathbf{b}_{a1})) \tag{7}$$

$$\mathbf{c}_t = \sum_{i=1}^{N} a_i^t \mathbf{h}_i^{(d)} \tag{8}$$

where $\mathbf{W}_{a1}$ and $\mathbf{U}_{a1}$ are the weight matrixes, $\mathbf{v}_{a1}$ is the weight vector and $\mathbf{b}_{a1}$ is the bias vector. During training, we learn the a_i^t supervised by the citations in target survey. During test, we generate the citations according the trained a_i^t (This will be discussed in "Multi-task Test" section).

Following above **(II)**, we add skip residual connections between two GRU hidden states with a wide range of distance l, such that $\forall t = \{1+l, 1+2l, \cdots, 1+[\frac{L-1}{l}]l\}$:

$$\mathbf{h}_t^{(s)} = \mathrm{GRU}(\mathbf{v}_{t-1}, \mathbf{h}_{t-1}^{(s)}, \mathbf{c}_{t-1}) + \mathbf{h}_{t-l}^{(s)} \tag{9}$$

Word Decoder. At each word decoding time step k for sentence t, the unidirectional GRU reads the previous word hidden state $\mathbf{h}_{t,k-1}^{(w)}$, previous word $\mathbf{y}_{t,k-1}$ (while training, this is the previous word of the target survey; at test time it is the previous word emitted by the word decoder) and previous context vector $\mathbf{c}_{t,k-1}$ to calculate current word hidden state $\mathbf{h}_{t,k}^{(w)}$. The $\mathbf{h}_{t,0}^{(w)}$ is initialized according to sentence representation $\mathbf{h}_t^{(s)}$ obtained by sentence decoder:

$$\mathbf{h}_{t,k}^{(w)} = \mathrm{GRU}(\mathbf{y}_{t,k-1}, \mathbf{h}_{t,k-1}^{(w)}, \mathbf{c}_{t,k-1}) \tag{10}$$

$$\mathbf{h}_{t,0}^{(w)} = \mathrm{relu}(\mathbf{W}_o\mathbf{h}_t^{(s)} + \mathbf{b}_o) \tag{11}$$

where $\mathbf{W}_o$ is weight matrix and $\mathbf{b}_o$ is bias vector.

For current step k, the context vector $\mathbf{c}_{t,k}$ is calculated by softmax attention score:

$$e_{i,j}^{t,k} = \mathbf{v}_{a2}^T \tanh(\mathbf{W}_{a2}\mathbf{h}_{t,k}^{(w)} + \mathbf{U}_{a2}\mathbf{h}_{i,j}^{(x)} + \mathbf{b}_{a2}) \tag{12}$$

$$p_{i,j}^{t,k} = a_i^t e_{i,j}^{t,k} \tag{13}$$

$$a_{i,j}^{t,k} = \frac{p_{i,j}^{t,k}}{\sum_{i=1}^{N}\sum_{j=1}^{N_i} p_{i,j}^{t,k}} \tag{14}$$

$$\mathbf{c}_{t,k} = \sum_{i=1}^{N} \sum_{j=1}^{N_i} a_{i,j}^{t,k} \mathbf{h}_{i,j}^{(x)} \tag{15}$$

where $\mathbf{W}_{a2}$ and $\mathbf{U}_{a2}$ are the weight matrixes, $\mathbf{v}_{a2}$ is the weight vector and $\mathbf{b}_{a2}$ is the bias vector.

Then the current word hidden state $\mathbf{h}_{t,k}^{(w)}$ and the context vector $\mathbf{c}_{t,k}$ are used to estimate the probability distribution of word $y_{t,k}$:

$$P(y_{t,k}|y_{t,1:k-1}) = \text{softmax}(\mathbf{V}[\mathbf{h}_{t,k}^{(w)}; \mathbf{c}_{t,k}] + \mathbf{b}) \tag{16}$$

where $\mathbf{V}$ is the vocabulary weight matrix, $\mathbf{b}$ is the vocabulary bias vector and $P(y_{t,k}|y_{t,1:k-1})$ is the probability distribution over all words in the vocabulary.

Copy Source. The unknown words (*UNK*) is a common problem in text generation task. A well-known work to address the *UNK* is Pointer-Generator Network [20]. Following that, we adopt a pointer-generator network to reduce the generation of unknown words. The pointer-generator network add a hybrid $p_{gen} \in [0,1]$ which controls ratio between copying words from original papers and generating words from a vocabulary, p_{gen} is defined as:

$$p_{gen} = \text{sigmoid}(\mathbf{w}_p^T \mathbf{c}_{t,k} + \mathbf{u}_p^T \mathbf{h}_{t,k}^{(w)} + \mathbf{v}_p^T \mathbf{y}_{t,k-1} + \mathbf{b}_p) \tag{17}$$

where $\mathbf{w}_p$, $\mathbf{w}_p$ and $\mathbf{v}_p$ are the weight vectors, and $\mathbf{b}_p$ is the learnable scalar, $\mathbf{y}_{t,k-1}$ is the word decoder input. Thus the probability of generating word w is defined as:

$$\begin{aligned} P_{copy}(y_{t,k} = w|y_{t,1:k-1}) = p_{gen} P(y_{t,k} = w|y_{t,1:k-1}) \\ + (1 - p_{gen}) \sum_{(i,j):x_{i,j}=w} a_{i,j}^{t,k} \end{aligned} \tag{18}$$

Coverage Mechanism. For a long text generation task, the neural network tends to generate phrases and words repeatedly. Pointer-Generator Network [20] used a coverage model where attention scores were tracked to avoid repeatedly attending to the same steps. We follow this work and maintain a coverage vector $\mathbf{cov}^{t,k}$ which is the sum of weight vectors for original words at previous time steps:

$$cov_{i,j}^{t,k} = \sum_{q=0}^{k-1} a_{i,j}^{t,q} \tag{19}$$

such that we change Eq. (12) to follows:

$$e_{i,j}^{t,k} = \mathbf{v}_{a2}^T \tanh(\mathbf{W}_{a2}\mathbf{h}_{t,k}^{(w)} + \mathbf{U}_{a2}\mathbf{h}_{i,j}^{(x)} + \mathbf{g}_{a2} cov_{i,j}^{t,k} + \mathbf{b}_{a2}) \tag{20}$$

where $\mathbf{g}_{a2}$ is the weight vector of same length as $\mathbf{v}_{a2}$.

3.3 Multi-task Learning

Multi-task learning [4] aims to solve the multiple learning tasks at the same time. Different from training the models separately, it exploits commonalities and differences across multiple learning tasks to improve learning efficiency and prediction accuracy for the task-specific models.

Automatic survey generation can be viewed as a multi-task learning in which generating citations and texts (i.e., words) are two inter-related tasks. The citations can guide to generate the more accurate texts and vice versa. So we can optimize the two learning tasks at the same time.

We first define the objective function of generating citations in each sentence:

$$\mathcal{L}^t_{citation} = -\sum_i \widetilde{a}^t_i \log(a^t_i) + (1 - \widetilde{a}^t_i)\log(1 - a^t_i) \tag{21}$$

where $\widetilde{a}^t$ is the binary vector representing citations of sentence s_t in a ground truth survey. The dimension size of $\widetilde{a}^t$ is N which is the number of original papers. We set the bits corresponding to citation indexes to "1" and the other bits to "0". For example, under the assumption that the number of original papers is "5" and the ground truth citations in s_t are "[1, 3]", we set $\widetilde{a}^t = [1, 0, 1, 0, 0]$.

Then we define the objective function of generating texts of each sentence, particularly, we add coverage loss to penalize the repetition:

$$\mathcal{L}^{t,k}_{words} = -\log P_{copy}(y_{t,k}) + \lambda \sum_{i,j} min(a^{t,k}_{i,j}, cov^{t,k}_{i,j}) \tag{22}$$

Finally, we add the $\mathcal{L}^t_{citation}$ into the $\mathcal{L}^{t,k}_{words}$ using weight β:

$$\mathcal{L} = \frac{1}{\sum_t L_t} \sum_{t=1}^{L} \sum_{k=1}^{L_t} \mathcal{L}^{t,k}_{words} + \beta \frac{1}{L} \sum_{t=1}^{L} \mathcal{L}^t_{citation} \tag{23}$$

where t is the t-th generated sentence in the survey and k is the k-th generated word in the current sentence.

3.4 Multi-task Test

During test, multiple papers are fed into trained HSDS model and the model generate citations and texts of a survey.

For generating citations of a survey, the a^t_i (as described in Eq. (7)) in sentence residual decoder is used to calculate the cited probability of original paper, furthermore, the citation $\widetilde{Idx}_t$ of sentence $\widetilde{s}_t$ in a generated survey are calculated according to the equation:

$$\widetilde{Idx}_t = \{i \mid a_{i,t} > 0.5\} \tag{24}$$

For generating texts of a survey, $P_{copy}(y_{t,k} = w | y_{t,1:k-1})$ (as described in Eq. (18)) is calculated to generate word w with the max probability.

4 Experiments

4.1 Dataset Construction

To train the proposed model, we need first to build a large dataset about surveys and their corresponding relevant papers. Fortunately, related work section in a paper is a survey for a specific research area, and cited papers in the related work section is relevant papers in the survey. Therefore, it is reasonable to build a dataset including related works and their respective cited papers.

Following the procedure in Fig. 3., we first crawl related work sections of numerous papers from IEEE Xplore Digital Library[2]. Each related work section contains multiple cited papers. Considering that the main idea of a cited paper is described in its abstract, title and authors, we use a combination of above three parts to represent full paper. The three parts can be obtained by searching the names of cited papers in Baidu Scholar[3]. For simplicity and not to cause confusion, we call the combination of three parts in a cited paper as a **pseudo reference paper**, thus the corresponding related work section is a **survey** of multiple pseudo reference papers. Finally, each survey and its corresponding multiple pseudo reference papers are regarded as a dataset of parallel pair. The dataset totally have 390,000 parallel pairs.

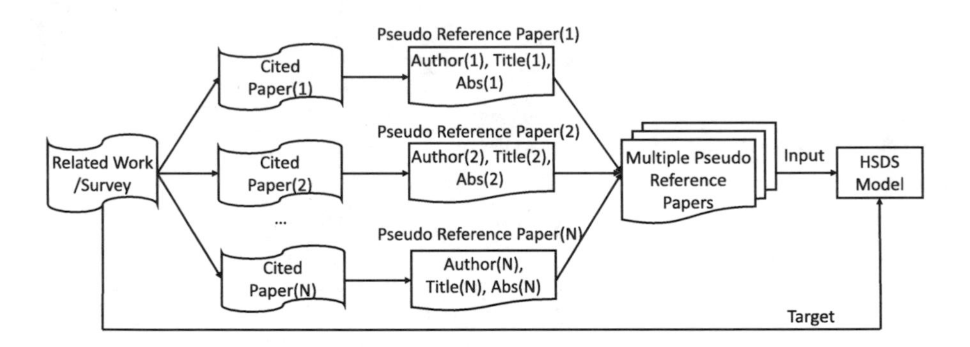

Fig. 3. The flowchart of dataset construction in HSDS model.

For further experiments, all words in dataset are lowercased. And the PTB tokenizer[4] is used as our word segmentation tool. Following the setting of automatic document summarization in [20], we save the top 50k most frequent words as vocabulary, and the other rare words are replaced with <*unk*>. Furthermore, for each survey, we use NLTK[5] to split it into sentences. Unfortunately, half of the sentences have no citations. For those sentences without citations, we take

[2] https://ieeexplore.ieee.org/search/searchresult.jsp?newsearch=true.
[3] http://xueshu.baidu.com/.
[4] https://nlp.stanford.edu/software/tokenizer.html.
[5] https://www.nltk.org/.

the citations of the most adjacent sentence as their citations, which is because the adjacent sentences may summarize the same papers. Particularly, if there are two most adjacent sentence located in front and back position respectively, we take the citations of front sentence, that is because when multiple sentences refer to the same citations, the citations usually will be marked in the front sentence. The dataset is divided into 312,000 training pairs, 39,000 validation pairs and 39,000 test pairs with the ratio of 8:1:1. More detailed statistics of our dataset are shown in Table 1.

Table 1. Statistics of our dataset. (avg.paper.num) indicates the average number of pseudo reference papers in a pair, (avg.paper.word.num) indicates the average number of words in a pseudo reference paper, (avg.sen.num) indicates the average number of sentences in a survey, (avg.sen.word.num) indicates the average number of words in a sentence of surveys.

	Avg.paper.num	Avg.paper.word.num	Avg.sen.num	Avg.sen.word.num
Train	9.89	187.36	25.75	21.45
Valid	9.87	190.78	25.71	21.36
Test	10.09	187.97	25.63	21.72

4.2 Implementation Details

We use Pytorch[6] for implementation. For all experiments, the GRU forward and backward hidden state size is 256 and the word embedding size is 128. In particular, the word embedding is pre-trained on our training set using word2vec[7] and that will be further trained in the model. The source and target have the same vocabulary including the top 50k most frequent words. According to the Table 1, we limit the number of pseudo reference papers in a pair to 10, length of each pseudo reference paper to 200, the number of sentences in a survey to 25, and length of each sentence to 20.

We use the Adam optimizer with a learning rate 0.001 and gradient clipping with range $[-5, 5]$. To speed up the training, we use layer normalization [1] and set mini-batch size 8. After the grid search of parameters, we set the coverage loss weight as $\lambda = 1$ (as described in Eq. (22)), multi-task learning loss weight as $\beta = 5$ (as described in Eq. (23)) to get the best performance. The residual connection distance l (as described in Eq. (9)) is empirically set as 2 (that will be further discussed in discussion section). Following the *PntrGen* [20], we first train the model without coverage and add it at the last two epochs. We use the loss on the validation set to implement early stopping. The model is trained for about 50 epochs in a NVIDIA 1080 Ti GPU with 12 days. During test, a beam search with beam size of 4 is used to generate survey.

[6] https://pytorch.org/.
[7] https://code.google.com/archive/p/word2vec/.

4.3 Baseline

We compare our proposed model with several well-known multi-document extractive methods and the state-of-the-art abstractive methods. For abstractive baselines, we regard multiple documents as a single document. Furthermore, to analyze the detailed effectiveness of three innovative components (described in "Our Proposed Model Section") in HSDS model, three additional baselines are introduced. The total compared methods are as follows:

- **Lead** is an extractive baseline which selects the lead several sentences.
- **TextRank** [15] proposed a graph-based method to calculate the similarity between two sentences and rank them using PageRank algorithm [18].
- **NeuralSum** [6] is a data-driven approach based on neural networks. They developed a hierarchical document encoder and an attention-extractor to extract sentences and words.
- **SummaRuNNer** [16] proposed a recurrent neural network (RNN) based sequence model for extractive summarization.
- **Attn+Seq2Seq** [17] proposed a single layer encoder-decoder model based on recurrent neural network (RNN). We regard it as our abstractive baseline.
- **PntrGen+Seq2Seq** [20] is a pointer-generator network proposed by [20] for the purpose of solving UNK and repetition.
- **HSDS-citation-residual** remove the citation supervision during training and residual connection in sentence decoder, then the HSDS model is only a hierarchical framework based on *PntrGen+Seq2Seq*.
- **HSDS-citation** only remove the citation supervision during training, thus our HSDS model is trained only supervised by text information, i.e., a single-task learning method. Through it, we can observe the effectiveness of citation supervision for generating better surveys.
- **HSDS-residual** only remove the residual mechanism in sentence decoder to observe whether the residual connection help generate the better citations.

4.4 Evaluation Metric

Evaluation for the quality of generated surveys contains following two aspects: the quality of generated citations and the quality of generated texts.

- **Evaluation for Quality of Generated Citation:** Idx_t is a set of indexes of the cited papers in a sentence s_t in ground truth survey, and $\widetilde{Idx_t}$ is a set of indexes of the cited papers in a sentence $\widetilde{s}_t$ in corresponding generated survey. The Precision is defined as $\frac{\widetilde{Idx_t} \cap Idx_t}{\widetilde{Idx_t}}$, Recall is defined as $\frac{\widetilde{Idx_t} \cap Idx_t}{Idx_t}$ and $F1$ score is defined as $\frac{2 * Precision * Recall}{Precision + Recall}$.
- **Evaluation for Quality of Generated Text:** We employ ROUGE-1.5.5 [12] which was widely used in text summarization task and report the F-measures of ROUGE-1 (unigram), ROUGE-2 (bigram) and ROUGE-SU4 (skip-gram plus unigram) to evaluate the overlap between generated text and target text. The formula of Rouge-N is defined in [12].

Table 2. Comparison results on our test set. The precision, recall and F1 are respectively the average results of all generated citations, '–' indicates not available. RG refers to ROUGE and all our ROUGE scores have a 95% confidence interval of at most ±0.25. The top half of the table are extractive methods and bottom half of the table are abstractive methods. The best results are shown in **bold face**.

Method	Precision	Recall	$F1$	RG-1	RG-2	RG-SU4
Lead	0.23	0.15	0.18	36.63	8.47	14.36
TextRank	0.46	0.32	0.38	38.13	**9.61**	16.04
NeuralSum	0.55	0.35	0.43	39.14	9.17	16.49
SummaRuNNer	0.59	0.36	0.45	39.58	9.51	17.12
Attn+Seq2Seq	–	–	–	33.05	7.19	13.10
PntrGen+Seq2Seq	–	–	–	35.91	8.29	14.24
HSDS-citation-residual	0.50	0.65	0.57	37.23	9.10	14.91
HSDS-citation	0.51	0.68	0.58	37.71	9.17	15.61
HSDS-residual	0.64	0.75	0.69	39.58	9.31	16.82
HSDS	**0.67**	**0.76**	**0.71**	**39.76**	9.49	**17.26**

4.5 Experimental Results

The experimental results contain the following two aspects:

- **Result of Generated Citation:** The results for citation generation contain precision, recall and $F1$ score and they are show in left half of Table 2. It is obvious that extractive methods assign only one citation for each generated sentence in a survey. Therefore the extractive methods have the high precision scores but the low recall scores. The traditional abstractive methods including *Attn+Seq2Seq* and *PntrGen+Seq2Seq* cannot assign any citations for generated sentences because of its single layer structure. Compared with above methods, *HSDS* model has two advantages: **(1)** The hierarchical structure overcomes the defect that traditional single-layer abstractive model cannot assign citations for sentences in a survey. It can be seen that *HSDS-citation-residual* has available evaluation results, i.e., 0.50 precision, 0.65 recall and 0.57 F1 score; **(2)** The dual supervised information including citation supervision and text supervision in *HSDS* model promote mutually to achieve the higher scores (+0.17 Precision, +0.11 Recall, +0.14 $F1$ score) than *HSDS-citation-residual.*
- **Result of Generated Text:** The results for text generation contain ROUGE-1, ROUGE-2 and ROUGE-SU4 and they are shown in right half of Table 2. Our *HSDS* model outperforms the state-of-the-art abstractive methods *PntrGen+Seq2Seq* with higher scores (+3.85 ROUGE-1, +1.20 ROUGE-2, +3.02 ROUGE-SU4), which is a significant improvements. What's more, comparing *PntrGen+Seq2Seq* (35.91 ROUGE-1, 8.29 ROUGE-2, 14.24 ROUGE-SU4) with *HSDS-citation-residual* (37.23 ROUGE-1, 9.10 ROUGE-2, 14.91 ROUGE-SU4), we observe that hierarchical structure indeed has a

good performance for the automatic survey generation. Furthermore, when adding citation supervision and residual mechanism into the hierarchical structure, our *HSDS* model has the higher scores (+2.53 ROUGE-1, +0.39 ROUGE-2 and +2.35 ROUGE-SU4). It demonstrates the citation supervision helps a lot not only in citation generation but also in text generation. However, our *HSDS* model do not always outperform extractive baselines. Our ROUGE-2 score (9.49 ROUGE-2) is a little lower than *TextRank* (9.61 ROUGE-2). One possible explanation is that the target surveys usually include some overlapping phrases with the abstract sections of pseudo reference papers, therefore, it is easier for extractive methods to achieve higher ROUGE scores.

5 Discussion

5.1 Citation Supervision Discussion

We divide dataset into several groups according to the number of multiple pseudo reference papers in pairs. For each group, different methods including *PntrGen+Seq2Seq*, *HSDS-citation* and *HSDS* generate different results. The comparisons are shown in Fig. 4. It shows that the ROUGE curves (ROUGE-1, ROUGE-2 and ROUGE-SU4) of *HSDS* always appear to be on the top of those of *PntrGen+Seq2Seq* and *HSDS-citation*. With the number of pseudo reference papers increasing, all curves have the lower ROUGE scores because of the longer text. However, our HSDS model decline more slowly than the other two models. It indicates that the citation supervision plays a more and more important role in generating the better survey when increasing the number of pseudo reference papers.

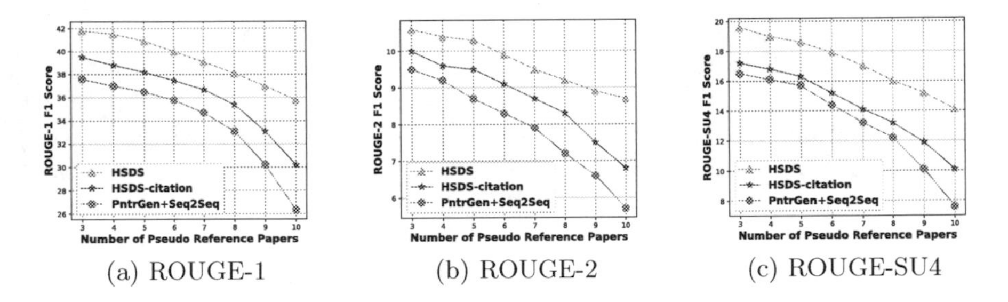

(a) ROUGE-1 (b) ROUGE-2 (c) ROUGE-SU4

Fig. 4. Rouge results of auto-evaluation of different number of pseudo reference papers on our dataset

5.2 Residual Hyper-parameter Discussion

The results of convergence for different residual connection distance l (as described in Eq. (9)) are shown in Fig. 5. It can be observed that a suitable value (e.g. $l = 2$) for l can accelerate the convergence, however, extreme value

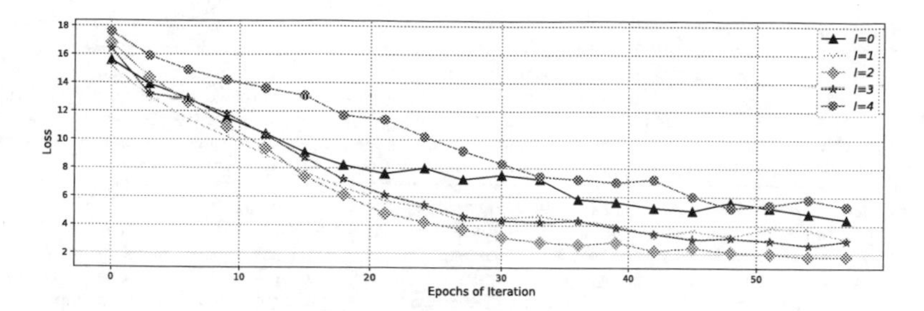

Fig. 5. Loss of training dataset with different l

(e.g. $l = 4$) for l may destroy the trend of gradient descent during training and lead to the worse performance. Intuitively, the residual connection distance is about the average interval between two adjacent and different citations.

5.3 The Area of Input Papers Discussion

Survey means making an overview for a specific area. Thus, there is a hidden assumption in our work, i.e., the input of our model is the relevant papers for a specific area, and these papers are required to be in the same area. Note that the area is not limited in theory. Of course, in practice, the effectiveness of the model depends not only the ability of the model, but also the quality of the input relevant paper. Table 3 shows the evaluation results of generated survey based on input papers in different areas. We found that if all input papers are in the same area, the generated survey has a better performance.

Table 3. Evaluated results of generated survey based on input papers in different areas. RG refers to ROUGE. NLP means natural language process, CV means computer vision, NLP+CV means that half of input papers is NLP and the other half of input papers is CV.

Area of input papers	Precision	Recall	$F1$	RG-1	RG-2	RG-SU4
NLP	0.67	0.74	0.70	39.64	9.49	17.18
CV	0.62	0.75	0.68	39.77	9.50	17.22
NLP+CV	0.57	0.54	0.55	37.24	9.09	15.83

5.4 Case Study

The examples of surveys generated by different methods are shown in Fig. 6. It can be observed that the state-of-the-art extractive model *SummaRuNNer* has a poor coherence between sentences and only contains one citation within a sentence. Therefore, the extractive survey has a lower average $F1$ score for

Ground Truth: In the literary of CAPTCHAs, most schemes were aimed at the Turing test that embeds characters in an image. However, *[1]* illustrated that computer vision techniques, by optical character recognition, have over 90\% accuracy to recognize the character in an image. Furthermore, *[2], [3], [4]* proposed alternative image question CAPTCHAs which does not have the above issue and *[5]* provided a combination of character and image CAPTCHA which possesses both of the above properties and users have to do simple mathematical computation in order to answer the question. Besides, the Turing test system of [6] is implemented in Linux kernel in order to improve the performance.
SummaRuNNer: We propose IMAGINATION (IMAge Generation for Internet AuthenticaTION), a system for the generation of attack-resistant, user-friendly, image-based CAPTCHAs [1]. CAPTCHA is now almost a standard security technology, and has found widespread application in commercial websites [3]. We propose and implement three CAPTCHAs based on naming images, distinguishing images, and identifying an anomalous image out of a set [4]. As a result, it improves performance, regardless of whether the server overload is caused by DDoS or a true Flash Crowd.
PntrGen+Seq2Seq: In the CAPTCHAs, produce controlled distortions on randomly chosed images and present them to the user. Usability and robustness are two fundamental issues with CAPTCHA. As a result some systems named CAPTCHA has been introduced. We present the Kill-Bots to protect against DDoS attacks in Linux kernel.
HSDS: In the CAPTCHAs, *[1]* produce controlled distortions on randomly chosen images and present them to the user for annotation from a given list of words. Furthermore, [2] [3] discusses usability issues that should be considered in the design of CAPTCHAs and a novel question framework is proposed. And *[5]* introduces a combination method and simple mathematical problem is generated according their images to answer the question. Besides, *[6]* design and implement Kill-Bots in Linux kernel extension to improve performance.

Fig. 6. Illustration of a case study. The green text indicates the concatenation words between two sentences. The red indexes indicates right citations generated by HSDS model. (Color figure online)

generated citations in each sentence than HSDS model. The survey generated by *PntrGen+Seq2Seq* is too short to describe the main ideas of multiple original papers roundly and lack the citations, while our HSDS model can generate a better survey.

6 Conclusion

In this paper, we investigate the features of automatic survey generation and propose a novel abstractive model named HSDS. Furthermore, we introduce the dual supervised information including citation supervision and text supervision in a survey to improve the performances of HSDS. Various experiments have demonstrated the effectiveness of our proposed method.

Acknowledgment. The work is supported by NSFC (No. 61772076 and 61751201), NSFB (No. Z181100008918002), BIGKE (No. 20160754021) and CETC (No. w-2018018).

References

1. Ba, J.L., Kiros, J.R., Hinton, G.E.: Layer normalization. arXiv preprint arXiv:1607.06450 (2016)
2. Barzilay, R., McKeown, K.R., Elhadad, M.: Information fusion in the context of multi-document summarization. In: Proceedings of the 37th Annual Meeting of the Association for Computational Linguistics on Computational Linguistics, pp. 550–557. Association for Computational Linguistics (1999)

3. Cao, Z., Wei, F., Dong, L., Li, S., Zhou, M.: Ranking with recursive neural networks and its application to multi-document summarization. In: AAAI, pp. 2153–2159 (2015)
4. Caruana, R.: Multitask learning. In: Thrun, S., Pratt, L. (eds.) Learning to Learn, pp. 95–133. Springer, Boston (1998). https://doi.org/10.1007/978-1-4615-5529-2_5
5. Chen, J., Zhuge, H.: Automatic generation of related work through summarizing citations. Pract. Exp. Concurr. Comput. **31**(3), e4261 (2017)
6. Cheng, J., Lapata, M.: Neural summarization by extracting sentences and words. arXiv preprint arXiv:1603.07252 (2016)
7. Gillick, D., Favre, B.: A scalable global model for summarization. In: Proceedings of the Workshop on Integer Linear Programming for Natural Langauge Processing, pp. 10–18. Association for Computational Linguistics (2009)
8. Goldstein, J., Mittal, V., Carbonell, J., Kantrowitz, M.: Multi-document summarization by sentence extraction. In: Proceedings of the 2000 NAACL-ANLP Workshop on Automatic Summarization, pp. 40–48. Association for Computational Linguistics (2000)
9. He, K., Zhang, X., Ren, S., Sun, J.: Deep residual learning for image recognition. In: Proceedings of the IEEE Conference on Computer Vision and Pattern Recognition, pp. 770–778 (2016)
10. Hoang, C.D.V., Kan, M.Y.: Towards automated related work summarization. In: Proceedings of the 23rd International Conference on Computational Linguistics: Posters, pp. 427–435. Association for Computational Linguistics (2010)
11. Hu, Y., Wan, X.: Automatic generation of related work sections in scientific papers: an optimization approach. In: Proceedings of the 2014 Conference on Empirical Methods in Natural Language Processing (EMNLP), pp. 1624–1633 (2014)
12. Lin, C.Y.: ROUGE: a package for automatic evaluation of summaries. In: Text Summarization Branches Out (2004)
13. Liu, H., Yu, H., Deng, Z.H.: Multi-document summarization based on two-level sparse representation model. In: AAAI, pp. 196–202 (2015)
14. Lopyrev, K.: Generating news headlines with recurrent neural networks. arXiv preprint arXiv:1512.01712 (2015)
15. Mihalcea, R., Tarau, P.: TextRank: bringing order into text. In: Proceedings of the 2004 Conference on Empirical Methods in Natural Language Processing (2004)
16. Nallapati, R., Zhai, F., Zhou, B.: SummaRuNNer: a recurrent neural network based sequence model for extractive summarization of documents. In: AAAI, pp. 3075–3081 (2017)
17. Nallapati, R., Zhou, B., Gulcehre, C., Xiang, B., et al.: Abstractive text summarization using sequence-to-sequence RNNs and beyond. arXiv preprint arXiv:1602.06023 (2016)
18. Page, L., Brin, S., Motwani, R., Winograd, T.: The PageRank citation ranking: bringing order to the web. Technical report, Stanford InfoLab (1999)
19. Rush, A.M., Chopra, S., Weston, J.: A neural attention model for abstractive sentence summarization. arXiv preprint arXiv:1509.00685 (2015)
20. See, A., Liu, P.J., Manning, C.D.: Get to the point: summarization with pointer-generator networks. arXiv preprint arXiv:1704.04368 (2017)
21. Shen, D., Sun, J.T., Li, H., Yang, Q., Chen, Z.: Document summarization using conditional random fields. In: IJCAI, vol. 7, pp. 2862–2867 (2007)

22. Takase, S., Suzuki, J., Okazaki, N., Hirao, T., Nagata, M.: Neural headline generation on abstract meaning representation. In: Proceedings of the 2016 Conference on Empirical Methods in Natural Language Processing, pp. 1054–1059 (2016)
23. Yasunaga, M., Zhang, R., Meelu, K., Pareek, A., Srinivasan, K., Radev, D.: Graph-based neural multi-document summarization. arXiv preprint arXiv:1706.06681 (2017)

PU-Shapelets: Towards Pattern-Based Positive Unlabeled Classification of Time Series

Shen Liang[1,2], Yanchun Zhang[1,2,3(✉)], and Jiangang Ma[4]

[1] School of Computer Science, Fudan University, Shanghai, China
sliang11@fudan.edu.cn
[2] Cyberspace Institute of Advanced Technology (CIAT), Guangzhou University, Guangzhou, China
[3] Institute for Sustainable Industries and Liveable Citie, Victoria University, Melbourne, Australia
Yanchun.Zhang@vu.edu.au
[4] School of Science, Engineering and Information Technology, Federation University Australia, Ballarat, Australia
j.ma@federation.edu.au

Abstract. Real-world time series classification applications often involve positive unlabeled (PU) training data, where there are only a small set PL of positive labeled examples and a large set U of unlabeled ones. Most existing time series PU classification methods utilize all readings in the time series, making them sensitive to non-characteristic readings. Characteristic patterns named *shapelets* present a promising solution to this problem, yet discovering shapelets under PU settings is not easy. In this paper, we take on the challenging task of shapelet discovery with PU data. We propose a novel pattern ensemble technique utilizing both characteristic and non-characteristic patterns to rank U examples by their possibilities of being positive. We also present a novel stopping criterion to estimate the number of positive examples in U. These enable us to effectively label all U training examples and conduct supervised shapelet discovery. The shapelets are then used to build a one-nearest-neighbor classifier for online classification. Extensive experiments demonstrate the effectiveness of our method.

Keywords: Time series · Shapelets · Positive unlabeled classification

1 Introduction

Time series classification (TSC) is an important research topic with applications to medicine [3,9], biology [4], electronics [17], etc. Conventional TSC [2] tasks are

This work is funded by NSFC grants 61672161 and 61332013. We sincerely thank Dr. Nurjahan Begum and Dr. Anthony Bagnall for granting us access to the code of [3] and [7], and all our colleagues who have contributed valuable suggestions to this work.

G. Li et al. (Eds.): DASFAA 2019, LNCS 11446, pp. 87–103, 2019.
https://doi.org/10.1007/978-3-030-18576-3_6

fully supervised. However, real-world TSC problems often fall into the category of **positive unlabeled (PU) classification** [8]. In such cases, only a small set PL of positive and labeled training examples and a large set U of unlabeled ones are available to help distinguish between two classes[1]. For example, in heartbeat classification for medical care, we may need to train a classifier based on a limited number of abnormal heartbeats and a large number of unlabeled (normal or abnormal) ones [3]. To the best of our knowledge, no conventional supervised TSC methods can be applied to such cases where only one class is labeled, thus specialized PU classification methods are required.

Most existing PU classification methods for time series [3,4,6,13,16,17] are *whole-stream* based, utilizing all readings in the training examples. This makes them sensitive to non-characteristic readings [15,19]. One effective solution to this problem is **time series shapelets** [7,10,15,18,19], which are characteristic patterns[2] that can effectively distinguish between different classes. For instance, consider three electrocardiography time series from the *TwoLeadECG* dataset [5]. Under whole-stream matching (Fig. 1 *left*) with the highly effective [4] DTW distance [6,13,16], ts_2 is incorrectly deemed to be more similar to ts_3 than ts_1. In contrast, with a shapelet (Fig. 1 *right*), we can obtain its best matching subsequence in each time series, and uncover the correct link.

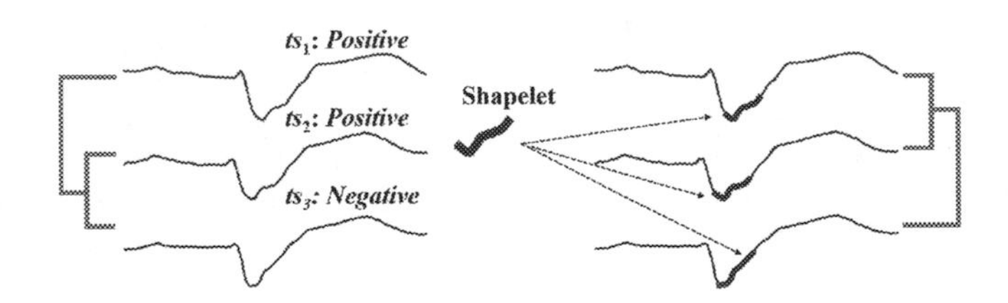

Fig. 1. A comparison of whole-stream based and shapelet-based methods. While the former incorrectly links ts_2 with ts_3 (*left*), the latter uncovers the correct link (*right*).

In this paper, we undertake the task of shapelet discovery with PU data. To the best of our knowledge, no previous work deals with this problem. Existing shapelet discovery methods are either supervised [7,10,18] or unsupervised [15,19]. Concretely, a classic framework [18,19] of shapelet discovery is to extract a pool of subsequences as shapelet candidates, rank them with an evaluation metric, and select the top-ranking ones as shapelets. For the choice of the evaluation metric, *supervised* metrics [7,10,18] can effectively discover high-quality shapelets. However, they need labeled examples from both classes, while

[1] The term *positive unlabeled* can be confusing, where *positive* actually means *positive labeled*. In this paper, we still use *positive unlabeled (PU)* to refer to what is actually *positive-labeled unlabeled*. However, in other cases, we use *positive/negative* to refer to all positive/negative examples, regardless of whether they are labeled or not. Positive examples that are labeled will be explicitly referred to as being *positive labeled (PL)*.

[2] In this paper, we use the terms *subsequence* and *pattern* interchangeably.

under PU settings, only one class is (partly) labeled. An *unsupervised* evaluation metric [15, 19] aims to maximize the inter-class gap and minimize the intra-class variance, yet this rationale often fails to hold, which is likely due to the typically noisy and high-dimensional nature of time series, and the sparsity of small datasets.

Faced with the difficulties of directly applying existing shapelet discovery methods, we propose our novel PU-Shapelets (PUSh) algorithm. To be specific, we opt to first label the unlabeled (U) examples, thus obtaining a fully labeled training set. This enables us to conduct supervised shapelet discovery. To label the U examples, we present a novel **Pattern Ensemble (PE)** technique that iteratively ranks all U examples by their possibilities of being positive. PE utilizes both *potentially* characteristic and *potentially* non-characteristic shapelet candidates, without the need to know their *actual* quality. We then develop a novel **Average Shapelet Precision Maximization (ASPM)** stopping criterion. Based on a novel concept called *shapelet precision*, ASPM determines the point where the PE iterations should stop [3, 4, 6, 13, 16, 17]. All U examples ranked before and at this point are labeled as being positive and the rest are considered negative. ASPM is essentially an estimation of the number of positive examples in U. Having labeled the entire training set, we select the shapelets with the supervised evaluation metric of *information gain* [10, 18]. The discovered shapelets are used to build a nearest-neighbor classifier for online classification. The complete workflow of PUSh is shown in Fig. 2.

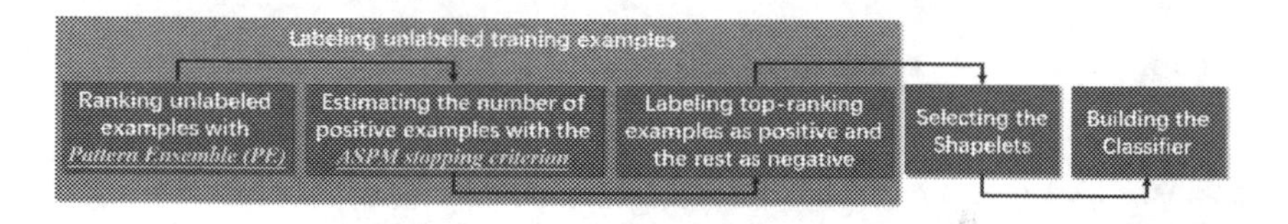

Fig. 2. The workflow of our PU-Shapelets (PUSh) algorithm.

Our main contributions in the paper are as follows.

- We present PU-Shapelets (PUSh), which addresses the challenging task of discovering time series shapelets [7, 10, 15, 18, 19] with positive unlabeled (PU) data. As far as we know, this is the first time this task has been undertaken.
- We develop a novel **Pattern Ensemble (PE)** technique to iteratively rank the unlabeled (U) examples by their possibilities of being positive. PE effectively utilizes both *potentially* characteristic and *potentially* non-characteristic patterns, without the need to know their *actual* quality.
- We present a novel **Average Shapelet Precision Maximization (ASPM)** stopping criterion. Based on a novel concept called *shapelet precision*, ASPM can effectively estimate the number of positive examples in U and determine when to stop the PE iterations. We combine PE and ASPM to label all U examples. We then conduct supervised shapelet selection and build a nearest-neighbor classifier for online classification.

- We conduct extensive experiments to demonstrate the effectiveness of our PUSh method.

The rest of the paper is organized as follows. Section 2 introduces the preliminaries. Section 3 presents our PUSh algorithm. Section 4 reports the experimental results. Section 5 reviews the related work. Section 6 concludes the paper.

2 Preliminaries

We now formally define several important concepts used in this paper. We begin with the concept of positive unlabeled classification [8].

Definition 1 ***Positive unlabeled (PU) classification.*** *Given a training set with a (small) set PL of positive labeled examples and a (large) set U of unlabeled examples, the task of positive unlabeled (PU) classification is to train a classifier with P and U and apply it to predicting the class of future examples.*

We move on to the definitions of time series and subsequence.

Definition 2 ***Time series and subsequence.*** *A time series is a sequence of real values in timestamp ascending order. For a length-L time series $T = t_1, \ldots, t_L$, a subsequence S of T is a sequence of contiguous data points in T. The length-l ($l \leq L$) subsequence that begins with the p-th data point in T is written as $S = t_p, \ldots, t_{p+l-1}$.*

We then introduce the concept of subsequence matching distance (SMD).

Definition 3 ***Subsequence matching distance (SMD).*** *For a length-l subsequence $Q = q_1, \ldots, q_l$ and a length-L time series $T = t_1, \ldots, t_L$, the subsequence matching distance (SMD) between Q and T is the minimum distance between Q and all length-l subsequences of T under some distance measure D, i.e. $SMD(Q,T) = min\{D(Q,S)|S = t_p, \ldots, t_{p+l-1}, \forall p, 1 \leq p \leq L-l+1\}$.*

For the choice of the distance measure, we apply the *length-normalized Euclidean distance* [10], which is the Euclidean distance between two equal-length subsequences divided by the square root of the length of the subsequences.

We now formally define the concept of orderline [10,15,18,19].

Definition 4 ***Orderline.*** *Given a subsequence S and a time series dataset DS, the corresponding orderline O_S is a sorted vector of SMDs between S and all time series in DS.*

We conclude with the definition of time series shapelets.

Definition 5 ***Time series shapelets.*** *Given a set DS of training time series, time series shapelets are characteristic subsequences that can distinguish between different classes in DS. Concretely, given a shapelet candidate set CS consisting of subsequences extracted from time series in DS, let m be the desired number of shapelets. Time series shapelets are the top-m ranking subsequences in CS under some evaluation matric E. E indicates how well separated different classes are on the orderline of a shapelet candidate.*

3 The PU-Shapelets Algorithm

We now present our PU-Shapelets (PUSh) algorithm. Following the workflow shown in Fig. 2, we will first elaborate on how to label the unlabeled (U) set, and then introduce the shapelet selection and classifier construction processes.

3.1 Labeling U Examples

We now introduce the process of labeling U examples. Our first step is to obtain a pool of patterns as shapelet candidates, which will also be useful when labeling the U set. Concretely, we set a range of possible shapelet lengths. For each length l, we apply a length-l sliding window to each training time series (regardless of whether it is labeled or not), extracting all length-l subsequences in the training set and adding them to the candidate pool. The final pool of shapelet candidates is obtained when all possible lengths are exhausted [7,10,15,18,19].

Having obtained all shapelet candidates, we now move on to labeling the U set. This is typically achieved by first rank the U examples by their possibilities of being positive, and then estimate the number np_u of positive examples in U [3,4,6,11–13,16,17], thus the top-ranking np_u examples in U are labeled as being positive and the rest are labeled as being negative. This workflow has been illustrated in Fig. 2. We now separately discuss how to rank the U examples, and how to estimate the number of positive examples.

Ranking U Examples with Pattern Ensemble (PE). We first discuss ranking the U examples. Previous works [3,4,6,13,16,17] have adopted the *propagating one-nearest-neighbor* (P-1NN) algorithm [21]. P-1NN works in an iterative fashion. In each iteration, the nearest neighbor of the positive labeled (PL) set in the unlabeled (U) set is moved from U to PL. The nearest neighbor of PL in U is defined as the U example with the minimum nearest neighbor distance to PL, i.e.

$$NN(PL, U) = \arg\min\{NNDist(u, PL) \mid u \in U\} \tag{1}$$

The iterations go on until U is exhausted. The order by which the U examples are added into PL is their rankings.

The problem with previous works is that when obtaining the nearest neighbors, they calculate the distances between entire time series, utilizing all the readings. This makes them susceptible to non-characteristic readings. In contrast, we attempt to actively minimize the interference from non-characteristic shapelet candidates. However, as was discussed in Sect. 1, no existing evaluation metric can effectively estimate the qualities of the candidates under PU settings. Without such prior knowledge, which candidates should we rely on? The answer is surprisingly simple: *All* of them.

To be specific, we develop the following **Pattern Ensemble (PE)** technique, whose workflow is shown in Fig. 3. PE adopts a similar iterative process as P-1NN. However, in each iteration of PE, we let each shapelet candidate individually identify the nearest neighbor of PL in U on its orderline (Fig. 4),

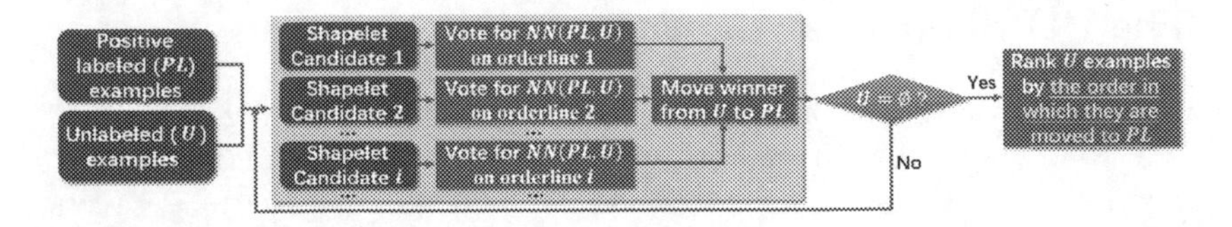

Fig. 3. The workflow of pattern ensemble (PE).

and vote for it. The U example receiving the most votes is moved to PL. The iterations stop when U is exhausted, and the order by which U examples are moved to PL is their rankings.

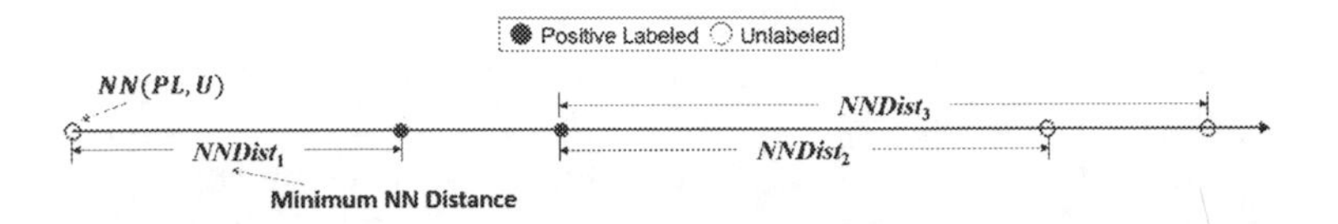

Fig. 4. An illustration of finding the nearest neighbor of PL in U on an orderline.

At first glance, PE seems highly unlikely to perform well, especially when the number of non-characteristic patterns significantly exceeds that of characteristic ones. However, note that in many cases, *while the non-characteristic patterns do not significantly favor the positive set, they are not significantly biased to the negative set either*. This is because non-characteristic readings exist not only in negative examples, but also positive ones. As a result, various non-characteristic patterns can vote for both negative and positive examples, thus cancelling out the effect of each other. On the other hand, the characteristic patterns strongly favor the positive class, ensuring that an actual positive example wins the vote. This effect is illustrated in Fig. 5.

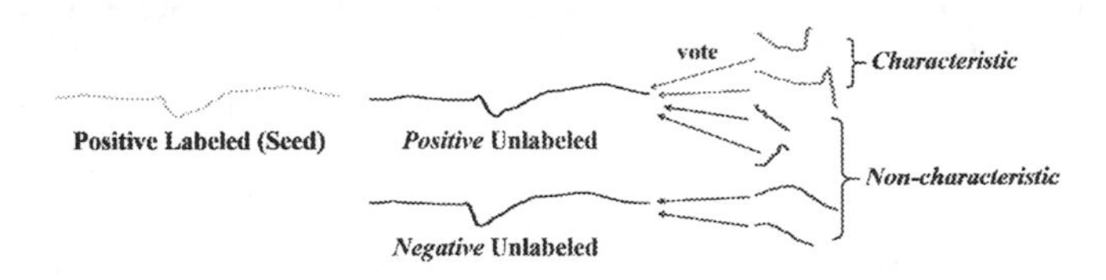

Fig. 5. An illustration of the rationale of pattern ensemble. Here PL contains only one example. While the votes from non-characteristic patterns cancel each other out, the characteristic patterns ensure the correct example is chosen.

Compared with previous works [3,4,6,13,16,17], our method also utilizes potentially non-characteristic readings. The critical difference is that for each time series, our method exploits *multiple* patterns. In contrast, previous works utilize *only one* pattern (i.e. the entire time series itself). This means the negative effect of one non-characteristic pattern cannot be cancelled out by the positive effect of another, making previous works less robust than our method.

Algorithm 1. PatternEnsemble(PL, U, CS)

Input : initial positive labeled examples PL, initial unlabeled examples U, shapelet candidate pool CS
Output: U example rankings R (by the U examples' possibilities of being positive)

```
1  np_0 = |PL|;
2  while U ≠ ∅ do
3      votes = zeros(1, |PL| + |U|);
4      foreach S ∈ CS do
5          u_s = FindNN(PL,U,S);
6          votes(u_s) + +;
7      nextP =argmax(votes);
8      PL = [PL, nextP]; U = U \ {nextP};
9  R = PL(np_0 + 1 : end);
10 return R;
```

The complete process of ranking U examples with PE is illustrated in Algorithm 1. To begin with, we cache the number of initial positive labeled examples (line 1), then iteratively take the following steps until U is exhausted (line 2): We first initiate a vote counter (line 3; note that among the $|PL| + |U|$ indices, only $|U|$ are valid. The others are simply used to avoid index mapping.). Then, we let every shapelet candidate S (line 4) identify the nearest neighbor u_s of PL in U on its orderline (line 5) and vote for it (line 6). The U example receiving the most votes (line 7) is moved to PL (line 8). The order by which the U examples are added into PL is their rankings. (lines 9–10).

The Average Shapelet Precision Maximization (ASPM) Stopping Criterion. With the U examples ranked, we can now move on to estimating the number np_u of positive examples in U. Note that for iterative algorithms such as the aforementioned P-1NN [21] and our PE, estimating np_u is essentially finding a *stopping criterion* to decide when to stop the iterations. All examples ranked before and at the stopping point is labeled as being positive, and the rest are considered negative. Previous works [3,6,13,16,17] have proposed several stopping criteria for whole-stream based P-1NN algorithms. However, these methods are susceptible to interference from non-characteristic readings, and some [6,13,17] are incompatible with our PE technique.

In light of these drawbacks, we present a brand new stopping criterion tailored to our PE technique. We first introduce the novel concept of *shapelet precision*. In a certain iteration of PE, for the current PL set and a pattern S, let LS and

RS be the sets of the leftmost and rightmost $|PL|$ examples on the orderline of S. The shapelet precision (SP) of S with respect to PL is

$$SP_S^{PL} = \frac{\max(|PL \cap LS|, |PL \cap RS|)}{|PL|} \tag{2}$$

For example, for the orderline in Fig. 4, we have $|PL| = 2$ in the current iteration. One of the two leftmost examples is in PL, and none of the two rightmost examples is in PL, thus the shapelet precision is $max(1, 0)/2 = 0.5$.

Note that SP is derived from the concept of *precision* in the classification literature. Essentially, we "classify" the leftmost (or rightmost) $|PL|$ examples on the orderline as being positive, and evaluate the "classification" performance with SP. Intuitively, at the best stopping point where PL is most similar to the *actual* positive set (which is unknown for U), the *average SP* (ASP) value of the top shapelet candidates should be the highest. Based on this intuition, we develop the following **Average Shapelet Precision Maximization (ASPM)** stopping criterion, whose workflow is illustrated in Fig. 6.

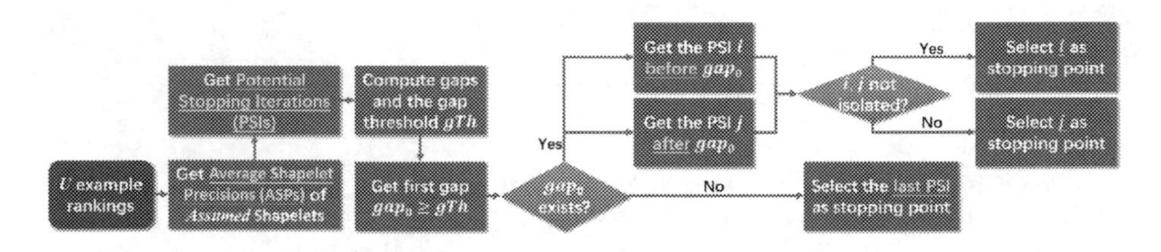

Fig. 6. The workflow of the Average Shapelet Precision Maximization (ASPM) stopping criterion.

To be specific, after each iteration of PE (lines 3–10 of Algorithm 1), we select the top-k *assumed* shapelets (rather than *actual* shapelets, since we do not know if they are actually the final shapelets yet) with the highest SP values and calculate their ASP score. Each iteration with the highest ASP value is considered to be a *potential stopping iteration* (PSI). Note that the ASP score of the last iteration is always 1, since at this point all examples are labeled as being positive and the SP scores of all shapelet candidates are 1. In response, we disregard this trivial case.

To break the ties between multiple PSIs, we consider their *gaps*. Suppose iterations i and j ($i < j$) are two consecutive PSIs (i.e. all iterations between them, if any, are non-PSIs with lower ASP scores), their gap is defined as

$$gap(i, j) = \begin{cases} j - i - 1, & \text{if } j - i - 1 > 1 \\ 0, & \text{otherwise} \end{cases} \tag{3}$$

Essentially, the gap between i and j is the number of non-PSIs between them. If the gap is 1, we consider it accidental and reset the gap to 0.

After getting all gaps, we set a gap threshold gTh that equals half the maximum gap between consecutive PSIs (except when the maximum gap is 0, where gTh is set to a random positive value). We find the first "large" gap $gap_0 \geq gTh$. Under the assumption that the positive class is relatively compact while the negative class can be diverse [4], gap_0 indicates a decision boundary between the positive and negative classes. Later "large" gaps may indicate boundaries between sub-classes of the negative class. At gap_0, we select the final stopping point in one of three cases (Fig. 7).

1. No gap_0 exists (namely the maximum gap is 0). Here we select the last PSI as the stopping point. The rationale is that on the orderlines of multiple *assumed* shapelets, the rankings of the negative examples are too diverse to yield a high ASP score, thus all PSIs correspond to the positive class.
2. Neither of the PSIs i before gap_0 and j after gap_0 is *isolated* (we say a PSI is isolated if there are no PSIs before and after it within the range of gTh). Here we select i as the stopping point. The rationale is that in the last few iterations before i, the rankings of the remaining positive unlabeled examples are relatively uniform on multiple orderlines, resulting in high ASPs before and at i. Similarly, the rankings of the first few negative unlabeled examples are relatively uniform, resulting in high ASPs at and after j.
3. At least one of i and j is isolated. Here we select j as the stopping point. Empirically, if i is isolated, i being a PSI is more likely a coincidence. If j is isolated, it is more likely that on multiple orderlines, the rankings are diverse for both the last few positive unlabeled examples (between i and j) and the first few negative unlabeled examples (after j), yet a clear decision boundary between the positive and negative classes is at j, resulting in an isolated point with a high ASP score.

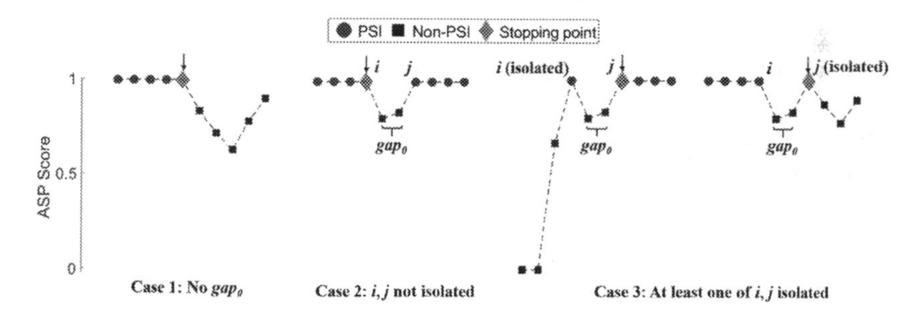

Fig. 7. Different strategies of stopping point selection in the three cases of ASPM. Note that we have left out the last iteration for its triviality.

To determine the number of *assumed* shapelets k, we set a largest allowed value $maxK$ and examine all $k \in [1, maxK]$. We find the stopping point for each k and pick the one with the maximum gap_0. Ties are broken by picking the one with the latest stop. This reduces the risk of false negatives, which is more troublesome than false positives in applications such as anomaly detection in medical care. Also, to prevent too early or too late a stop, we pre-set the lower and upper bounds of the stopping point. Note that we usually only need loose bounds to yield satisfactory performance, which are relatively easy to estimate in real applications.

Our ASPM stopping criterion is illustrated in Algorithm 2. After initiation (line 1), we examine each possible number k of *assumed* shapelets (line 2). We first calculate the ASP values for all iterations except the last (lines 3–6), and then obtain the PSIs (line 7). Next, we obtain the gap threshold gTh (line 8) and gap_0 along with the two PSIs before and after it (line 9). We then select the stopping point for the current k (lines 10–12), and update the best-so-far stopping point if the current k is the better than previous ones (lines 13–14). The best stopping point is obtained after examining all k values (line 15).

Algorithm 2. ASPM(R, CS, lb, ub, $maxK$)

Input : the U example rankings R, the shapelet candidate pool CS, the lower and upper bounds of the stopping point lb and ub, the maximum number of *assumed* shapelets $maxK$

Output: the stopping point $bestStop$

```
1  bestStop = INF; maxGap0 = −INF;
2  for k = 1 : maxK do
3      aspList = [];
4      for iter = 1 : |R| − 1 do
5          asp = getAvgShapeletPrecision(CS, R, iter, k);
6          aspList = [aspList, asp];
7      psiList = getPotentialStopIter(aspList, lb, ub);
8      maxGap = getMaxGap(psiList); gTh = ⌈maxGap/2⌉;
9      [gap0, i, j] = getGap0(psiList, gTh);
10     if gap0 == 0 then stop = psiList(end);
11     else if !(isIsolated(i) || isIsolated(j)) then stop = i;
12     else stop = j;
13     if maxGap0 < gap0 then maxGap0 = gap0; bestStop = stop;
14     else if maxGap0 == gap0 && bestStop < stop then bestStop = stop;
15 return bestStop;
```

Having obtained the stopping point, we label all U examples ranked before and at it as being positive and the rest as being negative. The newly labeled U examples and the initial PL examples make up a fully labeled training set.

3.2 Selecting the Shapelets and Building the Classifier

With a fully labeled training set, we can now select the shapelets using a supervised evaluation metric [7,10,18]. Concretely, we adopt the classic [18] *information gain* metric [10,18] to rank and select the top-m shapelet candidates as the final shapelets. Next, we conduct feature extraction with the shapelets. To be specific, we represent each training time series with an m-dimensional feature vector in which each value is the SMD between the time series and one of the shapelets. This representation is called *shapelet transformed representation* [7]. The feature vectors are used to train a one-nearest-neighbor classifier. To classify a future time series, we obtain its shapelet transformed representation and assign to it the label of its nearest neighbor in the training set.

4 Experiments

For experiments, we use 21 datasets from [5]. For brevity, we omit further description of the datasets. The names of the datasets will be presented along with the experimental results, and their detailed information can be found on [5].

All datasets have been separated into training and test sets by the original contributors [5]. We designate examples with the label "1" in each dataset as being positive, and all others as being negative. Let the number of positive training examples in each dataset be np, For datasets with $np \geq 10$, we randomly generate 10 initial PL sets for each dataset, each containing 10% of all positive training examples. For datasets with $np < 10$, we generate np initial PL sets, each containing one positive training example. All experimental results are averaged over the 10 (or np) runs.

Our baseline methods come from [3,4,6,13,17]. Like our PUSh method, they also label the U examples by first ranking them, and then find a stopping criterion. To rank the U examples, the baselines utilize the P-1NN algorithm [21] (see Sect. 3.1) on the original time series with one of three distance measures: Euclidean distance (**ED**) [3,4,17] **DTW** [6,13], and **DTW-D** [4]. As to stopping criteria, our baselines utilize eight stopping criteria: **W** [17], **R** [13], **B** [3] and **G1–G5** [6] which are a family of five stopping criteria. The description of criterion W in [17] is insufficient for us to accurately implement it. Luckily, another criterion is *implicitly* used by [17], which is the one we use. To make up for not testing the former, we first find a stopping point using the latter and then examine all iterations before and at this point, reporting only the best performance achieved. Also, criterion B [3] only supports initial PL sets with a single example. For initial PL sets with multiple examples, we use each example to individually find a stopping point and pick the one with the minimum RDL value (RDL is a metric used in [3] to determine the stopping point). We compare PUSh (i.e. PE+ASPM) against the combination of each of the three U example ranking methods with each of the eight stopping criteria, resulting in a total of 24 baseline methods. For all 25 methods being compared, we label all U examples before and at the stopping point as being positive, and the rest as being negative. The fully labeled training set is used for one-nearest-neighbor classification.

For parameter settings, we set the range of possible shapelet lengths to $10 : (L-10)/10 : L$, where L is the time series length. For Algorithm 2, we set the lower bound lb to 5 if the number of positive examples $np \geq 10$. Otherwise, it is set to 1 which is essentially no lower bound at all. The upper bound ub is set to $n \times 2/3 - np_0$, where n is the total number of training examples and np_0 is the size of the initial PL set. This means we assume that the positive class makes up no more than two thirds of all training data. Again, we stress that these settings are usually loose bounds than can be estimated relatively easily. For fairness, we apply the same lower and upper bounds to our baselines. The maximum number of *assumed* shapelets $maxK$ is set to 200. The number of final shapelets m is set to 10. As we will show later, our method is not sensitive to m. For DTW and DTW-D, we set the warping constraints as the values provided on [5], including the setting of no constraint if it yields better supervised performance. If this setting is 0, DTW is reduced to ED and DTW-D is ineffective. In such cases, we set the constraints to 1%, 2%, ..., 10% of the time series length L, and only report the best results. The parameters *cardinality* and β for criteria B [3] and G1–G5 [6] are set to 16 and 0.3 as suggested by the original authors.

For reproducibility, our source code and all raw experimental results can be found on [1]. All experiments were run on a laptop computer with Intel Core i7-4710HQ @2.50 GHz CPU, NVIDIA GTX850M graphics card (GPU acceleration was used to speed up DTW computation [14]), 12 GB memory and Windows 10 operating system.

4.1 Performance of Labeling the U Examples

We first look into the performance of labeling the U set. Note that this can be seen as classifying the U set, thus we can apply an evaluation metric for classification. Here we adopt the widely used [6,11–13,16] **F1-score**, which is defined as $F1 = 2 \times precision \times recall/(precision + recall)$.

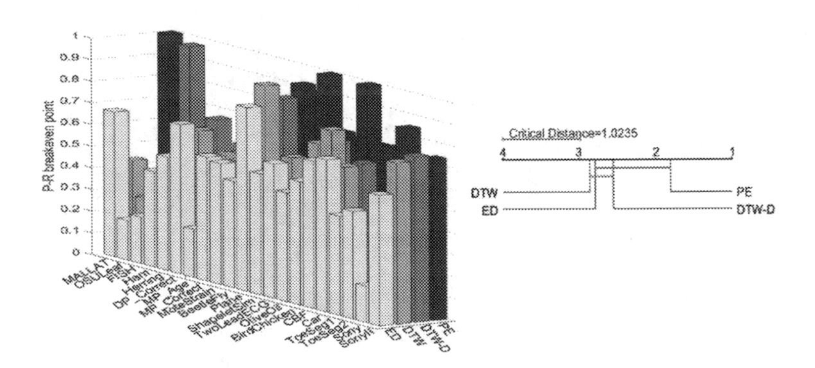

Fig. 8. Performances of ranking the U examples (disregarding the stopping criteria). *(left)* Precision-recall breakeven points. *(right)* Critical difference diagram for all four methods. PE outperforms all baseline methods and significantly outperforms DTW.

We first evaluate the performance of PE. In this case, we need to disregard the effect of the stopping criterion. Therefore, we assume the *actual* number of positive examples np is known, and the stopping point is where there are np examples in PL. At this point, precision, recall and F1-score share the same value, which is called *precision-recall (P-R) breakeven point* [17]. The P-R breakeven points for all methods are illustrated in Fig. 8. There are no significant differences among the performances of the three baseline methods. Our PE outperforms all baselines and significantly outperforms DTW.

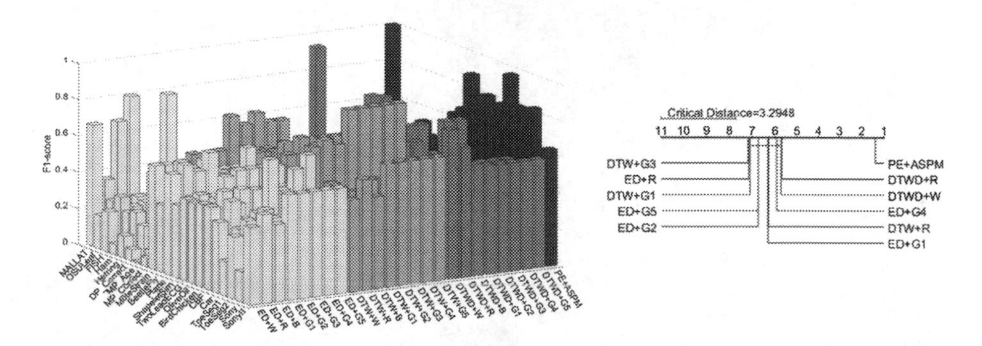

Fig. 9. Performances of labeling the U set (taking into account the stopping criteria). *(left)* F1-scores. *(right)* Critical difference diagram for PUSh (PE+ASPM) and the top-10 baselines. PUSh significantly outperforms the others.

We then take the stopping criteria into account. The F1-scores at the stopping points are shown in Fig. 9. Among the top-10 baselines, no significant difference in performance is observed. Most top ranking baselines utilize one of G1–G5. Their high performances is likely due to G1–G5's abilities to take into account long term trends in minimum nearest neighbor distances [6]. Our PUSh (PE+ASPM) significantly outperforms the top-10 baselines.

4.2 Performance of Online Classification

We now move on to classification performance. Once again we use the F1-score for evaluation. We need to first set the number of shapelets m for our PUSh. We have set $m = 10:10:50$ and performed pairwise Wilcoxon signed rank test on the performances of PUSh under these settings. The minimum p-value is 0.0766. With 0.05 as the significance threshold, there are no significant differences among these settings. We set m to 10 for shorter running time.

The classification performances are shown in Fig. 10. Not surprisingly, most of the top ranking methods in the U example labeling process (Fig. 9) remain highly competitive. This is because for online classification, the labels of the training examples are the labels obtained from the U example labeling process, *not* the *actual* labels (which are unknown for U). Therefore the performance of labeling U directly affects the classification performance. While no significant difference is observed among the top-10 baselines, our PUSh (PE+ASPM) significantly outperforms nine of them and is as competitive as DTWD-R.

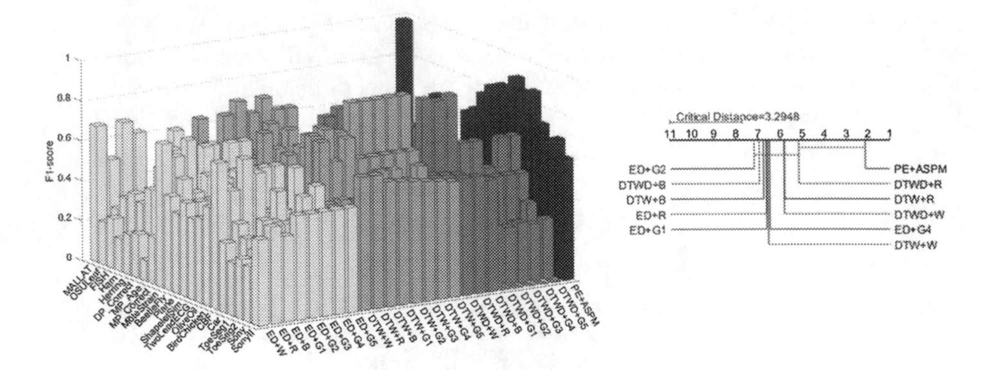

Fig. 10. Online classification performances. *(left)* F1-scores. *(right)* critical difference diagram for PUSh (PE+ASPM) and the top-10 baselines. PUSh is as competitive as DTWD-R and significantly outperforms the others.

4.3 Running Time

We now look into the efficiency aspect of PUSh. For the training step (from labeling the U examples to building the classifier, see Fig. 2), the computational bottlenecks are obtaining the orderlines and calculating the shapelet precisions. Let the number of training examples be N and the length of the time series be L, there are $O(NL)$ shapelet candidates. For each candidate, the amortized time to obtain its orderline is $O(NL)$ using the fast algorithm proposed by [10], and the time to calculate its SP values in all iterations is $O(N^2)$, thus the total time is $O(N^2L^2) + O(N^3L)$. As is shown in Fig. 11 (*left*), despite the relatively high time complexity, PUSh is able to achieve reasonable running time, with the longest average running time less than 1100 s.

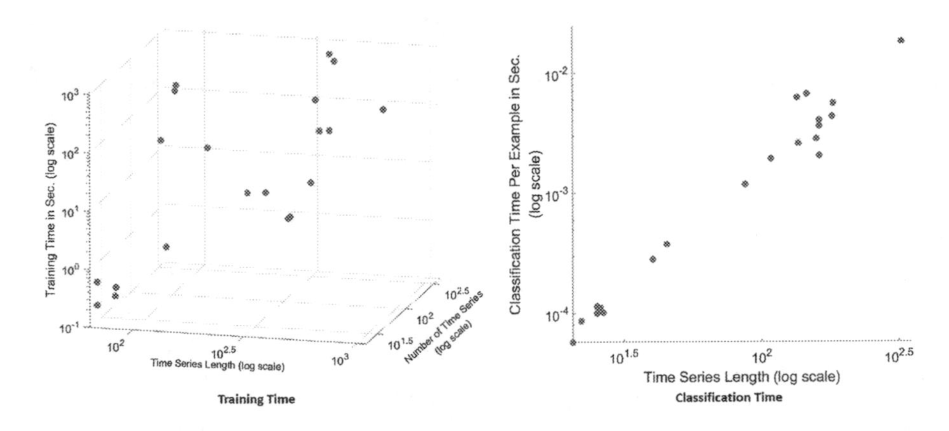

Fig. 11. Running time of PUSh. Note that all axes are in log scale. *(left)* Training time. *(right)* Online classification time per example.

For online classification, the bottleneck is to obtain the shapelet transformed representation [7] (see Sect. 3.2), whose time is $O(L^2)$ per test example. As is shown in Fig. 11 (*right*), for time series lengths in the order of 10^2 to $10^{2.5}$ (which is typical in applications such as heartbeat classification [3] in medicine), the average time is in the order of 10^{-3} to 10^{-2} s, which is sufficient for real-time processing.

5 Related Work

PU classification of time series [3,4,6,11–13,16,17] is a relatively less well-studied task in time series mining. Most existing works [3,4,6,13,16,17] are whole-stream based propagating one-nearest-neighbor [21] algorithms which tend to be sensitive to non-characteristic readings [15,19]. [11,12] selects local features from time series. However, the selected features are *discrete readings* that do not necessarily form *continuous patterns*, while the latter often contains valuable information on the trend of the data. In this work, we explicitly discover characteristic patterns called shapelets, which have been applied to supervised classification [7,10,18] and clustering [15,19,20]. Previous works utilize supervised [7,10,18] and unsupervised evaluation [15,19] metrics to assess shapelet candidates. However, both are not directly suitable for PU settings. Therefore we opt to first label the U set and then conduct supervised shapelet discovery.

Most previous works on PU classification of time series [3,4,6,13,16,17] iteratively rank the U examples by their possibilities of being positive. A stopping criterion is needed to determine where to stop the iterations. Existing stopping criteria can be divided into two types: Distance-based criteria [6,13,17] utilize distances between PL and U to decide the stopping point. Minimum description length based criteria [3,16] utilize the initial PL to encode the training set. The stopping point is where the encoding is most compact. Both types of criteria suffer from the interference from non-characteristic readings, and distance-based criteria are not compatible to our method. This has motivated us to develop our novel ASPM stopping criterion.

6 Conclusions and Future Work

In this paper, we have taken on the challenging task of positive unlabeled [8] discovery of time series shapelets [7,10,15,18,19]. To label the U set, we have developed a novel pattern ensemble (PE) method that ranks U examples with both *potentially* characteristic and *potentially* non-characteristic patterns, with no need to know their *actual* qualities. We have also developed a novel ASPM stopping criterion, which estimates the number of positive examples based on the novel concept of shapelet precision. After labeling the entire training set, we have conducted supervised shapelet selection and built a one-nearest-neighbor classifier. Extensive experiments have demonstrated the effectiveness and efficiency of our method. Currently, our method utilizes the orderlines of all shapelet candidates, which is highly costly in terms of space and time efficiency. For future work, we plan to develop heuristics for more efficient selection of shapelet candidates for the PE subroutine. We also plan to apply GPU acceleration to our method.

References

1. PU-Shapelets source code. https://github.com/sliang11/PU-Shapelets
2. Bagnall, A., Lines, J., Bostrom, A., Large, J., Keogh, E.: The great time series classification bake off: a review and experimental evaluation of recent algorithmic advances. Data Min. Knowl. Discov. **31**(3), 606–660 (2017)
3. Begum, N., Hu, B., Rakthanmanon, T., Keogh, E.: Towards a minimum description length based stopping criterion for semi-supervised time series classification. In: 2013 IEEE 14th International Conference on Information Reuse Integration, pp. 333–340 (2013)
4. Chen, Y., Hu, B., Keogh, E., Batista, G.: DTW-D: time series semi-supervised learning from a single example. In: Proceedings of the 19th ACM SIGKDD International Conference on Knowledge Discovery and Data Mining, pp. 383–391 (2013)
5. Chen, Y., et al.: The UCR time series classification archive, July 2015. www.cs.ucr.edu/~eamonn/time_series_data/
6. González, M., Bergmeir, C., Triguero, I., Rodríguez, Y., Benítez, J.: On the stopping criteria for k-nearest neighbor in positive unlabeled time series classification problems. Inf. Sci. **328**, 42–59 (2016)
7. Hills, J., Lines, J., Baranauskas, E., Mapp, J., Bagnall, A.: Classification of time series by shapelet transformation. Data Min. Knowl. Discov. **28**(4), 851–881 (2014)
8. Li, X.-L., Liu, B.: Learning from positive and unlabeled examples with different data distributions. In: Gama, J., Camacho, R., Brazdil, P.B., Jorge, A.M., Torgo, L. (eds.) ECML 2005. LNCS (LNAI), vol. 3720, pp. 218–229. Springer, Heidelberg (2005). https://doi.org/10.1007/11564096_24
9. Ma, J., Sun, L., Wang, H., Zhang, Y., Aickelin, W.: Supervised anomaly detection in uncertain pseudoperiodic data streams. ACM Trans. Internet Technol. **16**(1), 4:1–4:20 (2016)
10. Mueen, A., Keogh, E., Young, N.: Logical-shapelets: an expressive primitive for time series classification. In: Proceedings of the 17th ACM SIGKDD International Conference on Knowledge Discovery and Data Mining, pp. 1154–1162 (2011)
11. Nguyen, M.N., Li, X., Ng, S.: Positive unlabeled learning for time series classification. In: Proceedings of the Twenty-Second International Joint Conference on Artificial Intelligence, pp. 1421–1426 (2011)
12. Nguyen, M.N., Li, X.-L., Ng, S.-K.: Ensemble based positive unlabeled learning for time series classification. In: Lee, S., Peng, Z., Zhou, X., Moon, Y.-S., Unland, R., Yoo, J. (eds.) DASFAA 2012. LNCS, vol. 7238, pp. 243–257. Springer, Heidelberg (2012). https://doi.org/10.1007/978-3-642-29038-1_19
13. Ratanamahatana, C.A., Wanichsan, D.: Stopping criterion selection for efficient semi-supervised time series classification. In: Lee, R. (ed.) Software Engineering, Artificial Intelligence, Networking and Parallel/Distributed Computing. SCI, vol. 149, pp. 1–14. Springer, Heidelberg (2008). https://doi.org/10.1007/978-3-540-70560-4_1
14. Sart, D., Mueen, A., Najjar, W., Keogh, E., Niennattrakul, V.: Accelerating dynamic time warping subsequence search with GPUs and FPGAs. In: 2010 IEEE 10th International Conference on Data Mining, pp. 1001–1006 (2010)

15. Ulanova, L., Begum, N., Keogh, E.: Scalable clustering of time series with U-shapelets. In: Proceedings of the 2015 SIAM International Conference on Data Mining, pp. 900–908 (2015)
16. Vinh, V.T., Anh, D.T.: Two novel techniques to improve MDL-based semi-supervised classification of time series. In: Nguyen, N.T., Kowalczyk, R., Orłowski, C., Ziółkowski, A. (eds.) Transactions on Computational Collective Intelligence XXV. LNCS, vol. 9990, pp. 127–147. Springer, Heidelberg (2016). https://doi.org/10.1007/978-3-662-53580-6_8
17. Wei, L., Keogh, E.: Semi-supervised time series classification. In: Proceedings of the 12th ACM SIGKDD International Conference on Knowledge Discovery and Data Mining, pp. 748–753 (2006)
18. Ye, L., Keogh, E.: Time series shapelets: a novel technique that allows accurate, interpretable and fast classification. Data Min. Knowl. Discov. **22**(1–2), 149–182 (2011)
19. Zakaria, J., Mueen, A., Keogh, E.: Clustering time series using unsupervised-shapelets. In: 2012 IEEE 12th International Conference on Data Mining, pp. 785–794 (2012)
20. Zhou, J., Zhu, S., Huang, X., Zhang, Y.: Enhancing time series clustering by incorporating multiple distance measures with semi-supervised learning. J. Comput. Sci. Technol. **30**(4), 859–873 (2015)
21. Zhu, X., Goldberg, A.B.: Introduction to semi-supervised learning. In: Synthesis Lectures on Artificial Intelligence and Machine Learning, vol. 3, no. 1, pp. 1–130 (2009)

Clustering and Classification

Discovering Relationship Patterns Among Associated Temporal Event Sequences

Chao Han[1], Lei Duan[1,2(✉)], Zhangxi Lin[3], Ruiqi Qin[1], Peng Zhang[1], and Jyrki Nummenmaa[4]

[1] School of Computer Science, Sichuan University, Chengdu, China
scuhanchao@163.com, 18610660375@163.com, leiduan@scu.edu.cn, richforgood@163.com, zp_jy1993@163.com
[2] West China School of Public Health, Sichuan University, Chengdu, China
[3] Texas Tech University, Lubbock, Texas, USA
[4] Tampere University, Tampere, Finland
jyrki.nummenmaa@tuni.fi

Abstract. Sequential data mining is prevalent in many real world applications, such as gene sequence analysis, consumer shopping log analysis, social networking analysis, and banking transaction analysis. Contrast sequence data mining is useful in describing the differences between two sets (classes) of sequences. However, in prior studies, little work has been done in how to mine the patterns from sequences formed by associated temporal events, where there exist relationships in chronological order between any two events in a sequence. To fill this gap, we consider the problem of mining associated temporal relationship pattern (ATRP) and propose a method, called *ATTEND (AssociaTed Temporal rElationship patterN Discovery)*, to discover ATRPs with top contrast measure from two sets of associative temporal event sequences. Moreover, we design several heuristic strategies to improve the efficiency of *ATTEND*. Experiments on both real and synthetic data demonstrate that *ATTEND* is effective and efficient.

Keywords: Contrast sequence data mining · Relationship pattern · Temporal event sequence

1 Introduction

Sequential data exists in many fields including gene sequence analysis, shopping log analysis, social networking analysis, banking transaction analysis, etc. Mining sequential patterns from sequences can unveil useful hidden information and provide decision support. Distinguishing sequential patterns (DSPs) [1,2], based on the events frequently occurring in a class of sequences but infrequently in

This work was supported in part by NSFC 61572332, the Fundamental Research Funds for the Central Universities 2016SCU04A22, and the China Postdoctoral Science Foundation 2016T90850.

G. Li et al. (Eds.): DASFAA 2019, LNCS 11446, pp. 107–123, 2019.
https://doi.org/10.1007/978-3-030-18576-3_7

another, can describe the differences between two associated sequence classes. Consider the scenario where a male patient $\mathcal{A}$ and two female patients $\mathcal{B}$ and $\mathcal{C}$ all suffer the same disease, amyotrophic lateral sclerosis (ALS). By clinical records, it can be found that patients of different genders may have some distinguishable symptoms (a patient's record is a temporal event sequence). Moreover, relationships can be found between different symptoms such as the onset of heavy periods came *before* that of spotting menstrual for patients $\mathcal{B}$ and $\mathcal{C}$, while for $\mathcal{A}$, the symptom prostatic hyperplasia came *after* the symptom enlarged prostate. Such relationships could be clinically interesting, but hard to formulate with traditional DSPs. Intuitively, there are different relationships among events, such as the order and duration, which can provide informative characteristics of the events. Thus, it is necessary to take the relationships among events into consideration when mining associated data sequences.

However, there was no existing sequential data mining or DSP mining methods aiming at solving the above problem. To fill this gap, we propose a novel problem of finding DSPs with temporal relationships, i.e., we aim to discover relationship patterns from associated temporal event sequences, named as *associative temporal relationship patterns* (ATRPs). Once the relationship patterns are identified, they can be further applied for clustering even sequences in accordance with given criteria.

Despite there are existing studies on DSP mining and event temporal relationship mining respectively, none of these studies has ever focused on the DSP mining taking temporal relationships among events into consideration. Thus, without either the DSP or the temporal relationship, such methods cannot find the ATRPs.

Nevertheless, finding ATRPs is a challenging problem due to the following reasons. First, in order to guarantee the completeness of solution space, all possible candidate event patterns as well as various temporal relationships should be enumerated. However, as the number of all candidates is extremely large, reducing the enumeration space becomes a challenging task. Second, the time complexity is high because of the large number of candidates. Thus, it is important to develop adequate techniques to empower our method for the efficiency of discovering ATRPs.

Coping with the above challenges, we made the following contributions: (1) proposing a novel problem of mining top-k associated temporal relationship patterns; (2) designing the algorithm *ATTEND* (short for AssociaTed Temporal rElationship patterN Discovery) in conjunction with heuristic strategies to efficiently speed up the mining of top-k ATRPs; (3) evaluating the proposed method on both real and synthetic data with satisfactory outcomes of effectiveness and efficiency of the method.

Note that in our work, the sequence order is formed by the starting and end time of events. If the events are initially given in some other order (unordered), they can be sorted to form the sequence order. In this sense we could talk about event sets, but conceptually we focus on the sequential nature of the data.

The rest of the paper is organized as follows. We review related studies in Sect. 2, and formulate the problem of mining top-k associated temporal relationship patterns (ATRPs) in Sect. 3. The design of the proposed method *ATTEND* is detailed in Sect. 4. The experiments based on real-world data and synthetic data are presented in Sect. 5 and the paper is concluded in Sect. 6.

2 Related Work

Sequential pattern mining is an important task in data mining, which has long attracted wide attention in academic studies. Srikant *et al.* [3] firstly introduced the problem of sequential pattern mining and proposed a sequential pattern mining algorithm called *GSP*. Zaki *et al.* [4] presented *SPADE* for fast discovery of sequential patterns. Ayres *et al.* [5] used bitmaps to represent the sequential data, and proposed *SPAM* to mine sequential patterns. However, all the above methods can only deal with sequences of single class, let alone discover differences among sequence sets of different classes.

Distinguishing sequential pattern (DSP) mining has many meaningful applications, as it aims to discover patterns that best describe the significant differences between two classes of sequences. Dong *et al.* [6] proposed the minimal distinguishing sequential subsequence (MDS), and designed the algorithm *ConsGapMiner* to discover MDSs. Yang *et al.* [1] proposed an approach to find top-k minimal item-based distinguishing sequential patterns with largest contrast values. Duan *et al.* [7] studied the problem of mining distinguishing customer focus sets from customer reviews, which can be helpful for online shopping decision support. Wang *et al.* [2] introduced the concept of density-aware distinguishing sequential patterns. Zheng *et al.* [8] proposed a CSP-tree-based approach to client sequential behavior analysis. Zhao *et al.* [9] focused on the problem of discovering diagnostic gene patterns from microarray data. Zhu *et al.* [10] designed an approach to mining user-related rare sequential topic patterns from document streams on the Internet to characterize and detect personalized and abnormal behaviors of users.

Many studies have focused on the problem of mining sequential event patterns with time intervals. Allen *et al.* [11] first proposed thirteen event relationships and described the temporal representation, and there are studies to discover frequent patterns with temporal intervals [12–14]. Yang *et al.* [15] utilized an index structure to extract the time interval-based events with duration. Patel *et al.* [16] used frequent temporal patterns to build an interval-based classifier. Mörchen *et al.* [17] presented a method for the understandable description of local temporal relationships in multivariate data. Tang *et al.* [18] studied the problem of finding lag intervals for temporal dependency analysis. Duan *et al.* [19] proposed the distinguishing temporal event patterns (DTEP) and designed a method called DTEP-Miner to find DTEPs.

However, to the best of our knowledge, none of existing methods has ever focused on the problem of mining associated temporal relationship patterns (ATRPs) from associated temporal event sequences, which are meaningful and necessary in many real-life scenarios.

Table 1. A toy set of temporal event sequences ($\Omega = \{a, b, c\}$)

ID	Temporal event sequence	Dataset
S_1	<(a, 3, 8), (c, 4, 5), (b, 4, 9), (a, 9, 14), (c, 10, 11), (b, 14, 15), (c, 16, 17)>	
S_2	<(b, 1, 6), (c, 6, 7), (b, 8, 27), (a, 8, 12), (c, 10, 25), (a, 13, 14), (b, 37, 39)>	
S_3	<(b, 1, 5), (a, 2, 3), (c, 4, 10), (b, 6, 8)>	D_+
S_4	<(a, 2, 8), (b, 3, 5), (c, 6, 7), (b, 6, 9), (a, 10, 12), (c, 16, 18)>	
S_5	<(a, 3, 8), (b, 4, 5), (c, 8, 12), (b, 9, 12), (c, 13, 16), (b, 18, 20)>	
S_6	<(b, 2, 6), (b, 8, 10), (a, 9, 10), (b, 11, 14), (a, 12, 15), (b, 16, 19)>	
S_7	<(c, 2, 22), (a, 7, 8), (b, 9, 10), (a, 10, 12), (b, 11, 14), (a, 13, 14), (b, 22, 24)>	
S_8	<(a, 1, 5), (b, 4, 5), (b, 6, 10), (a, 6, 8), (a, 16, 18), (b, 16, 18)>	D_-
S_9	<(b, 3, 5), (a, 4, 7), (b, 8, 14), (a, 10, 12), (b, 22, 24)>	
S_{10}	<(b, 3, 6), (a, 7, 8), (b, 8, 11), (a, 9, 12), (b, 12, 16), (b, 22, 28)>	

3 Problem Definition

We start with some preliminaries. Let Ω be the set of all possible *events*. Examples of events include "shopping", "sleep" or "travel" etc. We use the symbol o, possibly with subscripts, to denote an event in Ω. We use a series of continuous non-negative integers starting from 0 to denote the time points of events. Without loss of generality, we assume that the smaller the value is, the earlier the time is.

An *event instance* e is a triplet (o, t^+, t^-) where $o \in \Omega$, t^+ denotes the starting time point of e, and t^- denotes the end time point of e. Naturally, $t^+ < t^-$. A *temporal event sequence* S is a list of event instances, ordered by their starting time points, of the form $S = <(o_1, t_1^+, t_1^-), (o_2, t_2^+, t_2^-), ..., (o_n, t_n^+, t_n^-)>$, where $o_i \in \Omega$, $0 \le t_i^+ \le t_j^+$ $(1 \le i < j \le n)$. The *length* of S is the number of event instances in S, denoted by $|S|$. We denote by $S[i]$ the i-th element in S $(1 \le i \le |S|)$. For $S[i]$, we use $S[i].o$ to denote the event, and use $S[i].t^+$ $(S[i].t^-)$ to denote the starting (end) time point of $S[i]$. Taking S_6 in Table 1 for instance, we have $|S_6| = 6$, $S_6[2].o = b$, $S_6[2].t^+ = 8$, $S_6[2].t^- = 10$.

Allen *et al.* pointed out that for any two event instances, there exist 7 temporal relations between them [11], which we include into a relation set denoted by $\mathcal{R}$. For the sake of clarity, we define 7 symbolic notations in Table 2.

For two events $o, o' \in \Omega$, we use oRo' $(R \in \mathcal{R})$ to describe the *temporal relationship* between o and o'. For example, $o \uparrow o'$ means that o happens before o'. Given a set of temporal relationships $P = \{oRo' \mid o, o' \in \Omega, R \in \mathcal{R}\}$, we denote by $\mathcal{E}(P)$ the events referred to in P. For a given temporal event sequence S, we say S *holds* P, denoted by $P \sqsubset S$, if there exist integers $1 \le k_1 < k_2 < \cdots < k_{|\mathcal{E}(P)|} \le |S|$ such that $\{S[k_i].o \mid 1 \le i \le |\mathcal{E}(P)|\} = \mathcal{E}(P)$ and time points of $S[k_i]$ satisfy the conditions of every temporal relation stated in P.

Example 1. Considering S_3 in Table 1, the temporal relationships contained in S_3 are $\{b_1 \Leftrightarrow a, b_1 \fallingdotseq c, b_1 \uparrow b_2, a \uparrow c, a \uparrow b_2, c \Leftrightarrow b_2\}$. For a temporal relationship set $P = \{a \uparrow c, a \uparrow b, c \Leftrightarrow b\}$, we have $P \sqsubset S_3$.

Table 2. Temporal relations between events e_i and e_j

Relation	Illustration	Notation	Condition
before	e_i e_j	$e_i \uparrow e_j$	$e_i.t^- < e_j.t^+$
overlaps	e_i e_j	$e_i \fallingdotseq e_j$	$e_i.t^+ < e_j.t^+, e_i.t^- > e_j.t^+, e_i.t^- < e_j.t^-$
during	e_i e_j	$e_i \Leftrightarrow e_j$	$e_i.t^+ < e_j.t^+, e_i.t^- > e_j.t^-$
equal	e_i e_j	$e_i = e_j$	$e_i.t^+ = e_j.t^+, e_i.t^- = e_j.t^-$
starts	e_i e_j	$e_i \leftarrow e_j$	$e_i.t^+ = e_j.t^+, e_i.t^- > e_j.t^-$
finishes	e_i e_j	$e_i \rightarrow e_j$	$e_i.t^+ < e_j.t^+, e_i.t^- = e_j.t^-$
meets	e_i e_j	$e_i \leftrightarrow e_j$	$e_i.t^- = e_j.t^+$

The *support* of a temporal relationship set P in a set of temporal event sequences D, denoted by $Sup(D, P)$, is

$$Sup(D, P) = \frac{|\{S \in D \mid P \sqsubset S\}|}{|D|} \tag{1}$$

Definition 1. *Given two sets of temporal event sequences D_+ and D_-, the contrast score of P targeting D_+ against D_-, denoted by $cScore(P)$, is*

$$cScore(P) = Sup(D_+, P) - Sup(D_-, P) \tag{2}$$

Example 2. For a temporal relationship set $P = \{a \uparrow c, a \uparrow b, c \Leftrightarrow b\}$, we have $|D_+| = |D_-| = 5$, $Sup(D_+, P) = 1/5 = 0.2, Sup(D_-, P) = 0$. Thus, $cScore(P) = 0.2 - 0 = 0.2$.

Definition 2. *Given two sets of temporal event sequences D_+ and D_-, a temporal relationship set P is an associative temporal relationship pattern (ATRP) targeting D_+, if $cScore(P) > 0$.*

To select top-k ATRPs, we first define a total order on all discovered ATRPs.

Definition 3. *Given two ATRPs P and P', $P \succ P'$ (called P precedes P' or P has a higher precedence than P') if:*

1. *$cScore(P) > cScore(P')$, or*
2. *$cScore(P) = cScore(P')$, but $|\mathcal{E}(P)| > |\mathcal{E}(P')|$, or*
3. *all of the above parameters are the same, but $\mathcal{E}(P)[i]$ is lexically smaller than $\mathcal{E}(P')[i]$, where $i = \min\{j \mid \mathcal{E}(P)[j] \neq \mathcal{E}(P')[j], 1 \leq j \leq |\mathcal{E}(P)|\}$.*

Table 3. Top-10 ATRPs discovered from Table 1

Rank	ATRP	*cScore*
1	$\{b_1 \uparrow c, b_1 \uparrow b_2, c \uparrow b_2\}$	0.8
2	$\{c_1 \uparrow c_2\}$	0.8
3	$\{b \uparrow c_1, b \uparrow c_2, c_1 \uparrow c_2\}$	0.6
4	$\{b_1 \uparrow c_1, b_1 \uparrow b_2, b_1 \uparrow c_2, b_2 \uparrow c_1, b_2 \uparrow c_2, c_1 \uparrow c_2\}$	0.4
5	$\{a \uparrow c_1, a \uparrow c_2, c_1 \uparrow c_2\}$	0.4
6	$\{a \Leftrightarrow c, a \uparrow b, c \uparrow b\}$	0.4
7	$\{c \uparrow a, c \uparrow b, a \uparrow b\}$	0.4
8	$\{c_1 \uparrow a, c_1 \uparrow c_2, a \uparrow c_2\}$	0.4
9	$\{a \Leftrightarrow b\}$	0.4
10	$\{a \risingdotseq c\}$	0.4

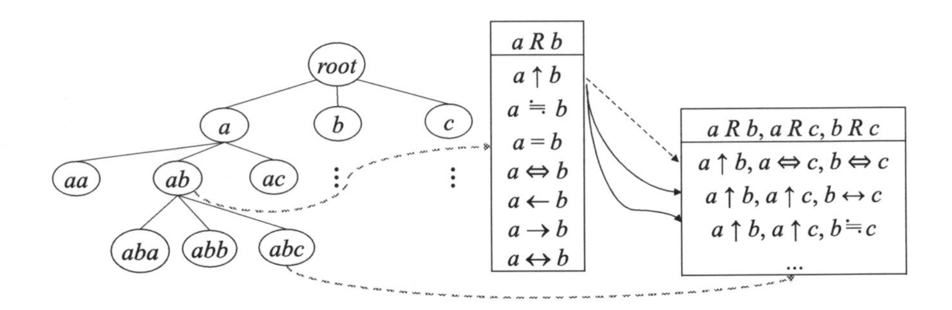

Fig. 1. An example of event set enumeration tree

Given k, the problem of mining top-k temporal relationship patterns is to find ATRPs with top-k precedence targeting D_+ against D_-.

Table 3 lists the top-10 ATRPs discovered from Table 1.

4 The Design of *ATTEND*

In this section, we present the details of *ATTEND* finding top-k ATRPs targeting D_+ against D_-, and discuss the key techniques including heuristic strategies to speed up our method.

In general, *ATTEND* consists of two main steps: (1) generating the set of candidate ATRPs (Sect. 4.1), and (2) evaluating the contrast score of each candidate to find the top-k ATRPs (Sect. 4.2).

4.1 Candidate ATRPs Generation

To generate all possible candidate ATRPs systematically, the set enumeration tree approach [20] is adopted by *ATTEND*. *ATTEND* firstly generates the events involved in the candidate. Then, for the generated events, *ATTEND* enumerates all possible pairwise temporal relationships between every two events.

Figure 1 illustrates an example of an event set enumeration tree generating candidate ATRPs over $\Omega = \{a, b, c\}$. Taking the candidate ATRPs containing two events for instance. We can see that there are 9 combinations of two events, i.e., $\{aa, ab, ac, ba, bb, bc, ca, cb, cc\}$. Furthermore, for ab, 7 temporal relationships can be generated, i.e., $\{a \uparrow b, a \fallingdotseq b, a \Leftrightarrow b, a = b, a \leftarrow b, a \rightarrow b, a \leftrightarrow b\}$.

Clearly, in the l-th ($l \geq 2$) level of an event set enumeration tree, there exist $|\Omega|^l |\mathcal{R}|^{|l-1|!}$ candidate ATRPs. A straightforward (time consuming) way is evaluating the *cScore* of each candidate on the two sets of temporal event sequences. Fortunately, Theorem 1 demonstrates the monotonicity of $Sup(D, P)$ with respect to P.

Theorem 1. *Given a temporal event sequence set D, for any two events temporal relationship sets P and P′, $Sup(D, P) \leq Sup(D, P')$ if $P' \subset P$.*

Proof (Outline). For given P and P' satisfying $P' \subset P$, we have $\{S \in D \mid P \sqsubset S\} \subseteq \{S \in D \mid P' \sqsubset S\}$. Then, by Eq. 1: $\frac{|\{S \in D | P \sqsubset S\}|}{|D|} \leq \frac{|\{S \in D | P' \sqsubset S\}|}{|D|}$. Thus, $Sup(D, P) \leq Sup(D, P')$. □

To minimize computation cost, we develop following pruning rule determining the candidate ATRPs that cannot be included in the top-k result based on Theorem 1.

Pruning Rule 1. *For a temporal relationship set P, if $Sup(D_+, P) < min$, all supersets of P can be pruned, where min (initialized as 0) is the k-th largest contrast score value of the candidate ATRPs searched so far.*

We also get the following observations.

Observation 1. *For any event $o \in \Omega$, there does not exist the following temporal relations: overlaps ($\fallingdotseq$), during ($\Leftrightarrow$), equal ($=$), starts ($\leftarrow$), finishes ($\rightarrow$), meets ($\leftrightarrow$) for o itself.*

Observation 2. *Some temporal relations are not compatible. For event $a, b, c \in \Omega$, if the temporal relations between a and b is before ($a \uparrow b$), and that between a and c is during ($a \Leftrightarrow c$), then the temporal relations $\{b \uparrow c, b \fallingdotseq c, b \Leftrightarrow c, b = c, b \leftarrow c, b \rightarrow c, b \leftrightarrow c\}$ do not hold (conflicting with $a \Leftrightarrow c$).*

To avoid generating a candidate ATRP containing conflicting temporal relations, *ATTEND* generates all logical relations among the starting time points and end time points of every event following the conditions listed in Table 2. If there are more than one illogical relations between the time points of two events, then we say this candidate ATRP is *conflicting*.

Example 3. Given a temporal relation set $P = \{a \uparrow b, a \Leftrightarrow c, b \Leftrightarrow c\}$. (I) From the conditions of $a \uparrow b$ and $a \Leftrightarrow c$, we have $a.t^+ < c.t^+, c.t^- < a.t^-, a.t^- < b.t^+$. As relation '↑' (*before*) is transitive, we can get $c.t^- < b.t^+$. (II) From the

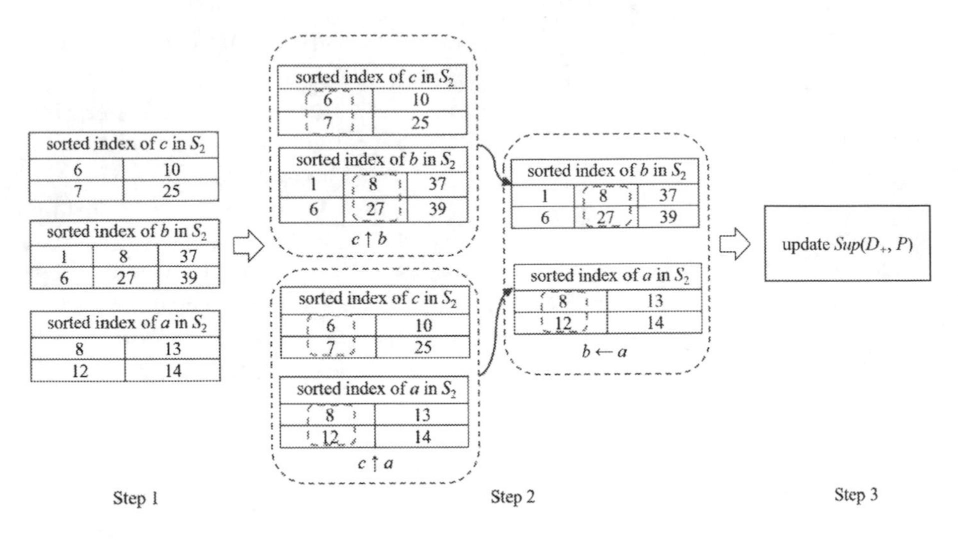

Fig. 2. Sorted index of S_2 in Example 4

conditions of $b \Leftrightarrow c$, we have $b.t^+ < c.t^+, c.t^- < b.t^-$. As $c.t^+ < c.t^-$, we can get $b.t^+ < c.t^-$. (I) conflicts with (II). Thus, P is conflicting. (There are more than one temporal relations between $b.t^+$ and $c.t^-$.).

Based on Observations 1 and 2, we design the following pruning rule to remove invalid candidate ATRPs.

Pruning Rule 2. *For a given temporal relationship oRo' in a temporal relationship set P,*

- *if o and o' are the same event, only $o \uparrow o'$ is generated;*
- *if $oR'o$ is conflicting, P and its supersets can be pruned.*

Observation 3. *The frequencies of different temporal relationships listed in Table 2 vary. Intuitively, the condition strength of the temporal relationship before is the weakest, while that of the relationship meets is the strongest.*

Based on Observation 3, we get following heuristic rule.

Heuristic Rule 1. *For an event set E,* ATTEND *enumerates the temporal relationships among every two events in E in the preference order of "before, overlaps, during, equal, starts, finishes, meets".*

By Heuristic Rule 1, *ATTEND* pretends to find temporal relationships with higher *cScore* value as early as possible, correspondingly the value of *min* can also be updated with a higher value quickly. As a result, more temporal relationships can be pruned by Pruning Rule 1.

Algorithm 1. $ATTEND(D_+, D_-, k)$

Input: D_+ and D_-: two sets of temporal event sequences, k: an integer
Output: $atrp$: the set of top-k ATRPs

```
1: initialize min ← 0, atrp ← ∅;
2: for each event set E searched by traversing the event set enumeration tree in a
   depth-first way do
3:    generate a candidate ATRP P ← {oRo′ | o, o′ ∈ E, o ≠ o′, R ∈ 𝓡};
4:    P ← P ∪ {o ↑ o | o ∈ E};
5:    if P is conflicting then
6:       perform Pruning Rule 2 and go to Step 2;
7:    end if
8:    compute Sup(D+, P) using Equation 1;
9:    if Sup(D+, P) < min then
10:      perform Pruning Rule 1 and go to Step 2;
11:   end if
12:   compute Sup(D−, P) using Equation 1;
13:   cScore ← Sup(D+, P) − Sup(D−, P);
14:   if cScore > min then
15:      if |atrp| < k then
16:         atrp ← atrp ∪ {P};
17:      else
18:         update atrp with P;
19:      end if
20:      min ← cScore;
21:   end if
22: end for
23: return atrp;
```

4.2 Contrast Score Calculation

According to Eq. 2, given a temporal event sequence set D and a candidate ATRP P, $ATTEND$ evaluates $Sup(D, P)$. For the sake of efficiency, we design a sorted index recording the starting/end time point of each event in S ($S \in D$). Thanks to the efficient operations on the sorted index, $ATTEND$ can quickly check whether or not S holds P. Specifically, the contrast score of P can be got by three steps: (1) building the sorted index of each event in S by the ascending order of starting time points; (2) searching each temporal relationship in P by the sorted index; (3) updating the value of $Sup(D, P)$.

Example 4. The sorted index of $S_2 \in D_+$ in Table 1 is illustrated in Fig. 2. For a given candidate ATRP $P = \{c \uparrow b, c \uparrow a, b \leftarrow a\}$, to check whether or not $P \subset S_2$, $ATTEND$ searches the sorted index of c, b and a one by one. For event c, $(c, 6, 7)$ is the first event instance stored in the sorted index. Then, for event b, $ATTEND$ searches the first event instance satisfying the temporal relationship $c \uparrow b$. As a result, event instance $(b, 8, 27)$ is got. Next, for event a, $(a, 8, 12)$ is the first event instance satisfying the temporal relationship $c \uparrow a$. Since $(b, 8, 27)$ and $(a, 8, 12)$ satisfy the temporal relationship $b \leftarrow a$, we have $P \subset S_2$. Correspondingly, the value of $Sup(D_+, P)$ is increased by 1.

Table 4. Characteristics of datasets

Datasets	D_+/D_-	$\|S\|_{avg}$	$\|S\|_{min}$	$\|S\|_{max}$	$\|D\|$	$\|\Omega\|$
ADL$_1$	*A*	18	12	27	16	22
	B	12	8	17	17	24
ADL$_2$	*Subject*1	145	64	210	58	33
	*Subject*2	176	90	280	61	33
ADL$_3$	h_1	23	30	17	17	26
	h_2	78	42	201	16	26
PRO-ACT	*male*	16	9	76	1862	3111
	female	17	9	78	1264	2594
	dead	18	10	77	1168	2512
	alive	18	10	71	277	1021

Based on the discussion above, we present the pseudo-code of *ATTEND* in Algorithm 1.

5 Empirical Evaluation

In this section, we evaluate the performance of *ATTEND* on both real and synthetic data. All experiments were conducted on a PC with an Intel Core i7-4790 3.60 GHZ and 16 GB main memory, running the Windows 10 operating system. All algorithms were implemented in Python 2.7.

5.1 Effectiveness

We apply *ATTEND* to three daily activity datasets and a disease dataset to test its effectiveness. Each of the three activity datasets, called ADL$_1$ [21], ADL$_2$[1] and ADL$_3$[2], respectively, records the daily activities of two users. Table 5 lists the abbreviations of the activities discovered in the experiment. PRO-ACT[3] is an open clinical trial dataset, collected from the clinical diagnosis and treatment data from more than 8500 patients suffering amyotrophic lateral sclerosis (*ALS*). Each PRO-ACT record collects the information of each patient suffering *ALS*, such as the basic statistical information, family history, onset time of symptoms, classification of symptoms, and the treatments used. Table 4 summarizes the characteristics of all datasets.

[1] http://courses.media.mit.edu/2004fall/.
[2] http://ailab.wsu.edu/casas/hh/.
[3] http://nctu.partners.org/ProACT.

Table 5. Activities and corresponding abbreviations

Bathing (BA)	Breakfast (BR)	Bed Toileting (BT)
Call (CA)	Doing Laundry (DL)	Dinner (DN)
Dressing (DR)	Going Entertainment (GE)	Grooming (GR)
Going Work (GW)	Leaving (LE)	Lunch (LU)
Morning Meds (MM)	Preparing Breakfast (PB)	Preparing Dinner (PD)
Personal Hygiene (PH)	Preparing Lunch (PL)	Preparing Snack (PS)
Read (RD)	Relax (RE)	Sleep (SE)
Shower (SH)	Snack (SN)	Spare_Time/TV (ST)
Toileting (TO)	Wash Breakfastdishes (WB)	Work on Computer (WC)
Wash Dishes (WD)	Wash Dinnerdishes (WI)	Work (WO)

Table 6. Top-10 ATRPs discovered from ADL_1 ($k = 10$)

Rank	*userA*(+)		*userB*(+)	
	ATRP	*cScore*	ATRP	*cScore*
1	{BR $\risingdotseq$ LE, BR $\uparrow$ TO, LE $\risingdotseq$ TO}	0.476	{BR $\risingdotseq$ LE, BR $\uparrow$ LU, LE $\risingdotseq$ LU}	0.476
2	{BR $\risingdotseq$ LE, BR $\uparrow$ TO, LE $\uparrow$ TO}	0.476	{SE_1 $\uparrow$ TO, SE_1 $\uparrow$ SE_2, TO $\leftrightarrow$ SE_2}	0.476
3	{BR $\risingdotseq$ LE_1, BR $\uparrow$ LE_2, LE_1 $\uparrow$ LE_2}	0.476	{BR $\uparrow$ TO}	0.476
4	{SH $\uparrow$ LE, SH $\uparrow$ BR, LE $\leftrightarrow$ BR}	0.476	{ST $\Leftrightarrow$ SN}	0.476
5	{DN $\uparrow$ LE}	0.476	{GR $\uparrow$ LE}	0.476
6	{SH $\uparrow$ BR}	0.476	{GR $\uparrow$ BR}	0.476
7	{BR $\uparrow$ GR}	0.476	{SH $\uparrow$ GR}	0.333
8	{SH $\uparrow$ ST}	0.476	{LE $\uparrow$ DN}	0.333
9	{BR $\uparrow$ LU}	0.476	{TO $\uparrow$ SH}	0.333
10	{TO $\uparrow$ ST}	0.333	{GR $\uparrow$ TO}	0.333

Table 6 lists the top-10 ATRPs discovered by *ATTEND* targeting user A and user B, respectively, where some interesting patterns characterizing the daily activities of each user can be found. For example, user A often ate breakfast *before* grooming (the top-7 ATRP targeting user A), while user B often groomed *before* eating breakfast (the top-6 ATRP targeting user B). Also, user A often had dinner *before* leaving from home (the top-5 ATRP targeting user A), while user B often left from home *before* dinner (the top-8 ATRP targeting user B). In addition, the result of top-6 ATRPs targeting user A indicated that user A often did three activities, that is, breakfast, leaving and toileting, with the corresponding ATRP as {Breakfast $\risingdotseq$ Leaving, Breakfast $\uparrow$ Toileting, Leaving $\risingdotseq$ Toileting}. While user B often had breakfast, leaving and having lunch, with the corresponding ATRP as {Breakfast $\risingdotseq$ Leaving, Breakfast $\uparrow$ Lunch, Leaving $\risingdotseq$ Lunch}.

Table 7. Top-10 ATRPs discovered from ADL_2 and ADL_3 ($k = 10$)

Rank	*Subject*1 (+)		*Subject*2 (+)		*h*1 (+)		*h*2 (+)	
	ATRP	*cScore*	ATRP	*Score*	ATRP	*cScore*	ATRP	*cScore*
1	{PL ↑ WD}	1.0	{PD ↑ GW}	1.0	{PH ↑ RD}	1.0	{PH ⇔ WC}	1.0
2	{PL ↑ GE}	1.0	{PD ↑ DL}	1.0	{PH ≒ RD}	1.0	{PH ≒ WC}	1.0
3	{PL ≒ WD}	1.0	{PL ↑ GW}	1.0	{WD ↑ RD}	1.0	{PD ⇔ WC}	1.0
4	{PL ⇔ WD}	1.0	{PL ↑ DL}	1.0	{DR ⇔ BT}	1.0	{BT ↑ MM}	1.0
5	{PL ≒ GE}	1.0	{TO ≒ DR}	1.0	{SE ⇔ BA}	1.0	{WC ↑ SE}	1.0
6	{PL ⇔ GE}	1.0	{PD ↑ GE}	1.0	{WI ⇔ PL}	1.0	{MM ≒ PB}	1.0
7	{TO ↑ WD}	1.0	{PB ≒ PL}	1.0	{PB ≒ RE}	1.0	{PD → DR}	1.0
8	{TO ≒ WD}	1.0	{PS ↑ GW}	1.0	{GR ≒ PD}	1.0	{PL ⇔ WC}	1.0
9	{TO ↑ GE}	1.0	{BA ≒ PD}	1.0	{WO ⇔ RE}	1.0	{PD ≒ CA}	1.0
10	{TO ⇔ WD}	1.0	{DR ↑ PL}	1.0	{WB ↔ WO}	1.0	{PD ↔ RE}	1.0

Table 8. Top-10 ATRPs discovered from PRO-ACT ($k = 10$)

Rank	ATRP	*cScore*	D_+
1	{Normocytic Anaemia ≒ Perleche}	0.0383	
2	{Tired Eyes ↑ Arcus Senilis}	0.0380	
3	{Contact Lens Intolerance → Arcus Senilis}	0.0380	
4	{PTT Prolonged ↑ Thrombocytopenia}	0.0380	
5	{Alkaline Phosphatase Increased ⇔ Blood Lactate Dehydrogenase Abnormal}	0.0380	
6	{Blood Lactate Dehydrogenase Abnormal ≒ Blood Alkaline Phosphatase High}	0.0380	*dead*
7	{Unattended Death ↑ Died in Sleep}	0.0380	
8	{T4 Abnormal ↑ Increased TSH}	0.0380	
9	{Increased TSH ⇔ Increased TSH}	0.0380	
10	{Blood Pressure Orthostatic ≒ Normocytic Anaemia}	0.0379	
1	{Pancytopenia ↑ Incomplete Bundle Branch Block}	0.0380	
2	{Tachycardia ≒ Atrial Fibrillation}	0.0380	
3	{Creatine Phosphokinase Abnormal ↑ Blood Alkaline Phosphatase Abnormal}	0.0380	
4	{Cataract ↑ Burning Oral Sensation}	0.0380	
5	{Cataract ↑ Graves-Basedow Disease}	0.0380	
6	{Bilateral Cataracts ≒ Bilateral cataracts}	0.0379	*alive*
7	{Normocytic Anaemia ↑ Supraventricular Arrhythmia NOS}	0.0337	
8	{Tachycardia Nervous ≒ Tachycardia Nervous}	0.0337	
9	{Nocturnal Dyspnea ↑ Activities of Daily Living Impaired}	0.0337	
10	{Joint Manipulation ↑ Hip Prosthesis Insertion}	0.0168	

Table 7 lists the top-10 ATRPs discovered by *ATTEND* targeting user *Subject*1 and user *Subject*2 and the top-10 ATRPs targeting user *h*1 and user *h*2, respectively, from which we can also deduce the habits of different users.

The dataset PRO-ACT is divided into two target classes according to the living status of the patient, i.e., dead or alive and Table 8 shows the top-10 ATRPs targeting the patients belonging to each of the classes, respectively. For the class *dead*, {Normocytic Anaemia ≒ Perleche} is the most distinguishing ATRP, where proofs can be found to support that Normocytic Anaemia and Perleche are two main symptoms accounting for the death of *ALS* patients. For the patients with Pancytopenia and are alive, some measures can be taken in advance to avoid the onset of Incomplete Bundle Branch Block. PRO-ACT is then divided into another two target classes according to the gender of patients. Table 9 shows the top-10 ATRPs targeting *male* patients and *female* patients. Take the class *female* for example, {Heavy Periods ≒ Bleeding Menstrual Heavy} got the highest *cScore*, and Heavy Periods, Absence of Menstruation, Premenstrual Syndrome, Breast Mass, Breast Cyst are some frequent symptoms for the *female* patients with *ALS*. Clearly, ATRPs can give some assistance to the treatment of patients with *ALS* and provide different treatment options for patients in different gender.

Figure 3 shows the contrast scores of top-k ATRPs with respect to k. As k grew larger, more ATRPs were found by *ATTEND* and ATRPs with lower contrast scores were got.

Table 9. Top-10 ATRPs discovered from PRO-ACT ($k = 10$)

Rank	ATRP	*cScore*	D_+
1	{Prostatic Hyperplasia ≒ Benign Prostatic Hyperplasia}	0.0380	*male*
2	{Prostatic Hyperplasia ↑ Benign Prostatic Hyperplasia}	0.0380	
3	{Enlarged Prostate ↑ Prostatic Hyperplasia}	0.0380	
4	{Genital Itching ↑ Genital Rash}	0.0380	
5	{Impotent ↑ Disorder Testicle}	0.0380	
6	{Impotence ≒ Erection Failure}	0.0300	
7	{Impotence ↑ Erection Failure}	0.0270	
8	{Pain in Testis ↑ Genital Itching}	0.0270	
9	{Penile Abrasion ≒ Sexual Dysfunction}	0.0270	
10	{Testicular Cyst ≒ Cystocele}	0.0240	
1	{Heavy Periods ≒ Bleeding Menstrual Heavy}	0.0420	*female*
2	{Heavy Periods ↑ Spotting Menstrual}	0.0380	
3	{Heavy Periods ≒ Spotting Menstrual}	0.0380	
4	{Primary Ovarian Failure ↑ Metrorrhagia}	0.0380	
5	{Absence of Menstruation ↑ Premenstrual Syndrome}	0.0378	
6	{Amenorrhea ↑ Premenstrual Syndrome}	0.0370	
7	{Menstrual Cycle Abnormal ↑ Menstrual Cycle Abnormal}	0.0370	
8	{Breast Mass ≒ Breast Cyst}	0.0337	
9	{Breast Mass ↑ Breast Cyst}	0.0210	
10	{Breast Tension ≒ Breast Cyst}	0.0120	

In addition, Fig. 4 shows the frequency of relationships of top-k ATRPs with respect to k. We can see that frequency of some relationships increased as k became larger, such as the relationships *before*, *overlaps* and *during* occurred frequently, while *starts*, *finishes* and *meets* occurred little. This demonstrates that Heuristic Rule 1 is effective.

5.2 Efficiency

To the best of our knowledge, there are no previous methods handling this problem. Thus we test the efficiency of *ATTEND* compared with two of its variations, that is, Baseline and Baseline*. Baseline does not adopt Pruning Rule 2 and Heuristic Rule 1, while Baseline* does not follow Pruning Rule 1. In

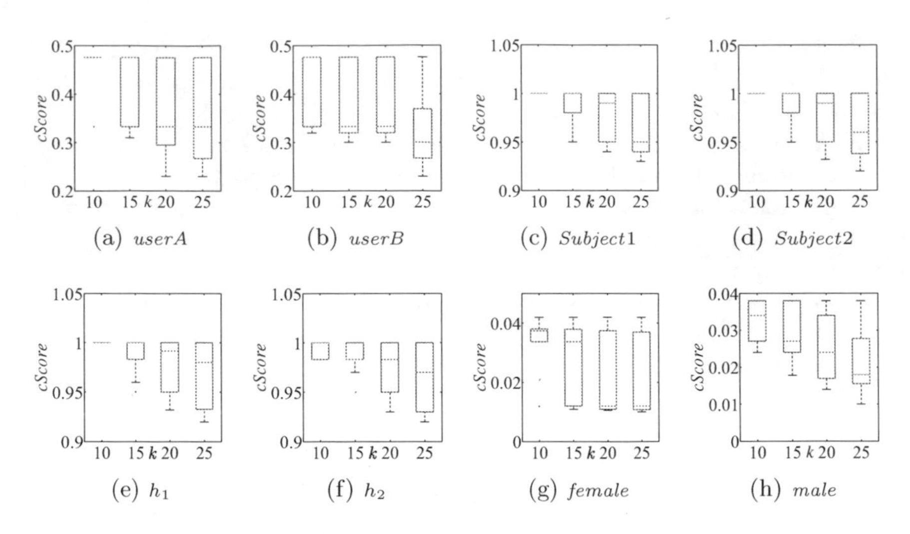

Fig. 3. Effectiveness test: *cScore* w.r.t. k

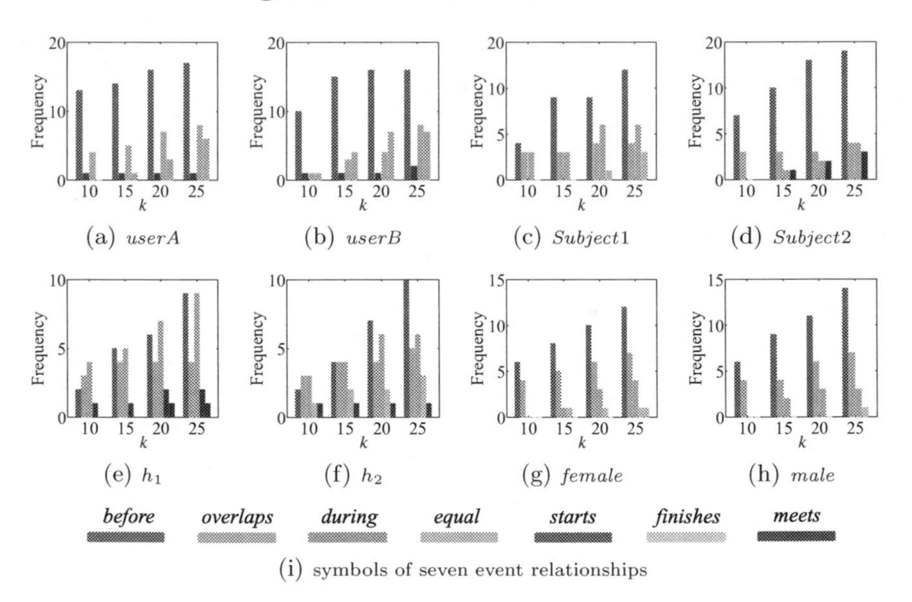

Fig. 4. Effectiveness test: the number of relationships w.r.t. k

all efficiency tests, we set $k = 10$, $|D| = 100$, $|S| = 100$, and $|\Omega| = 20$ in default, and the synthetic event sequences are generated randomly.

Figure 5 shows the runtime with respect to k. As k increased, all algorithms (*ATTEND*, Baseline and Baseline*) took more time to run. In particular, *ATTEND* ran much faster than both Baseline and Baseline*. The result indicates that *ATTEND* is insensitive to k.

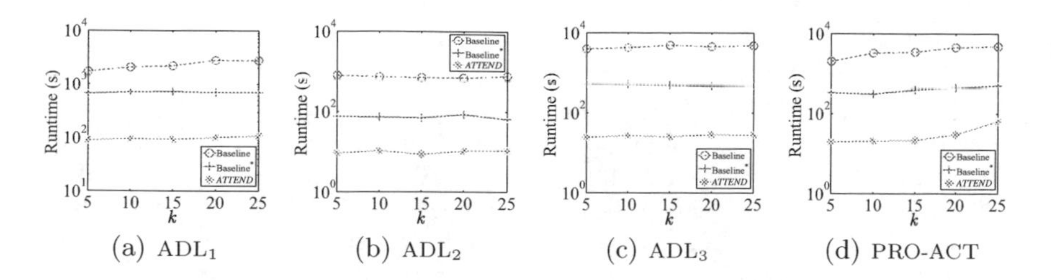

(a) ADL_1 (b) ADL_2 (c) ADL_3 (d) PRO-ACT

Fig. 5. Efficiency test: runtime w.r.t. k

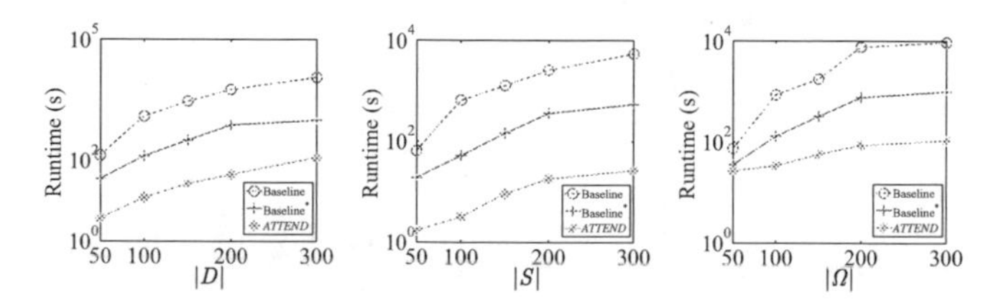

Fig. 6. Efficiency test: runtime w.r.t. $|D|$, $|S|$, $|\Omega|$

Figure 6 shows the impact of $|D|$, $|S|$ and $|\Omega|$ on the runtime of each algorithm. As $|D|$, $|S|$ and $|\Omega|$ grew larger, the runtime of Baseline and Baseline* both increased rapidly, while the runtime of *ATTEND* also increased but in a slow and steady way. The runtime of all algorithms increased because the generation of candidate ATRPs are positively correlated to $|\Omega|$, and the computation of contrast scores of candidate ATRPs also relies on both $|D|$ and $|S|$. Compared with the baseline methods, *ATTEND* ran much faster because it adopted all heuristic strategies, which helped to speed up the search of top-k ATRPs. The results indicate that the heuristic strategies are effective and *ATTEND* is scalable.

6 Conclusions

In this paper, we studied the novel problem of mining ATRPs from temporal event sequences. We proposed *ATTEND* with heuristic strategies to find top-k ATRPs. Experiments on both real and synthetic data demonstrate that our proposed *ATTEND* is effective and efficient.

In the future, *ATTEND* can be applied to different fields such as financial data and electric data to further verify its effectiveness. Considering the huge search space for candidate ATRP generation, we also plan to design an approximate algorithm to find ATRPs faster.

References

1. Yang, H., Duan, L., Dong, G., Nummenmaa, J., Tang, C., Li, X.: Mining itemset-based distinguishing sequential patterns with gap constraint. In: Renz, M., Shahabi, C., Zhou, X., Cheema, M.A. (eds.) DASFAA 2015. LNCS, vol. 9049, pp. 39–54. Springer, Cham (2015). https://doi.org/10.1007/978-3-319-18120-2_3
2. Wang, X., Duan, L., Dong, G., Yu, Z., Tang, C.: Efficient mining of density-aware distinguishing sequential patterns with gap constraints. In: Bhowmick, S.S., Dyreson, C.E., Jensen, C.S., Lee, M.L., Muliantara, A., Thalheim, B. (eds.) DASFAA 2014. LNCS, vol. 8421, pp. 372–387. Springer, Cham (2014). https://doi.org/10.1007/978-3-319-05810-8_25
3. Srikant, R., Agrawal, R.: Mining sequential patterns: generalizations and performance improvements. In: Apers, P., Bouzeghoub, M., Gardarin, G. (eds.) EDBT 1996. LNCS, vol. 1057, pp. 1–17. Springer, Berlin (1996). https://doi.org/10.1007/BFb0014140
4. Zaki, M.J.: SPADE: an efficient algorithm for mining frequent sequences. Mach. Learn. **42**(1/2), 31–60 (2001)
5. Ayres, J., Flannick, J., Gehrke, J., Yiu, T.: Sequential pattern mining using a bitmap representation. In: Proceedings of the 8th ACM SIGKDD International Conference on Knowledge Discovery and Data Mining, pp. 429–435 (2002)
6. Dong, G., Pei, J.: Sequence Data Mining. Advance in Database System, vol. 33. Springer, Boston (2007). https://doi.org/10.1007/978-0-387-69937-0
7. Duan, L., et al.: Mining distinguishing customer focus sets from online customer reviews. Computing **100**(4), 335–351 (2018)
8. Zheng, Z., Wei, W., Liu, C., Cao, W., Cao, L., Bhatia, M.: An effective contrast sequential pattern mining approach to taxpayer behavior analysis. World Wide Web **19**(4), 633–651 (2016)
9. Zhao, Y., Wang, G., Li, Y., Wang, Z.: Finding novel diagnostic gene patterns based on interesting non-redundant contrast sequence rules. In: Proceedings of the 11th IEEE International Conference on Data Mining, pp. 972–981 (2011)
10. Zhu, J., Wang, K., Wu, Y., Hu, Z., Wang, H.: Mining user-aware rare sequential topic patterns in document streams. IEEE Trans. Knowl. Data Eng. **28**(7), 1790–1804 (2016)
11. Allen, J.F.: Maintaining knowledge about temporal intervals. Commun. ACM **26**(11), 832–843 (1983)
12. Papapetrou, P., Kollios, G., Sclaroff, S., Gunopulos, D.: Discovering frequent arrangements of temporal intervals. In: Proceedings of the 5th IEEE International Conference on Data Mining, pp. 354–361 (2005)
13. Winarko, E., Roddick, J.F.: ARMADA - an algorithm for discovering richer relative temporal association rules from interval-based data. Data Knowl. Eng. **63**(1), 76–90 (2007)
14. Hui, L., Chen, Y., Weng, J.T., Lee, S.: Incremental mining of temporal patterns in interval-based database. Knowl. Inf. Syst. **46**(2), 423–448 (2016)
15. Yang, C., Jaysawal, B.P., Huang, J.: Subsequence search considering duration and relations of events in time interval-based events sequences. In: Proceedings of 2017 IEEE International Conference on Data Science and Advanced Analytics, pp. 293–302 (2017)
16. Patel, D., Hsu, W., Lee, M.: Mining relationships among interval-based events for classification. In: Proceedings of the ACM SIGMOD International Conference on Management of Data, pp. 393–404 (2008)

17. Mörchen, F., Ultsch, A.: Efficient mining of understandable patterns from multivariate interval time series. Data Min. Knowl. Discov. **15**(2), 181–215 (2007)
18. Tang, L., Li, T., Shwartz, L.: Discovering lag intervals for temporal dependencies. In: Proceedings of the 18th ACM International Conference on Knowledge Discovery and Data Mining, 633–641 (2012)
19. Duan, L., Yan, L., Dong, G., Nummenmaa, J., Yang, H.: Mining top-*k* distinguishing temporal sequential patterns from event sequences. In: Candan, S., Chen, L., Pedersen, T.B., Chang, L., Hua, W. (eds.) DASFAA 2017. LNCS, vol. 10178, pp. 235–250. Springer, Cham (2017). https://doi.org/10.1007/978-3-319-55699-4_15
20. Rymon, R.: Search through systematic set enumeration. In: Proceedings of the 3rd International Conference on Principles of Knowledge Representation and Reasoning, KR, pp. 539–550 (1992)
21. Lichman, M.: UCI machine learning repository (2013)

Efficient Mining of Event Periodicity in Data Series

Hua Yuan(✉), Yu Qian, and Mengna Bai

School of Management and Economics,
University of Electronic Science and Technology of China,
Chengdu 611731, China
{yuanhua,qiany}@uestc.edu.cn, mengnabai@163.com

Abstract. This paper investigates the problem of efficiently discovering periodicity of a certain event in data series. To that end, the current work argues firstly that the periodicity of an event in data series may be formalized as the distribution period, the structure period, or the both. Along this line, a partition method, $\pi(n)$, is proposed to divide the data series into length-equal and position-continuous segments. Based on the results of implementing $\pi(n)$ on a data series, we propose two new concepts of distribution periodicity and structure periodicity. Then, a cross-entropy-based method, namely CEPD, is proposed to mine the periodicity of data series. The experimental results show that CEPD can be used to mine feasible event periodicity in data series, especially, with very low level of time consumption and high capability of noise resilience.

Keywords: Data series · Cross entropy · Distribution periodicity · Structure periodicity

1 Introduction

Data series is commonly used in presentation of the events sequentially happened in real world, such as the weather data for a location [22], the gene expression data [12], the finance fluctuation data [20,32], the web site visiting traffic [5,29], and the consumption sequence of a user [1]. Data series is mostly characterized by being composed of repeating cycles [19], especially, for those data series generated by user behaviors [30]. For instance, "The vendors purchase *twice a month* from the suppliers," "Bob visits gym *every Tuesday*," and so on. Basically, such repeating patterns could reveal important observations about the behavior and future trends of the events represented by the data series, and hence would lead to more effective decision making [16]. These gave rise to an important process for mining regular patterns within a data series.

In general, event(s) may show three types of periodicity in a data series: the symbol periodicity, the partial (sequence) periodicity, and the full-cycle (segment) periodicity [19]. Given a periodicity mining task, the methods proposed in the literature would like to treat the task of periodicity detection as a process

G. Li et al. (Eds.): DASFAA 2019, LNCS 11446, pp. 124–139, 2019.
https://doi.org/10.1007/978-3-030-18576-3_8

of finding temporal regularities within the data series [8,13,19,28,30]. Although these methods had performed well for event periodicity mining in certain situations, nevertheless, they also face some technical challenges:

- First, the common problem for these approaches is their computational performance, especially in a big data environments. To address this issue, they would assume previously that users either know the value of the period beforehand or are willing to try various period values until satisfactory periodic patterns emerge [7]. However, if there are multiple events embedded in a data series, then more prior information is needed for event periodicity mining task, or it will make the mining methods present relative poor performance, both on efficiency and completeness [31].
- Second, these methods mainly identify the structural periodicity of the events over a set of periods (time intervals), i.e., those events which occurred at a fixed position in each period may be considered as having periodicity [13].
- Third, data collected from the real-world, which is the input of mining algorithms, are affected by several components; among them, noise is an unavoidable problem [25]. Therefore, the event periodicity mining methods are expected to provide better robustness to noise [11,18].

In this work, we will introduce the *distribution periodicity* and *structure periodicity* to measure the periodic information of an event in a data series. The main contributions of this paper lie in two aspects: first, to the best of our knowledge, it is the first time to distinguish the idea of distribution periodicity and structure periodicity; second, based on the minimum cross entropy principle, an efficient method is proposed to mine the periodicity of an event in data series, which also has a better performance on noise resilience.

2 Related Work

There are lots of studies proposed in the literature of data stream mining. In summary, they can be categorized into types of signal-processing-based, data-structure-based, and statistics-based method.

Signal-Processing-Based Method. The signal processing method in periodic pattern mining is mainly reflected in the data processing and transformation. [6] used the Haar Wavelet Transform and discrete fourier transform (DFT) for time series indexing. The algorithm presented by [24] is the first one that exploits the information in both periodogram and autocorrelation to provide accurate periodic estimates without upsampling. In this work, both DFT and power spectral density (PSD) estimation method are introduced to deal the time series data. A convolution-based algorithm is proposed for segment periodicity and symbol periodicity, and the periodic patterns of unknown periods are also discovered without affecting the time complexity [7]. As pointed out in [18] and [13], the fast Fourier transform (FFT) [3] can also be used to identify periodicity. However, there are two problems in the FFT method. First, it does not cope well with random off-segments in periodic patterns. Further, the computational efficiency is very complicate when events in data series are sparse [30].

Data-Structure-Based Method. In earlier studies, the work in [10] use a sliding window over the data sequence and extract its features, then [26] presented mining technology from a time series database based on a moving-window. Also, by using an expanding sliding windows, [9] improved the accuracy of the discovered periodicity rates. In recent, a pattern-growth approach which is based on a tree structure, called Periodic Frequent-tree (PF-tree) has been discussed for mining periodic patterns [23]. In the paper of [23], the authors use a so called Periodic-frequent pattern tree to capture the database contents and generate the complete set of periodic-frequent patterns.

Since partial periodicity is very common in practice, [13] studied an interesting data mining problem of searching for partial periodic patterns in time-series databases, their algorithm based on a max-subpattern tree offers excellent performance. Promoted by this research, [4] proposed a new structure, the abbreviated list table (ALT), and several efficient algorithms to compute the partial periods. Sheng et al. [21] developed an algorithm to utilize optimization steps to find dense periodic areas in the time series.

Statistics-Based Method. Some basic static methods such as autocorrelation and ranking are commonly used. [2] proposed an algorithm for finding approximate periodicities in large time series data, utilizing autocorrelation function and FFT. And it can discover weak periodic signals in time series databases. [14] investigated an interesting type of periodic pattern, called partial periodic (PP) correlation. Especially, a more suitable measurement, information, is introduced in [27] to naturally value the degree of surprise of the pattern within a data sequence. In [30], the authors presented a variance-based approach to model periodicity, which is to detect event periodicity basing on the statistical variance of the gaps at which a pattern occurs in data series (i.e., the variance of the interarrival times of the pattern). Ghosh et al. [11] have demonstrated the use of a sequential Monte Carlo method to detect and track the periodicity in discrete event streams. Unlike other methods, this technique does not rely on the underlying process sticking to a constant phase.

3 Concepts and Model Formulation

3.1 Data Series and Event Periodicity

A data series S is an ordered sequence of $|S|$ feature values:

$$S = (s_1 s_2 ... s_t ... s_{|S|}),\ s_t \in \mathbb{R}, \tag{1}$$

where s_t is the value of the feature at position t, for example, the feature might be the daily average stock price of a company.

In order to facilitate the calculation, we usually transfer S into a more easy-to-compute formation. If we discretize the feature values in S into nominal discrete events (e.g, stock price goes "up", "down" or "flat"), then the set of feature values can be denoted as $\Sigma = \{a, b, c, ...\}$[7] by representing each event as a symbol (i.e., $a=$ "goes up", $b=$ "goes down", $c=$ "goes flat", etc.). As a result, S can be viewed as a sequence of $|S|$ events (symbols) drawn from a finite event set of Σ_S. Further more, let e_t be any event occurred at position t in S, then a set of events happened sequentially over **continuous position space** of $[1, |S|]$ can be specified as follows:

$$S = (e_1 e_2 ... e_t ... e_{|S|}),\ e_t \in \Sigma_S. \tag{2}$$

Using x to denote the focal event in Σ_S, and $\tilde{x}$ denotes any event type in $\Sigma_S \setminus \{x\}$, if $e_t = x$, then we said that event x appears at position t in S. Especially, when we only concern about whether event x occurred at position t, BS_x is then can be encoded as a **binary data series**,

$$BS_x = (b_1 b_2 ... b_t ... b_{|S|}),\ b_t \in \{0, 1\}. \tag{3}$$

where b_t is specified as follows:

$$b_t = \begin{cases} 1 & \text{if } e_t = x \text{ (i.e., } x \text{ appears at position } t); \\ 0 & \text{otherwise (i.e., } \tilde{x} \text{ appears at position } t). \end{cases}$$

Definition 1. *An event $x \in \Sigma_S$ is said to have* **periodicity** *(or x is a periodic event) in data series S, if its appearances are shown repeated periodically in S.*

Apparently, if event x appears periodically in data series S, we can expect that the appearances of code "b_t" in BS_x are also periodically.

3.2 Data Series Partitioning

To mine the periodicity of event x in data series S, a feasible way is to divide S into segments [17].

Given a set of partition methods $\prod$ defined over S, if there always exists a partitioning scheme $\pi(n) \in \prod$ (where $n < |S|$) such that n is the (distribution or structure) period of event x, then $\prod$ is called a **complete partition set** with respect to the periodicity of x in S. Moreover, $\pi(n)$ is then called a **"good" partition** for detecting the periodicity of x in S.

Accordingly, we propose a simple and complete partitioning method, i.e., $\pi(n)$, $n \in [1, |S|]$, to divide the data series S into segments iteratively as follows:

Step 1: Begin with the first position $t = 1$;
Step 2: Every n position-continuous elements are partitioned into a same segment P, i.e., the first n events are in P_1, the second n events are in P_2, and so on. As a result, $\pi(n) = \{P_1 | P_2 | \cdots | P_{\lceil \frac{|S|}{n} \rceil}\}$ partitions S into $\lceil \frac{|S|}{n} \rceil$ length-equal[1] segments, and $P_j = \left(e_{(j-1)*n+1}, \cdots, e_{j*n}\right)$, where $j \in [1, \lceil \frac{|S|}{n} \rceil]$.

[1] $|P_{\lceil \frac{|S|}{n} \rceil}| \leq n$ is allowed.

For example, $BS = (00111000100110001000)$, then method $\pi(4)$ will partition BS into five segments as follows:

$$\pi(4) = \{\underbrace{0011}_{\mathcal{P}_1}|\underbrace{1000}_{\mathcal{P}_2}|\underbrace{1001}_{\mathcal{P}_3}|\underbrace{1000}_{\mathcal{P}_4}|\underbrace{1000}_{\mathcal{P}_5}\}.$$

By implementing $\pi(n) = \{P_1|P_2|\cdots|P_{\lceil\frac{|S|}{n}\rceil}\}$ on a binary time-series BS, we can obtain a **partition matrix** by rearranging all the segments P_j as follows:

$$\begin{Vmatrix} P_1 \\ P_2 \\ \vdots \\ P_j \\ \vdots \\ P_{\lceil\frac{|S|}{n}\rceil} \end{Vmatrix} = \begin{Vmatrix} b_1 & \dots & b_n \\ b_{n+1} & \dots & b_{2n} \\ \vdots & \dots & \vdots \\ b_{(j-1)*n+1} & \dots & b_{j*n} \\ \vdots & \dots & \vdots \\ b_{(\lceil\frac{|S|}{n}\rceil-1)*n+1} & \dots & b_{(\lceil\frac{|S|}{n}\rceil)*n} \end{Vmatrix}.$$

In general, the matrix has a total of $\lceil\frac{|S|}{n}\rceil$ rows and n columns. Row $j \in [1,\cdots,\lceil\frac{|S|}{n}\rceil]$ is just the contents of j-th segment, i.e., P_j. Column $\tau \in [1,\cdots,n]$ is corresponding to the appearances of event x at the τ-th position, which is referred to as $C_\tau = \left(b_\tau, ..., b_{(j-1)*n+\tau}, ..., b_{(\lceil\frac{|S|}{n}\rceil-1)*n+\tau}\right)^T$.

Definition 2. *The total appearances of event x in segment P_j ($j \in [1,\cdots,\lceil\frac{|S|}{n}\rceil]$) is called the* **support** *of x in P_j, it is defined as*

$$supp(x|P_j) = \sum_{b_t \in P_j} b_t. \tag{4}$$

Lemma 1. $supp(x|P_j) + supp(\tilde{x}|P_j) = n,\ j \in [1, ..., \lceil\frac{|S|}{n}\rceil - 1]$[2].

Definition 3. *The total appearances of event x in C_τ ($\tau \in [1,\cdots,n]$) is called the* **support** *of x at position τ, which is represented by:*

$$supp(x|C_\tau) = \sum_{b_t \in C_\tau} b_t. \tag{5}$$

Lemma 2. $supp(x|C_\tau) + supp(\tilde{x}|C_\tau) = \lceil\frac{|S|}{n}\rceil,\ \tau = 1, ..., n.$

Further more, we can define that the total appearances of event x in S is called the **support** of x, which is defined as

$$supp(x) = \sum_{b_t \in BS_x} b_t. \tag{6}$$

[2] $supp(x|P_{\lceil\frac{|S|}{n}\rceil}) + supp(\tilde{x}|P_{\lceil\frac{|S|}{n}\rceil})$ may less than n while incomplete partition happened in the last segment.

Accordingly, the **distribution** of x in P_j, i.e., $p(P_j)$, and the **distribution** of x in C_τ, i.e., $q(C_\tau)$, are defined as following respectively,

$$p(P_j) = \frac{supp(x|P_j)}{supp(x)}, \quad \text{and} \quad q(C_\tau) = \frac{supp(x|C_\tau)}{supp(x)}. \tag{7}$$

3.3 Distribution Periodicity and Structure Periodicity

An event x is said to have **distribution periodicity** in S with respect to "good" partition $\pi(n)$, if its *support* (appearance) in each segment is equal. For the distribution periodicity, we have the following theorem:

Theorem 1. *If event x has a distribution period of n in S, then the ideal distribution of x in $\lceil \frac{|S|}{n} \rceil$ segments is as*

$$p_n = \left\{ \frac{1}{\lceil \frac{|S|}{n} \rceil}, ..., \frac{1}{\lceil \frac{|S|}{n} \rceil}, ..., \frac{1}{\lceil \frac{|S|}{n} \rceil} \right\}. \tag{8}$$

Proof. The good partition $\pi(n) = \{P_1|P_2|\cdots|P_{\lceil \frac{|S|}{n} \rceil}\}$ divides S into $\lceil \frac{|S|}{n} \rceil$ equal length segments. Since x shows distribution periodicity with respect to partition $\pi(n)$, we can expect that $supp(x|P_i) \approx supp(x|P_j)$ for any $i \neq j$, where $i, j = \{1, .., \lceil \frac{|S|}{n} \rceil\}$. Moreover, with Lemma 1, we obtain:

$$supp(x) = \sum_{j=1}^{\lceil \frac{|S|}{n} \rceil} supp(x|P_j) \approx \lceil \frac{|S|}{n} \rceil \times supp(x|P_j).$$

That is, the distributions of x in $\{P_j\}$, i.e., $\frac{supp(x|P_j)}{supp(x)}$, are equally to $1/\lceil \frac{|S|}{n} \rceil$.

Definition 4. *An event x is said to have* **structure periodicity** *in S with respect to "good" partition $\pi(n)$, if its position (time point) in each segment is the same.*

For the structure periodicity, we have the following theorem:

Theorem 2. *If event x has a structure period n in S at position $\tau^\# \in [1, ..., n]$, then the ideal distribution of x on the n positions is as*

$$q_n = \{q(C_1), ..., q(C_{\tau^\#}), ..., q(C_n)\} = \{0, ..., 1, ..., 0\}. \tag{9}$$

Proof. If event x has structure periodicity in data series BS_x with respect to "good" partition $\pi(n)$, then

- $b_{i*n+\tau} = b_{j*n+\tau}$, here $i, j \in [0, \cdots, \lceil \frac{|S|}{n} \rceil - 1]$ and $\tau \in [1, ..., n]$; and
- $\exists\ \tau^\# \in [1, ..., n]$ such that $b_{\tau^\#} = 1$ and $b_\tau = 0$ $(\tau \neq \tau^\#)$.

We obtain $\forall\, j \in [0, \cdots, \lceil \frac{|S|}{n} \rceil - 1]$, $b_{j*n+\tau\#} = 1$ and $b_{j*n+\tau} = 0$ holds. Then,

$$supp(x|C_{\tau\#}) = \sum_{j=0}^{\lceil \frac{|S|}{n} \rceil - 1} b_{j*n+\tau\#} = \lceil \frac{|S|}{n} \rceil,$$

and $supp(x|C_\tau)_{\tau \neq \tau\#} = 0$. Based on Lemma 2, we know that $q(C_{\tau\#}) = 1$, and $\{q(C_\tau)\}_{\tau \neq \tau\#}$ are all 0.

If $\pi(n)$ is the "good" partition for the distribution periodicity of x in S, and $\pi(n)$ is also the "good" partition for the structure periodicity of x in S, then x is said to have a **perfect periodicity** in S.

3.4 Research Problem

Implementing partition method $\pi(n) \in \Pi$ on a binary time-series BS_x, it would generate two distributions for the appearances of x in data series, i.e.,

$$\hat{p}_n = \{\hat{p}(P_j)\}_{1 \leq j \leq \lceil \frac{|S|}{n} \rceil]} \text{ and } \hat{q}_n = \{\hat{q}(C_\tau)\}_{1 \leq \tau \leq n}. \tag{10}$$

If there exists a feasible measurement of $d(\cdot)$ that can be used to evaluate the distance between two distributions in (10), then $d(\hat{p}_n, p_n)$ and $d(\hat{q}_n, q_n)$ would show how close a real probability distribution $\hat{p}_n$ ($\hat{q}_n$) is to a candidate distribution of p_n (q_n). Without losing generality, it can be assumed that the more closer $\hat{p}_n$ ($\hat{q}_n$) to p_n (q_n), the more smaller the value of $d(\hat{p}_n, p_n)$ and $d(\hat{q}_n, q_n)$ would be. Along this line, the event periodicity detection is changed to find an optimal partition $\pi(n)$ on S to minimize the distance between the generated two distributions with two distributions respectively:

$$\begin{array}{l} \min_{\pi(n) \in \prod}\{d(\hat{p}_n, p_n)\} \text{ and } \min_{\pi(n) \in \prod}\{d(\hat{q}_n, q_n)\} \\ \text{st. } 2 \leqslant n \leqslant \lceil \frac{|S|}{2} \rceil. \end{array} \tag{11}$$

4 Mining Event Periodicity

4.1 Cross Entropy

In this work, we introduce the cross entropy [15] to measure the similarity between two distributions. Given two distributions of $\hat{p}_n$ and p_n, the **cross entropy** or the Kullback-Leibler (KL) divergence between $\hat{p}_n$ and p_n is defined by

$$KL(\hat{p}||p)_n = \sum_n \hat{p}_n \log \frac{\hat{p}_n}{p_n}. \tag{12}$$

The cross entropy determines the ability to discriminate between two states of the world, yielding sample distributions $\hat{p}_n$ and ideal distribution p_n.

Theorem 3. *$KL(\hat{p}||p)_n \geq 0$, and it is minimized if the distributions match exactly, i.e., $KL(\hat{p}||p)_n = 0$ if $\hat{p}_n = p_n$.*

Theorem 3 provides theoretical clues for finding a feasible n in task of event periodicity detection.

4.2 Identifying Distribution Periodicity

According to the definition of Theorem 1, if event x has distribution periodicity in S with respect to the **"good" partition** $\pi(n)$, then

$$p_n = \left\{ \frac{1}{\lceil \frac{|S|}{n} \rceil}, \cdots, \frac{1}{\lceil \frac{|S|}{n} \rceil} \right\}.$$

The appearances of x in each segment P_j is $supp(x|P_j)$ with respect to $\pi(n)$. Accordingly, the posterior probability distribution of $\hat{p}_n$ is calculated as:

$$\hat{p}_n = \left\{ \frac{supp(x|P_1)}{supp(x)}, \cdots, \frac{supp(x|P_{\lceil \frac{|S|}{n} \rceil})}{supp(x)} \right\}.$$

Known from Theorem 3, a smaller value of $KL(\hat{p}||p)_n$ means the posterior distribution p_n is more close to q_n, which indicates that $p_n \sim q_n$ means $KL(\hat{p}||p)_n \sim 0$ and then $\pi(n)$ may be a "good" partition for detecting the periodicity of event x. Thus, the task of detecting distribution periodicity of event x in S is equal to find a "good" partition $\pi(n^*)$ to minimizes the KL distance:

$$n^* = \arg \min_{\pi(n) \in \prod} \{KL(\hat{p}||p)_n\} \quad \text{st. } 2 \leqslant n \leqslant \lceil \frac{|S|}{2} \rceil. \tag{13}$$

We propose an Algorithm 1 to calculate the minimized $KL(\hat{p}||p)_n$.

Algorithm 1. Calculate $KL(\hat{p}||p)_n$

1: **Input**: Binary data series BS_x;
2: **Output**: $\mathbb{KL}$;
3: $\mathbb{KL} = \phi$;
4: **for** $n = 2$ to $\lceil \frac{|S|}{2} \rceil$ **do**
5: $\quad KL_n = 0$;
6: $\quad$ **for** $j = 1$ to $\lceil \frac{|S|}{n} \rceil$ **do**
7: $\quad\quad P_j = \{b_{(j-1)*n+1}, ..., b_{j*n}\}$;
8: $\quad\quad p_j = \frac{supp(x|P_j)}{supp(x)}$;
9: $\quad\quad KL_n = KL_n + p_j \log \left(p_j * \lceil \frac{|S|}{n} \rceil \right)$;
10: $\quad$ **end for**
11: $\quad \mathbb{KL} \leftarrow KL_n$;
12: **end for**
13: **return** $\mathbb{KL}$;

The proposed method traverses all the n in $[2, \lceil \frac{|S|}{2} \rceil]$ to find the most feasible $\pi(n)$ such that the value of $KL(\hat{p}||p)_n$ can be minimized. Based on the partition results provided by $\pi(n)$, we have to calculate $\lceil \frac{|S|}{n} \rceil$ values, i.e., $supp(x|P_j)_{j=\{1,...,\lceil \frac{|S|}{n} \rceil\}}$, which can be obtained in $O(1)$ time. Therefore, the overall complexity of Algorithm 1 is very efficient of $\sum_2^{\lceil \frac{|S|}{2} \rceil} \lceil \frac{|S|}{n} \rceil = O(|S| \ln |S|)$.

4.3 Identifying Structure Periodicity

We consider the opposite side of the above mentioned "good" partition, that is, the "worst" partition for showing the structure periodicity of event x. In such a poor case, the distribution of x would not obey the rule of Theorem 2, which means the distributions of x in $C_t, (t = 1, ..., n)$ are the same instead of a distribution shown in relation (9). Such a distributions of x can be referred as:

$$q_n = \left\{\frac{1}{n}, \cdots, \frac{1}{n}\right\}.$$

In real, the posterior probability distribution of x in C_τ is:

$$\hat{q}_n = \left\{\frac{supp(x|C_1)}{supp(x)}, \cdots, \frac{supp(x|C_n)}{supp(x)}\right\}.$$

Using $KL(\hat{q}||q)_n$ to measure the difference between two distributions of $\hat{q}_n$ and q_n, a bigger value of $KL(\hat{q}||q)_n$ indicates that the posterior distribution $\hat{q}_n$ is deviated much from the route of q_n, and thus $\pi(n)$ may be a "good" partition for detecting the structure periodicity of event x. Detecting structure periodicity of event x is thus equal to find a $\pi(n) \in \Pi$ to minimizes the value of $-KL(\hat{q}||q)_n$.

$$n^{\#} = \arg \min_{\pi(n) \in \prod} \{-KL(\hat{q}||q)_n\} \quad \text{st. } 2 \leqslant n \leqslant \lceil \frac{|S|}{2} \rceil. \tag{14}$$

Algorithm 2 is used to calculate all the value of $-KL(\hat{q}||q)_n$ for x.

Algorithm 2. Compute $-KL(\hat{q}||q)$

1: **Input**: Binary data series BS_x;
2: **Output**: $\mathbb{KL}$;
3: $\mathbb{KL} = \phi$;
4: **for** $n = 2$ to $\lceil \frac{|S|}{2} \rceil$ **do**
5: $\quad KL_n = 0$;
6: $\quad$ **for** $k = 1$ to n **do**
7: $\quad\quad C_k = \{b_k, b_{n+k}, ..., b_{(\lceil \frac{|S|}{n} \rceil - 1)*n+k}\}$;
8: $\quad\quad q_k = \frac{supp(x|C_k)}{supp(x)}$;
9: $\quad\quad KL_n = KL_n + q_k * \log(q_k * n)$;
10: $\quad$ **end for**
11: $\quad \mathbb{KL} \leftarrow -KL_n$;
12: **end for**
13: **return** $\mathbb{KL}$;

There are totally $supp(x)$ appearances of x in S and $\lceil \frac{|S|}{2} \rceil$ partition results, we then can calculate $supp(x|C_\tau)$ for all the partition results by traversing all the appearance of x in S. Along this way, the complexity of Algorithm 2 is $O(supp(x)|S|)$ and the operation time can be optimized, especially, when x is sparsely distributed in S. Note that, the basic operation for this method is to calculate $supp(x|C_\tau)$ with (5) for all the $\lceil \frac{|S|}{2} \rceil$ partition results. Therefore, the complexity of Algorithm 2 is characterized by $\max\{O(|S| \ln |S|), O(supp(x)|S|)\}$.

5 Experimental Results

5.1 Experimental Setup

We conduct a series of experiments to evaluate the performance of the proposed method, namely Cross-Entropy based Periodicity Detection (CEPD). To that end, three algorithms of WARP [8], CONV [7] and VAR [30] are selected for comparisive purpose. Given that the performance of VAR are affected heavily by a user-specified parameter of va, i.e., a bigger threshold value of va will result in an increased time consumption [30], the VAR method will be conducted 3 times with different parameter settings of $va = 0.01$, 0.1, and 0.2 respectively.

All these algorithms suffer from the poor performance to big data and poor resilience to noise. To support this claim and make the experimental results more clear, we conduct a series of experiments using synthetic data. The synthetic data have been generated by controlling parameters of data distribution (uniform or normal), alphabet size (number of unique symbols in the data), size of the data series (total number of symbols), period length, and the amount of noise in the data, which is the same way as done in [7,19]. In addition, each algorithm will be run 10 times, and then take the averaged value of these experimental results as the final result to avoid potential bias.

The confidence of a periodic event x occurring in data series S is the ratio of its actual periodicity to its expected periodicity. Formally, the periodicity confidence of x in S under partition $\pi(n)$ is defined as [13,19]:

$$conf(x)_n = \frac{Actual\_Periodicity(x)}{\lceil \frac{|S|}{n} \rceil}, \quad (15)$$

where $Actual\_Periodicity(x)$ is computed by counting the number of segments in which x is appearanced.

5.2 Efficiency of the Method

In the efficiency experiments, we test the time consumption of the four algorithms of CEPD, WARP, CONV and VAR under the impacts of following circumstance: the total data size $|S|$, the period length n, the alphabet size $|\Sigma_S|$ and the (replacement) noise ratio in S.

The first set of experiments is about the effect of data series size, $|S|$, on the efficiency of algorithms, all the methods will be conducted on a set of synthetic data series by varying data size from 100 to 500. These synthetic data series have been generated by following uniform distribution with alphabet size of 4 and embedded identical period of 5. The results are presented in Fig. 1(a) and (b). Obviously, the running time of all the four methods will go increasing dramatically while $|S|$ becomes bigger. WARP has the highest complexity (Fig. 1(a))[3]. Figure 1(b) is a locally magnified image of Fig. 1(a), which shows that both the efficiency of CEPD and CONV are better than that of VAR,

[3] In Fig. 1, symbol † means the experimental results without WARP.

this advantage becomes more obvious with the decrease of va. The experiments indicate that the efficiency of CEPD is superior to the others when data series becomes very large.

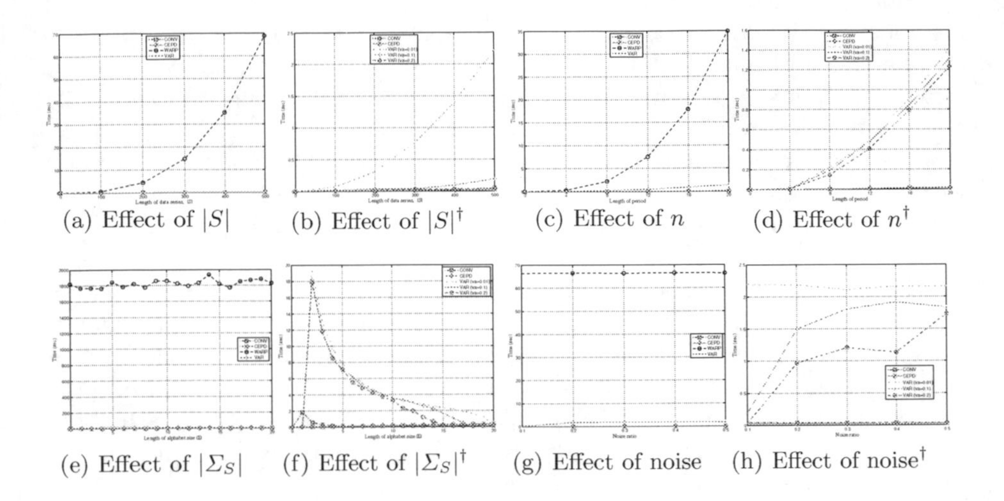

(a) Effect of $|S|$ (b) Effect of $|S|^\dagger$ (c) Effect of n (d) Effect of $n^\dagger$

(e) Effect of $|\Sigma_S|$ (f) Effect of $|\Sigma_S|^\dagger$ (g) Effect of noise (h) Effect of noise†

Fig. 1. Efficiency experiments

The next set of experiments are intended to show the efficiency of the algorithms by varying the embedded period size. In the experiment, we fixed the number of alphabets as 4, and embedded period are varied from 4 to 20. The curves of running time taken by CEPD, WARP, CONV and VAR have been plotted in Fig. 1(c), and an amplification version for the comaprison among CEPD, CONV and VAR is shown in Fig. 1(d). The results show that the time consumption of all the four algorithms are increased while the period length becomes longer. Again, WARP has the highest complexity which is followed by VAR, CONV and then CEPD.

The third set of experiments are intended to show the performances of all the four methods under effects of different alphabet size $|\Sigma|$ in a data series. The synthetic data series used in the experiments are embedded with a period of 32 and the number of alphabets are varied from 1 to 20 (smaller than the period value). The experimental results show that, along with the increasing of the alphabet size $|\Sigma|$, the running time of WARP is stable and significantly higher than the other methods. Interestingly, the running time of VAR and CONV increase dramatically when $|\Sigma|$ is relative small, and then fall down when $|\Sigma|$ goes bigger (Fig. 1(e) and (f)). The running time of CEPD increases slowly and lower than that of the other methods.

The fourth experiments are conducted to measure the impact of noise ratio on the time performance of the four methods. To that end, we fixed the length of the time series as 500, and replaced some regular symbols with noise symbols in the experiments. The noise ratio is varied from 0 to 0.5. As we can see, the running efficiency of CEPD, WARP and CONV perform stably under different noise ratio. In other words, the performance of these algorithms are not sensitive to replacement noise (Fig. 1(g) and (h)). This is similar to the results presented in [19]. As for VAR, it performs worse as the ratio noise increasing (Fig. 1(h)).

5.3 Accuracy of the Method

Two different ways are conducted to study the accuracy of the algorithms. The first compares the value of confidence for each method assuming that the period can be identified by all the methods. Whereas, the second compares the period identified by each method when the confidence is maximized. Five synthetic data sets are generated for the experiments (Table 1). In the experiments, the potential period length is set as 10, and the data series is generated by repeating the period 100 times.

Table 1. The generation of synthetic data set (denoted by D).

D	Generation rules	Sample data series
1	1 periodic event	$\underline{1}000\ \underline{1}000\ \underline{1}000\ \underline{1}000$
2	1 random event	1000 0010 0001 0100
3	2 periodic events	$\underline{1}0\underline{1}0\ \underline{1}0\underline{1}0\ \underline{1}0\underline{1}0\ \underline{1}0\underline{1}0$
4	1 periodic and 1 random event	$\underline{1}001\ \underline{1}100\ \underline{1}010\ \underline{1}010$
5	2 random events	0101 1001 1010 0011

For the experiments of confidence comparison, the results are listed in Table 2. As we can see, the CEPD can identify all the events with confidence 100% since all the events in the data series are uniformed distributed with respect to the period. WARP also show a better performance on accuracy (close to 100%) than CONV and VAR. However, CONV performs good on data set 1, 3 and 4, and bad on data set 2 and 5 (the events are randomly distributed). That is to say, CONV prefers to structure period. The VARs show good efficiency on the data series having only 1 event embedded (data set 1 and 2).

5.4 Noise Resilience

In data series, there are three types of noise: replacement, insertion, and deletion noise [8]. Accordingly, the purpose of the following experiments are to study the behavior of the different methods in periodicity detection with respect to toward these noise as well as some mixtures of them. In the experiments, we used a synthetic time series containing 4 symbols and period size of 10. The noise ratio increased gradually from 0.0 to 0.5. Finally, we report the averaged confidence level of all the symbols at which the actual period of 10 is detected.

Table 2. The comparison of confidence (the period is fixed).

Data Set	CEPD	CONV	WARP	VAR ($va = 0.2$)	VAR ($va = 0.1$)	VAR ($va = 0.01$)
1	1	1	1	1	1	1
2	1	0	0.9959	1	1	-
3	1	1	1	-	-	-
4	1	1	0.9989	-	-	-
5	1	0.0813	0.9975	-	-	-

In general, along with the increase of noise ratio, the accuracy of all the algorithms are in decline; especially when considering insertion, deletion or hybrid noise increasing, the accuracy of these algorithms will fall shapely (Figs. 2 and 3). In case that the noises are uniformly distributed in data series, CEPD has the best performance for the replacement noise (Fig. 2(a)). When the insertion, deletion and hybrid noises are embedded uniformly into a data series, WARP shows the best performance on noise resilience, followed by CEPD (Fig. 2(b), (c) and (d)). In case that the noises are normally distributed in data series, similarly, CEPD performances best under the situation that the replacement noises are embedded (Fig. 3(a)). However, for the situations of insertion, deletion and hybrid noises, the comparative results are mixed and no method has a significant advantage over the others in noise resilience (Fig. 3(b), (c) and (d)).

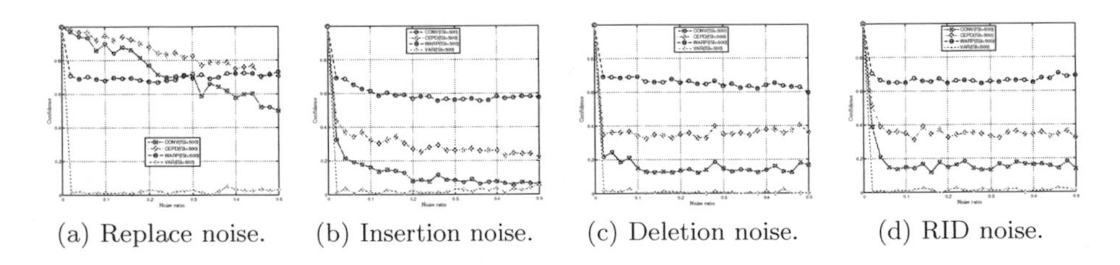

(a) Replace noise. (b) Insertion noise. (c) Deletion noise. (d) RID noise.

Fig. 2. Accuracy (uniformly distributed noise).

5.5 A Case Study on Real Dataset

A real-world data set, i.e., Amazon access samples data set (AASDS)[4], has been used in the experiments which was created and donated by Amazon.com in 2011 and has been cited for many times. AASDS contains 17612 users' access history from 2005.8 to 2010.8. To study the periodicity of each Amazon user, we take "Day" as the basic time unit, and all the accessing actions are then counted by 24-hours-day, for example, if a user had accessed Amazon.com more than 0

[4] http://archive.ics.uci.edu/ml/datasets/.

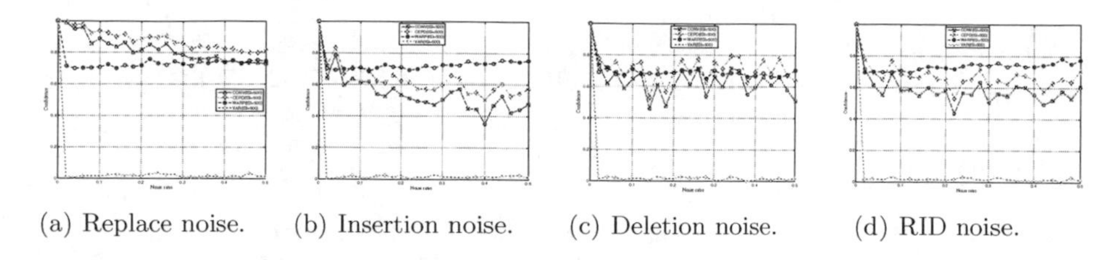

(a) Replace noise. (b) Insertion noise. (c) Deletion noise. (d) RID noise.

Fig. 3. Accuracy (normally distributed noise).

times in Sep. 2, 2005, then we marked the value of e_t at the position of day Sep. 2, 2005 as "1" in S_A, otherwise, "0" is marked. Finally, we can generate a data series of S_A for the targeted user.

Taking the No. #33400 user as an example, the results are shown in Fig. 4. As we can see that, the minimized value of KL shows that the distribution period of users #33400 accessing Amazon.com can be approximated as $n^* = 15$ days (two weeks). That is to say, the user trended to visit Amazon.com equal times every 15 days. Although the event of user #33400 accessing Amazon.com has a distribution of two weeks, but the multiple relationship between the positions of local minimized $-KL'$ for data set AASDS is weak, which means the user has no regular time point for visiting Amazon.com.

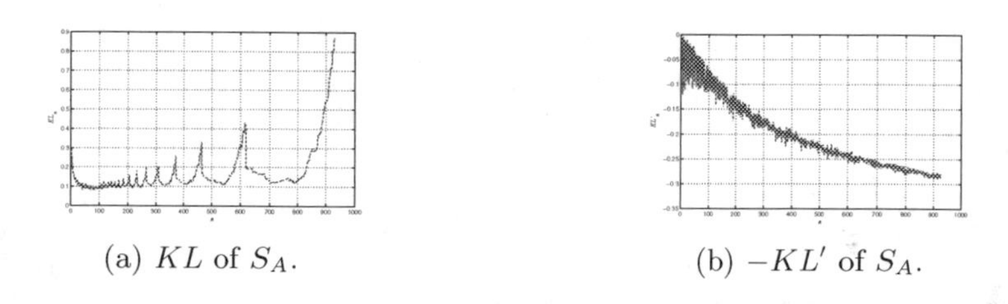

(a) KL of S_A. (b) $-KL'$ of S_A.

Fig. 4. KL and $-KL'$ in real data series.

6 Conclusion

Discovering the periodicity of event happened in sequential data series is a valuable work for data analyzing. In this paper, a novel and efficient method, namely CEPD is proposed to address the event-based periodicity mining problem in data series. The advantages of the proposed method are summarized as follows: first, it is the first time to distinguish the idea of distribution periodicity and structure periodicity. Second, a simple and complete partition method $\pi(n)$ is proposed. Third, basing on the minimum cross entropy principle and the property of periodic function, we present an efficient method to measure and determine the periodicity of an event. The experimental results show that CEPD has a best performance of running efficiency due to its less complexity.

Acknowledgments. The authors would like to thank the supports of the National Natural Science Foundation of China (71671027/91846105/71572029/71490723).

References

1. Benson, A.R., Kumar, R., Tomkins, A.: Modeling user consumption sequences. In: Proceedings of the 25th International Conference on World Wide Web, WWW 2016, pp. 519–529 (2016)
2. Berberidis, C., Vlahavas, I., Aref, W.G., Atallah, M., Elmagarmid, A.K.: On the discovery of weak periodicities in large time series. In: Elomaa, T., Mannila, H., Toivonen, H. (eds.) PKDD 2002. LNCS (LNAI), vol. 2431, pp. 51–61. Springer, Heidelberg (2002). https://doi.org/10.1007/3-540-45681-3_5
3. Brigham, E.: Fast Fourier Transform and Its Applications, 1st edn. Prentice Hall, Englewood (1988)
4. Cao, H., Cheung, D.W., Mamoulis, N.: Discovering partial periodic patterns in discrete data sequences. In: Dai, H., Srikant, R., Zhang, C. (eds.) PAKDD 2004. LNCS (LNAI), vol. 3056, pp. 653–658. Springer, Heidelberg (2004). https://doi.org/10.1007/978-3-540-24775-3_77
5. Cetintas, S., Chen, D., Si, L., Shen, B., Datbayev, Z.: Forecasting counts of user visits for online display advertising with probabilistic latent class models. In: Proceeding of the 34th International ACM SIGIR Conference, pp. 1217–1218 (2011)
6. Chan, K.P., Fu, A.W.C.: Efficient time series matching by wavelets. In: Proceedings of the 15th International Conference on Data Engineering, ICDE 1999, pp. 126–133 (1999)
7. Elfeky, M.G., Aref, W.G., Elmagarmid, A.K.: Periodicity detection in time series databases. IEEE Trans. Knowl. Data Eng. **17**(7), 875–887 (2005)
8. Elfeky, M.G., Aref, W.G., Elmagarmid, A.K.: WARP: time warping for periodicity detection. In: Proceedings of the Fifth IEEE International Conference on Data Mining, ICDM 2005, pp. 138–145 (2005)
9. Elfeky, M.G., Aref, W.G., Elmagarmid, A.K.: Stagger: periodicity mining of data streams using expanding sliding windows. In: Proceedings of the 6th IEEE International Conference on Data Mining, pp. 188–199 (2006)
10. Faloutsos, C., Ranganathan, M., Manolopoulos, Y.: Fast subsequence matching in time-series databases. In: Proceedings of the SIGMOD 1994, pp. 419–429. ACM (1994)
11. Ghosh, A., Lucas, C., Sarkar, R.: Finding periodic discrete events in noisy streams. Proc. CIKM **2017**, 627–636 (2017)
12. Glynn, E.F., Chen, J., Mushegian, A.R.: Detecting periodic patterns in unevenly spaced gene expression time series using lomb-scargle periodograms. Bioinformatics **22**(3), 310–316 (2006)
13. Han, J., Dong, G., Yin, Y.: Efficient mining of partial periodic patterns in time series database. In: Proceedings of International Conference on Data Engineering, pp. 106–115 (1999)
14. He, Z., Wang, X.S., Lee, B.S., Ling, A.C.H.: Mining partial periodic correlations in time series. Knowl. Inf. Syst. **15**, 31–54 (2008)
15. Kullback, S., Leibler, R.A.: On information and sufficienvy. Ann. Math. Stat. **22**, 79–86 (1951)
16. Li, Z., Ding, B., Han, J., Nye, R.K.P.: Mining periodic behaviors for moving objects. In: Proceedings of the 16th ACM SIGKDD International Conference on Knowledge Discovery and Data Mining, pp. 1099–1108 (2010)

17. Li, Z., Wang, J., Han, J.: Mining event periodicity from incomplete observations. In: Proceedings of the 18th ACM SIGKDD International Conference on Knowledge Discovery and Data Mining, pp. 444–452 (2012)
18. Ma, S., Hellerstein, J.L.: Mining partially periodic event patterns with unknown periods. In: Proceedings of the 17th International Conference on Data Engineering, pp. 205–214. IEEE (2001)
19. Rasheed, F., Alshalalfa, M., Alhajj, R.: Efficient periodicity mining in time series databases using suffix trees. IEEE Trans. Knowl. Data Eng. **23**(1), 79–94 (2011)
20. Ruiz, E.J., Hristidis, V., Castillo, C., Gionis, A., Jaimes, A.: Correlating financial time series with micro-blogging activity. In: Proceedings of the Fifth ACM International Conference on Web Search and Data Mining, WSDM 2012, pp. 513–522. ACM (2012)
21. Sheng, C., Hsu, W., Lee, M.L.: Mining dense periodic patterns in time series data. In: Proceedings of the 22nd International Conference on Data Engineering, ICDE 2006, p. 115. IEEE (2006)
22. Sripada, S.G., Reiter, E., Hunter, J., Yu, J.: Segmenting time series for weather forecasting. In: Macintosh, A., Ellis, R., Coenen, F. (eds.) Applications and Innovations in Intelligent Systems X, pp. 193–206. Springer, London (2003). https://doi.org/10.1007/978-1-4471-0649-4_14
23. Tanbeer, S.K., Ahmed, C.F., Jeong, B.-S., Lee, Y.-K.: Discovering periodic-frequent patterns in transactional databases. In: Theeramunkong, T., Kijsirikul, B., Cercone, N., Ho, T.-B. (eds.) PAKDD 2009. LNCS (LNAI), vol. 5476, pp. 242–253. Springer, Heidelberg (2009). https://doi.org/10.1007/978-3-642-01307-2_24
24. Vlachos, M., Yu, P.S., Castelli, V.: On periodicity detection and structural periodic similarity. In: SDM 2005, pp. 449–460 (2005)
25. Wang, R.Y., Storey, V.C., Firth, C.P.: A framework for analysis of data quality research. IEEE Trans. Knowl. Data Eng. **7**(4), 623–640 (1995)
26. Wang, X., Zhang, H., Zhang, D., Xiao, Y.: A moving-window based partial periodic patterns update technology in time series databases. In: 2008 International Symposium on Computational Intelligence and Design, ISCID 2008, vol. 2, pp. 98–101, October 2008
27. Yang, J., Wang, W., Yu, P.S.: Infominer: mining surprising periodic patterns. In: Proceedings of the seventh ACM SIGKDD International Conference on Knowledge Discovery and Data Mining, KDD 2001, pp. 395–400. ACM (2001)
28. Yang, J., Wang, W., Yu, P.S.: Mining asynchronous periodic patterns in time series data. IEEE Trans. Knowl. Data Eng. **15**(3), 613–628 (2003)
29. Yang, Y., Pan, B., Song, H.: Predicting hotel demand using destination marketing organization's web traffic data. J. Travel Res. **53**(4), 433–447 (2014)
30. Yang, Y.C., Padmanabhan, B., Liu, H., Wang, X.: Discovery of periodic patterns in sequence data: a variance-based approach. INFORMS J. Comput. **24**(3), 372–386 (2012)
31. Yuan, Q., Shang, J., Cao, X., Zhang, C., Geng, X., Han, J.: Detecting multiple periods and periodic patterns in event time sequences. Proc. CIKM **2017**, 617–626 (2017)
32. Ziegler, H., Jenny, M., Gruse, T., Keim, D.A.: Visual market sector analysis for financial time series data. In: IEEE VAST, pp. 83–90. IEEE (2010)

EPPADS: An Enhanced Phase-Based Performance-Aware Dynamic Scheduler for High Job Execution Performance in Large Scale Clusters

Prince Hamandawana[1], Ronnie Mativenga[1], Se Jin Kwon[2], and Tae-Sun Chung[1(✉)]

[1] Ajou University, Suwon 16499, South Korea
[2] Kangwon National University, Chuncheon, Gangwon 24341, South Korea
{phamandandawana,ronniematie,tschung}@ajou.ac.kr, sjkwon@kangwon.ac.kr

Abstract. The way in which jobs are scheduled is critical to achieve high job processing performance in large scale data clusters. Most existing scheduling mechanism employs a First-In First-Out, serialized approach encompassed with task straggler hunting techniques which launches speculative tasks after detecting slow tasks. This is often achieved through the instrumentation of processing nodes. Such node instrumentation incurs frequent communication overheads as the number of processing nodes increase. Moreover the sequential scheduling of job tasks and the straggler hunting approach fails to meet optimal performance as they increase job waiting time in queue and incurs delayed speculative execution of straggling tasks respectively. In this paper we propose an **E**nhanced **P**hase based **P**erformance **A**ware **D**ynamic **S**cheduler (EPPADS), which schedules job tasks without additional instrumentation modules. EPPADS uses a two staged scheduling approach, that is, the slow start phase (SSP) and accelerate phase (AccP). The SSP schedules the initial task in the queue in the normal FIFO way and records the initial execution times of the processing nodes. The AccP uses the initial execution times to compute the processing nodes task distribution ratio of the remaining tasks and schedules them using a single scheduling I/O. We implement EPPADS scheduler in Hadoop's MapReduce framework. Our evaluation shows that EPPADS can achieve a performance improvement on FIFO scheduler of 30%. Compared with existing Dynamic scheduling approach which uses node instrumentation, EPPADS achieves a better performance of 22%.

Keywords: Distributed processing · Scheduling · MapReduce

1 Introduction

For the past decade or so, we have experienced a data deluge [15] which keeps increasing at an exponential rate, doubling every two years and consequently

G. Li et al. (Eds.): DASFAA 2019, LNCS 11446, pp. 140–156, 2019.
https://doi.org/10.1007/978-3-030-18576-3_9

raising storage and computational concerns. With this massive amounts of data stored, the need for effective parallel processing in large scale data centers arises. Prior works have shown that in large parallel data processing clusters, job response time is a key performance factor that needs serious attention [4]. To achieve high job processing performance, most approaches split jobs into several smaller tasks which are then distributed to multiple processing nodes. However, job response time is often delayed by tasks running on slower nodes known as *stragglers*. The presence of stragglers is mainly due to the high contention or failures in processing nodes. It is therefore critical to mitigate the job execution slowdown caused by stragglers in order to achieve high processing performance. An example of a widely used parallel processing framework adopted by the cloud community is MapReduce [7]. One important benefit of MapReduce framework is that it abstracts the complications of handling stragglers, from the programmer by automatically invoking duplicate tasks of the stragglers onto idle slots of faster processing nodes. This phenomenon is known as *speculative execution.* Speculative execution of stragglers often results in reduced job completion time. Prior works [7] shows that speculative execution can reduce job execution time by 44%.

However, the default scheduler in MapReduce framework, splits the job tasks uniformly to the processing nodes called task trackers (TT) in a First-In First-Out (FIFO) manner for parallel task execution. Such task distribution assumes a homogeneous compute capability across all task trackers. However, this will not perform well when subjected to heterogeneous cluster setups, where some task trackers are faster than others. Previous work, LATE [19], proved that when the default FIFO scheduler is employed in a heterogeneous cluster environment, it frequently invokes unnecessary speculative tasks which often elongate the job completion time. To mitigate such unnecessary speculative task execution, a number of self-adaptive dynamic schedulers, LATE, LA, DDAS, ESAMR, COSSH [11,13,14,18] were proposed. These schedulers operate by regularly collecting and using system level information to dynamically make scheduling decisions in heterogeneous environments. This was achieved through employing some modules in the job tracker node (also called master node) which instruments the compute capabilities of all task trackers. This therefore enforced the use of always up to date system level information to make scheduling decisions. As a result, speculative tasks will be invoked only if their invocation can result in shorter job execution time. However the monitoring can contribute additional overhead as the number of nodes to be monitored increase above certain limits.

In summary, the existing approaches are coupled with some drawbacks that limits the attainment of maximum job execution performance in large scale clusters. firstly, sequential task scheduling, i.e., a successive task in the job queue is scheduled only if the predecessor task has completed, elongates the job completion time as all successive tasks has to wait until previously scheduled task is completed. Secondly, existing approaches uses the straggler "hunting" approach, which only invokes a speculative task in the middle of executing a straggling task. This speculative task invocation approach occurs when the job monitor notices that the progress of the job is well below average. Job completion performance is then degraded as speculative execution is often invoked late. lastly, frequent collection of processing nodes system level information to predict the time to complete a task and make scheduling decisions causes additional performance degradation. Such communication overheads increase dramatically as we scale up task trackers in the cluster.

To overcome the above mentioned problems in existing approaches, we propose an **E**nhanced **P**hase based **P**erformance **A**ware **D**ynamic **S**cheduler (EPPADS), which uses a rolling scheduling window in which incoming tasks are split into smaller partitions, called scheduling window instances. EPPADS launches the first tasks in the scheduling window instance in a normal FIFO way. Basing on the initial time to complete and progress rate of the initial tasks, straggler nodes are marked. This eradicates the need for system level communication modules used in previous works. This process happens in the first phase of EPPADS known as *Slow-Start Phase* (SSP). EPPADS resolves the drawback of sequential task scheduling by distributing the remaining job tasks within the scheduling window instance at one go. This is achieved in the second phase of EPPADS called the *Accelerate Phase* (AccP). Since the straggler nodes are marked during the SSP, tasks scheduled to straggler nodes are launched at the same time with their corresponding speculative tasks. This removes the drawback caused by late invocation of speculative tasks. The main contributions of our work are as follows;

- We design a lightweight dual-phase opportunistic job task scheduler, which amortizes the performance drawbacks caused by node monitoring and sequential task scheduling.
- We design and implement a sliding scheduling window technique that fuse job scheduling together with task pre-speculation based on the compute capabilities of each worker node. Task allocation decisions are made at each scheduling window instance.
- We implemented our EPPADS prototype in Yarn Scheduler Load Simulator (SLS) which has the capability to emulate a large scale production cluster. Our evaluation shows that EPPADS achieves a performance improvement of 30% and 22% as compared to default FIFO scheduler and existing dynamic self adaptive schedulers which uses node monitoring, respectively.

2 Background and Motivation

2.1 MapReduce and Job Scheduling

Figure 1 shows the general design of MapReduce framework. In a MapReduce cluster, clients submit job requests through the *job-tracker* node. The job-tracker splits the job requests into smaller tasks that are uniformly scheduled to multiple *task-tracker* nodes for parallel processing. The job-tracker is responsible for tracking the progress of each assigned task in all the task-trackers. Each MapReduce job has a map and reduce stage and it is the responsibility of the job-tracker

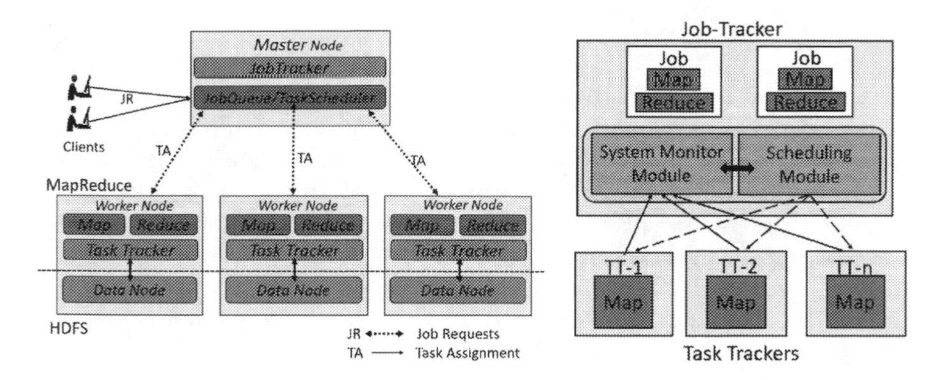

Fig. 1. MapReduce framework.

Fig. 2. Node instrumentation.

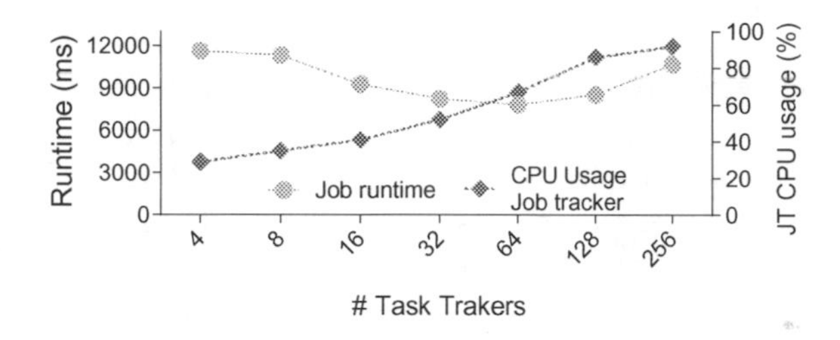

Fig. 3. An analysis of the default FIFO scheduler. JT in Y-axis means job tracker.

node to harmonize the map and reduce stages of the MapReduce job. The Map phase partitions the input data and outputs some intermediate key and value pairs which are sorted and combined so that similar key and value pairs are processed by the same reducer on the reduce phase. Finally the reducers will process the key, value pairs and output the final results.

It is important to note that the manner in which jobs are scheduled and processed is of significance importance to achieve high job execution performance in large scale data processing clusters. In MapReduce programming model, tasks are scheduled equally among task trackers for processing using the FIFO scheduler. The job-tracker monitors the progress of all tasks in all task tracker nodes. If a straggler task is detected, a speculative task is invoked on a faster task-tracker with an available processing slot. The results of the task which finishes first is then used, and the other task is destroyed immediately. This speculative task execution shortens the job completion time significantly [7]. However, the performance gain significantly decreases when the default FIFO scheduler is subjected to heterogeneous cluster environments [19]. This is because FIFO scheduler was designed with homogeneous environments in mind, where all nodes have equal processing capabilities. It is therefore vital to design effective performance-aware schedulers that improve the job execution performance in a heterogeneous large scale data clusters.

2.2 Motivation

Communication and Monitoring Overheads: Quite a number of previous works [8–11,13,14,18,19] has concentrated on enhancing job processing performance in heterogeneous clusters. Most large scale data processing clusters are commonly faced with frequently fluctuating system loading. Usually this might be caused by other non-MapReduce processes running on the cluster. Therefore, there is need for provisioning of a consistent and accurate speculative execution scheme that keeps track of task tracker nodes system level information [6,11,14]. Basing on these system level information, speculative task execution is invoked judiciously only on tasks that reduce the job completion time.

Figure 2 shows the widely used instrumentation approach in previous and existing works [6,11,14]. It uses additional modules for collecting system level information of all task-tracker nodes and make scheduling decisions based on the collected task trackers' system level information. However, additional instrumentation can cause system performance degradation as the number of tasks tracker nodes increases. Figure 3 shows an analysis of the default FIFO scheduler, and the job runtime performance when FIFO scheduler is adopted. The details of the testbed setup are the same as the ones in Sect. 4 - Table 1, using the Scan workload traces from the HiBench suite. We observed that when we initially increase the number of task trackers, the job runtime performance increases. However, it reaches a point (In our case above 64 task trackers) where increasing the number of task trackers starts to degrade the job execution performance. We observed also that the CPU utilization on the Job tracker node is very high when the number of task trackers are greater than 64. This shows that the communication and monitoring overheads are non negligible in large scale data processing clusters. This is because a large number of task trackers create significant contention on the job tracker node and consequently suppress the overall job execution performance.

Sequential Task Scheduling: All prior studies adopted the traditional way of scheduling tasks in a sequential manner. This implies that a successive task can only be scheduled only if one of the already scheduled tasks is completed. By so doing, it causes a major constraint in achieving maximum job completion throughput. Serialization of task scheduling usually result in increased scheduling time considering a large number of tasks that can be scheduled in large scale data processing clusters. Also, the average waiting time of successive tasks and jobs in the job tracker's job queue consequently increases, thereby affecting job processing performance.

Delayed Speculative Task Execution: Current scheduling techniques in MapReduce adopts a straggler "hunting" approach in-order to invoke speculative tasks of stragglers. In this approach, all scheduled tasks are monitored for their execution process. While in the middle of task execution, if the progress of any scheduled tasks is below the average progress rate of all processing tasks,

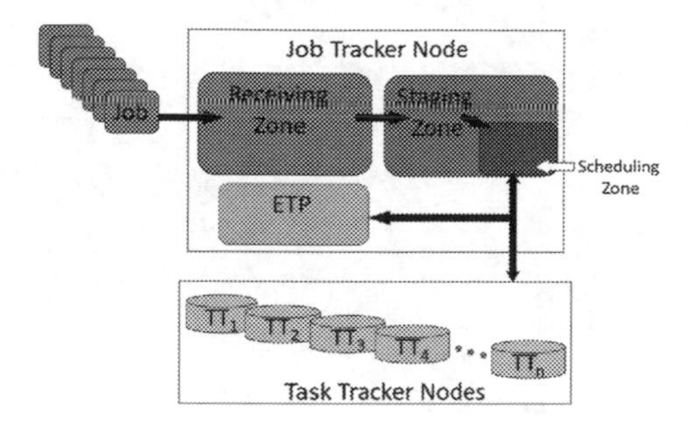

Fig. 4. An overview of EPPADS architecture.

a corresponding duplicate task are launched speculatively on other faster task trackers. However this approach waits for straggler task detection before invoking a speculative task. Invoking a speculative task at the time of straggler detection might be too late to limit the slowdown caused by the straggling tasks.

To amortize the above mentioned drawbacks we propose EPPADS, which achieves high performance job execution without the need for any additional instrumentation modules, schedules multiple tasks in the same scheduling window instance in a concurrent manner and incorporates job scheduling and task pre-speculation withing a scheduling window instance. In this case straggling tasks are predicted and their speculative tasks are invoked at the same time with their predicted straggler. The following section explains in detail the design of our proposed EPPADS scheduler.

3 EPPADS Design

3.1 Overview of EPPADS

Figure 4 depicts the overview of the EPPADS design. The design of EPPADS, splits the processing of incoming job requests into three memory regions, *receiving zone, staging zone and scheduling zone.* The receiving zone receives and splits the incoming job requests into smaller tasks. The staging zone is a fixed allocated buffer which prepares tasks for scheduling. Finally, the scheduling zone is a smaller buffer area inside the staging zone in which the scheduling occurs. Recall, when clients submit their job requests to the job tracker node, the job requests are split into smaller equal sized tasks. In the default scheduling in MapReduce, the smaller sized tasks are then distributed equally to the available task tracker nodes one at a time.

However, to achieve our goal in EPPADS when the job tracker receives job requests from clients, the job requests are initially split into smaller tasks as usual in the *receiving zone*. These smaller tasks are then forwarded to the *staging zone* in which the staged tasks are absorbed into the smaller scheduling region called scheduling zone. The scheduling zone uses a rolling scheduling window that schedules all the tasks in the staging zone. We discuss in detail the

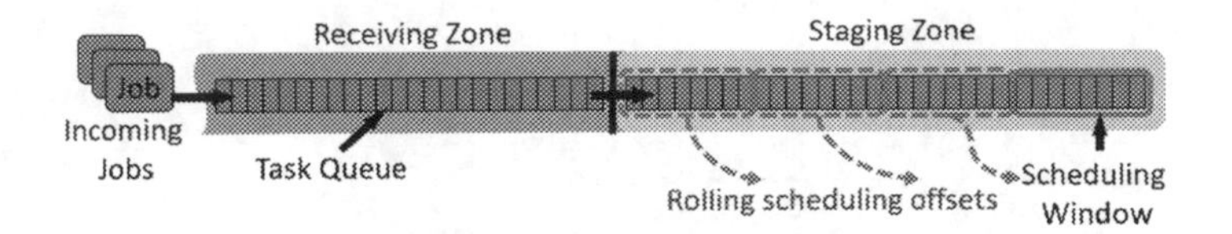

Fig. 5. Rolling scheduling window design in EPPADS

rolling scheduling window technique in Sect. 3.2. To reduce the impact caused by monitoring system level of task trackers, EPPADS uses a two phase scheduling approach. In the first phase, called the Slow Start Phase, the first tasks inside the current scheduling window instance, are scheduled in the usual FIFO manner to the available task trackers. The initial execution time for the first scheduled tasks are then recorded and stored in an *Execution Time Prediction Table* (ETP). We discuss the details of the ETP in Sect. 3.3

In the 2nd phase, called the Accelerate Phase, EPPADS uses the execution time ratios of the initial tasks and compute the task distribution of all the remaining tasks in the current scheduling window and distributes them to the task trackers via a single I/O operation. After the scheduled tasks finish execution, the window rolls to the next scheduling window and so on. Note that, during the Slow Start Phase, EPPADS also detect slow task tracker nodes and flag them as straggler nodes. We also explain in detail how the straggler node are detected in Sect. 3.3. When scheduling a task to a task tracker marked as a straggler node, EPPADS simultaneously schedules a redundant task to the nearest non-straggler node in the cluster. This amortize the overhead caused by the delay in speculative task invocation in prior works. Next we discuss in detail the scheduling in EPPADS.

3.2 Staging and Scheduling of Tasks

In order for tasks to be scheduled in EPPADS, they must be first moved onto the staging zone. This area is a pinned memory region which makes sure all task requests marked for scheduling are not swapped. The rolling scheduling window inside the staging zone will then start scheduling tasks per each scheduling window instance as depicted in Fig. 5. After all the scheduled tasks in the current scheduling window instance have finished execution, the scheduling window rolls to the next tasks in the staging zone by an offset equal to the scheduling window size. This offset shifting happens up to the end of the staging zone memory region and then reset back to the start of the staging zone region. We set the size of the staging zone to 4x the size of the scheduling zone. The reason is to reduce the memory copy time between the receiving zone and the staging zone and the waiting time to forward tasks in the staged area to the scheduling window area. Figure 9 shows that setting the staging zone size greater than 4× that of the scheduling zone didn't have any benefit. Every instance of a scheduling window, is composed of a dual phase scheduling approach which does the task scheduling in two stages (Slow Start phase and Acceleration phase). Before we detail the

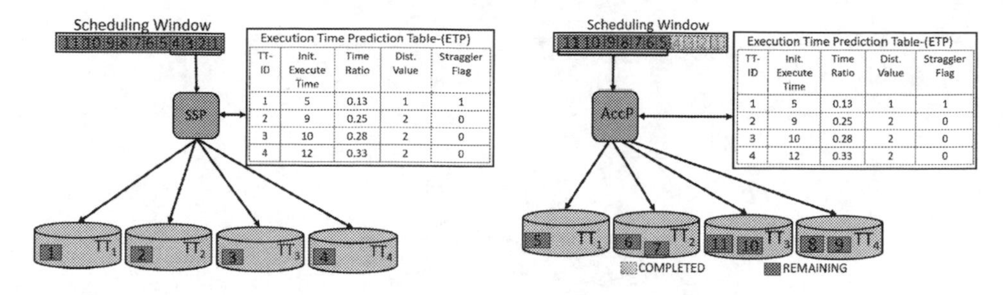

Fig. 6. Slow-start phase (SSP).

Fig. 7. Accelerate phase (AccP).

two phased scheduling approach, we will explain the execution time prediction table which is used in both the slow start phase and the accelerate phase.

3.3 Execution Time Prediction Table (ETP)

In the ETP, each task tracker ID (TT-ID) has a corresponding initial execution time, time ratio, distribution ratio and a straggler flag. The initial execution time is the time taken to compute the initial task in the slow start phase. The time ratio is the ratio of the task tracker execution time to the total initial execution time of all the first scheduled tasks, e.g., task-1 to task-4 in Fig. 6. The dist. value is the ratio used in the accelerate phase to distribute the remaining tasks, e.g., task-5 to tasks-11 in Fig. 7. This distribution value is calculated by multiplying the total number of remaining tasks in the scheduling window by the time ratio of the task tracker node and rounding the value to nearest whole number. For all task trackers, the straggler flag is set to either 0 when its not a straggler or to 1 when the node is a straggler. If the execution time of a task tracker is less by 8% or more from the average execution time, then its straggler flag is marked as 1, otherwise it is flagged to 0. To determine the 8% value, we measured job slow down caused by the increase in negative deviation from the average task on the straggler node. We increased the load in the straggler node at each experimental run as we measure the job slow down. Figure 10 shows that a negative deviation of less than 8% from the average progress will not cause significant slowdown. With the Dist. Value and straggler flag result we can schedule all the remaining tasks in the scheduling window at once rather than one at a time.

3.4 Slow-Start Phase

The scheduling of tasks inside the current scheduling window instance starts with the slow start phase. Figure 6 depicts an example of scheduling in the slow start phase. In this phase, the first tasks, e.g. tasks 1, 2, 3 and 4, at the front of the task queue in the scheduling window are scheduled to the available task tracker nodes sequentially just like in the default FIFO scheduler. The execution time of these first scheduled tasks are recorded in the ETP table. At this stage EPPADS also computes for each node, the time ratio, distribution value and sets the straggler

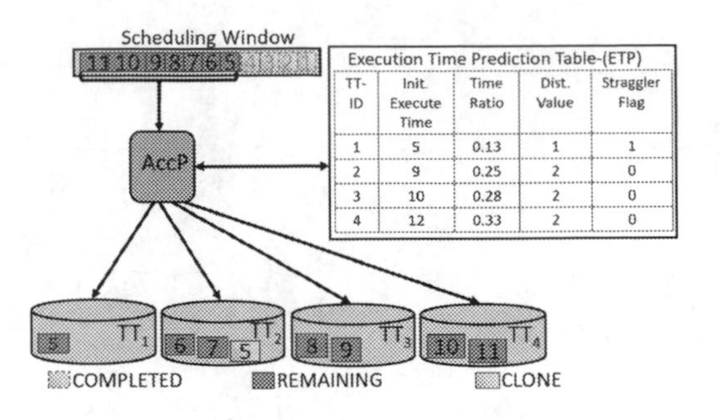

Fig. 8. Clone invocation in AccP.

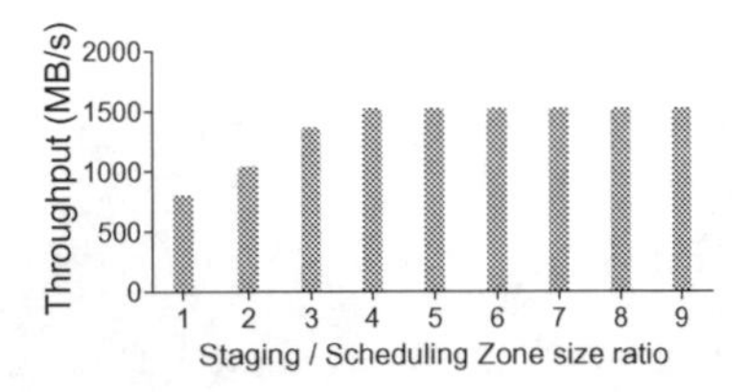

Fig. 9. Analysis of the staging zone buffer size.

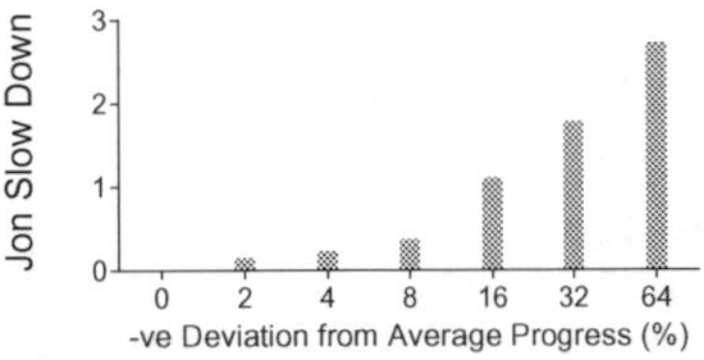

Fig. 10. Analysis of job slow down caused by stragglers.

flag to either 0 or 1. The reason to do this is to get an insight on a single task execution time of each task tracker node. Since the scheduling window is small we assume; (i) It is fast enough to execute all tasks in the current scheduling window before any significant change in task tracker node system load. (ii) Since it is fast enough to compute all tasks in a scheduling window instance, we assume that the time to compute each remaining task on the task-tracker is equal to the recorded time taken to compute the first issued task. By so doing, our key idea is to eliminate the overhead caused by task tracker system level instrumentation. Although the slow start phase of the EPPADS scheduling limits performance, the worst case scenario is equal to the performance of the default FIFO scheduler.

3.5 Acceleration Phase

With the knowledge of the time to compute a single task at each task tracker node (stored in the ETP table), EPPADS then distributes the remaining tasks using a single I/O. For example in Fig. 7 EPPADS uses the task tracker distribution value, which is computed from the initial time ratio, to schedule the remaining tasks 5, 6, 7, 8, 9, 10, and 11 at once. The key idea here is to achieve the best overall execution time by making sure that scheduled tasks in the AccP phase finish at almost same time on all nodes as they are scheduled based on the nodes execution capabilities. Therefore, for any scheduling window instance, we assume time to compute scheduled tasks on all task-trackers is almost equal. Consequently, the overall scheduling time is speed up as the remaining tasks are

Algorithm 1. Enhanced Phase-based Performance-Aware Dynamic Scheduler (EPPADS).

```
   Input: Ji, ith Requested job to cluster
   Ω, Rolling scheduling size;
   α, Initial sequentially scheduled tasks
   T, Total tasks for scheduling window
   ti, ith task in scheduling window
   M, Number of task-trackers in the cluster
   Wi, ith window of Ji
   Prate, Average Progress Rate
   prate, Task Progress Rate
   Cti, Clone of Task ti
   TSCi, , Completion time of Clone Cti
   TSti, Completion time of task Cti
   Output: RESULT, key, value pair of task ti;
 1 for Wi ≠ 0 do
 2   for 0 < ti <T /* begin Slow-Start Phase (SSP) */
 3   do
 4     Insert ti into scheduling window Wi
 5     Query progress for α tasks
 6     for all ti with prate < (Prate - 8 ) do
 7       Mark task-tracker as straggler
 8     end
 9     for Remaining tasks in Wi /* Initialize Accelerate Phase */
10     do
11       Preschedule all ti using the time from α tasks
12       Pre-clone tasks scheduled on stragglers
13     end
14     for All pre-scheduled Cti do
15       if slot_exists−true then
16         Launch Cti together with corresponding ti
17         if TSCi < TSTi then
18           RESULT = result of Cti
19         end
20         else
21           RESULT = result of ti
22         end
23       end
24       else
25         Execute Cti on next available slot
26       end
27     end
28   end
29   Slide by offset Ω to next scheduling window
30   return RESULT
31 end
```

scheduled at once in the accelerate phase, rather than in a serialized way. The main benefit of the acceleration phase is to amortize the drawback caused by sequential scheduling of tasks.

3.6 Optimizing Delay in Speculative Execution

In order to increase the job execution performance the manner in which straggler nodes within the cluster are detected is important. Two important questions to come up with a suitable solution arises; (i) At which stage during task execution is a straggling node detected? (ii) At the time of straggler node detection, does the launch of a speculative task abate the job execution delay or it actually causes more overheads? To answer these two questions, the dual-phase design of EPPADS assist in detecting straggler nodes early before serious degradation in job processing performance. At the slow start phase a straggling node is flagged. Later at the accelerate phase, if a task is scheduled to a straggler node a redundant task called a clone is concurrently scheduled to the nearest non-straggler node. We use a topology-aware mechanism that prioritizes a faster node within the same rack. The nearest node here refers to the shortest hop count, location in cabinet, etc, depending with cluster set-up. Figure 8 shows the invocation of a clone of task-5 which was scheduled to straggler task tracker-1. The benefit of the concurrent scheduling of a straggler task and its clone, is that we can fuse the task scheduling together with the task pre-speculation. This helps to amortize the delay in speculative execution caused in previous works where a straggler is detected at the middle of task execution. In the event that two or more tasks are scheduled to a straggler node, then the tasks are shared equally among the nearest non-straggler nodes. By so doing we improve the speculative task execution and consequently improve job processing time. A complete flow of our proposed approach is described in Algorithm 1.

We evaluate our proposed EPPADS scheduling scheme in the next section. We compare the performance with existing adaptive scheduling algorithm on top of the default FIFO schedulers.

4 Evaluation

4.1 Testbed Setup

To evaluate the behavior and performance of our proposed EPPADS scheduler on a large scale data cluster, we used a modified version of the Yarn Scheduler Load Simulator (SLS) [2] that fuses with our proposed scheduler. The reason for using the SLS simulator is because of its ability to simulate large scale MapReduce clusters and gives a near accurate analysis of our EPPADS prototype. We configured hadoop and the SLS simulator on a single powerful server running on Linux CentOS v7.3 and equipped with a dual CPU socket each with 32 core Intel Xeon Skylake-EP v5 processor, running at 2.30 GHz. The server is also equipped with 8 DIMM slots of 32 GB DRAM each, making a total of 256 GB of DRAM. The total amount of storage in the server is 24 TB made out of 6×4 TB HDDs. We physically define the size of containers and the amount of resources dedicated to each container using the YARN reservation system. We configure different range of containers (Task trackers) with different specification values to depict heterogeneity among task tracker nodes. For all experiments we fix the container specifications to those defined in Table 1. Each task tracker in the same allocation range is allocated the same amount of memory, e.g, TT-1 to TT-32 are all allocated with 1024 MB of RAM. We also allocate different numbers of processors to different task tracker ranges to enable variance in processing power. We then restrict containers in the same range to use only the allocated processors, e.g, TT-1 to TT-32 only utilize processors 0 to 13. Finally, On the hadoop configuration we set the replication factor of 3 and set the number of Map and reduce tasks per each node to 3.

Workloads: For all experiments irrespective of difference in workload type, we used a total of 10 TB input data. In cases where the input data is less than 10 TB we simply amplified the input data by a factor which makes it equal or close to 10 Terabytes. For experiments in Figs. 11, 12 and 14 we used the Scan workload traces with the number of processing nodes equal 256. For experiments in Fig. 13, we used a variety of workload traces from the HiBench suite and also with a fixed number of processing nodes equal 256.

To evaluate the effectiveness of the proposed EPPADS scheduler, we compare the following systems:

- ***FIFO:*** This is the default scheduler in MapReduce framework. We used this as the baseline in our evaluation. FIFO schedules tasks equally between task tracker nodes irrespective of any difference in processing capabilities that might exist between the nodes.
- ***DynMon:*** To show the impact of node instrumentation in large scale processing clusters, we implemented a Dynamic Scheduling Algorithm which uses some instrumentation modules to monitor task trackers. The instrumentation modules regularly collect task tracker system level information and compute some utility value which is used to make scheduling decisions. Like all the existing Dynamic approaches it follows a First-In First-Out approach of scheduling job tasks. We called this Approach DynMon.
- ***Proposed:*** This refers to our proposed EPPADS implementation approach refered to in Sect. 3.

Table 1. Testbed setup depicting task tracker (TT) heterogeneity in different processing node ranges. We configure the Job-Tracker (JT) with much higher specifications than TT nodes.

	JT-Node	TT 1–32	TT 33–64	TT 65–96	TT 97–128	TT 129–160	TT 161–192	TT 193–224	TT 225–256
RAM (MB)	4096	1024	768	1024	1280	1536	1024	512	1024
Storage (GB)	80	80	80	80	80	80	80	80	80
Processor no.	0–13	14–21	22–25	26–29	30–31	32–35	36–43	44–49	50–59

4.2 Scalability Analysis

To show the scalability effectiveness of our proposed EPPADS scheduler we first measure the performance effect caused by increasing the number of the nodes in the cluster. We start with a small cluster size and double the cluster processing nodes at each experimental run until the size depicts a large sized data processing cluster. Figure 11 shows the results of our experiment. The results shows that as the number of task-tracker nodes increases, the job execution performance increases for all algorithms. This is because the contention in the

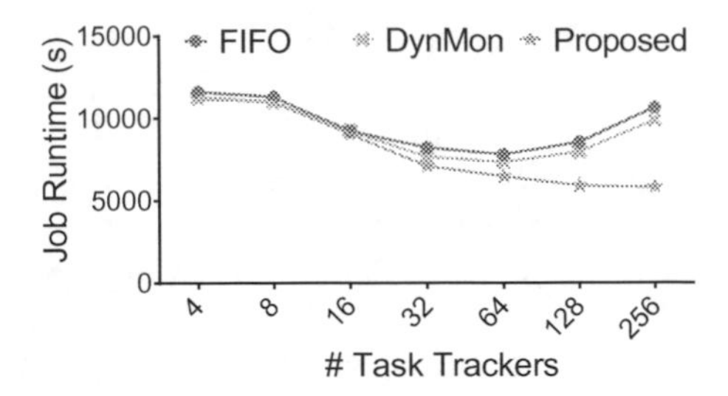

Fig. 11. Scalabilty analysis.

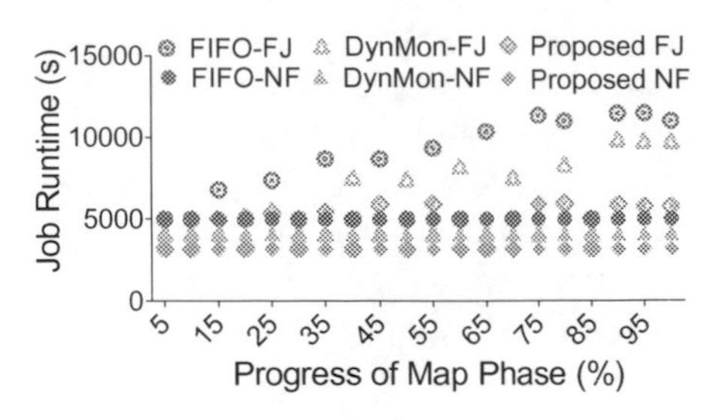

Fig. 12. Effects of job slow down.

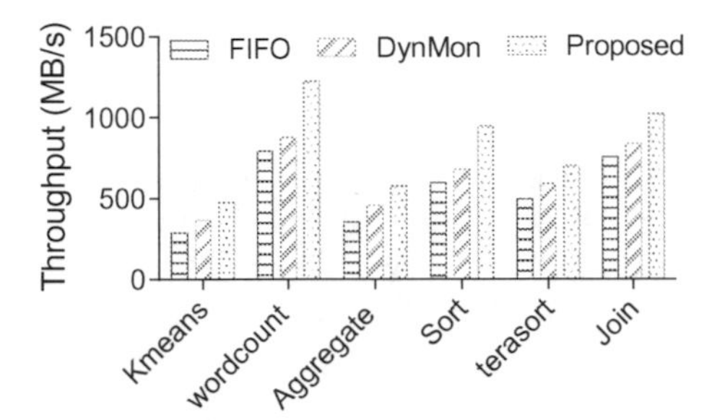

Fig. 13. Performance with HiBench suite.

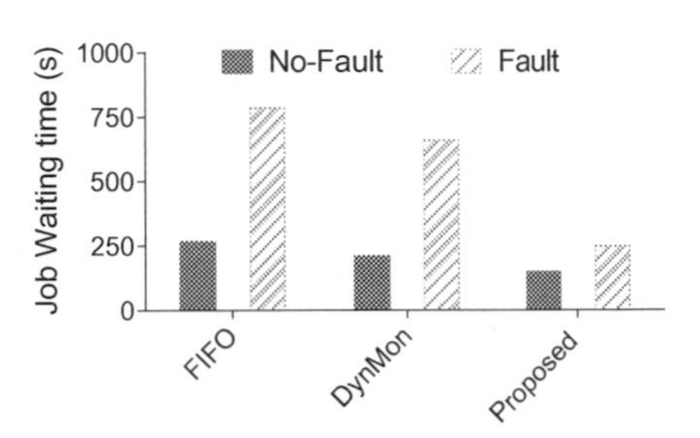

Fig. 14. Comparison of job waiting time.

cluster decreases with increasing number of task-trackers. However as we keep on adding the cluster processing nodes, we observe that the performance start to degrade at a certain point (after 64 nodes in our setup) in FIFO and the DynMon scheduling approaches. However, EPPADS performance does not suffer from the performance degradation due to continuous increase in cluster nodes. This is because of the fact that, though the node instrumentation forms an integral part of dynamic and self adaptiveness of scheduling algorithm, it incurs monitoring overheads as the number of nodes to be monitored increases above some certain limit. This is mostly due to the frequent communication overheads between the centralized job tracker node and the processing task tracker nodes. However due to the approach in which EPPADS eliminates the use of instrumentation modules, this communication overhead is greatly amortized, thus no degradation as the cluster size becomes huge.

4.3 Execution Time Perfomance

Figure 12 shows the job execution time comparison between FIFO, DynMon and proposed EPPADS, at different stages in their Map phases. **NF** and **FJ**, stands for No-Fault and Faulty jobs respectively. Each doted point on the plot depicts a single job and its execution time at different stages of Map phase. The solid dotted points represent jobs without failures and the non-solid dotted points represent jobs which incurred some failure or slow down during their processing. We observed a big job execution performance degradation of 2$\times$ or more when there is a job failure in the baseline FIFO scheduling approach and the DynMon scheduling approach. We attribute this to the delayed speculative task invocation caused by the straggler hunting approach. Mostly a speculative task is invoked at a later stage when the primary task has been running for quite a number of seconds. The Job tracker later observes that the task is straggling and decides to launch a redundant task. However this speculative task invocation is often too late to minimize the slowdown to job processing performance. However the job slow is significantly less in our proposed EPPADS scheduling approach. This is due to the dual-phase approach used in our approach. The straggler detection is determined in the Slow Start Phase and then any task which are scheduled to nodes marked as stragglers has a corresponding cloned task which is scheduled at the same time with the straggler task. This drastically reduces the delay in speculative task invocation caused in current scheduling approaches. EPPADS reduces the delayed speculative task invocation by almost 2x, thereby drastically improving the job processing performance.

4.4 Job Throughput Analysis

Figure 13 shows the job throughput comparison between the baseline FIFO scheduler, Adaptive Dynamic Approach with monitoring-DynMon, and the proposed EPPADS scheduler. To do an intensive evaluation of our proposed approach, we employed the HiBench benchmark suite which consists of a number of benchmark workloads. We observed that our proposed EPPADS scheduler have an average throughput performance improvement of 30% and 22%, over baseline and DynMon scheduling approach respectively. The reason behind this is the concurrent scheduling of job tasks in the accelerate phase (AccP) of EPPADS using a single scheduling I/O. One might argue that the scheduling in the slow-start phase degrades the job processing throughput, but the worst case scenario is equal with the best case of baseline FIFO and DynMon cases which all schedules the task in a sequential First-In First-Out Approach. However the AccP of EPPADS increases the job throughput significantly as multiple jobs are scheduled at one go. Furthermore the concurrent scheduling of straggling tasks with their clones increases the performance as they amortize the speculative execution delay.

4.5 Job Queuing Time

Another important factor in determining job processing performance is the amount of time a job spends in the queue before it can be scheduled for processing. For each approach we profiled the job waiting time from the time the job tracker node receive a job request until the first task of that job is scheduled for processing in any of the task tracker nodes. Figure 14 shows the results of this experiments. We observed that when there is no fault in the processing nodes, the job waiting time in EPPADS is less as compared to baseline FIFO and DynMon approaches. However this gap increases when there are job faults within the processing nodes. We observe that in the existence of job faults the job waiting time in FIFO and DynMon approaches increases dramatically.

This is mostly due to the following two factors; (i) The serialized scheduling of job tasks in FIFO and DynMon approaches increases the job waiting time. Furthermore, when multiple job failure occurs, other jobs has to wait whilst the failed jobs execute extra speculative tasks. (ii) To make it worse there is a delay in launching these extra speculative tasks in the FIFO and DynMon scheduling approaches. On the other hand the job waiting time in the proposed EPPADS scheduler does not increase significantly. This is because of the early detection of straggler nodes in EPPADS. Furthermore, the job waiting time is reduced significantly due to the scheduling of multiple job tasks in one go in the accelerate phase. The reduction of job waiting time in the queue further increases the job processing performance.

5 Related Works

Scheduling and processing mechanisms play a major role in determining the job processing efficiency. The default FIFO scheduler distributes job tasks in a uniform manner across available task tracker nodes [7]. However, when subjected to heterogeneous cluster environments, performance is heavily degraded due to the imbalance of processing power between the nodes. Many prior works on self adaptive schedulers [1,10–14,18,19] has been explored to minimize job response time in heterogeneous clusters. All these prior works achieved this through instrumentation of processing nodes, which frequently collects nodes load level information, which in turn is used for job tasks scheduling decisions and speculative tasks invocation. However, additional instrumentation can cause system performance degradation as the number of tasks tracker nodes increases. Furthermore, all these prior studies follows a sequential task scheduling approach in which a successive task can only be scheduled only if one of the already scheduled tasks is completed, which is a major constraint in achieving maximum job processing performance.

The authors of [4–6,10,16], proposed proactive straggler mitigation techniques to minimize job processing time in production clusters. The technique of straggler "hunting" was used in these works which often wait until tasks are straggling so as to invoke a speculative task. However, these solutions suffer from enough information needed to separate between slow nodes and faster nodes. These approaches are likely to cause some unnecessary over-utilization of resources without any improvement in job completion performance. This is likely due to scheduling of speculative tasks onto already slower nodes. To avoid such scenarios our proposed EPPADS scheduler detects the stragglers in an early stage of the slow-start phase and speculative tasks are launched on the nearest faster node using a topology aware policy (preference given to node in the same rack).

Other previous works [3,8,17] proposed an enhanced blacklisting instrumentation approach that avoids scheduling tasks on slower blacklisted nodes.

The scheduling node (Master), collects system level information from processing nodes (slaves) at finer time intervals. This help to provide warning about possible stragglers and scheduling is done whilst avoiding such blacklisted nodes. However, this blacklisting at finer time interviews can degrade the cluster performance due to frequent communication overheads between the scheduling node and processing nodes. Our approach eliminates this huge monitoring overhead by implementing a small time based scheduling window which amortizes significantly the communication costs caused by frequent node monitoring in large scale data processing clusters.

Overally, none of the previous works implemented a multi-task scheduling approach which is employed in our proposed approach. the phase based implementation in EPPADS improves greatly the job execution performance as evidenced by our evaluation results.

6 Conclusion

This paper presents EPPADS, a lightweight and high performance job scheduler for improving job processing performance in large scale data processing clusters. With its dual phased scheduling approach, EPPADS can drastically mitigate the job straggler problem in production clusters. This is achieved through the early detection of stragglers in the slow-start phase (SSP) and the speed-up in scheduling performance that occurs at the acceleration phase(AccP). The evaluation results confirms the effectiveness of EPPADS over existing approaches.

Acknowledgement. This research was supported by Basic Science Research Program through the National Research Foundation of Korea (NRF) funded by the Ministry of Education (2016R1D1A1B03934129).

References

1. Fair Scheduler. http://hadoop.apache.org/docs/r1.2.1/fair_scheduler.html
2. Yarn Scheduler Load Simulator (SLS). https://hadoop.apache.org/docs/stable/hadoop-sls/SchedulerLoadSimulator.html
3. Ananthanarayanan, G., Ghodsi, A., Shenker, S., Stoica, I.: Effective straggler mitigation: attack of the clones. In: Presented as part of the 10th USENIX Symposium on Networked Systems Design and Implementation (NSDI 13), pp. 185–198. USENIX, Lombard, IL (2013)
4. Ananthanarayanan, G., et al.: Reining in the outliers in Map-Reduce clusters using mantri. In: 9th USENIX Symposium on Operating Systems Design and Implementation (OSDI 10), Vancouver, BC (2010)
5. Chang, H., Kodialam, M., Kompella, R.R., Lakshman, T.V., Lee, M., Mukherjee, S.: Scheduling in MapReduce-like systems for fast completion time. In: 2011 Proceedings of IEEE INFOCOM, pp. 3074–3082 (2011)
6. Chen, Q., Liu, C., Xiao, Z.: Improving MapReduce performance using smart speculative execution strategy. IEEE Trans. Comput. **63**(4), 954–967 (2014)
7. Dean, J., Ghemawat, S.: MapReduce: simplified data processing on large clusters. Commun. ACM **51**(1), 107–113 (2008)

8. Fu, H., Chen, H., Zhu, Y., Yu, W.: FARMS: efficient MapReduce speculation for failure recovery in short jobs. Parallel Comput. **61**, 68–82 (2017)
9. Hamandawana, P., Mativenga, R., Kwon, S.J., Chung, T.: PADS: performance-aware dynamic scheduling for effective mapreduce computation in heterogeneous clusters. In: 2018 IEEE International Conference on Cluster Computing (CLUSTER), pp. 160–161 (2018)
10. Hsiao, J.H., Kao, S.J.: A usage-aware scheduler for improving MapReduce performance in heterogeneous environments. In: 2014 International Conference on Information Science, Electronics and Electrical Engineering, vol. 3, pp. 1648–1652 (2014)
11. You, H.-H., Yang, C.C., Huang, J.L.: A load-aware scheduler for MapReduce framework in heterogeneous cloud environments. In: Proceedings of the 2011 ACM Symposium on Applied Computing, SAC 2011, pp. 127–132. ACM, New York (2011)
12. Rasooli, A., Down, D.G.: A hybrid scheduling approach for scalable heterogeneous hadoop systems. In: 2012 SC Companion: High Performance Computing, Networking Storage and Analysis, pp. 1284–1291 (2012)
13. Rasooli, A., Down, D.G.: COSHH: a classification and optimization based scheduler for heterogeneous hadoop systems. Future Gener. Comput. Syst. **36**, 1–15 (2014)
14. Sun, X., He, C., Lu, Y.: ESAMR: an enhanced self-adaptive MapReduce scheduling algorithm. In: 2012 IEEE 18th International Conference on Parallel and Distributed Systems, pp. 148–155 (2012)
15. Swabey, P.: The data deluge: five years on. https://www.slideshare.net/economistintelligenceunit/the-data-deluge-five-years-on
16. Xu, H., Lau, W.C.: Task-cloning algorithms in a MapReduce cluster with competitive performance bounds. In: 2015 IEEE 35th International Conference on Distributed Computing Systems, pp. 339–348 (2015)
17. Yadwadkar, N.J., Ananthanarayanan, G., Katz, R.: Wrangler: predictable and faster jobs using fewer resources. In: Proceedings of the ACM Symposium on Cloud Computing, SOCC 2014, pp. 26:1–26:14 (2014). https://doi.org/10.1145/2670979.2671005
18. Yang, S.J., Chen, Y.R., Hsieh, Y.M.: Design dynamic data allocation scheduler to improve MapReduce performance in heterogeneous clouds. In: 2012 IEEE Ninth International Conference on e-Business Engineering, pp. 265–270 (2012)
19. Zaharia, M., Konwinski, A., Joseph, A.D., Katz, R., Stoica, I.: Improving MapReduce performance in heterogeneous environments. In: Proceedings of the 8th USENIX Conference on Operating Systems Design and Implementation, OSDI 2008, pp. 29–42. USENIX Association, Berkeley (2008)

Incremental Discovery of Order Dependencies on Tuple Insertions

Lin Zhu[1,2], Xu Sun[1,2], Zijing Tan[1,2(✉)], Kejia Yang[3], Weidong Yang[1,2], Xiangdong Zhou[1,2], and Yingjie Tian[4]

[1] School of Computer Science, Fudan University, Shanghai, China
zjtan@fudan.edu.cn
[2] Shanghai Key Laboratory of Data Science, Shanghai, China
[3] Ann Arbor EECS Department, University of Michigan, Ann Arbor, USA
[4] State Grid Shanghai Municipal Electric Power Company, Shanghai, China

Abstract. Order dependencies (ODs) are recently proposed to describe a relationship of ordering between lists of attributes. It is typically too costly to design ODs manually, since the number of possible ODs is of a factorial complexity in the number of attributes. To this end, automatic discovery techniques for ODs are developed. In practice, data is frequently updated, especially with tuple insertions. Existing techniques do not lend themselves well to these situations, since it is prohibitively expensive to recompute all ODs from scratch after every update. In this paper, we make a first effort to investigate incremental OD discovery techniques in response to tuple insertions. Given a relation D, a set Σ of valid and minimal ODs on D, and a set $\triangle D$ of tuple insertions to D, it is to find, changes $\triangle\Sigma$ to Σ that makes $\Sigma \oplus \triangle\Sigma$ a set of valid and minimal ODs on $D + \triangle D$. Note that $\triangle\Sigma$ contains both new ODs to be added to Σ and outdated ODs to be removed from Σ. Specifically, (1) We formalize the incremental OD discovery problem. Although the incremental discovery problem has a same complexity as its batch (non-incremental) counterpart in terms of traditional complexity, we show that it has good data locality. It is linear in the size of $\triangle D$ to validate on $D+\triangle D$ any OD φ that is valid on D. (2) We present effective incremental OD discovery techniques, leveraging an intelligent traversal strategy for finding $\triangle\Sigma$ and chosen indexes to minimize access to D. Our approach computes $\triangle\Sigma$ based on ODs in Σ, and is independent of the size of D. (3) Using real-life data, we experimentally verify that our approach substantially outperforms its batch counterpart by orders of magnitude.

1 Introduction

Data dependencies, *a.k.a.* integrity constraints, specify data semantics and inherent attribute relationships. They are widely employed in schema design, query optimization [15] and data cleaning [3,5,10,11], among other things. Recently, order dependencies (ODs) [15,18] are proposed to describe the relationship between two lexicographical ordering specifications on lists of attributes. ODs

G. Li et al. (Eds.): DASFAA 2019, LNCS 11446, pp. 157–174, 2019.
https://doi.org/10.1007/978-3-030-18576-3_10

properly subsume functional dependencies (FDs), and can define lexicographic orders used in the SQL order-by clause. Hence, ODs are proved to be useful in query optimizations concerning sorting [15,18]. Compared to traditional dependencies based on sets, *e.g.*, FDs, denial constraints (DCs) [5] and differential dependencies (DDs) [14], ODs are defined on lists, and are hence quite different. We will review formal definitions of OD in Sect. 2, and first provide an illustrative example to highlight features of OD.

		A	B	C	D	E	F	G
D	t_1:	1	2	3	4	1	1	1
	t_2:	1	2	3	4	2	1	2
	t_3:	2	1	4	2	4	1	3
	t_4:	2	4	5	1	4	1	4
$\triangle D$	t' :	1	2	3	5	4	2	5
	t'':	1	2	4	3	5	6	7

Fig. 1. An instance D, and $\triangle D$ of tuple insertions to D.

Example 1: Figure 1 shows an instance D with four tuples $\{t_1, t_2, t_3, t_4\}$. If we sort tuples by attribute A, and then break ties by attribute B, these tuples are also sorted by attribute C first and then by attribute D. This sorting specification is in accordance with the SQL order by clauses. With the notation of OD, this is written as $\mathsf{AB} \mapsto \mathsf{CD}$, *i.e.*, AB *orders* CD. Here AB and CD are lists of attributes. Leveraging this example, we illustrate several unique features of ODs.

(1) OD $\mathsf{AB} \mapsto \mathsf{CD}$ states that values on CD are monotonically non-decreasing with respect to values on AB [15]. Specifically, (a) $\mathsf{AB} \mapsto \mathsf{CD}$ implies an FD $\mathcal{AB} \rightarrow \mathcal{CD}$. Here set $\mathcal{AB}$ (resp. $\mathcal{CD}$) denotes the set of elements in list AB (resp. CD). To guarantee that when tuples are sorted by AB, they are also sorted by CD, tuples with a same value on AB must have a same value on CD, *e.g.*, t_1 and t_2. (b) ODs also impose order semantics on tuples with different values on AB. For example, t_3 has a larger AB's value than t_2, the CD's value of t_3 cannot be less than that of t_2, to satisfy $\mathsf{AB} \mapsto \mathsf{CD}$.
(2) Unlike other constraints, *e.g.*, FDs and DCs, ODs are specified on *lists* of attributes, and the order of attributes on the left-hand side (LHS) and right-hand side (RHS) matters. For example, neither $\mathsf{BA} \mapsto \mathsf{CD}$ nor $\mathsf{AB} \mapsto \mathsf{DC}$ holds.
(3) FDs can be always converted into the form with a single RHS attribute. For example, $\mathcal{AB} \rightarrow \mathcal{CD}$ can be expressed as $\mathcal{AB} \rightarrow \mathcal{C}$ and $\mathcal{AB} \rightarrow \mathcal{D}$. This simplifies FD discovery [9,13]. In contrast, RHS attributes of an OD are taken as a whole and *may not* be splitted. As an example, $\mathsf{AB} \mapsto \mathsf{D}$ does not hold. □

No matter how desirable, it is prohibitively time-consuming to design ODs manually, even by domain experts. Automatic discovery techniques for ODs [12, 16,17] are hence studied, just like those for FDs [9,13], DCs [2,4] and DDs [14], among others. The need for automatic OD discovery is even more evident, since the number of possible list-based ODs is of a factorial complexity in the number of attributes [12], much larger than that of FDs.

Discovering ODs is already shown to be a hard problem. Worse, Data in practice is typically dynamic, *i.e.,* frequently updated. Even if tuple deletions are not allowed, tuple insertions are generally supported on most data. It is too expensive to recompute all ODs, especially when data grow with tuple insertions. This highlights the quest for incremental OD discovery techniques, to update the set Σ of discovered (*valid* and *minimal*) ODs on an instance D as a set $\triangle D$ of tuple insertions is applied to D. Intuitively, when $\triangle D$ is small compared to D, it is more efficient to find update $\triangle\Sigma$ to Σ than the entire set of ODs on $D + \triangle D$ from scratch. However, the incremental OD discovery problem is very intricate, as illustrated by the example below.

Example 2: We illustrate two ways to find ODs in $\triangle\Sigma$. In Fig. 1, suppose $\triangle D$ with a single tuple t' is applied to D (neglecting t'' at this time).

(1) In instance D, the algorithm for OD discovery, *e.g.,* [12], finds $\mathsf{AB} \mapsto \mathsf{CD}$ and adds it into Σ. It can be verified that $\mathsf{AB} \mapsto \mathsf{CD}$ is no longer valid (holds) on $D + \triangle D$. As a violation, tuples t_1, t' agree on their AB's value, but have different CD's values. An incremental OD discovery algorithm then has to compute updates $\triangle\Sigma$ to Σ. It finds that $\mathsf{ABE} \mapsto \mathsf{CD}$, $\mathsf{ABF} \mapsto \mathsf{CD}$, $\mathsf{ABG} \mapsto \mathsf{CD}$ and $\mathsf{AB} \mapsto \mathsf{C}$ are all valid on $D+\triangle D$. Therefore, it adds all these ODs into $\triangle\Sigma^+$, the set of ODs to be added into Σ, and adds $\mathsf{AB} \mapsto \mathsf{CD}$ into $\triangle\Sigma^-$, the set of ODs to be removed from Σ. Note that none of $\mathsf{ABE} \mapsto \mathsf{CD}$, $\mathsf{ABF} \mapsto \mathsf{CD}$, $\mathsf{ABG} \mapsto \mathsf{CD}$ or $\mathsf{AB} \mapsto \mathsf{C}$ is in Σ, since they are not *minimal* (formalized in Sect. 2). Theoretically, they are not in Σ because $\mathsf{AB} \mapsto \mathsf{CD}$ is in Σ and $\mathsf{AB} \mapsto \mathsf{CD}$ logically implies them [12,15]: any instance that satisfies $\mathsf{AB} \mapsto \mathsf{CD}$ also satisfies them. However, when $\mathsf{AB} \mapsto \mathsf{CD}$ no longer holds, our incremental method has to discover them, since they are valid on $D + \triangle D$, and are *minimal* now.

(2) Both $\mathsf{E} \mapsto \mathsf{F}$ and $\mathsf{EB} \mapsto \mathsf{G}$ are in Σ. $\mathsf{EFB} \mapsto \mathsf{G}$ is not in Σ; it is valid but not minimal. Theoretically, $\mathsf{EFB} \mapsto \mathsf{G}$ is an "embedded" OD *w.r.t.* $\mathsf{EB} \mapsto \mathsf{G}$ according to $\mathsf{E} \mapsto \mathsf{F}$ [12,15]. Since the order of E determines that of F, adding F after E in a list does not impose any new order restrictions and is regarded as redundancy (formalized in Sect. 2). However, $\mathsf{E} \mapsto \mathsf{F}$ is not valid after t' is inserted. Our incremental algorithm has to check the validity of $\mathsf{EFB} \mapsto \mathsf{G}$. Indeed, although $\mathsf{EB} \mapsto \mathsf{G}$ is not valid on $D + \triangle D$, $\mathsf{EFB} \mapsto \mathsf{G}$ is valid. $\mathsf{EFB} \mapsto \mathsf{G}$ is hence put into $\triangle\Sigma^+$: it is both valid and minimal now. Note that even when $\mathsf{EB} \mapsto \mathsf{G}$ is valid on $D + \triangle D$ and is not removed from Σ, $\mathsf{EFB} \mapsto \mathsf{G}$ is still minimal. This is because EB is not a *prefix* of EFB. Also note that $\mathsf{EFB} \mapsto \mathsf{G}$ *cannot* be discovered following the strategy in (1), *i.e.,* adding attributes to the tail of LHS attribute list, or removing attributes from the tail of RHS attribute list. □

Contributions. We make a first effort to investigate incremental OD discovery.

(1) We formalize the incremental OD discovery problem (Sect. 3). We show that this problem has a same complexity as its batch (non-incremental) counterpart, in terms of the traditional complexity analysis. Nevertheless, we prove that the incremental problem has a good data locality. Specifically, given an OD φ valid on D and a set $\triangle D$ of tuple insertions to D, it is linear in the size of $\triangle D$ to check the validity of φ on $D + \triangle D$.
(2) We present efficient methods for incremental OD discovery (Sect. 4). We present an intelligent traversal strategy for finding new valid and minimal ODs on $D + \triangle D$, based on those already discovered ODs in Σ. We study techniques for choosing indexes, such that the access to D is minimized, and hence the required *local* data can be effectively fetched. Our approach has a desirable property that it is independent of the size of D.
(3) Using real-life data, we experimentally verify that our incremental algorithm outperforms the batch counterpart by orders of magnitude (Sect. 5).

Related Work. Dependency discovery is one of the most important aspects of data profiling. To alleviate the burden of users, automatic dependency discoveries are conducted for a host of different constraints; see *e.g.,* FDs [9,13], conditional FDs (CFDs) [6,8], DCs [2,4] and DDs [14].

Order dependencies (ODs) [15,18] state a relationship of order between lists of attributes. Theoretical foundations of ODs are well discussed in [15,18]. ODs properly subsume FDs, and are well employed in query optimizations concerning order. [12] presents the first approach for discovering ODs, with a level-wise bottom-up traversal of the lattice of permutations of attributes, an efficient OD validation method and some pruning rules to reduce the search space. Since ODs are defined on lists, the search space (the number of possible ODs) is factorial in the number of attributes. [16,17] present a polynomial mapping form ODs defined in lists to a canonical form of ODs in sets; this canonical form of ODs in sets has an advantage that the search space is exponential in the number of attributes. [16,17] then present discovery techniques for ODs via set-based axioms. In this paper, we follow the notation of list-based ODs. This is because (1) ODs defined in lists are preferable, since they naturally model the lexicographic orders employed in the SQL order-by clause; (2) each list-based OD of the form $\mathsf{X} \mapsto \mathsf{Y}$ has to be expressed in $|\mathsf{X}|\cdot|\mathsf{Y}|$ set-based ODs, where $|\mathsf{X}|$ (resp. $|\mathsf{Y}|$) is the number of attributes in list X (resp. Y). Therefore, the discovery of set-based ODs does not lead to the discovery of list-based ODs; and (3) we address the incremental OD discovery problem, to improve the efficiency from another perspective.

Incremental techniques are developed in different aspects of data quality. [3] discusses incremental data repairing for CFDs. Taking a clean relation D *w.r.t.* a set Σ of CFDs and a set $\triangle D$ of tuple insertions, [3] presents methods to repair tuples in $\triangle D$ such that Σ is satisfied. [7] investigates incremental detection of CFD violations in distributed data. Given a set V of violations *w.r.t.* a set Σ of CFDs on a distributed database D and updates $\triangle D$ to D, [7] aims to find changes $\triangle V$ to V, with minimum data shipment among sites. Our work considers incremental constraint discovery for ODs, and significantly differs from [3,7].

To our best knowledge, the only incremental algorithm for constraint discovery on dynamic data is [1], concerning unique column combinations (Uccs), *a.k.a.* candidate keys. [1] employs indexes on attributes to reduce access to old data, and expands attributes in old Uccs for new Uccs when old Uccs no longer hold. Note that OD subsumes FD, and FD subsumes Ucc. We consider ODs with multiple LHS and RHS attributes in lists. This makes the traversal for OD candidates, OD validations and index choice far more complicated, compared to [1].

2 Preliminaries

We review basic notations and formal definitions of ODs [12,15–18].

Relation. $R(A_1, \ldots, A_m)$ denotes a relation schema, where each $A_j (j \in [1, m])$ denotes a single attribute. D denotes a specific instance, and t, s denote tuples. For an attribute A_j, t_{A_j} denotes the value of attribute A_j in a tuple t.

Sets and Lists. $\mathcal{X}$ and $\mathcal{Y}$ denote sets of attributes, while X and Y denote lists of attributes. Specifically, $\{\}$ (resp. $[\,]$) denotes the empty set (resp. empty list). $\mathcal{XY}$ is a shorthand for $\mathcal{X} \cup \mathcal{Y}$, and XY is a shorthand for the concatenation of X and Y. For a list X, set $\mathcal{X}$ denotes the set of elements in X.

For an attribute list $\mathsf{X} = [A_1, \ldots, A_k]$, we say X *contains* another list Y, when there exists some $1 \leq i \leq j \leq k$ such that $\mathsf{Y} = [A_i, \ldots, A_j]$. We use $prefixes(\mathsf{X})$ to denote the set of all possible prefixes of X, *i.e.*, $[A_1, \ldots, A_i]$ for any $i \leq k$.

Order on Lists. For a tuple t and an attribute list $\mathsf{X} = [A_1, \ldots, A_k]$, we use t_{X} to denote the projection of tuple t on X, *i.e.*, $[t_{A_1}, \ldots, t_{A_k}]$. For two tuples t, s,

(1) $t \prec_{\mathsf{X}} s$ if there exists some $i \leq k$ such that $t_{A_i} < s_{A_i}$ and for all $j < i$, $t_{A_i} = s_{A_i}$.
(2) $t =_{\mathsf{X}} s$ when $t_{A_i} = s_{A_i}$ for all $i \in [1, k]$.
(3) $t \preceq_{\mathsf{X}} s$ if $t \prec_{\mathsf{X}} s$ or $t =_{\mathsf{X}} s$.

Order Dependency [12,15–18]**.** For attributes lists X, Y on schema R, $\mathsf{X} \mapsto \mathsf{Y}$ denotes an *order dependency.* An instance D of R satisfies an OD $\varphi = \mathsf{X} \mapsto \mathsf{Y}$, if for any two tuples $t, s \in D$, when $t \preceq_{\mathsf{X}} s$, $t \preceq_{\mathsf{Y}} s$. We say φ is valid on D and φ holds on D interchangeably. If $\mathsf{X} \mapsto \mathsf{Y}$ is not valid on D, we write $\mathsf{X} \not\mapsto \mathsf{Y}$.

Remark. As stated in [15,18], ODs *strictly generalize* FDs. Specifically, each OD $\mathsf{X} \mapsto \mathsf{Y}$ has an "embedded FD" $\mathcal{X} \to \mathcal{Y}$; if $\mathsf{X} \mapsto \mathsf{Y}$ holds, $\mathcal{X} \to \mathcal{Y}$ holds.

Violations of Order Dependency [15,18]**.** For an OD $\varphi = \mathsf{X} \mapsto \mathsf{Y}$, two sources of OD violations exist:

(1) A split *w.r.t.* φ is a pair of tuples t and s such that $t =_{\mathsf{X}} s$, but $t \neq_{\mathsf{Y}} s$.
(2) A swap *w.r.t.* φ is a pair of tuples t and s such that $t \prec_{\mathsf{X}} s$ but $s \prec_{\mathsf{Y}} t$.

Example 3: (1) A split is actually a violation of the "embedded" FD. In Fig. 1, t_1 and t_2 lead to a split *w.r.t.* $\mathsf{F} \mapsto \mathsf{G}$. $t_1 =_{\mathsf{F}} t_2$, but $t_1 \neq_{\mathsf{G}} t_2$. (2) t_1 and t_3 cause a swap *w.r.t.* $\mathsf{A} \mapsto \mathsf{B}$. $t_1 \prec_{\mathsf{A}} t_3$, but $t_3 \prec_{\mathsf{B}} t_1$. □

The number of ODs valid on an instance can be very large. Similar to discovery techniques for other constraints, it is more instructive to find *minimal* valid ODs than to find all valid ODs. List-based ODs lead to an intricate definition of minimality. We follow similar criteria as [12,15], formalized as follows.

Minimality of an Attribute List. An attribute list X is minimal, iff for any disjoint sub-lists Y and W in X: if W follows (maybe not directly) Y,$\mathsf{Y} \not\mapsto \mathsf{W}$.

Intuitively, when the order of Y determines that of W, adding W after Y in a list does not impose any new order restrictions. It is easy to see that an attribute A_i occurs at most once in any minimal attribute list X.

Minimality of ODs. An OD $\mathsf{X} \mapsto \mathsf{Y}$ is minimal, iff

(1) $\nexists\, \mathsf{X}' \mapsto \mathsf{Y}\mathsf{Y}'$, such that $\mathsf{X}' \in prefixes(\mathsf{X})$ and $\mathsf{X}' \mapsto \mathsf{Y}\mathsf{Y}'$ is valid (if $\mathsf{X}' = \mathsf{X}$, Y' is not empty; otherwise Y' can be empty); and
(2) both X and Y are minimal.

Example 4: Recall the instance D presented in Fig. 1. (1) $\mathsf{AB} \mapsto \mathsf{CD}$ is minimal, but $\mathsf{ABE} \mapsto \mathsf{CD}$ is not minimal. (2) Because $\mathsf{E} \mapsto \mathsf{F}$, EFB is not a minimal attribute list and hence $\mathsf{EFB} \mapsto \mathsf{G}$ is not a minimal OD. □

3 Data Locality for Incremental OD Discovery

We formalize the incremental OD discovery problem and show its complexity. We then justify that the incremental OD discovery problem has good data locality.

Incremental OD Discovery. Given a relation D of schema R, a set Σ of valid and minimal ODs on D, and a set $\triangle D$ of tuple insertions to D, incremental OD discovery is to find, changes $\triangle\Sigma$ to Σ that makes $\Sigma \oplus \triangle\Sigma$ a set of valid and minimal ODs on $D + \triangle D$; $\triangle\Sigma$ contains both new ODs to be added to Σ and outdated ODs to be removed from Σ.

Note that each batch (non-incremental) OD discovery problem on D can be directly modeled as an incremental OD discovery problem with inputs D', $\triangle D'$ and Σ, by setting $D' = \phi$, $\triangle D' = D$ and $\Sigma = \phi$. Since the incremental OD discovery includes the batch counterpart as a special case, the incremental problem at least has a same complexity as the batch one in terms of traditional complexity. In practice, $\triangle D$ is typically (much) smaller than D. In contrast to batch algorithms that recompute the output from scratch, an incremental algorithm can greatly improve efficiency if its cost is independent of D.

In light of this, we classify *local* data for an inserted tuple t' and an OD φ that is already valid on D, followed by computations concerning only local data.

Local Data of a Single Tuple Insertion. Given an OD $\mathsf{X} \mapsto \mathsf{Y}$ valid on D and a tuple t' inserted into D, we can find the following three sets of tuples on D, as *local data* of t' *w.r.t.* $\mathsf{X} \mapsto \mathsf{Y}$:

$equ(\mathsf{X}, t')$: tuple $s \in equ(\mathsf{X}, t')$, if $s =_{\mathsf{X}} t'$;

$\overline{low(\mathsf{X}, t')}$: tuple $s \in low(\mathsf{X}, t')$, if (1) $s \prec_{\mathsf{X}} t'$; and (2) there exists no s' such that $s \prec_{\mathsf{X}} s' \prec_{\mathsf{X}} t'$;

$high(\mathsf{X}, t')$: tuple $s \in high(\mathsf{X}, t')$, if (1) $t' \prec_{\mathsf{X}} s$; and (2) there exists no s' such that $t' \prec_{\mathsf{X}} s' \prec_{\mathsf{X}} s$;

Note that $equ(\mathsf{X}, t')$ (resp. $low(\mathsf{X}, t')$, $high(\mathsf{X}, t')$) may be empty. Incremental OD discovery takes as inputs D and the set Σ of ODs valid on D. Hence, some auxiliary data structure can be built to effectively obtain required local data.

Example 5: In Fig. 2(a), we show a simplified B+ tree built on D (Fig. 1) with AB as the key. For each key value in a leaf node, we store the set of tuple *ids*. In addition, we build a doubly linked list between successive leaf nodes. For t', $equ(\mathsf{AB}, t') = \{t_1, t_2\}$, $high(\mathsf{AB}, t') = \{t_3\}$ and $low(\mathsf{AB}, t') = \{\ \}$. With the B+ tree, it takes $O(\log |D|)$ to fetch $equ(\mathsf{AB}, t')$, and then $O(1)$ to fetch (non-empty) $low(\mathsf{AB}, t')$ (resp. $high(\mathsf{AB}, t')$), where $|D|$ is the number of tuples in D. □

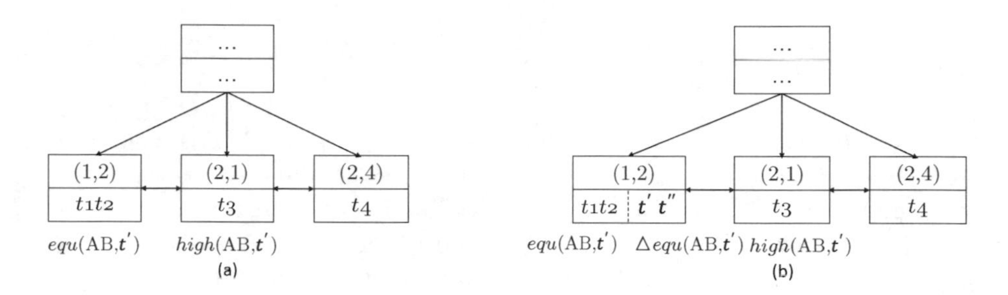

Fig. 2. Example local data

Note that $\forall s, t \in equ(\mathsf{X}, t')$ (resp. $low(\mathsf{X}, t')$, $high(\mathsf{X}, t')$), $t =_{\mathsf{Y}} s$ on instance D, since $\mathsf{X} \mapsto \mathsf{Y}$ is valid on D. We choose an arbitrary tuple from $equ(\mathsf{X}, t')$ (resp. $low(\mathsf{X}, t')$, $high(\mathsf{X}, t')$), denoted as $t_e'^{\mathsf{X}}$ (resp. $t_l'^{\mathsf{X}}$, $t_h'^{\mathsf{X}}$), when the set is not empty. The following theorem states that whether $\mathsf{X} \mapsto \mathsf{Y}$ is valid on $D + \{t'\}$ concerns computations only on t' and $t_e'^{\mathsf{X}}$, $t_l'^{\mathsf{X}}$, $t_h'^{\mathsf{X}}$.

Theorem 1: *$\varphi = \mathsf{X} \mapsto \mathsf{Y}$ is valid on $D + \{t'\}$, when (1) there is no* split *between t' and $t_e'^{\mathsf{X}}$; and (2) there is no* swap *between t' and $t_l'^{\mathsf{X}}$ (resp. $t_h'^{\mathsf{X}}$).* □

Local Data of $\triangle D$. We then consider $\triangle D$ with multiple tuple insertions. For a tuple t' in $\triangle D$, we still denote by $equ(\mathsf{X}, t')$, $low(\mathsf{X}, t')$ and $high(\mathsf{X}, t')$ local data on D, while denote by $equ'(\mathsf{X}, t')$, $low'(\mathsf{X}, t')$ and $high'(\mathsf{X}, t')$ local data on $D + \triangle D$. $equ'(\mathsf{X}, t') = equ(\mathsf{X}, t') \cup \triangle equ(\mathsf{X}, t')$, where $\triangle equ(\mathsf{X}, t')$ is the set of tuples $t \in \triangle D$ such that $t =_{\mathsf{X}} t'$. Obviously, $t' \in \triangle equ(\mathsf{X}, t')$; $equ'(\mathsf{X}, t') = equ'(\mathsf{X}, t'')$ if $t'' =_{\mathsf{X}} t'$. Similarly for $low'(\mathsf{X}, t')$ and $high'(\mathsf{X}, t')$.

Leveraging the auxiliary structure to effectively obtain required data, Theorem 2 shows the good data locality of incremental OD validation, as an important building block of incremental OD discovery.

Theorem 2: *For any OD φ that is valid on D, it is linear in $|\triangle D|$ to check the validity of φ on $D + \triangle D$, where $|\triangle D|$ is the number of tuples in $\triangle D$.* □

Algorithm. We prove Theorem 2 by providing algorithm $\triangle Check(\varphi)$ with the required property, for checking the validity of $\varphi = \mathsf{X} \mapsto \mathsf{Y}$ on $D + \triangle D$. $\triangle Check(\varphi)$ divides $\triangle D$ into k disjoint sets $= \{\triangle D_1, \ldots, \triangle D_k\}$ with hashing, such that $\forall t', t'' \in \triangle D_i$ ($i \in [1, k]$), $t' =_{\mathsf{X}} t''$. It then selects an arbitrary t'_i in each $\triangle D_i$.

(1) To check swap *w.r.t.* t'_i, it finds $t_e^{'max} = argmax_t(t_{\mathsf{Y}})$ and $t_e^{'min} = argmin_t(t_{\mathsf{Y}})$ in all $t \in equ'(\mathsf{X}, t'_i)$ (ties broken by an arbitrary one). It then finds $t_l^{'max}$ (resp. $t_h^{'min}$) in $low'(\mathsf{X}, t'_i)$ (resp. $high'(\mathsf{X}, t'_i)$) similarly. There is no swap iff $t_e^{'max} \preceq_{\mathsf{Y}} t_h^{'min}$ and $t_l^{'max} \preceq_{\mathsf{Y}} t_e^{'min}$.
(2) To check split *w.r.t.* t'_i, it suffices to check whether $t_e^{'max} =_{\mathsf{Y}} t_e^{'min}$.

Example 6: Consider $\triangle D$ with two tuples t', t'' (Fig. 1) and $\varphi = \mathsf{AB} \mapsto \mathsf{CD}$. As shown in Fig. 2(b), $equ'(\mathsf{AB}, t') = equ'(\mathsf{AB}, t'') = \{t_1, t_2, t', t''\}$: $equ(\mathsf{AB}, t') = \{t_1, t_2\}$, $\triangle equ(\mathsf{AB}, t') = \{t', t''\}$. $high'(\mathsf{AB}, t') = high(\mathsf{AB}, t') = \{t_3\}$.
$t_e^{'max} = t''$, $t_e^{'min} = t_1$ and $t_h^{'min} = t_3$. split exists since $t_1 \neq_{\mathsf{CD}} t''$, and swap exists since $t_3 \prec_{\mathsf{CD}} t''$. □

Complexity. It is easy to see the correctness of $\triangle Check(\varphi)$. We then prove $\triangle Check(\varphi)$ is linear in $|\triangle D|$. Observe that the total number of required $equ'(\mathsf{X}, t'_i)$, $low'(\mathsf{X}, t'_i)$ and $high'(\mathsf{X}, t'_i)$ for all t'_i is at most $3 \times |\triangle D|$. In $equ'(\mathsf{X}, t'_i)$, it takes $O(1 + |\triangle equ(\mathsf{X}, t'_i)|)$ to find $t_e^{'max}$ and $t_e^{'min}$, where $|\triangle equ(\mathsf{X}, t'_i)|$ is the number of tuples in $\triangle equ(\mathsf{X}, t'_i)$. This is because all tuples in $equ(\mathsf{X}, t'_i)$ agree on values of Y. Note that the sum of $|\triangle equ(\mathsf{X}, t'_i)|$ for all t'_i equals $|\triangle D|$.

If $\triangle low(\mathsf{X}, t'_i)$ is not empty for some t'_i, $low'(\mathsf{X}, t'_i) = equ'(\mathsf{X}, s')$ for any $s' \in \triangle low(\mathsf{X}, t'_i)$. In this case, no additional computation on $low'(\mathsf{X}, t'_i)$ is required. If $\triangle low(\mathsf{X}, t'_i)$ is empty, it takes $O(1)$ to find $t_l^{'max}$ in $low'(\mathsf{X}, t'_i)$ because all tuples in $low(\mathsf{X}, t'_i)$ have a same value on Y. Similarly for $high'(\mathsf{X}, t'_i)$.

To sum up, $\triangle Check(\varphi)$ is linear in $|\triangle D|$.

4 Incremental OD Discovery

We first discuss methods for finding $\triangle \Sigma$ on $D + \triangle D$, based on Σ. We then study techniques for choosing indexes, to minimize access to the original data.

4.1 Finding ODs in $\triangle \Sigma$

Note that $\triangle \Sigma$ consists of two disjoint sets $\triangle \Sigma^+$ and $\triangle \Sigma^-$; $\triangle \Sigma^+$ contains new valid and minimal ODs as additions to Σ, while $\triangle \Sigma^-$ contains non-valid ODs that should be removed from Σ. Taking as input the set Σ of minimal and valid ODs on D, it is relatively easy to compute $\triangle \Sigma^-$. As stated in Theorem 2, we can effectively check the validity of any OD φ in Σ on $D + \triangle D$ by $\triangle Check(\varphi)$. If φ no longer holds on $D + \triangle D$, we add it to $\triangle \Sigma^-$.

It is, however, much more intricate to compute $\triangle \Sigma^+$ as shown in Example 2. To fully take advantage of incremental computations, we should always leverage $\triangle Check(\varphi)$ in the validation of any OD φ; a prerequisite is that φ must hold on D. This is possible since any OD valid on $D + \triangle D$ must be valid on D, and hence

ODs in $\triangle\Sigma^+$ can be computed based on ODs in Σ. We present two ways to find ODs (candidates) in $\triangle\Sigma^+$, referred to as enrichment and expansion, respectively.

Enrichment of an Attribute List. Given an OD $\varphi = \mathsf{X} \mapsto \mathsf{Y} \in \Sigma$, but invalid on $D + \triangle D$, and an attribute list Z, when (1) Z contains X, and (2) Z and Y are disjoint, we "enrich" Z by φ, to generate a set of attribute lists, denoted by $enrich(\mathsf{Z}, \varphi)$. W.l.o.g., let $\mathsf{Z} = \mathsf{X'XA_{1'} \ldots A_{k'}}$. $enrich(\mathsf{Z}, \varphi) = \{\ \mathsf{Z}, \mathsf{X'XYA'_1 \ldots A'_k}, \ldots, \mathsf{X'XA'_1 \ldots A'_kY}\ \}$. If either condition (1) or (2) is false, $enrich(\mathsf{Z}, \varphi) = \{\ \mathsf{Z}\ \}$.

Intuitively, $enrich(\mathsf{Z}, \varphi)$ (excluding Z) is a set of *non-minimal* attribute lists due to the validity of $\mathsf{X} \mapsto \mathsf{Y}$ on D. When $\mathsf{X} \mapsto \mathsf{Y}$ no longer holds on $D + \triangle D$, these attribute lists become minimal. We then present our first way to generate candidates in $\triangle\Sigma^+$, referred to as enrichment.

Enrichment of an OD. Given an OD $\varphi = \mathsf{X} \mapsto \mathsf{Y}$ in Σ and a set Υ of ODs $\{\varphi_1, \ldots, \varphi_m\}$, where each φ_i ($i \in [1, m]$) is in Σ, but is invalid on $D + \triangle D$, enrichment of φ by Υ, denoted by $Enrich(\varphi, \Upsilon)$ is to generate a set of ODs = $\{\mathsf{U} \mapsto \mathsf{V}\ \}$, where $\mathsf{U} \in enrich(\mathsf{X}, \varphi_i)$, $\mathsf{V} \in enrich(\mathsf{Y}, \varphi_j)$, $\forall \varphi_i, \varphi_j \in \Upsilon$.

Example 7: Suppose $\varphi = \mathsf{AB} \mapsto \mathsf{CD}$ and $\Upsilon = \{\mathsf{B} \mapsto \mathsf{E}, \mathsf{C} \mapsto \mathsf{F}\ \}$, enrichment of φ by Υ is $\{\mathsf{AB} \mapsto \mathsf{CD}, \mathsf{AB} \mapsto \mathsf{CFD}, \mathsf{AB} \mapsto \mathsf{CDF}, \mathsf{ABE} \mapsto \mathsf{CD}, \mathsf{ABE} \mapsto \mathsf{CFD}, \mathsf{ABE} \mapsto \mathsf{CDF}\ \}$. □

Complexity. Observe that enrichment is conducted based on attributes in ODs. Its complexity is irrelevant of $|R|$, the number of attributes of the schema R, and it requires no data access to D.

Intuitively, enrichment of φ by Υ is to enrich the LHS and RHS attribute lists of φ by ODs in Υ respectively. It is easy to prove the following results. (1) Any OD in $Enrich(\varphi, \Upsilon)$ is valid on D; and (2) none of ODs in $Enrich(\varphi, \Upsilon)$ is minimal on D (excluding $\varphi = \mathsf{X} \mapsto \mathsf{Y}$). ODs generated by enrichment are only candidates in $\triangle\Sigma^+$, since some of them are invalid on $D + \triangle D$. Based on those invalid candidates and ODs in $\triangle\Sigma^-$, we provide another approach to computing ODs in $\triangle\Sigma^+$, referred to as expansion.

Algorithm. Algorithm Expand is presented to apply expansion to an OD $\varphi = \mathsf{X} \mapsto \mathsf{Y}$ valid on D but invalid on $D + \triangle D$. It produces a set Υ of ODs valid on $D + \triangle D$. Each OD $\in \Upsilon$ is of the form $\mathsf{XZ} \mapsto \mathsf{Y'}$, where $\mathsf{Y'} \in prefixes(\mathsf{Y})$. It is easy to see that these ODs are valid but not minimal on D.

(1) Expand first eliminates possible swap by removing attributes from the tail of Y one by one, until no swap exists or Y is empty (lines 2–4). Note that adding attributes to the tail of X does not help remove swap. If swap cannot be removed, Expand returns an empty set; no valid ODs can be generated based on $\mathsf{X} \mapsto \mathsf{Y}$.
(2) Expand then turns to eliminate split. In its loop (lines 6–11), Expand tries $\mathsf{X} \mapsto \mathsf{Y'}$ for each prefix $\mathsf{Y'}$ of Y (line 11). (a) If no split exists, Expand returns current results (line 8). This is because adding more attributes to the tail of X and (or) removing attributes from the tail of Y cannot further produce *minimal* ODs. (b) Otherwise, function *ExpandL* is called to remove split by

Algorithm 1: Expand

input : *a relation D of schema R, an OD $\varphi = \mathsf{X} \mapsto \mathsf{Y}$ valid on D, and a set $\triangle D$ of tuple insertions to D.*
output: *a set Υ of ODs valid on $D + \triangle D$, where each OD $\in \Upsilon$ is of the form $\mathsf{XZ} \mapsto \mathsf{Y}'$, where $\mathsf{Y}' \in prefixes(\mathsf{Y})$.*

```
1  Υ := {}; LHS_set := {}; V_set := {};
2  while Y is not empty do
3  |  if there is no swap detected by △Check(φ) then break;
4  |  Remove the last attribute from Y;
5  if Y is empty then return { };
6  while Y is not empty do
7  |  Check split by △Check(φ), and put violations into V_set;
8  |  if V_set is empty then return Υ∪{X ↦Y };
9  |  foreach X' ∈ ExpandL(V_set, X, Y, LHS_set) do
10 |  |  Add X' ↦Y to Υ; Add X' to LHS_set;
11 |  Remove the last attribute from Y;
12 return Υ;
13 Function ExpandL(V_set, U, W, LHS_set)
   input : a set V_set of split violations; each violation is a set of tuples with a
           same U's value but different W's values. LHS_set is a set of attribute
           lists: ∀ Z ∈ LHS_set, Z ↦WW' is valid on D + △D for some list W'.
   output: a set Ω of attribute lists, ∀U' ∈ Ω, U' ↦W is valid on D + △D.
14 Ω := {};
15 foreach attribute A ∈ R\U such that swap_free(V_set, A), and there is no Z in
   LHS_set, where Z ∈ prefixes(UA) do
16 |  V'_set := update(V_set, A);
17 |  if V'_set is empty then Ω := Ω ∪ { UA };
18 |  else Ω := Ω ∪ ExpandL(V'_set, UA, W, LHS_set);
19 return Ω;
```

adding attributes to the tail of X, and its results are kept (lines 9–10). Recall that $\triangle Check(\varphi)$ divides tuples in $\triangle D$ into $\{\triangle D_1, \ldots, \triangle D_k\}$ based on their X's values. It then detects split on $equ'(\mathsf{X}, t'_i)$ for a tuple t'_i in each $\triangle D_i$. All $equ'(\mathsf{X}, t'_i)$ with different Y's values are collected in V_{set} (line 7), as a parameter of *ExpandL* (line 9).

(3) Function *ExpandL* takes a set V_{set} of split violations, where each violation is a set of tuples that have a same value on U but different values on W. *ExpandL* returns a set Ω of attribute lists. Each $\mathsf{U}' \in \Omega$ is obtained from U by adding attributes to its tail, and $\mathsf{U}' \mapsto \mathsf{W}$ is valid on $D + \triangle D$. Instead of simply trying permutation of all attributes in $R \backslash \mathcal{U}$, *ExpandL* employs both instance-based and schema-based strategies to effectively prune the search space (line 15). (a) *ExpandL* only chooses attribute A that does not cause swap among tuples in a same set in V_{set} (checked by $swap\_free(V_{set}, A)$). Recall that in V_{set}, each set (violation) $vio = \{t_1, \ldots, t_m\}$, contains tuples that have a same value on U but different values on W. Adding A to the tail of U does not cause swap if for any two tuples t_i, t_j in vio, when $t_i \prec_{\mathsf{A}} t_j$,

$t_i \preceq_{\mathsf{W}} t_j$. (b) *ExpandL* avoids UA when $\mathsf{Z} \in prefixes(\mathsf{UA})$ for some Z in LHS_{set}; if $\mathsf{Z} \mapsto \mathsf{WW}'$ is valid, $\mathsf{UA} \mapsto \mathsf{W}$ is not minimal. Recall that LHS_{set} is maintained when new ODs are found (line 10).

(4) V_{set} is updated after A is added to the tail of U ($update(V_{set}, A)$) (line 16). Specifically, (a) it further divides sets in V_{set} based on values on A; tuples t', t'' are in a same set in V'_{set} when $t' =_{\mathsf{UA}} t''$; and (b) it discards set vio in V'_{set} when vio contains only one tuple, or $\forall t', t'' \in vio$, $t' =_{\mathsf{W}} t''$. If V'_{set} is empty, no further attribute additions are required (line 17). Otherwise, *ExpandL* is recursively called with updated violations and a lengthened LHS attribute list (line 18).

Example 8: Recall D and $\triangle D$ with two tuples t', t'' (Fig. 1) and $\varphi = \mathsf{AB} \mapsto \mathsf{CD}$. (1) Expand first tries to eliminate swap by removing attributes from the RHS of φ; this is done after removing D. (2) Expand detects split on $equ'(\mathsf{AB}, t')$, and hence the set $\{t_1, t_2, t', t''\}$ is put into V_{set}. (3) *ExpandL* tries to eliminate split by adding attributes to the end of AB. For example, adding E does not cause a swap. (4) After that, the only violation $\{t_1, t_2, t', t''\}$ in V_{set} is divided into four singleton sets $\{t_1\}$,$\{t_2\}$,$\{t'\}$,$\{t''\}$. Therefore, no split exists now. ABE is collected in Ω. There is no need for more attributes at the end of ABE, and step (3) is repeated by trying other attributes at the end of AB. (5) After *ExpandL* returns, all ODs of the form $\mathsf{ABS} \mapsto \mathsf{C}$ (resp. ABS) are collected in Υ (resp. LHS_{set}). Since no more attributes can be removed from the RHS of φ, Expand terminates. □

Complexity. (1) In terms of data complexity, recall that $\triangle Check(\varphi)$ is linear in $|\triangle D|$. On V_{set}, function $update(V_{set}, A)$ is linear in m on a set vio with m tuples (line 16). The most expensive part is function $swap\_free(V_{set}, A)$ (line 15). To check whether adding attribute A causes swap *w.r.t.* W, it takes $O(m \cdot logm)$ to sort tuples in vio based on values on A, followed by a linear scan to check values on W between successive tuples in $O(m)$. V_{set} is initialized with $equ'(\mathsf{X}, t'_i)$ with different Y's values (line 7). Hence, Expand is irrelevant of $|D|$. (2) Removing attributes from the tail of Y is linear in the size of Y, while adding attributes to the tail of X has a worst-case factorial complexity in the number of attributes in $R \backslash \mathcal{X}$. However, *ExpandL* is also bounded by the number of violations in V_{set}; the size of each violation monotonously decreases and all violations are eventually eliminated. Moreover, effective pruning rules are applied in lines 8, 15 and 17.

We are now ready to present the algorithm to compute $\triangle\Sigma$, by combining enrichment and expansion together.

Algorithm. Algorithm IncOD takes as inputs a relation D of schema R, a set Σ of valid and minimal ODs on D, and a set $\triangle D$ of tuple insertions to D. It computes $\triangle\Sigma$ such that $\Sigma \oplus \triangle\Sigma$ is a set of valid and minimal ODs on $D + \triangle D$.

(1) It initializes three empty sets Σ_{cand}, Σ_{valid} and Σ_{pre}, for OD candidates, new valid ODs in $D + \triangle D$, and ODs in Σ that are also valid on $D + \triangle D$, respectively. It validates every $\varphi \in \Sigma$ on $D + \triangle D$ by $\triangle Check(\varphi)$, and puts invalid (resp. valid) φ into $\triangle\Sigma^-$ (resp. Σ_{pre}) (lines 2–4).

(2) It applies enrichment to every $\varphi \in \Sigma$ by $\triangle\Sigma^-$, and collects results in Σ_{cand} (line 5). ODs in Σ_{cand} are then validated on $D + \triangle D$. Those valid ones

Algorithm 2: IncOD

input : *a relation D of schema R, a set Σ of valid and minimal ODs on D, and a set $\triangle D$ of tuples insertions to D.*

output: $\triangle\Sigma = \triangle\Sigma^+ \cup \triangle\Sigma^-$. *$\triangle\Sigma^+$ contains new valid and minimal ODs as additions to Σ, $\triangle\Sigma^-$ contains non-valid ODs to be removed from Σ.*

1 $\Sigma_{cand} := \{\}$; $\Sigma_{valid} := \{\}$; $\Sigma_{pre} := \{\}$;
2 **foreach** $\varphi \in \Sigma$ **do**
3 | **if** *φ is invalid by $\triangle Check(\varphi)$* **then** add φ into $\triangle\Sigma^-$;
4 | **else** add φ into Σ_{pre};
5 **foreach** $\varphi \in \Sigma$ **do** $\Sigma_{cand} := \Sigma_{cand} \cup Enrich(\varphi, \triangle\Sigma^-)$;
6 **foreach** $\varphi \in \Sigma_{cand}$ **do**
7 | **if** *φ is valid by $\triangle Check(\varphi)$* **then** move φ from Σ_{cand} to Σ_{valid};
8 **foreach** $\varphi \in \Sigma_{cand} \cup \triangle\Sigma^-$ **do** $\Sigma_{valid} := \Sigma_{valid} \cup Expand(D, \varphi, \triangle D)$;
9 $\triangle\Sigma^+ := Prune(\Sigma_{valid}, \Sigma_{pre})$;

are moved from Σ_{cand} to Σ_{valid} (lines 6–7). It applies expansion to ODs in $\Sigma_{cand} \cup \triangle\Sigma^-$, *i.e.*, ODs valid on D but invalid on $D + \triangle D$, and adds results to Σ_{valid} (line 8).

(3) It finally prunes non-minimal ODs in Σ_{valid} to get $\triangle\Sigma^+$ (line 9); Σ_{pre} is required in this step. (a) For each OD $\mathsf{X} \mapsto \mathsf{Y} \in \Sigma_{valid}$, it requires to check whether there exists some OD $\mathsf{U} \mapsto \mathsf{V}$ in $\Sigma_{valid} \cup \Sigma_{pre}$, such that $\mathsf{U} \in prefixes(\mathsf{X})$ and $\mathsf{Y} \in prefixes(\mathsf{V})$. It suffices to consider only those ODs $\mathsf{U} \mapsto \mathsf{V}$, whose $|\mathsf{U}| \leq |\mathsf{X}|$, and whose $|\mathsf{V}| \geq |\mathsf{Y}|$. (b) To verify whether X (resp. Y) is minimal, it requires to check whether there exists some OD $\mathsf{U} \mapsto \mathsf{V}$ in $\Sigma_{valid} \cup \Sigma_{pre}$, such that U is before V, both contained in X (resp. Y). It suffices to consider only those ODs $\mathsf{U} \mapsto \mathsf{V}$, whose $|\mathsf{U}| + |\mathsf{V}| \leq |\mathsf{X}|$ (resp. $|\mathsf{Y}|$).

Complexity. IncOD employs $\triangle Check(\cdot)$, $Enrich(\cdot)$ and $Expand(\cdot)$ in OD validations and computations of Σ_{cand}, Σ_{valid} and Σ_{pre}. Pruning of non-minimal ODs in Σ_{valid} concerns attributes of ODs in $\Sigma_{valid} \cup \Sigma_{pre}$, and requires no visits to D. To conclude, IncOD is irrelevant of $|D|$, and $\triangle\Sigma$ is computed based on ODs in Σ via enrichment and expansion only.

We provide insights into the interaction between enrichment and expansion, for developing optimization techniques.

Theorem 3: *On $D + \triangle D$, if* $\mathsf{W} \mapsto \mathsf{V}$ *does not cause a* split, *(1) when* $\mathsf{UWA_1 \ldots A_k} \mapsto \mathsf{Y}$ *is valid (resp. invalid),* $\mathsf{UWA_1 \ldots A_iVA_{i+1} \ldots A_k} \mapsto \mathsf{Y}$ *is valid (resp. invalid); and (2) when* $\mathsf{UWA_1 \ldots A_kZ} \mapsto \mathsf{Y'}$ *is valid for some* Z, *and some* $\mathsf{Y'} \in prefixes(\mathsf{Y})$, $\mathsf{UWA_1 \ldots A_iVA_{i+1} \ldots A_kZ} \mapsto \mathsf{Y'}$ *is valid.* □

Theorem 3 states that when an OD φ is invalid only due to swap (no split), (1) the enrichment of any valid OD ξ by φ also generates valid ODs; and (2) the enrichment of any invalid OD ξ by φ also generates invalid ODs, and any expansion of ξ that results in valid ODs also works for those ODs. We leverage these observations to avoid unnecessary expansion in our implementation. This optimization is proved to be very effective in our experimental studies, since expansion is the most expensive part of IncOD.

Algorithm 3: CoverIndex

---

input : *a set Σ of ODs*
output: *a set of attribute lists on which indexes to be built*
1 $U := \phi$; $output := \phi$;
2 **foreach** $\mathsf{X} \mapsto \mathsf{Y} \in \Sigma$ **do**
3 | **foreach** $\mathsf{X}' \in prefixes(\mathsf{X})$ **do**
4 | | **if** $\mathsf{X}' \notin U$ **then**
5 | | | add X' to U; $\mathsf{X}'.price := 0$; $\mathsf{X}'.weight := assignweight(\mathsf{X}')$;
6 **while** *there exists* $\mathsf{X} \mapsto \mathsf{Y}$ *such that* $\forall\, \mathsf{X}' \in prefixes(\mathsf{X})$, $\mathsf{X}'.price < \mathsf{X}'.weight$ **do**
7 | $\mathsf{Z} := \underset{\mathsf{X}' \in prefixes(\mathsf{X})}{\text{argmin}} (\mathsf{X}'.weight - \mathsf{X}'.price)$;
8 | **foreach** $\mathsf{X}' \in prefixes(\mathsf{X})$ **do** $\mathsf{X}'.price := \mathsf{X}'.price + \mathsf{Z}.weight - \mathsf{Z}.price$;
9 **foreach** $\mathsf{X}' \in U$ **do** **if** $\mathsf{X}'.weight = \mathsf{X}'.price$ **then** put X' into *output*;

4.2 Building Indexes

Only local data are required in IncOD. Our incremental OD discovery problem takes as inputs D and the set Σ of ODs valid on D, and hence some auxiliary structures can be built to help fetch those required data more efficiently.

We employ composite indexes (indexes on multiple attributes) as our auxiliary structure. In a composite index, tuples are sorted by concatenating values of the indexed attributes (see Example 5). Note that a composite index on attributes [$\mathsf{ABC}\ldots$] can be used when values of A, or AB or ABC are provided. We use memory-based B+ tree to implement composite indexes in this paper. Since B+ tree is well adopted in most commercial DBMS, our approach can be easily extended to handle data stored in DBMS as well.

To speed up data visits concerning $\mathsf{X} \mapsto \mathsf{Y}$, a straightforward way is to build a composite index ind_{X} on X. In practice when the number of ODs in Σ is large, building composite indexes on all distinct LHS attribute lists for ODs may become costly in terms of both computation and storage. We present another strategy that aims to build a *minimal* set of composite indexes and guarantees that for any OD at least one index is usable. We tackle this by relating the problem of building indexes to techniques for weighted vertex cover problems [19].

More specifically, for all $\mathsf{X} \mapsto \mathsf{Y}$ in Σ, (1) for any $\mathsf{X}' \in prefixes(\mathsf{X})$, we treat X' as a *vertex*, to build a set of vertices; and (2) we treat X as a *hyperedge*, with all $\mathsf{X}' \in prefixes(\mathsf{X})$ as its vertices. Then, our goal is to index at least one X' $\in prefixes(\mathsf{X})$ for any $\mathsf{X} \mapsto \mathsf{Y}$ in Σ, the same as the goal of vertex cover, to pick at least one vertex for any hyperedge. We also assign a weight to each prefix $\mathsf{X}' = [A_1, \ldots, A_k]$, based on its *selectivity*. The weight of X' is computed as $(1 - \frac{dist(A_1)}{|D|}) \cdot \ldots \cdot (1 - \frac{dist(A_k)}{|D|})$, where $dist(A_i)$ is the number of distinct values of A_i. We use uniform random sampling to estimate $dist(A_i)$ in our implementation. If the weight of some X' is zero, we assign a small number α as its weight.

Algorithm. Algorithm CoverIndex is to find a set of attribute lists on which we build indexes. It is an adaption of the "pricing" method for weighted vertex cover. It first initializes the set U of vertices (lines 2–5). It then continues to pick

X (hyperedge) when neither of its prefix X′ (vertex) is *tight* (line 6); a prefix X′ is tight when X′.price = X′.weight. It then increases the price of all X′ as much as possible, but guarantees that X′.*price* ≤ X′.*weight* (lines 7–8). Finally, all *tight* prefixes are collected as the output (line 9).

Complexity. CoverIndex terminates when at least one prefix is tight for each X, and all tight prefixes form a cover. CoverIndex is linear in the size of U, *i.e.*, the number of prefixes of all X ↦Y ∈ Σ. CoverIndex is a d-approximation algorithm where $d = \max(|X|)$ for all X ↦Y ∈ Σ in our setting, following [19].

Remark. The index built on a prefix X′ of X can be used for new ODs based on X ↦Y by both expansion and enrichment. We denote by $local'(\mathsf{X}, t') = equ'(\mathsf{X}, t')$ $\cup\, low'(\mathsf{X}, t') \cup high'(\mathsf{X}, t')$. Observe that (1) in expansion, we generate ODs of the form XZ ↦$prefixes$(Y), $local'(\mathsf{XZ}, t') \subseteq local'(\mathsf{X}', t')$; and (2) in enrichment, we generate ODs of the form X″Z ↦Y′, where X″ ∈$prefixes$(X). (a) If X′ is a prefix of X″, $local'(\mathsf{X}''\mathsf{Z}, t') \subseteq local'(\mathsf{X}', t')$. (b) If X″ is a prefix of X′, $local'(\mathsf{X}''\mathsf{Z}, t') \subseteq local'(\mathsf{X}'', t')$, and the index on X′ can be used when X″ value is available.

5 Experimental Study

Experimental Setting. We used one machine with Intel Xeon CPU E5-2640 and 32GB RAM, ran each experiment 3 times and report the average here.

Data. We used two real datasets that have been used to evaluate OD discovery algorithms [12,16,17]. FLI is about US flights information, with 500K tuples and 20 attributes (www.transtats.bts.gov). NC contains data of registered voters from North Carolina, with 1M tuples and 22 attributes (ncsbe.gov). To improve efficiency and avoid uninteresting ODs, we replaced attribute values with integers in a way that the ordering is preserved, and removed tuples with NULL values, similar to [12,16,17].

Algorithms. We implemented our algorithms in Java: IncOD for incremental OD discovery (with $\triangle Check(\cdot)$, $Enrich(\cdot)$ and $Expand(\cdot)$) and CoverIndex for choosing attributes on which to build minimal indexes. For comparison, we obtained a batch OD discovery implementation ORDER [12] from www.metanome.de. To our best knowledge, this is the only algorithm for list-based OD discovery.

All experiments are controlled by 3 parameters: (1) $|D|$: the number of original tuples; (2) $|\triangle D|$: the number of tuples inserted into D; and (3) $|R|$: the number of attributes. We vary $|R|$ by taking random projections of the dataset. We employ ORDER to compute Σ on D, as inputs of CoverIndex for index building. IncOD then computes $\triangle\Sigma$ with inputs D, $\triangle D$ and Σ, leveraging indexes. The correctness of IncOD is verified by checking whether $\Sigma \oplus \triangle\Sigma$ equals the results of ORDER on $D + \triangle D$. We report the time of ORDER on $D + \triangle D$, against the time of IncOD for updating indexes and computing $\triangle\Sigma$ on tuple insertions.

Exp-1. We compare IncOD against ORDER using FLI. We set $|D| = 300$K,$|\triangle D| =$ 90K and $|R| = 8$ by default, and vary one parameter in each of the experiments.

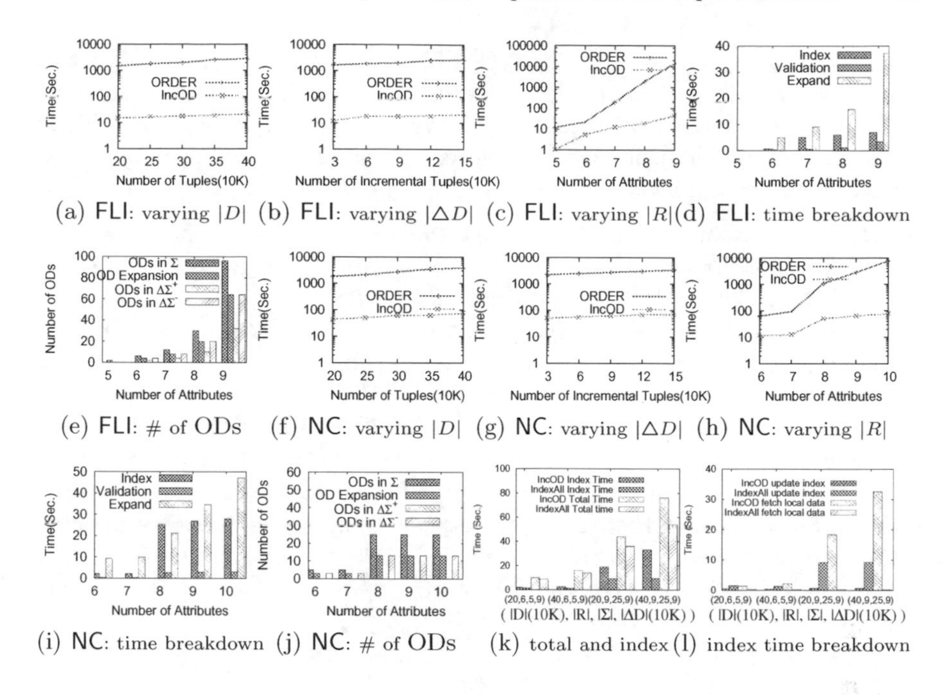

(a) FLI: varying $|D|$ (b) FLI: varying $|\triangle D|$ (c) FLI: varying $|R|$ (d) FLI: time breakdown

(e) FLI: # of ODs (f) NC: varying $|D|$ (g) NC: varying $|\triangle D|$ (h) NC: varying $|R|$

(i) NC: time breakdown (j) NC: # of ODs (k) total and index (l) index time breakdown

Fig. 3. Experimental results

Varying $|D|$. Fig. 3(a) shows results by varying $|D|$ from 200K to 400K. ORDER scales well with $|D|$, consistent with results in [12]. Times of IncOD increase slightly, due to more *local* data *w.r.t.* $|\triangle D|$ as $|D|$ increases. IncOD outperforms ORDER by two orders of magnitude on all sizes of D. As an example, ORDER takes more than 45 min when $|D|$ is 400K, while IncOD takes only 22 s.
Varying $|\triangle D|$. Fig. 3(b) shows results by varying $|\triangle D|$ from 30K to 150K. We find IncOD scales very well with $|\triangle D|$: the time increases from 12 s to 21 s, when the ratio of $|\triangle D|$ to $|D|$ increases from 10% to 50%. IncOD outperforms ORDER by two orders of magnitude even when $|\triangle D|$ is half of $|D|$.
Varying $|R|$. We vary $|R|$ from 5 to 9 in Fig. 3(c). $|R|$ has the most effect on the time of list-based OD discovery, since the number of possible list-based ODs is of a factorial complexity in $|R|$. ORDER does not scale well with $|R|$, consistent with results in [12]. The scalability of IncOD is far more better. As $|R|$ increases from 8 to 9, the time for ORDER increases from 33 min to more than 4 h, while the time for IncOD only increases from 19 s to 46 s.

In Fig. 3(d) we decompose the overall time into times for (i) updating indexes and obtaining local data via indexes for $\triangle D$, (ii) OD validations by $\triangle Check(\cdot)$, and (iii) OD expansion; other times are marginal. The times for (i) and (ii) are related to ODs in Σ, while time (iii) is related to ODs for expansion, whose numbers are shown in Fig. 3(e). We also report in Fig. 3(e) the number of ODs in $\triangle\Sigma$. We find time (i) is short, due to the fact that almost all of ODs on FLI contain a single LHS attribute, and hence local data *w.r.t.* $\triangle D$ can be directly fetched via indexes built by CoverIndex. Time (ii) is also short; $\triangle Check(\cdot)$ requires only local data of $\triangle D$ and is linear in $|\triangle D|$ (Theorem 2). The time for expansion (Time (iii)) governs the overall time. The search space of our approach is much smaller than its batch counterpart since $\triangle \Sigma$ is computed based on Σ, fully leveraging incremental computations. Moreover, instance-based pruning rules in expansion and optimizations by Theorem 3 are proved to be quite effective.

Exp-2. We then compare IncOD against ORDER using NC, with $|D| = 300\text{K}$, $|\triangle D| = 90\text{K}$ and $|R| = 9$ by default. We vary $|D|$ from 200K to 400K in Fig. 3(f), vary $|\triangle D|$ from 30K to 150K in Fig. 3(g), and vary $|R|$ from 6 to 10 in Fig. 3(h). In the same setting as Fig. 3(h), we report the time breakdown and number of related ODs in Figs. 3(i) and 3(j). The results confirm our observations on FLI. (1) IncOD significantly outperforms ORDER: IncOD is on average 48 and 51 times faster in Figs. 3(f) and 3(g), respectively. (2) IncOD scales much better with $|R|$. As $|R|$ increases from 6 to 10 in Fig. 3(h), the time for ORDER increases by more than two orders of magnitude, while the time for IncOD increases by less than 7 times. (3) Fig. 3(i) shows that more time is required in the index processing phase of NC. Most of ODs found on NC have multiple LHS attributes. Since CoverIndex may choose to build indexes on prefixes of LHS attributes, some post-processing after index visits is required to fetch local data of $\triangle D$. Specifically, to fetch local data $local'(\mathsf{X}, t')$ with an index $ind_{\mathsf{X}'}$ where X' is a prefix of X, we need to sort tuples in $local'(\mathsf{X}', t')$ on $\mathsf{X} \setminus \mathsf{X}'$; this incurs additional costs. Note that as $|R|$ increases from 8 to 10, the same number of ODs are found on D (Fig. 3(j)).

Exp-3. We evaluate different index strategies on NC. We denote by IndexAll when indexes are built on all distinct LHS attribute lists of ODs in Σ, and compare it against IncOD with CoverIndex. We denote by $(|D|,|R|,|\Sigma|,|\triangle D|)$ indexes with different settings: index building depends on D and Σ; Σ is determined by D and R; running times concern $\triangle D$. We report in Fig. 3(k) total running time and index processing time; index time is part of the running time and IndexAll differs from IncOD only in this time. We also show in Fig. 3(l) index time breakdown. We find IndexAll takes less time compared to IncOD, as expected. The total time of IndexAll is about [68%, 88%] of that of IncOD in Fig. 3(k). The efficiency of IndexAll comes at the cost of more indexes. For the case that $|\Sigma| = 25$ in Fig. 3(k), IndexAll has to build 25 indexes since each OD has a distinct LHS attribute list, while CoverIndex suffices to cover all ODs with 5 indexes. Hence, IndexAll takes more time to update indexes, shown in Fig. 3(l). IncOD takes more time for fetching local data of $\triangle D$ due to required post-processing, as illustrated before. We contend that CoverIndex is a better choice when index space is a major concern, *e.g.*, for large $|\Sigma|$ or $|D|$. IncOD already achieves very good performance. In practice if we can afford more space, we can combine some extra indexes with the indexes built by CoverIndex, to further improve the efficiency.

6 Conclusions

We have formalized the problem of incremental OD discovery, studied its computational complexity, discussed its data locality property, presented algorithms and optimizations, and experimentally demonstrated our approaches.

We are developing distributed techniques for incremental OD discovery to further enhance the scalability, and studying incremental discoveries for other constraints.

Acknowledgements. This work is supported by NSFC 61572135, NSFC 61370157, The Shanghai Innovation Action Project 17DZ1203600, China Grid (Shanghai) 52094017001x, Shanghai Science and Technology Project (No. 16DZ1110102), Airplane Research Project, Industry Internet Innovation Development Project.

References

1. Abedjan, Z., Quian-Ruiz, J., Naumann, F.: Detecting unique column combinations on dynamic data. In: ICDE (2014)
2. Bleifub, T., Kruse, S., Naumann, F.: Efficient denial constraint discovery with hydra. PVLDB **11**(3), 311–323 (2017)
3. Cong, G., Fan, W., Geerts, F., Jia, X., Ma, S.: Improving data quality: consistency and accuracy. In: VLDB (2007)
4. Chu, X., Ilyas, I., Papotti, P.: Discovering denial constraints. PVLDB **6**(13), 1498–1509 (2013)
5. Chu, X., Ilyas, I., Papotti, P.: Holistic data cleaning: putting violations into context. In: ICDE (2013)
6. Fan, W., Geerts, F., Li, J., Xiong, M.: Discovering conditional functional dependencies. TKDE **23**(5), 683–698 (2011)
7. Fan, W., Li, J., Tang, N., Yu, W.: Incremental detection of inconsistencies in distributed data. TKDE **26**(6), 1367–1383 (2014)
8. Golab, L., Karloff, H., Korn, F., Srivastava, D., Yu, B.: On generating near-optimal tableaux for conditional functional dependencies. PVLDB **1**(1), 376–390 (2008)
9. Huhtala, Y., Karkkainen, J., Porkka, P., Toivonen, H.: TANE: an efficient algorithm for discovering functional and approximate dependencies. Comput. J. **42**(2), 100–111 (1999)
10. Hao, S., Tang, N., Li, G., He, J., Ta, N., Feng, J.: A novel cost-based model for data repairing. TKDE **29**(4), 727–742 (2017)
11. Khayyat, Z., et al.: BigDansing: a system for big data cleansing. In: SIGMOD (2015)
12. Langer, P., Naumann, F.: Efficient order dependency detection. VLDB J. **25**(2), 223–241 (2016)
13. Papenbrock, T., Naumann, F.: A hybrid approach to functional dependency discovery. In: SIGMOD (2016)
14. Song, S., Chen, L.: Differential dependencies: reasoning and discovery. TODS **36**(3), 16:1–16:41 (2011)
15. Szlichta, J., Godfrey, P., Gryz, J.: Fundamentals of order dependencies. PVLDB **5**(11), 1220–1231 (2012)
16. Szlichta, J., Godfrey, P., Golab, L., Kargar, M., Srivastava, D.: Effective and complete discovery of order dependencies via set-based axiomatization. PVLDB **10**(7), 721–732 (2017)

17. Szlichta, J., Godfrey, P., Golab, L., Kargar, M., Srivastava, D.: Effective and complete discovery of bidirectional order dependencies via set-based axioms. VLDB J. **27**(4), 573–591 (2018)
18. Szlichta, J., Godfrey, P., Gryz, J., Zuzarte, C.: Expressiveness and complexity of order dependencies. PVLDB **6**(14), 1858–1869 (2013)
19. Vazirani, V.: Approximation Algorithms. Springer, Heidelberg (2003). https://doi.org/10.1007/978-3-662-04565-7

Multi-view Spectral Clustering via Multi-view Weighted Consensus and Matrix-Decomposition Based Discretization

Man-Sheng Chen[1,2], Ling Huang[1,2], Chang-Dong Wang[1,2](✉), and Dong Huang[3]

[1] School of Data and Computer Science, Sun Yat-sen University, Guangzhou, China
chenmsh27@mail2.sysu.edu.cn, huanglinghl@hotmail.com, changdongwang@hotmail.com
[2] Guangdong Province Key Laboratory of Computational Science, Guangzhou, China
[3] College of Mathematics and Informatics, South China Agricultural University, Guangzhou, China
huangdonghere@gmail.com

Abstract. In recent years, multi-view clustering has been widely used in many areas. As an important category of multi-view clustering, multi-view spectral clustering has recently shown promising advantages in partitioning clusters of arbitrary shapes. Despite significant success, there are still two challenging issues in multi-view spectral clustering, i.e., (i) how to learn a similarity matrix for multiple weighted views and (ii) how to learn a robust discrete clustering result from the (continuous) eigenvector domain. To simultaneously tackle these two issues, this paper proposes a unified spectral clustering approach based on multi-view weighted consensus and matrix-decomposition based discretization. In particular, a multi-view consensus similarity matrix is first learned with the different views weighted w.r.t. their confidence. Then the eigen-decomposition is performed on the similarity matrix and a set of c eigenvectors are obtained. From the eigenvectors, we first learn a continuous cluster label and then discretize it to build the final clustering label, which avoids the potential instability of the conventional k-means discretization. Extensive experiments have been conducted on multiple multi-view datasets to validate the superiority of our proposed approach.

Keywords: Multi-view spectral clustering · Weighted consensus · Matrix-decomposition · Discretization

1 Introduction

With the development of the information technology [1], a huge amount of multi-view data have emerged from various kinds of real-world applications [2–12].

G. Li et al. (Eds.): DASFAA 2019, LNCS 11446, pp. 175–190, 2019.
https://doi.org/10.1007/978-3-030-18576-3_11

Multi-view data can be captured from heterogenous views or sources, and these different views or sources reveal the distinct information of the same object. For instance, a YouTube video consists of text features, auditory features and visual features. A text news can be translated into different languages. In traditional multi-view clustering, a straightforward idea to deal with multi-view data is to concatenate all the features into a new feature vector, and then perform single-view clustering method on the new feature vector to obtain the clustering result. However, this simple strategy ignores the different characteristics as well as the correlation among multiple views. The features for multiple views are able to provide complementary information between views. To capture the diversity and correlation in multi-view data, many multi-view clustering algorithms have been developed to improve the robustness of the clustering by making full use of the information from multiple views [13–18].

In the past few years, many multi-view clustering algorithms have been proposed by considering the rich information of multiple views [19–24]. For example, Cai et al. [22] developed a multi-view spectral clustering framework to integrate heterogeneous image features. Kumar et al. [21] introduced the co-regularization technique in multi-view spectral clustering. These methods, however, may be affected by weak or poor views, and thereby result in degraded clustering performances. In multi-view clustering, different views may be associated with very different reliability and should be weighted accordingly. Inspired by the co-training technique [19], Kumar and Daumé III [20] exploited prior knowledge to decide the view weights, and designed a consensus cluster label matrix for multi-view spectral clustering. However, besides the view-weighting issue, another limitation to these existing multi-view spectral clustering methods [21,25,26] is that they mostly rely on the k-means algorithm to perform discretization on the continuous eigenvector domain, where the inherent instability of k-means may significantly affect the final clustering result after discretization.

To simultaneously deal with the issue of view weighting and the issue of potentially unstable discretization of k-means, in this paper, we propose a unified multi-view spectral clustering framework based on multi-view weighted consensus similarity and matrix-decomposition based discretization. Specifically, a consensus similarity matrix is first built with the multiple views evaluated and weighted. Then, a continuous cluster label is learned, from which the final discrete clustering label can be obtained in an optimization model. In the optimization model, we exploit an alternative iteration scheme to achieve an approximate solution. Extensive experiments have been conducted on multiple multi-view datasets, which demonstrate the superiority of our proposed method.

The following sections are organized as follows. In Sect. 2 we describe the proposed model in detail, and present an optimization algorithm to solve the model. Next in Sect. 3, extensive experiments are conducted on four real-world datasets to show the superiority of our method. Finally in Sect. 4, we conclude the whole paper.

Notations. In this paper, uppercase letters are used to represent the matrices. For a matrix M, its i-th row can be written as m_i whose j-th entry is denoted as m_{ij}. $Tr(M)$ stands for the trace of the matrix M. The v-th view of the matrix M is expressed as $M^{(v)}$. We use $\|M\|_2$ and $\|M\|_F$ to respectively represent the l_2-norm and the Frobenius norm of the matrix M. In addition, 1_n means the column vector whose length is n and the entries are all one.

2 The Proposed Algorithm

In this section, we introduce in detail the proposed Multi-view Spectral Clustering via Multi-view Weighted Consensus and Matrix-decomposition based Discretization (MvWCMD) algorithm. First of all, we will briefly introduce the preliminary knowledge. And then we will describe in detail the proposed model, the optimization problem of which will be solved by the alternative iteration scheme. Finally, we will summarize the entire algorithm and provide time complexity analysis.

2.1 Preliminary Knowledge

Graph-Based Clustering Description. Suppose there are n samples which can be partitioned into c categories. To well represent the affinities between these samples, a similarity matrix is supposed to be constructed in a graph-based clustering method. A decent graph plays a vital role therein, therefore it has been studied in many works [27]. When a similarity matrix is ideal, the number of its connected components must be c the same as the number of the final clusters, and it can be directly applied for clustering. Inspired by the idea above, Nie et al. [28] proposed a Constraint Laplacian Rank (CLR) method which aims to learn an ideal graph from the given similarity matrix. Given an arbitrary similarity matrix $A \in \mathbb{R}^{n \times n}$, the target graph can be solved by the following model

$$\min_{s_i 1_n = 1, s_{ij} \geq 0, S \in C} \|S - A\|_F^2, \tag{1}$$

where S is non-negative, and the entries of each row sum up to 1. C indicates a set of n by n square matrices whose connected components are c. In the light of the graph theory in [29], the connectivity constraint can be substituted for a rank constraint, and thus the problem (1) can be rewritten as

$$\min_{s_i 1_n = 1, s_{ij} \geq 0, rank(L) = n-c} \|S - A\|_F^2, \tag{2}$$

where $rank(L)$ stands for the rank of the Laplacian matrix L, and $L = D - \frac{(S^T + S)}{2}$. The n by n degree matrix D is a diagonal matrix, and $D(ii) = \frac{\sum_j (s_{ij} + s_{ji})}{2}$. In this way, the ideal similarity matrix S can be obtained, and thus it can be directly used in clustering. However, the CLR method is just applicable for single-view clustering.

Spectral Clustering Revisit. Looking back on the spectral clustering method [30], data points can be partitioned into different groups according to their similarities. Not requiring data is linearly separable, the method can explore the non-convex pattern. For spectral clustering, Laplacian matrix $L \in \mathbb{R}^{n\times n}$ is required as an input. To obtain the Laplacian matrix L, the similarity matrix $S \in \mathbb{R}^{n\times n}$ is firstly needed to be constructed in traditional spectral clustering methods by one of the three common strategies, such as the k-nearest-neighborhood (knn). Suppose in data X there are c clusters, the spectral clustering problem can be written as

$$\min_{Y} Tr\left(Y^T L Y\right), \qquad \text{s.t. } Y \in Ind, \tag{3}$$

where $Y = [y_1, y_2, ..., y_n]^T \in \mathbb{R}^{n\times c}$ is the cluster indicator matrix whose labels are discrete, and $Y \in Ind$ indicates that the cluster label vector of each point $y_i \in \{0,1\}^{c\times 1}$ only comprises one and only one element "1" to reveal the cluster membership of x_i. Actually, the problem (3) is an NP-hard problem according to the discrete constraint on Y. Thus, the matrix Y is usually relaxed to allow continuous values, and finally the problem becomes

$$\min_{F} Tr\left(F^T L F\right), \qquad \text{s.t. } F^T F = I, \tag{4}$$

where $F \in \mathbb{R}^{n\times c}$ is the relaxed continuous cluster label matrix, and the trivial solution can be avoided by the orthogonal constraint therein. And then the approximate solution of F can be achieved by the c eigenvectors of L corresponding to the c smallest eigenvalues. Subsequently, traditional clustering method such as k-means is applied to compute on F to get the final discrete cluster labels [31]. Nevertheless, there still exists potential instability. Due to the uncertainty of the post-processing step, the final solution may deviate from the real discrete labels unpredictably [32].

2.2 The Proposed Model

Motivated by the idea that the spectral embedding matrix F is spanned by the column vectors of the cluster indicator matrix $Y \in Ind$ [31] when the similarity matrix is ideal, we extend the CLR method mentioned above to the multi-view clustering. Despite of this idea, the spectral embedding matrix F is actually not equal to the cluster indicator matrix $Y \in Ind$. Thus, in this paper, not only the spectral embedding matrix can be focused on, but also the cluster indicator matrix can be solved finally without k-means discretization.

In multi-view clustering, the same object represented in different views is expected to be partitioned into the same group. Thus, the ground truth similarity matrix of each view is supposed to be the same. That is to say, there is a consensus similarity matrix among all the views. For multi-view data, suppose that there are m views, and $A^{(1)}$, $A^{(2)}$, ..., $A^{(m)}$ corresponding to the similarity matrix of each view, we aim to get the multi-view consensus similarity matrix S that can well approximate the original input similarity matrix $A^{(v)} \in \mathbb{R}^{n\times n} (1 \le v \le m)$. A straight-forward solution is to assign the same weight to every input similarity matrix and achieve an average similarity matrix by the equation $\overline{A} = \frac{1}{m}\sum_{v=1}^{m} A^{(v)}$. However, this simple way ignores the different contributions among views, leading to bad clustering performance when there are poor quality views. Accordingly, a group of meaningful weights are needed to be introduced to measure the importance of different views. In this paper, a trick idea [24] is followed by our algorithm to adaptively measure the weights of the views. Consequently, the target multi-view weighted consensus similarity matrix S with rank constraint is learned to approximate the similarity matrix of each view with different weights. To solve this problem, a linear combination of the reconstruction error $\|S - A^{(v)}\|_F^2$ for each view will be minimized [24]. Thus, the problem can be written as

$$\begin{aligned} &\min_{S} \sum_{v=1}^{m} w^{(v)} \|S - A^{(v)}\|_F^2, \\ &\text{s.t. } s_i 1_n = 1, s_{ij} \geq 0, rank(L) = n - c, \end{aligned} \tag{5}$$

where the constant $w^{(v)}$ is the optimal target function value of the following problem:

$$w^{(v)} \stackrel{def}{=} \min_{S} \frac{1}{\|S - A^{(v)}\|_F}. \tag{6}$$

We can obviously find that $w^{(v)}$ depends on S. If the view v is good, the value of $\|S - A^{(v)}\|_F$ should be small, and therefore $w^{(v)}$ is supposed to be large. Otherwise, a small weight is required to be assigned to a weak view.

Problem (5) is not easy to be solved, due to the rank constraint where $L = D - \frac{(S^T+S)}{2}$ and D is an n by n diagonal matrix whose diagonal elements $D(ii) = \frac{\sum_j (s_{ij}+s_{ji})}{2}$ also depend on the similarity matrix S. Here L is a positive semi-definite matrix, and thus $\sigma_i(L) \geq 0$, where $\sigma_i(L)$ corresponds to the i-th smallest eigenvalue of the Laplacian matrix L. Inspired by [29], $rank(L) = n - c$ is tantamount to $\sum_{i=1}^{c} \sigma_i(L) = 0$. To cope with the optimization question with rank constraint whose complexity analysis is combinatorial, the rank constraint is incorporated into the objective function as a regularizer term [28,33]. Therefore, the constraint is relaxed and our model is reformulated as

$$\begin{aligned} &\min_{S} \sum_{v=1}^{m} w^{(v)} \|S - A^{(v)}\|_F^2 + \alpha \sum_{i=1}^{c} \sigma_i(L), \\ &\text{s.t. } s_i 1_n = 1, s_{ij} \geq 0. \end{aligned} \tag{7}$$

If α is enough large, the minimization of Eq. (7) will make the regularizer term $\sum_{i=1}^{c} \sigma_i(L) \to 0$. And then the rank constraint $rank(L) = n - c$ will be solved.

Despite all this, problem (7) still remains a challenging problem as a result of the last term. Fortunately, the Ky Fan's Theorem [34] can be applied to solve the problem above, that is to say

$$\sum_{i=1}^{c} \sigma_i(L) = \min_{F^T F = I} Tr\left(F^T L F\right), \tag{8}$$

where $F \in \mathbb{R}^{n\times c}$ is a spectral embedding matrix, and the spectral embedding matrix F is actually not equal to the cluster indicator matrix $Y \in Ind$. To better achieve our clustering task, our multi-view spectral clustering via multi-view weighted consensus and matrix-decomposition based discretization (MvWCMD) model is proposed as follows:

$$\begin{aligned}
\min_{S,F,Y,Q} \quad & \underbrace{\sum_{v=1}^{m} w^{(v)} \|S - A^{(v)}\|_F^2}_{\text{multi-view weighted consensus similarity learning}} + \\
& \underbrace{\alpha Tr\left(F^T L F\right)}_{\text{continuous cluster label learning}} + \underbrace{\beta\|Y - FQ\|_F^2}_{\text{discrete cluster label learning}}, \\
\text{s.t. } & s_i 1_n = 1, s_{ij} \geq 0, F^T F = I, Q^T Q = I, Y \in Ind,
\end{aligned} \tag{9}$$

where α and β are the penalty parameters, and Q is a rotation matrix. Due to the invariance property of spectral solution [35], FQ is another solution for any solution F [36]. The last term expects to find an appropriate orthogonal rotation matrix Q so that the result of FQ is closely approaching to the ground truth discrete cluster label matrix Y. From Eq. (9), the multi-view weighted consensus similarity matrix S, the continuous cluster label matrix F and the final discrete cluster label matrix Y can be automatically learned from the data. Ideally, we must have $s_{ij} = 0$ if data point i and j belong to different groups and vice versa. That is to say, if and only if data point i and j belong to different groups, we have $s_{ij} = 0$ or $f_i \neq f_j$. Therefore, the correlation between the learned similarity matrix and the cluster labels can be exploited in our unified framework Eq. (9). In fact, there is a self-taught property in our clustering model because of the feedback of cluster labels to induce the ideal similarity matrix and vice versa.

2.3 Optimization

In this subsection, an alternative iteration scheme is utilized to solve the problem (9). When updating one variable, the remaining variables will be fixed in the alternative iteration scheme.

Computation of S. With F, Q and Y fixed, the problem is reduced to

$$\begin{aligned}
&\min_{S} \sum_{v=1}^{m} w^{(v)} \|S - A^{(v)}\|_F^2 + \alpha Tr\left(F^T L F\right), \\
&\text{s.t. } s_i 1_n = 1, s_{ij} \geq 0.
\end{aligned} \tag{10}$$

In particular, the problem (10) can be further written as

$$\min_{s_i 1_n = 1, s_{ij} \geq 0} \sum_{v=1}^{m} w^{(v)} \sum_{i,j=1}^{n} \left(s_{ij} - a_{ij}^{(v)}\right)^2 + \alpha \sum_{i,j=1}^{n} \|f_i - f_j\|_2^2 s_{ij}. \tag{11}$$

Due to the independence of the problem (11) for different i, it is equivalent to separately solving the following problem for each i

$$\min_{s_i 1_n=1, s_{ij}\geq 0} \sum_{j=1}^{n}\sum_{v=1}^{m} w^{(v)} \left(s_{ij} - a_{ij}^{(v)}\right)^2 + \alpha \sum_{j=1}^{n} \|f_i - f_j\|_2^2 s_{ij}. \tag{12}$$

For briefness, $v_{ij} = \|f_i - f_j\|_2^2$ is used, and v_i is a vector whose j-th entry is v_{ij}. s_i and a_i are in like manner. Thus, the problem (12) becomes

$$\min_{s_i 1_n=1, s_i\geq 0_n^T} \|s_i - \frac{\sum_{v=1}^{m} w^{(v)} a_i^{(v)} - \frac{\alpha}{2} v_i}{\sum_{v=1}^{m} w^{(v)}}\|_2^2. \tag{13}$$

The problem above can be addressed by an efficient iterative algorithm proposed in [37]. To rapidly obtain the totally sparse multi-view consensus similarity matrix S, the neighbors of the i-th data can be chosen to be updated, and exactly the neighbors can be set as a const, like 10 in our algorithm.

Computation of F. With S, Q and Y fixed, we have

$$\min_F \alpha Tr\left(F^T L F\right) + \beta \|Y - FQ\|_F^2, \qquad \text{s.t. } F^T F = I. \tag{14}$$

The problem (14) which is constrained by the orthogonal condition can be settled efficiently by the algorithm proposed by [38].

Computation of Q. With S, F and Y fixed, the problem becomes

$$\min_Q \|Y - FQ\|_F^2, \qquad \text{s.t. } Q^T Q = I. \tag{15}$$

This is an orthogonal Procrustes problem [39], which allows a closed-form solution, and the solution is as follows

$$Q = UV^T, \tag{16}$$

where U and V are the left and right components of the SVD decomposition of $Y^T F$.

Computation of Y. With S, F and Q fixed, it is equivalent to solving

$$\min_Y \|Y - FQ\|_F^2, \qquad \text{s.t. } Y \in Ind. \tag{17}$$

Knowing that $Tr\left(Y^T Y\right) = n$, the problem above can be reformulated as

$$\max_Y Tr\left(Y^T FQ\right), \qquad \text{s.t. } Y \in Ind. \tag{18}$$

Consequently, the optimal solution can be achieved from the following equation

$$Y_{ij} = \begin{cases} 1, & j = \arg\max_k (FQ)_{ik} \\ 0, & \text{otherwise.} \end{cases} \tag{19}$$

The variables S, F, Q and Y are separately initialized at first. And then they are updated iteratively in an interplay manner until convergence. In this way, an overall optimal solution can be achieved.

2.4 Algorithm Summary and Time Complexity Analysis

For clarity, the main procedure of the proposed MvWCMD method is summarized in Algorithm 1. In what follows, we will provide the time computational complexity analysis. With our optimization strategy, the computation of S requires $\mathcal{O}\left(n^3 + nv\right)$ complexity where $v \ll n$, since it needs to perform eigenvalue decomposition in every iterative step. SVD is involved in the updating of Q, and its computational complexity is $\mathcal{O}\left(nc^2 + c^3\right)$. The complexity for F is $\mathcal{O}\left(nc^2 + c^3\right)$. To update Y, $\mathcal{O}\left(nc^2\right)$ is needed. The number of clusters c is usually a small digit. Therefore, the main computational load of the model in Eq. (9) relies on obtaining the multi-view consensus similarity matrix S.

Algorithm 1. Multi-view Spectral Clustering via Multi-view Weighted Consensus and Matrix-decomposition based Discretization

---

Input: Similarity matrices for m views $A^{(1)}, A^{(2)}, ..., A^{(m)}$ and $A^{(v)} \in \mathbb{R}^{n \times n}$, number of clusters c, parameter $\alpha > 0$, $\beta > 0$

1: Initialize the weight of each view $w^{(v)} = \frac{1}{m}$, random matrices $F \in \mathbb{R}^{n \times c}$ and $Q \in \mathbb{R}^{n \times n}$, zero matrix $Y \in \mathbb{R}^{n \times c}$.
2: Let $A = \sum_{v=1}^{m} w^{(v)} A^{(v)}$.
3: Compute F, which is spanned by the c eigenvectors of $L = D - \frac{A^T + A}{2}$ corresponding to the c smallest eigenvalues.
4: **repeat**
5:     **repeat**
6:         For each i, update the i-th row of S by solving the problem of Eq. (13).
7:     **until** stopping criterion is met.
8:     Update F according to Eq. (14).
9:     Update Q by solving Eq. (16).
10:     $Y = 0$.
11:     Update Y by solving Eq. (19).
12: **until** stopping criterion is met.

Output: $S \in \mathbb{R}^{n \times n}$ with exactly c components, spectral embedding matrix $F \in \mathbb{R}^{n \times c}$, orthogonal rotation matrix $Q \in \mathbb{R}^{n \times n}$ and indicator matrix $Y \in Ind$.

---

3 Experiment

In this section, extensive experiments are conducted to verify the superiority of the proposed method on four real-world datasets. In our experiments, two common evaluation metrics, accuracy (ACC), and normalized mutual information (NMI) are used to estimate the clustering performance of our proposed method and baselines. For each measure, the value is higher, the clustering performance is better [40]. Readers can refer to [41] for further details of the two measures. In addition, parameter analysis, convergence analysis and comparison experiments are separately conducted on the four real-world datasets.

3.1 Real-World Datasets

In our experiment, the four benchmark datasets, UCI Handwritten digits, MSRCv1, Caltech101-7 and Caltech101-20 are used. In the following, we will introduce the details of these datasets.

1. **Handwritten digits dataset**
   Coming from UCI machine learning repository, multiple features (Mfeat) dataset is a handwritten digits dataset[1]. The dataset consists of 2000 samples in which there are 10 classes. In our experiment, three kinds of features, 216 profile correlations, 76 Fourier coefficients and 47 Zernike moments are used to represent images. Each type of features is considered as a view.
2. **MSRCv1 dataset**
   MSRCv1 dataset is an image dataset [42]. The dataset consists of 210 objects and 7 classes. In our experiment, four kinds of features, CM feature, GIST feature, LBP feature and GENT feature are used to represent images, and each type of features is regarded as a view.
3. **Caltech101 datasets**
   Consisting of 101 categories of images, caltech101 [43] is an image dataset. For experimental purpose, two subsets are chosen to represent two datasets following the previous work [25]. The one dataset is named Caltech101-7, and it has 1474 images and 7 widely used classes. The other dataset which is larger is called Caltech101-20, and it is made up of 2386 images and 20 classes. Three types of features, 1984-dimensional HOG feature, 512-dimensional GIST feature and 928-dimensional LBP feature from the images are selected to stand for three views.

The summarization of the four real-world datasets is shown in Table 1.

Table 1. Statistic of the four real-world datasets.

	Mfeat	MSRCv1	Caltech101-7	Caltech101-20
View1	fac(216)	cm(24)	hog(1984)	hog(1984)
View2	fou(76)	gist(512)	gist(512)	gist(512)
View3	zer(47)	lbp(256)	lbp(928)	lbp(928)
View4	-	gent(254)	-	-
# Size	2000	210	1474	2386
# Class	10	7	7	20

3.2 Parameter Analysis

There are two parameters in our model: α and β. In the following, parameter analysis is conducted to show the effect of the two parameters. There are

[1] http://archive.ics.uci.edu/ml/index.php.

different properties of different datasets, and thus different ranges of α and β are applied to different datasets. For example, the ranges of α and β are separately $10, 30, 50, 70, 90$ and $0.01, 0.03, 0.05, 0.07, 0.09$ in Mfeat dataset, while the ranges of α and β are separately $1, 3, 5, 7, 9$ and $0.001, 0.003, 0.005, 0.007, 0.009$ in Caltech101-7 dataset. The experimental results are respectively exhibited in Figs. 1, 2, 3 and 4. According to the figures, best results in different datasets can be obtained. For Mfeat dataset, there are the best results when α is 50 and β is 0.01. Similarly, when α is 1 and β is 0.009, best results are achieved for MSRCv1 dataset. In particular, when α is 7 and β is 0.003, the best ACC value can be obtained in Caltech101-7, but the NMI value is lower at this time. To be balanced, the comparatively better results are chosen when α is 9 and β is 0.007 for Caltech101-7. In Caltech101-20 dataset, α is 30 and β is 1 when there are the best results.

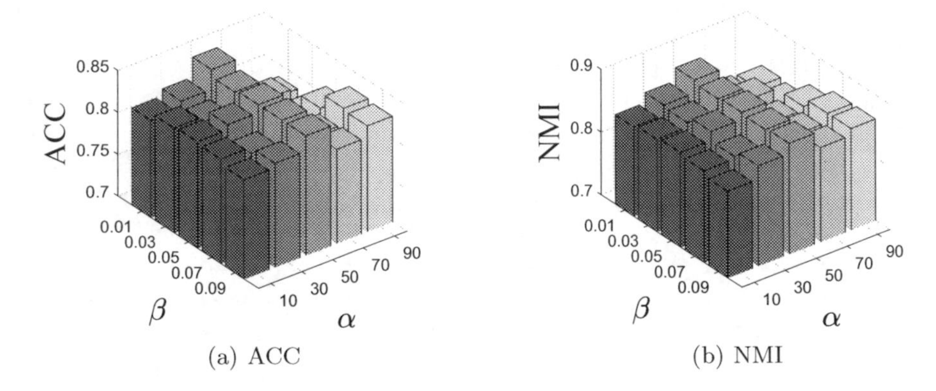

(a) ACC (b) NMI

Fig. 1. Parameter analysis on α and β on Mfeat.

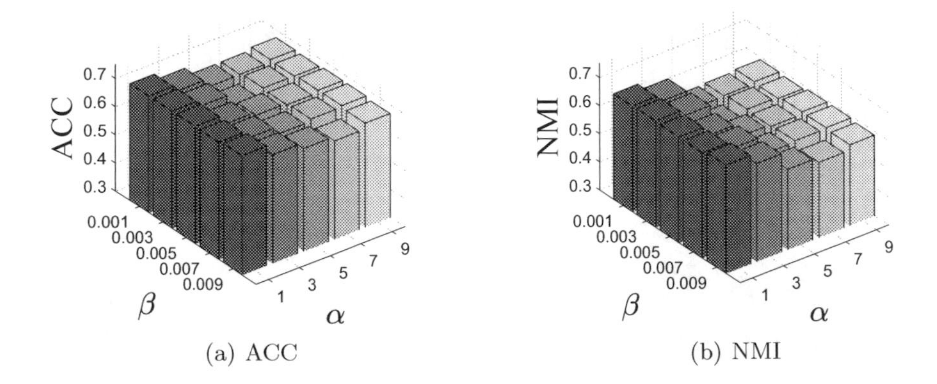

(a) ACC (b) NMI

Fig. 2. Parameter analysis on α and β on MSRCv1.

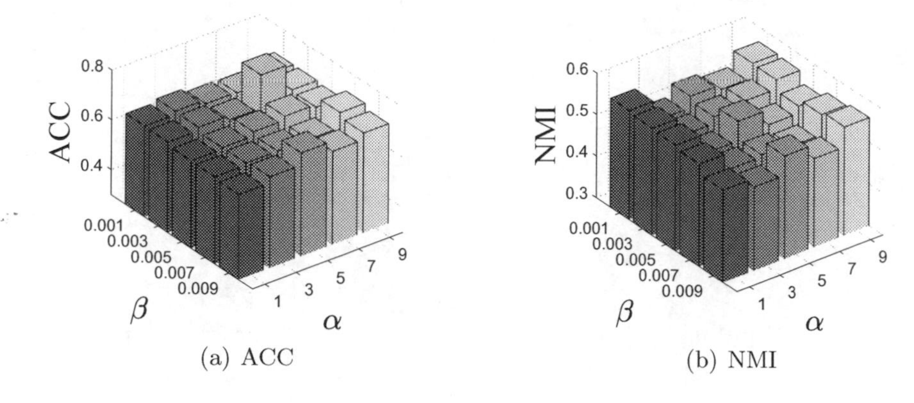

(a) ACC (b) NMI

Fig. 3. Parameter analysis on α and β on Caltech101-7.

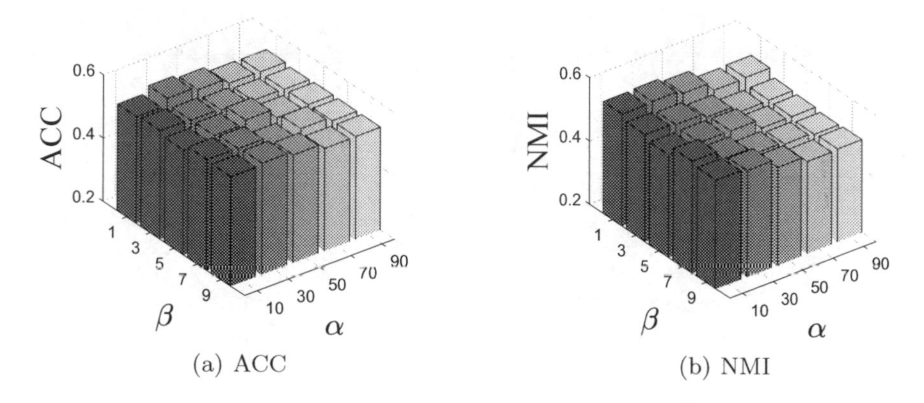

(a) ACC (b) NMI

Fig. 4. Parameter analysis on α and β on Caltech101-20.

3.3 Convergence Analysis

To verify the convergence property of the proposed method, convergence analysis is conducted. With the best results, the values of α and β from different datasets are set according to the parameter analysis. The experimental results are showed in Fig. 5. Obviously, we can generally conclude that the method will converge during the 30 times of iterations from the subfigures.

3.4 Comparison Experiment

To validate the superiority of the proposed MvWCMD method, we compare our algorithm with the following methods: Constraint Laplacian Rank [28] (CLR), Co-Regularized Spectral Clustering [21] (CoReg), Co-Training Multi-view Clustering [20] (CoTrn), Self-weighted Multi-view Clustering [24] (SwMC), Multi-View Spectral Clustering [22] (MVSC), Robust Multi-view Spectral Clustering [26] (RMSC) and Multi-view Learning with Adaptive Neighbors [44] (MLAN). Following the CLR method, an initial input similarity matrix $A^{(v)}$

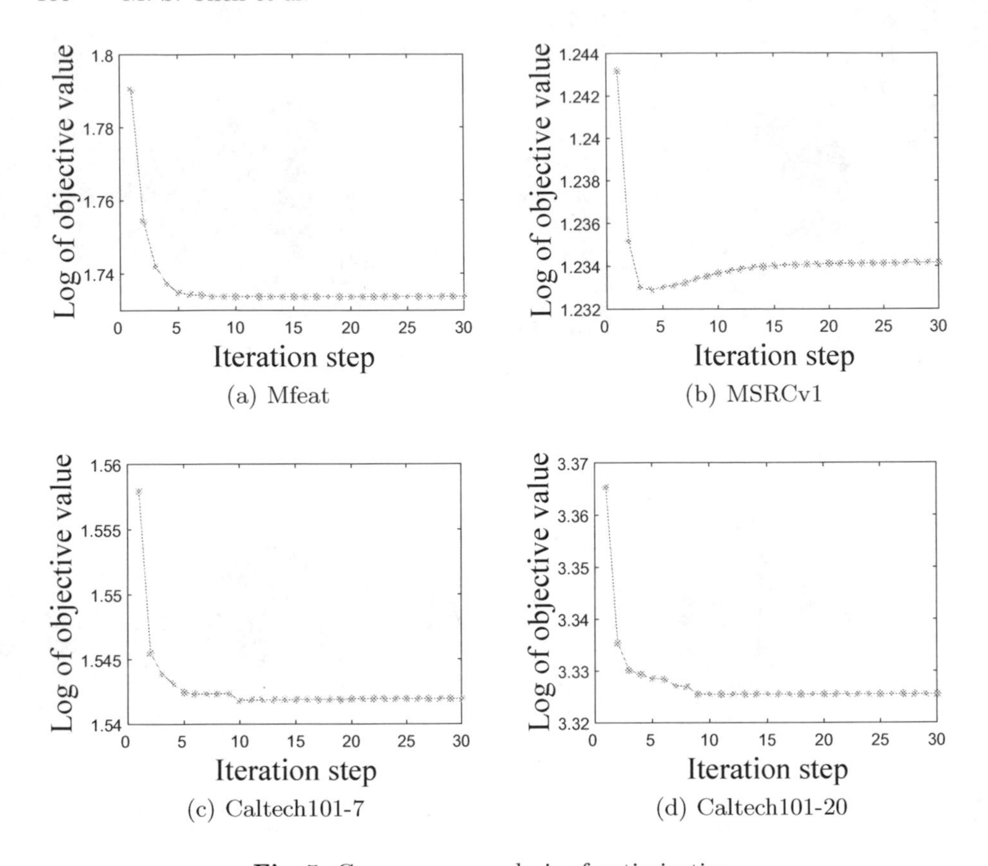

(a) Mfeat (b) MSRCv1

(c) Caltech101-7 (d) Caltech101-20

Fig. 5. Convergence analysis of optimization.

can be constructed for each view. Only a parameter k that means the number of neighbors is needed to be set in the construction method. For the proposed method, the k is fixed as 10. With the advantage of this graph construction method, the neat normalized similarity matrix of each view is achieved. For all the compared methods, the corresponding parameters are tuned to achieve better performance suggested by the authors. The number of clusters c is set to be equal to the number of the ground truth cluster labels. At the same time, all the methods are conducted for 20 times to avoid the randomness, and the average performance and their standard deviation (std) are computed. The best experimental results will be remarked in bold face.

Tables 2 and 3 show the ACC and NMI results of all algorithms on the four real-world datasets. From the two tables, the proposed algorithm can be seen to obtain the best results among all the state-of-the-art methods in comparison. Thus, our proposed method MvWCMD which jointly learn the multi-view weighted consensus similarity matrix and the cluster label matrix in a unified framework is preferred.

Table 2. Clustering results in terms of ACC on all datasets.

Method	Mfeat	MSRCv1	Caltech101-7	Caltech101-20
CLR	0.7385(0.0000)	0.5524(0.0000)	0.6859(0.0000)	0.4933(0.0000)
CoTrn	0.8002(0.0128)	0.6898(0.0139)	0.4184(0.0082)	0.4283(0.0055)
CoReg	0.7387(0.0095)	0.6241(0.0078)	0.4101(0.0033)	0.3796(0.0033)
SwMC	0.8300(0.0000)	0.5619(0.0000)	0.6635(0.0000)	0.4434(0.0000)
MLAN	0.7640(0.0000)	0.7098(0.0066)	0.6255(0.0000)	0.5270(0.0223)
MVSC	0.8224(0.0511)	0.7205(0.0452)	0.6197(0.0159)	0.4316(0.0329)
RMSC	0.5789(0.0126)	0.3246(0.0098)	0.5369(0.0047)	0.4909(0.0066)
MvWCMD	**0.8479(0.0092)**	**0.7243(0.0332)**	**0.7169(0.0633)**	**0.5585(0.0485)**

Table 3. Clustering results in terms of NMI on all datasets.

Method	Mfeat	MSRCv1	Caltech101-7	Caltech101-20
CLR	0.7609(0.0000)	0.4857(0.0000)	0.5112(0.0000)	0.3795(0.0000)
CoTrn	0.7494(0.0051)	0.6142(0.0090)	0.4145(0.0022)	0.5380(0.0015)
CoReg	0.6949(0.0044)	0.5088(0.0058)	0.4026(0.0027)	0.4884(0.0023)
SwMC	0.8542(0.0000)	0.5639(0.0000)	0.5251(0.0000)	0.4123(0.0000)
MLAN	0.8110(0.0005)	0.6007(0.0134)	0.5482(0.0002)	0.5478(0.0295)
MVSC	0.8393(0.0270)	0.6162(0.0198)	0.5256(0.0206)	0.5505(0.0111)
RMSC	0.5759(0.0142)	0.3103(0.0077)	0.5333(0.0055)	0.5021(0.0064)
MvWCMD	**0.8792(0.0124)**	**0.6868(0.0310)**	**0.5521(0.0178)**	**0.5569(0.0439)**

4 Conclusion

In this work, to eliminate the potential instability from the conventional k-means discretization, we have proposed a novel Multi-view Spectral Clustering via Multi-view Weighted Consensus and Matrix-decomposition based Discretization (MvWCMD) method aiming to jointly learn the multi-view weighted consensus similarity matrix, the continuous cluster label matrix and the final discrete cluster label matrix without k-means discretization. With the help of this framework, variables are updated iteratively in an interplay manner until convergence, so that an overall optimal solution can be achieved. Extensive experiments have been conducted on several real-world datasets to show the superiority of our proposed method.

Acknowledgments. This work was supported by NSFC (61876193, 61602189), Guangdong Natural Science Funds for Distinguished Young Scholar (2016A030306014), and Tip-top Scientific and Technical Innovative Youth Talents of Guangdong special support program (2016TQ03X542).

References

1. Bertino, E.: Introduction to data science and engineering. Data Sci. Eng. **1**(1), 1–3 (2016)
2. Cesa-Bianchi, N., Hardoon, D.R., Leen, G.: Guest editorial: learning from multiple sources. Mach. Learn. **79**(1–2), 1–3 (2010)
3. Chen, N., Zhu, J., Sun, F., Xing, E.P.: Large-margin predictive latent subspace learning for multiview data analysis. IEEE Trans. Pattern Anal. Mach. Intell. **34**(12), 2365–2378 (2012)
4. Gao, Y., Gu, S., Li, J., Liao, Z.: The multi-view information bottleneck clustering. In: Kotagiri, R., Krishna, P.R., Mohania, M., Nantajeewarawat, E. (eds.) DASFAA 2007. LNCS, vol. 4443, pp. 912–917. Springer, Heidelberg (2007). https://doi.org/10.1007/978-3-540-71703-4_78
5. Huang, L., Wang, C.D., Chao, H.Y.: A harmonic motif modularity approach for multi-layer network community detection. In: International Conference on Data Mining (ICDM 2018), pp. 1043–1048 (2018)
6. Zhang, H., Wang, C.D., Lai, J.H., Yu, P.S.: Community detection using multilayer edge mixture model. Knowl. Inf. Syst. (2018). (In press)
7. Li, J.H., Wang, C.D., Li, P.Z., Lai, J.H.: Discriminative metric learning for multi-view graph partitioning. Pattern Recognit. **75**, 199–213 (2018)
8. Sun, Z.R., et al.: Multi-view intact space learning for tinnitus classification in resting state EEG. Neural Process. Lett. **49**, 1–14 (2018)
9. Huang, L., Wang, C.D., Chao, H.Y.: Overlapping community detection in multi-view brain network. In: International Conference on Bioinformatics and Biomedicine (BIBM 2018), pp. 655–658 (2018)
10. Hu, Q.Y., Zhao, Z.L., Wang, C.D., Lai, J.H.: An item orientated recommendation algorithm from the multi-view perspective. Neurocomputing **269**, 261–272 (2017)
11. Hu, Q.Y., Huang, L., Wang, C.D., Chao, H.Y.: Item orientated recommendation by multi-view intact space learning with overlapping. Knowl. Based Syst. **164**, 358–370 (2018)
12. Huang, L., Wang, C.D., Chao, H.Y.: Higher-order multi-layer community detection. In: 33rd AAAI Conference on Artificial Intelligence (AAAI 2019) (2019)
13. Xu, C., Tao, D., Xu, C.: A survey on multi-view learning. CoRR abs/1304.5634 (2013)
14. Lin, K.Y., Wang, C.D., Meng, Y.Q., Zhao, Z.L.: Multi-view unit intact space learning. In: Proceedings of the 10th International Conference on Knowledge Science, Engineering and Management, pp. 211–223 (2017)
15. Zhang, G.Y., Wang, C.D., Huang, D., Zheng, W.S.: Multi-view collaborative locally adaptive clustering with Minkowski metric. Expert Syst. Appl. **86**, 307–320 (2017)
16. Tao, H., Hou, C., Liu, X., Liu, T., Yi, D., Zhu, J.: Reliable multi-view clustering. In: Proceedings of the Thirty-Second AAAI Conference on Artificial Intelligence, (AAAI-2018), the 30th Innovative Applications of Artificial Intelligence (IAAI-2018), and the 8th AAAI Symposium on Educational Advances in Artificial Intelligence (EAAI-2018), New Orleans, Louisiana, USA, 2–7 February 2018, pp. 4123–4130 (2018)
17. Zhang, G.Y., Wang, C.D., Huang, D., Zheng, W.S., Zhou, Y.R.: TW-Co-k-means: two-level weighted collaborative k-means for multi-view clustering. Knowl. Based Syst. **150**, 127–138 (2018)
18. Huang, L., Chao, H.Y., Wang, C.D.: Multi-view intact space clustering. Pattern Recognit. **86**, 344–353 (2019)

19. Blum, A., Mitchell, T.M.: Combining labeled and unlabeled data with co-training. In: Proceedings of the Eleventh Annual Conference on Computational Learning Theory, COLT 1998, Madison, Wisconsin, USA, 24–26 July 1998, pp. 92–100 (1998)
20. Kumar, A., Daumé III, H.: A co-training approach for multi-view spectral clustering. In: Proceedings of the 28th International Conference on Machine Learning, ICML 2011, Bellevue, Washington, USA, 28 June–2 July 2011, pp. 393–400 (2011)
21. Kumar, A., Rai, P., Daumé III, H.: Co-regularized multi-view spectral clustering. In: 25th Annual Conference on Neural Information Processing Systems, Advances in Neural Information Processing Systems 24, Granada, Spain, 12–14 December 2011, pp. 1413–1421 (2011)
22. Cai, X., Nie, F., Huang, H., Kamangar, F.: Heterogeneous image feature integration via multi-modal spectral clustering. In: The 24th IEEE Conference on Computer Vision and Pattern Recognition, CVPR 2011, Colorado Springs, CO, USA, 20–25 June 2011, pp. 1977–1984 (2011)
23. Xu, Y.M., Wang, C.D., Lai, J.H.: Weighted multi-view clustering with feature selection. Pattern Recognit. **53**, 25–35 (2016)
24. Nie, F., Li, J., Li, X.: Self-weighted multiview clustering with multiple graphs. In: Proceedings of the Twenty-Sixth International Joint Conference on Artificial Intelligence, IJCAI 2017, Melbourne, Australia, 19–25 August 2017, pp. 2564–2570 (2017)
25. Li, Y., Nie, F., Huang, H., Huang, J.: Large-scale multi-view spectral clustering via bipartite graph. In: Proceedings of the Twenty-Ninth AAAI Conference on Artificial Intelligence, Austin, Texas, USA, 25–30 January 2015, pp. 2750–2756 (2015)
26. Xia, R., Pan, Y., Du, L., Yin, J.: Robust multi-view spectral clustering via low-rank and sparse decomposition. In: Proceedings of the Twenty-Eighth AAAI Conference on Artificial Intelligence, Québec City, Québec, Canada, 27–31 July 2014, pp. 2149–2155 (2014)
27. Zelnik-Manor, L., Perona, P.: Self-tuning spectral clustering. In: Advances in Neural Information Processing Systems 17, Neural Information Processing Systems, NIPS 2004, Vancouver, British Columbia, Canada, 13–18 December 2004, pp. 1601–1608 (2004)
28. Nie, F., Wang, X., Jordan, M.I., Huang, H.: The constrained Laplacian rank algorithm for graph-based clustering. In: Proceedings of the Thirtieth AAAI Conference on Artificial Intelligence, Phoenix, Arizona, USA, 12–17 February 2016, pp. 1969–1976 (2016)
29. Mohar, B., Alavi, Y., Chartrand, G., Oellermann, O.: The Laplacian spectrum of graphs. Graph Theory Comb. Appl. **2**(871–898), 12 (1991)
30. von Luxburg, U.: A tutorial on spectral clustering. Stat. Comput. **17**(4), 395–416 (2007)
31. Huang, J., Nie, F., Huang, H.: Spectral rotation versus k-means in spectral clustering. In: Proceedings of the Twenty-Seventh AAAI Conference on Artificial Intelligence, Bellevue, Washington, USA, 14–18 July 2013 (2013)
32. Yang, Y., Shen, F., Huang, Z., Shen, H.T.: A unified framework for discrete spectral clustering. In: Proceedings of the Twenty-Fifth International Joint Conference on Artificial Intelligence, IJCAI 2016, New York, NY, USA, 9–15 July 2016, pp. 2273–2279 (2016)
33. Wang, X., Liu, Y., Nie, F., Huang, H.: Discriminative unsupervised dimensionality reduction. In: Proceedings of the Twenty-Fourth International Joint Conference on Artificial Intelligence, IJCAI 2015, Buenos Aires, Argentina, 25–31 July 2015, pp. 3925–3931 (2015)

34. Fan, K.: On a theorem of Weyl concerning eigenvalues of linear transformations I. Proc. Nat. Acad. Sci. **35**(11), 652–655 (1949)
35. Yu, S.X., Shi, J.: Multiclass spectral clustering. In: 9th IEEE International Conference on Computer Vision (ICCV 2003), Nice, France, 14–17 October 2003, pp. 313–319 (2003)
36. Kang, Z., Peng, C., Cheng, Q., Xu, Z.: Unified spectral clustering with optimal graph. In: Proceedings of the Thirty-Second AAAI Conference on Artificial Intelligence, (AAAI-2018), the 30th Innovative Applications of Artificial Intelligence (IAAI-2018), and the 8th AAAI Symposium on Educational Advances in Artificial Intelligence (EAAI-2018), New Orleans, Louisiana, USA, 2–7 February 2018, pp. 3366–3373 (2018)
37. Duchi, J.C., Shalev-Shwartz, S., Singer, Y., Chandra, T.: Efficient projections onto the l_1-ball for learning in high dimensions. In: Proceedings of the Twenty-Fifth International Conference on Machine Learning (ICML 2008), Helsinki, Finland, 5–9 June 2008, pp. 272–279 (2008)
38. Wen, Z., Yin, W.: A feasible method for optimization with orthogonality constraints. Math. Program. **142**(1), 397–434 (2013)
39. Schönemann, P.H.: A generalized solution of the orthogonal procrustes problem. Psychometrika **31**(1), 1–10 (1966)
40. Lin, K.-Y., Huang, L., Wang, C.-D., Chao, H.-Y.: Multi-view proximity learning for clustering. In: Pei, J., Manolopoulos, Y., Sadiq, S., Li, J. (eds.) DASFAA 2018. LNCS, vol. 10828, pp. 407–423. Springer, Cham (2018). https://doi.org/10.1007/978-3-319-91458-9_25
41. Wang, C.D., Lai, J.H., Yu, P.S.: Multi-view clustering based on belief propagation. IEEE Trans. Knowl. Data Eng. **28**(4), 1007–1021 (2016)
42. Winn, J.M., Jojic, N.: LOCUS: learning object classes with unsupervised segmentation. In: 10th IEEE International Conference on Computer Vision (ICCV 2005), Beijing, China, 17–20 October 2005, pp. 756–763 (2005)
43. Li, F.F., Fergus, R., Perona, P.: Learning generative visual models from few training examples: an incremental bayesian approach tested on 101 object categories. Comput. Vis. Image Underst. **106**(1), 59–70 (2007)
44. Nie, F., Cai, G., Li, X.: Multi-view clustering and semi-supervised classification with adaptive neighbours. In: Proceedings of the Thirty-First AAAI Conference on Artificial Intelligence, San Francisco, California, USA, 4–9 February 2017, pp. 2408–2414 (2017)

SIRCS: Slope-intercept-residual Compression by Correlation Sequencing for Multi-stream High Variation Data

Zixin Ye[1], Wen Hua[1](✉), Liwei Wang[2], and Xiaofang Zhou[1]

[1] School of Information Technology and Electrical Engineering, The University of Queensland, Brisbane, Australia
zixin.ye@uqconnect.edu.au, w.hua@uq.edu.au, zxf@itee.uq.edu.au
[2] International School of Software, Wuhan University, Wuhan, China
liwei.wang@whu.edu.cn

Abstract. Multi-stream data with high variation is ubiquitous in the modern network systems. With the development of telecommunication technologies, robust data compression techniques are urged to be developed. In this paper, we humbly introduce a novel technique specifically for high variation signal data: SIRCS, which applies linear regression model for slope, intercept and residual decomposition of the multi data stream and combines the advanced tree mapping techniques. SIRCS inherits the advantages from the existing grouping compression algorithms, like GAMPS. With the newly invented correlation sorting techniques: the correlation tree mapping, SIRCS can practically improve the compression ratio by 13% from the traditional clustering mapping scheme. The application of the linear model decomposition can further facilitate the improvement of the algorithm performance from the state-of-art algorithms, with the RMSE decrease 4% and the compression time dramatically drop compared to the GAMPS. With the wide range of the error tolerance from 1% to 27%, SIRCS performs consistently better than all evaluated state-of-art algorithms regarding compression efficiency and accuracy.

Keywords: High variation data · Multi-signal compression · Correlation mapping · Linear regression model · Error detection

1 Introduction

Multi-stream data is ubiquitous in the modern network systems [13]. With the development of telecommunication technologies, information is usually generated as a collective and multi-dimensional data stream from different sources. As the popularisation of the Internet of Things [17], the time-series group data compression is becoming more popular and important than ever before in both industry and academia. Meanwhile, in today's critical network systems, information with high variation is also frequently generated, such as in the stock

G. Li et al. (Eds.): DASFAA 2019, LNCS 11446, pp. 191–206, 2019.
https://doi.org/10.1007/978-3-030-18576-3_12

trade, traffic systems, massively distributed solar systems, etc. Such data usually preserves ambiguous variation pattern, big data range and high variance, and hence becomes a challenging data type to compress in the communication network. Therefore, current research needs to be widely extended to optimally encode and reconstruct the high variation data in a highly correlated multi-signal network system.

Previous work has been conducted for single-stream time-series data compression, such as APCA [2] and SF [6], to name a few. In a multi-signal environment, however, if we apply these methods directly to compress each single stream one by one without considering their correlation, it is highly possible to result in a small compression ratio.

To simultaneously handle all streaming data, multi-signal compression algorithms, such as GAMPS [7], are developed based on the data correlation information. Particularly, GAMPS first groups signals within spatial proximity into a cluster, and determines the best base signal in the cluster by iteratively checking the compression performance of using each stream as the base signal. For each data other than the base signal, it then constructs a ratio signal based on its difference with the base signal, called "cluster mapping". Finally, it applies APCA to compress both base signal and ratio signals. However, such methods still have some drawbacks especially when dealing with high variation data: (1) The brute-force search for the base signal is extremely time-consuming; (2) The correlation information is never fully utilised when we transform each signal only according to the base signal in the cluster mapping; (3) Ratio signal cannot comprehensively capture complex patterns in high variation data, leading to relatively large reconstruction error.

To address the above issues, we propose a novel algorithm, SIRCS (Slope-Intercept-Residual compression by Correlation Sequencing), for multi-stream compression with high variation data. We introduce decomposition-based compression and tree mapping techniques in this work, and SIRCS is a condign combination of these techniques, which demonstrates an overall improvement over current state-of-the-art compression methods in both efficiency and precision. Our major contributions can be summarised as follows:

1. We study the problem of multi-signal compression which has important applications in modern network systems. The problem is challenging due to various correlation levels, and variation patterns existed in the streaming signals.
2. We introduce the correlation tree mapping technique for data grouping to fully utilise the correlation information between signals efficiently. The mapping can efficiently configure a tree index with a selected base signal, and meanwhile, maximise the preservation of the highest correlation information in the index. We theoretically prove the improvability of the tree mapping technique over traditional cluster mapping.
3. We propose a regression-based decomposition technique for data-variation reduction, which results in smaller fluctuation in the residual signals and hence better compression performance.

4. We propose a new idea of residual compression with the guarantee of the worst-case maximum L_∞ error derived from the base signal error bound. This assures all signals to be perfectly reconstructed with a maximum error guarantee.
5. We empirically compare SIRCS with several state-of-the-art compression algorithms on a real-world dataset, and the experimental result demonstrates better performance achieved by SIRCS regarding compression ratio, reconstruction precision, and compression speed.

For the rest of the paper, we review the work of data compression in Sect. 2, then formulate the problem of multi-stream high variation data compression in Sect. 3. In Sect. 4, the SIRCS algorithm is introduced to solve the problem in Sect. 3 by integrating the tree mapping, regression-based decomposition, and residual compression. We report our empirical results in Sect. 5, followed by a brief conclusion in Sect. 6.

2 Related Work

Numerous state-of-art algorithms exist in the computing systems, usually classified into lossless and lossy compression schemes. Prevalent application of the lossless algorithms, such as Adaptive and Non-adaptive Huffman Coding [19], LZ77 [22], LZ78 [23], LZW [16], BWT [14] and PPM [4], remain robust and functional even in most of the modern operating systems. BWT-based compression reaches the optimised performance at $O(\frac{log(n)}{n})$, improving from $O(\frac{log(log(n))}{log(n)})$ from LZ77 [21] and $O(\frac{1}{log(n)})$ from LZ78 [12]. However, the compression ratio cannot be dramatically increased from a lossless algorithm, therefore, lossy compression is introduced for a better trade-off of the compression efficiency.

In lossy compression for single data, the piecewise approximation algorithms, in particular, are the most fundamental and can be furthermore classified into: piecewise constant approximation (eg., PCA [11], APCA [2], PAA [8], etc.), linear approximation (eg., SF [6], PWLH [1], PLA [3], etc.), and polynomial approximation (eg., CHEB [20], etc.). Another lossy compression type is the decomposition based algorithms, such as DWT [15], DCT [9], DFT [10], etc. Those compression algorithms usually preserves high compression ratio but longer compression time. However, in modern network systems, the correlation between multiple signals should also be considered to improve compression performance further.

Group data compression algorithms are introduced in the lossy compression domain. GAMPS is the first application using the data correlation. In GAMPS, ratio signals are introduced by dividing one signal value with the selected base signal value. Due to the signals similarity, the ratio signal from two highly correlated signals is much flatter than its original data, thus largely reducing the variation level. Compressing the low variation data by APCA, in turn, increases the total compression ratio. To select the proper base signal, GAMPS computes the compression ratio in every scenario with different signals as the base signal. Thorough iteration occurs to estimate the consumptions by summing the size of all compressed signals. The algorithm then picks the base signal leading to the smallest compressed file size. Consequently, GAMPS can lead to an excellent compression ratio but relatively large precision error and long compression time. Our work, on the contrary, aims to optimise all the three performance criteria in multi-stream compression.

3 Problem Definition

Definition 1. *(High Variation Time Series Data) The time-series data with high variation D is defined as a stream of data points (t_i, v_i) with a consecutive time index t_i (i.e., $D = [(t_1, v_1), (t_2, v_2), \ldots, (t_n, v_n)]$), where the standard deviation σD and the range $D_{max} - D_{min}$ are much higher than regular time-series data. The time index follows monotonicity: $\forall i < j, t_i < t_j$.*

We use $S = \{D_1, D_2, \ldots, D_n\}$ to denote a multi-signal time series dataset, i.e., a set of time-series data D which share the same time index with the length n (i.e., $D_i = [(t_1, v_1^i), (t_2, v_2^i), \ldots, (t_n, v_n^i)]$). The problem studied in this paper can be formulated as follows.

Definition 2. *(Group Compression of High Correlation Data with Max-error Precision) A dataset S formed by the high variation time-series data D_i, where $i \in [1, n]$, and an error bound ϵ are given. The problem is to compress all D_i in the dataset S so that the reconstructed signal D_i' suffice the equation: $\forall D_i \in S, (D_i'(t)) - D_i(t)) \leq \epsilon$.*

Intuitively, the hypothesis can be made that the higher the correlation, the higher the compression ratio will be obtained. We conduct an empirical evaluation to test the relationship between the correlation of a paired signal and their compression ratio, along with the precision. Assume two randomly picked signals from the signal network are D_i, D_j, and their correlation is $R_{i,j}^2$. In this two signals compression, we link the R^2 values to the CR and $NRMSE$ from the compression between D_i, D_j. The result of the evaluation shows the statistical significance in the positively associated relationship between the correlation and the compression ratio. The detailed information of the empirical evaluation will be reported in Sect. 5. This result validates the hypothesis that a high correlation between two signals can improve the compression performance. Therefore, we focus our study of multi-stream data compression in a highly correlated network system as defined below.

Definition 3. *(Correlated Multi-signal Network) The correlated multi-signal network is a system where any two randomly selected signals, $D_i(t)$ and $D_j(t)$, are correlated, thus similar in variation pattern, with a mathematical relation as a function of F, denoted as $D_i(t) = F(D_j(t)) + \delta_j(t)$.*

4 SIRCS: Slope-intercept-residual Compression by Correlation Sequencing

4.1 Overview

The algorithm consists of three main components: correlation tree sequencing, regression-based decomposition and the residual data compression. First, given the time-series dataset $S = \{D_1, D_2, \ldots, D_n\}$, the correlation tree sequencing is to create a compressing index I_{tree} and select the base signal D_{base}. Second, following I_{tree} and D_{base}, the regression-based decomposition dissemble D_i into its residual R_i and the regression coefficients. Finally, the residual data is compressed with a newly estimated error bound. This residual error bound assures that the recovered residuals and the regression coefficients can reconstruct the raw signal under the original maximum error guarantee. We will elaborate on the technical details of these three components in the following sections, respectively.

4.2 Correlation Sequencing Mechanism

According to our hypothesis, data correlation can effectively minimise the memory consumption of multi-stream data. In this section, we will introduce our method of correlation tree mapping and meanwhile theoretically prove the improvability of the tree mapping over the cluster mapping.

Technique 1. (Correlation Tree Mapping) The cluster mapping is based on a unique base signal, so its information index, $I_{cluster}$, processes only one pass to each of the child signal ($D_{base} \rightarrow D_i$, where $0 \leq i \leq n-1$). Replacing the cluster to tree mapping, whose information index, I_{tree}, processes multiple passes from one child signal to another child signal ($D_{base} \rightarrow D_i \rightarrow \cdots \rightarrow D_j$, where $0 \leq i, j \leq n-1$), we always have the compression ratio compared as

$$(CR)_{tree}(\sum_{i=0}^{n} D_i(t)) \geq (CR)_{cluster}(\sum_{i=0}^{n} D_i(t)). \tag{1}$$

Tree Components Formation

Definition 4. *Correlation pairs are the signal link between two signals; correlation branches are the signal link with multiple signals sharing one head node and; correlation twigs are the branch components which are different in length but share the same head node with their branch.*

Tree components formation aims to extract the high correlation pairs from signals m and n and arrange them in an ordered sequence. Considering the non-repetitive collection: if $r_{(m,n)}$ is chosen, the system will check if either m and n is already collected, and if not, the system will register $r_{(m,n)}$. Such correlation collection will not end until all signals are contained. During the signal collection, there will be three scenarios:

1. $r_{(m,n)}$ where m is in the list, and n is in the list. In this case, there will be no collecting operation occurred.
2. $r_{(m,n)}$ where m is in the list, but n is not in the list. In this case, only the signal n is collected. It further implies the node m is an intermediate connection between a collected signal and n.
3. $r_{(m,n)}$ where both m and n are not in the list. In this case, both signal ID m and n will be collected. It implies that the connecting pair m and n are isolated from the other signal nodes.

These three possibilities will impact on the branch creation in the later procedure: if there's a node acting as an intermediate connection with two other nodes, a branch will be created. Then the high correlation pairs obtained previously will be connected to several branches with longer connections in each segment. The connection starts with connecting one pair's head with the other pair's tail if the head and tail have the same signal index. To achieve the repetitive seeking for the same heads and tails, the recursion algorithm is implemented to keep connecting the previously and newly generated segments until no same heads and tails occur in the segments. As a special case of the branch, several twigs may be included in one branch. In this case, they will be encapsulated in one branch.

Example 1. In Fig. 1, $r_{11,12} \rightarrow r_{11,0} \rightarrow r_{1,2} \rightarrow \ldots \rightarrow r_{7,6}$ is sorted and there are 14 elements in total. All the 18 signals are just recovered from those 14 paired segments, where the signal 6 is the last selected element. The rest of the correlation pairs after $r_{7,6}$ will be ignored. In the left figure of Fig. 2, we find the repetition of the signals in the parental and child node position, such as the pairs $3 \leftrightarrow 15$ against $15 \leftrightarrow 17$ and $15 \leftrightarrow 7$. The connection will be ended with the segment $3 \leftrightarrow 15 \leftrightarrow 17$ and $3 \leftrightarrow 15 \leftrightarrow 7 \leftrightarrow 6$. After the connections, the branch with only one twig is $1 : [[2, 0]]$, and the branch with multiple twigs is $11 : [[0, 8], [12]]$, $3 : [[15, 7, 6], [15, 17], [9]]$, and $5 : [[13], [4, 16], [14]]$.

Base Signal Selection. In this step, we aim to find a common based signal for all branches by seeking the highest correlation pair between one branch's head node and any elements in the other branches. To assure the result of iteration is the highest correlation among all possibilities, the connecting candidates will not be defined until all the head nodes of the branches go through every element of the other branches and estimate their correlation level. The highest pair will be given the priority to connect and for each loop. As the plantation of branches is accomplished, there will be only one head node in the tree, which will be nominated as the base signal $D_{base}(t)$.

Example 2. Right figure in Fig. 2(a) demonstrates four branches with the head-node 3, 5, 1, and 11. The first highest correlation searching ends up with connecting the signal 12 with the head node 3 at $R^2 = 0.88$. The second searching follows up with the connection between signals 5 and 7 at $R^2 = 0.81$. The last searching ends up with connecting signals 1 and 10 at $R^2 = 0.78$. Finally, the tree index is constructed and $D_{base}(t)$ is $D_{11}(t)$, shown in Fig. 2(b).

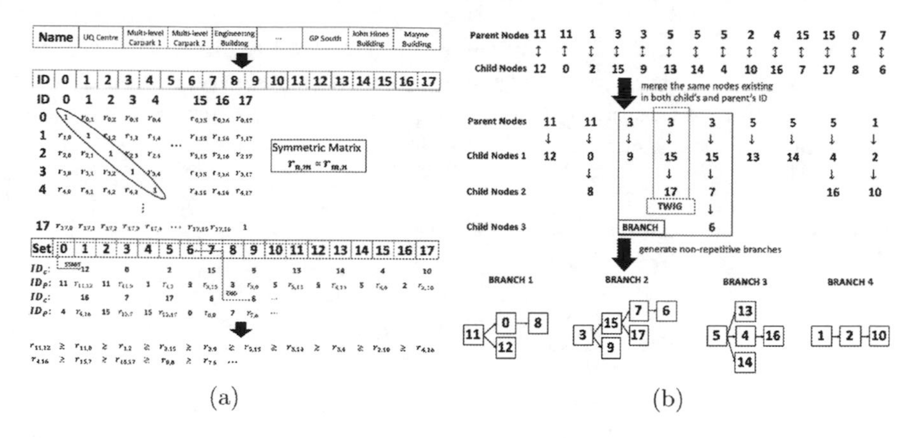

Fig. 1. (a) Shows the example of the correlation sequencing: the system will arrange those collected signal pair into a structure similar to: $D_{r^2} = r_{11,12} : [11, 12], r_{11,0} : [11, 0], \ldots, r_{7,6} : [7, 6]$. (b) demonstrates the example of connecting same ID of different pairs to branches.

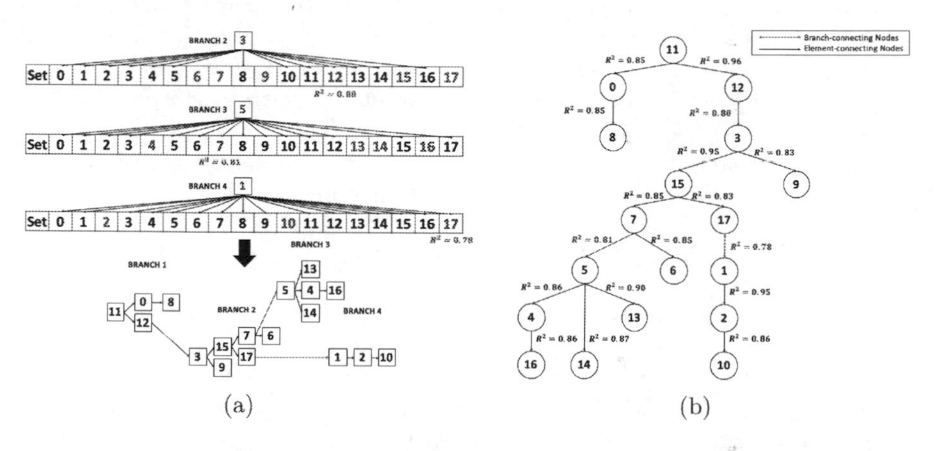

Fig. 2. (a) Demonstrates searching for the highest correlated pair: the searching process iterates three times in total, indicating the optimal connecting index among the branches, after which the correlation tree is eventually created. The whole steps guarantee that for each connection, the chosen correlation level remains the highest from the rest. (b) shows the result of the correlation tree mapping, the index of the signals is encoded in digital numbers as the header of the compressed file.

Proof of Improvability. The following theorem highlights the superiority of our proposed tree mapping over traditional cluster mapping.

Theorem 1. *If the information of the total correlation level from an index is given by I, the sum of correlation level from the cluster mapping is always less equal than the sum of correlation level from the tree mapping, denoted as* $I_{cluster} \leq I_{tree}$.

Proof. Assume the branch number of the index is m, and within a branch, if the connecting nodes number is greater than 2, assume the node connection number as n. The formulas of the total correlation level from both cluster index mapping and tree index mapping can be written as

$$I_{cluster} = \sum_{i=1}^{m} Cor(D_{base,0}(t), D_{i,1}(t)), \tag{2}$$

$$I_{tree} = \sum_{i=1}^{m} \sum_{j=1}^{n} Cor(D_{i,j}(t), D_{i,j+1}(t)). \tag{3}$$

In the cluster mapping formula, it is known there are only two signals in one branch: the base signal as $D_{base,0}(t)$ and child signal as $D_{i,1}(t)$, where the base signal is fixed once the index is created. Suppose the first component in the correlation calculation is a set of the possible parental signals, denoted as P, the set of the parental signals in the cluster mapping will then be $P_{cluster} = D_{base,0}(t)$. It can be observed that the total number of elements in the cluster mapping is unique, while in the tree mapping, multiple parental signals including that in the cluster mapping case can concurrently exist, denoted as $D_{base,0}(t) \in P_{tree}$. Therefore, the relation between the parental set from cluster and tree mapping will be $P_{cluster} \subset P_{tree}$. Since the parental-signal selection in the tree mapping has greater flexibility, a wider range of the correlation selection exists in the tree mapping than the cluster mapping, denoted as

$$Set(Cor(D_{base,0}(t), D_{i,1}(t))) \subset Set(Cor(D_{i,j}(t), D_{i,j+1}(t))). \tag{4}$$

More correlation selection in the tree mapping further implies the tree index can cover higher correlation information. After all, cluster mapping only manifests the correlation between the base signal and its child signals, while the in tree mapping, both correlation between two child signals are also free to choose. With a wider range of selection, total correlation from tree mapping is no less than that from cluster mapping, denoted as $I_{tree} \geq I_{cluster}$.

4.3 Regression-Based Decomposition Mechanism

The essential reason for using regression-based decomposition is to reduce the data variation from raw to residual signal compression. With one base signal D_{base} selected from the dataset S, other signals can be decomposed via the base signal and the correlation coefficients into another signal with a much smaller size $\hat{D}$. In this paper, $\hat{D}$ is the residual data $\hat{D}_i = R_i$, whose validity will be affirmed by proving $\sigma(R_i) < \sigma(D)$ in this section. Reversely, based on D_{base}, correlation coefficients and $\hat{D}_i$, the signals can be reconstructed with a given normalised error tolerance of ϵ. The problem can be formulated as follow.

Technique 2. (Reduction of Data Variation) Based on the preceding assumptions, if a child signal is given by $D_i(t)$ and its residual signal is denoted as $R_i(t)$, for the variation level represented by standard deviation of σ, they are always satisfying the following relation: $\sigma(R_i(t)) < \sigma(D_i(t))$.

We recall the definition of the correlated signal network that $D_i(t) = F(D_j(t)) + \delta_j(t)$, while we also assume the time lag between two randomly selected signals in the network cannot be too large compared to the signal period: $\Delta_i \ll T$. Then we configure the linear model (LM) as $\hat{y} = \beta_0\hat{x} + \beta_1$. To minimise the mean square error of the regression line from the real data: $E_i^k = \sum_{i=0}^{n}(y - \hat{y})^2$, the coefficients are adjusted to the least square estimates [18] as

$$\beta_0 = \frac{\sum_{i=0}^{N}(y_i - \hat{y_i})(x_i - \hat{x_i})}{\sum_{i=0}^{N}(x_i - \hat{x_i})^2}, \tag{5}$$

$$\beta_1 = \frac{\sum_{i=0}^{N}(y_i - \bar{y})(x_i - \bar{x})}{\sum_{i=0}^{N}(x_i - \bar{x})^2}. \tag{6}$$

Such regression model can extract the coefficient of slope, intercept, and the residual data with zero mean and lower variation level. This theorem of data variation reduction can be proved bellow and visually shown in Fig. 3.

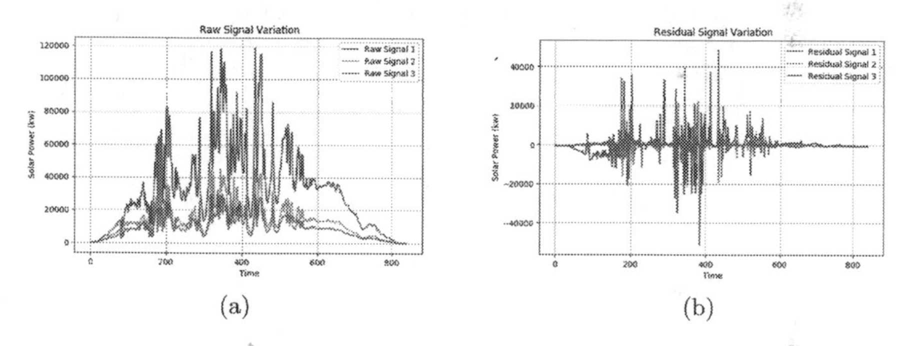

(a) (b)

Fig. 3. (a) Shows an example of three raw signals from the solar panel in St Lucia Campus. (b) Indicates the residual signals from the left figure have been visually reduced in data variation.

Theorem 2. *Assume the original data has high range and standard deviation, while their variation patterns are also similar. Given the child signal $S_i(t)$ and its parental signal $S_j(t) = \beta_0 S_i(t + \Delta) + \beta_1 + \delta_i(t)$ where $S_i(t), S_j(t) \geq 0$, we define $S_\Delta(t) = S_i(t) - S_j(t)$. If σ represent the standard deviation, we can always have $\sigma(S_\Delta(t)) \leq \sigma(S_i(t))$.*

Proof. If the coefficient of variation is defined as $\sigma(S(t)) = \sqrt{\frac{\sum_{i=0}^{n}(S(t)-\bar{S})^2}{n-1}}$. Taking $S_\Delta(t)$, we have

$$\sigma(S_\Delta(t)) = \sqrt{\frac{\sum_{i=0}^{n}(S_\Delta(t) - \bar{S}_\Delta)^2}{n-1}}. \tag{7}$$

Since $S_\Delta(t) = S_i(t) - S_j(t)$, we can deduce

$$\sigma(S_\Delta(t)) = \sqrt{\frac{\sum_{i=0}^{n}((S_i(t) - (\beta_0 S_i(t+\Delta) + \beta_1 + \delta_i(t))) - \bar{S}_\Delta)^2}{n-1}}. \quad (8)$$

As $S_i(t)$ is periodic, and apparently $\beta_0 S_i(t+\Delta)+\beta_1$ is also periodic, according to the linearity of Fourier Transform, $S_T(t) = S_i(t) - (\beta_0 S_i(t+\Delta)+\beta_1)$ is a periodic signal. Since $\bar{S}_\Delta = E(S_T(t)) + E(\delta_i(t))$ and from the linear regression model, we know $E(\delta_i(t)) = 0$, then $\bar{S}_\Delta = E(S_T(t))$. The formula can be rewritten as

$$\sigma(S_\Delta(t)) = \sqrt{\frac{\sum_{i=0}^{n}(S_T(t) - \delta_i(t) - \bar{S}_T)^2}{n-1}}. \quad (9)$$

Since the assumption of the distributed signals network is geographically closed, while the time latency Δ should be small enough for the similarity detection, which is formulated as $S_i(t) \approx S_i(t+\Delta)$.

We assume that the order of magnitude in delta signal and the error signal is much smaller than that of the original signal. This is normal to expect since $Power(S_i(t)) \approx \beta_0 S_i(t+\Delta) + \beta_1$ and, without bad leverages and outliers, $Power(\delta_i(t)) \ll S_i(t)$. Therefore, we have $S_T(t), \delta_i(t) \ll S_i(t))$.

Now that we want to compare the between $\sigma(S_i(t))$ and $sigma(S_\Delta(t))$. In the course, we can rely on the aforementioned assumptions to approximate $\sum_{i=0}^{n}(S_T(t) - \bar{S}_T)^2 \approx 0$, compared to the much larger value of $S_i(t)$. Therefore we have

$$\sigma(S_\Delta(t)) \approx \sqrt{\frac{\sum_{i=0}^{n}(\delta_i(t))^2}{n-1}}. \quad (10)$$

As one of the assumptions, $Var(S_i(t)) = \sigma^2(S_i(t)) \gg \sigma^2(S_\Delta(t))$, we can deduce

$$\sigma(S_i(t)) \geq \sqrt{\frac{\sum_{i=0}^{n}(\delta_i(t))^2}{n-1}} = \sigma(S_\Delta(t)). \quad (11)$$

4.4 Residual Data Compression

This section is proposed to compressed the $R_i(t)$ decomposed from the signals $D_i(t)$ based on the linear regression model with $D_{base}(t)$. The problem of the residual compression is shown as follow.

Technique 3. (Error Bound of Residual Compression) If the error precision of the raw signal is given by ϵ_{raw} and the corresponding error precision of the residual signal is given by ϵ_{res}, the algorithm needs to assure for any signal reconstruction from its residual data with ϵ_{res}, the precision of the reconstructed signal should fall in the range of ϵ_{raw}.

The problem is solved by the theorem of the residual error bound, in which the signals $D(t)$ are divided into direct parent signal $P(t)$ and its child signals $C(t)$. It implies that the error precision of the residual signal is equal to the error precision of the child signal, which we assumed to be the maximum error guarantee as ϵ_{raw}.

Theorem 3. *Suppose the error precision of the raw signal is given by* ϵ_{raw} *and the reconstructed parental and child signal is denoted as* $P_{rec}(t)$ *and* $C_{rec}(t)$*. If the linear regression model gives*

$$C_{raw}(t) = \beta_0 P_{rec}(t) + \beta_1, \tag{12}$$

the maximum error tolerance of the residual signal will be equal to that of its raw signal, denoted as $\epsilon_{res} = \epsilon_{raw}$*.*

Proof. The relationship between the raw signal data and the residual data can be denoted as

$$C_{raw}(t) = \beta_0 P_{raw}(t) + \beta_1 + R_{raw}(t). \tag{13}$$

Symbol C means the child signal, P means the parental signal, and R means the residual signal. We also recall the relation between the raw and reconstructed signal as $Rec(t) = Raw(t) + \delta(t)$. We can deduce that the coefficients β_0 and β_1 are from the raw child signal and raw parental signal. Let us redesign the linear model between the raw child signal and the recovered parental signal. The equation is reformatted:

$$C_{raw}(t) = \beta_0 P_{rec}(t) + \beta_1 + R_{raw}(t). \tag{14}$$

Let us assume that $P_{rec}(t)$ has high similarity with $P_{raw}(t)$. In decompression side, what are known are the values of β_0 and β_1, two reconstructed signals $P_{rec}(t)$ and $R_{raw}(t)$. The reconstructed child signal will be

$$C_{rec}(t) = \beta_0 P_{rec}(t) + \beta_1 + R_{rec}(t). \tag{15}$$

Taking the residual signal to the linear model, we have

$$C_{rec}(t) - C_{raw}(t) = \beta_0 P_{rec}(t) + \beta_1 + R_{rec}(t) - \beta_0 P_{rec}(t) - \beta_1 - R_{raw}(t). \tag{16}$$

Eventually, the formula can be rewritten as $\delta_R(t) = \delta_C(t) \leq \epsilon_{raw}$.

The theorem finalises the estimation of the residual data error bound, therefore, the final design of the SIRCS algorithm can be integrated in Algorithm 1. Here we assume the group dataset as S, single data stream as D, lists for signal collection as L, and encapsulate the tree index creation in the starting procedure of pseudo code.

5 Experiment and Results

5.1 Experiment Setup

In the experiment, we use the real world dataset of the solar network system of the University of Queensland. 26 historical solar data are used from three different campuses: St Lucia Campus (18 signals), Gatton Campus (6 signals), and Herston Campus (2 signals). The time range of the data is 20 days from 10th to 29th in November in 2017, with the data sampling period of 60 s.

Algorithm 1. SIRCS(S, ϵ)

```
1: procedure TREE(S)                                        ▷ Tree Configuration from S
2:     I_pairs ← sort(S)                                    ▷ correlation sequencing
3:     I_branches ← sort(I_pairs)                           ▷ R^2 to branches
4:     I_tree ← sort(I_branches)                            ▷ Plantation of branches
5: end procedure
6: I_tree ← Cor(S)
7: for b ← branch to last branch in tree do
8:     for D ← b to last element in current branch do
9:         if s ∈ compressedbucket then continue            ▷ skip shared-node signals
10:        else                                             ▷ start compression
11:            function get_lm_coefficient(last D_rec, currentsignal)
12:                β0 ← lmCoeff[0]                          ▷ function's returned list: lmCoeff
13:                β1 ← lmCoeff[1]
14:                residual ← lmCoeff[2]
15:            end function
16:            L_β0 ← append(β0)
17:            L_β1 ← append(β1)
18:            function residual_compression(residual, ε)
19:                D_com ← compression_algorithms           ▷ from single data compression
20:                D_rec ← recover_algorithms               ▷ for finding next LM coefficient
21:            end function
22:            L_com ← append(D_com)
23:            L_rec ← append(D_rec)
24:        end if
25:    end for
26: end for
```

The performance evaluation is mainly based on traditional compression benchmarks, including compression ratio, normalised root-mean-square error, and computational time. They are formulated as follow:

$$CR = \frac{Size(F_{raw}(t))}{Size(F_{compressed}(t))} \tag{17}$$

$$NRMSE = \frac{1}{norm}\sqrt{\frac{\sum_{i=0}^{N}(\hat{y} - y)^2}{N}} \tag{18}$$

Additional evaluation, nominated as the precision test, is introduced in RIDA [5]. The test demonstrates the compression precision in a given compression ratio, regardless of the error tolerance selection.

State-of-art algorithms are realised under Python Environment (3.6.4) in the operating system with a 2.2 GHz Intel Core i7 processor and a 16 GB 1600 MHz DDR3 memory. Particularly, APCA, SF and GAMPS are selected for the performance comparison against the SIRCS. Their algorithm realisation is slightly customised in favour of the maximum performance: we adjust the floating precision to 5 digits and the coefficient c as 0.4 in GAMPS.

5.2 Effect of Correlation Level

The linear model test shows that a positive association exists between compression ratio and the signal pairs correlation, with the p-value approaching zero. From LM test in Fig. 4(a), p-value approaches to 0. For every unit increase of the correlation, the compression ratio rises 1.42856. The linear model test also shows that a positive association exists between NRMSE and the signal pairs correlation, with the p-value approaching zero. From LM test in Fig. 4(b), p-value also approaches to 0. The outcome implies the higher correlation grouping between two data streams will statistically lead to a higher compression ratio, therefore we validate the statement that picking high correlation signal pairs can improve the total compression performance.

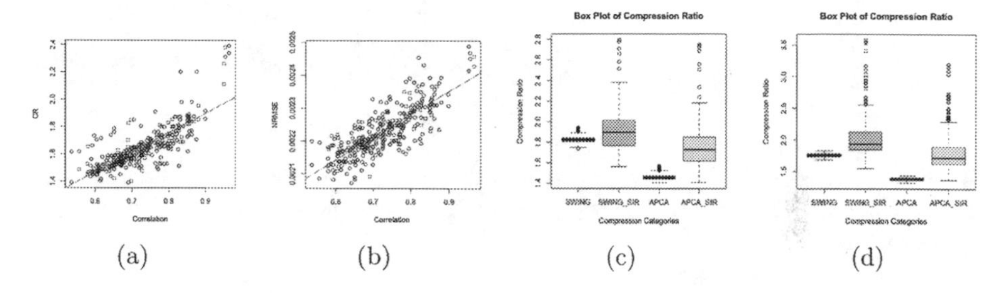

Fig. 4. (a) Implies that higher the correlation level, smaller the file will be compressed and (b) Implies that higher the correlation level, greater the compression error will be generated. (c) Shows the box plot of the two-sample t-test of the one-day dataset, and (d) shows that of twenty-day dataset, both of which manifests the improvement of the residual data compression.

5.3 Effect of Regression-Based Decomposition

We conduct two-sample t-tests between using and not using residual data compression for both the twenty days dataset and a one-day dataset on 21st of January 2018. In Fig. 4(c), practically significant increase can be observed in SIR algorithm with the corresponding state-of-art algorithms: SIR application on Swing Filter has average 0.27 increase in compression ratio, while on APCA also has average 0.42 increase. To consolidate the persuasiveness of the result, Fig. 4(d) shows the outcome of the compression ratio comparison over a one-day dataset and the improvement is similar to the twenty-day dataset scenario. Two-sample t-tests imply a strong evidence that using SIR algorithm can significantly improve the compression ratio based on the corresponding state-of-art algorithms.

5.4 Effect of Tree Mapping

First, we demonstrate the difference between the cluster mapping and the tree mapping in Fig. 5(a). The bar chart in Fig. 5 shows practical improvement, from

0.04 to 0.18, for all eighteen tested signals in both APCA and Swing Filter. The improvement in compression ratio in APCA is averagely 0.026 higher than the improvement in Swing Filter. From this outcome, the improvement of the tree mapping is practically significant. The increased level varies with the base signal selection, but the improvement applies in all circumstances. Therefore, from the empirical evaluation, the improvement from the tree mapping is practically significant over the cluster mapping.

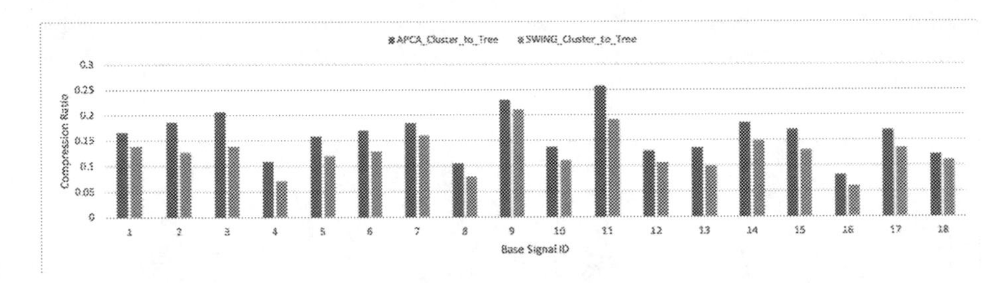

Fig. 5. Shows the compression ratio with or without the tree mapping in different base signal selection. In all situations, tree mapping improves the compression efficiency in various extent.

5.5 Effect of Error Tolerance

From the outcome of the evaluation, we estimate the percentage improvement of SIRCS based on the state-of-art algorithms. In the compression ratio performance, the SIRCS has averagely 15% of the increase from the compression ratio of APCA, shown in Fig. 6 (a). It can shoot up to 30% of increase with the error tolerance equal to 1% and also go up to 11% when the error tolerance is equal to 13%. The swing filter algorithm applying SIRCS can increase its compression ratio up to 14%, and averagely increase 5% for any error level, shown in Fig. 6(b). The compression time shows in the similar level except that of GAMPS, which shoots up to 103.67 s to compress the whole datasets, according to Fig. 6(c). The rest has similar computational time varying from 1.05 to 4.72 s. In the precision test, SIRCS also has the noticeable improvement in reducing the NRMSE-compression ratio trade-off, shown in Fig. 6(d). In the APCA scheme, the SIRCS can decrease almost 75% of NRMSE when the compression ratio is 1.11, and it can also reduce 15% more in most of the compression ratio level. The swing-filter-based algorithm can reduce its NRMSE by using SIRCS up to 50%. Even though such improvement differs from applying different error tolerance, the improvement is proved to be practically significant.

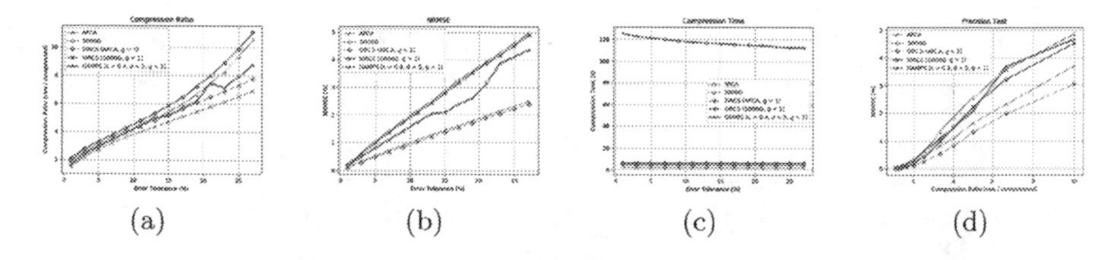

Fig. 6. Comparison of the performance between SIRCS and the other three state-of-art algorithms in terms of compression ratio in (a), NRMSE in (b), compression time in (c), and the precision level against a given compression ratio in (d).

6 Conclusion

In this paper, we have demonstrated the impact of data correlation level on the group compression performance. We proposed a new correlation grouping techniques: correlation tree mapping and developed a novel compression technique SIRCS for high variation data in the multi-signal network under a certain error bound. Conspicuous features of SIRCS include: (i) For high variation data, it improves the original algorithm's performance in both compression ratio and NRMSE. (ii) Tree index provides optimal solutions of preserving the highest correlation level of the signal network, taking less compression time than the traditional grouping techniques. In summary, SIRCS is the first algorithm providing maximum correlation preservation and effectively compressed the high variation data. The evaluation of SIRCS from the real world dataset shows the practical improvement from its existing counterparts.

Acknowledgement. This research is partially supported by the Australian Queensland Government (Grant No. AQRF12516).

References

1. Buragohain, C., Shrivastava, N., Suri, S.: Space efficient streaming algorithms for the maximum error histogram, pp. 1026–1035. IEEE (2007)
2. Chakrabarti, K., Keogh, E., Mehrotra, S., Pazzani, M.: Locally adaptive dimensionality reduction for indexing large time series databases. ACM Trans. Database Syst. (TODS) **27**(2), 188–228 (2002)
3. Chen, F., Deng, P., Wan, J., Zhang, D., Vasilakos, A.V., Rong, X.: Data mining for the Internet of Things: literature review and challenges. Int. J. Distrib. Sens. Netw. **11**(8) (2015)
4. Cleary, J., Witten, I.: Data compression using adaptive coding and partial string matching. IEEE Trans. Commun. **32**(4), 396–402 (1984)
5. Dang, T., Bulusu, N., Feng, W.: Robust data compression for irregular wireless sensor networks using logical mapping. Sens. Netw. **2013**, 18 (2013)

6. Elmeleegy, H., Elmagarmid, A.K., Cecchet, E., Aref, W.G., Zwaenepoel, W.: Online piece-wise linear approximation of numerical streams with precision guarantees. Proc. VLDB Endowment **2**(1), 145–156 (2009)
7. Gandhi, S., Nath, S., Suri, S., Liu, J.: GAMPS: compressing multi sensor data by grouping and amplitude scaling. In: Proceedings of the 2009 ACM SIGMOD International Conference on Management of Data, SIGMOD 2009, pp. 771–784. ACM (2009)
8. Keogh, E., Chakrabarti, K., Pazzani, M., Mehrotra, S.: Dimensionality reduction for fast similarity search in large time series databases. Knowl. Inf. Syst. **3**(3), 263–286 (2001)
9. Korn, F., Jagadish, H., Faloutsos, C.: Efficiently supporting ad hoc queries in large datasets of time sequences, vol. 26, pp. 289–300 (1997). http://search.proquest.com/docview/26522991/
10. Krause, A., Guestrin, C., Gupta, A., Kleinberg, J.: Near-optimal sensor placements: maximizing information while minimizing communication cost, vol. 2006, pp. 2–10. IEEE (2006)
11. Lazaridis, I., Mehrotra, S.: Capturing sensor-generated time series with quality guarantees (2003). http://handle.dtic.mil/100.2/ADA465863
12. Louchard, G., Szpankowski, W.: On the average redundancy rate of the Lempel-Ziv code. IEEE Trans. Inf. Theory **43**(1), 2–8 (1997)
13. McAnlis, C., Haecky, A.: Understanding Compression Data Compression for Modern Developers, 1st edn. O'Reilly Media, Sebastopol (2016)
14. Mochizuki, T.: WSJ.D technology: artificial intelligence gets a shake – tiny Japanese startup presses for gains in 'deep learning' efforts; a tech boon for Japan? Wall Street J. (2015). http://search.proquest.com/docview/1738468090/
15. Rafiei, D., Mendelzon, A.: Similarity-based queries for time series data, vol. 26, pp. 13–25 (1997). http://search.proquest.com/docview/23040591/
16. Sarlabous, L., Torres, A., Fiz, J.A., Morera, J., Jané, R.: Index for estimation of muscle force from mechanomyography based on the Lempel-Ziv algorithm. J. Electromyogr. Kinesiol. **23**(3), 548–547 (2013)
17. Sayood, K.: Introduction to Data Compression. The Morgan Kaufmann Series in Multimedia Information and Systems, 3rd edn. Elsevier Science, Amsterdam (2005)
18. Sheather, S.: A Modern Approach to Regression with R. Springer Texts in Statistics, vol. 02. Springer, New York (2009). https://doi.org/10.1007/978-0-387-09608-7
19. Uthayakumar, J., Vengattaraman, T., Dhavachelvan, P.: A survey on data compression techniques: from the perspective of data quality, coding schemes, data type and applications. J. King Saud Univ. Comput. Inf. Sci. (2018)
20. Wang, W., Liu, G., Liu, D.: Chebyshev similarity match between uncertain time series. Math. Prob. Eng. **2015**, 13 (2015). http://search.proquest.com/docview/1722855792/
21. Wyner, A., Wyner, A.: Improved redundancy of a version of the Lempel-ziv algorithm. IEEE Trans. Inf. Theory **41**(3), 723–731 (1995)
22. Ziv, J., Lempel, A.: A universal algorithm for sequential data compression. IEEE Trans. Inf. Theory **23**(3), 337–343 (1977)
23. Ziv, J., Lempel, A.: Compression of individual sequences via variable-rate coding. IEEE Trans. Inf. Theory **24**(5), 530–536 (1978)

Crowdsourcing

Fast Quorum-Based Log Replication and Replay for Fast Databases

Donghui Wang[1], Peng Cai[1,2](✉), Weining Qian[1], and Aoying Zhou[1]

[1] School of Data Science and Engineering, East China Normal University, Shanghai 200062, People's Republic of China
donghuiwang@stu.ecnu.edu.cn, {pcai,wnqian,ayzhou}@dase.ecnu.edu.cn
[2] Guangxi Key Laboratory of Trusted Software, Guilin University of Electronic Technology, Guilin 541004, People's Republic of China

Abstract. The modern In-Memory Database (IMDB) can support highly concurrent OLTP workloads and generate massive transactional logs per second. Quorum based replication protocols such as Paxos or Raft have been widely used in distributed databases. However, it's non-trivial to replicate IMDB because high transaction rate has brought new challenges. First, the leader node in quorum replication should have adaptivity by considering various transaction arrival rates and the processing capability of follower nodes. Second, followers are required to replay logs to catch up the state of the leader in the highly concurrent setting to reduce visibility gap. To this end, we built QuorumX, an efficient quorum-based replication framework for IMDB under heavy OLTP workloads. QuorumX combines critical path based batching and pipeline batching to provide an adaptive log propagation scheme to obtain a stable and high performance at various settings. Further, we propose a safe and coordination-free log replay scheme to minimize the visibility gap between the leader and follower IMDBs. Our evaluation results with the YCSB and TPC-C benchmarks demonstrate that QuorumX achieves the performance close to asynchronous primary-backup replication without sacrificing the data consistency and availability.

Keywords: Log replication · Log replay · High performance · Quorum

1 Introduction

Replication is the technique used for a traditional DBMS or fast, multi-core scalable In-Memory Database (IMDB) to support high-availability. In this work, we assume a full database copy is held on a single IMDB node, and each backup node has the full replication. In replicated IMDBs, the execution of a transaction is completely in the primary IMDB. Primary-backup replication is the well-known replication method in database community. The asynchronous primary-backup replication used in traditional database systems [3,4] trades consistency for performance and availability. The synchronous primary-backup replication trades performance and availability for consistency.

G. Li et al. (Eds.): DASFAA 2019, LNCS 11446, pp. 209–226, 2019.
https://doi.org/10.1007/978-3-030-18576-3_13

Today's mission-critical enterprise applications in Banking or E-commerce require the back-end database system to provide high-performance and high-availability without sacrificing consistency. Compared with primary-backup replication, the quorum-based replication (e.g. Multi-Paxos [9], Raft [19], etc.) can guarantee strong consistency, tolerate up to F out of 2F+1 fail-stop failures, and achieve good performance because it only requires the majority of replicas to response to the leader. The quorum-based replication adopts consensus protocols to take more reasonable trade-off among performance, availability and consistency, and thus it has been regarded as a practical and efficient replication protocol for large scale datastores [14,22,23].

Quorum-based replication protocols are the natural choice for replicating IMDB as a highly available and strongly consistent OLTP datastore. However, it's non-trivial to translate the quorum-based replication protocol into a pragmatic implementation for industrial use. The basic principle of various quorum-based protocols is that committing a transaction requires its log to be replicated and flushed on non-volatile storage on the majority of follower replicas. A transaction may take extremely short time to complete its execution in the leader IMDB. But, committing this transaction may take more time to wait its log replicated to the majority of followers. As a result, the performance of replicated IMDBs significantly depends on the quorum-based log replication which is influenced by many factors.

To achieve read scalability, the followers need to replay committed logs at a fast speed to keep up with the leader's state. The classic quorum-based replication needs the leader to send followers the maximal committed log sequence number (MaxComLSN), and then follower can commit and replay these logs with LSN smaller or equal to MaxComLSN. *Replaying logs after receiving the specified MaxComLSN leads to that the committed data on followers are visible at a later time than that on the leader, referred to as visibility gap (VGap).* Without careful design, VGap would be larger when the leader IMDB is running under a heavy OLTP workload, and generates transactional logs at a high rate.

In this paper, we present an efficient quorum-based replication framework, called as QuorumX, to optimize log replication and replay for IMDB under highly concurrent OLTP workloads. Main contributions are summarized as follows:

- QuorumX combines critical path based batching and pipeline based batching to adaptively replicate transactional logs, which takes into account various factors including the characteristics of transactional workloads and the processing capability of follower.
- We introduce a fast and coordination-free log replay scheme without waiting for the MaxComLSN, which applies logs to memory ahead of time in parallel to reduce the risk of increased VGap.
- QuorumX has been implemented in Solar [10], an in-memory NewSQL database system that has been successfully deployed on Bank of Communications, one of the biggest commercial banks in China. Extensive experiments are conducted to evaluate QuorumX under different benchmarks.

2 Preliminary

Overview of Quorum-Based Log Replication. Figure 1 shows the overall architecture of replicating an IMDB. The replicated IMDB cluster contains one primary IMDB as a leader and more than two replica IMDBs as followers. All requests of read/write transactions are routed to the leader IMDB. Transactions are concurrently executed on the leader. When a transaction completes all transactional logics and starts to execute the COMMIT statement, the leader generates its transactional logs and appends them to log buffer (at steps 1 and 2 in the left side of Fig. 1). Then this transaction enters the commit phase, waits to be committed (at step 3) and finally responses to the client (at step 6). The single commit thread in the leader sends these logs to all followers and flush them to local disks (at steps 4 and 5). A transaction can be committed only after the leader receives more than half responses from followers. After that, leader will asynchronously send the latest committed log sequence number (MaxComLSN) to followers. Follower replicas then replay committed logs less than the latest received MaxComLSN. It should be noted that the execution worker is multi-threaded. The new arrived transaction requests from clients can be processed in parallel although previous transactions have not been committed. The new transactions cannot be committed until the previous ones have been committed. That means the commit order is sequential.

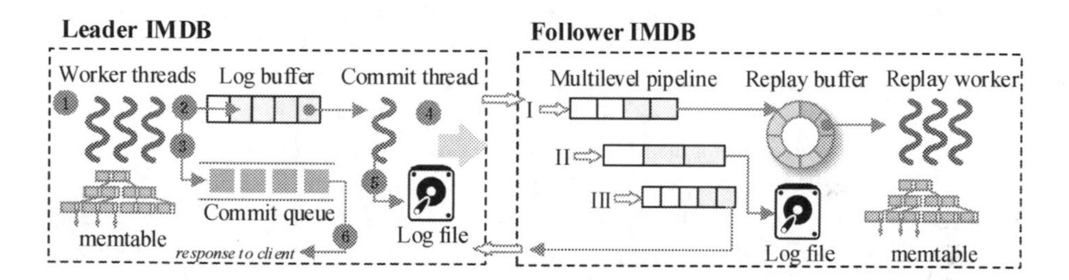

Fig. 1. Overall architecture of replicating an IMDB.

The follower replica who receives logs will first parse it into entries with log format and check the integrity, then write it to non-volatile storages and send a response message to leader. Under a heavy OLTP workload, if followers use a single thread to process received logs in a sequential manner, the replication latency would be unacceptable in practical settings. Pipeline and batching are general methods used to improve the performance of log replication.

- **Pipeline parallelism in a follower replica.** The basic steps for processing a received log by replica can be divided into three relatively independent stages: parsing logs, flushing logs and sending response to leader. The pipeline of processing logs in follower is that: the parsing thread gets network packets, parses them to log entries and appends these logs to the *replay buffer* (at steps

I in the right side of Fig. 1). At the back-end, the single persistence thread reads logs from the replay buffer and flushes them to log files (at step II). When finishing writing a batch of logs, the persistence thread notifies the reply thread to send a response to the leader (at step III). The replication latency introduced by follower replicas is hidden through pipelined log processing.
- **Batching logs in the leader.** Pipeline and batching are often used together [22,25]. Without batching, the pipeline will be hard to work effectively. Basically, batching several requests into a single instance allows the overhead to be amortized over per-request. The systems built over quorum-based replication can adopt the batching method to boost the throughput. However, the parameters such as batch size have greater impacts on the performance of batching method. The manual configurations for these parameters are proved to be time consuming and can not adapt to different settings. Existing works on automatic batching are limited in replicated IMDBs. For example, the factor on processing capability of follower has not been fully considered in the log replication.

In this work, we investigated several batching methods and found that they were not always effective under the context of replicating a fast IMDB. Quorum-based replication needs an adaptively self-tuning batching mechanism that is not only parameter-free but also considers: (1) the capacity of follower; and (2) the workload characteristics (e.g. the arrival rate).

Log Replay. To avoid the follower lagging behind the leader too much, follower requires a fast mechanism of replaying committed logs. On the back-end, follower IMDBs replay logs to memtables (which is often implemented by B+ Tree or SkipList in IMDB) to provide read-only transaction requests. The maximal committed log sequence number (MaxComLSN) is piggybacked on logs to notify the follower the latest committed point. Conventional quorum replication schemes only allows logs with LSN less than MaxComLSN to be replayed. In the case of highly concurrent workloads, this principle of relaying logs by follower causes a challenge in visibility gap. In this paper, visibility gap is defined as the time difference between leader and follower for making the committed data be visible. Real Applications such as HTAP often take real-time OLAP analysis over follower nodes [20], and it's expected that there is a as small as possible VGap between leader and followers.

Recently proposed solutions to VGap aim at resolving the problem in the asynchronous primary-backup replication, which can not be applied to the quorum-based replication [16,21]. In the asynchronous primary-backup replication, follower could replay the received logs immediately without any coordination with leader. However, in the quorum-based replication, it is leader that notifies followers the consensus decision of committing transactions by sending the current MaxComLSN. After receiving MaxComLSN, follower nodes are agreed to replay logs with LSN no larger than MaxComLSN. Since it's expensive to read logs from disk for replay, the replicated and uncommitted logs need to reside in the memory for a period of time before being replayed. The structure holding un-replayed logs is the replay buffer. However, in the case where the leader generates logs at a high speed, e.g SiloR could produce logs at gigabytes-per-second rates [27], un-replayed logs in the buffer can be soon erased by the new arrivals if the size of replay buffer is insufficient. Followers still needs to read flushed logs from disk for replay, and would definitely lag behind the leader and produces larger and larger VGap. Therefore, to achieve read scalability for IMDB replicated by quorum based protocols, VGap of a follower should be minimized in order to keep up with the state of leader.

3 Adaptively Self-tuning Batching Scheme

The design objectives of batching scheme have three aspects. First, no parameters are required to be calculated offline and then manfully tune system configurations. Because once the environment settings are changed, these parameters need to be calculated again. It should be totally automatic to cope with uncertainty without manual intervention. Second, workloads in real setting are often dynamically changed and have an important effect on the performance of batching scheme. For instance, if the transaction arrival rate becomes low, a batch should be constructed by a small number of transactional logs. Last but not least, considering the processing capacity of follower is essential for adaptively tuning algorithm, especially in the case where the whole performance relies on the processing speed of followers under heavy workloads. Follower replicas may be overloaded if logs are replicated with a wrong batch size (Table 1).

Table 1. Features of batching algorithms.

Batching scheme	Parameter-free	Workload-adaptive	Replica-friendly
JPaxos [13]	×	✓	×
Nuno Santos [25]	×	✓	×
Paolo Romano [24]	×	✓	×
AB [11]	✓	×	×
TAB [11]	×	✓	×
QumrumX	✓	✓	✓

3.1 Batching Scheme

Based on the above design objectives, we propose to combine critical-path-based batching (CB) [11] and pipeline-based batching (PB). CB automatically adjusts the batch size according to workload characteristics. PB is complementary to CB by considering the processing capability of follower, which can adaptively tune the frequency of sending logs to avoid followers being overloaded in highly concurrent workloads.

The CB mechanism operates as following: as shown in Fig. 2, after finishing processing transaction logics, each worker thread will enter a global common code

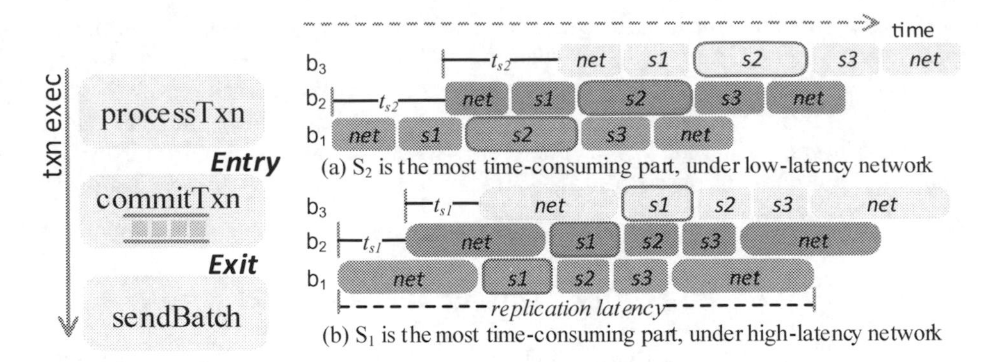

Fig. 2. Critical-path-based batching (CB).

Fig. 3. Pipeline-based batching (PB).

fragment, that is `commitTxn`. The entry code is used for registering the commit queue as the task is inside. Similarly, the exit code deregisters the task and appends it to the sending batch. The intuition behind CB is that multiple tasks should be included in the same batch only if they arrive "close together" to the `sendBatch`. When implementing CB, we treat the commit queue as a doorway. A batch is complete and sent to follower when the commit queue is empty, since the next task is too far behind to join into the current batch. Compared with batching with a fixed time or a fixed size, CB could adjust sending frequency according to the arrival rate. When the arriving rate is high, CB gathers a lot of close tasks and achieves good throughput. And if the arrival rate is low, CB will not waste a long time for waiting for more tasks. The disadvantage of CB is that when the arrival rate stays constantly high, CB will continue to gather too many tasks without sending a batch in a proper size. We combine PB with CB mechanisms to resolve this issue.

PB takes a full consideration of the pipelined replication scheme in follower. As described above, pipelined replication scheme in follower consists of three stages (s_1, s_2, s_3). *It should be noted that an optimal performance can be achieved if the slowest pipeline stage handles tasks all the time and has no idle time.* Taking Fig. 3(a) for example, suppose that s_2 is the most time-consuming stage, and the optimal send interval for a batch should be t_{s2}. Upon this sending rate, every batch could get a smallest replication latency and next batches would not be blocked by the previous ones. As a result, during the pipeline replication, QuorumX collects the consumed time of each stage by followers for each batch, and embedded them into the response to be sent to the leader. QuorumX requires the time interval of sending two batches should not to be less than it. If logs are sent with a interval larger than that value, the resources cannot be utilized sufficiently. On the contrary, if the sending interval is less than that value, congestion should happen during replication.

3.2 Discussion

We demonstrate that network latency between leader and follower has no effect to set sending frequency with Fig. 3(a) and (b). No matter how the network latency changes, the optimal frequency is always restricted by the most time-consuming stage of follower. However, we need to point out that network bandwidth can affect the sending frequency. Under complicated network environment especially the wide area network, bandwidth is often limited and may be occupied by some unknown applications. Here, log replication is constrained by the limited network bandwidth of the leader. As a result, the sending frequency should be lowered properly. How to automatically adjust the frequency of sending logs over complicated, unreliable networks is still an open question, and we will study this problem in our future work.

4 Coordination-Free Log Replay

4.1 Design Choices for Replay Buffer

The replay buffer in follower is an important structure which is responsible for caching the received logs from the leader. The persistence thread can flush a batch of buffered logs at one time, and the replay thread can directly replay the buffered log to keep in sync with the primary. The design of replay buffer should guarantee replicated logs are replayed from memory most of the time and avoid re-loading them from HDD/SDD.

The size of the replay buffer is a key design consideration. IMDB such as SiloR could generate logs at gigabytes-per-second rates. Caching all logs in the replay buffer leads to excessive memory consumption. If the size is set to a small value, the buffered and non-replayed logs would be covered by the new arrivals under heavy workload. Reading the received yet covered logs from disk for log replay would introduce extra disk I/O latency. This causes the risk of cascading latency as more non-replayed logs continue to be covered by newly arrived logs. Finally, it will make the follower nodes never catch up with the leader. The basic idea of determining the buffer size is that it should be greater than the rate of log generation on the leader.

In order to provide read services on fresh data by followers, they need to replay received logs to memory as soon as possible. However, as discussed above, different from asynchronous replication, the time to replay a log entry is restricted by the quorum-based replication scheme. A follower is only allowed to replay logs with LSN not larger than MaxComLSN for guaranteeing consistency. However, wait-for-replay logs residing in the memory may cause the replay buffer overwhelmed. To this end, we design a coordination-free log replay (CLR) scheme which directly applies the received logs to the memtable without waiting for the MaxComLSN. CLR ensures consistency by separating the replay procedure into two phases. The first phase converts logs into uncommitted cell lists of memtable in parallel, where the applied data are invisible. The second phase sequentially installs them into memtable according their LSNs, where the consistency is guaranteed. *It's should be emphasized that the second phase is extremely lightweight without introducing overhead as the installation only contains a few pointer manipulations.*

4.2 Mechanism of Coordination-Free Log Replay

Basically, different from transaction execution in the leader, there has **NO** rollback when replaying logs in follower. That means all of the logs must be replayed successfully in order. We choose to replicate *value logs* instead of operation logs, which could promise a lock-free replay strategy. When CLR begins to replay a batch of logs, in the first phase, multiple threads (replay workers) works in parallel. Replay worker first starts a transaction for each log entry. Then it looks up the memtable to find the node that the transaction wants to modify. After that, logs are translated into several uncommitted cell informations in which each cell has a pointer pointing to the actual node in the memtable. Translating won't directly installed modifications into the memtable and therefore has no need to acquire any locks. The uncommitted cell informations are stored in the transaction context.

Algorithm 1. QuorumX commit algorithm of replaying

```
   /* Commit transactions according to log sequence                */
   Input: MaxComLSN
 1 while !thread_.stop() do
       /* Get a transaction from commit queue sequentially.        */
 2     log_id = commit_queue.seq_;
 3     while true do
 4         if log_id > MaxComLSN then
 5             wait(wait_time_ms);
 6             continue;
 7         txn_ctx = commit_queue.get(log_id);
 8         if NULL == txn_ctx then
 9             wait(wait_time_ms);
10             continue;
11         __sync_bool_compare_and_swap(&commit_queue.seq_, log_id, log_id + 1);
12         break;
       /* Install the modification into memtable.                  */
13     for cell_info in txn_ctx.uc_info do
14         memnode = cell_info → node;
15         exclusive_lock(memnode.rowlock);
16         memnode.value_list.append(cell_info);
17         exclusive_unlock(memnode.rowlock);
```

After completing the above procedures, transaction will be pushed into the commit queue of a single commit thread. It was the single commit thread that ensures the safety and consistency of quorum-based replication. In the second phase of CLR, commit thread sequentially pops transaction whose log id is smaller than the MaxComLSN, and does the commit transaction operation. As shown in the Algorithm 1, transactions with log id less than MaxComLSN will be committed and their uncommitted cell informations will be directly append to the value list in the memtable. Locks are necessary in this part, but as we can see, the duration is short (lines 13–17).

The main processing flow of CLR can be processed totally in parallel and only the commit part is done sequentially in order to promise transaction modifications are installed into memtable by the LSN order. CLR immediately replays the received logs without waiting for MaxComLSN. One advantage is to avoid the risk of reading flushed logs from disk and the memory resources consumed by the replay buffer has a minor risk of being excessive. Besides, since CLR performs replaying ahead of time, the VGap can be minimized compared with scheme waiting for MaxComLSN.

4.3 Discussion

Nevertheless, there are additional demands on fault handing introduced by our proposed replay strategy. Suppose such a scenario in Fig. 4, five replicas ($R_1 - R_5$) form a cluster and R_1 is the initial leader. Before crashed, R_1 has generated five log entries and committed four log entries. Log five has flushed to disk and entered into the first phase of CLR in R_2 while the other three followers haven't received log five. According to election algorithm, R_4 is elected as the new leader. It generates a different log five and replicates it, and there is growing problem that R_2 has began to replay a log five from the old leader. Although the modification has not been installed in the memtable, the transaction context with log five still reside in the memory (dirty contents). If R_2 begins to replay another log five, there may be some checksum errors. If similar situations arise when we don't adopt CLR, there are no dirty contents in R_2's memory, R_2 only needs to rewrite log five to its disk.

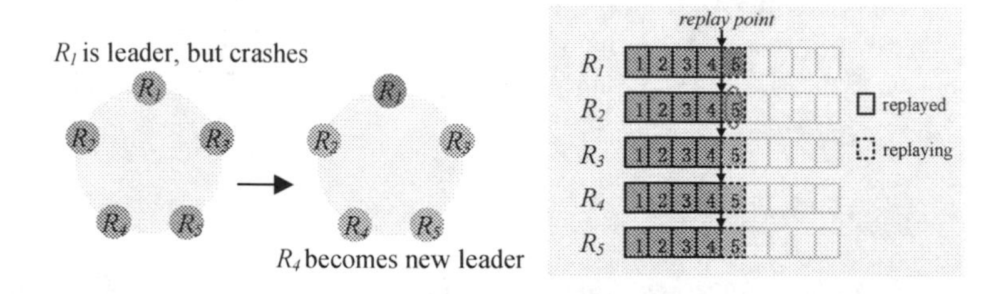

Fig. 4. An example illustrates a fault caused by CLR.

Based on above description and discussion, when introducing CLR, we also refine the fault handing algorithm. More concretely, when role change happens, each node will firstly perform replay-revoking operation before actually getting into working. CLR ensures that dirty contents can be easily erased since it neither modify any structure that stores data nor hold any locks. The commit thread pops all tasks from its commit queue, cleaning uncommitted cell informations and ending these transactions.

5 Evaluation

In this section, we evaluate the performance of QuorumX for answering the following questions:

- The first question is whether QuorumX could support a high performance replication for fast IMDB, and how much additional performance is sacrificed by QuorumX through comparing it with the asynchronous primary-backup replication and the single replica without replication.
- Another question is that whether QuorumX can be self-tuning to workloads. We evaluate its performance under different concurrency by comparing batching methods include AB [11] and JPaxos [13]. Since the calculation of offline models in [24,25] requires a lot of additional parameters which are difficult to collect, we didn't implement them in QumrunX.
- The final question is that how much VGap can be reduced by the CLR of QuorumX in contrast with asynchronous primary-backup replication. Besides, CLR replays logs without waiting for MaxComLSN in order to avoid reading logs from disk and thus reduces the VGap. We also measured how much VGap could be reduced by CLR even if QuorumX replays logs after receiving the MaxComLSN.

Experiment Setup. We have implemented QuorumX in Solar [10], an open-source, scalable IMDB. We implement QumrumX by adding or modifying 31282 lines of C++ code on the original base. Therefore, Solar is a completely functional and high available in-memory database system. It has also been deployed on Bank of Communications, one of the biggest commercial banks in China. The default cluster consists of three replicas and the leader has the full-copy of data. We also evaluate performance of different number of replicas. Each server is equipped with *two 2.3* GHz *20-core E5-2640 processors*, *504* GB DRAM, and connected by a *10 Gigabit Ethernet*.

5.1 Workloads

In the following experiments, we use three benchmarks that allows us to measure how QuorumX performs in specific aspects.

YCSB. The Yahoo! Cloud Serving Benchmark (YCSB) [6] is designed to evaluate large-scale Internet applications. The scheme contains a single table (`usertable`) which has one primary key (INT64) and 9 columns (VARCHAR). The usertable is initialized to consist of 10 million records. A transaction in YCSB is simple and only includes one read/write operation. The record is accessed according to an uniform distribution.

TPC-C. This benchmark models a warehouse ordering processing which simulates an industry OLTP application. We use a standard TPC-C workload and populated 200 warehouses in the database by default. The transaction parameters are generated according to the TPC-C specification.

Micro-benchmark. As a fully functional database, Solar requires to interact with clients, interpret SQL statements and translate them into physical execution plans, so it could not achieve a similar performance like Silo. Therefore, we build a write-intensive micro-benchmark, which originates from a realistic bank application used for importing massive data into databases everyday. Instead of sending the leader IMDB transaction requests coded by SQL statements, this micro-benchmark directly issues raw write operations to the leader. As a result, the micro-benchmark makes leader IMDB is running under extremely high-concurrent, write-intensive workloads. By default, the micro-benchmark contains 10 GB data modifications, which could produce gigabyte of logs per second.

5.2 Replication Performance

We firstly measure the throughput and latency under the YCSB workload with 100% write operations and the complicated TPC-C workload. The comparing methods include QuorumX with three replicas one of which servers as the leader (abbr. QuorumX), asynchronous primary-backup replication (abbr. AsynR) with three replicas and a single replica without replication (abbr. NR).

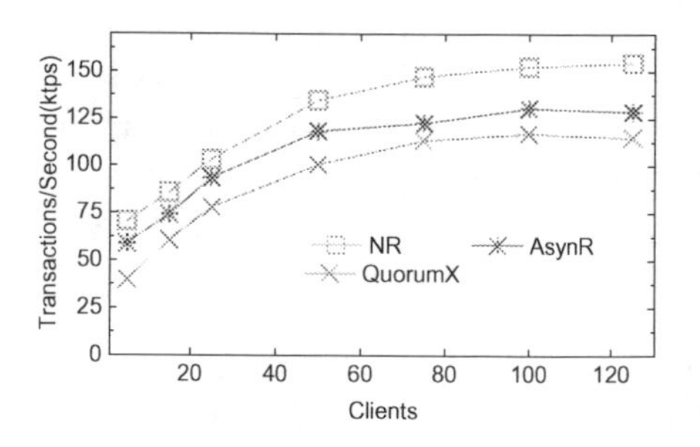

Fig. 5. Throughput of YCSB.

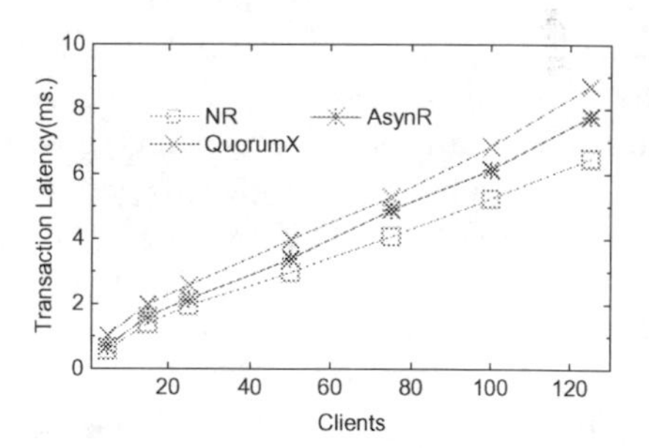

Fig. 6. Latency of YCSB.

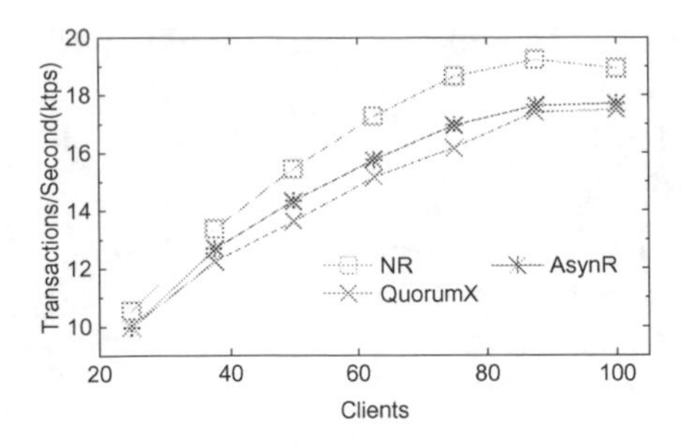

Fig. 7. Throughput of TPC-C.

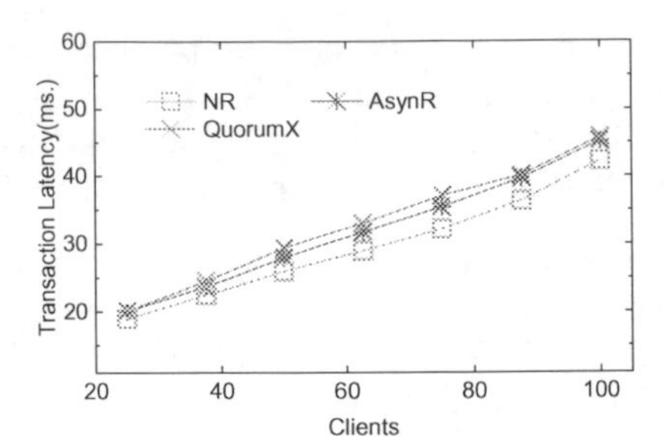

Fig. 8. Latency of TPC-C.

Experimental results of YCSB are shown in Figs. 5 and 6. We can observe that the throughput trend of all replication scheme is increasing firstly and then remaining at a high level. In general, QuorumX sacrifices about 11% performance compared with AsynR and 26% compared with NR to provide data consistency and high availability. As for latency, QuorumX produces about 0.6 more milliseconds than AsynR and 1.1 ms than NR in average. Figures 7 and 8 illustrates the performance under the TPC-C workload. We find that the throughput gap among QuorumX and AsynR and NR reaches to 2% and 8% respectively, which is smaller than that in YCSB. The reason is that a transaction in TPC-C contains more read/write operations than that in YCSB so the leader takes more time to execute a TPC-C transaction. As a result, the percentage of replication latency is relatively small in the whole transaction latency.

5.3 The Ability of Adaptive Self-tuning

To compare the performance of self-tuning batching scheme of QuorumX with other batching algorithms, we implemented a parameter-free method—AB, which adopts critical-path-based batching. We choose AB instead of TAB since the batching method of AB is totally parameter-free. Besides, JPaxos, which needs manually set the parameter of batch size, is also compared with QuorumX to evaluate their effectiveness under various number of concurrent clients. JPaxos is configured to two *batchsize* values: 32 and 256 respectively, referred to as JPaxos-32 and JPaxos-256. Experiments are run over YCSB workloads with 100% write requests.

Figure 9 illustrates the experimental results on different client concurrency. It is clear that QuorumX performs best under all concurrency. We can observe that the performance of AB is close to that of QuorumX when the concurrency is low. However, as the number of clients increases, AB could not achieve good performance. Recall from Sect. 3, under a light workload, critical path based batching works well. But, under a highly concurrent workload, the throughput of the system would be determined by the slowest stage in the pipelined processing on followers. In this case, the pipeline batching mechanism in QuorumX can adaptively tune the interval of sending logs.

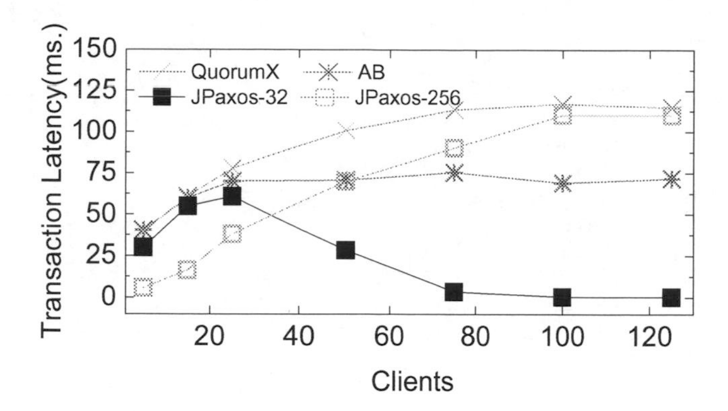

Fig. 9. Throughput under different client concurrency.

The trend of JPaxos-32 increases firstly and could stay at a similar throughput to QuorumX, but decreases sharply when the number of client exceeds 25. This is because, when the client number is small, the arrival rate of transactions is slow, and waiting 32 requests to generate a batch is relatively reasonable. However, when the arrival rate rises, sending batches with size of 32 exceeds the processing capacity of followers. Follower cannot process as many as batches produced by JPaxos-32 in time and these received batches would be blocked. So there is a sudden drop of the performance. On the contrary, JPaxos-256 performs badly when the client concurrency is low and gradually close to QuorumX with the increasing of the number of client. It is clear that, sending batches with size of 256 is too slowly for followers when the arrival rate is low. The leader wastes too much time on waiting for enough requests. Under the high concurrency, collecting 256 requests for a batch becomes easier, and the sending frequency can match the processing capacity of follower.

5.4 VGap Results

We measure the VGap between the leader and followers to explore the effectiveness of CLR under a continued, write-intensive micro-benchmark. Assuming that the leader l and the follower f commit the same transaction at physical time t_l and t_f, we use the value $t_f - t_l$ to donate the VGap between the same visible state of leader l and the follower f. We compare VGap of three methods: QuorumX, QuorumX without CLR and AsynR.

Figure 10 shows the VGap results over 60 s. The number of client is fixed to 800. Results shows that QuorumX could gain the lowest and most stable VGap among three methods. The VGap of AsynR exceeds 200 ms, which suggests that follower in AsynR lags far behind the leader. And the VGap of QuorumX without CLR remains about 100 ms at beginning, but it suddenly increases sharply at time 45. By our analysis, the replica may perform disk-read operations for getting logs to replay, and the trace log also proved that. QuorumX with CLR has a stable VGap and most of it is under 60 ms. Using CLR could achieve a 3.3x

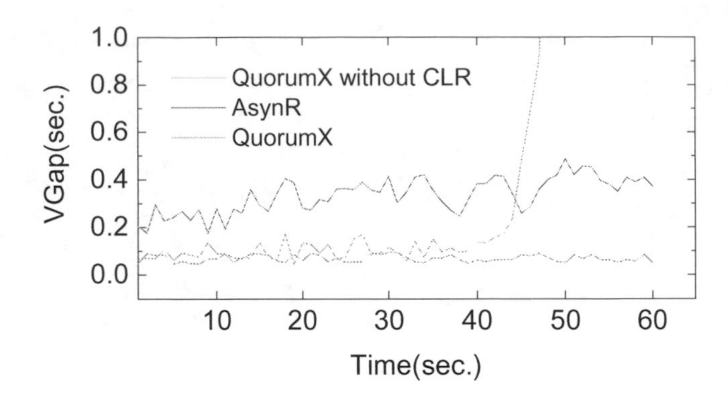

Fig. 10. VGap under write-heavy load.

lower VGap than AyncR and 1.67x than not using CLR. Therefore, in the case of heavy workload, reading from follower under QuorumX with CLR could get a fresher and more stable state.

5.5 The Number of Replicas

To investigate the scalability of QuorumX, we evaluate the performance over different number of replicas under two YCSB workloads of different write/read ratios: 100/0 and 50/50. Experimental results are shown in Fig. 11. The number of clients is fixed to 125. We can see that under workload with 100% writes, the performance decreased most significantly when the number of replicas is changed from one to three, dropped about 26%. This is because transaction processed under three-replica cluster has obviously longer latency than under single server. When the number of replicas keeps increasing, the throughput decline is not intense, performance under five replicas only decreases 9% than three replicas. This is acceptable since logs have to be replicated to more replicas. Under the workload with 50/50 write/read ratio, the performance decline is even less obvious. As more replicas could provide scalable read service, we can see that with the number of replicas increase, the performance could achieve a sustainable growth. After all, QuorumX has a good scalability with more replicas.

6 Related Work

Replication is an important research topic across database and distributed system communities for decades [15,18]. In this section, we review relevant works mainly on two widely used replication schemes, i.e. primary-backup replication and quorum based replication.

Primary-Backup Replication. Asynchronous primary-backup replication [26], proposed by Michael Stonebraker in 1979, has been implemented in

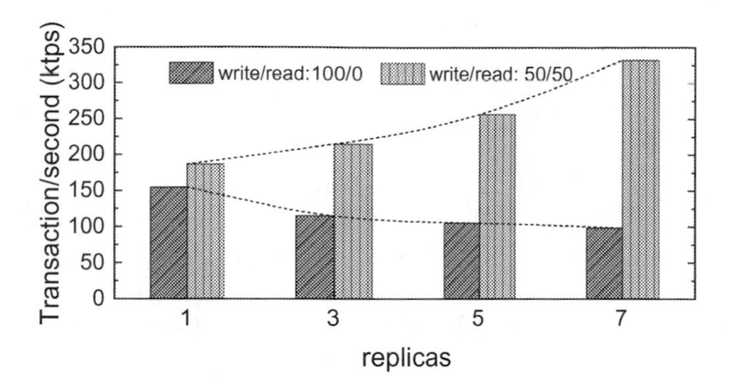

Fig. 11. Throughput over replicas number.

many traditional database systems. In most typical deployment scenarios, asynchronous primary-backup replication is used to transfer recovery logs from a master database to a standby database. The standby database is usually set up for fault tolerance, and not required to provide the query on the latest data. The performance of log replication and replay have not received much attentions in the last several decades. Recently, the researchers [12,16] suggest that serial log replay in the primary-backup replication can cause the state of replica is far behind that of the primary with modern hardware and under heavy workloads. KuaFu [12] constructs a dependency graph based on tracking write-write dependency in transactional logs, and it enables logs to be replayed concurrently. The dependency tracking method works well for traditional databases under normal workloads, and it might introduce overheads for IMDB under highly-concurrent workloads. [16] proposed a parallel log replay scheme for SAP HANA to speed up log replay in the scenario where logs are replicated from an OLTP node to an OLAP node. Qin et al. [21] proposed to add the transactional write-set into its log in SQL statement formats, which can reduce the logging traffics. Log replay in classical quorum-based replication has different logics to primary-backup replication. Followers using quorum-based replication cannot replay received logs to memtable immediately, and they need to wait for MaxComLSN from the primary. Due to this difference, these works that optimize log replay for primary-backup replication can not be directly applied to the quorum-based replication.

Despite the low transaction latency, the asynchronous primary-backup replication cannot guarantee high availability and causes data loss when the primary is crashed. PacificA [17] resolves these problems by requiring the primary to commit transactions only after receiving persistence responses from all replicas. The introduced synchronous replication latency depends on the slowest server in all replicas. Kafka [5] reduces replication latency by maintaining a set of in-sync replicas (ISR) in the primary. Here ISR indicates the set of replicas that keep the same states with the primary. A write request is committed until all replicas in ISR reply. Kafka uses the high watermark (HW) to mark the offset of the last committed logs. The replicas in ISR need to keep the same HW with the primary. When the offset of a replica is less than HW, it would be removed from ISR. Through ISR, Kafka can reduce negative impact on performance caused by the network dithering.

Quorum-Based Replication. Replication based on consensus protocols is referred to as quorum-based replication, which is also called as state machine replication in the community of distributed system. Paxos based replication ensures all replicas to execute operations in their state machines with the same order [9]. Paxos variants such as Multi-Paxos used by Spanner [7] are designed to improve the performance. Raft [19] is a consensus algorithm proposed in recent years. One of its design goals is more understandable than Paxos. For this reason, Raft separates log replication from the consensus protocol. Many systems such as AliSQL [1] and etcd [2] adopt Raft to provide high availability. However, these systems use Paxos or Raft to replicate meta data, where replication performance is not a serious problem. Spanner as a geo-distributed database system supports distributed transactions, and each partitioned database node is not designed to handle highly concurrent OLTP workloads. AliSQL only uses Raft to elect leader in the occurrence of system failures. Etcd is a distributed, reliable key-value store that uses the Raft for log replication. Similar to Zookeeper [8], these kinds of datastore are designed to provide high availability for meta data management and are not suitable for highly concurrent OLTP workloads.

There are a few works on tuning replication performance of Paxos with batching and pipeline [13,25]. Nuno Santos et al. [25] provide an analytical model to determine batch size and the pipeline size through gathering a lot of parameters, like bandwidth and the application properties. [13] proposed to generate batches and instances according to three input parameters: the maximum number of instances that can be executed in parallel, the maximum batch size, and the batch timeout. These parameters need to be calculated offline and set manually which can not adapt to various environments.

7 Conclusion and Future Work

In this paper, we built QuorumX, an efficient quorum-based replication framework for replicating fast IMDB. We propose an adaptive batching scheme which could self-tuning sending frequency and could adapt to both light and heavy workloads. In order to produce a minimal and stable visibility gap between leader and follower, we design a fast and coordinate-free log replay mechanism to replay logs without waiting for MaxComLSN. Experimental results show that QuorumX supports strong data consistency and high availability by sacrificing only 8%–25% performance than single IMDB replica and has a 2%–11% decline than asynchronous primary-backup replication. The batching scheme always performs better than existing methods. Also, the visibility gap produced by QuorumX can reach to a low level.

QuorumX is designed for fast IMDBs without harsh assumptions, so it is also applicable to NoSQL systems. In our future work, we will erect a more general, pluggable quorum-based replication framework that not only provides high replication performance for these high-throughput systems but also takes many complicated factors like network bandwidth into consideration.

Acknowledgement. This work is partially supported by National Key R&D Program of China (2018YFB1003404), NSFC under grant numbers 61432006, and Guangxi Key Laboratory of Trusted Software (kx201602). We thank anonymous reviewers for their very helpful comments.

References

1. AliSQL. https://github.com/alibaba/AliSQL
2. etcd. https://coreos.com/etcd/
3. IBM DB2. https://www.ibm.com
4. Oracle Corporation and/or its affiliates. MySQL Cluster (2017)
5. W. contributors. Apache kafka (2018). https://en.wikipedia.org/w/index.php?title=Apache_Kafka&oldid=831864654
6. Cooper, B.F., Silberstein, A., Tam, E., Ramakrishnan, R., Sears, R.: Benchmarking cloud serving systems with YCSB. In: SoCC (2010)
7. Corbett, J.C., Dean, J., Epstein, M., Fikes, A., et al.: Spanner: Google's globally distributed database. ACM Trans. Comput. Syst. **31**(3), 8:1–8:22 (2013)
8. Hunt, P., et al.: ZooKeeper: wait-free coordination for internet-scale systems. In: USENIX ATC (2010)
9. Chandra, T.D., et al.: Paxos made live: an engineering perspective. In: PODC (2007)
10. Zhu, T., et al.: Towards a shared-everything database on distributed log-structured storage. In: ATC (2018)
11. Friedman, R., Hadad, E.: Adaptive batching for replicated servers. In: 25th IEEE Symposium on Reliable Distributed Systems, pp. 311–320 (2006)
12. Hong, C., Zhou, D., Yang, M., Kuo, C., Zhang, L., Zhou, L.: KuaFu: closing the parallelism gap in database replication. In: ICDE (2013)
13. Kończak, J., de Sousa Santos, N.F., et al.: JPaxos: state machine replication based on the Paxos protocol. Technical report (2011)
14. Zheng, J., et al.: PaxosStore: high-availability storage made practical in WeChat. PVLDB **10**(12), 1730–1741 (2017)
15. Kemme, B., Alonso, G.: Don't be lazy, be consistent: Postgres-R, a new way to implement database replication. In: VLDB, pp. 134–143 (2000)
16. Lee, J., Moon, S., et al.: Parallel replication across formats in SAP HANA for scaling out mixed OLTP/OLAP workloads. PVLDB **10**, 1598–1609 (2017)
17. Lin, W., Yang, M., Zhang, L., Zhou, L.: PacificA: replication in log-based distributed storage systems (2008)
18. Wiesmann, M., Pedone, F., et al.: Database replication techniques: a three parameter classification. In: SRDS, pp. 206–215 (2000)
19. Ongaro, D., Ousterhout, J.K.: In search of an understandable consensus algorithm. In: ATC, pp. 305–319 (2014)
20. Özcan, F., Tian, Y., Tözün, P.: Hybrid transactional/analytical processing: a survey. In: SIGMOD Conference, pp. 1771–1775. ACM (2017)
21. Qin, D., Goel, A., Brown, A.D.: Scalable replay-based replication for fast databases. PVLDB **10**(13), 2025–2036 (2017)

22. Rao, J., Shekita, E.J., Tata, S.: Using paxos to build a scalable, consistent, and highly available datastore. PVLDB **4**, 243–254 (2011)
23. Liu, Y.A., Chand, S., Stoller, S.D.: Moderately complex Paxos made simple: high-level specification of distributed algorithm. CoRR abs/1704.00082 (2017)
24. Romano, P., Leonetti, M.: Self-tuning batching in total order broadcast protocols via analytical modelling and reinforcement learning. In: ICNC, pp. 786–792 (2012)
25. Santos, N., Schiper, A.: Tuning paxos for high-throughput with batching and pipelining. In: Bononi, L., Datta, A.K., Devismes, S., Misra, A. (eds.) ICDCN 2012. LNCS, vol. 7129, pp. 153–167. Springer, Heidelberg (2012). https://doi.org/10.1007/978-3-642-25959-3_11
26. Stonebraker, M.: Concurrency control and consistency of multiple copies of data in distributed INGRES. IEEE Trans. Softw. Eng. **5**(3), 188–194 (1979)
27. Zheng, W., Tu, S., et al.: Fast databases with fast durability and recovery through multicore parallelism. In: USENIX OSDI (2014)

PDCS: A Privacy-Preserving Distinct Counting Scheme for Mobile Sensing

Xiaochen Yang[1], Ming Xu[1], Shaojing Fu[1,2(✉)], and Yuchuan Luo[1]

[1] College of Computer, National University of Defense Technology, Changsha, China
shaojing1984@163.com
[2] State Key Laboratory of Cryptology, Beijing, China

Abstract. Mobile sensing mines group information through sensing and aggregating users' data. Among major mobile sensing applications, the distinct counting problem aiming to find the number of distinct elements in a data stream with repeated elements, is extremely important for avoiding waste of resources. Besides, the privacy protection of users is also a critical issue for aggregation security. However, it is a challenge to meet these two requirements simultaneously since normal privacy-preserving methods would have negative influence on the accuracy and efficiency of distinct counting. In this paper, we propose a Privacy-preserving Distinct Counting Scheme (PDCS) for mobile sensing. By integrating the basic idea of homomorphic encryption into Flajolet-Martin (FM) sketch, PDCS allows an aggregator to conduct distinct counting over large-scale data sets without knowing privacy of users. Moreover, PDCS supports various forms of sensing data, including camera images, location data, etc. PDCS expands each bit of the hashing values of users' original data, FM sketch is thus enhanced for encryption to protect users' privacy. We prove the security of PDCS under known-plaintext model. The theoretic and experimental results show that PDCS achieves high counting accuracy and practical efficiency with scalability over large-scale data sets.

Keywords: Distinct counting · Privacy-preserving · Mobile sensing · Flajolet-Martin sketch · Secure bitwise XOR

1 Introduction

With the rapid development of information technology and modern manufacturing, mobile devices have occupied an indispensable position in daily life. Especially, those devices, like smartphones, which are equipped with CPUs and a variety of sensors such as GPS and camera, are used not only for their traditional functions, but also for sensing and calculation. These features make these devices ideal mobile carriers favored by researchers as they study many issues. The mobile sensing problem is one of them. Recent years, an amount of mobile sensing projects have been developed with different mobile devices [18,21,23].

G. Li et al. (Eds.): DASFAA 2019, LNCS 11446, pp. 227–243, 2019.
https://doi.org/10.1007/978-3-030-18576-3_14

The process of mobile sensing can be described as: the aggregator issues tasks to users with mobile devices, then mobile devices of users collect sensing data and send them to the aggregator, after that, the aggregator processes all the data to draw valid conclusions. Generally, the aggregator needs to collect and monitor users' data continuously, meaning a considerable scale of collected sensing data.

There are two essential challenges in actual mobile sensing projects. One is that whether users are willing to give the original sensing data to the aggregator. As original data may contain users' private information such as physical location, consumption habits, physical health, etc. Most users would give a negative reaction to such a mobile sensing application lacking reliable privacy protection. The other one is, for the aggregator, how to solve the distinct counting problem [3] when facing the huge sensing data set with a large amount of duplicate data in various forms. If the aggregator fails to get the cardinality of users' data, then lots of computing resources will be wasted to handle duplicate data. In addition, excessive repetitive elements in an aggregated data set may result in characteristics of data being inconspicuous. In other words, a solution which can ensure users' privacy safety as well as solve distinct counting problem is in urgent need.

Exiting studies about distinct counting problem in mobile sensing mainly focus on various algorithms (such as Flajolet-Martin sketch [13], LogLog [24]), while few works have considered users' privacy during data aggregation. Han et al. [8] propose a secure data aggregation scheme to enable the traffic monitoring center to verify whether an aggregate sensing report is correct. Their security refers to the reliability of aggregated data rather than the user's privacy protection.

In this paper, we propose a scheme, **P**rivacy-preserving **D**istinct **C**ounting **S**cheme (PDCS), to solve the distinct counting problem with privacy protection of users. PDCS is based on a semi-honest model and it can complete distinct counting over large data sets with various forms of elements in the mobile sensing scenario. Through expanding each bit of the hashing values of users' original data added to the FM sketch, PDCS enhances FM sketch to apply the bitwise XOR homomorphic encryption algorithm as an encryption method, so that users' privacy gets protected even under known plaintext model. We conduct theoretical analysis and experiments, and the results show that our scheme achieves practical counting accuracy and efficiency.

The remainder of this paper is organized as follows. Section 2 discusses the related work. Section 3 defines related models and introduces several necessary preliminaries. Section 4 presents main idea and essential module of PDCS and analyze the correctness and security. Section 5 evaluates the accuracy rate and efficiency of PDCS. We conclude the paper in Sect. 6.

2 Related Work

2.1 Privacy Preserving in Mobile Sensing Applications

In terms of the applications about mobile sensing, most works focus on researching the various methods of user's privacy protection or discussing the operation of aggregated data, like [1,2,15,16,19].

Both [10] and [11], consider about the protection of user's privacy and then to seek the minimum computation in the aggregated data set. They study how an untrusted aggregator can periodically obtain desired statistics over the data contributed by multiple mobile users, without compromising the privacy of each user. Their scheme [11], which is based on [10], utilizes the redundancy in security to decrease the communication cost caused by each users joining and/or leaving activities. Their protocol traverses the entire data space to find the minimum value based on summation protocols rather than bitwise XOR operations.

In [25], Zhang and Chen propose semi-honest protocols to calculate the minimum and kth minimum values in mobile sensing systems. The data can be time-series. By using probabilistic coding schemes and a cipher system, they construct two protocols that allows homomorphic bitwise XOR computations for their problems. And the homomorphic bitwise XOR algorithm ensures privacy during the whole process. As the interaction times increase, the bits sent or received by users and the aggregator are much more.

2.2 Distinct Counting

On the other hand, distinct counting is also interested by a lot of researches. However, it is mainly discussed in Vehicular Ad Hoc Networks [8,13], rather than a more general mobile sensing scenario. Meanwhile, there are a group of works adopting different algorithms including FM sketch to solve this problem [5,14].

Considine et al. in [4] use FM sketches to accomplish a kind of robust in-network aggregation in sensor networks. The application situation is believed to exist packet loss or node failures. They consider about the coordinated collection of information towards a sink in the sensor network. However, the security problem is overlooked during the entire aggregation process. In [22], FM sketch is used to integrated with spatio-temporal indexes to solve the problem: "How many objects were in region x over the time interval t?". Like [4], Tao et al. in [22] do not mention the privacy protection of user's data.

Han et al. propose a secure data aggregation scheme in Vehicular Ad Hoc Networks which is based on FM sketch in [8]. And they also consider about the security problem, while their security goal is to enable the traffic monitoring center to verify whether an aggregate sensing report is correct or not. Their security refers to the aggregator's aggregated security rather than the user's privacy protection.

Remark: In general, there are few works that have studied on privacy protection of users and distinct counting for mobile sensing simultaneously. PDCS reduces the consumption of storage space and other resources, and has practicality due to guaranteeing user's privacy, which makes it worthy to be studied.

3 Problem Statements and Preliminaries

3.1 System and Security Model

System Model. We consider the system model in this paper as follows: there is a group of users with mobile devices who are providing data to an aggregator to do some sensing task. Assume that sensing data of each user is a set of data in various forms. The aggregator needs to find the cardinality of all users data, a big data set composed by plenty of sub data sets. When transmitting sensing data, all users would not reveal their original data to the aggregator. We discuss a general network model in mobile sensing, which exists a direct communication channel between the aggregator and every user. That is to say, the aggregator and all users form a star network topology. The communication channels could be 3G/4G, WiFi, or others supported by mobile devices and the aggregator in practical applications. Besides, as for each device, it can do the hash and bitwise XOR operation on its sensing data and transmit them to the aggregator.

Security Model. In this paper, we assume that it is a semi-honest model. All the aggregators and users observe the data transmission and collection process described above. However, they may attempt to derive extra information about other participators' private inputs, which they should not know. Therefore, the scheme is believed to be secure if it guarantees that every participator can learn no more information from the process than that this participator is entitled to know. For the users, they should not be able to get any data of each other without a permission. While for the aggregator, except for the encrypted data from users and the calculating result of these aggregated data, no extra knowledge about users ought to be acquired or speculated from the data he aggregates.

3.2 XOR Homomorphic Encryption

We choose the bitwise XOR homomorphic encryption as the encryption algorithm in this paper. A trusted third party, the authority, is needed during the process of key generation. Let $f_{m,\alpha,\beta}()$ denote a function in the pseudo-random function family $F_{m,\alpha,\beta} = \{f_{m,\alpha,\beta} : \{0,1\}^\alpha \rightarrow \{0,1\}^\beta\}_{m\in\{0,1\}^\gamma}$, where $\alpha, \beta, \gamma \in \mathbb{N}$. Denote $t \in \{0, ..., 2^v - 1\}$ the nonce information. The details are shown as follows.

a. **Key generation:**
   (1) The trusted authority uniformly and independently picks $m_1, ..., m_n \in \{0,1\}^\gamma$. Then the authority computes $M_a^i = m_i$ and $M_b^i = m_{(i\ mod\ n)+1}$ for each user $i (i = 1, ..., n)$, and sends them to user i.

(2) For each data set with the nonce information t which is different in each time of transmission, user i computes its secret key by

$$k_i = F_{M_a^i,v,l}(t) \oplus F_{M_b^i,v,l}(t) \quad (1)$$

b. **Encryption:** Denote by $x_i \in \{0,1\}^l$ a bit-string. The user i encrypts it by computing

$$\overline{x_i} = x_i \oplus k_i \quad (2)$$

c. **Decryption:** Denote by $\overline{x_i}$ a ciphertext of user i. The user i decrypts it by computing

$$x_i = \overline{x_i} \oplus k_i \quad (3)$$

d. **Aggregation:** Anyone can decrypt the bitwise XOR of all users plaintexts without any user's secret key by computing

$$x_1 \oplus ... \oplus x_n = \overline{x_1} \oplus ... \oplus \overline{x_n} \quad (4)$$

From Eq. (4), it is straightforward to see that the bitwise XOR of all users' keys equals to 0. As a result, the bitwise XOR of all users' ciphertexts is equal to the bitwise XOR of all users' plaintexts. In other words, this encryption algorithm is homomorphic on the bitwise XOR computation.

In this paper, the aggregator does not have any user's private key, so that it cannot decrypt any user's plaintext. Instead, it decrypts the bitwise XOR of all users' plaintexts and uses this information to solve the distinct counting problem. Therefore, when we talk about the aggregator's decryption operation, it means the decryption of the bitwise XOR of all users' plaintexts.

3.3 FM Sketch

A FM sketch is a data structure for probabilistic counting of distinct elements that has been introduced in [6]. It is widely used in network applications, such as data dissemination [12] and probabilistic aggregation [7,17].

FM sketch represents an approximation of a positive integer by a bit field $S = s_1, s_2, ..., s_w$ of length w, where $w \geq 1$. The bit field is initialized to zero at all positions. To add an element x to the sketch, it is hashed by a hash function h with geometrically distributed positive integer output, where the probability is $P(h(x) = i) = 2^{-i}$. The entry $s_{h(x)}$ is then set to 1. With probability 2^{-w}, we have $h(x) > w$ and no operation is performed in this case. A hash function with the necessary properties can easily be derived from a common hash function with equidistributed bit string output by using the position of the first 1-bit in the output string as the hash value.

According to [6], an approximation $C(S)$ of the number of distinct elements added to the sketch can be obtained by locating the end of the initial, uninterrupted sequence of ones.

$$Z(S) := \min(\{i \in \mathbb{N}_0 \mid i < w \wedge s_{i+1} = 0\} \cup \{w\}) \quad (5)$$

$$C(S) := \frac{2^{Z(S)}}{\varphi}, \varphi \approx 0.775351 \tag{6}$$

Since the variance of $Z(S)$ is pretty significant, the approximation $C(S)$ in Eq. (6) is not very accurate. To avoid this situation, a set of sketches will be used to represent a single value instead of only one sketch. [6] proposes the respective technique called Probabilistic Counting with Stochastic Averaging (PCSA). With PCSA, before being added there, each element is first mapped to one of the sketches by using an equidistributed hash function. If d sketches are used, denoted by $S_1, ..., S_d$, the estimation for the total number of distinct elements added is then calculated through

$$C(S_1, ..., S_d) := d \cdot \frac{2^{\sum_{i=1}^{m} \frac{Z(S_i)}{d}}}{\varphi} \tag{7}$$

However, [6] also points out that Eq. (7) is rather inaccurate as long as the number of elements is below approximately $10 \cdot d$. According to [20], we modify Eq. (7) in the following way:

$$C(S_1, ..., S_d) := d \cdot \frac{2^{\sum_{i=1}^{d} \frac{Z(S_i)}{d}} - 2^{-\kappa \cdot \sum_{i=1}^{d} \frac{Z(S_i)}{d}}}{\varphi}, \kappa \approx 1.75 \tag{8}$$

This alleviates the initial inaccuracies, while otherwise being asymptotically equivalent to Eq. (7). PCSA with d sketches yields a standard error of approximately $0.78/\sqrt{d}$ [6,9]. For many mobile sensing projects, it can achieve sufficiently good approximations when the sizes of data set are reasonable.

The FM sketch can be merged to obtain the total number of distinct elements added to any of them by a simple bitwise OR. Important here is that, by their construction, repeatedly combining the same sketches or adding already present elements again will not change the results, no matter how often or in which order these operations occur. This makes FM sketches ideally suited for the distinct counting scheme in mobile sensing.

4 Privacy-Preserving Distinct Counting Computation

In this section, we describe a specific operation on the sensing data of users based on FM sketch. Here we employ a knack to greatly reduce the overall computing time. And then, the important part in PDCS, operations of encryption and decryption(i.e. calculation based on ciphertexts), are presented in detail. After that, in Sect. 4.3, the correctness and the security of PDCS will be discussed. Assume that the space of users' data is $[0, N-1] (N \geqslant 2)$, and $w = \lceil \log_2 N \rceil$.

4.1 Main Idea

From Sect. 3.3, it is obvious that while dealing with the distinct counting problem, FM sketch cannot provide the protection of users' privacy during the transmission and calculation process. Therefore we provide a method that expands each bit of the string to make the sketch suitable for encryption operation, where the string is the calculating result of each user's data set.

As mentioned above, there are n users in total. In the FM sketch, the bit field $S = s_1, s_2, ..., s_w$ of length w is initialized to zero at all positions. In the meanwhile, each user's sensing data is a data set with various forms of elements, ranging in size from small to large. Assume that user $i(i = 1, ..., n)$ has L_i elements in his sensing data set. Let the set $\{x_l^{(i)}\}$ denote the original sensing data set of user i, where $l = 1, ..., L_i$. While putting user i's sensing data $\{x_l^{(i)}\}$ into the FM sketch, PDCS determines the bit field S bit by bit, from the Most Significant Bit (MSB) to the Least Significant Bit (LSB). The MSB refers to the last bit of S, and the LSB is the first bit relatively.

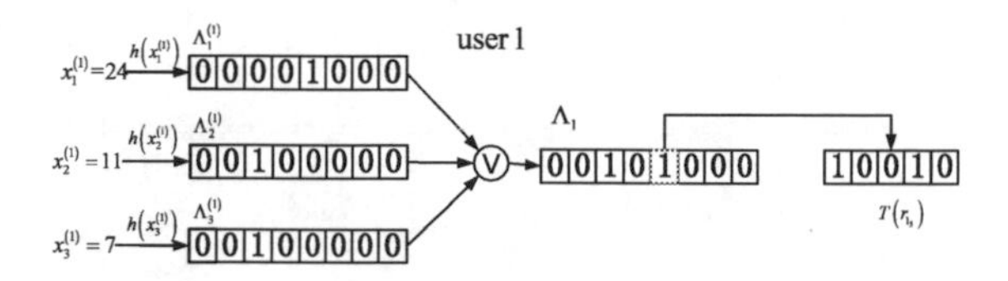

Fig. 1. An example of the process that user 1 deals with his original data set. Here, for example, the elements in the data set are all integers. User 1 hashes elements of his data set into 8-bit strings and do the bitwise OR operation to get a string as the representation of his data set. Then each bit of this string is coded to a 5-bit string.

Step 1

This step is taken by users. For a user $i(i = 1, ..., n)$, every element $x_l^{(i)}(l = 1, ..., L_i)$ in his original data set $\{x_l^{(i)}\}$ is hashed by a hash function h with a w-bit string output. Let $\Lambda_l^{(i)} = (r_l^{(i)})_1, ..., (r_l^{(i)})_w$ denote this bit string output of length w, where $(r_l^{(i)})_j$ is the jth bit in the string and the probability is $P((r_l^{(i)})_j = 1) = 2^{-j}$. That is to say, $\Lambda_l^{(i)} = h(x_l^{(i)})$. Then a bitwise OR operation is taken to get

$$\Lambda_i = \Lambda_1^{(i)} \vee ... \vee \Lambda_{L_i}^{(i)} \tag{9}$$

The string $\Lambda_i = r_{i_1}, ..., r_{i_w}$ represents all elements in user i's original data set.

According to the knowledge mentioned in Sect. 3.3, Eq. (10) should be correct.

$$S = \Lambda_1 \vee ... \vee \Lambda_n \tag{10}$$

However, S should not be calculated out straightforward. Because according to Eq. (9), if the aggregator could receive Λ_i directly, Λ_i would reveal the original data of user i, especially when the size of his data set is small. Therefore, a series of operations should be carried on the Λ_i.

Step 2

This step is also done on the user's side. The user i operates on each bit of the string Λ_i in order to avoid any damage caused on the privacy. In PDCS, we design a kind of specific coding scheme for these bits. Let $T(r_{i_j})$ denote the corresponding code of r_{i_j} in the coding scheme, where $j = 1, ..., w$. The coding scheme is defined as follows:

$$T(r_{i_j}) \begin{cases} = 0^q, & if\ r_{i_j} = 0 \\ \stackrel{\circ}{=} \{0,1\}^q \setminus 0^q, & if\ r_{i_j} = 1 \end{cases} \tag{11}$$

where $q \in N$ is the accuracy controlling parameter and $\stackrel{\circ}{=}$ denotes to sample uniformly at random. Figure 1 shows an example of the process user 1 deals with his original data set.

Step 3

The aggregator takes this step after aggravation all users' coded data. Let $G(j) = T(r_{1_j}) \oplus ... \oplus T(r_{n_j})$ with bitwise XOR operation. Then there is a judgement rule designed to determine each bit of FM sketch S, corresponding to the coding scheme (11). We define the rule as follows:

$$s_j = \begin{cases} 0, if\ G(j) = 0^q \\ 1, if\ G(j) \neq 0^q \end{cases} \tag{12}$$

where s_j is the jth-LSB, or $(w-j)$th-MSB in the bit field S. Notice here that when PDCS judging each bit in FM sketch, it starts from the MSB to the LSB.

Step 4

The calculation work is done by the aggregator. Based on the FM sketch S, the aggregator can get a significant parameter $Z(S)$, the position of the last bit in S that is 1, according to Eq. (5). As mentioned in Sect. 3.3, the approximation of distinct counting needs several more FM sketches in which the hash functions are different. After taking Step 1 to Step 3 for d times and according to Eq. (8), the aggregator can get the final result $C(S)$, the number of distinct elements in the sensing data set.

Remark 1: In Step 2, it is worth noting that there is a probability of $1 - 1/2^q$ to occur such a situation, where r_{i_j} equals to 1 but $T(r_{i_j})$ is coded to be 0^q. Thus in Eq. (11), the coding scheme requires that if this situation happened, r_{i_j} should be recoded until $T(r_{i_j})$ is not 0^q. In that step, each bit of the string Λ_i, from MSB to LSB, would be expanded into a q-bit string $T(r_{i_j})$ under the action of our coding scheme (11), which is suitable for encryption operation.

Remark 2: Notice that during the whole process, in order to reduce the computing time, we employ a knack here which is that PDCS determines bits of FM sketch S from the last bit to the first bit. When the aggregator applies FM sketches, the purpose is to find out the position of the last bit in S that is 1 and regard it as an index. And this purpose is equal to find out the position of the first bit in S that is 1, when PDCS starts finding from the last bit of S. This transformation means the aggregator does not have to determine all bits in S, after all, what the aggregator needs is the index to calculate the number of distinct counting of the data set rather than the whole S. Through this knack, PDCS can leave out a lot of computing steps, thus improving the efficiency.

4.2 Privacy-Preserving Distinct Counting Scheme

In Sect. 4.1, we can calculate the number of distinct counting through PDCS. The specific operation towards users' data is prepared for the homomorphic encryption to protect users' privacy. In this subsection, we highlight the modules of encryption and decryption(i.e. calculation based on ciphertexts) in PDCS that allow the aggregator to solve the distinct counting problem and to avoid acquiring each user's data privacy at the same time. In our assumption, there is a trusted authority as a third party who helps users and the aggregator to establish a key system each time.

(1) Setup

The protection mechanism of PDCS is based on the bitwise XOR homomorphic encryption introduced in Sect. 3.2. The trusted authority has $m_1, ..., m_n \in \{0,1\}^\gamma$ privately and he computes $M_a^i = m_i$ and $M_b^i = m_{(i \ mod \ n)+1}$ for each user $i(i = 1, ..., n)$. Then the two seeds are sent to the corresponding user. The user i does a bitwise XOR operation on the seeds as well as a nonce number t according to Eq. (1) to acquire his own key k_i. Notice that the nonce number t used for calculating k is different in each transmission.

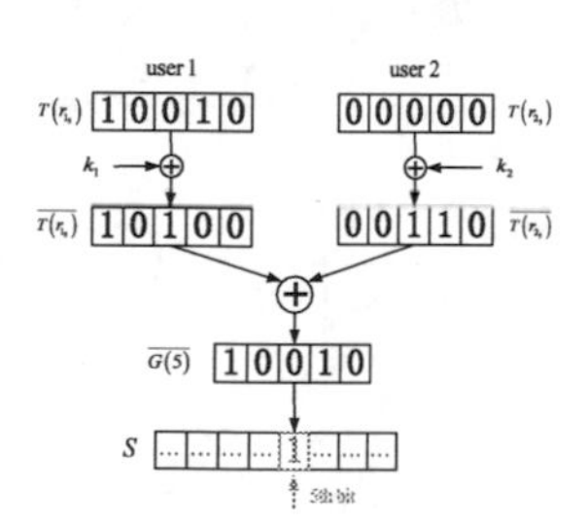

Fig. 2. An example of determining the 5th bit in FM sketch S with the bitwise XOR homomorphic encryption, where the coding scheme defined the length of the bit string $T(r_{i_j})$ is 5. The result shows that the aggregation result is not influenced by the operations of user's privacy protection.

(2) Encrypt

The data encryption is operated on the user's side. The user i regards the coding string $T(r_{i_j})$ for jth bit of his data representation Λ_i as the plaintext and encryptes it with the bitwise XOR homomorphic encryption algorithm to get the ciphertext

$$\overline{T(r_{i_j})} = T(r_{i_j}) \oplus k_i, (j = 1, ..., w) \tag{13}$$

where the user i's key k_i is generated as introduced above. Then the ciphertext $\overline{T(r_{i_j})}$ is sent to the aggregator as user i's sensing data.

(3) Aggregate

On the side of the aggregator, he collects all the n users' data about the jth-LSB and then does the bitwise XOR computations. Denote by $\overline{G(j)}$ the bit string result. According to Eq. (4) it can be drawn that

Algorithm 1. Privacy-preserving Distinct Counting Scheme

Input:
$\{x_i^l\}$: User i's data set with various elements, $l \in [0, L_i], x_i^l \in [0, N-1], i = 1, ..., n$;
h: a hash function with a w-bit string output;
d: the number of FM sketches;
M_a^i and M_b^i: two secret seeds of User i;
$t \in [0, 2^v - 1]$: a public known nonce number;
$q \in \mathbb{N}$: an accuracy controlling parameter.

Output: The number of distinct elements in $\{x_i\}$, $i = 1, ..., n$.

1: **for** $k = 1$ to d **do**
2:     **for** $i = 1$ to n **do**
3:         User i: $\Lambda_i \leftarrow h(x_i)$, $len(\Lambda_i) = w$;
4:         User i: $k_i = F_{M_a^i,v,q}(t) \oplus F_{M_b^i,v,q}(t)$;
5:     **end for**
6:     **for** $j = w$ to 1 **do**
7:         **for** $i = 1$ to n **do**
8:             User i: $T(r_{i_j}) \leftarrow r_{i_j}$ in Λ_i, $len(T(r_{i_j})) = q$;
9:             User i: $\overline{T(r_{i_j})} = T(r_{i_j}) \oplus k_i$;
10:        **end for**
11:        Aggregator P: $\overline{G(j)} = \overline{T(r_{1_j})} \oplus ... \oplus \overline{T(r_{n_j})}$.
12:        **if** $\overline{G(j)} = \{0\}^q$ **then**
13:            continue;
14:        **else**
15:            break;
16:        **end if**
17:    **end for**
18: **end for**
19: **return** $C(S) = d \cdot \frac{2^{\sum_{k=1}^{d} \frac{Z(S_k)}{d}} - 2^{-\kappa \cdot \sum_{k=1}^{d} \frac{Z(S_k)}{d}}}{\varphi}$;

$$\overline{G(j)} = \overline{T(r_{1_j})} \oplus ... \oplus \overline{T(r_{n_j})} \tag{14}$$

It is easy to see if the jth-LSBs of the n users are all 0, then the bitwise XOR of the corresponding strings $\overline{G(j)}$ is always a q-bit string of 0s. If there is any user whose data Λ is 1 on the jth-bit, the bitwise XOR of all reports corresponding strings is not a q-bit string of 0s with a probability of $1 - 1/2^q$. However this situation has little damage on the accuracy which will be proved in Sect. 4.3.

(4) Judge

Just like the rule mentioned above, we define the rule as follows:

$$s_j = \begin{cases} 0, if\ \overline{G(j)} = 0^q \\ 1, if\ \overline{G(j)} \neq 0^q \end{cases} \tag{15}$$

where s_j is the jth-LSB in the bit field S.

In Fig. 2, a detailed example of aggregation in the FM sketch is shown. And we present the formal description of PDCS in Algorithm 1.

Remark 1: The operation of homomorphic encryption causes no damage on the accuracy of PDCS. According to the property of the bitwise XOR homomorphic encryption, there is $\overline{G(j)} = \overline{T(r_{1_j})} \oplus ... \oplus \overline{T(r_{n_j})} = T(r_{1_j}) \oplus ... \oplus T(r_{n_j}) = G(j)$. Thus, we can say that Eq. (15) is equal to Eq. (12), which means that the calculated aggregation result is not influenced by the encryption and decryption operations for user's privacy protection. Then the final result of distinct counting problem is calculated by Step 4 in Sect. 4.1.

Remark 2: The correctness and security of PDCS are credible. According to Eq. (10), we have

$$s_j = r_{i_1} \vee ... \vee r_{i_w} \tag{16}$$

In PDCS, Eq. (15) is equal to Eq. (16) with a probability of $1-1/2^q$, which means that the operations in PDCS have nearly no effect on the final result when the parameter q is appropriate. We will prove it in Theorem 1. The security of PDCS will later be formally proved in Theorem 2.

4.3 Scheme Analysis

Theorem 1. *(Correctness). The probability that the result of Eq. (15) equals to that of Eq. (16) is greater or equal to* $1-1/2^q$. *The correctness of PDCS is greater or equal to* $1-(w/2^q)^d$.

Proof. According to the definition in Sect. 3.3, it is obvious that the sketch constructed by Eq. (10) is the correct result of our problem. Equation (16) is one of Eq. (10)'s mutually independent w parts to determine the jth-LSB bit. While in PDCS, Eq. (15) represents the result. Actually, Eq. (15) is equal to Eq. (16) with a probability of $1 - 1/2^q$ on the calculation.

On the basis of the regular of the bitwise OR, only when all the numbers on that bit are 0, the result bit is 0, which is $0 \vee ... \vee 0 = 0$. Otherwise, that bit should be 1. If our scheme is 100 percent accurate, when $s_j = 0$, it means $G(j)$ should be 0^q where all users' r_{i_j} should be 0. However there is a special case that a user y whose r_{y_j} is 1 while y's encrypted bit string equals the bitwise XOR of all other users' encrypted strings. Since the encoding function in our scheme is random and the encoding string has 2^q different choices, the probability for the result of our scheme being not accurate is $1/2^q$. Therefore, we have:

$$\begin{aligned} & P(the\ jth\ bit\ is\ accurate) \\ & = P(all\ users'\ jth\ bits\ are\ 0) \times 1 + P(any\ user's\ jth\ bit\ is\ 0) \times (1-1/2^q) \\ & \geqslant 1-1/2^q \end{aligned}$$

Because Eq. (15) has to be independently calculated for w times to achieve the goal of Eq. (10) and there are d FM sketches used, it is obvious that the correctness of PDCS is greater or equal to $1-(w/2^q)^d$.

As in most cases, a malicious aggregator can only have the knowledge of ciphertext in privacy-preserving mobile sensing schemes. This belongs to the ciphertext-only attacks, which corresponds to an attaker of minimal capability. However, we still analyze the security of PDCS under a more stronger model,

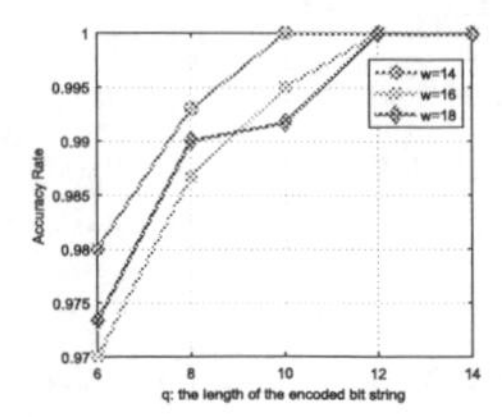

Fig. 3. Accuracy rate influenced by the value of q.

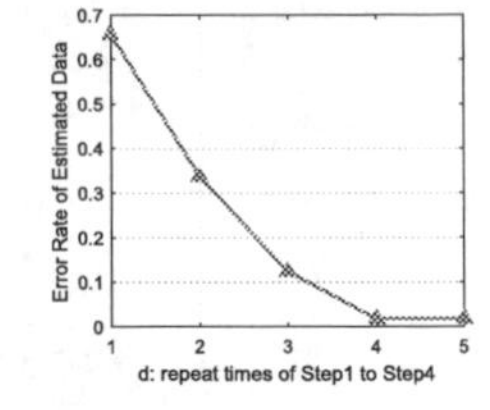

Fig. 4. Error rate affected by the value of d.

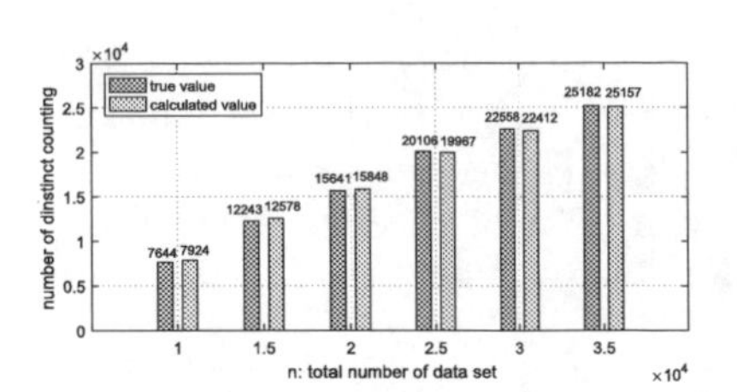

Fig. 5. The contrast of the numbers of distinct counting coming from the true value and the calculated value.

known plaintext model, which assumes that the attackers may obtain a certain number of plaintext-ciphertext pairs through extra channels. The security of the proposed PDCS is summarized in the following theorem.

Theorem 2. *(Security). or the homomorphic operations in PDCS, there is no probabilistic polynomial time (P.P.T.) adversary that can break the data confidentiality of user's data under known plaintext model.*

Proof. In PDCS, for the aggregator, we prove that there is no extra knowledge revealed to him in PDCS. We consider the situation that the aggregator could acquire most information. To calculate out the final result, all the w bits in S should be confirmed, which means that there must be w times communication between each user and the aggregator and each time the aggregator could get n cipertexts from all users. Let $I = (I_1, ..., I_w)$ denote the aggregator's information received from all users for w times, where $I_j = (\overline{T(r_{1_j})}, ..., \overline{T(r_{n_j})})$ $(j = 1, ..., n)$ is the ciphertexts of q-bit bit strings from all users to decide the jth bit in S. The aggregator calculates $G(j)$ according to Eq. (14) and then determines $s(j)$ in S by Eq. (15).

If the result is $s_j = 1$, then $G(j) \neq 0^q$, which can help the aggregator speculate that in I_j there is at least one bit string $\overline{T}$ from some user which is not 0^q. The aggregator wants to speculate T from $\overline{T}$. Since the aggregator has no corresponding key, the probability that he guesses the correct plaintext is $1/2^q$. Under known plaintext model, the aggregator could get a plaintext-ciphertext pairs to calculate a keys. Then the probability of knowing a certain user's original data rises to $1/A_n^a$. However, the fact that users' keys are different each time leads to low probability and the attacking time $O(nq)$ is over P.P.T. Therefore, the aggregator could not conjecture any extra knowledge about the users.

If $s_j = 0$, then I_j are all 0^q or it happens to such a case that some bit strings, $\{0,1\}^q \setminus 0^q$, equal to 0^q under the effect of XOR operations. However the aggregator could not distinguish these two situations by calculation in P.P.T.

Moreover, because the keys are pseudo-random for each user and each time, $I_1, ..., I_w$ are independent. As a result, there is no probabilistic polynomial time (P.P.T.) adversary that can break the data confidentiality of user's data under known plaintext model. PDCS ensures the privacy protection of users.

Table 1. Communication cost and computation complexity

	A user		The aggregator		R. complexity
	Comm. cost	Comp. complexity	Comm. cost	Comp. complexity	
Baseline	$N\lceil\log_2 n\rceil$	$O(N\lceil\log_2 n\rceil)$	$nN\lceil\log_2 n\rceil$	$O(nN\lceil\log_2 n\rceil)$	1
PDCS	$q\lceil\log_2 N\rceil$	$O(q\lceil\log_2 N\rceil)$	$nq\lceil\log_2 N\rceil$	$O(nq\lceil\log_2 N\rceil)$	d

5 Performance Evaluation

In this section, we conduct experiments to evaluate the accuracy of PDCS and its efficiency compared with the situation lacking of privacy protection.

5.1 Accuracy Evaluation of PDCS

In PDCS, q is an accuracy controlling parameter, the length of encoded bit string in *Step 2* of Sect. 4.1. Figure 3 shows the relationship between the value of q and the accuracy rate when the length w of FM sketch is changing, where the total amount of users is $n = 15000$. It can be concluded that no matter what value w is, as the value of q approaches w, the accuracy rate gradually increases to nearly 100%, which is in accord with theoretical analysis above.

Since there is a significant error in applying only one FM sketch, d, the number of FM sketches, must be discussed. Figure 4 shows that the error rate of estimated data decreases dramatically along with the increase of repeat times in the beginning, then keeps relatively stable after a specific threshold, like $d = 4$ in this experiment. For different sizes of data sets, the threshold will be different.

We set experiments with different number of users participating in the program. The corresponding number of distinct counting in each data set is independent and irregular since the elements which present users' sensing data are generated entirely randomly. The calculated values of PDCS are contrasted with true values in Fig. 5. There is difference between the two values, and as the size of the data set improves, the overall difference tends to decrease but still exists fluctuation. The fluctuation is associated with the nature result of FM sketch which has a close relationship with the multiply of 2.

According to the results in Fig. 5, we calculate out the accuracy rate of PDCS presented in Fig. 6 to evaluate the correctness of PDCS more intuitively. It is obvious that the accuracy rate of PDCS is gradually raising to close 100% along with the increase of the size of data sets, where even in the case of a small data set the accuracy rate can still reach 97%. When the amount of users are huge and corresponding cardinal number is big, PDCS can perform much better.

5.2 Efficiency Evaluation of PDCS

We explore efficiency of PDCS by testing the communication cost and computing time.

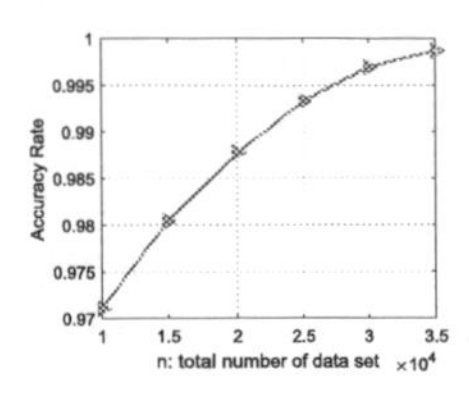

Fig. 6. Accuracy rate of PDCS on different size of data sets.

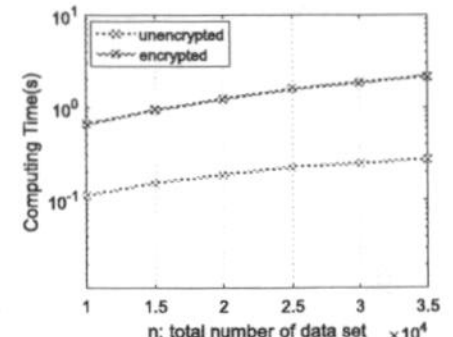

Fig. 7. Computing time of the unencrypted process and PDCS.

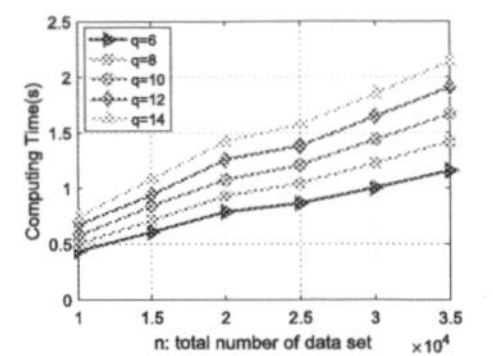

Fig. 8. Computing time of PDCS with different value of q.

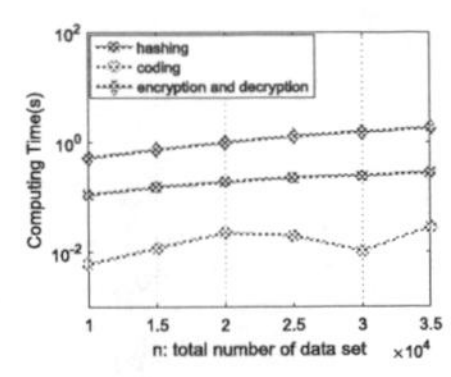

Fig. 9. Computing time spent by different steps in PDCS.

(1) Communication Overhead

Table 1 shows the comparison of communication cost between baseline method without privacy protection and PDCS. In the Table 1, the total bits sent by a user, as the communication cost of a user, and the total bits received by the aggregator, as the communication cost of the aggregator, are the measured standards, as well as the computation complexity and round complexity of two schemes. Mentioned parameters includes: n is the total number of users and the range of users' data is $[0, N-1]$, and $w = \lceil \log_2 N \rceil$ is the length of each user's bit string, and d is the number of FM sketches we applied. As the proof of Theorem 1 shows, $1 - (w/2^q)^d$ is the upper limit of the correctness of PDCS. When q is approximately equal to w and not too small, the error rate of PDCS will decrease to an acceptable level(for example, less than 0.001). At the meanwhile, the communication cost of PDCS affected by q would also be reduced. Besides, PDCS can achieve to send or receive less data than the baseline method when n is not too much greater than N, i.e. $n = O(N)$.

However, the total communication cost is also influenced by the round complexity. The round complexity refers to the amount of time a user has to keep communicating online. Notice that the baseline method needs only one round communication which is its most significant advantage. Therefore, the cases PDCS performs better are that the network connection is stable, while when the network connection cannot stay reliable, the baseline method is more suitable.

(2) Computation Overhead

We discuss the computing time spent during the whole process. Here the computing time of PDCS includes the time of hash operation and coding, encryption time for each user, and the time of decryption to determine the final FM sketch S and calculating results for the aggregator. The data used as a comparison is the computing time of the number of distinct counting calculated without privacy protection in Fig. 7. It can be seen that it takes more time for PDCS to calculate out the results. Since there are more processes like encoding, encrypting, decrypting and formula calculating than the general method, PDCS is relatively more time-consuming. However, this consumption is within an acceptable range, as shown in Fig. 7 that with the data set expanding, the trend of the increase in the consumption time is slower than the linear increase. Besides, due to solving the distinct counting problem, the following other operations on the aggregated data set will reduce resources consumption of repetitive process. Furthermore, when the size of data set is huge, the probability of the index $Z(S)$ approaching the end of FM sketch S is high. With our knack in Sect. 4.1, the computing time will gradually decrease accordingly. Therefore, on the whole, PDCS does not waste computing time. This conclusion proves that the efficiency of PDCS is appropriate for large-scale data aggregation processing.

About the computing time, we also conducted assessments of PDCS under different factors. Firstly, the value of q is obvious to have affection on the computing time. Figure 8 shows the variation trend of computing time of PDCS with different value of q. As the q is bigger, the corresponding time is much more and the difference caused between contiguous different q is rising when the size of data set increases. Therefore, an appropriate value of q is needed. Next, we evaluate the computing time spent by each step in PDCS. The corresponding results are present in Fig. 9. In Fig. 9, the encryption and decryption step which is the most important part of achieving users' privacy-preserving in PDCS, is the most costly compared with other steps. Due to the bitwise XOR operation, such result is reasonable. And as the size of data set raises, the increase of encryption and decryption time will slow down, since we employ the knack on FM sketches in PDCS. Besides, the increasing curves of all steps in Fig. 9 tend to be lower than the linear increase, which is in accord with the result in Fig. 7.

6 Conclusion

Both privacy protection of users and distinct counting problem on large data sets are essential issues in the mobile sensing applications. In this paper, we propose a privacy-preserving distinct counting scheme, PDCS, to solve these two problems simultaneously. PDCS expands each bit of the hashing values of users' original data, so that FM sketch is enhanced for encryption to protect user's privacy. And we choose the bitwise XOR encryption algorithm as the encryption algorithm. According to the theoretical analysis, PDCS causes little damage on the accuracy of FM sketch. Moreover, a set of experiments demonstrates that with appropriate value of several parameters, PDCS achieves high counting accuracy and practical efficiency with scalability over large-scale data sets.

Acknowledgement. This work is supported in part by the National Natural Science Foundation of China (NSFC) under grants 61872372, 61572026, 61672195, and Open Foundation of State Key Laboratory of Cryptology (No:MMKFKT201617).

References

1. Au, M.H., Liang, K., Liu, J.K., Lu, R., Ning, J.: Privacy-preserving personal data operation on mobile cloud—chances and challenges over advanced persistent threat. Future Gener. Comput. Syst. **79**, 337–349 (2018)
2. Bae, M., Kim, K., Kim, H.: Preserving privacy and efficiency in data communication and aggregation for AMI network. J. Netw. Comput. Appl. **59**, 333–344 (2016)
3. Bar-Yossef, Z., Jayram, T.S., Kumar, R., Sivakumar, D., Trevisan, L.: Counting distinct elements in a data stream. In: Rolim, J.D.P., Vadhan, S. (eds.) RANDOM 2002. LNCS, vol. 2483, pp. 1–10. Springer, Heidelberg (2002). https://doi.org/10.1007/3-540-45726-7_1
4. Considine, J., Li, F., Kollios, G., Byers, J.: Approximate aggregation techniques for sensor databases. In: Proceedings of the 20th International Conference on Data Engineering, pp. 449–460. IEEE (2004)
5. Dietzel, S., Bako, B., Schoch, E., Kargl, F.: A fuzzy logic based approach for structure-free aggregation in vehicular ad-hoc networks. In: Proceedings of the Sixth ACM International Workshop on VehiculAr InterNETworking, pp. 79–88. ACM (2009)
6. Flajolet, P., Martin, G.N.: Probabilistic counting algorithms for data base applications. J. Comput. Syst. Sci. **31**(2), 182–209 (1985)
7. Garofalakis, M., Hellerstein, J.M., Maniatis, P.: Proof sketches: verifiable in-network aggregation. In: IEEE 23rd International Conference on Data Engineering, ICDE 2007, pp. 996–1005. IEEE (2007)
8. Han, Q., Du, S., Ren, D., Zhu, H.: SAS: a secure data aggregation scheme in vehicular sensing networks. In: 2010 IEEE International Conference on Communications (ICC), pp. 1–5. IEEE (2010)
9. Kirschenhofer, P., Prodinger, H., Szpankowski, W.: How to count quickly and accurately: a unified analysis of probabilistic counting and other related problems. In: Kuich, W. (ed.) ICALP 1992. LNCS, vol. 623, pp. 211–222. Springer, Heidelberg (1992). https://doi.org/10.1007/3-540-55719-9_75
10. Li, Q., Cao, G.: Efficient and privacy-preserving data aggregation in mobile sensing. In: 2012 20th IEEE International Conference on Network Protocols (ICNP), pp. 1–10. IEEE (2012)
11. Li, Q., Cao, G., La Porta, T.F.: Efficient and privacy-aware data aggregation in mobile sensing. IEEE Trans. Dependable Secure Comput. **11**(2), 115–129 (2014)
12. Lochert, C., Rybicki, J., Scheuermann, B., Mauve, M.: Scalable data dissemination for inter-vehicle-communication: aggregation versus peer-to-peer (skalierbare informationsverbreitung für die fahrzeug-fahrzeug-kommunikation: aggregation versus peer-to-peer). IT-Inf. Technol. **50**(4), 237–242 (2008)
13. Lochert, C., Scheuermann, B., Mauve, M.: Probabilistic aggregation for data dissemination in VANETs. In: Proceedings of the Fourth ACM International Workshop on Vehicular ad Hoc Networks, pp. 1–8. ACM (2007)
14. Lochert, C., Scheuermann, B., Mauve, M.: A probabilistic method for cooperative hierarchical aggregation of data in VANETs. Ad Hoc Netw. **8**(5), 518–530 (2010)
15. Lu, R., Liang, X., Li, X., Lin, X., Shen, X.: EPPA: an efficient and privacy-preserving aggregation scheme for secure smart grid communications. IEEE Trans. Parallel Distrib. Syst. **23**(9), 1621–1631 (2012)
16. Ma, R., Cao, Z.: Serial number based encryption and its application for mobile social networks. Peer-to-Peer Netw. Appl. **10**(2), 332–339 (2017)

17. Nadeem, T., Dashtinezhad, S., Liao, C., Iftode, L.: TrafficView: traffic data dissemination using car-to-car communication. ACM SIGMOBILE Mob. Comput. Commun. Rev. **8**(3), 6–19 (2004)
18. Rana, R.K., Chou, C.T., Kanhere, S.S., Bulusu, N., Hu, W.: Ear-phone: an end-to-end participatory urban noise mapping system. In: Proceedings of the 9th ACM/IEEE International Conference on Information Processing in Sensor Networks, pp. 105–116. ACM (2010)
19. Samanthula, B.K., Jiang, W., Madria, S.: A probabilistic encryption based min/max computation in wireless sensor networks. In: 2013 IEEE 14th International Conference on Mobile Data Management (MDM), vol. 1, pp. 77–86. IEEE (2013)
20. Scheuermann, B., Mauve, M.: Near-optimal compression of probabilistic counting sketches for networking applications. In: DIALM-POMC. Citeseer (2007)
21. Tan, X., et al.: An autonomous robotic fish for mobile sensing. In: IEEE/RSJ International Conference on Intelligent Robots and Systems, pp. 5424–5429. IEEE (2006)
22. Tao, Y., Kollios, G., Considine, J., Li, F., Papadias, D.: Spatio-temporal aggregation using sketches. In: Proceedings of the 20th International Conference on Data Engineering, pp. 214–225. IEEE (2004)
23. Thiagarajan, A., et al.: VTrack: accurate, energy-aware road traffic delay estimation using mobile phones. In: Proceedings of the 7th ACM Conference on Embedded Networked Sensor Systems, pp. 85–98. ACM (2009)
24. Wang, L., et al.: Fine-grained probability counting: refined loglog algorithm. In: IEEE Bigcomp (2018)
25. Zhang, Y., Chen, Q., Zhong, S.: Efficient and privacy-preserving min and k th min computations in mobile sensing systems. IEEE Trans. Dependable Secure Comput. **14**(1), 9–21 (2017)

Reinforced Reliable Worker Selection for Spatial Crowdsensing Networks

Yang Wang, Junwei Lu, Jingxiao Chen, Xiaofeng Gao(✉), and Guihai Chen

Shanghai Key Laboratory of Scalable Computing and Systems,
Department of Computer Science and Engineering,
Shanghai Jiao Tong University, Shanghai, China
{y_wang,luke_13,timemachine}@sjtu.edu.cn, {gao-xf,gchen}@cs.sjtu.edu.cn

Abstract. Spatial Crowdsensing Networks limit the sensing tasks in some special places where workers should sense data for them. Due to the lack of a priori information about quality of workers, guaranteeing the quality of the sensing tasks remains a key challenge. In this paper, we model the quality of workers through two factors, namely bias and variance, which describe the continuous value feature of sensing tasks. After calibrating the bias, we should iteratively estimate worker variances more and more accurately. Meanwhile, we should select more reliable workers with low variances to finish sensing tasks. This is a classic exploration and exploitation dilemma. Therefore, to overcome the dilemma, we design a novel Multi-Armed Bandit (MAB) algorithm which is based on Upper Confidence Bounds (UCB) scheme and combined with a weighted data aggregation scheme to calculate a better ground truth of a sensing task. Then, we prove the expected sensing error of sensing tasks can be bounded according to the regret bound of the MAB in our setting. In simulation experiments, we use a real world data set to validate the theoretical results of our algorithm and it outperforms two baselines significantly in different settings.

Keywords: Crowdsensing · Worker selection · Multi-Armed Bandit · Data aggregation

1 Introduction

Crowdsourcing (CSo) is becoming an important outsourcing way in ubiquitous mobile wireless networks. There are many crowdsourcing platforms to release tasks, e.g., *Amazon Mechanical Turk*, *Upwork*, *CrowdFlower*, and *uTest*. In these systems, guaranteeing the completion quality of tasks remains a main challenge, due to the limited a priori information about the quality of workers. Therefore, data management becomes interest of many industrial and academic communities. Meanwhile, many techniques in CSo, such as data aggregation [26], worker selection [14], and task assignment [29] (online version of worker selection) are developed to improve the completion quality of the tasks.

G. Li et al. (Eds.): DASFAA 2019, LNCS 11446, pp. 244–259, 2019.
https://doi.org/10.1007/978-3-030-18576-3_15

Most crowdsourcing tasks require workers to upload their answers for question tasks or choices for labelling tasks. The metrics for evaluating quality of workers in question or labelling tasks are normally error ratio or accuracy which are commonly discrete values. From the perspective of data type, Crowdsensing (CSe) can be viewed as a special kind of CSo, which brings in continuous value. In order to seek for high-quality sensing data, we consider using worker bias and variance to model quality of workers in CSe, because worker bias and variance can provide a continuous and uncertain description for quality of workers.

In this paper, we propose a spatial CSe framework which aims to infer the truth value of some physical measures such as temperature, humidity, noise and so on, at different points of interest (PoIs). Opportunistic workers with mobile devices, who go by one of PoIs, will be candidates to execute sensing tasks for CSe requesters. However, we do not know the quality of these workers in advance. Therefore, we should select them to do sensing tasks for many times and estimate their quality according to their sensing data. It means that the more times a worker is selected to do tasks, the more accurate the estimation of his/her quality is. Additionally, a budget is given by platform and limited for selecting finite high-quality workers. Obviously, the exploration and exploitation procedure appropriately matches the feature of Multi-Armed Bandit (MAB) which expects to reduce the regret of multi-decision procedure under the budget. We should define a suitable reward for the MAB in CSe, which involves worker bias and variance together, to implement a more accurate measure. Authors in [19] also use MAB to solve the worker selection problem. However, their task type and data aggregation method are different from ours in this paper.

Due to lack of Ground Truth (GT) of physical measure, we should exploit the sensing data from workers to infer it and apply it in the estimation of quality of workers. Therefore, estimating the GT well not only helps estimate quality of workers more accurately but also accelerates the convergence of MAB method. We will introduce a Bayesian estimation method for GT which outperforms maximum likelihood method in our simulation experiments.

The framework of reinforced reliable worker selection for Crowdsensing is illustrated in Fig. 1. The circles in this figure are the main operations of the framework. As we can see from the top left of the figure, the tasks requester will send requests associated with several PoIs for sensing tasks via the crowd platform. Then, the crowd platform divides the workers into groups according to the current locations of workers and the locations of PoIs. Each group of workers finish one sensing task at a specific PoI. Note that some of the workers that are far away from the PoIs will not be selected as candidates in this round, however, they might still be selected in future rounds.

After the process of worker groups division, the processes of workers calibration and selection are described as follows. Firstly, the platform will calibrate the newly coming workers whose biases are unknown. Under the constraint of the sensing budget, the platform will select workers who have as high quality as possible to produce sensing data for each task (PoI). However, the worker variance is unknown, which makes it difficult to select the best one. Therefore,

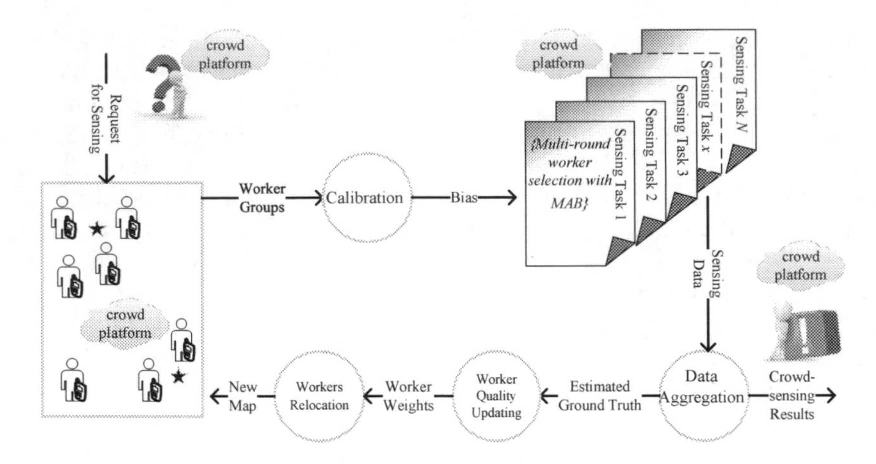

Fig. 1. The framework of reinforced reliable workers selection for Crowdsensing.

the platform introduces one MAB to make the decision of worker selection for each one of tasks. As a result, there are N tasks to be done and N MABs to perform the worker selection procedure. Moreover, workers can be selected several times to do the same sensing task, hence their estimated variance will be updated to approach the true variances gradually. In data aggregation stage, the platform should infer a good estimation of GT. The estimations of GTs as the results for sensing task requests will be output. The worker sensing quality and their locations should be updated for next requests in following time. As the process continues, the knowledge about the quality of workers becomes clearer and clearer. Due to the fact that the regrets of MABs are bounded, the expected sensing error of the tasks will become smaller and smaller, which will be theoretically analyzed in Sect. 4.

The contributions of this paper have several folds and are summarized as follows.

1. We model the worker quality with two continuous factors: bias and variance. To the best of our knowledge, it is a first try to combine them to model quality of worker in CSe. We also propose an expected sensing error minimization problem based on the quality of workers.
2. We address a worker selection problem without the quality of workers in CSe tasks, which is very rare in research literature in comparison with worker selection in question tasks or labeling tasks.
3. A novel MAB is proposed to solve the exploration and exploitation dilemma of worker selection according to the quality of workers in multi-rounds. Besides, it is also novel to use the bayesian estimation for GT estimation which brings a significant improvement of performance as shown in our experiments on the real world data.

The rest of this paper is organized as follows. Some related work is reviewed in Sect. 2. The framework of reinforced reliable worker selection is introduced in Sect. 3. Expected error minimization problem is proposed and analyzed in Sect. 4. We propose a Multi-Armed Bandit based estimation method in Sect. 5. Finally experiments and conclusions are discussed in Sects. 6 and 7 respectively.

2 Related Work

This section discusses related work mainly about three topics in crowdsourcing, namely task allocation, truth discovery, and exploration and exploitation which are wildly used in reinforcement learning.

2.1 Task Allocation

To achieve better data quality in crowdsourcing, workers and tasks should be matched properly. There are two perspectives of settings: task assignment and worker selection. In the task assignment setting, when a worker comes, the focus is on studying which subset of tasks should be assigned to the coming worker, namely online task assignment problem [4,9,11,12,29]. In the worker selection setting, given a task and a set of candidate workers, the focus is on studying which subset of workers should be selected to answer the task, namely jury selection problem [6,27,28].

In this paper, our problem is the worker selection setting but we consider the worker selection without knowledge of worker quality which is a more open and more challenging problem. Moreover, authors in [28] uses the Bayesian voting which is the optimal strategy. We also consider the Bayesian method to estimate the ground truth of a task. However, we exceed the scope of [28] because we take estimating numerical sensing values into consideration instead of discrete decision-making values.

2.2 Truth Discovery

In order to identify reliable answers from workers' noisy or even conflicting sensing data in mobile crowdsourcing systems, truth discovery methods [13,15–17,25] have drawn significant attentions. Researches on truth discovery jointly estimate data quality of workers and the underlying truths through quality-aware data aggregation. Besides of sensing quality, there are some researches discussing different application scenarios in crowdsourcing, e.g. [8,18,24]. Authors in [8] proposed a general truth discovery method for human answer tasks by combining reliability model and pattern of task clusters and source clusters. Research in [18] exploited cognitive psychology studies on dynamic memory structures and cognitive heuristics by tagging a place of the crowd participant. Authors in [24] proposed a graph-based PageRank-HITS Hybrid model to distinguish authoritative workers from unreliable ones in machine translation.

However, none of above research proposes similar truth discovery for numerical sensing value. Although Researches in [3,16,20] also considered numerical tasks in crowdsourcing, we innovatively considers both data bias, variance together as factors of quality of workers. Moreover, we borrow a reinforcement learning idea to solve a expected error minimization problem.

2.3 Exploration vs. Exploitation

The quality of workers is commonly unknown to crowdsourcing platform. Therefore, the trade-off between exploration and exploitation in reinforcement learning can be borrowed as a strategy for worker selection. The Multi-Armed Bandit (MAB) model gives a description of the sequential decision-making problem under incomplete information [21]. Here, we will discuss the allocation of budget for exploration and exploitation.

A variety of budget constrained models have been studied in the MAB setup [1,5,10]. These works considered a budget-limited exploration in the initial phase followed by a cost-free exploitation phase. However, in a real world setting such as the one considered in budget-aware crowdsourcing, the exploitation phase is not free of cost. This limitation is addressed in the budget-limited MAB problem, where both the exploration and exploitation phase are limited by a single budget. This model also considers different costs for arm selection. Two different policies were proposed in this setting, called ε-first policy and KUBE [22,23].

In our research, we do not consider the budget allocation directly for exploration and exploitation and use an UCB based MAB algorithm to solve the exploration and exploitation tradeoff. Authors in [19] also have this settings to optimize the discrete label annotation task but they do not focus on a numerical value Crowdsensing task, while we model worker quality with bias and variance which are suitable for numerical value Crowdsensing tasks.

3 Reinforced Reliable Worker Selection Framework

3.1 Overview

In this section, we propose the reinforced reliable worker selection framework, to iteratively estimate workers' sensing quality and select workers to execute regular spatial sensing tasks. The framework is a multi-round process. In each round, there is a set of workers available to execute sensing tasks. As the quality of some newly coming workers is unknown and the quality of existing workers might change, the framework should estimate the quality of workers while allocate spatial sensing tasks to workers with high quality.

In the following subsections, we will firstly model the factors that influence the sensing quality of workers. Then the process of worker selection is introduced. At last we provide a scheme of worker sensing quality estimation and weighted ground truth aggregation.

3.2 Model Establishment

The sensing quality of a worker could be modeled by two factors, namely bias and variance. Bias is a factor caused by the device which represents a constant distance between the mean of the sensing data and the ground truth. Variance reflects the stability of a worker to accurately estimate the sensing target.

Definition of Worker Attributes. Typically, a data sample X of a worker u_j taken for task t_i could be regarded as the result of a bias and a noise added to the ground truth of the task. The sensor on worker's device might have a natural bias which is introduced by device model and the calibration accuracy of the sensor producer. The noise is generated due to the distortion in the mobile device within usage environment. Therefore, suppose the ground truth of the task is GT_i, then the sample X is drawn a from Gaussian distribution $\mathcal{N}(GT_i + b_j, \sigma_j^2)$, where b_j is worker bias and σ_j^2 is worker variance.

As σ_j^2 is a nature of the mobile device of u_j, its value is unavailable for the platform and is required to be estimated. The number of samples utilized to estimate σ_j^2 has an impact on the reliability of the estimation on σ_j^2. That is to say, the larger size of data used, the more accurate we can estimate σ_j^2.

Definition 1 (Worker Attributes). *A worker $u_j \in \mathcal{K}$ is associated with b_j, σ_j^2 to represent her bias and variance respectively.*

Impact of Worker Attributes. On the one hand, the bias of a worker is used to adjust the value of data samples so that the ground truth could be estimated accurately (without bias). On the other hand, The variance of a worker reflects how stable her data sample is. That is to say, worker variance could be referred to as a weight to the worker's data samples when we aggregate her and other workers' data samples to estimate the ground truth. This is an intuition of the impacts of worker attributes, a mathematical analysis of how these attributes are used to estimate the ground truth is shown in Sect. 3.5.

3.3 Process of Reliable Worker Selection

Initially, the platform does not know the sensing quality of workers. Therefore, our framework employs a multi-round reinforced worker selection scheme to both estimate the sensing quality of workers and allocate sensing tasks to workers.

A round in our framework is a period of time when the workers might not change their positions and are available to repeatedly execute sensing tasks in the same place. For example, a couple of office hours could be regarded as the time range of a round and the positions of workers in this round are stable. On the other hand, we also assume that during a round, the ground truth of the sensing tasks will not change as the time range of a round will not be too long.

As the bias is a nature of the mobile devices of workers, we assume that in different rounds of our framework, the bias of workers remains unchanged. However, the variance of a device is influenced by both the device and environment. As a result the variance of workers is different in different rounds.

In a round r, the platform will receive a set of sensing tasks $\mathcal{T}^{(r)}$ and a budget $B^{(r)}$ from requesters. Suppose $|\mathcal{T}^{(r)}| = N^{(r)}$. Our framework will firstly check the availability of workers and obtain $N^{(r)}$ sets of candidate workers for the tasks. The candidate set of task $t_i \in \mathcal{T}^{(r)}$ is denoted by $\mathcal{K}_i^{(r)}$. Denote the number of workers in $\mathcal{K}_i^{(r)}$ as $M_i^{(r)}$.

As both the bias and variance of workers are unknown, the framework is supposed to estimate these factors. For the new coming workers who the platform has no prior knowledge of, our framework will allocate calibration tasks to these workers to obtain some knowledge of her bias and variance. In real situations, we may not be able to get the ground truth, so a method in Sect. 5.2 is proposed to get a approximation of ground truth for calibration tasks.

Suppose a subset $\mathcal{K}_i^{'(r)}$ of workers in $\mathcal{K}_i^{(r)}$ are new to the platform. Then we split $B_c^{(r)} = M_i^{\prime(r)}$ units of total budgets to allocate calibration tasks to workers in $\mathcal{K}_i^{\prime(r)}$. In a calibration task, both the bias and variance of workers are estimated according to the known ground truth of the calibration tasks. The rest of budgets are equally split to each task, namely budget for $t_i \in \mathcal{T}^{(r)}$ is $B_i^{(r)} = (B^{(r)} - B_c^{(r)})/N^{(r)}$.

The confidence level of estimation of variance in the calibration task is not a hundred percent and the variance of workers might be changed in later rounds. Hence the variance of workers should be further explored. The exploration of variance is considered in our framework by formulating the problem as a MAB problem, which will be discussed in Sect. 4.

3.4 Estimation of Worker Attributes

The estimation of worker attributes takes place in two cases. Firstly, in the calibration phase, the bias and variance of workers are estimated. Secondly, we update the value of variance according to the data samples of sensing task.

The bias of workers might be estimated in two ways. The first method is used when we have some ground truth of PoIs where workers might visit. In this case, the bias could be directly estimated by the mean of sampled data and the ground truth. The second method is used when the ground truth is not available. This method of estimation will be introduced in Sect. 5.2.

Notations for Sensing Data. Suppose the data sample of the τ-th time that worker u_j execute the task t_i is $\mathbf{X}_{i,j}^{(\tau)} = \{X_{i,j,1}^{(\tau)}, \ldots, X_{i,j,n}^{(\tau)}\}$, where n is the number of samples generated in a task. The sample mean and sample variance of data are denoted as $\bar{X}_{i,j}^{(r)}$ and $S_{i,j}^{(r)}$ respectively. For the calibration task, the notation is defined as $\mathbf{X}_j^{(0)}, \bar{X}_j^{(0)}$ and $S_j^{(0)}$.

Bias Estimation in Calibration Phase. As the ground truth of the calibration task is known, denoted by GT, the bias of u_j could be estimated by

$$\widehat{b}_j = \bar{X}_j^{(0)} - GT \tag{1}$$

The sample variance could be regarded as an unbiased estimation of σ_j^2. Hence, $(\widehat{\sigma}_j^2)^{(0)} = S_j^{(0)}$.

Variance Updating in a Spatial Sensing Task. For a sensing task t_i, suppose worker u_j is allocated to execute this task. At the τ-th execution of task t_i for u_j, the variance of this worker could be updated by

$$\left(\widehat{\sigma}_j^2\right)^{(\tau)} = \frac{1}{(\tau+1)n-1}\left(S_{i,j}^{(\tau)}\cdot(n-1)+\left(\widehat{\sigma}_j^2\right)^{(\tau-1)}\cdot(\tau n-1)\right) \tag{2}$$

3.5 Estimation of Ground Truth

We use a weighted combination of the estimated results provided by $\mathcal{K}_i^{(r)}$ to estimate the ground truth of task $t_i \in \mathcal{T}^{(r)}$, denoted by $\widehat{GT}_i^{(r)}$. As a worker might be selected multiple times to execute t_i, we denote the number of times that our allocation scheme A has selected u_j to execute t_i as $N_j^A(B_i^{(r)})$.

The estimated result of worker u_j is

$$\widehat{X}_{i,j}^{(r)} = \frac{1}{N_j^A(B_i^{(r)})}\sum_{\tau=1}^{N_j^A(B_i^{(r)})} \bar{X}_{i,j}^{(\tau)} \tag{3}$$

The estimated ground truth could be expressed as

$$\widehat{GT}_i^{(r)} = \frac{\sum_{j=1}^{M_i^{(r)}} w_j^{(r)} N_j^A(B_i^{(r)})(\widehat{X}_{i,j}^{(r)} - \widehat{b}_j)}{\sum_{j=1}^{M_i^{(r)}} w_j^{(r)} N_j^A(B_i^{(r)})} \tag{4}$$

Without ambiguity, we omit superscript r in this section. To analyze the case without considering the influence of allocation scheme, we assume we are aggregating the result using one execution of task t_i. The problem is to determine the weight for each worker u_j so that the ground truth of t_i is well estimated. For a task t_i, the error of the worker is modeled by a Gaussian distribution $\varepsilon_j \sim \mathcal{N}(b_j, \sigma_j^2)$. Then the total error for estimated ground truth after each worker executes once is

$$\varepsilon \sim \mathcal{N}\left(\frac{\sum_{j=1}^{M_i^{(r)}} w_j(b_j - \widehat{b}_j)}{\sum_{j=1}^{M_i^{(r)}} w_j}, \frac{\sum_{j=1}^{M_i^{(r)}} w_j^2\sigma_j^2}{n(\sum_{j=1}^{M_i^{(r)}} w_j)^2}\right) \tag{5}$$

To find a weight of workers that minimize the variance of the total error, the following problem is solved.

$$\min_{\{w_j\}} \sum_{j=1}^{M_i^{(r)}} w_j^2\sigma_j^2 \qquad s.t. \sum_{j=1}^{M_i^{(r)}} w_j = 1 \tag{6}$$

Solving this convex optimization problem, we can obtain that the optimal weight is $w_j \propto 1/\sigma_j^2$. Therefore, in our framework, the weight is set as

$$w_j^{(\tau)} = \frac{1}{\left(\widehat{\sigma}_j^2\right)^{(\tau)}} \tag{7}$$

Interestingly, the same conclusion that the weight could be set as Eq. (7) could be drawn by bayesian estimation. This is because bayesian estimation minimizes the combined posterior probabilistic distribution of data samples from different workers and Eq. (7) coincidentally is the solution of the minimization problem of bayesian estimation of ground truth. The proof is ignored due to space limit. Besides, if all w_j are equal, the estimation of ground truth is maximal likelihood estimation. In the experiments, we will show the bayesian estimation outperforms the maximal likelihood estimation.

4 Problem Formulation

4.1 Objective Description

From the perspective of requesters in many scenarios, they wish to obtain accurate data on some POIs. Therefore, the sensing quality is highly dependent on the expectation of error between the estimated result and the ground truth. Therefore, we define the following expected sensing error as our objective. Note that in this section we only analyze the result of one task in one round. Hence we omit the superscript r.

Definition 2 (Expected Sensing Error). *The expected sensing quality $\mathcal{E}_i$ of a task t_i is defined as*

$$\begin{aligned} \mathcal{E}_i &= \mathbb{E}_{\{N_j^A(B_i)\}}\mathbb{E}_{\{\mathbf{X}_{i,j}\}}\left[(\widehat{GT}_i - GT_i)^2\right] \\ &= \mathbb{E}_{\{N_j^A(B_i)\}}\mathbb{E}_{\{\mathbf{X}_{i,j}\}}\left[\left(\frac{\sum_{j=1}^{M_i} w_j N_j^A(B_i)(\widehat{X}_{i,j} - \widehat{b}_j)}{\sum_{j=1}^{M_i} w_j N_j^A(B_i)} - GT_i\right)^2\right] \end{aligned} \tag{8}$$

4.2 Expected Error Minimization Problem

Definition 3 (Expected Error Minimization Problem). *Given a sensing task t_i, sets of candidate workers $\mathcal{K}_i$ available for task t_i, and a total budget B_i, select workers from $\mathcal{K}_i$ to execute task t_i such that $\mathcal{E}_i$ is minimized and the budget B_i is not exhausted.*

As the bias and variance of workers are unknown to the platform, the selection scheme of workers could not be directly obtained by minimizing the function of expected sensing error. The bias of workers could be calibrated via calibration tasks. However, the variance of workers are to be explored after we allocate some tasks to workers. This meets an *exploration-exploitation* dilemma. In this paper, we employ an algorithm for MAB problem to depict the process of exploring the workers' variance while selecting low variance workers to perform sensing tasks. Meanwhile, we provide a theoretical bound to show that when selecting low variance workers using the MAB, the expected error could be bounded by a value which decreases as the total budget increases.

In this setting, workers are regarded as the arms of a bandit. The reward of a worker is set according to her variance. As shown in Definition 4, we set the reward of a worker u_j to be $1 - \frac{\widehat{\sigma}_j^2}{C}$ where C is a sufficiently large constant, which means that when we select workers with lower variance, the reward will be higher.

Definition 4 (Reward). *The reward of a worker u_j in a step τ is defined by $1 - \frac{(\widehat{\sigma}_j^2)^{(\tau)}}{C}$.*

4.3 Theoretical Bound for Expected Sensing Error

Suppose our algorithm A for the MAB will allocate $N_j^A(B_i)$ times of the task t_i to worker w_j given a total budget B_i. Note that $B_i = \sum_{j=1}^{K} N_j^A(B_i)$. The regret of the MAB is described in Eq. (9), where A^* is an optimal algorithm for the MAB.

$$\begin{aligned} R(A) &= \mathbb{E}\left[G(A^*)\right] - \mathbb{E}\left[G(A)\right] \\ &= \sum_{j=1}^{M_i} \mathbb{E}\left[N_j^{A^*}(B_i)\right]\mu_j - \sum_{j=1}^{M_i} \mathbb{E}\left[N_j^{A}(B_i)\right]\mu_j \\ &= \sum_{j=1}^{M_i} \mathbb{E}\left[N_j^{A^*}(B_i) - N_j^{A}(B_i)\right]\left(1 - \frac{\sigma_j^2}{C}\right) \end{aligned} \tag{9}$$

Theorem 1. *Suppose $R(A)$ could be bounded by $O(F(B_i))$ where $F(B_i)$ is a function of budget B_i, then $\mathcal{E}_i$ could be bounded by*

$$\mathcal{E}_i \leq \frac{C \cdot O(F(B_i)) + \mathbb{E}\left[\sum_{j=1}^{M_i} N_j^{A^*}(B_i)\sigma_j^2\right]}{B_i^2} + \max_{j' \in [M_i]} \{b_{j'} - \widehat{b}_{j'}\}^2 \tag{10}$$

The proof of Theorem 1 could be seen in our technical report[1].

As we can see in Eq. (10), the bound of $\mathcal{E}_i$ is related to both the accuracy of estimated bias (i.e. $\max_{j' \in [M_i]}\{b_{j'} - \widehat{b}_{j'}\}$) and the budget for the task. Note that an optimal algorithm solving the MAB problem as a regret of $O(\log B_i)$. Hence the first term of Eq. (10) is bounded by $O(1/B_i)$. As a result, both providing more budget and obtaining a better estimation of worker bias could lead to a smaller bound of expected sensing error, which improves the accuracy of estimated result.

5 Algorithm

5.1 Multi-Armed Bandit Based Estimation Method

Our algorithm is depicted in Algorithm 1. In each round of task assignment, we use $|\mathcal{T}^{(r)}|$ MAB with Upper Confidence Bound (UCB) algorithm to assign tasks,

[1] https://www.dropbox.com/s/3643ygf0jvu11vs/Technical_Report.pdf?dl=0.

in order to select workers with high sensing quality. The reason why we use UCB algorithm is that the quality of workers is unstable between different rounds.

In order to select the optimal worker, we estimate the expected rewards of all the the workers. Both estimating the qualities of workers and exploiting the optimal worker consume budget, and the UCB algorithm can help us effectively balance these trade-off between estimation and exploitation.

In each step of the UCB algorithm, we select the worker u_j with maximum trade-off reward $\hat{r}_j = \overline{r}_j + \sqrt{\frac{2\ln n}{n_j}}$, where $\overline{r}_j$ is the estimate of expected real reward $\mathbb{E}(r_j)$ of worker u_j. In this formulation, n is total number of steps, and n_j is the number of times that worker u_j has been selected.

In each round of our algorithm, we set up p bandits in p places, separately. At one step, every bandit selects one worker u_j to sense data $(X_1, \cdots, X_n)$. Then from the data, we can update the estimated quality of the worker u_j, by using Eqs. (1) and (2).

With the goal to maximize the total quality of all the worker we chosen, reward r_j was formulated as:

$$r_j = 1 - \frac{\hat{\sigma}_j^2}{C} \tag{11}$$

Note that C is a constant which satisfy $C \geq \hat{\sigma}_j^2, \forall j$ in the equation above. With the action-reward pair, UCB algorithm can update its estimate of reward $\overline{r}$.

Repeat the process until all the budget runs out. After the task assignment, we can estimate the ground truth $\widehat{GT}$ in all the places for this round and update our estimate of all worker's bias $\hat{b}_j$, according to Eq. (1).

For each step, bandit i infers worker and updates estimate in $O(|\mathcal{K}_i^{(r)}|)$. In a round, the upper bound of total number of steps for task t_i equals to the upper bound of budget, which denoted as B. With the assumption that one worker only belongs to one place in a round, the num of all over the worker $|\mathcal{K}^{(r)}| = \sum_i^{|\mathcal{T}^{(r)}|} |\mathcal{K}_i^{(r)}|$. Therefore, the time complexity is bounded by $O(B|\mathcal{K}_i^{(r)}|)$.

The regret of UCB algorithm is proven to be $O(\log B)$ [2].

5.2 Estimate Bias with Calibration

Without given ground truth, we need to assume the expected value of all the bias $\mathbb{E}[b_j] = 0$. To get the estimation of bias $\hat{b}_j$, we need to use Eq. (1). In fact, we actually do not have the ground truth GT, and the estimation, $\widehat{GT}$ need use the bias b_j of every worker u_j. To solve this dilemma, we use the approximations of $\widehat{GT}$ for r-th round and rewrite Eq. (1) as:

$$\hat{b}_j = \frac{1}{\sum_{r,i} N_j^A(B_i^{(r)})} \sum_{r,i} \bar{X}_{i,j}^{(r)} - \widehat{GT}_i^{(r)} \tag{12}$$

Bias $\hat{b}_j$ update after getting the estimation of ground truth $\widehat{GT}_i^{(r)}$ at the end of round r, and in this way, the estimation for bias will become more and more

Algorithm 1. The weighted MAB algorithm in one round

```
1: % Initialize MABs
2: for task t_i in T^(r) do
3:     Set up the i-th bandit for task t_i
4: end for
5: % Make task assignment until no budget left
6: while budget left do
7:     for task t_i in T^(r) do
8:         if Remaining budget B̂_i^(r) for task t_i large than 0 then
9:             Infer action j from i-th MAB with UCB algorithm
10:            Let worker u_j sense new data (X_1, ···, X_n) from environment
11:            Get reward r_j of action j using Equation (11).
12:            Update the i-th MAB with reward r
13:            B̂_i^(r) ← B̂_i^(r) − 1
14:        end if
15:    end for
16: end while
17: % Update values for every place
18: for task t_i in T^(r) do
19:     Update ground truth of task t_i with weighted estimation method according to
        Equation (4)
20:     Estimate each worker's bias (b̂_1, ···, b̂_n)
21: end for
```

accurate. However, the approximation may bring some error and let the estimation for GT become harder, and what is more, there is not a $\hat{b}$ to estimate $\widehat{GT}$ for the first round.

A straightforward method is to set the default initialization as 0 for every $\hat{b}_i$. When the value of bias is far away from 0, this initialization is obviously inappropriate. With this assumption, another method is to randomly select the worker and simply estimate the ground truth by taking average. Interestingly, this method can find good estimations without $\hat{b}_j$.

Therefore, an intuitive approach is to use this trivial method as calibration. In the first L rounds, we only use the random method to get a preliminary estimation of b_j, and then use these estimations in the latter rounds.

6 Simulation Experiments

6.1 Experimental Setup

Data Description. We apply our algorithm to a synthetic data set based on the real world trace data and temperature data. We use real world data provided by the trace data set to obtain available workers and use temperature data set to obtain the ground truth of each sensing task.

The trace data is a set of real GPS data in DataTech Modeling Contest Samples [7] from ChinaMobile. They collected daily GPS track logs from Hang

Zhou in Zhe Jiang province. We select 50 locations that are frequently visited by people as PoIs. All the people staying at a PoI within the time range of a task are regarded as available workers.

The temperature of a PoI is obtained from Dark Sky API[2] which can retrieve observed temperature from a couple of reliable data sources. Whenever a task is generated in our simulations, the temperature on PoI within the time range of the task is queried via Dark Sky API.

Environment Simulation. Without loss of generality, we use the temperature queries of 50 fixed places as the spatial task set $\mathcal{T}^{(r)}$. In a single round, tasks need to find the ground truths of these places. Each round lasts one hour.

We extract a possible worker set from people once passed that place according to the trace data above. At the beginning of every round, the 40–50 workers in set $\mathcal{K}_i^{(r)}$ are randomly picked from the available worker set.

The variance of each worker is randomly chosen from the interval $[0.1, \sigma_{max}^2]$, and the bias of each worker is randomly chosen from the interval $[-b_{max}, +b_{max}]$, where σ_{max}^2, b_{max} is the maximum of the variance and bias for all the workers, separately. When a worker sense data, her sensing samples are generated from a Gaussian distribution $\mathcal{N}(GT_i + b_j, \sigma_j^2)$, and the number of samples for an execution is $n = 50$.

Baselines. To understand the properties of our methods and their performance, we consider the following the baselines:

Random selector. This method simply selects one worker at each step and estimates the ground truth by taking average value of all the samples. In fact, with the assumption that the expected average of samples equal to the ground truth, this method is enough powerful.

Naive MAB. This baseline uses UCB algorithm for MAB in order to select workers with high quality. The difference between this baseline and our method is that it finally estimate the ground truth by simply setting all the w_j in Eq. (4) equally. This weights setting leads to a maximal likelihood estimation of the ground truth.

6.2 Performance Evaluation

The first part of experiments shows the advantage of our algorithm. By the assumption that the bias of workers $\hat{b}$ is well estimated, we set $b_{max} = 0$. Our algorithm selects the workers with high quality, therefore, the more diverse the quality of workers are, the better performance our algorithm will receive. On the other hand, if the budget is infinity, all the methods will get good performance. It means that our method should get better result with limited budget.

[2] https://darksky.net/dev.

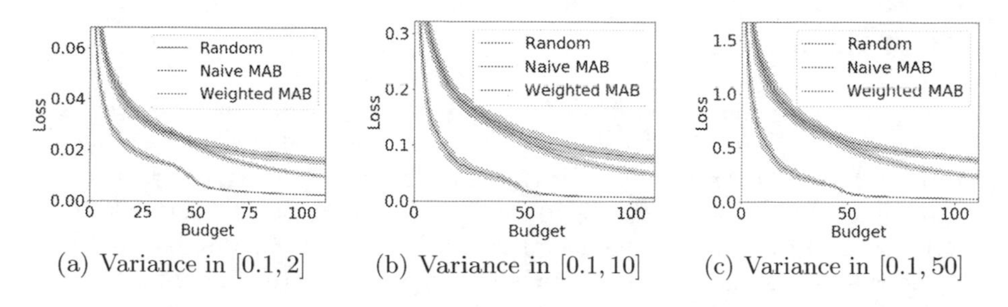

(a) Variance in $[0.1, 2]$ (b) Variance in $[0.1, 10]$ (c) Variance in $[0.1, 50]$

Fig. 2. Performance of methods with different budgets in one round.

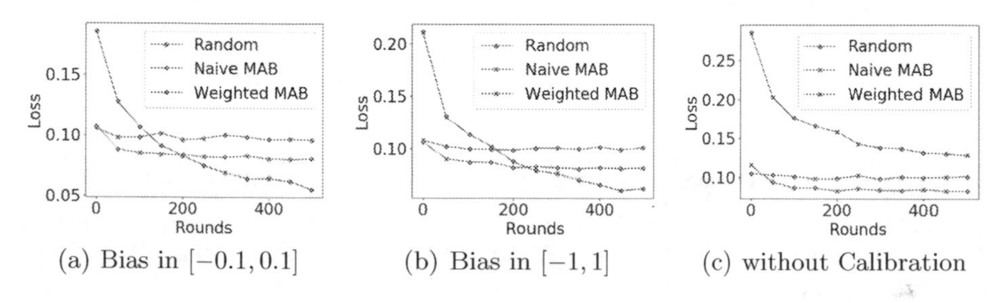

(a) Bias in $[-0.1, 0.1]$ (b) Bias in $[-1, 1]$ (c) without Calibration

Fig. 3. Performance of methods in several rounds.

Therefore, we evaluate these methods in different σ^2_{max}, the results are showed in Fig. 2. The loss in these figures is calculated by $\sqrt{\frac{1}{N^{(r)}}\sum_i^{N^{(r)}}(GT_i - \widehat{GT_i})^2}$. The max budget is almost two times of the number of workers. These results show that our algorithm has a good performance with limited budget, and both the MAB algorithm and the weighted estimation can improve the performance.

With only 40–50 budgets, MAB algorithm tries all the worker one by one, which has the same effect with Random method. Hence the Naive MAB get a better performance than the Random method with more than 50 budgets. With the budget increasing, the difference between different methods becomes smaller. Comparing these sub-figure (a), (b), and (c) in Fig. 2, it shows that the more diverse workers are, the better our method works significantly.

At the second part of experiments, we take the estimation of bias into consideration. Therefore, we fix $\sigma^2_{max} = 10$, change the value of b_{max} and add $L = 10$ rounds of calibration in the beginning. The results are shown as Fig. 3.

In sub-figure (a) and (b), the estimation of ground truth finally converges, and workers with small bias can slightly decrease the difficulty of convergence.

Finally, we fix $\sigma^2_{max} = 10, b_{max} = 1$, and try to run our algorithm with $\hat{b} = 0$ in the first round to show the effects of calibration. The result in sub-figure (c) shows that it can hardly find a correct estimation of bias.

7 Conclusion

In this paper, we model the quality of workers in Crowdsensing with two factors: bias and variance. For the classic exploration and exploitation dilemma, we introduce a novel Multi-Arms Bandit (MAB) to solve it. A weighted data aggregation (based on bayesian estimation) scheme is proposed which is shown to be a better way to calculate the ground truth of a sensing task. Besides, we prove that the expected sensing error can be bounded according to the bounded regret of the MAB. In simulation experiments, we use a real world data set to validate the theoretical results of our algorithm and it outperforms random and naive MAB baselines significantly in different settings. In the future, we will consider the partition method to allocate different budgets for different tasks for a global optimization instead of minimizing expected sensing error of each sensing task.

Acknowledgment. This work was supported by the National Key R&D Program of China (2018YFB1004703), the National Natural Science Foundation of China (61872238, 61672353), the Shanghai Science and Technology Fund (17510740200), the Huawei Innovation Research Program (HO2018085286), and the State Key Laboratory of Air Traffic Management System and Technology (SKLATM20180X).

References

1. Antos, A., Grover, V., Szepesvári, C.: Active learning in multi-armed bandits. In: Freund, Y., Györfi, L., Turán, G., Zeugmann, T. (eds.) ALT 2008. LNCS (LNAI), vol. 5254, pp. 287–302. Springer, Heidelberg (2008). https://doi.org/10.1007/978-3-540-87987-9_25
2. Auer, P., Cesa-Bianchi, N., Fischer, P.: Finite-time analysis of the multiarmed bandit problem. Mach. Learn. **47**(2–3), 235–256 (2002)
3. Aydin, B.I., Yilmaz, Y.S., Li, Y., Li, Q., Gao, J., Demirbas, M.: Crowdsourcing for multiple-choice question answering. In: AAAI, pp. 2946–2953 (2014)
4. Boim, R., Greenshpan, O., Milo, T., Novgorodov, S., Polyzotis, N., Tan, W.C.: Asking the right questions in crowd data sourcing. In: ICDE, pp. 1261–1264 (2012)
5. Bubeck, S., Munos, R., Stoltz, G.: Pure exploration in multi-armed bandits problems. In: Gavaldà, R., Lugosi, G., Zeugmann, T., Zilles, S. (eds.) ALT 2009. LNCS (LNAI), vol. 5809, pp. 23–37. Springer, Heidelberg (2009). https://doi.org/10.1007/978-3-642-04414-4_7
6. Cao, C.C., She, J., Tong, Y., Chen, L.: Whom to ask? Jury selection for decision making tasks on micro-blog services. PVLDB **5**(11), 1495–1506 (2012)
7. DataTech: Datatech modeling contest public data sample, Zhejiang Division Co. China Mobile Co. (2017). http://datatech.zjdex.com
8. Du, Y., Xu, H., Sun, Y.-E., Huang, L.: A general fine-grained truth discovery approach for crowdsourced data aggregation. In: Candan, S., Chen, L., Pedersen, T.B., Chang, L., Hua, W. (eds.) DASFAA 2017. LNCS, vol. 10177, pp. 3–18. Springer, Cham (2017). https://doi.org/10.1007/978-3-319-55753-3_1
9. Fan, J., Li, G., Ooi, B.C., Tan, K., Feng, J.: iCrowd: an adaptive crowdsourcing framework. In: SIGMOD, pp. 1015–1030 (2015)
10. Guha, S., Munagala, K.: Approximation algorithms for budgeted learning problems. In: STOC, pp. 104–113 (2007)

11. Ho, C., Jabbari, S., Vaughan, J.W.: Adaptive task assignment for crowdsourced classification. In: ICML, pp. 534–542 (2013)
12. Ho, C., Vaughan, J.W.: Online task assignment in crowdsourcing markets. In: AAAI (2012)
13. Jin, H., Su, L., Nahrstedt, K.: Theseus: incentivizing truth discovery in mobile crowd sensing systems. In: MobiHoc, pp. 1:1–1:10 (2017)
14. Lappas, T., Liu, K., Terzi, E.: Finding a team of experts in social networks. In: SIGKDD, pp. 467–476 (2009)
15. Li, Q., et al.: A confidence-aware approach for truth discovery on long-tail data. PVLDB **8**(4), 425–436 (2014)
16. Li, Q., Li, Y., Gao, J., Zhao, B., Fan, W., Han, J.: Resolving conflicts in heterogeneous data by truth discovery and source reliability estimation. In: SIGMOD, pp. 1187–1198 (2014)
17. Meng, C., et al.: Truth discovery on crowd sensing of correlated entities. In: SenSys, pp. 169–182 (2015)
18. Mordacchini, M., et al.: Crowdsourcing through cognitive opportunistic networks. TAAS **10**(2), 13:1–13:29 (2015)
19. Rangi, A., Franceschetti, M.: Multi-armed bandit algorithms for crowdsourcing systems with online estimation of workers' ability. In: AAMAS, pp. 1345–1352 (2018)
20. Raykar, V.C., et al.: Learning from crowds. J. Mach. Learn. Res. **11**, 1297–1322 (2010)
21. Robbins, H.: Some aspects of the sequential design of experiments. Bull. Am. Math. Soc. **55**, 527–535 (1952)
22. Tran-Thanh, L., Chapman, A., FloresLuna, J.E.M.D., Rogers, A., Jennings, N.R.: Epsilon-first policies for budget-limited multi-armed bandits. In: NCAI, pp. 1211–1216 (2010)
23. Tran-Thanh, L., Chapman, A.C., Rogers, A., Jennings, N.R.: Knapsack based optimal policies for budget-limited multi-armed bandits. In: AAAI, pp. 1134–1140 (2012)
24. Yan, R., Song, Y., Li, C., Zhang, M., Hu, X.: Opportunities or risks to reduce labor in crowdsourcing translation? Characterizing cost versus quality via a pagerank-HITS hybrid model. In: IJCAI, pp. 1025–1032 (2015)
25. Yao, S., et al.: Recursive ground truth estimator for social data streams. In: IPSN, pp. 14:1–14:12 (2016)
26. Yin, X., Han, J., Yu, P.S.: Truth discovery with multiple conflicting information providers on the web. TKDE **20**(6), 796–808 (2008)
27. Zhao, Z., Wei, F., Zhou, M., Chen, W., Ng, W.: Crowd-selection query processing in crowdsourcing databases: a task-driven approach. In: EDBT, pp. 397–408 (2015)
28. Zheng, Y., Cheng, R., Maniu, S., Mo, L.: On optimality of jury selection in crowdsourcing. In: EDBT, pp. 193–204 (2015)
29. Zheng, Y., Wang, J., Li, G., Cheng, R., Feng, J.: QASCA: a quality-aware task assignment system for crowdsourcing applications. In: SIGMOD, pp. 1031–1046 (2015)

SeqST-ResNet: A Sequential Spatial Temporal ResNet for Task Prediction in Spatial Crowdsourcing

Dongjun Zhai[1,2,4], An Liu[1(✉)], Shicheng Chen[3], Zhixu Li[1], and Xiangliang Zhang[2]

[1] Soochow University, Suzhou, China
{anliu,zhixuli}@suda.edu.cn
[2] King Abdullah University of Science and Technology, Thuwal, Saudi Arabia
{Dongjun.Zhai,Xiangliang.Zhang}@kaust.edu.sa
[3] National University of Singapore, Singapore, Singapore
coder.chen.shi.cheng@gmail.com
[4] Blockshine Technology Corp., Shanghai, China

Abstract. Task appearance prediction has great potential to improve task assignment in spatial crowdsourcing platforms. The main challenge of this prediction problem is to model the spatial dependency among neighboring regions and the temporal dependency at different time scales (e.g., hourly, daily, and weekly). A recent model ST-ResNet predicts traffic flow by capturing the spatial and temporal dependencies in historical data. However, the data fragments are **concatenated** as one tensor fed to the deep neural networks, rather than learning the temporal dependencies in a sequential manner. We propose a novel deep learning model, called SeqST-ResNet, which well captures the temporal dependencies of historical task appearance in **sequences** at several time scales. We validate the effectiveness of our model via experiments on a real-world dataset. The experimental results show that our SeqST-ResNet model significantly outperforms ST-ResNet when predicting tasks at hourly intervals and also during weekday and weekends, more importantly, in regions with intensive task requests.

Keywords: Task prediction · Spatial crowdsourcing · Deep neural network

1 Introduction

Spatial Crowdsourcing (SC) [20] has attracted lots of attentions in recent years. A typical SC system consists of a platform that is responsible for releasing location-based tasks and a crowd of workers that can physically move to specified locations to perform tasks. Emerging SC applications include taxi-hailing service such as Didi and Uber, meal order and delivery service such as Eleme, and dynamic information collection service such as Gigwalk. In practice, hundreds

G. Li et al. (Eds.): DASFAA 2019, LNCS 11446, pp. 260–275, 2019.
https://doi.org/10.1007/978-3-030-18576-3_16

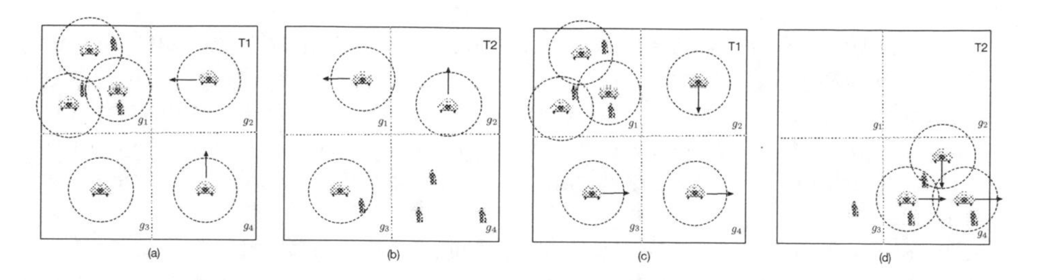

Fig. 1. Task assignment with/without task prediction in spatial crowdsourcing. Without task prediction, idle workers at T_1 in (a) make aimless movement, then at T_2 in (b) only one task-worker pair can be matched. With the guidance of task prediction, idle workers at T_1 in (c) move to g_4 which is predicted to have several task requests. Then at time T_2 in (d), three task-worker pairs are matched.

or thousands of workers and tasks could be online simultaneously, so an important problem in SC is to assign tasks to workers with the aim to maximize the number of total assigned task-worker pairs. In previous studies the online task assignment problem is usually decomposed into a series of offline task assignment problems, each of which is further reduced to be a maximum matching problem in a weighted bipartite graph [13,22]. While this reduction can generate feasible solutions, it does not consider the dynamic feature of workers and tasks. After one round of offline task assignment, idle workers are assumed to stay in place or just move around aimlessly, waiting for the next round of offline task assignment with new coming tasks. In fact, these idle workers can move towards some places or areas where new tasks are likely to arrive, so that more task-worker pairs can be assigned. This motivation can be further illustrated by the following example.

Figure 1 demonstrates a real time taxi-hailing service where tasks are passengers and workers are taxis. To facilitate later discussion, the whole area is divided into 4 grids and each taxi has a service range shown by dotted circles. At current time T_1, the platform can generate three task-worker pairs as three passengers are just in the service range of three taxis in g_1, as shown in Fig. 1(a). For taxis in other grids, if they stay at place (e.g., the taxi in g_3) or move around aimlessly (e.g., the taxis in g_2 and g_4), at next time T_2 shown in Fig. 1(b), only one task-worker pair can be made as most new tasks are appeared in g_4 but no workers are there. However, if the platform can guide idle workers to move to grid g_4 as shown in Fig. 1(c), then at time T_2 in Fig. 1(d), three task-worker pairs can be matched. From this toy example, we observe that it is beneficial to let idle workers move to areas that new tasks may appear, but the challenge here is how to predict accurately the time and the location of new coming tasks.

However, since tasks are always published individually, it is challenging to predict the specific position and time of the new coming tasks. This important problem recently has caused increasing attention. Several studies relax the problem to predict the future distribution of the new coming tasks using grid-based methods. Cheng et al. [4] used linear regression to predict the future number of workers/tasks of each grid cell. Tong et al. [23] compared the performance of several traditional statistical methods (e.g., ARIMA [3]) and traditional machine learning methods (e.g., GBRT [7]) on this problem. However, these methods are too simple and not good enough to model both spatial and temporal dependencies in task prediction problem, and the experimental results in this paper show that these methods have close performance and it is difficult to improve one over another.

Our study in this paper, for the first time, attempts to apply deep neural networks on addressing the task prediction problem. Deep neural networks have shown its success on diverse applications fields, and outperform traditional methods on modeling complex temporal and spatial feature dependency. For example, deep spatio-temporal residual network (ST-ResNet) in [29] is proposed to predict inflow and outflow of crowds in grid regions of a city by using convolution-based residual networks to model nearby and distant spatial dependencies between any two regions in a city, and the temporal properties of flows regarding temporal closeness, period, and trend.

Our problem of task appearance prediction shares similar challenges in traffic flow prediction on modeling the spatial dependency among regions and temporal dependency on dates and trend. However, the ST-ResNet model has a major limitation on learning the temporal dependency in historical sequences. It **concatenates data in one sequence as one tensor**, which weakens the sequential dependency in successive time units. Thus, we design a spatial-temporal residual network that learns from the **sequential historical data**, in a manner like Recurrent neural networks (RNN) and Long-short term memory (LSTM) learn from time series and word sequence.

The main contributions in this paper are summarized as follows:

- We analyze the spatial dependencies and temporal dependencies in task prediction problem, and targets on modeling the temporal dependencies in the sequences of historical task appearances at different scales, e.g., on the time interval level (e.g., half-hour) from $t-2$ to $t-1$ to t, on the daily level from one day to another, and on the weekly level from one week to another.
- We proposed a model called SeqST-ResNet, which well captures the temporal dependencies of historical task appearance in **sequences**, and the spatial dependencies among neighboring regions.
- We validate the effectiveness of our model on a real-world dataset. The experimental results show that our SeqST-ResNet model significantly outperforms the most competitive baseline ST-ResNet when predicting tasks at different time scales, and more importantly, in regions with intensive task requests.

The remainder of this paper is organized as follows. We give the definitions of task prediction problem and then analyze the spatial and temporal dependencies in Sect. 2. Section 3 presents the details of the proposed model. Experimental results are presented in Sect. 4 to demonstrate the performance of our methods. Related work and conclusion are discussed in Sects. 5 and 6, respectively.

2 Problem Definition and Analysis

2.1 Problem Definition

Following the previous work in [4,23], we address the task prediction problem by relaxing it to predict the number of task appearances in a specific spatio-temporal scope. In a city or region of interest, we divide its map into $M \times N$ grids, by partitioning along the longitude and latitude. The resolution of grid partition is controlled by M and N according to the application need. This practical grid-based map partition has been widely used in many spatio-temporal problems [4,23,29–31].

Based on the grid partition, at each time moment t, the task distribution can be considered as an image. We formally define it as:

Definition 1. *(Task image at time t) Task image of the whole area ($M \times N$ grids) at interval t can be defined as:*

$$\boldsymbol{X}_t = \begin{bmatrix} X_t^{(0,0)} & X_t^{(0,1)} & \cdots & X_t^{(0,N)} \\ X_t^{(1,0)} & X_t^{(1,1)} & \cdots & X_t^{(1,N)} \\ \vdots & \vdots & \ddots & \vdots \\ X_t^{(M,0)} & X_t^{(M,1)} & \cdots & X_t^{(M,N)} \end{bmatrix} \tag{1}$$

The element of task image matrix at the i-th row and the j-th column is

$$X_t^{(i,j)} = \sum_{P \in \mathbb{P}} |\{P_{loc} \in grid(i,j) \wedge P_{pt} = t\}| \tag{2}$$

which is the count of tasks P published at time P_{pt} at location P_{loc} belonging to grid cell $grid(i,j)$. Here, $\mathbb{P}$ denotes the task set in the spatial crowdsourcing system.

Then, our task appearance prediction problem is defined as

Definition 2. *(Task Prediction Problem) Given the sequence of historical task images $\{\boldsymbol{X}_0, \boldsymbol{X}_1, \ldots, \boldsymbol{X}_{t-1}\}$, the goal is to predict $\boldsymbol{X}_t$ at next time t.*

2.2 Temporal and Spatial Dependency Analysis

To address the task appearance prediction problem, we discuss two important types of dependencies that should be included in prediction model design.

- **Temporal Dependencies:** at a given location, the task image value at a time t depends on the values before t. Such dependencies are often observed due to the time dependent properties of tasks. For example, at a central business district (CBD), the number of lunch orders on Grubhub or Eleme will have little differences between 11:30am–12:00pm and 12:00pm–12:30pm because these time intervals are both in lunch time. And the orders quantity

will be similar from Monday to Friday as they are workdays. Also, similar order numbers can be found from one weekend to another. More importantly, the dependency between t and $t-1$ is stronger than that between t and $t-2$. Therefore, the temporal dependencies should be modeled in the **sequence of task images at different scales**, e.g., on the time interval level (e.g., hourly or half-hourly) from $t-2$ to $t-1$ to t, on the daily level from $day1$ to $day2$ to $day3$, and on the weekly level from 1-st Monday to 2-nd Monday to 3-rd Monday and so on.
- **Spatial Dependencies:** the task image value at one location depends on the values at neighboring locations, because the neighboring grid cells may cover a same urban functional region. For example, at different grid cells around a commercial center. there will be similar number of taxi orders. Moreover, similar task appearance can even exist in distant grid cells when they cover the same type of functional region, e.g., train stations distributed in a big city.

3 The Proposed Approach, SeqST-ResNet

In this section, we will introduce our proposed model, SeqST-ResNet, which is a deep neural network model capturing the **sequential dependencies** in temporal features and spatial features of task appearance predication problem. We first present the network architecture and then discuss the network learning process, and its relevance to the ST-ResNet in [29].

3.1 SeqST-ResNet Architecture

As we discussed in the previous section, the sequential dependency on time-line is important for making correct prediction. Like in language models, video processing models and time-series models, the historical sequence is memorized for making predicting at the next time moment. Recurrent neural networks (RNN), Long-short term memory (LSTM), and all their variants have been widely used for this purpose. To incorporate the temporal dependency on different levels, we learn from the three types of historical sequences of task images:

$$\{\boldsymbol{X}_{t-n}, \boldsymbol{X}_{t-n+1}, \ldots, \boldsymbol{X}_{t-2}, \boldsymbol{X}_{t-1}\} \quad \text{interval-level}$$
$$\{\boldsymbol{X}_{t-n*p}, \boldsymbol{X}_{t-(n-1)*p}, \ldots, \boldsymbol{X}_{t-2*p}, \boldsymbol{X}_{t-p}\} \quad \text{day-level when } p = \text{one day}$$
$$\{\boldsymbol{X}_{t-n*q}, \boldsymbol{X}_{t-(n-1)*q}, \ldots, \boldsymbol{X}_{t-2*q}, \boldsymbol{X}_{t-q}\} \quad \text{week-level when } q = \text{one week}$$

where p and q are parameters controlling the different level of temporal dependency to be modeled.

The architecture of our model, SeqST-ResNet, is shown in Fig. 2. The time axis on the top indicates the three types of dependency we modeled, denoting recent time (interval-level, in color red), near history (day-level, in color green) and distant history (week-level, in color blue). The sequence of each level (task images in the same color) goes through the same multi-layer deep network, including a convolution layer followed by several ResUnits, which are designed

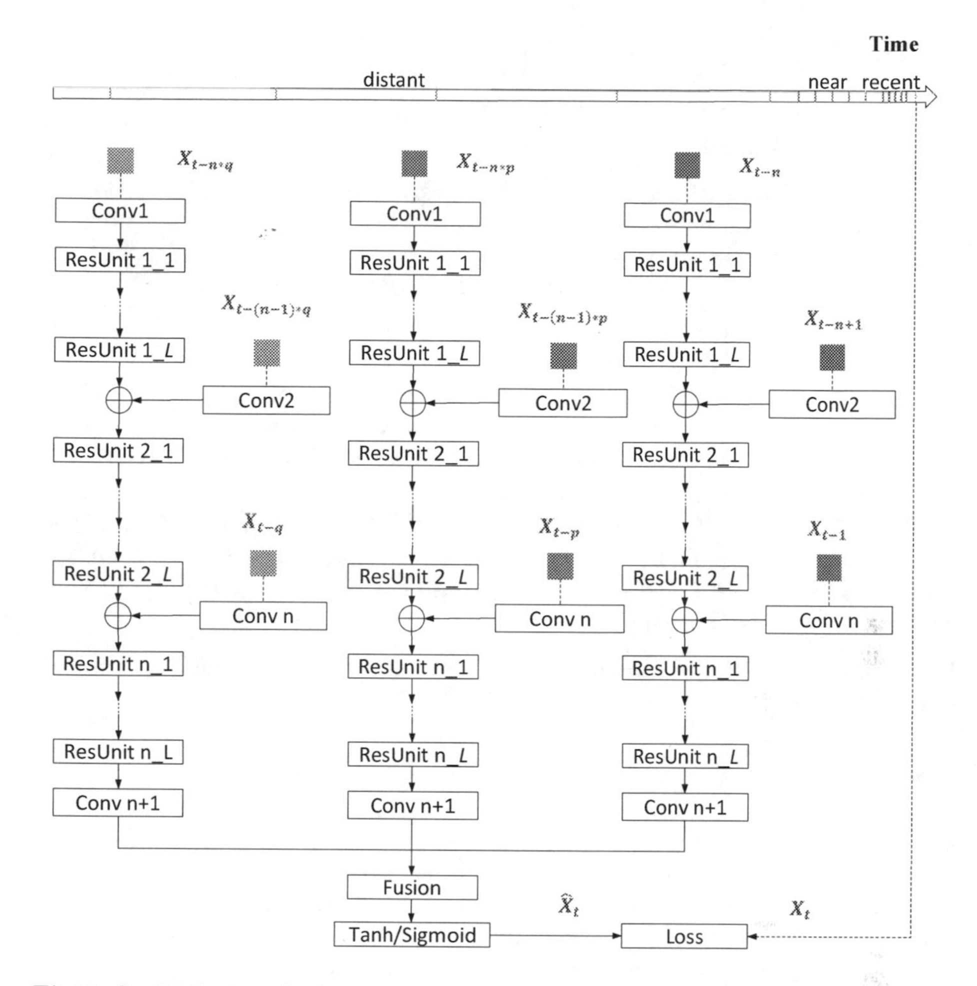

Fig. 2. SeqST-ResNet Architecture. The time axis on the top indicates three types of dependency, denoting recent time (interval-level, in color red), near history (day-level, in color green) and distant history (week-level, in color blue). The sequence of each level (task images in the same color) goes through the same multi-layer deep network, including a convolution layer followed by several ResUnits. The outputs from each sequence learning component are then combined by a fusion layer and followed by a final activation layer, as shown in bottom. (Color figure online)

following the Residual Network (ResNet) [9]. ResNet has been a great achievement in deep learning for addressing the gradients vanish problem and makes the deep network trainable even with over 1000 layers [10]. The outputs from each sequence learning component are then combined by a fusion layer and followed by a final activation layer, as shown in bottom of Fig. 2.

We next discuss the details of each sequence learning component, which consists of four parts: inputs, convolution layers, residual units, and addition layers.

Inputs. The input sequence of task images can be an interval-level sequence $\{\boldsymbol{X}_{t-n}, \boldsymbol{X}_{t-n+1}, \ldots, \boldsymbol{X}_{t-1}\}$, or a day-level sequence $\{\boldsymbol{X}_{t-n*p}, \boldsymbol{X}_{t-(n-1)*p}, \ldots, \boldsymbol{X}_{t-p}\}$, or a week-level sequence $\{\boldsymbol{X}_{t-n*q}, \boldsymbol{X}_{t-(n-1)*q}, \ldots, \boldsymbol{X}_{t-q}\}$, as we discussed previously.

Convolution Layers. The convolution operation captures the spatial dependency. Taking a task image, it outputs[1]:

$$\boldsymbol{X}_{t-n}^{(1)} = f(\boldsymbol{W}^{(1)} * \boldsymbol{X}_{t-n}^{(0)} + b^{(1)}) \tag{3}$$

where * denote the convolution operation and f is an activation function (e.g., ReLu, Tanh, or Sigmoid), $\boldsymbol{W}^{(1)}$ and $b^{(1)}$ are the learnable network parameters. With the increase of the number of the convolution layers, each unit of the feature map in the output of the component can cover more grids in input (task images). We use a 3×3 kernel as the filter. After one convolution layer, each unit of the output feature map can catch 9 pixels' information of the input. After two convolution layers, each unit can catch 27 pixels' information as the image is big enough. Thus, to capture the whole task image's information we use several convolution layers. To reduce training time, most convolution layers are contained in residual unit.

Residual Unit. The residual unit we use is shown in Fig. 3. Formally, a residual unit is defined as:

$$\boldsymbol{X}_l = \boldsymbol{X}_{l-1} + F(\boldsymbol{X}) \tag{4}$$

where $\boldsymbol{X}_l$ and $\boldsymbol{X}_{l-1}$ denote the input and output of a residual unit, respectively. Function F is a residual function which consists of multiple convolution layers and Batch Normalization [12] layers also with several ReLu transition function [17]. Batch Normalization is known by a lot of advantages, such as faster training, allowing, higher learning rate, easy to initialize network weights, regularization and improvement of network performance.

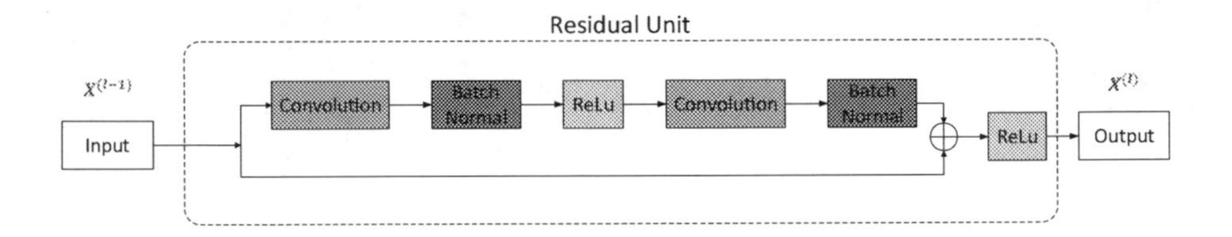

Fig. 3. Residual unit

Addition Layer. The addition layer, denoted by $\bigoplus$ serves for absorbing the task image into the sequence modeling component. Taking the first addition layer as an example, it operates

$$\boldsymbol{X}_{t-n+1}^{(1)} : \boldsymbol{X}_{t-n+1}^{(1)} + \boldsymbol{X}_{t-n}^{(l+1)} \tag{5}$$

[1] We show conv1 as an example. Other convolution layers follow the same function.

where $\boldsymbol{X}_{t-n+1}^{(1)}$ on the right is the output of conv2, and the second term $\boldsymbol{X}_{t-n}^{(l+1)}$ is the output of *ResUnit 1_L*. After merging $\boldsymbol{X}_{t-n}^{(l+1)}$, $\boldsymbol{X}_{t-n+1}^{(1)}$ is sent forward to next layers.

To demonstrate the end-to-end SeqST-ResNet model, we take a simple sequence of two task images, $\{\boldsymbol{X}_0, \boldsymbol{X}_1\}$, as input to one sequence modeling component (e.g., the interval-level model). The feed-forward procedure will be:

$$\begin{aligned}
\boldsymbol{X}_0^{(1)} &= f_0(\boldsymbol{W}_0^{(1)} * \boldsymbol{X}_0^{(0)} + b_0^{(1)}) \\
\boldsymbol{X}_1^{(1)} &= f_1(\boldsymbol{W}_1^{(1)} * \boldsymbol{X}_1^{(0)} + b_1^{(1)}) \\
\boldsymbol{X}_0^{(l+1)} &= \boldsymbol{X}_0^{l} + F_0^{(l)}(\boldsymbol{X}_0^{(l)}) \text{ (for } l \text{ from 1 to } L) \\
\boldsymbol{X}_1^{(1)} &= \boldsymbol{X}_1^{(1)} + \boldsymbol{X}_0^{(l+1)} \\
\boldsymbol{X}_1^{(l+1)} &= \boldsymbol{X}_1^{l} + F_1^{(l)}(\boldsymbol{X}_1^{(l)}) \text{ (for } l \text{ from 1 to } L)
\end{aligned}$$

The fusion layer will coalesce the output of three components with a simple parameter-matrix-based method. Suppose $\boldsymbol{X}_c$, $\boldsymbol{X}_p$, $\boldsymbol{X}_q$ are the output of each component respectively, the output of the fusion layer is:

$$\boldsymbol{X}_{out} = \boldsymbol{W}_c * \boldsymbol{X}_c + \boldsymbol{W}_p * \boldsymbol{X}_p + \boldsymbol{W}_q * \boldsymbol{X}_q \tag{6}$$

where the symbol $*$ is the hadamard product. The final activation function is Sigmoid, which predicts the output $\widehat{\boldsymbol{X}_t} = sigmoid(\boldsymbol{X}_{out})$. The loss function MSE (Mean Square Error) is defined to measure the difference between the predicted task image at t and the ground truth $\boldsymbol{X}_t$:

$$loss = ||\boldsymbol{X}_t - \widehat{\boldsymbol{X}_t}||_2^2 \tag{7}$$

3.2 Network Training

We use Adam algorithm as the optimizer. The learning rate is set as 0.003, and batch size is set as 16. The convolution kernels are with size of 3×3 both in convolution layers and residual blocks.

To reduce the influence of anomalies in training task image sequences, we apply moving average (e.g., with window length 3) on the input sequences. Min-max normalization is used for data pre-processing. The length of the interval-level and day-level sequences is set as 3, and the length of the week-level sequences is set to be 1 unless specified differently. The influence of these length parameters will be analyzed in the experimental result section.

3.3 Discussion

Our proposed model SeqST-ResNet shares the similar architecture of using three components of ResNet in ST-ResNet [29]. Our interval-level component corresponds to the closeness component in ST-ResNet, while day-level and week-level component corresponds to the period and trend component in ST-ResNet, respectively. The key difference lies in how to model the historical sequences. In each component of ST-ResNet, the fragments in one sequence are **concatenated as one tensor**, and then modeled by convolution layer and residual units. In our proposed SeqST-ResNet, **the task images in a sequence are modeled by considering their time order, and thus captures the temporal dependency in a more reasonable way than taking concatenation as a tensor**. The evaluation results in next section also verify that our sequential deep network architecture can better model the temporal dependency in task image sequences and achieves better prediction results than ST-ResNet in [29].

4 Experiments

In this section, we evaluate our proposed model SeqST-ResNet on a real-world dataset and compare it with the state-of-the-art models.

4.1 Dataset

The dataset we use is the taxi request data in Chengdu, China, provided by Didi GAIA Open Dataset [1]. The detail information of this dataset is shown in Table 1. The whole area covered by the dataset in Chengdu is divided by a 10×10 grid-based map partition. We set time interval as 30 min, and then get 2928 intervals totally from Oct. 1st to Nov. 30th, 2016.

Table 1. Dataset information

City	Chengdu, China
Time span	10/1/2016–11/30/2016
# taxi orders	11779076
Time interval	30 min
# intervals	2928
Grid-based map partition	10×10
Area of each grid	Nearly 800×800 m^2

4.2 Baselines and Settings

We compare our method with the following baselines including both traditional methods and deep learning methods:

- **HA:** Historical Average. This naive approach uses the historical average value in the same interval and same grid as the prediction.
- **ARIMA:** Auto-Regressive Integrated Moving Average [3]. It is a well-known time series prediction model.

Table 2. Prediction errors of different models

	Model	RMSE
Baselines	HA	22.99
	ARIMA	17.40
	LSTM-48	39.87
	LSTM-96	39.97
	DeepST	18.52
	ST-ResNet	16.28
Our methods	SeqST-ResNet-1AVG	14.00
	SeqST-ResNet-3AVG	**12.95**
	SeqST-ResNet-5AVG	12.97
	SeqST-ResNet-7AVG	13.61

- **LSTM:** Long-Short-Term-Memory Network [11]. Its chain like neural network structure is capable of learning long-term dependencies.
- **ST-ResNet:** Spatio-Temporal Residual Network [29]. The deep learning model is designed for predicting traffic flow based on historical data.
- **DeepST:** Deep Spatio-Temporal Network [30]. It is similar to ST-ResNet, but without using residual units.

For all methods, we select the data in the last 7 days (nearly 10%) for evaluating the prediction accuracy, while all data before that are used for training. The parameters of deep learning models (DeepST, ST-ResNet and our SeqST-ResNet) are set to the same values: $p = 48$ intervals (24 h) and $q = 48 \times 7$ intervals (one week), such that day-level and week-level sequences are fed with interval-level sequences to the corresponding components in the deep networks. The parameter settings in other baselines methods are tuned for achieving their best performance.

Evaluation Metrics. We use the most common metrics in this paper: Root Mean Squared Error (RMSE) for evaluation:

$$RMSE = \sqrt{\frac{1}{n}\sum_{i=1}^{n}\left(\boldsymbol{X}_i - \widehat{\boldsymbol{X}_i}\right)} \tag{8}$$

The results reported next are the average of 5 independent runs.

4.3 Results

Overall Prediction Performance. Table 2 presents the prediction results of baseline methods and our proposed models with different smoothing window sizes (SeqST-ResNet-τAVG represents our model is trained by smoothed sequences with window size τ). The lower RMSE value indicates more accurate prediction. We can see that our SeqST-ResNet models consistently outperform the baseline methods. Especially SeqST-ResNet-3AVG shows the best performance with RMSE of 12.95. Obviously, smoothing of sequences can improve the results due to the elimination of anomalies (SeqST-ResNet-1AVG without smoothing is not as good as other SeqST-ResNet settings). One interesting observation is that SeqST-ResNet-5AVG has the closest RMSE (12.97) to the best result, and SeqST-ResNet-7AVG has higher RMSE (13.61). This is due to the over-smoothing of the sequences with a large window (length of 7 intervals means 3.5 h). Over all, the results in Table 2 verify that our proposed SeqST-ResNet model can capture well the sequential dependency in time order and among spatial locations, and thus make more accurate prediction of future task appearance than other models.

To demonstrate the model training and testing errors in the learning process, the training error curve and test curve of epochs are shown in Figs. 4 and 5. We can see clearly that SeqST-ResNet models always have lower error than DeepST and ST-ResNet in both training and testing. Moreover, SeqST-ResNet-7AVG is worse than SeqST-ResNet-5AVG and SeqST-ResNet-3AVG, but still better than SeqST-ResNet-1AVG without using smoothing. The training error curve and test error curve have the same tendency and similar value, which indicates that the parameters we use in these models are suitable and do not result in over-fitting or under-fitting issues.

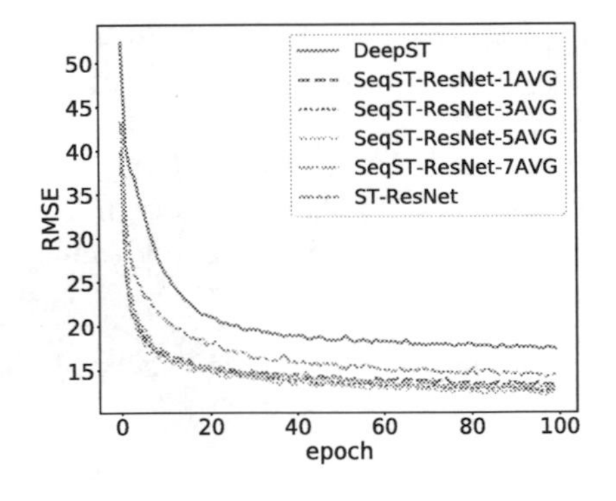

Fig. 4. Training error

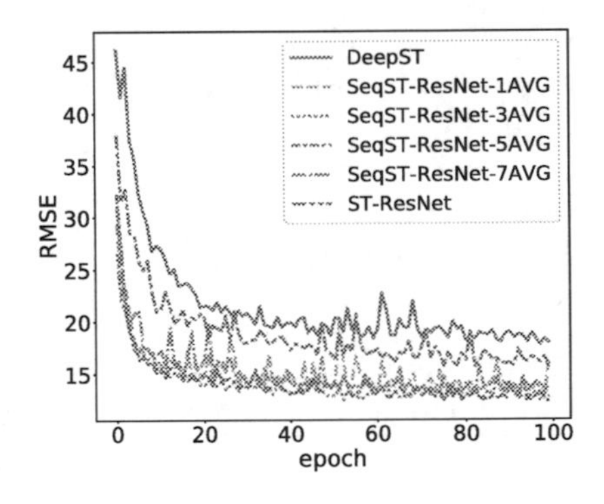

Fig. 5. Test error

Prediction Performance with Respect to Time. To evaluate the prediction accuracy at different time moments during a day, and on weekdays and weekends, we compare our best model SeqST-ResNet-3AVG and the most competitive baseline ST-ResNet. Figure 6(a) and (b) show that our model has lower prediction errors at the half-hour resolution prediction during a day, and also much lower prediction errors during weekends and weekdays. That is to say, our model can predict the task appearance at different time scales more accurately than ST-ResNet. This is mainly because our model learns from the sequences of task images at different time scales, rather than taking concatenation of tasks images as done in ST-ResNet.

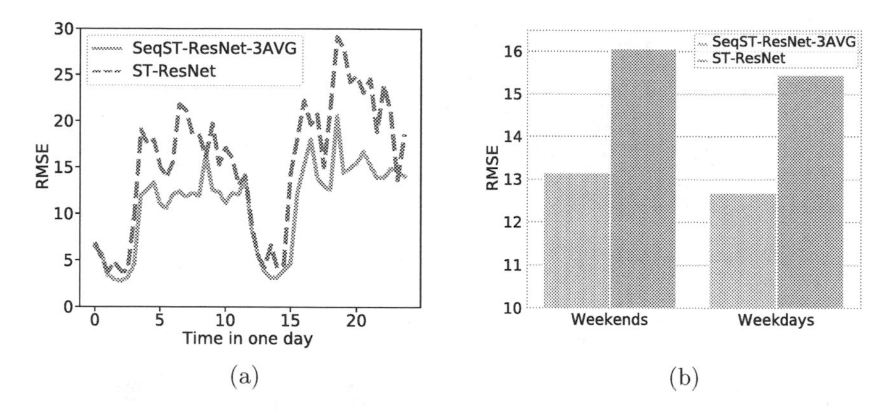

Fig. 6. Prediction at different time in one day (a) and at weekends and weekdays (b)

Prediction Performance on Regions with Different Request Intensity. We are also interested in evaluating the prediction performance of our model on regions with different true task frequency, e.g., request intensive regions vs mild regions. We calculate the average of ground truth task appearance frequency in each grids (10×10 grids), and then sort these grid cells by the average frequency with ascending order (from the most idle region to the most busy region). Figure 7 shows the prediction performance on these grid cells ordered on x-axis. In regions with intensive task requests, our SeqST-ResNet-3AVG model has much lower error than the baseline ST-ResNet model, while they have similar performance in mild regions. This is an important advantage, because correct prediction of tasks in request intensive regions (e.g., around commercial centers, central transportation stations) will highly improve the task assignment efficacy in spatial crowdsourcing platform.

Influence of the Sequence Length. Our model learns from three types of sequences. We also evaluate how the sequence length can affect the prediction performance. Let l_i, l_d, l_w denote the length of interval-level sequence, day-level sequence and week-level sequence, respectively. In Fig. 8(a), we show the prediction error when changing l_i from 1 to 5, while fixing $l_d = 1$, $l_w = 1$. The result shows that with l_i increasing, the RMSE decreases. This indicates longer interval-level sequences can help on improving prediction accuracy, but also takes more training time. An appropriate setting is $l_i = 3$ because the error decreases slowly after l_i is larger than 3. Figure 8(b) shows the influence of l_d when fixing $l_i = 3$, $l_w = 1$. The results show that the model performs best when $l_d = 3$. Neither too long nor too short day-level sequence is helpful. Then we set $l_i = 3$ and $l_d = 3$, and show the impact of l_w in Fig. 8(c) where we find $l_w = 1$ is the best setting.

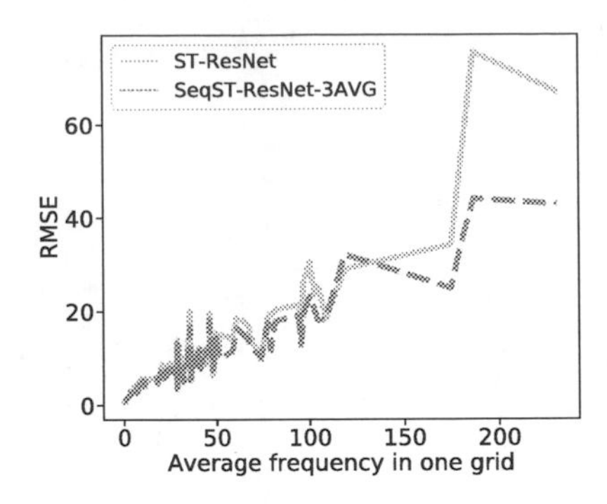

Fig. 7. Prediction on regions with different request intensity

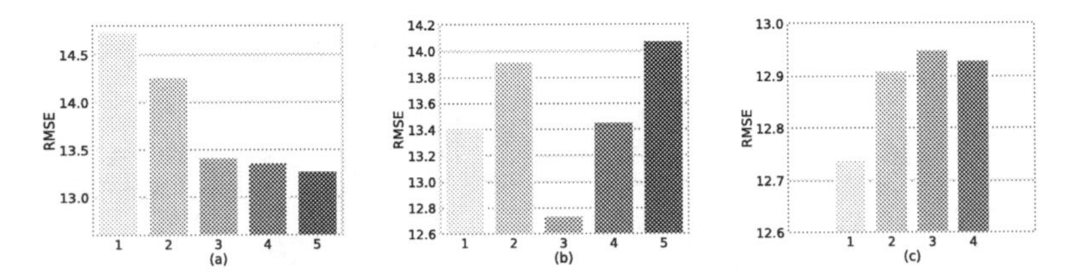

Fig. 8. Impact of sequence length at different levels. In (a), the length of interval-level sequence l_i varies when $l_d = 1$, $l_w = 1$. In (b), the length of day-level sequence l_d varies when $l_i = 3$, $l_w = 1$. In (c) the length of week-level sequence l_w varies when $l_i = 3$ and $l_d = 3$.

5 Related Work

In this section we briefly review some recent advances on task assignment in spatial crowdsourcing and deep learning since the focus of this paper is to apply deep neural networks on addressing the task prediction problem during task assignment in spatial crowdsourcing.

In [13], Kazemi and Shahabi propose several heuristics to maximize the number of assigned tasks in a given time interval while meeting the constraints specified by workers. In practice, tasks often arrives dynamically. This kind of online scenarios is more challenging and has been addressed in [20,21,23] where efficient algorithms with provable competitive ratio are proposed. Song et al. in [18] extend conventional task assignment from two objects matching problem to trichromatic matching problem. In [15,16,25,27,28], privacy of user or platform are considered when making the task assignment. In [4], spatial distribution of workers and tasks are taken into account when maximizing a global assignment quality score. In [14,24], travel time, as an important factor, the prediction of which has drawn some attentions recently. [32] tackles the problem of assigning tasks to workers such that mutual benefit are maximized. All these studies fail to model the spatial dependency among regions and temporal dependency on successive time units, which is the focus of our work in this paper. To the best of our knowledge, there is no deep learning based method to solve the task prediction problem. Thus, it is necessary to design a deep neural network based method.

On the other hand, recent years have witnessed the big success of deep learning in a variety of application domains. Specifically, there are lots of achievements in catching spatial or temporal properties. For temporal property, recurrent neural networks (RNN) [6] is designed to make use of sequential information, and has been shown great success in many NLP tasks [19]. However, vanishing gradient problem causes it to be difficult to capture the long-term dependency [2]. Long short term memory networks (LSTM) [11] improved RNN by using "gates" which control what to forget or remember. Gated Recurrent Unit (GRU) [5] simplifies LSTM by reducing the "gates" from 3 to 2. Bidirection LSTM (BiLSTM) [8] not only considers the forward data flow but also backward. For spatial property, convolution neural networks (CNN) can effectively capture the spatial property from near to distant with the depth deeper. Residual Networks (ResNet) [9] makes the deep network realize really "deep" even over 1000 layers. For combination of spatial and temporal properties, convolution LSTM (ConvLSTM) [26] and deep spatial temporal networks (DeepST) [29] are proposed to capture the two properties. However none of them can model long dependency as the training cost is really huge in practice.

6 Conclusion and Future Work

In this paper, we study the problem of predicting the task appearance in a spatial crowdsourcing platform. To take advantage of the temporal dependency of the historical sequential task request and the spatio-dependency in neighboring regions, we proposed a novel deep network model that learns from the sequences of task image data at different time scales. Experimental results on a real dataset demonstrated that our methods can significantly improve the prediction accuracy when comparing with the baselines. In future, we will consider to add the attention mechanism to further reduce the prediction error, and extend the application to other spatial crowdsourcing data such as meal order data.

Acknowledgements. This research was supported in part by National Natural Science Foundation of China (NSFC) (Grant No. 61572336, 61632016, 61572335, 61772356), and Natural Science Research Project of Jiangsu Higher Education Institution (No. 18KJA520010, 17KJA520003), and the King Abdullah University of Science and Technology (KAUST), and the Blockshine Technology corp. Data source: Didi Chuxing.

References

1. Didi Chuxing. https://gaia.didichuxing.com
2. Bengio, Y., Simard, P., Frasconi, P.: Learning long-term dependencies with gradient descent is difficult. IEEE Trans. Neural Netw. **5**(2), 157–166 (1994)

3. Box, G.E., Pierce, D.A.: Distribution of residual autocorrelations in autoregressive-integrated moving average time series models. J. Am. Stat. Assoc. **65**(332), 1509–1526 (1970)
4. Cheng, P., Lian, X., Chen, L., Shahabi, C.: Prediction-based task assignment in spatial crowdsourcing. In: 33rd IEEE International Conference on Data Engineering, ICDE 2017, San Diego, CA, USA, 19–22 April 2017, pp. 997–1008 (2017)
5. Cho, K., et al.: Learning phrase representations using RNN encoder-decoder for statistical machine translation. arXiv preprint arXiv:1406.1078 (2014)
6. Elman, J.L.: Finding structure in time. Cogn. Sci. **14**(2), 179–211 (1990)
7. Friedman, J.H.: Greedy function approximation: a gradient boosting machine. Ann. Stat. **29**(5), 1189–1232 (2001)
8. Graves, A., Schmidhuber, J.: Framewise phoneme classification with bidirectional LSTM and other neural network architectures. Neural Netw. **18**(5–6), 602–610 (2005)
9. He, K., Zhang, X., Ren, S., Sun, J.: Deep residual learning for image recognition. In: Proceedings of the IEEE Conference on Computer Vision and Pattern Recognition, pp. 770–778 (2016)
10. He, K., Zhang, X., Ren, S., Sun, J.: Identity mappings in deep residual networks. In: Leibe, B., Matas, J., Sebe, N., Welling, M. (eds.) ECCV 2016. LNCS, vol. 9908, pp. 630–645. Springer, Cham (2016). https://doi.org/10.1007/978-3-319-46493-0_38
11. Hochreiter, S., Schmidhuber, J.: Long short-term memory. Neural Comput. **9**(8), 1735–1780 (1997)
12. Ioffe, S., Szegedy, C.: Batch normalization: accelerating deep network training by reducing internal covariate shift. arXiv preprint arXiv:1502.03167 (2015)
13. Kazemi, L., Shahabi, C.: GeoCrowd: enabling query answering with spatial crowdsourcing. In: Proceedings of the 20th International Conference on Advances in Geographic Information Systems, pp. 189–198. ACM (2012)
14. Li, Y., Fu, K., Wang, Z., Shahabi, C., Ye, J., Liu, Y.: Multi-task representation learning for travel time estimation. In: International Conference on Knowledge Discovery and Data Mining, KDD (2018)
15. Liu, A., et al.: Privacy-preserving task assignment in spatial crowdsourcing. J. Comput. Sci. Technol. **32**(5), 905–918 (2017)
16. Liu, A., Wang, W., Shang, S., Li, Q., Zhang, X.: Efficient task assignment in spatial crowdsourcing with worker and task privacy protection. GeoInformatica **22**(2), 335–362 (2018)
17. Nair, V., Hinton, G.E.: Rectified linear units improve restricted Boltzmann machines. In: Proceedings of the 27th International Conference on Machine Learning, ICML 2010, pp. 807–814 (2010)
18. Song, T., et al.: Trichromatic online matching in real-time spatial crowdsourcing. In: ICDE, pp. 1009–1020 (2017)
19. Sutskever, I., Vinyals, O., Le, Q.V.: Sequence to sequence learning with neural networks. In: Advances in Neural Information Processing Systems, pp. 3104–3112 (2014)
20. Tong, Y., Chen, L., Shahabi, C.: Spatial crowdsourcing: challenges, techniques, and applications. PVLDB **10**(12), 1988–1991 (2017)
21. Tong, Y., She, J., Ding, B., Chen, L., Wo, T., Xu, K.: Online minimum matching in real-time spatial data: experiments and analysis. PVLDB **9**(12), 1053–1064 (2016)
22. Tong, Y., She, J., Ding, B., Wang, L., Chen, L.: Online mobile micro-task allocation in spatial crowdsourcing. In: 32nd IEEE International Conference on Data Engineering, ICDE 2016, Helsinki, Finland, 16–20 May 2016, pp. 49–60 (2016)

23. Tong, Y., et al.: Flexible online task assignment in real-time spatial data. Proc. VLDB Endow. **10**(11), 1334–1345 (2017)
24. Wang, Z., Fu, K., Ye, J.: Learning to estimate the travel time. In: Proceedings of the 24th ACM SIGKDD International Conference on Knowledge Discovery & Data Mining, pp. 858–866. ACM (2018)
25. Xiao, M., et al.: SRA: secure reverse auction for task assignment in spatial crowdsourcing. IEEE Trans. Knowl. Data Eng. 1 (2019). https://doi.org/10.1109/TKDE.2019.2893240
26. Shi, X., Chen, Z., Wang, H., Yeung, D.Y., Wong, W., Woo, W.: Convolutional LSTM network: a machine learning approach for precipitation nowcasting. In: Advances in Neural Information Processing Systems, pp. 802–810 (2015)
27. Zhai, D., et al.: Towards secure and truthful task assignment in spatial crowdsourcing. World Wide Web 1–24 (2018). https://doi.org/10.1007/s11280-018-0638-2
28. Zhang, D., Chow, C.Y., Liu, A., Zhang, X., Ding, Q., Li, Q.: Efficient evaluation of shortest travel-time path queries through spatial mashups. GeoInformatica **22**(1), 3–28 (2018)
29. Zhang, J., Zheng, Y., Qi, D.: Deep spatio-temporal residual networks for citywide crowd flows prediction. In: AAAI, pp. 1655–1661 (2017)
30. Zhang, J., Zheng, Y., Qi, D., Li, R., Yi, X.: DNN-based prediction model for spatio-temporal data. In: Proceedings of the 24th ACM SIGSPATIAL International Conference on Advances in Geographic Information Systems, p. 92. ACM (2016)
31. Zhao, J., Xu, J., Zhou, R., Zhao, P., Liu, C., Zhu, F.: On prediction of user destination by sub-trajectory understanding: a deep learning based approach. In: Proceedings of the 27th ACM International Conference on Information and Knowledge Management, CIKM 2018, Torino, Italy, 22–26 October 2018, pp. 1413–1422 (2018)
32. Zheng, L., Chen, L.: Mutual benefit aware task assignment in a bipartite labor market. In: ICDE, pp. 73–84 (2016)

Towards Robust Arbitrarily Oriented Subspace Clustering

Zhong Zhang, Chongming Gao, Chongzhi Liu, Qinli Yang, and Junming Shao(✉)

School of Computer Science and Engineering,
University of Electronic Science and Technology of China, Chengdu 611731, China
{zhongzhang,chongming.gao,liuchongzhi}@std.uestc.edu.cn
{qinli.yang,junmshao}@uestc.edu.cn

Abstract. Clustering high-dimensional data is challenging since meaningful clusters usually hide in the arbitrarily oriented subspaces, and classical clustering algorithms like k-means tend to fail in such case. Subspace clustering has thus attracted growing attention in the last decade and many algorithms have been proposed such as ORCLUS and 4C. However, existing approaches are usually sensitive to global and/or local noisy points, and the overlapping subspace clusters are little explored. Beyond, these approaches usually involve the exhaustive local search for correlated points or subspaces, which is infeasible in some cases. To deal with these problems, in this paper, we introduce a new subspace clustering algorithm called RAOSC, which formulates the Robust Arbitrarily Oriented Subspace Clustering as a group structure low-rank optimization problem. RAOSC is able to recover subspace clusters from a sea of noise while noise and overlapping points can be naturally identified during the optimization process. Unlike existing low-rank based subspace clustering methods, RAOSC can explicitly produce the subspaces of clusters without any prior knowledge of subspace dimensionality. Furthermore, RAOSC does not need a post-processing procedure to obtain the clustering result. Extensive experiments on both synthetic and real-world data sets have demonstrated that RAOSC allows yielding high-quality clusterings and outperforms many state-of-the-art algorithms.

Keywords: Subspace clustering · Correlation clustering

1 Introduction

In high-dimensional data set, meaningful clusters usually hide in the arbitrarily oriented subspaces, i.e., subsets of points showing linear correlations among subsets of dimensions [11]. To explain this idea, consider a real-world example illustrated in Fig. 1. The 2D plot represents a Height/Weight Standard[1], which consists of two subspace clusters of male and female, respectively. In contrast to

[1] http://www.angelo.edu/dept/rotc/height_weight_chart.php.

G. Li et al. (Eds.): DASFAA 2019, LNCS 11446, pp. 276–291, 2019.
https://doi.org/10.1007/978-3-030-18576-3_17

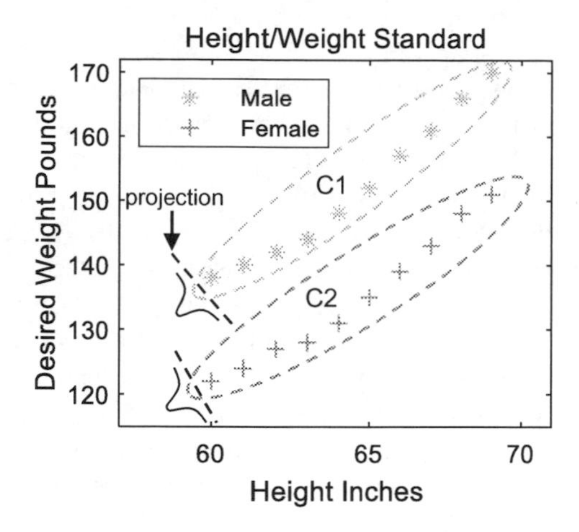

Fig. 1. A real-world example of subspace clustering. C1 and C2 are two subspace clusters.

the classical clustering, points are grouped into the same cluster because they exhibit high correlation (i.e., they locate near the same line or plane), rather than closeness. Only when projecting the points into the subspace orthogonal to the plane where they are lying in, they will exhibit high density. Since each cluster lies in an arbitrarily oriented subspace, it is referred to as arbitrarily oriented subspace clustering or correlation clustering.

Clustering a 2D data is just a piece of cake, but how about a high-dimensional data having dozens or hundreds of attributes? In such case, detecting subspace clusters is a challenging task since many dimensions are irrelevant and only a few of dimensions truly contribute to the cluster structure. The word "Relevant" means that a cluster shows high correlation in and only in these relevant dimensions. More importantly, the relevant dimensions often differ largely for different clusters [11]. Therefore, global dimensionality reduction methods like Principal Component Analysis (PCA) cannot be used to preserve the subspace cluster structure. To tackle this problem, most approaches adopt certain assumptions/heuristics and start from a local search of subspaces and clusters.

During the past decade, many subspace clustering approaches have been proposed from various perspectives. The earliest attempt is to heuristically examine all possible axis-parallel subspaces and identify clusters, algorithms include CLIQUE [4], ENCLU [7], PROCLUS [2], SUBCLU [10], DUSC [5], to name a few. However, these algorithms can only find axis-parallel subspace clusters. Afterwards, arbitrarily oriented subspace clustering emerges. Most of these algorithms rely on the search of local correlated points to identify clusters and subspaces. Algorithms include, for examples, ORCLUS [3], 4C [6], CURLER [21], SSCC [9], FOSSCLU [8], ORSC [18] and CoSync [19]. However, most previous solutions of finding suitable subspaces work well if and only if subspace clusters are locally well separated and no noise/outlier points exist. In the presence of noise/outliers in the local neighbourhood of cluster points or cluster representatives in the entire feature space, most previous methods fail to detect meaningful subspace clusters. Besides, due to the (exhaustive) heuristic local search, they are generally time consuming especially when the dimensionality is high.

In contrast to previous methods that use certain heuristic ways to search subspace clusters, we turn to formulate the subspace clustering task as a group structure low-rank optimization problem. The key idea is to assign data points to clusters to meet the **correlation** and **closeness** criteria. Specifically, if we examine an arbitrarily oriented subspace cluster w.r.t. its relevant dimensions, we find these cluster objects exhibit a high correlation. Meanwhile, when projecting these cluster objects into the orthogonal complementary space spanned by the irrelevant dimensions, they exhibit a high closeness. We argue that taking both of the correlation and closeness into account improves the robustness. Motivated by the observations, in this paper, we propose a new subspace clustering algorithm called RAOSC, which formulates the Robust Arbitrarily Oriented Subspace Clustering as a group structure low-rank optimization problem. It has several attractive properties. Firstly, since the optimization does not rely on the local search, it is more efficient. Furthermore, the optimization problem well characterizes the two criteria of correlation and closeness simultaneously where most previous methods only consider the correlation criterion. This makes RAOSC more robust to noisy objects or outliers. Inspired by [17], we develop an effective and efficient optimization algorithm to solve RAOSC. During the optimization process, noise and overlapping points can be naturally identified. Last but not least, unlike the previous low-rank representation (LRR) based subspace clustering methods [12,13] that cannot give explicit subspaces of clusters and need a two-step algorithm to do clustering, RAOSC is able to find explicit subspace for each cluster without knowing the subspace's dimensionality. It directly obtains the discrete cluster membership indicators by the optimization, no further post-processing procedure is needed. In summary, the main contributions of our work are listed as follows.

- We formulate the identification of arbitrarily oriented subspace clustering as an optimization problem by exploiting two intrinsic properties of a subspace cluster: **correlation** and **closeness**. We integrate the two properties and formulate a group structure low-rank model for subspace clustering.
- We develop an effective and efficient optimization algorithm for RAOSC. During the optimization process, noise and overlapping points can be naturally identified. To the best of our knowledge, for arbitrarily oriented subspace clustering problem, we are the first to handle both noise and overlapping points in one unified framework.
- We perform extensive experiments on synthetic and real-world data sets and compare with the state-of-the-art algorithms to demonstrate the effectiveness of our approach.

2 The Proposed Method

2.1 Problem Formulation

Formally, let $\mathbf{X} \in \mathbb{R}^{m \times n}$ be a data matrix of n instances with m dimensions, the objective of this study is to find k overlapping arbitrarily oriented subspace clusters $\{\mathcal{C}_1, \mathcal{C}_2, \ldots, \mathcal{C}_k\}$, and a noise point set $\mathcal{C}_0$. $\mathbf{X}_i$ is a matrix containing data points in cluster $\mathcal{C}_i$. For each cluster, we use a diagonal indicator matrix $\mathbf{P}_i \in \{0,1\}^{m \times m}$ to indicate the relevant dimensions in the original space, and use $\bar{\mathbf{P}}_i = \mathbf{I} - \mathbf{P}_i$ to indicate the irrelevant dimensions, where $\mathbf{I}$ denotes an identity matrix. Specifically, $\mathbf{P}_i^T \mathbf{X}_i$ sets the rows of $\mathbf{X}_i$ corresponding to the irrelevant dimensions to zero, and leaves the rest of the rows corresponding to the relevant dimensions untouched, which extracts the axis-parallel relevant dimensions of $\mathbf{X}_i$. Since subspace clusters may accommodate in arbitrarily oriented subspaces, thereby, inspired by [14], an orthonormal rigid rotation matrix $\mathbf{S}_i \in \mathbb{R}^{m \times m}$ is further introduced. $\mathbf{S}_i$ rotates the i-th cluster so that its relevant dimensions align to the parallel axes. Combine these two matrices, $\mathbf{P}_i^T \mathbf{S}_i^T \mathbf{X}_i$ is thus used to characterize the relevant subspace for the cluster $\mathcal{C}_i$. Finally, we use a diagonal matrix $\mathbf{G}_i \in \{0,1\}^{n \times n}$ to indicate the corresponding cluster membership, where $\mathbf{G}_i(j,j) = 1$ if the j-th data point is grouped into the i-th subspace cluster, and 0 otherwise. Thus $\mathbf{X}\mathbf{G}_i$ leaves the columns corresponding to the i-th cluster points untouched and sets the others to zero. Therefore, for a given data set, the subspace clustering is conducted by learning the three matrices for each cluster.

2.2 Clustering via Correlation and Closeness

In this study, we consider that a subspace cluster should satisfy two criteria: **correlation** and **closeness**. Specifically, we first refer to the subspace spanned by the relevant dimensions as the correlation space, the subspace spanned by the irrelevant dimensions as the cluster space. Note the two subspaces are orthogonal complementary to each other w.r.t. the full space. Data points in a subspace cluster should show high correlation in the correlation space. Meanwhile, they should be as close as possible when projecting them into the cluster space. By contrast, most existing approaches only consider the correlation for subspace clustering. To illustrate this basic idea, Fig. 2 gives a toy example. Figure 2(a) shows that data points in a subspace cluster should locate near an arbitrarily oriented 2D plane in the full 3D space (i.e., strong correlation). In addition, data points are close to each other when projecting them into the cluster space (i.e., high closeness) (see Fig. 2(b)). In the following, we will formulate our objective function in terms of the two criteria.

For the i-th cluster, we assume that we can find an orthonormal rigid rotation matrix $\mathbf{S}_i$ [14], which rotates the original space and thus the first d_i dimensions span the correlation space accommodating all data points in the i-th cluster (e.g., S in Fig. 2). And the last $(m - d_i)$ dimensions span the cluster space (e.g., $R^3 \backslash S$ in Fig. 2). Note that the dimensionality d_i is not a parameter that needs to be manually set. It is automatically determined in the optimization process.

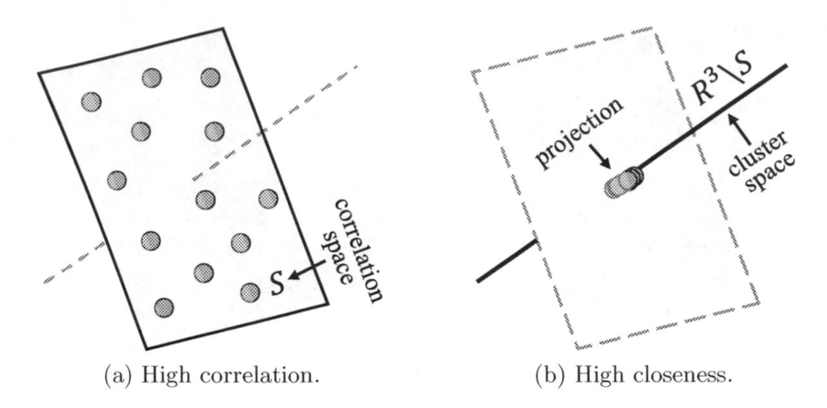

(a) High correlation.

(b) High closeness.

Fig. 2. Illustration of subspace clustering with criteria of correlation and closeness in correlation space $\mathcal{S}$ and cluster space $R^3\backslash\mathcal{S}$.

Accordingly, we use two diagonal indicator matrices $\mathbf{P}_i$ and $\bar{\mathbf{P}}_i$ defined as follows to split the two subspaces.

$$\mathbf{P}_i = \begin{bmatrix} \mathbf{I}_{d_i} & \\ & \mathbf{0}_{m-d_i} \end{bmatrix}, \bar{\mathbf{P}}_i = \begin{bmatrix} \mathbf{0}_{d_i} & \\ & \mathbf{I}_{m-d_i} \end{bmatrix}. \tag{1}$$

A data point $\mathbf{x}$ can project into the correlation space and the cluster space by $\mathbf{P}_i^T\mathbf{S}_i^T\mathbf{x}$ and $\bar{\mathbf{P}}_i^T\mathbf{S}_i^T\mathbf{x}$, respectively.

Since we require closeness of a cluster in the cluster space, one intuitive option is to minimize the pairwise distance of all cluster points, which can be formulated as $\sum_{\mathbf{x},\mathbf{y}\in\mathcal{C}_i}||\bar{\mathbf{P}}_i^T\mathbf{S}_i^T\mathbf{x}-\bar{\mathbf{P}}_i^T\mathbf{S}_i^T\mathbf{y}||$. To characterize the correlation in the correlation space, we consider a group structure low-rank model, i.e., a group of data points in the same cluster are low-rank in the full-dimensional space. We can minimize the ranks of clusters to find subspace clusters. That is, $\min\limits_{\mathcal{C}}\sum_{i=1}^k rank(\mathbf{X}_i)^2$. Since multiplying an orthogonal matrix, and discarding the irrelevant dimensions of a subspace cluster in the full-dimensional space do not change the rank, we can equivalently write it as $\min\limits_{\mathbf{P},\mathbf{S},\mathcal{C}}\sum_{i=1}^k rank(\mathbf{P}_i^T\mathbf{S}_i^T\mathbf{X}_i)^2$. Note the square on the rank function is used to avoid trivial solution as discussed in [17]. We formulate the problem as the following objective function.

$$\begin{aligned} &\min_{\mathbf{P},\mathbf{S},\mathcal{C}}\sum_{i=1}^{k}\Big(rank(\mathbf{P}_i^T\mathbf{S}_i^T\mathbf{X}_i)^2+\sum_{\mathbf{x},\mathbf{y}\in\mathcal{C}_i}||\bar{\mathbf{P}}_i^T\mathbf{S}_i^T(\mathbf{x}-\mathbf{y})||\Big) \\ &s.t.\ \forall\, 1\le i\le k,\ \ \mathbf{S}_i^T\mathbf{S}_i=\mathbf{I}. \end{aligned} \tag{2}$$

We use an alternative diagonal indicator matrix $\mathbf{G}_i$ to rewrite Eq. (2) as follows.

$$\min_{\mathbf{P},\mathbf{S},\mathbf{G}} \sum_{i=1}^{k} \left(rank(\mathbf{P}_i^T \mathbf{S}_i^T \mathbf{X} \mathbf{G}_i)^2 + \sum_{\mathbf{x},\mathbf{y} \in \mathcal{C}_i} ||\bar{\mathbf{P}}_i^T \mathbf{S}_i^T (\mathbf{x} - \mathbf{y})|| \right)$$
$$s.t. \sum_{i=1}^{k} \mathbf{G}_i \succeq \mathbf{I}, \sum_{i=1}^{k} Tr(\mathbf{G}_i) = kl, \tag{3}$$
$$\forall\, 1 \leq i \leq k, \quad \mathbf{G}_i \subseteq \{0,1\}^{n \times n}, \mathbf{S}_i^T \mathbf{S}_i = \mathbf{I},$$

where the term $\sum_{i=1}^{k} \mathbf{G}_i \succeq \mathbf{I}$ and $\sum_{i=1}^{k} Tr(\mathbf{G}_i) = kl$ ensure that each data point has to be assigned to at least one group, and some data points can be assigned to multiple groups to allow overlapping clustering. l is an integer parameter that controls the degree of overlapping, which is within the range of $[n/k, n]$. $l = n/k$ means no overlapping clusters, $l = n$ is complete overlapping (i.e., each data point belongs to all clusters). Since l depends on the number of data points so that it is not convenient to set, we use a variable substitution trick which replaces l with $\tilde{l} \in [0,1]$. It stands for the proportion of the overlapping data points, where $l = \left\lfloor \frac{\tilde{l}n(k-1)+n}{k} \right\rfloor$.

Since the rank minimization problem is NP-hard, the common practice is to relax the rank function to the nuclear norm. However, inspired by [17], we use the Schatten-1 norm for relaxation. Schatten-1 norm is numerically equal to the nuclear norm, but it can be more efficiently optimized by adopting certain strategy. In addition, finding an optimal cluster assignment on the minimization of all pairwise distances (i.e., $\sum_{\mathbf{x},\mathbf{y} \in \mathcal{C}_i} ||\bar{\mathbf{P}}_i^T \mathbf{S}_i^T \mathbf{x} - \bar{\mathbf{P}}_i^T \mathbf{S}_i^T \mathbf{y}||$) is also computationally infeasible. We relax it by minimizing distances between data points and cluster centroids. Thereby, the second term can be relaxed as a k-means style term. The overall objective function is given as follows.

$$\min_{\mathbf{P},\mathbf{S},\mathbf{G},\mathbf{M}} \sum_{i=1}^{k} \Big((||\mathbf{P}_i^T \mathbf{S}_i^T (\mathbf{X} - \mathbf{M}_i) \mathbf{G}_i||_{Sp}^p)^2$$
$$+ \alpha ||\bar{\mathbf{P}}_i^T \mathbf{S}_i^T (\mathbf{X} - \mathbf{M}_i) \mathbf{G}_i||_F^2 \Big) \tag{4}$$
$$s.t. \sum_{i=1}^{k} \mathbf{G}_i \succeq \mathbf{I}, \sum_{i=1}^{k} Tr(\mathbf{G}_i) = kl,$$
$$\forall\, 1 \leq i \leq k, \quad \mathbf{G}_i \subseteq \{0,1\}^{n \times n}, \mathbf{S}_i^T \mathbf{S}_i = \mathbf{I},$$

where $||\mathbf{X}||_{Sp}^p = Tr((\mathbf{X}\mathbf{X}^T)^{\frac{p}{2}})$ is the Schatten-p norm of matrix $\mathbf{X}$, and we take $p = 1$ in this study. $\mathbf{M}_i = \boldsymbol{\mu}_i \mathbf{1}_n^T$, $\boldsymbol{\mu}_i$ is the mean vector of the i-the cluster. Using $\mathbf{X} - \mathbf{M}_i$ to replace $\mathbf{X}$ at the first term is to make the rank approximation more robust [17]. $\alpha > 0$ is a real number parameter to balance the dimensionality of the correlation space and the cluster space, which will be discussed later.

2.3 Optimization Algorithm

To solve the optimization problem, we use an iterative algorithm to update $\mathbf{P}_i$, $\mathbf{S}_i$, $\mathbf{G}_i$ and $\mathbf{M}_i$ one at a time while fixing the others to achieve a local optimum.

Updating G. Let $\mathcal{O}$ denote the objective function Eq. (4), we write the Lagrange function of Eq. (4) w.r.t. variables $\mathbf{G}$ as follows.

$$L(\mathbf{G}, \Lambda) = \mathcal{O}(\mathbf{G}) + g(\Lambda, \mathbf{G}), \tag{5}$$

where Λ is the Lagrange dual variable, $g(\Lambda, \mathbf{G})$ encodes the constraints on $\mathbf{G}$ in problem Eq. (4).

By taking the derivative of Eq. (5) w.r.t. $\mathbf{G}$, and setting it to zero, we have:

$$\sum_{i=1}^{k} 2\mathbf{A}_i\mathbf{G}_i + \frac{\partial g(\Lambda, \mathbf{G}_i)}{\partial \mathbf{G}_i} = 0, \tag{6}$$

where $\mathbf{A}_i$ and $\mathbf{B}_i$ are:

$$\mathbf{A}_i = \tilde{\mathbf{X}}_i^T\mathbf{S}_i\mathbf{P}_i\mathbf{B}_i\mathbf{P}_i^T\mathbf{S}_i^T\tilde{\mathbf{X}}_i + \alpha\tilde{\mathbf{X}}_i^T\mathbf{S}_i\bar{\mathbf{P}}_i^2\mathbf{S}_i^T\tilde{\mathbf{X}}_i, \tag{7}$$

$$\mathbf{B}_i = p||\mathbf{P}_i^T\mathbf{S}_i^T\tilde{\mathbf{X}}_i\mathbf{G}_i||_{Sp}^p(\mathbf{P}_i^T\mathbf{S}_i^T\tilde{\mathbf{X}}_i\mathbf{G}_i^2\tilde{\mathbf{X}}_i^T\mathbf{S}_i\mathbf{P}_i)^{\frac{p-2}{2}}. \tag{8}$$

$\tilde{\mathbf{X}}_i = \mathbf{X} - \mathbf{M}_i$. $\mathbf{A}_i$ and $\mathbf{B}_i$ depend on $\mathbf{G}_i$. It can be solved via an iteration based re-weighted algorithm [16,17]. We first calculate $\mathbf{A}_i$ and $\mathbf{B}_i$ based on the current $\mathbf{G}_i$. After $\mathbf{A}_i$ and $\mathbf{B}_i$ are fixed and treated as constants, we can solve the following problem which satisfies Eq. (6) to update $\mathbf{G}_i$.

$$\begin{aligned} &\min_{\mathbf{G}} \sum_{i=1}^{k} Tr(\mathbf{G}_i^T\mathbf{A}_i\mathbf{G}_i) \\ &s.t. \sum_{i=1}^{k}\mathbf{G}_i \succeq \mathbf{I}, \sum_{i=1}^{k} Tr(\mathbf{G}_i) = kl, \\ &\forall\, 1 \le i \le k, \quad \mathbf{G}_i \subseteq \{0,1\}^{n\times n}. \end{aligned} \tag{9}$$

Due to the diagonality and the discrete constraints of $\mathbf{G}_i$, Eq. (9) can be equivalently written as:

$$\begin{aligned} &\min_{g} \sum_{i=1}^{k}\sum_{j=1}^{n} a_{ij}g_{ij} \\ &s.t. \sum_{i=1}^{k} g_{ij} \ge 1, \sum_{i=1}^{k}\sum_{j=1}^{n} g_{ij} = kl, \\ &\forall\, 1 \le i \le k,\ 1 \le j \le n,\ g_{ij} \in \{0,1\}, \end{aligned} \tag{10}$$

where g_{ij} is the j-th diagonal element of matrix $\mathbf{G}_i$, a_{ij} is the j-th diagonal element of matrix $\mathbf{A}_i$. The objective function above derives a 0-1 integer programming problem, which is usually NP-hard. Fortunately, the constraints imposed on g_{ij} significantly reduce the searching space, we can still obtain an efficient algorithm to solve this problem. Firstly, for every data point, we assign cluster i which has the minimum a_{ij} value to the j-th data point. This ensures every data point has been assigned to one cluster. Afterwards, we deal with the remaining $(kl - n)$ overlapping data points. We sort the remaining a_{ij} in an ascending order, then record the first $(kl - n)$ subscripts (ij) of a_{ij} and set the corresponding $g_{ij} = 1$.

Meanwhile, noise can be naturally identified during the cluster membership assigning process. We additionally let $\mathbf{G}_0$ denote the noise indicator matrix, $\mathbf{G}_0(j, j) = 1$ if data point j is a noise, otherwise 0. Given the current cluster indicator vector $\mathbf{g}_i$ for the i-th cluster, we first collect all a_{ij} which satisfy $g_{ij} = 1$, then find their median m_i. A data point j in the i-the cluster is a noise, if $a_{ij} > \lambda m_i$, where $\lambda > 0$ is a real number parameter. The rationale is, if we closely look into a_{ij}, it consists of two parts: the first part can be regarded as a weighted Mahalanobis-like distance between $\mathbf{x}_j$ and the group mean of the i-th cluster in the correlation space. If data point j has been assigned to the i-the subspace cluster but deviates from the principal component directions of the i-th subspace cluster (i.e., deviates from the plane where the subspace cluster is lying in), it tends to produce a large value of a_{ij}. The second part represents the Euclidean distance between $\mathbf{x}_j$ and the group mean in the cluster space. Noise points are far away from the cluster center in the cluster space, which also produce large a_{ij}. In summary, a noise point can be pinpointed by checking the anomaly large a_{ij}. Note that the overlapping points are identified by fulfilling the constraint of $\sum_{i,j} g_{ij} = kl$. We need to recalculate $l = \left\lfloor \frac{\tilde{l}(n - Tr(\mathbf{G}_0))(k-1) + (n - Tr(\mathbf{G}_0))}{k} \right\rfloor$ at each iteration otherwise noise points will be assigned to the overlapping clusters. We summarize the algorithm to solve problem Eq. (10) in Algorithm 1.

Updating S and P. Adopting the similar derivation of updating $\mathbf{G}_i$, we can use the iteration based re-weighted method to solve $\mathbf{S}_i$ by optimizing the following objective function.

$$\begin{aligned} &\min_{\mathbf{S}} \sum_{i=1}^{k} Tr(\mathbf{S}_i^T \mathbf{C}_i \mathbf{S}_i \mathbf{P}_i \mathbf{P}_i^T) + Tr(\mathbf{S}_i^T \mathbf{D}_i \mathbf{S}_i \bar{\mathbf{P}}_i \bar{\mathbf{P}}_i^T) \\ &s.t.\ \forall\, 1 \leq i \leq k,\ \ \mathbf{S}_i^T \mathbf{S}_i = \mathbf{I}, \end{aligned} \tag{11}$$

where $\mathbf{C}_i$, $\mathbf{D}_i$ and $\mathbf{E}_i$ are:

$$\mathbf{C}_i = \tilde{\mathbf{X}}_i \mathbf{G}_i \mathbf{E}_i \mathbf{G}_i^T \tilde{\mathbf{X}}_i^T, \tag{12}$$

$$\mathbf{D}_i = \alpha \tilde{\mathbf{X}}_i \mathbf{G}_i^2 \tilde{\mathbf{X}}_i^T, \tag{13}$$

$$\mathbf{E}_i = p \left|\left| \mathbf{G}_i^T \tilde{\mathbf{X}}_i^T \mathbf{S}_i \mathbf{P}_i \right|\right|_{Sp}^{p} (\mathbf{G}_i^T \tilde{\mathbf{X}}_i^T \mathbf{S}_i \mathbf{P}_i^2 \mathbf{S}_i^T \tilde{\mathbf{X}}_i \mathbf{G}_i)^{\frac{p-2}{2}}. \tag{14}$$

Note that $\mathbf{P}_i \mathbf{P}_i^T$ leaves the upper left $d_i \times d_i$ matrix untouched and sets the other elements to zero, thus the value of $Tr(\mathbf{S}_i^T \mathbf{C}_i \mathbf{S}_i \mathbf{P}_i \mathbf{P}_i^T)$ is equal to the summation of the eigenvalues of that upper left matrix. It is similar for $\bar{\mathbf{P}}_i \bar{\mathbf{P}}_i^T$.

To solve problem Eq. (11), we first calculate the eigenvalues of $\mathbf{C}_i$ and $\mathbf{D}_i$ and put them in an array as $\boldsymbol{\delta}_i = \{\delta_{Ci}^{(1)}, \ldots, \delta_{Ci}^{(m)}, \delta_{Di}^{(1)}, \ldots, \delta_{Di}^{(m)}\}$, then sort $\boldsymbol{\delta}_i$ in

Algorithm 1. Algorithm to solve problem Eq. (10).

Input:

Variable $\mathbf{A}_i(1 \leq i \leq k)$, number of clusters k, parameters $\tilde{l}$, λ.

Output:

Cluster and noise indicators $\mathbf{G}_i(0 \leq i \leq k)$.

1: Initialize $\mathbf{G}_0 = \mathbf{0}_{n \times n}$.
2: **for** $i = 1$ to k **do**
3: // Make sure every point has been assign to one cluster.
4: **for** $j = 1$ to n **do**
5: $\mathbf{G}_i(j,j) = \begin{cases} 1, \, i = \underset{i}{\text{argmin}} \, \mathbf{A}_i(j,j) \\ 0, \quad \text{otherwise} \end{cases}$
6: **end for**
7: // Handle the noise points.
8: $\mathcal{J} = \{j \mid \mathbf{G}_i(j,j) = 1\}$.
9: $m_i =$ median of $\mathbf{A}_i(\mathcal{J}, \mathcal{J})$.
10: **for** j in $\mathcal{J}$ **do**
11: **if** $\mathbf{A}_i(j,j) > \lambda \times m_i$ **then**
12: $\mathbf{G}_0(j,j) = 1$.
13: $\mathbf{G}_i(j,j) = 0$.
14: **end if**
15: $\mathbf{A}_i(j,j) = \text{Inf}$.
16: **end for**
17: **end for**
18: // Handle the overlapping points.
19: $l = \left\lfloor \frac{\tilde{l}(n - Tr(\mathbf{G}_0))(k-1) + (n - Tr(\mathbf{G}_0))}{k} \right\rfloor$.
20: $\mathcal{I} = \{(i,j) \mid$ sort $\mathbf{A}_i(j,j)$ in an ascending order then leave the first $kl - (n - Tr(\mathbf{G}_0))$ entries$\}$.
21: $\mathbf{G}_i(j,j) = 1$, if $(i,j) \in \mathcal{I}$.

an ascending order. Without loss of generality, we assume that d_i eigenvalues of $\mathbf{C}_i$ and $m - d_i$ eigenvalues of $\mathbf{D}_i$ are in the m-smallest set of $\boldsymbol{\delta}_i$. Then we permute the m-smallest set so that the first d_i and the last $m - d_i$ entries are the eigenvalues of $\mathbf{C}_i$ and $\mathbf{D}_i$, respectively. The optimal solution of problem Eq. (11) can be obtained by putting d_i eigenvectors of $\mathbf{C}_i$ and $m - d_i$ eigenvectors of $\mathbf{D}_i$ corresponding to their smallest eigenvalues into $\mathbf{S}_i$'s columns. Note that d_i is automatically determined by the sorting rather than a parameter. Then we can update $\mathbf{P}_i$ and $\bar{\mathbf{P}}_i$ by using Eq. (1).

Updating M. $\mathbf{M}_i$ is a matrix whose columns are identical, every column is the mean vector of the i-th cluster. Thus the actual variable that needs to be solved is $\boldsymbol{\mu}_i$. We solve the following problem to update $\mathbf{M}_i$.

$$\begin{aligned} &\min_{\boldsymbol{\mu}} \sum_{i=1}^{k} Tr(\mathbf{G}_i^T (\mathbf{X} - \mathbf{M}_i)^T \mathbf{F}_i (\mathbf{X} - \mathbf{M}_i) \mathbf{G}_i) \\ &s.t. \; \forall \, 1 \leq i \leq k, \;\; \mathbf{M}_i = \boldsymbol{\mu}_i \mathbf{1}_n^T, \end{aligned} \tag{15}$$

Algorithm 2. Algorithm to solve RAOSC

Input:

Data matrix $\mathbf{X}$, the number of clusters k, parameters $\tilde{l}$, λ, α.

Output:

Cluster and noise indicators $\mathbf{G}_i(0 \leq i \leq k)$, subspace indicators $\mathbf{P}_i(1 \leq i \leq k)$ and the orthonormal rigid rotation matrix $\mathbf{S}_i(1 \leq i \leq k)$.

1: Initialize all $\mathbf{G}_i$ such the constraints in Eq. (10) are satisfied. $\mathbf{S}_i$, $\mathbf{P}_i$, $\bar{\mathbf{P}}_i = \mathbf{I}$. $\mathbf{M}_i = \mathbf{0}$.
2: **repeat**
3: Calculate $\tilde{\mathbf{X}}_i|_{i=1}^k = \mathbf{X} - \mathbf{M}_i$.
4: Calculate $\mathbf{A}_i|_{i=1}^k$ and $\mathbf{B}_i|_{i=1}^k$ using Eq. (7-8).
5: Update $\mathbf{G}_i|_{i=0}^k$ using Algorithm 1.
6: Calculate $\mathbf{C}_i|_{i=1}^k$, $\mathbf{D}_i|_{i=1}^k$, $\mathbf{E}_i|_{i=1}^k$ using Eq. (12-14).
7: Update $\mathbf{S}_i|_{i=1}^k$ by putting m eigenvectors of $\mathbf{C}_i|_{i=1}^k$ or $\mathbf{D}_i|_{i=1}^k$ corresponding to the m-smallest eigenvalues in $\boldsymbol{\delta}_i|_{i=1}^k = \{\delta_{Ci}^{(1)}, ..., \delta_{Ci}^{(m)}, \delta_{Di}^{(1)}, ..., \delta_{Di}^{(m)}\}$ into its columns.
8: Permute columns of $\mathbf{S}_i|_{i=1}^k$ so that the first d_i and the last $m - d_i$ columns are the eigenvectors of $\mathbf{C}_i|_{i=1}^k$ and $\mathbf{D}_i|_{i=1}^k$, respectively.
9: Update $\mathbf{P}_i|_{i=1}^k$ and $\bar{\mathbf{P}}_i|_{i=1}^k$ using Eq. (1).
10: Update $\mathbf{M}_i|_{i=1}^k$ using Eq. (16).
11: **until** convergence or max no. iterations reached.

where $\mathbf{F}_i = \mathbf{S}_i\mathbf{P}_i\mathbf{B}_i\mathbf{P}_i^T\mathbf{S}_i^T + \alpha\mathbf{S}_i\bar{\mathbf{P}}_i^2\mathbf{S}_i^T$, though it is irrelevant for updating $\mathbf{M}_i$. By substituting the variable and calculating the derivative w.r.t. $\boldsymbol{\mu}_i$ and setting it to zero, it is easy to obtain the update rule of $\mathbf{M}_i$ as follows.

$$\mathbf{M}_i = \frac{1}{Tr(\mathbf{G}_i)}\mathbf{X}\mathbf{G}_i\mathbf{1}_n\mathbf{1}_n^T. \tag{16}$$

It can be seen that updating $\mathbf{M}_i$ is just simply calculating the mean of a cluster in the original space.

Finally, we summarize the overall optimization procedure in Algorithm 2.

2.4 Relationship to Existing Clustering Paradigms

For α, when setting it to a relative small value, step 7 in Algorithm 2 will put all $\mathbf{D}_i$'s eigenvectors into $\mathbf{S}_i$'s columns, so that the correlation space is vanished ($\mathbf{P}_i = \mathbf{0}$) and the cluster space gains full dimensionality ($\bar{\mathbf{P}}_i = \mathbf{I}$). This yields the problem Eq. (4) without the first term, which is the ordinary k-means algorithm (suppose no overlapping or noise). Similarly, if we set α to a large value, it yields the problem Eq. (4) without the second term, which degenerates to a generalized version of the LRS model [17]. When setting α to a medium value, it balances the dimensionality between the correlation space and the cluster space, thus the correlation and closeness of a cluster are both taken into consideration.

2.5 Time Complexity

Without loss of generality, we assume $m < n$ in the following analysis. The computational bottleneck of RAOSC lies in the SVD decomposition at step 4, 6 and 7 of Algorithm 2. In step 4, computing $\mathbf{B}_i$ needs to compute $||\mathbf{P}_i^T\mathbf{S}_i^T\tilde{\mathbf{X}}_i\mathbf{G}_i||_{Sp}^p$ and $(\mathbf{P}_i^T\mathbf{S}_i^T\tilde{\mathbf{X}}_i\mathbf{G}_i^2\tilde{\mathbf{X}}_i^T\mathbf{S}_i\mathbf{P}_i)^{\frac{p-2}{2}}$, both rely on the SVD of $\mathbf{P}_i^T\mathbf{S}_i^T\tilde{\mathbf{X}}_i\mathbf{G}_i$, which costs $O(m^2n)$. Thus computing all $\mathbf{B}_i$ costs $O(m^2nk)$. Computing $\mathbf{A}_i$ costs $O(m^2nk)$. Thus step 4 costs $O(m^2nk)$. Similar to step 4, step 6 costs $O(mn^2k)$. Step 7 needs to compute SVD of $\mathbf{C}_i$ and $\mathbf{D}_i$, which costs $O(m^3k)$. In summary, the time complexity of Algorithm 2 is $O((mn^2 + m^2n + m^3)kt)$, where k is usually a small constant that can be ignored, t is the number of iterations. Usually, the algorithm converges in a few iterations, e.g., 50 iterations.

3 Experiments

In this section, we evaluate our method with respect to its clustering results on both synthetic data and real-world data. We start with the synthetic data to show a proof-of-concept of finding arbitrarily oriented subspace clusters in the presence of noise and overlapping points. Afterwards, we compare our method with six state-of-the-art algorithms on nine real-world data sets obtained from the UCI and UCR repositories. For real-world data, we have no prior knowledge about the noise nor overlapping, and most of the typical comparison algorithms cannot handle such case. So we only perform clustering on real-world data sets with an assumption that there are no noise or overlapping points, since we have demonstrated it on the synthetic data.

3.1 Evaluation on Synthetic Data

We start with the synthetic data to demonstrate the effectiveness of finding arbitrarily oriented subspace clusters in the presence of noise and overlapping points. Here three synthetic data sets are generated. In details, synthetic data 1 consists of three subspace clusters in 3D space, where one of the subspace cluster forms a 2D plane and the other two subspace clusters form two cross lines passing through the plane's origin. Each subspace cluster contains 500 points with 5% level perturbation added to deviate from the subspace. Synthetic data 2 consists of two subspace clusters forming two perpendicular 2D planes in 3D space, each subspace cluster contains 500 uniformly distributed points with 5% level perturbation added. Synthetic data 3 consists of two perpendicular subspace clusters in 2D space. Each subspace cluster contains 500 uniformly distributed points forming a long and narrow rectangle shape. Note that synthetic data 2 and 3 have approximate 10% overlapping points. Finally, we add uniform noise points into all synthetic data set with noise level ranging from 0.1 to 0.8.

For comparison, we select three state-of-the-art subspace clustering algorithms and compare the clustering performance while varying the noise level. The comparison algorithms are NrKmeans [15], ISAAC [22] and ORSC [19]. The reason we select these algorithms is that they can handle noise and/or overlapping to a certain extent. We select the parameters according to the true statistics of the data and from a wide tuning range to obtain the best result. For numerical evaluation, we use the pair-counting F1-measure [1], which is commonly used when encountering overlapping clustering.

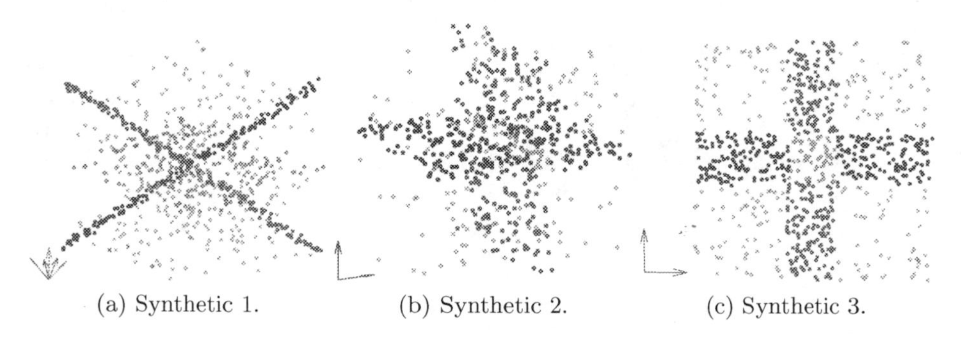

(a) Synthetic 1. (b) Synthetic 2. (c) Synthetic 3.

Fig. 3. Visualization of the clustering results of RAOSC on synthetic data sets. Colored points are the found subspace clusters. The orange points are the found overlapping points. Noise are plotted with gray points. Arrows at the bottom-left represent the found subspaces of the corresponding clusters. (Color figure online)

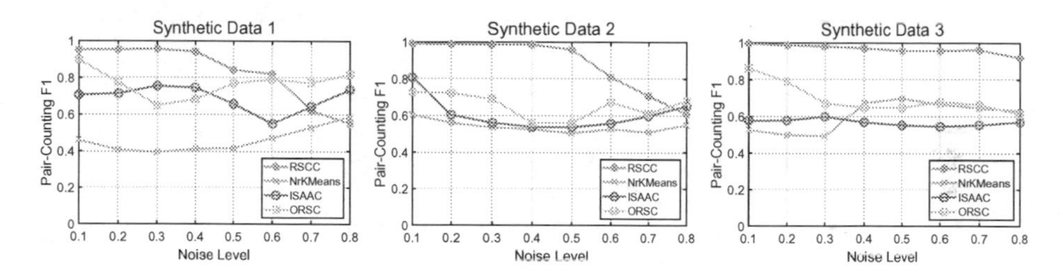

Fig. 4. Clustering results while varying noise levels on synthetic data sets in the presence of noise and overlapping points.

Figure 3 visualizes the clustering results of RAOSC. Note that we only plot the low noise level (about 0.3) results for legibility reason, though it can correctly find the subspace clusters and identify noise points at a higher level of noise. As we can see, RAOSC successfully assigns data points into the correct clusters in a sea of noise, and the noise and overlapping points are all correctly identified. Besides, it obtains the corresponding subspaces as well (indicated by the arrows at the bottom left corner). Figure 4 shows the numerical evaluation of the comparison algorithms. RAOSC achieves promising results and outperforms other algorithms in most cases. When the noise level is extremely high, the performance drops sharply, because the extreme noise points have covered the cluster structure and meaningful clusters no longer exist. We observe that the performance of comparison algorithms drop down at first and then go up. This might be because they are mild arbitrators. With the increase of noise level, they tend to assign noise into multiple overlapping clusters. This can still increase the score w.r.t. the pair-counting F1-measure though they actually produce the wrong assignment.

3.2 Evaluation on Real-World Data

Next we compare with extensive clustering algorithms on the real-world data sets. Nine real-world data sets are used in this study, which include Pendigits, Seeds, Soybean, Spam, Wine and Zoo from the UCI repository, OliveOil, Plane and Symbols from the UCR repository. The statistics of these data sets are given in Table 1.

Table 1. Statistics of the real-world data sets.

Name	#Classes	#Dim.	#Inst.	Name	#Classes	#Dim.	#Inst.
Wine	3	13	178	Seeds	3	7	210
Pendigits	10	16	10092	Soybean	4	35	47
Zoo	7	16	101	Spam	2	57	4601
Plane	7	144	210	Symbol	6	398	1020
Olive	4	570	60				

Here, six arbitrarily oriented subspace clustering algorithms are selected. ORCLUS [3] and 4C [6] are the two most typical arbitrarily oriented subspace clustering algorithms. SubKmeans [14] and FOSSCLU [8] are two recent subspace clustering algorithms. Different from ORCLUS and 4C, they only find one optimal subspace for clustering. LRR [12] and LRS [17] are two low-rank based subspace clustering algorithm. Though they are not considered as the classical subspace clustering algorithm in the data mining community, they are still closely related to our method. Source codes of all algorithms are downloaded from the authors' websites. The source code of RAOSC can be downloaded from Dropbox[2].

For a comprehensive evaluation, we tune all the algorithms' parameters from wide ranges while being compatible with the original papers. We search LRR's parameter λ within the set of $\{10^{-5}, 10^{-4}, \ldots, 10^{5}\}$. For LRS, we select its parameter p from $\{0.1, 0.2, \ldots, 1\}$, and set $K = 2$. We use PCA and k-means to do initialization as described in its paper. We search ORCLUS's parameter l in the range of $[2; \min(20, m)]$, and run 4C for $\epsilon \in [2; 20]$, $minPts \in [1; 15]$ and $\lambda \in [2; \min(20, m)]$. SubKmeans has no additional parameters except for k. FOSSCLU determines parameters automatically. For RAOSC, we search parameter α from $\{10^{-5}, 10^{-4}, \ldots, 10^{5}\}$. In addition, we set parameter l to zero, λ to a large number, which gives no overlapping nor noise result. Similar to LRS, we use PCA and k-means to initialize the cluster indicator matrices and the cluster centroids. For all experiments, we standardize all data so that all features have zero mean and unit variance. Since some algorithms may run into a local optimum and produce insufficient outcomes, we run all algorithms for 10 times and

[2] https://www.dropbox.com/s/7csm3itojmb5glh/RAOSC_code.rar?dl=0.

Table 2. Clustering results in terms of NMI (%) on the real-world data sets. Results marked with † were aborted due to memory limit or convergence issues, or cannot obtain results in reasonable running time. In either case, we report the best results that have been achieved.

	Pendigits	OliveOil	Seeds	Soybean	Spam	Symbol	Wine	Zoo	Plane
RAOSC	**73.68**	**76.30**	**73.69**	**100.00**	**41.05**	**81.56**	**88.26**	**89.11**	91.23
SubKmeans	67.94	75.42	72.79	**100.00**	2.17	79.48	87.59	83.39	91.23
ORCLUS	68.32	75.86	72.69	96.89	36.13	65.60†	87.79	87.33	70.84
4C	69.99	63.96	16.50	**100.00**	11.78	81.41	48.11	85.48	88.08
FOSSCLU	70.20	0.00†	63.95	37.02	0.00†	0.00†	84.68	0.00†	0.00†
LRS	65.13	44.16	73.30	69.10	27.04	60.72	63.22	67.03	57.37
LRR	70.85	40.12	21.58	81.49	4.67	78.09	41.36	76.02	**93.29**

sort the results by their costs and remove the half with higher costs. Then we report the average NMI for evaluation.

Table 2 summarizes the clustering results. As shown in Table 2, SubKmeans, ORCLUS, and 4C achieve comparable results. However, SubKmeans only finds one optimal subspace for clustering rather than distinct subspaces for all clusters. ORCLUS and 4C are time consuming especially when the number of dimensions is high. Besides, it is hard to find the optimal parameters for 4C. FOSSCLU also only finds one optimal subspace for clustering, however, there seems to be convergence issue in the author provided implementation. In general, the two low-rank based algorithms LRR and LRS perform poorly on these data sets. Neither LRR nor LRS can give the explicit subspaces of clusters, they all find implicit subspace clusters in the full-dimensional space, where the low-rank structure is seem to be dim. By contrast, RAOSC outperforms other algorithms on eight data sets, only slightly lags behind on the Plane data set. RAOSC naturally characterizes the intrinsic correlation and closeness properties of subspace cluster, which accounts for the promising clustering results.

4 Conclusion

In this paper, towards the arbitrarily oriented subspace clustering problem, we propose a novel algorithm called RAOSC. RAOSC formulates the task as a group structure low-rank optimization problem, which well characterizes the intrinsic correlation and closeness properties of subspace cluster. RAOSC can not only recover the subspace clusters from a sea of noise points but also explicitly obtains the corresponding subspaces. It can naturally identify the noise and overlapping points during the optimization process. Empirical experiments on both synthetic data sets and real-world data sets have demonstrated its effectiveness. In future work, we would like to reduce the computational complexity. One potential route is to incorporate accelerated SVD, another is to develop data parallelism at the algorithmic level [20].

Acknowledgments. This work is supported by the National Natural Science Foundation of China (61403062, 61433014, 41601025), Science-Technology Foundation for Young Scientist of SiChuan Province (2016JQ0007), Fok Ying-Tong Education Foundation for Young Teachers in the Higher Education Institutions of China (161062) and National key research and development program (2016YFB0502300).

References

1. Achtert, E., Goldhofer, S., Kriegel, H.P., Schubert, E., Zimek, A.: Evaluation of clusterings–metrics and visual support. In: Proceedings of the 28th IEEE International Conference on Data Engineering, pp. 1285–1288 (2012)
2. Aggarwal, C.C., Wolf, J.L., Yu, P.S., Procopiuc, C., Park, J.S.: Fast algorithms for projected clustering. In: Proceedings of the 1999 ACM SIGMOD International Conference on Management of Data, vol. 28 (1999)
3. Aggarwal, C.C., Yu, P.S.: Finding generalized projected clusters in high dimensional spaces. In: Proceedings of the 2000 ACM SIGMOD International Conference on Management of Data, vol. 29 (2000)
4. Agrawal, R., Gehrke, J., Gunopulos, D., Raghavan, P.: Automatic subspace clustering of high dimensional data for data mining applications. In: Proceedings of the 1999 ACM SIGMOD International Conference on Management of Data, vol. 27 (1998)
5. Assent, I., Krieger, R., Emmanuel, M., Seidl, T.: DUSC: dimensionality unbiased subspace clustering. In: Proceedings of the 7th IEEE International Conference on Data Mining, pp. 409–414 (2008)
6. Böhm, C., Kailing, K., Kröger, P., Zimek, A.: Computing clusters of correlation connected objects. In: Proceedings of the 2004 ACM SIGMOD International Conference on Management of Data, pp. 455–466 (2004)
7. Cheng, C.H., Fu, A.W., Zhang, Y.: Entropy-based subspace clustering for mining numerical data. In: Proceedings of the 5th ACM SIGKDD International Conference on Knowledge Discovery and Data Mining, pp. 84–93 (1999)
8. Goebl, S., He, X., Plant, C., Böhm, C.: Finding the optimal subspace for clustering. In: Proceedings of the 14th IEEE International Conference on Data Mining, pp. 130–139 (2014)
9. Günnemann, S., Färber, I., Virochsiri, K., Seidl, T.: Subspace correlation clustering: finding locally correlated dimensions in subspace projections of the data. In: Proceedings of the 18th ACM SIGKDD International Conference on Knowledge Discovery and Data Mining, pp. 352–360 (2012)
10. Kailing, K., Kriegel, H.P., Kröger, P.: Density-connected subspace clustering for high-dimensional data. In: Proceedings of the 2004 SIAM International Conference on Data Mining, pp. 246–256 (2004)
11. Kriegel, H.P., Kröger, P., Zimek, A.: Clustering high-dimensional data: a survey on subspace clustering, pattern-based clustering, and correlation clustering. ACM Trans. Knowl. Discov. Data **3**(1), 1 (2009)
12. Liu, G., Lin, Z., Yan, S., Sun, J., Yu, Y., Ma, Y.: Robust recovery of subspace structures by low-rank representation. IEEE Trans. Pattern Anal. Mach. Intell. **35**(1), 171–184 (2013)
13. Liu, G., Lin, Z., Yu, Y.: Robust subspace segmentation by low-rank representation. In: Proceedings of the 27th International Conference on Machine Learning, pp. 663–670 (2010)

14. Mautz, D., Ye, W., Plant, C., Böhm, C.: Towards an optimal subspace for k-means. In: Proceedings of the 23rd ACM SIGKDD International Conference on Knowledge Discovery and Data Mining, pp. 365–373 (2017)
15. Mautz, D., Ye, W., Plant, C., Böhm, C.: Discovering non-redundant k-means clusterings in optimal subspaces. In: Proceedings of the 24th ACM SIGKDD International Conference on Knowledge Discovery and Data Mining, pp. 1973–1982 (2018)
16. Nie, F., Yuan, J., Huang, H.: Optimal mean robust principal component analysis. In: Proceedings of the 31st International Conference on International Conference on Machine Learning, vol. 32, pp. 1062–1070 (2014)
17. Nie, F., Huang, H.: Subspace clustering via new low-rank model with discrete group structure constraint. In: Proceedings of the 25th International Joint Conference on Artificial Intelligence, pp. 1874–1880 (2016)
18. Shao, J., Gao, C., Zeng, W., Song, J., Yang, Q.: Synchronization-inspired co-clustering and its application to gene expression data. In: 2017 IEEE International Conference on Data Mining, pp. 1075–1080 (2017)
19. Shao, J., Wang, X., Yang, Q., Plant, C., Böhm, C.: Synchronization-based scalable subspace clustering of high-dimensional data. Knowl. Inf. Syst. **52**(1), 83–111 (2017)
20. Shao, J., Yang, Q., Dang, H.V., Schmidt, B., Kramer, S.: Scalable clustering by iterative partitioning and point attractor representation. ACM Trans. Knowl. Discov. Data **11**(1), 5 (2016)
21. Tung, A.K.H., Xu, X., Ooi, B.C.: CURLER: finding and visualizing nonlinear correlation clusters. In: Proceedings of the 2005 ACM SIGMOD International Conference on Management of Data, pp. 467–478 (2005)
22. Ye, W., Maurus, S., Hubig, N., Plant, C.: Generalized independent subspace clustering. In: Proceedings of the 2016 IEEE International Conference on Data Mining, pp. 569–578 (2016)

Truthful Crowdsensed Data Trading Based on Reverse Auction and Blockchain

Baoyi An[1], Mingjun Xiao[1(✉)], An Liu[2], Guoju Gao[1], and Hui Zhao[1]

[1] School of Computer Science and Technology, Suzhou Institute for Advanced Study, University of Science and Technology of China, Hefei, China
xiaomj@ustc.edu.cn

[2] School of Computer Science and Technology, Soochow University, Suzhou, China

Abstract. Crowdsensed Data Trading (CDT) is a novel data trading paradigm, in which each data consumer can publicize its demand as some crowdsensing tasks, and some mobile users can compete for these tasks, collect the corresponding data, and sell the results to the consumers. Existing CDT systems either depend on a trusted data trading broker or cannot ensure sellers to report costs honestly. To address this problem, we propose a Reverse-Auction-and-blockchain-based crowdsensed Data Trading (RADT) system, mainly containing a smart contract, called RADToken. We adopt a greedy strategy to determine winners, and prove the truthfulness and individual rationality of the whole reverse auction process. Moreover, we exploit the smart contract with a series of devises to enforce mutually untrusted parties to participate in the data trading honestly. Additionally, we also deploy RADToken on an Ethereum test network to demonstrate its significant performances. To the best of our knowledge, this is the first CDT work that exploits both auction and blockchain to ensure the truthfulness of the whole data trading process.

1 Introduction

Owing to the huge potential economic value of data resources, many online data trading systems [12] have emerged in recent years, such as CitizenMe, DataExchange, Datacoup, Factual, and Terbine, etc., whereby data consumers can search and purchase their interested data. However, most data in the real world are preserved by few research institutions or companies only for their own analysis purposes rather than sharing them with others who have data needs but cannot afford to collect data by themselves, which causes trouble to the availability of data. Consequently, the volumes of data in trading systems are still very limited, which has significantly suppressed the increasing market demand for data [24]. To tackle this problem, a novel data trading paradigm, called Crowdsensed Data Trading (CDT), is proposed, in which the mobile crowdsensing technology is adopted to provide data resources for trading, i.e., a large crowd of mobile users are leveraged to collect data with their smart phones [17,24].

G. Li et al. (Eds.): DASFAA 2019, LNCS 11446, pp. 292–309, 2019.
https://doi.org/10.1007/978-3-030-18576-3_18

In general, a typical CDT system (e.g., Thingful [1], Thingspeak [2]) includes a data trading broker, some data consumers, and data sellers (a.k.a., crowdsensing workers). The broker employs a large amount of sellers to collect data according to some requirements, and then sells the sensed data to the consumers who are interested in. So far, there have been a few works focusing on the CDT system design. For example, [24] proposes a profit-driven data collection framework for crowd-sensed data markets, called VENUS, in which a data procurement auction is adopted to determine the minimum payment for each data collection. A data sharing market is introduced in [11], where the sensed data is saved and processed in user devices locally and shared among users in a P2P manner. A brokerage-based market is launched in [23], where sellers and consumers propose their selling and buying quantities, respectively, to match the market supply and demand in the trading platform. However, these CDT systems have to depend on a data trading broker, making those data consumers worry about the truthfulness and even unwilling to use the systems.

On the other hand, blockchain [16], a newly-emerging decentralized transaction recording technology, shows a glimpse of solutions to fairness and transparency issues which is resistant to modification of the data. In addition, blockchains allow mutually distrusted users to complete data exchange or transaction securely without a centralized truthful intermediary, avoiding high legal and transactional costs [13]. Smart contracts [5] are some complex programs deployed on blockchains which can automatically execute operations according to treaty conditions. An important advantage of smart contracts is that they can enforce the participants, who might not trust each other, to fulfill their obligations. Due to this characteristic, smart contracts are introduced into CDT systems as a truthful broker to conduct the data trading between sellers and consumers. For instance, [6] proposes a CDT framework which enables efficient truth discovery over encrypted crowdsensed data streams and knowledge monetization on smart contract. A reliable mechanism that allows consumers to search directly over encrypted data is also implemented on blockchain [10].

Although blockchain-based CDT systems can create trust between sellers and consumers of data trading, they cannot guarantee the truthfulness of individuals, i.e., sellers might report fake data collection costs to achieve more rewards. As we know, auction mechanisms can ensure the participants to report their bids honestly, which have been widely used in crowdsensing systems [8,21]. Hence, to construct fully truthful data trading systems on blockchain, we propose a Reverse-Auction-and-blockchain-based crowdsensed Data Trading (RADT) system in this paper, which mainly contains a smart contract, called RADToken. Each consumer can start a RADT by issuing the data demand via RADToken (e.g., report traffic conditions of multiple locations), and sellers who have registered in RADToken can bid for their interested tasks. RADToken will automatically execute to determine the winners and payments. After sellers complete the data collection and submit the results, the consumer will pay some rewards to sellers via RADToken. Since bids on blockchain is transparent and public visible, RADToken takes a two-stage bidding strategy to ensure the security of bids.

Moreover, both consumer and sellers are asked for a deposit which keeps them comply with the prescribed rules. Once they finish their obligations, RADToken will automatically and transparently transfer their deposits and payments. We summarize the contributions of this paper as follows:

1. We propose a CDT system based on reverse auction and blockchain, i.e., RADT. It does not need any truthful broker but can provide truthful data trading between mutually untrusted consumers and sellers. To the best of our knowledge, this is the first CDT system that exploits both auction theory and blockchain technology to ensure the truthfulness of the whole data trading.
2. We design a reverse-auction-based smart contract for the RADT system, i.e., RADToken, where the reliability of each seller is taken into consideration and a greedy strategy is used to determine winners. Moreover, we prove the truthfulness and individual rationality of the whole auction process, which implies that all sellers will report their data collection costs honestly.
3. To ensure the truthfulness of the data trading process, we devise some modifiers and set life span limitations for each procedure to filter the illegal invokes which can resist certificate forgery and tamper attacks effectively. Meanwhile, we adopt a two-stage bid strategy to protect bid privacy in auction procedure and use symmetric and asymmetric encryptions for data delivery procedure which can protect the confidentiality of the sensed data.
4. We implement a prototype of the RADT system and deploy RADToken to an Ethereum test network. Extensive simulations are conducted to demonstrate the significant performances and the practicability of RADToken.

The paper is organized as follows. In Sect. 3, we present a system model for RADT and analyze the security in Sect. 4. The reverse auction is elaborated in Sect. 5. We carry on the theoretical analysis in Sect. 6. We present simulations and evaluations in Sect. 7 and review related works in Sect. 8. Finally, we conclude in Sect. 9.

2 Preliminaries

We first introduce the background of Ethereum before the system overview.

2.1 Account Types

- **Externally Owned Accounts (EOAs).** An EOA only has a balance which can send transactions either to transfer ether or to trigger contract code.
- **Smart Contract Accounts.** A smart contract has a balance and associated code. Code execution is triggered by transactions from EOAs.

A contract is invoked by a transaction and is run by Ethereum Virtual Machine on each node participating in the network as part of their verification of new blocks. Contracts like *autonomous agents* have direct control over their balances and dictionary (key/value-datatype) storage.

2.2 Transaction

The term *transaction* is used in Ethereum to refer to the signed data package. Its *VALUE* field is the amount of **wei** to be transferred from the sender to the recipient. All values are denominated in units of wei: 1 **ether** is 10^{18} **wei**. Once a smart contract receives a transaction (*msg*), it can obtain two parameters:

- **msg.sender** is sender's account.
- **msg.value** is the amount of **ether** that sender transfer to it.

2.3 Gas System

Ethereum charges a fee (gas) per computational step to prevent deliberate attacks and abuse on Ethereum. Each transaction is required to include a gas limit and a fee that it is willing to pay per gas. If the total gas used for the computational steps spawned by the transaction is not greater than the gas limit, the transaction will be processed. Otherwise, all modifications are reverted. The excess gas is reimbursed to the sender. The total cost involves two aspects:

- **gasUsed** is the total gas which is consumed by the transaction.
- **gasPrice** is the price of one unit gas that is specified in the transaction.

Hence, the total cost = gasUsed × gasPrice.

3 System Overview

3.1 The RADT System

Figure 1 illustrates the RADT system including 4 major entities, i.e., *consumer*, RADToken, *Secure Cloud Server (SCS)* and *seller*. The consumer sends data collection job requirements to RADToken, a smart contract on Ethereum. The job includes a set of Points of Interest (PoIs) which correspond to the location-sensitive sensing tasks, denoted as $\mathcal{T} = \{t_1, t_2, \cdots, t_l\}$. Sellers who participate in the job, denoted by $\mathcal{W} = \{w_1, w_2, \cdots, w_n\}$. Each seller w_i can submit a bid β_i for their interested tasks $\mathcal{T}_i$ $(\subseteq \mathcal{T})$. Moreover, we denote the sellers who can execute t_j by $\mathcal{W}_j$ $(\subseteq \mathcal{W})$. RADToken as a broker then executes a reverse auction to select winners $\mathcal{S}(\subseteq \mathcal{W})$ and determine payments $\mathcal{P} = \{p_i | s_i \in \mathcal{S}\}$ for the winners. The winners upload the encrypted sensed data (*EnData*) to SCS where the consumer can download and decrypt *EnData* to obtain the real sensed data (*Data*). The consumer and sellers can get the refund and payments respectively after the job ends. We specify two key features of a smart contract [13]:

- *Timing.* A smart contract has a time clock which is modeled as a continuously increasing variable *now*. *now* is an alias for a timestamp of the blockchain.
- *Function Modifier.* Modifiers are inheritable properties of contracts which are used to automatically check a condition prior to executing the function.

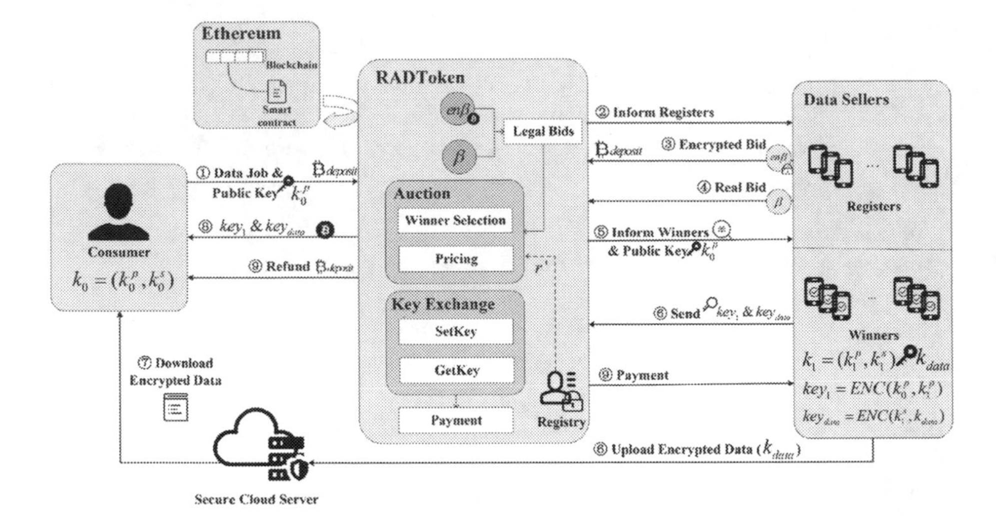

Fig. 1. The RADT system

3.2 The Workflow of the RADT System

Sellers register in RADToken to be informed once a data collection job appears. RADToken maintains **Registry** which contains a dictionary $\mathcal{R}$ to record sellers' reliabilities and $\mathcal{R}(w_i) = r_i$. The system works as follows:

Initiate($\mathcal{T}, \tau_{bid}, \tau_{reveal}, \tau_{exec}, k_0^p$):

1. require \$deposit($msg.value$) $\geq$ $guaranty$.
2. set $\mathcal{Q} = msg.sender$ and store k_0^p.
3. set $\mathcal{T} = \mathcal{T}$, $t_{bid} = now + \tau_{bid}$, $t_{reveal} = t_{bid} + \tau_{reveal}$, $t_{exec} = t_{reveal} + \tau_{exec}$.
4. trigger $Notify$ event to inform the registered data sellers.

Fig. 2. The initiate function

Step 1: Job Initialization. The consumer describes a data collection job with requirements: $\mathcal{T}$, bid commitment duration τ_{bid}, bid reveal duration τ_{reveal} and job execution duration τ_{exec}. The consumer generates a pair of keys $k_0 = (k_0^p, k_0^s)$, where k_0^p is his public key and k_0^s is his private key. Then, he sends a transaction which contains the job description and the public key k_0^p to RADToken. The detailed function is given in Fig. 2.

Step 2: Notify Data Providers. *Initiate()* notifies all registered sellers of the new job. A seller provides a deposit (\$$deposit_i$) for participating in the job.

To prevent the bids from being intercepted, a two-stage bidding strategy is adopted in Sect. 4.2. Each seller has an encrypted bid $enB_i = (\mathcal{T}_i, en\beta_i)$ and a real bid $B_i = (\mathcal{T}_i, \beta_i)$. RADToken uses two dictionaries *EnBids* and *Bids* to record them respectively, where $EnBids(w_i) = en\beta_i$ and $Bids(w_i) = (\mathcal{T}_i, \beta_i, \$deposit_i)$.

CommitBid($\mathcal{T}_i, en\beta_i$) payable:

1. require now $\leq t_{bid}$ and $msg.value \geq guaranty$.
2. add $(msg.sender, \mathcal{T}_i, msg.value)$ to $Bids$. $\triangleright w_i = msg.sender,\ deposit_i = msg.value$
3. add $(msg.sender, en\beta_i)$ to $EnBids$.

RevealBid($\mathcal{T}_i, \beta_i, nonce_i$):

1. require $t_{bid} \leq$ now $\leq t_{reveal}$.
2. require $EnBids(msg.sender)$=sha3$(\mathcal{T}_i, \beta_i, nonce_i)$.
3. require $Bids(msg.sender).T = \mathcal{T}_i$.
4. add β_i to $Bids(msg.sender)$.

Fig. 3. The two-stage bidding procedure

Step 3: Commit Encrypted Bid. A seller computes his encrypted bid $en\beta_i$ using Secure Hash Algorithm-3 (SHA-3) [4]. SHA-3 takes as input his account w_i, his bid β_i and a randomly selected $nonce_i$, i.e., $en\beta_i = sha3(w_i, \beta_i, nonce_i)$. *CommitBid()* in RADToken takes as input a binary tuple enB_i and records it.
Step 4: Reveal Real Bid. RADToken is invoked by a seller w_i to send $\mathcal{T}_i$, β_i and $nonce_i$. *RevealBid()* will check the bid's authenticity and record the legal bid. If w_i sends illegal bid, he is untruthful and his $\$deposit_i$ will be forfeited. The detailed functions of *CommitBid()* and *RevealBid()* are given in Fig. 3.

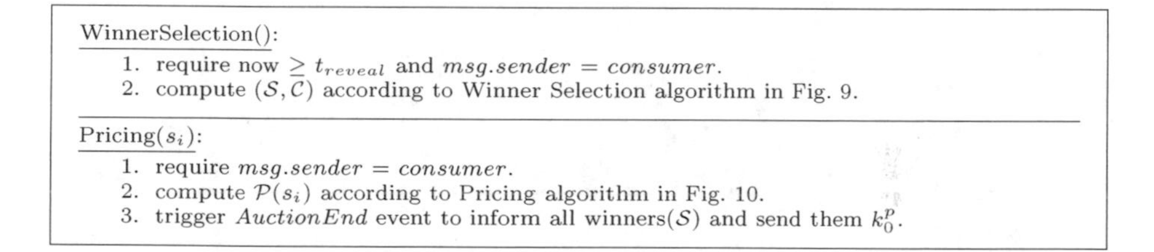
WinnerSelection():

1. require now $\geq t_{reveal}$ and $msg.sender = consumer$.
2. compute $(\mathcal{S}, \mathcal{C})$ according to Winner Selection algorithm in Fig. 9.

Pricing(s_i):

1. require $msg.sender = consumer$.
2. compute $\mathcal{P}(s_i)$ according to Pricing algorithm in Fig. 10.
3. trigger $AuctionEnd$ event to inform all winners($\mathcal{S}$) and send them k_0^p.

Fig. 4. The reverse auction procedure

Step 5: Inform Winners. RADToken executes a reverse auction in Fig. 4 to select winners $\mathcal{S}$ and determine payments for all winners. All winners will get informed the auction results and public key k_0^p. The detailed reverse auction process is in Sect. 5.
Step 6: Send Keys. To ensure the confidentiality of the sensed data, w_i generates a symmetric key $k_{data,i}$ and encrypts $Data_i$ with $k_{data,i}$. After uploading the encrypted data ($Endata_i$) to SCS, w_i generates a pair of asymmetric keys $k_{1,i} = (k_{1,i}^p, k_{1,i}^s)$. And he obtains $key_{data,i}$ and $key_{1,i}$ by encrypting $k_{data,i}$ and $k_{1,i}^p$ as described in Sect. 4.3. Then the seller invokes $SetKey()$ to delivery

$key_{data,i}$ and $key_{1,i}$. RADToken maintains two dictionaries KDs and $K1s$ to store $key_{data,i}$ and $key_{1,i}$ respectively, as shown in Fig. 5.

SetKey($key_{data,i}$, $key_{1,i}$):

1. add ($msg.sender$, $key_{data,i}$) to KDs.
2. add ($msg.sender$, $key_{1,i}$) to $K1s$.

GetKey(s_i):

1. require $msg.sender = consumer$.
2. require $\$rewards \geq \sum_{s_i \in \mathcal{S}} \mathcal{P}(s_i)$. ▷ consumer transferred \$rewards before
3. send ($KDs(s_i), K1s(s_i), \mathcal{R}(s_i)$) to $consumer$.

Fig. 5. The key exchange procedure

Step 7: Download ***EnData*****.** The consumer downloads *EnData* from SCS.
Step 8: Get Keys. In Fig. 5, the consumer can invoke *GetKey*() to get keys for decrypting *EnData*. Before the invocation, he should transfer some ethers ($\$rewards$) to RADToken which are no less than the total payments.

Refund():

1. require now $\geq t_{exec}$.
2. compute untruthful sellers' total deposits which is \$fines.
3. compute the remaining rewards $\$rewards = \$rewards - \sum_{s_i \in \mathcal{S}} \mathcal{P}(s_i)$.
4. transfer \$deposit+\$fines+\$rewards to consumer.

Payment():

1. require $Bids(msg.sender).\beta \neq 0$. ▷ untruthful according to step 4 in Fig. 3
2. if $msg.sender \in \mathcal{S}$, require KDs(msg.sender)$\neq$ 0 and K1s(msg.sender)$\neq$ 0.
3. transfer the final payment $\$P = \mathcal{P}(msg.sender) + \$deposit_i$+\$fine to $msg.sender$

Fig. 6. The refund and payment procedure

Step 9: Payment and Refund. Upon receiving the total rewards, RADToken will return the deposit to the consumer. Each winner can get his payment only when he had sent his keys. The functions are detailed in Fig. 6.

4 Security Analysis

4.1 Robustness of RADToken

We use some modifiers introduced at the beginning of Sect. 4 in Fig. 7 to ensure that RADToken can run steadily. The keyword *require* can roll back all states without deducting gas when encountering some invalid codes. The robustness of RADT is guaranteed by the following points:

```
Modifiers:
  1. modifier onlyBefore(uint time)  require(now < time);
  2. modifier onlyAfter (uint time)  require(now > time);
  3. modifier onlyTrue (uint flag)  require(flag == true);
  4. modifier onlyFalse (uint flag)  require(flag == false);
```

Fig. 7. Some modifiers

1. **Only one job in one round.** Once a consumer invokes RADToken to launch a job like in Fig. 2, others cannot invoke it until the job ends. If there exists an active job, the value of flag *jobEnded* whose default is *true* would be *false*, which cannot pass the check of *onlyTrue(jobEnded)* in *Initiate*, so that other consumers will be rejected.
2. **Each participant should offer deposit.** We set *Initiate()* and *CommitBid() payable*, a keyword of smart contract, which requires invokers to transfer ethers to RADToken. We assume that the consumer will transfer rewards and will not quit the RADT system midway. Otherwise his deposit offered in *Initiate()* will be fined as a compensation for sellers and he even cannot get keys to decrypt *EnData*. Each seller should invoke *CommitBid()* with his deposit, which urges him to reveal his bid truthfully.

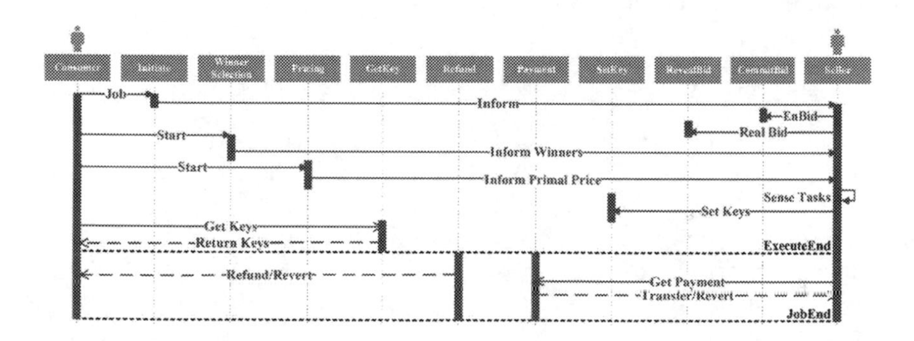

Fig. 8. The sequence diagram of RADT system

3. **Each procedure only be executed orderly.** Every function in our RADToken is an independent entry through separate calls. We also set some time modifiers to ensure the safety. We illustrate the sequence diagram of RADT in Fig. 8. For example, *RevealBid()* should be invoked after t_{bid} and before t_{reveal} where the auction only can be executed after *RevealBid()*.

4.2 Security and Truthfulness of Bid

Everything on blockchain is transparent, which enables a potential adversary to reconstruct the entire transaction history and find out the meanings and logic behind it. That means the adversary can grab other bids and make a bid in his favour. Thus, we should consider how to guarantee the security of each bid.

Security. Since smart contracts cannot handle complex encryption and decryption, we adopt a two-stage bidding strategy to protect bids. At the first stage, a seller w_i can encrypt his bid β_i using SHA-3 as follows:

$$en\beta_i = \text{“}0x\text{”} + ethereumjs.ABI.\ soliditySHA3([\text{“}address\text{”}, \text{“}uint256\text{”}, \text{“}uint256\text{”}], [w_i, \beta_i, nonce_i]).toString(\text{“}hex\text{”})$$

Then, he sends enB_i to RADToken but other users can only read a hash value $en\beta_i$ rather than the real bid β_i. Moreover, a wise adversary will not try to crack the hash value which is impossible to succeed. So β_i is secure until t_{bid}.

Truthfulness. Even though an adversary can obtain other bids which will be revealed at the second stage, he cannot adjust his bid which will make him untruthful for the bid check of *RevelBid()* in Fig. 3. First, it recomputes the hash value $sha3(msg.sender, \beta_i, nonce_i)$ of w_i and compares it with his recorded $en\beta_i$ (step 2). RADToken also checks if his new submitted tasks are same with $\mathcal{T}_i$ (step 3). w_i will be accepted as a truthful seller only when he passes the check.

4.3 Confidentiality of Data

Symmetric encryption has more efficiency and less computation overhead. So we use it to encrypt data and asymmetric encryption to protect keys as below:

1. A winner s_i encrypts his sensed data with a symmetric key k_{data}, i.e., $EnData_i = ENC(Data_i, k_{data,i})$. Then he uploads the $EnData_i$ to SCS.
2. s_i uses his private key to encrypt k_{data} and get $key_{data,i} = ENC(k^s_{1,i}, k_{data,i})$.
3. s_i encrypts $k^p_{1,i}$ with k^p_0, i.e., $key_{1,i} = ENC(k^p_0, k^p_{1,i})$.
4. s_i invokes $SetKey()$ to send $key_{data,i}$ and $key_{1,i}$ to RADToken.
5. The consumer downloads $EnData_i$ and invokes $GetKey()$ to get $(key_{data,i}, key_{1,i})$.
6. The consumer decrypts $key_{1,i}$ with k^s_0 to get $k^p_{1,i} = DEC(key_{1,i}, k^s_0)$.
7. The consumer then decrypts key_{data} with $k^p_{1,i}$ to get $k_{data,i} = DEC(key_{data,i}, k^p_{1,i})$.
8. Finally, the consumer can obtain the real data $Data_i = DEC(EnData_i, k_{data,i})$.

5 The Reverse Auction Mechanism of RADT

5.1 Problem Formulation

As more sellers take part in the job, the actual sensed data would exceed the reliability requirements $\epsilon = \{\epsilon_j | t_j \in \mathcal{T}\}$, whereas it also increases the total costs $\mathcal{C}$. Furthermore, we adopt σ^S_j to denote the overall reliability that all winners contribute to $t_j \in \mathcal{T}$. In order to estimate $\sigma_j^{\mathcal{S}}$, we compute the sum of the reliabilitics of the winners who process t_j as follows:

$$\sigma_j^{\mathcal{S}} = \sum\nolimits_{s_i \in \mathcal{S} \cap \mathcal{W}_j} r_i \tag{1}$$

Due to the truthfulness of sellers, we regard a seller w_i's bid β_i as his cost (i.e., $c_i = \beta_i$). The goal of the auction is to find a subset of sellers that minimize the overall cost while satisfying the reliability requirements of data. Hence, the RADT problem can be formulated as follows:

$$Minimize : C(\mathcal{S}) = \sum\nolimits_{s_i \in \mathcal{S}} \beta_i, \; //c_i = \beta_i \tag{2}$$

$$Subject\ to : \quad \mathcal{S} \subseteq \mathcal{W} \tag{3}$$

$$\sigma_j^{\mathcal{S}} \geq \epsilon_j, \; 1 \leq j \leq l \tag{4}$$

Here, Eq. 4 indicates that the total reliability of task t_j is no less than ϵ_j.

The auction first selects winners who minimize the total costs under reliability constraints. Then, it determines payments for winners so that the whole auction satisfies truthfulness and individual rationality which are defined as follows:

Definition 1 ***(Truthfulness).*** *Let B_i be the truthful bid and B_i' be the untruthful bid where the payments are $p_i(B_i)$ and $p_i(B_i')$ respectively. Then, if*

$$p_i(B_i) - c_i \geq p_i(B_i') - c_i, \tag{5}$$

we say that the auction mechanism is truthful.

Definition 2 ***(Individual Rationality).*** *The payoff for B_i is non-negative,*

$$p_i(B_i) - c_i \geq 0. \tag{6}$$

5.2 The Winner Selection Algorithm of RADT

WinnerSelection():

1. repeat until $G(\mathcal{S}) = \sum_{j=1}^{l} \epsilon_j$:
   (a) for $\forall B_i$, compute $\rho_i = \frac{v_i(\mathcal{S})}{\beta_i}$;
   (b) record the index of the maximum ρ_i as i^*;
   (c) add s_{i^*} to $\mathcal{S}$, set $\mathcal{C} = \mathcal{C} + \beta_{i^*}$;
2. return $(\mathcal{S}, \mathcal{C})$.

Fig. 9. The reliability-aware winner selection algorithm

The RADT problem is NP-hard, because the minimum weight set cover problem is to find a subset that minimize the total weight which can be polynomial-time reducible to the RADT problem according to [8]. So we propose a greedy algorithm to solve RADT. A seller who has the largest reliability to execute the most tasks with the least cost will be selected and added into the set $\mathcal{S}$ first.

To design an appropriate approximation algorithm, we first define a reliability contribution function $G(\mathcal{S})$. $G(\mathcal{S})$ indicates the current total reliability of winners who process $\mathcal{T}$ under the constraint of ϵ, defined as follows:

$$G(S) = \sum_{j=1}^{l} \min\{\sigma_j^S, \epsilon_j\} \tag{7}$$

Based on $G(S)$, the marginal reliability contribution $v_i(\mathcal{S})$ is the marginal reliability that $w_i \in \mathcal{W} - \mathcal{S}$ can contribute to the whole job, defined as follows:

$$v_i(\mathcal{S}) = G(\mathcal{S} \cup \{w_i\}) - G(\mathcal{S}) \tag{8}$$

Based on Eq. 8, we illustrate the winner selection algorithm in Fig. 9. The algorithm begins from an empty set $\mathcal{S}$. In each iteration, it adds the winner who has the maximum weight $\rho_i = \frac{v_i(\mathcal{S})}{\beta_i}$ into $\mathcal{S}$. The algorithm terminates when $G(\mathcal{S}) = \sum_{j=1}^{l} \epsilon_j$. The computation overhead of the algorithm is $O(n^2 l)$, where n is the number of sellers and l is the number of tasks.

5.3 The Critical Pricing Algorithm of RADT

Pricing(s_i):

1. create a empty winner set $\mathcal{S}'$.
2. record $bid = Bids(s_i)$ and set $Bids(s_i) = 0$. ▷ remove B_i from $\mathcal{B}$
3. repeat until $G(\mathcal{S}') = \sum_{j=1}^{l} \epsilon_j$:
   (a) compute $\rho_k = \frac{v_i(\mathcal{S}')}{\beta_k}$, for $\forall s_k$ is not in $\mathcal{S}'$ and satisfies $Bids(s_k) \neq 0$;
   (b) record the index of the maximum ρ_k as k^*;
   (c) if $\mathcal{P}(s_i) < \frac{\beta_{k^*} v_i(\mathcal{S}')}{v_{k^*}(\mathcal{S}')}$, set $\mathcal{P}(s_i) = \frac{\beta_{k^*} v_i(\mathcal{S}')}{v_{k^*}(\mathcal{S}')}$;
   (d) add s_{k^*} to $\mathcal{S}'$.
4. set $Bids(s_i) = bid$, delete $\mathcal{S}'$.

Fig. 10. The reliability-aware pricing algorithm

The pricing algorithm in Fig. 10 is to determine payments for winners. We consider that each winner is priced at p_i for his winning bid B_i. Let $\mathcal{B}_{-i}$ denote all bids except B_i. Then, we conduct the greedy winner selection over $\mathcal{B}_{-i}$ to get a solution, denoted by $\mathcal{S}'$. We assume that the bid B_k is the winning bid in the k_{th} iteration, where $G(\mathcal{S}')$ is the utility before adding w_k into $\mathcal{S}'$. So the payment of B_i must be no more than $\frac{\beta_k v_i(\mathcal{S}')}{v_k(\mathcal{S}')}$. Otherwise, the weight ρ_i is not the largest. So the critical payment of winner s_i is the maximum critical value:

$$p_i = \max\{\frac{\beta_k\, v_i(\mathcal{S}')}{v_k(\mathcal{S}')} | k = 1, 2, \cdots\} \tag{9}$$

which terminates at $G(\mathcal{S}') = \sum_{j=1}^{l} \epsilon_j$. The total time complexity of pricing is $O(n^3 l)$, where n is the number of sellers and l is the number of tasks.

6 Theoretical Analysis

To prove the truthfulness of our RADT system, we should ensure some properties of the RADT auction. First, we simply define a notation:

$$G(j|\mathcal{S}) = \min\{\sigma_j^{\mathcal{S}}, \epsilon_j\}.$$

Lemma 1. *$G(\mathcal{S})$ is an increasing function.*

Proof. Considering two arbitrary winner sets $\mathcal{S}_1$ and $\mathcal{S}_2$, $\mathcal{S}_1 \subseteq \mathcal{S}_2 \subseteq \mathcal{S}$. According to Eq. 1, we have $\sigma_i^{\mathcal{S}_1} \leq \sigma_i^{\mathcal{S}_2}$ and $G(j|\mathcal{S}_1) \leq G(j|\mathcal{S}_2)$ for $\forall t_j \in \mathcal{T}$. Then $G(\mathcal{S}_1) \leq G(\mathcal{S}_2)$ when $\mathcal{S}_1 \subseteq \mathcal{S}_2$, which implies $G(\mathcal{S})$ is increasing. ∎

Theorem 1. *$G(\mathcal{S})$ is submodular.*

Proof. Without loss of generality, we assume that for two arbitrary winner sets $A, B \subseteq \mathcal{S}$. For $\forall t_i \in \mathcal{T}$, we have the conclusion that $\sigma_j^A + \sigma_j^B = \sigma_j^{A\cap B} + \sigma_j^{A\cup B}$ which indicates $\sigma_j^{\mathcal{S}}$ is submodular. Since $G(j|\mathcal{S})$ is the cut-off function of $\sigma_j^{\mathcal{S}}$, we can prove that $G(j|\mathcal{S})$ is submodular according to [8]. Hence, $G(\mathcal{S})$ is submodular because of the fact that $G(\mathcal{S}) = \sum_{j=1}^{l} G(j|\mathcal{S})$. ∎

Lemma 2 ***(Bid monotonicity).*** *Each seller w_i who wins by bidding $(\mathcal{T}_i, \beta_i)$ still wins by biding any $\beta_i' < \beta_i$ and any $\mathcal{T}_i' \supset \mathcal{T}_i$ given that other bids are fixed.*

Proof. Let ρ_i, $v_i(\mathcal{S})$ denote the weight and marginal reliability of seller w_i who bids $(\mathcal{T}_i, \beta_i)$, where $\rho_i = \frac{v_i(\mathcal{S})}{\beta_i}$. Let ρ_i', $v_i(\mathcal{S})'$ denote the weight and marginal reliability respectively if w_i bids $(\mathcal{T}_i', \beta_i)$ or $(\mathcal{T}_i, \beta_i')$. Either in $(\mathcal{T}_i', \beta_i)$ or in $(\mathcal{T}_i, \beta_i')$, it is clear that $v_i(\mathcal{S}) \geq v_i(\mathcal{S})'$ and $\rho_i \geq \rho_i'$ because of the submodularity of $G(\mathcal{S})$ according to Theorem 1. Moreover, if w_i has not been selected by bidding $(\mathcal{T}_i, \beta_i)$, he will not be selected by bidding $(\mathcal{T}_i, \beta_i')$ or $(\mathcal{T}_i', \beta_i)$. ∎

Lemma 3 ***(Critical payment).*** *Each seller s_i is paid a critical value p_i.*

Proof. We assume s_i wins in the k_{th} iteration, so the set $\mathcal{S}$ of winner selection and $\mathcal{S}'$ of pricing is same from the 0_{th} to the $(k-1)_{th}$ iterations. If s_i reports a bid β_i' instead of β_i. We need to prove that β_i' will fail if $\beta_i' > p_i$, otherwise he still wins when $\beta_i' \leq p_i$. Then we consider these two cases in the k_{th} iteration:

Case 1: $\beta_i' > p_i$. According to Fig. 9, we can derive that $\frac{v_i(\mathcal{S}_{k-1})}{\beta_i'} = \frac{v_i(\mathcal{S}_{k-1}')}{\beta_i'} < \frac{v_i(\mathcal{S}_{k-1}')}{p_i} \leq \frac{v_i(\mathcal{S}_{k-1}')}{\beta_k}$, where s_k is a winner, so that $\mathcal{S}' = \mathcal{S}' \cup \{s_k\}$. The first equation holds because $\mathcal{S}_{k-1} = \mathcal{S}_{k-1}'$. And the last inequation makes sense due to $p_i \geq \frac{\beta_k v_i(\mathcal{S}_{k-1}')}{v_k(\mathcal{S}_{k-1}')}$ according to Eq. 9. Hence, s_k is selected as a winner instead of s_i in the k_{th} iteration. So, $\mathcal{S}_k' = \mathcal{S}_{k-1} \cup \{s_k\} = \mathcal{S}_k$. Based on the above analysis, we can conclude that s_i will fail in all iterations of the winner selection in Fig. 9.

Case 2: $\beta_i' \leq p_i$. Assume that the winner selection runs over $\mathcal{B}_{-i}$ which is the process for pricing s_i. According to Eq. 9, we assume that $p_i = \frac{\beta_k v_i(\mathcal{S}_{k'-1})}{v_k(\mathcal{S}_{k'-1})}$, where s_k is the winner in the k_{th}' iteration. Now we run the winner selection again with the input set $\mathcal{B}$. We discuss two subcases of this process: (1) s_i wins before the k'_{th} iteration; (2) s_i does not win before the k'_{th} iteration. In the k'_{th} iteration: $\frac{v_i(\mathcal{S}_{k'-1})}{\beta'_i} \geq \frac{v_i(\mathcal{S}_{k'-1})}{p_i} \geq \frac{v_k(\mathcal{S}_{k'-1})}{\beta_k}$. Therefore, s_i wins in this iteration.

In Summary, the payments for all winners are critical. ■

Theorem 2. *The RADT auction is truthful.*

Proof. Lemmas 2 and 3 prove that the winner selection is monotonic and all the payments are critical respectively. So the auction is truthful according to [15]. ■

Theorem 3. *The RADT auction is individually rational.*

Proof. We consider that a seller w_i probably encounters these two situations, $w_i \in \mathcal{S}$ and $w_i \notin \mathcal{S}$. If $w_i \notin \mathcal{S}$, his payment will be zero. Otherwise, he wins the auction and his payment is p_i. According to Lemma 3, w_i will always be paid with the critical value p_i when he bids any $\beta_i < p_i$. Each seller bids his truthful cost due to the truthfulness in Theorem 2. Apparently, $p_i - c_i \geq 0$ holds. ■

Theorem 4. *The RADT system is truthful.*

Proof. We guarantee the truthfulness of RADT from three aspects.

1. **Deposit.** The consumer and sellers are asked for deposits which enforce them not to deviate from RADT, i.e., quit midway.
2. **Truthful auction.** The two-stage bidding strategy in Sect. 4.2 requires sellers to reveal bids truthfully. And the reverse auction can make sellers bid their truthful cost according to Theorem 2.
3. **Modifier.** The consumer must offer rewards to sellers to get keys for decryption, guaranteed by the modifiers in $GetKey()$. Moreover, some time modifiers are used to ensure that each procedure is invoked orderly.

Hence, the whole data trading process of RADT is truthful. ■

7 Implementation and Evaluations

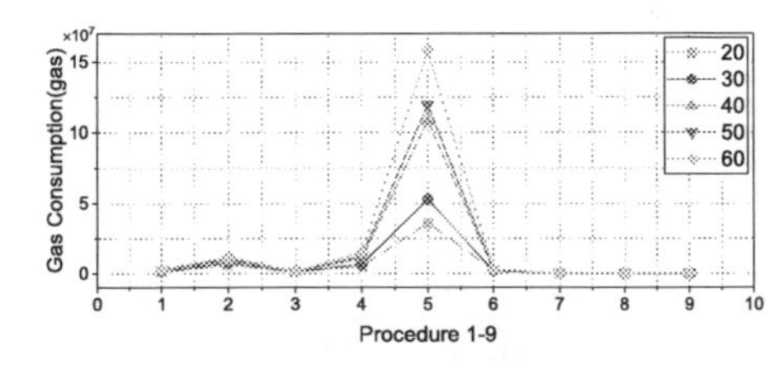

Fig. 11. Gas consumption of each procedure in Ganache Cli (with 20 sellers)

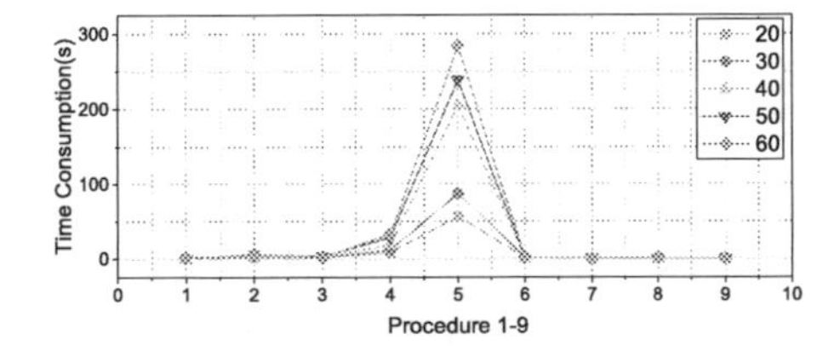

Fig. 12. Time consumption of each procedure in Ganache Cli (with 20 sellers)

We implement a prototype of RADT including the RADToken, the consumers and sellers. RADToken is deployed to a local simulated network TestRPC using Ethereum development tool Ganache Cli which is realized in the programming language Solidity with JavaScript (JS) as the intermediate interactive language. The consumer side is written in Python who needs to complete data decryption. And seller side is written in JS and Python who should finish the bid encryption and data encryption.

Due to difference of SHA-3 between JS and Solidity, we implement a custom SHA-3 in JS to make $en\beta_i$ have the identical value with sha3 $(w_i, \beta_i, nonce_i)$ in Solidity. We employ the standard cryptographic toolkit in Python, where we use AES for symmetric encryption and RSA for asymmetric encryption. Before the evaluation, we set some major parameters of RADT. The number of tasks l varies in [20, 30, 40, 50, 60] while the number of sellers n is fixed at 20. For each seller, his reliability is randomly generated from 0.6 to 1 and his bid is from 10 to 20. The reliability requirements ranges from 1 to 2.

7.1 Evaluations on Simulated Network at System Level

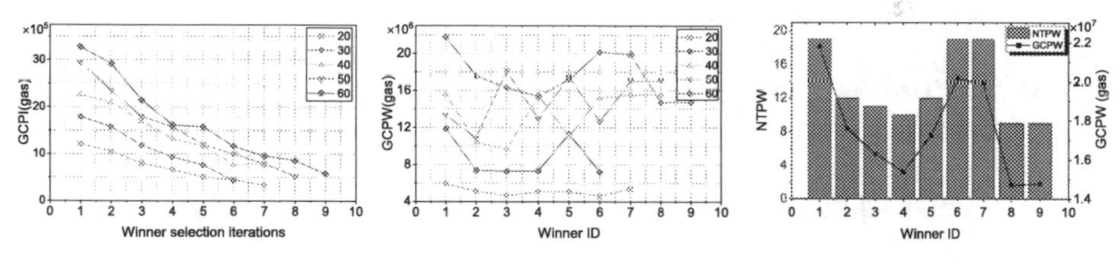

Fig. 13. GCPI vs. iterations

Fig. 14. GCPW vs. ID

Fig. 15. NTPW vs. ID

To demonstrate the practicality of our system, we first use Ganache Cli to construct a simulated network which is much like the real Ethereum environment except for its automatically mining mechanism in the background. This allows us to focus on the performance of RADToken, irrespective of time-consuming mining process and complex network circumstances in Ethereum. Our RADToken consists of nine main functions which correspond to **Procedures 1–9** that reflect the functionalities of **Initiate**, **CommitBid**, **RevealBid**, **WinnerSelection**, **Pricing**, **SetKey**, **GetKey**, **Refund** and **Payment** in Sect. 3.2.

To evaluate the unique performance of RADToken at the system level, we use two metrics for each procedure: *gas consumption* and *time consumption* which are depicted in Figs. 11 and 12 respectively. Since each computational step will be charged some gas, the more complicated the procedure is, the more gas and time it will consume. The operations to create and write storage data are relatively expensive [20], as we can see, Procedure 2, 4 and 5 use more gas and time. Procedure 4 need execute a nontrivial set of add, subtract, multiply, divide, compare and write operations, and there is a positive correlation between the number of winners and gas consumption. Procedure 5 is roughly equivalent to execute Procedure 4 $|\mathcal{S}|$ times. On the other hand, we may traverse more iterations than the entire Procedure 4 to find the critical payment for a winner, which uses more gas and time accordingly. Procedure 3 uses much gas because each encrypted bid is a 32-bytes hash value which will take up more storage. Other procedures use less gas due to their most read operations and smaller data length. Notice that here we use gas in **wei** where 1 ether $= 10^{18}$ wei.

7.2 Evaluations of Auction Mechanism

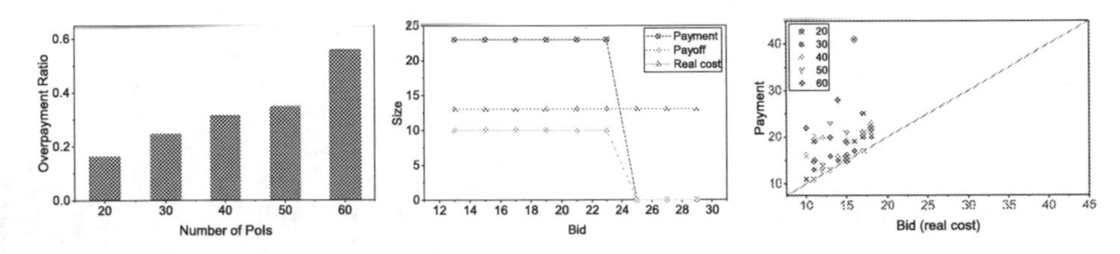

Fig. 16. Overpayment ratio

Fig. 17. Payoff of a bid

Fig. 18. Payments vs. bids

Since the **auction** including **WinnerSelection** and **Pricing** is the core mechanism of RADToken, we explicitly evaluate the performance of **WinnerSelection** and **Pricing** respectively. To avoid exceeding **gasLimit**, we divide the winner selection and pricing procedures into multiple iterations and repeatedly send a transaction to RADToken to select a winner and price the winner.

WinnerSelection. We give an example in Fig. 13 to compare the *gas consumption per iteration* (GCPI) under different number of tasks from 20 to 60. We notice a gradual decline of GCPI. We explain that by the constant cost of loading past mined blocks from storage into memory before each selection [10].

Pricing. We use *gas consumption per winner* (GCPW) and *number of transactions per winner* (NTPW) as two metrics in Figs. 14 and 15 respectively. We figure out that determining the payment for each winner will consume how much gas and need how many transactions. Figure 14 shows that the total gas consumption increases as the increasing number of tasks from an overall perspective. The GCPW has nothing to do with the iteration sequence which is only related to the number of PoIs and the number of traverse times to obtain its critical payment, which we can see in Fig. 15. The NTPW represents traverse times needed to price a winner and the corresponding GCPW shows that more gas will be used if more transactions are needed when the number of PoIs is 60.

Beyond valuating the performance on blockchain, we should ensure that the properties of our auction mechanism holds. we use the following metrics: overpay ratio, truthfulness and individual rationality. The overpay ratio is defined as:

$$\lambda = (\mathcal{P} - \mathcal{C}(\mathcal{S}))/\mathcal{C}(\mathcal{S}) \tag{10}$$

where $\mathcal{P}$ is the total payment and $\mathcal{C}(\mathcal{S})$ is the total cost. It measures the cost paid by the consumer to induce the truthfulness overall. Ensuring truthfulness means that no sellers can improve his payment by committing a different bid from the real one. Individual rationality ensures that each payoff is non-negative.

Overpayment Ratio: Figure 16 plots the overpayment ratio λ when l changes from 20 to 60. The results show that λ is always less than 0.6, which means that the consumer does not have to pay much extra money to induce truthfulness. λ increases monotonously with increasing l because more sellers will be selected and the increments of the payments are greater than those of the costs.

Truthfulness: To verify the truthfulness, we randomly pick a winner and change its claimed bid, then recalculate the payments as well as the payoffs. The results illustrated in Fig. 17 show that when the truthful bid (real cost) is 13, the payment is 23 and the payoff is 10. The payoff remains unchanged when the bid is no more than 23. However, if the bid is larger than the critical payment 23, the payoff becomes zero which means that the winner loses the auction.

Individual Rationality: We demonstrate individual rationality in Fig. 18. Each payment is greater than the related bid when l varies from 20 to 60.

8 Related Works

We review related works from the following two aspects:

Trading on Blockchain: Blockchain offers users new options for managing their holdings and their trading intentions which can ensure the data integrity. Due to the honest-but-curious property of secure third party, a few works resort to blockchain to build trading systems. [3] implements a decentralized energy trading system using blockchain to address the problem of providing transaction security. [12] proposes AccountTrade for big data trading which can achieve book-keeping ability and accountability against dishonest consumers. In addition to P2P trading, blockchain is fit for the crowd trading. For instance, [14] conceptualizes a blockchain-based decentralized framework named CrowdBC, which does not depend on any central third party to accomplish crowdsourcing process. However, the high storage requirement prevents the wide usage of blockchains on mobile phones. A novel concept, Consensus Unit (CU) [22], organizes different nodes into one unit and lets them to store at least one copy of blockchain data, which can be applied in more application scenarios. Blockchains conduct trading will consume resources, so BLOCKBENCH is designed in [7] to understand the performance of blockchains against data processing workloads.

Incentive Data Trading Mechanism: In order to improve the repetitive use rate of data, [18] designs a DataMart to determine the pricing and consumer-seller matching in distributed fashion which is suited for highly dynamic and heterogeneous market environment and ad-hoc setting. It adopts double auction for pricing to ensure the truthfulness. However this cannot satisfy some consumers' needs who want buy large volumes of data which is not easy to collect. [24] designs a practical data collection scheme leveraging mobile crowdsensing and proposes VENUS-PRO for profit maximization and VENUS-PAY for payment minimization which is a data procurement auction in Bayesian setting. The above incentive data trading which adopts auction is executed by a third-party which may disclose privacy or data information. A novel distributed agent-based privacy-preserving framework DADP proposed in [19] enables real-time crowd-sourced statistical data publishing with strong privacy protection under an untrusted server. [9] considers the introduction of homomorphic cryptography to allow the auctions to be processed using only encrypted bids. Moreover, it uses the digital signature to ensure that data has not been manipulated in transmission or by a compromised entity in network.

9 Conclusion

In this paper, we first propose a Reverse-Auction-and-blockchain-based crowdsensed Data Trading system. Different from the existing CDT, we use a meticulous designed smart contract to replace a third-party data broker which can ensure the truthfulness of data consumers and data sellers. In order to incentivize more sellers to participate in the crowdsensing data collection, we propose a reverse auction mechanism to prompt sellers to provide high quality sensing data and claim truthful bids. Meanwhile, we protect sellers' bids by leveraging a two-stage bidding strategy which can blame untruthful sellers and ensure the immutability of bids. The confidentiality of data is preserved by the introduction of symmetric and asymmetric cryptography where the keys cannot be grabbed in transmission. Finally, we implement a prototype on an Ethereum test network and the evaluations demonstrate its practicability.

Acknowledgment. This research was supported in part by National Natural Science Foundation of China (NSFC) (Grant No. 61872330, 61572336, 61572457, 61632016, 61379132, U1709217), Natural Science Foundation of Jiangsu Province in China (Grant No. BK20131174, BK2009150), Anhui Initiative in Quantum Information Technologies (Grant No. AHY150300), and Natural Science Research Project of Jiangsu Higher Education Institution (No. 18KJA520010).

References

1. Thingful. https://www.thingful.net/
2. ThingSpeak. https://thingspeak.com/
3. Aitzhan, N.Z., Svetinovic, D.: Security and privacy in decentralized energy trading through multi-signatures, blockchain and anonymous messaging streams. IEEE Trans. Dependable Secure Comput. **15**(5), 840–852 (2018)
4. Aumasson, J.P., Henzen, L., Meier, W., Phan, R.C.W.: SHA-3 proposal BLAKE. Submission to NIST (2008)
5. Buterin, V.: A next-generation smart contract and decentralized application platform. White paper (2014)

6. Cai, C., Zheng, Y., Wang, C.: Leveraging crowdsensed data streams to discover and sell knowledge: a secure and efficient realization. In: IEEE ICDCS (2018)
7. Dinh, T.T.A., Liu, R., Zhang, M., Chen, G., Ooi, B.C., Wang, J.: Untangling blockchain: a data processing view of blockchain systems. IEEE Trans. Knowl. Data Eng. **30**(7), 1366–1385 (2018)
8. Gao, G., Xiao, M., Wu, J., Huang, L., Hu, C.: Truthful incentive mechanism for nondeterministic crowdsensing with vehicles. IEEE Trans. Mob. Comput. **17**(12), 2982–2997 (2018)
9. Gao, W., Yu, W., Liang, F., Hatcher, W.G., Lu, C.: Privacy-preserving auction for big data trading using homomorphic encryption. IEEE Trans. Netw. Sci. Eng. (2018)
10. Hu, S., Cai, C., Wang, Q., Wang, C., Luo, X., Ren, K.: Searching an encrypted cloud meets blockchain: a decentralized, reliable and fair realization. In: IEEE INFOCOM (2018)
11. Jiang, C., Gao, L., Duan, L., Huang, J.: Scalable mobile crowdsensing via peer-to-peer data sharing. IEEE Trans. Mob. Comput. **17**(4), 898–912 (2018)
12. Jung, T., et al.: AccountTrade: accountable protocols for big data trading against dishonest consumers. In: IEEE INFOCOM (2017)
13. Kosba, A., Miller, A., Shi, E., Wen, Z., Papamanthou, C.: Hawk: the blockchain model of cryptography and privacy-preserving smart contracts. In: IEEE S&P (2016)
14. Li, M., et al.: CrowdBC: a blockchain-based decentralized framework for crowdsourcing. IEEE Trans. Parallel Distrib. Syst. (2018)
15. Myerson, R.B.: Optimal auction design. Math. Oper. Res. **6**(1), 58–73 (1981)
16. Nakamoto, S.: Bitcoin: a peer-to-peer electronic cash system. https://bitcoin.org/bitcoin.pdf
17. Niu, C., Zheng, Z., Wu, F., Gao, X., Chen, G.: Trading data in good faith: integrating truthfulness and privacy preservation in data markets. In: ICDE (2017)
18. Susanto, H., Zhang, H., Ho, S., Liu, B.: Effective mobile data trading in secondary ad-hoc market with heterogeneous and dynamic environment. In: IEEE ICDCS (2017)
19. Wang, Z., et al.: Privacy-preserving crowd-sourced statistical data publishing with an untrusted server. IEEE Trans. Mob. Comput. (2018)
20. Wood, G.: Ethereum: a secure decentralised generalised transaction ledger (2014). https://gavwood.com/paper.pdf
21. Xiao, M., Wu, J., Huang, L., Cheng, R., Wang, Y.: Online task assignment for crowdsensing in predictable mobile social networks. IEEE Trans. Mob. Comput. **16**(8), 2306–2320 (2017)
22. Xu, Z., Han, S., Chen, L.: CUB, a consensus unit-based storage scheme for blockchain system. In: ICDE (2018)
23. Yu, J., Cheung, M.H., Huang, J., Poor, H.V.: Mobile data trading: behavioral economics analysis and algorithm design. IEEE J. Sel. Areas Commun. **35**(4), 994–1005 (2017)
24. Zheng, Z., Peng, Y., Wu, F., Tang, S., Chen, G.: Trading data in the crowd: profit-driven data acquisition for mobile crowdsensing. IEEE J. Sel. Areas Commun. **35**(2), 486–501 (2017)

Data Integration

Selective Matrix Factorization for Multi-relational Data Fusion

Yuehui Wang[1], Guoxian Yu[1,3](✉), Carlotta Domeniconi[2], Jun Wang[1], Xiangliang Zhang[3], and Maozu Guo[4]

[1] College of Computer and Information Science, Southwest University, Chongqing, China
{yuehuiwang,gxyu,kingjun}@swu.edu.cn
[2] Department of Computer Science, George Mason University, Fairfax, USA
carlotta@cs.gmu.edu
[3] King Abdullah University of Science and Technology, Thuwal, Saudi Arabia
xiangliang.zhang@kaust.edu.sa
[4] College of Electrical and Information Engineering, Beijing University of Civil Engineering and Architecture, Beijing, China
guomaozu@bucea.edu.cn

Abstract. Matrix factorization based data fusion solutions can account for the intrinsic structures of multi-relational data sources, but most solutions equally treat these sources or prefer sparse ones, which may be irrelevant for the target task. In this paper, we introduce a Selective Matrix Factorization based Data Fusion approach (SelMFDF) to collaboratively factorize multiple inter-relational data matrices into low-rank representation matrices of respective object types and optimize the weights of them. To avoid preference to sparse data matrices, it additionally regularizes these low-rank matrices by approximating them to multiple intra-relational data matrices and also optimizes the weights of them. Both weights contribute to automatically integrate relevant data sources. Finally, it reconstructs the target relational data matrix using the optimized low-rank matrices. We applied SelMFDF for predicting inter-relations (lncRNA-miRNA interactions, functional annotations of proteins) and intra-relations (protein-protein interactions). SelMFDF achieves a higher AUROC (area under the receiver operating characteristics curve) by at least 5.88%, and larger AUPRC (area under the precision-recall curve) by at least 18.23% than other related and competitive approaches. The empirical study also confirms that SelMFDF can not only differentially integrate these relational data matrices, but also has no preference toward sparse ones.

Keywords: Matrix factorization · Data fusion · Multi-relational data · Association prediction

---

This work is supported by NSFC (61872300, 61873214 and 61871020), Natural Science Foundation of CQ CSTC (cstc2018jcyjAX0228 and cstc2016jcyjA0351).

G. Li et al. (Eds.): DASFAA 2019, LNCS 11446, pp. 313–329, 2019.
https://doi.org/10.1007/978-3-030-18576-3_19

1 Introduction

With the rapid growth of Internet and modern technologies, we can obtain various data sources that are directly related to the main task, and also other data sources indirectly related to the task, which can still facilitate the completion of this task. For example, the accuracy of gene function prediction can be improved by integrating the gene-level data (gene expression, gene-gene interactions), and also by fusing transcript-level data (miRNA-gene interactions, miRNA-miRNA interactions) that convey complementary information about gene functions [7,26]. The ever-increasing heterogeneous data sources make data fusion approaches increasingly popular over the past decade, which aim to collectively explore interesting patterns from multiple data sources, and to reduce the impact of noisy or irrelevant ones [7,15].

An intuitive solution to fuse multiple data sources is concatenating the feature vectors of the same object across different data sources into a longer feature vector, and then applying off-the-shelf learners on this long vector. But this concatenation ignores the intrinsic characteristics of these feature vectors and may (and often does) suffer from the issue of curse of dimensionality and of missing features. Another intuitive solution is to train a classifier on each feature view and then combine these classifiers for ensemble prediction [21], but this ensemble solution may be impacted by low-quality base classifiers independently trained on individual views, which can not ensure a base classifier with sufficient accuracy. Furthermore, the early fusion (feature concatenation) and late fusion (classifier ensemble) can not capture heterogeneous relations between different object types. For these reasons, many *inter-median* data fusion solutions have been proposed in recent years [6,13,23].

Inter-median data fusion methods can be generally divided into three categories: *multiple kernel(network) learning*-based (MKL), *Bayesian network*-based (BN) and *matrix factorization*-based (MF) [7]. MKL methods firstly transform multi-relational data matrices onto the homologous data matrices that are directly related with the target task, and then applies different techniques to combine these transformed data matrices for prediction [8,13,23]. These MKL-based methods can selectively integrate multiple homologous data matrices. However, they have to transform heterogeneous features or project multi-relational data into a common feature space before fusion. This hand-crafted transformation and projection may enshroud the intrinsic structure of multi-relational data, and thus does not make full usage of them [26]. BN-based approaches combine the concepts from probability and graph theory to represent and model causal relations between random variables [17]. BN was initially applied to gene function prediction [18] and also shows the potentiality in patient-specific data integration [25]. Although BN-based solution can capture conditional dependence between data sources and variables, it suffers from a heavy computational limitation and asks for sufficient training data with labels.

MF-based solutions generally factorize multiple data matrices into low-rank matrices to explore latent relationships between objects across different data sources. Solutions in this type do not need to project multi-relational data matrices into the common feature space and thus can account for the intrinsic structure of these information sources. To name a few, Ding *et al.* [5] extended the classical nonnegative matrix factorization (NMF) [14] to nonnegative matrix tri-factorization (NMTF) to co-cluster heterogeneous data, but NMTF can only fuse inter-relational data matrices and ignore the intra-relational ones. Wang *et al.* [19] proposed the symmetric nonnegative matrix tri-factorization (SNMTF) to simultaneously cluster different types of objects, and incorporates the intra-relational ones through manifold regularization [1]. However, SNMTF has a heavy computational complexity and large runtime, because it performs matrix factorization on a big matrix, whose block matrices embody inter-relations between objects. Zitnik and Zupan [26] developed a penalized matrix tri-factorization based model (DFMF) to jointly factorize multiple relational data matrices for predicting gene functions and pharmacologic actions.

These aforementioned MF-based solutions show great potential in exploring the underlying relations between objects, but they ignore the different relevances of multi-relational data sources, since they implicitly assume each source having equal relevance toward the target prediction task, while they may not (and often does). To overcome this problem, Fu *et al.* [6] introduced a MF-based model (MFLDA) to predict lncRNA-disease associations by assigning different weights to multiple inter-relational data matrices for objects of different types and by jointly factorizing these matrices into low-rank ones. MFLDA then uses the optimized matrices to reconstruct the target matrix to predict new inter-relations between objects of different types. However, MFLDA does not account for the different relevances of intra-relational data matrices for objects of the same types, and thus its performance may be compromised by the low-quality or irrelevant data sources. To simultaneously account for the different relevances of multiple intra-relational data matrices, MFLDA was further extended to WMFLDA, which can selectively fuse multiple intra-relation matrices [24]. However, these extended solutions *prefer* to assigning larger weights to *sparse* data matrices, or have a priority toward sparser ones, which may be irrelevant (or even harmful) for the target task. In fact, this preference is also suffered by many MKL-based solutions [9,20,23].

To address these issues, we propose a *Sel*ective *M*atrix *F*actorization based *D*ata *F*usion (SelMFDF) solution for integrating multi-relational data. SelMFDF can avoid preferring the sparse relational data matrices during the fusing process. It performs collaborative matrix tri-factorization to optimize the low-rank representation matrices of respective object types and the weights of inter-relational data matrices. To selectively integrate multiple intra-relational data matrices, it further optimizes these low-rank matrices by approximating them to multiple intra-relational data matrices and the weights of these matrices. These two types of weights contribute to identify relevant data sources and remove irrelevant ones. After that, it approximates the target relational data matrix using the optimized low-rank matrices. The main contributions of this paper are summarized as follows:

(i) Our introduced SelMFDF can respect and explore the intrinsic structures of multi-relational data matrices to simultaneously predict inter(intra)-relation between objects of different (same) types, automatically discard irrelevant data sources and credit larger weights to the more relevant ones.

(ii) An alternative optimization procedure is developed to jointly optimize the low-rank matrix approximations and weights of multi-relational data matrices for the target prediction task.

(iii) Empirical study on predicting lncRNA-miRNA associations, gene functions and protein-protein interactions shows that SelMFDF significantly outperforms the related and competitive methods NMTF [5], SNMTF [19], DFMF [26], MFLDA [6], and WMFLDA [24].

2 Methodology

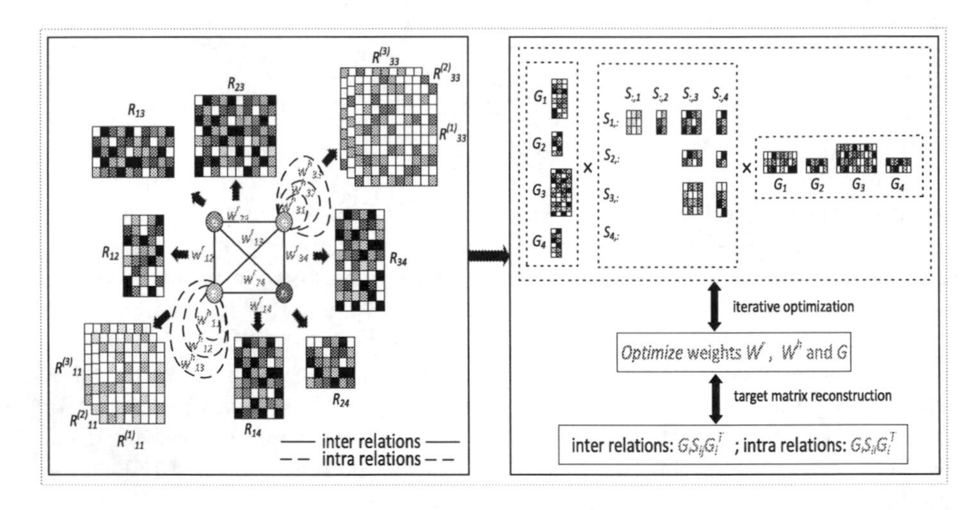

Fig. 1. The operating principle of SelMFDF. In the left figure, $\mathbf{R}_{ij}$ is the inter-relational data matrix between object type i and j, $\mathbf{R}_{ii}^{(v)}$ is the v-th intra-relational matrix of the i-th object type; in the right figure, $\mathbf{G}_i$ is the low-rank representation matrix of the i-th object type. $\mathbf{W}_{ij}^r$ and $\mathbf{W}_{iv}^h$ are the weights assigned to respective inter-relational and intra-relational data matrices.

The operating principle of SelMFDF is illustrated in Fig. 1. SelMFDF presets weights for inter-relational and intra-relational data matrices, and performs collaborative low-rank matrix factorization. It then jointly optimizes the weights and the low-rank matrix approximations of these relational matrices. After that, it reconstructs the target relational data matrix based on the product of optimized low-rank matrices.

2.1 Matrix Factorization Model for Multiple Relational Data

The relationships between multi-type objects can be divided into inter-relations and intra-relations, both of which can be encoded by relational data matrices. To fuse these relational data sources, various solutions follow different principles to transform these data matrices toward the target relational matrix using the inter-relations between objects [7,22]. However, this transformation often overrides or even distorts the intrinsic structures of multi-relational data. To avoid this issue, Zitnik and Zupan [26] introduced a penalized matrix factorization based data fusion framework (DFMF). The objective function of DFMF is:

$$\begin{aligned}\min_{G\geq 0}\mathcal{L}(\mathbf{G},\mathbf{S}) &= \sum_{\mathbf{R}_{ij}\in\mathcal{R}} \left\|\mathbf{R}_{ij}-\mathbf{G}_i\mathbf{S}_{ij}\mathbf{G}_j^T\right\|_F^2 \\ &+\sum_{i=1}^{m}\sum_{t=1}^{\max_i t_i} tr(\mathbf{G}^T\mathbf{\Theta}_i^{(t)}\mathbf{G})\end{aligned} \tag{1}$$

where $\|\cdot\|_F^2$ and $tr(\cdot)$ are the Frobenius norm of a matrix and the matrix trace operator. DFMF simultaneously considers m object types and fuses a collection of relational data sources ($\mathcal{R}$). The inter relations between n_i objects of type i and n_j objects of type j are stored in $\mathbf{R}_{ij}\in\mathbb{R}^{n_i\times n_j}$, $\mathbf{G}_i\in\mathbb{R}^{n_i\times k_i}$ is the low-rank representation of object type i, $\mathbf{S}_{ij}\in\mathbb{R}^{k_i\times k_j}$ encodes the latent relationship between $\mathbf{G}_i$ and $\mathbf{G}_j$, $k_i\ll n_i$ is the low-rank size of the respective object type, $\mathbf{G}=diag(\mathbf{G}_1,\mathbf{G}_2,\cdots,\mathbf{G}_m)$. Without loss of generality, suppose the i-th object type has t_i intra relational data matrices and $\mathbf{\Theta}_i^{(t)}$ is the t-th one. $\mathbf{\Theta}^{(t)}$ collectively contains all the following block diagonal matrices: $\mathbf{\Theta}^{(t)}=diag(\mathbf{\Theta}_1^{(t)},\mathbf{\Theta}_2^{(t)},\cdots,\mathbf{\Theta}_m^{(t)})$, $t\in\{1,2,\cdots,\max_i t_i\}$, and the i-th block matrix along the main diagonal of $\mathbf{\Theta}^{(t)}$ is zero if $t\geq t_i$.

Equation (1) can respect and explore the intrinsic structure of multiple relational data matrices, since it does not project these matrices onto the same space for fusion. However, it *equally* treats all the relational matrices and ignores the different relevances of them toward the target task. As a result, its performance may be dragged down by noisy or irrelevant data sources. To address this issue, Fu *et al.* [6] extended DFMF by optimizing the weights assigned to inter-relational data matrices. However, it still can not differentiate noisy intra-relational matrices during the fusing process. Given that, Yu *et al.* [24] further specified weights to different intra-relational matrices. The theoretical analysis and experimental results show that these two extensions can indeed selectively fuse multiple relational data sources. However, they are inclined to select sparse ones with more zero elements, since the sparse data matrices generally have a smaller approximate loss ($\left\|\mathbf{R}_{ij}-\mathbf{G}_i\mathbf{S}_{ij}\mathbf{G}_j^T\right\|_F^2$) or smoothness loss ($tr(\mathbf{G}_i^T\mathbf{\Theta}_i^{(t)}\mathbf{G}_i)$). In practice, a too sparse data matrix often cannot encode sufficient information for the target task, and thus is irrelevant for the task.

2.2 Objective Function of SelMFDF

Based on the above analysis, to reduce the impact of noisy data sources and to avoid inclined to sparse ones, we define the objective function of SelMFDF as follows:

$$\begin{aligned}\min_{\mathbf{G}\geq 0} \mathcal{L}(\mathbf{G}, \mathbf{S}, \mathbf{W}^r, \mathbf{W}^h) = &\sum_{R_{ij}\in\mathcal{R}} \mathbf{W}^r_{ij} \left\|\mathbf{R}_{ij} - \mathbf{G}_i\mathbf{S}_{ij}\mathbf{G}_j^T\right\|_F^2 \\ &+ \sum_{i=1}^{m}\sum_{t=1}^{\tau} \mathbf{W}^h_{it} \left\|\mathbf{R}^{(t)}_{ii} - \mathbf{G}_i\mathbf{S}_{ii}\mathbf{G}_i^T\right\|_F^2 \\ s.t. \quad \mathbf{W}^r \geq 0, \mathbf{W}^h \geq 0 &\end{aligned} \tag{2}$$

where $\mathbf{W}^r \in \mathbb{R}^{m\times m}$, $\mathbf{W}^h \in \mathbb{R}^{m\times\tau}$, $\tau = \max_i t_i$, $\mathbf{W}^r$ contains the weights assigned to different inter-relational data matrices, if $\mathbf{R}_{ij} \notin \mathcal{R}$, $\mathbf{W}^r_{ij} = 0$. $\mathbf{W}^h$ contains the weights assigned to different intra-relational data matrices. If $\mathbf{R}^{(t)}_{ii} \notin \mathcal{R}$ or $t > t_i$, $\mathbf{W}^h_{it} = 0$. Unlike Eq. (1), our objective function utilizes the shared low-rank matrices $\mathbf{G}_i$ and $\mathbf{S}_{ii} \in \mathbb{R}^{k_i\times k_i}$ across t_i intra-relational data matrices to approximate $\mathbf{R}^{(t)}_{ii}$. In this way, a data matrix inconsistent with other intra-relational data matrices of the same objects will be assigned with a lower weight. Particularly, for a sparse data matrix $\mathbf{R}^{(t)}_{ii}$, $\left\|\mathbf{R}^{(t)}_{ii} - \mathbf{G}_i\mathbf{S}_{ii}\mathbf{G}_i^T\right\|_F^2$ results in a large loss, because $\mathbf{R}^{(t)}_{ii}$ encodes much fewer relations between objects than its cousin matrices ($\{\mathbf{R}^{(t')}_{ii}\}_{t'=1}^{t_i}, t' \neq t\}$) and the loss is dominated by $tr(\mathbf{G}_i\mathbf{S}_{ii}\mathbf{G}_i^T)$. Similarly, for a dense matrix with noisy entries, $\left\|\mathbf{R}^{(t)}_{ii} - \mathbf{G}_i\mathbf{S}_{ii}\mathbf{G}_i^T\right\|_F^2$ also results in a big loss. To minimize the above objective function, a smaller weight will be automatically assigned to these two types of data matrices. Since $\mathbf{G}_i$ is also shared by the inter-relational data matrices, the first term in Eq. (2) can also avoid preferring to sparse ones. As a result, Eq. (2) can avoid the preference toward the sparse data matrices. We want to remark that low-rank matrix approximation can also reduce the impact of noises to some extent [4,16].

However, Eq. (2) may only set $\mathbf{W}^r_{ij} = 1$ to $\mathbf{R}_{ij}$ if $\mathbf{R}_{ij}$ has the smallest approximation loss ($\|\mathbf{R}_{ij} - \mathbf{G}_i\mathbf{S}_{ij}\mathbf{G}_j\|_F^2$) among all the inter-relational matrices, and the other inter-relational ones will be discarded. Equation (2) may also assign $\mathbf{W}^h_{it} = 1$ to $\mathbf{R}^{(t)}_{ii}$, if $\mathbf{R}^{(t)}_{ii}$ has the smallest approximation loss among all the intra-relational matrices. As a result, the contribution of other intra-relational ones will be disregarded. To remedy this issue, we add two l_2-norm based regularizations on $\mathbf{W}^r$ and $\mathbf{W}^h$, and update the objective function as follows:

$$\begin{aligned}\min_{G\geq 0} \mathcal{L}(\mathbf{G}, \mathbf{S}, \mathbf{W}^r, \mathbf{W}^h) = &\sum_{R_{ij}\in\mathbb{R}} \mathbf{W}^r_{ij} \left\|\mathbf{R}_{ij} - \mathbf{G}_i\mathbf{S}_{ij}\mathbf{G}_j^T\right\|_F^2 \\ &+ \sum_{i=1}^{m}\sum_{t=1}^{\tau} \mathbf{W}^h_{it} \left\|\mathbf{R}^{(t)}_{ii} - \mathbf{G}_i\mathbf{S}_{ii}\mathbf{G}_i^T\right\|_F^2 \\ &+ \alpha \left\|vec(\mathbf{W}^r)\right\|_F^2 + \beta \left\|vec(\mathbf{W}^h)\right\|_F^2 \\ s.t. \quad \mathbf{W}^r \geq 0, \mathbf{W}^h \geq 0, &\sum vec(\mathbf{W}^r) = 1, \sum vec(\mathbf{W}^h) = 1\end{aligned} \tag{3}$$

where $vec(\mathbf{W}^r)$ and $vec(\mathbf{W}^h)$ are the vectorization operator that stacks the rows of $\mathbf{W}^r$ and $\mathbf{W}^h$, $\alpha > 0$ and $\beta > 0$ are used to control the complexity of $vec(\mathbf{W}^r)$ and $vec(\mathbf{W}^h)$. By adding these two regularization terms, SelMFDF can selectively integrate several relevant data matrices, and automatically remove irrelevant ones. Our following optimization procedure for $\mathbf{W}^h$ and $\mathbf{W}^r$ will theoretically confirm this advantage.

$\tilde{\mathbf{G}}$ can be viewed as the optimized low-rank matrices of these object types, we can approximate the target inter-relational data matrix between object type i and j as Eq. (4). Similarly, we can also approximate the intra-relational data matrix as Eq. (5).

$$\widehat{\mathbf{R}}_{ij} = \tilde{\mathbf{G}}_i \tilde{\mathbf{S}}_{ij} \tilde{\mathbf{G}}_j^T \tag{4}$$

$$\widehat{\mathbf{R}}_{ii} = \tilde{\mathbf{G}}_i \tilde{\mathbf{S}}_{ii} \tilde{\mathbf{G}}_i^T \tag{5}$$

In this way, SelMFDF can not only predict the inter-relations between different types of objects, but also the intra-relations between objects of the same type.

2.3 Optimization of SelMFDF

The optimization problem in Eq. (3) is non-convex with respect to $\mathbf{G}$, $\mathbf{S}$, $\mathbf{W}^r$ and $\mathbf{W}^h$ simultaneously. It is difficult to seek the global optimal solutions for all the variables at the same time. Here, we follow the idea of alternating direction method of multipliers (ADMM)[2] and DFMF [26] to alternatively optimize one variable by fixing other three of these four variables in an iterative way.

To account for $\mathbf{G}_i \geq 0$, we import the Lagrangian multipliers $\{\lambda_i\}_{i=1}^m$ and reformulate Eq. (3) as follows:

$$\begin{aligned}
&\min_{\mathbf{G}\geq 0} \tilde{\mathcal{L}}(\mathbf{G}, \mathbf{S}, \mathbf{W}^r, \mathbf{W}^h, \lambda) = \\
&\sum_{R_{ij}\in\mathcal{R}} \mathbf{W}_{ij}^r tr(\mathbf{R}_{ij}^T \mathbf{R}_{ij} - 2\mathbf{R}_{ij}^T \mathbf{G}_i \mathbf{S}_{ij} \mathbf{G}_j^T + \mathbf{G}_j^T \mathbf{G}_j \mathbf{S}_{ij}^T \mathbf{G}_i^T \mathbf{G}_i \mathbf{S}_{ij}) \\
&+ \sum_{i=1}^{m} \sum_{t=1}^{\tau} \mathbf{W}_{it}^h tr(\mathbf{R}_{ii}^{(t)^T} \mathbf{R}_{ii}^{(t)} - 2\mathbf{R}_{ii}^{(t)^T} \mathbf{G}_i \mathbf{S}_{ii} \mathbf{G}_i^T + \mathbf{G}_i^T \mathbf{G}_i \mathbf{S}_{ii}^T \mathbf{G}_i^T \mathbf{G}_i \mathbf{S}_{ii}) \\
&+ \alpha \left\| vec(\mathbf{W}^r) \right\|_F^2 + \beta \left\| vec(\mathbf{W}^h) \right\|_F^2 - \sum_{i=1}^{m} tr(\lambda_i \mathbf{G}_i^T) \\
&s.t. \quad \mathbf{W}^r \geq 0, \mathbf{W}^h \geq 0, \sum vec(\mathbf{W}^r) = 1, \sum vec(\mathbf{W}^h) = 1
\end{aligned} \tag{6}$$

Next, we goto the alternative optimization procedure.

Optimizing $\mathbf{S}_{ij}$: Suppose $\mathbf{G}$, $\mathbf{W}^r$ and $\mathbf{W}^h$ are known and fixed, and let the partial derivative of Eq. (6) with respect to $\mathbf{S}_{ij}$ and $\mathbf{S}_{ii}$ equal to 0, we can obtain the explicit solution of $\mathbf{S}_{ij}$ and $\mathbf{S}_{ii}$ as follows:

$$\mathbf{S}_{ij} = (\mathbf{G}_i^T \mathbf{G}_i)^{-1} \mathbf{G}_i^T \mathbf{R}_{ij} \mathbf{G}_j (\mathbf{G}_j^T \mathbf{G}_j)^{-1} \tag{7}$$

$$\mathbf{S}_{ii} = (\mathbf{G}_i^T\mathbf{G}_i)^{-1}\frac{\sum_{t=1}^{\tau}\mathbf{W}_{it}^h(\mathbf{G}_i^T\mathbf{R}_{ii}^{(t)}\mathbf{G}_i)}{\sum_{t=1}^{\tau}\mathbf{W}_{it}^h}(\mathbf{G}_i^T\mathbf{G}_i)^{-1} \tag{8}$$

Optimizing $\mathbf{G}_i$: Similar as the optimization of $\mathbf{S}$, we also take the partial derivative of Eq. (6) with respect to $\mathbf{G}_i$ with known $\mathbf{S}$, $\mathbf{W}^r$ and $\mathbf{W}^h$:

$$\begin{aligned}\frac{\partial\tilde{\mathcal{L}}}{\mathbf{G}_i} &= \sum_{j:\mathbf{R}_{ij}\in\mathcal{R}}\mathbf{W}_{ij}^r(-2\mathbf{R}_{ij}\mathbf{G}_j\mathbf{S}_{ij}^T + 2\mathbf{G}_i\mathbf{S}_{ij}\mathbf{G}_j^T\mathbf{G}_j\mathbf{S}_{ij}^T)\\ &+ \sum_{j:\mathbf{R}_{ji}\in\mathcal{R}}\mathbf{W}_{ji}^r(-2\mathbf{R}_{ji}^T\mathbf{G}_j\mathbf{S}_{ji} + 2\mathbf{G}_i\mathbf{S}_{ji}^T\mathbf{G}_j^T\mathbf{G}_j\mathbf{S}_{ji})\\ &+\sum_{t=1}^{\tau}\mathbf{W}_{it}^h2\mathbf{R}_{ii}^{(t)}\mathbf{G}_i - \lambda_i\end{aligned} \tag{9}$$

Multipliers λ_i can be obtained from Eq. (9) by letting $\frac{\partial\tilde{\mathcal{L}}}{\mathbf{G}_i} = 0$ and the KKT (Karush-Kuhn-Tucker) complementary condition [2] for nonnegativity of $\mathbf{G}_i$ as:

$$0 = \lambda_i \circ \mathbf{G}_i \tag{10}$$

where $\circ$ denotes the Hadamard product. Equation (10) is a fixed point equation and the solution must satisfy it at convergence. Thus, we can obtain:

For $\mathbf{R}_{ij} \in \mathcal{R}$:

$$\begin{aligned}\mathbf{G}_i^{(e)} +&= \mathbf{W}_{ij}^r(\mathbf{R}_{ij}\mathbf{G}_j\mathbf{S}_{ij}^T)^+ + \mathbf{W}_{ij}^r\mathbf{G}_i(\mathbf{S}_{ij}\mathbf{G}_j^T\mathbf{G}_j\mathbf{S}_{ij}^T)^-\\ \mathbf{G}_i^{(d)} +&= \mathbf{W}_{ij}^r(\mathbf{R}_{ij}\mathbf{G}_j\mathbf{S}_{ij}^T)^- + \mathbf{W}_{ij}^r\mathbf{G}_i(\mathbf{S}_{ij}\mathbf{G}_j^T\mathbf{G}_j\mathbf{S}_{ij}^T)^+\\ \mathbf{G}_j^{(e)} +&= \mathbf{W}_{ij}^r(\mathbf{R}_{ij}^T\mathbf{G}_i\mathbf{S}_{ij})^+ + \mathbf{W}_{ij}^r\mathbf{G}_j(\mathbf{S}_{ij}^T\mathbf{G}_i^T\mathbf{G}_i\mathbf{S}_{ij})^-\\ \mathbf{G}_j^{(d)} +&= \mathbf{W}_{ij}^r(\mathbf{R}_{ij}^T\mathbf{G}_i\mathbf{S}_{ij})^- + \mathbf{W}_{ij}^r\mathbf{G}_j(\mathbf{S}_{ij}^T\mathbf{G}_i^T\mathbf{G}_i\mathbf{S}_{ij})^+\end{aligned} \tag{11}$$

For $t = 1, 2, \ldots, \tau$:

$$\begin{aligned}\mathbf{G}_i^{(e)} +&= 2\mathbf{W}_{it}^h(\mathbf{R}_{ii}^{(t)}\mathbf{G}_i\mathbf{S}_{ii}^T)^+ + 2\mathbf{W}_{it}^h(\mathbf{G}_i\mathbf{S}_{ii}\mathbf{G}_i^T\mathbf{G}_i\mathbf{S}_{ii}^T)^-\\ \mathbf{G}_i^{(d)} +&= 2\mathbf{W}_{it}^h(\mathbf{R}_{ii}^{(t)}\mathbf{G}_i\mathbf{S}_{ii}^T)^- + 2\mathbf{W}_{it}^h(\mathbf{G}_i\mathbf{S}_{ii}\mathbf{G}_i^T\mathbf{G}_i\mathbf{S}_{ii}^T)^+\end{aligned} \tag{12}$$

where the matrices with positive and negative symbols are defined as $\mathbf{A}^+ = \frac{|\mathbf{A}|+\mathbf{A}}{2}$ and $\mathbf{A}^- = \frac{|\mathbf{A}|-\mathbf{A}}{2}$, respectively. Then we can construct $\mathbf{G}$ as:

$$\mathbf{G} \leftarrow \mathbf{G} \circ diag(\sqrt{\frac{\mathbf{G}_1^{(e)}}{\mathbf{G}_1^{(d)}}}, \sqrt{\frac{\mathbf{G}_2^{(e)}}{\mathbf{G}_2^{(d)}}}, \ldots, \sqrt{\frac{\mathbf{G}_m^{(e)}}{\mathbf{G}_m^{(d)}}}) \tag{13}$$

Optimizing $\mathbf{W}^r$: After updating $\mathbf{S}$ and $\mathbf{G}$, we view them as known and take the partial derivative of Eq. (6) with respect to $\mathbf{W}^r$. In this case, the second, fourth and fifth terms on the right of Eq. (6) are irrelevant to $\mathbf{W}^r$. Then we have:

$$\begin{aligned}\widetilde{\mathcal{L}}(\mathbf{G}, \mathbf{S}, \mathbf{W}^r) = &\sum_{\mathbf{R}_{ij} \in \mathcal{R}} \mathbf{W}^r_{ij} \left\|\mathbf{R}_{ij} - \mathbf{G}_i \mathbf{S}_{ij} \mathbf{G}_j^T\right\|_F^2 \\ &+ \alpha \left\|vec(\mathbf{W}^r)\right\|_F^2 \\ s.t. \quad &\mathbf{W}^r_{ij} \geq 0, \sum vec(\mathbf{W}^r) = 1\end{aligned} \tag{14}$$

Let $\mathbf{L}_{ij} = \left\|\mathbf{R}_{ij} - \mathbf{G}_i \mathbf{S}_{ij} \mathbf{G}_j^T\right\|_F^2$ be the reconstruction loss for $\mathbf{R}_{ij}$, then Eq. (14) can be updated as:

$$\begin{aligned}\widetilde{\mathcal{L}}(\mathbf{L}, \mathbf{W}^r, \delta, \gamma) = &vec(\mathbf{W}^r)^T vec(\mathbf{L}) + \alpha vec(\mathbf{W}^r)^T vec(\mathbf{W}^r) \\ &- \sum_{i,j=1}^{m} \delta_{ij} \mathbf{W}^r_{ij} - \gamma(\sum_{i,j=1}^{m} \mathbf{W}^r_{ij} - 1)\end{aligned} \tag{15}$$

Equation (15) is a quadratic optimization problem with respect to $vec(\mathbf{W}^r)$ and the Lagrangian multipliers (δ and γ) are the two constraints of $\mathbf{W}^r$.

Base on the KKT conditions, the optional $\mathbf{W}^r$ should satisfy the following four conditions:

(i) Stationary condition: $\frac{\partial \widetilde{\mathcal{L}}}{\partial \mathbf{W}^r} = \mathbf{L} + 2\alpha \mathbf{W}^r - \delta - \gamma = 0$
(ii) Feasible condition: $\mathbf{W}^r_{ij} > 0, \sum_{i,j=1}^{m} \mathbf{W}^r_{ij} - 1 = 0$
(iii) Dual feasibility: $\delta_{ij} \geq 0, \forall \mathbf{R}_{ij} \in \mathcal{R}$
(vi) Complementary slackness: $\delta_{ij} \mathbf{W}^r_{ij} = 0, \forall\ \mathbf{R}_{ij} \in \mathcal{R}$

From the stationary condition, $\mathbf{W}^r_{ij}$ can be computed as follows:

$$\mathbf{W}^r_{ij} = \frac{\delta_{ij} + \gamma - \mathbf{L}_{ij}}{2\alpha} \tag{16}$$

We can find that $\mathbf{W}^r_{ij}$ depends on the specification of δ_{ij} and γ, and the specification of δ_{ij} and γ can be analyzed in the following three cases:

(i) If $\gamma > \mathbf{L}_{ij}$, then $\mathbf{W}^r_{ij} > 0$, because of the complementary slackness $\delta_{ij} \mathbf{W}^r_{ij} = 0, \delta_{ij} = 0$ and $\mathbf{W}^r_{ij} = \frac{\gamma - \mathbf{L}_{ij}}{2\alpha}$
(ii) If $\gamma = \mathbf{L}_{ij}$, because of $\delta_{ij} \mathbf{W}^r_{ij} = 0$ and $\mathbf{W}^r_{ij} = \frac{\delta_{ij}}{2\alpha}$, then $\delta_{ij} = 0$ and $\mathbf{W}^r_{ij} = 0$
(iii) If $\gamma < \mathbf{L}_{ij}$, since $\mathbf{W}^r_{ij} \geq 0$, it requires $\delta_{ij} > 0$; because $\delta_{ij} \mathbf{W}^r_{ij} = 0$, then $\mathbf{W}^r_{ij} = 0$

From the above analysis, we can set $\mathbf{W}^r_{ij}$ as:

$$\mathbf{W}^r_{ij} = \begin{cases} \frac{\gamma - \mathbf{L}_{ij}}{2\alpha} & \text{if } \gamma > \mathbf{L}_{ij} \text{ and } \mathbf{R}_{ij} \in \mathcal{R} \\ 0 & \text{if } \gamma \leq \mathbf{L}_{ij} \text{ or } \mathbf{R}_{ij} \notin \mathcal{R} \end{cases}, \tag{17}$$

Let $\mathbf{v}_L \in \mathbb{R}^{|\mathcal{R}|}$ store the entries of vector $vec(\mathbf{L})$ in ascending order with entries corresponding to $\mathbf{R}_{ij} \notin \mathcal{R}$ removed. Accordingly, $\mathbf{v}^r \in \mathbb{R}^{|\mathcal{R}|}$ stores the corresponding entries of $vec(\mathbf{W}^r)$. For a not too big predefined α, there exists $p \in \{1, 2, \ldots, |\mathcal{R}|\}$ with $\mathbf{v}_L(p) < \gamma$ and $\mathbf{v}_L(p+1) \geq \gamma$, satisfying $\sum \mathbf{v}_L = \sum_{\mathbf{v}_L(p)<\gamma} \frac{\gamma - \mathbf{v}_L(p)}{2\alpha} = 1$. Then $\mathbf{v}^r(p')$ has the following explicit solution:

$$\mathbf{v}^r(p') = \begin{cases} \frac{\gamma - \mathbf{v}_L(p')}{2\alpha} & \text{if } p' \leq p \\ 0 & \text{if } p' > p \end{cases}, \tag{18}$$

From $\sum_{p'=1}^{|\mathcal{R}|} \mathbf{v}^r(p') = \sum_{p'=1}^{p} \frac{\gamma - \mathbf{v}_L(p')}{2\alpha} = 1$, we can get the value for γ as:

$$\gamma = \frac{2\alpha + \sum_{p'=1}^{p} \mathbf{v}_L(p')}{p} \tag{19}$$

To search the optimal p, we initialize $p = |\mathcal{R}|$ and decrease it step by step. In each step, we repeatedly refer to Eqs. (18–19) and stop the search once a feasible p is obtained. From Eq. (19), we can observe that for a nonnegative γ, at least one inter-relational data matrix can be selected.

Optimizing $\mathbf{W}^h$: When $\mathbf{G}$, $\mathbf{S}$ and $\mathbf{W}^r$ are fixed, the first, the third and fifth terms on the right of Eq. 3 are irrelevant to $\mathbf{W}^h$, and can be ignored. Then we can follow the similar procedure as that of $\mathbf{W}^r$ to obtain the explicit solution of $\mathbf{W}^h$:

$$\mathbf{W}^h_{it} = \begin{cases} \frac{\mu - \mathbf{O}_i^{(t)}}{2\beta} & \text{if } \mu > \mathbf{O}_i^{(t)} \text{ and } t \leq \max_i t_i \\ 0 & \text{if } \mu \leq \mathbf{O}_i^{(t)} \text{ and } t > \max_i t_i \end{cases}, \tag{20}$$

where $\mathbf{O}_i^{(t)} = \left\| \mathbf{R}_{ii}^{(t)} - \mathbf{G}_i \mathbf{S}_{ii} \mathbf{G}_i^T \right\|_F^2$, $\mu = \frac{2\beta + \sum_{h'=1}^{h} \mathbf{v}_O(h')}{h}$, $\mathbf{v}_O$ stores the entries of vector $vec(\mathbf{O})$ in ascending order with entries corresponding to $\{\mathbf{R}_{ii}^{(t)}\}_{t=1}^{t_i}$ $(i = 1, 2, \cdots, m)$, and h can also be sought in the similar way as p in Eq. (18). We can see from Eq. (20) that if $\mathbf{O}_i^{(t)}$ is larger, $\mathbf{W}^h_{it}$ will be smaller. Once $\mathbf{G}_i \mathbf{S}_{ii} \mathbf{G}_i^T$ is a fixed appropriation, a sparser (or denser) $\mathbf{R}_{ii}^{(t)}$ causes a larger reconstruction loss ($\mathbf{O}_i^{(t)}$). As a result, the explicit solution of $\mathbf{W}^h$ can also avoid the preference toward the 'sparse' data matrices.

3 Experiments

3.1 Experimental Setup

To investigate the effectiveness of SelMFDF, we apply it for inter-relation and intra-relation prediction tasks. The inter-relation prediction tasks include lncRNA-miRNA associations and Gene Ontology (GO) annotations of genes, where the target relational matrix is a binary matrix, representing associations between lncRNAs and miRNAs or between genes and GO terms (labels). The intra-relation prediction task is to predict protein-protein interactions by reconstructing the target adjacent matrix of proteins. We collect five object types: lncRNA, genes, miRNA, diseases and Gene Ontology, and adopt eight inter-relational data sources and twelve intra-relational data sources between these objects for experiments. The details of these sources are provided in Table 1.

Table 1. Details on the collected inter-relations and intra-relations from different data sources

Datasets	Size	#Associations		Sources
LncRNA-Gene	240 × 15527	6186	$\mathbf{R}_{12}$	http://www.lncrna2target.org/
LncRNA-miRNA	240 × 495	1002	$\mathbf{R}_{13}$	http://starbase.sysu.edu.cn/mirLncRNA.php/
LncRNA-Disease	240 × 412	2697	$\mathbf{R}_{14}$	http://www.cuilab.cn/lncrnadisease/
LncRNA-GO	240 × 6428	3094	$\mathbf{R}_{15}$	ftp://ftp.ncbi.nih.gov/gene/GeneRIF/
Gene-Disease	15527 × 412	115317	$\mathbf{R}_{24}$	http://www.disgenet.org/
Gene-GO	15527 × 6428	1191503	$\mathbf{R}_{25}$	http://geneontology.org/
miRNA-Gene	495 × 15527	135852	$\mathbf{R}_{32}$	http://mirtarbase.mbc.nctu.edu.tw/
miRNA-Disease	495 × 412	13562	$\mathbf{R}_{34}$	http://www.cuilab.cn/hmdd/
Gene-Gene	2719 × 2719	4551	$\mathbf{R}_{22}^{(1)}$	http://dip.doe-mbi.ucla.edu/dip/Main.cgi
	7898 × 7898	32097	$\mathbf{R}_{22}^{(2)}$	http://hprd.org/index_html
	13106 × 13106	283306	$\mathbf{R}_{22}^{(3)}$	http://ophid.utoronto.ca/ophidv2.204/index.jsp
	11778 × 11778	113973	$\mathbf{R}_{22}^{(4)}$	http://www.ebi.ac.uk/intact/
	7898 × 7898	32097	$\mathbf{R}_{22}^{(5)}$	http://mint.bio.uniroma2.it/
	13086 × 13086	223546	$R_{22}^{(6)}$	http://thebiogrid.org/
miRNA-miRNA	239 × 239	57121	$\mathbf{R}_{33}^{(1)}$	https://doi.org/10.1186/1471-2164-8-166
	443 × 443	196249	$\mathbf{R}_{33}^{(2)}$	https://doi.org/10.1093/bioinformatics/btx019
	495 × 495	225645	$\mathbf{R}_{33}^{(3)}$	http://www.cuilab.cn/hmdd/
	495 × 495	202833	$\mathbf{R}_{33}^{(4)}$	http://mirtarbase.mbc.nctu.edu.tw/
	495 × 495	42723	$\mathbf{R}_{33}^{(5)}$	http://starbase.sysu.edu.cn/mirLncRNA.php/
	22 × 22	32	$\mathbf{R}_{33}^{(6)}$	https://doi.org/10.1016/j.gene.2012.09.066

To comparatively study the performance of SelMFDF, we compare it against five matrix factorization based data fusion methods, including NMTF [5], S-NMTF [19], DFMF [26], MFLDA [6] and WMFLDA [24]. The first three comparing methods equally treat inter-relational matrices or intra-relational matrices during the fusion process. MFLDA optimizes weights to different inter-relational ones and WMFLDA further assigns weights to different intra-relational ones. The input parameters of these methods are set as specified by the authors in the code, or optimized in the suggested ranges. We use the area under the receiver operating characteristic curve (AUROC) and the area under the precision recall curve (AUPRC) to quantify the overall performance. We run five fold cross validation for ten independent rounds, and report the average results.

3.2 Results of Inter-relation Prediction Tasks

For this investigation, we randomly divide the original lncRNA-miRNA associations ($\mathbf{R}_{13}$) into five folds for cross validation. Next, we plot the ROC curves of the comparing methods and report their corresponding AUROCs in Fig. 2(a). We can see that SelMFDF always has the highest TPRs (true positive rates) under the same FPRs (false positive rates), and achieves the highest AUROC among these methods. Figure 2(b) plots the PR curves and reports the AUPRCs, we can also observe that SelMFDF consistently outperforms these comparing methods.

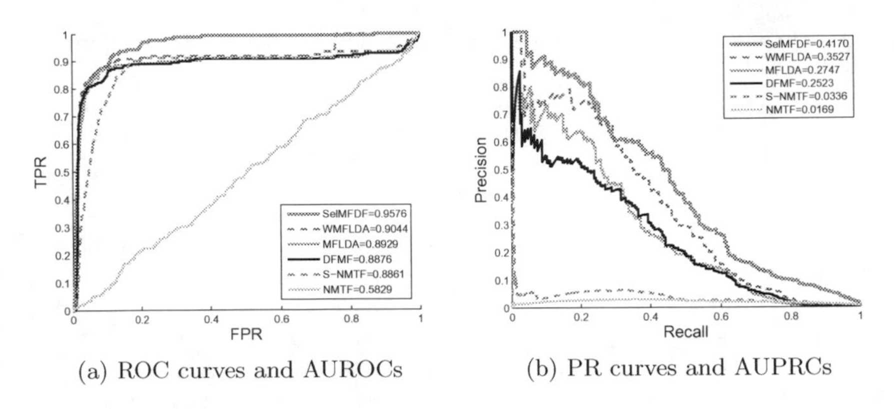

(a) ROC curves and AUROCs (b) PR curves and AUPRCs

Fig. 2. Results of lncRNA-miRNA association prediction. (a) ROC curve and AUROCs. (b) PR curve and AUPRCs.

SelMFDF performs significantly better than WMFLDA and MFLDA, although the latter two also account for the different relevances of multiple relational data matrices. This is because they both use the manifold regularization and approximation loss to determine the relevance of these matrices. As such, they prefer sparse data matrices during the fusion process. However, those sparse matrices may be irrelevant for the target task. SelMFDF does not have such preference, and thus it obtains better results than WMFLDA and MFLDA. The other comparing methods equally treat all the data matrices. As expected, they have much lower AUROC and AUPRC than those of WMFLDA and MFLDA, and say nothing of SelMFDF. In practice, S-NMTF costs the largest runtime costs and memory, since it performs matrix factorization on a big adjacency matrix of all objects. NMTF only fuses inter relational data matrix and thus loses to all the comparing methods.

To investigate whether SelMFDF has the capability to identify relevant data matrices and avoid too sparse ones, we report the weights assigned to different intra-relational matrices. The weights assigned to $\mathbf{R}_{22}^{(i)}, (i = 1, \ldots, 6)$ are (0, 0.0946, 0, 0.0537, 0.0946, 0.1084) and the weights assigned to $\mathbf{R}_{33}^{(i)}, (i = 1, \ldots, 6)$ are (0.1033, 0.1572, 0.1133, 0.1092, 0.1568, 0.0089). We can see SelMFDF assigns a zero weight to the sparsest $\mathbf{R}_{22}^{(1)}$ and $\mathbf{R}_{33}^{(6)}$, and it also assigns a zero weight to the densest $\mathbf{R}_{22}^{(3)}$. The sparsity of these data matrices is included in Table 1 (column '#Associations'). These two assignments are expected from Eq. (20) that SelMFDF can avoid preferring to too sparse and too dense data matrices by crediting lower weights to them. In contrast, these comparing methods either equally integrate them or prefer the sparse ones.

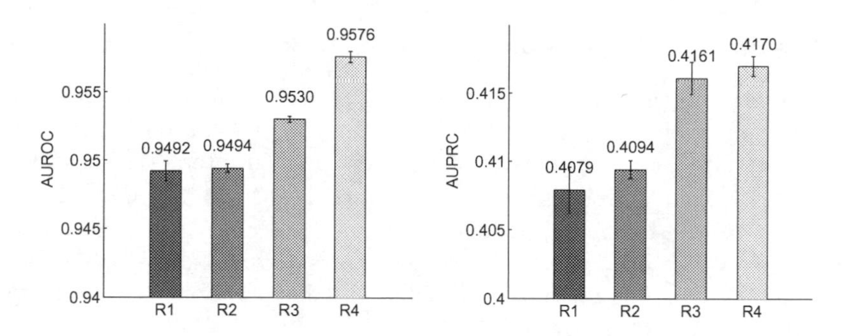

Fig. 3. The AUROCs and AUPRCs of SelMFDF with different collections of intra-relational data matrices. $\mathcal{R}1 = \{\mathbf{R}_{22}^{(1)}, \mathbf{R}_{22}^{(2)}, \mathbf{R}_{22}^{(3)}, \mathbf{R}_{22}^{(4)}, \mathbf{R}_{22}^{(5)}, \mathbf{R}_{22}^{(6)}\}$, $\mathcal{R}2 = \mathcal{R}1 - \mathbf{R}_{22}^{1}$, $\mathcal{R}3 = \mathcal{R}1 - \mathbf{R}_{22}^{3}$, $\mathcal{R}4 = \mathcal{R}1 - \mathbf{R}_{22}^{1} - -\mathbf{R}_{22}^{3}$.

To prove these discarded matrices are indeed irrelevant, we further report the results of SelMFDF by discarding $\mathbf{R}_{22}^{(1)}$ and $\mathbf{R}_{22}^{(3)}$, in Fig. 3. SelMFDF obtains the highest AUROC and AUPRC when $\mathbf{R}_{22}^{(1)}$ and $\mathbf{R}_{22}^{(3)}$ are excluded. We also see that $\mathbf{R}_{22}^{(1)}$ has little contribution. This observation confirms the sparse data matrix has a tiny impact on the target prediction task, since it is too sparse to encode sufficient information for the target task. In addition, SelMFDF has an increased performance when $\mathbf{R}_{22}^{(3)}$ is discarded. That is possible because $\mathbf{R}_{22}^{(3)}$ is a dense matrix with many noisy entries.

We further apply these comparing methods to predict GO annotations of genes (the target relational matrix is $\mathbf{R}_{25}$) in five-fold cross validation. The AUROCs and AUPRCs of these comparing methods are revealed in Fig. 4. We can clearly see that SelMFDF again performs consistently better than the other five approaches and the results give the similar conclusions as those on predicting lncRNA-miRNA associations.

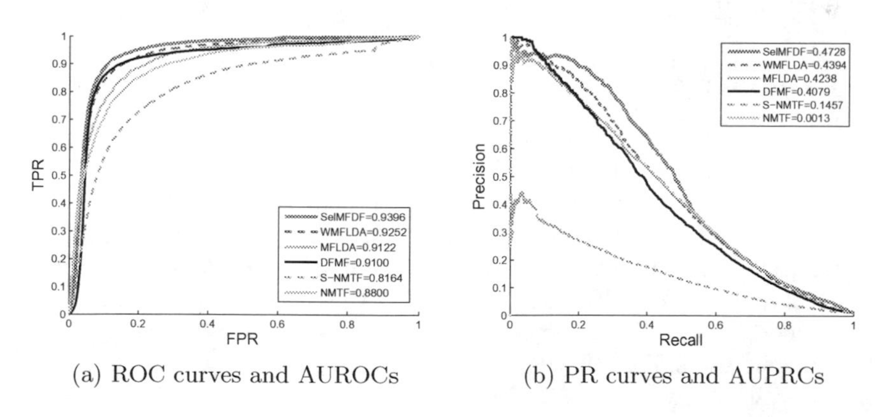

(a) ROC curves and AUROCs (b) PR curves and AUPRCs

Fig. 4. Results of predicting GO annotations of proteins. (a) ROC curve and AUROCs. (b) PR curve and AUPRCs.

3.3 Results of Intra-relation Prediction Task

To further explore the usage of SelDFMF in predicting intra-relations between the same type of objects, we apply SelMFDF to predict protein-protein interactions. For this study, we pick out the protein-protein interaction matrix collected from BioGrid [3] from the collection of intra relational matrices $\{\mathbf{R}_{22}^{(t)}\}_{t=1}^{6}$, and then use $\mathbf{G}_2\mathbf{S}_{22}\mathbf{G}_2^T$ to approximate the target intra-relational data matrix. Next, we select the top K predicted interact-pairs and check them by referring to available interactions in BioGrid [3]. The number of confirmed interactions under different K is reported in Table 2.

Table 2. Number of confirmed PPIs (from BioGrid) predicted by comparing methods.

Methods	Confirmed interactions					
	K = 20	K = 50	K = 100	K = 500	K = 1000	K = 10000
SelMFDF	5	9	17	56	118	879
WMFLDA	2	5	10	31	69	521
MFLDA	2	4	13	26	52	511
DFMF	2	4	10	23	43	482
S-NMTF	2	4	4	9	24	140
NMTF	0	0	0	1	1	9

From Table 2, we can clearly see that SelMFDF always more accurately predicts protein-protein interactions than other methods. In addition, from the remaining 15 interactions (not recorded in the BioGrid) in top 20 predicted by SelMFDF, we further find 6 interactions confirmed by HRPD [11], IntAct [10] and I2D [12] databases. These results indicate SelMFDF can be more reliably applied for the intra-relations prediction.

3.4 Parameter Analysis

The low-rank size k_i is an important parameter for low-rank matrix approximation based solutions. To study the sensitivity of k_i, we fix all $k_i = k$ across these five types of objects for simplicity, and then increase k from 10 to 200. Figure 5 reports the AUROC and AURPC under different input values of k_i in predicting lncRNA-miRNA associations in five-fold cross validation. Both the AUROC value and AUPRC value increase as the increase of k and reach to a highest when $k \approx 20$. Then the AUROC value has a slight decrease and keeps stable after $k > 100$. The AURPC value nearly keeps stable when $k \geq 20$. Given these observations, we adopt $k = 20$ for experiments.

From Eqs. (18) and (20), we can find that once the input value of α or β is specified, the weights $\mathbf{W}^r$ and $\mathbf{W}^h$ assigned to the relational data matrices are also determined. Thus, we further conduct five-fold validation to evaluate the

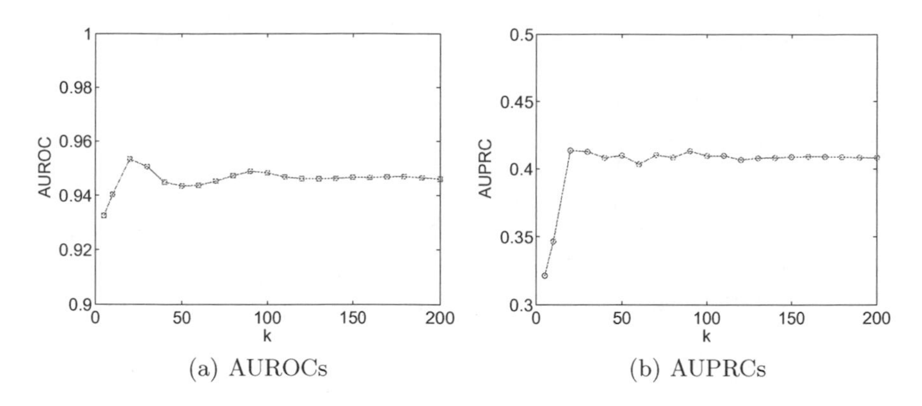

(a) AUROCs (b) AUPRCs

Fig. 5. The AUROC and AUPRC of SelMFDF under different low-rank sizes k.

performance of SelMFDF under different combinations of α and β. We vary α and β in $\{10^{-2}, 10^{-1}, \cdots, 10^{10}\}$ and report the average AUROC and AUPRC in Fig. 6. We can clearly see that when $\alpha = 10^7$ and $\beta = 10^4$, SelMFDF achieves the highest AUPRC. The input value of α significantly affects the performance; the AUROC value and AUPRC value increase as α increase, and reach a plateau when $\alpha > 10^7$. The input value of β also affects the performance; the AUPRC value increases as β get larger, and then it slightly decreases when $\beta > 10^4$.

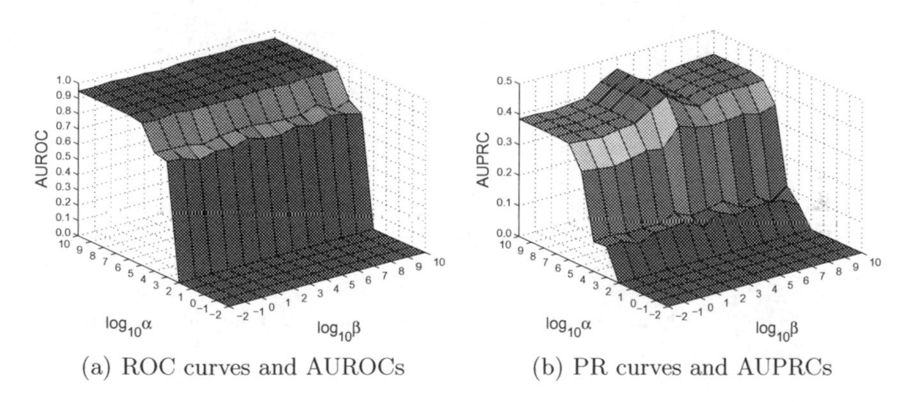

(a) ROC curves and AUROCs (b) PR curves and AUPRCs

Fig. 6. AUROC and AUPRC of SelMFDF under different input values of α and β. (a) AUROCs. (b) AUPRCs.

From these results, we can conclude that SelMFDF can automatically identify irrelevant relational data matrices, and achieve a more prominent performance on predicting the inter- and intra-relations between multiple object types. In addition, it is effective in a wide combination of α and β values, and low-rank sizes.

4 Conclusion

We introduced a selective matrix factorization based solution (SelMFDF) to fuse multi-relational data matrices. Unlike existing matrix factorization based data fusion approaches, SelMFDF can not only selectively integrate multi-relational data matrices, but also avoid preferring to sparse ones and dense ones. Extensive experimental results show that SelMFDF achieves a much better performance than the state-of-the-art solutions in predicting inter-relations and intra-relations between objects. In our future work, we will extend SelMFDF for large scale heterogeneous data fusion. The code and datasets are available at http://mlda.swu.edu.cn/codes.php?name=SelMFDF.

References

1. Belkin, M., Niyogi, P., Sindhwani, V.: Manifold regularization: a geometric framework for learning from labeled and unlabeled examples. JMLR **7**(11), 2399–2434 (2006)
2. Boyd, S., Vandenberghe, L.: Convex Optimization. Cambridge University Press, Cambridge (2004)
3. Chatr-Aryamontri, A., Oughtred, R., et al.: The biogrid interaction database: 2017 update. Nucleic Acids Res. **45**(D1), D369–D379 (2017)
4. Chen, X., Yu, G., Domeniconi, C., Wang, J., Zhang, Z.: Matrix factorization for identifying noisy labels of multi-label instances. In: Geng, X., Kang, B.-H. (eds.) PRICAI 2018. LNCS (LNAI), vol. 11013, pp. 508–517. Springer, Cham (2018). https://doi.org/10.1007/978-3-319-97310-4_58
5. Ding, C., Li, T., Peng, W., Park, H.: Orthogonal nonnegative matrix t-factorizations for clustering. In: KDD, pp. 126–135 (2006)
6. Fu, G., Wang, J., Domeniconi, C., Yu, G.: Matrix factorization-based data fusion for the prediction of lncRNA-disease associations. Bioinformatics **34**(9), 1529–1537 (2018)
7. Gligorijević, V., Pržulj, N.: Methods for biological data integration: perspectives and challenges. J. Roy. Soc. Interface **12**(112), 20150571 (2015)
8. Gönen, M., Alpaydın, E.: Multiple kernel learning algorithms. JMLR **12**(7), 2211–2268 (2011)
9. Karasuyama, M., Mamitsuka, H.: Multiple graph label propagation by sparse integration. TNNLS **24**(12), 1999–2012 (2013)
10. Kerrien, S., et al.: The intact molecular interaction database in 2012. Nucleic Acids Res. **40**(D1), D841–D846 (2011)
11. Keshava Prasad, T., et al.: Human protein reference database–2009 update. Nucleic Acids Research **37**(S1), D767–D772 (2008)
12. Kotlyar, M., Pastrello, C., Sheahan, N., Jurisica, I.: Integrated interactions database: tissue-specific view of the human and model organism interactomes. Nucleic Acids Res. **44**(D1), D536–D541 (2015)
13. Lanckriet, G.R., De Bie, T., Cristianini, N., Jordan, M.I., Noble, W.S.: A statistical framework for genomic data fusion. Bioinformatics **20**(16), 2626–2635 (2004)
14. Lee, D.D., Seung, H.S.: Algorithms for non-negative matrix factorization. In: NIPS, pp. 556–562 (2001)
15. Li, Y., Wu, F.X., Ngom, A.: A review on machine learning principles for multi-view biological data integration. Brief. Bioinf. **19**(2), 325–340 (2016)

16. Meng, D., De La Torre, F.: Robust matrix factorization with unknown noise. In: ICCV, pp. 1337–1344 (2013)
17. Nielsen, T.D., Jensen, F.V.: Bayesian Networks and Decision Graphs. Springer, Heidelberg (2009)
18. Troyanskaya, O.G., Dolinski, K., Owen, A.B., Altman, R.B., Botstein, D.: A Bayesian framework for combining heterogeneous data sources for gene function prediction (in Saccharomyces cerevisiae). PNAS **100**(14), 8348–8353 (2003)
19. Wang, H., Huang, H., Ding, C.: Simultaneous clustering of multi-type relational data via symmetric nonnegative matrix tri-factorization. In: CIKM, pp. 279–284 (2011)
20. Wang, M., Hua, X.S., Hong, R., Tang, J., Qi, G.J., Song, Y.: Unified video annotation via multigraph learning. TCSVT **19**(5), 733–746 (2009)
21. Yu, G., Domeniconi, C., Rangwala, H., Zhang, G., Yu, Z.: Transductive multi-label ensemble classification for protein function prediction. In: KDD, pp. 1077–1085 (2012)
22. Yu, G., Fu, G., Lu, C., Ren, Y., Wang, J.: BRWLDA: bi-random walks for predicting lncRNA-disease associations. Oncotarget **8**(36), 60429 (2017)
23. Yu, G., Rangwala, H., Domeniconi, C., Zhang, G., Zhang, Z.: Predicting protein function using multiple kernels. TCBB **12**(1), 219–233 (2015)
24. Yu, G., Wang, Y., Wang, J., Fu, G., Guo, M., Domeniconi, C.: Weighted matrix factorization based data fusion for predicting lncRNA-disease associations. In: BIBM, pp. 1–6 (2018)
25. Yuan, Y., Savage, R.S., Markowetz, F.: Patient-specific data fusion defines prognostic cancer subtypes. PLoS Comput. Biol. **7**(10), e1002227 (2011)
26. Žitnik, M., Zupan, B.: Data fusion by matrix factorization. TPAMI **37**(1), 41–53 (2015)

Selectivity Estimation on Set Containment Search

Yang Yang[1,2](✉), Wenjie Zhang[1,2], Ying Zhang[3], Xuemin Lin[1,2,4], and Liping Wang[4]

[1] Guangzhou University, Guangzhou, China
[2] UNSW, Sydney, Australia
{yang.yang,zhangw,lxue}@cse.unsw.edu.au
[3] University of Technology Sydney, Sydney, Australia
ying.zhang@uts.edu.au
[4] East China Normal University, Shanghai, China
lipingwang@sei.ecnu.edu.cn

Abstract. In this paper, we study the problem of selectivity estimation on set containment search. Given a query record Q and a record dataset $\mathcal{S}$, we aim to accurately and efficiently estimate the selectivity of set containment search of query Q over $\mathcal{S}$. The problem has many important applications in commercial fields and scientific studies.

To the best of our knowledge, this is the first work to study this important problem. We first extend existing distinct value estimating techniques to solve this problem and develop an inverted list and *G-KMV* sketch based approach *IL-GKMV*. We analyse that the performance of *IL-GKMV* degrades with the increase of vocabulary size. Motivated by limitations of existing techniques and the inherent challenges of the problem, we resort to developing effective and efficient sampling approaches and propose an ordered trie structure based sampling approach named *OT-Sampling*. *OT-Sampling* partitions records based on element frequency and occurrence patterns and is significantly more accurate compared with simple random sampling method and *IL-GKMV*. To further enhance performance, a divide-and-conquer based sampling approach, *DC-Sampling*, is presented with an inclusion/exclusion prefix to explore the pruning opportunities. We theoretically analyse the proposed techniques regarding various accuracy estimators. Our comprehensive experiments on 6 real datasets verify the effectiveness and efficiency of our proposed techniques.

1 Introduction

Set-valued attributes are ubiquitous and play an important role in modeling database systems in many applications such as information retrieval, data cleaning, machine learning and user recommendation. For instance, such set-valued attributes may correspond to the profile of a person, the tags of a post, the domain information of a webpage, and the tokens or q-grams of a document. In

G. Li et al. (Eds.): DASFAA 2019, LNCS 11446, pp. 330–349, 2019.
https://doi.org/10.1007/978-3-030-18576-3_20

the literature, there has been a variety of interests in the computation of set-valued records including set containment search (e.g., [6,18,24,32]), set similarity joins (e.g., [27,29]), and set containment joins (e.g., [10,21,22,30]).

In this paper, we focus on the problem of *selectivity estimation of set containment search*. Considering a query record Q and a collection of records $\mathcal{S}$ where a record consists of an identifier and a set of elements (i.e., terms), a set containment search retrieves records from $\mathcal{S}$ which are contained by Q, i.e., $\{X|X \in \mathcal{S} \wedge Q \supseteq X\}$, where Q *contains* X ($Q \supseteq X$) if all the elements in X are also in Q. Figure 1 shows an example with eight records in a dataset and a query record Q where Q contains X_2, X_3 and X_5. Selectivity (cardinality) of a query refers to the size of the query result size. For instance, the selectivity of Q in Fig. 1 is 3.

Selectivity estimation on set containment search aims at estimating the cardinality of the containment search. As an essential and fundamental tool on massive collections of set-values, the problem has a wide spectrum of applications because it can provide users with fast and useful feedback. As a simple example, when introducing a new product to the market, its characteristics and features could be described as a set of keywords. Assume a preference dataset consists of such characteristics and features desired by users from online survey. Size estimation of the new product descriptions on the preference dataset estimates the total number of users who may be interested in the product and could serve as a prediction of the product's market potential. In another example, companies may post positions in an online job market website where a position description contains a set of required skills. A job-seeker may want to have a basic understanding of the job market by obtaining the total number of active job vacancies that he/she perfectly matches (i.e., the skill set of the job-seeker contains the required skills of the job).

id	record	id	record
X_1	$\{e_1, e_2, e_3, e_4, e_7\}$	X_5	$\{e_1, e_3, e_5, e_7\}$
X_2	$\{e_2, e_3, e_5\}$	X_6	$\{e_2, e_6, e_7, e_8\}$
X_3	$\{e_2, e_5, e_7\}$	X_7	$\{e_4, e_8\}$
X_4	$\{e_1, e_2, e_6, e_{10}\}$	X_8	$\{e_4, e_{10}\}$
Q	$\{e_1, e_2, e_3, e_5, e_7, e_9\}$		

Fig. 1. A record dataset with eight records and a query Q

Challenges. The key challenges of selectivity estimation on set containment search come from the following three aspects. *Firstly*, the dimensionality (i.e., the number of distinct elements) is high. As shown in our empirical studies, the vocabulary size in real-world dataset could reach more than 3 million when the high-order shingles are used. This makes the selectivity estimation techniques which are sensitive to dimensionality inapplicable to our problem. *Secondly*, the number of records in the dataset could be very large. Moreover, the length of query and data record may also be large. To deal with the sheer volume of the data, it is desirable to efficiently and effectively provide approximate solutions. *Thirdly*, the distribution of element frequency may be highly skewed in real applications. It is desirable to devise sophisticated data-dependent techniques to properly handle the skewness of data distribution to boost accuracy.

Even though selectivity estimation has been widely explored, most of the existing techniques cannot be trivially applied to handle the problem studied in this paper. We discuss two categories of techniques which can be extended to support the selectivity estimation problem, range counting estimating (e.g., [12], [5]) and distinct value estimating [9,13].

Given the element universe (vocabulary) $\mathcal{E}$, a record X_i can be regarded as an $|\mathcal{E}|$-dimensional binary vector, where $X_{ij} = 1$ if element e_j appears in X_i ($e_j \in X_i$) and $X_{ij} = 0$ otherwise, for $1 \leq j \leq |\mathcal{E}|$. Let n denote the vocabulary size $|\mathcal{E}|$. Under this context, the dataset $\mathcal{S}$ can be modeled as a set of points in $\{0,1\}^n$ where each record corresponds to an n-dimensional point and the query is a hypercube in $\{0,1\}^n$. Thus, we can rewrite the selectivity estimation problem as the approximate range counting problem in computational geometry. However, the approximate range counting problem suffers from the curse of dimensionality where the computing cost is exponentially dependent on dimensionality n [13, 23]. As the vocabulary size is usually large, applying range counting estimating methods to our problem is not applicable.

Distinct value estimators (e.g., KMV [9], bottom-k, min-hash [13]) can effectively support size estimation for set operations (e.g., union and intersection) and are widely used for problems of size estimation under different context. In Sect. 3.2, we show how to extend the distinct value based estimator to the problem studied in this paper combining with inverted list techniques. We also analyse that the performance of distinct value estimators based approach degrades when the vocabulary size is large due to the inherent *superset* containment semantics of the problem studied in this paper. [28] studies selectivity estimation on streaming spatio-textual data where the textual data is a set of keywords/terms (i.e., elements). However, the query semantic is different as it specifies a *subset containment search* on the textual data, i.e., the keywords (elements) in the query should be *contained by* the keywords from spatial objects. This is different from the *superset* query semantic in our problem which is more challenging to handle using distinct value estimators as discussed in Sect. 3.2.

Contributions. Motivated by the challenges and limitations of existing techniques, in the paper we aim to develop efficient and effective sampling based approaches to tackle the problem. Naively applying random sampling over the dataset ignores the element frequency distribution and results in compromised performance. Intuitively, combinations of high-frequency elements (i.e., frequent patterns) occur among data records with high frequency, and records with similar frequent patterns are more likely to be contained by the same query. Thus, we use the frequent patterns as labels and partition records by these labels to boost efficiency and accuracy. Moreover, assume that the elements are ordered based on frequency, we use ordered trie structure to maintain partitions of the dataset and present *OT-Sampling* method. This ordered trie based approach, though demonstrated to be highly efficient and accurate, does not consider element distribution of the query Q. Inspired by the observation that query Q must include a subset of record X in order to contain X, efficient pruning techniques are developed on the partitions of dataset. We further propose a divide-and-conquer based sampling approach named *DC-Sampling* which only conducts sampling on the qualified partitions surviving from the pruning.

The principle contributions of this paper are summarized as follows.

- This is the first work to systematically study the problem of selectivity estimation on set containment search, which is an essential tool for set-valued attributes analyses in a wide range of applications.
- Two baseline algorithms are devised. The first algorithm is based on random sampling. We also extend distinct value estimator *G-KMV* sketch and propose an inverted list based approach *IL-GKMV*. Insights about the limitations of the two baseline approaches are theoretically analysed and empirically studied.
- We develop two novel sampling based techniques *OT-Sampling* and *DC-Sampling*. *OT-Sampling* integrates ordered trie index structure to group the dataset and achieves higher accuracy by capturing the element frequency and frequent patterns. *DC-Sampling* employs divide-and-conquer philosophy and an exclusion/inclusion-set prefix to further improve performance by exploring pruning opportunities and skipping sampling on pruned partitions of the dataset.
- Comprehensive experiments on a variety of real-life datasets demonstrate superior performance of the proposed techniques compared with baseline algorithms.

2 Preliminary

In this section, we first formally present the problem of containment selectivity estimation, and then give some preliminary knowledge. The notations used throughout this paper are summarized in Table 1.

2.1 Problem Definition

Suppose the element universe is $\mathcal{E} = \{e_1, e_2, ..., e_n\}$. Each record X consists of a set of elements from domain $\mathcal{E}$. Let $\mathcal{S}$ be a collection of records $\{X_1, X_2, ..., X_m\}$. Given two records X and Y, we say X contains Y, denoted as $X \supseteq Y$, if all elements of Y can be found in X. In the paper, we also say X is a superset of Y or Y is a subset of X. Given a query record Q and a dataset $\mathcal{S}$, a set containment search of Q over $\mathcal{S}$ returns all records from $\mathcal{S}$ which are contained by Q, i.e., $\{X | X \in \mathcal{S}, Q \supseteq X\}$. We use t to denote the selectivity (cardinality) of the set containment search. The selectivity of Q measures the number of records returned by the search, namely, $t = |\{X | X \in \mathcal{S}, Q \supseteq X\}|$.

Table 1. The summary of notations

Notation	Definition	Notation	Definition
X, Q, $\mathcal{S}$	A record, a query record, a set of records	P_i, $\mathcal{P}$	A partition of dataset, all partitions
e, $\mathcal{E}$	An element, element domain (vocabulary)	m_i	Size of partition P_i
m	Number of records in $\mathcal{S}$	m'_i	Sampling size in partition P_i
n	Number of distinct elements (vocabulary size)	p_i	Sampling probability in P_i
t ($\hat{t}$)	Containment selectivity (estimation of t)	t_i	Containment selectivity of Q in P_i

Considering the containment relationship between a given query Q and a record $X_i \in \mathcal{S}$ $(1 \leq i \leq m)$, let $\mathbf{n}_i$ be the indicator function such that

$$\mathbf{n}_i := \begin{cases} 1 & \text{if } Q \supseteq X_i, \\ 0 & \text{otherwise} \end{cases} \tag{1}$$

then the *selectivity of the set containment search* on dataset $\mathcal{S}$ with respect to the query Q can also be calculated as $t = \sum_{X_i \in \mathcal{S}} \mathbf{n}_i$.

Problem Statement. In this paper, we investigate the problem of selectivity estimation on set containment search. Given a query record Q and a dataset $\mathcal{S}$, we aim to accurately and efficiently estimate the selectivity of the set containment search of Q on $\mathcal{S}$.

Hereafter, whenever there is no ambiguity, selectivity estimation on set containment search is abbreviated to containment selectivity estimation.

Estimation Measure. In order to evaluate the accuracy of containment selectivity estimation, we apply the *mean square error* (MSE) to measuring the expected difference between an estimator and the true value. The MSE formula is as follows,

$$E(\hat{t} - t)^2 = Var(\hat{t}) + (E(\hat{t}) - t)^2 \tag{2}$$

where $\hat{t}$ is an estimator for t. If $\hat{t}$ is an unbiased estimator, the MSE is simply the variance.

2.2 KMV Synopses

The k minimum values (**KMV**) technique first introduced in [8] is to estimate the number of distinct elements in a large dataset. Given a no-collision hash function h which maps elements to range $[0, 1]$, a *KMV* synopses of a record (set) X, denoted by $\mathcal{L}_X$, is to keep k minimum hash values of X. Then the number of distinct elements $|X|$ can be estimated by $\widehat{|X|} = \frac{k-1}{U_{(k)}}$ where $U_{(k)}$ is

k-th smallest hash value. [9] also methodically analyses the problem of distinct element estimation under set operations. As for union operation, consider two records X and Y with corresponding *KMV* synopses $\mathcal{L}_X$ and $\mathcal{L}_Y$ of size k_X and k_Y, respectively. In [9], $\mathcal{L}_X \oplus \mathcal{L}_Y$ represents the set consisting of the k smallest hash values in $\mathcal{L}_X \cup \mathcal{L}_Y$ where $k = min(k_X, k_Y)$. Then the *KMV* synopses of $X \cup Y$ is $\mathcal{L} = \mathcal{L}_X \oplus \mathcal{L}_Y$. An unbiased estimator for the number of distinct elements in $X \cup Y$, denoted by $D_\cup = |X \cup Y|$, is as follows.

$$\hat{D}_\cup = \frac{k-1}{U_{(k)}} \quad (3)$$

The variance of $\hat{D}_\cup$, as shown in [9], is

$$Var[\hat{D}_\cup] = \frac{D_\cup(D_\cup - k + 1)}{k-2} \quad (4)$$

As shown in [9], Eq. 3 can be modified to compound set operation where $\mathcal{L} = \mathcal{L}_{A_1} \oplus ... \oplus \mathcal{L}_{A_n}$ and $k = min(k_{A_1}, ..., k_{A_n})$.

An improved *KMV* sketch, named *G-KMV*, is proposed to estimate the multi-union size in [28]. *G-KMV* imposes a global threshold and ensures that all hash values smaller than the threshold will be kept. Considering a union operation $\bigcup X_i$ with the sketch as $\mathcal{L} = \mathcal{L}_{X_1} \cup \mathcal{L}_{X_2} ... \cup \mathcal{L}_{X_n}$, the sketch size k for the union is $k = |\mathcal{L}_{X_1} \cup \mathcal{L}_{X_2} ... \cup \mathcal{L}_{X_n}|$. The estimation variance by *G-KMV* method is smaller than that of simple *KMV* method under reasonable assumptions as analysed in [31].

3 Baseline Solutions

In this section, we introduce two baseline solutions following simple random sampling and *G-KMV* sketching techniques, respectively.

3.1 Random Sampling Approach

A simple way to tackle the set containment estimation problem is to adopt the random sampling techniques and conduct set containment search over a sampled dataset $\mathcal{S}'$ which is usually much smaller compared with the original dataset $\mathcal{S}$. After getting the selectivity of Q on sampled dataset $\mathcal{S}'$, we scale it up to get an estimation of containment selectivity regarding $\mathcal{S}$.

Given sampling size budget b in terms of number of records, we describe the random sampling based approach in the following two steps: (1) uniformly at random sample b ($b \ll m$) records $X_1, X_2, ..., X_b$ from $\mathcal{S}$; (2) compare each sampled record X_i ($1 \leq i \leq b$) with the query Q and assign $\mathbf{n}_i$ accordingly. Recall that $\mathbf{n}_i$ is the containment indicator for a record X_i as shown in Eq. 1. Based on this, the containment selectivity estimator ($\hat{t_\mathcal{R}}$) of the random sampling approach is:

$$\hat{t_\mathcal{R}} = \frac{m}{b} \sum_{i=1}^{b} \mathbf{n}_i \quad (5)$$

Note that $\mathbf{n}_i$ is a binary random variable because of the random sampling on records. Next we show that the estimator for baseline solution $\hat{t_\mathcal{R}}$ is an unbiased estimator and then derive its variance. We first compute the probability of the event $\{\mathbf{n}_i = 1\}$. Let t denote the containment selectivity over dataset $\mathcal{S}$ with respect to query Q, i.e., $t = |\{X|X \in \mathcal{S}, Q \supseteq X\}|$, then $Pr[\mathbf{n}_i = 1] = \frac{t}{m}$ where m is total number of records, and thus the expectation of $\mathbf{n}_i$ is $E[\mathbf{n}_i] = \frac{t}{m}$. By the linearity of expectation, we get the expectation of the estimator for baseline solution in Eq. 5 is $E[\hat{t_\mathcal{R}}] = t$, and the variance is

$$Var[\hat{t_\mathcal{R}}] = \frac{t(m-t)}{b}. \tag{6}$$

3.2 IL-GKMV: Inverted List and G-KMV Sketch Based Approach

The random sampling method, which is very efficient, may result in poor accuracy because it ignores the data distribution information, e.g., the distribution of element frequency or record length. In this section, we develop containment selectivity estimation techniques which are data-dependent by utilizing the inverted list and *G-KMV* sketch techniques.

In the first step, we build an inverted index $\mathcal{I}$ on the dataset $\mathcal{S}$ where an element (token) e_i is associated with a list of record identifiers such that the corresponding records contain the element e_i [7]. For instance, in Fig. 1, the inverted list of element e_3 is $\{X_1, X_2, X_5\}$. Let f_i denote the frequency of an element e_i, i.e., the size of the inverted list I_{e_i}; let $Pr[e_i = 1]$ denote the probability that a record in a dataset contains the element e_i, then we have $Pr[e_i = 1] = \frac{f_i}{m}$. Similarly, given a record $X = \{e_1, e_2, ..., e_{|X|}\}$, the probability of X appearing in the dataset is

$$Pr[X = 1] = Pr[\bigcap_{e \in X} \{e = 1\}, \bigcap_{e \in \mathcal{E} \setminus X} \{e = 0\}].$$

Note that record X can be duplicated in the dataset $\mathcal{S}$; given a query Q, the containment selectivity t of Q is calculated as

$$\hat{t} = \sum_{X \in 2^Q} m * Pr[X = 1] \tag{7}$$

where the sum is over all subsets of Q. The above equation enumerates every subset of the query Q to check if it appears in the dataset. In order to compute Eq. 7, we need to compute the joint probability $Pr[X = 1]$ for each subset X of Q. Clearly, the complexity in Eq. 7 is exponentially dependent on the query size $|Q|$ which is not acceptable when $|Q|$ is large. Furthermore, the joint probability computation of $Pr[X = 1]$ is complicated and expensive.

Given the difficulty of directly computing the containment selectivity, we consider the complement version of set containment search. It is easy to see that $X_i \subseteq Q$ if and only if $\mathcal{E} \setminus X_i \supseteq \mathcal{E} \setminus Q$; this implies that, if an element $e \in \mathcal{E} \setminus Q$ and there exists a record X with $e \in X$, then record X is definitely not a subset of the query Q. Thus, if we exclude all the records that contain any element in $\mathcal{E}\backslash Q$, the remaining records in dataset $\mathcal{S}$ are all subsets of Q, namely, satisfying the set containment search. Given that, the containment selectivity t of query Q can be computed as

$$t = m - m * Pr[\bigcup_{e \in \mathcal{E}\backslash Q} e = 1] \tag{8}$$

where $Pr[e = 1]$ denotes the probability that some record in the dataset $\mathcal{S}$ contains the element e. Remind that the event $\{e = 1\}$ corresponds to all the records containing element e in dataset $\mathcal{S}$, i.e., the inverted list $I_e = \{X|e \in X\}$, we can rewrite Eq. 8 as

$$t = m - |\bigcup_{e \in \mathcal{E}\backslash Q} I_e| \tag{9}$$

The key point in the above equation is to calculate the union size of the inverted lists, which has the time complexity of $\sum_{e \in \mathcal{E}\backslash Q} |I_e|$ by merge-join. Since the set of $\mathcal{E}\backslash Q$ and the inverted list I_e could both be very large, directly computing the multi-union operation could result in unaffordable time consumption. Based on this, we adopt approximate methods (e.g., *G-KMV* sketch) to estimate the union size of the inverted lists.

For each element $e \in \mathcal{E}$, $\mathcal{L}_e$ denotes the *G-KMV* synopsis of its inverted list with k $(=|\mathcal{L}_e|)$ smallest hash values. Considering the union of inverted lists in Eq. 9, we have the sketch $\mathcal{L} = \bigcup_{e \in \mathcal{E}\backslash Q} \mathcal{L}_e$ and $k = |\mathcal{L}|$ as introduced in Sect. 2.2, then the size $D_\cup$ of the multi-union set $\bigcup_{e \in \mathcal{E}\backslash Q} I_e$ can be estimated as $\hat{D}_\cup = \frac{k-1}{U_{(k)}}$, where $U_{(k)}$ is the k-th smallest hash value in the synopsis $\mathcal{L}$. Thus the containment selectivity of *G-KMV* sketch based method is computed as $\hat{t}_\mathcal{G} = m - \hat{D}_\cup$. Furthermore, the variance can be calculated as $Var[\hat{t}_\mathcal{G}] = \frac{D_\cup(D_\cup - k + 1)}{k-2}$ by Eq. 4.

Analysis. Given the space budget b in terms of number of records, the sketch size of *IL-GKMV* method is $|\mathcal{L}| \approx b * \bar{d}$ where $\bar{d}$ denotes the average record length. By *G-KMV* sketch, the budget size is proportionally assigned to each inverted list. Apparently, with the very large vocabulary size, the performance significantly deteriorates since each inverted list receives little sampling space. Remark that the *time complexity* for simple random sampling method is $O(b*C)$ where C is the time cost for set comparison. The time cost of *IL-GKMV* is $O(|\mathcal{L}|)$ which is comparable with $O(b * C)$ since $|\mathcal{L}| \approx b * \bar{d}$.

4 Our Approach

As analysed in the previous section, the random sampling approach fails to capture the element frequency distribution. *IL-GKMV* approach, on the other hand, considers data distribution by utilizing the inverted lists (i.e., frequent elements are associated with longer inverted lists) and *G-KMV* sketch (i.e., inverted lists with larger size keep more hashing values) techniques. However, due to the inherent superset query semantics studied in this paper, the number of inverted lists involved in *IL-GKMV* method linearly depends on the vocabulary size which leads to compromised accuracy. In this section, we aim to develop sophisticated sampling approaches which strike a balance between accuracy and efficiency.

4.1 Trie-Structure Based Stratified Sampling Approach

Trie is a widely used tree data structure for storing a set of records (i.e., dataset). Observing that combinations of high-frequency elements (i.e., frequent patterns) occur among records with high frequency and records with similar frequent patterns are more likely to be included by the same query, we adopt the trie structure to partition the dataset using the combinations of high-frequency elements as *labels*. Assume that elements of the vocabulary $\mathcal{E}$ are ordered based on decreasing frequency in the underlying dataset. For example, the most frequent element in Fig. 1 is e_2 as it appears 5 times; e_7 appears 4 times and is ranked 2nd place. Based on this ordering, we refer the top-k high-frequency elements as $\mathcal{E}_k$, and adopt the combination of high-frequency elements within $\mathcal{E}_k$ as label. The choice of k will be discussed later in Sect. 5.

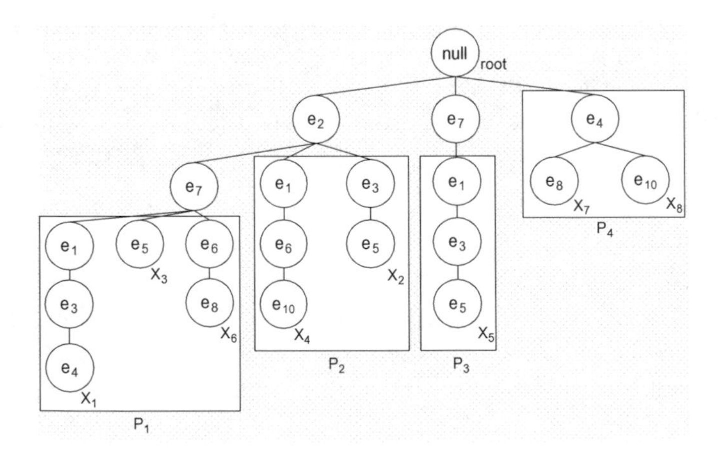

Fig. 2. Trie structure

Figure 2 illustrates an ordered trie T built on dataset in Fig. 1. It is easy to see that each record in the trie is stored in a top-to-down manner with a start node as *null*. Next we give an example about the *labels*.

Example 1. Consider the top-2 elements $\mathcal{E}_2$ in Fig. 2; $\{e_2, e_7\}$ is the label for records X_1, X_3, X_6, $\{e_2\}$ is for records X_4, X_2 and $\{e_7\}$ is for X_5.

It is interesting to notice that the left and upper part of the trie encompasses most of the dataset, since this part is made up of high-frequency elements in the

Algorithm 1. Ordered Trie Structure Based Estimation

Input : Q, a query set; b, sample size budget
$\mathcal{S}$, a dataset; k, top-k high-frequency elements
Output : $\hat{t}$: estimation of containment selectivity under query Q
1 $\mathcal{E}_k \leftarrow$ the top-k high-frequency elements;
2 construct a trie T on dataset $\mathcal{S}$;
3 $L \leftarrow$ all labels in trie T w.r.t $\mathcal{E}_k$;
4 **for each** label $L_i \in L$ **do**
5 $\quad P_i \leftarrow$ records with L_i as the prefix in trie T;
6 $\quad P_i' \leftarrow$ sample m_i' records from P_i based on sample size budget b;
7 $\quad$ conduct containment search regarding Q over sampled records P_i';
8 $\hat{t} \leftarrow$ estimator based on each partition $\mathcal{P} = \{P_1, ..., P_{|\mathcal{P}|}\}$;
9 **return** $\hat{t}$

dataset. Based on this, there is a natural partition strategy generated by the trie T. Namely, from the root node along the high-frequency part (left and upper of trie), each path (label for records) comprises a partition of the dataset since records in the corresponding partition are all made up of this path as prefix. Note that all the remaining records that do not share any high-frequency element are accumulated as a partition by themselves, and we set the label of this partition as ϕ. Here is an example about the partition on trie.

Example 2. In Fig. 2, there are four partitions as $\{X_1, X_3, X_6\}$, $\{X_2, X_4\}$, $\{X_5\}$ and $\{X_7, X_8\}$ with labels $\{e_2, e_7\}$, $\{e_2\}$, $\{e_7\}$ and ϕ, respectively.

Next, we propose an approximate method to compute the containment selectivity based on the partition $\mathcal{P} = \{P_1, ..., P_{|\mathcal{P}|}\}$. Given a query record Q and sample size budget b (number of sampled records), we allocate the sample size budget proportionally to the size $m_i = |P_i|$ of each partition in $\mathcal{P}$ (i.e., stratified sampling). Namely, for partition P_i, there are $m_i' = \frac{|P_i|}{m} * b$ records uniformly at random sampled from P_i. Let P_i' denote these sampled records, i.e., $P_i' = \{X_{i1}, ..., X_{im_i'}\}$, then in each partition, the query Q is compared with each sampled records X_{ij}; let $\mathbf{n}_{ij}$ be the indicator such that

$$\mathbf{n}_{ij} := \begin{cases} 1 & \text{if } X_{ij} \subseteq Q, \\ 0 & \text{otherwise}, \end{cases} \tag{10}$$

then an estimator of the containment selectivity is

$$\hat{t}_{\mathcal{P}} = \sum_{P_i \in \mathcal{P}} \frac{m_i}{m_i'} \sum_{j=1}^{m_i'} \mathbf{n}_{ij} \tag{11}$$

Algorithm 1 illustrates the ordered trie based sampling approach (*OT-Sampling*). Line 1 collects the k most frequent elements $\mathcal{E}_k$ and Line 2 constructs the ordered trie structure based on the dataset $\mathcal{S}$, followed by obtaining the labels according to $\mathcal{E}_k$ (Line 3). Lines 4–7 groups the dataset based on the labels, and conduct the set containment search over each sampled P_i' from individual partitions regarding Q. Line 8 retrieves the final selectivity estimation.

Analysis. Next we show that the estimator $\hat{t}_{\mathcal{P}}$ in Eq. 11 is unbiased, followed by an analysis of the variance $Var[\hat{t}_{\mathcal{P}}]$. Recall that the containment selectivity is $t = |\{X|X \subseteq Q \ and \ X \in \mathcal{S}\}|$; for each partition P_i, let t_i be the size of subsets of Q in partition P_i, i.e., $t_i = |\{X|X \subseteq Q \ and \ X \in P_i\}|$, and $t = \sum_{P_i \in \mathcal{P}} t_i$, then we have $Pr[\mathbf{n}_{ij} = 1] = \frac{t_i}{m_i}$ which means that the probability of a sampled record X_{ij} in partition P_i being the subset of Q is $\frac{t_i}{m_i}$; the expectation of $\mathbf{n}_{ij}$ is $E[\mathbf{n}_{ij}] = \frac{t_i}{m_i}$ and variance is $Var[\mathbf{n}_{ij}] = \frac{t_i(m_i - t_i)}{m_i^2}$. Let $\hat{t}_i = \frac{m_i}{m_i'} \sum_{j=1}^{m_i'} \mathbf{n}_{ij}$, then $E[\hat{t}_i] = t_i$ and $Var[\hat{t}_i] = \frac{t_i(m_i - t_i)}{m_i'}$ by linearity of expectation, thus the expectation of Eq. 11 is

$$E[\hat{t}_{\mathcal{P}}] = \sum_{P_i \in \mathcal{P}} E[\hat{t}_i] = t$$

which proves that $\hat{t}_{\mathcal{P}}$ is an unbiased estimator of containment selectivity. Similarly, the variance of $\hat{t}_{\mathcal{P}}$ is

$$Var[\hat{t}_{\mathcal{P}}] = \sum_{P_i \in \mathcal{P}} Var[\hat{t}_i] = \sum_{P_i \in \mathcal{P}} \frac{t_i(m_i - t_i)}{m_i'} \tag{12}$$

Compare with Random Sampling (RS) Approach. Comparing the variance of *OT-Sampling* in Eq. 12 with that of *RS-Sampling* in Eq. 6, we show that $Var[\hat{t}_{\mathcal{P}}] \leq Var[\hat{t}_B]$ as follows. Let p_i denote the sampling probability in partition P_i, and there is $p_i = \frac{m_i'}{m_i} = \frac{b}{m}$ by the stratified sampling strategy. Suppose that the number of partitions is $q = |\mathcal{P}|$, then we have $Var[\hat{t}_{\mathcal{P}}] - Var[\hat{t}_B] = -\sum_{(i,j) \in \binom{q}{2}} \frac{\prod_{k=1}^{q} m_k}{m_i m_j} (t_i m_j - t_j m_i)^2 \leq 0$.

Time Complexity. The time complexity of the *OT-Sampling* method is $O(b * C) + O(P)$ where C is the containment check cost and $O(P)$ is the pre-process time on trie partition. As demonstrated in our empirical studies, $O(b * C)$ is the dominating cost and $O(P)$ is negligible since we only consider top-k (small k).

4.2 Divide-and-Conquer Based Sampling Approach

In *OT-Sampling*, the sampling strategy is independent of query workload; that is, we do not distinguish the data information (e.g., labels) of each partition with respect to the query. In this section, we propose a query-oriented sampling approach to improve the estimation accuracy.

Consider the records X's in a dataset as binary vectors with respect to the element universe $\mathcal{E} = \{e_1, ..., e_n\}$, i.e., each record is regarded as a size-n vector with i-th position as 1 if $e_i \in X$ and 0 otherwise; *divide* the element universe $\mathcal{E}$ into two disjoint parts as $\mathcal{E}_1$ and $\mathcal{E}_2$, then each record X can be written as two parts X_1 and X_2 corresponding $\mathcal{E}_1$ and $\mathcal{E}_2$ respectively, and we have

$X = \{X_1; X_2\}$ where X_1 is concatenated with X_2. We give a lemma based on the division.

Lemma 1 ***(Subset Inclusion).*** *Given a query record Q and a record X from the dataset $\mathcal{S}$, Q and X are under the same division strategy described above and let $Q = \{Q_1; Q_2\}$ and $X = \{X_1; X_2\}$. We have $X \subseteq Q$ if and only if $X_1 \subseteq Q_1$ and $X_2 \subseteq Q_2$.*

The proof of the lemma is straightforward. From this lemma, a simple pruning technique can be derived such that if $X_1 \nsubseteq Q_1$ then $X \nsubseteq Q$.

Recall the tire-based partition method; we partition the dataset into several groups by the labels of records, where the label can be regarded as the *representative* for each partition. Before drawing samples from a partition with label X_1, we can calculate if X_1 is a subset of query Q. If not, we can skip sampling from that collection of records with X_1 as a label. In order to specify the grouping of records, we give a definition as follows.

Definition 1 ***((E_1, E_2)-Prefix Collection).*** *Given E_1 and E_2 as the subsets of element universe $\mathcal{E}$, the (E_1, E_2)-prefix collection of records denoted as $\mathcal{S}(E_1, E_2)$ consists of all records X's from dataset $\mathcal{S}$ such that all elements of E_1 are contained in X while no element of E_2 appears in X, that is, $\mathcal{S}(E_1, E_2) = \{X \in \mathcal{S} | E_1 \subseteq X \text{ and } E_2 \cap X = \Phi\}$.*

Note that E_1 and E_2 are respectively named as inclusion element set and exclusion element set.

Example 3. An $(\{e_2\}, \{e_7\})$-prefix collection in Fig. 1 is $\{X_2, X_4\}$.

Recall that in Sect. 3.2 we model the record X as a random variable and give the probability that X appears in dataset $\mathcal{S}$. Similarly, we compute the *generating probability* of the prefix collection $\mathcal{S}(E_1, E_2)$ as follows:

$$Pr[\mathcal{S}(E_1, E_2)] = Pr[\bigcap_{e \in E_1} \{e = 1\}, \bigcap_{e \in E_2} \{e = 0\}]. \tag{13}$$

Next we compute the number of subsets of a given query Q within the prefix collection $\mathcal{S}(E_1, E_2)$, i.e., the containment selectivity in regard to $\mathcal{S}(E_1, E_2)$. Let $\mathbf{n}_X$ denote the indicator function such that

$$\mathbf{n}_X := \begin{cases} 1 & \text{if } Q \supseteq X, \\ 0 & \text{otherwise} \end{cases}$$

then the containment selectivity of Q with respect to $\mathcal{S}(E_1, E_2)$ is

$$t_{\mathcal{S}(E_1, E_2)} = \sum_{X \in \mathcal{S}(E_1, E_2)} \mathbf{n}_X * \frac{Pr[X]}{Pr[\mathcal{S}(E_1, E_2)]} \tag{14}$$

Now we can present the lemma which lay the foundation of the divide-and-conquer algorithm.

Algorithm 2. Divide-And-Conquer Exact Algorithm

Input : $\mathcal{S}$, a collection of records as dataset; Q, a query set
E_1 (E_2), elements included (excluded) in the prefix collection
Output : $\hat{t}$: containment selectivity of query Q within $\mathcal{S}(E_1, E_2)$
1 **procedure** $\mathbf{T}(\mathcal{S}, E_1, E_2, Q)$
2 **if** $E_1 \nsubseteq Q$ **then**
3   **return** 0
4 choose an element $e \notin E_1 \cup E_2$;
5 **return** $Pr[e = 1|\mathcal{S}(E_1, E_2)] * \mathbf{T}(\mathcal{S}, E_1 \cup \{e\}, E_2, Q) + Pr[e = 0|\mathcal{S}(E_1, E_2)] * \mathbf{T}(\mathcal{S}, E_1, E_2 \cup \{e\}, Q)$

Lemma 2. *Considering a prefix collection $\mathcal{S}(E_1, E_2)$ and an element e which does not belong to $E_1 \cup E_2$, the containment selectivity of a given query Q within $\mathcal{S}(E_1, E_2)$ can be calculated as*

$$t_{\mathcal{S}(E_1,E_2)} = Pr[e = 1|\mathcal{S}(E_1, E_2)] * t_{\mathcal{S}(E_1 \cup \{e\}, E_2)} + Pr[e = 0|\mathcal{S}(E_1, E_2)] * t_{\mathcal{S}(E_1, E_2 \cup \{e\})}.$$

The key point in the proof of Lemma 2 is to consider the conditional probability. We omit the detailed proof here due to space limitation.

Based on Lemma 2, we propose the divide-and-conquer algorithm illustrated in Algorithm 2. We can calculate the containment selectivity of Q within dataset $\mathcal{S}$ by invoking procedure $\mathbf{T}(\mathcal{S}, \phi, \phi, Q)$; by Lemma 2, the dataset is partitioned into two groups of records by choose an element $e \in \mathcal{E}$ and we have

$$t_{\mathcal{S}(\phi,\phi)} = Pr[e = 1|\mathcal{S}(\phi, \phi)] * t_{\mathcal{S}(\{e\},\phi)} + Pr[e = 0|\mathcal{S}(\phi, \phi)] * t_{\mathcal{S}(\phi,\{e\})}$$

then compute the containment selectivity in each of the two groups recursively as shown in Line 4–5. When there is $E_1 \nsubseteq Q$, we can prune this collection of records $\mathcal{S}(E_1, E_2)$ by Lemma 1. Obviously, the time complexity of the exact divide-and-conquer algorithm is $O(C * 2^n)$ where n is the size of the element universe $\mathcal{E}$ and C is the cost of set comparison. Recall that the element frequency distribution is usually skew in real dataset, and we can arrange the elements by decreasing frequency order when choosing the element e in Line 4 of Algorithm 2, which can accelerate the computation by pruning more records corresponding to the high-frequency elements.

Approximate Divide-And-Conquer Algorithm. Next we propose an approximate method based on the exact divide-and-conquer algorithm. In Algorithm 2, the dataset $\mathcal{S}$ is recursively partitioned into two collection of records by choosing an element $e \notin E_1 \cup E_2$. In addition, we can order the elements by decreasing element frequency to boost the computation efficiency. However, the complexity is still $O(C * 2^n)$. In this section, we only consider the top-k high-frequency elements $\mathcal{E}_k$, from which the element is selected to partition the dataset. After finishing all the elements in $\mathcal{E}_k$, we end up with 2^k prefix collections of records $\mathcal{S}_i(E_1, E_2)$, $i = 1, 2, ..., 2^k$, which is much smaller than 2^n. Note that (E_1, E_2) can be regarded as the label for each prefix collection.

Recall Lemma 1, all the records X's can be described as the binary vector with $X = \{X_1; X_2\}$ where X_1 corresponds to the top-k high-frequency elements part $\mathcal{E}_k$ and X_2 is the rest part concatenated with X_1. Similarly, when a query record Q arrives, let Q be $Q = \{Q_1; Q_2\}$ following the same manner; then by Lemma 1, we can exclude all the prefix collections $\mathcal{S}(E_1, E_2)$ with $E_1 \nsubseteq Q_1$. For the remaining prefix collections, we sample some records from each group and conduct containment search of Q over sampled records. Let $X = \{X_1; X_2\}$ be a sampled record, it is only required to test if $X_2 \subseteq Q_2$ since $X_1 \subseteq Q_1$. In the following part, we formally demonstrate how to estimate the containment selectivity of Q by the divide-and-conquer method.

Let $\mathbf{I}_i$ denote the indicator function for prefix collection $\mathcal{S}_i(E_1, E_2)$ ($\mathcal{S}_i$ for short) such that $\mathbf{I}_i = 1$ when $E_1 \subseteq Q_1$ otherwise 0. The size of prefix collection $\mathcal{S}_i(E_1, E_2)$ can be computed as $m_i = |\mathcal{S}_i(E_1, E_2)| = m * Pr[\mathcal{S}_i(E_1, E_2)]$ by Eq. 13. Let p_i be the sampling probability in $\mathcal{S}_i$, then the sample size is $m'_i = m_i * p_i$. For any sampled record $X_j = \{X_1; X_2\}$ in this prefix collection $\mathcal{S}_i$, let $\mathbf{n}_{ij}$ be the indicator for which $\mathbf{n}_{ij} = 1$ if $X_2 \subseteq Q_2$ otherwise 0. Then an estimator for the containment selectivity of Q by divide-and-conquer algorithm can be expressed as

$$\hat{t}_{\mathcal{D}} = \sum_{\mathcal{S}_i} \mathbf{I}_i \sum_{j=1}^{m'_i} \frac{\mathbf{n}_{ij}}{p_i} \tag{15}$$

It can be verified that $\hat{t}_{\mathcal{D}}$ is an unbiased estimator and the variance of $\hat{t}_{\mathcal{D}}$ is

$$Var[\hat{t}_{\mathcal{D}}] = \sum_{\mathcal{S}_i} \mathbf{I}_i * \frac{t_i(m_i - t_i)}{p_i m_i} \tag{16}$$

where t_i is the number of records satisfying $X_2 \subseteq Q_2$ in $\mathcal{S}_i$. Let $\mathcal{S}_i$, $i = 1, 2, ..., l$ be all the prefix collections with $E_1 \subseteq Q_1$ for a given query Q, then the variance can be written as $Var[\hat{t}_{\mathcal{D}}] = \sum_{i=1}^{l} \frac{t_i(m_i - t_i)}{p_i m_i}$.

Compare with *OT-Sampling*. Obviously, in *DC-Sampling* method, we avoid allocating the space budget to unqualified partitions compared with *OT-Sampling*. In formal, assume there are q partitions (corresponding to prefix collections) in total with $\{P_1, ..., P_q\}$; after pruning, there remains l partitions, w.l.o.g, $\{P_1, ..., P_l\}$. Then for *DC-Sampling*, the sampling probability is $p_i = \frac{b}{\sum_{i=1}^{l} m_i}$ where $m_i = |P_i|$ and b is space budget, and the sampling probability of *OT-Sampling* is $p'_i = \frac{b}{\sum_{i=1}^{q} m_i}$. Thus we have $Var[\hat{t}_{\mathcal{P}}] - Var[\hat{t}_{\mathcal{D}}] = \sum_{i=1}^{l} (\frac{1}{p'_i m_i} - \frac{1}{p_i m_i}) t_i(m_i - t_i) + \sum_{i=l+1}^{q} \frac{1}{p'_i m_i} t_i(m_i - t_i) \geq 0$ since $p'_i \leq p_i$.

Time Complexity. The time complexity of *DC-Sampling* method is $O(b * \tilde{C}) + O(P)$ where $\tilde{C}$ is the cost for two-record containment check. After pruning the unqualified partitions, we can skip comparing the prefix part of a record with the query by our algorithm, thus the time cost of $\tilde{C}$ is smaller than that of *OT-sampling*, which leads to better efficiency than *DC-Sampling*.

5 Experimental Evaluation

In this section, we evaluate the estimation accuracy and computation efficiency of different strategies on a variety of real-life datasets. All experiments are conducted on PCs with Intel Xeon 2x2.3 GHz CPU and 128 GB RAM running Debian Linux.

5.1 Experimental Setting

Algorithms. Since there exists no previous work for tackling the problem of set containment selectivity estimation, we evaluate the following estimation methods introduced in this paper.

- **RS.** Direct random sampling method in Sect. 3.1.
- **IL-GKMV.** Inverted lists and G-KMV sketch based method in Sect. 3.2.
- **OT-Sampling.** Ordered trie structure based sampling method in Sect. 4.1.
- **DC-Sampling.** The divide-and-conquer based sampling method in Sect. 4.2.

The above algorithms are implemented in C++. In verifying the inclusion relationship between the query and records, we apply the merge-join method. For records with large size, we utilize the prefix-tree structure to boost the computation efficiency.

Datasets. We deploy 6 real-life datasets which are chosen from various domains with different data properties. In Table 2, we illustrate the characteristics of these 6 datasets in details. For each dataset, we show the representations of record and element, the number of records, the average record length, and the number of distinct elements in dataset.

Table 2. Characteristics of datasets

Dataset	Abbreviation	Record	Elements	#Records	AvgLength	#Elements
Bookcrossing [1]	BOOKC	Book	User	340,523	3.38	105,278
Delicious [2]	DELIC	User	Tag	833,081	98.42	4,512,099
Livejournal [3]	LIVEJ	User	Group	3,201,203	35.08	7,489,073
Netflix [10]	NETFLIX	Movie	Rating	480,189	209.25	17,770
Sualize [4]	SUALZ	Picture	Tag	495,402	3.63	82,035
Twitter [19]	TWITTER	Partition	User	371,586	65.96	1,318

Workload. The workload for the selectivity estimation of set containment search is made up of 10000 queries, each of which is uniformly at random selected from the dataset. Note that we exclude the queries with size smaller than 10 in order to evaluate the accuracy properly.

Measurement. In the following part, we use relative error to measure the accuracy. Let t be the exact result and $\hat{t}$ be the estimation one, then the relative error denoted by ϵ is calculated as $\epsilon = \frac{|t-\hat{t}|}{t}$. The sampling size is in terms of the number of records. For *IL-GKMV* approach, the space budget is allocated as discussed at the end of Sect. 3.

Tuning k. In order to evaluate the impact of the high-frequency elements in *OT-Sampling* and *DC-Sampling*, we first tune the number of highest-frequency elements, i.e., top-k. By experimental study, we set the k value as 12 which can well balance the trade-off between accuracy and efficiency.

5.2 Overall Performance

Figure 3(a) compares the estimation accuracy and time cost of the four algorithms on 6 datasets. The sample size is set as 1000 in terms of number of records; for trie-structure based approach and divide-and-conquer algorithm, the k-value is 12 as mentioned above. Overall, we can see that the divide-and-conquer (*DC-Sampling*) algorithm achieves the best performance in accuracy on all datasets, which can reduce the relative error of the random sampling (*RS*) method by around 60% and cut the relative error of *IL-GKMV* method by more than 80%. Also, the ordered-trie structure-based approach (*OT-Sampling*) can diminish the relative error of *RS* by around 40% for most datasets and narrow the relative error of *IL-GKMV* by about 70%. Moreover, divide-and-conquer (*DC-Sampling*) algorithm outperforms the ordered tire structure based approach (*OT-Sampling*) by decreasing the relative error about half.

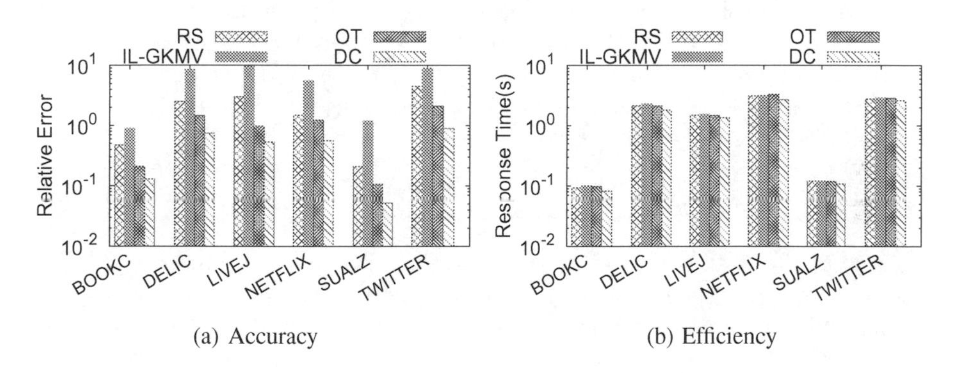

(a) Accuracy (b) Efficiency

Fig. 3. Overall performance

Figure 3(b) reports the query response time on 6 datasets with 10000 queries, where *DC-Sampling* method consumes less time than the other three because of the pruning techniques. It is remarkable that for each dataset, the time costs of the four algorithms are comparable since we keep the same sample size in every algorithm. Meanwhile, the response time varies among different datasets because of the diverse average record lengths, and datasets with larger average length, e.g., NETFLIX with AvgLength 209.25, consume more query time.

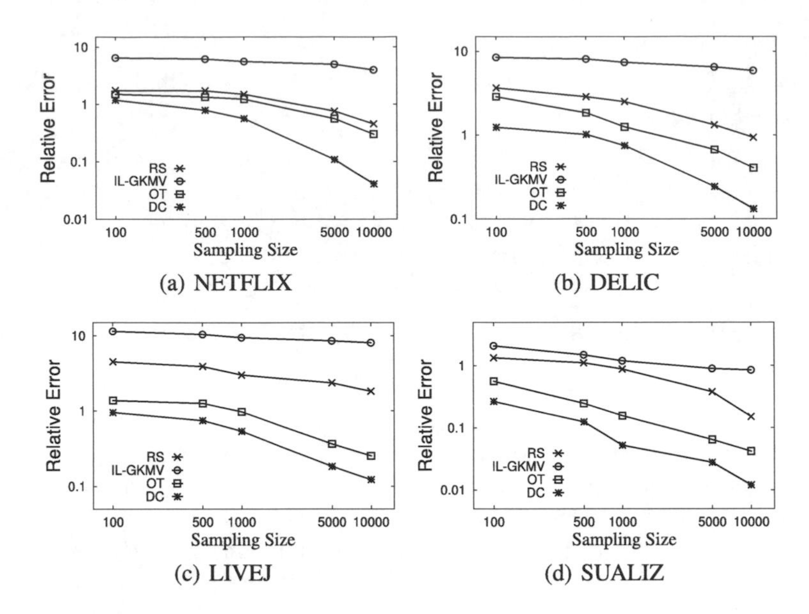

Fig. 4. Accuracy vs space

5.3 Estimation Accuracy Evaluation

In this section, we assess the effectiveness of the four methods in terms of relative error. We consider the effect of space budget on the estimation accuracy by changing the sampling size. Figure 4 illustrates superior accuracy achievement of *DC-Sampling* against the other three by varying the space budget. As anticipated, the accuracy performance of all algorithms is ameliorated when more sampling size is provided.

5.4 Computation Efficiency Evaluation

In the last part of experiment, we evaluate the efficiency of the four algorithms in terms of query response time with 10,000 queries. Figure 5 demonstrates the response time of four algorithms with different space budget. Obviously, the query response time increases as the sampling size grows. The *DC-Sampling* method outperforms the other three algorithms because of the pruning techniques.

6 Related Work

To the best our knowledge, there is no existing work on selectivity estimation of set containment search. In this section, we review two important directions closely related to the problem studied in this paper.

Searching Set-Valued Data. The study of set-valued data has attracted great attention from research communities and industrial organizations due to an ever

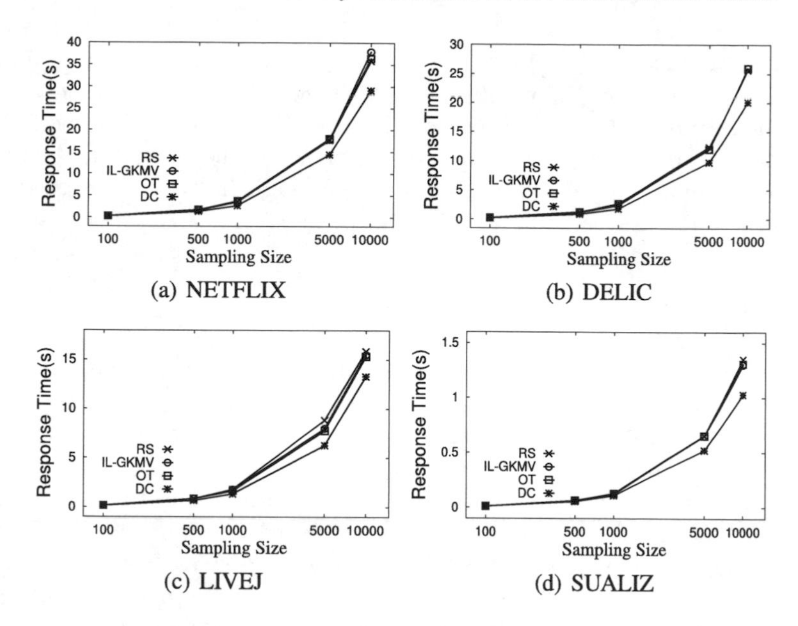

Fig. 5. Efficiency vs space

increasing prevalence of set-valued data in a wide range of applications. The research in this area focuses on set containment search [15,16,24], set similarity and set containment joins [17,19,20,26]. In one of the representative work on set containment search [24], Terrovitis *et. al* introduce a OIF index combined the inverted index with B-tree to tackle three kinds of set-containment queries: subset queries, equality queries and superset queries. In a recent work [30], Yang *et. al* propose a TT-join method for the set containment join problem, which is based on prefix tree structure and utilize the element frequency information; they also present a detailed summary of the existing set-containment join methods. The containment queries can also be modeled as range searching problem in computational geometry [5]; nevertheless, the performance is exponentially dependent on dimension n which is unsuitable in practice for our problem.

Selectivity Estimation. The problem of selectivity estimation has been studied for a large variety of queries and over a diverse range of data types such as range queries (e.g., [13]), boolean queries (e.g., [11]), relational joins (e.g., [25]), spatial join (e.g., [14]), and set intersection (e.g., [13]). Nevertheless, many of the techniques developed above are sensitive to the dimension of data and not applicable to the problem studied in this paper. Moreover, the superset containment semantics brings in extra challenges in adopting existing techniques. Although the set containment search query can be naturally modeled as range counting problem as discussed in Sect. 1, existing range counting techniques are exponentially dependent on the dimensionality (i.e., number of distinct elements in our problem) and not applicable to solving the containment selectivity estimation problem in our problem [13,23]. Distinct value estimators (e.g., *KMV*

[9], bottom-k, min-hash [13]) are adopted in [28] to solve subset containment search (i.e., query record is a subset of data record). We also extend the distinct value estimator *KMV* and develop the *IL-GKMV* approach in Sect. 3 and demonstrate theoretically and through extensive experiments that distinct value estimators cannot efficiently and accurately support the superset containment semantics studied in this paper.

7 Conclusion

The prevalence of set-valued data generates a wide variety of applications that call for sophisticated processing techniques. In this paper, we investigate the problem of selectivity estimation on set containment search and develop novel and efficient sampling based techniques, *OT-Sampling* and *DC-Sampling*, to address the inherent challenges of set containment search and the limitations of existing techniques. Simple random sampling techniques and a *G-KMV* sketch based estimating approach *IL-GKMV* are also devised as baseline solutions. We theoretically analyse the accuracy of the proposed techniques by means of expectation and variance. Our comprehensive experiments on 6 real-life datasets empirically verify the effectiveness and efficiency of the sampling based approaches.

References

1. http://www.informatik.uni-freiburg.de/~cziegler/BX/
2. http://dai-labor.de/IRML/datasets
3. http://socialnetworks.mpi-sws.org/data-imc2007.html
4. http://vi.sualize.us/
5. Agarwal, P.K.: Range searching. Technical report, Duke University Durham NC Dept of Computer Science (1996)
6. Agrawal, P., Arasu, A., Kaushik, R.: On indexing error-tolerant set containment. In: SIGMOD, pp. 927–938 (2010)
7. Baeza-Yates, R., Ribeiro-Neto, B., et al.: Modern Information Retrieval, vol. 463. ACM Press, New York (1999)
8. Bar-Yossef, Z., Jayram, T.S., Kumar, R., Sivakumar, D., Trevisan, L.: Counting distinct elements in a data stream. In: Rolim, J.D.P., Vadhan, S. (eds.) RANDOM 2002. LNCS, vol. 2483, pp. 1–10. Springer, Heidelberg (2002). https://doi.org/10.1007/3-540-45726-7_1
9. Beyer, K., Haas, P.J., Reinwald, B., Sismanis, Y., Gemulla, R.: On synopses for distinct-value estimation under multiset operations. In: SIGMOD, pp. 199–210 (2007)
10. Bouros, P., Mamoulis, N., Ge, S., Terrovitis, M.: Set containment join revisited. Knowl. Inf. Syst. 1–28 (2015)
11. Chen, Z., Korn, F., Koudas, N., Muthukrishnan, S.: Selectivity estimation for boolean queries. In: PODS, pp. 216–225 (2000)
12. Cohen, E., Cormode, G., Duffield, N.G.: Structure-aware sampling on data streams. In: SIGMETRICS, pp. 197–208 (2011)

13. Cohen, E., Cormode, G., Duffield, N.G.: Is min-wise hashing optimal for summarizing set intersction? In: PODS, pp. 109–120 (2014)
14. Das, A., Gehrke, J., Riedewald, M.: Approximation techniques for spatial data. In: SIGMOD, pp. 695–706 (2004)
15. Goldman, R., Widom, J.: WSQ/DSQ: a practical approach for combined querying of databases and the web. In: ACM SIGMOD Record, vol. 29, pp. 285–296. ACM (2000)
16. Helmer, S., Moerkotte, G.: A performance study of four index structures for set-valued attributes of low cardinality. VLDB J. **12**(3), 244–261 (2003)
17. Jampani, R., Pudi, V.: Using prefix-trees for efficiently computing set joins. In: Zhou, L., Ooi, B.C., Meng, X. (eds.) DASFAA 2005. LNCS, vol. 3453, pp. 761–772. Springer, Heidelberg (2005). https://doi.org/10.1007/11408079_69
18. Li, C., Lu, J., Lu, Y.: Efficient merging and filtering algorithms for approximate string searches. In: ICDE, pp. 257–266 (2008)
19. Luo, Y., Fletcher, G.H., Hidders, J., De Bra, P.: Efficient and scalable trie-based algorithms for computing set containment relations. In: ICDE, pp. 303–314. IEEE (2015)
20. Mamoulis, N.: Efficient processing of joins on set-valued attributes. In: SIGMOD, pp. 157–168. ACM (2003)
21. Melnik, S., Garcia Molina, H.: Adaptive algorithms for set containment joins. TODS **28**(1), 56–99 (2003)
22. Ramasamy, K., Patel, J.M., Naughton, J.F., Kaushik, R.: Set containment joins: the good, the bad and the ugly. In: VLDB, pp. 351–362 (2000)
23. Suri, S., Toth, C., Zhou, Y.: Range counting over multidimensional data streams. Discrete Comput. Geometry **36**(4), 633–655 (2006)
24. Terrovitis, M., Bouros, P., Vassiliadis, P., Sellis, T., Mamoulis, N.: Efficient answering of set containment queries for skewed item distributions. In: Proceedings of the 14th International Conference on Extending Database Technology, pp. 225–236. ACM (2011)
25. Tzoumas, K., Deshpande, A., Jensen, C.S.: Efficiently adapting graphical models for selectivity estimation. PVLDB **22**(1), 3–27 (2013)
26. Vernica, R., Carey, M.J., Li, C.: Efficient parallel set-similarity joins using MapReduce. In: SIGMOD, pp. 495–506. ACM (2010)
27. Wang, J., Li, G., Feng, J.: Can we beat the prefix filtering?: an adaptive framework for similarity join and search. In: SIGMOD, pp. 85–96 (2012)
28. Wang, X., Zhang, Y., Zhang, W., Lin, X., Wang, W.: Selectivity estimation on streaming spatio-textual data using local correlations. PVLDB **8**(2), 101–112 (2014)
29. Xiao, C., Wang, W., Lin, X., Yu, J.X.: Efficient similarity joins for near duplicate detection. In: WWW, pp. 131–140 (2008)
30. Yang, J., Zhang, W., Yang, S., Zhang, Y. Lin, X.: TT-join: efficient set containment join. In: ICDE, pp. 509–520 (2017)
31. Yang, Y., Zhang, Y., Zhang, W., Huang, Z.: GB-KMV: an augmented kmv sketch for approximate containment similarity search. arXiv preprint arXiv:1809.00458 (2018)
32. Zhu, E., Nargesian, F., Pu, K.Q., Miller, R.J.: LSH ensemble: internet scale domain search. In: VLDB, pp. 1185–1196 (2016)

Typicality-Based Across-Time Mapping of Entity Sets in Document Archives

Yijun Duan[1(✉)], Adam Jatowt[1], Sourav S. Bhowmick[2], and Masatoshi Yoshikawa[1]

[1] Graduate School of Informatics, Kyoto University, Kyoto, Japan
{yijun,adam}@dl.kuis.kyoto-u.ac.jp, yoshikawa@i.kyoto-u.ac.jp
[2] Nanyang Technological University, Singapore, Singapore
assourav@ntu.edu.sg

Abstract. News archives constitute a rich source of knowledge about the past societies. In order to effectively utilize such large and diverse accounts of the past, novel approaches need to be proposed. One of them is comparison of the past and present entities which can lay grounds for better comprehending the past and the present, as well as can support forecasting techniques. In this paper, we propose a novel research task of *automatically generating across-time comparable entity pairs given two sets of entities*, as well as we introduce an effective method to solve this task. The proposed model first applies the idea of typicality analysis to measure the representativeness of each entity. Then, it learns an orthogonal transformation between temporally distant entity collections. Finally, it generates a set of typical comparables based on a concise integer linear programming framework. We experimentally demonstrate the effectiveness of our method on the New York Times corpora through both qualitative and quantitative tests.

Keywords: Comparable entity mining · Temporal embeddings alignment · Integer linear programming

1 Introduction

Comparison is an effective strategy extensively adopted in practice to discover commonalities and differences between two or more objects. Users can benefit from comparison for a myriad of needs such as understanding complex concepts, gaining insights about similar objects/situations, making better decisions and so on. Entity comparison has been studied in the past (e.g., [31]). For an input entity pair the task was to find their differences and similarities. Other work focused on finding a comparable entity for a given input entity [15,17].

Sometimes, however, what users want is to compare different entity sets across time. For example, a journalist or historian may be interested in the comparison of contemporary politicians with ones of 30 years ago. Another example could be a comparison of electronic gadgets used in 1980s–1990s with those

G. Li et al. (Eds.): DASFAA 2019, LNCS 11446, pp. 350–366, 2019.
https://doi.org/10.1007/978-3-030-18576-3_21

used at present, i.e., 2000s–2010s, that a student or scholar in the history of science/technology might wish to conduct. This kind of temporal analogy determination could be beneficial for general understanding of the relation of and the similarity between the past and present. Furthermore, it could then lead to not only improved comprehension of the past but could also complement and support our forecasting abilities. Manual comparison of entity sets across time is however non-trivial due to the following reasons: (1) In the current fast-paced world, people tend to possess limited knowledge about things from the past. In other words, it is difficult for average users to find temporal comparable entities (e.g. to know that music device like *Walkman* was playing a similar role 30 years ago as *iPod* does nowadays.); (2) such comparison involves entire entity sets, which is not an easy task given their diversity and complexity, thus it would require much cognitive effort. Note that entity sets are quite common in the real world and can be massive. For example, Wikipedia, which is considered to be the most comprehensive encyclopedia, contains over 1.13 million categories grouping numerous entities and concepts in many diverse ways [1].

A natural method of comparison is to find pairs of corresponding entities (e.g., finding a set of representative pairs of similar politicians or corresponding technological devices from among temporally distant time periods). Indeed, learning from examples is regarded an effective strategy extensively adopted in daily life. Good examples are often easier to be understood for learning concepts or categories of entities than high-level feature descriptions. Therefore, given two collections of entities from different historical times (e.g. the lists of contemporary politicians and the one of politicians active 30 years ago), it would be useful to automatically find a diverse set of corresponding entity pairs (e.g., U.S. Presidents: *Donald Trump* and *Ronald Reagan*, Russian Presidents: *Vladimir Putin* and *Mikhail Gorbachev*) as such pairs do not only provide contrasting information, but can be also understandable and intuitive.

Users can benefit from our study with respect to many needs. First of all, our study paves the way for the automatic discovery of mapping relationships between exemplars, which gives rise to the entity analogy solving task. Solving analogy tasks and generating analogical examples can be then enhanced using our method. Besides, finding typical comparables is a natural prerequisite step of discovering the commonalities and differences.

The problem of automatically detecting comparable entity pairs is however non-trivial due to the following reasons: (1) To measure across-time entity correspondence is a difficult task. The general context of the two compared entity collections which originate from different time periods may be fairly different. Intuitively, the correspondence of entities in different contexts cannot be computed properly without a solid understanding of the connection (analogies) between their contexts. Moreover, it is difficult to collect training data for learning such connections. (2) Naturally, only typical entities should be chosen for comparison. This is because typical instances are usually associated with more representative features and thus are less likely to cause misunderstanding. For instance, to compare mammals with another animal class, typical examples of mammals such as lions should be preferred rather than atypical instances like platypuses (which lay eggs instead of giving birth). (3) The input sets of entities can be very diverse and may cover multiple latent subgroups. Thus instead of a single output entity pair, a set of pairs that represent latent subgroups within the input entity sets should be returned. Given the limitation on the size of output, selecting a subset of optimal pairs is a challenging problem since they should contain both typical and temporally comparable entities.

In view of the above-mentioned challenges, we propose a novel method to address the task of generating *typical comparables*. First of all, we formulate the measurement of *entity typicality* inspired by research in psychology and cognitive science [6,14,38]. In particular, for an entity to be typical in a diverse set it should be representative within a significant subset of that set. Moreoever, we formulate the measurement of across-time entity comparability by aligning different vector spaces and finding corresponding terms. We first adopt the distributed vector representation [27] to represent the context vectors of entities; then we learn linear and orthogonal transformations between two vector spaces of input collections for establishing across-time entity correspondence. Finally, inspired by the popular Affinity Propagation algorithm (AP) [11], we propose a concise joint integer linear programming framework (J-ILP) which detects typical entities (which we call exemplars) and, at the same time, generates comparable pairs from the detected exemplars. Based on this formulation, the optimal solution can be obtained.

To sum up, we make the following contributions: (1) We introduce a new research problem of automatically discovering comparable entity pairs from two across-time collections of entities. (2) We develop a novel method to address this task based on an efficient entity typicality estimation, an effective across-time entity comparability measurement, and a concise integer linear programming framework. (3) Finally, we perform extensive experiments on the New York Times Annotated Corpus, which demonstrates the effectiveness of our approach.

2 Problem Definition

Formally, given two sets of entities denoted by D_A and D_B, where D_A and D_B come from different time periods T_A and T_B, respectively ($T_A \cap T_B = \emptyset$ and, typically T_A represents some period in the past while T_B represents more present time period), the task is to discover m comparable entity pairs $P = [p_1, p_2, ..., p_m]$ to form a concise subset conveying the most important comparisons, where $p_i = (e_i^A, e_i^B)$. e_i^A and e_i^B are entities from D_A and D_B, respectively.

3 Estimation of Entity Typicality

Learning from examples is an effective strategy extensively adopted in cognition and education [14]. Good examples should be however typical. In this work, we apply the strategy of using typical examples for discovering comparable entity pairs. We denote the typicality of an entity e with regard to a set of entities S as

$Typ(e, S)$. The entities to be selected for comparison should be typical in their sets, namely, $Typ(e_i^A, D_A)$ and $Typ(e_i^B, D_B)$ should be as high as possible when $p_i = (e_i^A, e_i^B)$ is a selected entity pair.

As suggested by the previous research in typicality analysis [14], an entity e in a set of entities S is typical, if it is likely to appear in S. We denote the likelihood of an entity e given a set of entities S by $L(e|S)$ (to be defined soon). However, it is not appropriate to simply use $L(e|S)$ as an estimator of typicality $Typ(e, S)$ considering the characteristics of our task. First of all, the collections of entities for comparison can be very complex, thus they may cover many different kinds of entities. For example, if we want to compare US scientists across time, each of entity collections will include multiple kinds of entities such as mathematicians, physicists, chemists and so on. It is then very difficult for a single entity to represent all of them. In addition, different entity kinds vary in their significance. For instance, "physicists" are far more common than "entomologists". Naturally, entities typical in a salient entity subset should be more important than those belonging to small subsets.

Given a set S including k mutually exclusive latent subgroups $[S^1, S^2, ..., S^k]$, let e_i^t denote the ith entity in the tth subgroup of S. We state two criteria required for e_i^t to be typical in the entire set S:

Criterion 1 *e_i^t should be representative in S^t.*
Criterion 2 *The significance of S^t in S should be high.*

The typicality of e_i^t with respect to S is then defined as follows:

$$Typ(e_i^t, S) = L(e_i^t|S^t) \cdot \frac{|S^t|}{|S|} \tag{1}$$

where $L(e_i^t|S^t)$ measures the representativeness of e_i^t with regard to the subgroup S^t. In addition, $\frac{|S^t|}{|S|}$ indicates the relative size of S^t regarded as an estimator of significance. e_i^t is more typical when the number of entities in its subgroup is large.

The likelihood $L(e|S)$ of an entity e given a set of entities S is the posterior probability of e given S, which can be computed using probability density estimation methods. Many model estimation techniques have been proposed including parametric and non-parametric density estimations. We use kernel estimation [3] as it does not require any distribution assumption and can estimate unknown data distributions effectively. Moreover, we choose the commonly used Gaussian kernels. We set the bandwidth of the Gaussian kernel estimator $h = \frac{1.06s}{\sqrt[5]{n}}$ as suggested in [35], where n is the size of the data and s is the standard deviation of the data set. Formally, given a set of entities $S = (e_1, e_2, ..., e_n)$, the underlying likelihood function is approximated as:

$$L(e|S) = \frac{1}{n}\sum_{i=1}^{n} G_h(e, e_i) = \frac{1}{n\sqrt{2\pi}}\sum_{i=1}^{n} e^{-\frac{d(e,e_i)^2}{2h^2}} \tag{2}$$

where $d(e, e_i)$ is the cosine distance between e and e_i, and $G_h(e, e_i)$ is a *Gaussian kernel.*

4 Measurement of Temporal Comparability

In this section, we describe the method for measuring temporal comparability between an entity e_A in set D_A and an entity e_B in the other set D_B. Intuitively, if e_A and e_B comparable to each other, then e_A and e_B contain comparable aspects. For instance, (*iPod*, *Walkman*) could be regarded as comparable based on the observation that *Walkman* played the role of a popular portable music player 30 years ago same as *iPod* does nowadays. The key difficulty comes from the fact that there is low overlap between terms' contexts across time (e.g., the set of top co-occurring words with *iPod* in documents published in 2010s has typically little overlap with the set of top co-occurring words with *walkman* that are extracted from documents in 1980s). Thus our task is then to build the connection between semantic spaces of D_A and D_B.

Let transformation matrix W map the entities from D_A into D_B, and transformation matrix Q map the entities in D_B back into D_A. Let a and b be normalized entity representations from D_A and D_B, respectively. The comparability between entities a and b can be evaluated as the similarity between vectors b and Wa, i.e., $Comp(a,b) = b^T Wa$. However we could also form this correspondence as $Comp'(a,b) = a^T Qb$. To be self-consistent, we require $Comp(a,b) = Comp'(a,b)$, thus the linear transformations W and Q between entity collections D_A and D_B should be orthogonal [36,40], i.e., $W^T W = I$ (where I denotes the identity matrix).

Our task is then to train the transformation matrix W to automatically align the semantic vector space across time. We adopt here a technique proposed by [42] for preparing sufficient training data. Namely, we use so-called Common Frequent Terms (CFT) as the training term pairs. CFT are very frequent terms in both dates T_A and T_B, which the compared entity collections originate from (e.g. man, woman, sky, water). Such frequent terms tend to change their meanings only to a small extent across time. The phenomenon that words which are intensively used in everyday life evolve more slowly has been reported in several languages including English, Spanish, Russian and Greek [13,23,29]. We first train the time-aware distributional vectors of CFTs using the New York Times Corpus [32] published within T_A and T_B, respectively. Given L pairs composed of normalized vectors of CFTs trained in both news corpora $[(a_1,b_1), (a_2,b_2), ..., (a_L,b_L)]$ (where a_i and b_i denote the vector of i-th CFT in T_A and T_B, respectively), we should learn the transformation W by maximizing the accumulated cosine similarity of CFT pairs,

$$\underset{W}{max} \sum_{i=1}^{L} b_i^T W a_i, s.t. W^T W = I \tag{3}$$

The solution corresponds to the best rotational alignment [34] and can be obtained efficiently using an application of SVD. By computing the SVD of $M = A^T B = U \Sigma V^T$, the optimized transformation matrix W^* satisfies $W^* = U \cdot V^T$.

Based on it, we measure the temporal comparability between an entity e_A in set D_A and an entity e_B in the other set D_B as follows:

$$Comp(e_A, e_B) = Sim_{cosine}(W^* \cdot e_A, e_B) \tag{4}$$

5 ILP Formulation for Detecting Comparables

In this section, we describe our method for discovering comparable entity pairs. Given two sets of entities D_A and D_B the output are m comparable entity pairs $[p_1, p_2, ..., p_m]$, where each pair contains an entity from D_A and an entity from D_B. Inspired by AP algorithm [11], we formulate our task as a process of identifying a subset of typical comparable entity pairs. It has been empirically found that using AP for solving objectives such as in our case (see Eq. (5)) suffers considerably from convergence issues [41]. Thus, we propose a concise integer linear programming (ILP) formulation for discovering comparable entities, and we use the *branch-and-bound* method to obtain the optimal solution.

Specifically, we formulate the task as a process of selecting a subset of k_A and k_B exemplars for each set respectively and choosing m entity pairs based on the identified exemplars. Each non-exemplar entity is assigned to an exemplar entity based on a measure of similarity, and each exemplar e represents a subgroup comprised of all non-exemplar entities that are assigned to e. On the one hand, we wish to maximize the overall typicality of selected exemplars w.r.t. their representing subgroups. On the other hand, we expect to maximize the overall comparability of the top m entity pairs, where each pair consists of two exemplars from different sets.

We next introduce some notations used in our method. Let e_i^A denote the ith entity in D_A. $M_A = [m_{ij}]^A$ is a $n_A \times n_A$ binary square matrix such that n_A is the number of entities within D_A. m_{ii}^A indicates whether entity e_i^A is selected as an exemplar or not, and $m_{ij:i \neq j}^A$ represents whether entity e_i^A votes for entity e_j^A as its exemplar. Similar to M_A, the $n_B \times n_B$ binary square matrix M_B indicates how entities belonging to D_B choose their exemplars, where n_B is the number of entities within D_B. m_{ii}^B indicates whether entity e_i^B is selected as an exemplar or not, and $m_{ij:i \neq j}^B$ represents whether entity e_i^B votes for entity e_j^B as its exemplar. Different from M_A and M_B, $M_T = [m_{ij}]^T$ is a $n_A \times n_B$ binary matrix whose entry m_{ij}^T denotes whether entities e_i^A and e_j^B are paired together as the final result. Then the following ILP problem is designed for the task of selecting k_A and k_B exemplars for each set respectively and for selecting m comparable entity pairs:

$$\begin{aligned} max \;\; & \lambda \cdot m \cdot [T'(M_A) + T'(M_B)] \\ & + (1 - \lambda) \cdot (k_A + k_B) \cdot C'(M_T) \end{aligned} \tag{5}$$

$$T'(M_X) = \sum_{i=1}^{n_X} m_{ii}^X \cdot Typ(e_i^X, G(e_i^X)), X \in \{A, B\} \tag{6}$$

$$C'(M_T) = \sum_{i=1}^{n_A} \sum_{j=1}^{n_B} m_{ij}^T \cdot Comp(e_i^A, e_j^B) \tag{7}$$

$$G(e_i^X) = \{e_j^X | m_{ji}^X = 1, j \in \{1, ..., n_X\}\}, \quad i \in \{1, ..., n_X\}, X \in \{A, B\} \tag{8}$$

$$s.t. \quad m_{ij}^X \in \{0, 1\}, i \in \{1, ..., n_X\}, \quad j \in \{1, ..., n_X\}, X \in \{A, B\} \tag{9}$$

$$m_{ij}^T \in \{0, 1\}, i \in \{1, ..., n_A\}, j \in \{1, ..., n_B\} \tag{10}$$

$$\sum_{i=1}^{n_X} m_{ii}^X = k_X, X \in \{A, B\} \tag{11}$$

$$\sum_{j=1}^{n_X} m_{ij}^X = 1, i \in \{1, ..., n_X\}, X \in \{A, B\} \tag{12}$$

$$m_{jj}^X - m_{ij}^X \geq 0, i \in \{1, ..., n_X\}, \quad j \in \{1, ..., n_X\}, X \in \{A, B\} \tag{13}$$

$$\sum_{i=1}^{n_A} \sum_{j=1}^{n_B} m_{ij}^T = m \tag{14}$$

$$m_{ii}^A - m_{ij}^T \geq 0, i \in \{1, ..., n_A\}, j \in \{1, ..., n_B\} \tag{15}$$

$$m_{jj}^B - m_{ij}^T \geq 0, i \in \{1, ..., n_A\}, j \in \{1, ..., n_B\} \tag{16}$$

$$\sum_{j=1}^{n_B} m_{ij}^T \leq 1, i \in \{1, ..., n_A\} \tag{17}$$

$$\sum_{i=1}^{n_A} m_{ij}^T \leq 1, j \in \{1, ..., n_B\} \tag{18}$$

We now explain the meaning of the above formulas. First, Eq. (11) forces that k_A and k_B exemplars are identified for both sets D_A and D_B, respectively, and Eq. (14) guarantees that m entity pairs are selected as the final result. The restriction given by Eq. (12) means each entity must choose only one exemplar. Equation (13) enforces that if one entity e_j^X is voted by at least one other entity, then it must be an exemplar (i.e., $m_{jj}^X = 1$). The constraint given by (15) and (16) jointly guarantees that if an entity is selected in any comparable entity pair (i.e., $m_{ij}^T = 1$), then it must be an exemplar in its own subgroup (i.e., $m_{ii}^A = 1$ and $m_{jj}^B = 1$). Restricted by Eqs. (17) and (18), each selected exemplar in the result is only allowed to appear once to avoid redundancy. $T'(M_X)$ represents the overall typicality of selected exemplars in both sets D_A and D_B, and $G(e_i^X)$ denotes the representing subgroup for entity e_i^X (if e_i^X is not chosen as an exemplar, its representing subgroup will be null). $C'(M_T)$ denotes the overall comparability of generated entity pairs. In view of the fact that there are $(k_A + k_B)$ values (each value is in [0,1]) in the typicality component $T'(M_A) + T'(M_B)$, and m numbers (each number is in [0,1]) in the comparability part $C'(M_T)$, we add the coefficients m and $(k_A + k_B)$ in the objective function to avoid suffering from skewness problem. Finally, the parameter λ[1] is used to balance the weight of the two parts. Our proposed ILP formulation guarantees to achieve the optimal solution by using *branch-and-bound method.*

6 Experiments

6.1 Datasets

We perform the experiments on the New York Times Annotated Corpus [32]. This corpus is a collection of 1.8 million articles published by the New York Times between January 01, 1987 and June 19, 2007 and has been frequently used to evaluate different researches that focus on temporal information processing or extraction in document archives [4]. For the experiments, we first divide the corpus into four parts based on article publication dates: [1987, 1991], [1992, 1996], [1997, 2001] and [2002, 2007]. The vocabulary size of each time period is around 300k. We then set on comparing the pair of time periods which are separated by the longest time gap, [1987, 1991] (denoted as T_A) and [2002, 2007] (denoted as T_B). We assume here that the more the two time periods are farther apart, the stronger is the context change, which increases the difficulty of finding corresponding entity pairs. We obtain the distributed vector representations for time period T_A and ones for T_B by training the Skip-gram model using the gensim Python library [30]. The number of dimensions of word vectors is experimentally set to be 200.

To prepare the entity sets for each period, we retain all unigrams and bigrams which appear more than 10 times in the collection of news articles within that period, excluding stopwords and all numbers. We then adopt spaCy[2] for recognizing named entities based on all unigrams and bigrams. In total we extract 68,872 entities and 34,151 entities in T_A ([1987, 1991]) and T_B ([2002, 2007]), respectively. The details of identified entities are shown in Table 1. The meaning of sub-categories can be found at spaCy website[3]. Note that some sub-categories of entities were not used due to their weak significance, e.g., TIME/DATE.

Table 1. Summary of datasets.

Period	LOC	PRODUCT	NORP	WOA	GPE	PERSON	FACT	ORG	LAW	EVENT	TOTAL
T_A	427	87	2,959	129	7,810	33,127	328	23,775	18	212	68,872
T_B	304	44	1,573	91	4,460	16,103	221	11,215	11	129	34,151

[1] We experimentally set the value of λ to be 0.4 in Sec "Experiments".
[2] https://github.com/explosion/spaCy.
[3] https://spacy.io/api/annotation#named-entities.

6.2 Test Sets

As far as we know, there is no ground truth data available for the task of identification of across-time comparable entities. Hence, we then apply pooling technique for creating test sets. In particular, we have leveraged the pooling technique by pulling the resulting comparable entity pairs from all the proposed methods and baselines as listed in Sect. 6.4). Three annotators then judged every result in the pool based on the following steps: firstly highlight all the typical entities in the results, then create reference entity pairs based on the highlighted entities. There was no limit on the number of highlighted entities nor chosen entity pairs. The annotators did not know which systems generated which answers. They were allowed to utilize any external resources or use search engines in order to verify the correctness of the results. In total, 447 entities and 315 entities were chosen as typical exemplars for periods T_A and T_B, respectively. Among them, 168 pairs were constructed.

6.3 Evaluation Criteria

Criteria for Quantitative Evaluation. Given the human-labeled typical entity set and the comparable entity pairs' set, we compare the generated results with the ground truth. We compute *precision*, *recall* and F_1-*score* to measure the performance of each method.

Criteria for Qualitative Evaluation. To further evaluate the quality of the results we also conducted user-based analysis. In particular, 3 subjects were invited to annotate the results generated by each method using the following quality criteria: (1) *Correctness* - it measures how sound the results are. (2) *Comprehensibility* - it measures how easy it is to understand and explain the results. (3) *Diversity* - it quantifies how varying and diverse information the annotators could acquire. All the scores were given in the range from 1 to 5 (1: not at all, 2: rather not, 3: so so, 4: rather yes, 5: definitely yes). We averaged all the individual scores given by the annotators to obtain the final scores per each comparison. During the assessment, the annotators were allowed to utilize any external resources including the Wikipedia, Web search engines, books, etc.

6.4 Baselines

We prepare different methods to select temporally comparable entity pairs. We first compare our model with three widely-used clustering methods: K-Means clustering, DBSCAN clustering [7] and aforementioned AP clustering [11]. Besides, we also adopt the mutually-reinforced random walk model [5] (denoted as MRRW) to judge entity typicality based on the hypothesis that typical exemplars are those who are similar to the other members of its category and dissimilar to members of the contrast categories. Finally, we also test a limited version of our approach called Independent ILP (denoted as I-ILP) that separately identifies exemplars of each input sets based on our proposed ILP framework. I-ILP aims to maximize the overall typicality of selected exemplars for each set respectively without considering whether chosen exemplars are comparable or not. In this study we use the Gurobi solver [12] for solving the proposed ILP framework. After the exemplars have been selected by the above methods, we construct the entity pairs which have the maximal comparability based on identified exemplars as follows.

$$\mathrm{P} \equiv \mathrm{argmax} \sum_{i=1}^{m} Comp(e_i^A, e_i^B) \tag{19}$$

where $P = [p_1, p_2, ..., p_m]$ are expected comparables, and $p_i = (e_i^A, e_i^B)$. e_i^A and e_i^B are chosen exemplars from the compared sets.

Besides, we also test effectiveness of orthogonal transformation for computing across-time comparability. To this end, we test the method which directly compares the vectors trained in different time periods separately without performing any transformation (denoted as Embedding-S + Non-Tran). Moreover, we also analyze the methods which utilize the distributional entity representation trained on the combination of news articles from two compared periods jointly (denoted as Embedding-J). We denote the proposed transformation-based methods as Embedding-S + OT.

6.5 Experiment Settings

We set the parameters as follows:

(1) **number of subgroups of each input set:** Following [39] we set the number k of latent subgroups of each input set as:

$$k = \lceil \sqrt{n} \rceil \tag{20}$$

where n is the number of entities in the set.

(2) **number of generated pairs for comparison:** In view of the fact that the number of counterparts for each entity is at most one in the output, we set the number of generated pairs m to be its upper bound $min\{k_A, k_B\}$, where k_A and k_B are the numbers of identified exemplars of two compared entity sets.

(3) **number of used CFTs:** Following [42] we utilize the top 5% ($\approx$18k) of *Common Frequent Terms* to train the orthogonal transformation in Sect. 3.

6.6 Evaluation Results

Results of Quantitative Evaluation. Table 2 shows the performance of all the analyzed methods in terms of *Precision*, *Recall* and F_1*-score*, while we show the detailed results for a few examples in Table 3. We first notice that the performance is extremely poor without transforming the contexts of entities. Only very few results in *Non-Tran* approaches are judged as correct. On the other hand, although methods based on the jointly-trained word embeddings perform better than *Non-Tran*, the performance increase is quite limited. It can be observed that the across-time orthogonal transformation is quite helpful since it exhibits significantly better effectiveness in terms of all the metrics than the other two types of methods. This observation suggests little overlap in the contexts of news articles separated by longer time gaps, and that the task of identifying temporal analogous entities is quite difficult.

Moreover, a closer look at Table 2 reveals that regardless of the type of evaluation metric, J-ILP improves the performance of the other models under transformation. From Table 2, it can be seen that 27.3% entity pairs generated by J-ILP model are judged as correct by human annotators, and that 29.0% of ground truth entity pairs are discovered. Specifically, J-ILP improves the baselines by 87.3% when measured using the main metric F_1-*score* on average. These results are observed because the proposed J-ILP formulation takes both necessary factors (typicality and comparability) into consideration. Based on this formulation, the optimal solution can be obtained using the *branch-and-bound* method.

We also investigate the possible reasons for the poor performance of baselines. K-Means suffers from strong sensitivity to outliers and noise, which leads to a varying performance. On the other hand, although AP shares many similar characteristics with J-ILP, its belief propagation mechanism does not guarantee to find the optimal solution, hence its lower performance. DBSCAN relies on the concept of "core point" for identifying exemplars with high density, however it is possible that a typical point does not have many points lying close to it, and a "core point" may not be typical in the scenarios of unbalanced clusters. Finally, MRRW tends to select entities that contain more discriminative features rather than common traits, which can explain why it has worse performance.

Table 2. Performance of models in terms of *Precision*, *Recall* and F_1-*score*. The best results of each setting are indicated in bold, while the best overall results are underlined.

Method	Embedding-S+Non-Tran			Embedding-J			Embedding-S+OT		
	Precision	Recall	F_1-score	Precision	Recall	F_1-score	Precision	Recall	F_1-score
K-Means	**0.027**	**0.030**	**0.028**	0.081	0.089	0.085	0.186	0.195	0.190
DBSCAN	0.000	0.000	0.000	0.000	0.000	0.000	0.105	0.106	0.105
AP	0.016	0.018	0.017	0.049	0.054	0.051	0.154	0.160	0.156
MRRW	0.000	0.000	0.000	0.027	0.030	0.028	0.132	0.136	0.133
I-ILP	0.016	0.018	0.017	0.049	0.054	0.051	0.165	0.171	0.167
J-ILP	0.000	0.000	0.000	**0.124**	**0.137**	**0.130**	**0.273**	**0.290**	**0.281**

Results of Qualitative Evaluation. Figure 1 shows the evaluation scores in terms of *Correctness*, *Comprehensibility* and *Diversity* judged by annotators, respectively. We first note that our J-ILP model achieves better results than the baselines based on both *Correctness* and *Comprehensibility* criteria. On average,

Table 3. Example results where entity pairs are ground truth. The entity on the left in parentheses is from period [1987, 1991] while the entity on the right is from [2002, 2007]. The tags (0,1) shown in parentheses denote the appearance of ground truth entity in results (1 means the entity matches the ground truth exemplars, while 0 means otherwise). Note that only the tag (1,1) indicates the ground truth entity pair was identified correctly, while (1,1)* denotes that although both entities are recognized as exemplars, they are not paired together in the results.

Entity pair	K-Means	DBSCAN	AP	MRRW	I-ILP	J-ILP
(iraq, syria)	(0,0)	(1,1)*	(1,0)	(1,0)	(1,1)*	(1,1)
(president_reagan, george_bush)	(1,1)*	(0,1)	(0,1)	(0,1)	(1,0)	(1,1)
(american_express, credit_card)	(1,0)	(0,0)	(0,0)	(0,0)	(1,1)	(1,1)
(macintosh, pc)	(1,1)	(1,0)	(0,0)	(0,0)	(1,0)	(1,0)
(salomon, morgan_stanley)	(0,1)	(0,0)	(1,0)	(1,0)	(0,1)	(1,1)
(national_basketball, world_series)	(1,1)	(0,0)	(0,1)	(0,0)	(0,1)	(0,1)
(european_community, china)	(0,1)	(0,0)	(0,1)	(1,0)	(0,0)	(0,1)
(pan_am, american_airlines)	(1,1)*	(1,0)	(1,1)*	(0,0)	(1,1)	(1,0)
(mario_cuomo, george_pataki)	(0,1)	(1,0)	(0,1)	(0,0)	(1,1)*	(1,1)
(bonn, berlin)	(0,0)	(0,0)	(1,0)	(1,0)	(1,1)	(1,1)
(sampras, federer)	(0,0)	(1,1)	(0,0)	(0,0)	(0,1)	(0,0)
(saddam, al_qaeda)	(1,1)	(1,0)	(1,0)	(0,1)	(0,0)	(1,0)

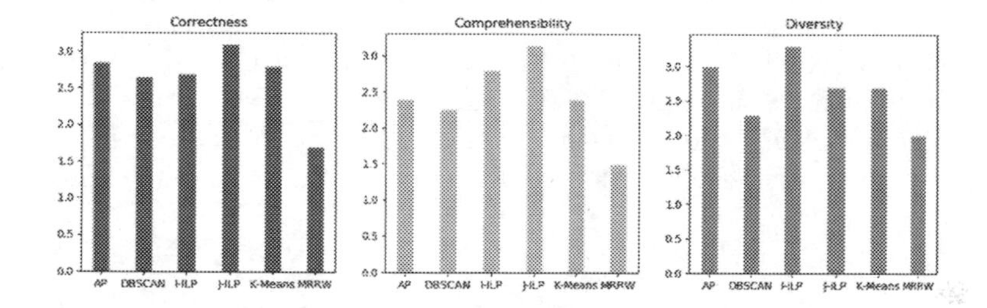

Fig. 1. Qualitative evaluation of results.

J-ILP outperforms baselines by 20.2% and 28.0% in terms of *Correctness* and *Comprehensibility*, respectively. This observation proves that J-ILP has relatively good performance in detecting dominant and reasonable entity pairs, which tend to be highly scored by annotators. On the other hand, J-ILP underperforms two baselines AP algorithm and I-ILP in terms of diversity by 10.0% and 18.2%, respectively. It may be because AP algorithm and I-ILP are intrincically better in capturing representative and diverse exemplars, while J-ILP aims to balance the entity typicality and comparability simultaneously.

6.7 Additional Observations

Effects of Trade-Off Parameter. We perform a grid search to find the best trade-off parameter λ. We set λ in the range [0.0, 1.0] with a step of 0.1. Note that when $\lambda = 1.0$, the J-ILP formulation degenerates into the aforementioned I-ILP model. From Fig. 2, we see that when λ is within the range [0.0, 0.4], the performance of J-ILP reaches its maximal value and remains stable. On the other hand, the values of all metrics degrade when increasing the value of λ after $\lambda = 0.4$. In general, we can see that λ needs to be fine-tuned to achieve an optimal performance. In this study we set λ as 0.4 based on the observations received from Fig. 2.

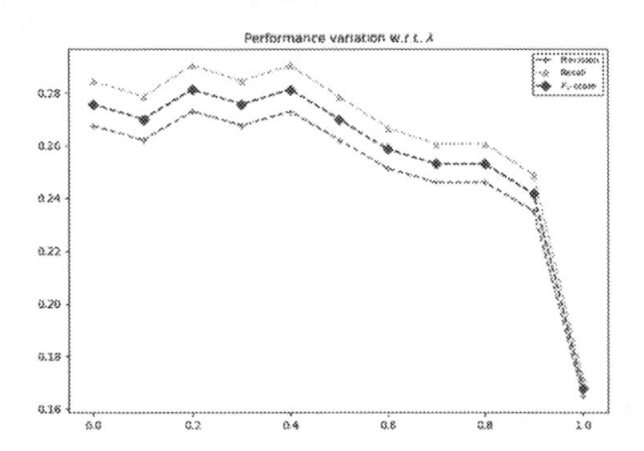

Fig. 2. Performance variation of *precision*, *recall* and F_1-*score* w.r.t. λ.

Sensitivity to Kernel Choice. In this work we adopt Gaussian kernel function for computing entity typicality. Let the generated pairs returned by using Gaussian kernel be P_G and the results generated by other popular kernel functions be P_O. The difference of P_G and P_O is measured as the difference rate d as follows.

$$d = \frac{|P_G - PO|}{|PG|} \cdot 100\% \tag{21}$$

Table 4. Difference rate vs. kernel function.

Kernel function	Quatic	Triweight	Epanechnikov	Cosine
Difference rate	15.9	10.3	5.5	15.9

Table 4 shows that the exemplars identified by different kernels are in general consistent, as the difference rate d is low.

7 Related Work

Comparable Entity Mining. The task of comparable entity mining has attracted much attention in the NLP and Web mining communities [15–19, 22].

Approaches to this task include hand-crafted extraction rules [8], supervised machine-learning methods [26,33] and weakly-supervised methods [17,22]. Jindal *et al.* [18,19] was the first to propose a two-step system in finding comparable entities which first tackles a classification problem (i.e., whether a sentence is comparative) and then a labeling problem (i.e., which part of the sentence is the desideratum). Later work refined that system by using a bootstrapping algorithm [22], or extended the idea of mining comparables to different types of corpora including query logs [16,17] and comparative questions [22]. In addition, comparable entity mining is strongly related to the problem of automatic structured information extraction, comparative summarization and named entity recognition. Some work lies in the intersection of these tasks [10,24].

Temporal Analog Detection and Embeddings Alignment. A part of our system approaches the task of identifying temporally corresponding terms across different times. The related work to this subtask include computing term similarity across time [2,20,21,37]. In this study we represent terms using the distributed vector representation [27]. Thus the problem of connecting news articles' context across different time periods can be approached by aligning pre-trained word embeddings in different time periods. Mikolov *et al.* proposed a linear transformation aligning bi-lingual word vectors for automatic text translation such as translation from Spanish to English [28]. Faruqui *et al.* obtained bi-lingual word vectors using CCA [9]. More recently, Xing *et al.* argued that the linear matrix adopted by Mikolov *et al.* should be orthogonal [40]. Similar suggestion has been given by Samuel *et al.* [36]. Besides linear models, non-linear models such as "deep CCA" has also been introduced for the task of mapping multi-lingual word embeddings [25]. In this study we adopt the orthogonal transformation for computing across-time entity correspondence due to its high accuracy and efficiency.

To the best of our knowledge, we are the first to focus on the task of automatically generating across-time comparable entity pairs given two entity sets, and on using the notion of typicality analysis from cognitive science and psychology.

8 Conclusions and Future Work

Entity is an evolving construct. This fact is nicely portrayed by the Latin proverb: *"omnia mutantur, nihil interit"* (in English: "everything changes, nothing perishes") which indicates that there are no completely static things [42]. Across-time comparison based on typical exemplars is an effective strategy used by humans for obtaining contrastive knowledge or for understanding unknown entity groups by their comparison to familiar groups (e.g., entities from the past compared to ones from present). In this work, we propose a novel research problem of automatically detecting across-time typical comparable entity pairs from two input sets of entities and we introduce effective method for solving it. We adopt a concise ILP model for maximizing the overall representativeness and comparability of the selected entity pairs. The experimental results demonstrate the effectiveness of our model compared to several competitive baselines.

In future we plan to test our model on more heterogeneous datasets where contexts of entities are more difficult to be compared. We will also modify our model for query-sensitive comparative summarization tasks benefiting from high flexibility of the proposed ILP framework.

Acknowledgements. This research has been supported by JSPS KAKENHI grants (#17H01828, #18K19841). We thank the anonymous reviewers for their insightful comments.

References

1. Bairi, R.B., Carman, M., Ramakrishnan, G.: On the evolution of Wikipedia: dynamics of categories and articles. In: AAAI (2015)
2. Berberich, K., Bedathur, S.J., Sozio, M., Weikum, G.: Bridging the terminology gap in web archive search. In: WebDB (2009)
3. Breiman, L., Meisel, W., Purcell, E.: Variable kernel estimates of multivariate densities. Technometrics **19**(2), 135–144 (1977)
4. Campos, R., Dias, G., Jorge, A.M., Jatowt, A.: Survey of temporal information retrieval and related applications. ACM Comput. Surv. (CSUR) **47**(2), 15 (2015)
5. Chen, Y.N., Metze, F.: Two-layer mutually reinforced random walk for improved multi-party meeting summarization. In: 2012 IEEE SLT, pp. 461–466. IEEE (2012)
6. Dubois, D., Prade, H., Rossazza, J.P.: Vagueness, typicality, and uncertainty in class hierarchies. Int. J. Intell. Syst. **6**(2), 167–183 (1991)
7. Ester, M., Kriegel, H.P., Sander, J., Xu, X., et al.: A density-based algorithm for discovering clusters in large spatial databases with noise. In: KDD, vol. 96, pp. 226–231 (1996)
8. Etzioni, O., et al.: Web-scale information extraction in knowitall: (preliminary results). In: Proceedings of the 13th WWW, pp. 100–110. ACM (2004)
9. Faruqui, M., Dyer, C.: Improving vector space word representations using multilingual correlation. In: EACL, pp. 462–471 (2014)
10. Feldman, R., Fresco, M., Goldenberg, J., Netzer, O., Ungar, L.: Extracting product comparisons from discussion boards. In: Data Mining, ICDM 2007, pp. 469–474. IEEE (2007)
11. Frey, B.J., Dueck, D.: Clustering by passing messages between data points. Science **315**(5814), 972–976 (2007)
12. Gurobi Optimization, Inc.: Gurobi optimizer reference manual (2016). http://www.gurobi.com
13. Hamilton, W.L., Leskovec, J., Jurafsky, D.: Diachronic word embeddings reveal statistical laws of semantic change. arXiv preprint arXiv:1605.09096 (2016)
14. Hua, M., Pei, J., Fu, A.W., Lin, X., Leung, H.F.: Efficiently answering top-k typicality queries on large databases. In: Proceedings of VLDB, pp. 890–901. VLDB Endowment (2007)
15. Huang, X., Wan, X., Xiao, J.: Learning to find comparable entities on the web. In: Wang, X.S., Cruz, I., Delis, A., Huang, G. (eds.) WISE 2012. LNCS, vol. 7651, pp. 16–29. Springer, Heidelberg (2012). https://doi.org/10.1007/978-3-642-35063-4_2
16. Jain, A., Pantel, P.: Identifying comparable entities on the web. In: Proceedings of the 18th ACM CIKM, pp. 1661–1664. ACM (2009)
17. Jiang, Z., Ji, L., Zhang, J., Yan, J., Guo, P., Liu, N.: Learning open-domain comparable entity graphs from user search queries. In: Proceedings of the 22nd ACM CIKM, pp. 2339–2344. ACM (2013)

18. Jindal, N., Liu, B.: Identifying comparative sentences in text documents. In: Proceedings of ACM SIGIR, pp. 244–251. ACM (2006)
19. Jindal, N., Liu, B.: Mining comparative sentences and relations. In: AAAI, vol. 22, pp. 1331–1336 (2006)
20. Kaluarachchi, A.C., Varde, A.S., Bedathur, S., Weikum, G., Peng, J., Feldman, A.: Incorporating terminology evolution for query translation in text retrieval with association rules. In: CIKM, pp. 1789–1792. ACM (2010)
21. Kanhabua, N., Nørvåg, K.: Exploiting time-based synonyms in searching document archives. In: JCDL, pp. 79–88. ACM (2010)
22. Li, S., Lin, C.Y., Song, Y.I., Li, Z.: Comparable entity mining from comparative questions. IEEE TKDE **25**(7), 1498–1509 (2013)
23. Lieberman, E., Michel, J.B., Jackson, J., Tang, T., Nowak, M.A.: Quantifying the evolutionary dynamics of language. Nature **449**(7163), 713 (2007)
24. Liu, J., Wagner, E., Birnbaum, L.: Compare&contrast: using the web to discover comparable cases for news stories. In: Proceedings of the 16th WWW, pp. 541–550. ACM (2007)
25. Lu, A., Wang, W., Bansal, M., Gimpel, K., Livescu, K.: Deep multilingual correlation for improved word embeddings. In: NAACL HLT, pp. 250–256 (2015)
26. McCallum, A., Jensen, D.: A note on the unification of information extraction and data mining using conditional-probability, relational models (2003)
27. Mikolov, T., Chen, K., Corrado, G., Dean, J.: Efficient estimation of word representations in vector space. arXiv preprint arXiv:1301.3781 (2013)
28. Mikolov, T., Le, Q.V., Sutskever, I.: Exploiting similarities among languages for machine translation. arXiv preprint arXiv:1309.4168 (2013)
29. Pagel, M., Atkinson, Q.D., Meade, A.: Frequency of word-use predicts rates of lexical evolution throughout indo-European history. Nature **449**(7163), 717 (2007)
30. Řehůřek, R., Sojka, P.: Software framework for topic modelling with large corpora. In: Proceedings of the LREC 2010 Workshop on New Challenges for NLP Frameworks, pp. 45–50. ELRA, Valletta, May 2010. http://is.muni.cz/publication/884893/en
31. Rodríguez, M.A., Egenhofer, M.J.: Determining semantic similarity among entity classes from different ontologies. IEEE TKDE **15**(2), 442–456 (2003)
32. Sandhaus, E.: The new york times annotated corpus overview, pp. 1–22. The New York Times Company, Research and Development (2008)
33. Sarawagi, S., Cohen, W.W.: Semi-markov conditional random fields for information extraction. In: NIPS, pp. 1185–1192 (2005)
34. Schönemann, P.H.: A generalized solution of the orthogonal procrustes problem. Psychometrika **31**(1), 1–10 (1966)
35. Scott, D.W., Sain, S.R.: 9-multidimensional density estimation. Handb. Stat. **24**, 229–261 (2005)
36. Smith, S.L., Turban, D.H., Hamblin, S., Hammerla, N.Y.: Offline bilingual word vectors, orthogonal transformations and the inverted softmax. arXiv preprint arXiv:1702.03859 (2017)
37. Tahmasebi, N., Gossen, G., Kanhabua, N., Holzmann, H., Risse, T.: NEER: an unsupervised method for named entity evolution recognition. COLING, pp. 2553–2568 (2012)
38. Tamma, V., Bench-Capon, T.: An ontology model to facilitate knowledge-sharing in multi-agent systems. Knowl. Eng. Rev. **17**(1), 41–60 (2002)
39. Wan, X., Yang, J.: Multi-document summarization using cluster-based link analysis. In: Proceedings of ACM SIGIR, pp. 299–306. ACM (2008)

40. Xing, C., Wang, D., Liu, C., Lin, Y.: Normalized word embedding and orthogonal transform for bilingual word translation. In: NAACL HLT, pp. 1006–1011 (2015)
41. Yu, H.T., et al.: A concise integer linear programming formulation for implicit search result diversification. In: Proceedings of the Tenth ACM WSDM, pp. 191–200. ACM (2017)
42. Zhang, Y., Jatowt, A., Bhowmick, S., Tanaka, K.: Omnia mutantur, nihil interit: Connecting past with present by finding corresponding terms across time. In: ACL, vol. 1, pp. 645–655 (2015)

Unsupervised Entity Alignment Using Attribute Triples and Relation Triples

Fuzhen He[1,2], Zhixu Li[1,3(✉)], Yang Qiang[4], An Liu[1], Guanfeng Liu[5], Pengpeng Zhao[1], Lei Zhao[1], Min Zhang[1], and Zhigang Chen[3]

[1] School of Computer Science and Technology, Soochow University, Suzhou, China
fzhe@stu.suda.edu.cn
{zhixuli,anliu,ppzhao,zhaol,minzhang}@suda.edu.cn
[2] Neusoft Corporation, Shenyang, China
[3] IFLYTEK Research, Suzhou, China
zgchen@iflytek.com
[4] King Abdullah University of Science and Technology, Jeddah, Saudi Arabia
qiangyanghm@hotmail.com
[5] Department of Computing, Macquarie University, Sydney, Australia
guanfeng.liu@mq.edu.au

Abstract. Entity alignment aims to find entities referring to the same real-world object across different knowledge graphs (KGs). Most existing works utilize the relations between entities contained in the relation triples with embedding-based approaches, but require a large number of training data. Some recent attempt works on using types of their attributes in attribute triples for measuring the similarity between entities across KGs. However, due to diverse expressions of attribute names and non-standard attribute values across different KGs, the information contained in attribute triples can not be fully used. To tackle the drawbacks of the existing efforts, we novelly propose an unsupervised entity alignment approach using both attribute triples and relation triples of KGs. Initially, we propose an interactive model to use attribute triples by performing entity alignment and attribute alignment alternately, which will generate a lot of high-quality aligned entity pairs. We then use these aligned entity pairs to train a relation embedding model such that we could use relation triples to further align the remaining entities. Lastly, we utilize a bivariate regression model to learn the respective weights of similarities measuring from the two aspects for a result combination. Our empirical study performed on several real-world datasets shows that our proposed method achieves significant improvements on entity alignment compared with state-of-the-art methods.

Keywords: Unsupervised entity alignment · Interactive model · Bivariate regression model · Relation triples · Attribute triples

G. Li et al. (Eds.): DASFAA 2019, LNCS 11446, pp. 367–382, 2019.
https://doi.org/10.1007/978-3-030-18576-3_22

1 Introduction

Entity alignment is a core task in knowledge graphs (KGs) integration, which aims to find entities referring to the same real-world object across different KGs. It is pretty challenging due to the diverse expressions and structures of knowledge in different KGs. An example of entity alignment scenario is depicted in Fig. 1, where both e_2^1 from the KG_1 and e_2^2 from the KG_2 refer to *Steven Jobs*.

Most existing approaches for entity alignment tend to utilize the relations between entities on different KGs. Typically, they encode these entities and their relations with the other entities on KGs into a semantic space, such that the similarity between entities could be measured. Various embedding models, TransE [2] and its variants [12,13,19], have been proposed to perform the encoding. However, building the relation embedding model requires a large number of aligned entity pairs for training, which may not always be easily available. In addition, entity relations do not always have high quality (such as incompleteness or inaccuracy), which may harm the accuracy of the entity alignment results. Out of these reasons, some researchers turn to explore how to leverage crowdsourcing to achieve large-scale annotated data for training [21].

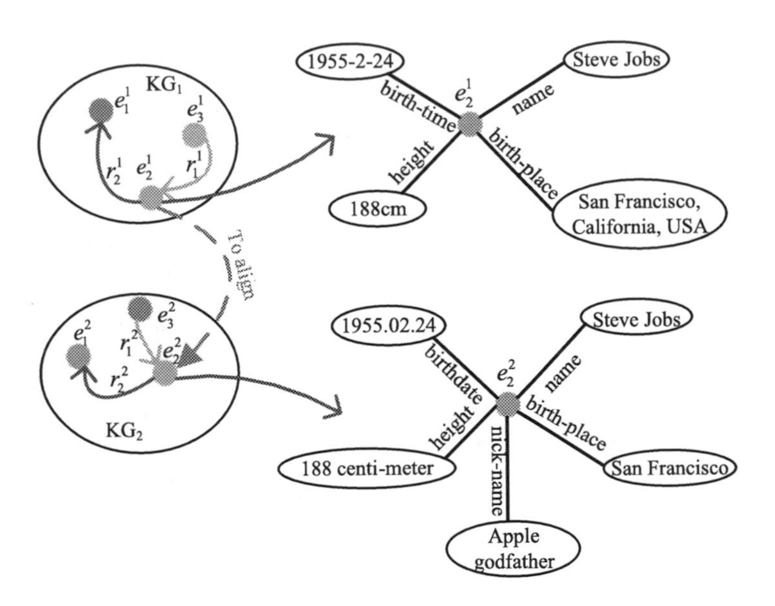

Fig. 1. An example for entity alignment between two KGs

Some recent attempt utilizes the attributes of entities in entity alignment. An alignment model called JAPE (Joint Attribute-Preserving Embedding) [17] is proposed, which jointly embeds the structures of two KGs into a unified vector space and further refines it by leveraging the attribute correlations between KGs. To avoid from tackling the problem of diverse expressions of attribute values, JAPE does not fully use the information of attribute values, but simplifies the attribute values into data types, e.g., (Steven Jobs, birthdate, 1955-02-24) is simplified into (Steven Jobs, birthdate, Datetime). So far, no existing work considers to fully use attribute names and values for entity alignment given that both the expressions of attributes and attribute values are diverse across KGs. As can be observed in Fig. 1, an attribute "birth time" could be denoted by "birthtime" in KG_1 but "birthdate" in KG_2. There also exist a lot of non-standard attribute values in KGs, e.g., a date time could be represented in several different formats such as "1955-02-24", "02/24/1955", and "24th Feb. 1955", etc.

To tackle the drawbacks of the existing approach, we propose an unsupervised entity alignment approach using both attribute triples and relation triples of KGs. Initially, we propose to fully use attribute triples for entity alignment, which will generate a lot of high-quality aligned entity pairs. We then use these aligned entity pairs to train a relation embedding model such that we could use relation triples to further align the remaining entities.

The challenge lies on how to fully use attribute triples for entity alignment. So as to tackle the two challenges mentioned above in using attribute triples for entity alignment, we novelly propose an interactive approach for entity alignment by performing entity alignment and attribute alignment alternately. It is worth noting that instead of performing entity alignment based on attribute names and values in one round, which deprives us further chances to update the alignment results, we make full use of the interaction between them to benefit each other in an iterative way. That is, we first perform entity alignment using values of common attributes and then do attribute alignment based on the matched entities, which could mutually promote the two alignment results iteratively.

Lastly, we utilize a bivariate regression model to learn respective weights of the similarity results, measured from the proposed iteration model using attribute triples and the relation embedding model using relation triples. In this way, we could fit the weights between the leveraging of relations and attributes. That is, on one hand, it is definitely beneficial for the alignment of entities with few relations in KGs to use attributes. On the other hand, taking advantage of relations could greatly reduce the harm brought by the inconsistent expressions of attributes names and values across different KGs.

We summarize our contributions as follows:

- We propose an unsupervised entity alignment approach using both attribute triples and relation triples of KGs, where the entity alignment results based on attribute triples could provide the training data for learning the relation embedding model based on relation triples.
- To deal with the challenges of diverse expressions of attributes names, we novelly propose an interactive approach for entity alignment by performing entity alignment and attribute alignment alternately.
- We utilize a bivariate regression model to fit the weights between the leveraging of relations and attributes in merging the alignment results of the two models.

Our empirical study conducted on several real-world data sets demonstrates that our entity alignment approach reaches a higher precision than the state-of-the-art relationship-embedding and attribute-embedding approaches.

Roadmap. The rest of the paper is organized as follows: We define the entity alignment problem in Sect. 2, and then present our approach in Sect. 3. After reporting our empirical study in Sect. 4, we cover the related work in Sect. 5. We finally conclude in Sect. 6.

2 Problem Definition

A typical *KG* consists of a number of *facts*, usually in the form of *triples* denoted by *(subject, predicate, object)*, where the *subject* is an *entity*, and the *object* can be either another *entity* or an attribute *value* of the subject entity. We call a triple as a *Relation Triple* if the *object* of the triple is also an entity and the *predicate* denotes the relation between the two entities. We call a triple as an *Attribute Triple* if the *predicate* denotes an attribute of the entity and the *object* of the triple is the corresponding attribute value of the entity.

Given two KGs, the task of entity alignment aims at identifying all pairs of entities referring to the same real-world objects between the two KGs. More formally, we define the entity alignment problem as follows.

Definition 1 (Entity Alignment). *Given two knowledge graphs denoted by* $KG_1 = \{E_1, RT_1, AT_1\}$ *and* $KG_2 = \{E_2, RT_2, AT_2\}$*, where* E_i*,* RT_i *and* AT_i *are the set of entities, relation triples and attribute triples of* KG_i *respectively* ($i = \{1, 2\}$)*, the task of* ***Entity Alignment*** *aims at finding every entity pair* $\{(e^1_m, e^2_n)|m \in [1, |E_1|], n \in [1, |E_2|], e^1_m \in E_1, e^2_n \in E_2, e^1_m \circeq e^2_n\}$*, where* $e^1_m \circeq e^2_n$ *means* e^1_m *and* e^2_n *refer to the same real-world object.*

3 Our Approach

The architecture of our approach is given in Fig. 2: The inputs are two KGs, KG_1 and KG_2, and the outputs are aligned entity pairs. Our approach uses both relation triples and attribute triples for entity alignment, and then combines the two alignment results by using a bivariate regression model to learn the respective weights. In our approach, we propose to fully use attribute triples for entity alignment, which will generate a lot of high-quality aligned entity pairs. We then use these aligned entity pairs to train a relation embedding model (called structure embedding) such that we could use relation triples to further align the remaining entities. For using attribute triples for entity alignment, we novelly propose an interactive approach for entity alignment by performing entity alignment and attribute alignment alternately.

In the following of this section, we first introduce the interactive model using attribute triples for entity alignment in Sect. 3.1, and then employ the results of the interactive model on the structure embedding for the training set in Sect. 3.2. We finally present how we use the bivariate regression model for alignment results combination in Sect. 3.3.

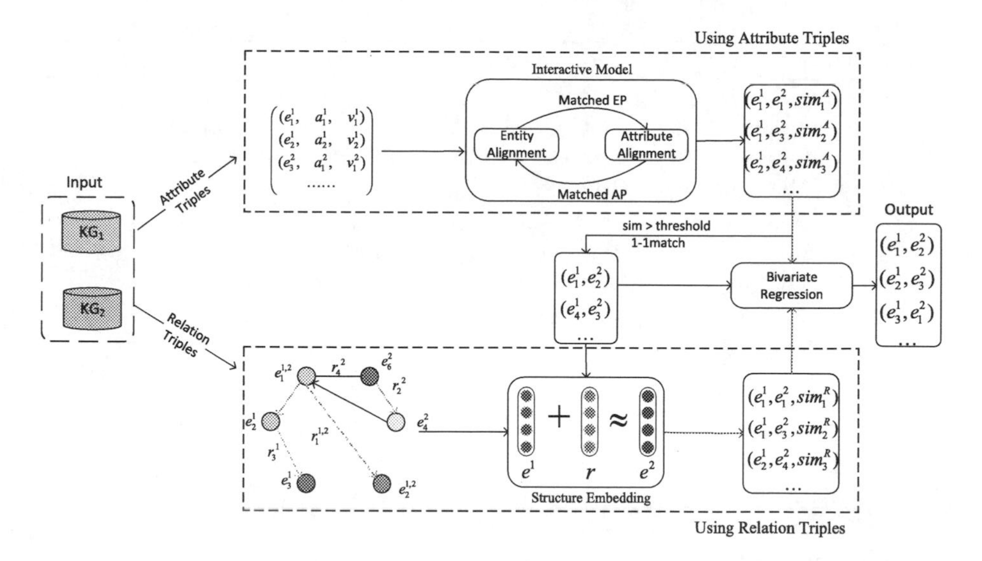

Fig. 2. Architecture of our approach

3.1 An Interactive Model Using Attribute Triples

A basic way to use attribute triples for entity alignment is to identify the percentage of common attributes and attribute values between the attribute triples of two entities, according to which we could measure a similarity between entities. However, the number of the common attributes with the same names across different KGs might be small. The diverse expressions of attribute values also aggravate the case. Thus it is very intractable for entity alignment with the basic way.

In this approach, we would like to perform attribute alignment together with entity alignment, based on the observation that the two tasks could be mutually reinforced by each other. That is, the matched entities are helpful to find more matched attributes and vice versa. Therefore, we propose an interactive model to make the two processes perform alternately.

Specifically, in each iteration, we first do entity alignment based on attribute values and build the matching set of entity pairs $\mathcal{O}_E$ increasingly and then employ the results to do attribute alignment and establish the matching set of attribute pairs $\mathcal{O}_A$ cumulatively. We iteratively repeat the above process until there are no more new common attributes or matched entities generated. The algorithm of interactive model is introduced in Algorithm 1. Next, we introduce the details of the interactive model as follows.

(1) Entity Alignment Based on Aligned Attributes. As mentioned before, we do entity alignment based on the common attributes between KGs. However, we find that there exists a huge gap in the same entities' attributes when analyzing data extracted from Baidu Encyclopedia and Wikipedia. This results

Algorithm 1. An Interaction Algorithm

Input : $KG_i = \{E_i, RT_i, AT_i\}, i = 1, 2$, where $E_i = \{e_1^i, e_2^i, ..., e_{|E^i|}^i\}$, RT_i, AT_i denote the Relation Triples and Attribute Triples. T_e, T_a denote the threshold of the accepted Entity Pairs and Attribute Pairs.

Output: Matching Entity Pairs

1. Set $\mathcal{O}_E = \mathcal{O}_A = \mathcal{O}_A^{iter} = \mathcal{O}_E^{iter} = \oslash$, where $\mathcal{O}_E, \mathcal{O}_A$ stand for all accepted matching Entity Pairs and Attribute Pairs, $\mathcal{O}_E^{iter}, \mathcal{O}_A^{iter}$ are the results of iter-th iteration. And let $iter = 0$;

2. **while** $\mathcal{O}_E^{iter} \neq \oslash$ *or* $\mathcal{O}_A^{iter} \neq \oslash$ *or* $iter = 0$ **do**

  3. $\mathcal{O}_E^{iter} = \mathcal{O}_A^{iter} = \oslash$;

  4. $\forall ep = (e_i^1, e_j^2), e_i^1 \in E_1, e_j^2 \in E_2$;

  5. **if** $Sim(e_i^1, e_j^2) \geq T_e$ **then**

    6. Add (e_i^1, e_j^2) to $\mathcal{O}_E^{iter}$,$\mathcal{O}_E$;

  end

  7. Extract two attribute sets $Attr_1, Attr_2$ from KG_1 and KG_2, among which are all belonging to the entities in $\mathcal{O}_E$;

  8. $\forall ap = (attr_i^1, attr_j^2), attr_i^1 \in Attr_1, attr_j^2 \in Attr_2$;

  9. **if** $Sim_A(attr_i^1, attr_j^2) \geq T_a$ **then**

    10. Add $(attr_i^1, attr_j^2)$ to $\mathcal{O}_A^{iter}$,$\mathcal{O}_A$;

    11. Replace $attr_i^1$ in KG_1 to $attr_j^2$;

  end

  12. $iter + +$;

end

return $\mathcal{O}_E, \mathcal{O}_A$;

from the following reasons: low coverage of attributes, various values of common attributes, and diversity of attribute names.

First, when using the values of attributes to do entity alignment, we find that the coverage of attributes is very low such that we cannot give the attributes different weights from a holistic view according to their importance, just like the way we do in relational databases. Thus, we have to repute that all common attributes of two entities share the equal weights. Based on this idea, we define the following function to calculate the similarity between two entities:

$$Sim_A(e_1, e_2) = \frac{1}{n} \sum_{k=1}^{n} Sim_V(v_k^1, v_k^2) \tag{1}$$

where $Sim_A(e_1, e_2)$ denotes the similarity between entities e_1 and e_2, v_k^1 represents the e_1's value of the $k-th$ attribute owned by both e_1 and e_2, and n is the size of all same attributes. Sim_V denotes the similarity between attribute values, v_k^1 and v_k^2, which is defined as the following equation:

$$Sim_V(v_k^1, v_k^2) = \frac{lcsSim(v_k^1, v_k^2)}{levenshteinSim(v_k^1, v_k^2) + lcsSim(v_k^1, v_k^2)} \tag{2}$$

where $levenshteinSim(\cdot,\cdot)$ is Levenshtein distance [8], a string metric for measuring the difference between two sequences. $lcsSim(\cdot,\cdot)$ measures the similarity through two strings' Longest Common Substring [9]. As a result, we can make full use of entities' common attributes to represent the entities' similarity.

Second, we also find that although entities have the common attributes, the values often vary greatly, especially for numeric attributes, e.g., the form of a person's birth time may show as "1955-02-24", "24/02/1955", or "24th Feb. 1955", etc. If calculating their similarity through $Sim_V(\cdot,\cdot)$, the results will be rather inaccurate. Moreover, due to different statistical time, the values of the number vary a lot, e.g., the values of attribute "population" for the same city in different datasources are different in different years. So as to tackle these problems, we normalize values into the most popular ones, e.g., "yyyy-mm-dd" for date. Then we extract all numbers in string through regular expression and define the following function to calculate their values' similarity through extracting number from the normalized values:

$$Sim(num_1, num_2) = \frac{|num_1 - num_2|}{max(num_1, num_2) + 1} \tag{3}$$

And only if $Sim(num_1, num_2) \leq 0.01$, num_1 will be seen as equivalence to num_2. Furthermore, we normalize some other attributes like area, length, height, population.

Last but not the least, considering the case when two entities share no common attributes, we are unable to calculate their similarity even if they refer to the same real-world object. In addition, the diversity of attribute names across KGs makes the phenomenon quite serious, such as the attribute "birth time" displaying as "birthdate" in Baidu Encyclopedia but presenting as "birth-time" in Wikipedia. We cannot use this kind of attributes well for entity alignment. Therefore, we decide to utilize the aligned entities to align this kind of attributes. The details will be introduced in the following part.

(2) Attribute Alignment Based on Aligned Entities. As mentioned before, we can find more possibly aligned attribute pairs by leveraging the set of aligned entity pairs. Supposing that in a certain iteration $iter$, we have a set of aligned entity pairs, $\mathcal{O}^E_{iter} = \{ep_1, ep_2, ep_3, ..., ep_{|\mathcal{O}^E_{iter}|}\}$ where ep_i is the aligned entity pair (e^1_i, e^2_j). Next, for the candidate attribute pairs (a^1_m, a^2_n) which are possessed by the subset of $\mathcal{O}^E_{iter}$ denoted as $EP_{(a^1_m,a^2_n)} = \{ep_1, ep_4, ep_{10}, ..., ep_k\}$. We leverage all values of attributes owned by entity pairs to represent the similarity of attribute pairs. For each $ep_l (1 \leq l \leq k)$ whose attribute values are v^i_m and v^i_n for attribute pair (a^1_m, a^2_n), the similarity can be calculated with the following equation:

$$Sim_A(a^1_m, a^2_n) = \frac{1}{Z}\sum_{i=1}^{Z} Sim_V(v^i_m, v^i_n) \tag{4}$$

where $Z = |EP_{\{a^1_m,a^2_n\}}|$ represents the size of $EP_{(a^1_m,a^2_n)}$ and $SimV(\cdot,\cdot)$ denotes the similarity between attribute values. Here we take a running example to illustrate our model as follows.

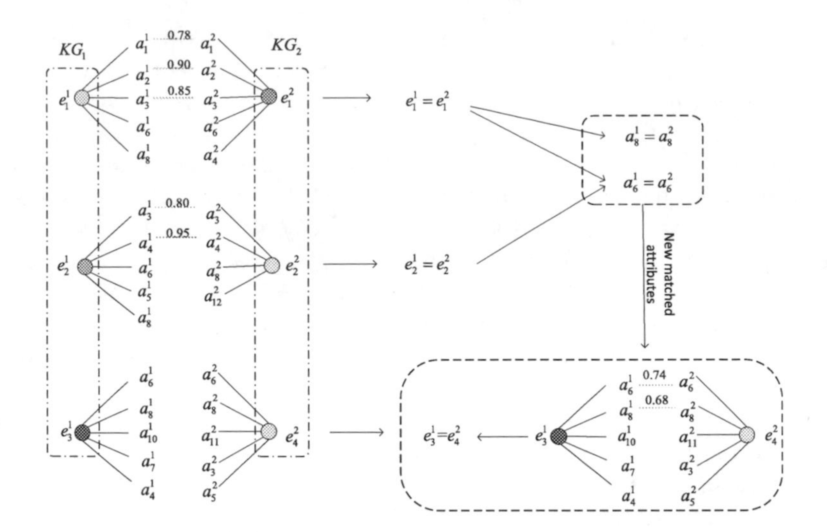

Fig. 3. An example of interactive alignment model

Example 1. As we can see in Fig. 3, we suppose the precondition is $a_i^1 = a_i^2, i \in [1, 5]$ *and we do not know the matching results of the remaining attributes. Under the precondition, each entity pair has at least one common attribute except* (e_3^1, e_4^2)*. Through leveraging these attribute alignment seeds, we can get two aligned entity pairs* (e_1^1, e_1^2) *and* (e_2^1, e_2^2)*, but they are not able to determine whether* e_3^1 *and* e_4^2 *are aligned. Then, based on the two aligned entity pairs, we do attribute alignment. Two more attribute matching pairs* (a_6^1, a_6^2) *and* (a_8^1, a_8^2) *are generated. In the next iteration, leveraging the attribute alignment results generated in the first iteration, we can determine the matching result of the entity pair* (e_3^1, e_4^2)*, which is aligned.*

3.2 Structure Embedding

We select entity pairs with high confidence (larger than a predefined threshold) from the results of interactive model as the training set for structure embedding. Then we use relation triples and the training set to do structure embedding (SE), aiming to model the geometric structures of two KGs and learning approximate representations for entities and relations. Formally, given a relation triple $tr = (h, r, t)$ where h is the head entity and r is the tail entity, we expect $h + r = t$. To measure the plausibility of tr, SE model optimizes the margin-based ranking loss to make the scores of positive triples lower than negative ones [17]:

$$\mathcal{O}_{SE} = \sum_{tr \in Tr} \sum_{tr' \in Tr'} (f(tr) - \alpha(tr')) \tag{5}$$

where $f(tr) = ||h+r-t||_2^2$ is the score function, Tr and Tr' are the sets of all positive triples and the associated negative triples, and α is a ratio hyper-parameter that weights positive and negative triples and its range is [0, 1]. Through the above embedding process, we could learn approximate vector representations of the entities across KGs, e.g., in order to get the similarity between an entity pair (e_i, e_j), we can calculate its similarity with entities' embedding results $(\boldsymbol{e_i}, \boldsymbol{e_j})$ according to the following equation:

$$simR(e_i, e_j) = \cos(\boldsymbol{e_i}, \boldsymbol{e_j}) \tag{6}$$

3.3 A Bivariate Regression Model for Weights Allocation

As we said before, we represent the similarity between entities both from the aspects of relations and attributes. Specifically, we incorporate the similarities calculated by the relation triples and attribute triples in linear weighting. We utilize the following formulation to represent the final similarity of any entity pair $e_i \in KG_1$ and $e_j \in KG_2$:

$$Sim(e_i, e_j) = \lambda \cdot simR(e_i, e_j) + (1 - \lambda) \cdot simA(e_i, e_j) \tag{7}$$

where $simR(\cdot, \cdot)$ is the similarity measure to calculate the embedding similarity using relation triples, and $simA(\cdot, \cdot)$ represents the similarity of entities calculated through their attribute triples. λ is the parameter to balance the importance of left part and right part.

Instead of setting parameter λ artificially, we take the aligned entity pairs as training data to learn λ through a bivariate regression model. It comes from the fact that for different datasets, the respective importance of relations and attributes should be different, i.e., for a data set full of entities with high-quality relations, the weight of relations should have a higher confidence. On the contrary, if the number of entities' relations is quite small, attributes should be assigned with a higher weight. Specifically, we leverage those training entity pairs, also treating as the training set of *Unsupervised SE* model, to construct the input of our regression model. e.g., for a matched entity pair (e_i, e_j), we suppose it's final similarity to be 1.0. So we let the input form be $(simR(e_i, e_j), simA(e_i, e_j), 1.0)$. We want the final similarities $Sim(e_i, e_j)$ of these matching entity pairs close to 1.0 as much as possible. Besides, we use MSE (Mean Squared Error) [20] as our loss function and SGD [3] as the optimizer.

4 Experiments

We first introduce our datasets and the metrics we would use for evaluation, and then represent the existing state-of-the-art approaches that we would compare with. Finally, we evaluate our proposed approach on several metrics.

4.1 Datasets and Metrics

We collect our data from Baidu Encyclopedia[1], Wikipedia[2] and Hudong Encyclopedia[3]. We extract triples from the infobox, if a value in the infobox is a link or contains several links, we will save it as a relation triple or separate it into several relation triples. Otherwise, we take it as an attribute triple. Then, we do object linking for all triples to replace the values with the URIs of the entities that they actually refer to in the KG. Specially, for some relation triples, whose objects are not in our KG, we treat them as attribute triples, aiming to construct a small but complete KG. Here, we have two datasets where one is about persons entities, places entities and others, while the other is about natural creatures, like animals and plants. As we can see in Table 1, the number of attribute triples are nearly 4 times of relation triples in the first dataset while the number of relation triples and attribute triples are almost equal in the second dataset.

Table 1. Statistics of the datasets

Datasets		Entities	Relations	Attributes	Rel. triples	Attr. triples
Dataset1	Baidu	12,647	715	5,166	29,373	108,052
	Wiki	8,218	231	1,904	14,351	66,249
Dataset2	Baidu	13,983	470	2,322	79,025	79,658
	Hudong	10,263	79	870	44,613	56,270

As for the evaluation metrics, we use $Hits@k$ and $Mean$ to assess the performance of the approaches. $Hits@k$ reflects the proportion of correctly aligned entities ranked in the top k. And $Mean$ calculates the mean of these ranks. Naturally, a higher $Hits@k$ and a lower $Mean$ indicate a better performance. We also evaluate the influence on the precision of the iterative time in the interactive model.

4.2 Approaches for Comparison

In this section, we briefly introduce five comparative methods including *SE* [2], *JAPE* [17], our proposed *Baseline Model*, the *Interactive Model* and *Interactive Model + Unsupervised SE*.

- The *SE* method aims to learn representations of all entities and relations in KG_1 and KG_2. In order to model the geometric structures of two KGs, *SE* serves the seed alignments as bridge to build an overlay relationship graph, essentially encoding the entities and relationships of various KGs into a semantic space to measure the similarity between entities.

[1] https://baike.baidu.com.
[2] https://zh.wikipedia.org.
[3] http://www.baike.com.

- The *JAPE* method jointly embeds the structures of two KGs into a unified vector space and further refines it by leveraging attribute correlations in the KGs. It employs two models, namely structure embedding (SE) and attribute embedding (AE), to learn embedding of KGs where SE models relation structures of two KGs from the aspect of relation triples and AE attempts to encode the entities more accurately leveraging attribute triples.
- The *Baseline* only leverages the attribute values to do entity alignment just one time and attribute alignment is not involved.
- The *Interactive Model* (*IM* for short) also just considers the attribute values, but different from *Baseline*, it also lets entity alignment and attribute alignment benefit each other iteratively.
- The *Interactive Model + Unsupervised SE* (*IMUSE* for short) method takes some high-quality entity pairs generated by the *Interactive Model* as the training data of *Unsupervised SE* model, and then incorporates results of two models by linear weighting.

4.3 Experimental Results

(1) Top-K and Mean Results. For *SE* and *JAPE*, we use 60% entity pairs of the groundtruth as alignment seeds while the rest is the testing data. Note that the other three methods do not need these seeds for training. The predefined thresholds are 0.89 and 0.92 respectively for dataset1 and dataset2. As can be seen from Table 2, when varying $hits@k$, *IMUSE* always largely outperforms the other methods on the two datasets in that it incorporates relation embedding method with attribute-value based method which can find more aligned entities. And *JAPE* performs better than *SE* since it encodes entities through jointly embedding the relation and attribute triples. And we can see that the iterative process enables *IM* perform better than *Baseline*, because *IM* can benefit from the interaction to align more entities. We can also see that in Table 2, *IMUSE* has the smallest Mean values among *IM* and *Baseline* in two datasets while *SE* and *JAPE* have the highest Mean values.

Table 2. Comparison with Hits@k and Mean

Approaches	Dataset1				Dataset2			
	Hits@1	Hits@5	Hits@10	Mean	Hits@1	Hits@5	Hits@10	Mean
SE	0.5825	0.6539	0.7464	29	0.6331	0.6570	0.6982	58
JAPE	0.6317	0.7182	0.8157	17	0.7029	0.7362	0.7451	30
Baseline	0.7162	0.9448	0.9682	8	0.6102	0.8943	0.9270	14
IM	0.7901	0.9590	0.9747	5	0.6809	0.9274	0.9416	10
IMUSE	**0.8232**	**0.9593**	**0.9755**	**3**	**0.7336**	**0.9328**	**0.9560**	**7**

It is noticeable that the results of these models on the two datasets perform a little difference on $Hits@1$, which results from the following reasons. (1) Entities in the first dataset have an average of 2 relations and 8 attributes while entities in the second dataset have 5 relations and 5 attributes averagely. Therefore, *SE*, only using entities' relation information, performed worse than *Baseline* and *IM* on the dataset1 but better on the dataset2. (2) As is shown in Table 1, the kinds of relations and attributes are much smaller in the dataset2 than in the dataset1. So, these methods performed a little worse on the dataset2. (3) Entities in the dataset2 are mainly about animals and plants, whose properties (relations and attributes) are almost "Kingdom", "Phylum", "Classes", "Family", etc, and the values under these properties are very similar. So, the final result of dataset2 is not so good as that in the dataset1.

(2) The Effectiveness of Iterations. We also evaluate the influence of iterations on *Hits@1* for our proposed method *IMUSE*. As shown in Fig. 4, with the increase of iterative time, *Hits@1* also went up with it since new aligned attributes contribute to aligning more entities and vice versa. In addition, when the iterative time increases to 4 and 3 for the dataset1 and the dataset2 respectively, *Hits@1* of them keep stable in that no more entities or attributes can be aligned.

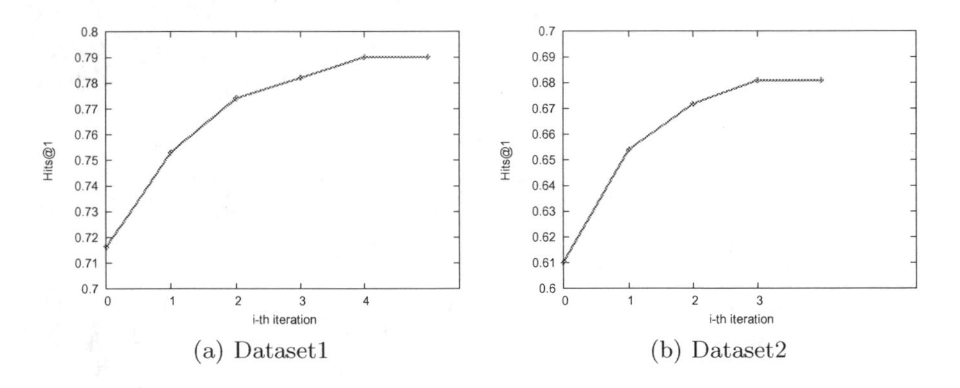

Fig. 4. The effectiveness of iteration

Table 3. The effectiveness of dynamic combination

	SE	IM	Static combination	Dynamic combination
Dataset1	0.5825	0.7901	0.8127	**0.8232**
Dataset2	0.6331	0.6809	0.7110	**0.7336**

(3) Dynamic Combination (Bivariate Regression) v.s. Static Combination. Instead of combining the results generated by *Interactive Model* and *Unsupervised SE* in a static way, we choose to learn respective weights of similarities measuring from the aspects of relations and attributes through bivariate regression, called *Dynamic Combination*. On the contrary, we manually give the

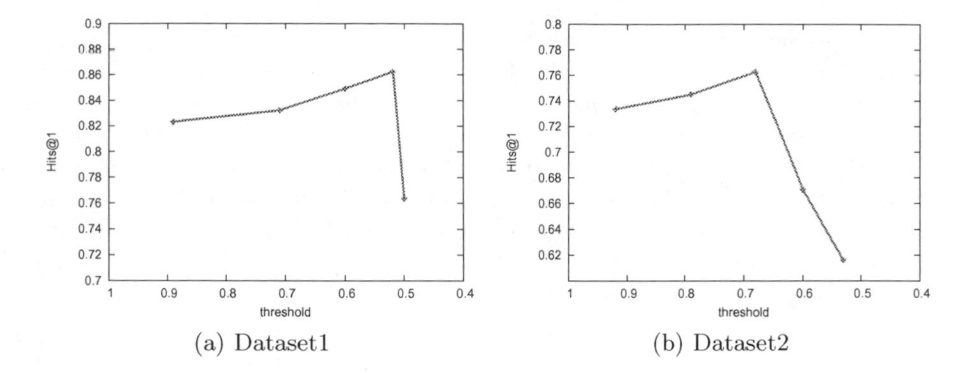

(a) Dataset1 (b) Dataset2

Fig. 5. The effectiveness of different thresholds

two results the same weights to combine them together to represent the process of *Static Combination.* As is shown in Table 3, *Dynamic Combination* truly performs better than *Static Combination* because learning weights contain the information of different importance of the entities' relations and attributes when doing entity alignment.

(4) The Effectiveness of Different Predefined Thresholds. As mentioned before, the predefined threshold enables us to select entity pairs with high confidence as the training set for structure embedding. Obviously, the higher the threshold is, the higher accuracy that the training set is, but the less training entity pairs it gets. Intuitively, there may be a balance between the accuracy and quantity. In order to find the optimal result, according to the size of training set, we set five different thresholds, corresponding to 30%, 40%, 60%, 80% and 90% of the size of SE's training set, to illustrate the idea on two datasets.

As we can see in Fig. 5, on dataset1 when the threshold decreases, the value of hit@1 goes up. When the threshold is 0.52, the accuracy is the highest, 86.25%. After that, it begins to drop. Similarly, on the dataset2 the accuracy reaches the best result, 76.27%, when the threshold is 0.68. Therefore, we can conclude that the predefined threshold truly plays an important role to the final results.

5 Related Work

Entity alignment is a sub-problem of KG integration. In this section, we first give some introduction to a similar problem of KG integration called database integration, and then cover some mainstream methods on entity alignment.

5.1 Database Integration v.s. KG Integration

Data integration in relational databases has gained lots of attentions [11], which consists of two main tasks, i.e., Record Matching [5] and Schema Matching [16]. Record Matching aims at identifying records in the same or different databases that refer to the same real-world objects.

There are several typical approaches commonly used to detect approximately duplicate records such as probabilistic methods [14], supervised or unsupervised machine learning techniques [15,18], variations based on active learning [4], etc. Schema Matching refers to matching combinations of elements that appear together in a structure. In current implementations, schema matching is typically performed manually, which has significant limitations. Of course, some automatic matching algorithms have been proposed based on name similarity [1], description similarity [10], etc. Besides, some other methods [6] incorporate crowdsourcing to improve the accuracy of automatic algorithms.

Compared to the relational database integration, KG integration is more challenging due to the complex structures of KGs and more diverse expressions of properties and values. In relational databases, entities belonging to the same field always appear in the same table. Even if some difficulties, like transcription errors and incomplete information, actually exist, these entities' properties are all same to each other. However, due to diverse expressions of properties and values in KGs, only a few entities have identical properties. As a result, entity alignment in KGs are faced with more challenges.

5.2 Entity Alignment

Entity alignment is pretty challenging due to the diverse expressions and structures of knowledge in different KGs. So far, plenty of work has been done on this problem, most of which are based on embedding methods. Some of them only leverage relation triples, commonly called KG Embedding, while the others jointly embed both relation and attribute triples, like JAPE [17].

KG Embedding. Recently, a lot of researchers have been trying every effort to learning and improving KG embedding. For example, TransE [2], the basic of all embedding based methods, treats a relation triple as (h, r, t) where h is head entity, r is relation, and t is tail entity. TransE tries to embed entities and relations of multi-relational data to low-dimensional vector spaces. So as to decrease the costs of training and reduce the number of parameters, Bordes et al. interpret relations as translating operations from head to tail entities, which can be expressed as $\mathbf{h} + \mathbf{r} = \mathbf{t}$. Since then, a lot of embedding models have been proposed to improve TransE, such as TransR [13] and PTransE [12]. Different from TransE, they project both entities and relations into a continuous vector space. Specifically, TransR model maps entities and relations into separated entity space and relation space, and performs translation in relation space. Considering that TransE only leverages individual triples and ignores multi-step relation paths, PTransE [12] encodes multi-step paths to address this issue. However, most of these methods just focus on how to encode relation triples in better ways, neglecting those attribute triples. Especially for the entities lacking in relations, it does not work well if only leveraging relations.

JAPE. In order to take advantage of attribute triples, Sun et al. propose an alignment model called JAPE (Joint Attribute-Preserving Embedding) which focuses on the cross-lingual entity alignment [7]. It jointly embeds the structures of two KGs and further refines it by leveraging attribute correlations in the KGs. Specifically, it employs two models, namely structure embedding (SE) and attribute embedding (AE), to learn embedding of KGs where SE models relation structures of two KGs from the aspect of relation triples and AE attempts to cluster those attributes often used together to describe an entity. The average of attributes embedding results will be used to encode entities. As a result, those entities which possess similar attributes will be close to each other. Finally, it combines SE and AE to jointly embed all the entities in two KGs into a unified vector space. However, JAPE does not use the information of attribute values well. Alternatively, so as to tackle the problem of diverse expression of values, they reduce the attribute values to data type, e.g., (Barack Obama, birthdate, 1961-08-04) is replaced by (Barack Obama, birthdate, Datetime). Obviously, it is just a coarse-grain way to deal with attribute values where the importance of attribute values is ignored. In order to overcome the heterogeneity between different KGs, we propose an interactive approach to integrate entity alignment and attribute alignment together.

6 Conclusions and Future Work

We work on leveraging both relation and attribute triples to do entity alignment between two KGs through our proposed model. In order to make full use of attribute triples, we propose an interactive method to do entity alignment and attribute alignment to calculate the similarities. And then we utilize the results of the interaction as the training set to embed relation triples for computing the similarities in an unsupervised manner. Last, we incorporate the two kinds of results in linear weights to represent the final similarities of entity pairs through a bivariate regression model. Our experiments on real-world datasets demonstrate that our approach performs much better than the state-of-the-art methods.

In the future work, we look forward to decreasing the influence of non-standard attribute values by learning patterns to build standard data forms. In addition, we would like to turn to crowdsourcing to help decide whether attribute pairs with great difference in expressions should be aligned.

Acknowledgments. This research is partially supported by National Natural Science Foundation of China (Grant No. 61632016, 61572336, 61572335, 61772356), the Natural Science Research Project of Jiangsu Higher Education Institution (No. 17KJA520003, 18KJA520010), and the Open Program of Neusoft Corporation (No. SKLSAOP1801).

References

1. Bell, G.B., Sethi, A.: Matching records in a national medical patient index. Commun. ACM **44**(9), 83–88 (2001)
2. Bordes, A., Usunier, N., Garcia-Duran, A., Weston, J., Yakhnenko, O.: Translating embeddings for modeling multi-relational data. In: Advances in Neural Information Processing Systems, pp. 2787–2795 (2013)

3. Bottou, L.: Large-scale machine learning with stochastic gradient descent. In: Saporta, G., Lechevallier, Y. (eds.) COMPSTAT 2010, pp. 177–186. Springer, Heidelberg (2010). https://doi.org/10.1007/978-3-7908-2604-3_16
4. Cohn, D., Atlas, L., Ladner, R.: Improving generalization with active learning. Mach. Learn. **15**(2), 201–221 (1994)
5. Elmagarmid, A.K., Ipeirotis, P.G., Verykios, V.S.: Duplicate record detection: a survey. IEEE Trans. Knowl. Data Eng. **19**(1), 1–16 (2007)
6. Fan, J., Lu, M., Ooi, B.C., Tan, W.C., Zhang, M.: A hybrid machine-crowdsourcing system for matching web tables. In: 2014 IEEE 30th International Conference on Data Engineering (ICDE), pp. 976–987. IEEE (2014)
7. Fu, B., Brennan, R., O'Sullivan, D.: Cross-lingual ontology mapping – an investigation of the impact of machine translation. In: Gómez-Pérez, A., Yu, Y., Ding, Y. (eds.) ASWC 2009. LNCS, vol. 5926, pp. 1–15. Springer, Heidelberg (2009). https://doi.org/10.1007/978-3-642-10871-6_1
8. Heeringa, W.J.: Measuring dialect pronunciation differences using Levenshtein distance. Ph.D. thesis. Citeseer (2004)
9. Hirschberg, D.S.: A linear space algorithm for computing maximal common subsequences. Commun. ACM **18**(6), 341–343 (1975)
10. Larson, J.A., Navathe, S.B., Elmasri, R.: A theory of attributed equivalence in databases with application to schema integration. IEEE Trans. Softw. Eng. **15**(4), 449–463 (1989)
11. Lenzerini, M.: Data integration: a theoretical perspective. In: Proceedings of the Twenty-First ACM SIGMOD-SIGACT-SIGART Symposium on Principles of Database Systems, pp. 233–246. ACM (2002)
12. Lin, Y., Liu, Z., Luan, H., Sun, M., Rao, S., Liu, S.: Modeling relation paths for representation learning of knowledge bases. arXiv preprint arXiv:1506.00379 (2015)
13. Lin, Y., Liu, Z., Sun, M., Liu, Y., Zhu, X.: Learning entity and relation embeddings for knowledge graph completion. In: AAAI, vol. 15, pp. 2181–2187 (2015)
14. Palopoli, L., Saccá, D., Terracina, G., Ursino, D.: A unified graph-based framework for deriving nominal interscheme properties, type conflicts and object cluster similarities. In: Proceedings of 1999 IFCIS International Conference on Cooperative Information Systems, CoopIS 1999, pp. 34–45. IEEE (1999)
15. Perkowitz, M., Doorenbos, R.B., Etzioni, O., Weld, D.S.: Learning to understand information on the internet: an example-based approach. J. Intell. Inf. Syst. **8**(2), 133–153 (1997)
16. Rahm, E., Bernstein, P.A.: A survey of approaches to automatic schema matching. VLDB J. **10**(4), 334–350 (2001)
17. Sun, Z., Hu, W., Li, C.: Cross-lingual entity alignment via joint attribute-preserving embedding. In: d'Amato, C., et al. (eds.) ISWC 2017. LNCS, vol. 10587, pp. 628–644. Springer, Cham (2017). https://doi.org/10.1007/978-3-319-68288-4_37
18. Verykios, V.S., Elmagarmid, A.K., Houstis, E.N.: Automating the approximate record-matching process. Inf. Sci. **126**(1–4), 83–98 (2000)
19. Wang, Z., Zhang, J., Feng, J., Chen, Z.: Knowledge graph embedding by translating on hyperplanes. In: AAAI, vol. 14, pp. 1112–1119 (2014)
20. Wang, Z., Bovik, A.C.: Mean squared error: love it or leave it? a new look at signal fidelity measures. IEEE Sig. Process. Mag. **26**(1), 98–117 (2009)
21. Yang, J., Fan, J., Wei, Z., Li, G., Liu, T., Du, X.: Cost-effective data annotation using game-based crowdsourcing. Proc. VLDB Endow. **12**(1), 57–70 (2018)

Combining Meta-Graph and Attention for Recommendation over Heterogenous Information Network

Chenfei Zhao(✉), Hengliang Wang, Yuan Li, and Kedian Mu

School of Mathematical Sciences, Peking University,
Beijing 100871, China
zhaochenfei@pku.edu.cn

Abstract. Recently heterogeneous information network (HIN) has gained wide attention in recommender systems due to its flexibility in modeling rich objects and complex relationships. It's still challenging for HIN based recommenders to capture high-level structure and fuse the mined features of users and items effectively. In this paper, we propose an approach for the recommendation over HIN, called MGAR, which combines Meta-Graph and Attention to address the challenge. Informally speaking, meta-graph is applied to feature extraction, so as to capture more semantic information, while the attention mechanism is used to fuse the features arising from different meta-graphs. MGAR can be divided into two stages. In the first stage, we apply the matrix factorization technique to generate latent factors based on predefined meta-graphs. In the second stage, the embeddings of users and items are fused with the neural attention mechanism. And then the deep neural network is employed to make recommendations by modeling complicated interactions. Experiments over two real datasets indicate MGAR achieves state-of-the-art performance.

Keywords: Heterogeneous information network · Meta-graph · Attention · Recommender system

1 Introduction

Recommender system gains extensive attention with the widespread use of the Internet. Users' preferences can be mined from the feedback of items that were rated or clicked in recommender systems. For example, on Yelp[1], users are provided a 5-point rating scale to express their likes and dislikes of each item. Existing models mainly utilized the rating information to make recommendations. But they can not address the sparsity and cold start issues. Actually, there is rich semantic information in different types of data, which can be considered as a starting point to alleviate the above two issues. But it is not yet easy to incorporate the heterogeneity of data to the recommender system.

[1] http://www.yelp.com.

G. Li et al. (Eds.): DASFAA 2019, LNCS 11446, pp. 383–397, 2019.
https://doi.org/10.1007/978-3-030-18576-3_23

Heterogeneous information network (HIN) [20] is proposed to capture rich semantic information in data, which is promising to solve the challenges of sparsity and cold start. Figure 1(a) shows a example of HIN schema on Yelp dataset. Comparing with the homogeneous information network, HIN has multiple objects and relations. To capture the semantic information of HIN, a network schema, called meta-path, has been proposed in [19]. Meta-path based methods over HIN have been applied in many different domains, such as social network analysis [10] and relation graph [8]. Recently, meta-path is utilized in the recommender system as a powerful representation of heterogeneous types of data [26]. Two objects in HIN can be connected via different meta-paths and these paths have different meanings. From HIN's schema on Yelp dataset, we can build a meta-path $User \rightarrow Item \rightarrow User$ to capture the semantic relation that users provide ratings to the same item. However, meta-path may not suitable to incorporate complex relationships. For example, if we want to obtain user's similarity by user's similar reviews on items, a meta-path $User \rightarrow Review \rightarrow Aspect \rightarrow Review \rightarrow Item$ can be defined to capture this similarity according to users' reviews on the same aspect. But if we want to capture the semantic that users have similar review content, and at the same time, they rate on the same item, none of single meta-path can model such complex information.

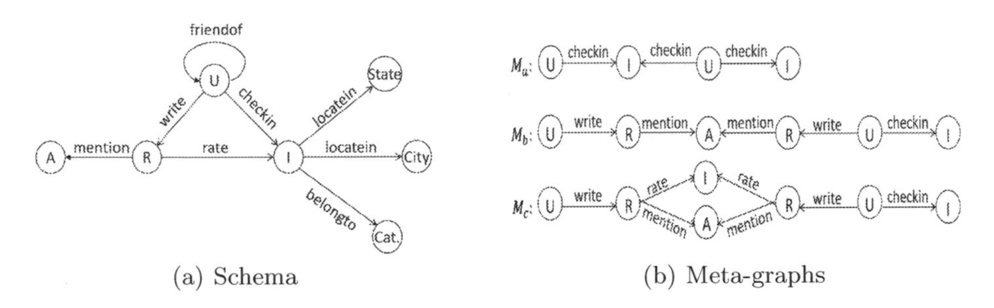

Fig. 1. Example of HIN schema and meta-graphs on Yelp dataset. A: aspects extracted from reviews; R: reviews; U: users; I: items; Cat: category of item.

As an extension of meta-path, a new network schema called meta-graph has been proposed to model complex information and incorporate richer semantics [6]. M_c presented in Fig. 1(b) is a meta-graph, which describes that users not only rated on the same item and they have also mentioned the same aspect. The meta-path is a special case of meta-graph, such as M_a and M_b presented in Fig. 1(b). As such, introducing meta-graph in the HIN-based recommender system can allow us to consider the relation between users and items meticulously. Then, we can capture more semantic information in HIN.

Nevertheless, different meta-graphs have different effects on recommendation. For example, $User \rightarrow Review \rightarrow Aspect \rightarrow Review \rightarrow Item$ counts more when we compare some similar products of the different brand, while

$User \rightarrow Item \rightarrow User$ matters when we want to compare user's preference on the same item. Thus, each meta-graph should be attached a different weight. However, as far as we know, meta-graph based works, such as [27], did not consider that different meta-graphs may have different contributions to the recommendation procedure explicitly. Actually, attention mechanism presented in [12] provides a promising way to address such an issue. Note that a prominent advantage of attention mechanism is the ability to attach attentive weights to the obtained latent features. The attention mechanism has made a significant improvement in many different domains, like computer vision [22,23] and natural language processing [1,18]. Recently, attention mechanism is applied to recommender system [2,4].

In this paper, we propose a two-stage method for HIN-based recommendation. Our contributions are listed as follows.

- We propose an approach for recommendation over HIN, called MGAR, which combines Meta-Graph and Attention.
- Predefined meta-graphs are utilized to model the multiple types of nodes and links, while attention mechanism is used to integrate the mined features from different meta-graphs.
- Experiments are conducted on Amazon and Yelp datasets. The experimental results validate the effectiveness of our model.

The rest of the paper is organized as follows: Sect. 2 presents a brief overview of related work. We introduce our model in detail in Sect. 3. And we analyze the experimental results in Sect. 4. At last, a conclusion is presented in Sect. 5.

2 Related Work

In this section, we will briefly introduce heterogeneous information network (HIN) and recommendations in HIN.

2.1 Heterogeneous Information Network

HIN has rapidly become a hot research topic, which contains multiple types of objects and relationships. Meta-path based methods over HIN have been widely applied in data mining tasks over HIN, including classification [7], clustering [21], especially recommendation [17,26], etc. For example, the graph-based ranking model established on meta-path was built to iteratively computing the ranking distribution of the objects within each class [7]. Sun et al. [21] integrated the meta-path selection problem with the user-guided clustering problem in HIN to generate the clusters under the learned weights of meta-paths. Yang et al. [24] proposed a semantic path-based similarity measure for weighted HIN, which can capture the semantics of weighted meta-path. However, meta-path fails to describe complex relations over HIN. Recently, meta-graph was proposed to model complicated information and incorporate the richer semantics in HIN [6].

Huang et al. [6] computed the relevance between the same type of two objects based on meta-graphs in large HIN. Jiang et al. [9] proposed a meta-graph guided random walk ensemble method in HIN for classification. In this paper, we exploit the rich information provided by meta-graphs. And we utilize user-item matrices based on meta-graphs to generate latent features for the prediction.

2.2 Recommendations in HIN

Recommendation over HIN has been paid much attention recently. Yu et al. [25] proposed a matrix factorization method with similarity regularization for recommendations in HIN. In [26], different types of meta-paths were used to generate user and item latent features, then a recommendation model was designed for both global and personalized levels. SemRec [17] proposed by Shi et al., calculated similarities between users with same ratings based on meta-paths, then combined the similarities with all meta-paths by a weight mechanism. Han et al. [4] proposed a neural network model based on meta-path to exploit different aspect latent factors, such as the brand-aspect and category-aspect of items. All of the above methods for recommendations are based on meta-paths. In [27], FMG model proposed by Zhao et al. used Factorization Machine with Group lasso based on latent features generated by matrix factorization on meta-graphs. FMG did not yet consider the explicit difference of meta-graphs for the recommendation. To address this problem, our approach utilizes the attention mechanism to make recommendations on meta-graphs with different weights.

3 Meta-Graph Based Neural Network Model

In this section, we present an approach for recommendation over HIN, called MGAR, by combining Meta-Graph and Attention. For capturing high-level structure and fusing the mined features of users and items effectively, the proposed MGAR method can be divided into two stages. In the first stage, matrix factorization (MF) based method is used to obtain the embeddings of users and items. Then, we aggregate the embedding of users and items with the attention mechanism and use DNN to make a prediction of ratings in the second stage.

3.1 Node Embedding Based on Meta-Graph

Definition 1. ***Meta-Graph*** [27]. *A meta-graph $\mathcal{M}$ is a directed acyclic graph (DAG) with a single source node n_s and a single sink node n_t, defined on a HIN $\mathcal{G} = (\mathcal{V}, \mathcal{E})$ with schema $\mathcal{T}_G = (\mathcal{A}, \mathcal{R})$, where $\mathcal{V}$ is the node set, $\mathcal{E}$ is the edge set, $\mathcal{A}$ is the node type set, and $\mathcal{R}$ is the edge type set. Then we define a meta-graph as $\mathcal{M} = (\mathcal{V}_M, \mathcal{E}_M, \mathcal{A}_M, \mathcal{R}_M, n_s, n_t)$, where $\mathcal{V}_M \subset \mathcal{V}, \mathcal{E}_M \subset \mathcal{E}$ constrained by $\mathcal{A}_M \subset \mathcal{A}$ and $\mathcal{R}_M \subset \mathcal{R}$, respectively. A* ***meta-path*** *is a special case of meta-graph when the schema $\mathcal{T}_G$ is a path.*

The definition of HIN and HIN schema have been introduced in [19]. From the definition we can know that each meta-graph contains explicit semantic information in HIN. As chosen in [27], we use the meta-graphs presented in Fig. 2 on Yelp and Amazon datasets without $U \xrightarrow{buy} I \xrightarrow{belongto} Brand \xleftarrow{belongto} I$. The reason is that we can only extract little information from $U \xrightarrow{buy} I \xrightarrow{belongto} Brand \xleftarrow{belongto} I$ on Amazon dataset.

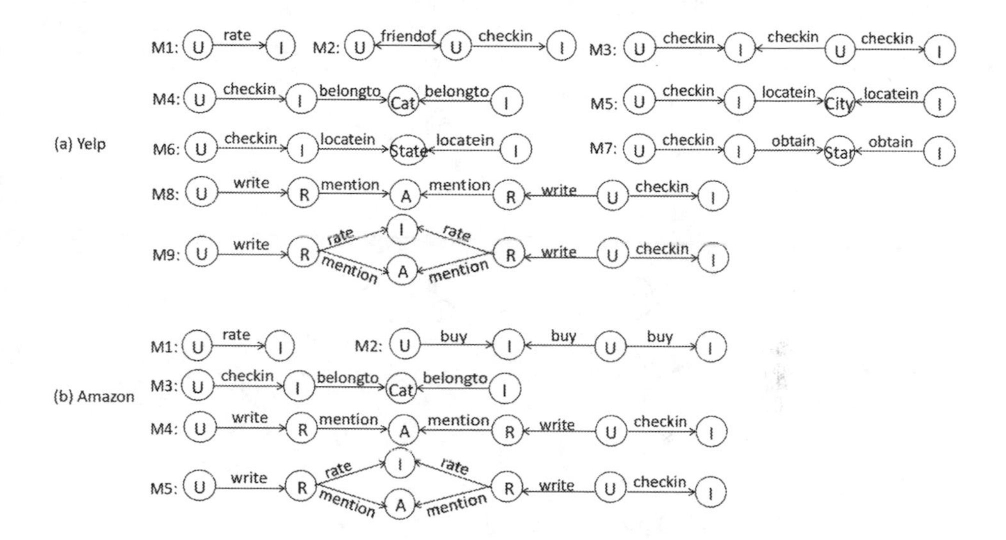

Fig. 2. Meta-graphs used for Yelp and Amazon datasets

In order to obtain embeddings of users and items, we need to calculate the commuting matrix between users and items. For a given HIN $\mathcal{G}$ and a meta-path $P = (A_1, A_2, \cdots, A_n)$, the commuting matrix of the meta-path is calculated by Eq. 1:

$$\boldsymbol{C}_P = \boldsymbol{M}_{A_1,A_2}\boldsymbol{M}_{A_2,A_3}\cdots\boldsymbol{M}_{A_{n-1},A_n}, \tag{1}$$

where $A_i \in \mathcal{A}$ denotes a node type, for $i = 1, 2\cdots n$ and n is the length of meta-path. $\boldsymbol{M}_{A_i,A_j}$ denotes the adjacency matrix between type A_i and type A_j.

Next, we extend the meta-path based method to meta-graph based method. As presented in [6], a meta-graph can be spilt into many meta-paths. The method provided by [27] to calculate similarity matrices of meta-graphs can only be used to calculate similarity matrix like M_9 in Fig. 2. Our method is a generalization of the previous method presented in [27]. A meta-graph can be seen as a network with many layers, where each layer refers to the objects from one or more types. As shown in Fig. 3, a meta-graph can be treated as a meta-path from layers' perspective. For a meta-graph $G = (L_1, L_2, \cdots, L_n)$ represented in layers' perspective, the commuting matrix of G can be calculated by Eq. 2, which is an extension of Eq. 1:

$$C_G = \boldsymbol{M}_{L_1,L_{j_1}} \boldsymbol{M}_{L_{j_1},L_{j_2}} \cdots \boldsymbol{M}_{L_{j_m},L_n}, \tag{2}$$

where $1 < j_1 < j_2 < \cdots < j_m < n$, and for all $k = 1, 2, \cdots, m$, L_{j_k} only have one node type. The adjacency matrix $\boldsymbol{M}_{L_{j_k},L_{j_{k+1}}}$ between $L_{j_k}, L_{j_{k+1}}$ is defined by the paths between two layers. We denote the commuting matrix of paths between $L_{j_k}, L_{j_{k+1}}$ as $\boldsymbol{M}_1^k, \cdots, \boldsymbol{M}_{k_l}^k$. The adjacency matrix is defined as follows:

$$\boldsymbol{M}_{L_{j_k},L_{j_{k+1}}} = \boldsymbol{M}_1^k \odot \boldsymbol{M}_2^k \cdots \odot \boldsymbol{M}_{k_l}^k, \tag{3}$$

where $\odot$ denotes the Hadamard product. For example, in Fig. 3, paths between L_1 and L_3 are $P_1 = (A, B_1, C)$ and $P_2 = (A, B_2, C)$. The commuting matrices are $\boldsymbol{M}_1^1 = \boldsymbol{M}_{A,B_1}\boldsymbol{M}_{B_1,C}$ and $\boldsymbol{M}_2^1 = \boldsymbol{M}_{A,B_2}\boldsymbol{M}_{B_2,C}$. So the commuting matrix $\boldsymbol{M}_{L_1,L_3} = \boldsymbol{M}_1^1 \odot \boldsymbol{M}_2^1$.

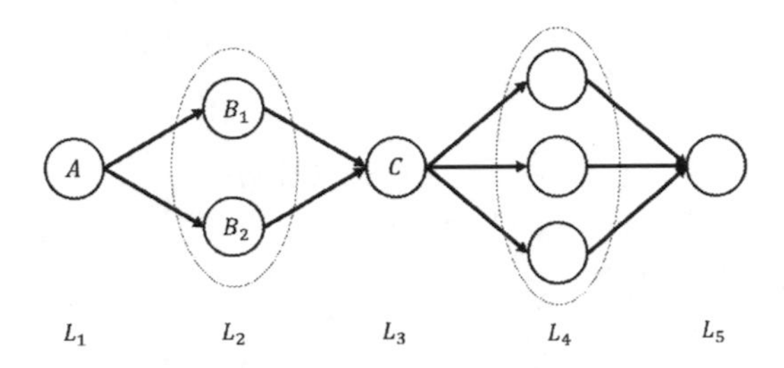

Fig. 3. Layers' perspective of a meta-graph

Given a specific meta-graph, a commuting matrix can be used to calculate similarities of users and items over HIN. Here we employ matrix factorization [16] to extract features on the obtained commuting matrices, because it can reduce noise and solve the sparsity problem. The method of matrix factorization is to factorize the commuting matrix into two low-rank matrices, the users' embedding matrix $\boldsymbol{U}$ and items' embedding matrix $\boldsymbol{I}$. The embedding matrix of users (items) is designed to capture the abstract feature of users (items) in a low dimensional vector space. With the Frobenius norm regularization to avoid overfitting, the embedding problem is converted to optimize the following objective function:

$$\min_{\boldsymbol{U},\boldsymbol{I}} \frac{1}{2}||\boldsymbol{\Omega} \odot (\boldsymbol{U}\boldsymbol{I}^T - \boldsymbol{R})||_F^2 + \frac{\lambda_u}{2}||\boldsymbol{U}||_F^2 + \frac{\lambda_i}{2}||\boldsymbol{I}||_F^2, \tag{4}$$

where $\boldsymbol{\Omega}$ is the indicator matrix that Ω_{ui} is equal to 1 if user u rated the item i and equal to 0 otherwise, and λ_u, λ_i represent the regularization parameters.

3.2 Neural Network Recommendation

Once the embeddings of users and items for different meta-graphs are obtained from the above procedure, aggregator function can be employed to aggregate all the embeddings based on meta-graphs to make a more representative embedding.

There are some commonly used aggregation techniques such as concatenating [27] and averaging [3] the latent factors. Concatenation is a straightforward and high-dimensional way, but the dimension of embedding increases a lot when adding more meta-graphs. In contrast, averaging all embeddings is not influenced by the dimension, but this way will lose something useful. Intuitively, users have different preferences on different meta-graphs. Then in a user-item interaction, we should make a distinction between these different meta-graphs. We consider the attention mechanism to integrate meta-graph embeddings, which can assign the attentive weights to the meta-graphs. We apply this method to HIN based recommender system, called attention aggregator. The weight obtained for a meta-graph represents the importance of the meta-graph, which can provide the choice and suggestion for recommender system agent. Attention is implemented as follows.

Attention Aggregator. In this paper, we adopt a two-layer architecture to implement the attention. We denote the embedding of user and item as $\boldsymbol{u}^l, \boldsymbol{i}^l$ under the meta-graph l. The attentive score α_u^l is calculated by Eq. 5:

$$\alpha_u^l = \boldsymbol{W}_2^T f(\boldsymbol{W}_1^T \boldsymbol{u}^l + \boldsymbol{b}_1) + b_2, \tag{5}$$

where $\boldsymbol{W}_1$ and $\boldsymbol{W}_2$ denote the weight matrix, $\boldsymbol{b}_1$ and b_2 denote the bias. $f(\cdot)$ is set to the ReLU function [13]. By normalizing the above attentive scores, user's final attention weight is:

$$w_u^l = \frac{exp(\alpha_u^l)}{\sum_{p \in L} exp(\alpha_u^p)}. \tag{6}$$

Then the aggregated latent factor representation of user is calculated by Eq. 7. L means the number of meta-graphs. A means the attention aggregator. Item's final embedding is also obtained by the same way:

$$\boldsymbol{u}^A = \sum_{l \in L} w_u^l \cdot \boldsymbol{u}^l, \quad \boldsymbol{i}^A = \sum_{l \in L} w_i^l \cdot \boldsymbol{i}^l. \tag{7}$$

To verify the effectiveness of the attention mechanism in our model, we compare concatenation and mean aggregators. C denotes the concatenation aggregator. M denotes the mean aggregator. The specific formalizations are given as follows, respectively:

$$\boldsymbol{u}^C = [\boldsymbol{u}^1, \boldsymbol{u}^2, \cdots, \boldsymbol{u}^L], \quad \boldsymbol{i}^C = [\boldsymbol{i}^1, \boldsymbol{i}^2, \cdots, \boldsymbol{i}^L]. \tag{8}$$

$$\boldsymbol{u}^M = \sum_{l \in L} \boldsymbol{u}^l / L, \quad \boldsymbol{i}^M = \sum_{l \in L} \boldsymbol{i}^l / L. \tag{9}$$

Deep Neural Network. With the embeddings of user $\boldsymbol{u}$ and item $\boldsymbol{i}$, the rating of each pair is denoted as $R_{u,i}$. We use a 3-layer DNN to predict the rating of each pair, the entire model represents in Fig. 4. The neural network can extract more attractive features of the provided data.

For each (user, item) pair $(\boldsymbol{u}, \boldsymbol{i})$, we concatenate the features of user and item as the input of DNN. The input vector of DNN is $\boldsymbol{x} = [\boldsymbol{u}, \boldsymbol{i}]$ and the output of hidden layer is denoted as $\boldsymbol{h}_l$ where l is the l-th layer, $l = 0, 1, 2, \cdots, n-1$ and n means the total number of layers. The predicted rating $R_{u,i}$ can be learned by multi-layer functions. The formalizations are as follows:

$$
\begin{aligned}
\boldsymbol{h}_0 &= \boldsymbol{x} \\
\boldsymbol{h}_1 &= f(\boldsymbol{W}_1^T \boldsymbol{h}_0 + \boldsymbol{b}_1) \\
&\cdots \\
\boldsymbol{h}_l &= f(\boldsymbol{W}_l^T \boldsymbol{h}_{l-1} + \boldsymbol{b}_l) \\
&\cdots \\
R_{u,i} &= f(\boldsymbol{W}_n^T \boldsymbol{h}_{n-1} + b_n)
\end{aligned}
\tag{10}
$$

where $\boldsymbol{W}_j$ is the weight matrix and $\boldsymbol{b}_j$ is the bias for the j-th layer, $j = 1, 2, \cdots, n$. $f(\cdot)$ is a nonlinear activation function. In experiment, we use ReLU [13] as the activation function.

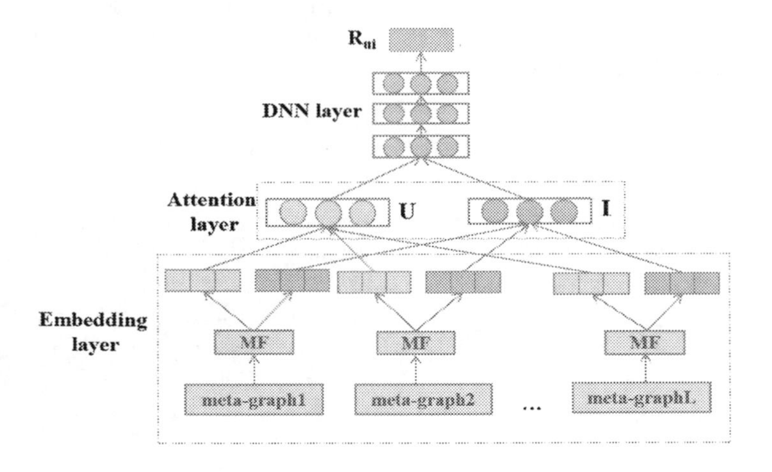

Fig. 4. Combining meta-graph and attention for recommendation over HIN

The entire method called MGAR, which combines meta-graph and attention for the recommendation over HIN. In addition, we apply concatenation aggregator and mean aggregator to our model, called MGCR and MGMR respectively.

4 Experiments

In this section, we conduct experiments on two real datasets. We demonstrate the ability of MGAR on recommender system by comparing with some baseline methods. Then we analyze the effect of MGAR and other two aggregate functions applied on our model, respectively. Finally, we explore the differences between meta-graphs and the effects of the dimension of latent features on MGAR.

4.1 Datasets

In experiments, we adopt Yelp[2] and Amazon[3] datasets with rich heterogeneous information. Yelp dataset is provided for recommending businesses to users. We randomly extract subsets from Yelp dataset containing 18465 users, 536 businesses, 20k ratings from 1 to 5. The higher rating means the higher degree of user's preference on business. Amazon dataset is provided by [5,11] for recommending multi-categories items to users who are surfing on the Amazon website, including 16970 users, 336 businesses with 20k ratings from 1 to 5. The detailed information of datasets is shown in Table 1.

Review-Aspect in Table 1 means the aspects extracted from review text. We use the Gensim [14] model to extract 10 topics, from the review text on Yelp and Amazon datasets. Each row of this Review-Aspect matrix refers to the probability of the 10 topics corresponding to each review.

Table 1. Statistics of Yelp/Amazon datasets

Dataset	Relations (A-B)	Number of A	Number of B	Number of (A-B)	Ave. Degrees of A/B
Yelp	User-User	18454	18454	125223	6.79/6.79
	User-Business	18454	576	20000	1.08/34.7
	User-Review	18454	20000	20000	1.08/1.0
	Business-Star	576	9	576	1.0/64.0
	Business-State	576	51	576	1.0/11.29
	Business-Category	576	1237	1827	3.17/1.48
	Business-City	576	1010	576	1.0/0.57
	Review-Business	20000	576	20000	1.0/34.72
	Review-Aspect	20000	10	172349	8.62/17234.9
Amazon	User-Business	16970	336	19287	1.14/57.40
	User-Review	16970	18331	18198	1.07/0.99
	Business-Category	336	16	323	0.96/20.19
	Review-Business	18331	336	20000	1.09/59.52
	Review-Aspect	18331	10	162407	8.86/16240.7

[2] http://www.yelp.com/dataset/.
[3] http://jmcauley.ucsd.edu/data/amazon/.

4.2 Metric

In order to evaluate the performance of models for recommendations, we utilize Root-Mean-Square-Error which is commonly used as metric on recommender system. A smaller RMSE means better performance. The formula is as follows:

$$RMSE = \sqrt{\frac{\sum_{(u,i)\in\mathcal{R}_{test}}(R_{u,i} - \hat{R}_{u,i})^2}{|\mathcal{R}_{test}|}}$$

where $\mathcal{R}_{test}$ denotes the test set, $\hat{R}_{u,i}$ represents the predicted rating of user u on item i and $R_{u,i}$ is the real rating of the user u on item i.

4.3 Baseline Methods

To verify the effectiveness of MGAR model, we compare baseline methods which are quite typical and perform well on recommender system. The baseline methods are introduced following.

• **PMF** [16]: Probabilistic Matrix Factorization (PMF) model without considering meta-graphs only uses user-item rating matrix $\boldsymbol{R}$ to product two lower-rank users matrix $\boldsymbol{U}$ and items matrix $\boldsymbol{V}$, and then predicts the ratings with $\hat{\boldsymbol{R}} = \boldsymbol{U}^T\boldsymbol{V}$.

• **FM** [15]: Factorization Machine (FM) only uses rating matrix to predict the rating of users on items. FM not only works with any real-valued feature vector, it also uses decomposition parameters to model all the interactions between variables. FM works on recommender system without considering meta-graphs.

• **SemRec** [17]: The recommendation method SemRec on weighted HIN defined the similarity measure by the same rating of users on items under given meta-paths. Different similarities between users can be learned from different meta-graphs. By the method of weight learning, SemRec can predict the ratings according to the weighted sum of $\hat{R}^l_{u,i}$ under each path l.

• **FMG** [27]: FMG is a meta-graph based method for recommendations, which is modeled by matrix factorization and factorization machine. The method adopts MF to generate latent features for users and items, and then use FM with Group lasso (FMG) to learn from the observed ratings to make predictions.

• **NeuACF** [4]: NeuACF is a neural network model based on meta-paths to exploit different aspect latent factors. The method applied PathSim [19] to compute similarity matrix on meta-paths.

• **MGMR**: MGMR denotes that mean aggregator by the average of latent features from users and items replaced attention aggregator on our model.

• **MGCR**: MGCR denotes that concatenation aggregator connecting the embeddings of users and items directly replaced attention aggregator on our model.

In experiments, we use the meta-graphs from Yelp and Amazon datasets shown in Fig. 2. The similarity measure of SemRec and NeuACF is built between users under given meta-paths, then we select $U \rightarrow I \rightarrow U$, $U \rightarrow U$, $U \rightarrow I \rightarrow * \rightarrow I \rightarrow U$, where $*$ includes categories, cities, and states of business. The parameters on all methods are set with the best performance.

4.4 Experimental Results

The results are shown in Table 2. In experiments, we use the 80% of a dataset as the training dataset, which means utilizing 80% rating data of rating matrix to predict the remaining 20%. The percentage in Table 2 means the gain of MGAR on two datasets comparing with each method. The time here presented in Table 2 only includes the training time of models after obtaining matrices or embeddings of users and items.

Table 2. Experimental results on Yelp and Amazon datasets

RMSE on Yelp and Amazon datasets								
	PMF	FM	SemRec	FMG	NeuACF	MGMR	MGCR	MGAR
Yelp	1.8765 +69.1%	1.2917 +55.0%	1.2732 +54.4%	1.1238 +48.3%	0.8023 +27.6%	1.0604 +45.2%	0.6716 +13.5%	**0.5807**
Amazon	1.9513 +64.1%	1.1575 +39.5%	1.2038 +41.9%	1.1323 +38.2%	0.9451 +26.0%	1.1112 +37.0%	0.8565 +18.3%	**0.6998**
Average training time on Yelp and Amazon datasets								
Time	97.45 s	107.26 s	167.29 s	181.51 s	3231.44 s	160.83 s	168.64 s	303.94 s

Comparing with the Former Five Baseline Methods. For Yelp and Amazon dataset, MGAR always achieves the best performance. As compared to PMF, MGAR improves the recommendation performance by 69.1% of RMSE on Yelp. MGAR outperforms FMG 48.3% on Yelp, validating that fusing the meta-graph-based features with weights is significant. The MGAR method does not perform as well as the Yelp dataset on the Amazon dataset. It's because the Amazon dataset has only 5 meta-graphs. There is not enough information, but it also performs better than other baseline methods, which fully demonstrates the effectiveness of the MGAR method. MGAR outperforms SemRec 48.2% on average of two datasets. NeuACF preforms only inferior to MGCR and MGAR, but costing much time. It's mainly because the obtained feature by PathSim with a high dimension. MGAR model uses the Matrix Factorization to obtain low-dimensional features to avoid the problem of long time caused by high dimensionality. The above phenomenons can indicate the information contained in the meta-graph is much more than which contained in the meta-path. And combining meta-graph and attention can improve the recommendation performance effectively.

The Effect of Three Aggregator Functions. Experimental results of three aggregator functions on our model can be seen in Table 2. As compared to MGCR, MGAR improves by an average of 15.9% on RMSE for recommendations. MGMR performs not well, from Table 2, we can know that the performance gap between MGMR and FMG is small. Because averaging the latent features will lose some useful information. We can get a conclusion that using attention aggregator function on our model can improve the learning performance in recommendation tasks.

4.5 Single Meta-Graph

In this section, we compare different meta-graphs on our model respectively. The RMSEs of all single meta-graphs on two datasets are shown in Fig. 5. M-A-all means the RMSE of MGAR using all meta-graphs.

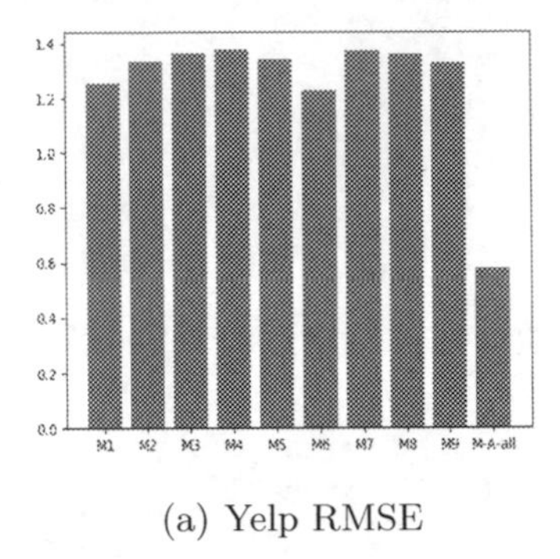

(a) Yelp RMSE

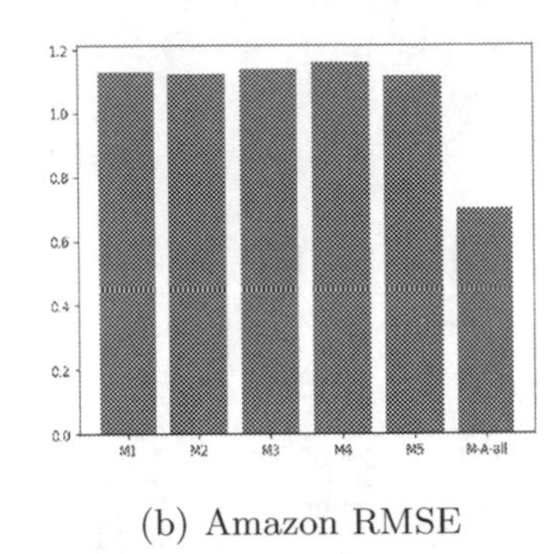

(b) Amazon RMSE

Fig. 5. Single meta-graph

Single meta-graph always performs not well. In Fig. 5(a), $M6$ defined as $U \rightarrow I \rightarrow State \rightarrow I \rightarrow U$ performs better than other meta-graphs. The result indicates that the states where businesses are located in provide more important information. In Fig. 5(b), $M5$ on Amazon performs best compared with other single meta-graphs. $M5$ contains more types of objects and relations which are helpful to capture critical information. In Fig. 5, MGAR with all meta-graphs always performs best, indicating that more meta-graphs used will make better recommendations. From the results, we can get some differences between different meta-graphs, but we can't find obvious rules. In the future, we will extract more meta-graphs from datasets to explore the difference.

4.6 The Parameter K

In this section, we conduct experiments with varying parameter K in the range of {2, 5, 10, 20, 30, 40, 50, 100} on Yelp and Amazon datasets. K is set as the dimension of latent features to obtain the embedding of users and items. The results can be seen in Fig. 6. After the threshold value 10, with the increase of parameter K, RMSE will not change significantly, but the training time will increase a lot. When setting K = 10, MGAR obtains the better performance without costing much time, indicating the latent features can provide enough information. So we set K = 10 on experiments in Sects. 4.4 and 4.5.

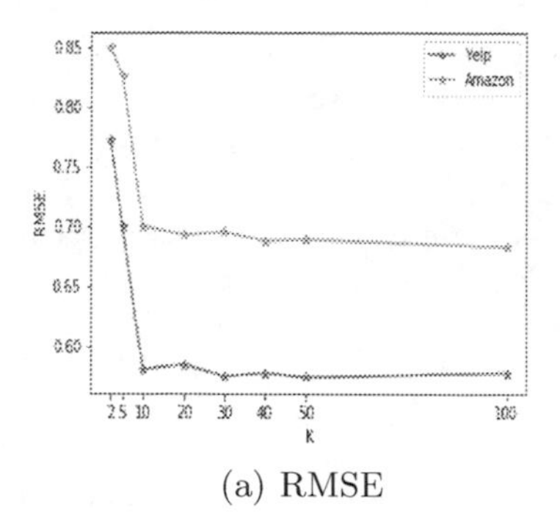

(a) RMSE

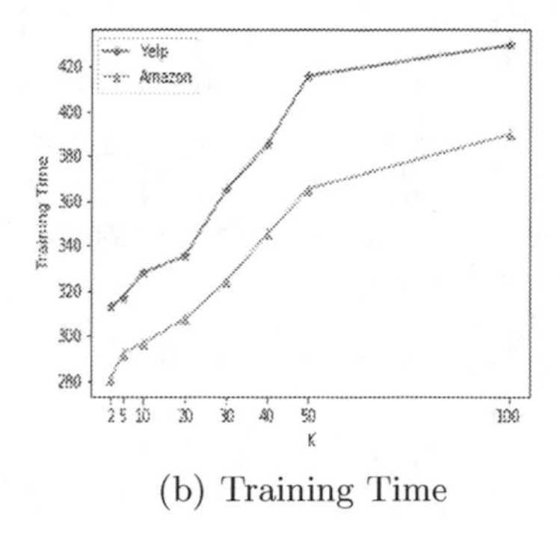

(b) Training Time

Fig. 6. RMSE and training time with varying K

5 Conclusion

In this paper, we proposed an approach for recommendation over HIN, called MGAR, by combining Meta-Graph and Attention. MGAR can effectively model the complicated information in HIN for the recommendation. We used meta-graphs to capture the complex relationships. The attention mechanism was applied to fuse the mined features of uscs and items from different meta-graphs. Our model contains two stages. In the first stage, matrix factorization was used to generate latent features which were based on many pre-defined meta-graphs to capture semantic features of multiple types of objects. In the second stage, the embeddings of users and items were aggregated by the attention mechanism, then we utilized a DNN method to predict the rating of users on items. Also, we considered the other two types of alternative aggregator functions. Experimental results validated the effectiveness of MGAR on the recommendation tasks, which achieved state-of-the-art performance on Yelp and Amazon datasets. In addition, we provided a way to choose the optimal dimension of latent features, which can achieve the balance of effect and training time.

In this paper, meta-graphs we applied were predefined, which was restricted by the quantity and length of meta-graphs in complex scenarios. It is interesting and natural to automatically mine meta-graphs according to different scenarios. Combining our model with meta-graph mining will be the next step.

Acknowledgements. This work was partly supported by the National Natural Science Foundation of China under Grant No. 61572002, No. 61170300, No. 61690201, and No. 61732001.

References

1. Bahdanau, D., Cho, K., Bengio, Y.: Neural machine translation by jointly learning to align and translate. CoRR abs/1409.0473 (2014)
2. Devooght, R., Bersini, H.: Long and short-term recommendations with recurrent neural networks. In: Bieliková, M., Herder, E., Cena, F., Desmarais, M.C. (eds.) Proceedings of the 25th Conference on User Modeling, Adaptation and Personalization, UMAP 2017, 09–12 July 2017, Bratislava, Slovakia, pp. 13–21. ACM (2017)
3. Hamilton, W.L., Ying, Z., Leskovec, J.: Inductive representation learning on large graphs. In: Advances in Neural Information Processing Systems 30: Annual Conference on Neural Information Processing Systems 2017, 4–9 December 2017, Long Beach, CA, USA, pp. 1025–1035 (2017)
4. Han, X., Shi, C., Wang, S., Yu, P.S., Song, L.: Aspect-level deep collaborative filtering via heterogeneous information networks. In: Lang, J. (ed.) Proceedings of the Twenty-Seventh International Joint Conference on Artificial Intelligence, IJCAI 2018, 13–19 July 2018, Stockholm, Sweden, pp. 3393–3399 (2018)
5. He, R., McAuley, J.: Ups and downs: modeling the visual evolution of fashion trends with one-class collaborative filtering. In: WWW, pp. 507–517. ACM (2016)
6. Huang, Z., Zheng, Y., Cheng, R., Sun, Y., Mamoulis, N., Li, X.: Meta structure: Computing relevance in large heterogeneous information networks. In: KDD, pp. 1595–1604. ACM (2016)
7. Ji, M., Han, J., Danilevsky, M.: Ranking-based classification of heterogeneous information networks. In: KDD, pp. 1298–1306. ACM (2011)
8. Ji, M., Sun, Y., Danilevsky, M., Han, J., Gao, J.: Graph regularized transductive classification on heterogeneous information networks. In: Balcázar, J.L., Bonchi, F., Gionis, A., Sebag, M. (eds.) ECML PKDD 2010. LNCS (LNAI), vol. 6321, pp. 570–586. Springer, Heidelberg (2010). https://doi.org/10.1007/978-3-642-15880-3_42
9. Jiang, H., Song, Y., Wang, C., Zhang, M., Sun, Y.: Semi-supervised learning over heterogeneous information networks by ensemble of meta-graph guided random walks. In: IJCAI, pp. 1944–1950 (2017)
10. Kong, X., Zhang, J., Yu, P.S.: Inferring anchor links across multiple heterogeneous social networks. In: CIKM, pp. 179–188. ACM (2013)
11. McAuley, J.J., Targett, C., Shi, Q., van den Hengel, A.: Image-based recommendations on styles and substitutes. In: SIGIR, pp. 43–52. ACM (2015)
12. Mnih, V., Heess, N., Graves, A., et al.: Recurrent models of visual attention. In: Advances in Neural Information Processing Systems, pp. 2204–2212 (2014)
13. Nair, V., Hinton, G.E.: Rectified linear units improve restricted Boltzmann machines. In: ICML, pp. 807–814. Omnipress (2010)
14. Rehurek, R., Sojka, P.: Software framework for topic modelling with large corpora. In: Proceedings of the LREC 2010 Workshop on New Challenges for NLP Frameworks. Citeseer (2010)
15. Rendle, S.: Factorization machines. In: ICDM, pp. 995–1000. IEEE Computer Society (2010)
16. Salakhutdinov, R., Mnih, A.: Probabilistic matrix factorization. In: NIPS, pp. 1257–1264. Curran Associates, Inc. (2007)
17. Shi, C., Zhang, Z., Luo, P., Yu, P.S., Yue, Y., Wu, B.: Semantic path based personalized recommendation on weighted heterogeneous information networks. In: CIKM, pp. 453–462. ACM (2015)

18. Sudhakaran, S., Lanz, O.: Attention is all we need: nailing down object-centric attention for egocentric activity recognition. In: British Machine Vision Conference 2018, BMVC 2018, 3–6 September 2018, Northumbria University, Newcastle, UK, p. 229. BMVA Press (2018)
19. Sun, Y., Han, J., Yan, X., Yu, P.S., Wu, T.: Pathsim: meta path-based top-k similarity search in heterogeneous information networks. PVLDB **4**(11), 992–1003 (2011)
20. Sun, Y., Han, J., Zhao, P., Yin, Z., Cheng, H., Wu, T.: Rankclus: integrating clustering with ranking for heterogeneous information network analysis. In: EDBT. ACM International Conference Proceeding Series, vol. 360, pp. 565–576. ACM (2009)
21. Sun, Y., Norick, B., Han, J., Yan, X., Yu, P.S., Yu, X.: Integrating meta-path selection with user-guided object clustering in heterogeneous information networks. In: KDD, pp. 1348–1356. ACM (2012)
22. Wang, F., et al.: Residual attention network for image classification. In: 2017 IEEE Conference on Computer Vision and Pattern Recognition, CVPR 2017, 21–26 July 2017, Honolulu, HI, USA, pp. 6450–6458. IEEE Computer Society (2017)
23. Xu, K., et al.: Show, attend and tell: neural image caption generation with visual attention. In: Bach, F.R., Blei, D.M. (eds.) Proceedings of the 32nd International Conference on Machine Learning, ICML 2015, 6–11 July 2015, Lille, France. JMLR Workshop and Conference Proceedings, vol. 37, pp. 2048–2057 (2015)
24. Yang, C., Zhao, C., Wang, H., Qiu, R., Li, Y., Mu, K.: A semantic path-based similarity measure for weighted heterogeneous information networks. In: Liu, W., Giunchiglia, F., Yang, B. (eds.) KSEM 2018. LNCS (LNAI), vol. 11061, pp. 311–323. Springer, Cham (2018). https://doi.org/10.1007/978-3-319-99365-2_28
25. Yu, X., Ren, X., Gu, Q., Sun, Y., Han, J.: Collaborative filtering with entity similarity regularization in heterogeneous information networks. In: IJCAI HINA (2013)
26. Yu, X., et al.: Personalized entity recommendation: a heterogeneous information network approach. In: WSDM, pp. 283–292. ACM (2014)
27. Zhao, H., Yao, Q., Li, J., Song, Y., Lee, D.L.: Meta-graph based recommendation fusion over heterogeneous information networks. In: KDD, pp. 635–644. ACM (2017)

Efficient Search of the Most Cohesive Co-located Community in Attributed Networks

Jiehuan Luo[1], Xin Cao[2], Qiang Qu[1(✉)], and Yaqiong Liu[3]

[1] Shenzhen Institutes of Advanced Technology, Chinese Academy of Sciences, Beijing, China
jiehuan.luo1@gmail.com, qiang.qu@siat.ac.cn
[2] School of Computer Science and Engineering, University of New South Wales, Sydney, Australia
xin.cao@unsw.edu.au
[3] Beijing University of Posts and Telecommunications, Beijing, China
liuyaqiong@bupt.edu.cn

Abstract. Attributed networks are used to model various networks, such as social networks, knowledge graphs, and protein-protein interactions. Such networks are associated with rich attributes such as spatial locations (e.g., check-ins from social network users and positions of proteins). The community search in attributed networks have been intensively studied recently due to its wide applications in recommendation, marketing, biology, etc. In this paper, we study the problem of searching the most cohesive co-located community (MC^3), which returns communities that satisfy the following two properties: (i) structural cohesiveness: members in the community are connected the most intensively; (ii) spatial co-location: members are close to each other. The problem can be used for social network user behavior analysis, recommendation, disease predication etc. We first propose an index structure called DkQ-TREE to integrate the spatial information and the local structure information together to accelerate the query processing. Then, based on this index structure we develop two efficient algorithms. The extensive experiments conducted on both real and synthetic datasets demonstrate the efficiency and effectiveness of the proposed methods.

Keywords: Community search · Attributed networks · Co-located community

1 Introduction

Attributed networks are used to model various networks, including social networks, knowledge graphs, and protein-protein interactions [2,11,27]. The increasing data volume and rich attributes of such networks pose great challenges to community search, which have attracted much attention in recent years (e.g.,

G. Li et al. (Eds.): DASFAA 2019, LNCS 11446, pp. 398–415, 2019.
https://doi.org/10.1007/978-3-030-18576-3_24

[12,19]). The research studies on finding communities can be categorized into community detection (e.g., [14,26,30]) and community search (e.g., [20]). Community detection methods are often used to discover communities in social networks based on the predefined implicit criteria, e.g., modularity [14]. Differently, community search is to find cohesive communities satisfying a given set of explicit criteria in an online manner, such as k-core [28] and k-truss [8].

Spatial attribute is one of the most important and useful feature in attributed networks. In a spatial-aware network, each node is attached with the spatial information. For example, the social networks such as Twitter and Foursquare can be modeled by such networks where each node (i.e., user) has one or more locations (e.g., the current position or check-in histories) [12]. Searching communities by taking into the account of the users' location information can bring the understanding of user activities from the virtual world to reality.

Various structure cohesiveness metrics have been used to find densely-connected communities, including k-core [4,10,29], k-clique [9,24], k-truss [8,18], densest subgraph [31], connectivity [17], etc. Most of existing studies only consider non-attributed networks and overlook the rich information of vertices in attributed networks. The studies on spatial-aware communities [7,12] look for communities such that the vertices are densely and closely connected in terms of both social and spatial proximity. In these works, the structure cohesiveness is a query constraint. For example, the users need to specify a k value in community search for the k-core [12] or k-truss [7] measure.

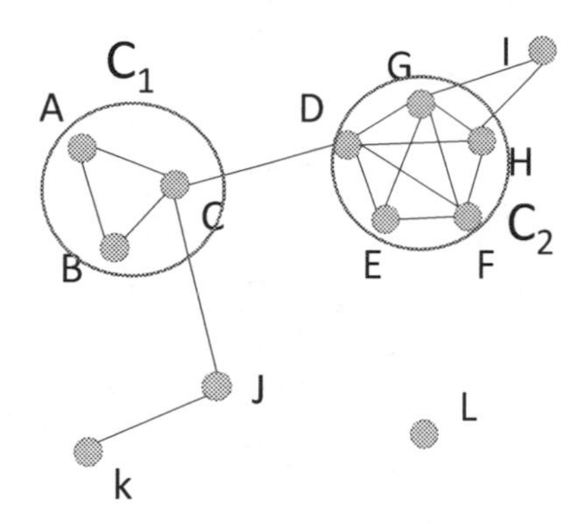

Fig. 1. An example of attributed networks and MC^3

In this work, we propose to study the problem of finding the most cohesive co-located communities (MC^3) from attributed networks in which each node has one or several spatial locations. Specifically, given a spatial-aware network and a value D, we aim to find the most cohesive connected subgraph (a community) that can be enclosed by a circle with diameter D. In this paper, we apply the widely-used k-core [4,12,29] to measure the structural cohesiveness, and thus we finally find a core with the largest value of k enclosed by a circle with diameter D. Different from existing work, the structure cohesiveness is the optimization objective in our problem rather than a constraint. Our framework works generally for other cohesive measures such as k-clique [24] and k-truss [8].

Example 1. For example, Fig. 1 shows a small attributed network, in which each vertex in the network is associated with a location in the space. C_2 is the MC^3 for the given circle size, since it encloses a 3-core which is the core with the largest k value.

The problem of finding MC^3 can find applications in many fields. For example, it can be used in social network users behavior analysis. Assume Fig. 1 represented a geo-social network, where each location is the position of a user at a certain time point. By detecting MC^3 from the network, we can further analyze the group of users' behavior for public security (e.g., detect emerging community gathering). If the attributed network represent protein-protein interactions, the proteins interact with each other the most intensively in a small area represent some abnormal functions. The MC^3 can help predict the diseases or protein abnormal functionalities.

The key challenge of finding MC^3 is that the global structure cohesiveness does not consist with the local structure cohesiveness. The members in a structure cohesive community may be faraway form each other, while the nearby vertices may not be connected in the network. In order to deal with this problem, we propose an index structure called DkQ-TREE (distance-aware k-core quadtree) based on the Quadtree. We integrate the location information and the structure information together for each tree node, which enable us to obtain the local structure cohesiveness at query time. Based on this index structure, we first develop an algorithm MC^{3}ALG which involve two iterative steps: pruning the vertices and verifying the candidate vertices. We also develop an enhanced algorithm MC^{3}ALG+ to further improve the verification cost by utilizing a two-phase binary search.

In summary, the contribution of this work is as follows: (1) we first propose the most cohesive co-located community search, in which the structure cohesiveness is the optimization objective; (2) we propose an index structure DkQ-TREE to organize the attributed network data; (3) we develop two efficient algorithms based on DkQ-TREE to find MC^3; (4) we conducted experiments on both real-world and synthetic data sets to demonstrate the performance of our methods.

2 Related Work

The research studies on finding communities can be categorized into community detection (e.g., [14,26,30]) and community search (e.g., [10,20,29]). Community detection methods are often used to discover communities in social networks based on the predefined implicit criteria. Differently, community search (CS) is to find cohesive and densely communities satisfying a given set of query request in an online manner. Various structure cohesiveness metrics have been used to find densely-connected communities, including k-core [4,10,22,29], k-clique

[9,24], k-truss [1,8,18,21], densest subgraph [31], connectivity [17], etc. These studies only consider non-attributed networks, and overlook the rich information of vertices in attributed networks.

The studies on spatial-aware communities [7,12,23,33,34] look for communities such that the vertices are densely and closely connected in terms of both social and spatial proximity. For instance, Fang et al. [12] studied searching the minimum spatial-aware community, in which the vertices form a k-core and their locations are very close. Chen et al. [7] proposed to search co-located community with the maximum number of nodes by using k-truss. The main difference between [7] and our work is the optimization objective. We aim to find communities with the most cohesive structure rather than the community with the maximum number of nodes. However, there exist no studies on finding the most cohesive co-located/spatial-aware communities. In these works, the structure cohesiveness is a constraint rather than the optimization objective.

In spatial databases, several works studied the group objects retrieval problem based on users' spatial locations such as [16,25,32]. Guo et al. [16] studied the m-closest keywords query which retrieves a group of objects close to each other and cover a set of keywords together. Wu et al. [32] adapted the densest subgraph model to the spatial community search problem on dual networks. Qu et al. [25] proposed localitySearch which retrieves top-k sets of spatial web objects by integrating spatial distance, textual relevance, and a "co-locality" measure into one ranking function. These studies did not consider the structure cohesiveness, and thus are different from our problem.

The densest subgraph search is also relevant to our work. In its basic form, the problem is to find the subgraph with maximum average edge weight. This problem can be solved in polynomial time [15]. For large graphs, efficient approximation algorithms have been developed. A 2-approximation algorithm is proposed in studies [3,6]. This problem is different from MC^3 because the cohesiveness measure is different and more importantly they ignore the spatial information in the subgraph search.

3 Problem Definition

Our problem is defined over an undirected attributed network $G = (V, E, S)$ with vertex set V, edge set E, and spatial location set S. The degree of a vertex v (e.g., a user in social networks) in G is denoted by $deg_G(v)$. Each vertex v has a spatial location $v.l = (x, y) \in S$ (e.g., the check-in of users), where x and y are the coordinates along x- and y-axis in a two-dimensional space. Table 1 lists the notations used throughout the paper.

We aim to find a community represented by a connected subgraph satisfying: (1) ***structure cohesiveness:*** the vertices in the subgraph are connected the most intensively; (2) ***spatial cohesiveness:*** the vertices in the subgraph are highly compact in the space.

Before formally defining the problem, we first introduce the following important concepts.

Table 1. The summary of notations

Notation	Definition
$G(V,E,S)$	A geo-social graph with vertex set V and edge set E and spatial location set S
$(v.x, v.y)$	The position of vertex v along $x-$ and $y-$axis
$deg_G(v)$	The degree of vertex v in G
$\gamma(N)$	The side length of node N in index structure

Definition 1 (k-core [28]). *Given a non-negative integer k, the k-core of G is the largest subgraph of G in which the degree of each vertex v is no less than k.*

The concept of k-core has been widely used for structure cohesiveness in many applications [10,12] due to its effectiveness. A k-core may not be connected. In this paper, we use a connected k-core in graph G (denoted by G_k) to represent a communities. We say that G_k has an order of k. Given a graph, the k-cores can be obtained by a linear core decomposition algorithm proposed by Batagelj and Zaversnik [5] with complexity $O(|E|)$.

Note that our proposed solutions can be easily adapted to other cohesive structure concepts (e.g., k-truss [8] and clique [24]) which can be used to capture the social cohesiveness from different perspectives.

Definition 2 (Core Number). *Given a vertex v in G, its core number is the highest order of a k-core that contains v, denoted by $C_G[v]$.*

Definition 3 (Co-located Community). *A co-located community is a connected subgraph (k-core) G_k such that the locations of the vertices in this subgraph can be enclosed by a circle with a pre-defined diameter D.*

We expect that the vertices in a co-located community have nearby locations. This reflect the "co-location" of this community.

Now we are ready to formally define the Most Cohesive Co-located Community (MC^3) search problem:

Problem 1. Given an undirected attributed graph G and a diameter D, the MC^3 query returns any groups of vertices with their locations, satisfying the following constraints: (a) Locations of vertices can be enclosed in a circle with diameter D; (b) Vertices form a k-core with the highest order.

More generally, we also consider the MC^3 problem when each vertex could have multiple locations (e.g., the check-in histories). The following example explains what is a MC^3.

Example 2. As shown in Fig. 1, both C_1 and C_2 are two co-located communities (i.e., members can be enclosed by a circle with diameter D). C_2 is a 3-core, which is a core with the highest order among all co-located communities w.r.t. D, and thus C_2 is the MC^3 for the given attributed network.

4 Baseline Approaches

4.1 A Spatial-First Approach

The MC^3 problem aims to find the most cohesive community that can be enclosed in a circle with diameter D. Hence, a naive approach is to check all possible circles in the space and then check the largest core order of the vertices enclosed in this circle. Finally, after all circles are checked, we return the largest core order across all circles.

To enumerate all possible circles, we can fix two locations in S whose distance is smaller than or equal to D, and then we can obtain at most two circles with diameter D passing by the two locations in the space. Next, we get the vertices from the graph to which these locations belong, and we apply the well-known liner core decomposition algorithm [5] to compute the largest core order. This methods require us to check $O(|V|^2)$ circles in the worst case. Thus, it is obvious that this method is quite time-consuming.

4.2 A Social-First Approach

Another approach is to perform the social-first search, and the idea is to utilize the structure of the network to accelerate the search. We can first do the core decomposition to compute the core number on each vertex. Next, we search in the core with the largest k value (we denote this as k_{max}). We get the locations from the vertex in this k_{max}-core, and then perform a search over these locations by enumerating all possible circles (similar to the spatial-first approach). After this step, we can obtain the current best core order (we denote this as k_{cur}). Then, we move to (k_{max}-1)-cores for further checking. We repeat this process until we reach the k_{cur}-cores, and we can terminate the algorithm. This can help reduce the number of circle verifications. However, it is still not efficient since the globally cohesive subgraph may not be locally cohesive.

5 The Distance-Aware *k*-Core Quadtree

It is shown that both the spatial first and social first approaches cannot achieve good performance, because the MC^3 problem considers both the spatial and structure cohesiveness, but the two methods ignore either the spatial or structure features of the data. This inspires us to design an index that can pre-compute the local structure cohesiveness, and this would enable us to accelerate the search and prune the search space. We first introduce the proposed tree index structure named D*k*Q-TREE (Distance-aware *k*-core Quadtree) based on the well-known spatial index Quadtree [13], and then we present two algorithms based on this index for solving the MC^3 problem.

5.1 Index Overview

The linear k-core decomposition algorithm [5] can only compute the global core number of the vertices, and thus the local cohesive information is unknown during query time. The key challenge of developing such an index is how to integrate both the structure and spatial information together for computing the local cohesiveness w.r.t. the user provided D. To this end, we utilize the Quadtree and pre-compute the local cohesiveness and other useful information for each tree node, based on the following property of the local cohesiveness.

Property 1. **Spatial-monotonicity:** Given a spatial region R (e.g., a square), if the vertices in this region can form a k-core with an order at most h, for any region R' inside R, the order of the k-core formed by the vertices in R' is no larger than h.

This property is quite straightforward since it is obvious that a smaller region has fewer vertices. In each node N of the DkQ-TREE, we pre-compute the core number of each vertex in the node within the subgraph extracted from this area and record the largest core number of vertices in the node, denoted by LC_N. We perform this computation is due to the following lemma:

Lemma 1. *Given a query diameter D and a tree node N that can be enclosed by a circle with diameter D, the order of* MC^3 *is no smaller than LC_N.*

Since N can be enclosed by a circle with diameter D, it can be easily proved according to Property 1. Hence, we are able to get a lower bound estimation of the order of MC^3 from the DkQ-TREE with this pre-computed information.

However, this is still not enough to obtain the local cohesiveness. We only can get the largest core number in each node. From Fig. 2, we can see that the vertices could form a k-core with the other vertices not in this node. Thus, we cannot get a bound for the core number of these nodes for a given D. Thus, we further compute a distance map $DistMap$ in each tree for the vertices. The idea is that, given a node N, for each value $k > LN_C$, we expand the node to a vertex with the smallest distance d such that the vertices involved during the expansion can form a k-core. We record this distance d and k in the distance map. The distance map can help us to prune the search space according to the following lemma:

Lemma 2. *Assume the current best order of* MC^3 *is k_{cur}. Given a query diameter D and a node N, if $N.DistMap[k_{cur}] > D$, N cannot contribute any vertex to* MC^3.

Again this lemma can be proved by using Property 1. If $N.DistMap[k_{cur}] > D$, it means that when we expand the border of N with length D we still cannot find a k_{cur}-core in this area. Hence, the core number of any node in this area is smaller than k_{cur}, and thus can be pruned.

In addition, in order to get vertices quickly from locations, we also use a vertex map to organize the mapping information when vertices have multiple

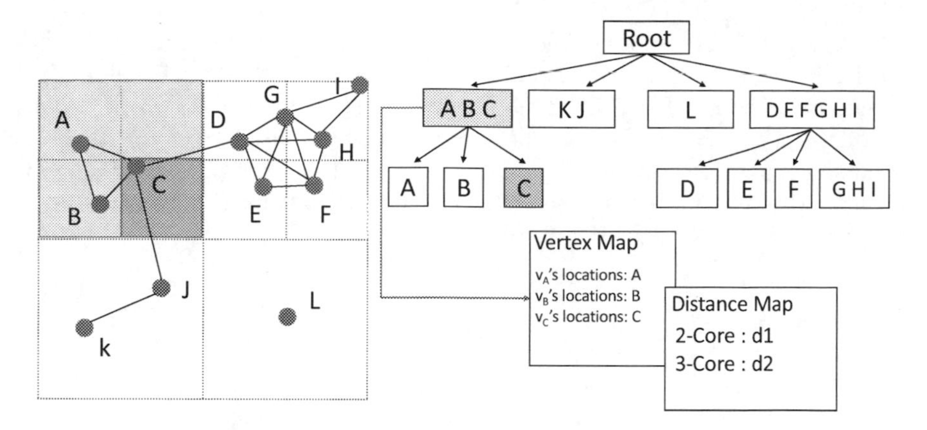

Fig. 2. An example of DkQ-TREE (Color figure online)

locations. In summary, in each node of DkQ-TREE, we store: (1) vertices in this node; (2) the largest core number in this node; (3) a vertex map; (4) a distance map. The tree structure is as shown in the Fig. 2.

5.2 Index Construction

We proceed to explain how to build the DkQ-TREE. The root node is the entire space. Then, we repeatedly partition each node into four child nodes. When we obtain a new node, we first do a core decomposition using the vertices within the node and store the largest core number. If it is smaller than a certain value k_ϵ, we do not further split the node. For example, in Fig. 2, in the blue area, the vertices {A, B, C} form a 2-core, so we split this area. After splitting, any sub area cannot form a 2-core, and we stop splitting these nodes.

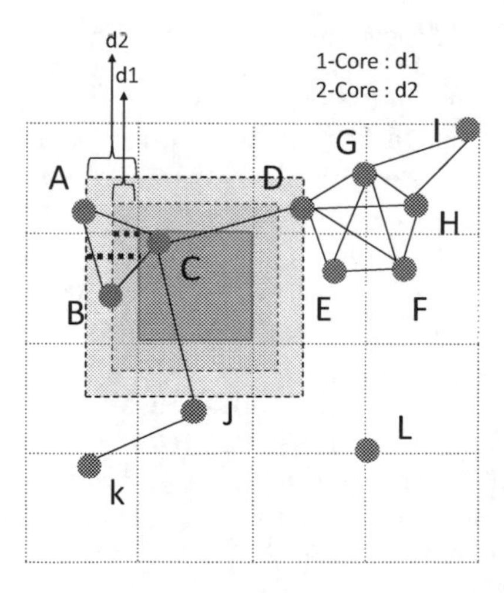

Fig. 3. An example of building distance map

Next, when we obtain a new node, we also build its distance map and vertex map. Building the vertex map is quit straightforward. The idea of building the distance map is that, For each value k, we perform a binary search to expand the node to a vertex with the smallest distance such that the vertices involved during the expansion can form a k-core. For example, in Fig. 3, the node only has one vertex C. When it expands to vertex A, it first form a 2-core, and we store the distance $d2$ to the distance map.

6 Algorithms

In this section, we propose two efficient algorithms by utilizing the DkQ-TREE. First, we propose an algorithm MC^{3}ALG, which involve two major iterative steps: one step is to get candidate nodes from DkQ-TREE and prune the nodes we do not need to check; the other step is to find MC^3 from the candidate nodes. In our experiments, MC^{3}ALG is more efficient than baseline algorithms. However, the verification of MC^{3}ALG still has high complexity. To address this issue, we further develop a more efficient algorithm MC^{3}ALG+. It improves the deficiencies of MC^{3}ALG's verification step and achieves excellent performance.

6.1 Algorithm MC^{3}ALG

The MC^{3}ALG algorithm involves two iterative steps: (1) prune nodes in DkQ-TREE; and (2) find MC^3 from nodes cannot be pruned. We proceed to present this algorithm.

We first get a lower bound for the order of MC^3 based on Lemma 1. Given a diameter D, we traverse the DkQ-TREE from top to bottom and get all nodes whose side length is smaller than D and whose parent nodes' side length is larger than D. We store these nodes in a node list $nodeList$. Next, we get the maximal core number from these nodes, which serves as a lower bound denoted by k_{cur}. Using this lower bound of the order of MC^3, we can further prune nodes in $nodeList$ according to Lemma 2.

For the remaining nodes in $nodeList$, we order them according to their core number upper bound obtained from their distance maps, and we start the verification on the best node N. Specifically, given a node N, if $N.distMap[k_1] \leq D \leq N.distMap[k_2]$, we can know that k_1 is the core number upper bound of vertices in N. We first extend N with D length and do a core decomposition in the extended square region. Then, we can safely ignore the vertices whose core number is smaller than k_{cur}, since they cannot be contained in MC^3. To verify if there exits a k-core with higher order on the remaining vertices in the extended area, instead of checking all the possible circles as we did in the spatial-first baseline method, we utilize the rotating circle method [16]. The idea is that, for each vertex in node N, we make it on the boundary of a circle with diameter D. Then, we rotate the circle clockwise. When a vertex enter the circle, we check if there is a k-core with order higher than k_{cur}. If so, we record the k-core and update k_{cur}. For example, in Fig. 4, we make vertex G on the boundary of circle and rotate the circle clockwise, when F enters the circle (circle with solid blue line), we can find a 2-core formed by $\{G, F, H, I\}$.

Since k_{cur} may be updated after verifying N, we can further prune more nodes in *nodeList*, and then perform verification from the next best node. We repeat the two steps until all nodes in *nodeList* are processed. Algorithm 1 presents the framework of MC^{3}ALG. First, it gets *nodeList* from DkQ-TREE (line 1). Then, we obtain the lower bound of the order of MC^3 and use φ to store the best k-core in N_{max} (lines 2–4). For each node in *nodeList*, we get its distance map *DistMap* and check how far we need to expand it to include a k-core. We safely remove nodes by Lemma 2 (lines 5–8). We get the core number upper bound of vertices in this node (line 9). Next, we sort *nodeList* in ascending order of the nodes' upper bounds (line 10). For each node, we extend it with D length and prune vertices as mention before. For each vertices not pruned in N, we use the rotating circle method to check k-cores and update φ (lines 11–15). The k-cores with the highest order is finally stored in φ (line 16).

Algorithm 1. MC^{3}ALG(G, D, *root*)

```
1  nodeList ← getnodes(root,D);
2  initialize k_cur ← 0,φ ← ∅;
3  N_max ← find a node in nodeList which has maximal value k;
4  k_cur ← N_max.k, φ ← get maximal kcore in N_max;
5  foreach node in nodeList do
6  |   DistMap ← node.distMap;
7  |   if DistMap[k_cur] > D then
8  |   |   remove node from nodeList;
9  |   node.upper ← getUpper(DistMap);
10 nodes ← sort nodeList according to node.upper;
11 foreach N in nodes do
12 |   S ← SearchNode(root, N, D, G, k_cur);
13 |   if S ≠ ∅ & S.k >= k_cur then
14 |   |   k_cur ← S.k;
15 |   |   update φ;
16 return φ
```

Example 3. Given a candidate node which contains G, H, I, we make G on the boundary of the circle, and I, H, F, E, D are in the circle's rotating area. We get ordered list $\{I, H, F, E, D\}$ according to the order in which they enter the circle. Then, we rotate the circle clockwise. Each time, when a vertex in $\{I, H, F, E, D\}$ enters the circle (on its boundary), the rotation stops and check if there is a k-core inside it. when F enters the circle (solid line), we can get a 2-core ($\{G, I, H, F\}$) inside it. When circle rotates to D, we can get a 3-core ($\{G, H, F, E, D\}$). After processing H and I in the same way, we find $\{G, H, F, E, D\}$ is the k-core with the highest order in this node.

Complexity. We assume that in average each unit space area contains n vertices and m edges, and we get X nodes from DkQ-TREE for the given D. We first sort the nodes according to their core number upper bounds with complexity $O(X \log X)$. Then, for each node N with $\gamma(N) = l$, we extend N by length D, i.e., $\gamma(N_{ex}) = 2D + l$, and do a core decomposition in this square area. In the extended square, there are $(2D + l)^2 m$ edges, and thus the core decomposition costs $O((2D+l)^2 m)$. Next, we rotate the circle on each vertex in N. In each circle, there are $\pi(\frac{D}{2})^2 n$ vertices and $\pi(\frac{D}{2})^2 m$ edges. Note that, the k-core verification we perform in the circle can be divided into 3 steps: (1) The degree check costs $O((\frac{D}{2})^2 n)$; (2) The core decomposition costs $O((\frac{D}{2})^2 m)$; and (3) The BFS check costs $O((\frac{D}{2})^2 m)$. Hence, the k-core verification costs at most $O(\pi(\frac{D}{2})^2(n+2m))$. In the worst case, we do at most $\pi D^2 n$ times for each vertex in N (the number of verteces in N is $l^2 n$). Thus, the total complexity of MC^{3}ALG is $O(X \log X + X((2D + l)^2.m + l^2 n \times \pi D^2 n \times \pi(\frac{D}{2})^2(n + 2m)))$.

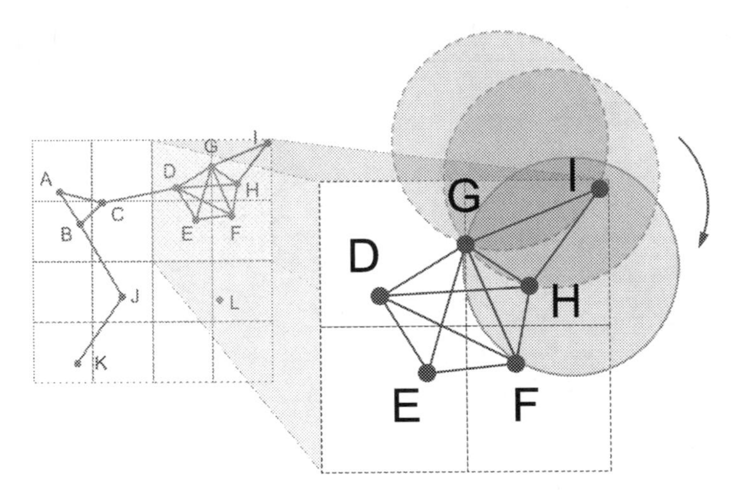

Fig. 4. An example of MC^{3}ALG (Color figure online)

6.2 Enhanced Algorithm MC^{3}ALG+

MC^{3}ALG is still not efficient enough in large attributed networks and has its limitations. First, in each node to be checked, there are lots of vertices and on each of them we need to apply the rotating circle method. Second, there are many vertices in the extended area of a node, and thus during rotating the circles we need to verify the k-cores for many times. To alleviate these issues, we develop a more efficient algorithm MC^{3}ALG+. The main difference between MC^{3}ALG+ and MC^{3}ALG is the verification cost on a node (prune nodes in DkQ-TREE the same as MC^{3}ALG).

In MC^{3}ALG+, for each node N to be checked, we perform a binary search to find the largest core number in this node. The upper bound of the core number is obtained from N's distance map as we did in MC^{3}ALG, and the lower bound is the current best order. During the binary search process, we check if there is a k-core in the extended region of N with the current core number k_c. This strategy can quickly get a larger k_c, and we can benefit from this in two ways: first, it reduces the vertices in N to be checked since; second, it also reduces the number of vertices in the extended area involved in the circle rotation.

Next, to further reduce the vertices in N to be checked, we divide the extended square region into $m \times m$ cells and use a small square to filter out the vertices which cannot form the solution. The idea is that, instead of checking the vertices one by one directly, we use a square covering $s \times s$ cells that can enclose a circle with diameter D to search all k-cores in the extended square region. We move the $(s \times s)$ square from left top corner of the extended square region to the right bottom corner, and we check if there exists a k_c-core at each position of the square. We record all the squares that contain k_c-cores, and we only do circle rotation on vertices in both N and such squares. The verification granularity is cells rather than vertices, and thus it is much faster.

Finally, we propose a binary rotating circle method to check the candidate vertices to improve the verification costs. The main difference with MC^{3}ALG is that when we rotate the circle, we do not stop rotating when a new vertex enters the circle. Instead, we use the binary search strategy to deal with this. We stop when we reach a vertex such that from the starting entering vertex to this vertex, a k_c-core is firstly met. Then, we check the circle with this vertex on the boundary. If there exist a k_c-core, we record it and stop rotating; otherwise, we start from the checked circle, and find the next vertex that we can meet a k_c-core. This technology is very efficient since we can skip a large region containing no cores.

Example 4. Figure 5 shows an example of the binary searching process (i.e.,). Given the same candidate node as in Example 3, we perform the binary search on the core number. We first have $upper = 3$ (from distance map) and $lower = 2$ (the current best value), so k_c is $\lfloor \frac{3+2}{2} \rfloor = 2$. Then, we set vertices G, H, I as the boundary vertex in. During the rotation process, we consider a binary strategy. First, we get a ordered list (denoted by $InAngleList$) $\{I, H, F, E, D\}$ according to the order that they enter the search circle. Next, we perform the binary search over $InAngleList$ to find the vertex such that a 2-core is first met, which is H since the rotation area (light blue area) form a 2-core (i.e., $\{G, H, I\}$). We rotate the circle to H (blue solid circle) and find a 2-core (i.e., $\{G, H, I\}$). We record it and update $lower = 2 + 1 = 3$. k_c now is 3, and we set vertex G as the boundary vertex and repeat above process. When D is on the boundary of search circle, the rotation area (light red area) form a 3-core (i.e., $\{G, D, E, F, H\}$). Rotate the circle to D directly (red solid circle), we can find a 3-core inside the circle. Finallly, we find $\{G, D, E, F, H\}$ as the best core in this node.

Complexity. We do the same assumption as MC^{3}ALG. We perform the binary search on each extended node N_{ex} ($\gamma(N_{ex}) = 2D + l$) that needs to be checked. Assume that the largest core number obtained from the distance map is k_{max}, and the binary search on k is at most $\log k_{max}$ times. We divide the extended square region into $T \times T$ cells and use a small square covering $s \times s$ to filter out some vertices. The small square covers $(\frac{s(2D+l)}{T})^2 n$ vertices and $(\frac{s(2D+l)}{T})^2 m$ edges, and we need to move the small square $(T-s)^2$ times. Thus the moving process costs at most $O((T-s)^2(\frac{s(2D+l)}{T})^2(n+2m))$. For each vertex in N, during the process of binary circle rotation, it costs at most $O(\log(\pi D^2 n)\pi(\frac{D}{2})^2(n+2m))$ (each circle covers $\pi(\frac{D}{2})^2 n$ vertices and $\pi(\frac{D}{2})^2 m$ edges). Thus, in the worst case, the total complexity of MC^{3}Alg+ is $O(X\log X + X(\log k_{max}((T-s)^2(\frac{s(2D+l)}{T})^2(n+2m) + l^2 n \times \log(\pi D^2 n) \times \pi(\frac{D}{2})^2(n+2m))))$.

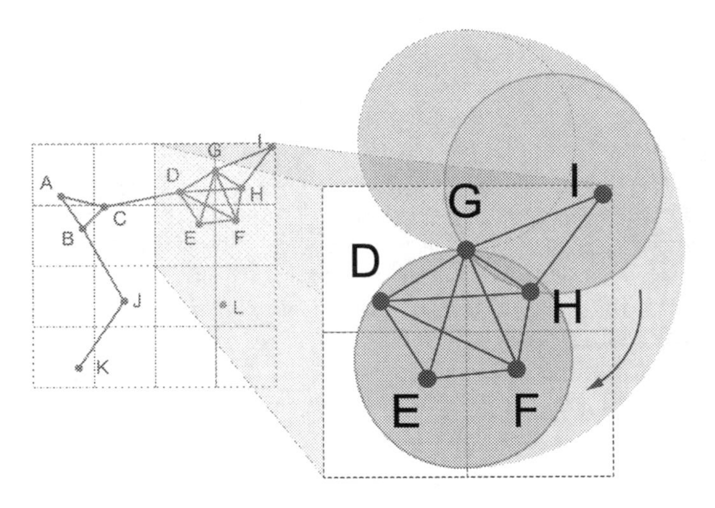

Fig. 5. An example of MC^{3}Alg+ (Color figure online)

7 Experiments

7.1 Experimental Setting

Algorithms. We evaluate our 2 algorithms MC^{3}Alg and MC^{3}Alg+, as well as the two baseline methods. However, since the two baselines runs extremely slow, we only report their performance on one set of experiments.

Datasets. We consider four datasets in our experiment, three real datasets (`Gowalla`, `FourSquare`, `Flickr`) and one synthetic datasets (`YoutubeSyn`). In `Gowalla`[1], each vertex is a user in Gowalla, and each edge represents the friendship between two users. Each user has many checkins, we choose the most frequent one as his location. Note that, we also do an experiment for the situation that a user has many checkins in this dataset. In `FourSquare`[2], each vertex is

[1] http://snap.stanford.edu/data/index.html.

[2] https://foursquare.com/.

a user of Foursquare website, and each edge represents the social relationship between two users. For each user, we choose his most frequent checkin information as his location. In `Flickr`[3], a vertex is a user and an edge represents the "follow" relationship between two users. We mark a user's location where she has token the most photos. In `YoutubeSyn`[4], each vertex represents a user of Youtube, each edge is the"follow" relationship between two users. However, there is no location information of the users. So, we generate a location for each user. In addition, in our experiment, we also use two distribution methods to generate locations, i.e., random distribution and gaussian distribution. Detailed information of the datasets is summarized in Table 2, where $\widehat{deg}$ is the average degree, max_k is the maximum number of locations on a node.

Table 2. Dataset properties

Type	Name	Vertices	Edges	$\widehat{deg}$	max_k
Real	`Gowalla`	107092	456830	4.50	43
Real	`Flickr`	214600	2168900	19.73	50
Real	`FourSquare`	1050000	3362325	7.40	130
Synthetic	`YoutubeSyn`	550000	1952308	7.09	50

Parameters and Query Generation. We set the number of m (the number of grid cells in an extended search area) to 10. We have conducted experiments on this parameter, and it does not affect the performance much. We achieve the best runtime when $m = 10$, and thus we use 10 as the default value on all experiments. We ignore the details due to the space limitation. In the experiment of multiple locations for a user, for `Gowalla`, a user's locations are all the checkins of this user. For `YoutubeSyn`, we randomly generate the locations of a user. In the different distribution experiment, we generate locations satisfying two distribution requirement, i.e., random distribution and gaussian distribution. For all the datasets, we put the locations in a square with size [0, 100] × [0, 100].

Setup. We run experiments on a machine having a Intel i7-6700 3.40 GHz processor and 16 GB of memory, with Windows 10 installed, and all algorithms were implemented in Java.

7.2 Experimental Results

Varying the Given Diameter. The value of the given diameter D affects the search region and efficiency of baselines, MC^{3}Alg and MC^{3}Alg+. In this experiment, we vary the given diameter from 2.5 to 12.5. Figure 6(a) to (c) shows the runtime of algorithms. It can be observed that MC^{3}Alg+ always

[3] https://www.flickr.com/.
[4] http://snap.stanford.edu/data/index.html.

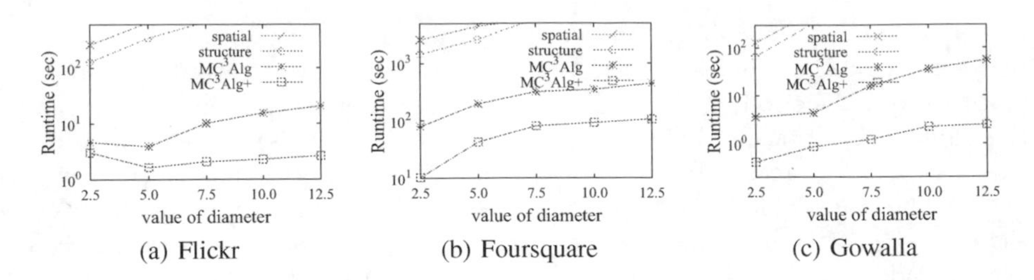

(a) Flickr (b) Foursquare (c) Gowalla

Fig. 6. Varying the diameter

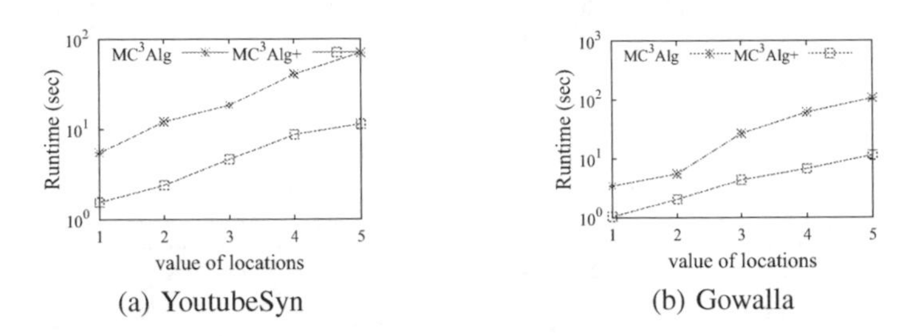

(a) YoutubeSyn (b) Gowalla

Fig. 7. Runtime of multiple checkins

outperforms than other algorithms, since it has the most prune strategies and optimization strategies. The two baselines spatial-first and structure-first are very time consuming, and thus we ignore them in the subsequent experiments.

Effectiveness When a User has Mutiple Checkins. In this experiment, we test the algorithms' performance when a vertex has multiple checkins. Recall the process of rotating circle, more checkins will cause more checks of k-core. Hence, the number of checkins affects the performance of both MC^{3}Alg and MC^{3}Alg+. Figures 7(a) to (b) show the results. We can observe that MC^{3}Alg+ are less affected by multiple checkins, the reason is we perform a binary search to accelerate the rotating process of MC^{3}Alg+. Observing that the runtime of MC^{3}Alg+, it is about 7 times faster than MC^{3}Alg.

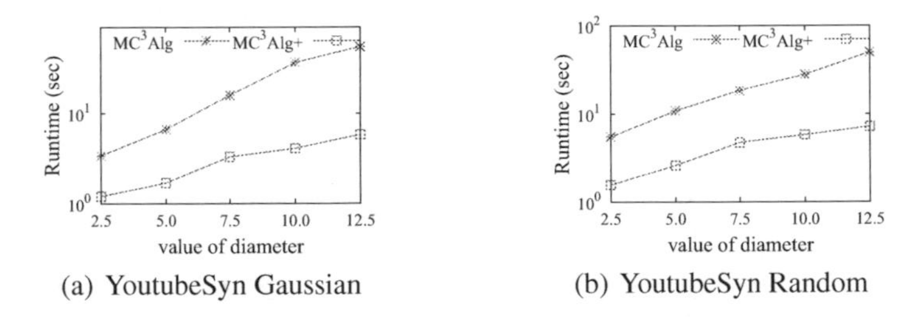

(a) YoutubeSyn Gaussian (b) YoutubeSyn Random

Fig. 8. Runtime of varying the distribution of checkins

Varying the Distribution of Locations. Figures 8(a)–(b) show the results of varying the location distribution of locations. We consider using two classical distributions, Gaussian distribution and Random distribution to evaluate the performance of the two algorithms. It can be observed that MC^{3}ALG+ outperforms MC^{3}ALG consistently. Note that, the superiority of MC^{3}ALG+ is more obvious in Gaussian distribution. The reason is that some nodes contain very large number of vertices, and this leads to a higher complexity in searching these nodes for MC^{3}ALG.

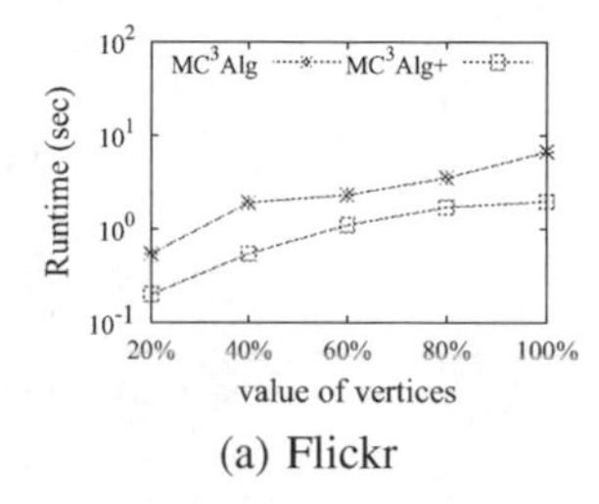

(a) Flickr

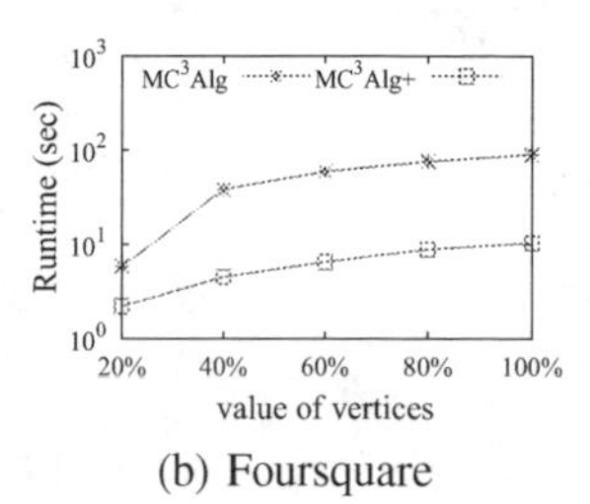

(b) Foursquare

Fig. 9. Runtime of scalability

Scalability. By following the work [12], we vary the percentage of vertices on two datasets, i.e., `Flickr` and `FourSquare`. The results are reported in Figs. 9(a) (b). We can observe that both algorithms scale well with the data set size and MC^{3}ALG+ also runs fastest as it has more pruning strategies.

8 Conclusion

In this paper, we study the most cohesive co-located communities search problem. We first propose an index structure, i.e., DkQ-TREE, to integrate the spatial information and the local structure information together to accelerate the query processing. Then, based on DkQ-TREE, we develop two efficient algorithms. Extensive experiments conducted on both real and synthetic datasets demonstrate the efficiency and effectiveness of our proposed algorithms.

Acknowledgments. Xin Cao is supported by ARC DE190100663. Qiang Qu is supported by the CAS Pioneer Hundred Talents Program, China (grant number Y84402, 2017). Yaqiong Liu is supported by the Fundamental Research Funds for the Central Universities (Grant No. 2018RC03) and 111 Project of China (B17007).

References

1. Akbas, E., Zhao, P.: Truss-based community search: a truss-equivalence based indexing approach. PVLDB **10**(11), 1298–1309 (2017)
2. Altaf-Ul-Amine, M., et al.: Prediction of protein functions based on k-cores of protein-protein interaction networks and amino acid sequences. Genome Inform. **14**, 498–499 (2003)

3. Asahiro, Y., Iwama, K., Tamaki, H., Tokuyama, T.: Greedily finding a dense subgraph. In: Karlsson, R., Lingas, A. (eds.) SWAT 1996. LNCS, vol. 1097, pp. 136–148. Springer, Heidelberg (1996). https://doi.org/10.1007/3-540-61422-2_127
4. Barbieri, N., Bonchi, F., Galimberti, E., Gullo, F.: Efficient and effective community search. DMKD **29**(5), 1406–1433 (2015)
5. Batagelj, V., Zaversnik, M.: An O (m) algorithm for cores decomposition of networks. arXiv preprint arXiv:cs/0310049 (2003)
6. Charikar, M.: Greedy approximation algorithms for finding dense components in a graph. In: Jansen, K., Khuller, S. (eds.) APPROX 2000. LNCS, vol. 1913, pp. 84–95. Springer, Heidelberg (2000). https://doi.org/10.1007/3-540-44436-X_10
7. Chen, L., Liu, C., Zhou, R., Li, J., Yang, X., Wang, B.: Maximum co-located community search in large scale social networks. PVLDB **11**(9), 1233–1246 (2018)
8. Cohen, J.: Trusses: cohesive subgraphs for social network analysis. NSATR **16**, 3 (2008)
9. Cui, W., Xiao, Y., Wang, H., Lu, Y., Wang, W.: Online search of overlapping communities. In: SIGMOD, pp. 277–288 (2013)
10. Cui, W., Xiao, Y., Wang, H., Wang, W.: Local search of communities in large graphs. In: SIGMOD, pp. 991–1002 (2014)
11. Doulkeridis, C., Vouros, G.A., Qu, Q., Wang, S. (eds.): Mobility Analytics for Spatio-Temporal and Social Data. LNCS. Springer, Cham (2018). https://doi.org/10.1007/978-3-319-73521-4
12. Fang, Y., Cheng, R., Li, X., Luo, S., Hu, J.: Effective community search over large spatial graphs. PVLDB **10**(6), 709–720 (2017)
13. Finkel, R.A., Bentley, J.L.: Quad trees: a data structure for retrieval on composite keys. Acta Inf. **4**, 1–9 (1974)
14. Fortunato, S.: Community detection in graphs. Phys. Rep. **486**(3–5), 75–174 (2010)
15. Goldberg, A.V.: Finding a maximum density subgraph. Technical report (1984)
16. Guo, T., Cao, X., Cong, G.: Efficient algorithms for answering the m-closest keywords query. In: SIGMOD, pp. 405–418 (2015)
17. Hu, J., Wu, X., Cheng, R., Luo, S., Fang, Y.: Querying minimal Steiner maximum-connected subgraphs in large graphs. In: CIKM, pp. 1241–1250 (2016)
18. Huang, X., Cheng, H., Qin, L., Tian, W., Yu, J.X.: Querying k-truss community in large and dynamic graphs. In: SIGMOD, pp. 1311–1322 (2014)
19. Huang, X., Lakshmanan, L.V.: Attribute-driven community search. PVLDB **10**(9), 949–960 (2017)
20. Huang, X., Lakshmanan, L.V., Xu, J.: Community search over big graphs: models, algorithms, and opportunities. In: ICDE, pp. 1451–1454 (2017)
21. Huang, X., Lakshmanan, L.V., Yu, J.X., Cheng, H.: Approximate closest community search in networks. PVLDB **9**(4), 276–287 (2015)
22. Li, R.H., Qin, L., Yu, J.X., Mao, R.: Influential community search in large networks. PVLDB **8**(5), 509–520 (2015)
23. Li, Y., Chen, R., Xu, J., Huang, Q., Hu, H., Choi, B.: Geo-social k-cover group queries for collaborative spatial computing. TKDE **27**(10), 2729–2742 (2015)
24. Luce, R.D., Perry, A.D.: A method of matrix analysis of group structure. Psychometrika **14**(2), 95–116 (1949)
25. Qu, Q., Liu, S., Yang, B., Jensen, C.S.: Efficient top-k spatial locality search for co-located spatial web objects. In: 2014 IEEE 15th International Conference on Mobile Data Management (MDM), vol. 1, pp. 269–278. IEEE (2014)
26. Ruan, Y., Fuhry, D., Parthasarathy, S.: Efficient community detection in large networks using content and links. In: WWW, pp. 1089–1098 (2013)

27. Scott, J.: Social Network Analysis. Sage, Thousand Oaks (2017)
28. Seidman, S.B.: Network structure and minimum degree. Soc. Netw. **5**(3), 269–287 (1983)
29. Sozio, M., Gionis, A.: The community-search problem and how to plan a successful cocktail party. In: SIGKDD, pp. 939–948 (2010)
30. Wang, M., Wang, C., Yu, J.X., Zhang, J.: Community detection in social networks: an in-depth benchmarking study with a procedure-oriented framework. PVLDB **8**(10), 998–1009 (2015)
31. Wu, Y., Jin, R., Li, J., Zhang, X.: Robust local community detection: on free rider effect and its elimination. PVLDB **8**(7), 798–809 (2015)
32. Wu, Y., Jin, R., Zhu, X., Zhang, X.: Finding dense and connected subgraphs in dual networks. In: ICDE, pp. 915–926. IEEE (2015)
33. Yang, D.N., Shen, C.Y., Lee, W.C., Chen, M.S.: On socio-spatial group query for location-based social networks. In: SIGKDD, pp. 949–957 (2012)
34. Zhu, Q., Hu, H., Xu, C., Xu, J., Lee, W.C.: Geo-social group queries with minimum acquaintance constraints. VLDB J. **26**(5), 709–727 (2017)

Embedding

A Weighted Word Embedding Model for Text Classification

Haopeng Ren[1], ZeQuan Zeng[1], Yi Cai[1(✉)], Qing Du[1], Qing Li[2], and Haoran Xie[3]

[1] School of Software Engineering, South China University of Technology, Guangzhou, China
ycai@scut.edu.cn

[2] Department of Computing, Hong Kong Polytechnic University, Hung Hom, Kowloon, Hong Kong

[3] The Education University of Hong Kong, Hong Kong, China

Abstract. Neural bag-of-words models (NBOW) have achieved great success in text classification. They compute a sentence or document representation by mathematical operations such as simply adding and averaging over the word embedding of each sequence element. Thus, NBOW models have few parameters and require low computation cost. Intuitively, considering the important degree of each word and the word-order information for text classification are beneficial to obtain informative sentence or document representation. However, NBOW models hardly consider the above two factors when generating a sentence or document representation. Meanwhile, term weighting schemes assigning relatively high weight values to important words have exhibited successful performance in traditional bag-of-words models. However, it is still seldom used in neural models. In addition, n-grams capture word-order information in short context. In this paper, we propose a model called weighted word embedding model (WWEM). It is a variant of NBOW model introducing term weighting schemes and n-grams. Our model generates informative sentence or document representation considering the important degree of words and the word-order information. We compare our proposed model with other popular neural models on five datasets in text classification. The experimental results show that our proposed model exhibits comparable or even superior performance.

Keywords: Neural bag-of-words models · Term weighting schemes · N-grams · Text classification

1 Introduction

Text categorization (TC) is a fundamental and traditional problem in natural language processing (NLP), which automatically classifies sentences or documents into some predefined categories. In a TC task, text representation is an important step. In recent years, neural models have been employed to learn text

G. Li et al. (Eds.): DASFAA 2019, LNCS 11446, pp. 419–434, 2019.
https://doi.org/10.1007/978-3-030-18576-3_25

representation because of the effectiveness of word embeddings learned from massive unstructured text data [1,6,16,25]. Compositional function is used in neural models to represent a text as a low dimensional vector. It is a mathematical process for combining multiple word embeddings into a single vector. Compositional functions fall into two classes: syntactic and unordered [5]. Syntactic functions take word-order and sentence structure information into account dubbed "sophisticated models". These models have shown impressive results in text classification, such as Recurrent Neural Networks (RNNs) [20,21], Recursive Neural Networks [17] and Convolutional Neural Networks (CNNs) [2,7,27]. However, there is a drawback: they are typically computationally expensive or limited computing resources, due to the need to estimate hundreds of thousands, if not millions, of parameters [15]. In contrast, unordered functions treat texts as bags of word embeddings dubbed "neural bag-of-words models" (NBOW). They compute a sentence or document representation by simply adding, or averaging, over the word embedding of each sequence element obtained via, e.g., word2vec [13]. Although NBOW models hardly consider the word-order information, they still exhibit comparable or even superior performance in the text classification, comparing with sophisticated models [16]. In this paper, our works are focus on NBOW models.

In addition, term weighting schemes are shown to bring significant improvements over raw Bag-Of-Words representation in text classification. They are used in traditional Vector Space Model to assign a reasonable weight value for each token [12,23]. For instance, in sentiment classification, the word *'excellent'* in sentence *"I have been here twice and it was excellent both times"* are much more important than other words like *'I'* and *'it'*. Thus, the weight value of the word 'excellent' is higher in the Bag-of-Words representation. Intuitively, considering the important degree of each word is also beneficial to neural models. Note that word embeddings cannot explicitly reflect the important degree of words in a sentence or document representation. Therefore, most NBOW models based on word embeddings ignore that the different tokens have different importance degrees to text classification. Meanwhile, most NBOW models cannot take the word-order information into consideration. For example, *"not really good"* and *"really not good"* convey different levels of negative sentiment, while being different only by their word orderings [16].

Thus, both the **important degree of words** and **word-order** information should not be ignored in a sentence or document representation. In this paper, we propose a model called weighted word embedding model (WWEM). WWEM is a variant of neural bag-of-words models. It combines term weighting schemes and n-grams with the NBOW model. Different with most current NBOW models, a new sentence or document representation method is applied in our proposed model. It takes the important degree of each word and word-order information into consideration. Meanwhile, term weighting schemes explicitly reflect the important degree of words and just require a small amount of computing resources. Therefore, we initialize the weight value of each token through term weighting schemes in our model. The weight values of tokens are fine-tune during training. After training, they become more reasonable. What's more, n-grams can capture the word-order information in the short context, thus we exploit it to enrich the semantic of a sentence or document representation [11]. We conduct experiments on five datasets in text classification. The experimental results show the effectiveness of our proposed model. We also find that the weight value of tokens become more reasonable after training.

The rest of our paper is organized as follows: Sect. 2 introduces the related work about our proposed model. Section 3 details relative models and our proposed model WWEM. Section 4 conducts five datasets to verify the effectiveness of our model. Meanwhile, two cases are given to interpret our proposed model intuitively. We draw a conclusion in Sect. 5.

2 Related Work

Bag-of-words models treat each word in text as an independent token, which ignore the fact that texts are essentially sequential data [14]. Thus, the word-order and syntax information is discarded. Due to the advantages of being effective, efficient and robust, they are still widely used in various kinds of NLP tasks such as information retrieval, question answering and text classification [11]. Usually, term weighting schemes are used to obtain better performance in bag-of-words models. These schemes can be interpreted as methods to measure the utility of a token in discriminating different categories. According to whether the categories information is used, term weighing schemes can be divided into unsupervised ones and supervised ones. TF, IDF, and TF·IDF [18] are the unsupervised ones while RF [9], DC and BDC [24] belong to supervised ones. BDC scheme measures the discrimination power of a token based on its global distributional concentration in the categories of a corpus [24]. Important words are given higher weight values while unimportant words are given less weight values. Currently, term weighting schemes have been successfully applied in bag-of-words models, but it is still seldom used in neural models. To the best of our knowledge, there is a few works in combining term weighting schemes and n-grams with neural models [11]. [11] exploit term weighting schemes and n-grams in objective function based on the Paragraph Vector (PV) model [10].

In recent years, many deep neural models with expressive compositional functions (e.g. CNNs or RNNs) are proposed and have achieved impressive results. [7] proposes convolutional neural networks (CNNS) to extract text features and generate a sentence or document representation. Meanwhile, Recurrent Neural Networks (RNNs) is suitable for the sequence text due to its special network structure [20,21]. Although these models have achieved impressive performance, they are typically computationally expensive. Comparing with CNNs and RNNs models, neural bag-of-words models with simple compositional functions are effective and only need low computational cost such as SWEM [16] and LEAM [22]. SWEM is a variant of NBOW model with simple compositional functions, which gets comparable results and even obtains better performance in some NLP tasks. At the same time, LEAM is another variant of NBOW model which introducing an attention between words and labels [22]. In this paper, we propose a variant of NBOW model introducing term weighting schemes and n-grams. Our proposed model generates informative sentence or document representation considering the important degree of words and the word-order information.

3 Model

In order to consider the important degree of each word and the word-order information for text classification, we propose a variant of NBOW model called Weighted Word Embedding Model (WWEM). The overall architecture of Weighted Word Embedding Model (WWEM) is shown in Figs. 1 and 2. The intuitive of our model is that not all words in the sentences or documents are equally important for text classification. Thus, our model introduces term weighting schemes to initialize the weight values of words. The more important the word, the higher the weight values. Note that the weight values of words are fine-tune in training process. What's more, n-grams information is considered in our proposed model to capture word-order information in short context shown in Fig. 2. The last layer of our model is a classifier which a sentence or document representation will be fed into it. It is implemented as a 300-dimensional Multilayer Perceptron (MLP) layer followed by a sigmoid or softmax function depending on the specific task.

In this section, we first introduce the neural bag-of-words model which performs simple operations over word embeddings to get a sentence or document representation. Then, we introduce term weighting schemes which give each word a reasonable weight value in a sentence or document representation. Finally, our proposed model which combines term weighting schemes and n-grams with neural bag-of-words model will be described in detail. For convenience, we define the mathematical symbols shown in Table 1.

Table 1. The description of some mathematical symbols

Symbol	Description
$X = \{x_1, x_2, ..., x_L\}$	A text sequence
$x_i \, (i \in [1, L])$	The ith token in X
L	The number of tokens in X
$V = \{v_1, v_2, ..., v_L\}$	A word embedding sequence
$v_i \in R^{\lvert e \rvert} \, (i \in [1, L])$	The word embedding of ith word $x_i \, (i \in [1, L])$
$\overrightarrow{z}$	A sentence or document representation vector
$\mathbf{g}\,(\mathbf{V})$	The compositional function

3.1 Neural Bag-of-Words (NBOW) Model

Neural Bag-of-Words (NBOW) models take each token in the text sequence X as independent and unordered. They treat the text sequence as bags of word embeddings. Although NBOW models compute a sentence or document representation by simply adding or averaging over the word embeddings, they have exhibited comparable or even superior performance in some NLP tasks [16]. Commonly, most NBOW models map the text sequence X into a low-dimensional vector $\overrightarrow{z}$ using the simple compositional function $\mathbf{g}(\mathbf{V})$ over the word embedding sequence V:

$$\overrightarrow{z} = \mathbf{g}(\mathbf{V}) \tag{1}$$

When $\mathbf{g}(\mathbf{V})$ is a linear function, a sentence or document representation $\overrightarrow{z}$ can be viewed as the weighted summation of word embeddings. Thus, a sentence or document representation $\overrightarrow{z}$ and the word embedding have the same number of dimension. The linear function can be uniformly described as follows:

$$\overrightarrow{z} = w_1 \cdot v_1 + w_2 \cdot v_2 + ... + w_L \cdot v_L \tag{2}$$

where $w_i \in [1, L]$ is a scalar value used to measure the important degree of each word in the text sequence. If w_i is equal to $\frac{1}{L}$, the model can be viewed as the simplest NBOW model [16] which takes each word equally important in the text sequence. The simplest NBOW model can be described as follows:

$$\overrightarrow{z}_{avg} = \frac{1}{L} \cdot \sum_{i=1}^{L} v_i \tag{3}$$

At the same time, $\mathbf{g}(\mathbf{V})$ can also be a non-linear function. For instance, to extract the most salient features, each dimension of $\overrightarrow{z}$ can be obtained by taking the maximum value over same dimension of the word embedding sequence. Note that a sentence or document representation and the word embedding also have the same number of dimension:

$$\overrightarrow{z}_{max} = \text{Max-pooling}(v_1, v_2, ..., v_L) \tag{4}$$

Meanwhile, $\mathbf{g}(\mathbf{V})$ can also be a composite function. For example, a common operation is to concatenate $\overrightarrow{z}_{avg}$ and $\overrightarrow{z}_{max}$ to get a new sentence or document representation $\overrightarrow{z}_{concat}$. Note that the dimension of $\overrightarrow{z}_{concat}$ is twice the dimension of $\overrightarrow{z}_{avg}$ or $\overrightarrow{z}_{max}$:

$$\overrightarrow{z}_{concat} = concat(\overrightarrow{z}_{max}, \overrightarrow{z}_{avg}) \tag{5}$$

3.2 Term Weighting Schemes

In the text classification task, term weighting schemes are often used to extract text features in traditional bag-of-words models. It is a strategy to measure the important degree of words. We can divide term weighting schemes into an unsupervised weighting schemes and supervised weighting schemes according to whether the label information is used. TF, IDF, and TF·IDF [18] belong to the unsupervised ones. RF [9], DC and BDC [24] can be divided into supervised ones. BDC scheme is based on information theory and measures the discriminating power of a token based on its global distributional concentration in the categories of a corpus [24]. BDC scheme is used to initialize the weight value of tokens in our proposed model and is described in the following subsection.

$$bdc\left(t\right) = 1 - \frac{BH\left(t\right)}{log\left(\left|C\right|\right)} = 1 + \frac{\sum_{i=1}^{|C|} f\left(t|c_i\right) logf\left(t|c_i\right)}{log\left(\left|c\right|\right)} \tag{6}$$

$$f\left(t|c_i\right) = \frac{p\left(t|c_i\right)}{\sum_{i=1}^{|C|} p\left(t|c_i\right)} \tag{7}$$

$$p\left(t|c_i\right) = \frac{num\left(t|c_i\right)}{num\left(c_i\right)}, i \in [1, |C|] \tag{8}$$

where $|C|$ denotes the number of the categories, $num\left(t|c_i\right)$ denotes the frequency of token t in category c_i, and $num\left(c_i\right)$ represents the frequency sum of all terms in category c_i. We can conclude that the larger the value of $bdc\left(t\right)$, the greater the term's discriminating power.

3.3 Weighted Word Embedding Model

Word embedding is a kind of word representation obtained by training a large amount of corpus in an unsupervised manner. Thus, word embedding can be viewed as a spatial representation of a word. However, it cannot explicitly express the important degree of each word in a text sequence. On the contrary, term weighting schemes can explicitly measure the important degree of each word. Therefore, as shown in Figs. 1 and 2, we propose a model which combines the term weighting schemes and n-grams with the neural bag-of-words model, termed Weighted Word Embedding Model (WWEM). For each ngram, we can get a weighted token representation. Then, a sentence or document representation can be obtained with compositional function over the weighted token representations. Meanwhile, according to the difference of n-grams and the compositional function, our proposed models can be divided into two types, namely WWEM-uni_gram models (WWEM-uni_avg, WWEM-uni_max) and WWEM-bi_gram models (WWEM-bi_avg and WWEM-bi_max). Then, we will introduce these four models separately.

WWEM model initializes the weight value of each token by BDC scheme. We normalize them using the softmax function:

$$w_i^{'} = bdc\left(x_i\right), i \in [1, L], x_i \in X \tag{9}$$

$$w_i = \frac{e^{w_i^{'}}}{\sum_{i=1}^{L} e^{w_i^{'}}}, i \in [1, L] \tag{10}$$

At the same time, the weight $w_i, i \in [1, L]$ calculated by Eq. (10) is the initial weight value of each token in WWEM model. The sentence or document representation can be obtained by Eq. (2). Thus, both the word embeddings and term weight values are the input of our proposed model. They are set as a variable respectively. Meanwhile, the weight value of each token would be fine-tune to get more reasonable. It can be shown in our experiments.

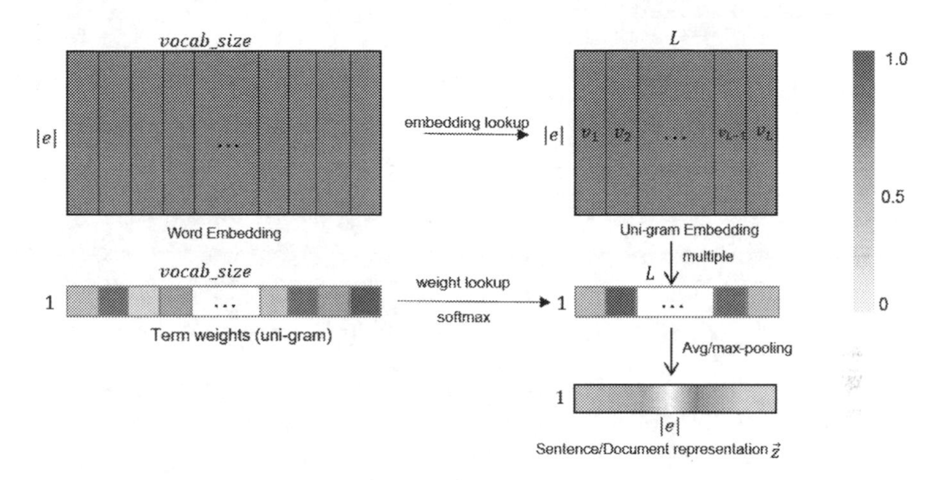

Fig. 1. Uni-gram weighted word embedding model (WWEM-uni_avg, WWEM-uni_max). (1) The shade of the blue color is used to measure the weight value of tokens. (2) The value of term weight range from 0 to 1

As shown in Fig. 1, to extract the most salient features from every word-embedding dimension, our proposed model take the maximum value over each dimension of the weighted word embeddings. Then, the sentence or document representation of WWEM-uni_max model and WWEM-uni_avg model can be separately obtained by Eqs. (11) and (12).

$$\overrightarrow{z}_{uni\_max} = \text{Max-pooling}\,(w_1 \cdot v_1, w_2 \cdot v_2, ..., w_L \cdot v_L) \tag{11}$$

$$\overrightarrow{z}_{uni\_avg} = \sum_{i=1}^{L} (w_i \cdot v_i) \tag{12}$$

Figure 2 shows the framework of WWEM-bi_avg model or WWEM-bi_max model. In order to take account of the word-order information, we combine n-grams information with our model. In WWEM-bi_avg model, the sentence or document representation can be obtained by Eqs. (13), (14) and (15).

$$\overrightarrow{z}_{WWEM\text{-}ngram\_avg} = concat(\overrightarrow{z}_{WWEM\_avg}, \overrightarrow{z}_{ngram\_avg}) \tag{13}$$

$$\overrightarrow{z}_{WWEM\text{-}avg} = w_1 \cdot v_1 + w_2 \cdot v_2 + ... + w_L \cdot v_L \tag{14}$$

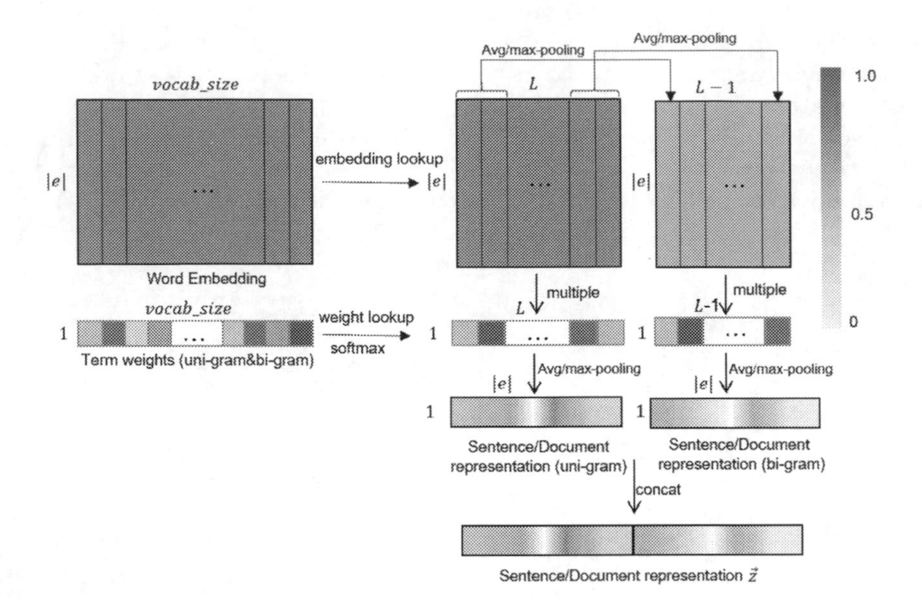

Fig. 2. Bi-gram weighted word embedding model (WWEM-bi_avg, WWEM-bi_max). (1) The shade of the blue color is used to measure the weight value of tokens. (2) The value of term weight range from 0 to 1 (Color figure online)

$$\vec{z}_{ngram\_avg} = w_{ngram,1} \cdot \sum_{i=1}^{n} w_i v_i + w_{ngram,2} \cdot \sum_{i=2}^{n+1} w_i v_i + \\ ... + w_{ngram,L-n+1} \cdot \sum_{i=L-n+1}^{L} w_i v_i \,(n \geq 2) \tag{15}$$

where $w_i, i \in [1, L]$ denotes the weight values of uni-grams and $w_{ngram,i}$, $i \in [1, L-n+1]$ denotes the weight values of n-grams. Both of them are initialized by BDC scheme. In our proposed model, n is set to 2. The sentence or document representation in WWEM-bi_max model can be calculated by Eqs. (16), (17), (18) and (19).

$$\vec{z}_{WWEM\text{-}ngram\_max} = concat\,(\vec{z}_{WWEM\_max}, \vec{z}_{ngram\_max}) \tag{16}$$

$$\vec{z}_{WWEM\_max} = \text{Max-pooling}\,(w_1 \cdot v_1, w_2 \cdot v_2, ..., w_L \cdot v_L) \tag{17}$$

$$\vec{z}_{ngram\_max} = w_{ngram,1} \cdot \vec{z}_{win\_max}\,(1, n) + w_{ngram,2} \cdot \vec{z}_{win\_max}\,(2, n+1) + \\ ... + w_{ngram,L-n+1} \cdot \vec{z}_{win\_max}\,(L-n+1, L) \tag{18}$$

$$\vec{z}_{win\_max}\,(i, j) = \text{Max-pooling}\,(w_i v_i, w_{i+1} v_{i+1}, ..., w_j v_j) \tag{19}$$

4 Experiments

Experimental Setting. We conduct the experiments on five datasets to evaluate the effectiveness of our proposed model. In our experiments, we use Google 300-dimensional word2vec as initialization for our model's word embeddings. Out-of-Vocabulary (OOV) words are initialized from a uniform distribution with range $[-0.01, 0.01]$. The final layer of our proposed model is a classifier which is implemented as an MLP layer followed by a sigmoid or softmax function depending on the special task. The hyper-parameter settings of the neural network are as follows:

(1) We train our model's parameters with the Adam Optimizer [8] with an initial learning rate of 0.001 and a minibatch size of 100.
(2) The MLP layer's dimension is selected from the set [100, 150, 200, 300].
(3) Dropout regularization [19] is employed on the MLP layer, with dropout rate 0.8. The model is implemented using Tensorflow and is trained on GTX 1080Ti.

Table 2. Summary statistics of five datasets. #classes denotes the number of classes, #Training: the number of training samples. #Testing: the number of testing samples. $|V|$ denotes the vocabulary size.

Dataset	#classes	#Training/#Testing	$\|V\|$
AGNews	4	120k/7.6k	90k
Yelp binary	2	560k/38k	314k
Yelp full	5	650k/38k	340k
DBPedia	14	560k/70k	667k
Yahoo	10	1400k/60k	990k

4.1 Datasets

We evaluate our proposed model on the same five benchmark datasets as in [26]. All the datasets use accuracy as the metric. The summary statistic of the datasets is shown in Table 2 and the simple descriptions of each dataset are shown as follows:

AGNews. The dataset is obtained from Internet news articles [4]. Each article consists of news title and the description fields. The articles are classified into four topics: Word, Entertainment, Sports and Business.

Yelp Review Polarity. The dataset is obtained from the Yelp Dataset Challenge in 2015. It is used for sentiment classification task, predicting a polarity label by considering stars 1 and 2 negative, and 4 and 5 positive.

Yelp Review Full. The Yelp Review Full dataset is also obtained from the Yelp Dataset Challenge in 2015. It is used for sentiment classification task, predicting full number of stars the user has given.

DBPedia. DBPedia is an ontology dataset, and is constructed by picking 14 non-overlapping classes from DBPedia 2014 (Wikipedia). Each article consists of the title and abstract of the Wikipedia article.

Yahoo! Answers Topic. The dataset is obtained from Yahoo! Answers Comprehensive Questions and Answers version 1.0 dataset. It is used for the topic classification task. Each article consists of question title, question content and best answer.

We compare with different types of models. According to the complexity of model, the models can be summarized into three types as follows:

(1) The traditional bag-of-words model in [26] denoted as Bag-of-words.
(2) Sophisticated deep CNN/RNN models: Small word CNN, Large word CNN, LSTM reported in [26] and SA-LSTM (word-level) [3]
(3) Simple neural bag-of-words models: simple word embedding models (SWEM-aver, SWEM-max) [16] and Label-Embedding Attentive Model (LEAM) [22].

4.2 Result Analysis

Accuracy Analysis. Compared with sophisticated deep learning models (such as CNN and RNN models), simple neural network structure models also achieve comparable performance and even get higher accuracy on some datasets (Table 3). Then, we analyse the experimental results from two perspectives as follows.

Comparing with Different Models

The traditional bag-of-words models perform worse than other methods. We analyse that bag-of-words models may not take the contextual, syntax and more semantic information of the sentence or document into account. In traditional bag-of-words method, term weighting scheme is a significant method to extract features. To some extent, our proposed model combining term weighting scheme with word embedding may obtain semantic and contextual information. From the experimental results, our proposed model gets higher test accuracy rate than traditional bag-of-words methods. We argue that our model may make use of the information of the word embedding.

Meanwhile, models with expressive compositional function (e.g. RNNs or CNNs), have demonstrated impressive results; however, they are typically computationally expensive [22]. On the contrary, our proposed model with simple compositional functions achieves comparable results or even exhibits stronger performance on some datasets. For example, in dataset Yahoo! Answers Topic, DBPedia, and Yelp Review Full, the test accuracy of our proposed model is about 1–2% higher than that of the CNN or LSTM models.

Table 3. Test accuracy on classification tasks, in percentage; the results of the other models are directly cited from the respective papers.

Model	Yahoo	DBPedia	AGNews	Yelp P.	Yelp F.
Bag-of-words [26]	68.90	96.60	88.80	92.20	58.00
Small word CNN [26]	69.98	98.15	89.13	94.46	58.59
Large word CNN [26]	70.94	98.25	91.45	**95.11**	59.48
LSTM [26]	70.84	98.55	86.06	94.74	58.17
SA-LSTM (word-level) [3]	-	98.60	-	-	-
SWEM-aver [16]	73.14	98.42	91.71	93.59	60.66
SWEM-max [16]	72.66	98.24	91.79	93.25	59.63
LEAM [22]	**75.22**	98.32	92.45	93.43	61.03
WWEM-uni_avg	72.34	98.65	93.08	93.20	59.73
WWEM-uni_max	72.79	98.11	92.54	93.11	60.00
WWEM-bi_avg	73.50	98.72	92.86	94.50	**61.35**
WWEM-bi_max	73.49	**98.73**	**93.20**	94.20	61.03

Finally, simple word embedding models (SWEM) [16] with parameters-free pooling operation and other simple compositional functions also exhibit comparable or even superior performance. However, SWEM ignores a point that not all the words are equally important for the text classification task. Meanwhile, word-order information has not been considered in SWEM model. At the same time, the LEAM [22] model obtains the important degree of words by measuring the similarity between the label embedding and the word embedding. However, word-order information is ignored. In some cases, there is no explicit corresponding word embedding available for the label embedding initialization during learning. Thus, it may give an unreasonable weight to words in a sentence or document representation. Intuitively, considering the important degree of each word and the word-order information are beneficial to obtain informative a sentence or document representation. Our proposed model takes account of the above factors and get better performance than SWEM model and LEAM model in some datasets.

Comparing with Different Type Tasks

On topic classification task (e.g. Yahoo dataset and Agnew's dataset), our proposed model exhibits stronger performances relative to both LSTM and CNN sophisticated architectures. On the ontology classification problem (DBPedia dataset), we find the same trend that our proposed model obtains comparable or even superior results relative to CNN or LSTM models.

On the sentiment analysis tasks, several deep learning models based on CNN or LSTM perform better than our proposed model. There is probably due to two reasons: (1) Word-order information may be required for predict sentiment orientations. For instance, two phrases *"not really good"* and *"really not good"* convey different levels of negative sentiment, while being different only by their word orderings [16]. (2) Syntax information also affects the sentiment orientations. The examples are shown in Table 4. For the first case, the first part of the sentence conveys positive sentiment. However, the latter part of the sentence is what the reviewer really want to express. The word *"But"*, a transitional word, conveys a signal that the latter part of sentence should be give more attention. On the contrary, we consider that word-order and syntax information may be less useful for the topic classification tasks.

Table 4. Case samples extracted from dataset Yelp Review Polarity.

Label	Sentence
Negative	*Breakfast is always good.* **But** *too much Pamela's seems greasy*
Negative	*I like domino's pizza.../The location* **however** *is awful*

4.3 Case Study

In order to validate that our proposed model is able to select informative words and generate the informative sentence or document representation, we analyze two cases. In Tables 5 and 6, we give two cases and visualize the change in term weights based on the Yelp Review Polarity dataset. In both two cases, we normalize the term weights: $w_{normalize} = \frac{w-min}{max-min}$, where w denotes the term weight vector of the sentence and min denotes the minimum of the term weight vector, while max is the maximum of the term weight vector.

Table 5. Case samples extracted from dataset Yelp Review Polarity.

Model type	Label	Predict result	Sentence
WWEM_uni	Positive	Positive	*"I love old navy clothing it's vintage styling with good prices. Great jeans too I love the painter's jeans."*

Case One

As shown in Table 5, the category of the sample sentence can be predicted correctly by our proposed model. In Fig. 3(a) and (b), our model can assign the words carrying strong sentiment like *"great"*, *"vintage"*, *"love"* and *"old"* to high weight. On the Contrary, the words hardly expressing sentiment obtain low weights, like *"with"*, *"it"* and the punctuation.

However, the word *"good"* carrying strong sentiment get a low initialization weight in Fig. 3(a). We analysis that the word *"good"* appears frequently in both categories. BDC schemes based on information theory give the word *"good"* low discrimination power. We argue that term weighting schemes hardly take the semantic information into account.

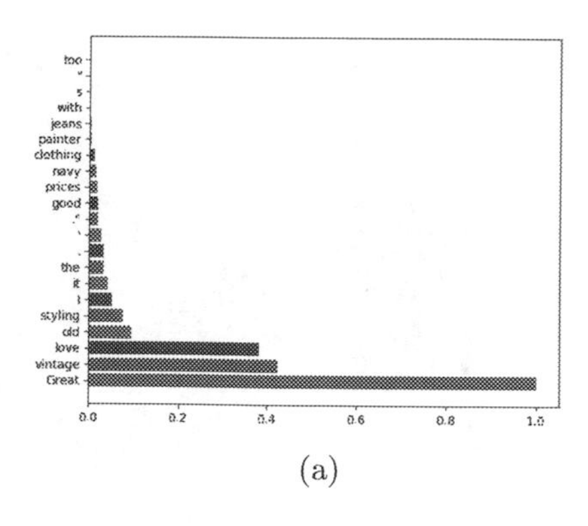

(a)

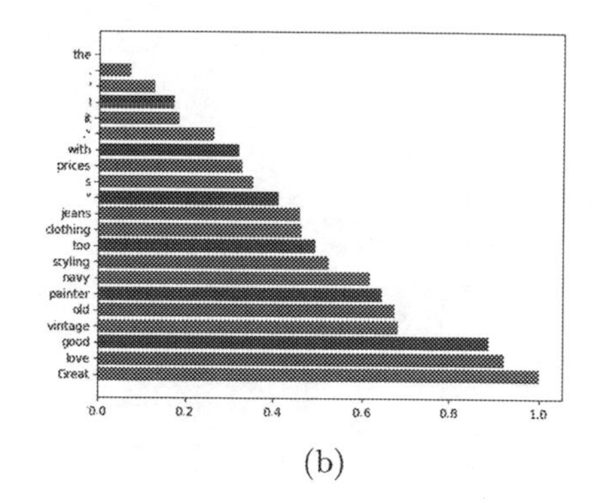

(b)

Fig. 3. Term weight values in WWEM-uni_avg model. (a) Term weights before model training; (b) Term weights after model training

In order to get informative sentence or document representation, the term weights can be set to fine-tune during model training. As shown in Fig. 3(b), the word *"good"* can be given high weight after training the model. To some extent, our proposed model may make good use of the semantic information. Meanwhile, the words, expressing strong sentiment, also get high weight values, like *"great"*, *"love"*, *"vintage"* and *"old"*.

Table 6. Case samples extracted from dataset Yelp Review Polarity

Model type	Label	Predict result	Sentence
WWEM-uni_avg	Positive	Negative	*"Pretty good food fantastic atmosphere* ***slightly overpriced*** *but* ***not unreasonable*** *for the quality and atmosphere"*
WWEM-bi_avg	Positive	Positive	

Case Two

As shown in Table 6, the WWEM-uni_avg model gives a wrong prediction for the sample sentence. Instead, the model WWEM-bi_avg correctly predicts the category of the sample sentence. In the following content, we analyze this case. In Fig. 4(a) and (b), Our proposed model can pay more attention to the sentiment words like *"overprice"*, *"fantastic"*, *"unreasonable"* and *"good"*. From these key words expressing strong sentiment, the WWEM-uni_gram models would be difficult to make a correct prediction. We analysis that the WWEM-uni_gram model ignores the word-order information. Thus, the bigram features cannot be extracted, like "slightly overpriced" and "not reasonable".

N-grams can also reflect semantic information that cannot be obtained by considering the words individually. The unigram *"overprice"* and *"unreasonable"*

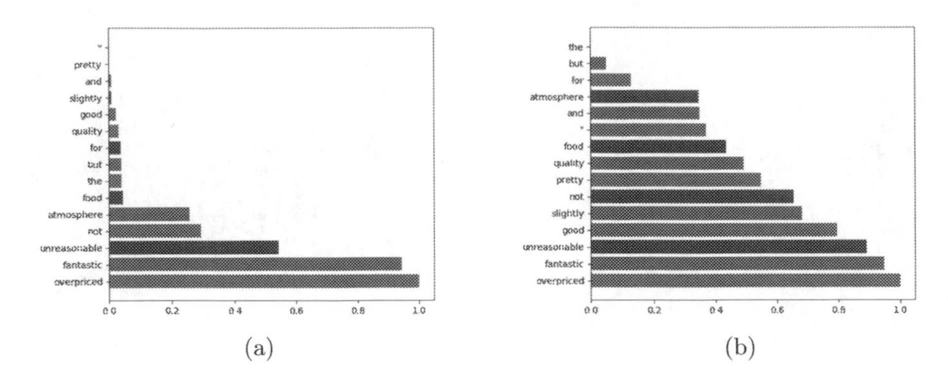

Fig. 4. Term weight values in WWEM-uni_avg model. (a) Term weights before model training; (b) Term weights after model training

are given high weights in WWEM-uni_avg models. Thus, the WWEM-uni_avg model is difficult to give a correct prediction. On the contrary, both bi-gram *"slightly overprice"* and *"not unreasonable"* are given relatively low weight in WWEM-bi_avg model. Meanwhile, the bigram *"good food"* and *"fantastic atmosphere"* are given more attention (Fig. 5). Finally, the WWEM-bi_avg model can give a correct prediction. Thus, WWEM-bi_gram models can take the word-order information into account and capture more significant semantic information.

Through the analysis of the two cases above, we can conclude that our proposed model can capture more semantic information by combining term weight schemes and n-gram information.

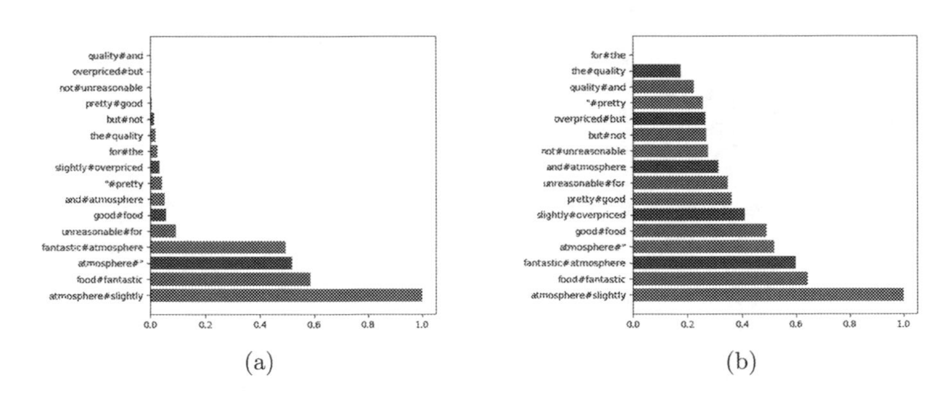

Fig. 5. Term weight values in WWEM-bi_avg model. (a) Term weights before model training; (b) Term weights after model training

5 Conclusion

In this work, we propose a variant of neural bag-of-works model combining with term weighting schemes and n-grams information. The model has few parameters and requires much lower computational cost. It is tested on several large public datasets. According to our experimental results, we find that our proposed model exhibits comparable or even superior performance in the text classification compared with the state-of-the-art. The weight values of words are highly interpretable. It is beneficial to generate an informative sentence or document representation. Meanwhile, n-gram information is significant for the model to achieve better performance. We also find that sentiment analysis tasks are more sensitive to the word-order information than the topic classification.

Acknowledgement. This work was supported by the Fundamental Research Funds for the Central Universities, SCUT (No. 2017ZD048, D2182480), the Tiptop Scientific and Technical Innovative Youth Talents of Guangdong special support program (No. 2015-TQ01X633), the Science and Technology Planning Project of Guangdong Province (No. 2017B050506004), the Science and Technology Program of Guangzhou International Science & Technology Cooperation Program (No. 201704030076). The research described in this paper has been supported by a collaborative research grant from the Hong Kong Research Grants Council (project No. C1031-18G).

References

1. Arora, S., Liang, Y., Ma, T.: A simple but tough-to-beat baseline for sentence embeddings (2016)
2. Blunsom, P., Grefenstette, E., Kalchbrenner, N.: A convolutional neural network for modelling sentences. In: Proceedings of the 52nd Annual Meeting of the Association for Computational Linguistics (2014)
3. Dai, A.M., Le, Q.V.: Semi-supervised sequence learning. In: Advances in Neural Information Processing Systems, pp. 3079–3087 (2015)
4. Del Corso, G.M., Gulli, A., Romani, F.: Ranking a stream of news. In: Proceedings of the 14th International Conference on World Wide Web, pp. 97–106. ACM (2005)
5. Iyyer, M., Manjunatha, V., Boyd-Graber, J., Daumé III, H.: Deep unordered composition rivals syntactic methods for text classification. In: Proceedings of the 53rd Annual Meeting of the Association for Computational Linguistics and the 7th International Joint Conference on Natural Language Processing (Volume 1: Long Papers), vol. 1, pp. 1681–1691 (2015)
6. Joulin, A., Grave, E., Bojanowski, P., Mikolov, T.: Bag of tricks for efficient text classification. arXiv preprint arXiv:1607.01759 (2016)
7. Kim, Y.: Convolutional neural networks for sentence classification. arXiv preprint arXiv:1408.5882 (2014)
8. Kingma, D.P., Ba, J.: Adam: A method for stochastic optimization. arXiv preprint arXiv:1412.6980 (2014)
9. Lan, M., Tan, C.L., Su, J., Lu, Y.: Supervised and traditional term weighting methods for automatic text categorization. IEEE Trans. Pattern Anal. Mach. Intell. **31**(4), 721–735 (2009)

10. Le, Q., Mikolov, T.: Distributed representations of sentences and documents. In: International Conference on Machine Learning, pp. 1188–1196 (2014)
11. Li, B., Zhao, Z., Liu, T., Wang, P., Du, X.: Weighted neural bag-of-n-grams model: new baselines for text classification. In: Proceedings of COLING 2016, the 26th International Conference on Computational Linguistics: Technical Papers, pp. 1591–1600 (2016)
12. Martineau, J., Finin, T., et al.: Delta TFIDF: an improved feature space for sentiment analysis. In: ICWSM, vol. 9, p. 106 (2009)
13. Mikolov, T., Sutskever, I., Chen, K., Corrado, G.S., Dean, J.: Distributed representations of words and phrases and their compositionality. In: Advances in Neural Information Processing Systems, pp. 3111–3119 (2013)
14. Pang, B., Lee, L., Vaithyanathan, S.: Thumbs up?: sentiment classification using machine learning techniques. In: Proceedings of the ACL-02 Conference on Empirical Methods in Natural Language Processing, vol. 10, pp. 79–86. Association for Computational Linguistics (2002)
15. Parikh, A.P., Täckström, O., Das, D., Uszkoreit, J.: A decomposable attention model for natural language inference. arXiv preprint arXiv:1606.01933 (2016)
16. Shen, D., et al.: Baseline needs more love: On simple word-embedding-based models and associated pooling mechanisms. arXiv preprint arXiv:1805.09843 (2018)
17. Socher, R., Lin, C.C., Manning, C., Ng, A.Y.: Parsing natural scenes and natural language with recursive neural networks. In: Proceedings of the 28th International Conference on Machine Learning (ICML 2011), pp. 129–136 (2011)
18. Sparck Jones, K.: A statistical interpretation of term specificity and its application in retrieval. J. Documentation **28**(1), 11–21 (1972)
19. Srivastava, N., Hinton, G., Krizhevsky, A., Sutskever, I., Salakhutdinov, R.: Dropout: a simple way to prevent neural networks from overfitting. J. Mach. Learn. Res. **15**(1), 1929–1958 (2014)
20. Sutskever, I., Vinyals, O., Le, Q.V.: Sequence to sequence learning with neural networks. In: Advances in Neural Information Processing Systems, pp. 3104–3112 (2014)
21. Tai, K.S., Socher, R., Manning, C.D.: Improved semantic representations from tree-structured long short-term memory networks. arXiv preprint arXiv:1503.00075 (2015)
22. Wang, G., et al.: Joint embedding of words and labels for text classification. arXiv preprint arXiv:1805.04174 (2018)
23. Wang, S., Manning, C.D.: Baselines and bigrams: simple, good sentiment and topic classification. In: Proceedings of the 50th Annual Meeting of the Association for Computational Linguistics: Short Papers-Volume 2, pp. 90–94. Association for Computational Linguistics (2012)
24. Wang, T., Cai, Y., Leung, H., Cai, Z., Min, H.: Entropy-based term weighting schemes for text categorization in VSM. In: 2015 IEEE 27th International Conference on Tools with Artificial Intelligence (ICTAI), pp. 325–332. IEEE (2015)
25. Wieting, J., Bansal, M., Gimpel, K., Livescu, K.: Towards universal paraphrastic sentence embeddings. arXiv preprint arXiv:1511.08198 (2015)
26. Zhang, X., Zhao, J., LeCun, Y.: Character-level convolutional networks for text classification. In: Advances in Neural Information Processing Systems, pp. 649–657 (2015)
27. Zhang, Y., et al.: Adversarial feature matching for text generation. arXiv preprint arXiv:1706.03850 (2017)

Bipartite Network Embedding via Effective Integration of Explicit and Implicit Relations

Yaping Wang[1], Pengfei Jiao[2(✉)], Wenjun Wang[1], Chunyu Lu[1], Hongtao Liu[1], and Bo Wang[1]

[1] College of Intelligence and Computing, Tianjin University, Tianjin, China
{yapingwang,wjwang,tjulcy,htliu,bo_wang}@tju.edu.cn
[2] Center for Biosafety Research and Strategy, Tianjin University, Tianjin, China
pjiao@tju.edu.cn

Abstract. Network representation learning, or network embedding, aims at mapping the nodes of the network to low-dimensional vector space, in which the learned node representations can be used for a variety of tasks, such as node classification, link prediction, and visualization. As a special class of complex networks, the bipartite network is composed of two different types of nodes in which the links only exist among different types of nodes, has important applications in the recommendation system, link prediction, and disease diagnosis. However, most existing methods for network representation learning are aimed at homogeneous networks in general, while the special properties of bipartite networks are not taken into account, such as the implicit relations (i.e., unobserved links) between nodes of the same type. In this paper, we propose a novel deep learning framework for bipartite networks, which integrates the explicit and implicit relations, while preserving the local and global structure, to learn the highly non-linear representations of nodes. Extensive experiments conducted on several real-world datasets, based on the link prediction, recommendation, and visualization, demonstrate the effectiveness of our proposed method compared with state-of-the-art network representation learning based methods.

Keywords: Bipartite networks · Network representation · Deep learning

1 Introduction

Many biological, social, and information systems in the world can be modeled as complex networks, which has received extensive attention and research in the past few years. The bipartite network is an important manifestation of the complex network, which is composed of two different types of nodes, and nodes of the same type are not connected. In the real world, the user-item network, author-paper collaboration network, and disease-gene interaction network are all

G. Li et al. (Eds.): DASFAA 2019, LNCS 11446, pp. 435–451, 2019.
https://doi.org/10.1007/978-3-030-18576-3_26

belong to bipartite networks. Modeling and predicting bipartite networks have widely applications, such as the link prediction, question answering systems and recommender systems.

For analysis on the bipartite network, traditional methods usually extract features by feature engineering [17], but searching for valuable features in large-scale data is a time-consuming and costly effort, and can lead to poor accuracy. Recently, network representation learning can solve this challenge, which enable learning meaningful feature representations from the network data automatically. It aims at mapping each node of the network into a low-dimensional vector space, where similar nodes are closer. These representations can be used as features of nodes in a variety of network application tasks, such as node classification, link prediction, and visualization [8,19].

Recently, a variety of representation learning methods for complex networks (we call homogeneous networks) have been proposed, such as DeepWalk [12], LINE [16], Node2vec [6] and so on. However the applications of these methods in the bipartite network face the following two major challenges: (1) Ignore implicit relations between the same type of nodes. (2) The highly non-linear structure of the bipartite network cannot be captured. Taking the LINE model as an example, it learns the first-order similarity and second-order similarity of the homogeneous network respectively, which doesn't consider the type of nodes in the bipartite network. Although there are no links between nodes of the same type, there will be some implicit relations between them. For example, in a user-item network, there will be some kind of friendship between the users, and similar attributes between the items. Meanwhile, LINE only leverages a single-layer nonlinear function, which can't capture the highly non-linear structure of the bipartite network.

At the same time, the bipartite network can be regarded as a special heterogeneous network. Metapath2vec++ is proposed as one of the pioneers of the heterogeneous network representation learning [3]. It applies the meta-path-guided random walk to capture the semantic and structural correlations among different types of nodes. Although metapath2vec++ can be applied to bipartite networks, its inherent structural properties are not taken into account. To our knowledge, Gao et al. [5] proposed a BiNE model for bipartite networks, which is the only specially proposed method for bipartite network representation learning. It models the explicit relations between different types of nodes and the implicit relations between nodes of the same type simultaneously. However, the BiNE model learns the explicit and implicit relations in two steps without integrating them, and cannot capture the high non-linear structure of the network.

In order to solve the above limitations of network representation learning methods for bipartite network, in this paper, we propose a novel Bipartite Network Embedding model Integrating the Explicit and Implicit relations, named **BiNE-IEI**. First, we apply a method based on projection to obtain the implicit relations between nodes of the same type, and model the explicit relations between different types of nodes by the observed links. For each node of bipartite networks, we use a uniform representation of the explicit and implicit relations.

Second, to integrate explicit and implicit relations effectively of each node while capturing the highly non-linear structure of bipartite networks, we propose a novel deep learning framework based on two parallel deep autoencoders which can be considered as extensions to the multi-layer autoencoder proposed by Salakhutdinov et al. [14]. It can reconstruct the neighborhood structure of each node through the deep autoencoders to preserve the global structure of bipartite networks. Besides, to preserve the local structure of networks, we model two directly connected nodes by learning the node representations. Last, we evaluate our model compared with a variety of baselines on three datasets, which show the effectiveness of our model on several data mining tasks, including the link prediction, recommendation, and visualization.

The contributions and advantages of this paper can be summarized as follows:

- We propose a novel deep learning framework, named **BiNE-IEI**, which integrates explicit and implicit relations to learn the node representations of the bipartite network.
- Our deep learning framework can capture the highly non-linear structure of the bipartite network while preserving local and global network structure.
- The experimental results show the effectiveness of our proposed model which outperform other state-of-the-art methods.

2 Related Work

Here, we introduce the traditional methods for link prediction in the bipartite network, the node representation learning methods based on the homogeneous network and heterogeneous network.

2.1 Traditional Bipartite Network Link Prediction

We divide these methods into two categories, projection based and topological similarity-based. The nature of the first type is mapping the bipartite network into the unipartite network for link prediction, such as the methods proposed in ProbS [22] and HeatS [21]. The idea of the latter is to directly calculate the similarity between the two kinds of nodes based on the observed networks. Several classical similarity indices based on local network topological structure have been proposed [11], including Common Neighbors (CN), Jaccard's index (JC), Adamic Adar (AA), allocation of resources (RA) and Preferential Attachment (PA). Compared with the link prediction methods of local network topological structure, the global network topological structure similarity methods [11] take the structure of the whole network into account, including Katz Index (Katz) and Leicht Holme-Newman (LHN2).

2.2 Representation Learning of the Homogenous Network

The network representation learning aims to learn a low-dimension representation of nodes. The methods and models for homogeneous networks are mainly divided into the following two categories: (1) Linear dimension reduction: Principal component analysis (PCA) [9] is one of the common methods for linear dimension reduction, while singular value decomposition (SVD) contains the kernel of PCA. (2) Nonlinear dimension reduction: Perozzi et al. [12] proposed a method DeepWalk, which generates a large number of random walk sequences by truncated random walk. After that, a simple neural network based model LINE [16] was proposed to learn the network representations separately through first-order and second-order similarity. Node2vec [6] is another improved algorithm for DeepWalk, which adjusts the process of random walk by introducing a depth-first and breadth-first strategy. Wang et al. [18] proposed a model named SDNE, which exploits deep learning to learn node representations. The first and second order similarity of networks are preserved while the highly non-linear structure is captured. Detailed reviews can be seen in [1,2]. However, these methods are designed for homogeneous networks without considering the type of nodes in bipartite networks, which are not optimal or apply to node representations of the bipartite network.

2.3 Representation Learning of the Heterogeneous Network

The representation learning on heterogeneous networks can well describe the complex relations among different types of nodes. Dong et al. [3] extracted the node structure information by performing the random walk based on the meta-paths in the heterogeneous network and applied the skip-gram algorithm to learn the node representations. On the basis of this work, the author proposed Metapath2Vec++ [3], which considers the node type information in softmax. The HIN2Vec proposed by Fu et al. [4] can not only learn the representations of nodes but also learn the vector representation of meta-paths. HINE [7] first calculates the similarity between nodes based on the random walk of the meta-paths and uses it as the supervised information to guide the vector representations of the nodes. Meanwhile, Gao et al. [5] proposed a BiNE model for bipartite network embedding, which models the explicit relations between different types of nodes by the first-order similarity and captures the implicit relations between the same nodes by performing a biased and self-adaptive random walk. However, BiNE does not consider the integration of the explicit and implicit relations, nor does it capture the highly non-linear structure of networks.

3 Methodology

In this section, we first describe the notations used in this paper and define our problem. Then we present our proposed model and give our designed loss functions for optimization. Note that the bold letters used in mathematics that appear in this paper represent vector or matrix.

3.1 Definition and Notation

Definition 1 (Bipartite Graph). *Let $G = (U, V, E)$ denote a general undirected bipartite graph, where U and V are two sets of different types of nodes* *and E is the set of links. Suppose there are n nodes of type $U = \{u_i, ..., u_n\}$ and m nodes of type $V = \{v_i, ..., v_m\}$. For a bipartite graph, we denote its weight matrix as $\mathbf{A} \in R^{n \times m}$. Each edge $e_{ij} \in E$ is associated with a weight a_{ij}, where the weight $a_{ij} > 0$ if nodes u_i and v_j are connected and $a_{ij} = 0$ otherwise.*

Definition 2 (Bipartite Network Embedding). *Given a bipartite graph denoted as $G = (U, V, E)$, the bipartite network embedding aims to learn a function $f : U \cup V \to R^d$, which maps each node $u_i \in U$ (or $v_j \in V$) into a low-dimensional embedding space. In the embedding space, both the local and global structure are preserved while integrating explicit and implicit relations.*

Definition 3 (Local Network Structure). *The local network structure can be described by local similarity between directly connected nodes. For any node pair (u_i, v_j), the local structure can be defined as: $\{a_{ij} | a_{ij} > 0, \forall u_i \in U, \forall v_j \in V\}$.*

Definition 4 (Global Network Structure). *The global network structure can be described by the similarity of the node pair's neighborhood structure. For any node $u_i \in U$, it's global structure is formulated as: $\{a_{i*} | a_{i*} > 0, \forall v_* \in V, u_i \in U\}$. Similarly, for any node $v_j \in V$, the global structure is defined as: $\{\mathbf{a}_{*j} | \mathbf{a}_{*j} > 0, \forall u_* \in U, v_j \in V\}$.*

3.2 Modeling Implicit Relations

The recommendation system is an important application of the bipartite network. Some researches have shown that implicit social relations can help improve the performance of recommendation systems. The correlation between users or items can be treated as an implicit relation. For example, users are more likely to purchase items based on their friends' recommendations, and if two users have purchased the same item, it means they have some kind of similarity. The more the same items they have purchased, the more similar they are. Similarly, if two items are purchased by many users, then they are similar. In this paper, we apply the bipartite network projection technique based on Cosine similarity to obtain implicit relations between nodes of the same type. The Cosine distance between two nodes of type U is defined as:

$$S_{ij}^U = \frac{|N(u_i) \times N(u_j)|}{\sqrt{k_i k_j}}, u_i, u_j \in U, \tag{1}$$

where $N(u_i) = \{v_j \in V | a_{ij} > 0,\ u_i \in U\}$ denote the set of neighbors of node u_i, and k_i is the degree of node u_i. The same similarity measure is also applicable to the nodes of the V type, which can be defined as:

$$S_{ij}^V = \frac{|N(v_i) \times N(v_j)|}{\sqrt{k_i k_j}}, v_i, v_j \in V, \tag{2}$$

In fact, not all nodes of the same type have significant implicit relations. To solve this problem, we introduce a threshold γ :

$$S_{ij}^U = \begin{cases} S_{ij}^U, & S_{ij}^U \geq \gamma, \\ 0, & S_{ij}^U < \gamma, \end{cases} \tag{3}$$

The same process is applied to S_{ij}^V.

3.3 Deep Learning Framework

In a bipartite network, we take the observed links as explicit relations, which can be represented by the weight matrix **A**. For unobserved links between nodes of the same type, we regard them as implicit relations and represent them with the similarity matrix $\mathbf{S}^U$ (or $\mathbf{S}^V$) defined in Eqs. 2 and 3. In order to obtain a uniform representation of the explicit and implicit relations of nodes in bipartite networks, we introduce an extended weight matrix $\mathbf{A}^{'} \in R^{(n+m)\times(n+m)}$ in Eq. 4, which contains both the explicit and implicit relations.

$$\mathbf{A}^{'} = \left[\begin{array}{c|c} \mathbf{S}^U & \mathbf{A} \\ \hline \mathbf{A}^{\mathsf{T}} & \mathbf{S}^V \end{array}\right], \tag{4}$$

The process for calculating the extended weight matrix $\mathbf{A}^{'}$ is shown in Algorithm 1. Next, we propose a deep learning framework to integrate explicit and implicit relations while preserving the highly non-linear structure of the bipartite network, whose framework is shown in Fig. 1.

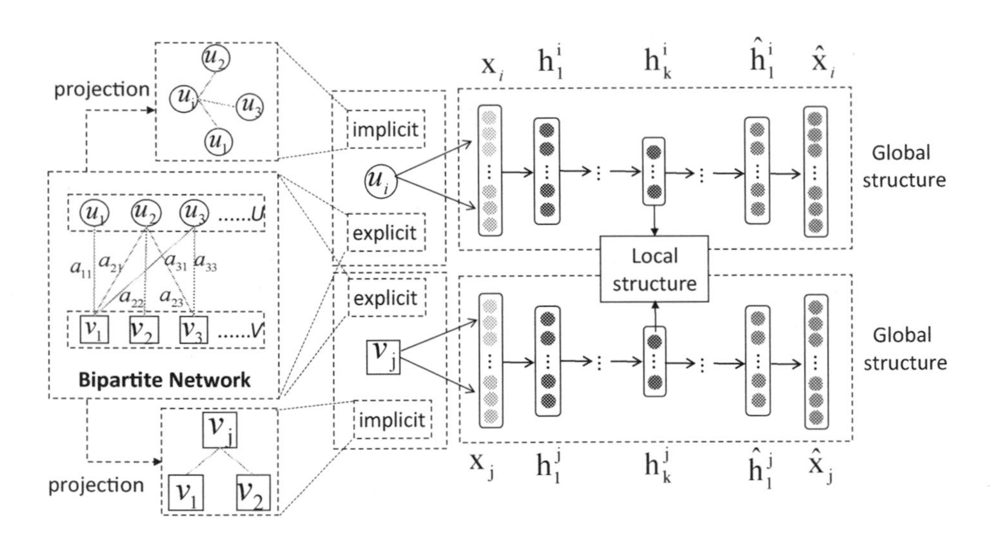

Fig. 1. The framework of the deep model of BiNE-IEI, which the blue solid line represents the explicit relations between different types of nodes, and the blue dotted line represents the implicit relations between the same type of nodes. (Color figure online)

In detail, in order to integrate the explicit and implicit relations in the weight matrix $\mathbf{A}^{'}$, We use two parallel deep autoencoders to capture the highly non-linear structure of the bipartite network. For each node in the bipartite network,

Algorithm 1. Extend_Adj

Input: Bipartite network $G = (U, V, E)$
1: $\mathbf{A} \leftarrow$ convert G to adjacency matrix
2: **for** $i \in \{1, 2, \ldots, | U |\}$ **do**
3:     **for** $j \in \{1, 2, \ldots, | U |\}$ **do**
4:         $S_{ij}^{U} \leftarrow$ apply Eq. 1
5:     **end for**
6: **end for**
7: **for** $i \in \{1, 2, \ldots, | V |\}$ **do**
8:     **for** $j \in \{1, 2, \ldots, | V |\}$ **do**
9:         $S_{ij}^{V} \leftarrow$ apply Eq. 2
10:     **end for**
11: **end for**
12: $\mathbf{A}' \leftarrow$ Based on A, S_{ij}^{U} and S_{ij}^{V}, apply Eq.4
13: **return** an extended weight matrix $\mathbf{A}^{'}$

we can obtain its explicit and implicit neighborhood structure according to the weight matrix $\mathbf{A}^{'}$, and then reconstruct the neighborhood structure of each node in the network through the deep autoencoders, which preserves the global structure of the bipartite network. Capturing the local structure of the network is equally important, we use the hidden feature representation of two directly connected nodes learned from two parallel deep autoencoders to preserve the local similarity structure of networks.

3.4 Loss Functions

From the extended weight matrix $\mathbf{A}^{'}$, each node x_i ($x_i \in U \cup V$) is represented as a high-dimensional vector as $\mathbf{a}_i^{'}$ (i.e., $\mathbf{x}_i$), which represents the i-th node's explicit and implicit neighborhood structure (the i-th row of the adjacency matrix $\mathbf{A}^{'}$). We reconstruct the neighborhood structure of each node by deep autoencoders, which preserves the global structure of the bipartite network. A deep autoencoder consists of two parts: the encoder and the decoder. Formally, we denote the input data and reconstructed data as $\mathbf{X} = \{\mathbf{x}_i\}_{i=1}^{n+m}$ and $\hat{\mathbf{X}} = \{\hat{\mathbf{x}}_i\}_{i=1}^{n+m}$, then given an input data $\mathbf{x}_i = \mathbf{a}_i^{'}$ and the number of layers K of the encoder, the hidden feature representations of each layer in the encoding process are as follows:

$$\begin{aligned} \mathbf{h}_1^i &= f(\mathbf{W}_1\mathbf{x}_i + \mathbf{b}_1), \\ \mathbf{h}_k^i &= f(\mathbf{W}_k\mathbf{h}_{k-1} + \mathbf{b}_k), k = 2, ..., K, \end{aligned} \tag{5}$$

where $\mathbf{h}_k^i$ denotes the representation of the k-th hidden layer, $\mathbf{W}_k$ and $\mathbf{b}_k$ denote the k-th hidden layer's weight matrix and bias, and f is the sigmoid function.

In reverse, we denote the hidden representations for a node in each layer in the decoding process as:

$$\begin{aligned} \hat{\mathbf{h}}_{k-1}^i &= f(\hat{\mathbf{W}}_k\hat{\mathbf{x}}_i + \hat{\mathbf{b}}_k), k = K, ..., 2, \\ \hat{\mathbf{x}}_i &= f(\hat{\mathbf{W}}_1\hat{\mathbf{h}}_1 + \hat{\mathbf{b}}_1), \end{aligned} \tag{6}$$

where $\hat{\mathbf{h}}_{k-1}^{i}$, $\hat{\mathbf{W}}_k$, and $\hat{\mathbf{b}}_k$ denote the hidden representations, weight matrix, and bias term of the $(k-1)$-th reconstruction layer, respectively.

The objective of the loss function is to minimize the reconstruction loss of $\mathbf{X}$ and $\hat{\mathbf{X}}$ to optimize the parameters and learn the latent representations, which can be formulated as:

$$\mathcal{L}_{global} = \|(\mathbf{X} - \hat{\mathbf{X}}) \odot \mathbf{Z}\|_F^2 \tag{7}$$

where $\odot$ denotes the Hadamard product, and $\mathbf{Z}$ is the weight matrix, which contains $n+m$ weight vectors $\mathbf{z}_1, \ldots, \mathbf{z}_{n+m}$. For each weight vector $\mathbf{z}_i = \{z_{ij}\}_{j=1}^{n+m}$, the detailed definition is as follows:

$$z_{ij} = \begin{cases} \alpha > 1, & a_{ij}^{'} > 0, \\ 1, & a_{ij}^{'} = 0, \end{cases} \tag{8}$$

where $a_{ij}^{'}$ is the j-th elements of $\mathbf{a}_i^{'}$ and α is the hyper-parameter. Due to the sparsity of the network, we introduce a weight matrix $\mathbf{Z}$ to impose more penalty to the reconstruction loss of the non-zero elements than that of zero elements in input data. Intuitively, minimizing the loss function $\mathcal{L}_{global}$ will make two nodes with similar neighborhood structure in the original network also similar in the embedding space, which is able to preserve the global network structure as desired.

We preserve local proximity of the network by modeling explicit link relations between two directly connected nodes. The loss function for capturing the local structure of a bipartite network can be formulated as:

$$\mathcal{L}_{local} = \sum_{i=1}^{n} \sum_{j=1}^{m} a_{ij} \|\mathbf{h}_k^i - \mathbf{h}_k^j\|_2^2, \tag{9}$$

Minimizing the loss function $\mathcal{L}_{local}$ makes two nodes with direct links to be mapped close in the embedding space, which can preserve the local network structure. In order to integrate the explicit and implicit relations while preserving the local and global structure of bipartite networks, the joint objective function of our proposed framework is defined as follows:

$$\mathcal{L} = \mathcal{L}_{global} + \lambda_1 \mathcal{L}_{local} + \lambda_2 \mathcal{L}_{reg}, \tag{10}$$

where λ_1 and λ_2 are balancing parameters and $\mathcal{L}_{reg}$ is the regularization term that prevents overfitting, which is formulated as follows:

$$\mathcal{L}_{reg} = \sum_{k=1}^{K} (\|\mathbf{W}_k\|_2^2 + \|\hat{\mathbf{W}}_k)\|_2^2 + \|\mathbf{b}_k\|_2^2 + \|\hat{\mathbf{b}}_k\|_2^2), \tag{11}$$

Our method is summarized in Algorithm 2.

4 Experiments

In this section, we conduct experiments on several real-world datasets to systemically evaluate the effectiveness of BiNE-IEI.

Algorithm 2. Training algorithm for the deep model of BiNE-IEI

Input: Bipartite network $G = (U, V, E)$, the parameters α, λ_1 and λ_2.
Output: The network representations $\mathbf{H}$

1: Initialize $\mathbf{W} = \{\mathbf{W}_1, \dots, \mathbf{W}_K\}$, $\mathbf{b} = \{\mathbf{b}_1, \dots, \mathbf{b}_K\}$, $\hat{\mathbf{W}} = \{\hat{\mathbf{W}}_1, \dots, \hat{\mathbf{W}}_K\}$, $\hat{\mathbf{b}} = \{\hat{\mathbf{b}}_1, \dots, \hat{\mathbf{b}}_K\}$
2: $A' \leftarrow Extend\_Adj(G)$
3: $X = A'$
4: **repeat**
5: $\hat{\mathbf{x}} \leftarrow$ apply Eqs. 5 and 6
6: $\mathcal{L}_{global} \leftarrow \|(X - \hat{X}) \odot \mathbf{Z}\|_F^2$
7: $\mathcal{L}_{local} = \sum_{i=1}^{n} \sum_{j=1}^{m} a_{ij} \|\mathbf{h}_k^i - \mathbf{h}_k^j\|_2^2$
8: $\mathcal{L} = \mathcal{L}_{global} + \lambda_1 \mathcal{L}_{local} + \lambda_2 \mathcal{L}_{reg}$
9: Back-propagate to get updated parameters $\mathbf{W}, \mathbf{b}, \hat{\mathbf{W}}$ and $\hat{\mathbf{b}}$
10: **until** converge
11: **return** The network representations $\mathbf{H} = \{\mathbf{h}_1, \dots, \mathbf{h}_{|U|+|V|}\}$

4.1 Datasets

We evaluate our proposed model BiNE-IEI on the following three datasets to demonstrate it's effectiveness.

- DBLP[1]: The DBLP dataset is a weighted bipartite publish network, which depicts the author's publishing relationship on the venues, where the edge weight means the number of papers an author has been published on a venue.
- VisualizeUs[2]: The VisualizeUs dataset is the bipartite picture tagging network consisting of two types of nodes, where the nodes represent tags and pictures respectively, and the edge weight indicates the number of times a tag has been tagged on a picture.
- Wikipedia[3]: It is an unweighted Wikipedia dataset. The nodes in this dataset contain authors and pages, in which the edge indicates that the authors have edited a page in Wikipedia.

The detailed statistics of the above datasets are summarized in Table 1.

Table 1. Statistics of the dataset.

Name	DBLP	VisualizeUs	Wikipedia
$\mid U \mid$	6001	6009	15000
$\mid V \mid$	1308	3355	3214
$\mid E \mid$	29256	38780	172426
Density	0.4%	0.2%	0.4%

1 http://dblp.uni-trier.de/xml.
2 http://konect.uni-koblenz.de/networks/pics_ti.
3 http://konect.uni-koblenz.de/networks/wikipedia_link_en.

4.2 Baselines

For comparison, we introduced the benchmark methods of three categories. The first categories are the benchmark methods only for link prediction tasks, The second categories are the benchmark methods for all tasks, while the third category is the benchmark approach specifically for recommended tasks.

(1) Methods based on topological structure [11]: We compare with several traditional link prediction methods based on the topological structure in the network, including Common Neighbors (CN), Jaccard's index (JC), Adamic Adar (AA), Preferential Attachment (PA).
(2) Methods based on network embedding: We compare with the following several methods based on network embedding.
   - DeepWalk [12]: DeepWalk leverages skip-gram model to learn node embedding with truncated random walks.
   - LINE [16]: LINE preserves both the first-order and second-order proximity to learn representations of nodes.
   - Node2vec [6]: Node2vec designs a biased random walk by introducing two hyper-parameters to balance the sampling process and generate the corpus of node sequences in the network, which improves the performance of the node representation.
   - Metapath2vec++ [3]: As a heterogeneous network embedding method, Metapath2vec++ performs the meta-path-guided random walk, which is able to capture the semantic and structural correlations among different types of nodes.
   - BiNE [5]: BiNE models the explicit relations by the first-order similarity, and captures the implicit relations by performing a biased and self-adaptive random walk in the bipartite network.
   - BiNE-IEI-I: It's the variant of BiNE-IEI, which removes the modeling of implicit relations.
   - BiNE-IEI-L: BiNE-IEI-L is similar to BiNE-IEI except that it removes the component which preserves the local structure.
(3) To evaluate the performance of the Top-N recommendations, we compare BiNE-IEI with the following three methods:
   - BPR [13]: It presents an optimization criterion BPR-Opt with a pairwise ranking loss for personalized ranking, which aims to learn from implicit feedback. It is a classic benchmark for item recommendations.
   - RankALS [15]: This method directly minimizes the ranking objective function without sampling, and can cope with the case of implicit feedback.
   - FISM [10]: FISM is a method based on item similarity for the top-N recommendation tasks, which is able to handle sparse datasets effectively.

4.3 Parameter Settings

BiNE-IEI contains a multi-layer deep neural network and we use a 2-layer auto-encoders by default, and the embedding layer or the dimension d of node representation is set to 120. Besides, the weight α of reconstruction loss for non-zero elements is set to 10, the threshold γ is set to 0.2, and the hyper-parameters to balance the loss function is set to $\lambda_1 = 10$, $\lambda_2 = 0.02$, respectively. The gradient is calculated using back-propagation and optimization using Adadelta [20] algorithm. And the parameters for all the baseline methods are set to the optimal value for the model mentioned in their own paper.

4.4 Link Prediction

For the link prediction task, we randomly sample a node pair which is not connected as a negative instance for each edge, while the links are considered positive instances. Then we randomly split 60% of instances as training set and the remaining instances as the test set. Then we learn the node embedding on the training set and generate edge embedding by concatenating the two node embedding of links. Finally, the embeddings of edges are treated as features and whether or not a node pair has edges as the ground truth. We train a simple logistic regression classifier on the training set and adopt area under the ROC curve (AUC-ROC) and Precision-Recall curve (AUC-PR), which have been used in the previous work [5] to evaluate the performance on the test set. The results of the experiment are shown in Table 2 compared with the baseline methods in the link prediction task. Note that the "N/A" in Table 2 represents that the result could not be computed for the corresponding method which cannot apply to large-scale networks. The main observations we made are as follows:

- BiNE-IEI obviously outperforms those methods based on network topological structure, which only consider the local or global network structure.

Table 2. Link prediction on DBLP and Wikipedia.

	DBLP		Wikipedia	
	AUC-ROC	AUC-PR	AUC-ROC	AUC-PR
CN	82.85%	N/A	86.85%	90.68%
JC	81.05%	N/A	63.90%	73.04%
AA	82.70%	N/A	87.37%	91.12%
PA	81.05%	N/A	90.71%	93.37%
DeepWalk	66.94%	71.51%	89.71%	91.20%
LINE	69.36%	73.64%	91.62%	93.28%
Node2vec	63.24%	67.69%	89.93%	91.23%
Metapath2vec++	71.61%	66.78%	89.56%	91.72%
BiNE	84.48%	86.21%	92.91%	94.45%
BiNE-IEI-I	79.98%	79.65%	93.53%	94.22%
BiNE-IEI-L	84.76%	85.95%	93.40%	94.50%
BiNE-IEI	**85.46%**	**86.50%**	**93.62%**	**95.19%**

- DeepWalk, LINE and Node2vec perform worse than BiNE-IEI, because our embedding method is aimed at bipartite networks, which considers the special properties of bipartite networks compared with those methods based on homogeneous networks.
- The performance of BiNE-IEI is better than Metapath2vec++ and BiNE, which is due to our method integrates implicit and explicit relations. Although Metapath2vec++ and BiNE consider both the implicit and explicit relations, they don't consider the integration of them.
- BiNE-IEI-I and BiNE-IEI-L are two variants of BiNE-IEI, which outperform poor than our method on the dataset of DBLP and Wiki. This is due to the two variants not consider the integration of implicit and explicit relations or preserve local network structure which is helpful for link prediction.

4.5 Recommendation

The performance of recommendation can reveal the quality of learned node representations. Specially, we randomly split 60% of the links as training set and remaining links as test set, for a user and an item in training set, we use the inner product of their embedding to evaluate the user's preference for the item, and for each user, we select $n = 10$ items with a largest preference scores for recommendation. All the results of three data sets on F1@10 and MAP@10 [5] are listed in Table 3. From these results, we have the following insightful observations.

- Obviously, BiNE-IEI outperforms all baseline methods in three data sets, indicating the effectiveness of integrating explicit and implicit relationships and preserving the local and global structures.

Table 3. Top-10 recommendation on VisualizeUs, DBLP, and Wikipedia.

	VisualizeUs		DBLP		Wikipedia	
	F1@10	MAP@10	F1@10	MAP@10	F1@10	MAP@10
BPR	6.22%	5.51%	8.95%	13.55%	14.12%	17.20%
RankALS	2.72%	1.50%	7.62%	7.52%	9.70%	14.05%
FISM	10.25%	8.86%	9.81%	7.38%	16.03%	16.74%
DeepWalk	5.82%	4.28%	8.50%	19.71%	2.28%	1.20%
LINE	9.62%	7.81%	8.99%	9.62%	5.52%	14.93%
Node2vec	6.73%	6.25%	8.54%	19.44%	3.83%	2.59%
Metapath2vec++	5.92%	5.35%	8.65%	19.06%	2.05%	1.26%
BiNE	13.63%	16.46%	11.79%	20.62%	13.67%	19.66%
BiNE-IEI-I	9.85%	12.00%	11.60%	12.78%	14.45%	13.30%
BiNE-IEI-L	10.33%	21.67%	15.38%	29.82%	18.97%	35.27%
BiNE-IEI	**20.21%**	**49.18%**	**16.81%**	**31.75%**	**19.45%**	**35.91%**

- Overall, the three state-of-the-art homogeneous network embedding methods, including DeepWalk, LINE and Node2vec show relatively poor performance compared with our method, which indicate the methods based on the homogeneous network does not take into account the node type information of bipartite network and are not the optimal methods for the representation learning for bipartite network.
- BiNE-IEI outperforms Metapath2vec++ significantly, which is a method based on the heterogeneous network. This is due to the factors that Metapath2vec++ models explicit and implicit relations equally. The bipartite network embedding method BiNE is also shows poor performance compared with BiNE-IEI. Although BiNE considers the different importance of the implicit and explicit relations, it do not integrate them in an efficient way.
- BiNE-IEI outperforms it's variants BiNE-IEI-I and BiNE-IEI-L, removing the modeling of implicit relations and local network structure, respectively. The results show the effectiveness of our method which integrates explicit and implicit relations while preserving the local and global network structure.

4.6 Parameters Sensitivity

In this subsection, we investigate the impact of the dimension d, the hyper-parameters α and λ_1 for our model. we use AUC-ROC and AUC-PR to evaluate the performance of the link prediction on the dataset of Wikipedia in Fig. 2. Figure 2(a) demonstrates the impact of the embedding dimension d. At first, as the dimension increases, the effect raises significantly, this is due to the increase of d can embed more information. However, the performance does not increase and even decrease slightly when d continues to increase, which shows that too large a dimension d cannot embed more information and even introduce noises.

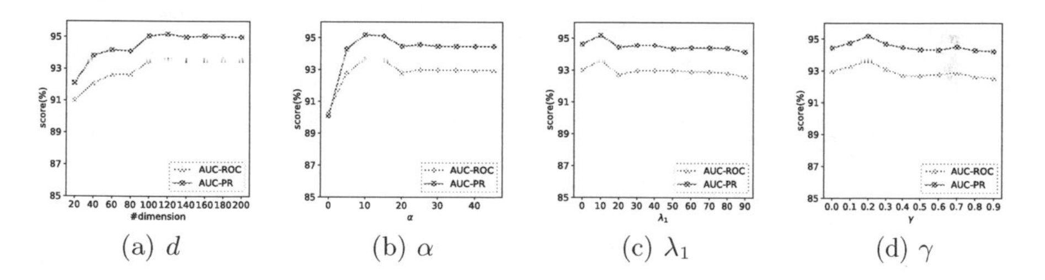

(a) d (b) α (c) λ_1 (d) γ

Fig. 2. Parameter w.r.t. dimension d, hyper-parameters α, λ_1 and γ

The hyper-parameter α can impact the reconstruction loss for non-zero elements. From Fig. 2(b), we can see that introducing parameter α is effective. On the other hand, when α is large enough, the performance remains stable. The reason is that too much reconstruction loss is useless in the learning.

The parameter sensitivity analysis for λ_1 is shown in Fig. 2(c). The hyper-parameter λ_1 is used to balance the loss of reconstruction and the loss of local network structure, when $\lambda_1 = 0$, the model cannot preserve the local network structure. From Fig. 2(c), we can see that the performance of $\lambda_1 > 0$ is better than that of $\lambda_1 = 0$, which demonstrates that preserving both the local and global network structure are essential for our model.

Figure 2(d) shows the impact of the threshold γ, which is used to preserve significant implicit relations between nodes of the same type. We can see that with the increase of γ, the experimental performance first increases and then decreases, and the best experimental result is obtained when $\gamma = 0.2$.

4.7 Visualization

Due to the lack of ground truth in the above dataset nodes, we conduct a visualization task on a subset of Aminer dataset[4] which is a heterogeneous collaboration network. The subset of Aminer dataset consists of 981 authors and 28 venues which are from the research field of Theoretical Computer Science or Computer Science Databases & Information Systems. A link will be constructed between an author and a venue if the author published a paper in this venue. In addition, we select the research field in which the author publishes the most papers as the author's ground truth. We leverage the t-SNE tool to reduce the embedding of authors to 2 dimensions, the visualization results are shown in Fig. 3. The color of the vertex represents the author's research field, where blue represents the research field of Theoretical Computer Science and red represents the research field of Computer Science Databases & Information Systems. DeepWalk, Node2vec and metapath2vec++ which based on random walks perform

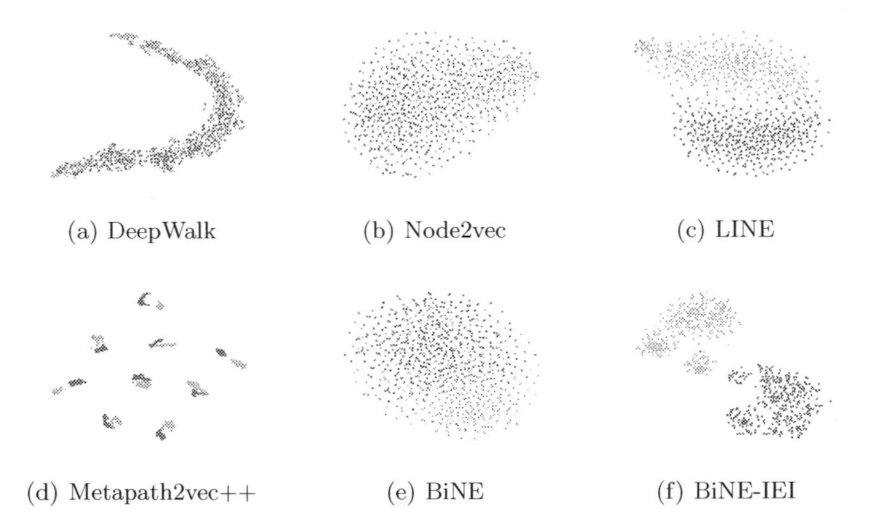

Fig. 3. Visualization of authors in the subset of Aminer, where blue represents the research field of Theoretical Computer Science and red represents the research field of Computer Science Databases & Information Systems. (Color figure online)

[4] https://www.aminer.cn/data.

worst because the two types of nodes are mixed together. For BiNE which captures the implicit relations by performing a biased and self-adaptive random walk in the bipartite network perform slightly better. It is surprising that LINE can distinguish well the category of authors, this may be due to the second-order proximity of LINE helps to distinguish the authors in different research fields. However, BiNE-IEI obviously performs best, because the authors of the same category are closely gathered together, and the different types of nodes are clearly distinguished. In summary, Fig. 3 proves the superiority of our model once again.

5 Conclusions

In this paper, we propose BiNE-IEI for bipartite networks embedding, which integrates explicit and implicit relations to learn the representations of nodes. Besides, our model can capture the highly non-linear structure while preserving the local and global structure of the network. We conduct extensive experiments on some widely used datasets compared with several state-of-the-art baselines and show the superior performance of our method on link prediction, recommendation, and visualization tasks. Intuitively, SDNE [18] and our model are partially similar in form but there are mainly the following differences. Firstly, SDNE is the representation learning method of the homogeneous network, and cannot model the bipartite network compared with our method; Secondly, SDNE preserves the local structure between the nodes of the same type, while our proposed method BiNE-IEI preserves the local structure between different types of nodes in the bipartite network; Finally, BiNE-IEI can integrate the explicit and implicit relations of the bipartite network. Because of the sparseness and large missing observed data, in the future, we will extend BiNE-IEI to model auxiliary information, such as images, textual descriptions or other attributes of the node in bipartite networks for embedding.

Acknowledgments. This work was supported by the National Key R&D Program of China (2018YFC0809800, 2016QY15Z2502-02, 2018YFC0831000), the National Natural Science Foundation of China (91746205, 51438009, U1736103), Tianjin Science and Technology Development Strategic Research Project (17ZLZ DZF00430) and the Key R&D Program of Tianjin (18YFZCSF01370).

References

1. Cai, H., Zheng, V.W., Chang, K.: A comprehensive survey of graph embedding: problems, techniques and applications. IEEE Trans. Knowl. Data Eng. **30**, 1616–1637 (2018)
2. Cui, P., Wang, X., Pei, J., Zhu, W.: A survey on network embedding. IEEE Trans. Knowl. Data Eng. **31**(5), 833–852 (2019)
3. Dong, Y., Chawla, N.V., Swami, A.: metapath2vec: scalable representation learning for heterogeneous networks. In: Proceedings of the 23rd ACM SIGKDD International Conference on Knowledge Discovery and Data Mining, pp. 135–144. ACM (2017)

4. Fu, T., Lee, W.C., Lei, Z.: Hin2vec: explore meta-paths in heterogeneous information networks for representation learning. In: Proceedings of the 2017 ACM on Conference on Information and Knowledge Management, pp. 1797–1806. ACM (2017)
5. Gao, M., Chen, L., He, X., Zhou, A.: Bine: bipartite network embedding. In: Proceedings of the 41th International ACM SIGIR Conference on Research and Development in Information Retrieval, pp. 715–724. ACM (2018)
6. Grover, A., Leskovec, J.: node2vec: scalable feature learning for networks. In: Proceedings of the 22nd ACM SIGKDD International Conference on Knowledge Discovery and Data Mining, pp. 855–864. ACM (2016)
7. Huang, Z., Mamoulis, N.: Heterogeneous information network embedding for meta path based proximity. arXiv preprint arXiv:1701.05291 (2017)
8. Jin, D., Ge, M., Yang, L., He, D., Wang, L., Zhang, W.: Integrative network embedding via deep joint reconstruction. In: Proceedings of the 27th International Joint Conference on Artificial Intelligence, pp. 3407–3413. AAAI Press (2018)
9. Jolliffe, I.: Principal component analysis. In: Lovric, M. (ed.) International Encyclopedia of Statistical Science. Springer, Heidelberg (2011)
10. Kabbur, S., Ning, X., Karypis, G.: Fism: factored item similarity models for top-n recommender systems. In: Proceedings of the 19th ACM SIGKDD International Conference on Knowledge Discovery and Data Mining, pp. 659–667. ACM (2013)
11. Lü, L., Zhou, T.: Link prediction in complex networks: a survey. Phys. A: Stat. Mech. Appl. **390**(6), 1150–1170 (2011)
12. Perozzi, B., Al-Rfou, R., Skiena, S.: Deepwalk: online learning of social representations. In: Proceedings of the 20th ACM SIGKDD International Conference on Knowledge Discovery and Data Mining, pp. 701–710. ACM (2014)
13. Rendle, S., Freudenthaler, C., Gantner, Z., Schmidt-Thieme, L.: BPR: Bayesian personalized ranking from implicit feedback. In: Proceedings of the Twenty-Fifth Conference on Uncertainty in Artificial Intelligence, pp. 452–461. AUAI Press (2009)
14. Salakhutdinov, R., Hinton, G.: Semantic hashing. Int. J. Approximate Reasoning **50**(7), 969–978 (2009)
15. Takács, G., Tikk, D.: Alternating least squares for personalized ranking. In: Proceedings of the Sixth ACM Conference on Recommender Systems, pp. 83–90. ACM (2012)
16. Tang, J., Qu, M., Wang, M., Zhang, M., Yan, J., Mei, Q.: Line: large-scale information network embedding. In: Proceedings of the 24th International Conference on World Wide Web, pp. 1067–1077. International World Wide Web Conferences Steering Committee (2015)
17. Turner, C.R., Wolf, A.L., Fuggetta, A., Lavazza, L.: Feature engineering. In: Proceedings of the 9th International Workshop on Software Specification and Design, p. 162. IEEE Computer Society (1998)
18. Wang, D., Cui, P., Zhu, W.: Structural deep network embedding. In: Proceedings of the 22nd ACM SIGKDD International Conference on Knowledge Discovery and Data Mining, pp. 1225–1234. ACM (2016)
19. Xu, H., Liu, H., Wang, W., Sun, Y., Jiao, P.: NE-FLGC: network embedding based on fusing local (first-order) and global (second-order) network structure with node content. In: Phung, D., Tseng, V.S., Webb, G.I., Ho, B., Ganji, M., Rashidi, L. (eds.) PAKDD 2018. LNCS (LNAI), vol. 10938, pp. 260–271. Springer, Cham (2018). https://doi.org/10.1007/978-3-319-93037-4_21
20. Zeiler, M.D.: Adadelta: an adaptive learning rate method. arXiv preprint arXiv:1212.5701 (2012)

21. Zhou, T., Kuscsik, Z., Liu, J.G., Medo, M., Wakeling, J.R., Zhang, Y.C.: Solving the apparent diversity-accuracy dilemma of recommender systems. Proc. Natl. Acad. Sci. **107**(10), 4511–4515 (2010)
22. Zhou, T., Ren, J., Medo, M., Zhang, Y.C.: Bipartite network projection and personal recommendation. Phys. Rev. E **76**(4), 046115 (2007)

Enhancing Network Embedding with Implicit Clustering

Qi Li[1], Jiang Zhong[1(✉)], Qing Li[1], Zehong Cao[2], and Chen Wang[1]

[1] College of Computer Science, Chongqing University, Chongqing, China
zhongjiang@cqu.edu.cn

[2] Discipline of ICT, School of Technology, Environments and Design, University of Tasmania, Hobart, Australia

Abstract. Network embedding aims at learning the low dimensional representation of nodes. These representations can be widely used for network mining tasks, such as link prediction, anomaly detection, and classification. Recently, a great deal of meaningful research work has been carried out on this emerging network analysis paradigm. The real-world network contains different size clusters because of the edges with different relationship types. These clusters also reflect some features of nodes, which can contribute to the optimization of the feature representation of nodes. However, existing network embedding methods do not distinguish these relationship types. In this paper, we propose an unsupervised network representation learning model that can encode edge relationship information. Firstly, an objective function is defined, which can learn the edge vectors by implicit clustering. Then, a biased random walk is designed to generate a series of node sequences, which are put into Skip-Gram to learn the low dimensional node representations. Extensive experiments are conducted on several network datasets. Compared with the state-of-art baselines, the proposed method is able to achieve favorable and stable results in multi-label classification and link prediction tasks.

Keywords: Network embedding · Feature learning · Edge representation · Network mining

1 Introduction

Social networks, paper citation networks, semantic networks and other large-scale networks have penetrated into all aspects of our real life [14]. These networks usually have complex structure and large scale. Moreover, the high dimensional and sparse characteristics of networks have brought unprecedented challenges to existing network mining technologies. To solve these problems, network embedding is designed to learn the low dimensional representation of nodes, while preserving the structure and inherent characteristics of the network. It can be effectively used by vector-based machine learning models for mining tasks,

G. Li et al. (Eds.): DASFAA 2019, LNCS 11446, pp. 452–467, 2019.
https://doi.org/10.1007/978-3-030-18576-3_27

including node classification, personalized recommendation, and link prediction, etc. [2,13,18,23].

Following the initial ideas in network embedding [18,24], recent techniques such as DeepWalk [18] and node2vec [13] learn node representation using random walks sampled in the network. Thereafter, Cao et al. [5] developed a GraRep model, which integrates the global structure information of the network into the learning process. They adopt the idea of matrix decomposition, achieve the dimensionality reduction by decomposing the relationship matrix, and thus obtain the network representation of nodes. Tang et al. [20] proposed a large-scale information network embedding method called LINE that preserves both the first-order and second-order proximity. Wang et al. [21] designed a Structural Deep Network Embedding (SDNE) model, which maintains the proximity between 2-hop neighbors through deep automatic encoders. Recently, Ribeiro et al. [6] developed a novel and flexible model, called Struc2vec, which uses hierarchical structure to measure the similarity of nodes at different scales, and constructs a multi-layer network to encode the similarity of nodes and generate the structure context for nodes.

However, most of the aforementioned methods mainly focus on the existence of edges between nodes and ignore the differences between edges. A node may be connected with other nodes for different relationship types. The edges with the same relationship types can form a cluster. These clusters hide abundant information. For example, the similarity between the inner vertices within the same cluster is relatively higher than that within the different clusters. The clusters reflect auxiliary information for network representation learning, and contribute to the generation of more accurate node vectors. In this paper, we propose an unsupervised model for network representation learning, which can strengthen the use of the first-order proximity of network structure and improve accuracy of preserving two-order proximity. Our main contributions can be summarized as follows:

- We propose an unsupervised model for network representation learning, called NEWEE, which can utilize the information of node neighbors as well as the information of the relationship types between nodes.
- We propose a new way to distinguish the relationship types for edges, which can learn similar vectors from similar relationship types without labeling data, and only use the structure information of the network itself.
- Extensive experiments on several datasets demonstrate that our proposed method produces significantly increased performance over the current state-of-the-art network embedding methods in most cases.

The rest of this paper is organized as follows; Sect. 2, briefly outlines a list of related works and our motivation. Detailed steps of the proposed method are presented in Sect. 3. Section 4, presents the experiment results, and comparison with completing algorithms. Finally, this paper is concluded in Sect. 5.

2 Related Work

Traditional network representation learning methods include network representation learning based on spectral method, such as locally linear embedding (LLE) [19,25] and laplacian eigenmaps based on manifold assumption [3]. In addition, there is optimization based representation learning for networks, a low-dimensional representation of network can be grasped by optimizing an objective function. Representative algorithms include mapping to homogeneous model (MTH) [9], content diffusion kernel (CDK) [4], and content-based source diffusion kernel (CSDK) [4]. Some scholars improve the description of node content by introducing network information based on subject probability model. Representative methods include Link-PLSA-LDA [16], relational topic model (RTM) [26] and probabilistic latent document network embedding (PLANE) [10]. These methods cannot be applied to generalized node feature representation. Besides, most of above-mentioned methods are expensive in calculation and non-expandable for large networks.

Nowadays, representation learning methods are widely applied in the field of natural language processing (NLP), among which, a representative one is word embed-ding [15]. The researchers believe that words with similar contexts should also have similar semantics. The word vectors obtained through unsupervised learning method have achieved excellent performance in many tasks.

Inspired by the above method, the researchers began to apply word embedding into feature learning of network nodes [7,8]. Perozzi [18] discovered that the number of words appearing in text corpus and the number of visits for nodes by random walk from network obey exponential distribution. Therefore, Perozzi [18] considered that the Skip-Gram model could be transplanted to representation learning of network as well, and DeepWalk model was proposed [18]. The similar method is Node2vec [1], a process of adjusting random walk by introducing depth first and breadth first strategies based on DeepWalk. Struc2vec [6], another type of node embedding strategy, is based on random walk, which finds similar embedding on nodes that are structurally similar. Wang et al. [22] developed an innovative network representation learning framework, called GraphGAN, which unifies *generative* models and d*iscriminative* models. The LINE [20] method combined first-order proximity with second-order proximity, which was as the final representation of nodes.

Although these methods are fast and effective, all existing methods mainly consider the existence of a link between nodes instead of the difference between these links. Therefore, we propose a new way to distinguish these relationship types, which can encode the edges to update network by implicit clustering, and without labeling data. Then a biased random walk from the updated network can generate more accurate node sequences.

3 The Proposed Model

The problem in this paper is how to construct a suitable model for network representation learning, which can map the networks data to a low-dimensional vector space. Each low-dimensional vector represents one node, and the relationships between these vectors reflect the first-order and the second-order proximity between nodes.

3.1 Framework

In this section, we describe the main steps of NEWEE model. The flow-graph of the proposed model is shown in Fig. 1. The NEWEE model is divided into three phases, which are described in the following procedure:

1. The edge sampling is used to optimize an objective function, and to learn a low-dimensional representation for each edge in the network. If the relationship type of two edges is similar, their vectors are similar as well;
2. By learning the edge vectors from the first phase, a biased random walk is adopted, which can increase the similarity of the two edge vectors before and after walking;
3. The node sequences are obtained from the second phase as the input of Skip-Gram. The original Skip-Gram model only indirectly preserves part of the first-order proximity. Therefore, the improvement of the original Skip-Gram model is made to enhance the similarity between directly connected nodes.

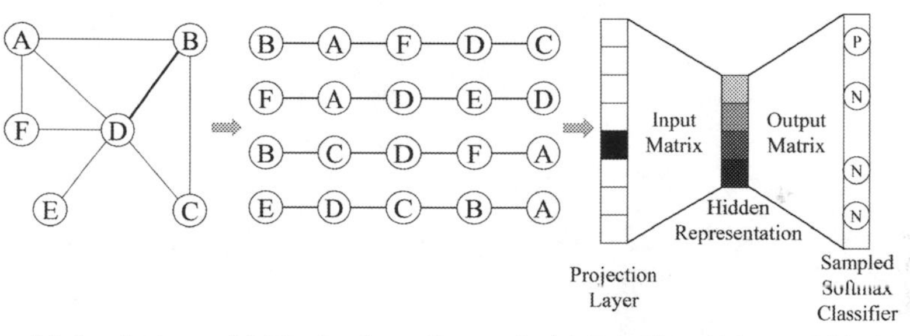

Fig. 1. Overview of NEWEE model: (a) Encode edge: Reconstruct the network and learn a low-dimensional representation for each edge in network. If the relationship type of two edges is similar, their vectors are similar as well; (b) The node sequences generated by biased random walk from the network; (c) Embedding with improved Skip-Gram.

3.2 Encode Edge

The purpose for encoding edges is to learn a low-dimensional representation for each edge of the network. If the relationship types between two edges are similar, their vectors are also similar. We have noticed that a node can be clustered with other nodes due to different relationship types, so the node neighbors can be divided into different neighbor clusters, and the relationship types among different neighbor clusters are different. That means we only need to train one model, which ensures the similarity of the inner edges of the same neighbor clusters, is higher than that of the outer edges of the clusters. Here, we first introduce the concept of self-centered network.

Definition 1 ***(self-centered network).*** *Given a network $G = (E, V)$. For any node v_i in G, its self-centered network is $G' = (E', V')$. The node set V' includes the node v_i and its neighbors, and E' represents the set of edges between all nodes in V'.*

Each node has its own self-centered network. Figure 2 (left) shows the self-centered networks of node a. The neighbors of node a are divided into two neighbor clusters of C_1 and C_2, b and c belong to C_1. We have also noticed that most of the edges in C_1 also exist in the self-centered networks of b and c, as shown in Fig. 2 (middle and right). In general, the closer the cluster is, the more edges exist simultaneously in the self-centered networks of the multiple nodes within the cluster; conversely, if multiple edges exist simultaneously in the self-centered networks of multiple nodes, the multiple edges should belong to the same cluster. Thus, we cannot only avoid explicitly calling clustering algorithm to cluster the node neighbors, but also use the nature of the network itself to implicit clustering.

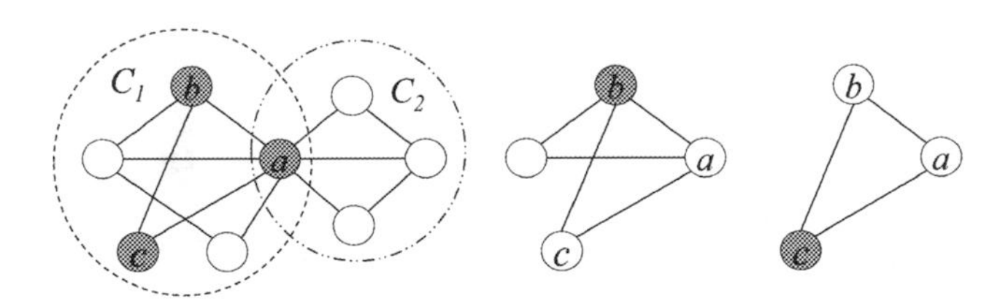

Fig. 2. The self-centered networks of nodes a (left), b (middle) and c (right).

In order to make the similarity of edge vectors of the same self-centered networks higher than that of the other self-centered networks, the objective function is defined as follows:

$$\max \sum_{v \in V'} \sum_{e \in E'} \log P(v|e) \tag{1}$$

Where $P(v|e)$ is the probability that the network is the self-centered network of node v when the edge is e. To achieve the purpose of making the edge vectors of the same self-centered networks similar, we regard it as a binary classification problem, and use the logical regression as the classification method to reconstruct the probability function. The negative sampling technique [17] is used to speed up the training. For $\forall u \in V$, we first define the following indication function:

$$I^v(u) = \begin{cases} 1, & u = v \\ 0, & u \neq v \end{cases} \tag{2}$$

For a given node v, the set of negative sampling is $NST(v)$. The probability function (1) is reconstructed by using negative sampling technique as follows:

$$\begin{aligned} P(v|e) &= \prod\nolimits_{u \in \{v\} \cup NST^e(v)} P(u|e) \\ P(u|e) &= \begin{cases} \sigma\left(e^T \theta^u\right), & I^v(u) = 1 \\ 1 - \sigma\left(e^T \theta^u\right), & I^v(u) = 0 \end{cases} \end{aligned} \tag{3}$$

Where σ is the sigmoid function. The parameter θ^u is the vector of node u. e is the vector of edge e, and is the final output. It is obtained by bitwise operations of the vectors of two ends of edge. In order to adapt to both the directed networks and the undirected networks, the average operation is used. That is, two ends of edge e are respectively v_i and v_j. The edge vector $e_{i,j}$ is denoted as follows:

$$e_{i,j} = \frac{v_i + v_j}{2} \tag{4}$$

The final objective function is:

$$\begin{aligned} &\max \sum_{v \in V'} \sum_{e_{i,j} \in E'} \sum_{u \in \{v\} \cup NST^{e_{i,j}}(v)} L(v, e, u) \\ &L(v, e, u) = I^v(u) \cdot \log\left[\sigma\left(e_{i,j}^T \theta^u\right)\right] + \left[1 - I^v(u)\right] \cdot \log\left[1 - \sigma\left(e_{i,j}^T \theta^u\right)\right] \end{aligned} \tag{5}$$

We use gradient descent method to optimize the formula (5). First, we consider the gradient of $L(v, e, u)$ on θ^u.

$$\begin{aligned} \frac{\partial L(v, e_{i,j}, u)}{\partial \theta^u} &= \frac{\partial}{\partial \theta^u} \left\{ I^v(u) \cdot \log\left[\sigma\left(e_{i,j}^T \theta^u\right)\right] + \left[1 - I^v(u)\right] \cdot \log\left[1 - \sigma\left(e_{i,j}^T \theta^u\right)\right] \right\} \\ &= I^v(u)\left[1 - \sigma\left(e_{i,j}^T \theta^u\right)\right] e_{i,j} - \left[1 - I^v(u)\right] \sigma\left(e_{i,j}^T \theta^u\right) e_{i,j} \\ &= \left[I^v(u) - \sigma\left(e_{i,j}^T \theta^u\right)\right] e_{i,j} \end{aligned} \tag{6}$$

The update formula of θ^u is:

$$\theta^u = \theta^u + \eta\left[I^v(u) - \sigma\left(e_{i,j}^T \theta^u\right)\right] e_{i,j} \tag{7}$$

Where η is the learning rate. Then, we consider the gradient of $L(v, e, u)$ about $e_{i,j}$. Because $e_{i,j}$ and θ^u are symmetrical in $L(v, e, u)$, it is easy to obtain the following formula:

$$\frac{\partial L(v, e_{i,j}, u)}{\partial e_{i,j}} = \left[I^v(u) - \sigma\left(e_{i,j}^T \theta^u\right)\right] \theta^u \tag{8}$$

According to the continuous derivation rule and the symmetry of v_i and v_j in $e_{i,j}$.

$$\frac{\partial L(v, e_{i,j}, u)}{\partial v_i} = \frac{1}{2}\left[I^v(u) - \sigma\left(e_{i,j}^T \theta^u\right)\right] \theta^u \tag{9}$$

The update formula of v_i is:

$$v_i = v_i + \frac{\eta}{2} \sum_{u \in \{v\} \cup NST^{e_{i,j}}(v)} \left[I^v(u) - \sigma\left(e_{i,j}^T \theta^u\right) \right] \theta^u \tag{10}$$

The update formula of v_i is same to v_j. If inputting the self-centered networks of multiple nodes, the following situations will occur:

- The similarity between the inner edges within the same self-centered network will be higher than that within the different clusters. For example, when inputting the self-centered networks of the nodes a, b, c, and the edges similarity in clusters C_1 and C_2 will be constantly strengthened;
- The similarity between edges within the different self-centered networks will be weaken. For example, when inputting the self-centered networks of the nodes a, b, c, and the similarity between edges in clusters C_1 and C_2 constantly weaken.

3.3 Learning Node Features

This section mainly describes how to use the edge vectors obtained in the first phase to train nodes. Like the article [1,18], which first obtain a series of node sequences by random walk from the network, but we adopt a biased random walk. In particular, by learning the edge vectors from the first stage, the similarity of the two edge before and after walking can be increased, so that the preservation accuracy of the second-order proximity of the network structure can be improved. Then, the node sequences are as the input of Skip-Gram model.

Biased Random Walk. After the first phase, we get a network with edge vectors, which preserves the relationship types information. Then, a series of node sequences are obtained by a biased random walk from the network. If the started node is v_0, the next walk node is randomly selected from its neighbors as v_1. If the current walk node is v_k $(k \geq 1)$, the selection of the next walk node v_{k+1} follows the following probability distribution:

$$P\left(v_{k+1} = x | v_k = v, v_{k-1} = t\right) = \begin{cases} \frac{\pi(t,v,x)}{Z}, e_{v,x} \in E \\ 0, \quad otherwise \end{cases} \tag{11}$$

Where Z is a normalization constant. $\pi(t, v, x)$ is a transition probability of walking from node t to node v and then walking from node v to node x:

$$\pi(t, v, x) = \begin{cases} \mu, & if\ x = t \\ similarity\left(e_{t,v}, e_{v,x}\right), & if\ x \neq t\ and\ e_{v,x} \in E \\ 0, & otherwise \end{cases} \tag{12}$$

Where μ is a return parameter and set to 0.5. In addition, we use cosine similarity to calculate *similarity*.

$$similarity\left(e_{t,v}, e_{v,x}\right) = \frac{e_{t,v} \cdot e_{v,x}}{\|e_{t,v}\| \cdot \|e_{v,x}\|} \tag{13}$$

Where $e_{t,v}$ and $e_{v,x}$ are the vectors of edge $e_{t,v}$ and $e_{v,x}$ respectively. They are learned from the first phase. Each node in the network is taken as the walk started node of the sequence in turn, and sampling the neighbors[1] according to the selection probability distribution of neighbors. For each walk started node v_0, we do biased random walk from the network to get a node sequence with length l. After repeating the above operation r times, a series of node sequences are obtained.

Example. There are two node sequences of $\langle v_1, v_2, v_3, v_4, v_5\rangle$ and $\langle v_1, v_2, v_6, v_4, v_5\rangle$ (Fig. 3). The nodes v_3 and v_6 have similar contexts, so they can learn the similar learning representations. In order to get the node sequences of (1) and (2), the edge vector $e_{1,2}$ should be similar to $e_{2,3}$, and the edge vector $e_{2,6}$ should be similar to $e_{1,2}$. That is, the edge vector $e_{2,3}$ should be similar to $e_{2,6}$. If adopting the random walk method of DeepWalk, it may get the node sequences of (2), (3), and (4). v_3 and v_6 may have similar left and right neighbors v_2 and v_4, but due to the uncertain relationship type, it is difficult to have the opportunity to reappear both v_1 and v_5 in the nodes extending forward and backward, which greatly reduce the context similarity of v_3 and v_6. On the contrary, if the relationship types between nodes (v_3 and v_8) and their neighbors (v_2 and v_4) are not similar, the conclusion of v_3 similar to v_8 is not credible even their contexts are similar.

According to the rule of NEWEE model for generating node sequences, any two connected edges have a high similarity in the sequence. If the two node sequences are similar, the edges of the two sequences are also similar. Conversely, If the nodes have different relationship types with their neighbors. As shown (2) and (3), with the sequence extends the similarity of the learned vector representation of v_6 and v_8 is decreased.

Improved Skip-Gram. As mentioned above, we have enhanced the utilization of first-order proximity. The objective function of Skip-Gram can achieve similar

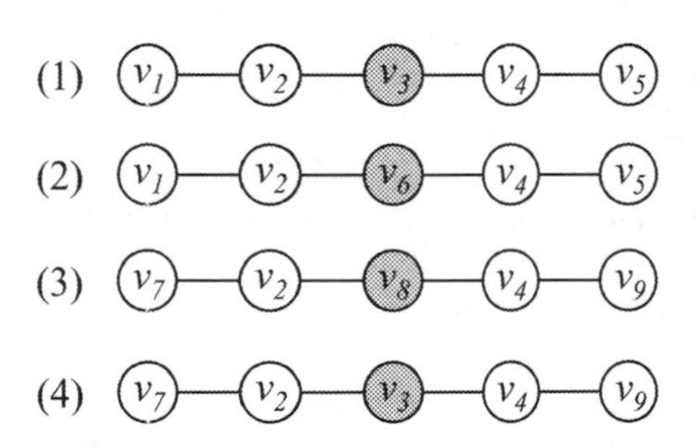

Fig. 3. An example of the influence of relationship type information on the node sequences (same type lines mean similarity relationship type).

[1] The alias sampling algorithm [12] method can be used to complete the sampling process in the time complexity of $O(1)$.

vectors from nodes with similar contexts. We make the improvement to it as follows:

$$\prod_{\tilde{w}\in C(w)} p\left(w|\tilde{w}\right) \rightarrow \prod_{e_{i,j}\in E} p\left(v_i, v_j\right) \prod_{\tilde{w}\in C(w)} p\left(w|\tilde{w}\right) \quad (14)$$

Where $p(v_i,\ v_j)$ is used to preserve first-order proximity, and defined as:

$$p\left(v_i, v_j\right) = \sigma\left(v_i^T v_j\right) \quad (15)$$

Where v_i and v_j are vector representations of node v_i and v_j as context nodes respectively. When the sequences are put into the improved Skip-Gram model, the nodes with similar contexts will be similar.

3.4 Complexity Analysis

In the first phase, the time complexity of training the edge vectors is $O(|V| \cdot kndi)$, where $|V|$ is the number of nodes in the network, k is the average degree of nodes, n is the number of negative sampling, d is the dimension of edge vectors, and i is the number of iterations. The parameters n, d and i are constants. The time complexity of the first phase is linear correlation with the number of nodes $|V|$.

The second phase includes random walk and training a Skip-Gram model. The time complexity of random walk is $O(|V| \cdot kdrl)$, where r is walk times, l is the length of the node sequence, these parameters are all constants. The time complexity of the random walk is also linear correlation with the number of nodes $|V|$. As for training a Skip-Gram model, its time complexity is $O(swndi)$, where s is the number of nodes in the input document and w is the size of the context window. The time complexity of training a Skip-Gram model is linear correlation with the number of nodes s. Therefore, the overall computational time complexity of NEWEE is $O\ (|V| \cdot kndi + |V| \cdot kdrl + swndi)$.

4 Experiments

In this section, we mainly consider the method of quantitative analysis for the NEWEE model. In order to fully describe the effectiveness of our model, the experiments are conducted on the two tasks of link prediction and multi-label classification. For the sake of verifying the robustness and efficiency, the experiments are performed from the perspectives of parameters sensitivity and the running time for learning different size networks. Furthermore, we also apply the same networks in the competing algorithms, including DeepWalk [18], LINE [20], AANE [8], Stru2vec [6], GraphSAGE [7] and Node2Vec [1]. The parameters of the six comparison algorithms are set in such a way that they either take advantage of the default settings suggested by the authors or adjust them experimentally to find the best Settings. After applying these network embedding algorithms, the representation of low-dimensional nodes can be obtained respectively. The hardware environment of the experiment is a PC with a stand-alone Intel Xeon processor with 2.67 GHz and 16 G memory. The software platform is python 2.7 in Windows.

4.1 Parameter Settings

The default settings of our parameters are mostly consistent with those in article [21]: the negative sampling parameters n_1 and n_2 are both set to 5. The vector dimensions d_1 and d_2 are both set to 128. The number of walks started per node r is 10. Each sequence length l and the size of context window w is set to 80 and 10 respectively.

4.2 Evaluation Metrics

For link prediction, we use *precision@k* and *Mean Average Precision* (MAP) to evaluate the performance. Their definitions are listed as follows:

precision@k is a metric, which gives equal weight to the returned instance. It is defined as follows:

$$Precision@k = \frac{|\{e_{i,j}|v_i, v_j \in V, index(e_{i,j}) < k, \triangle_{i,j} = 1\}|}{k} \tag{16}$$

Where $E^{''}$ is a hidden edge set hidden in the network. $e_{i,j}$ represents an edge between nodes v_i and v_j. index($e_{i,j}$) is the ranked index of an edge $e_{i,j}$ in prediction results. $\triangle_{i,j} = 1$ indicates an edge $e_{i,j}$ exists in $E^{''}$.

Mean Average Precision (***MAP***) is a metric with good discrimination and stability. Compared with *Precision@k*, *MAP* pays more attention to the instances of ranked ahead in prediction results. It is defined as follows:

$$AP = \frac{\sum_{i=1}^{|E^{''}|} Precision@i \cdot \triangle_i}{|E''|}, \quad MAP = \frac{\sum_{j=1}^{Q} AP(j)}{Q} \tag{17}$$

Where $\triangle i$ is an indicator function. When the i-th prediction result is hit, the value $\triangle i$ is 1, otherwise, it is zero. Q is query times.

For multi-label classification, we adopt *Macro-F1* and *Micro-F1* as evaluation indexes. Specifically, Suppose C is a label set and A is a label. We denote $TP(A)$, $FP(A)$ and $FN(A)$ as the number of true positives, false positives and false negatives in the instances which are predicted as A, respectively. $F1(A)$ is the *F1-measure* for the label A. *Micro-F1* and *Macro-F1* are defined as follows:

$$\begin{aligned} Pr &= \frac{\sum_{A \in C} TP(A)}{\sum_{A \in C}(TP(A) + FP(A))}, \quad R = \frac{\sum_{A \in C} TP(A)}{\sum_{A \in C}(TP(A) + FN(A))} \\ Macro-F1 &= \frac{\sum_{A \in C} F1(A)}{|C|}, \quad Micro-F1 = \frac{2 \cdot Pr \cdot R}{Pr + R} \end{aligned} \tag{18}$$

4.3 Multi-label Classification

Multi-label classification is an important task to measure the effectiveness of network representation. We select three social networks to perform multi-label classification task in this experiment. The detailed statistics of datasets can be summarized in Table 1. For Blogcatalog, we randomly select 10% to 90% of nodes as training data. For Flickr and Youtube, we randomly select 1% to 10% of nodes as training data. We run 5 times for each algorithm and recorded the mean values in our results.

Table 1. Statistics of the dataset.

Dataset	$\|V\|$	$\|E\|$	Average degree	Label number
Blogcatalog	10,312	333,983	64.9	39
Flickr	80,513	5,899,882	146.7	195
Youtube	1,138,499	2,945,443	5.25	47

The results are shown in Fig. 4. For the Blogcatalog dataset, when the ratios of training data are 10% and 20%, the *Micro-F1* value of NEWEE is slightly lower than the values of other models. For other ratios of training data, NEWEE and Stru2vec perform well, especially when setting 50% of nodes as training data, our model is 10% higher than Stru2vec on *Macro-F1*.

Node2Vec is superior to DeepWalk, but it has no advantage only on the Youtube dataset. The *Micro-F1* value of Node2Vec is lower than DeepWalk. Because Youtube network is relatively sparse and the randomness of sampling neighbor nodes is reduced, therefore, the walk strategy of Node2Vec cannot bring obvious improvement. On the contrary, LINE performs well on the sparsest Youtube network, but not on other datasets. Because LINE preserves the first-order proximity well. Our model not only controls the way of walks, but also strengthens the utilization of first-order proximity. Therefore, the performance of NEWEE is superior to Node2Vec and LINE.

The performance of DeepWalk and GraphSAGE is the worst among the network embedding methods. The reason is that they do not well capture the network structure. Based on the above results, although the proposed method does not perform best on different types of networks, overall, compared with the other six algorithms, our model shows good performance.

4.4 Link Prediction

We conduct the link prediction task on arXiv GR-QC [11] to test our model. The dataset arXiv GR-QC is a collaboration network of papers. It has 5,242 nodes and 14,490 edges. Each node represents an author. If two authors cooperate to write a paper, there is an undirected edge between the two nodes. We randomly hide some edges from the network as test samples, and the remaining part of

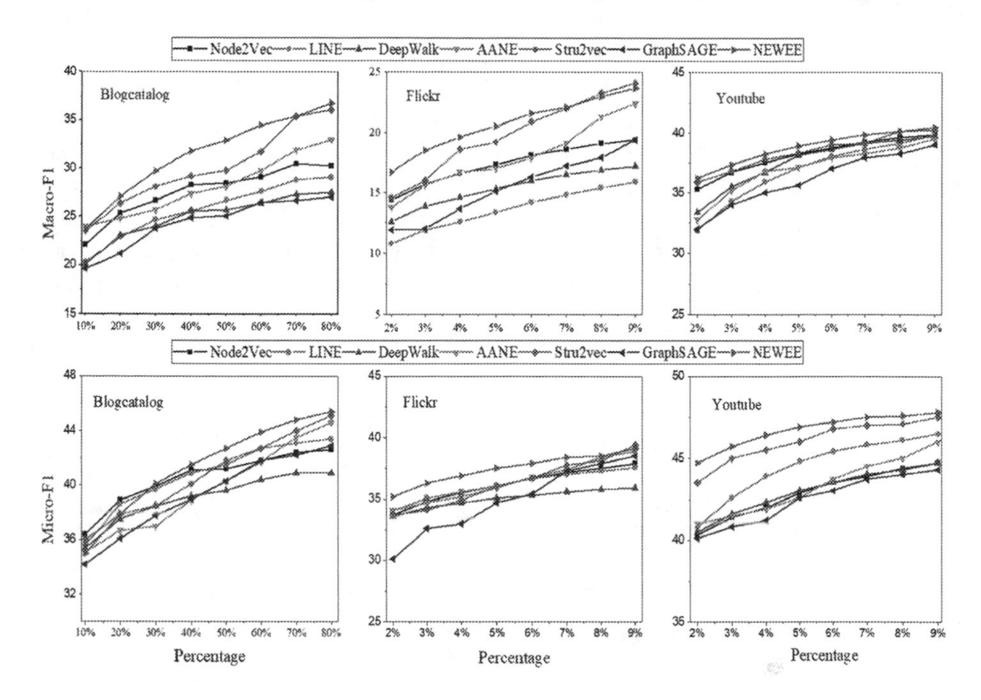

Fig. 4. *Macro-F1* scores and *Micro-F1* scores on Blogcatalog, Flickr, and Youtube.

the network as training samples. The nodes vectors are obtained after training, and the cosine similarity between the two nodes is calculated. We consider that there may be an edge between the two nodes with larger similarity. We conduct two experiments: The first evaluates the performance; the second evaluates the performance impact of different sparsity of networks on link prediction.

Table 2. *precision@k* values of arXiv GR-QC on link prediction task.

Method	*P@10*	*P@100*	*P@200*	*P@300*	*P@500*	*P@800*	*P@1000*
Node2vec	0.51	0.42	0.36	0.31	0.26	0.25	0.24
LINE	0.43	0.22	0.17	0.15	0.19	0.21	0.21
DeepWalk	0.42	0.27	0.31	0.31	0.26	0.24	0.25
AANE	0.65	0.48	0.31	0.37	0.31	0.27	0.30
Stru2vec	0.61	0.41	0.34	0.36	0.35	0.31	0.29
GraphSAGE	0.39	0.35	0.28	0.20	0.21	0.29	0.23
NEWEE	**0.71**	0.45	**0.35**	**0.40**	**0.38**	**0.34**	**0.31**

For the first experiment, we extract 15% of edges from the network, and use *Precision@k* as evaluation criterion. The value k increased from 2 to 1,000. The results are shown in Table 2. NEWEE is slightly better than other models in most cases. For the second experiment, we change the ratio of edges extracted from the network and use MAP as evaluation criterion. The experimental results are shown in Fig. 5. The results show that NEWEE is always better than the other six models. The performance of LINE and GraphSAGE is poor, because the LINE method relies more on first-order proximity. When the ratio of edges extracted reaches 80%, the damage to first-order proximity is more serious, so the effect of LINE has been greatly reduced. In addition, we find that with the increase of the ratio of edges extracted from the network, the effect of the seven models increases first and then decreases. This is because an increase in the ratio of edges extracted means an increase in the set of test samples. Therefore, the probability hitting the correct edge is decreased. On the other hand, as the ratio of edges extracted increases, the less information is provided for training. When the benefit of increasing the test samples can no longer offset the loss caused by the reduction of training samples, the effect of model begins to decline.

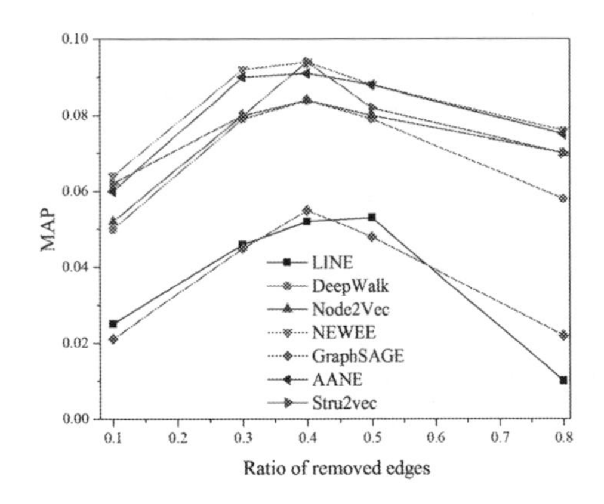

Fig. 5. Influence of ratio of removed links.

4.5 Parameter Sensitivity

In this section, the sensitivity of our model to parameters is tested. In addition to the parameters currently being tested, other parameters keep the default value. Multi-label classification task on Blogcatalog is performed to show the effect.

Firstly, the effect of the edge vector dimension and the node vector dimension on NEWEE model are evaluated respectively. The results are shown in Fig. 6((a) and (b)). Along with the increase of dimension, the performance of the model is slightly improved since the larger dimension can store more information. Especially for the edge vectors, they contain more information than node vectors. Therefore, the influence of edge vector dimension on NEWEE model is slightly more obvious than that of node vector dimension. In addition, the effect of random walk parameters (walk times r and walk length l) on the model is tested.

The results are shown in Fig. 6((c) and (d)). With the increase of r and l value, the performance of the model is improved rapidly and then became relatively stable. The two parameters can also improve the performance of NEWEE model due to that the random walk can traverse more paths from the network to provide more useful information. However, when the two values increase to a certain value, the provision of information becomes redundant.

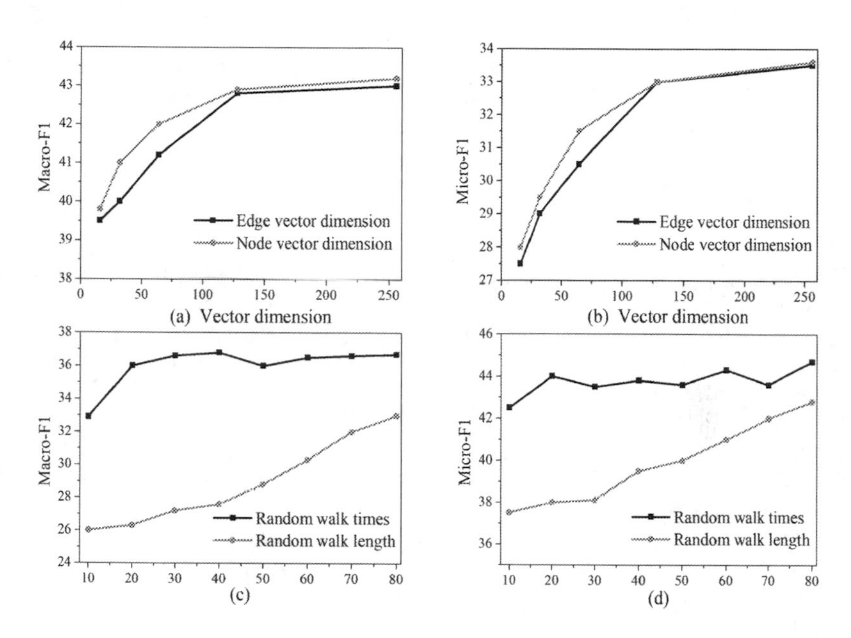

Fig. 6. Effect of different parameters on performance of NEWEE model.

5 Conclusion

This paper presents an unsupervised network representation learning model, called NEWEE, which can not only preserve the information of neighbor nodes, but also preserve the information of the relationship types between nodes and their neighbors. By performing multi-label classification and link prediction tasks on several real-world networks, our model can achieve excellent performance. Moreover, we provide a new way to distinguish relationship types without labeling data, and it is scalable, and can be applied to large-scale real-world networks.

Acknowledgements. We are grateful to the anonymous reviewers for their valuable comments on this manuscript. The research has been supported by the National Key Research and Development Program of China under Grant 2017YFB1402400, in part by the Frontier and Application Foundation Research Program of CQ CSTC under Grant cstc2017jcyjAX0340, in part by the Key Industries Common Key Technologies Innovation Projects of CQ CSTC under Grant cstc2017zdcy-zdyxx0047, in part by the Chongqing Technological Innovation and Application Demonstration Project under Grant cstc2018jszx-cyzdX0086, in part by the Social Undertakings and Livelihood Security Science and Technology Innovation Fund of CQ CSTC under Grant cstc2017shmsA0641, in part by the Fundamental Research Funds for the Central Universities under Grant 2018CDYJSY0055, and in part by the Graduate Research and Innovation Foundation of Chongqing under Grant CYB18058.

References

1. Aditya Grover, J.L.: node2vec: scalable feature learning for networks. In: ACM SIGKDD International Conference on Knowledge Discovery and Data Mining, pp. 855–864 (2016)
2. Bandyopadhyay, S., Kara, H., Biswas, A., Murty, M.N.: SaC2Vec: information network representation with structure and content (2018)
3. Belkin, M., Niyogi, P.: Laplacian eigenmaps and spectral techniques for embedding and clustering. Adv. Neural Inf. Process. Syst. **14**(6), 585–591 (2001)
4. Bourigault, S., Lagnier, C., Lamprier, S., Denoyer, L., Gallinari, P.: Learning social network embeddings for predicting information diffusion, pp. 393–402 (2014)
5. Cao, S., Lu, W., Xu, Q.: Grarep: learning graph representations with global structural information. In: ACM International on Conference on Information and Knowledge Management, pp. 891–900 (2015)
6. Figueiredo, D.R., Ribeiro, L.F.R., Saverese, P.H.P.: struc2vec: learning node representations from structural identity, pp. 385–394 (2017)
7. Hamilton, W.L., Ying, R., Leskovec, J.: Inductive representation learning on large graphs (2017)
8. Huang, X., Li, J., Hu, X.: Accelerated Attributed Network Embedding (2017)
9. Jacob, Y., Denoyer, L., Gallinari, P.: Learning latent representations of nodes for classifying in heterogeneous social networks, pp. 373–382 (2014)
10. Le, T.M.V., Lauw, H.W.: Probabilistic latent document network embedding, pp. 270–279 (2014)
11. Leskovec, J., Kleinberg, J., Faloutsos, C.: Graph evolution: densification and shrinking diameters. ACM Trans. Knowl. Discov. Data **1**(1), 2 (2007)
12. Li, A.Q., Ahmed, A., Ravi, S., Smola, A.J.: Reducing the sampling complexity of topic models, pp. 891–900 (2014)
13. Li, J.-H., Wang, C.-D., Huang, L., Huang, D., Lai, J.-H., Chen, P.: Attributed network embedding with micro-meso structure. In: Pei, J., Manolopoulos, Y., Sadiq, S., Li, J. (eds.) DASFAA 2018. LNCS, vol. 10827, pp. 20–36. Springer, Cham (2018). https://doi.org/10.1007/978-3-319-91452-7_2
14. Li, Q., Zhong, J., Li, Q., Wang, C., Cao, Z.: A community merger of optimization algorithm to extract overlapping communities in networks. IEEE Access **7**, 3994–4005 (2019)
15. Mikolov, T., Chen, K., Corrado, G.S., Dean, J.: Efficient estimation of word representations in vector space. Computation and Language arXiv:1301.3781 (2013)
16. Nallapati, R.M., Ahmed, A., Xing, E.P., Cohen, W.W.: Joint latent topic models for text and citations. In: ACM SIGKDD International Conference on Knowledge Discovery and Data Mining, August, Las Vegas, Nevada, USA, pp. 542–550 (2008)
17. Neelakantan, A., Shankar, J., Passos, A., Mccallum, A.: Efficient non-parametric estimation of multiple embeddings per word in vector space. Comput. Sci. (2015)
18. Perozzi, B., Alrfou, R., Skiena, S.: Deepwalk: online learning of social representations. In: ACM SIGKDD International Conference on Knowledge Discovery and Data Mining, pp. 701–710 (2014)

19. Roweis, S.T., Saul, L.K.: Nonlinear dimensionality reduction by locally linear embedding. Science **290**(5500), 2323–2326 (2000)
20. Tang, J., Qu, M., Wang, M., Zhang, M., Yan, J., Mei, Q.: Line: large-scale information network embedding, vol. 2, no. 2, pp. 1067–1077 (2015)
21. Wang, D., Cui, P., Zhu, W.: Structural deep network embedding. In: ACM SIGKDD International Conference on Knowledge Discovery and Data Mining, pp. 1225–1234 (2016)
22. Wang, H., et al.: GraphGAN: graph representation learning with generative adversarial nets (2017)
23. Wang, S., Chang, X., Li, X., Sheng, Q.Z., Chen, W.: Multi-task support vector machines for feature selection with shared knowledge discovery. Sig. Process. **120**, 746–753 (2016)
24. Yoshua, B., Aaron, C., Pascal, V.: Representation learning: a review and new perspectives. IEEE Trans. Pattern Anal. Mach. Intell. **35**(8), 1798–1828 (2013)
25. Yue, L., Chen, W., Li, X., Zuo, W., Yin, M.: A survey of sentiment analysis in social media. Knowl. Inf. Syst. 1–47 (2018)
26. Zhang, A., Zhu, J., Zhang, B.: Sparse relational topic models for document networks. In: Blockeel, H., Kersting, K., Nijssen, S., Železný, F. (eds.) ECML PKDD 2013. LNCS (LNAI), vol. 8188, pp. 670–685. Springer, Heidelberg (2013). https://doi.org/10.1007/978-3-642-40988-2_43

MDAL: Multi-task Dual Attention LSTM Model for Semi-supervised Network Embedding

Longcan Wu, Daling Wang(✉), Shi Feng, Yifei Zhang, and Ge Yu

School of Computer Science and Engineering,
Northeastern University, Shenyang, China
longcanwu@gmail.com, {wangdaling, fengshi, zhangyifei, yuge}@cse.neu.edu.cn

Abstract. In recent years, both the academic and commercial communities have paid great attentions on embedding methods to analyze all kinds of network data. Despite of the great successes of DeepWalk and the following neural models, only a few of them have the ability to incorporate contents and labels into low-dimensional representation vectors of nodes. Besides, most network embedding methods only consider universal representations and the optimal representations could not be learned for specific tasks. In this paper, we propose a **M**ulti-task **D**ual **A**ttention **L**STM model (dubbed as MDAL), which can capture structure, content, and label information of network and adjust representation vectors according to the concrete downstream task simultaneously. For the target node, MDAL leverages Tree-LSTM structure to extract structure, text and label information from its neighborhood. With the help of dual attention mechanism, the content related and label related neighbor nodes are emphasized during embedding. MDAL utilizes a multi-task learning framework that considering both network embedding and downstream tasks. The appropriate loss functions are proposed for task adaption and a joint optimization process is conducted for task-specific network embedding. We compare MDAL with the state-of-the-art and strong baselines for node classification, network visualization and link prediction tasks. Experimental results show the effectiveness and superiority of our proposed MDAL model.

Keywords: Dual attention · Network embedding · Multi-task learning

1 Introduction

At present, analyzing network data have drawn extensive attentions from research communities for a wide range of applications, such as node classification [18, 29], link prediction [8], community discovery [7], anomaly detection [10]. Now network representation learning (NRL) has emerged as the primary method for modeling the network structures and paved the way for the downstream application task [1, 18, 30]. NRL aims to map node into low-dimensional vector, which ideally should retain all the node information in the network, such as structure, content, and label information.

Inspired by DeepWalk [18], a large number of network embedding algorithms based on neural models have been proposed. Some of these algorithms use network

G. Li et al. (Eds.): DASFAA 2019, LNCS 11446, pp. 468–483, 2019.
https://doi.org/10.1007/978-3-030-18576-3_28

topological structure [8, 18, 21], and some algorithms consider both network structure and node content [24, 31]. But only a handful of existing literature take into account the node label information [17, 27] partially because of the sparse labels. Typically, only a small subset of nodes in the network had labels. Moreover, compared with text, node labels are all high-level concepts with limited semantics, so it is usually not easy to model this kind of information. Finally, structure, content and label information are three different kinds of information sources, and the heterogeneity between these sources make it difficult to embed all the three information into the same vector space. Therefore, how to properly embed three aspects of node still remains as a major challenge.

The vector representations learned by network embedding algorithm are usually evaluated by two classic downstream tasks: node classification and link prediction. For node classification, learning representations of nodes and training node classifier are usually separated. That means we need firstly to obtain vector representation by specific network embedding algorithm, and then fed the vectors into the subsequent classifier. This two stage approach will make the node embeddings insensitive to node classification task. To tackle this problem, previous literature have attempted to learn network representation and node classifier jointly, and the experimental results show that the joint training methods have achieved better results than the two stage approach [15, 22]. However, the existing joint learning framework does not fully utilize the label information for network embedding, because the node label is only regarded as the standard of classifier during joint training.

For link prediction, the learned node vectors are directly fed into a function to obtain the features of node pair, then the AUC metric is used to evaluate the performance [2]. Similar to node classification task, the learned node embeddings are task-insensitive for link prediction, which will decrease the performance of prediction model. As far as we known, only limited literature have been published for jointly optimizing network presentation and link prediction task [25].

In this paper, we propose a **M**ulti-task **D**ual **A**ttention **L**STM model (dubbed as MDAL) for semi-supervised network embedding. MDAL employs a multi-task deep learning framework that can jointly optimize the network representation learning and downstream task. The core of MDAL is a proposed Dual Attention Tree-LSTM network (dubbed as DAL), which can capture structure, content, and label information of node simultaneously. The Tree-LSTM naturally represents the network structure between nodes and the dual attention mechanism that considering both text contents and labels is able to obtain the relatedness between neighborhood and target node. Furthermore, with the help of MDAL framework, the vector representation of target node is fine-tuned by the downstream tasks. As a result, MDAL can generate task-sensitive network embeddings and further improve the performance of the specific tasks.

In addition, MDAL model has good extensibility and interpretability. On one hand, based on MDAL framework we can easily incorporate auxiliary information, such as community structure and user profile, from nodes into vector representations. On the other hand, the dual attention mechanism gives us reasonable explanation for the weights of text and label when we analyze the downstream task. We evaluate MDAL by node classification, network visualization, and link prediction tasks on three real-world datasets. The experimental results show that the proposed model not only outperforms the state-of-the-art baseline algorithms, but also has strong adaptability to different tasks.

The main contributions of our paper are three-fold: (1) We propose a semi-supervised network embedding model MDAL with a multi-task deep learning framework, which can jointly optimize the network representation learning and specific downstream tasks. (2) The proposed MDAL model can integrate structure, text and label information into the low-dimensional vector representation of nodes, and generate better task-sensitive embeddings by using dual attention mechanism in multi-task framework. (3) We conduct extensive experiments on three real-world benchmark datasets. The results confirm the superiority of our proposed approach over the state-of-the-art baseline methods.

The rest of this paper is organized as follows. We briefly review semi-supervised NRL models and the Tree-LSTM in the Sect. 2. In Sect. 3, we introduce our MDAL model in detail, and discuss two specific tasks, i.e. node classification and link prediction in Sect. 4. Extensive experiments are conducted in Sect. 5 to show the effectiveness of MDAL model. Finally, Sect. 6 summarizes the work of this paper.

2 Related Work

In this paper, the proposed model leverages semi-supervised learning strategy and Tree-LSTM for modeling the network. Therefore, we briefly introduce a variety of semi-supervised NRL algorithms and Tree-LSTM models. A detailed summary of the semi-supervised NRL is made in the survey [30]. In general, such algorithms can be divided into two categories depending on whether node content is considered: the first category is semi-supervised structure preserving NRL, and the second is semi-supervised content augmented NRL. The structure preserving NRL mainly considers the structure and label information of nodes. The existing algorithm such as DeepWalk [18], node2vec [8] and LINE [21], only consider the structure information. Getting vector representation of nodes and training subsequent classifiers for classification of nodes are separated. This two-stage model does not obtain discriminative vector representations and good classification result. To solve this problem, DDRW [15], MMDW [22] and TLINE [32] add classifier optimization target into the NRL objective function. The experimental results show that these methods are better than the above two-stage model in terms of node classification. Other semi-supervised structure preserving NRL models, such as GENE [3], LENE [5], and PNE [4], incorporate label information into the vector representation of node by maximizing node and label co-occurring probability. Then these model trains another classifier for the node classification task. In general, this kind of NRL models does not take full advantage of node content.

Semi-supervised content augmented NRL uses textual information as complement of structure and label information. The following models are the most common. TriDNR [17] and LDE [24] incorporate label and content information into node embedding by maximizing the node-text-label co-occurring conditional probability. LANE [11] enriches the vector representation of nodes by embedding the structure, text and the label information into the same low-dimensional space. In addition, GCN [13] and GraphSAGE [9] spread information of each node to its neighbors through multiple iterations and aggregations and use the label information of nodes through the classifier. The most relevant work with this paper is AGRNN model [27]. AGRNN first obtains neighbors of target node by constructing a subtree with target node as root, then applies the Tree-LSTM structure to obtain the vector representation of target node. AGRNN takes into account network structure and text information, and jointly optimizes node embedding model and node classifier so that the label information can be used. In our paper, the proposed MDAL model is different from AGRNN. Firstly, MDAL uses label attention to directly fuse label information into node embedding; secondly, in MDAL model, we design different loss functions according to different downstream tasks and jointly optimize the network embedding and specific downstream task under a multi-task learning framework.

Tree-structured neural networks, also known as recursive neural networks, were first used in the NLP domain [19]. If Tree-structured neural network is very deep, then gradient exploding or vanishing problem will damage the model's effect. Therefore, Tai et al. [20] first introduced Long Short-Term Memory (LSTM) unit to tree-structured neural networks and proposed a Tree-LSTM model. The special structure of Tree-structured neural networks allows this model to be well fit for network data. Kim et al. [12] firstly used this structure to classify users in Twitter space. Then AGRNN model [27] added the attention mechanism into this structure. Based on the achievements of above work, our MDAL model also employs Tree-structured neural networks, and adopts LSTM as recursive neural unit.

3 The Proposed Model

Assume information network $G = \{V, E, X, L\}$, V is the node set and $V = \{v_1, v_2, \ldots, v_n\}$, E is the edge set and each edge represents a relationship between nodes; $X = \{x_1, x_2, \ldots, x_n\}$ is the text set of nodes and x_t is the text vector of node v_t; L is the label set of nodes and $L = \{l_1, l_2, \ldots, l_n\}$. If node v_t has a label, then l_t is the corresponding label vector value, otherwise l_t is zero vector. For a given target node v_t, our goal is to learn a low-dimensional vector representation of $h_t \in R^k$ ($k \ll n$). Here h_t should not only contain the structure, text, and label information of v_t in the original network G, but also be suitable for specific task such as node classification and link prediction.

The general framework of our proposed model MDAL is presented as Fig. 1. Firstly, for a given target node v_t (red node) in the network, we need to obtain its neighbors (green node) and the first order and second order neighbors are elected here. Then the target node and its neighbors are fed into Tree-LSTM model to obtain the hidden vector representation h_t of target node. Finally, the hidden vector is fed into different loss functions for specific downstream tasks aiming to get target node' vector representation for a specific task. We will explain details of the MDAL model next.

3.1 Dual Attention Tree-LSTM Model

The nature of node in network is not only related to its text and labels, but also to the properties of its surrounding neighbors. Further, for the target node v_t, the influence of its neighbors is different depending on distance between neighbors and target node. For

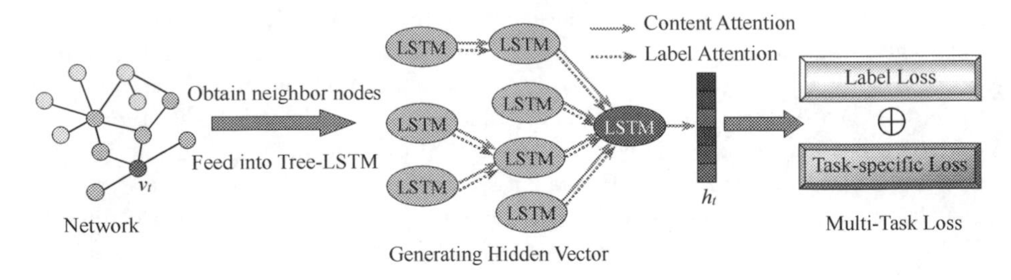

Fig. 1. MDAL framework (Color figure online).

example, first order neighbor information has a great influence on v_t, and second order neighbors and third order neighbors have smaller impact. So we need a special model to consider effect of neighbor on target node embedding. Tree-LSTM model is a common network model, which can continuously and recursively learn vector representation of parent node by child nodes. Based on its essential characteristic, Tree-LSTM model is a well fit for network data.

With Tree-LSTM model, the hidden vector representation of target node can naturally contains information from itself and surrounding neighbors. Besides, in order to better consider the influence of surrounding neighbor information on target node, we propose a dual attention mechanism, namely, text attention and label attention. Through text attention, vector representation of target node can contain more information of neighbors that have more similar text to target node. Identically, label attention can also make target node more focus on neighbors with similar label information. Because of the fact that only some of the nodes in the network have labels, in order to use label attention mechanism, we need to get label vectors for all the nodes. In this paper, the label vectors of those nodes without tags are obtained by bootstrap algorithm. In brief, we use traditional node classification algorithm ICA [16] to train a classifier and then this classifier is applied to unlabeled nodes to obtain the pseudo-label.

Given a target node v_t, in order to use Tree-LSTM model, we first need to obtain a subtree T_t with v_t being root node. As shown in the left part of Fig. 1, we can take v_t as root node (red node) in original network to obtain a subtree of depth d (here $d = 2$) by breadth-first search, which contains v_t and its neighbors. Unlike the traditional LSTM model, as shown in the middle part of Fig. 1, Tree-LSTM model starts with leaf nodes of subtree and uses LSTM unit to obtain hidden vector representation of each node in a bottom-up manner. In this way, the information of neighbors is gradually converging to the root node, i.e. target node v_t. The Tree-LSTM model obtains structure information of v_t through subtree, and uses the LSTM unit to obtain text and label of each node in the subtree. So the Tree-LSTM model can incorporate both structure, text and label information of nodes. For any node v_p in the subtree, the computation equations of hidden vector h_p are as follows:

$$\tilde{h}_p = mean_{v_r \in C(v_p)}\{h_r\} \tag{1}$$

$$i_p = \sigma(W^{(i)}x_p + \lambda_2 V^{(i)}l_p + U^{(i)}\tilde{h}_p + b^{(i)}) \tag{2}$$

$$f_{pr} = \sigma(W^{(f)}x_p + \lambda_2 V^{(f)}l_p + U^{(f)}h_r + b^{(f)}) \tag{3}$$

$$o_p = \sigma(W^{(o)}x_p + \lambda_2 V^{(o)}l_p + U^{(o)}\tilde{h}_p + b^{(o)}) \tag{4}$$

$$u_p = \tanh(W^{(u)}x_p + \lambda_2 V^{(u)}l_p + U^{(u)}\tilde{h}_p + b^{(u)}) \tag{5}$$

$$c_p = i_p \Theta u_p + \sum\nolimits_{v_r \in C(v_p)} f_{pr} \Theta c_r \tag{6}$$

$$h_p = o_p \, \Theta \tanh(c_p) \tag{7}$$

where x_p and l_p are text vector and label vector of node v_p; $C(v_p)$ is the set of child of v_p and v_r is a concrete child; h_r and c_r are hidden vector and cell vector of child v_r; $\widetilde{h}_p$ is the summation of child hidden vector; $W^{(t)}$ denotes the weight matrix and $b^{(t)}$ is bias with $t \in \{i, f, o, u\}$; σ denotes the logistic sigmoid function and Θ denotes element-wise multiplication; λ_2 is a hyper-parameter that controls the weight of label vectors of node v_p and belongs to numerical interval [0, 1], because for many nodes their label vectors are pseudo-labels, so we need this parameter. According to the above formulas, we can obtain the hidden vector h_p of node v_p. In this way, we can obtain the hidden vector of root node v_t, in which we consider it as the node embedding. It is important to note that there is no child node for the leaf node, so h_r and c_r of child node of leaf node are set to zero vector. In addition, in order to prevent overfitting, h_r and c_r of child node are regularized with zoneout [14].

To better consider influence of neighbor information on root node embedding, we propose a dual attention mechanism to focus on nodes that are more relevant to root node. As shown in middle part of Fig. 1, for subtree with v_t being root node, the hidden vector h_p of any node v_p in the subtree should contain information more relevant to root node by content attention and label attention. To achieve this, we need to reconsider formula of $\widetilde{h}_p$ as follow:

$$\alpha_\gamma^c = softmax(x_T W^c x_r) \tag{8}$$

$$\alpha_\gamma^l = softmax(l_T W^l l_r) \tag{9}$$

$$\tilde{h}_p = \sum\nolimits_{v_r \in C(v_p)} (\alpha_\gamma^c + \lambda_1 \alpha_\gamma^l) h_r \tag{10}$$

First we select child nodes that have more similar text with respect to root node content. As shown in Formula (8), we use a parameter matrix W^c to determine the content attention score α_γ^c between root node text x_T and child node text x_r. Secondly, we select children whose label is more similar to root node label. As shown in Formula (9), we use a parameter matrix W^l to get the label attention score α_γ^l. For a child node, if it has not a label, it is assigned the corresponding pseudo-label. Finally, $\widetilde{h}_p$ is the summation of child hidden vectors with corresponding text and label attention score in Formula (10), in which λ_1 is a hyper-parameter that controls the proportion of text attention and label attention, and belongs to numerical interval [0, 1].

3.2 Parameters Learning

When we obtain the hidden vector h_t of target node v_t, h_t is fed into subsequent fully connected layer classifier to train the parameters of LSTM unit. It is important to note that we use the same LSTM unit parameters for the entire Tree-LSTM model. Finally the predicted probability distribution p_t is obtained by softmax function:

$$p_t = softmax(W^t h_t + b^t) \tag{11}$$

Here cross-entropy loss with L_2 regularization is used as cost function. The goal of model training is to minimize the cross-entropy J_1 between predicted probability distribution and label vectors for all labeled nodes:

$$E(l_t, p_t) = -\sum\nolimits_{i=1}^{k} l_t^i * \log p_t^i \tag{12}$$

$$J_1 = -\frac{1}{N}\sum\nolimits_{t=1}^{N} E(l_t, p_t) + \frac{\lambda}{2}\|\theta\|^2 \tag{13}$$

where k is number of categories, N is number of labeled nodes, all the model parameters including W^c and W^d are denoted as θ, λ denotes a regularization coefficient. The vector representation of node obtained through the objective function J_1 is universal, which can be applied to any downstream tasks. We call this Dual Attention Tree-LSTM model as DAL. Note that DAL is a universal model and the core of MDAL framework. Finally, the algorithm of DAL model is presented in Algorithm 1.

Algorithm 1. DAL

Input: $G=\{V, E, X, L\}$
Output: vector representations h_t for all $v_t \in V$
1) obtain pseudo label for unlabeled nodes by ICA algorithm;
2) construct a subtree T_t with v_t being root node for all $v_t \in V$;
3) **repeat**
  for all labeled node v_t **do**:
  {obtain h_t of v_t by feeding T_t into Tree-LSTM structure;
  send h_t to cross-entropy J_1 to train the parameters of model}
 until convergence;
4) **for** all nodes $v_t \in V$ **do**:
 obtain node embedding h_t according to send T_t into Tree-LSTM structure;

---

4 MDAL Variants for Specific Downstream Tasks

In this section, MDAL integrates DAL into a multi-task learning framework, which can simultaneously learn task-sensitive embeddings and fulfill downstream task. We introduce two MDAL variant models for node classification and link prediction tasks respectively.

4.1 Node Classification

When we obtain hidden vector h_t of target node v_t, we need to design a specific objective function, which lets h_t be adapted to node classification task. For node classification, we can train the parameters of Tree-LSTM model by feeding h_t into the objective function J_1 as described in Sect. 3.2. However, for the practical application, the node label is not easy to be obtained, namely, only a small number of nodes have tags. If only the actual labeled nodes are calculated to obtain hidden vector and then fed into loss function J_1, the information of untagged nodes is not fully exploited. Based on the characteristics of network data, we use the link between nodes to conduct unsupervised learning. For the target node v_p, we use the first-order proximity of nodes [8, 17] to enforce v_p and its neighbors with similar hidden vectors, and v_p and its non-neighbors with distinct hidden vectors. Based on the above description, for the node classification task, we define graph-based loss function J_2 as follow:

$$J_2 = -\sum\nolimits_{(i,j)\in E} (\log \sigma(h_i^T h_j) + \sum\nolimits_{k=1}^{K} E_{v_n \sim P_n(v)}[\log \sigma(-h_i^T h_n)]) \tag{14}$$

where h_i is hidden vector of node v_i learned from Tree-LSTM model and (i, j) is an edge; v_n is obtained by negative sampling; K is the number of negative edges; P_n is a negative sampling distribution.

Therefore, for the node classification task, we can minimize the label loss function J_1 and graph-based loss function J_2 in the multi-task learning framework, and obtain the final node classification loss J_c by a weighted combination between J_1 and J_2 as shown in Formula (15), where λ_c is used to balance the relative proportions of two losses. We call this variant model MDAL-C, where C stands for classification.

$$J_c = \lambda_c J_1 + J_2 \tag{15}$$

4.2 Link Prediction

After getting hidden vectors of nodes, we can use these vectors for link prediction. The common approach is to get edge features by binary operators [8], and then train a classifier with aim to classify node pairs that have link or no link. Through the classifier, for any node pairs we can get the probability of existence of link, and finally use AUC indicator to measure node embedding performance in the link prediction task. The above approach has the following problems. First, for the network data, the node pair with a link can be used as a positive example and the node pair with unknown link state cannot be used as a negative example, because these node pairs may have a link that is not observed or what we need to predict. Therefore, it is unreasonable to directly take these node pairs as negative examples. So the link prediction problem itself is a Positive-Unlabeled learning problem [23]. Secondly, embedding of nodes obtained by model has task independence, and it is less effective to measure the performance of model in the link prediction task if we directly use these nodes embedding. In order to solve the second question, we use the AUC indicator as part of loss function. However, the direct optimization of AUC also need to pre-select positive and negative examples and as mentioned above directly selecting the node pair with unknown link state as a negative example is unreasonable. To overcome this obstacle, we optimize the AUC with node pairs that has a link (that is, positive examples) or have unknown link state. The specific reason is as follows: research work has demonstrated that, as shown in Formula (16), PU-AUC risk R_{PU} is equivalent to supervised AUC risk R_{PN} with a linear transformation [26]. So even if we do not have negative data, we can optimize R_{PU} instead of optimizing R_{PN}.

$$R_{PU} = \frac{1}{2}\theta_P + \theta_N R_{PN} \tag{16}$$

where θ_P and θ_N are percentage of positive data and negative data. Based on the above conclusions, for link prediction tasks, we need to include PU-AUC risk J_3 in the loss function of model, where J_3 is defined as follows:

$$J_3 = \sum\nolimits_{(i,j)\in E_p} \sum\nolimits_{(m,n)\in E_U} g(S(h_i, h_j), S(h_m, h_n)) \tag{17}$$

where E_p represents a set of node pairs having link and E_U is a set of node pairs having unknown link state; g denotes loss function and here we adopt hinge loss function; $S(h_i, h_j)$ represents how we get the similarity of node pairs for v_i and v_j, and we use the L_2 norm. Because many nodes have unknown link in the network, in order to reduce the computational cost, we use the idea of negative sampling to construct a set of triplets P, where $(i, j, k) \in P$ and v_i has link with v_j and has unknown link state with v_k. Based on the above description, J_3 is redefined as follows:

$$J_3 = \sum\nolimits_{(i,j,k)\in P} \max(0, \delta + S(h_i, h_j) - S(h_i, h_k)) \tag{18}$$

where δ is a threshold to regulate the similarity.

For the link prediction task, we can not only optimize the AUC index directly by minimizing the loss function J_3, but also use label information in the graph. Therefore, we minimize the loss function J_1 and J_3 in the multi-task learning framework. Through a weighted combination, we obtains the final link prediction loss function J_l as shown in Formula (19), where λ_l is used to balance the relative proportions of two losses. We call this variant model MDAL-L, where L stands for link prediction.

$$J_l = \lambda_l J_1 + J_3 \tag{19}$$

5 Experiments

Our MDAL model can be applied to many network data related tasks. In this paper, we compare MDAL model with other baseline models in the real-world datasets with respect to node classification, network visualization and link prediction tasks. Here we first introduce the benchmark datasets and baselines. Then we analyze the performance of the MDAL model and its variants on three classical tasks.

5.1 Experiment Setup

Datasets. There are three datasets for this experiment, two of which are citation networks: Cora and Citeseer [22]. These two citation networks use papers as nodes and reference relationships between papers as edges (undirected). Each of paper contains a set of keywords as attributes of node, and the corresponding attribute features are represented by a 0/1-valued word vector. The research field of each paper is used as a label. In order to verify the effectiveness of proposed model on social network da-ta, we use WebKB dataset [27]. In WebKB dataset, each node represents a website, each link represents a hyperlink between web pages, node attribute represents contents of web page and the node label represents the department to which website belongs. For the sparsity of WebKB dataset, we use the WebKB-sim dataset [27], which adds additional three edges for each node on the basis of WebKB dataset. And extra three edges are connected to each node most similar to attribute of each node. An overview about above three datasets is given in Table 1.

Table 1. Datasets statistics

Dataset	Cora	Citeseer	WebKB-sim
Nodes	2708	3312	877
Edges	5429	4732	2631
Features	1433	3703	1703
Labels	7	6	5

Baselines. For comprehensive and comparative analysis of MDAL model, we utilize the following methods as strong baselines. DeepWalk [2] directly leverages the network topology. TADW [28] model uses structure and attribute information in the form of matrix decomposition. ICA [16] algorithm is a classical network data classification algorithm and we use its variant, ICA-count, which uses the number of labels as label information. Considering the fact that the number of tagged nodes is sparse, the ICA model approximately can be regarded as a content-only baseline. GraphSAGE-sup [9] obtains the vector representation of node by aggregator functions, which can take advantage of structure, label and attribute of nodes. It is important to note that we use GCN aggregator for the GraphSAGE-sup model. Similar to our MDAL model, AGRNN [27] model can obtain the vector representation of target node through attribute attention, and also make use of three aspects of node information.

Parameter Setup. For all datasets and model, the dimension of node represents is set to 128. For baselines we refer to the parameter settings in the corresponding paper. For the model proposed in this paper, unless otherwise specified, for each node we build a subtree with depth d being 2. For the balance parameter λ_1 (controlling label attention) and λ_2 (controlling label vector), we conduct a parameter sweep on {0.01, 0.001, 0.0001}. The number of negative edges is 5; regularization coefficient λ is 1e–4; threshold δ is 1. In terms of the optimization process, for the DAL model, the learning rate is set to 0.01; for the MDAL-C and MDAL-L model, the learning rate is set to 1e–5. For the loss balance parameter λ_c and λ_l, we conduct grid-search on numerical interval (0, 2).

5.2 Node Classification

In this section, we first verify the performance of MDAL model in terms of multi-class node classification task. First for the network dataset, we select some nodes to give the real label, then select 10% as the validation set from unlabeled data, and the remaining 90% as the test set. Before conducting the experiment, we used the ICA model to obtain pseudo-label of unlabeled node through preprocessing, so as to make the label attention mechanism can be implemented. Then the node embedding is obtained using a specific model on the network dataset. We use representation vector as feature vector for classification tasks. In order to eliminate the effect of classifier on the performance of multi-label classification, if not explained, we send representation vector of node into single-layer neural network model to obtain the label.

Classification performance is measured by Micro F1-score metric. We use 5-fold cross validation to evaluate Micro F1-score. For node classification tasks, we use the baselines mentioned in Sect. 5.1, as well as two proposed models: DAL and MDAL-C. It is important to note that DAL represents a model not designed for any task, and MDAL-C is a model designed for node classification tasks. Table 2 shows the performance of different models on different datasets, and the best result is boldfaced. From these three tables, we have the following observations and analysis.

Compared with DeepWalk model, which only utilizes structure information of network, TADW and ICA can make use of node attribute information, so these two models get better results on three datasets across all training ratios. The reason is that the three datasets are very sparse and DeepWalk model cannot give full play to its own advantages. Therefore, in this case, the text information can be more helpful to node classification. Secondly, for GraphSAGE, AGRNN and MDAL variant models, all these models can take advantage of structure, label and text information of nodes. So compared with TADW and ICA, these models can achieve obvious improvements on Cora and Citeseer datasets. However, it is important to note that in WebKB-sim dataset, the GraphSAGE model does not achieve satisfactory results. The reason may be that the WebKB-sim dataset is a semi-synthetic dataset and additionally added links disrupt the aggregator functions. Comparatively speaking, AGRNN and MDAL variant models can still be effective for semi-synthetic dataset WebKB-sim, which shows the effectiveness of Tree-LSTM structure on modeling network data.

For AGRNN, DAL takes label information of node fully into account in the vector representation learning process, so DAL obtains better results on three datasets.

Table 2. Node classification Micro-F1 score (%) with various methods on different datasets

Dataset	%Labeled nodes	Deep walk	TADW	ICA	Graph SAGE	AGRNN	DAL	MDAL-C
Cora	5%	71.73	70.89	74.23	80.06	78.32	78.40	**80.64**
	10%	74.33	77.89	75.10	82.86	81.24	82.13	**82.95**
	15%	76.20	81.71	76.29	84.21	84.59	85.26	**85.51**
	20%	77.25	83.94	77.38	85.01	85.23	85.69	**86.66**
	25%	78.24	84.17	78.76	85.19	85.65	86.15	**87.51**
Citeseer	5%	50.58	62.04	65.53	70.90	69.38	69.84	**70.90**
	10%	52.45	66.81	68.61	69.76	69.68	70.17	**71.02**
	15%	53.76	69.26	69.33	71.02	71.45	72.65	**72.85**
	20%	54.73	71.06	68.79	70.47	72.74	73.20	**74.21**
	25%	55.28	72.46	69.44	71.49	73.16	73.82	**74.63**
WebKB-sim	5%	46.40	46.64	48.60	55.15	50.80	55.06	**64.66**
	10%	47.72	53.03	67.79	56.43	68.07	70.45	**73.13**
	15%	50.34	57.51	70.34	56.71	72.87	74.96	**75.41**
	20%	51.56	58.40	73.05	58.01	74.32	75.11	**76.70**
	25%	53.79	60.63	73.64	59.62	75.01	76.68	**78.88**

Because there are relatively few tagged nodes, the label information is limited. In this case, the information of unlabeled nodes is fully taken into account in the MDAL-C model through graph-based loss function J_2, so that MDAL-C achieves better results than AGRNN and DAL. For example, on Cora and Citeseer dataset, MDAL-C obtains improvement from 0.5% to 2% over AGRNN and DAL under all training ratio settings. For WebKB-sim, MDAL-C achieves about average 3.3% gain over DAL. In general, DAL outperforms or competitive with baselines, which proves that the proposed model DAL effectively fuses structure, content and label information of nodes. For node classification tasks, the proposed task-specific model MDAL-C is consistently better on all datasets, which effectively shows that the proposed model MDAL has good task adaptability.

5.3 Network Visualization

In order to further illustrate that vector representation of node is discriminative, we map those representations for node classification task into the 2-D space and display the distribution of different categories of nodes in the planar axis using t-SNE visualization tool with its default parameter values [6]. Here we take Cora dataset as an example, first select 20% of data set as a labeled dataset, and then get vector representation of node through the corresponding model. As shown in Fig. 2, each dot in figure represents a node, and each color represents a category. The good vector representations should enable nodes with similar tags to gather together in space.

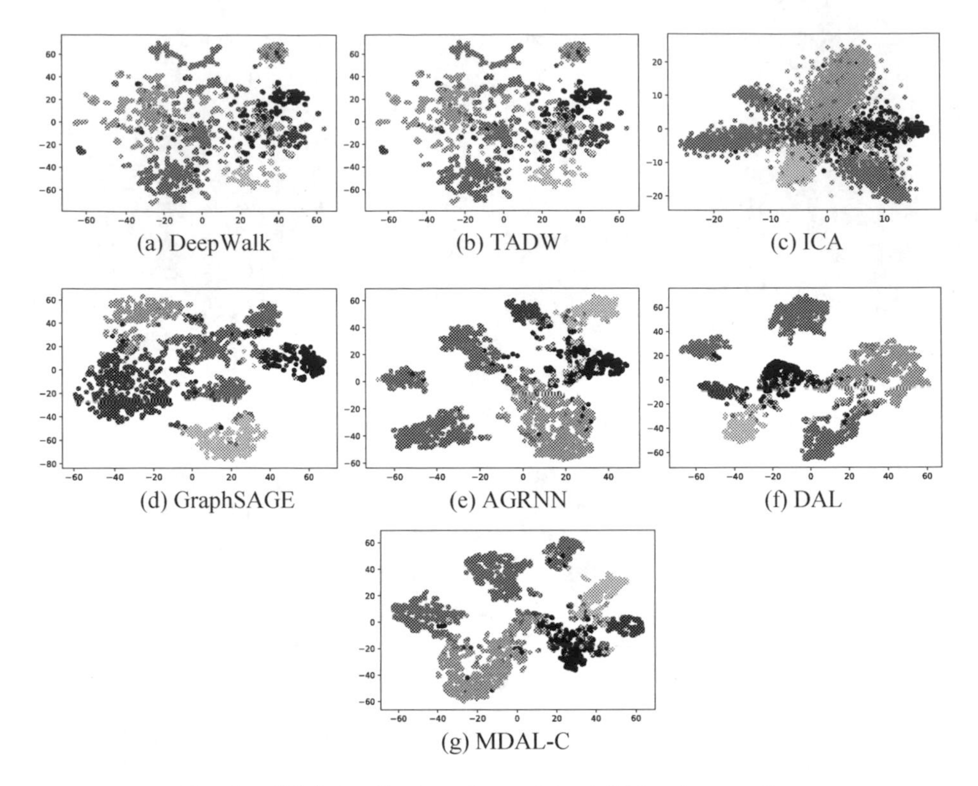

Fig. 2. t-SNE visualization of cora dataset (Color figure online)

For Cora datasets, nodes represent papers, link represents references between papers, and labels represent the research field to which paper belongs. Because the correlation between these research fields is relatively large, a paper may have many references bibliography belong to many research fields. So using only the reference relationship between papers does not distinguish the label of paper very well, which can be seen from the visualization results of the DeepWalk. As shown in Fig. 2(a), DeepWalk model uses only the structure information of network, so the visualization result is not good and we can find that different types of nodes are mixed together. As far as the citation dataset, the content information of paper can better express the domain of paper. The TADW model incorporates content information on the basis of DeepWalk, so visualization results of TADW in Fig. 2(b) are slightly better than DeepWalk. As mentioned earlier, ICA model approximately can be seen as a content-only model. From Fig. 2(c), the visualization result of ICA is better than DeepWalk and TADW. After incorporating the label information of nodes, GraphSAGE, AGRNN, DAL and MDAL-C get better visualization results (Fig. 2(d)–(g)). It is clear that the visualization of DAL and MDAL-C performs best in both models. From Fig. 2 (g), we can find that different types of nodes are clustered together, and the boundary between different groups is obvious. From the above analysis and Fig. 2, we can find the effectiveness of the proposed MDAL-C model in network visualization.

5.4 Link Prediction

In this section, we evaluate MDAL model on the link prediction task. In terms of link prediction task, our goal is to predict the unobserved edge based on existing network structure. In order to carry out the link prediction task, first we need to randomly select $p\%$ node pairs with unknown link state as negative samples, and then randomly hide $p\%$ linked node pairs as positive samples. It is important to note that after hiding $p\%$ linked nodes, the residual network needs to be connected. Secondly, we use corresponding model to get vector representation of nodes on the residual network. Finally, we calculate similarity between node pairs and use AUC index to measure the performance of model on link prediction task. Here for three datasets, we select 20% nodes to assign labels and conduct experiment after respectively hiding 10%, 20% and 30% links. For link prediction task, we use baselines mentioned in Sect. 5.1, as well as two models proposed: DAL and MDAL-L. DAL represents a model not designed for any task and MDAL-L is designed for link prediction tasks. It is important to note that for Cora and Citeseer datasets, for each node we build a subtree with depth 2 to achieve better results; for WebKB-sim datasets, the depth is 1. Table 3 shows the results of link prediction on three datasets.

Table 3. Link prediction AUC with various methods on different datasets

Dataset	%Linked nodes	Deep walk	TADW	ICA	Graph SAGE	AGRNN	DAL	MDAL-C
Cora	10%	0.904	0.919	0.909	0.922	0.889	0.902	**0.932**
	20%	0.885	0.903	0.904	0.914	0.878	0.896	**0.917**
	30%	0.865	0.898	0.897	0.908	0.869	0.887	**0.911**
Citeseer	10%	0.887	0.956	0.925	0.939	0.944	0.951	**0.960**
	20%	0.836	0.944	0.904	0.935	0.941	0.949	**0.958**
	30%	0.822	0.942	0.894	0.932	0.933	0.948	**0.954**
WebKB-sim	10%	0.768	0.675	0.533	0.663	0.656	0.656	**0.942**
	20%	0.781	0.648	0.544	0.655	0.596	0.636	**0.937**
	30%	0.771	0.628	0.536	0.617	0.584	0.605	**0.926**

Overall, as the percentage of hidden node pairs increases, the performances of all models decrease. Compared with AGRNN and DAL, MDAL-L model which is designed for link prediction tasks, improves the AUC score by at least 2.4% in Cora, 0.6% in Citeseer and 28% in WebKB-sim. This phenomenon shows that the proposed model MDAL has good task adaptability. Compared to other baselines, MDAL-L still outperforms these baselines to all datasets, because the loss function of MDAL-L contains the loss items that directly optimize the AUC index. It is important to note that for WebKB-sim datasets, all baselines do not achieve good results on account of some links are manually added to this network, which makes the dataset full of noise. However, MDAL-L still achieves very satisfactory results on this dataset by directly optimizing the AUC indicator.

6 Conclusion

In this paper, we proposed a semi-supervised network embedding model MDAL, which is based on Tree-LSTM structure. For target node, MDAL model can recursively obtain information from its surrounding neighbors. In order to better obtain the effective information contained in surrounding neighborhood, we use dual attention, i.e. content-attention and label-attention to choose neighbor nodes more related to target node. For the purpose of making embedded vector more suitable for downstream task, we design corresponding objective functions for node classification and link prediction, and optimize the network representation learning task and specific downstream tasks jointly under the multi-task learning framework. A large number of experimental results show the validity of MDAL model on node classification, link prediction and network visualization tasks. Future research work can introduce MDAL model into the recommendation system.

Acknowledgement. The work was supported by the National Key R&D Program of China under grant 2018YFB1004700, and National Natural Science Foundation of China (61772122, 61872074, 61602103, U1435216).

References

1. Bhuiyan, M., Hasan, M.A.: Representing graphs as bag of vertices and partitions for graph classification. Data Sci. Eng. **3**(2), 150–165 (2018)
2. Cai, H., Zheng, V.W., Chang, K.C.: A comprehensive survey of graph embedding: problems, techniques, and applications. IEEE Trans. Knowl. Data Eng. **30**(9), 1616–1637 (2018)
3. Chen, J., Zhang, Q., Huang, X.: Incorporate group information to enhance network embedding. CIKM **2016**, 1901–1904 (2016)
4. Chen, W., Mao, X., Li, X., Zhang, Y., Li, X.: PNE: label embedding enhanced network embedding. In: Kim, J., Shim, K., Cao, L., Lee, J.-G., Lin, X., Moon, Y.-S. (eds.) PAKDD 2017. LNCS (LNAI), vol. 10234, pp. 547–560. Springer, Cham (2017). https://doi.org/10.1007/978-3-319-57454-7_43
5. Chen, Y., Qian, T., Zhong, M., Li, X.: Exploit label embeddings for enhancing network classification. In: Benslimane, D., Damiani, E., Grosky, W.I., Hameurlain, A., Sheth, A., Wagner, R.R. (eds.) DEXA 2017. LNCS, vol. 10439, pp. 450–458. Springer, Cham (2017). https://doi.org/10.1007/978-3-319-64471-4_36
6. Der Maaten, L.V., Hinton, G.E.: Visualizing data using t-SNE. J. Mach. Learn. Res. **9**, 2579–2605 (2008)
7. Du, L., Lu, Z., Wang, Y., Song, G., Wang, Y., Chen, W.: Galaxy network embedding: a hierarchical community structure preserving approach. In: IJCAI, pp. 2079–2085 (2018)
8. Grover, A., Leskovec, J.: node2vec: scalable feature learning for networks. In: KDD, pp. 855–864 (2016)
9. Hamilton, W.L., Ying, Z., Leskovec, J.: Inductive representation learning on large graphs. In: NIPS, pp. 1025–1035 (2017)
10. Hu, R., Aggarwal, C.C., Ma, S., Huai, J.: An embedding approach to anomaly detection. In: ICDE, pp. 385–396 (2016)

11. Huang, X., Li, J., Hu, X.: Label informed attributed network embedding. In: WSDM, pp. 731–739 (2017)
12. Kim, S.M., Xu, Q., Qu, L., Wan, S., Paris, C.: Demographic inference on twitter using recursive neural networks. In: ACL, pp. 471–477 (2017)
13. Kipf, T.N., Welling, M.: Semi-supervised classification with graph convolutional networks. arXiv preprint arXiv:1609.02907 (2016)
14. Korbak, T., Żak, P.: Fine-tuning Tree-LSTM for phrase-level sentiment classification on a polish dependency treebank. arXiv preprint arXiv:1711.01985 (2017)
15. Li, J., Zhu, J., Zhang, B.: Discriminative deep random walk for network classification. In: ACL, pp. 1004–1013 (2016)
16. Lu, Q., Getoor, L.: Link-based classification. In: ICML, pp. 496–503 (2003)
17. Pan, S., Wu, J., Zhu, X., Zhang, C., Wang, Y.: Tri-party deep network representation. In: IJCAI, pp. 1895–1901 (2016)
18. Perozzi, B., Al-Rfou, R., Skiena, S.: Deepwalk: online learning of social representations. In: The ACM SIGKDD International Conference, pp. 701–710. ACM (2014)
19. Socher, R., Lin, C.C., Manning, C.D., Ng, A.Y.: Parsing natural scenes and natural language with recursive neural networks. In: ICML, pp. 129–136 (2011)
20. Tai, K.S., Socher, R., Manning, C.D.: Improved semantic representations from tree-structured long short-term memory networks. In: ACL, pp. 1556–1566 (2015)
21. Tang, J., Qu, M., Wang, M., Zhang, M., Yan, J., Mei, Q.: Line: large-scale information network embedding. In: WWW, pp. 1067–1077 (2015)
22. Tu, C., Zhang, W., Liu, Z., Sun, M.: Max-margin deepwalk: discriminative learning of network representation. IJCAI **2016**, 3889–3895 (2016)
23. Wang, J., Shen, J., Li, P., Xu, H.: Online matrix completion for signed link prediction. In: WSDM, pp. 475–484 (2017)
24. Wang, S., Tang, J., Aggarwal, C.C., Liu, H.: Linked document embedding for classification. In: CIKM, pp. 115–124 (2016)
25. Wang, Z., Chen, C., Li, W.: Predictive network representation learning for link prediction. In: SIGIR, pp. 969–972 (2017)
26. Xie, Z., Li, M.: Semi-supervised AUC optimization without guessing labels of unlabeled data. In: AAAI, pp. 4310–4317 (2018)
27. Xu, Q., Wang, Q., Xu, C., Qu, L.: Attentive graph-based recursive neural network for collective vertex classification. In: CIKM, pp. 2403–2406 (2017)
28. Yang, C., Liu, Z., Zhao, D., Sun, M., Chang, E.Y.: Network representation learning with rich text information. In: IJCAI, pp. 2111–2117 (2015)
29. Ye, Q., Zhu, C., Li, G., Liu, Z., Wang, F.: Using node identifiers and community prior for graph-based classification. Data Sci. Eng. **3**(1), 68–83 (2018)
30. Zhang, D., Yin, J., Zhu, X., et al.: Network representation learning: a survey. arXiv preprint arXiv:1801.05852 (2018)
31. Zhang, D., Yin, J., Zhu, X., Zhang, C.: Homophily, structure, and content augmented network representation learning. In: ICDM, pp. 609–618 (2016)
32. Zhang, X., Chen, W., Yan, H.: TLINE: scalable transductive network embedding. In: Ma, S., et al. (eds.) AIRS 2016. LNCS, vol. 9994, pp. 98–110. Springer, Cham (2016). https://doi.org/10.1007/978-3-319-48051-0_8

Net2Text: An Edge Labelling Language Model for Personalized Review Generation

Shaofeng Xu[1,2], Yun Xiong[1,2(✉)], Xiangnan Kong[3], and Yangyong Zhu[1,2]

[1] Shanghai Key Laboratory of Data Science, School of Computer Science, Fudan University, Shanghai, China
{sfxu16,yunx,yyzhu}@fudan.edu.cn

[2] Shanghai Institute for Advanced Communication and Data Science, Fudan University, Shanghai, China

[3] Worcester Polytechnic Institute, Worcester, MA, USA
xkong@wpi.edu

Abstract. Writing an item review for online shopping or sharing the dining experience of a restaurant has become major Internet activities of young people. This kind of review system could not only help users express and exchange experience but also prompt business to improve service quality. Instead of taking time to type in the review, we would like to make the review process more automated. In this work, we study an edge labelling language model for personalized review generation, *e.g.*, the problem of generating text (*e.g.*, a review) on the edges of the network (*e.g.*, online shopping). It is related to both network structure and rich text semantic information. Previously, link prediction models have been applied to recommender system and event prediction. However, they could not migrate to text generation on the edges of networks since most of them are numerical prediction or tag labelling tasks. To bridge the gap between link prediction and natural language generation, in this paper, we propose a model called Net2Text, which can simultaneously learn the structural information in the network and build a language model over text on the edges. The performance of Net2Text is demonstrated in our experiments, showing that our model performs better than other baselines, and is able to produce reasonable reviews between users and items.

Keywords: Link prediction · Language model · Graph mining · Data mining

1 Introduction

Under the wave of mobile Internet development, we witness and engage more and more online review system. For example, before we purchase a product on Amazon we always check reviews from other buyers to determine if buy it or

G. Li et al. (Eds.): DASFAA 2019, LNCS 11446, pp. 484–500, 2019.
https://doi.org/10.1007/978-3-030-18576-3_29

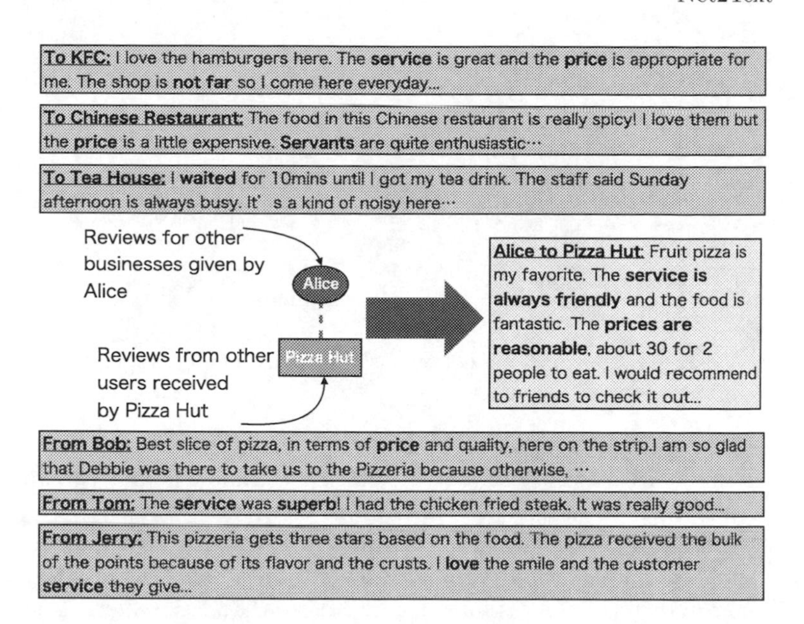

Fig. 1. Our personalized review generation task. Top blocks are reviews that Alice gave to other restaurants, and bottom blocks are reviews that Pizza Hut received from other customers. Our goal is to help Alice generate review automatically for Pizza Hut (see the text in the green box). We provide detail explanation in Sect. 1. (Color figure online)

not. Meanwhile, we write our own using experience for the products purchased as well. In addition to online shopping, we also focus on movie reviews and restaurant reviews. Review system influences various aspects of our life.

Sometimes, writing a review is a trivial work for customers, and most customers share thoughts on fixed aspects (*e.g.*, quality, service). In this work, we would like to help customers generate accurate reviews automatically given historical reviews between users and items. Descriptive text can encode rich semantic information and structural relationship between user-item network.

Figure 1 describes our task of personalized review generation. Top blocks are reviews that Alice gave to other restaurants, and bottom blocks are reviews that Pizza Hut received from other customers. After analyzing the reviews given by Alice, we can conclude that he concerns more about price, position and service (emphasized in red bold). Pizza Hut has received some reviews from other customers and some of them mentioned the attributes that Alice cares about. Since most of reviews to Pizza Hut are positive, our generated review for Alice is friendly and reasonable. Consequently, it will help users to alleviate the burden of typing by generating accurate reviews automatically.

Essentially, it is a kind of edge labelling task for network, which can be recognized as a link prediction problem. In previous researches, link prediction models are mainly applied to recommender system. For example, in social networks, we usually predict the evolution of networks like recommending new friends [1] by

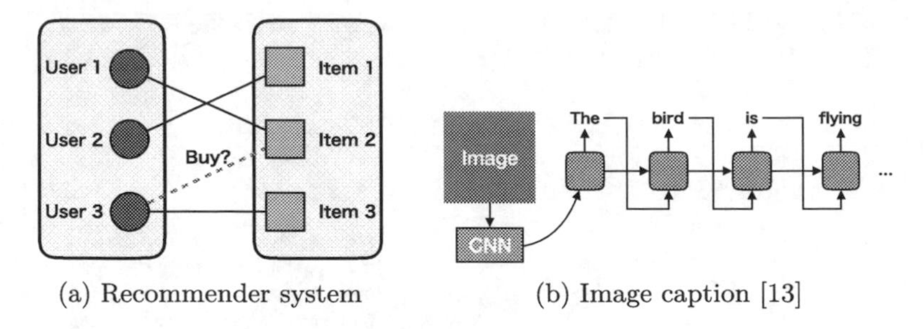

(a) Recommender system (b) Image caption [13]

Fig. 2. Previous models on link prediction task and multimodel language generation task.

calculating similarities based on user activities. In online e-commerce websites, instead of just recommending new links, link prediction models also predict events like buying an item (Fig. 2a). In this area, collaborative filtering (CF) is a successful recommendation paradigm that employs transaction information to enrich user and item features for recommendation [12]. However, these models could not migrate to text generation on the edges since most of them are numerical prediction or tag labelling tasks.

From the aspect of language modeling, although we have a lot of sophisticated language models nowadays, it is not a trivial task to employ it into network structure. Common language generation models, including n-gram models [7] and deep learning models (*e.g.*, LSTM [24]), always generate text without considering additional features of network structures. Text produced by these models cannot reflect the specialization between vertices. Karpathy et al. proposed an image caption model (Fig. 2b) and it is a kind of language model that considers image features [13]. This model cannot be applied in network structure but inspired us to build a multimodel language model over a network structure.

To our best of knowledge, there are no previous models on personalized review generation. Therefore, in this paper, we propose a model, Net2Text, based on network structure with text on the edges. Our model innovatively solves the task of generating personalized text on network edges automatically and makes up the gap in the study of edge labelling language model.

Our method proposed in this paper has following three-fold contributions:

1. It is the first work to propose automated personalized review generation task, which is an extensive work of image caption to network structure. It builds a bridge between network structure and text generation.
2. Our Net2Text is a novel model on edge labelling for personalized comment generation by learning the structural information in the network and language modeling in the text.
3. We demonstrate the performance of our model in experiments on real datasets. By comparing Net2Text to other baselines, we show that our model performs better under machine translation metrics. We also present some generated reviews to prove the readability and personality.

2 Problem Definition

In this section, we first introduce related concepts and notations, then define our problem.

Table 1. Important Notations.

Symbol	Definition
$G = (\mathcal{V}_1, \mathcal{V}_2, \mathcal{E})$	A network with text on edges
$\mathcal{V}_1 = \{1, \cdots, n_1\}, \mathcal{V}_2 = \{1, \cdots, n_2\}$	Two sets of vertices, n_1 and n_2 are vertex number of each set
$\mathcal{E} = \{(l_i, r_i, \mathbf{s}_i)\}_{i=1}^m$	The set of m links in network G
l_i, r_i	Two connected vertices of i-th link in $\mathcal{E}$
$\mathbf{s}_i = (s_i^1, \cdots, s_i^{T_i})$	Sequence of i-th edge text. $\forall t \in [1..T_i]$, s_i^t correspondes to a word of text $\mathbf{s}_i$ at position t. T_i is the number of words on the i-th sequence
$\mathbf{x}_i = (s_i^0, \cdots, s_i^{T_i})$	Text sequence on i-th edge $\mathbf{s}_i$ with an additional pre tag *START*
$\mathcal{X} = \{\mathbf{x}_1, \cdots, \mathbf{x}_m\}$	Input vectors for all instances
$\mathcal{F} = \{\mathbf{f}_1, \cdots, \mathbf{f}_m\}$	Link features for all instances
$\mathbf{f}_i = [\Phi(l_i), \Phi(r_i)] \in \mathbb{R}^{2 \times d}$	i-th link features combined by vector representation of vertex $l_i \in \mathcal{V}_1$ and vertex $r_i \in \mathcal{V}_2$. d is the dimension of each vertex vector
$\mathcal{Y} = \{\mathbf{y}_1, \cdots, \mathbf{y}_m\}$	Output vectors for all instances
$\mathbf{y}_i = (s_i^1, \cdots, s_i^{T_i+1})$	i-th edge text sequence $\mathbf{s}_i$ with an additional post tag *END*
$\Phi \in \mathbb{R}^{(n_1+n_2) \times d}$	Representation vectors for all the vertices in network G
$\mathcal{L}$ and $\mathcal{U}$	The training set and testing set

In our task, we have two types of vertices in network. One is the user (Alice), the other is the item or business (Pizza Hut). Therefore, we denote a network, $G = (\mathcal{V}_1, \mathcal{V}_2, \mathcal{E})$, where $\mathcal{V}_1 = \{1, \cdots, n_1\}$ and $\mathcal{V}_2 = \{1, \cdots, n_2\}$ are two partitions of this network, and $\mathcal{E} = \{(l_i, r_i, \mathbf{s}_i)\}_{i=1}^m$ denotes the edges of the network. m is the total number of edges in G. For the i-th edge, $l_i \in \mathcal{V}_1$ and $r_i \in \mathcal{V}_2$ are two connected vertices. $\mathbf{s}_i = (s_i^1, \cdots, s_i^{T_i})$ denotes the text sequence on i-th edge. $\forall t \in [1..T_i]$, s_i^t correspondes to a word in sequence $\mathbf{s}_i$ at position t. T_i is the number of words of the sequence on the i-th edge.

Now we formally define our problem as follow:

Definition 1. *Given a network $G = (\mathcal{V}_1, \mathcal{V}_2, \mathcal{E})$ with text on its edges, for each edge $(l, r, \mathbf{s}) \in \mathcal{E}$, our goal is to learn an edge labelling language model $g(l, r) \approx \mathbf{s}$,* *that can map all the edges in the network into personalized reviews. For a new edge* (l_{new}, r_{new})*, we could generate a personalized review* $\mathbf{s}_{new} = g(l_{new}, r_{new})$.

Since we take network structure information into consideration in our task, for i-th edge $(l_i, r_i, \mathbf{s_i})$ we begin with learning vector representation $\Phi(l_i) \in \mathbb{R}^d$ of vertex $l_i \in \mathcal{V}_1$ and $\Phi(r_i) \in \mathbb{R}^d$ of vertex $r_i \in \mathcal{V}_2$, and then we denote a vector $\mathbf{f}_i = [\Phi(l_i), \Phi(r_i)] \in \mathbb{R}^{2\times d}$ by concatenating representation of two vertices as features of i-th edge. d is the embedding dimensions of each vertex.

Now we create a new edge set $\mathcal{E}' = (\mathcal{X}, \mathcal{F}, \mathcal{Y})$, where $\mathcal{X} = \{\mathbf{x}_1, \cdots, \mathbf{x}_m\}$ represents input vectors, $\mathcal{F} = \{\mathbf{f}_1, \cdots, \mathbf{f}_m\}$ represents edge features and $\mathcal{Y} = \{\mathbf{y}_1, \cdots, \mathbf{y}_m\}$ represents output vectors.

For the sequence $\mathbf{s}_i$, we add a special tag *START* as s_i^0 to the beginning, which forms our input vector $\mathbf{x}_i = (s_i^0, \cdots, s_i^{T_i})$ for i-th edge. We add another special tag *END* as $s_i^{T_i+1}$ to the end, which forms our output vector $\mathbf{y}_i = (s_i^1, \cdots, s_i^{T_i+1})$.

Our instances in $\mathcal{E}'$ are then divided into a training set $\mathcal{L} \subset \{\mathcal{E}'_1, \cdots, \mathcal{E}'_m\}$ and test set $\mathcal{U} \subset \{\mathcal{E}'_1, \cdots, \mathcal{E}'_m\}$, where $\mathcal{L} \bigcup \mathcal{U} = \mathcal{E}'$ and $\mathcal{L} \bigcap \mathcal{U} = \emptyset$. For $\forall i \in \mathcal{L}$, $\mathbf{x}_i$, $\mathbf{f}_i$ and $\mathbf{y}_i$ are fully observed, while $\forall i \in \mathcal{U}$, we only observe x_i^0, which is tag *START*, and feature $\mathbf{f}_i$ for i-th edge. Since we use previous output as new input in text generation process, we denote $\mathbf{y}_i^o$ as previously generated word sequence. Here, $\mathbf{y}_i^o = (y_i^0, \cdots, y_i^{t-1})$ changes as the step goes to t-th position.

Our task on the i-th instance is to predict $\mathbf{y}_i = (y_i^j)_{j=1}^{T_i}$ based on $\mathbf{y}_i^o$ and $\mathbf{f}_i$ until tag *END* appears or designated text length arrives.

We use $\mathcal{Y}_\mathcal{U} = \{\mathbf{y}_i | \forall i \in \mathcal{U}\}$ to denote all the target text on unconnected vertices for prediction. In addition, we define $\mathcal{Y}_\mathcal{U}^o = \{\mathbf{y}_i^o | \forall i \in \mathcal{U}\}$ to denote all the generated outputs during the test phase. $\mathcal{F}_\mathcal{U} = \{\mathbf{f}_i | \forall i \in \mathcal{U}\}$ is the collection of edge features in test set $\mathcal{U}$. The overall goal of Net2Text model is to estimate a probability distribution:

$$\Pr(\mathcal{Y}_\mathcal{U} | \mathcal{F}_\mathcal{U}, \mathcal{Y}_\mathcal{U}^o) \propto \prod_{i \in \mathcal{U}} \Pr(\mathbf{y}_i | \mathbf{y}_i^o, \mathbf{f}_i) \tag{1}$$

Table 1 explains details of our important notations.

3 Proposed Method

3.1 Overview

In this section we will explain details of our method.

In general, our model integrates the process of vertex embedding and text generation. We first learn representations of vertices with random-walk-based method, then train a language generation model over existing text on the edges by incorporating vertical features. Figure 3 describes the architecture of our model Net2Text.

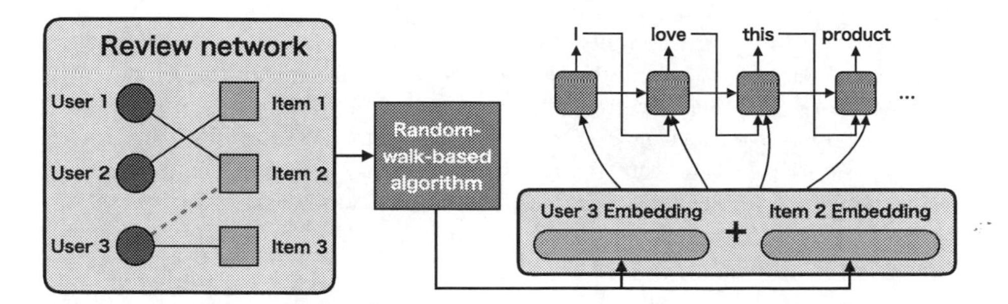

Fig. 3. Architecture of Net2Text. We first learn vertex representations from review network. For unconnected user 3 and item 2, we combine their embeddings into an edge feature, together with each word in the review sequence, as the input to the nGRU.

Algorithm 1. Learning vector representations.

Input: vertices set $\mathcal{V}$, edge set $\mathcal{E}$, windows size w
embedding size d, number of walks per vertex b
Output: matrix of vector representations $\Phi \in \mathbb{R}^{N \times d}$
1 Initialize Φ
2 **for** $i = 1$ **to** b **do**
3 Reorder the vertices set $\mathcal{V}$
4 **foreach** $v_i \in \mathcal{V}$ **do**
5 $\mathcal{W}_{v_i} = RandomWalk(\mathcal{V}, \mathcal{E}, v_i)$
6 $SkipGram(\Phi, \mathcal{W}_{v_i}, w)$
7 **end**
8 **end**

3.2 Learning Representation of Vertices

To learn latent representations of vertices in a network, we use local information obtained from truncated random walks by treating walks as the equivalent of sentences [20]. We merge vertices from $\mathcal{V}_1$ and $\mathcal{V}_2$ into $\mathcal{V} = \{1, \cdots, N\}$ where $N = n_1 + n_2$, then we define $\Phi \in \mathbb{R}^{N \times d}$ as vector representations of all the vertices.

We initialize the mapping function Φ by uniformly sampling a random walk. We choose a random vertex v_i as the root of the walk, then sample uniformly from the neighbors of previous vertices visited. Therefore, the objective function of this optimization problem is:

$$\min_{\Phi} -log \Pr(\{v_{i-w}, \cdots, v_{i-1}, v_{i+1}, \cdots, v_{i+w}\}|\Phi(v_i)) \tag{2}$$

where w is the context windows size for each vertex in the walk.

In each iteration, we start a new random walk at each vertex and update the objective function Φ.

For each vertex v_i in the inner loop, we generate a random walk $\mathcal{W}_{v_i}$, which starts from vertex in one side to the vertex in the other side back and forth in the network. We then use the generated walk to update our representations Φ and our objective function Eq. 3.2 with SkipGram [17] loop. For each vertex $v_j \in \mathcal{W}_{v_i}$ we map it to current representation vector $\Phi(v_j) \in \mathbb{R}^d$, and maximize the probability of its neighbors $\{u_k\}$ with context window size w in the walk.

3.3 Text Generation on Edges

Next step after learning vertex representations is to incorporate them into our edge labelling language model.

Formally, for each edge $(l_i, r_i, \mathbf{s}_i)$ in the network G, we first construct edge features $\mathbf{f}_i = [\Phi(l_i), \Phi(r_i)]$ by combining vector representations of vertex l_i and vertex r_i, then convert text $\mathbf{s}_i$ on the edge into a sequence of words $\mathbf{x}_i = (x_i^0, \cdots, x_i^{T_i})$.

Recurrent Neural Network (RNN) [16] is one of frequently used deep neural language model. It defines a probability distribution of the next word given the current word and previous context sequence generated. Simple RNN unit in vanilla version could not capture long dependency due to the gradient explosion/vanishing problem, therefore, we replace it with Gated Recurrent Unit [8]. Since current GRU model just tries to generate text that has correct spelling and grammar while ignoring our network structural information, we define a new unit called *nGRU*. nGRU makes an improvement to the original GRU, that it can condition text generation on vector representations Φ of all the vertices learned from previous process to build our edge labelling language model for our personalized review generation task.

The t-th nGRU unit has two gates: an update gate $\mathbf{z}_t$ and a reset gate $\mathbf{r}_t$, and two states: a candidate state $\tilde{\mathbf{h}}_t$ and a hidden state $\mathbf{h}_t$. For each step t, we compute input vector $\mathbf{u}_t$ which is a linear transformation of edge features f_i^t and current t-th word x_i^t, and outputs the next word $\mathbf{o}_t$. Update rules for our recurrent units are below:

$$\mathbf{u}_t = W_{uf} f_i^t + W_{ux} embed(x_i^t) + \mathbf{b}_u \quad (3)$$

$$\mathbf{z}_t = \sigma(W_{zu}\mathbf{u}_t + W_{zh}\mathbf{h}_{t-1}) \quad (4)$$

$$\mathbf{r}_t = \sigma(W_{ru}\mathbf{u}_t + W_{rh}\mathbf{h}_{t-1}) \quad (5)$$

$$\tilde{\mathbf{h}}_t = \tanh(W_{hu}\mathbf{u}_t + W_{hh}(\mathbf{r}_t \odot \mathbf{h}_{t-1})) \quad (6)$$

$$\mathbf{h}_t = (1 - \mathbf{z}_t) \odot \mathbf{h}_{t-1} + \mathbf{z}_t \odot \tilde{\mathbf{h}}_t \quad (7)$$

$$\mathbf{o}_t = softmax(W_{oh}\mathbf{h}_t + \mathbf{b}_o) \quad (8)$$

Here, function *embed*() turns a word into a low dimension embedding vector. It can be assigned with fixed global word vector or trained by the model itself. $\mathbf{W}^{(\cdot)}$ are weight matrix and $\mathbf{b}^{(\cdot)}$ are bias vectors. σ is the logistic function and the operator $\odot$ means element-wise product between two vectors. We initialize $\mathbf{h}_0$ as a vector of all zeros.

Until now, we build the basic architecture of our model as shown in Fig. 3, in next part we will introduce the procedure of our algorithm.

Algorithm 2. Net2Text Algorithm

Input: Network $G = (\mathcal{V}, \mathcal{E})$, training set $\mathcal{L}$, test set $\mathcal{U}$
windows size w, embedding size d, walks per vertex b, max sentence length p_{max}
Output: $\mathcal{Y}_{\mathcal{U}} = \{\mathbf{y}_i | \forall i \in \mathcal{U}\}$, each $\mathbf{y}_i$ is a generated link text in test set $\mathcal{U}$.
Data Preparing:
- Generate vector representations:
Φ = *Learning vector representations*$(\mathcal{V}, \mathcal{E}, w, d, b)$
- Create new link dataset: $\mathcal{E}' = (\mathcal{X}, \mathcal{F}, \mathcal{Y})$
Training Phase:
- Construct extended dataset $\mathcal{X}' = (\mathbf{x}'_1, \cdots, \mathbf{x}'_m)$:
for $i = 1$ **to** m **do**
Extend i-th link $\mathbf{x}'_i = ([x_i^0, f_i], \cdots, [x_i^T, f_i])$ by concatenating link features to each timestep over the whole text.
end
- Let g = Net2Text$(\mathcal{X}')$ be our model trained on $\mathcal{X}'$
Test:
foreach $i \in \mathcal{U}$ **do**
- Use START as our first token: y_i^0=(START)
- Initialize our test parameters: $len = 1$, $\mathbf{x}_i = ([y_i^0, f_i])$
repeat
- Predict next character with our trained model: token $= g(\mathbf{x}_i)$
- Append newly generated character to existed text: $\mathbf{y}_i$.append(token)
- Update input vector with the new token: $\mathbf{x}_i$.append($[token, f_i]$)
- Move to next character: $len = len + 1$
until $token = END$ *or* $len > p_{max}$
end

3.4 Net2Text Algorithm

We conclude a procedure of our method as shown in Algorithm 2. According to Sect. 2, the input to our algorithm will be a network $G = (\mathcal{V}, \mathcal{E})$ with random splitted training set $\mathcal{L}$ and test set $\mathcal{U}$. The algorithm will output generated text that belongs to edges from test set $\mathcal{U}$.

The algorithm has following steps:

1. **Data Preparing.** First, we use random-walk-based algorithm to extract the representations of all the vertices Φ, and generate a new edge set $\mathcal{E}' = (\mathcal{X}, \mathcal{F}, \mathcal{Y})$ through the problem definition mentioned in Sect. 2.
2. **Model Training.** We extend text vectors $\mathcal{X}$ to a new combined vectors $\mathcal{X}'$ by concatenating edge features to each word over the whole sequence. With this operation, we can take local network features into consideration while modeling languages. Finally we train a language model g on dataset $\mathcal{X}'$.

Table 2. Experiment datasets.

Dataset	Average words/review	Number of reviews
Amazon - clothing & jewelry	57	263890
Amazon - health care	72	301284
Amazon - sports	67	263212
Amazon - video games	81	151018
Yelp	98	517729

3. **Text Prediction.** In the testing phase, we first set tag *START* as the initial word to our model. For each generation step, we update input vectors as in the training phase, and repeat predicting new word until the tag *END* appears or maximum length of text reaches.

4 Experiments

In order to validate the performance of our model, we applied Net2Text to several real world review datasets in our experiments.

4.1 DataSets

- Amazon[1] is the world leading e-ecommerce platform, and customers often write down their reviews for their purchases. Amazon datasets [11] we used in this experiment contains product reviews, including 142.8 million reviews spanning May 1996 - July 2014. We choose 4 categories of items, and each of which has been reduced to extract the 5-core, such that each of the remaining users and items has 5 reviews each.
- Yelp[2] is one of the biggest online restaurant reservation service provider. It has a huge amount of data, including customer profile, resturant information, customer review and so on. They open-sourced their dataset on the official website[3]. We reduced the dataset to the 30-core, such that each of the remaining customers and businesses has 30 reviews each.

All the datasets are splitted into training sets and test sets. Statistical details on our datasets are presented in Table 2.

4.2 Experiment Setup

We have some basic configurations in our experiments. The vector representation dimension for user and item vertices is set to 256. For language modelling part, we use nGRU units with 512 dimension hidden nodes. We choose cross-entropy [23]

[1] https://www.amazon.com/.
[2] https://www.yelp.com/.
[3] https://www.yelp.com/dataset/challenge.

Table 3. Algorithms for comparison.

Method	Content	Network	Publication
wordRNN	✓	⊘	[25]
Random-walk-based	⊘	✓	[10,20]
Net2Text	✓	✓	This paper

Table 4. Experiments results under different datasets and algorithms.

Dataset	BLEU		
	wordRNN	Random-walk-based	Net2Text
Amazon - clothing & jewelry	0.192	0.173	**0.345**
Amazon - health care	0.185	0.165	**0.332**
Amazon - sports	0.181	0.168	**0.357**
Amazon - video games	0.184	0.153	**0.323**
Yelp	0.19	0.17	**0.382**

as our loss function. In Sect. 4.4 we compare performance of different dimensions for nGRU units and vector representation. For comparison algorithms we have the same configurations as ours. All the experiments were conducted under a Linux workstation with a Nvidia GTX 1080 GPU (8 G graphic memory).

4.3 Algorithms for Comparison

To our best of knowledge, there are no comparable models since we are the first to propose text generation on network edges. In addition to various versions of Net2Text, we evaluated the performance against the following baseline algorithms (Summarized in Table 3):

- wordRNN [25]: This method does not consider the local network structure between user and item but only concentrates on review content. It only builds word level generation language model for all the reviews and generates review based on the first few words. We use the same language model structure as Net2Text but remove edge features.
- Random-walk-based algorithm [10,20]: We compare to a baseline with the idea of edge similarity. It chooses a review from the training set that expresses the most similar edge features. In other words, it only considers network structure information and neglects text content.
- Net2Text: This is our proposed method mentioned in Sect. 3, which generates personalized review for users conditioned on both historical review content and network relationships between users and items.

Table 5. Experiments results under different embedding methods.

Dataset	BLEU	
	Deepwalk	Node2vec
Amazon - clothing & jewelry	0.202	**0.345**
Amazon - health care	0.211	**0.332**
Amazon - sports	0.302	**0.357**
Amazon - video games	0.296	**0.323**
Yelp	0.247	**0.382**

Table 6. Experiments results under different vertex embedding dimension.

Dataset	BLEU			
	Dim. 64	Dim. 128	Dim. 256	Dim. 512
Amazon - clothing & jewelry	0.205	0.168	0.229	**0.345**
Amazon - health care	0.194	0.221	0.214	**0.332**
Amazon - sports	0.252	0.212	0.241	**0.357**
Amazon - video games	0.187	0.199	0.291	**0.323**
Yelp	0.239	0.253	0.343	**0.382**

4.4 Evaluation

BLEU (Bilingual Evaluation Understudy) score [18] is a commonly used metric in the area of language generation. We apply it to measure the quality of our generated reviews. Since the length of our generated reviews are longer than that of image captions and we have only one reference review per case for evluation, the BLEU score of our experiments will be lower than other language generation tasks (*e.g.* image caption).

We compared our model Net2Text with the other two baseline algorithms. As the results showed in Table 4, our method Net2Text considers not only review semantic information, but also local network structure features, and has significant improvement than baseline algorithms. As a contrast, the baseline random-walk-based algorithm has the worst performance under BLEU score. One explanation is that random-walk-based algorithm only selects a similar existing review without concerning its content, it may be totally different from what the review really is, which leads to a low BLEU score. Instead, wordRNN tries to generate reviews based on first few words from real reviews, and it helps establish a similar context (quality, service) and predicts in the right direction. Therefore, wordRNN performs better than random-walk-based algorithm under BLEU score.

We will show some cases in Sect. 4.6.

Table 7. Experiments results under different nGRU dimension.

Dataset	BLEU		
	Dim. 128	Dim. 256	Dim. 512
Amazon - clothing & jewelry	0.225	0.243	**0.345**
Amazon - health care	0.282	0.304	**0.332**
Amazon - sports	0.221	0.289	**0.357**
Amazon - video games	0.224	0.25	**0.323**
Yelp	0.282	0.319	**0.382**

4.5 Model Selection

In this section, we will conduct some analysis on model parameters. We mainly consider three aspects of parameter influence. First two are embedding method and embedding dimension for the network. Another one is hidden dimension of nGRU units. All of them have effects on performance according to our experiments.

Influence of Embedding Method. Our model use random-walk-based algorithms as our vertex embedding method, therefore, we compare experiments performance between Deepwalk [20] and node2vec [10]. Table 5 tells us node2vec leads to a better performance. Since node2vec defines a flexible notion of a node's network neighborhood and design a biased random walk procedure, which efficiently explores diverse neighborhoods, and this kind of flexibility in exploring neighborhoods is the key to learning richer representations [10]. We choose the flexible parameters that applicable in our experiments.

Influence of Vertex Embedding Dimensions. In accordance with [20], network embedding dimensional representations are distributed, meaning each user or item is expressed by a subset of the dimensions and each dimension contributes to a subset of social concepts expressed by the space. We used the default experiment setup, and changed the variable embedding size d in vector representations learning phase. Table 6 shows that the dimension 512 performs the best, which means larger dimension could express complex social concepts and that could help make personalized review more accurate.

Influence of nGRU Parameters. In our experiment we chose nGRU as the base language modeling method. To improve the performance of Net2Text, we adjust hidden dimension of nGRU units. Table 7 shows that 512 hidden dimension has significant improvement than other small dimension.

Table 8. Comparison of generated reviews with ground-truth reviews.

Case 1: Amazon - clothing & jewelry	
Source	Crocs are a really comfortable slippers. Looks great and you can make your activities with comfort and good looking
Generated	I really love them! They are perfect and very comfortable. They look good and I will recommend this item
Case 2: Amazon - health care	
Source	I take vitamins three times a day. This product fits the bill perfectly when I travel. I like it so much! I bought two of them. I first went to a chain pharmacy and discount store but their offerings were too expensive and were too small. This product was far cheaper and works for me
Generated	I love these vitamins. Easy small pills to take and a good quality product. Great price on Amazon! Will order again
Case 3: Amazon - sports	
Source	It is very easy to use good equipment. It came with in a short time. Really happy to have it
Generated	It's great and easy and it is not too big to install. It will work. I have a good purchase
Case 4: Amazon - video games	
Source	This game is awesome! Beautiful landscapes and adventures! It puts you right in the fantasy world. Very addictive!
Generated	Great graphics! I would love playing the series and it was very well! Great price and I recommend this
Case 5: Yelp	
Source	Always great sushi no matter if you are doing AYCE or individual orders. Always a wait for this place
Generated	I enjoy the panko a lot. I always love family style ramen but always looking for a spot

4.6 Case Study

Our goal is that the predicted reviews can reflect suggestions and feelings existed in the real reviews as much as possible, since it could alleviate the burden of user's typing modification.

To gain a better understanding of our method, we explain more details on some example reviews we generated. Table 8 lists cases of our results selected from each of five datasets compared to original reviews.

Case 1 is a perfect generated review example. The customer mentions good feedback about the looking and the feeling of purchased product slippers (*a really comfortable slippers* and *Looks great*). In our generated review, we predict precisely the same keywords (*very comfortable* and *They look good*) as the original one.

In case 2, the customer originally gave a long review to the item. Although long sentence generation is more challenging, we still have the ability to capture the important information for this purchase relationship. For example, we produce the name of this product (*vitamins*), and mention the product quality (*works for me* and *good quality product*) and the price (*far cheaper* and *Great price*) as well.

Case 3 is the case from sports shopping in Amazon. We generate the word *easy* as the same from source review. The customer also wrote the joy feeling to have this item, and we generate it in a different expression (*happy to have it* v.s. *good purchase*).

Review in cases 4 looks like a general comment to video games. We emphasize graphical performance that summarize description in source review.

After comparing the generated review with the original review in case 5, we find that this customer went to a Japanese food resturant and he/she praised the food (*great sushi*, *enjoy the panko a lot* and *love family style ramen*). This customer also complained about the waiting time (*Always a wait* and *always looking for a spot*).

According to above cases, our model could not only capture real facts about items/restaurants but also express original meanings for users. This work will greatly help users alleviate the trouble of typing text by filling the review content automatically.

5 Related Work

As we mentioned in Sect. 1, our task is a variant to common link prediction problems.

Most link prediction models study the task of recommender system. For example, in social networks (*e.g.* Facebook, Twitter, Instagram, *etc.*) we can recommend new friends based on current relationships [1]. In some domain specific social networks like academic social network, link prediction can help find domain experts or co-authors [19]. Large online shopping platforms could form a behaviour network based on user behaviours as well, and link prediction models can recommend personalized items for individual customers [14]. In the advertisement recommendation area, Wang et al. propose a framework SHINE to predict possibly existing sentiment links in the presence of heterogeneous information, which could generate sentiment tags between users and advertisers [26]. Link prediction also predicts the missing links on an incomplete observed networks [15]. For dynamic networks such as P2P lending networks, it can infer the future newly added links and evolution process [27,30,31] as well. In other domains like bioinformatics area, there exists networks such as gene expression networks. We can use link prediction to predict new protein-protein interactions [2], which has great significance to human life health and disease treatment. All the problems mentioned above are numerical prediction or tag labelling tasks on links, and they cannot be applied to our problem.

Statistical language modeling (SLM) is an important research topic all the time in natural language processing field. It is commonly used in speech recognition [9], machine translation [4], handwriting recognition [29] and other applications. N-gram model [7] was the earliest technique for language modeling. It models the probability of a word appears next given a sequence of context words. Later, models based on decision tree [21] and maximum entropy techniques [6] appeared. They build models with consideration of other features such as part of speech tags. Bengio et al. [5] first applied neural networks to language model domain in 2003. He proposed a Feed-forward Neural Network (FNN) which can predict the next word given the fixed size of the previous sequence of words and learn the word representation in the vocabulary at the same time. This model greatly improves the performance of speech recognition [22]. However, previous models still have the shortcoming that long range dependencies cannot be captured due to the fixed context constrain. Mikolov et al. proposed a Recurrent Neural Network (RNN) which allows context information passing all through [16]. To address the issue of gradient vanishing, a variant of RNN called Long-Short Term Memory (LSTM) [24] is presented later. In recent years, generating natural language text for image (Image Caption) has been researched in depth. Karpathy et al. proposed an image caption model and it builds a multimodel RNN that considers image features [13]. Anderson et al. propose a combined bottom-up and top-down attention mechanism that can enable deeper image understanding for image caption and visual question answering [3]. However, image caption models cannot be migrated to network because they have different structures. Yao et al. also study the problem of generating fake reviews for specific restaurant automatically [28]. Their reviews cannot reflect the personalization because they do not concern about the network structure even the customers.

6 Conclusion

In this paper, we propose a novel application of language modelling on network structure, which is generating text on each edge of a network. It is a challenging work because text on edges encodes not only the relationship between two network entities but also rich semantic information. We build an edge labelling language model Net2Text that considers both network structure information and text on edges, to generate personalized reviews for users. The performance of Net2Text is demonstrated in experiments on real world datasets, showing that our model performs better than other baselines, and is able to generate reasonable reviews for customers.

Acknowledgment. This work is supported in part by the National Natural Science Foundation of China Projects No. U1636207, No. 91546105, the Shanghai Science and Technology Development Fund No. 16JC1400801, No. 17511105502, No. 17511101702.

References

1. Aiello, L.M., Barrat, A., Schifanella, R., Cattuto, C., Markines, B., Menczer, F.: Friendship prediction and homophily in social media. ACM Trans. Web **6**(2), 9 (2012)
2. Almansoori, W., et al.: Link prediction and classification in social networks and its application in healthcare and systems biology. Netw. Model. Anal. Health Inf. Bioinf. **1**(1–2), 27–36 (2012)
3. Anderson, P., et al.: Bottom-up and top-down attention for image captioning and visual question answering. In: CVPR 2018, p. 6 (2018)
4. Bahdanau, D., Cho, K., Bengio, Y.: Neural machine translation by jointly learning to align and translate. In: ICLR (2015)
5. Bengio, Y., Ducharme, R., Vincent, P., Jauvin, C.: A neural probabilistic language model. J. Mach. Learn. Res. **3**(Feb), 1137–1155 (2003)
6. Berger, A.L., Pietra, V.J.D., Pietra, S.A.D.: A maximum entropy approach to natural language processing. Comput. Linguist. **22**(1), 39–71 (1996)
7. Brown, P.F., Desouza, P.V., Mercer, R.L., Pietra, V.J.D., Lai, J.C.: Class-based n-gram models of natural language. Comput. Linguist. **18**(4), 467–479 (1992)
8. Chung, J., Gulcehre, C., Cho, K., Bengio, Y.: Empirical evaluation of gated recurrent neural networks on sequence modeling. arXiv preprint arXiv:1412.3555 (2014)
9. Graves, A., Mohamed, A.r., Hinton, G.: Speech recognition with deep recurrent neural networks. In: ICASSP 2013, pp. 6645–6649. IEEE (2013)
10. Grover, A., Leskovec, J.: node2vec: Scalable feature learning for networks. In: KDD 2016, pp. 855–864. ACM (2016)
11. He, R., McAuley, J.: Ups and downs: modeling the visual evolution of fashion trends with one-class collaborative filtering. In: WWW 2016, pp. 507–517. International World Wide Web Conferences Steering Committee (2016)
12. Herlocker, J.L., Konstan, J.A., Terveen, L.G., Riedl, J.T.: Evaluating collaborative filtering recommender systems. ACM Trans. Inf. Syst. **22**(1), 5–53 (2004)
13. Karpathy, A., Fei-Fei, L.: Deep visual-semantic alignments for generating image descriptions. In: CVPR 2015, pp. 3128–3137 (2015)
14. Li, X., Chen, H.: Recommendation as link prediction in bipartite graphs: a graph kernel-based machine learning approach. Decis. Support Syst. **54**(2), 880–890 (2013)
15. Marchette, D.J., Priebe, C.E.: Predicting unobserved links in incompletely observed networks. Comput. Stat. Data Anal. **52**(3), 1373–1386 (2008)
16. Mikolov, T., Karafiát, M., Burget, L., Cernockỳ, J., Khudanpur, S.: Recurrent neural network based language model. In: InterSpeech 2010, vol. 2, p. 3 (2010)
17. Mikolov, T., Sutskever, I., Chen, K., Corrado, G.S., Dean, J.: Distributed representations of words and phrases and their compositionality. In: NIPS 2013, pp. 3111–3119 (2013)
18. Papineni, K., Roukos, S., Ward, T., Zhu, W.J.: Bleu: a method for automatic evaluation of machine translation. In: ACL 2002, pp. 311–318. Association for Computational Linguistics (2002)
19. Pavlov, M., Ichise, R.: Finding experts by link prediction in co-authorship networks. In: FEWS 2007, pp. 42–55 (2007)
20. Perozzi, B., Al-Rfou, R., Skiena, S.: DeepWalk: online learning of social representations. In: KDD 2014, pp. 701–710. ACM (2014)
21. Potamianos, G., Jelinek, F.: A study of n-gram and decision tree letter language modeling methods. Speech Commun. **24**(3), 171–192 (1998)

22. Schwenk, H., Gauvain, J.L.: Training neural network language models on very large corpora. In: HLT/EMNLP 2005, pp. 201–208. Association for Computational Linguistics (2005)
23. Shore, J., Johnson, R.: Axiomatic derivation of the principle of maximum entropy and the principle of minimum cross-entropy. TIT **26**(1), 26–37 (1980)
24. Sundermeyer, M., Schlüter, R., Ney, H.: LSTM neural networks for language modeling. In: InterSpeech (2012)
25. Sutskever, I., Martens, J., Hinton, G.E.: Generating text with recurrent neural networks. In: ICML 2011, pp. 1017–1024 (2011)
26. Wang, H., Zhang, F., Hou, M., Xie, X., Guo, M., Liu, Q.: SHINE: signed heterogeneous information network embedding for sentiment link prediction. In: WSDM 2018, pp. 592–600. ACM (2018)
27. Xiong, Y., Zhang, Y., Kong, X., Zhu, Y.: NetCycle+: a framework for collective evolution inference in dynamic heterogeneous networks. IEEE Trans. Knowl. Data Eng. **30**(8), 1547–1560 (2018)
28. Yao, Y., Viswanath, B., Cryan, J., Zheng, H., Zhao, B.Y.: Automated crowdturfing attacks and defenses in online review systems. In: CCS 2017, pp. 1143–1158. ACM (2017)
29. Zamora-Martinez, F., Frinken, V., España-Boquera, S., Castro-Bleda, M.J., Fischer, A., Bunke, H.: Neural network language models for off-line handwriting recognition. Pattern Recogn. **47**(4), 1642–1652 (2014)
30. Zhang, Y., Xiong, Y., Kong, X., Li, S., Mi, J., Zhu, Y.: Deep collective classification in heterogeneous information networks. In: WWW 2018, pp. 399–408. International World Wide Web Conferences Steering Committee (2018)
31. Zhang, Y., Xiong, Y., Kong, X., Zhu, Y.: NetCycle: collective evolution inference in heterogeneous information networks. In: KDD 2016, pp. 1365–1374 (2016)

Understanding Information Diffusion via Heterogeneous Information Network Embeddings

Yuan Su[1], Xi Zhang[1(✉)], Senzhang Wang[2], Binxing Fang[1,3], Tianle Zhang[4], and Philip S. Yu[5,6]

[1] Key Laboratory of Trustworthy Distributed Computing and Service, Ministry of Education, Beijing University of Posts and Telecommunications, Beijing, China
{timsu,zhangx,fangbx}@bupt.edu.cn
[2] The Hong Kong Polytechnic University, Kowloon, Hong Kong
csszwang@comp.polyu.edu.hk
[3] Institute of Electronic and Information Engineering of UESTC in Guangdong, Dongguan, China
[4] Guangzhou University, Guangzhou, China
tlezhang@gzhu.edu.cn
[5] University of Illinois at Chicago, Chicago, USA
psyu@cs.uic.edu
[6] Institute for Data Science, Tsinghua University, Beijing, China

Abstract. Predicting information diffusion in social networks has attracted substantial research efforts. For a specific user in a social network, whether to forward a contagion is impacted by complex interactions from both her neighboring users and the recent contagions she has been involved in, which is difficult to be modeled in a unified model. To address this problem, we investigate the contagion adoption behavior under a set of interactions among users and contagions, which are learned as latent representations. Instead of learning each type of representations separately, we try to jointly encode the users and contagions into the same latent space, where their complex interaction relationships can be properly incorporated. To this end, we construct a heterogeneous information network consisting of users and contagions as two types of objects, and propose a novel random walk algorithm by using meta-path-based proximity as a guide to learn the representations of heterogeneous objects. In the end, to predict contagion adoption, we judiciously design an effective neural network model to capture the interactions based on the representations. The evaluation results on a large-scale Sina Weibo dataset demonstrate our proposal can outperform the competing baselines. Moreover, the latent representations are also suitable for multi-class classification of contagions.

G. Li et al. (Eds.): DASFAA 2019, LNCS 11446, pp. 501–516, 2019.
https://doi.org/10.1007/978-3-030-18576-3_30

1 Introduction

The online social networks such as Twitter and Weibo have become the fundamental platforms for information and contagion diffusion. The opinions of the social networking users tend to be influenced by their neighbors [11], and such influence can propagate through links and spread over the network.

Massive efforts have been devoted to understanding the information dynamics in social networks [3,4,11,12,20,30,31,33]. However, most previous works focus on studying how a single piece of information spreads, but the interactive effects among multiple contagions and users are not fully explored. The interactions among users and contagions have impacts on the diffusion of contagions. For example, if two users u_1 and u_2 have common interests and are close friends, a contagion posted by u_1 will be more likely adopted by u_2. In addition to user interactions, interactions among contagions, i.e., competition or cooperation, may also affect the adoption of a contagion [1,7,16,18,19,23,28,32], which should also be considered for making the decision for contagion adoption. Therefore, there is a clear motivation to extend the understanding of information diffusion by studying the joint interactions among the contagions and users. Though recent works begin to make initial discoveries [16,23,24], how to effectively capture and use their comprehensive interactions remains an open problem.

In this paper, we study the problem that when a user is exposed to a set of contagions posted or forwarded by her neighbors, whether she would like to forward them (as shown in Fig. 1). Our aim is to effectively capture the interactions among users and contagions, and use such interactions to predict the user's adoption behavior more accurately. It is challenging due to the interactions are rather complex and often present strong coupling relations. Instead of exploiting each kind of interactions separately, it is necessary to incorporate all the interactions in a unified way.

We approach this problem by introducing the representation learning techniques, which have been widely exploited in natural language processing [14,15], into our interaction-aware diffusion analysis. Although existing methods can produce user embeddings given the social structures (e.g., DeepWalk [17], node2vec [10], and Line [27]) and contagion embeddings given the text information (e.g., LDA [2]), they learn each kind of embeddings separately, ignoring the coupling effects between users and contagions in the information diffusion process, which may limit the predictive capability. To address this challenge, we develop a novel framework, termed *Heterogeneous integrated Users and Contagions Embedding* (HUCE), to (1) jointly learn the representations of users and contagions based on a heterogeneous information network (HIN) consisting of two types of objects, users and contagions and (2) judiciously integrate the user and contagion representations while capturing the various types of interactions among users and contagions. Specifically, we first construct the HIN and then propose HWalk, a novel random walk algorithm designed for the HIN, where meta-path-based proximity is used to guide the transitions between neighboring nodes. With HWalk, a set of random walk paths is generated with semantic meanings. Then we extend the traditional Skip-gram model [15] to process these paths to encode different types of nodes in a low-dimensional continuous vector space. Finally, to predict contagion adoption, we design an effective neural network model to capture the interactions based on the latent representations.

Although several recent works have also tried to use representation learning techniques for information diffusion, they either only embed the users into a latent space without considering any contagion information [4], or fall short in modeling the comprehensive interactions among users and contagions [3]. For some works [3,9], the prediction goals and the required prior inputs for the models are also different from our work. Moreover, compared to some heterogeneous representation learning techniques in [6,21,26,35,36], the proposed representation learning algorithm is suitable for large-scale dense heterogeneous network, doesn't require external networks and labeled data, and is task-guided for information diffusion. The obtained representations in our proposal can also be applied to other applications, e.g., the multi-class classification for contagions, sentiment analysis [34].

The major contributions can be summarized as follows:

(1) A HUCE framework based on representation learning is proposed to model the forwarding contagion behavior of users by jointly incorporating the characteristics of users and contagions, and their interactions.
(2) To obtain the embeddings of users and contagions in the same latent space, a HIN is constructed, and a meta-path guided random walk algorithm HWalk is developed to capture the comprehensive neighborhood information for various types of nodes. An effective neural network model is designed based on the latent representations for capturing the various types of interactions and predicting contagion adoption.
(3) Experiments on a large-scale Weibo dataset demonstrate that HUCE outperforms the state-of-the-art works for predicting the contagion adoption behavior. Moreover, we also observe superior performance when using the latent representations for the task of contagion classification.

2 Overview

In this section, we first provide the statement of the problem, and then describe the framework of the proposed model.

2.1 Problem Statement

In a social network, when a contagion is posted or forwarded by a user, the contagion is exposed to her neighbors, and the contagion is called an exposure. If one neighbor forwards a contagion, we call that the neighbor adopts this contagion, and this contagion gets propagated. We make the same assumption as [16] that at a specific time, the user has read through all the contagions that forwarded by her neighbors. We study the problem that whether a user will adopt a contagion that she has just read, and the contagion is called the

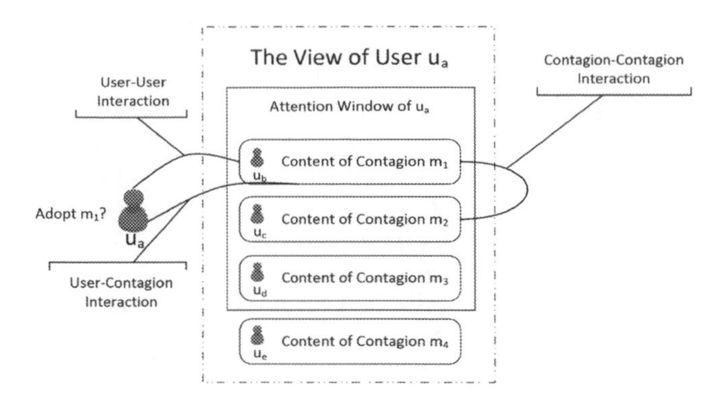

Fig. 1. An illustration of the interacting scenario that we model. User u_a has been exposed to a sequence of contagions $\{m_1, m_2, m_3\}$ in her attention window, and is examining whether to adopt m_1. u_a's decision is not only influenced by her own preference and the content of m_1, but also impacted by three types of interactions: the interaction between u_a and m_1, the interaction between u_a and user u_b, and the interactions among m_1 and other contagions (m_2 and m_3).

examined contagion. There is an attention window of the user, which contains the examined contagion and the other most recent K exposures kept in her mind. An example of the interacting scenario in Weibo that we study is shown in Fig. 1, where m_1 is the contagion which is previously forwarded by u_b, and now is examined by u_a. $M = \{m_1, m_2, m_3\}$ is the attention window, and it implies $K = 2$ for this case.

The decision made by u_a is not only affected by the inherent characteristics of m_1 and u_a, but also by a few interaction factors among users and contagions, including: (1) the *User-Contagion Interaction*, which is the interaction between the examining user u_a and the examined contagion m_1, reflecting the preference of u_a over m_1; (2) the *User-User Interaction*, which is the interaction between the examining user u_a and the neighbor u_b who has forwarded the examined contagion, reflecting the influence between them; and (3) the *Contagion-Contagion Interaction*, which is the interaction between the examined contagion m_1 and the other contagions in the attention window (e.g., m_2), reflecting their competing or cooperating relationships. Given an interacting scenario, our task is to predict the user's adoption behavior by incorporating the above factors.

2.2 Framework

In this paper, we model users' adoption behavior by jointly considering the characteristics of both users and contagions, as well as their interaction effects. These factors are commonly learned through Bayesian methods in existing studies [23]. We devise here an alternative approach based on representation learning techniques, which use the representations of users and contagions as input features for the prediction task. Since the users and contagions are tightly correlated, rather than learning their representations separately, we develop the HUCE framework that learns their embeddings simultaneously by encoding the users and contagions into the same latent space, in order to achieve more effective embeddings. Moreover, a task-guided neural network is devised to capture the various interactions among users and contagions for predicting contagion adoption.

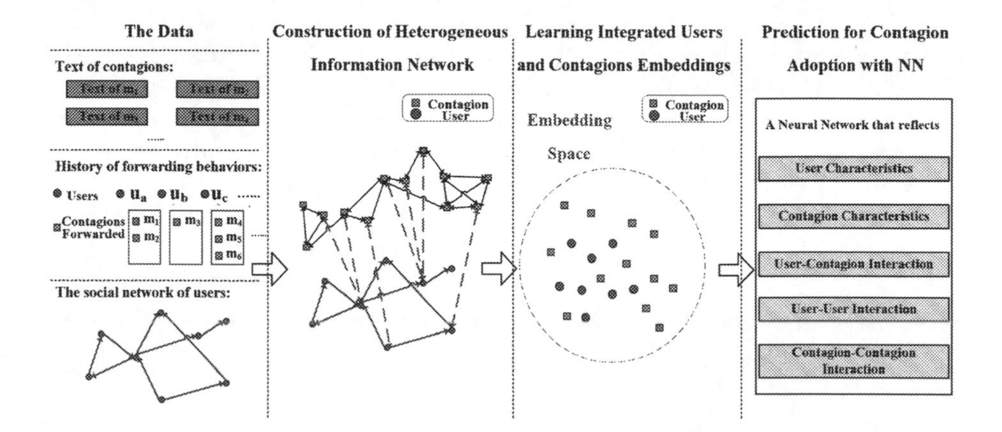

Fig. 2. HUCE framework.

The overall framework is depicted in Fig. 2, which consists

1. It first takes the social structures, texts of contagions and the historical forwarding behaviors as input, and construct the HIN that consists of two kinds of vertices, i.e., users and contagions. This HIN can describe the relations among users and contagions in the diffusion process.
2. To encode two types of objects into one latent space, a random walk algorithm called HWalk is proposed, which is guided by meta paths to generate sequences of heterogeneous vertices for node representation learning.
3. Given the representations, an effective neural network is designed to capture the necessary interaction characteristics required for contagion adoption.

3 Construction of User-Contagion HIN

A HIN is defined as a network with multiple types of objects and/or multiple types of links. Given the social relations between users, contagions and the historical adoption behaviors, we construct a HIN involving all these elements. Formally, we define a directed graph $G = (V, E)$, where V is the set of vertices and $E \subseteq (V \times V)$ is the set of edges. There is a network schema $S = (R, L)$, where R and L are the sets of node types and edge types, respectively. More specifically, $R = \{R_u, R_m\}$, where R_u denotes the type of user nodes and R_m denotes the type of contagion nodes. $L = \{L_{uu}, L_{um}, L_{mm}\}$, where L_{uu}, L_{um} and L_{um} represent the types of user-user links, the user-contagion links and the contagion-contagion links, respectively.

The overall HIN can be viewed as a combination of three subnetworks, and each subnetwork consists of only one link type. We introduce each subnetwork in detail as follows. The **User-User Following Subnetwork** is the social network structure indicating the follower and followee relationships among users. The **User-Contagion Adoption Subnetwork** is built based on historical adoption behaviors. Specifically, if a user u_a has forwarded a contagion m_i, we will set up two opposite directed edges between them. For the **Contagion-Contagion Correlation Subnetwork**, there are no explicit relationships between contagions in the social network. To set up such a subnetwork, we define the similarity of contagions according to their contents. If two contagions are similar, there will be two directed links between them, with opposite directions.

Here we explain how to measure the contagion-contagion similarity. LDA [2] algorithm is exploited to extract the latent topic distribution of each contagion. For one contagion, if the distribution value of one topic is larger than 0.5, then we define that this contagion belongs to this topic. Each pair of contagions belonging to the same topic would have two directed links between them. Besides, not every contagion belongs to a specific topic, to measure the contagion-contagion similarity, we also propose a clustering method. The similarity between two contagions is calculated according to the distance of their topic distributions. Given the similarity metric and the topic distribution of contagions, we can cluster the contagions with Mini Batch K-means [22], which is effective and fast to handle large-scale dataset. Now the contagions are clustered into T clusters, and each pair of contagions belonging to the same cluster would also have two directed links between them.

With the three subnetworks, the HIN can be constructed by combining them.

4 Learning Integrated Users and Contagions Embeddings

In this section, we present how to embed users and contagions of the HIN into a common latent space. We first introduce the HWalk algorithm to obtain the node sequences, and then give the optimization method to get the embeddings.

4.1 Random Walks on the HIN

Motivated by the idea of random walks in word2vec [15], DeepWalk [17] and node2vec [10], we extend this technique to HINs by considering the network heterogeneity. The premise of our method is to preserve the proximity of nodes in HINs. To perform this, we present a new random walk algorithm for HINs termed HWalk. Different from traditional random walk algorithms, HWalk is guided by meta paths in HIN that involves the rich semantics among different types of relationships among nodes in the context of information diffusion, which facilitates our adoption prediction problem. Compared to the recent random walk algorithm that is constrained to step over meta paths only [8], HWalk is able to walk through any path, and thus can incorporate richer semantics.

Algorithm 1. TransProbGeneration (G)

```
   Input: HIN G = (V, E)
   Output: transition probabilities T_P (P ∈ {UUU, UMU, MUM})
 1 for each node x ∈ V do
 2     for each node y ∈ ne(x) do
 3         if x and y are both users then
 4             Calculate S_P(x, y) (P ∈ {UUU, UMU});
 5         end
 6         if x and y are both contagions then
 7             Calculate S_P(x, y) (P = MUM);
 8         end
 9     end
10     for each node y ∈ ne(x) do
11         if x and y are nodes of the same type then
12             Calculate Prob_P(y|x);
13             T_P(x, y) = Prob_P(y|x);
14         end
15     end
16 end
```

Meta paths in HINs have been proven to be beneficial to a lot of data mining tasks, and here we show how to define meta paths in our application and how to use them to guide the random walkers. Given the network schema $S = (R, L)$, where R and L are the sets of node types and edge types respectively, a meta path P is defined in the form of $R_1 \xrightarrow{L_1} R_2 \xrightarrow{L_2} \ldots \xrightarrow{L_n} R_{n+1}$. Thus, a meta path defines a composite relation between node types R_1 and R_{n+1}. In our HIN, similar users are measured based on their common followees and common contagions they adopted, and similar contagions can be measured based on their contents and common adopters. The corresponding meta paths are defined as

$$R_u \xrightarrow{follow} R_u \xrightarrow{followed-by} R_u, \text{ for short as } UUU;$$
$$R_u \xrightarrow{adopt} R_m \xrightarrow{adopted-by} R_u, \text{ for short as } UMU;$$
$$R_m \xrightarrow{adopted-by} R_u \xrightarrow{adopt} R_m, \text{ for short as } MUM.$$

Here, U denotes the user type and M denotes the contagion type. Please note that these meta paths are symmetric as the first node and the last node are the same type. Then we use PathSim [25], a meta-path-based similarity measure to capture the subtlety of peer similarity between two nodes of the same type. The intuition behind is that two similar objects should not only be strongly connected, but also share comparable visibility. Here the visibility is defined as the number of path instances between them. Given a symmetric meta path P, PathSim similarity between objects x and y is defined as:

$$S_P(x, y) = \frac{2 \times |\{p_{x \rightsquigarrow y} : p_{x \rightsquigarrow y} \in P\}|}{|\{p_{x \rightsquigarrow x} : p_{x \rightsquigarrow x} \in P\}| + |\{p_{y \rightsquigarrow y} : p_{y \rightsquigarrow y} \in P\}|} \tag{1}$$

Algorithm 2. HWalk (G, T_P, v, l)

Input: HIN $G = (V, E)$, transition probabilities T_P, source node v, walk length l
Output: Random walk path W
1 Initialization: the walk $W = [v]$, $cur = v$;
2 **while** $W.length < l$ **do**
3 Randomly choose a node type $r(r \in \{R_u, R_m\})$ from the types of cur's neighbors;
4 **if** r *equals to cur's node type* **then**
5 **if** $r = R_u$ **then**
6 Randomly choose a meta path $P \in \{UUU, UMU\}$;
7 **else**
8 $P = MUM$;
9 **end**
10 $s = AliasSample(ne_r(cur), T_P)$;
11 Append s into W;
12 **else**
13 $s = Random(ne_r(cur))$;
14 Append s into W;
15 **end**
16 $cur = s$;
17 **end**

Here $p_{x \rightsquigarrow y}$ is a path instance between x and y. $p_{x \rightsquigarrow x}$ is that between x and x, and $p_{y \rightsquigarrow y}$ is that between y and y. If both x and y are users, the meta path P is either UUU or UMU. If both x and y are contagions, P is MUM.

We denote the set of x's outgoing neighbors of node type r as $ne_r(x)$. Given a node x, a node type r which is the same as that of x, and a meta path P, the similarity $S_P(x, y)$ between x and each node $y \in ne_r(x)$ can be calculated according to Eq. (1). In each step, if the current node is x and the type of the next node is r, the transition probability to $y \in ne_r(x)$ is proportional to

$$Prob_P(y|x) = \frac{S_P(x, y)}{\sum_{z \in ne_r(x)} S_P(x, z)} \tag{2}$$

The transition probabilities between neighboring nodes of the same type can be precomputed in Algorithm 1, where $T_p(x, y)$ is the transition probability between nodes x and y with respect to meta path P. If two connected nodes are both users, two transition probabilities can be derived with respect to $P = UUU$ and $P = UMU$ respectively. If they are both contagions, meta path $P = MUM$ is used to derive the transition probabilities.

Please note that PathSim is only able to calculate the similarity between two nodes of the same type, and thus is only used for transition to a node with the same type as the current node. If the walker is stepping to a node with a different type, it just randomly chooses a neighboring node of that type. The overall walking procedure is described in Algorithm 2. The walk starts from the source node v, and in each step, a node type r is first chosen randomly from

Algorithm 3. HEmbedding

Input: HIN $G = (V, E)$, window size w, embedding size d, walks per node γ, walk length l
Output: Node representations $\Phi \in \mathbb{R}^{|V| \times d}$
1 $T_P = TransProbGeneration(G)$;
2 **for** *iter=1 to* γ **do**
3   **for** *each node* $v \in V$ **do**
4         $W_v = HWalk(G, T_P, v, l)$;
5   Skip-gram(Φ, W_v, w);
6   **end**
7 **end**

the types of the current node's neighbors (line 3). If r is the same as the type of current node (line 4), we sample one neighbor of type r to walk according to the transition probabilities generated in Algorithm 1 (line 10). This sampling process can be done efficiently using alias sampling [13,29], which takes only $O(1)$ time when repeatedly drawing a sample from the same discrete probability distribution. If r is different from the current node's type, we randomly choose a node from the current node's neighbors of type r (line 13). The walk continues until the length of the path achieves the pre-defined walk length l. Thus the time complexity of HWalk is $O(l)$.

4.2 Heterogeneous Node Embedding

Based on HWalk, a sequence of node paths can be obtained. Then we learn representations of heterogeneous nodes based on these random paths. Our goal is to learn a mapping function $\Phi : v \in V \rightarrow \in \mathbb{R}^{|V| \times d}$, where d is the dimension of the representations. The mapping Φ represents the representation associated with each user and contagion in the HIN. We proceed by extending the Skip-gram [14] to HINs, with the corpus of node paths are taken as input. The algorithm of heterogeneous node embedding (HEmbedding) is shown in Algorithm 3.

The generated representations are then used for the contagion adoption problem as well as contagion classification.

5 Capturing Interactions for Predicting Contagion Adoption

After generating the latent representations of multiple types of objects of the HIN, conventional methods commonly measure the similarity between a user and a contagion by their representations, and use this similarity to make the prediction for contagion adoption. However, this may limit its performance as it discards the complex interactions from other users and contagions. In contrast, we delicately design an effective neural network model to capture the interactions based on the representations.

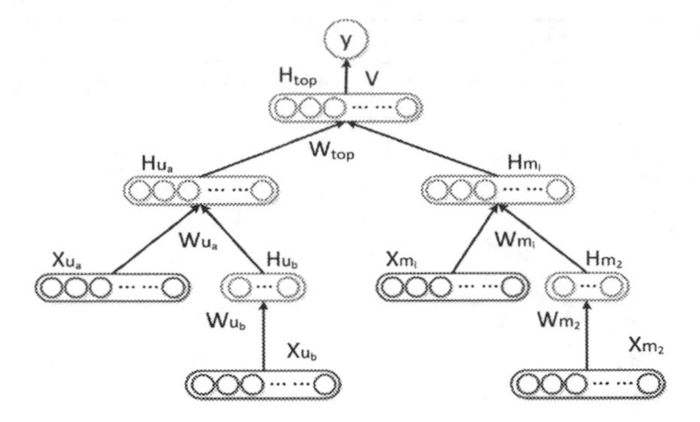

Fig. 3. NN for capturing interactions on contagion adoption. Here u_a is the examining user; m_i is the examined contagion; u_b is the user who has forwarded m_i; m_2 is the contagion in the examining user's attention window. We assume $K = 1$, that is, there is one contagion in the examining user's attention window. The inputs X_{u_a}, X_{m_i}, X_{u_b} and X_{m_2} are embeddings of u_a, m_i, u_b and m_2. H_{u_a}, H_{m_i}, H_{u_b}, H_{m_2} and H_{top} in hidden layers are real-valued vectors. All the W as well as V are parameters in this neural network, and y is the output.

As shown in Fig. 3, the neural network contains five layers. In this neural network, X_{u_a} reflects the user characteristic and X_{m_i} reflects the contagion characteristic. X_{u_a} is integrated with H_{u_b} to generate the vector H_{u_a}, and this vector indicates the impact of u_b on u_a's adoption behavior, i.e., the User-User Interaction. Similarly, the generated vector H_{m_i} of the integration of X_{m_i} and H_{m_2} indicates the impact of m_2 on the adoption of m_i, i.e., the Contagion-Contagion Interaction. In the upper layer, H_{u_a} and H_{m_i} are integrated to generate H_{top} that directly indicates the preference of u_a over m_i, i.e., the User-Contagion Interaction, and indirectly indicates the impact of u_b on u_a and the impact of m_2 on m_i. Based on H_{top}, whether u_a will adopt m_i can be predicted in the form of output layer y. The neural network not only considers the intrinsic characteristics of the examining user and the examined contagion, but also takes three types of interactions into account, which provides more informative prediction.

The calculations in all the layers are shown as follows:

$$
\begin{aligned}
H_{u_b}(j) &= sig(\sum_i X_{u_b}(i) W_{u_b}(j,i)) \\
H_{m_2}(j) &= sig(\sum_i X_{m_2}(i) W_{m_2}(j,i)) \\
H_{u_a}(j) &= sig(\sum_i (X_{u_a} \oplus H_{u_b})(i) W_{u_a}(j,i)) \\
H_{m_i}(j) &= sig(\sum_i (X_{m_i} \oplus H_{m_2})(i) W_{m_i}(j,i)) \\
H_{top}(j) &= sig(\sum_i (H_{m_i} \oplus H_{u_a})(i) W_{top}(j,i)) \\
y &= sig(\sum_j H_{top}(j) V(j))
\end{aligned}
\tag{3}
$$

Here ⊕ represents a concatenation operator and sig is the sigmoid function. Backpropagation algorithm is used to train the parameters of the neural network.

6 Experiments

In this section, we conduct experiments on a public Sina Weibo dataset. Then, we discuss various qualitative insights obtained by applying our framework to the dataset.

6.1 Experimental Settings

Dataset. The Weibo dataset [37] provides the friendship links of Weibo users (1,776,950 users, 308,489,739 links), the contagion adoption behaviors of the users (23,755,810 retweets), as well as the textual information of the contagions (300,000 original microblogs). Due to the crawling strategy, the amounts of retweet behaviors in different months are highly imbalanced. Thus, we select the diffusion data from July 2012 to December 2012, in which the retweet counts are large enough and more balanced. Consequently, we get 19,388,727 retweets on 140,400 popular microblogs. To construct the HIN, we use the retweets from July 2012 to November 2012 to generate the user-contagion links. To set up the contagion-contagion links, the contagions are clustered into 3,000 sets. To balance the effectiveness and efficiency of the model, we choose the parameters $d = 30$, $l = 15$, $\gamma = 10$ and $w = 5$ in the experiments. We set $K = 1$ and $K = 2$ as the size of contagions a user kept in mind besides the examined contagion.

For each user, when she reads a newly posted contagion, an interacting scenario occurs. If the examined contagion is adopted, the instance is positive, otherwise, it is negative. To train and test our model, we do statistical work to extract interacting scenarios from the data in December 2012, and observe that the positive and negative instances are highly unbalanced in the dataset, so we sampled the positive and negative instances with the equal number. In total, 2,000,000 interacting scenarios are used for training and testing. We randomly sample 90% of the instances as the training set, and the left 10% as the testing set. The performance is evaluated in terms of F1-score and Accuracy.

Baselines. We compare our proposal with:

- **DeepWalk** [17] is a representative representation learning method for homogeneous networks. We provide two versions of DeepWalk with different inputs. One is called **uDeepWalk** where we only feed the social network structure for embedding, and thus it only produces user representations. The other is called **hDeepWalk**, with the constructed HIN as input. Thus, hDeepWalk learns the representations of both users and contagions.
- **metapath2vec** [8] is a meta-path-based representation learning method for heterogeneous networks, in which the random walkers traverse through predefined meta-paths. Since it can only support one meta-path scheme, we choose UMU.

- **LDA** [2] learns contagion representations as the latent distributions over topics.
- **IAD** [23] is a state-of-the-art information diffusion prediction framework based on Bayesian theory, which also takes the interactions among contagions and users into account.

For all the representation learning methods that are compared, we adopt the same parameter settings as HUCE. Given the representations learned by these methods, we apply logistic regression (LR) as the off-the-shelf classification algorithm for prediction. For uDeepWalk, the input feature for LR is the representation of the examining user. For LDA, the input feature for LR is the representation of the examined contagion. As hDeepWalk and metapath2vec all produce representations of both users and contagions, their input features are concatenated using the representation of the examining user, the representation of the examined contagion, together with the representations of three kinds of interactions. Here the vector of User-User Interaction is established by the difference in each dimension of the representation of the examining user and that of the user who has forwarded the examined contagion; the vector of Contagion-Contagion Interaction is established by the difference in each dimension of the representation of the examined contagion and that of the contagion which is in the examining user's attention window; the vector of User-Contagion Interaction is established by the difference in each dimension of the representation of the examining user and that of the examined contagion. Besides, to analyze the effectiveness of the proposed neural network, we propose a variation of the HUCE, named *HUCE-LR*, which applies logistic regression for prediction instead, and the input features are constructed in the same way as those in hDeepWalk and metapath2vec.

6.2 Results

Table 1. Performance of HUCE compared to baselines (%)

Models	Accuracy ($K = 1$)	F1-score ($K = 1$)	Accuracy ($K = 2$)	F1-score ($K = 2$)
uDeepWalk	68.95	61.79	68.95	61.79
LDA	64.16	59.13	64.16	59.13
IAD	73.50	65.58	74.03	66.52
hDeepWalk	73.49	66.74	73.38	66.65
metapath2vec	71.47	61.37	71.61	61.21
HUCE-LR	77.58	72.42	77.53	72.32
HUCE	**78.76**	**73.87**	**78.62**	**73.75**

Table 1 summarize the experimental results. Overall, HUCE consistently and significantly outperforms all the baselines. Generally, uDeepWalk and LDA perform worse than HUCE, hDeepWalk and metapath2vec, indicating that considering only the characteristic of users or contagions is not enough to make accurate predictions. HUCE outperforms IAD by 5.26% and 8.29% in terms of accuracy and F1-score respectively when $K = 1$, and by 4.59% and 7.23% in terms of accuracy and F1-score respectively when $K = 2$, showing that HUCE can model the interactions more effectively than IAD. HUCE outperforms hDeepWalk by 5.27% and 7.13% in terms of accuracy and F1-score respectively when $K = 1$, and by 5.24% and 7.10% in terms of accuracy and F1-score respectively when $K = 2$, demonstrating that considering the heterogeneity of nodes in walking can facilitate the embedding. However, though metapath2vec is also designed for HINs, it performs worse than HUCE and hDeepWalk. The possible reason is that metapath2vec can only traverse through a pre-defined meta-path schema, which works well on some HINs such as academic networks, but is difficult to adapt to other scenarios such as information diffusion which requires more comprehensive neighborhood information. Moreover, the jointly derived representations on users and contagions are more effective, since HUCE-LR is better than all the baselines. It can also be observed that HUCE-LR is worse than HUCE, which indicates that neural network is effective in capturing the various types of interactions (user-user, contagion-contagion, user-contagion) and fusing the heterogeneous representations together.

To analyze the importance of different features, we propose several variations of the HUCE framework as follows:

- *HUCE-U*: Only the user representations learned by HUCE are fed into logistic regression.
- *HUCE-M*: Only the contagion representations learned by HUCE are fed into logistic regression.
- *HUCE-UM*: Both the representations of the users and contagions learned by HUCE are fed into logistic regression. Compared with HUCE, HUCE-UM lacks the three interaction features among users and contagions.

Table 2. Results with different input features (%)

Models	Accuracy ($d = 20$)	F1-score ($d = 20$)	Accuracy ($d = 30$)	F1-score ($d = 30$)
HUCE-U	70.90	65.75	70.85	65.63
HUCE-M	63.70	53.29	70.58	63.63
HUCE-UM	71.88	65.90	72.38	66.09
HUCE	**77.96**	**73.02**	**78.76**	**73.87**

Table 2 shows the comparison results, while we set $K = 1$ for HUCE-LR and HUCE. When $d = 30$, in terms of F1-score, HUCE achieves an improvement of 8.24%, 10.24% and 7.78% over HUCE-U, HUCE-M and HUCE-UM. When $d = 20$, in terms of F1-score, HUCE achieves an improvement of 7.27%, 19.73% and 7.12% over HUCE-U, HUCE-M and HUCE-UM, respectively. In terms of accuracy, HUCE also outperforms the variations when $d = 30$ and $d = 20$. It can be concluded that all the factors including the user's characteristic, the contagion's characteristic and three kinds of interactions among users and contagions, are judicious for the contagion adoption problem, and ignoring any one of them would limit the predictive capability.

6.3 Contagion Classification

Table 3. Multi-class contagion classification results (%)

Models	Macro F1-score	Micro F1-score
LDA	23.03	26.78
HUCE	**31.15**	**32.33**

As the contagion representation is learned jointly with users, it is in fact more effective than doing it alone even for contagion classification. For the Weibo dataset, we manually define 15 categories of contagions, that is, *advertisement, constellation, culture, economy, food, health, history, life, movie, music, news, politics, sports, technology and traffic*. We randomly labeled 10,000 contagions in the dataset by hand to these categories, while each category has around 600 to 700 contagions. The labeled dataset is publicly available[1]. We randomly sample 90% of these labeled contagions for training, and the left 10% for testing. For the multi-class classification problem, the performance is evaluated in terms of Macro F1-score as well as Micro F1-score.

We perform the multi-class classification with the contagion representations learned by HUCE and LDA, using C-Support Vector Classification (SVC) implemented in scikit-learn based on libsvm [5]. It can be observed in Table 3 that HUCE performs much better than LDA. In terms of Macro F1-score, our proposal outperforms LDA by an impressive 8.12% (relatively 35.26%), and in terms of Micro F1-score, the gain achieves 5.55% (relatively 20.72%). This represents that involving the user-contagion relationships can learn better representations than only considering the contents.

7 Conclusion

We propose HUCE, a novel approach for modeling the contagion adoption behavior in information diffusion by considering the complex interactions among users

[1] https://www.dropbox.com/s/b0ym8cmyzp5gpyx/InforClass.zip?dl=0.

and contagions. At first, a HIN consisting of users and contagions is constructed, and then we propose a new algorithm that uses the meta-path-based proximity to guide the transitions between nodes. After jointly learning the representations of various types, to predict contagion adoption, we delicately design an effective neural network model to capture the interactions based on the representations. Although our framework is designed for information diffusion prediction, the learned representations are general and can be used for other tasks such as contagion classification. Experiments on a large-scale Weibo dataset demonstrate the effectiveness of our framework over the state-of-the-art baselines.

Acknowledgments. This work has been supported in part by the National Key Research and Development Program of China (No. 2017YFB0803301), the Natural Science Foundation of China (No. U1836215, 61602237, 61672313), NSF through grants IIS-1526499, IIS-1763325, and CNS-1626432, and DongGuan Innovative Research Team Program (No. 201636000100038).

References

1. Bi, Y., Wu, W., Zhu, Y.: CSI: charged system influence model for human behavior prediction. In: ICDM (2013)
2. Blei, D., Ng, A., Jordan, M.: Latent Dirichlet allocation. J. Mach. Learn. Res. **3**, 993–1022 (2003)
3. Bourigault, S., Lagnier, C., Lamprier, S., Denoyer, L., Gallinari, P.: Learning social network embeddings for predicting information diffusion. In: WSDM (2014)
4. Bourigault, S., Lamprier, S., Gallinari, P.: Representation learning for information diffusion through social networks: an embedded cascade model. In: WSDM (2016)
5. Chang, C., Lin, C.: LIBSVM: a library for support vector machines. ACM Trans. Intell. Syst. Technol. **2**, 27:1–27:27 (2011)
6. Chen, W., Liu, C., Yin, J., Yan, H., Zhang, Y.: Mining E-commercial data: a text-rich heterogeneous network embedding approach. In: ISNN (2017)
7. Coscia, M.: Competition and success in the meme pool: a case study on quickmeme.com. In: ICWSM (2013)
8. Dong, Y., Chawla, N.V., Swami, A.: Metapath2vec: scalable representation learning for heterogeneous networks. In: KDD (2017)
9. Gao, S., Pang, H., Gallinari, P., Guo, J., Kato, N.: A novel embedding method for information diffusion prediction in social network big data. IEEE Trans. Ind. Inf. **13**(4), 2097–2105 (2017)
10. Grover, A., Leskovec, J.: node2vec: scalable feature learning for networks. In: KDD (2016)
11. Kempe, D., Kleinberg, J., Tardos, E.: Maximizing the spread of influence through a social network. In: KDD (2003)
12. Li, C., Ma, J., Guo, X., Mei, Q.: DeepCas: an end-to-end predictor of information cascades. In: WWW (2017)
13. Marsaglia, G., Tsang, W.W., Wang, J., et al.: Fast generation of discrete random variables. J. Stat. Softw. **11**(3), 1–11 (2004)
14. Mikolov, T., Chen, K., Corrado, G., Dean, J.: Efficient estimation of word representations in vector space. arXiv preprint arXiv:1301.3781 (2013)
15. Mikolov, T., Sutskever, I., Chen, K., Corrado, G., Dean, J.: Distributed representations of words and phrases and their compositionality. In: NIPS (2013)

16. Myers, S.A., Leskovec, J.: Clash of the contagions: cooperation and competition in information diffusion. In: ICDM (2012)
17. Perozzi, B., Al-Rfou, R., Skiena, S.: DeepWalk: online learning of social representations. In: KDD (2014)
18. Prakash, B.A., Beutel, A., Rosenfeld, R., Faloutsos, C.: Winner takes all: competing viruses or ideas on fair-play networks. In: WWW (2012)
19. Rong, X., Mei, Q.: Diffusion of innovations revisited: from social network to innovation network. In: CIKM (2013)
20. Rotabi, R., Kamath, K., Kleinberg, J., Sharma, A.: Cascades: a view from audience. In: WWW (2017)
21. Santos, L.D., Piwowarski, B., Denoyer, L., Gallinari, P.: Representation learning for classification in heterogeneous graphs with application to social networks. ACM Trans. Knowl. Discov. Data **12**(5), 62 (2018)
22. Sculley, D.: Web-scale k-means clustering. In: WWW (2010)
23. Su, Y., Zhang, X., Yu, P.S., Hua, W., Zhou, X., Fang, B.: Understanding information diffusion under interactions. In: IJCAI (2016)
24. Su, Y., Zhang, X., Liu, L., Song, S., Fang, B.: Understanding information interactions in diffusion: an evolutionary game-theoretic perspective. Front. Comput. Sci. **10**(3), 518–531 (2016)
25. Sun, Y., Han, J., Yan, X., Yu, P.S., Wu, T.: PathSim: meta path-based top-k similarity search in heterogeneous information networks. Very Large Data Bases **4**(11), 992–1003 (2011)
26. Tang, J., Qu, M., Mei, Q.: PTE: predictive text embedding through large-scale heterogeneous text networks. In: Knowledge Discovery and Data Mining, pp. 1165–1174 (2015)
27. Tang, J., Qu, M., Wang, M., Zhang, M., Yan, J., Mei, Q.: LINE: large-scale information network embedding. In: WWW (2015)
28. Valera, I., Gomez-Rodriguez, M.: Modeling adoption and usage of competing products. In: ICDM (2015)
29. Walker, A.J.: An efficient method for generating discrete random variables with general distributions. ACM Trans. Math. Softw. (TOMS) **3**(3), 253–256 (1977)
30. Wang, S., Hu, X., Yu, P.S., Li, Z.: MMRate: inferring multi-aspect diffusion networks with multi-pattern cascades. In: KDD (2014)
31. Wang, S., Yan, Z., Hu, X., Yu, P.S., Li, Z.: Burst time prediction in cascades. In: AAAI (2015)
32. Weng, L., Flammini, A., Vespignani, A., Menczer, F.: Competition among memes in a world with limited attention. Sci. Rep. **2**(1), 335–335 (2012)
33. Yang, J., Leskovec, J.: Modeling information diffusion in implicit networks. In: ICDM (2010)
34. Yue, L., Chen, W., Li, X., Zuo, W., Yin, M.: A survey of sentiment analysis in social media. Knowl. Inf. Syst. 1–47 (2018)
35. Zhang, D., Yin, J., Zhu, X., Zhang, C.: MetaGraph2Vec: complex semantic path augmented heterogeneous network embedding. In: Phung, D., Tseng, V.S., Webb, G.I., Ho, B., Ganji, M., Rashidi, L. (eds.) PAKDD 2018. LNCS (LNAI), vol. 10938, pp. 196–208. Springer, Cham (2018). https://doi.org/10.1007/978-3-319-93037-4_16
36. Zhang, J., Xia, C., Zhang, C., Cui, L., Fu, Y., Yu, P.S.: BL-MNE: emerging heterogeneous social network embedding through broad learning with aligned autoencoder. In: ICDM (2017)
37. Zhang, J., Liu, B., Tang, J., Chen, T., Li, J.: Social influence locality for modeling retweeting behaviors. In: IJCAI (2013)

Graphs

Distributed Parallel Structural Hole Detection on Big Graphs

Faming Li[1], Zhaonian Zou[1(✉)], Jianzhong Li[1], Yingshu Li[2], and Yubiao Chen[1]

[1] Harbin Institute of Technology, Harbin, China
{lifaming2016,znzou,lijzh,chenyubiao}@hit.edu.cn
[2] Georgia State University, Atlanta, USA
yili@gsu.edu

Abstract. Structural holes in social networks are vertices that serve as gateways for information exchange between communities. Although many algorithms have been proposed to detect structural holes, they are not scalable to big graphs. This paper proposes a structural hole detection algorithm ESH based on distributed parallel graph processing frameworks. Instead of using substructures in social networks, the algorithm exploits a factor diffusion process in structural hole detection. The algorithm naturally fits the vertex-centric programming models and can be easily implemented on the graph-parallel processing frameworks. Extensive experiments show that ESH can handle social networks with billions of links and produce structural holes of higher quality than the existing algorithms.

1 Introduction

Structural hole is an important concept in social network analysis. Informally, the structural holes in a social network refer to the vertices (i.e. persons) that bridge a number of communities in the network. Note that a community in a social network represents a group of persons who share common interests, and therefore the structural holes in the network serve as the gateways between communities, via which information exchanges between different communities. For illustration, the vertices in red corlor in Fig. 1 are the structural holes in the network, which bridge two non-overlapping densely communities. For another practical example, Lou and Tang [1] studied the AMiner academic network in computer science and found 107 researchers as structural holes who served as program committee members for conferences in different fields.

Structural holes play an important role in social network analysis and have attracted considerable research interests in recent years. For example, influence maximization can be benefited when structural holes are selected as seeds in information diffusion models [2]. Community detection can also be benefited as the removal of structural holes can cut off the ties between communities [3,4]. Monitoring the information that structural holes receive and publish can help grasp popular topics in social networks [5].

G. Li et al. (Eds.): DASFAA 2019, LNCS 11446, pp. 519–535, 2019.
https://doi.org/10.1007/978-3-030-18576-3_31

As social networks are becoming increasingly larger, the existing structural hole detection algorithms [1,2,6–9], are faced with a series of challenges, especially the scalability issue. In this paper, we propose a new structural hole detection algorithm ESH based on the distributed parallel graph processing framework PowerGraph. To this end, we first introduce a new concept of *entropy-based structural hole*. The concept is formulated based on a *factor diffusion process* rather than substructures in graphs. The process imitates the propagation of information on social networks. In the beginning, each vertex creates a unique factor (a user's interest or a topic). In the diffusion process, a factor is propagated from a vertex to its neighbors. After a vertex receives factors from its neighbors, the vertex selects one factor from the received ones and adds it to the factor list attached with the vertex. After many iterations of the process, the factor list attached with a vertex can be seen as the "fingerprint" of the vertex. The distribution of the factors in the factor list characterizes the role of the vertex in the graph. A structural hole is likely to collect a number of heterogeneous factors propagated from various communities that the structural hole links to; while an interior vertex in a community is likely to collect homogeneous factors propagated from other vertices within the same community. Therefore, we evaluate the likelihood of a vertex being a structural hole by the *entropy* of the distribution of the factors in the factor list attached with the vertex.

The proposed algorithm ESH is substantially different from the existing algorithms. First, our concept of structural hole is defined based on a factor diffusion process rather than the substructures or communities in a graph. Second, the algorithm is simple but effective and naturally fits to be implemented on the distributed parallel graph processing frameworks. Although some of the existing algorithms can also be implemented on the distributed parallel graph processing frameworks, they are really more complex than our algorithm. For example, the 2-Step algorithm needs to collect the 2-hop neighbors of each vertex and therefore yields high communication cost between computers. The other existing algorithms are not suitable to be implemented on the distributed parallel graph processing frameworks because they either need to know the global structure of the graph (such as PathCount [6]) or must do expensive matrix computation (such as HAM [2]).

The main contributions of this paper are listed as follows.

- We formulate the new concept of entropy-based structural hole based on a factor diffusion process.
- We propose a new structural hole detection algorithm ESH based on the distributed parallel graph processing framework PowerGraph.
- We compare our algorithm ESH with the existing algorithms by an extensive set of experiments and verify that the ESH algorithm can efficiently detect high-quality structural holes on graphs with billions of edges.

The rest of this paper is organized as follows. Section 2 presents the fundamental concepts and explains the rationale behind the new concept. Section 3 proposes the structural hole detection algorithm ESH based on the distributed parallel graph processing framework PowerGraph. Section 4 reports the experimental evaluation results. The related work is reviewed in Sect. 5. Lastly, the paper is concluded in Sect. 6.

2 Entropy-Based Structural Holes

In this section, we propose a new notion of entropy-based structural hole.

2.1 Formulation

In the literature, the notion of structural hole is defined based on static structural properties of a graph. In this paper, we give a new formulation of structural hole based on an information diffusion process. Given a graph $G = (V, E)$, let F be a set of *factors* of the vertices in V. Specifically, a factor in a social network is a user's interest; a factor in a citation network is the topic of a paper. Each vertex has a distinct factor, and factors can propagate within a graph.

We formulate a stochastic process of factor propagation over graph G as follows. Each vertex $v \in V$ maintains a factor list L_v, which initially contains the single factor f_v that uniquely corresponds to v. In the beginning of the process, each vertex v sends the only factor $f_v \in L_v$ to all its neighbours. Then, the factor propagation process iterates according to the *bulk synchronization protocol* (BSP) [10]. In each iteration of BSP, we carry out the following steps for each vertex $v \in V$.

1. Gather the factors sent from all neighbors of v in the previous iteration and add one of the factors to L_v.
2. Update the distribution of the factors stored in L_v.
3. Pick a factor f from L_v at random according to the distribution of the factors in L_v.
4. Send the picked factor f to all neighbors of v.

After executing a large number of iterations, we evaluate the *entropy* $H(v)$ of the factors in L_v, that is,

$$H(v) = -\sum_{f \in L_v} P(f) \log P(f), \tag{1}$$

where $P(f)$ is the probability of factor f in L_v. Factor list L_v can be seen as the "fingerprint" of vertex v, and entropy $H(v)$ evaluates the "skewness" of the factor distribution of L_v. Intuitively, a structural hole vertex is likely to collect a number of heterogeneous factors propagated from several different communities, and therefore, the entropy of its factors tends to be high; while an interior vertex within a community is likely to collect homogeneous factors, and thus, its entropy tends to be low. Hence, we have the following definition of structural holes in a graph.

Definition 1. *Given a graph G and an integer k, the top-k structural holes in G are the k vertices with the highest entropy of factor distribution.*

2.2 Rationale

In this subsection, we discuss the rationale behind the definition of entropy-based structural holes. In Step 1 of the factor propagation process, each vertex $v \in V$ gathers the factors sent from all its adjacent vertices and adds one of them to the factor list L_v associated with v. The description of Step 1 is too general. We have three specific strategies to pick one factor from the collected factors:

- **Strategy 1 (Random):** Select a factor uniformly at random from the collected factors.
- **Strategy 2 (MostFrequent):** Select the factor that occurs the most frequently from the collected factors. Ties are broken arbitrarily.
- **Strategy 3 (LeastFrequent):** Select the factor that occurs the least frequently from the collected factors. Ties are broken arbitrarily.

We analyze the factor propagation process and the entropy of a factor distribution as follows.

Factor Propagation. Let T_f^n be the number of propagation of factor f within graph G in the nth iteration of the factor propagation process. T_f^n is a random variable over $\{0, 1, 2, \ldots, n\}$. The probability distribution of T_f^n is essential for our study.

Recall that f_v is the unique factor corresponding to vertex v. The following theorem shows that $T_{f_v}^n$ is expected to be $d(v)$ when the factor diffusion process uses the Random strategy.

Theorem 1. *When the Random strategy is applied, we have $E[T_{f_v}^n] = d(v)$ for all vertices $v \in V$ and $n > 0$.*

Proof. Let $X_{v,f}^n$ be a 0–1 random variable which indicates whether vertex v sends factor f to its neighbors in the nth iteration. Particularly, $X_{v,f}^n = 1$ if and only if v sends f to its neighbors in the nth iteration. Let $Y_{v,f}^n$ be a 0–1 random variable which indicates whether vertex v adds factor f to its factor list L_v in the nth iteration. Particularly, $Y_{v,f}^n = 1$ if and only if v adds f to its factor list L_v in the nth iteration. Let $P_v^n(f)$ represent the probability of factor f occurring in the factor list L_v in the nth iteration. We have

$$E[X_{v,f}^n | Y_{v,f}^n = 1] = P(X_{v,f}^n = 1 | Y_{v,f}^n = 1) = \frac{(n-1)P_v^{n-1}(f) + 1}{n},$$
$$E[X_{v,f}^n | Y_{v,f}^n = 0] = P(X_{v,f}^n = 1 | Y_{v,f}^n = 0) = \frac{(n-1)P_v^{n-1}(f)}{n}.$$

Then,

$$\begin{aligned} E[X_{v,f}^n] &= P(Y_{v,f}^n = 1)E[X_{v,f}^n | Y_{v,f}^n = 1] + P(Y_{v,f}^n = 0)E[X_{v,f}^n | Y_{v,f}^n = 0] \\ &= \frac{P(Y_{v,f}^n = 1) + (n-1)P_v^{n-1}(f)}{n} \\ &= \frac{\frac{1}{d(v)} \sum_{u \in N(v)} P_u^{n-1}(f) + (n-1)P_v^{n-1}(f)}{n}, \end{aligned}$$

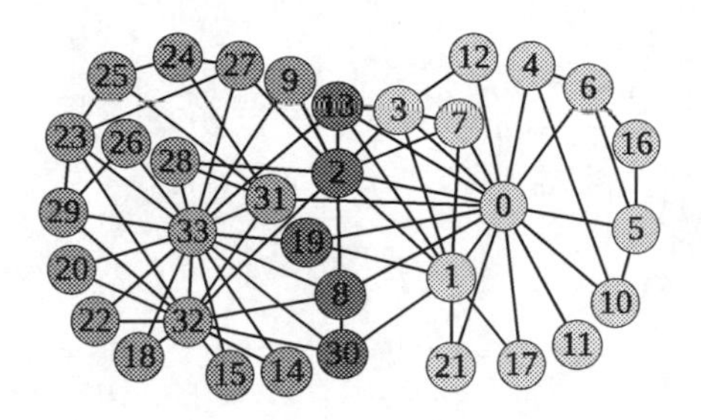

Fig. 1. The karate club graph. (Color figure online)

where $N(v)$ denotes the set of neighbors of vertex v. Thus,

$$\begin{aligned} E[T_f^n] = \sum_{v \in V} d(v) E[X_{v,f}^n] &= \frac{1}{n}\left(\sum_{v \in V}\sum_{u \in N(v)} p_u^{n-1}(f) + \sum_{v \in V}(n-1)d(v)p_v^{n-1}(f)\right) \\ &= \frac{1}{n}\left(\sum_{v \in V} d(v)p_v^{n-1}(f) + \sum_{v \in V}(n-1)d(v)p_v^{n-1}(f)\right) \\ &= \sum_{v \in V} d(v)p_v^{n-1}(f) = E[T_f^{n-1}]. \end{aligned}$$

Finally, we have $E[T_f^n] = E[T_f^{n-1}] = \cdots = E[T_f^1]$. In the first iteration, each vertex v sends its own distinguished factor f_v to exactly $d(v)$ neighbors, so $T_{f_v}^1 = d(v)$. Hence, $E[T_{f_v}^n] = E[T_{f_v}^1] = d(v)$, and thus the theorem holds. □

It is not easy to formulate the closed form of $E[T_{f_v}^n]$ when the MostFrequent strategy or the LeastFrequent strategy is applied. Instead, we empirically study the number of propagation of a factor in one iteration under these strategies. We first carried out experiments on the "karate club" graph [11], which is illustrated in Fig. 1. The graph can be conveniently visualized and has two significant communities, which are highlighted in green and yellow, respectively. It is convenient to reveal the concept of structural holes with the graph. The red vertices are the structural holes. We ran the factor propagation process for 150 iterations. Figure 2 depicts the relationship between the degree $d(v)$ of vertex v and the average number of propagation $\bar{T}_{f_v}$ of factor f_v in the middle 50 and the last 50 iterations. We have the following observations.

- As shown in Figs. 2a and b, when the Random strategy is applied, $\bar{T}_{f_v}$ is very close to $d(v)$. The Pearson's correlation coefficient between $d(v)$ and $\bar{T}_{f_v}$ is 0.92 and 0.95 in Figs. 2a and b, respectively, which means that $d(v)$ and $\bar{T}_{f_v}$ are strongly positively correlated. This experimental result is consistent with Theorem 1.
- As shown in Fig. 2c, when the MostFrequent strategy is applied, we have $\bar{T}_{f_6} = 70$ for vertex 6 whose degree is 4 and $\bar{T}_{f_{32}} = 64$ for vertex 32 whose degree is 16. As shown in Fig. 2d, when the MostFrequent strategy is applied,

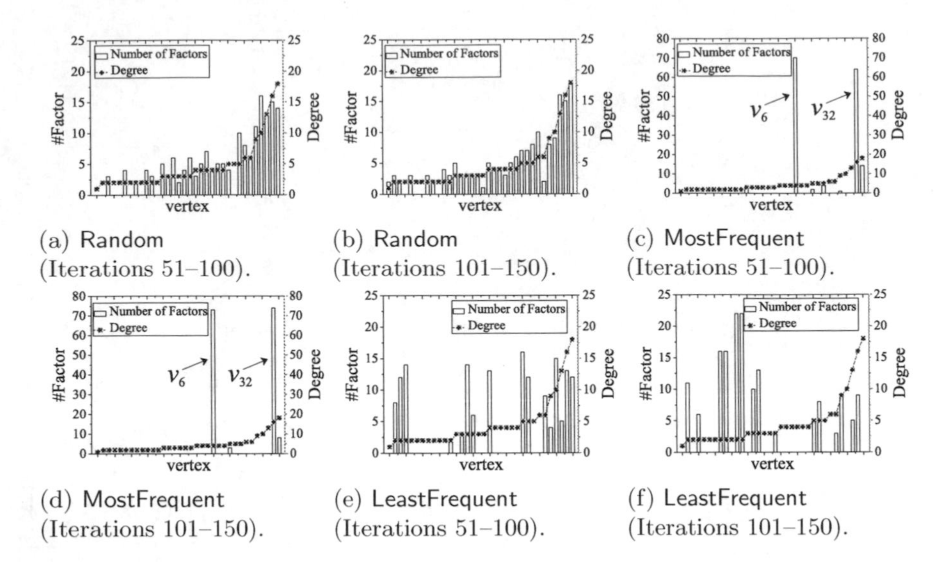

(a) Random (Iterations 51–100). (b) Random (Iterations 101–150). (c) MostFrequent (Iterations 51–100). (d) MostFrequent (Iterations 101–150). (e) LeastFrequent (Iterations 51–100). (f) LeastFrequent (Iterations 101–150).

Fig. 2. The degree $d(v)$ of vertex v and the average number $\bar{T}_{f_v}$ of propagation of factor f_v on the karate club graph in the last 100 iterations.

we have $\bar{T}_{f_6} = 73$ and $\bar{T}_{f_{32}} = 74$. Except vertices 6 and 32, $\bar{T}_{f_v} < d(v)$ holds for all vertices v. Therefore, factors f_6 and f_{32} dominate the factor propagation process because more than 40% of factors propagated in an iteration are f_6 and f_{32}. Interestingly, vertices 6 and 32 are in the interior of the two communities of the karate club graph, respectively. Moreover, the Pearson's correlation coefficient between $d(v)$ and $\bar{T}_{f_v}$ is 0.32 and 0.28 in Figs. 2c and d, respectively, which means that the correlation between $d(v)$ and $\bar{T}_{f_v}$ is very weak.

- As shown in Fig. 2e, when the LeastFrequent strategy is applied, we have $\bar{T}_{f_v} \leq 16$ for all vertices v. As shown in Fig. 2f, when the LeastFrequent strategy is applied, we have $\bar{T}_{f_v} \leq 23$ for all vertices v. The Pearson's correlation coefficient between $d(v)$ and $\bar{T}_{f_v}$ is 0.21 and -0.11 in Figs. 2e and f, respectively, which means that $d(v)$ and $\bar{T}_{f_v}$ are nearly uncorrelated. Figures 2e and f also show that the factor propagation process under the LeastFrequent strategy is unstable.

Furthermore, we performed the same experiments on the DBLP graph [12] which contains 317,080 vertices and 1,049,866 edges. We use the same experimental setting as in the previous experiments on the karate club graph. Figure 3 shows the experimental results. Overall, the experimental results obtained on the DBLP graph exhibit similar patterns as on the karate club graph. Notably,

- When the Random strategy is applied, $\bar{T}_{f_v}$ is very close to $d(v)$, so it is consistent with Theorem 1.
- When the MostFrequent strategy is applied, about 4% of factors dominate the factor propagation process. In particular, 78.3% of factors propagating in the

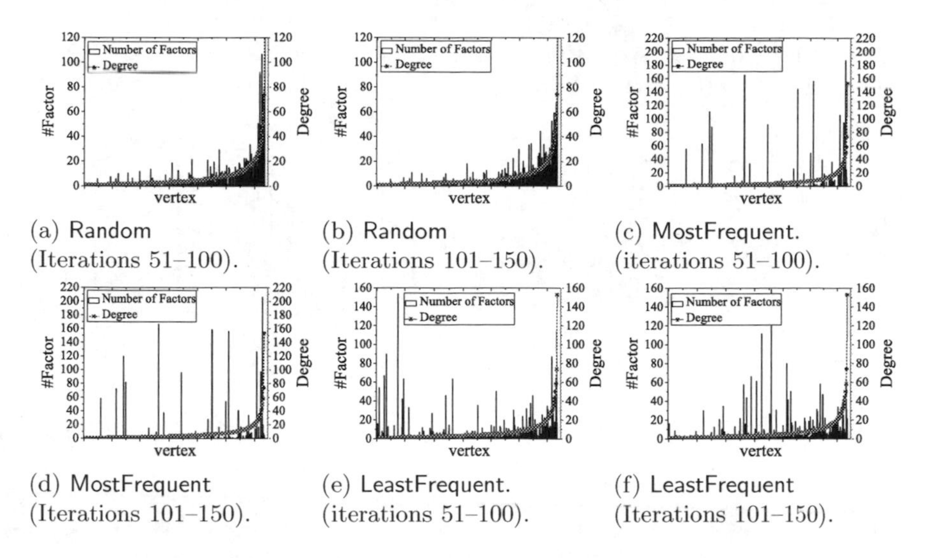

(a) Random (Iterations 51–100).
(b) Random (Iterations 101–150).
(c) MostFrequent. (iterations 51–100).
(d) MostFrequent (Iterations 101–150).
(e) LeastFrequent. (iterations 51–100).
(f) LeastFrequent (Iterations 101–150).

Fig. 3. The degree $d(v)$ of vertex v and the average number $\bar{T}_{f_v}$ of propagation of factor f_v on the DBLP graph in the last 100 iterations.

middle 50 iterations correspond to the interior vertices of the communities, and the proportion increases to 81.5% in the last 50 iterations.

- When the LeastFrequent strategy is applied, $d(v)$ and $\bar{T}_{f_v}$ are uncorrelated, and the factor propagation process is unstable.

Results. According to the experimental results shown above, the MostFrequent strategy is preferable since a fraction of factors dominate the factor propagation process after a long run of the factor propagation process, and these dominating factors correspond to the vertices in the interior of different communities. Therefore, in the rest of this paper, we only use the MostFrequent strategy in the factor propagation process.

Entropy of Factor Distribution. In order to explain the rationale behind the concept of entropy-based structural hole, we also study the entropy of the distribution of the factors in a vertex's factor list.

Figure 4 illustrates the changes of entropy $H(v)$ for all vertices v in the karate club graph in the 10th, 50th, 100th and 500th iterations. The red bars correspond to the structural hole vertices 2, 8, 13, 19 and 30 (the red ones in Fig. 1). In the 10th iteration, $H(v)$ is large for almost all vertices v in the graph. As the factor propagation process proceeds, the entropy $H(u)$ for most of the vertices u in the communities decreases more significantly than the entropy $H(v)$ for the structural hole vertices v. Therefore, in terms of entropy, the contrast between the structural holes and the vertices in the communities become more and more significant.

The entropy of the distribution of the factors in a vertex's factor list of DBLP graph is similar to the karate club graph, so we don't show the result repeatedly.

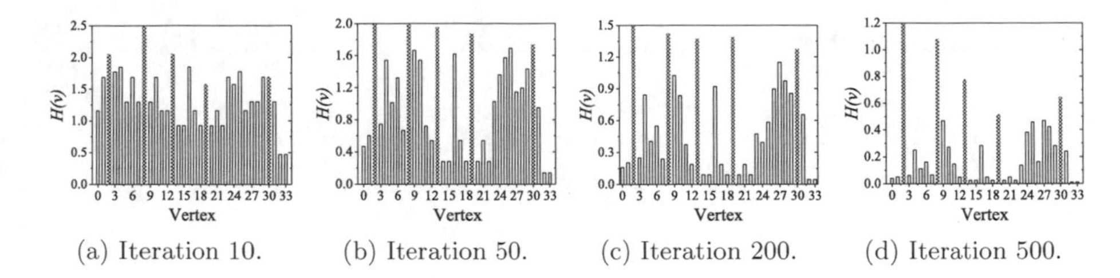

(a) Iteration 10. (b) Iteration 50. (c) Iteration 200. (d) Iteration 500.

Fig. 4. Variantion of entropy $H(v)$ of all vertices v in the karate club graph.

Results. The factor list L_v attached with vertex v can be seen as the "fingerprint" of v. A structural hole is likely to collect a number of heterogeneous factors propagated from various communities that the structural hole links to; while an interior vertex in a community is likely to collect a few homogeneous factors propagated from other vertices within the same community. The distribution of the factors in L_v characterizes the "role" of vertex v in the graph.

3 Structural Hole Detection Algorithm

The definition of entropy-based structural hole naturally implies a distributed parallel algorithm for structural hole detection on a big graph. In this section, we propose an algorithm called ESH based on the popular graph-parallel computing framework PowerGraph [13]. The ESH algorithm is described as a vertex-centric program that will be executed on each vertex in the graph according to the bulk synchronization protocol (BSP) [10].

A vertex-centric program running on PowerGraph basically needs to implement three functions `gather`, `apply` and `scatter`. For a vertex v, these functions work as follows.

- The `gather` function gathers and preprocesses data sent from the vertices adjacent to v in the graph.
- The `apply` function updates the data attached to v using the data obtained from the `gather` function.
- The `scatter` function scatters the updated data attached to v to all neighbours of v in the graph.

The ESH algorithm is described in Algorithm 1. The input is a graph $G = (V, E)$, the number k of structural hole vertices to be discovered from G and three integers n_1, n_2 and n_3 which indicate the numbers of iterations when to stop the three stages of the algorithm, respectively. In this paper, we assume that G has been partitioned into subgraphs and the subgraphs have already been placed on the nodes of a cluster.

Algorithm 1. ESH

```
Input: a graph G = (V, E) and integers k, n1, n2, n3
Output: k structural holes
 1: for all v ∈ V do
 2:     v.t ← 0; v.f ← v.id; v.L ← [ ]
 3:     v.L[0] ← v.f
 4:     scatter(v, v.f)
 5: for all v ∈ V do
 6:     while v.t ≤ n3 do
 7:         v.t ← v.t + 1
 8:         FP(v)
 9:     H(v) ← the entropy of the distribution of the factors v.L[n1], ..., v.L[n3]
10: return k vertices with the largest entropy H(v)

    Procedure FP(v)
11: f ← gather(v)
12: apply(v, f)
13: scatter(v)

    Procedure gather(v)
14: S ← ∅
15: for u ∈ N(v) do
16:     S.add(u.f)
17: return the most frequently occurred factor in S

    Procedure apply(v, f)
18: v.L[t_v] ← f
19: if v.t ≤ n2 then
20:     v.f ← the factor selected from v.L[0], ..., v.L[v.t] uniformly at random
21: else
22:     v.f ← the factor selected from v.L[n1], ..., v.L[n2] uniformly at random

    Procedure scatter(v)
23: active all the neighbors of v to gather v.f from v
```

4 Experimental Evaluation

This section evaluates the proposed structural hole detection algorithm ESH by experiments.

4.1 Experimental Setting

We implemented ESH on PowerGraph 2.2 and complied it by gcc 4.8.5. For comparisons, we implemented a bunch of existing algorithms, whose details are presented in Table 1. Except HAM which was implemented in MatLab, all the algorithms were implemented in C++. Since PageRank and 2-Step only require local neighborhood information of a vertex, we parallelized PageRank and 2-Step and implemented them on PowerGraph 2.2, which are called PageRank-P and 2-Step-P, respectively. The other algorithms are not parallelized because they require global information of a graph and are not suitable to be parallelized. For example, BICC must access l-hop neighbors of a vertex, and AP_BICC must know the articulation points of a graph. Specifically, the serial structural hole detection algorithms PathCount, PageRank, 2-Step, ICC, BICC, AP_BICC, HIS, MaxD and HAM were executed on a PC with 3.40 GHz Intel Core i7 CPU and 32 GB of DDR3 RAM, running Ubuntu 16.04. The distributed parallel algorithms ESH, PageRank-P and 2-Step-P were run on a cluster of 30 Dell PowerEdge 370 servers.

Every node is installed with two Intel Xeon E5-2609 V3 CPUs (1.9 GHz and 12 cores) and 32 GB of DDR3 RAM.

Table 1. Comparisons between structural hole detection algorithms.

Algorithm	Time complexity*	Communities
PathCount [6]	$O(\|V\|\|E\|)$	Not required
PageRank [7]	$O(T\Delta)$	Not required
2-Step [8]	$O(\|E\|^{1.5})$	Not required
HIS [1]	$O(T2^c\|E\|)$	Required
MaxD [1]	$O(4^c t(\|V\|))$	Required
ICC [9]	$O(\|V\|\|E\|)$	Not required
BICC [9]	$O(l(\|V\| + \|E\|))$	Not required
AP_BICC [9]	$O(\|V\| \log k + \|E\|)$	Not required
HAM [2]	$O(\|V\|^3)$	Not required
ESH (this paper)	$O(T\Delta)$	Not required

*Notation: $|V|$—number of vertices, $|E|$—number of vertices, T—number of iterations, Δ—maximum vertex degree, c—number of communities, l—radius of neighborhood, and k—number of structural holes.

Table 2. Statistics of graph datasets.

Dataset	Type	#Vertices	#Edges	#Communities	Diameter	Size
DBLP	Collaboration network	317,080	1,049,866	13,477	21	13.9 MB
Orkut	Social network	3,072,441	117,185,083	6,288,363	9	1.8 GB
Friendster	Social network	65,608,366	1,806,067,135	957,154	32	32.4 GB

In order to show that the vertices with the largest degrees may not be good structural holes, we implemented a distributed parallel algorithm called k-Hub that returns k vertices with the largest degrees as structural holes.

The experiments were carried out on three real graphs *DBLP*, *Orkut* and *Friendster* obtained from the SNAP project of Stanford University [12]. The statistics of these graphs are listed in Table 2.

By default, we use the following parameters in the experiments. For ESH, we set the numbers of end iterations of three stages to $n_1 = 33, n_2 = 66, n_3 = 100$ and set the number k of detected structural holes to 20. For PageRank and PageRank-P, the number T of iterations is set to 100 as well. For AP_BICC, the neighbors of a vertex are bounded to within $l = 3$ hops.

4.2 Evaluation Measures

We evaluate the quality of a set S of structural holes by three measures described below. The measures assess different aspects of the detected structural holes.

- **Average weighted number of connected communities (AWCC).** The AWCC measure was originally proposed in [9]. Let $C(v)$ be the set of communities adjacent to structural hole vertex $v \in S$ and $d(v)$ be the degree of v. Intuitively, the larger $|C(v)|$ is and the smaller $d(v)$ is, the more likely v is a structural hole. Thus, the quality of S is measured by

$$AWCC(S) = \frac{1}{|S|} \sum_{v \in S} \frac{|C(v)|}{d(v)}.$$

  We further revise this measure. Let $C_\alpha(v)$ be the set of communities that connect to structural hole vertex v and consist of more than α vertices. By setting threshold α, small communities are filtered out to eliminate noises. Thus, the quality of S is measured by

$$AWCC_\alpha(S) = \frac{1}{|S|} \sum_{v \in S} \frac{|C_\alpha(v)|}{d(v)}. \quad (2)$$

  If not otherwise stated in the rest of the paper, the AWCC measure refers to our revised AWCC measure.
- **Average community size (ACS) connected to structural holes.** Intuitively, the more and the larger communities a vertex connects to, the more important role it plays in information diffusion over the graph. Thus, we design the following measure to assess the quality of S.

$$ACS(S) = \frac{1}{|S|} \sum_{C \in \bigcup_{v \in S} C(v)} |C|. \quad (3)$$

4.3 Competing Algorithms

As the first step of our experimental evaluation, we select a list of competitors for ESH from the 12 algorithms mentioned earlier. First, we ran the serial algorithms on the *DBLP* graph, the smallest one in our experiments, on the PC that we used in the experiments. We have the following observations:

- MaxD returns the result in more than 20 h.
- ICC and PathCount return the result in more than 11 h.
- HIS fails to complete in 24 h due to its extremely high time complexity.
- HAM, which is implemented in MatLab, fails to complete in 24 h.
- Although BICC completes in 192 s on the *DBLP* graph, it takes more than 12 h on the *LiveJournal* graph.

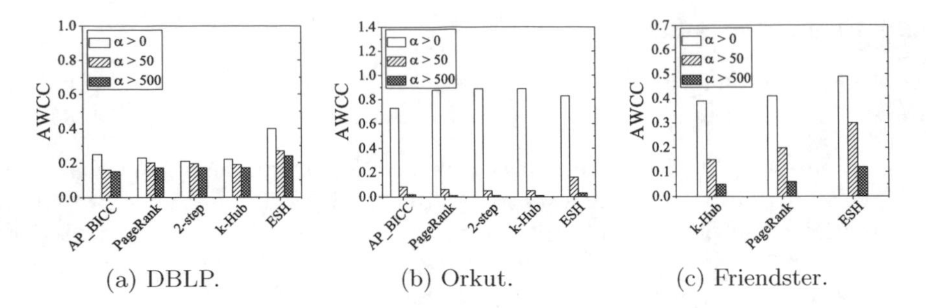

Fig. 5. AWCC values of structural holes detected by AP_BICC, PageRank, 2-Step, k-Hub and ESH.

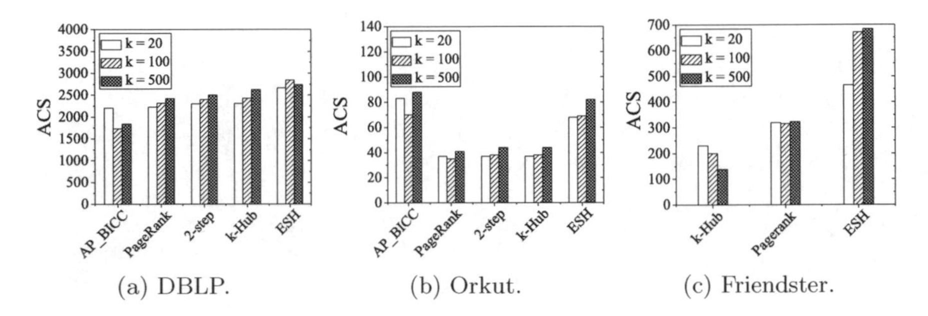

Fig. 6. ACS values of structural holes detected by AP_BICC, PageRank, 2-Step, k-Hub and ESH.

In addition, HIS and MaxD must be provided with the communities of a graph as additional input, so they differ from the other algorithms that only utilize graph structures. Thus, we can eliminate MaxD, ICC, PathCount, HIS, HAM and BICC from the list of competitors, so we select PageRank, 2-Step, AP_BICC, k-Hub, PageRank-P and 2-Step-P as the competitors for ESH in our experiments.

4.4 Quality Evaluation

First, we evaluate the quality of the structural holes detected by ESH using the measures described in Sect. 4.2.

Evaluation by the AWCC Measure. Figure 5 shows the AWCC values (Eq. (2)) of the structural holes detected by PageRank, 2-Step, AP_BICC, k-Hub and ESH on graphs *DBLP*, *Orkut* and *Friendster*. In the experiment, we varied α, the threshold on community size, from 0 to 500. Except on *Orkut* and at $\alpha = 0$, the AWCC value of the structural holes detected by ESH is much larger than the AWCC values of the structure holes produced by the other algorithms. It verifies that the structural holes detected by ESH connect to more important communities. Therefore, ESH outperforms the other algorithms in terms of AWCC. Notably, on *Orkut* and at $\alpha = 0$, the AWCC value for ESH is less

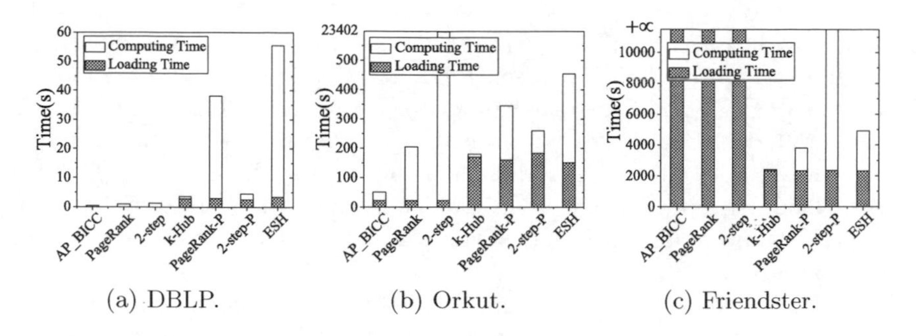

(a) DBLP. (b) Orkut. (c) Friendster.

Fig. 7. Execution time of structural hole detection algorithms.

than the AWCC values for PageRank, 2-Step and k-Hub because *Orkut* contains a very large number of communities, and the large amount of small communities become noises and have a negative effect on the AWCC measure when $\alpha = 0$. For this reason, we propose the enhanced AWCC measure to evaluate the quality of structural holes.

Evaluation by the ACS Measure. Figure 6 shows the ACS values of the structural holes (Eq. (3)) detected by PageRank, 2-Step, AP_BICC, k-Hub and ESH on the four graphs. In the experiment, we varied k, the number of structural holes, from 30 to 500. For any k, we find that the ACS value of the structural holes detected by ESH is larger than the ACS values of the structural holes detected by the other algorithms. It verifies that the structural holes detected by ESH connect to more and larger communities. Therefore, ESH outperforms the other algorithms in terms of ACS.

Summary. The ESH algorithm is able to detect the structural holes in a graph that connect to more and larger communities and that are separated from each other in the graph. It verifies the effectiveness of ESH. The most interesting finding in this experiment is that the simple factor diffusion process adopted by ESH can yield better structural holes in practice.

4.5 Efficiency Evaluation

This subsection evaluates the efficiency of ESH. Figure 7 shows the execution time on the three graphs. The serial algorithms AP_BICC, PageRank and 2-Step were executed on the PC, and the distributed parallel algorithms PageRank-P, 2-Step-P and ESH were run on the computer cluster. The executing time shown in Fig. 7 consists of the data loading time and the computing time.

The serial algorithms AP_BICC and PageRank run faster than the distributed parallel algorithms PageRank-P, 2-Step-P and ESH on DBLP and Orkut. This is because the distributed parallel algorithms spend much time in loading data and communication between the cluster nodes. It is not worth when the graph is small. However, as the graph size increases, the computing time of AP_BICC and PageRank increases significantly comparing with their data loading time, which indicates that AP_BICC and PageRank are not scalable to large graphs. As verified by Fig. 7c, when the graph is too large to fit into the main memory of the PC, AP_BICC and PageRank cannot terminate in reasonable time.

The serial algorithm 2-Step only runs fast on the smallest DBLP graph but performs worse than all the other algorithms on all the other graphs. The k-Hub algorithm shows good efficiency on all graphs. However, as verified by the previous experiments, the quality of the structural holes detected by this algorithm are not good.

The distributed parallel algorithms PageRank-P and ESH demostrate good scalability. The ratio between the data loading time and the computing time decreases as the graph becomes larger. In more details, PageRank-P is faster than ESH because PageRank-P only performs simple mathematical computation in its vertex-centric program. However, the quality of the structural holes detected by this algorithm is not as good as ESH as verified by the previous experiment.

The distributed parallel algorithm 2-Step-P cannot work on Friendster and always make one node of the cluster halt whenever we try to run the algorithm. This is because the algorithm needs to know the 2-hop neighbors of a vertex, which are too many to fit into the main memory for the Friendster graph.

5 Related Work

This section reviews the related work on structural hole detection and information diffusion.

5.1 Structural Hole Detection

Since Burt first introduced the concept of structural hole in [14], many algorithms have been proposed to detect structural holes in graphs in recent years. Many structural hole detection algorithms are based on *centrality* measures. Goyal and Vega-Redondo [6] evaluate the *betweenness centrality* [15] (the number of shortest paths passing a vertex) of the vertices in a graph and select k vertices with the largest betweenness centrality as structural holes. In this paper, we call this algorithm PathCount. Rezvani *et al.* [9] propose the concept of *inverse closeness centrality (ICC)* of a vertex. The basic idea is that the average distance between the vertices in a graph increases the most when the structural holes are removed. The time complexity of ICC is $O(mn)$, where m and n are the number of edges and vertices in the input graph, respectively. As the time complexity is too high, the BICC algorithm is proposed by limiting the computation of the ICC of a vertex in its l-hop neighborhood, i.e. *bounded inverse closeness centrality (BICC)*. To further improve efficiency, the AP_BICC algorithm is proposed, which first computes the BICC of the articulation points in a graph to filter out the vertices that need not to be considered. The 2-Step algorithm [8] proposed by Tang *et al.* introduces a concept that is a special case of BICC. *PageRank* [7] is a crucial technique to measure the importance of the vertices in a graph. Although its original use is different from the centrality measures, PageRank can also be applied in structural hole detection.

The algorithms above only make use of the topological structure of a graph in structural hole detection. Different from these work, Lou and Tang [1] find structural holes using prior knowledge about communities. They propose two algorithms, namely HIS and MaxD, whose time complexity are exponential to the number of cummunities. Moreover, the algorithms require information about communities in prior, which is often unavailable in data, and needs extra time to be discovered by the community detection algorithms.

In most work on structural hole detection, the structural holes in a graph are detected separately from the communities. Recently, He *et al.* [2] proposed a new approach to find structural holes and communities at the same time. They define a vertex to be a structural hole if the vertex connects to many vertices that belong to other communities. They detect communities through intra-community neighbors and structural holes through inter-community neighbors using a harmonic function. However, the algorithm must do matrix computation, and the time complexity is $O(n^3)$, where n is the number of vertices.

5.2 Information Diffusion

The research on information diffusion in online social networks mainly focus on three aspects [5], namely popular topic detection, information diffusion models and influential spreader identification. Popular topic detection aims to discover topics that are currently popular or will become popular in future [16]. Information diffusion models are used to understand how information diffuses within a network [17]. Influential spreader identification aims to find the most influential vertices in a network to ensure efficient diffusion of information [18].

The label propagation algorithm LPA [19] also applies the idea of information diffusion to community detection. Each vertex initializes itself with a distinct community label. Then, each vertex updates its community label based on its neighbors' community labels in each iteration. The community label is set to the most label of its neighbors. If two or more labels has the same occurrence number, one is selected randomly; while the others are broken down. Both LPA and ESH select labels (or factors) with the highest frequency. However, ESH maintains a list to store gathered factors. The list rather than using a single label is used as the fingerprint of the vertex. Moreover, LPA identifies a community by grouping vertices with same label; while ESH detects structural holes according to the distributions of the factors stored in the vertices' factor lists. Not surprisingly, ESH can be modified to identify communities as the information entropy of the interior vertex in a community is small. This will be our future work.

6 Conclusions

We address the structural hole detection problem on big graphs by proposing the ESH algorithm based on the graph-parallel processing framework PowerGraph.

The factor diffusion process adopted by ESH is very simple yet very effective too. The experimental results verify that ESH is able to detect structural holes of even better quality than the state-of-the-art algorithms. The ESH algorithm naturally fits the vertex-centric programming paradigm and is very easy to be implemented on PowerGraph. The experimental results verify that ESH is scalable to graphs consisting of billions of edges that are difficult to be handled by the state-of-the-art algorithms. These interesting findings make ESH an ideal choice for structural hole detection on big graphs.

Acknowledgements. This work was partially supported by the NSF of China (No. 61532015, No. 61672189). J. Li was supported by the NSF of China (No. 61832003, No. U1811461, No. 61732003); Y. Li was supported by the NSF of USA (No. 1741277 and No. 1829674); Y. Chen was supported by the NSF of China (No. 61602129 and No. 61872106).

References

1. Lou, T., Tang, J.: Mining structural hole spanners through information diffusion in social networks. In: WWW, pp. 825–836 (2013)
2. He, L., Lu, C., Ma, J., Cao, J., Shen, L., Yu, P.S.: Joint community and structural hole spanner detection via harmonic modularity. In: SIGKDD, pp. 875–884 (2016)
3. Andersen, R., Lang, K.J.: Communities from seed sets. In: WWW, pp. 223–232 (2006)
4. Wang, L., Lou, T., Tang, J., Hopcroft, J.E.: Detecting community kernels in large social networks. In: ICDM, pp. 784–793 (2011)
5. Guille, A., Hacid, H., Favre, C., Zighed, D.A.: Information diffusion in online social networks: a survey. SIGMOD Rec. **42**(2), 17–28 (2013)
6. Goyal, S., Vega-Redondo, F.: Structural holes in social networks. J. Econ. Theory **137**(1), 460–492 (2007)
7. Page, L., Brin, S., Motwani, R., Winograd, T.: The PageRank citation ranking: bringing order to the web. In: Stanford Digital Libraries Working Paper, vol. 9, no. 1, pp. 1–14 (1998)
8. Tang, J., Lou, T., Kleinberg, J.M.: Inferring social ties across heterogenous networks. In: WSDM, pp. 743–752 (2012)
9. Rezvani, M., Liang, W., Xu, W., Liu, C.: Identifying top-k structural hole spanners in large-scale social networks. In: CIKM, pp. 263–272 (2015)
10. Malewicz, G., et al.: Pregel: a system for large-scale graph processing. In: SIGMOD, pp. 135–146 (2010)
11. Zachary, W.W.: An information flow model for conflict and fission in small groups. J. Anthropol. Res. **33**(4), 452–473 (1977)
12. Leskovec, J., Krevl, A.: SNAP datasets: Stanford large network dataset collection, June 2014. http://snap.stanford.edu/data
13. Gonzalez, J.E., Low, Y., Gu, H., Bickson, D., Guestrin, C.: Powergraph: distributed graph-parallel computation on natural graphs. In: OSDI, pp. 17–30 (2012)
14. Burt, R.S.: Structural Holes: The Social Structure of Competition. Harvard University Press, Cambridge (1992)
15. Brandes, U.: A faster algorithm for betweenness centrality*. J. Math. Sociol. **25**(2), 163–177 (2010)

16. Rong, L., Qing, Y.: Trend analysis of news topics on twitter. Int. J. Mach. Learn. Comput. **2**(3), 327 (2012)
17. Gomez-Rodriguez, M., Leskovec, J., Schölkopf, B.: Structure and dynamics of information pathways in online media. In: WSDM, pp. 23–32 (2013)
18. Goldenberg, J., Libai, B., Muller, E.: Talk of the network: a complex systems look at the underlying process of word-of-mouth. Mark. Lett. **12**(3), 211–223 (2001)
19. Raghavan, U.N., Albert, R., Kumara, S.: Near linear time algorithm to detect community structures in large-scale networks. Phys. Rev. E Stat. Nonlinear Soft Matter Phys. **76**(3), 036106 (2007)

DynGraphGAN: Dynamic Graph Embedding via Generative Adversarial Networks

Yun Xiong[1,2(✉)], Yao Zhang[1], Hanjie Fu[1], Wei Wang[3], Yangyong Zhu[1,2], and Philip S. Yu[1,2,4]

[1] Shanghai Key Laboratory of Data Science, School of Computer Science, Fudan University, Shanghai, China
{yunx,yaozhang18,16210240052,yyzhu}@fudan.edu.cn

[2] Shanghai Institute for Advanced Communication and Data Science, Fudan University, Shanghai, China

[3] University of California, Los Angeles, Los Angeles, CA, USA
weiwang@cs.ucla.edu

[4] Computer Science Department, University of Illinois at Chicago, Chicago, IL, USA
psyu@uic.edu

Abstract. Graphs have become widely adopted as a means of representing relationships between entities in many applications. These graphs often evolve over time. Learning effective representations preserving graph topology, as well as latent patterns in temporal dynamics, has drawn increasing interests. In this paper, we investigate the problem of *dynamic graph embedding* that maps a time series of graphs to a low dimensional feature space. However, most existing works in the field of dynamic representation learning either consider temporal evolution of low-order proximity or treat high-order proximity and temporal dynamics separately. It is challenging to learn one single embedding that can preserve the high-order proximity with long-term temporal dependencies. We propose a Generative Adversarial Networks (GAN) based model, named DynGraphGAN, to learn robust feature representations. It consists of a generator and a discriminator trained in an adversarial process. The generator generates connections between nodes that are represented by a series of adjacency matrices. The discriminator integrates a graph convolutional network for high-order proximity and a convolutional neural network for temporal dependency to distinguish real samples from fake samples produced by the generator. With iterative boosting of the performance of the generator and discriminator, node embeddings are learned to present dynamic evolution over time. By jointly considering high-order proximity and temporal evolution, our model can preserve spatial structure with temporal dependency. DynGraphGAN is optimized on subgraphs produced by random walks to capture more complex structural and temporal patterns in the dynamic graphs. We also leverage sparsity and temporal smoothness properties to further improve the model efficiency. Our model demonstrates substantial gains over several baseline models in link prediction and reconstruction tasks on real-world datasets.

G. Li et al. (Eds.): DASFAA 2019, LNCS 11446, pp. 536–552, 2019.
https://doi.org/10.1007/978-3-030-18576-3_32

Keywords: Dynamic graph embedding · Adversarial Network · Data mining · Representation learning

1 Introduction

Graph embedding is known as learning graph representation in a latent low-dimensional space, in which each node can be represented as a vector. Graph embedding has demonstrated its utilities in a wide range of applications, such as merchant advertisements [1], friend recommendation in social networks [21] and gene expression analysis in biological networks [20]. Most existing works focus on analyzing a static graph. These include random walk based methods such as DeepWalk [26] and Node2vec [12], matrix factorization based methods such as HOPE [24], and deep autoencoder based methods such as SDNE [29].

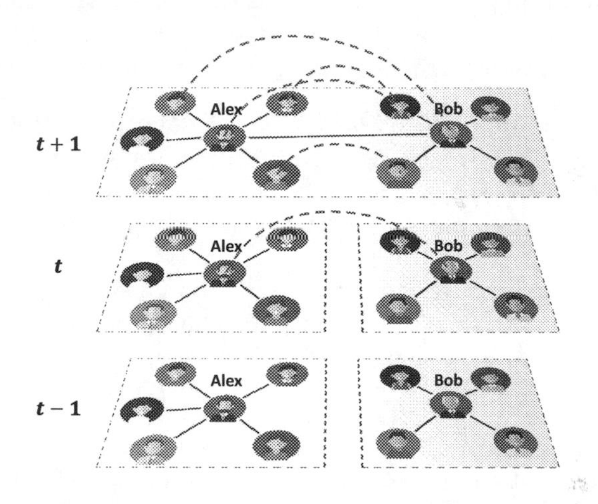

Fig. 1. An example of a dynamic social network.

However, many graphs in real world evolve over time. For example, new friendships may be formed in a social network [28]; and new co-authorships may emerge in a citation network [19]. We take the evolution of a social network as an example shown in Fig. 1, where each node indicates a user and each edge represents a friendship between two users. The black solid edges are existing relationships at time $t-1$ and the red dashed edges indicate newly formed relationships at time $\geq t$. At time $t-1$, users Alex and Bob are two core users that connect to most other users. They do not know each other until time t. At time t, a new friendship is built between Alex and Bob but their friends are still separated with no other connections yet. At time $t+1$, several new friendships are established between other users. Analyzing these temporal evolutions requires a graph embedding that can capture these changes as a function of time[1], while still preserves the structural feature over the entire graph. Traditional embedding techniques designed for static graphs fail to meet this requirement. Because they can only be applied to each graph independently, graphs at different time are represented in different latent spaces so that we cannot directly compare the embeddings of a given node across multiple time points to model the evolution of its relationships to others.

Various attempts have been made to jointly model spatial topology and temporal evolution. Matrix factorization based approaches such as TNE [36] and LIST [33] consider temporal evolution but limit the scope of spatial topology to first-order proximity. Features concerning high-order proximity are largely ignored in these models. Even though DHPE [35] preserves the high-order proximity, it does so by applying a static model to each graph and thus has limited capacity to capture temporal evolution. In this paper, we consider the task of learning node embeddings of dynamic graphs that can adequately capture graph topological features and temporal smoothness simultaneously. Despite the significance of it, this task is highly challenging, as summarized below.

- **High-order proximity** is often modeled by the powers of the adjacency matrix of a graph [3,32]. However, on a large graph, especially a dynamic one containing many snapshots, computing cost of high powers of adjacency matrices is expensive. It is a challenge to model high-order proximity efficiently.
- **Temporal evolution** is another key property of dynamic graphs. Most existing methods examine this property by modeling evolution at node- or edge-levels [33,36]. However, dynamics of a graph not only includes evolution of nodes and edges, but also the evolution of (sub-)structures.

While Zhou et al. [34] considered triads in graphs, it is still challenging to model temporal evolution at a higher level. To tackle the aforementioned challenges, instead of explicitly formulating the high-order proximity and temporal evolutions, we leverage the Generative Adversarial Network (GAN) [10] to capture the essence of real dynamic graphs and learn their distributions. This is inspired by the recent success of GAN (e.g., GraphGAN [30], NetGAN [2] and AIDW [5]) in learning embeddings for static graphs.

Our proposed model, DynGraphGAN, includes a generator and a discriminator. The generator generates a sequence of adjacency matrices mimicking the evolution of a subgraph (i.e., a graph connecting a subset of nodes). The generated subgraph evolutions and the subgraph evolutions from real data are fed to the discriminator whose job is to assess the probability of which a given subgraph evolution is real. The discriminator needs to be carefully designed. If the discriminator is weak and converged quickly, there will be no longer useful

[1] Following settings in related work, we only consider discrete-time dynamic graphs since we can take discrete snapshots from a continuously varying graph. This is also the case of many real applications where recording every changes is expensive or unnecessary, e.g., brain networks and bibliographic networks.

Table 1. Important notations.

Symbol	Definition
$\mathcal{V}$	The node set of a graph. Specially, a subset of nodes is defined as $\mathcal{V}'$
$\mathcal{E}$	The edge set of a graph
$\mathcal{G} = (\mathcal{V}, \mathcal{E})$	A static graph with a node set $\mathcal{V}$ and an edge set $\mathcal{E}$
$\mathcal{S}_{\mathcal{V}}$	The adjacency matrix representing the connections between nodes in $\mathcal{V}$
T	The number of timestamps of a dynamic graph ranging from 1 to T
$\mathbb{G} = \{\mathcal{G}^1, \ldots, \mathcal{G}^T\}$	The definition of a dynamic graph with a set of graph snapshots along time
$\mathbb{S}_{\mathcal{V}} = \{\mathcal{S}_{\mathcal{V}}^1, \ldots, \mathcal{S}_{\mathcal{V}}^T\}$	The structure of a dynamic graph represented by a sequence of adjacency matrices
G	The generator function
D	The discriminator function
$V(G, D)$	The value function with respect to G and D
$\{X^1, \ldots, X^T\}$	The node embedding matrices in the discriminator
$\{A^1, \ldots, A^T\}$	An input sample to the discriminator D, which is a sequence of adjacency matrices
$\tilde{A}^t$	The adjacency matrix with self-connections
$\hat{A}^t$	The reconstruction of A^t
H_l^t	The embeddings at the l-th layer at snapshot t in the graph convolutional network
$\mathbf{u_i^t}$, $\widetilde{\mathbf{u}}_\mathbf{i}^\mathbf{t}$	Embeddings of node u_i at time t, differing only in their random initialization

information to propagate to the generator, in which case, the generator fails to capture the true data distribution. Thus in our design, the discriminator employs two components to learn graph embeddings: a multi-layer Graph Convolutional Network (GCN) [17] for modeling high-order proximity, followed by a Convolutional Neural Network (CNN) for capturing temporal evolution. Similar to WGAN-GP [13], we use a gradient penalty to ensure smooth convergence.

It is worth noting that different from previous GAN-based models [2,5,30] that directly use nodes, edges, or random walks, we feed sequences of subgraphs to the discriminator. Subgraphs contain richer topological information that can be further revealed by graph convolutional layers, where high-order proximities are extracted. The following CNN module is then able to examine community-level temporal evolutions. Moreover, instead of sampling subgraphs fully randomly, we use random walks with different starting nodes to sample sets of nodes and then extract the corresponding subgraphs. This procedure ensures that each node can be covered, and meaningful structural and temporal patterns are contained in sampled subgraphs.

Our main contributions are three-folded. We propose DynGraphGAN, an effective GAN-based approach for dynamic graph embedding, which is suited for most common tasks such as link reconstruction and prediction; we apply a multi-layer GCN to model high-order proximity based on subgraphs. The following CNN module is then used to capture temporal evolution over time; we conduct experiments on four real-world datasets and compare DynGraphGAN with five baseline methods on two tasks. Our model is able to achieve significant improvements over baseline methods in all settings.

2 Problem Statement

A graph $\mathcal{G}$ can be considered as $(\mathcal{V}, \mathcal{E})$, where $\mathcal{V}$ is a set of nodes and $\mathcal{E} \subseteq \mathcal{V} \times \mathcal{V}$ contains edges between nodes. In this paper, we only consider undirected unweighted graphs. Let $\mathcal{S}_{\mathcal{V}}$ be the adjacency matrix of nodes $\mathcal{V}$. We have $(\mathcal{S}_{\mathcal{V}})_{ij} = (\mathcal{S}_{\mathcal{V}})_{ji} = 1$ if $(u_i, u_j) \in \mathcal{E}$, and 0 otherwise.

Definition 1. Dynamic graph. A dynamic graph is considered as a series of T graph snapshots $\mathbb{G} = \{\mathcal{G}^1, \ldots, \mathcal{G}^T\}$ defined over a given node set $\mathcal{V}$ with evolving edge sets $\mathcal{E}^t$ at time t. Here $\mathcal{G}^t = (\mathcal{V}, \mathcal{E}^t)$ is the graph snapshot at time t, whose adjacency matrix is denoted by $\mathcal{S}_{\mathcal{V}}^t$. Then the evolution of the dynamic graph can be represented by a sequence of adjacency matrices $\mathbb{S}_{\mathcal{V}} = \{\mathcal{S}_{\mathcal{V}}^1, \ldots, \mathcal{S}_{\mathcal{V}}^T\}$ over node set $\mathcal{V}$.

Definition 2. Dynamic graph embedding. Dynamic graph embedding aims to find a low-dimensional latent space to represent a dynamic graph. This can be represented by a mapping function from a node set to a low-dimensional space at time t, $f^t : u_i \mapsto \Re^d$, where $u_i \in \mathcal{V}$ and d is the dimension of the latent space.

Ideally, nodes sharing similar topological features at any time t should be close to each other in this latent space, and a node should not "move" too much in this latent space between two consecutive graph snapshots. For simplicity, $f^t(u_i)$ is denoted as $\mathbf{u_i^t}$ in the remainder of this paper. The notations are summarized in Table 1.

3 Methodology

In this section, we introduce the framework of DynGraphGAN, a Generative Adversarial Network based model for dynamic graph embedding. As shown in Fig. 2, our model is composed of a generator G and a discriminator D. During each iteration of training, instead of simply providing random noises to the generator, we sample a sizeable subset $\mathcal{V}'$ of nodes as the input and use the induced subgraph snapshots $\mathbb{S}_{\mathcal{V}'}$ as the "real sample" in our training. Using subgraphs will not only provide an effective means to capture proximity features [7,18] but also enable efficient computation. There exist many sampling algorithms to generate $\mathcal{V}'$. We adopt the random walk approach of Node2vec [12] here, because

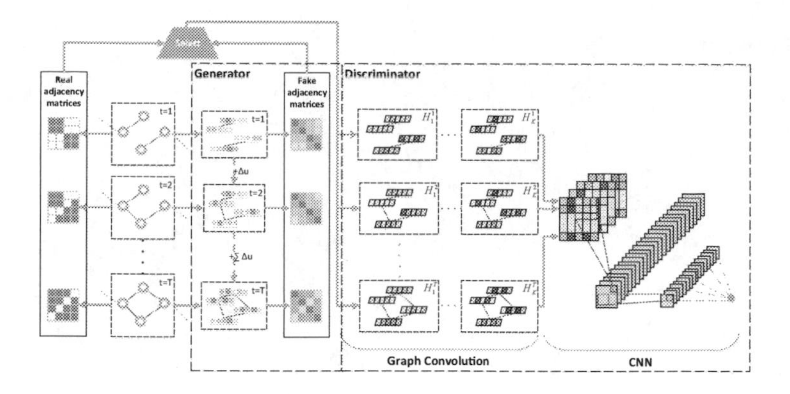

Fig. 2. Framework of DynGraphGAN

subgraphs induced by random walk contain richer structural information, e.g., denser connections. In the experimental section, we find a subgraph size around 80 generally leads to good performance.

Given $\mathcal{V}'$, the generator G returns a sequence of T adjacency matrices, denoted by $\hat{\mathbb{S}}_{\mathcal{V}'} \triangleq G(\mathcal{V}')$, representing the evolution of subgraph formed by nodes in $\mathcal{V}'$, which will be used as the "fake samples" in the training. The discriminator D is then trained to distinguish the real samples $\mathbb{S}_{\mathcal{V}'}$ from the fake samples $\hat{\mathbb{S}}_{\mathcal{V}'}$. Concretely, D and G play the minimax game with value function $V(G, D)$:

$$\begin{aligned}\min_G \max_D V(D,G) = & \mathop{\mathbb{E}}_{\mathcal{V}' \sim \mathbb{P}_{\mathcal{V}}} \left[D\left(\mathbb{S}_{\mathcal{V}'}\right) - D\left(G\left(\mathcal{V}'\right)\right)\right] \\ & + \alpha \mathop{\mathbb{E}}_{\mathcal{V}' \sim \mathbb{P}_{\mathcal{V}}} \left(\left\|\nabla_{\hat{\mathbb{S}}_{\mathcal{V}'}} D\left(\hat{\mathbb{S}}_{\mathcal{V}'}\right)\right\|_2 - 1\right)^2 \\ & + \lambda \left\|\Delta_G\right\|_F^2,\end{aligned}$$

where $\mathbb{P}_{\mathcal{V}}$ is a distribution of $\mathcal{V}$ from which $\mathcal{V}'$ is sampled by Node2vec. We adopt a linear gradient penalty function (defined on $\hat{\mathbb{S}}_{\mathcal{V}'}$ sampled uniformly along straight lines between $\mathbb{S}_{\mathcal{V}'}$ and $G\left(\mathcal{V}'\right)$) to provide stable convergence [13]. A $\|\Delta_G\|_F^2$ regularization term is used to preserve embedding smoothness between adjacent snapshots, where $Delta_G$ is the union of all $\left\{\mathbf{u_i^t} - \mathbf{u_i^{t-1}}\right\}_{u_i \in \mathcal{V}, t \leq T}$. α and λ are hyperparameters for tuning the gradients of the discriminator and the temporal smoothness. Both G and D are optimized in the adversarial process. In the rest of this section, we describe details of DynGraphGAN design and different strategies for adjacency matrix generation.

3.1 Generator

Given a subset of nodes $\mathcal{V}'$, G generates T adjacency matrices to represent the evolution of a subgraph over time:

$$\hat{\mathbb{S}}_{\mathcal{V}'} \triangleq G(\mathcal{V}') = \left\{G^0(\mathcal{V}'), G^1(\mathcal{V}'), \ldots, G^T(\mathcal{V}')\right\} \quad (1)$$

$$= \left\{\left(p_{ij}^0\right)_{|\mathcal{V}'|^2}, \left(p_{ij}^1\right)_{|\mathcal{V}'|^2}, \ldots, \left(p_{ij}^T\right)_{|\mathcal{V}'|^2}\right\}, \quad (2)$$

where in Eq. (1), $G^t(\mathcal{V}') \subseteq \Re^{|\mathcal{V}'|\times|\mathcal{V}'|}$ is the adjacency matrix at time t. The probability of whether there is an edge between two nodes $u_i, u_j \in \mathcal{V}'$ at time t is defined by p_{ij}^t in Eq. (2). Intuitively, p_{ij}^t should depend on the similarity or distance between the embeddings of u_i and u_j at time t. Thus we employ a sigmoid function of their learned embeddings, that is, $p_{ij}^t = \sigma(\mathbf{u_i^t}, \mathbf{u_j^t})$, which defines a mapping $\mathcal{V}' \times \mathcal{V}' \rightarrow (0, 1)$. The closer the two embeddings $\mathbf{u_i^t}$ and $\mathbf{u_j^t}$, the higher the probability that u_i and u_j are connected by an edge at time t. In this paper, we investigate two popular choices of the sigmoid function.

Inner Product Similarity: We use the sigmoid function of the inner product of two nodes' embeddings:

$$\sigma_{inner}\left(\mathbf{u_i^t}, \tilde{\mathbf{u}}_\mathbf{j}^\mathbf{t}\right) = \left[1 + \exp\left(-\left\langle \mathbf{u_i^t}, \tilde{\mathbf{u}}_\mathbf{j}^\mathbf{t}\right\rangle\right)\right]^{-1}. \quad (3)$$

$\langle\cdot,\cdot\rangle$ denotes the operation of the inner product. The sigmoid function squeezes node similarity into edge probability in $(0, 1)$. Instead of using the inner product of $\mathbf{u_i^t}$ and $\mathbf{u_j^t}$, we choose to replace $\mathbf{u_j^t}$ by $\tilde{\mathbf{u}}_\mathbf{j}^\mathbf{t}$ from another node embedding. Here, $\mathbf{u_i^t}$ and $\tilde{\mathbf{u}}_\mathbf{i}^\mathbf{t}$ are two node embeddings learned in the same way but with different random initializations. This is because that it has been observed that the inner product of $\mathbf{u_i^t}$ and $\mathbf{u_j^t}$ (learned from the same initialization) may result in unexpected artifact. Replacing $\mathbf{u_j^t}$ by $\tilde{\mathbf{u}}_\mathbf{j}^\mathbf{t}$ may avoid such artifact and make the training more stable and converge faster [4,25]. Therefore, in our model, we train two graph embeddings $\mathbf{u_i^t}$ and $\tilde{\mathbf{u}}_\mathbf{i}^\mathbf{t}$ in parallel using different random initializations. The final output of the graph embedding is the average of the two.

Euclidean Distance: The Euclidean distance indicates the difference between two nodes in the graph embedding space. We use a sigmoid function of negative Euclidean distance to define the edge probability.

$$\sigma_{diff}\left(\mathbf{u_i^t}, \mathbf{u_j^t}\right) = \left[1 + \exp\left(\left(\mathbf{u_i^t} - \mathbf{u_j^t}\right)^2\right)\right]^{-1}. \quad (4)$$

Because only matrices subtraction is used in this function, we do not need to use a second embedding $\tilde{\mathbf{u}}_\mathbf{i}^\mathbf{t}$.

3.2 Discriminator

Given a subset of nodes $\mathcal{V}'$, the discriminator D aims to distinguish real subgraph samples from fake subgraph samples produced by generator G. To facilitate the discussion, we let $\{A^1, \ldots, A^T\}$ denote the sequence of adjacency matrices of a sample (which could be either a real sample $\mathbb{S}_{\mathcal{V}'}$ or a fake one $\hat{\mathbb{S}}_{\mathcal{V}'}$), where $A^t \subseteq R^{|\mathcal{V}'| \times |\mathcal{V}'|}$ is the adjacency matrix at time t. $\tilde{A}^t = A^t + I$ represents the adjacency matrix with self-connections added, and $\tilde{D}^t = diag\left(\{\sum_j \tilde{A}^t_{ij}\}_i\right)$ the diagonal node degree matrix. The objective of the discriminator D is to maximize the probability of assigning the correct label to each sample.

As discussed in Sect. 1, the discriminator should be carefully designed so that it can provide useful information to the generator. To achieve this goal, we use the following components in the discriminator of DynGraphGAN in succession: (1) a graph convolutional network (GCN) [17] that encodes neighborhood features of nodes and captures high-order proximity through multi-layer propagation; (2) a convolutional neural network (CNN) that learns temporal graph evolution along the time dimension.

The GCN treats each snapshot separately. At time t, it takes A^t and $X^t_{\mathcal{V}'}$ as inputs through a layered propagation. Here $X^t_{\mathcal{V}'}$ is the embeddings of nodes in $\mathcal{V}'$ at time t, learned by the discriminator. At the l-th layer, we apply the forward propagation of graph convolutional network [17] defined as $H^t_{l+1} = \varphi\left((\tilde{D}^t)^{-\frac{1}{2}}\tilde{A}^t(\tilde{D}^t)^{-\frac{1}{2}}H^t_l W^t_l\right)$. Here, $(\tilde{D}^t)^{-\frac{1}{2}}\tilde{A}^t(\tilde{D}^t)^{-\frac{1}{2}}$ is the approximate Laplacian produced by the matrix multiplication of an adjacency matrix with self-connections $\tilde{A}^t$ and the diagonal node degree matrix $\tilde{D}^t$. $W^t_l \in R^{d_l \times d_{l+1}}$ is a matrix of filter parameters, $H^t_l \in R^{|\mathcal{V}'| \times d_l}$ is an embedding matrix of nodes and d_l is the embedding dimension at the l-th layer. φ is an activation function. Specifically, H^t_0 is set to node embedding $X^t_{\mathcal{V}'}$. After K convolutional operations, H^t_K captures the high-order proximity by convolutions of neighborhood which are K distance from the central node. The output of the GCN is $\{H^t_K\}^T_{t=1}$.

To capture temporal dependencies over time, we apply a CNN to the node embeddings generated by the GCN. For each node v, we concatenate the node embeddings in ascending order of time into a matrix of size $T \times d_K$. We call this matrix the *image* of node v.

$$image_v = stack\left(\left\{H^t_K(v)\right\}_{t \leq T}\right),$$

where $H^t_K(v)$ is the embedding vector of node v produced by the graph convolutional network at time t. As shown in Fig. 2, T vectors in green can be found in the output of the graph convolutional network. We stack these vectors along the time dimension and get a green matrix. Similarly, we can generate a yellow matrix, a blue matrix and a grey matrix for the remaining three nodes. Next, these images are stacked together to form a tensor $tensor \in R^{T \times d_K \times |\mathcal{V}'|}$:

$$tensor = stack\left(\{image_v\}_{v \in \mathcal{V}'}\right).$$

In the example in Fig. 2, the four matrices are stacked to form a tensor with four channels.

In the CNN, multiple kernels are used as the convolution filters. Specially, the last kernel is used to combine all features together. Finally, a fully-connected layer is used to produce a discriminative probability for a given $\{A^1, \ldots, A^T\}$:

Algorithm 1. DynGraphGAN training with batch size 64, penalty coefficient $\alpha = 0.1$ and learning rate $\tau = 0.001$, optimized by Adam with hyperparameters $\beta_1 = 0.5, \beta_2 = 0.9, \epsilon = 0.0001$.

Input: Dynamic graph, $\mathbb{G}$; Model parameters, α, λ, γ; Learning rate τ; Node2vec hyperparameters, p, q, nw, nl; Adam hyperparameters, $\beta_1, \beta_2, \epsilon$;
Output: Node embedding;
1: $walks \leftarrow$ node2vecWalk($\mathcal{G}_T$,p,q,nw,nl)
2: **for** $i = 1$ to n **do**
3: **for** node set $\mathcal{V}' \in walks$ **do**
4: $x \leftarrow \mathcal{S}_{\mathcal{V}'}$ // real sample
5: $\hat{x} \leftarrow (\gamma \cdot (1 - \mathcal{S}_{\mathcal{V}'}) + \mathcal{S}_{\mathcal{V}'}) \circ G(\mathcal{V}')$ // fake sample
6: $\mu \sim U[0, 1]$ // random value
7: $\tilde{x} \leftarrow \mu \cdot x + (1 - \mu) \cdot \hat{x}$
8: $\mathcal{L}_{penalty} \leftarrow (\|\nabla_{\tilde{x}} D(\tilde{x})\|_2 - 1)^2$
9: $\mathcal{L}_D \leftarrow D(\hat{x}) - D(x) + \alpha \mathcal{L}_{penalty}$
10: $\mathcal{L}_G \leftarrow -D(\hat{x}) + \lambda \|\Delta_G\|_F^2$
11: $\theta_D \leftarrow Adam(\nabla_{\theta_D} \mathcal{L}_D, \theta_D, \tau, \beta_1, \beta_2, \epsilon)$
12: $\theta_G \leftarrow Adam(\nabla_{\theta_G} \mathcal{L}_G, \theta_G, \tau, \beta_1, \beta_2, \epsilon)$
13: **end for**
14: **end for**
15: **return** $\left\{\frac{\mathbf{u}_i^t + \tilde{\mathbf{u}}_i^t}{2}\right\}_{u_i \in |V|, t \leq T}$

$$D(\{A^1, \ldots, A^T\}) = full(cnn(tensor)).$$

Here, cnn is a CNN function and $full$ is a fully-connected operation.

In both GCN and CNN, we use LeakyReLU [22,31] as the activation function. We demonstrate in Sect. 4 that combining GCN and CNN can capture the temporal dependency and high-order proximity.

3.3 Implementation Details

Node Sampling. The generator G simulates subgraph evolution on a subset of nodes $\mathcal{V}'$ selected by Node2vec [12]. The random walk approach of Node2Vec is a 2nd order random walk that explores network neighborhood.

Attention Balance. Most graphs we study are sparse in that each node only connects to a small number of other nodes at any time. This is particularly true for graphs representing social networks. This results in an overwhelming percentage of 0s in the adjacency matrices. In the generative adversarial network, if the model pays equal attention to each 0 and 1, these 0s will dominate the gradient descent procedure and produce a suboptimal solution. To overcome this sparsity challenge, it is necessary to balance the attention by reconstructing the input of discriminator D. Inspired by [29], a hyperparameter γ is used to reduce the attention of 0s in adjacency matrices. The reconstruction of the adjacency matrix at time t is denoted as $\hat{A}^t$:

$$\hat{A}^t = \begin{cases} (\gamma \cdot (1 - \mathcal{S}^t_{\mathcal{V}'}) + \mathcal{S}^t_{\mathcal{V}'}) \circ A^t & \{A^1, \ldots, A^T\} \sim \mathbb{P}_G, \\ A^t & otherwise, \end{cases}$$

where $\circ$ is the Hadamard product, i.e., element-wise multiplication, and $\gamma \leq 1$ is the hyperparameter for adjusting the attention. All operations are element-wise product applied to $\mathcal{S}^t_{\mathcal{V}'}$ and A^t. Reconstruction is only applied to the adjacency matrices produced by generator G because it does not affect the adjacency matrices of real samples.

Algorithm 1 summarizes the training procedure. All networks are optimized by Adam [16] with batch size 64 and an initial learning rate $\tau = 0.001$.

4 Experiments

In this section, we demonstrate the performance of our model DynGraphGAN on four real-world datasets, and compare with five baseline models.

4.1 Datasets

We use four dynamic graph datasets in our experiments, of which three are generated from social networks, and one is from co-authorship networks. All graphs are undirected and unweighted. Detailed descriptions are summarized below and statistics are described in Table 2:

- **Facebook (wall)** [28]**, Facebook (friendship)** [28]**, Digg** [15]**:** These three datasets are collected from the Facebook or Digg's front page. The nodes represent users and edges indicate friendships between users or posting behaviors. Based on the time at which a friendship was established or a post was made, we aggregate all such events in one year to construct a graph snapshot. This results in five graph snapshots in every datasets. For those events do not have any timestamps, we assume that they exist at the initialization and belong to the first snapshot.
- **DBLP**[2]**:** This dynamic graph represents the co-authorships between users from the DBLP computer science bibliography from 2000 to 2015. We aggregate all co-authorships in one year into a graph snapshot. This results in 16 snapshots of graphs.

The partition of training and test sets need to be carefully designed to avoid potential information leaking. To avoid the case that an edge in the test set may have already been observed in earlier graph snapshots in training set, we start with the last graph snapshot (at time T) and randomly sample 20% edges, denoted as $\mathcal{G}_{hide}$. To create the training set, we remove these edges from all earlier snapshots, $\mathbb{G}_{train} = \{\mathcal{G}^t_{train} = \mathcal{G}^t - \mathcal{G}_{hide}\}_{t \leq T}$. The test set consists of edges in $\mathcal{G}_{hide}$ and their occurrences in all earlier snapshots, $\mathbb{G}_{test} = \{\mathcal{G}^t_{test} = \mathcal{G}^t \cap \mathcal{G}_{hide}\}_{t \leq T}$. This will guarantee that each edge in $\mathbb{G}_{test}$

[2] http://projects.csail.mit.edu/dnd/DBLP/.

Table 2. Statistics of datasets

Dataset	#V	#E	#Snapshots
Facebook (wall)	46,952	876,993	5
Facebook (friendship)	63,731	817,035	5
Digg	279,630	1,731,653	5
DBLP	1,482,029	10,615,809	16

does not appear in any snapshots of $\mathbb{G}_{train}$. Edges in $\mathcal{G}^t_{train}$ are treated as positive samples. We also randomly select an equal number of node pairs without any edges to add to $\mathcal{G}^t_{train}$ as negative samples, resulting in a balanced training set.

4.2 Baseline Algorithms

We compare our model with several baselines:

- **Node2vec**[3] [12]**:** This is a random walk based method on static graphs, and therefore can only be applied to each graph snapshot independently.
- **TNE**[4] [36]**:** This is a dynamic graph embedding method based matrix factorization with temporal smoothness. We set the type of method to be "*global auto*". We run a grid search on parameter λ from 0 to 10 and report the best performance.
- **LIST** [33]**:** This model captures dynamic graph properties by decomposing the graph adjacency matrices into time-dependent matrices.
- **DHPE** [35]**:** This model learns graph embeddings with high-order proximity preserved and updates the embedding of nodes by an acceleration technique.
- **DynamicTriad**[5] [34]**:** A triadic closure process is considered for preserving both structural information and evolution patterns of dynamic graph. We run a grid search on parameters β_0 and β_1 from 0.01 to 10 and report the best performance.
- **DynGraphGAN (euclid):** This is the proposed method using the Euclidean distance (defined by Eq. (4)) in the generator. Subgraphs are generated by setting parameters $num\_walks = 1$ and $walk\_length = 80$.
- **DynGraphGAN (inner):** This is the proposed method using the inner similarity (defined by Eq. (3)) in the generator. Parameters of subgraph sampling are the same as DynGraphGAN (euclid).

The embedding dimension is set to 50 for all datasets except the DBLP dataset, for which it is set to 20 (in order to fit in the limited GPU memory). We also notice that different baseline methods choose different functions to infer

[3] https://github.com/aditya-grover/node2vec.
[4] https://github.com/linhongseba/Temporal-Network-Embedding.
[5] https://github.com/luckiezhou/DynamicTriad.

the likelihood of an edge. For example, TNE reconstructs each edge by the inner product of node embeddings. Therefore, we use a sigmoid function of the inner product of node embeddings in TNE. The sigmoid function of similarity $\langle \mathbf{u_i^t}, \mathbf{u_j^t} \rangle$ is used in Node2vec, LIST, DHPE and DynGraphGAN (inner). The sigmoid function of negative euclidean distance $(\mathbf{u_i^t} - \mathbf{u_j^t})^2$ is used in DynamicTriad and DynGraphGAN (euclid).

4.3 Evaluation Tasks

Two tasks are considered for evaluating the performance of the graph embeddings:

- *Link reconstruction:* The aim of this task is to infer whether an edge in $\mathcal{G}_{test}^t$ exists at time t using the embeddings $\mathbf{u}_i^t$. We do not use any information from earlier snapshots. The goal is to evaluate whether a dynamic graph embedding is equally effective on static settings.
- *Link prediction:* The aim of this task is to predict whether an edge in $\mathcal{G}_{test}^{t+1}$ exists in a future snapshot at time $t+1$ using the embeddings $\mathbf{u}_i^t$ of time t. This task can demonstrate whether dynamic graph embeddings can capture temporal evolutions, offering advantages over static methods.

For both link reconstruction and link prediction tasks, we use AUC [8] score and F1 score [27] to evaluate the performance. Because edges are not uniformly distributed across all snapshots in each dataset, we report both micro-averaging and macro-averaging.

- *Micro:* In micro-averaging, metrics are averaged globally over all edge decisions. Snapshots having more edges will contribute more to the overall performance.
- *Macro:* In macro-averaging, the performance is first aggregated for each snapshot and the metrics is averaged over all graph snapshots. This gives equal weights to each snapshot instead of edges.

4.4 Link Reconstruction and Link Prediction

Table 3 demonstrates the comparison of our model versus baseline models. LIST and DynamicTriad have no result on DBLP because of large data size. From Table 3, we have following observations and analysis:

- Our model DynGraphGAN achieves significant improvements in AUC and F1 scores over the baseline models. In the link reconstruction task, the average improvements of mi-AUC, ma-AUC, mi-F1 and ma-F1 are 0.0314, 0.035, 0.0409 and 0.0425 respectively. In the link prediction task, we observe 0.0671, 0.0616, 0.0755 and 0.0692 improvements of mi-AUC, ma-AUC, mi-F1 and ma-F1 on average respectively.
- As a static graph embedding model, Node2vec shows substantial performance loss in link prediction task. This is largely due to the inability of capturing temporal evolution by Node2Vec.

Table 3. Link reconstruction and link prediction.

Dataset	Algorithm	Link reconstruction				Link prediction			
		mi-AUC	ma-AUC	mi-F1	ma-F1	mi-AUC	ma-AUC	mi-F1	ma-F1
Facebook (wall)	Node2vec	0.8923	0.8217	0.8273	0.7644	0.6171	0.6237	0.5852	0.5896
	TNE	0.6872	0.6972	0.6412	0.6730	0.5461	0.5574	0.5373	0.5860
	LIST	0.7375	0.7428	0.6858	0.7428	0.7251	0.7234	0.6739	0.6706
	DHPE	0.8371	0.8302	0.7747	0.7643	0.8089	0.8071	0.7353	0.7404
	DynamicTriad	0.8321	0.8329	0.7669	0.7651	0.7798	0.7890	0.7158	0.7264
	DynGraphGAN (euclid)	**0.9196**	**0.9249**	**0.8546**	**0.8611**	**0.8885**	**0.8840**	**0.8193**	**0.8130**
	DynGraphGAN (inner)	0.8530	0.8575	0.7860	0.7928	0.8300	0.8250	0.7569	0.7571
Facebook (friendship)	Node2vec	0.9198	0.9169	0.8547	0.8517	0.8859	0.8941	0.8216	0.8284
	TNE	0.7892	0.7916	0.7191	0.7182	0.6854	0.6932	0.6197	0.6284
	LIST	0.8553	0.9139	0.7699	0.8491	0.8734	0.8717	0.8073	0.8051
	DHPE	0.9032	0.9046	0.8471	0.8491	0.8715	0.8618	0.8077	0.8061
	DynamicTriad	0.9412	0.9450	0.8788	0.8841	0.9079	0.8888	0.8414	0.8305
	DynGraphGAN (euclid)	0.9491	0.9493	0.8926	0.8928	**0.9291**	**0.9261**	0.8656	0.8602
	DynGraphGAN (inner)	**0.9506**	**0.9518**	**0.9002**	**0.9021**	0.9271	0.9189	**0.8675**	**0.8615**
Digg	Node2vec	0.8565	0.7939	0.7939	0.6923	0.5057	0.5043	0.5082	0.5041
	TNE	0.8279	0.7562	0.7575	0.7333	0.6003	0.5862	0.5679	0.6368
	LIST	0.8729	0.7646	0.8118	0.7113	0.8852	0.8366	0.8247	0.7778
	DHPE	0.7915	0.7034	0.7523	0.6686	0.7991	0.7626	0.7641	0.7252
	DynamicTriad	0.8824	0.7233	0.8264	0.6829	0.8861	0.8219	0.8332	0.7671
	DynGraphGAN (euclid)	**0.9021**	**0.8037**	**0.8301**	**0.7376**	0.8875	0.8412	0.8210	0.7712
	DynGraphGAN (inner)	0.8791	0.7737	0.8237	0.7119	**0.8948**	**0.8452**	**0.8431**	**0.7832**
DBLP	Node2vec	0.8660	0.8764	0.7873	0.8002	0.5789	0.5783	0.5524	0.5521
	TNE	0.5406	0.5860	0.5101	0.5673	0.5222	0.5614	0.4953	0.5514
	LIST	-	-	-	-	-	-	-	-
	DHPE	0.7844	0.7731	0.7151	0.7061	0.7856	0.7783	0.7134	0.7071
	DynamicTriad	-	-	-	-	-	-	-	-
	DynGraphGAN (euclid)	**0.9415**	**0.9407**	**0.8815**	**0.8827**	**0.9124**	**0.9211**	**0.8512**	**0.8663**
	DynGraphGAN (inner)	0.8386	0.8410	0.7761	0.7751	0.7979	0.8088	0.7371	0.7469

- TNE is an algorithm designed for dynamic graph embedding but still performs poorly on several datasets. One possible reason is that TNE is more sensitive to the edge sparsity than other models.
- Our model DynGraphGAN achieves better results than LIST, DHPE and DynamicTriad. This is because that LIST only captures first-order proximity. DHPE considers the spatial structure and temporal dependence separately. DynamicTriad only makes use of triads in graphs and ignores more globally information. DynGraphGAN uses convolutional networks to make full use of the graph structure along time.

4.5 Effect of Parameters

- **The Effect of Temporal Smoothness.** The temporal smoothness of dynamic graph embedding is controlled by parameter λ in DynGraphGAN. The higher the value of λ, the smoother the node transitions in the embedding space. We vary λ from 0 to 10 and report the mi-AUC scores in Figs. 3(a) and (b). Obviously, if we do not consider the temporal smoothness ($\lambda = 0$),

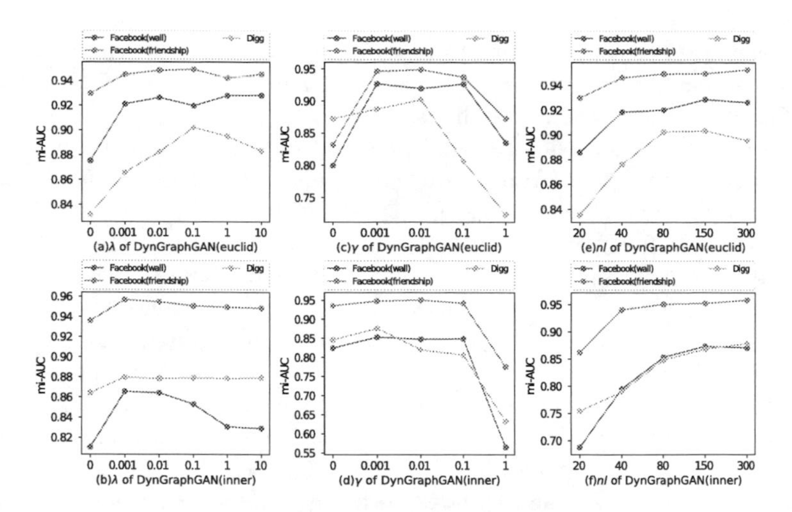

Fig. 3. Hyperparameter analysis

a low mi-AUC score is observed. It is evidenced that the temporal smoothness is necessary in dynamic graph embedding. Our model cannot score more improvement when $\lambda \geq 1$, because the strong temporal smoothness may put too much constraint on the embedding. Therefore, λ is chosen between 0 and 1 in the experiments.

- **The Effect of Attention Balance.** The attention is controlled by parameter γ. We vary γ from 0 to 1 and report the mi-AUC scores in Figs. 3(c) and (d). When presence and absence of an edge receives equal attention ($\gamma = 1$), our model performs substantially worse on these sparse graphs. This proves the importance of attention balance.
- **The Effect of Subgraph Size.** A larger subgraph may contain richer structural information but entail higher computational complexity. We vary the subgraph size nl from 20 to 300 and report the mi-AUC scores in Figs. 3(e) and (f). The mi-AUC score is positively correlated to nl and converges to a plateau when $nl \geq 80$. Thus, we use $nl = 80$ in most experiments, as the additional benefit of using larger subgraphs diminishes when $nl > 80$.

5 Related Work

Graph embedding models aim to learn low-dimensional features of nodes in a graph. On static graphs, the random walk based models, such as DeepWalk [26] and Node2vec [12], seek to maximize the log-likelihood of observing network neighborhoods for the given nodes. DeepWalk firstly uses random walk to generate node sequences from the original graph, and then learns node embeddings via Skip-gram [23]. As a variant of DeepWalk, Node2vec designs a biased random walk procedure to explores diverse neighborhoods. Matrix factorization based methods such as HOPE [24] approximate the high-order proximity measurements on directed graphs to capture asymmetric transitivity property. Autoencoder based method SDNE [29] uses deep autoencoders to code non-linear dependencies to enrich the embeddings.

All aforementioned methods do not perform well on dynamic graphs as they are not able to represent graph evolution. This inspired the development of dynamic graph embedding models. LIST [33] uses the time-dependent matrix factorization to express the network structure as a function of time. However, it only focuses on modeling the first-order proximity and assumes linear dependencies between nodes and between consecutive snapshots. DynGEM [11] designs a deep autoencoder framework for dynamic graph embedding, which combines first-order proximity and second-order proximity to preserve graph structure. DynamicTriad [34] imposes triad to model dynamic changes of graph structures. These two methods consider more complex relations between nodes, but high-order proximity is still largely ignored. DHPE [35] is the first model to consider high-order proximity by a GSVD-based static model, but it consider the spatial structure and temporal evolution separately.

Recently, many generative adversarial models [6,9] show promises in learning robust feature representations: NetGAN [2] employs the Long short-term memory (LSTM) [14] architecture for learning the distribution of biased random walks. GraphGAN [30] proposes a novel graph softmax and approximates the connectivity distribution by sampling node pairs. AIDW [5] uses the IDW model first to generate node embeddings, and then applies an adversarial learning component to regularize the embeddings. These models unfortunately cannot be easily adapted for dynamic graphs.

6 Conclusions

In this paper, we propose DynGraphGAN, a GAN based model that learns feature representations of dynamic graphs. Specifically, to capture the high-order proximity, a multi-layer graph convolutional network is used to preserve structure features of each graph snapshot in the discriminator. Then a convolutional neural network follows to capture the temporal dependency in a community-level. This convolutional architecture considers the high-order proximities and the temporal dependency jointly, from which the generator can model properties of the dynamic graph efficiently. Moreover, we further improve the model efficiency by exploiting sparsity and temporal smoothness. By simultaneously optimizing the generator and the discriminator, our model DynGraphGan is robust and scalable to large dynamic graphs. Experimental results on various real-world datasets demonstrate its effectiveness and advantages over alternative graph embedding models.

Acknowledgements. This work is supported in part by the National Natural Science Foundation of China Projects No. U1636207, No. 91546105, the Shanghai Science and Technology Development Fund No. 16JC1400801, No. 17511105502, No. 17511101702.

References

1. Barkan, O., Koenigstein, N.: ITEM2VEC: neural item embedding for collaborative filtering. In: IEEE International Workshop on Machine Learning for Signal Processing, pp. 1–6 (2016)
2. Bojchevski, A., Shchur, O., Zügner, D., Günnemann, S.: NetGAN: generating graphs via random walks. arXiv preprint arXiv:1803.00816 (2018)
3. Cao, S., Lu, W., Xu, Q.: GraREP: learning graph representations with global structural information, pp. 891–900 (2015)
4. Ciresan, D.C., Giusti, A., Gambardella, L.M., Schmidhuber, J.: Deep neural networks segment neuronal membranes in electron microscopy images. In: Neural Information Processing Systems, pp. 2852–2860 (2012)
5. Dai, Q., Li, Q., Tang, J., Wang, D.: Adversarial network embedding. In: AAAI Conference on Artificial Intelligence, pp. 2167–2174 (2018)
6. Donahue, J., Krähenbühl, P., Darrell, T.: Adversarial feature learning. arXiv preprint arXiv:1605.09782 (2016)
7. Faloutsos, C., McCurley, K.S., Tomkins, A.: Fast discovery of connection subgraphs. In: ACM SIGKDD International Conference on Knowledge Discovery and Data Mining, pp. 118–127 (2004)
8. Fawcett, T.: An introduction to ROC analysis. Pattern Recogn. Lett. **27**, 861–874 (2006)
9. Glover, J.: Modeling documents with generative adversarial networks. arXiv preprint arXiv:1612.09122 (2016)
10. Goodfellow, I., et al.: Generative adversarial nets. In: Neural Information Processing Systems, pp. 2672–2680 (2014)
11. Goyal, P., Kamra, N., He, X., Liu, Y.: DynGEM: deep embedding method for dynamic graphs. In: IJCAI International Workshop on Representation Learning for Graphs (2017)
12. Grover, A., Leskovec, J.: Node2vec: scalable feature learning for networks. In: ACM SIGKDD International Conference on Knowledge Discovery and Data Mining, pp. 855–864 (2016)
13. Gulrajani, I., Ahmed, F., Arjovsky, M., Dumoulin, V., Courville, A.C.: Improved training of Wasserstein GANs. In: Neural Information Processing Systems, pp. 5769–5779 (2017)
14. Hochreiter, S., Schmidhuber, J.: Long short-term memory. Neural Comput. **9**, 1735–1780 (1997)
15. Hogg, T., Lerman, K.: Social dynamics of digg. arXiv preprint arXiv:1202.0031 (2012)
16. Kingma, D.P., Ba, J.: Adam: a method for stochastic optimization. arXiv preprint arXiv:1412.6980 (2014)
17. Kipf, T.N., Welling, M.: Semi-supervised classification with graph convolutional networks. arXiv preprint arXiv:1609.02907 (2016)
18. Koren, Y., North, S.C., Volinsky, C.: Measuring and extracting proximity graphs in networks. ACM Trans. Knowl. Discov. Data **1**, 12 (2007)
19. Leskovec, J., Kleinberg, J.M., Faloutsos, C.: Graphs over time: densification laws, shrinking diameters and possible explanations. In: ACM SIGKDD International Conference on Knowledge Discovery and Data Mining, pp. 177–187 (2005)
20. Li, B., Zheng, C., Huang, D., Zhang, L., Han, K.: Gene expression data classification using locally linear discriminant embedding. Comput. Biol. Med. **40**, 802–810 (2010)

21. Liao, L., He, X., Zhang, H., Chua, T.S.: Attributed social network embedding. arXiv preprint arXiv:1705.04969 (2017)
22. Maas, A.L., Hannun, A.Y., Ng, A.Y.: Rectifier nonlinearities improve neural network acoustic models. In: ICML Workshop on Deep Learning for Audio, Speech and Language Processing, p. 3 (2013)
23. Mikolov, T., Sutskever, I., Chen, K., Corrado, G.S., Dean, J.: Distributed representations of words and phrases and their compositionality. In: Neural Information Processing Systems, pp. 3111–3119 (2013)
24. Ou, M., Cui, P., Pei, J., Zhang, Z., Zhu, W.: Asymmetric transitivity preserving graph embedding. In: ACM SIGKDD International Conference on Knowledge Discovery and Data Mining, pp. 1105–1114 (2016). https://doi.org/10.1145/2939672.2939751
25. Pennington, J., Socher, R., Manning, C.: Glove: global vectors for word representation. In: Empirical Methods in Natural Language Processing, pp. 1532–1543 (2014)
26. Perozzi, B., Al-Rfou, R., Skiena, S.: DeepWalk: online learning of social representations. In: ACM SIGKDD International Conference on Knowledge Discovery and Data Mining, pp. 701–710 (2014)
27. Sasaki, Y., et al.: The truth of the F-measure. Teach Tutor mater **1**, 1–5 (2007)
28. Viswanath, B., Mislove, A., Cha, M., Gummadi, K.P.: On the evolution of user interaction in Facebook. In: ACM Workshop on Online Social Networks, pp. 37–42 (2009)
29. Wang, D., Cui, P., Zhu, W.: Structural deep network embedding. In: ACM SIGKDD International Conference on Knowledge Discovery and Data Mining, pp. 1225–1234 (2016)
30. Wang, H., et al.: GraphGAN: graph representation learning with generative adversarial nets. In: AAAI Conference on Artificial Intelligence, pp. 2508–2515 (2018)
31. Xu, B., Wang, N., Chen, T., Li, M.: Empirical evaluation of rectified activations in convolutional network. arXiv preprint arXiv:1505.00853 (2015)
32. Yang, C., Sun, M., Liu, Z., Tu, C.: Fast network embedding enhancement via high order proximity approximation. In: International Joint Conference on Artificial Intelligence, pp. 3894–3900 (2017)
33. Yu, W., Cheng, W., Aggarwal, C.C., Chen, H., Wang, W.: Link prediction with spatial and temporal consistency in dynamic networks. In: International Joint Conference on Artificial Intelligence, pp. 3343–3349 (2017)
34. Zhou, L., Yang, Y., Ren, X., Wu, F., Zhuang, Y.: Dynamic network embedding by modelling triadic closure process. In: AAAI Conference on Artificial Intelligence, pp. 571–578 (2018)
35. Zhu, D., Cui, P., Zhang, Z., Pei, J., Zhu, W.: High-order proximity preserved embedding for dynamic networks. IEEE Trans. Knowl. Data Eng. **30**, 2134–2144 (2018)
36. Zhu, L., Guo, D., Yin, J., Steeg, G.V., Galstyan, A.: Scalable temporal latent space inference for link prediction in dynamic social networks. IEEE Trans. Knowl. Data Eng. **28**, 2765–2777 (2016)

Evaluating Mixed Patterns on Large Data Graphs Using Bitmap Views

Xiaoying Wu[1(✉)], Dimitri Theodoratos[2], Dimitrios Skoutas[3], and Michael Lan[2]

[1] School of Computer, Wuhan University, Wuhan, China
xiaoying.wu@whu.edu.cn
[2] New Jersey Institute of Technology, Newark, USA
{dth,ml122}@njit.edu
[3] IMSI, R.C. Athena, Athens, Greece
dskoutas@imis.athena-innovation.gr

Abstract. Developing efficient and scalable techniques for pattern queries over large graphs is crucial for modern applications such as social networks, Web analysis, and bioinformatics. In this paper, we address the problem of efficiently finding the homomorphic matches for tree pattern queries with child and descendant edges (mixed pattern queries) over a large data graph. We propose a novel type of materialized views to accelerate the evaluation. Our materialized views are the sets of occurrence lists of the nodes of the pattern in the data graph. They are stored as compressed bitmaps on the inverted lists of the node labels in the data graph. Reachability information between occurrence list nodes is provided by a node reachability index. This technique not only minimizes the materialization space but also reduces CPU and I/O costs by translating view materialization processing into bitwise operations. We provide conditions for view usability using the concept of pattern node coverage. We design a holistic bottom-up algorithm which efficiently computes pattern query matches in the data graph using bitmap views. An extensive experimental evaluation shows that our method evaluates mixed patterns up to several orders of magnitude faster than existing algorithms.

1 Introduction

Graphs are used as the underlying data structure in many modern applications ranging from chemical, medical, and bioinformatics to health informatics, social networks, and Web analysis applications. A fundamental operation in big data graph management and analysis is the evaluation of pattern queries to a data graph. This operation involves finding all the matches of a given pattern to the graph structure. As this is a central issue for graph analysis, it has been the focus of previous attention [2–4,6,8,13,17]. Existing approaches are characterized by: (a) the type of edges the patterns have, and (b) the type of morphism used to

The research of the first author was supported by the National Natural Science Foundation of China under Grant No. 61872276.

G. Li et al. (Eds.): DASFAA 2019, LNCS 11446, pp. 553–570, 2019.
https://doi.org/10.1007/978-3-030-18576-3_33

map the pattern to the graph structure. An edge in a pattern can be either a *child* edge, in which case it is mapped to an *edge* in the graph, or a *descendant* edge, in which case it is mapped to a *path* in the graph. The morphism determines how a pattern is mapped to the data graph and, in this context, it can be an *isomorphism* or a *homomorphism*.

Earlier works have considered isomorphisms for matching patterns with child edges [10,12] while more recent ones focus on homomorphisms for matching pattern queries with descendant edges [2–4,8,20,21]. By allowing edge-to-path mapping on graphs, homomorphisms are able to extract matches "hidden" deeply within large graphs which might be missed by isomorphims. On the other hand, the patterns with child edges can discover important parent-child relationships in the data graph which can be missed by patterns with descendant edges. In this paper, we consider *mixed patterns* which generalize the other two types of patterns by allowing both child and descendant relationships and can extract detailed information which cannot be extracted by either one of the other patterns. We focus, in particular, on tree-structured mixed patterns on data graphs as they are the building blocks for general graph patterns. Mixed pattern queries have been ignored in existing evaluation algorithms which allow edge-to-path mapping in graphs by considering only patterns with descendant edges [2,3,8,20,21][1]. Mixed pattern queries cannot be simply evaluated by post-filtering descendant-edge only pattern queries as this would generate a large number of redundant intermediate results.

Since answering pattern queries on massive graphs can be very time consuming, devising techniques to improve their response time is of great importance. A powerful query optimization technique in current database systems consists in materializing (i.e., precomputing and storing) views. The idea is that storing these materializations in a view pool is beneficial for query evaluation. In this paper, we adopt a novel approach for materializing views in the context of pattern queries on large data graphs. Unlike other materialized view approaches which store the matches of the patterns, our approach sees materialized views as sets of occurrence lists of the pattern query nodes.

Contribution. The main contributions of the paper are the following:

- We address the problem of efficiently finding the homomorphic matches of tree-pattern queries on large data graphs. Previous approaches consider isomorphic pattern matches or adopt simulation-based matches to cope with the hardness of the graph pattern matching problem. Our pattern queries involve both child and descendant edges (mixed pattern queries), which generalize child- or descendant-edge-only patterns, allowing for more specific information to be extracted from the data.
- We study how to efficiently evaluate mixed pattern queries on large data graphs using materialized views. We adopt bitmap views, a type of view which materializes the occurrence lists of the pattern nodes, stored as compressed bitmaps on the inverted lists of the data graph labels. Bitmap views

[1] [2] defines patterns which involve both child and descendant edges, but provides an algorithm for patterns with only descendant edges.

consume very little space. They further result in CPU cost savings by translating materialized view processing into bitwise operations over bitmaps.

- We provide conditions for answering pattern queries using views through view homomorphisms. We also show how multiple bitmap views can be exploited for answering a query by intersecting their node occurrence lists.
- We design a bitmap-view-based approach for evaluating pattern queries on data graphs. Our approach first identifies bitmap views in the view pool that can be used for answering the query. It then uses these views to generate occurrence lists for the query nodes excluding inverted list data graph nodes that do not contribute to the query answer, thus narrowing down the search space. Finally, it employs a bottom-up algorithm on the reduced size occurrence lists to compute the query answer.
- We run extensive experiments to evaluate the efficiency and scalability of our bitmap view-based approach. We also compare with previous approaches which do not use materialized views. Our results show that our approach outperforms previous approaches by orders of magnitude in terms of both time efficiency and scalability.

2 Data Graph, Pattern Queries and Bitmap Views

Next, we present the data model, pattern queries, and the concept of bitmap view.

Definition 1 (Data Graph). *A data graph is a directed node-labeled graph $G = (V, E)$ where V denotes the set of nodes and E denotes the set of edges (ordered pairs of nodes). Let $\mathcal{L}$ be a finite set of node labels. Each node v in V has a label $\tau(v) \in \mathcal{L}$ associated with it. Given a label x in $\mathcal{L}$, the inverted list I_x is the list of nodes in G whose label is x.*

Definition 2 (Reachability). *A node u is said to* reach *node v, denoted by $u \prec v$, if there exists a path from u to v in G. Clearly, if $(u, v) \in E$, then $u \prec v$. Abusing tree notation, we refer to v as a* child *of u (or u as a* parent *of v) if $(u, v) \in E$, and v as a* descendant *of u (or u is an* ancestor *of v) if $u \prec v$.*

Given two nodes u and v in G, in order to efficiently check whether $u \prec v$, graph pattern matching algorithms use some kind of reachability indexing scheme. Most reachability indexing schemes associate with every graph node a label which is an entry in the index for the data graph [11]. Our approach can flexibly use any labeling scheme to check node reachability. In order to check if v is a child of u, the basic access information of the graph G can be used; for example, the adjacency lists.

Definition 3 (Tree Pattern Query). *We focus on tree-pattern queries and views. Every node x in a tree pattern Q has a label $\tau(x)$ from $\mathcal{L}$. There can be two types of edges in Q. A* child *(resp.* descendant*) edge denotes a child (resp. descendant) structural relationship between the respective two nodes.*

A view is a named query. The class of views is not restricted; any query can be a view.

Definition 4 (Homomorphism). *Given a tree pattern Q and a data graph G, a* homomorphism *from Q to G is a function m mapping the nodes of Q to nodes of G, such that: (1) for any node $x \in Q$, $\tau(x) = \tau(m(x))$; and (2) for any edge $(x, y) \in Q$, if (x, y) is a child edge, $(m(x), m(y))$ is an edge of G, while if (x, y) is a descendant edge, $m(x) \prec m(y)$ in G.*

We call *occurrence* of a pattern query Q on a data graph G a tuple indexed by the nodes of Q whose values are the images of the nodes in Q under a homomorphism from Q to G.

Definition 5 (Query Answer). *The* answer *of Q on G is a relation whose schema is the set of nodes of Q, and whose instance is the set of occurrences of Q under all possible homomorphisms from Q to G.*

If x is a node in Q labeled by label a, the *occurrence list of x on G* is a sublist L_x of the inverted list I_a containing only those nodes that occur in the answer of Q on G for x (that is, nodes that occur in the column x of the answer). We introduce *answer graphs* to compactly encode all possible homomorphisms of a query in a data graph.

Definition 6 (Answer Graph). *The answer graph G_A of a pattern query Q is a k-partite graph. Graph G_A has an independent node set for every node $q \in Q$ which is equal to the occurrence list L_q of q. There is an edge (v_x, v_y) in G_A between a node $v_x \in L_x$ and a node $v_y \in L_y$ if and only if there is an edge (x, y) in Q and a homomorphism from Q to G which maps x to v_x and y to v_y.*

The answer graph G_A losslessly summarizes all the occurrences of Q on G. Similar to factorized representations of query results studied in the context of classical databases and probabilistic databases [9], G_A exploits computation sharing to reduce redundancy in the representation and computation of query results. A useful property of G_A is that through a top-down traversal, the answer of Q on G can be obtained in time linear to the total number of occurrences of Q on G; also, the cardinality of the query answer can be calculated without explicitly enumerating the occurrences of Q on G.

For evaluating pattern queries and views on a graph G, we use the inverted lists of the node labels that appear in G. Reachability information for a pair of nodes is provided by a reachability index. In order to evaluate a pattern query Q on G only the inverted lists of the query node labels are needed. The input to the query pattern evaluation algorithms on a graph G is the set I of the inverted lists of the node labels in G. In the following, we might refer to a graph G and to its inverted list set I interchangeably.

Materialized Views. We materialize views on a data graph by storing only the occurrence list of its nodes.

Definition 7 (View Materialization). *The* materialization $V(I)$ of a view V *on the inverted list set* I *of a graph is the set of occurrence lists of the nodes of* V *on* I *along with a function that maps each node* x *in* V *to its occurrence list* L_x. *A view is characterized as* materialized *if it has a materialization.*

Bitmap Views. The occurrence list L_x of a view node x labeled by a on I can be represented by a bitmap on I_a that has a '1' bit at position i iff L_x comprises the graph node at position i of I_a. We refer to the materialized views whose occurrence lists are bitmaps as *bitmap views.* The bitmaps are stored compressed to even further reduce the materialization space. Clearly, storing the materialization of multiple views as compressed bitmaps results in important space savings.

3 Answering Pattern Queries Using Bitmap Views

We show next how a pattern query can be answered using one or multiple materialized views. We first provide necessary and sufficient conditions for answering a pattern query using inclusively or exclusively one or multiple materialized views.

Let x be a node in a pattern query Q labeled by a. In order to evaluate Q on an inverted list set I of a data graph, an algorithm iterates over the inverted list I_a in I. If there is a sublist, say I_x, of I_a such that Q can be computed on I by iterating over I_x instead of I_a, we say that node x *can be computed using* I_x *on* I. Let now y be a node in a view V and let I_y be the occurrence list of y on I. If node x can be computed using I_y for every I, we say that node x is *covered* by node y or that y is a *covering node* for x. The idea of our approach for answering Q using V on I is to identify covering nodes in Q. The occurrence lists of these nodes on I can then be used to compute the answer of Q on I instead of using the corresponding inverted lists in I.

Answering a Pattern Query Using a View. We start by defining what answering a pattern query using a view means in our framework of view materialization.

Definition 1. *A pattern query* Q can be answered using a view V *if a node in* Q *is covered by a node in* V. *If every node in* Q *is covered by a node in* V, *we say that* Q *can be answered* exclusively *using* V. *Otherwise, we say that* Q *can be answered* inclusively *using* V.

When the answer of a query is computed using a view, a node of the query that is covered by a view node uses only the occurrence list of this view node. Since the occurrence list of the view node is usually much smaller than the inverted list of the corresponding node label, the cost for computing the answer of the query is reduced.

View Usability Conditions. The following theorem provides necessary and sufficient conditions for node coverage.

Theorem 1. *Let x be a node in a pattern query Q and y be a node in a view V. Node x in Q is covered by node y in V iff there is a homomorphism from V to Q that maps y to x.*

As a consequence, pattern query Q can be answered using V iff there is a homomorphism from V to Q. Further, query Q can be answered using V exclusively iff there are homomorphisms from V to Q such that every node of Q is the image of a node in V under some homomorphism.

Computing the Answer of a Pattern Query Using a View. In order to compute the answer of a query using a view what is needed is an association of the query nodes with covering view nodes. The set of covering view nodes of a given query node is determined as follows: let $h_1, \ldots, h_k$ be the homomorphisms from a view V to a query Q and $y_i^1, \ldots, y_i^{m_k}$ be the nodes in V whose image under h_i is x. Then, the set $m(x)$ of covering nodes for x in V is $m(x) = \cup_{i\in[1,k],\, j\in[1,m_k]}\{y_i^j\}$.

The occurrence list on I of any node in $m(x)$ can be used for computing x on I. However, we might also use the occurrence lists of multiple (or all the) nodes in $m(x)$ by computing and using the intersection of their occurrence lists.

Note that a view V can have a number of homomorphisms to a query which is exponential in the number of view nodes. Nevertheless, the number of covering nodes in $m(x)$ is bounded by the number of nodes in V. In the next section, we present a procedure which computes the covering nodes in $m(x)$ in polynomial time.

Answering a Pattern Query Using Multiple Views. In order to compute the answer of the query using a set of materialized views we need to associate query nodes with covering nodes in the views. Let x be a node in query Q, and $m_1(x), \ldots, m_n(x)$ be the sets of covering nodes of x in $V_1, \ldots, V_n$, respectively. Then, the set $m(x)$ of covering nodes of x in $V_1, \ldots, V_n$ is $m(x) = \bigcup_{i\in[1,n]} m_i(x)$.

As with the case of a single view, we might also use the occurrence lists of some (or all the) nodes in $m(x)$: during the computation of the answer, node x will be computed using the intersection of the occurrence lists of these view nodes in $m(x)$. As the occurrence lists are stored as bitmaps, their intersection can be computed efficiently by applying bit wise operations. Of course, the higher the number of covering nodes from $m(x)$ used for computing x, the smaller the size of the employed inverted list will be.

4 Algorithm for Evaluating Pattern Queries Using Bitmap Views

In this section, we present the bitmap view approach for optimizing pattern queries. We assume that a set of views are materialized as compressed bitmaps in a view pool. In order to compute the answer of a query Q, our approach computes, for every query node x, all the covering view nodes in the view pool. Then, it intersects the occurrence lists of these covering view nodes. The resulting sublist is used for the computation of x. If x does not have covering nodes, the corresponding inverted list is used for its computation. Next, we describe each step in more detail.

4.1 Computing Covering Nodes

As discussed in Sect. 3, given a query Q and a view V, the covering nodes for a node of Q in V are defined in terms of the homomorphisms of V to Q. A brute-force method for computing covering nodes computes all the possible homomorphisms from V to Q. Unfortunately, the number of these homomorphisms can be exponential on the size of V. Therefore, we have designed an algorithm which, given two patterns P and Q, compactly represents all the homomorphisms from P to Q in polynomial time and space $O(|P| \times |Q|)$. The algorithm employs a standard dynamic programming technique for computing a Boolean matrix $\mathcal{M}(p, q)$, $p \in \text{nodes}(P)$, $q \in \text{nodes}(Q)$, such that $\mathcal{M}(p, q)$ is true if: (1) there exists a homomorphism from the subpattern rooted at p to the subpattern rooted at q; and (2) there exists a homomorphism from the prefix path of p to the prefix path of q, where *prefix path* of a node is the path from the pattern root to that node. A detailed description is omitted here in the interest of space.

4.2 Computing Query Node Inverted Sublists with Bitwise Operations

The intersection of the occurrence lists of the covering view nodes can be implemented by a bitwise operation on the corresponding bitmaps: first, the bitmaps of the operand view nodes are fetched into memory and bitwise AND-ed. Then, the target inverted sublist is constructed by fetching into memory the inverted list nodes indicated by the resulting bitmap.

Besides space savings, exploiting bitmaps and bitwise operations results in time saving for two reasons. First, bitwise AND-ing bitmaps incurs less CPU cost than intersecting the corresponding inverted sublists, especially if there are many views. Second, fetching into memory the bitmaps of the operand view nodes and the target inverted sublist nodes indicated by the resulting bitmap incurs less I/O cost than fetching the entirety of the inverted sublist for the operand view nodes as this is required for performing a regular intersection operation.

4.3 Bottom-Up Evaluation of Pattern Queries

We present now a mixed pattern query evaluation algorithm called *BUP* (Fig. 1). Before describing the algorithm, we introduce the terminology and notation used.

Notation. Given query Q, let I^Q denote the set of inverted sublists obtained by the inverted sublists generation procedure (Sect. 4.2). Each node q of Q is associated with a sublist I^Q_q in I^Q. A candidate occurrence list CL_q of a query node q in Q is the occurrence list of the root of the subquery of Q rooted at q. Clearly, $CL_q \subseteq I^Q_q$ and $L_q \subseteq CL_q$. Let (q_i, q_j) be a query edge in Q, and v_i and v_j be two nodes in G, such that $\tau(q_i) = \tau(v_i)$ and $\tau(q_j) = \tau(v_j)$. The pair

Input: Tree pattern query Q, and inverted sublists I^Q
Output: Answer graph G_A of Q on G

1. G_A := a k-partite graph without edges having one data node set CL_q for every node $q \in Q$;
2. Let initially $CL_q = \emptyset$ for every node $q \in Q$;
3. traverse($root(Q)$);
4. **for** (every $q \in Q$, $q \neq root(Q)$, in a top-down manner) **do**
5. Remove the data nodes of CL_q which do not have an incoming edge;

Procedure traverse(q)

1. **if** ($isLeaf(q)$) **then**
2. $CL_q := I_q^Q$;
3. **return**
4. **for** ($q_i \in children(q)$) **do**
5. traverse(q_i);
6. **for** ($v_q \in I_q$) **do**
7. expand(q, v_q);

Procedure expand(q, v_q)

1. Append v_q to CL_q;
2. **for** ($q_i \in children(q)$) **do**
3. **for** ($v_{q_i} \in CL_{q_i}$) **do**
4. **if** ((v_q, v_{q_i}) is an occurrence of the query edge (q, q_i)) **then**
5. Add the edge (v_q, v_{q_i}) to G_A;
6. **if** (no match to (q, q_i) is found) **then**
7. remove v_q from CL_q;
8. **return**

Fig. 1. Algorithm *BUP*.

(v_i, v_j) is called *occurrence* of the query edge (q_i, q_j) if: (a) (q_i, q_j) is a child edge in Q and (v_i, v_j) is an edge in G, or (b) (q, q_i) is a descendant edge in Q and $v_i \prec v_j$ in G.

The Algorithm. Given query Q and its associated inverted sublists I^Q, Algorithm *BUP* builds Q's answer graph G_A by doing a postorder traversal on Q. Specifically, it first generates the candidate occurrence lists for the query nodes in Q and links their nodes with edges: a data node $v \in I_q^Q$ is put in the candidate occurrence list CL_q of a query node $q \in Q$ if there are data nodes $v_1, \ldots, v_k$ for the child query nodes $q_1, \ldots q_k$ of q such that: (a) for every $i \in [1, k]$, $v_i \in CL_{q_i}$, and (b) for every $i \in [1, k]$, (v, v_i) is an occurrence in G of the edge $(q, q_i) \in Q$. For every $i \in [1, k]$, an edge is added in G_A from a node $v \in CL_q$ to a node $v_i \in CL_{q_i}$ if and only if (v, v_i) is an occurrence in G of the edge $(q, q_i) \in Q$. Due to the bottom up traversal of Q, the candidate occurrence lists of the child nodes q_i of a q are available from the previous iteration of the algorithm. At the end

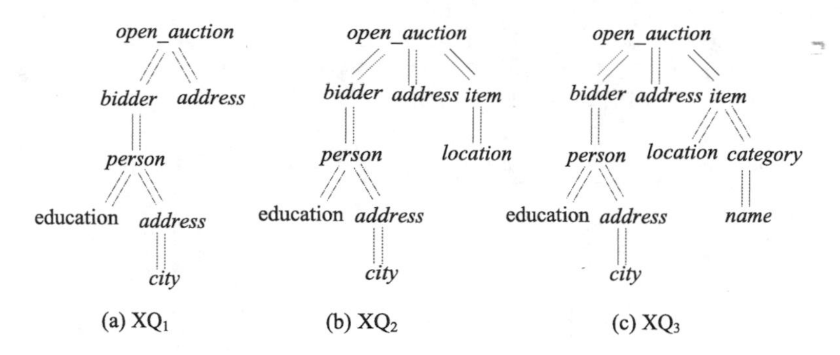

Fig. 2. Queries on the XMark graph.

Table 1. Parameters for query generation.

Parameters	Range	Description
Q	300 to 1800	Number of queries
D	6 to 16	Maximum depth of queries
DS	0 to 1	Probability of setting an edge to be a descendant edge ('//')
NP	1 to 3	Number of branches per query node

of the process, a top-down traversal of the answer graph G_A under construction eliminates nodes in the CL_q's (and their incident edges) which are not in L_q.

The bottom-up processing of *BUP* is realized by procedure *traverse*. Let q be the current query node under consideration. For each node v_q in I_q^Q, procedure *traverse* invokes procedure *expand* to potentially expand G_A by putting v_q into the candidate occurrence list CL_q and by adding incident edges to G_A (lines 6–7 in *traverse*).

When *traverse* terminates after processing the root of Q, $CL_{root(Q)}$ is $L_{root(Q)}$. The candidate occurrence lists CL_q for other nodes q of Q might contain nodes that are not in L_q. To discard these nodes, a breadth first traversal of Q is performed. For every node q of Q encountered (other than the root node), all the data nodes which do not have an incoming edge are removed from G_A along with their incident outgoing edges (lines 6–7 in the main procedure). The resulting graph is the answer graph G_A of Q.

The following theorem shows the correctness of Algorithm *BUP*. We omit its proof here in the interest of space.

Theorem 2. *Algorithm* BUP *puts a data node* $v_q \in I_q$ *in the list* CL_q *under construction if and only if it is in the candidate occurrence list of the root of the subquery of* Q *rooted at* q.

Complexity. The time complexity of *BUP* is determined by the total number of pattern edge matches checked, which is $|Q| \times |I_{max}^Q|^2$, where $|Q|$ is the number

Table 2. Dataset statistics. $|V|$, $|E|$ and $|L|$ are the number of nodes, edges and distinct labels, respectively. *maxout* and *maxin* are the maximum out-degree and in-degree of the graph. d_{avg} $(=|E|/|V|)$ denotes the average degree of a graph.

Dataset	|V|	|E|	|L|	Maxout	Maxin	d_{avg}
xm03	50266	57992	92	765	152	1.15
xm05	83533	96484	92	1275	162	1.16
xm07	118670	136828	92	1785	151	1.15
xm09	151289	174737	92	2295	154	1.15
xm5	832911	960941	92	12750	168	1.15
acm	629814	631215	12610	195	816	1.00
cite-lb10000	6540401	15011260	8343	181247	203695	2.30
cite-lb8000	6540401	15011260	6124	181247	203695	2.30
cite-lb7000	6540401	15011260	5662	181247	203695	2.30
cite-lb6000	6540401	15011260	4969	181247	203695	2.30
cite-lb5000	6540401	15011260	3970	181247	203695	2.30
rand-v300k	300000	518528	3799	10	12	1.73
rand-v500k	500000	864057	4038	10	11	1.73
rand-v700k	700000	1208868	3826	11	11	1.73
rand-v1000k	1000000	1727123	3721	12	11	1.73
rand-v1500k	1500000	2590639	4048	12	12	1.73

of pattern nodes and $|I^Q_{max}|$ is the size of the largest inverted list in I^Q. The time R for checking if a pair of data nodes is an edge occurrence is bound by the time for checking reachability for a pair of nodes in the data graph. Therefore, the time complexity of *BUP* is $O(|Q| \times |I^Q_{max}|^2 \times R)$. The memory consumption is determined by the size of the answer graph which is bound by $|Q| \times |I^Q_{max}|^2$.

5 Experimental Evaluation

We present an experimental evaluation of our bitmap view approach by comparing it with other previous approaches in terms of time performance and scalability.

5.1 Experimental Setting

Algorithms. We implemented and compared the following algorithms. The first includes two versions of the stack-based algorithm *TwigStackD* [2,21], using two different reachability index schemes: *TSD-SSPI* uses the *Surrogate Surplus Predecessor Index* (SSPI) [2], while *TSD-BFL* uses the *Bloom Filter Labeling* (BFL) index [11]. We modified *TwigStackD* to output the query answer graph, instead of enumerating solution tuples using expensive merge-join operations over query

Table 3. Query set statistics. '//' denotes the descendant pattern edge.

Query set	# queries	Avg. |V|	Avg. height	% of '//'	Maxout	Avg. # solutions
XQ1	10	7	4	100.00	2	19842.4
XQ2	10	9	4	100.00	3	2001.5
XQ3	10	11	4	100.00	3	7913.5
acm.qry	100	7.16	2.69	16.20	2.13	6804.59
cite-lb10000.qry	10	3.5	1.4	65.71	1.9	219.9
cite-lb8000.qry	10	3.2	1.2	65.63	2	93.4
cite-lb7000.qry	10	3.1	1.1	67.74	2	1859.1
cite-lb6000.qry	10	3.3	1.2	63.64	2	760.9
cite-lb5000.qry	10	3.2	1.1	65.63	2	582.9
rand-v300k.qry	10	4.5	2.5	60.00	1.7	46769.4
rand-v500k.qry	10	4.9	2.3	55.10	1.8	978.2
rand-v700k.qry	10	6.9	3.8	40.58	1.9	6375.1
rand-v1000k.qry	10	4.9	2.8	59.18	1.7	264620.5
rand-v1500k.qry	10	6.2	3.4	50.00	1.8	33960.2

path solutions. The second is our proposed bitmap materialized views approach, denoted as *BUP-MV*, presented in Sect. 4. The third is the bottom-up pattern evaluation algorithm without using materialized views, denoted as *BUP*. The last is the algorithm *BUP* with an node pre-filtering technique introduced in [2,21], denoted as *BUP-FLT*. All implemented algorithms, except *TSD-SSPI*, use BFL for reachability checking. We present no performance comparison with the decomposition-based algorithms R-Join [3] and HGJoin [13] because their design is closely tied to some specific reachability indexing schemes, whose overall performance has already been shown to be much worse than BFL [11]. Also, the performance of HGJoin has been reported in [8] to be even worse than *TSD-SSPI* for queries similar to those used in our experiments. We used roaring bitmaps [1] for the implementation of compressed bitmap view materializations. Roaring bitmaps have been shown to outperform conventional compressed bitmaps such as WAH, EWAH or Concise [14]. Our implementation was coded in Java. All the experiments were performed on a workstation having an Intel Xeon CPU 1240V5@3.50 GHz processor with 32 GB memory.

Datasets. We ran experiments on four types of graph datasets with different structural properties. Their main characteristics are summarized in Table 2.

XMark[2] is a synthetic benchmark dataset modeling an auction website. We generated five XMark datasets using scaling factors, 0.3, 0.5, 0.7, 0.9, 5, named xm03, xm05, xm07, xm09, and xm5, respectively. For each dataset, we generated a graph by treating internal links (parent-child) and ID/IDREF links as edges. As in [8,20], we randomly classified *person* and *item* elements of the XMark graphs into ten groups and assigned a distinct label to each group.

[2] xml-benchmark.org.

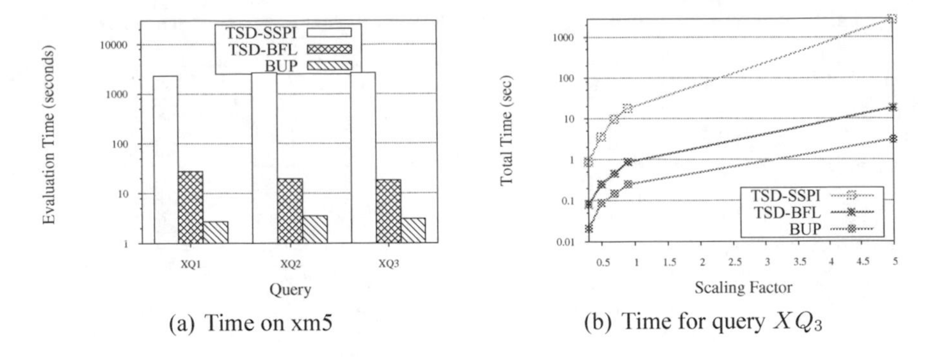

(a) Time on xm5

(b) Time for query XQ_3

Fig. 3. Performance evaluation on the XMark dataset.

acm[3] models citations of the ACM publications and consists of a directed graph with 615K nodes and 616K edges. Nodes represent papers while edges represent citations.

citeseerx[4] represents a directed graph consisting of 6.3M publications (nodes) and 14.3M citations between them (edges). The original graph does not have labels. We wrote a label assignment program which randomly adds a specified number of distinct labels to graph nodes, following a Gaussian distribution. Using this program, we generated five labeled citeseerx graphs whose number of labels ranges from 5,000 to 10,000. Each graph is named as cite-lb$x$, where x is the number of labels in the graph.

We also implemented a random graph generator which creates a random graph based on the Erdos-Renyi model. Given three input parameters n, m, and l, the generator first creates a random graph with n nodes and m edges; then it calls the label assignment program to randomly add l distinct labels to the nodes. Using the graph generator, we generated five graphs varying the size n of the nodes from 300,000 to 1,500,000. The size m of the edges was $2n$, and l was fixed to 5000. These five graphs are named rand-vx, where x is 300K, 500K, 700K, 1000K, or 1500K.

Our materialized view approach works on general graphs. Since most of the reachability indexing schemes including BFL and $SSPI$ work only with dags, in the experiments, we converted directed graphs with cycles to dags, by removing back edges. All the statistics shown in Table 2 are for dags.

Queries. For the XMark dataset, we used the three query templates XQ_1–XQ_3 shown in Fig. 2. These are descendant-only tree patterns which were also used in [8,20] and are useful for comparing the performance of algorithms *TSD-SSPI* and *TSD-BFL* which are not designed for mixed pattern queries. For each query template, we generated 10 queries by randomly choosing the labels on *person* and *item* nodes. For the experiments on the acm, citeseerx, and random datasets, we used randomly generated queries. We implemented a query generator that

[3] www.aminer.cn/citation.
[4] citeseerx.ist.psu.edu.

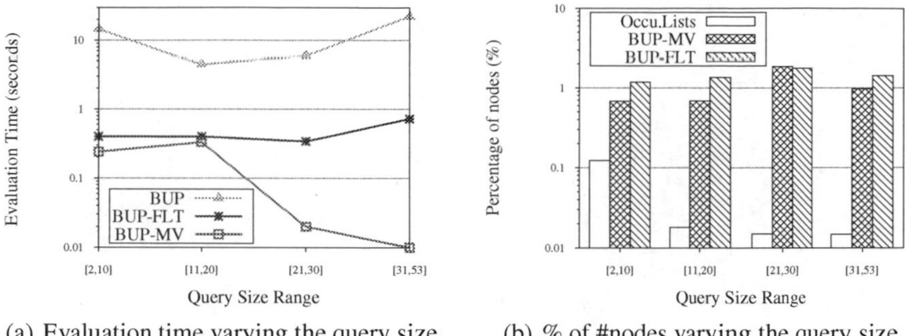

(a) Evaluation time varying the query size (b) % of #nodes varying the query size

Fig. 4. Performance evaluation on the *acm* graph.

creates a set of tree pattern queries based on the parameters listed in Table 1. Random queries are generated according to the input data graph and these parameters. For each data graph, we first generated a number of queries (in the range of 300 to 1800) using different value combinations of the parameters listed in Table 1, and then we formed a query set by randomly selecting 10 to 100 among them. Table 3 summarizes the statistics of the query sets.

Views. In the experiments, we used single edge (child or descendant) path patterns as views. The views cover in each dataset all the queries generated.

5.2 Algorithm TSD vs. BUP on XMark Graphs

We first compared the time performance of the three pattern matching algorithms *TSD-SSPI*, *TSD-BFL* and *BUP*. Figure 3(a) shows the execution time of the three algorithms for the query templates XQ_1–XQ_3 of Fig. 2 over the XMark graph dataset xm5. We can see that in all the cases, *TSD-BFL* outperforms *TSD-SSPI* by about two orders of magnitude, whereas *BUP* is at least six times faster than *TSD-BFL*. We also studied the scalability of the three algorithms for evaluating XQ_1–XQ_3 over XMark graphs of various sizes. In Fig. 3(b), we present the results for XQ_3 obtained for graphs whose scaling factors vary from 0.5 to 5. The increase in the execution time of *TSD-SSPI* is much sharper than that of *TSD-BFL*, which in turn grows faster than *BUP*. Therefore, *BUP* scales better than the other two when the input size increases.

Note that the underlying graph matching process of *TSD-SSPI* and *TSD-BFL* is stack-based [2, 20], which is different from the bottom-up technique used by *BUP*. The only difference between *TSD-SSPI* and *TSD-BFL* lies in the reachability index scheme used.

5.3 Results on *acm*, *citeseerx*, and *random* Graphs

In the experiments below, we use *BUP* as the baseline algorithm, and compare its performance with the two algorithms *BUP-MV* and *BUP-FLT*. The difference

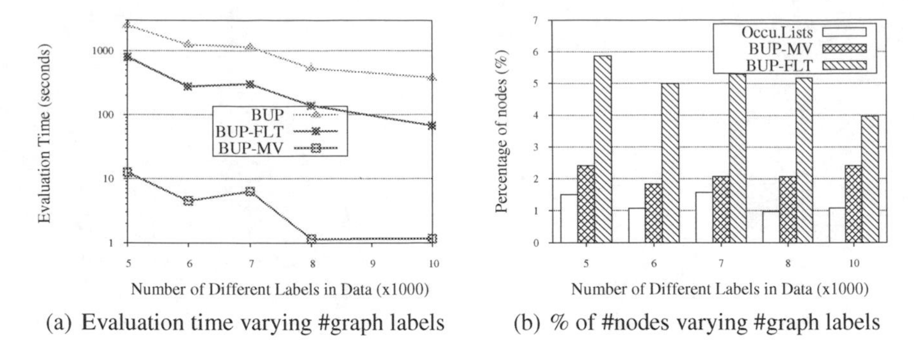

(a) Evaluation time varying #graph labels (b) % of #nodes varying #graph labels

Fig. 5. Performance evaluation on the *Citeseer* graph.

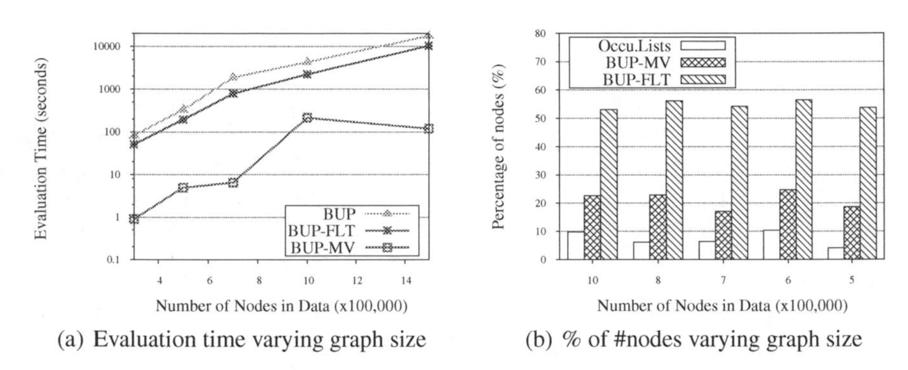

(a) Evaluation time varying graph size (b) % of #nodes varying graph size

Fig. 6. Performance evaluation on the *randomly* generated graphs.

of these algorithms is that, while *BUP* computes the answer over the original graph inverted lists, both *BUP-MV* and *BUP-FLT* employ some technique to filter out, in advance, nodes of the inverted lists that do not participate in the answer of the query.

Query Evaluation Time. The query *evaluation time* of *BUP-MV* consists of the *optimization time* and the query *execution time*. The query execution time is the time needed for computing the query using the view materializations. The optimization time consists of the time needed for finding the covering view nodes of the query nodes and the time needed for disk loading, decompressing and bitwise ANDing the bitmaps of the node materializations. Unlike the materialized view approach, the pre-filtering technique [2, 20] does not exploit pre-computation, but repeats the filtering process for every incoming query. Specifically, it conducts two traversals on the data graph: one for pruning nodes that violate the downward structural constraints and one for checking upward constraints. The pre-filtering technique prunes data nodes for descendant-only pattern queries. Our materialized view approach does not have such restriction. As *BUP-MV*, the mixed pattern query evaluation time of *BUP-FLT* has two parts: optimization time (for node punning) and query execution time.

Varying Query Sizes. We designed experiments to examine the impact of query size on the performance of the three algorithms. In order to do so, we used our query generation tool to generate a random query set called *acm.qry* for the acm graph (Table 3). Queries in this set were grouped by query size (i.e., the number of nodes in the query) into the following four groups: 2–10, 11–20, 21–30, 31–53. Figure 4(a) compares the average evaluation time of the three algorithms when the query size increases on the acm data graph. Figure 4(b) shows the average percentage of the number of inverted list nodes appearing in the query occurrence lists. It also shows the average percentage of the number of inverted list nodes accessed by *BUP-MV* and *BUP-FLT* during the matching process. *BUP* accesses all the nodes (100%) of the inverted lists during the matching process since it does not have a filtering phase.

We observe that: (1) *BUP* is more than one order of magnitude slower than both *BUP-MV* and *BUP-FLT*; this can be explained by the fact that the latter two algorithms access about 2% of the inverted list nodes accessed by *BUP* (Fig. 4(b)) during the matching process. (2) *BUP-MV* is faster than *BUP-FLT* since it only needs to access about 76% of the nodes accessed by *BUP-FLT*. Their performance gap widens when the query size becomes larger than 20 (Fig. 4(a)). We found that in this experiment, *BUP-FLT* spent most time on optimization, which increases when the query size increases. In contrast, the optimization time of *BUP-MV* was too small to even be noticed in all cases. This explains the performance difference between *BUP-MV* and *BUP-FLT*.

Varying the Number of Graph Data Labels. We examine the impact of the total number of distinct graph labels on the performance of the three algorithms. We used the aforementioned five labeled *citeseerx* graphs *cite-lbx* (Table 2), where the number of labels x increased from 5,000 to 10,000. For each graph *cite-lbx*, we generated a query set *cite-lbx.qry* with 10 distinct queries (Table 3). As shown in Fig. 5(a), the execution time for all algorithms decreases with the increase of the total number of graph labels. This is reasonable, since the average size of the input inverted list per graph node label is reduced when more labels are added. This confirms the complexity results of Sect. 4.3 that show dependency of the execution time on the input size. As with the previous experiment, *BUP-MV* has the best performance overall; it is more than one and two orders of magnitudes faster than *BUP-FLT* and *BUP*, respectively. It accesses around 2% and 41% of inverted list nodes visited by *BUP* and *BUP-FLT*, respectively, during the matching process (Fig. 5(b)). Its optimization time revolves around 1.5 ms in all cases. By accessing 5% of the number of nodes visited by *BUP*, *BUP-FLT* achieves a speedup of 2.6 on average. Its optimization time is around 6 s on average, i.e. about 2% of the average evaluation time.

Varying Random Graph Sizes. We evaluate the performance of the three algorithms on randomly generated graphs. We used the five random graphs *rand-vx* (Table 2), where the number of nodes x was increased from 300,000 to 1,500,000, while the total number of labels is fixed to 5,000. For each graph *rand-vx*, we generated a random query set *rand-vx.qry* with 10 distinct queries (Table 3).

The results are shown in Fig. 6(a). The execution time for all algorithms goes up when the total number of graph nodes increases. *BUP-MV* displays significantly better performance than the other two algorithms, achieving a speedup of at least one order of magnitude across the entire range of random graphs. During the filtering phase it eliminates on average 95% of the inverted list nodes accessed by *BUP* (Fig. 6(b)). Its average optimization time was 7 ms. The evaluation time of *BUP-FLT* is 56% of that of *BUP* on average over all random graphs. *BUP-FLT* accesses about 56% of inverted list nodes visited by *BUP* (Fig. 6(b)). Its average optimization time is 1.6 s.

5.4 Summary

The experiments reported here have examined the performance of four pattern query evaluation algorithms on graphs. The results can be summarized as follows:

- The performance of a graph pattern matching algorithm is affected significantly by both the graph matching process and the reachability indexing scheme used.
- The total number of graph nodes accessed by each algorithm for computing a pattern query is the principal determining factor of the query evaluation time.
- The bitmap view materialization approach has better graph node pruning power than the pre-filtering technique for optimizing mixed pattern queries over large graphs.
- *BUP-MV* shows the best efficiency and scalability performance among all tested algorithms, while displaying a negligible optimization cost.

6 Related Work

A number of papers have recently addressed the important problem of answering pattern queries on graphs using views [5,7,15,16]. Two types of approaches, namely, equivalent answering and approximate answering have been considered. The equivalent pattern query answering aims at producing all the results to the query using the given view materialization(s), whereas the approximate pattern query answering may produce a subset of the results. Previous work on this line has also considered two kinds of pattern matching models: subgraph isomorphism and graph simulation. Subgraph isomorphism retrieves the exact topological matches. However, subgraph isomorphism is an NP-complete problem. Graph simulation relaxes the restrictions enforced by subgraph isomorphism and can be computed in polynomial time.

The main problem studied in the simulation-based approach is pattern containment. It determines the conditions under which the results of a pattern query are contained in the results of a set of views. Fan et al. [5] proposed an equivalent pattern containment scheme based on (bounded) simulation model to characterize pattern matching using views. They also developed efficient algorithms for pattern containment checking. Equivalent pattern containment based on subgraph isomorphism was investigated in [16]. Wang et al. [15] studied the problem of reusing the results of previously executed queries to answer subgraph/supergraph queries based on subgraph isomorphism.

The approximate graph pattern answering, a.k.a maximally contained pattern rewriting, aims at identifying a maximal part of the pattern query that can be answered using a set of views. A contained rewriting helps to answer approximately the original pattern query on a data graph. Both [5] and [16] studied maximally contained pattern rewriting albeit under the two different pattern matching models described above (graph simulation and subgraph isomorphism). Li et al. [7] studied the problem of computing upper and lower approximations from views for pattern queries in both models.

Our work differs from prior work [5,7,15,16] in the following: (1) We address the problem of tree-pattern matching on large data graphs using homomorphisms. Homomorphisms relax the strict one-to-one mapping to a more general many-to-one matching, and allow a pattern edge to be matched to a path in the data graph. It is therefore more flexible than subgraph isomorphism. In addition, homomorphic tree-pattern matching on graphs can be computed in polynomial time. Homomorphism preserves the topology of data graphs. In contrast, simulation (and its variants) may match together a data graph and a pattern with drastically different structures. (2) We study how to efficiently evaluate mixed pattern queries on large data graphs using materialized views. Instead of materializing pattern matches, our approach materializes the occurrence lists of the pattern nodes, stored as compressed bitmaps on the inverted lists of the data graph labels. Materializing views as bitmapped occurrence lists are also adopted in [18,19]. However, the focus of those papers was on the problem of answering XML queries using materialized views over XML data.

7 Conclusion

In this paper we have addressed the problem of efficiently evaluating mixed pattern queries on large data graphs using materialized views. Unlike other approaches which store the matches of the patterns, our approach sees materialized views as sets of occurrence lists of the pattern query nodes and stores them as compressed bitmaps. We have provided conditions for pattern query answerability using bitmap views through view homomorphisms. We have also designed a bitmap-view-based approach for evaluating pattern queries on data graphs. An extensive experimental evaluation has verified the efficiency and scalability of our approach and has shown that it largely outperforms other approaches that do not use views. We are currently working on developing elaborate techniques for selecting views for bitmap materialization.

References

1. Chambi, S., Lemire, D., Kaser, O., Godin, R.: Better bitmap performance with roaring bitmaps. Softw. Pract. Exper. **46**(5), 709–719 (2016)

2. Chen, L., Gupta, A., Kurul, M.E.: Stack-based algorithms for pattern matching on DAGs. In: VLDB (2005)
3. Cheng, J., Yu, J.X., Yu, P.S.: Graph pattern matching: a join/semijoin approach. IEEE Trans. Knowl. Data Eng. **23**(7), 1006–1021 (2011)
4. Fan, W., Li, J., Ma, S., Wang, H., Wu, Y.: Graph homomorphism revisited for graph matching. PVLDB **3**(1), 1161–1172 (2010)
5. Fan, W., Wang, X., Wu, Y.: Answering pattern queries using views. IEEE Trans. Knowl. Data Eng. **28**(2), 326–341 (2016)
6. Gallagher, B.: Matching structure and semantics: a survey on graph-based pattern matching. AAAI FS **6**, 45–53 (2006)
7. Li, J., Cao, Y., Liu, X.: Approximating graph pattern queries using views. In: CIKM, pp. 449–458 (2016)
8. Liang, R., Zhuge, H., Jiang, X., Zeng, Q., He, X.: Scaling hop-based reachability indexing for fast graph pattern query processing. IEEE Trans. Knowl. Data Eng. **26**(11), 2803–2817 (2014)
9. Olteanu, D., Schleich, M.: Factorized databases. SIGMOD Rec. **45**(2), 5–16 (2016)
10. Shang, H., Zhang, Y., Lin, X., Yu, J.X.: Taming verification hardness: an efficient algorithm for testing subgraph isomorphism. PVLDB **1**(1), 364–375 (2008)
11. Su, J., Zhu, Q., Wei, H., Yu, J.X.: Reachability querying: can it be even faster? IEEE Trans. Knowl. Data Eng. **29**(3), 683–697 (2017)
12. Ullmann, J.R.: An algorithm for subgraph isomorphism. J. ACM **23**(1), 31–42 (1976)
13. Wang, H., Li, J., Luo, J., Gao, H.: Hash-base subgraph query processing method for graph-structured XML documents. PVLDB **1**, 478–489 (2008)
14. Wang, J., Lin, C., Papakonstantinou, Y., Swanson, S.: An experimental study of bitmap compression vs. inverted list compression. In: SIGMOD, pp. 993–1008 (2017)
15. Wang, J., Ntarmos, N., Triantafillou, P.: Indexing query graphs to speedup graph query processing. In: EDBT, pp. 41–52 (2016)
16. Wang, X.: Answering graph pattern matching using views: a revisit. In: Benslimane, D., Damiani, E., Grosky, W.I., Hameurlain, A., Sheth, A., Wagner, R.R. (eds.) DEXA 2017. LNCS, vol. 10438, pp. 65–80. Springer, Cham (2017). https://doi.org/10.1007/978-3-319-64468-4_5
17. Wu, X., Souldatos, S., Theodoratos, D., Dalamagas, T., Sellis, T.K.: Efficient evaluation of generalized path pattern queries on XML data. In: WWW (2008)
18. Wu, X., Theodoratos, D., Wang, W.H.: Answering XML queries using materialized views revisited. In: CIKM (2009)
19. Wu, X., Theodoratos, D., Wang, W.H., Sellis, T.: Optimizing XML queries: bitmapped materialized views vs. indexes. Inf. Syst. **38**(6), 863–884 (2013)
20. Zeng, Q., Jiang, X., Zhuge, H.: Adding logical operators to tree pattern queries on graph-structured data. PVLDB **5**(8), 728–739 (2012)
21. Zeng, Q., Zhuge, H.: Comments on "stack-based algorithms for pattern matching on dags". PVLDB **5**(7), 668–679 (2012)

Heterogeneous Information Network Hashing for Fast Nearest Neighbor Search

Zhen Peng[1], Minnan Luo[1](✉), Jundong Li[2], Chen Chen[2], and Qinghua Zheng[1,3]

[1] School of Electronic and Information Engineering, Xi'an Jiaotong University, Xi'an, China
zhenpeng27@outlook.com, {minnluo,qhzheng}@xjtu.edu.cn
[2] Computer Science and Engineering, Arizona State University, Tempe, USA
{jundongl,chen_chen}@asu.edu
[3] National Engineering Lab for Big Data Analytics, Xi'an Jiaotong University, Xi'an, China

Abstract. Heterogeneous information networks (HINs) are widely used to model real-world information systems due to their strong capability of capturing complex and diverse relations between multiple entities in real situations. For most of the analytical tasks in HINs (*e.g.*, link prediction and node recommendation), network embedding techniques are prevalently used to project the nodes into real-valued feature vectors, based on which we can calculate the proximity between node pairs with nearest neighbor search (NNS) algorithms. However, the extensive usage of real-valued vector representation in existing network embedding methods imposes overwhelming computational and storage challenges, especially when the scale of the network is large. To tackle this issue, in this paper, we conduct an initial investigation of learning binary hash codes for nodes in HINs to obtain the remarkable acceleration of the NNS algorithms. Specifically, we propose a novel heterogeneous information network hashing algorithm based on collective matrix factorization. Through fully characterizing various types of relations among nodes and designing a principled optimization procedure, we successfully project the nodes in HIN into a unified Hamming space, with which the computational and storage burden of NNS can be significantly alleviated. The experimental results demonstrate that the proposed algorithm can indeed lead to faster NNS and requires lower memory usage than several state-of-the-art network embedding methods while showing comparable performance in typical learning tasks on HINs, including link prediction and cross-type node similarity search.

1 Introduction

A heterogeneous information network (HIN) is a logical network that contains multiple types of entities and various relations between them [22]. Since it can capture rich information in a wide diversity of realistic scenarios via characterizing the heterogeneity of entities and their interactions, HINs become increasingly

G. Li et al. (Eds.): DASFAA 2019, LNCS 11446, pp. 571–586, 2019.
https://doi.org/10.1007/978-3-030-18576-3_34

important and have widespread usage in many high-impact domains, such as the critical infrastructure systems, the knowledge graphs and the social media networks. Over the past few decades, many prevalent network learning tasks have been studied in HINs, most of which heavily depend on the nearest neighbor search (NNS) algorithms, examples include network clustering [16], link prediction [14], personalized recommendation [15], chemical similarity search [29], to name a few. Therefore, it is essential to provide an effective and trustworthy search function to advance these tasks in HINs. More specifically, we need a measure (*e.g.*, Euclidean distance or cosine distance) to quantify the node closeness in HINs such that given a query node in the HIN, we can retrieve the sets of nodes that are most similar to the query node by computing their similarity. Moreover, it should be noted that for the NNS algorithms, their utility is normally determined by two leading factors: (1) time complexity - the retrieval time for the nearest neighbors; and (2) space complexity - the storage costs of the used data structures. Typically, network embedding [9] is often regarded as a popular way to tackle the NNS problem by projecting nodes into a low-dimensional real-valued space while maximally preserving the node proximity in the HIN.

In fact, a vast majority of existing network embedding approaches such as PTE [23], metapath2vec [5] and HIN2Vec [7] generate real-valued node embeddings, which brings formidable computational and storage challenges to the NNS algorithms. The main reason is that a large number of indispensable similarity calculations in the NNS algorithms are directly performed on the learned real-valued node vectors, while the storage and the associated floating-point arithmetic operations could be very expensive. For instance, recommending a movie of interest for all users in a social media network including n users and m movies has a time complexity of $O(nmd)$ and a space complexity of $O((n+m)d)$, where d is the dimensionality of the node embeddings. Considering that the real-world social media networks may contain millions or billions of users and movies, the computational and storage costs of the NNS algorithms will be extremely high and are difficult to scale. Fortunately, hashing has been widely recognized as a promising technique to accelerate the similarity search [26]. Through transforming the real-valued vectors to compact binary codes consisting of a sequence of bits, hashing not only gains the advantage of reducing the memory storage costs but also lowers the time complexity of the nearest neighbor retrieval, as the similarity of two binary hash codes can be easily obtained with fast Hamming distance calculations.

The success of the hashing technique in many applications motivates our initial investigation of learning binary hash codes for nodes in HINs. However, it is a nontrivial problem with the following two unique challenges. The first core problem lies in how to elaborately characterize the diversity of connectivity patterns observed in HINs for the binary hash code learning. Secondly, after imposing the binary constraints on the learning model, designing a principled optimization algorithm is imperative as binary constraints can make the learning procedure NP-hard. To tackle these challenges, in this paper, we develop a novel heterogeneous information network hashing algorithm, called *HIN2Hash*.

Benefiting from the collective matrix factorization (CMF) [21] technique which not only supports multi-relational learning but also improves the prediction accuracy, *HIN2Hash* models the diverse relations among nodes in HINs delicately and then learns unified binary hash codes for each node via a well-designed optimization framework. The superiority of *HIN2Hash* over other conventional methods rests with lower time cost and memory usage, and meanwhile, *HIN2Hash* exhibits competitive performance in some typical learning tasks on HINS, including link prediction and cross-type node similarity search. The main contributions of our work are as follow:

- Studying the problem of heterogeneous information network hashing which aims at (1) speeding up the downstream graph mining tasks on HINs that heavily depend on the NNS algorithms; and meanwhile (2) reducing the memory storage usage.
- Proposing a novel CMF based heterogeneous information network hashing algorithm *HIN2Hash* which carefully addresses the diversity of node types and relations among nodes, and encodes nodes into binary hash codes.
- Evaluating the performance of the proposed algorithm on three real-world datasets and the results illustrate that *HIN2Hash* can indeed accelerate similarity calculation and reduce storage with little sacrifice of performance.

2 Related Work

With the rise of representation learning research on networked data, a wide range of techniques have been developed in the past few years, especially for homogeneous network embedding. For instance, DeepWalk [17] and node2vec [8] are two typical algorithms based on the skip-gram model by using uniform and parameterized random walks, respectively. LINE [24] preserves 1st-order and 2nd-order proximities separately, to learn two different types of node embeddings. NetMF [18] is a matrix factorization based framework for network embedding. Different from the above methods, heterogeneous network embedding approaches fully consider the heterogeneity of nodes and edges for representation learning. For example, metapath2vec [5] learns node embeddings based on meta-path-based random walks and the heterogeneous skip-gram model.

Learning to hash (*a.k.a.* Hashing) [26], a widely studied solution to the approximate nearest neighbor search, attracts an enormous number of research efforts due to its comparable performance and superiority in time and storage. Typically, hashing algorithms can be classified into two classes: (1) two-stage approaches which first derive real-valued vectors and then convert them into binary codes, and (2) discrete hashing learning for binary hash codes directly. Recently, a number of network embedding algorithms adopt this idea to learn binary hash codes of nodes and are referred to as network hashing. However, these methods [13,20] only focus on homogeneous networks, and ignore the various types of nodes and edges in HINs. Different from existing works, our algorithm is proposed for heterogeneous information network hashing and derive binary hash codes straightly through joint optimization of hashing and network embedding.

3 The Proposed Methodology

We first introduce the notations used in this paper. Following the standard notation, we use bold uppercase letters (e.g., $\mathbf{A}$) for matrices and bold lowercase letters (e.g., $\mathbf{b}$) for vectors. Scalars are indicated as normal lowercase letters (e.g., c) and uppercase italic letters (e.g., V) represent sets. Also, we denote the i-th row of the matrix $\mathbf{A} \in \mathbb{R}^{m \times n}$ as $\mathbf{A}(i,:)$, the j-th column as $\mathbf{A}(:,j)$, and the (i,j)-th entry as $\mathbf{A}(i,j)$. As for the matrix norms, the only used matrix norm is the Frobenius norm, written as $\|\mathbf{A}\|_F = \sqrt{\sum_{i=1}^{m}\sum_{j=1}^{n}\mathbf{A}(i,j)^2}$. We denote $tr(\cdot)$ as the matrix trace, and $sign(\cdot): \mathbb{R} \rightarrow \{\pm 1\}$ as the round-off function.

3.1 Problem Formulation

With the above-mentioned notations, we now formally define the studied problem of heterogeneous information network hashing. It should be noted that we follow the definition of HINs as in [5,7].

Definition 1. ***Heterogeneous Information Networks.*** *A heterogeneous information network is defined as a graph $G = (V, E, \Phi, \Psi)$ in which each node $v \in V$ and each edge $e \in E$ is associated with its own mapping function $\Phi(v): V \rightarrow \mathcal{E}$ and $\Psi(e): E \rightarrow R$,* i.e., *each node is mapped to one particular node type in $\mathcal{E}$ and each edge belongs to a specific edge type in R. Furthermore, we have $|\mathcal{E}| + |R| > 2$.*

According to the definition, it is obvious that an HIN G contains two different types of links, one is the relations among nodes of the same type (intra-type relations), and the other is the relations among nodes across two different node types (inter-type relations). Formally, suppose $\mathcal{E} = \{\mathcal{E}_1, \ldots, \mathcal{E}_g\}$ indicate a set of g node types, then there are g intra-type relations and C_g^2 inter-type relations theoretically. We adopt a set of matrices $A = \{\mathbf{A}_1, \ldots, \mathbf{A}_g\}$ to describe the proximity among nodes within each node type, where $\mathbf{A}_i \in \{0,1\}^{n_i \times n_i}$ $(i = 1, \ldots, g)$; and a set of matrices $X = \{\mathbf{X}_{ij}, (i,j = 1, \ldots, g)(i \neq j)\}$ to represent interactions among nodes of different types, where $\mathbf{X}_{ij} \in \{0,1\}^{n_i \times n_j}$ denotes the inter-type relations between $\mathcal{E}_i$ and $\mathcal{E}_j$. Taking an HIN represented by A and X as an input, the problem of heterogeneous information network hashing is formally defined as follows.

Problem 1. ***Heterogeneous Information Network Hashing.*** *Given an HIN G, the problem of heterogeneous information network hashing aims to learn a function $\mathcal{F}: V \rightarrow \{\pm 1\}^d$ that projects each node $v \in V$ to a low-dimensional Hamming space $\{\pm 1\}^d$ shown in $U = \{\mathbf{U}_1, \ldots, \mathbf{U}_g\}$, where $\mathbf{U}_i \in \{\pm 1\}^{d \times n_i}$ $(i = 1, \ldots, g)$ and $d \ll |V|$. In the learned hashing space $\{\pm 1\}^d$, two additional constraints on the binary codes introduced by [28],* i.e., *bit uncorrelation and bit balance (more details later), should be satisfied as much as possible.*

3.2 Heterogeneous Information Network Hashing

As shown in [18], matrix factorization provides a principled framework to unify the prevalent network embedding methods such as DeepWalk, node2vec, LINE and achieves comparable performance. Hence, in the following context, we resort to the matrix factorization to elaborate our developed framework for heterogeneous information network hashing. Distinct from conventional homogeneous networks, in HINs, multiple types of nodes and edges are presented together, and a particular node type may be involved in more than one types of relations. In order to fully characterize these complex and diverse connectivity patterns among nodes, we expand our view from separately factoring the individual connectivity matrix of each node type to collective matrix factorization as it is able to take full advantage of different types of connections among nodes in HINs for representation learning.

Mathematically, we first define a pairwise relational schema $S = \{(i,j)|\mathcal{E}_i \sim \mathcal{E}_j, i < j\}$ to include all possible C_g^2 inter-type relations. Then the objective function is formulated as:

$$\min_{\mathbf{U}_i \in U_*} \sum_{(i,j)\in S} \alpha_{ij} \|\mathbf{X}_{ij} - \mathbf{U}_i^T \mathbf{U}_j\|_F^2 + \sum_{i=1}^{g} \beta_i \|\mathbf{A}_i - \mathbf{U}_i^T \mathbf{U}_i\|_F^2, \tag{1}$$

where $\mathbf{U}_i$ denotes the binary hash codes of all n_i nodes belonging to $\mathcal{E}_i$. It should be noted that $\mathbf{U}_i$ has to satisfy the bit uncorrelation and balance constraints such that $U_* = \{\mathbf{U}_i \in \{\pm 1\}^{d \times n_i} | \mathbf{U}_i \mathbf{U}_i^T = n_i \mathbf{I}_d, \mathbf{U}_i \mathbf{1}_{n_i} = \mathbf{0}\}$. The constraints in U_* maximize the information encoded in the binary embedding space. To be more specific, the bit uncorrelation constraint ensures that each bit should be as independent as possible. Meanwhile, the bit balance constraint makes each bit of almost equal chance of being 1 or -1. In other words, it maximizes the entropy of each bit. By simultaneously decomposing the relation matrices $\mathbf{X}_{ij}$ and $\mathbf{A}_i$ into a product of two matrices in a low-dimensional Hamming space, we obtain unified binary codes in which different types of relations among nodes can be captured. The trade-off parameters $\alpha_{ij}, \beta_i \geq 0$ measure the importance of each relation matrix in the reconstruction process.

Admittedly, solving Eq. (1) is a challenging task since the existence of binary constraints makes the optimization process of learning binary hash codes NP-hard [10] and may cause Eq. (1) infeasible to solve. For the sake of solving it in a computationally tractable manner, we reformulating Eq. (1) by softening the bit balance and uncorrelation constraints [13,30]. Considering that the bit balance condition is easier to deal with, we intend to handle these two constraints separately by splitting the constraint set U_* into $U_0 = \{\mathbf{U}_i \in \{\pm 1\}^{d \times n_i} | \mathbf{U}_i \mathbf{1}_{n_i} = \mathbf{0}\}$ and $U_\perp = \{\mathbf{U}_i \in \{\pm 1\}^{d \times n_i} | \mathbf{U}_i \mathbf{U}_i^T = n_i \mathbf{I}_d\}$. As for the uncorrelation constraint, we approximate it by introducing $Z = \{\mathbf{Z}_i \in \mathbb{R}^{d \times n_i} | \mathbf{Z}_i \mathbf{Z}_i^T = n_i \mathbf{I}_d\}$ and a penalty term recording the deviation between a feasible $\mathbf{U}_i$ and the set Z. Hence, the original objective function can be softened as:

$$\min_{\mathbf{U}_i \in U_0, \mathbf{Z}_i \in Z} \sum_{(i,j)\in S} \alpha_{ij} \|\mathbf{X}_{ij} - \mathbf{U}_i^T \mathbf{U}_j\|_F^2 + \sum_{i=1}^{g} \beta_i \|\mathbf{A}_i - \mathbf{U}_i^T \mathbf{U}_i\|_F^2 + \sum_{i=1}^{g} \varphi_i \|\mathbf{U}_i - \mathbf{Z}_i\|_F^2. \tag{2}$$

The above equation allows a certain discrepancy between the binary hash codes (e.g., $\mathbf{U}_i$) and their delegate continuous values (e.g., $\mathbf{Z}_i$), which makes these constraints computationally tractable. In fact, we can impose a very large parameter φ_i to force the matrix $\mathbf{U}_i$ to be equal to $\mathbf{Z}_i$ such that the constraints in Eq. (1) become feasible. By jointly optimizing the binary codes and the delegate real variables, we can obtain nearly balanced and uncorrelated Hamming codes for each node v in the HIN G.

4 The Optimization Algorithm

Due to the existence of binary constraints, we employ the alternating direction method of multipliers (ADMM) [1] to solve Eq. (2). First, we introduce variables $\widetilde{\mathbf{U}}_i$ $(i = 1, \ldots, g)$ to convert Eq. (2) to the following equivalent objective function:

$$\begin{aligned}\min_{\mathbf{U}_i \in U_0, \mathbf{Z}_i \in Z} & \sum_{(i,j)\in S} \alpha_{ij} \|\mathbf{X}_{ij} - \mathbf{U}_i^T \mathbf{U}_j\|_F^2 + \sum_{i=1}^{g} \beta_i \|\mathbf{A}_i - \widetilde{\mathbf{U}}_i \mathbf{U}_i\|_F^2 \\ & + \sum_{i=1}^{g} \varphi_i \|\mathbf{U}_i - \mathbf{Z}_i\|_F^2 \qquad s.t.\, \mathbf{U}_i^T = \widetilde{\mathbf{U}}_i. \end{aligned} \tag{3}$$

The augmented Lagrange function of Eq. (3) is:

$$\begin{aligned} & \mathcal{L}_{\{\rho_i\}_{i=1}^g}(\{\mathbf{U}_i, \widetilde{\mathbf{U}}_i, \mathbf{Z}_i, \Lambda_i\}_{i=1}^g) \\ = & \sum_{(i,j)\in S} \alpha_{ij} \|\mathbf{X}_{ij} - \mathbf{U}_i^T \mathbf{U}_j\|_F^2 + \sum_{i=1}^{g} \beta_i \|\mathbf{A}_i - \widetilde{\mathbf{U}}_i \mathbf{U}_i\|_F^2 + \sum_{i=1}^{g} \varphi_i \|\mathbf{U}_i - \mathbf{Z}_i\|_F^2 \\ + & \sum_{i=1}^{g} \langle \Lambda_i, \mathbf{U}_i^T - \widetilde{\mathbf{U}}_i \rangle + \sum_{i=1}^{g} \frac{\rho_i}{2} \|\mathbf{U}_i^T - \widetilde{\mathbf{U}}_i\|_F^2 \qquad s.t.\, \mathbf{U}_i \in U_0, \mathbf{Z}_i \in Z, \end{aligned} \tag{4}$$

where Λ_i is Lagrange multiplier, and ρ_i is the constraint violation penalty parameter. Normally, in ADMM, the basic Gauss-Seidel structure in $(t+1)$-th iteration is as follows:

$$\begin{cases} \mathbf{U}_i^{t+1} = \arg\min \mathcal{L}(\mathbf{U}_i, \{\widetilde{\mathbf{U}}_i^t, \mathbf{Z}_i^t, \Lambda_i^t\}_{i=1}^g) \ \ (i = 1, \ldots, g), \\ \widetilde{\mathbf{U}}_i^{t+1} = \arg\min \mathcal{L}(\widetilde{\mathbf{U}}_i, \{\mathbf{U}_i^{t+1}, \mathbf{Z}_i^t, \Lambda_i^t\}_{i=1}^g) \ \ (i = 1, \ldots, g), \\ \mathbf{Z}_i^{t+1} = \arg\min \mathcal{L}(\mathbf{Z}_i, \{\mathbf{U}_i^{t+1}, \widetilde{\mathbf{U}}_i^{t+1}, \Lambda_i^t\}_{i=1}^g) \ \ (i = 1, \ldots, g), \\ \Lambda_i^{t+1} = \Lambda_i^t + \rho_i((\mathbf{U}_i^{t+1})^T - \widetilde{\mathbf{U}}_i^{t+1}) \ \ (i = 1, \ldots, g). \end{cases} \tag{5}$$

Next, we give the details to show how to solve above-mentioned subproblems.

$\mathbf{U}_i^{t+1}$**-subproblem.** When the other variables are fixed, via the basic algebraic operations, the optimization problem for $\mathbf{U}_i^{t+1}$ is formulated as:

$$\mathbf{U}_i^{t+1} = \arg\max_{\mathbf{U}_i \in U_0} \sum_{(i,j)\in S_i} \alpha_{ij} tr(\mathbf{X}_{ij}(\mathbf{U}_j^t)^T \mathbf{U}_i) + \sum_{(k,i)\in S_i} \alpha_{ki} tr(\mathbf{X}_{ki}^T (\mathbf{U}_k^{t+1})^T \mathbf{U}_i)$$
$$+ \beta_i tr(\mathbf{A}_i^T \widetilde{\mathbf{U}}_i^t \mathbf{U}_i) + \frac{\rho_i}{2} tr(\widetilde{\mathbf{U}}_i^t \mathbf{U}_i) - \frac{1}{2} tr(\Lambda_i^t \mathbf{U}_i) + \varphi_i tr((\mathbf{Z}_i^t)^T \mathbf{U}_i), \qquad (6)$$

where S_i is a subset of S which contains all inter-type relations involving $\mathbf{U}_i$. Mathematically, Eq. (6) is also equivalent to:

$$\mathbf{U}_i^{t+1} = \arg\min_{\mathbf{U}_i \in U_0} \|\mathbf{U}_i - \Pi_i\|_F^2, \qquad (7)$$

where $\Pi_i = (\sum_{(i,j)\in S_i} \alpha_{ij}\mathbf{X}_{ij}(\mathbf{U}_j^t)^T + \sum_{(k,i)\in S_i} \alpha_{ki}\mathbf{X}_{ki}^T(\mathbf{U}_k^{t+1})^T + \beta_i \mathbf{A}_i^T \widetilde{\mathbf{U}}_i^t + \frac{\rho_i}{2}\widetilde{\mathbf{U}}_i^t - \frac{1}{2}\Lambda_i^t + \varphi_i(\mathbf{Z}_i^t)^T)^T$. It can also be interpreted as projecting Π_i onto a balanced Hamming space and the optimal solution can be easily obtained by:

$$\mathbf{U}_i^{t+1} = sign(\Pi_i - \lambda \mathbf{1}_{n_i}^T), \qquad (8)$$

where $\lambda = median(\Pi_i)$ is the row median of Π_i and can be viewed as a multiplier of the bit balance constraint.

$\widetilde{\mathbf{U}}_i^{t+1}$**-subproblem.** To achieve $\widetilde{\mathbf{U}}_i^{t+1}$, we need to solve:

$$\widetilde{\mathbf{U}}_i^{t+1} = \arg\min \beta_i \|\mathbf{A}_i - \widetilde{\mathbf{U}}_i \mathbf{U}_i^{t+1}\|_F^2 + \frac{\rho_i}{2}\|(\mathbf{U}_i^{t+1})^T - \widetilde{\mathbf{U}}_i + \frac{\Lambda_i^t}{\rho_i}\|_F^2. \qquad (9)$$

By setting the derivative of Eq. (9) w.r.t. $\widetilde{\mathbf{U}}_i$ to zero, we get a closed-form solution of $\widetilde{\mathbf{U}}_i^{t+1}$ as follows:

$$\widetilde{\mathbf{U}}_i^{t+1} = \frac{1}{2n_i\beta_i + \rho_i}(2\beta_i \mathbf{A}_i(\mathbf{U}_i^{t+1})^T + \rho_i(\mathbf{U}_i^{t+1})^T + \Lambda_i^t). \qquad (10)$$

$\mathbf{Z}_i^{t+1}$**-subproblem.** The subproblem regarding $\mathbf{Z}_i^{t+1}$ is:

$$\mathbf{Z}_i^{t+1} = \arg\min_{\mathbf{Z}_i \in Z} \varphi_i \|\mathbf{U}_i^{t+1} - \mathbf{Z}_i\|_F^2. \qquad (11)$$

It is easy to derive the update rule of $\mathbf{Z}_i$ which is equivalent to projecting $\mathbf{U}_i^{t+1}$ to the Stiefel manifold [6] and an analytical can be derived. Specifically, on the basis of Von Neumann's trace inequality [11], we have $tr(\mathbf{U}_i^{t+1}\mathbf{Z}_i^T) \leq \sum_{k=1}^d \sigma_k(\mathbf{Z}_i)\sigma_k(\mathbf{U}_i^{t+1})$, where $\sigma_k(\mathbf{Z}_i)$ is the k-th largest singular value of $\mathbf{Z}_i$. Assume that $\mathbf{U}_i^{t+1} = \mathbf{P}_i \Sigma \mathbf{Q}_i^T$ is the thin SVD decomposition of $\mathbf{U}_i^{t+1}$, then $\mathbf{Z}_i^{t+1}$ can be updated according to:

$$\mathbf{Z}_i^{t+1} = \sqrt{n_i}\mathbf{P}_i \mathbf{Q}_i^T. \qquad (12)$$

The key steps of the proposed algorithm are summarized in Algorithm 1. By optimizing the network embedding and hashing in a joint fashion, *HIN2Hash* generates compact binary hash codes for nodes in HINs. After analysis, the total time complexity is $\#iterations * O((g-1)n^2d + n^2d + nd^2)$, *i.e.*, $\#iterations * O(n^2d)$, where $n = \max\{n_i\}_{i=1}^g$ (detailed analysis omitted for brevity).

Algorithm 1. *HIN2Hash*: Heterogeneous information network hashing for fast nearest neighbor search

Input: An HIN G including intra-type relations A and inter-type relations X, the embedding dimension d, parameters α_i, β_i and φ_i $(i = 1, \ldots, g)$.
Output: The binary hash codes U of all nodes.
1: Initialize $\mathbf{U}_i$, $\widetilde{\mathbf{U}}_i$ and Λ_i $(i = 1, \ldots, g)$ to be zero matrices, $\rho_i = 10^{-6}$ $(i = 1, \ldots, g)$, $\max_\rho = 10^{10}$, $\tau = 1.1$, $\varepsilon = 10^{-3}$, $t = 0$.
2: Initialize $\mathbf{Z}_i$ $(i = 1, \ldots, g)$ by Eq. (12).
3: **while** objective function in Eq. (2) not converge **do**
4: Update $\mathbf{U}_i^{t+1}$ $(i = 1, \ldots, g)$ by Eq. (8);
5: Update $\widetilde{\mathbf{U}}_i^{t+1}$ $(i = 1, \ldots, g)$ by Eq. (10);
6: Update $\mathbf{Z}_i^{t+1}$ $(i = 1, \ldots, g)$ by Eq. (12);
7: Update Lagrange multipliers Λ_i^{t+1} $(i = 1, \ldots, g)$ via $\Lambda_i^{t+1} = \Lambda_i^t + \rho_i((\mathbf{U}_i^{t+1})^T - \widetilde{\mathbf{U}}_i^{t+1})$;
8: Update the parameters ρ_i $(i = 1, \ldots, g)$ by $\rho_i = \min(\tau\rho_i, \max_\rho)$;
9: $t = t + 1$;
10: Check the convergence conditions $\|(\mathbf{U}_i^t)^T - \widetilde{\mathbf{U}}_i^t\|_\infty < \varepsilon$ $(i = 1, \ldots, g)$ and $|\frac{\mathcal{F}^t - \mathcal{F}^{t-1}}{\mathcal{F}^{t-1}}| < \varepsilon$, where $\mathcal{F}^t$ is the value of Eq. (2) at the t-th iteration.
11: **end while**
12: Output the learned hash embedding U for the HIN G.

5 Experiments

In this section, we empirically verify the effectiveness of the proposed algorithm *HIN2Hash* on two fundamental tasks on HINs, including link prediction and cross-type node similarity search [3]. In particular, we attempt to answer the following three research questions:

Q1 How effective is the proposed algorithm in predicting missing links and performing cross-type node similarity search?
Q2 Will the binary codes reduce the processing time of the task that rely on the NNS algorithms?
Q3 Will the binary codes reduce the memory storage costs?

5.1 Datasets

We perform evaluations on three datasets that have been widely used in the previous research [2,12], including one academic network - AMINER, one infrastructure system network - INFRA, and one comparative toxicogenomics database (CTD) network in the biological domain - BIO. The statistics of the datasets are listed in Table 1, and Fig. 1 is a schematic diagram of their respective link relations.

AMINER [25] is an academic network in the domain of computer science[1] which contains three node types, *i.e.*, papers, authors and conference venues.

[1] https://aminer.org/.

Table 1. Statistics of the used datasets.

	AMINER	INFRA	BIO
#node types	3	3	3
#inter-type relations	2	3	3
#nodes	17,504	8,325	35,631
#intra-type links	107,466	15,138	253,827
#inter-type links	35,229	23,897	75,456

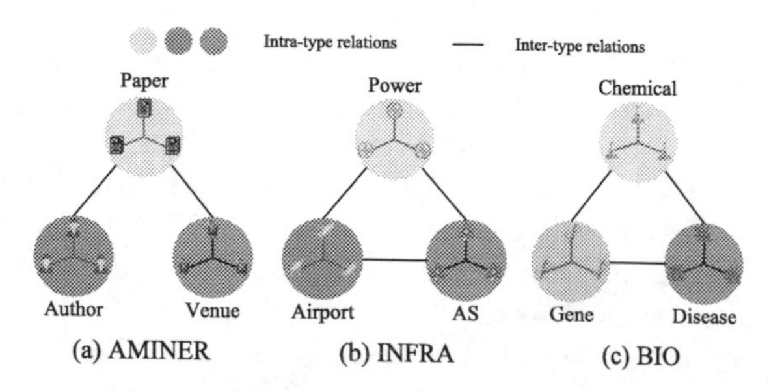

Fig. 1. Abstract network structure diagram.

Specifically, there are three intra-type relations: (1) a paper-paper citation network, (2) an author-author co-authorship network, and (3) a venue-venue citation network; and two inter-type relations: (1) the paper-author dependency, and (2) the paper-venue dependency. The abstract network structure diagram of AMINER is shown in Fig. 1(a).

INFRA [27] involves three critical infrastructure networks: (1) an autonomous system network[2], (2) an airport network[3], and (3) a power grid networks. The above three networks record the intra-type relations, and they are functionally dependent on each other and form a triangle-shaped inter-type dependency network as shown in Fig. 1(b).

BIO is a CTD network that is constructed based [4,19]. It includes three intra-type relations which are chemical, disease and gene similarity networks. Meanwhile, as shown in Fig. 1(c), interactions in the form of inter-type relations also exist among three types of nodes.

5.2 Compared Methods

We compare *HIN2Hash* with six state-of-the-art network embedding methods which can be roughly categorized into two classes: matrix factorization based methods and skip-gram based methods. The details of compared baseline methods are given as follow.

[2] http://snap.stanford.edu/data/.
[3] http://www.levmuchnik.net/Content/Networks/NetworkData.html.

- **CMF** [21]: The collective matrix factorization (CMF) approach can be regarded as a variant of *HIN2Hash*. Specifically, it removes the binary constraints imposed on the objective function, and encodes nodes into a continuous low-dimensional feature space.
- **DeepWalk** [17]: DeepWalk learns low-dimensional node representations via truncated random walks and the skip-gram model.
- **LINE** [24]: LINE is a large-scale information network embedding method which preserves the 1st-order and the 2nd-order node proximity. In the experiments, we concatenate both the 1st and the 2nd order representations together.
- **Node2vec** [8]: With parameterized random walks and the skip-gram model, node2vec explores neighborhood structure around each node in a more flexible way for embedding representation learning.
- **DCF** [30]: The discrete collaborative filtering (DCF) algorithm is originally applied to the recommendation problem. In the experiments, we ignore the node and edge types, and transform the heterogeneous network G into a large flattened homogeneous network.
- **Metapath2vec** [5]: Metapath2vec learns node embeddings in HINs through meta-path-guided random walks and a heterogeneous skip-gram model.

Among them, DeepWalk, LINE and node2vec are designed for homogeneous networks. In the experiments, they are performed in a way by treating different types of nodes and edges as the same type. According to the form of derived embeddings, the above algorithms can fall into two categories: algorithms for generating binary hash codes including DCF and *HIN2Hash*, and the remaining algorithms for generating real-valued embedding representations.

5.3 Settings and Evaluation Metrics

In terms of the default parameter settings, the dimension of node vectors d is set to 128 for all approaches. For DeepWalk and node2vec, the context window size, walk length and the number of walks per node are set to 8, 80, and 10, respectively. For LINE, the size of negative samples is set to 5. The two parameters p and q for parameterized random walks in node2vec are set to 1 and 4, respectively. As for metapath2vec, the context window size, walk length, the number of walks per node and the number of negative examples are set as 5, 80, 600, and 5, separately. Regarding the regularization parameters in the proposed framework, we tune them in the range of $\{10^{-3}, 10^{-2}, 10^{-1}, 10^{0}, 10^{1}, 10^{2}, 10^{3}\}$.

For the link prediction task, we refer to the experimental settings and the evaluation metrics in [8]. To be more specific, given a network with a certain fraction of edges removed, the task is to predict these missing edges. In particular, we randomly remove 10%, 20% and 50% neighbors of each node and use the rest for training, while ensuring that the residual network obtained after the edge removal is fully connected. The criteria we adopted is AUC which is the area under the ROC curve. Additionally, the negative examples involved in AUC are built by randomly sampling an equal number of node pairs with no connections in the original network.

For the cross-type node similarity search task, we also remove a portion of edges with the goal of detecting the robustness of *HIN2Hash*. Since our work is on heterogeneous network hashing, we focus more on the similarity search across different types of nodes. For instance, given a particular disease as a query node, the task can be finding the genes that highly associated with this disease. In order to quantitatively evaluate the performance of our method, we leverage MAP to measure the mean average precision over all nodes in the inter-type relations and the other types of nodes that are connected to the query node are taken as the ground truth.

5.4 Experimental Results

Quantitative Results. Tables 2 and 3 summarize the AUC and MAP scores of all the methods, respectively. We have the following observations from these tables:

- The AUC scores of *HIN2Hash* are comparable to the best baseline method on all datasets, regardless of the removal ratio of edges. The MAP scores of *HIN2Hash* is slightly worse than DeepWalk and node2vec on INFRA and BIO. While on the AMINER dataset, our approach achieves the best performance.
- In the link prediction task, CMF outperforms *HIN2Hash* in most cases but is not as good as ours on the AMINER dataset. In the cross-type node similarity search task, *HIN2Hash* outperforms CMF on all datasets. Hence, it is reasonable to conclude that even though transforming the node embeddings from real-valued vectors to binary hash codes will result in certain information loss, but the sacrifice is not much. In some certain cases, it may even be helpful to filter out the noise in the embedding such that a number of tasks that heavily reply on the NNS algorithms can be even slightly improved.
- The DCF algorithm, the other algorithm which learns the binary hash codes for nodes in the network, is almost the worst among all the methods. Their inferior performance can be attributed to the following two reasons: (1) it does not differentiate different inter-type and intra-type connections among nodes, and (2) the optimization algorithm used in DCF loses a lot of information as its subproblems are NP-hard to solve, while the subproblems in our algorithm have closed optimal solutions.
- For metapath2vec, it often requires prior human knowledge to define the most appropriate metapaths. Even though the high-quality matapaths have been well discovered in the academic datasets such as AMINR, it is difficult to achieve desirable results on INFRA and BIO with limited human experiences of the optimal metapaths.

Time Cost. Since the time cost of similarity calculation for CMF, DeepWalk, LINE, node2vec and metapath2vec (algorithms for generating real-valued embeddings) are quite close, Table 4 only illustrates the results of *HIN2Hash*

Table 2. AUC scores for link prediction on three datasets.

Algorithm	AMINER			INFRA			BIO		
	10%	20%	50%	10%	20%	50%	10%	20%	50%
CMF	0.9216	0.9074	0.8807	**0.8899**	0.8651	0.7612	**0.9452**	**0.9401**	**0.9314**
DeepWalk	0.9624	**0.9575**	0.9238	0.8764	**0.8676**	**0.8306**	0.9241	0.9200	0.9051
LINE	0.9424	0.9348	0.9035	0.8122	0.7244	0.5332	0.6048	0.5963	0.5789
Node2vec	**0.9684**	0.9568	**0.9354**	0.8753	0.8662	0.8281	0.9165	0.9076	0.9048
DCF	0.8495	0.8431	0.8055	0.6960	0.6850	0.5555	0.5681	0.5536	0.5337
Metapath2vec	0.9110	0.9008	0.8735	0.8615	0.7820	0.6762	0.8206	0.8027	0.7728
HIN2Hash	**0.9426**	**0.9316**	**0.9260**	**0.8313**	**0.8258**	**0.7950**	**0.9104**	**0.9101**	**0.8871**

The percentage denotes the removal ratio of edges.

Table 3. MAP scores for cross-type node similarity search on three datasets.

Algorithm	AMINER			INFRA			BIO		
	10%	20%	50%	10%	20%	50%	10%	20%	50%
CMF	0.6480	0.5941	0.5431	0.5914	0.5783	0.5080	0.6446	0.6158	0.5124
DeepWalk	0.5260	0.4949	0.4660	**0.8711**	**0.8436**	**0.8039**	**0.7544**	0.7439	**0.7248**
LINE	0.4370	0.4098	0.3875	0.4549	0.4271	0.3867	0.3678	0.3766	0.3601
Node2vec	0.6147	0.5704	0.5285	0.8579	0.8384	0.7927	0.7538	**0.7517**	0.7223
DCF	0.4083	0.3875	0.3624	0.4016	0.3921	0.3701	0.3445	0.3368	0.3154
Metapath2vec	0.4906	0.4780	0.4666	0.4665	0.4342	0.3980	0.5676	0.5493	0.5152
HIN2Hash	**0.7809**	**0.6978**	**0.6303**	**0.8147**	**0.7760**	**0.7138**	**0.7056**	**0.7006**	**0.6947**

The percentage denotes the removal ratio of edges.

and node2vec in node similarity computation. As can be seen, the similarity calculation on binary codes is significantly faster than that on real-valued vectors in all cases, which substantiates our motivation. We can attribute this superiority to the fact that similarity calculation in the real-valued vector space is replaced by the bit operations in a low-dimensional Hamming space. It is well known that computers are better at performing bit operations than floating-point arithmetic. Therefore, hashing is a promising and impactful technique in accelerating computing on large-scale networks.

Memory Usage. This part is carried out on the MATLAB platform. As algorithms for generating real-valued embeddings included in baselines use *double* data type to store vector representations, occupying the same storage space, we show the memory usage of node2vec and ours for storing network embedding/hashing results in Table 5. In theory, only 1 bit is enough to represent binary numbers. Hence, we design an algorithm to compress the binary codes and store them with *uint8* data type instead of *double*. As for external memory, we save the above embeddings as MAT files, where MATLAB compresses the data using HDF5-variant format. It is obvious to see that compared with node2vec, *HIN2Hash* remarkably reduces the memory usage of embeddings with the same dimensionality. Besides, *HIN2Hash* only needs to store less than 1 MB

Table 4. Time cost for similarity calculation of node2vec and *HIN2Hash*.

Metric	Dataset	*HIN2Hash*	Node2vec	
		Time (s)	Time (s)	Speedup
AUC	AMINER	0.013 ± 0.001	0.079 ± 0.004	6.08×
	INFRA	0.011 ± 0.001	0.074 ± 0.001	6.73×
	BIO	0.021 ± 0.001	0.149 ± 0.013	7.10×
MAP	AMINER	49.342 ± 0.196	404.118 ± 1.617	8.19×
	INFRA	22.241 ± 0.114	205.192 ± 0.791	9.23×
	BIO	355.383 ± 2.834	3077.695 ± 11.079	8.66×

Table 5. Memory usage of embeddings derived by node2vec and *HIN2Hash*.

Computer memory	Dataset	*HIN2Hash*	Node2vec	
		Size (KB)	Size (MB)	Reduction
Internal memory	AMINER	273.50	17.09	64×
	INFRA	130.08	8.13	64×
	BIO	556.73	34.80	64×
External memory	AMINER	274.03	15.76	59×
	INFRA	130.43	7.52	59×
	BIO	557.61	32.27	59×

of binary codes for all three datasets, which further demonstrates the superiority of hashing in reducing storage costs.

5.5 Parameter Analysis

Dimensionality and Training Time. Due to space limitation, only the results of link prediction are given here. Figure 2(a) shows how the AUC scores vary with different dimensionality d. In AMINER, the best performance is achieved when d is 128. In INFRA and BIO, the performance enhances with the increasing d up to 256, but only a little. Therefore, setting d to 128 is reasonable as it is sufficient to capture the node proximity among nodes, and too large d cannot gain desired benefits but greatly increases the burden of storage. Also, we have an interesting observation on BIO as shown in Fig. 2(b). *HIN2Hash* requires much less training time (657 s) to obtain competitive results with node2vec, while node2vec takes about three times (1,711 s). The main reason is that *HIN2Hash* benefits from the characteristics of ADMM that produces acceptable results for practical use within tens of iterations. The reduction of walk length in node2vec can accelerate training but has the side effect of reduced performance since walk length has a relatively high impact on the performance of skip-gram based network embedding algorithms.

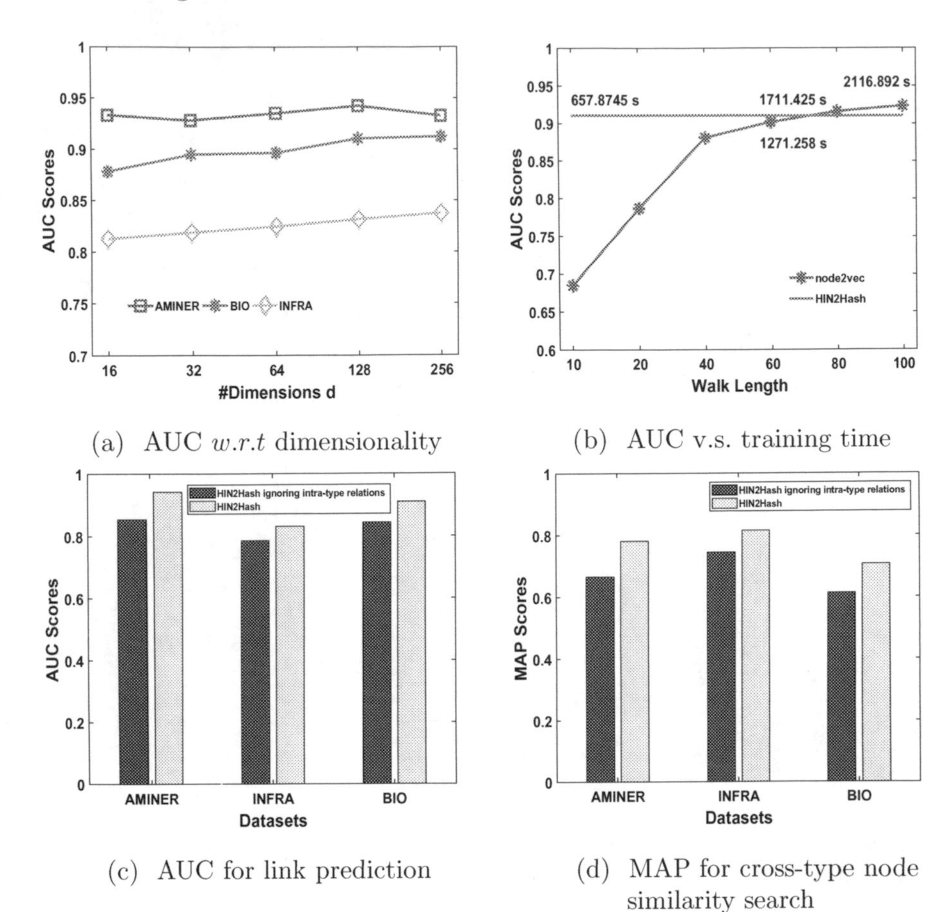

(a) AUC $w.r.t$ dimensionality (b) AUC v.s. training time

(c) AUC for link prediction (d) MAP for cross-type node similarity search

Fig. 2. (a) AUC scores in different dimensions. (b) AUC scores v.s. training time on BIO dataset. (c) Effects of intra-type relations on link prediction. (d) Effects of intra-type relations on cross-type node similarity search.

Effects of Intra-type Relations. Among the parameters appeared in Eq. (2), α_i and $\beta_i(i = 1, \ldots, g)$ are relatively more important since they control the participation of inter-type and intra-type relations for binary hash codes learning, respectively. In this section, we attempt to explore the impact of intra-type relations on the quality of heterogeneous network hashing. By setting β_i to be zero, only inter-type relations are taken into consideration in the learning process, while two kinds of relations are both used when β_i does not equal to zero. Figure 2(c) and (d) shows the performance of link prediction and cross-type node similarity search in these two scenarios. As can be observed, when the intra-type relations are ignored, the AUC and MAP scores are lower on all the three datasets. Hence, we can conclude that the intra-type relations play a vital role in heterogeneous network hashing and it is necessary to consider them in our framework.

6 Conclusions and Future Work

In this work, we study the issue of learning to hash HINs and propose a novel CMF based heterogeneous information network hashing algorithm *HIN2Hash* that projects nodes into a low-dimensional Hamming space instead of Euclidean space. Experiments on three real-world datasets indicate that *HIN2Hash* can achieve comparable results in most cases *w.r.t.* the link prediction and the cross-type node similarity search tasks, and even outperforms other prevalent network embedding techniques in certain cases. Meanwhile, *HIN2Hash* is superior to other embedding methods in both computation time and memory usage, due to the characteristics of the learned discrete binary hash codes. Future work will concentrate on attributed network hashing and task-oriented network hashing.

Acknowledgements. This work is supported by National Key Research and Development Program of China (2016YFB1000903), National Nature Science Foundation of China (61872287, 61532015 and 61672418), Innovative Research Group of the National Natural Science Foundation of China (61721002), Innovation Research Team of Ministry of Education (IRT_17R86), Project of China Knowledge Center for Engineering Science and Technology.

References

1. Boyd, S., Parikh, N., Chu, E., Peleato, B., Eckstein, J., et al.: Distributed optimization and statistical learning via the alternating direction method of multipliers. Found. Trends® Mach. Learn. **3**(1), 1–122 (2011)
2. Chen, C., Tong, H., Xie, L., Ying, L., He, Q.: FASCINATE: fast cross-layer dependency inference on multi-layered networks. In: KDD (2016)
3. Cui, P., Wang, X., Pei, J., Zhu, W.: A survey on network embedding. arXiv preprint arXiv:1711.08752 (2017)
4. Davis, A.P., et al.: The comparative toxicogenomics database's 10th year anniversary: update 2015. Nucleic Acids Res. **43**(D1), D914–D920 (2014)
5. Dong, Y., Chawla, N.V., Swami, A.: Metapath2vec: scalable representation learning for heterogeneous networks. In: KDD (2017)
6. Eldén, L., Park, H.: A Procrustes problem on the Stiefel manifold. Numerische Mathematik **82**(4), 599–619 (1999)
7. Fu, T., Lee, W.C., Lei, Z.: HIN2Vec: explore meta-paths in heterogeneous information networks for representation learning. In: CIKM (2017)
8. Grover, A., Leskovec, J.: Node2vec: scalable feature learning for networks. In: KDD (2016)
9. Hamilton, W.L., Ying, R., Leskovec, J.: Representation learning on graphs: methods and applications. arXiv preprint arXiv:1709.05584 (2017)
10. Håstad, J.: Some optimal inapproximability results. J. ACM **48**(4), 798–859 (2001)
11. Horn, R.A., Johnson, C.R.: Matrix Analysis. Cambridge University Press, Cambridge (1990)

12. Li, J., Chen, C., Tong, H., Liu, H.: Multi-layered network embedding. In: SDM (2018)
13. Lian, D., et al.: High-order proximity preserving information network hashing. In: KDD (2018)
14. Liben-Nowell, D., Kleinberg, J.: The link-prediction problem for social networks. J. Am. Soc. Inf. Sci. Technol. **58**(7), 1019–1031 (2007)
15. Ma, H., Zhou, D., Liu, C., Lyu, M.R., King, I.: Recommender systems with social regularization. In: WSDM (2011)
16. Opsahl, T., Panzarasa, P.: Clustering in weighted networks. Soc. Netw. **31**(2), 155–163 (2009)
17. Perozzi, B., Al-Rfou, R., Skiena, S.: DeepWalk: online learning of social representations. In: KDD (2014)
18. Qiu, J., Dong, Y., Ma, H., Li, J., Wang, K., Tang, J.: Network embedding as matrix factorization: unifying DeepWalk, LINE, PTE, and node2vec. In: WSDM (2018)
19. Razick, S., Magklaras, G., Donaldson, I.M.: iRefIndex: a consolidated protein interaction database with provenance. BMC Bioinform. **9**(1), 405 (2008)
20. Shen, X., Pan, S., Liu, W., Ong, Y.S., Sun, Q.S.: Discrete network embedding. In: IJCAI (2018)
21. Singh, A.P., Gordon, G.J.: Relational learning via collective matrix factorization. In: KDD (2008)
22. Sun, Y., Han, J., Yan, X., Yu, P.S., Wu, T.: PathSim: meta path-based top-k similarity search in heterogeneous information networks. Proc. VLDB Endow. **4**(11), 992–1003 (2011)
23. Tang, J., Qu, M., Mei, Q.: PTE: predictive text embedding through large-scale heterogeneous text networks. In: KDD (2015)
24. Tang, J., Qu, M., Wang, M., Zhang, M., Yan, J., Mei, Q.: LINE: large-scale information network embedding. In: WWW (2015)
25. Tang, J., Zhang, J., Yao, L., Li, J., Zhang, L., Su, Z.: ArnetMiner: extraction and mining of academic social networks. In: KDD (2008)
26. Wang, J., Zhang, T., Sebe, N., Shen, H.T., et al.: A survey on learning to hash. IEEE Trans. Pattern Anal. Mach. Intell. **40**(4), 769–790 (2018)
27. Watts, D.J., Strogatz, S.H.: Collective dynamics of 'Small-World' networks. Nature **393**(6684), 440 (1998)
28. Weiss, Y., Torralba, A., Fergus, R.: Spectral hashing. In: NIPS (2009)
29. Willett, P., Barnard, J.M., Downs, G.M.: Chemical similarity searching. J. Chem. Inf. Comput. Sci. **38**(6), 983–996 (1998)
30. Zhang, H., Shen, F., Liu, W., He, X., Luan, H., Chua, T.S.: Discrete collaborative filtering. In: SIGIR (2016)

Learning Fine-Grained Patient Similarity with Dynamic Bayesian Network Embedded RNNs

Yanda Wang[1(✉)], Weitong Chen[2], Bohan Li[1], and Robert Boots[2]

[1] Nanjing University of Aeronautics and Astronautics, Nanjing, China
{yandawang,bhli}@nuaa.edu.cn
[2] The University of Queensland, Brisbane, Australia
{w.chen9,r.boots}@uq.edu.cn

Abstract. The adoption of Electronic Health Records (EHRs) enables comprehensive analysis for robust clinical decision-making in the rapidly changing environment. Therefore, using historical and similar patient records, we investigate how to utilize EHRs to provide effective and timely treatments and diagnoses for them under the circumstances that our patients are likely to respond to the therapy. In this paper, We propose a novel framework that embeds the Markov decision process into the multivariate time series analysis to research the meaningful distance among patients in Intensive Care Units (ICU). Specifically, we develop a novel deep learning model TDBNN that employs Triplet architecture, Dynamic Bayesian Network (DBN), and Recurrent Neural Network (RNN). Causal correlations among medical events are firstly obtained by the conditional dependencies in DBN, and to transmit this kind of correlations over time as temporal features, conditional dependencies in DBN are used to construct extra connections among RNN units. With specially-designed connections, the RNN is further utilized as fundamental components of the Triplet architecture to study the fine-grained similarities among patients. The proposed method has been applied to a real-world ICU dataset MIMIC-III. The experimental results between our approach and several existing baselines demonstrate that the proposed approach outperforms those methods and provides a promising direction for the research on clinical decision support.

1 Introduction

Health Information Technology (HIT) has improved the efficiency and quality of modern healthcare systems while the accumulation of EHRs provides strong supporting evidence for clinical decision making in an ICU, including lab measurements and vital signs. However, most of the existing methods fail to explicitly obtain the causal correlations among medical events or transmit such correlations over time as temporal features to obtain underlying changing patterns of patient conditions, which could help search for similar patients who will support evidence-based decision making in the ICU. As a key and fundamental

G. Li et al. (Eds.): DASFAA 2019, LNCS 11446, pp. 587–603, 2019.
https://doi.org/10.1007/978-3-030-18576-3_35

component, the study of patient similarity aims at deriving a measurement of distance in the clinical field to measure the similarity among patients which is based on their multiple types of medical information. And an effective similarity metric plays an important part in obtaining accurate alternative treatments and diagnoses based on historical and similar cases, which could assist in improving chance of survival in the ICU. Moreover, case-based decision supports employ similar patients as explanations, making the decisions much more convincing.

Motivation. Explanations of the diagnoses and treatments given by a clinical decision-making system are crucial for convincing physicians and patients the reliability of the system. Although multiple strategies for decision support in the ICU have been proposed to handle tasks such as prediction of disease onsets and mortality [6,9,22,37], most of them focus on providing effective pattern recognition solutions and final results directly. When these applications are employed to provide assessments and diagnoses in the ICU, the explanations become critical and are mostly offered by physicians wielding individual experiences. However, it is common that a decision made by machine learning algorithms is not explainable to humans. Meanwhile, despite the fact that some proposed methods try to capture the correlation among medical variables by deriving DBN from EHRs [2,20], they ignore that EHRs could hardly meet the conditional independence assumption, an essential property for learning accurate DBN from data. Therefore, case-based explanations plus the reasons for the similarity of these cases are indispensable for a reliable decision support system.

Driven by providing similar patients for case-based explanations, we believe that an effective patient similarity metric is crucial for developing clinical decision-making system in the ICU. To demonstrate the importance of similar patients in case-based decision-making system, we illustrate how the illness severity of two similar patients shown by Sequential Organ Failure Assessment (SOFA) change over time in Fig. 1. These two patients are similar to each other according to their International Classification of Diseases, Ninth Revision (ICD-9) codes, but while one's SOFA score tends to be stable, the other patient still

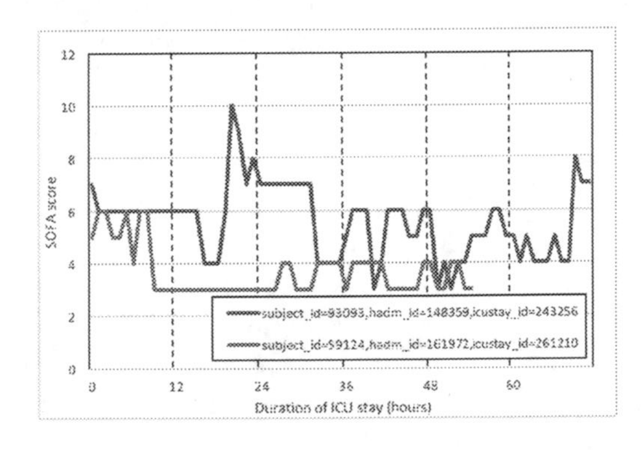

Fig. 1. Comparison of SOFA scores change in two similar patients.

suffers from worrying health status shown by the dramatically changing SOFA score. With patient similarity metric, there would be effective treatment recommendations and the deterioration of the second patient's condition could be avoided.

Challenge. Numerous deep learning models have been proposed in recent years to tackle difficulties in the study of the similarities of patients. Distributional representation models from Natural Language Processing (NLP) field are commonly adopted for deriving clinically meaningful representations for medical events [8,16], which are furthermore employed for patient profiling. Meanwhile, both sequence modeling methods [7] non-sequence methods [6,25] have been applied to mine sequential information from comprehensive representations of patients, while Siamese architecture [21], and Triplet architecture [34], are the two common choices for studying the similarities of patients.

However, most of the existing methods either combine all the medical events to treat them as a single time series, or process different variables separately without developing extra mutual connections, which leads to an obvious neglect of causal correlations among variables that reveal the fundamental mechanism of how different systems of the human bodies interact with each other. Meanwhile, they could only handle similarity in category-level, where two patients are considered to be similar as long as they suffer from a particular disease that chosen as the class label. This could hardly achieve good performance for healthcare applications that require the distinction of difference among patients within the same category, i.e., fine-grained patient similarity. Moreover, similar cases provided by previous models are complicated to be explained without the causal correlations among medical indicators revealing the dynamic trends of how changes in one indicator are affected by changes in others.

Solution. To achieve a better performance in the study of patient similarity, we propose the Triplet architecture based on Dynamic Bayesian Neural Network (TDBNN). The main challenges for this framework include capturing causal dependencies from heterogeneous EHRs regarding data types, data categories, and frequency of sampling, and transferring the time-invariant dependencies into correlated temporal features to study long-term memories contained in multivariate time series. To capture the causal correlations among various heterogeneous medical indicators, DBNs are derived from medical indicators and the conditional dependencies among indicators are then employed to construct specially-designed connections among Gated Recurrent Units (GRU) at two adjacent time stamps of RNN, so that the temporal correlations among medical indicators could be utilized and transmitted over time to research long-term memories while overcoming the constraint of conditional independence assumption in the DBN. The RNN with specially-designed connections is further employed as fundamental components of a Triplet architecture, which simultaneously utilize three patients to study how they are similar to each other.

Contributions. Our contributions could be summarized as follows:

- To obtain causal correlations among various medical events, the DBN model is applied to study conditional dependencies among variables, while overcome multiple challenges brought about by data heterogeneity in EHRs.
- Conditional dependencies in DNBs are employed to develop specially-designed connections among GRU units in RNN to transmit temporal correlations among medical indicators over time. The combination of DBN and RNN provides a promising way for capturing temporal causal correlations.
- The Triple architecture is used for fine-grained patient similarity learning, providing an effective basic component for case-based clinical decision making and explainable treatment recommendation.
- Comparative experiments are conducted between the proposed method and some other baselines on a real-world ICU dataset, and the experimental results have demonstrated the effectiveness of our method.

2 Related Work

Concept Embedding. Deriving meaningful representation vectors that capture the latent similarities and context of discrete medical events is the primary objective of medical concept embedding. Choi et al. [9] use skip-gram to project medical codes into real-value vectors since the chronologically arranged clinical concepts of a patient are similar to sentences. However, the definition of context is not as clear as in NLP when it comes to EHRs, so Choi et al. [10] choose to partition a patient's event sequence into small chunks, and randomize the order of events within each chunk to treat them as separate sequences. The Med2Vec model proposed by Choi et al. [8] takes both code-level and visit-level information into account for representation learning. In addition to NLP techniques, Tran et al. [28] apply Restricted Boltzmann machine (RBM) to increase representation interpretation, and Lv et al. [19] use Autoencoders (AEs) to generate concept vectors from word-based concepts that extracted from the clinical free text.

Extra Entity Connection. Constructing extra specially-designed connections in standard neural networks is a promising direction for a better understanding of the raw data. Santoro et al. [26] introduce a new Relational Memory Core (RMC) which uses multi-head dot product attention to allow memories to interact with each other. Li et al. [17] introduce an extension of Graph Neural Networks and construct special connections among entities according to graph-based inputs, presenting a novel framework on feature learning for graph-structured inputs. Beck et al. [1] make further improvements by incorporating the full graph structure to develop an encoder to encode the complete structural information that contained in the graph. Oord et al. [23] present the PixelRNNs for large-scale modeling of natural images, which adopts a convolution to compute at once all the states along one of the spatial dimensions of the data.

Sequence Modeling. The RNN is a typical technique for sequence modeling, and multiple variants such as LSTM and GRU have been proposed for long-term memories learning. Chen et al. [4] proposed a approach that based on multi-task RNN to predict the illness severity of patients in the ICU, and Lipton et al. [18] use LSTM to recognize latent patterns in multivariate time series of clinical measurements to classify diagnoses. Chen et al. [5] use RNN for EEG-based motion intention recognition. Aside from RNN, many non-sequence models are presented and used in sequence modeling problem. Van Den Oord et al. [29] use the dilated convolution skips input values with a certain step to enable the network to operate on sequential data while Vaswani et al. [30] apply convolutional neural networks with attention mechanism to sequence modeling tasks.

Deep Metric Learning. Deep metric learning aims to learn nonlinear meaningful representations of the raw data which could act as inputs of models for various different tasks. Muller et al. [21] present a Siamese adaptation of LSTM for labeled data comprised of pairs of variable-length sequences to learn sentence similarity. But Siamese architecture treats similarity learning as a classification problem and is not able to deal with fine-grained similarity. Wang et al. [34] propose a Triplet architecture for fine-grained image similarity learning, which characterizes the fine-grained image similarity relationship with a set of triplets. Ni et al. [22] further refine the problem and present a deep metric learning framework by optimizing quadruplet loss for fine-grained patient similarity.

Patient Similarity. Learning patient similarity has received enormous attention in recent years. Chan et al. [3] uses Support Vector Machine (SVM) to weight the similarity measure, while Wang et al. [32] uses a Local Spline Regression based method to embed medical events into an intrinsic space to measure patient similarity. Sun et al. [27] proposed the Locally Supervised Metric Learning (LSML) that is tailored toward physician feedback, which also combine multiple similarity measures from multiple physicians. Ni et al. [22] proposed a deep metric learning framework with a quadruple loss objective function for fine-grained patient similarity learning. But these methods do not take temporal information in EHRs into consideration, so Wang et al. [33] present a covolutional matrix factorization for detection of temporal patterns, and Cheng et al. [6] proposed an adjustable temporal fusion scheme using CNN extracted features.

3 Proposed Method

In this section, we describe the details of our framework for fine-grained patient similarity learning in the ICU. Firstly, we introduce data preprocessing and how to obtain meaningful embedding vectors for medical events based on Med2Vec. Then, we introduce the DBN structure learning process and how to employ the DBN to develop specially-designed connections among GRU units. At last, we describe how to apply the Triplet architecture for fine-grained patient similarity learning. The whole workflow of the proposed method is illustrated in Fig. 2.

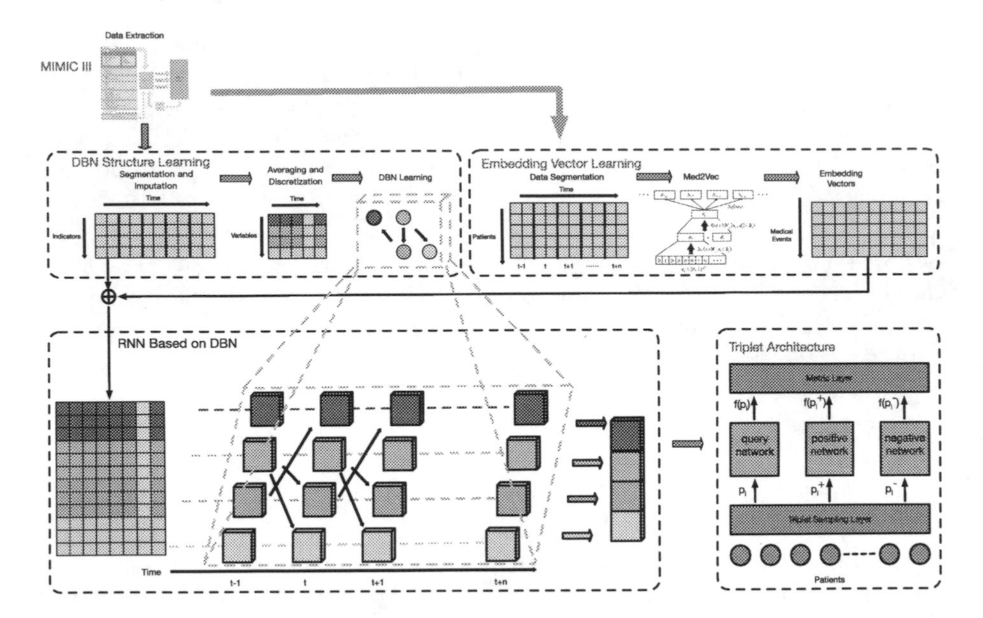

Fig. 2. Workflow of TDBNN. Medical indicators will be firstly extracted from MIMIC III fed into the **DBN Structure Learning** module, whose outputs would be used for special connection design in **RNN Based on DBN** module, while sequences of medical events from MIMIC III would act as input to the **Embedding Vector Learning** module, whose output would be combined with medical indicators to act as inputs of the RNN in the **RNN Based on DBN** module, which would act as the query network components in the **Triplet architecture**.

Data Acquisition. Medical Information Mart for Intensive Care (MIMIC) III [14] is a real-world clinical database comprising health data relating to over 40,000 patients admitted to ICU at the Beth Israel Deaconess Medical Center. We have applied the latest version MIMIC III v1.4 and carried out a selection of patient cohort to exclude those patients under the age of 15 or stay in the ICU for less than 48 h. Children are excluded since the definition of a normal range for medical indicators are different between adults and children, and the requirement for 48 h in ICU guarantees enough data for analysis. Meanwhile, since too much imputation for missing data may introduce variances with negative effect, patients with a large amount of missing data are excluded. Totally 3251 patients are finally selected for modeling and analysis.

To obtain the physiological features of patients, we select 21 medical variables from EHRs as shown in Table 1 according to studies that make use of MIMIC III and require comprehensive physiological characteristics of patients [16,18,25]. To make use of the supervised feedback to the physiological changes of patients from physicians, we also extract the prescriptions and procedures of these patients, which are presented as distinct medical events in the form of National Drug Code (NDC) and Current Procedural Terminology (CPT) code respectively.

Table 1. List of selected medical indicators from MIMIC III

Id	Vital signs	Id	Lab measurements
1	Systolic blood pressure (BPs)	11	Bicarbonate
2	Mean blood pressure (BPm)	12	Bilirubin
3	Arterial pH	13	Blood urea nitrogen
4	Fractional inspired oxygen (FiO2)	14	Calcium
5	Glasgow coma scale (GCS)	15	Chloride
6	Heart rate (HR)	16	Creatinine
7	Respiratory Rate (RR)	17	Glucose
8	Body temperature	18	Hematocrit
9	Blood oxygen saturation (SpO2)	19	Potassium
10	Partial pressure of carbon dioxide (PaCO2)	20	Sodium
		21	White blood cells

Embedding Vector Learning. Low-dimensional and clinically meaningful representation vectors of medical events are crucial for efficient similarity learning, and we employ Med2Vec to obtain embedding vectors for the selected vital signs, lab measurements, prescriptions, and procedures. Firstly, the numerical values of vital signs and lab measurements are discretized into different clinical states based on cut-offs from medical knowledge [24], which represent different physiological states, including unusually low, low, normal, high, and unusually high. Then these discrete states, as well as prescriptions and procedures, are further arranged chronologically to form a sequence of medical events, which would be divided into different time windows later. Thus, each patient can be represented by $X = \{x_1, x_2, ..., x_n\}$, where x_i is the sequence of the medical events within time window i, and n is the total number of time windows. Different sizes of time window are considered here, including 6 h, 12 h, 18 h and 24 h. Finally, these sequences of medical events are fed into Med2Vec to obtain embedding vectors for medical events. Totally 2868 events are taken into account, and the length of embedding vectors is set to 200.

DBN Learning. Deriving causal correlations among medical variables as conditional dependencies is the primary purpose of DBN structure learning, and the causal correlation is the relationship that indicates how changes in a medical variable is affected by the others, which reveals how different organ systems interact with each other.

To learn the DBN structure, those medical variables arranged chronologically are firstly divided into different time windows, and the numerical values of each vital signs and lab measurements within a time window will be averaged as the new value in that time slot, which would be further discretized into different physiological states based on clinically meaningful cut-offs known from medical knowledge. Due to different frequency of sampling, vital signs and lab

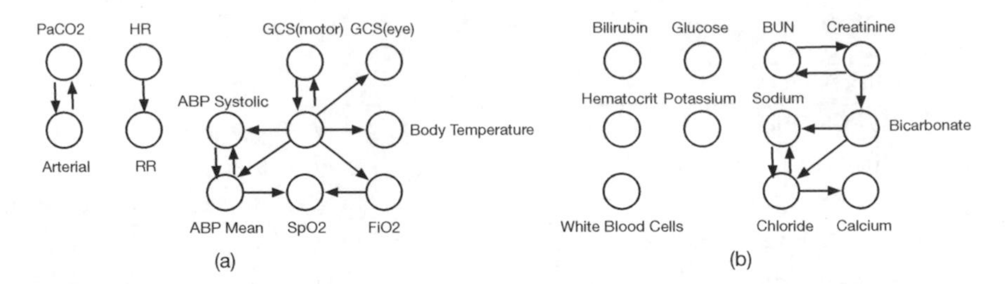

Fig. 3. DBN structure derived from vital signs (a) and lab measurements (b). Each dynamic variable is represented by a node, and each edge in the graphs represents a conditional dependency between two variables existing in adjacent time windows. If a node is not connected to any edge, then the corresponding variable is not affected by the other variables in previous time windows.

measurements are handled respectively, and the size of time windows for vital signs is 6 h while it is 12 h for lab measurements. To tackle the missing value problem, we have imputed those windows without values with the last value from previous windows, and if no value is sampled for a variable throughout the entire ICU admission, all the windows will be imputed with a 'normal' state. At last, greedy search and BDe score [13] are employed to learn DBN structures from the processed dataset. The DBN structures for selected vital signs and lab measurements are illustrated in Fig. 3(a) and (b) respectively.

Meanwhile, since prescriptions and procedures belong to static variables while vital signs and lab measurements are dynamic variables [20], these prescriptions and procedures would be represented by two distinct nodes respectively, which would act as parent nodes of all the other variables.

RNN Connection Design. How physicians make diagnoses on patients could be regarded as a Markov decision process, and a DBN could represent the process by decomposing the temporal evaluation of medical variables into local transition

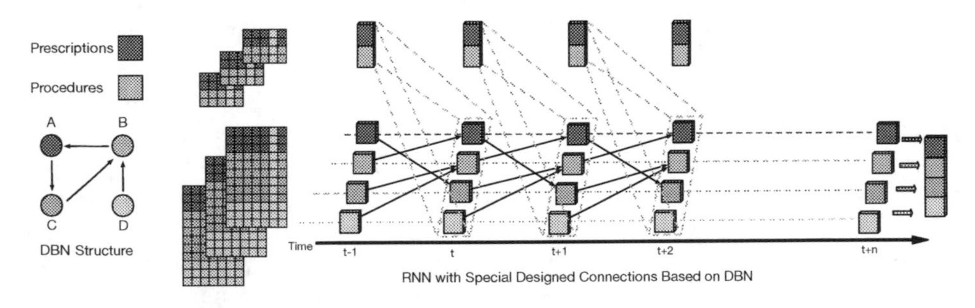

Fig. 4. The RNN with specially-designed connections among GRU units based on DBN derived from medical indicators. Extra connections are developed among GRU units based on the DBN, with two data nodes representing information from static variables. At last, The outputs of GRU units are flattened to form the final output.

probability. However, applying a DBN alone can not model the decision process accurately due to the conditional independence assumption, which EHRs can hardly meet. Besides, DBN can not learn long-term dependencies either.

Thus, we propose to develop extra connections among GRU units in two adjacent time stamps wielding DBN to learn long-term memories that reveal underlying temporal patterns in EHRs, so that the Markov decision process is embedded into RNN as specially-designed connections. For simplicity, suppose there are four selected medical variables A, B, C, and D, with a DBN indicating their conditional dependencies as illustrated in Fig. 4, and two static variables *Prescription* and *Procedures*. Accordingly, there are four GRU units at each time stamp that represent those four variables, and two data nodes which represent information that derived from prescriptions and procedures.

To propagate information over time, a GRU unit firstly receives the hidden state $h^{(t)}$ from previous time window t, then it combines information from four sources to form the input $x^{(t+1)}$ of the current time window using Eq. (1). The $e^{(t+1)}$ as calculated in Eq. (2) is the mean of all the n embedding vectors corresponding to the n different states of a variable in time window $t+1$, which indicate those discrete states assigned to the variable based on medical cut-offs, and $e_i^{(t+1)}$ is the embedding vector of the ith status in time window $t+1$. Similarly, there may also be multiple prescriptions and procedures in a time window, so the embedding vectors of those prescriptions and procedures are averaged as the second and third sources based on Eqs. (3) and (4) respectively, where p and q are the numbers of prescriptions and procedures while $e_{Pre,u}^{(t)}$ and $e_{Pro,v}^{(t)}$ are the embedding vectors for the uth and vth prescriptions and procedures in time window t. The final source comes from the variables on which a variable is conditionally depended in the DBN, and those m outputs from these units are added up according to Eq. (5). With hidden states from previous time windows and inputs for current time windows defined, the framework could now work as a normal GRU-based RNN. At last, the outputs of GRU units are flattened to form a final output vector.

$$x^{(t+1)} = e^{(t+1)} + Pre^{(t)} + Pro^{(t)} + C^{(t)} \tag{1}$$

$$e^{(t+1)} = \frac{\sum_{i=1}^{n}(e_i^{(t+1)})}{n} \tag{2}$$

$$Pre^{(t)} = \frac{\sum_{u=1}^{q}(e_{Pre,u}^{(t)})}{q} \tag{3}$$

$$Pro^{(t)} = \frac{\sum_{v=1}^{p}(e_{Pro,v}^{(t)})}{p} \tag{4}$$

$$C^{(t)} = \sum_{j=1}^{m}(h_j^{(t)}) \tag{5}$$

Since vital signs and lab measurements are processed respectively due to different frequency of sampling, two individual RNNs will be developed based on the derived DBN structures in Fig. 3(a) and (b), which will run simultaneously to form a single component of the Triplet architecture.

Triplet Architecture. The Triplet architecture is a commonly used framework for fine-grained similarity learning tasks. The framework takes information from three patients of a triplet as inputs, and uses three identical neural networks to transfer patients' information into effective representations in the feature space, which would act as inputs to the distance metric learning layer for distance calculation. Meanwhile, a triplet of patients contains three patients, namely a query patient p_i, a positive patient p_i^+ and a negative patient p_i^-. With a function $\mathcal{S}(p_i, p_j)$ measuring the similarity between two patients, these three patients should meet the constraint $\mathcal{S}(p_i, p_i^+) > \mathcal{S}(p_i, p_j^-)$.

There are three layers in the Triplet architecture. The sampling layer aims at sampling effective triplets of patients and providing inputs to the model. Pairwise supervised similarity of patients for triplet sampling and model training are calculated according to Eq. (6), where $\mathcal{LCP}()$ is the function that obtains the longest common prefix, and L_a and L_b are the ICD code sets belonging to patient a and b, while $L_{a,i}$ and $L_{b,j}$ are the ith and jth icd code in L_a and L_b.

$$\mathcal{S}(L_a, L_b) = \frac{\sum_{i=1}^{m} \sum_{j=1}^{n} (|\mathcal{LCP}(L_{a,i}, L_{b,j})|)}{|L_a| \times |L_b|} \tag{6}$$

The representation layer feeds medical information of triplets to three identical RNNs respectively. In our framework, RNNs with specially-designed connections will act as the three identical components. After obtaining the representations $f(p_i)$, $f(p_i^+)$ and $f(p_i^-)$ for patients of a triplet, the metric layer calculates the square Euclidean distance between two patients to get $\mathcal{D}(f(p_i), f(p_i^+))$ and $\mathcal{D}(f(p_i), f(p_i^-))$ respectively. Then the metric layer employs these two distances to define hinge loss as in Eq. (7), where m represents the preset threshold of the margin between the distances of two pairs of patients.

$$\mathcal{L}(p_i, p_i^+, p_i^-) = \max\{0, m + \mathcal{D}(f(p_i, p_i^+)) - \mathcal{D}(f(p_i), f(p_i^-))\} \tag{7}$$

4 Experiments

To evaluate the performance of TDBNN, we have conducted comparative experiments on a publicly available clinical dataset MIMIC III v1.4, and compared our model with several baselines in terms of different evaluation metrics. We also investigate the influence of different sizes of time windows based on which the embedding vectors for medical events are obtained, in order to reveal the period of change of patient's physiological features.

Dataset. We have conducted extensive experiments on a real-world ICU dataset MIMIC III v1.4. A patient cohort selection is carried out to select adult patients (age > 15) staying in the ICU for more than 48 h. We do not take those patients who entered ICU more than once in a single hospitalization or died into account either. Totally 3251 patients are finally selected and divided into training set (80%), test set (10%), and validation set (10%). The number of triples generated in these three datasets are 1,000,000 for the training set, 10,000 for the testing set, and 10,000 for the validation set.

The neural network model was implemented with PyTorch and trained on 2 Nvida 1080 Ti GUPs in a fully-supervised manner and the purpose is to minimize the hinge loss in Eq. 7, which was achieved by employing stochastic gradient descent with Adma update rule [15] with a learning rate of 5×10^{-4}. The implementation is available at Github[1].

Comparison Methods. To evaluate the effectiveness of the proposed TDBNN, we compare the method with the following baselines and approaches in terms of different performance metrics. We implemented the first three baselines with Python ourselves while the rest four methods were implemented with scikit-learn.

1. **T-SRNN:** A Triplet architecture based on standard RNNs, namely no extra connections among units in two adjacent time stamps are developed.
2. **T-MLP:** A Triplet architecture based on Multi-Layer Perceptron (MLP) inspired by Patient Similarity Deep Metric Learning Framework (PSDML).
3. **RV** and **dCor:** A unsupervised method proposed by Zhu et al. [37] that adopts RV and dCor coefficient to measure linear and non-linear relations between pairwise patients based on their temporal embedding matrices.
4. **ITML:** A framework proposed by Eavis et al. [11] that minimizes the differential relative entropy between two multivariate Gaussians under constraints on the distance function.
5. **LMNN:** A method presented by Weingerger et al. [35] that learns a Mahanalobis distance metric in the kNN classification setting to keep k-nearest neighbors in the same class, while keep examples from different classes separated by a large margin.
6. **NCA:** A supervised learning method introduced by Goldberger et al. [12] for classifying multivariate data into distinct classes according to a given distance metric over the data.
7. **MMC:** A framework introduced by Xing et al. [36] that minimizes the sum of squared distances between similar examples, while enforcing the sum of distances between dissimilar examples to be greater than a certain margin.

Different performance metrics are employed to compared TDBNN with those baselines in terms of different potential clinical applications.

Triplet Similarity Ranking: Given a large number of triplets consisting of patients with different similarities, we calculate the proportion of correctly ranked triplets as a performance evaluation. Specifically, given a triplet $t_i = (p_i, p_i^+, p_i^-)$ and the embedding representations $f(p_i), f(p_i^+), f(p_i^-)$, we say the triplet is correctly ranked if the the Euclidean distance $d_i^+ = distance(f(p_i), f(p))$ is smaller than $d_i^- = distance(f(p_i), f(p_i^-))$.

[1] https://github.com/vpccw152c/TDBNN.

Classification Based on SOFA: SOFA is a morbidity severity score and mortality estimation tool developed from a large sample of ICU patients throughout the world, and it can be divided into two groups: stable stage (SOFA score < 9) and critical stage (SOFA score ≥ 9) [31]. Thus, classification based on SOFA could indicate whether the obtained embedding vectors could represent the overall severity of patients. Given a visit, we use Logistic Regression (LR) to predict the binary SOFA class associated with the visit, and use Area Under The Curve (AUC) to measure the classification accuracy.

Multi-Lable K-Nearest-Neighbors (MLKNN) Classifier: MLKNN classifier based on patients' embedding representations could be used to indicate whether these representations truly reflect the distance among patients, while the classification result is a reasonable reference for diagnosis. We take the ten ICD-9 codes that appears most frequently in the selected cohort as class labels, and observe how Hamming loss changes with the number of neighbors.

Evaluation. Checking whether a model could accurately rank the patients of a triplet by their pairwise similarities is a sensible strategy to verify its performance. Table 2 shows the accuracy of similarity ranking of different methods, and it is clear that TDBNN achieves the best performance with an accuracy of 82.99%, followed by the figure of T-SRNN at 76.35%. Meanwhile, T-MLP, RV, and dCor get close accuracy ranging from 62.11% to 64.39%, which show significant gaps with those of TDBNN and T-SRNN. The remaining three methods, namely ITML, NCA, and MMC, show another decline in accuracy compared to all those previous methods and MMC achieves the lowest global accuracy. Overall, the proposed TDBNN surpasses all the other baselines.

Different frameworks are then evaluated using binary classification of SOFA score to see whether these representations truly reflect the overall severity of patients. Note that RV and dCor coefficients have not been taken into account here, since they do not generate embedding representations for patients that

Table 2. Accuracy of triplet ranking of different frameworks

Index	Method	Accuracy
1	T-SRNN	76.35%
2	T-MLP	62.11%
3	RV	63.02%
4	dCor	64.39%
5	ITML	55.8%
6	NCA	54.27%
7	MMC	53.84%
8	TDBNN	**82.99%**

Table 3. AUC of binary SOFA classification of different methods. Note that RV and dCor are not compared here since they do not generate patient representations that used for classification

Index	Method	Accuracy
1	T-SRNN	70.59%
2	T-MLP	51.85%
3	ITML	70.77%
4	NCA	71.44%
5	MMC	77.11%
6	TDBNN	**82.29%**

Table 4. Hamming loss of different frameworks

Neighbors K	T-SRNN	T-MLP	ITML	NCA	MMC	TDBNN
5	0.2858	0.2856	0.2874	0.2780	0.2752	0.2858
7	0.2743	0.2778	0.2787	0.2749	0.2651	0.2770
9	0.2722	0.2652	0.2739	0.2811	0.2621	0.2628
11	0.2694	0.2652	0.2682	0.2808	0.2608	0.2616
13	0.2698	0.2620	0.2661	0.2779	0.2618	0.2564
15	0.2720	0.2595	**0.2621**	0.2768	0.2617	0.2530
17	0.2616	**0.2593**	0.2629	**0.2676**	**0.2600**	**0.2464**
19	**0.2613**	0.2579	0.2651	0.2737	0.2644	0.2523

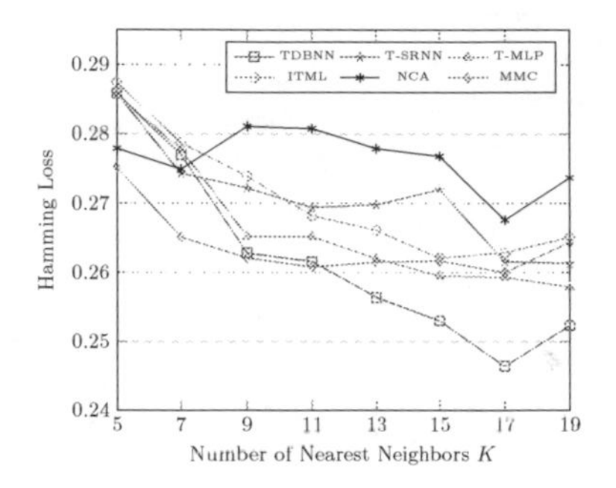

Fig. 5. Hamming loss of different frameworks.

could be used for classification. We convert patients' SOFA scores into binary labels with a cutoff value of 9, and use LR to predict the binary SOFA labels while the performance is evaluated with AUC. The corresponding AUC are shown in Table 3, and it is obvious that TDBNN achieves a higher AUC than all the other methods, indicating the effectiveness of the embedding representations obtained by TDBNN on representing overall severity of patients.

Next, we use MLKNN classification to compare the performance of different methods. The experimental results are shown in Table 4 and Fig. 5 illustrates trends of Hamming loss of those methods over different numbers of nearest neighbors K. The Hamming loss of TDBNN is lower than those of all the other methods for most of the values that K takes, decreasing modestly from 0.2858 to 0.2464 before getting to the minimum value when $K = 17$. Meanwhile, the values of MMC, T-MLP and ITML show a similar trend that decrease steadily with the increase in the number of nearest neighbors K before they remain roughly unchanged. By contrast, the losses of NCA and T-SRNN fluctuate throughout the whole period, showing no tendency to stabilize. Overall, most of the compared methods show a downward trend in Hamming loss except for NCA and T-SRNN, and the proposed TDBNN outperforms all the other methods.

Finally, to investigate the period of change of patient's status, we compare the performance of TDBNN using embedding vectors that obtained from 6-h, 12-h, 18-h, and 24-h time windows according to all the three performance metrics. Table 5 presents the results, and it is clear that when the TDBNN model carries out triplet similarity ranking using embedding vectors derived from 24-h time window, it achieves a much higher accuracy. The 24-h time window also leads to the lowest Hamming loss in MLKNN classification and the highest AUC in the binary classification of SOFA score. Therefore, it is reasonable to regard 24 h as the period of change of patient's conditions.

Table 5. Performance comparison of TDBNN based on different time window sizes

Index	Time window size (hours)	Triplet similarity ranking (Accuracy)	MLKNN classification (Hamming loss)	Binary classification of SOFA (AUC)
1	6	54.68%	0.2614	76.25%
2	12	57.68%	0.2612	78.86%
3	18	49.15%	0.261	77.23%
4	24	**82.99%**	**0.2464**	**82.29%**

The experimental results demonstrate that TDBNN achieves better performance than all the other methods on patient similarity learning in terms of different performance evaluation metrics. The comparison between TDBNN and T-SRNN implies that the specially-designed connections among RNN units help to transmit underlying dependencies among medical variables, which have a positive effect on learning patient similarity. Furthermore, both TDBNN and T-SRNN employ RNN to process sequential information obtained from EHRs, which achieve far better performances than those methods that do not consider temporal information. Thus, it is sensible to think that temporal information plays an important role in deriving meaningful information from EHRs. Meanwhile, the experiment on length of time window for embedding vector learning shows that 24 h could be the period of change of patient's conditions.

5 Conclusion

Providing physicians with historical and similar cases assist in making timely medical decisions for better clinical cares as well as predicting possible changing trends of patients' conditions. Causal correlations among medical events in the form of DBN are employed to develop specially-designed connections among RNN units, so that both such causal correlations and temporal features could be used for fine-grained patient similarity. Meanwhile, the changes in medical indicators following causal correlations could be used to explain the similarities among patients, making the recommended medical decisions much more convincing. The TDBNN achieves better performance in terms of multiple performance metrics, and indicates a promising approach for fine-grained similarity learning.

To sum up, taking causal correlations among medical indicators into account and applying such correlations to construct specially-designed connections in RNN are the most significant advantages of our method. The proposed TDBNN provides a promising direction for the study of fine-grained patient similarity, which acts as an important role in medical decision support system.

References

1. Beck, D., Haffari, G., Cohn, T.: Graph-to-sequence learning using gated graph neural networks. arXiv preprint arXiv:1806.09835 (2018)
2. Cai, X., et al.: Real-time prediction of mortality, readmission, and length of stay using electronic health record data. J. Am. Med. Inform. Assoc. **23**(3), 553–561 (2015)
3. Chan, L., Chan, T., Cheng, L., Mak, W.: Machine learning of patient similarity: a case study on predicting survival in cancer patient after locoregional chemotherapy. In: 2010 IEEE International Conference on Bioinformatics and Biomedicine Workshops (BIBMW), pp. 467–470. IEEE (2010)
4. Chen, W., Wang, S., Long, G., Yao, L., Sheng, Q.Z., Li, X.: Dynamic illness severity prediction via multi-task rnns for intensive care unit. In: 2018 IEEE International Conference on Data Mining (ICDM), pp. 917–922. IEEE (2018)
5. Chen, W., et al.: EEG-based motion intention recognition via multi-task RNNs. In: Proceedings of the 2018 SIAM International Conference on Data Mining, pp. 279–287. SIAM (2018)
6. Cheng, Y., Wang, F., Zhang, P., Hu, J.: Risk prediction with electronic health records: a deep learning approach. In: Proceedings of the 2016 SIAM International Conference on Data Mining, pp. 432–440. SIAM (2016)
7. Choi, E., Bahadori, M.T., Schuetz, A., Stewart, W.F., Sun, J.: Doctor AI: predicting clinical events via recurrent neural networks. In: Machine Learning for Healthcare Conference, pp. 301–318 (2016)
8. Choi, E., et al.: Multi-layer representation learning for medical concepts. In: Proceedings of the 22nd ACM SIGKDD International Conference on Knowledge Discovery and Data Mining, pp. 1495–1504. ACM (2016)
9. Choi, E., Schuetz, A., Stewart, W.F., Sun, J.: Medical concept representation learning from electronic health records and its application on heart failure prediction. arXiv preprint arXiv:1602.03686 (2016)
10. Choi, Y., Chiu, C.Y.I., Sontag, D.: Learning low-dimensional representations of medical concepts. AMIA Summits Transl. Sci. Proc. **2016**, 41 (2016)
11. Davis, J.V., Kulis, B., Jain, P., Sra, S., Dhillon, I.S.: Information-theoretic metric learning. In: ICML, Corvalis, Oregon, USA, pp. 209–216, June 2007
12. Goldberger, J., Hinton, G.E., Roweis, S.T., Salakhutdinov, R.R.: Neighbourhood components analysis. In: Saul, L.K., Weiss, Y., Bottou, L. (eds.) Advances in Neural Information Processing Systems 17, pp. 513–520. MIT Press, Cambridge (2005). http://papers.nips.cc/paper/2566-neighbourhood-components-analysis.pdf
13. Heckerman, D., Geiger, D., Chickering, D.M.: Learning bayesian networks: the combination of knowledge and statistical data. Mach. Learn. **20**(3), 197–243 (1995)
14. Johnson, A.E., et al.: MIMIC-iii, a freely accessible critical care database. Sci. Data **3**, 160035 (2016)
15. Kingma, D.P., Ba, J.: Adam: a method for stochastic optimization. arXiv preprint arXiv:1412.6980 (2014)

16. Lee, J., Maslove, D.M., Dubin, J.A.: Personalized mortality prediction driven by electronic medical data and a patient similarity metric. PloS one **10**(5), e0127428 (2015)
17. Li, Y., Tarlow, D., Brockschmidt, M., Zemel, R.: Gated graph sequence neural networks. arXiv preprint arXiv:1511.05493 (2015)
18. Lipton, Z.C., Kale, D.C., Elkan, C., Wetzel, R.: Learning to diagnose with LSTM recurrent neural networks. arXiv preprint arXiv:1511.03677 (2015)
19. Lv, X., Guan, Y., Yang, J., Wu, J.: Clinical relation extraction with deep learning. IJHIT **9**(7), 237–248 (2016)
20. Marini, S., et al.: A dynamic bayesian network model for long-term simulation of clinical complications in type 1 diabetes. J. Biomed. Inf. **57**, 369–376 (2015)
21. Mueller, J., Thyagarajan, A.: Siamese recurrent architectures for learning sentence similarity. In: AAAI, vol. 16, pp. 2786–2792 (2016)
22. Ni, J., Liu, J., Zhang, C., Ye, D., Ma, Z.: Fine-grained patient similarity measuring using deep metric learning. In: Proceedings of the 2017 ACM on Conference on Information and Knowledge Management, pp. 1189–1198. ACM (2017)
23. van den Oord, A., Kalchbrenner, N., Kavukcuoglu, K.: Pixel recurrent neural networks. arXiv preprint arXiv:1601.06759 (2016)
24. Pagana, K.D., Pagana, T.J.: Mosby's Diagnostic and Laboratory Test Reference-E-Book. Elsevier, Amsterdam (2012)
25. Razavian, N., Marcus, J., Sontag, D.: Multi-task prediction of disease onsets from longitudinal laboratory tests. In: Machine Learning for Healthcare Conference, pp. 73–100 (2016)
26. Santoro, A., et al.: Relational recurrent neural networks. arXiv preprint arXiv:1806.01822 (2018)
27. Sun, J., Wang, F., Hu, J., Edabollahi, S.: Supervised patient similarity measure of heterogeneous patient records. ACM SIGKDD Explor. Newslett. **14**(1), 16–24 (2012)
28. Tran, T., Nguyen, T.D., Phung, D., Venkatesh, S.: Learning vector representation of medical objects via EMR-driven nonnegative restricted boltzmann machines (eNRBM). J. Biomed. Inf. **54**, 96–105 (2015)
29. Van Den Oord, A., et al.: Wavenet: a generative model for raw audio. In: SSW, p. 125 (2016)
30. Vaswani, A., et al.: Attention is all you need. In: Advances in Neural Information Processing Systems, pp. 5998–6008 (2017)
31. Vincent, J.L., et al.: Use of the SOFA score to assess the incidence of organ dysfunction/failure in intensive care units: results of a multicenter, prospective study. Crit. Care Med. **26**(11), 1793–1800 (1998)
32. Wang, F., Hu, J., Sun, J.: Medical prognosis based on patient similarity and expert feedback. In: Proceedings of the 21st International Conference on Pattern Recognition (ICPR2012), pp. 1799–1802. IEEE (2012)
33. Wang, F., Lee, N., Hu, J., Sun, J., Ebadollahi, S.: Towards heterogeneous temporal clinical event pattern discovery: a convolutional approach. In: Proceedings of the 18th ACM SIGKDD International Conference on Knowledge Discovery and Data Mining, pp. 453–461. ACM (2012)
34. Wang, J., et al.: Learning fine-grained image similarity with deep ranking. In: Proceedings of the IEEE Conference on Computer Vision and Pattern Recognition, pp. 1386–1393 (2014)

35. Weinberger, K.Q., Blitzer, J., Saul, L.K.: Distance metric learning for large margin nearest neighbor classification. In: Weiss, Y., Schölkopf, B., Platt, J.C. (eds.) Advances in Neural Information Processing Systems 18, pp. 1473–1480. MIT Press, Cambridge (2006). http://papers.nips.cc/paper/2795-distance-metric-learning-for-large-margin-nearest-neighbor-classification.pdf
36. Xing, E.P., Ng, A.Y., Jordan, M.I., Russell, S.: Distance metric learning, with application to clustering with side-information. In: Proceedings of the 15th International Conference on Neural Information Processing Systems, NIPS 2002, pp. 521–528. MIT Press, Cambridge (2002). http://dl.acm.org/citation.cfm?id=2968618.2968683
37. Zhu, Z., Yin, C., Qian, B., Cheng, Y., Wei, J., Wang, F.: Measuring patient similarities via a deep architecture with medical concept embedding. In: 2016 IEEE 16th International Conference on Data Mining (ICDM), pp. 749–758. IEEE (2016)

Towards Efficient k-TriPeak Decomposition on Large Graphs

Xudong Wu[1], Long Yuan[2(✉)], Xuemin Lin[3], Shiyu Yang[1], and Wenjie Zhang[3]

[1] East China Normal University, Shanghai, China
xdwu@stu.ecnu.edu.cn, syyang@sei.ecnu.edu.cn
[2] Nanjing University of Science and Technology, Nanjing, China
longyuan@njust.edu.cn
[3] The University of New South Wales, Sydney, Australia
lxue@cse.unsw.edu.au, wenjie.zhang@unsw.edu.au

Abstract. Analyzing the structure of real-world networks has attracted much attention over years and cohesive subgraph models are commonly used to characterize the structure of a network. Recently, a model named k-Peak is proposed to address the issue failing to detect sparser regions if the network contains distinct regions of different densities in the cohesive subgraph models. However, k-Peak only considers the edge connection (i.e., degree) in the network and the loose structure restricts the effectiveness of the k-Peak. On the other hand, triangles are fundamental building blocks of a network and are widely used in the literature. Motivated by this, in this paper, we propose the k-TriPeak model based on the triangles and study the problem of k-TriPeak decomposition that computes the k-TriPeak for all possible k values to understand the structure of a network. Through investigating the drawbacks of the baseline algorithm following the idea of k-Peak decomposition, we devise a new efficient algorithm to perform the k-TriPeak decomposition. Our new algorithm adopts a top-down decomposition paradigm and integrates two novel upper bounds with which large unnecessary computation can be pruned. We conduct extensive experiments on several large real-world datasets and the experimental results demonstrate the efficiency and effectiveness of our proposed algorithm.

1 Introduction

Due to the rapid development of information technology, we are witnessing the proliferation of graph data based applications over recent years. This has led to huge research efforts devoted to real-world network analytics [2,4–6,10,16,23, 27]. Among them, identifying cohesive subgraphs to characterize the structure of real-world networks has been extensively studied. Observing that the cohesive subgraph models are often computed globally and fail to detect sparser regions if the network contains distinct regions of different densities, Govindan et al. proposed a new model named k-Peak recently [7]. By conducting the k-Peak decomposition (compute the k-Peak for all possible k values in the graph), [7]

G. Li et al. (Eds.): DASFAA 2019, LNCS 11446, pp. 604–621, 2019.
https://doi.org/10.1007/978-3-030-18576-3_36

can divide the graph into separate 'mountains' and can find the centers of distinct regions in the graph.

Motivation. The k-Peak decomposition addresses the issue of neglecting sparser regions in the existing cohesive subgraph model [7], however, since k-Peak only considers the edge connection (i.e., degree) between nodes in the subgraph, the returned results are often not that cohesive [5,14,28]. The returned loose structure restricts the effectiveness of the k-Peak decomposition. On the other hand, triangles are higher-order connectivity structure than degree [3,19] and are known as fundamental building blocks of a network [12,17,18]. Therefore, triangles are commonly treated as the building blocks for the cohesive subgraph model in the literature [5,8,16,27]. Motivated by this, in this paper, we propose a new model named k-TriPeak based on the triangles and study the k-TriPeak decomposition problem. The model inherits the ability to find the centers of distinct regions of k-Peak model and avoids the problem of incohesiveness for the returned result. Formally, given a graph G, the support of an edge is the number of triangles containing it. A k-TriContour of G is the largest subgraph of G such that (i) the support of edges in it is at least $k-2$; (ii) the k-TriContour does not include edges from a higher TriContour. The k-TriPeak of G is the union of j-TriContours, where $j \geq k$. And k-TriPeak decomposition computes the k-TriPeak for all possible k values.

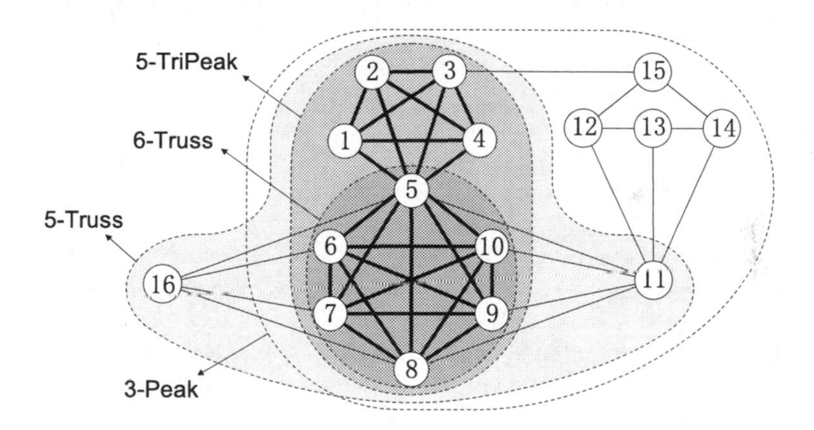

Fig. 1. k-Peak vs k-TriPeak

Example 1. Consider the graph in Fig. 1, we show the 3-Peak and 5-TriPeak of G (The counterpart of 3-Peak is 5-TriPeak since the support of edges is at least $k-2$ in the definition of k-TriContour). As shown in Fig. 1, since 3-Peak only considers the edge connection between nodes, the incohesive subgraph induced by $v_{11}, \ldots, v_{15}$ is also returned. On the contrary, this incohesive subgraph is filtered out by the 5-TriPeak. In the literature, another similar cohesive subgraph model defined based triangles, k-truss, is also studied [5]. It is defined as the largest subgraph of G in which every edge is contained in at least $(k-2)$ triangles within the subgraph. The problem of k-truss is that it is unable to find the centers of distinct regions in the graph. As shown in Fig. 1, two centers of regions exist in G, namely $\{v_1, \ldots, v_5\}$, $\{v_5, \ldots, v_{10}\}$. For k-truss model, if we return the 6-truss, the center $\{v_1, \ldots, v_5\}$ will be missed; if we return 5-truss, the two centers are returned but the returned result also contains node v_{11} and v_{16} and these two nodes are loosely connected with each other and are not what we want. On the contrary, 5-TriPeak can find these two centers.

Applications. k-TriPeak decomposition can be used in many applications. For example, in the community detection, since k-TriPeak model can find central regions with different densities, those sparser communities in the graph will not be missed if the k-TriPeak model is adopted [8,14]. Similar to k-Peak decomposition, k-TriPeak decomposition can also be used to visualize the graph through the mountain plot technique presented in [7]. Moreover, understanding the hierarchical structure facilitates graph-topology analysis [1,22]. The k-TriPeaks of a graph for all k values form a hierarchical structure. It is clear that k-TriPeak decomposition is helpful for understanding the hierarchical structure in a graph.

Our Approach. To perform the k-TriPeak decomposition, a direct approach is following the idea of k-Peak decomposition in [7]. In [7], the k-Peak decomposition is achieved by iteratively computing the k-core with maximum k value in the graph through the k-core decomposition algorithm [2] and removing the computed k-core until the graph is empty. Following the idea, we can implement the k-TriPeak decomposition through k-truss decomposition algorithm [15] in a similar way. However, as analysed in Sect. 3, lots of unnecessary edges will involve in the expensive k-truss decomposition procedure, which leads to the inefficiency of this direct approach. To address the drawback of this approach, in this paper, we propose a new algorithm for the k-TriPeak decomposition. Our new algorithm adopts a top-down decomposition paradigm in which k is explored in decreasing order. Based on this top-down decomposition paradigm, we design two effective but lightweight upper bounding techniques. Using these two upper bounding techniques, we can prune the unpromising edges involving in the expensive k-truss decomposition procedure and the unnecessary computation in the direct approach can be significantly reduced.

Contribution. In this paper, we make the following contributions:

(1) The k-TriPeak model to find the centers of distinct regions in the graph. We investigate the drawbacks of existing k-Peak model and propose a new model, namely k-TriPeak. Based on the k-TriPeak model, we study the problem of k-TriPeak decomposition. To the best of our knowledge, this is the first work to study the problem of efficient k-TriPeak decomposition.

(2) An efficient algorithm for k-TriPeak decomposition. We present an efficient algorithm to perform the k-TriPeak decomposition. In our algorithm, we adopt a top-down decomposition paradigm and devise a static upper bound

and a dynamic upper bound to reduce the unnecessary computation. Moreover, we also explore efficient techniques with which we can maintain the dynamic upper bound in $O(1)$ time for each update during the decomposition process.

(3) *Extensive performance studies on large real-world datasets.* We conduct extensive performance studies using large real-world datasets. The experimental results demonstrate the effectiveness of our proposed model and the efficiency and scalability of the devised decomposition algorithm.

2 Preliminaries

We model a undirected graph as $G(V, E)$, where $V(G)$ represents the set of nodes and $E(G)$ represents the set of edges in G. We denote the number of nodes as n and the number of edges as m, i.e., $n = |V(G)|$ and $m = |E(G)|$. We define the size of G, denoted by $|G|$, as $|G| = m+n$. For a node $u \in V(G)$, we use $\mathsf{nbr}(u, G)$ to denote the neighbor set of u in G, i.e., $\mathsf{nbr}(u, G) = \{v \in V(G)|(u, v) \in E(G)\}$. The degree of a node $u \in V(G)$, denoted by $\mathsf{deg}(u, G)$, is the number of neighbors of u, i.e., $\mathsf{deg}(u, G) = |\mathsf{nbr}(u, G)|$. A triangle in G is a cycle of length 3. In this paper, we omit G in the notations when it is explicit in context.

Definition 1 (Support). *Given a graph G, the support of an edge $e \in E(G)$, denoted by $\mathsf{sup}(e, G)$, is the number of triangles that contain e in G.*

Definition 2 (k-TriContour). *Given a graph G, a subgraph S is the k-TriContour of G, denoted by $C_k(G)$, if (i) $\mathsf{sup}(e, S) \geq k - 2$ for every edge $e \in S$; (ii) the k-TriContour does not include edges from a higher TriContour; (iii) S is maximal, i.e., any subgraph $S' \supset S$ is not a k-TriContour.*

Definition 3 (k-TriPeak). *Given a graph G, a k-TriPeak, denoted by $P_k(G)$, is the union of j-TriContours, where $j \geq k$.*

Definition 4 (TriPeak Number). *The TriPeak number of an edge e in G, denoted by $\kappa(e, G)$, is the value k such that e is contained in the k-TriContour.*

Problem Statement. In this paper, we study the problem of *k-TriPeak decomposition* that computes the k-TriPeak for all possible k values in the given graph. Since the k-TriPeak consists of the edges with TriPeak number at least k. The k-TriPeak decomposition problem equals to compute the TriPeak number for each edge in the given graph. Therefore, in this paper, we aim to design an efficient algorithm to perform the assignment of TriPeak number to each edge.

Example 2. Consider the graph G illustrated in Fig. 2, we also show its corresponding k-TriContour and k-TriPeak in Fig. 2. For example, for the edge (v_8, v_9), its support is 3 since it is contained in triangles $\{v_8, v_9, v_6\}$, $\{v_8, v_9, v_{10}\}$, $\{v_8, v_9, v_{11}\}$. The 5-TriContour of G is the subgraph induced by nodes $\{v_5, v_6, v_8 \dots v_{15}\}$ except edge (v_5, v_6) and (v_{10}, v_{13}) as this is the maximal subgraph such that the support for each edge in it is 3 and it does not contain any

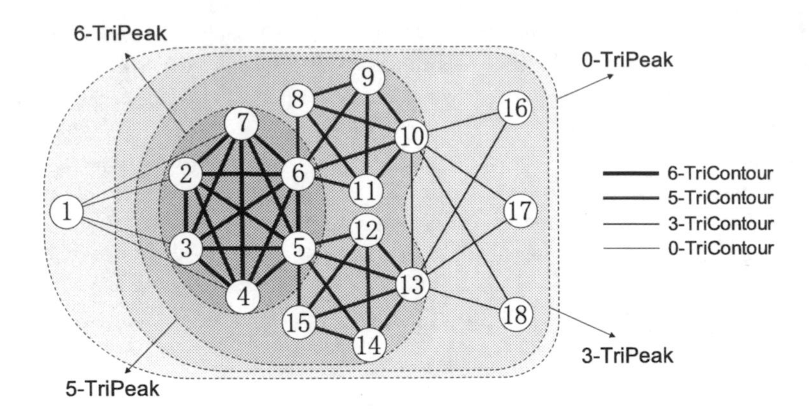

Fig. 2. An example graph G

edges from a k-TriContour with $k > 5$. Note that although (v_1, v_7) has support 3, it is not in the 5-TriContour of G. This is because (v_1, v_7) is contained in $\{v_1, v_7, v_2\}, \{v_1, v_7, v_3\}, \{v_1, v_7, v_4\}$ but edges (v_7, v_2), (v_7, v_3) and (v_7, v_4) are in the 6-TriContour of G. After the TriPeak decomposition, all the TriPeak number of edge can be obtained. For example, $\kappa((v_6, v_7)) = 6$ as it is in the 6-TriContour and $\kappa((v_{10}, v_{13}))$ is 3 as it is in the 3-TriContour.

3 Baseline k-TriPeak Decomposition Algorithm

Inspired by the solution proposed in [7] to perform the Peak decomposition, we present a baseline solution for the TriPeak decomposition problem in this section. In [7], to conduct the Peak decomposition, it iteratively computes the k-core with maximum k value in the graph by the k-core decomposition algorithm [2] and removing the computed k-core until the graph is empty. Following the same idea, we can perform the TriPeak decomposition through the k-truss decomposition [15] based on the following lemma:

Lemma 1. *Given a graph G, let k_{max} be the maximum value such that the corresponding k-truss in G, denoted by $T_{k_{max}}(G)$, is not empty, then $T_{k_{max}}(G) = P_{k_{max}}(G)$.*

Proof. According to Definition 2, $C_{k_{max}}(G) = T_{k_{max}}(G)$ since there doesn't exist $C_k(G)$ with $k > k_{max}$, otherwise such $C_k(G)$ is also a $T_k(G)$ that $k > k_{max}$. And by Definition 3, $P_{k_{max}}(G)$ is the union of $C_j(G)$ where $j \geq k_{max}$. Thus $P_{k_{max}}(G) = C_{k_{max}}(G) = T_{k_{max}}(G)$.

Based on Lemma 1, for a given graph G, the k_{max}-truss and k_{max}-TriPeak in G are the same, which means the TriPeak number of edges in $T_{k_{max}}(G)$ equals to k_{max} exactly. Moreover, based on Definitions 2 and 4, the edges with TriPeak number k have no impact on the TriPeak number of edges whose TriPeak number

Algorithm 1. Baseline(Graph G)

```
1: while not all edges in G are removed do
2:     C_{k_max} ← maxTruss(G);
3:     for each edge e ∈ C_{k_max} do
4:         κ(e) ← k_max;
5:         remove e from G;
6: procedure maxTruss(Graph G)
7:     compute sup(e) for each edge e ∈ E(G) using the triangle counting algorithm [9];
8:     sort all edges in ascending order of their support;
9:     k_max ← 2;
10:    while not all edges in G are removed do
11:        let e = (u, v) be the edge with the lowest support in G; (assume deg(u) ≤ deg(v))
12:        k ← sup(e) + 2;
13:        k_max ← max(k_max, k);
14:        Φ_{k_max} ← Φ_{k_max} ∪ {e};
15:        for each w ∈ nbr(u) do
16:            if (v, w) ∈ G_k then
17:                sup((u, w)) ← sup((u, w)) − 1;
18:                sup((v, w)) ← sup((v, w)) − 1;
19:                update the new positions of (u, w) and (v, w) in the sorted edge array;
20:        remove e from G;
21:    return Φ_{k_max};
```

is k', where $k' < k$. In other words, for a graph G, if we remove the edges with TriPeak number k from G, the edges with TriPeak number k' in G and the new generated graph G' after the edge removal are the same. Therefore, we can conduct the TriPeak decomposition by iteratively computing the k-truss with the maximum k value and removing the edges in the graph until the graph is empty.

Algorithm. Based on the above analysis, the baseline algorithm, Baseline, is shown in Algorithm 1. The baseline algorithm iteratively computes and removes the k-truss with the maximum k value of G at each iteration (line 2–5). If an edge is contained in k-truss with the maximum k value of G at current iteration, the TriPeak number of it will be assigned (line 4) and it will be removed from G then (line 5). This process is carried out until all edges in G are removed (line 1).

Procedure maxTruss computes the k-truss with the maximum k value in G. It first computes the support of each edge in G by the triangle counting algorithm [9] (line 7). Then it sorts all the edges in ascending order of their supports and keep them in an array (line 8). After that, the algorithm iteratively removes the edge e with the lowest support, which is the first edge in the sorted edge array, and add e into the result set of $\Phi_{k_{max}}$ (line 11–14). When removing e, the supports of all other edges that form a triangle with e should be decreased, and their new positions in the sorted edge array should be updated (line 15–20). This algorithm terminates after all edges in G are removed (line 10) and returns the k-truss with the maximum k value (line 21). [15] shows that the time complexity of procedure maxTruss is $O(m^{1.5})$.

Drawbacks of the Baseline Solution. In the baseline algorithm, we conduct the TriPeak decomposition through k-truss decomposition iteratively. In each iteration, the edges with $\kappa(e) = k_{max}$ are assigned by computing the k-truss of G with the maximum k value in current iteration (line 4). For a specific iteration, an ideal algorithm is that the computation in this iteration only involves the edges with $\kappa(e) = k_{max}$. However, in the baseline algorithm, all the edges in the remaining graph are taken as the input for the k-truss decomposition algorithm (line 2). Assume that there is an edge e with a small TriPeak number k' in G, it will participate in all iterations computing k-truss where $k \geq k'$ in the baseline algorithm. Therefore, lots of redundant computation exist in the baseline algorithm and it is inefficient to conduct TriPeak decomposition considering the time complexity of k-truss decomposition is $O(m^{1.5})$.

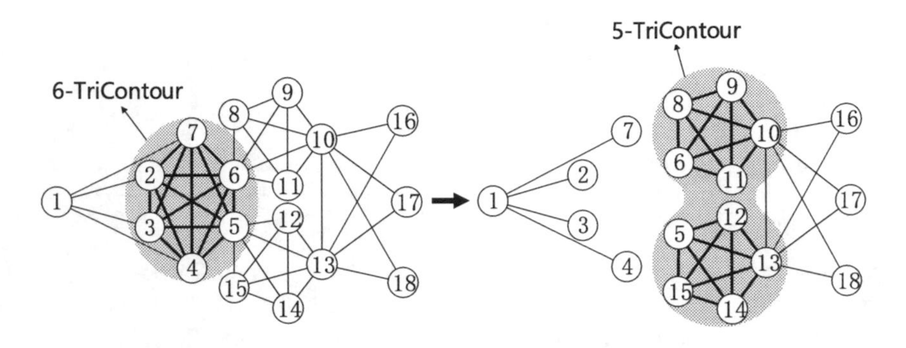

Fig. 3. A running example of Algorithm 1

Example 3. Figure 3 shows a running example of Algorithm 1 on the graph G in Fig. 2. It first performs k-truss decomposition on the whole graph and finds 6-TriPeak. Then, it removes edges in the 6-TriPeak and performs k-truss decomposition on all remaining edges to find the 5-TriContour on the remaining graph. The procedure terminates when all the edges are removed. As shown in Fig. 3, although the TriPeak number of edges incident to v_1, v_{16}, v_{17}, v_{18} and (v_{10}, v_{13}) is not 5, all of these edges involve the k-truss decomposition to compute 5-TriContour on the remaining graph, which leads to the inefficiency of Algorithm 1.

4 Our New Approach

To overcome the drawbacks of the baseline solution, we propose a new paradigm for the TriPeak decomposition problem. In this section, we first present an overview of the new paradigm in Sect. 4.1. Then, we show our concrete techniques in Sects. 4.2 and 4.3, respectively.

4.1 A New Top-Down Decomposition Paradigm

In the baseline algorithm (Algorithm 1), in a specific iteration, we compute the TriPeak number for the edges with $\kappa(e) = k_{max}$, where k_{max} is the maximum k value such that the corresponding k-truss exists in the remaining graph of current iteration. Since the k_{max} for current iteration cannot be determined in advance, it has to conduct the truss decomposition on the graphs consisting of the edges with $\kappa(e) \leq k_{max}$, which leads to the inefficiency of the baseline algorithm for the TriPeak decomposition problem. On the other hand, based on Lemma 1, we know the maximum TriPeak number for all edges in G equals to the maximum k value such that the k-truss exists in the original input graph. For brevity, we denote it as κ_{max}. In other words, we know the TriPeak number for all edges of G is in the range from 1 to κ_{max}. According to the definition of TriPeak decomposition, the essence of the problem is to determine the TriPeak number for each edge. Therefore, to perform the TriPeak decomposition, we can iterate all the possible TriPeak number of the graph in decreasing order based on their values and compuate the edges whose TriPeak number equals to the specific TriPeak number. The benefit of this paradigm is that it is possible to prune the edge with $\kappa(e) < k$ in a specific iteration as we know the TriPeak number k to be handled in each iteration in advance. In this way, we can reduce the redundant computation in the baseline algorithm caused by edges with small TriPeak number involving truss decomposition many times.

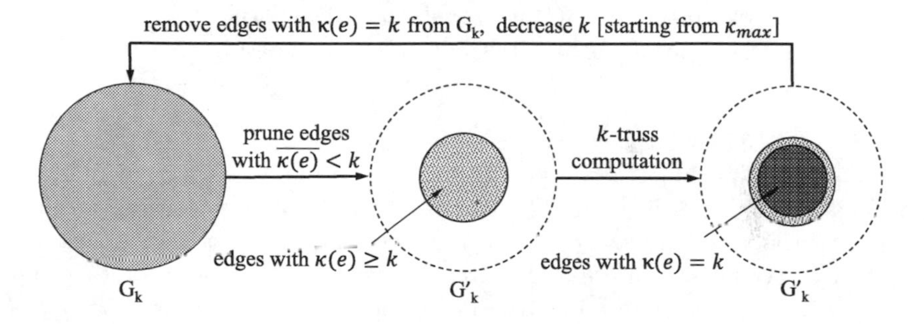

Fig. 4. The new paradigm

Algorithm Framework. Following the above analysis, the new TriPeak decomposition paradigm is illustrated in Fig. 4. Staring from κ_{max}, the paradigm computes the TriPeak number for the edges in decreasing order of k. For a specific k, the edges with $\kappa(e) = k$ are computed. As analysed above, we aim to limit the edges involving the truss computation in this step to the edges with $\kappa(e) = k$. However, this goal is hard to achieve. Therefore, we compute upper bounds of TriPeak number for the edges and use these upper bounds to prune the useless edges. Specifically, when processing a specific k (without loss of generality, we denote the input graph regarding k as G_k), it first prunes the edges whose upper bound of TriPeak number is less than k, i.e., $\overline{\kappa(e)} < k$. We denote the pruned graph as G'_k. Then, we determine the edges with $\kappa(e) = k$ by computing the k-truss on G'_k. After that, the edges with $\kappa(e) = k$ are removed from G_k and the remaining graph are treated as the input graph for the next iteration. The process terminates when G_k is empty.

4.2 Upper Bounding Techniques

As analysed in Sect. 4.1, the key point for the efficiency of the new paradigm is tight upper bounds of $\kappa(e)$. To achieve this goal, in this part, we will introduce two kinds of upper bounding techniques for $\kappa(e)$.

A Static Upper Bound. Based on the definition of TriPeak number, a direct upper bound of the TriPeak number of the edges in G can be obtained by the truss number:

Definition 5 (Truss Number). *Given a graph G, the truss number of an edge e in G, denoted by $\phi(e)$, is the maximal number of k such that e is contained in a k-truss.*

Lemma 2. *Let e be an edge in G, $\phi(e)$ is the truss number of e, and $\kappa(e)$ is the TriPeak number of e, then $\phi(e) \geq \kappa(e)$.*

Proof. By the definition of k-TriContour, every edge in k-TriContour has no less than $k - 2$ triangles inside k-TriContour, which indicates that the k-TriContour is a part of k-truss. Thus if e is in k-TriContour, it must also be in k-truss. Hence $\phi(e) \geq \kappa(e)$ holds.

The truss number for each edge can be easily obtained through the k-truss decomposition algorithm. However, since k-TriContour does not consider the support from triangles in higher TriContours, as our decomposition paradigm progresses, the pruning power of truss number weakens and the edges with $\phi(e) > \kappa(e)$ accumulate more and more. Therefore, we propose another tight but lightweight upper bound for $\kappa(e)$. The upper bound is defined based on G_k (the input graph of our paradigm when processing a specific k, i.e., the graph after removing all the edges with $\kappa(e) > k$) and is dynamically maintained as our decomposition paradigm progresses.

A Dynamic Upper Bound. Given an edge $(u, v) \in E(G_k)$, $\mathsf{sup}((u, v), G_k)$ be the support of $e = (u, v)$ in G_k. For a node $u \in V(G_k)$, let $h(u, G_k)$ returns the maximum value h such that there exist at least h neighbours v of u with $\mathsf{sup}((u, v), G_k) \geq h$. We define $\lambda(e, G_k) = \min\{\mathsf{sup}(e, G_k), h(u, G_k), h(v, G_k)\} + 2$. And we can prove that for any arbitrary valid k in our paradigm, $\lambda(e, G_k)$ is an upper bound of $\kappa(e, G)$, which is shown in the following lemma:

Lemma 3. *Let e be an edge in G, then $\lambda(e, G_k) \geq \kappa(e, G)$.*

Algorithm 2. TriPeakDecom(Graph G)

```
1: compute κ_max and sup(e), φ(e) for all edges by maxTruss in Algorithm 1;
2: compute h(u) for each node u ∈ V(G);
3: for each e = (u, v) ∈ E(G) do
4:     λ(e) ← min{sup(e), h(u), h(v)} + 2;
5:     κ̄(e) ← min{λ(e), φ(e)};
6: k ← κ_max; G_k ← G;
7: while G_k ≠ ∅ do
8:     G'_k ← {e|e ∈ E(G_k), κ̄(e) ≥ k};
9:     C_k' ← maxTruss (G'_k)
10:    if k' = k then
11:        S ← ∅;
12:        for each edge e = (u, v) ∈ C_k' (assume deg(u) ≤ deg(v)) do
13:            κ(e) ← k; remove e from G_k;
14:            update h(u), h(v);
15:            add u (or v) into S if h(u) (or h(v)) is changed;
16:            for each w ∈ nbr(u) do
17:                if (v, w) ∈ G_k then
18:                    sup((u, w)) ← sup((u, w)) − 1; sup((v, w)) ← sup((v, w)) − 1;
19:                    update h(w), h(u), h(v) and λ((u, w)) and λ((v, w));
20:                    add u(v or w) into S if h(u) (h(v) or h(w)) is changed;
21:        for each node w in S do
22:            update κ̄(e) for each edge e incident to w as line 4-5;
23:    G_{k-1} ← G_k; k ← k − 1;
```

Proof. Since $e = (u, v)$ is still remained in G_k, we know $C_{\kappa(e,G)}(G) \subseteq G_k$. And within $C_{\kappa(e,G)}(G)$, we know that $\mathsf{sup}(e, C_{\kappa(e,G)}(G)) + 2 \geq \kappa(e, G)$, $h(u, C_{\kappa(e,G)}(G)) + 2 \geq \kappa(e, G)$ and $h(v, C_{\kappa(e,G)}(G)) + 2 \geq \kappa(e, G)$. Thus $\lambda(e, G_k) = \min\{\mathsf{sup}(e, G_k),\ h(u, G_k), h(v, G_k)\} + 2 \geq \min\{\mathsf{sup}(e, C_{\kappa(e,G)}(G)), h(u, C_{\kappa(e,G)}(G)), h(v, C_{\kappa(e,G)}(G))\} + 2 \geq \kappa(e, G)$.

4.3 Our k-TriPeak Decomposition Algorithm

In this part, we present our algorithm to conduct the TriPeak decomposition. With the new decomposition paradigm and upper bounding techniques, the only challenge is integrating the upper bounding techniques into decomposition paradigm efficiently, especially the maintenance of upper bound $\lambda(e, G_k)$. This part addresses this challenge.

Algorithm. Our algorithm, TriPeakDecom, is shown in Algorithm 2. It first computes the κ_{max} and initializes the auxiliary information for the upper bounds (line 1–5). Then, it conduct the TriPeak decomposition following the new top-down paradigm until the graph is empty (line 6–23).

Specifically, it first invokes procedure maxTruss in Algorithm 1 to compute κ_{max}, $\mathsf{sup}(e)$ and $\phi(e)$ for each edge e ($\phi(e)$ equals to k_{max} when e is removed from G in line 20 of Algorithm 1). Then it computes $h(u)$ for each node u based on the supports of edges incident to u (line 2). At last, $\lambda(e)$ for each edge e is assigned according to its definition and $\overline{\kappa(e)}$ of each edge e is initialized as $\min\{\lambda(e), \phi(e)\}$ (line 4–5).

Then, it conducts the TriPeak decomposition iteratively starting with $k = \kappa_{max}$ and $G_k = G$ (line 6) and the decomposition terminates when G_k is empty (line 7). In a specific iteration processing k, it first extracts G'_k from G_k with edges of which the upper bound $\overline{\kappa(e)}$ is no less than k (line 8). This step filters out the unpromising edges. Then it computes the $C_{k'}$ of G'_k by the $\mathsf{maxTruss}$ procedure (line 9). If $k' = k$, for each edge $e = (u, v) \in C'_k$, it assigns $\kappa(e) = k$ and removes the edge from G_k (line 13). The remaining work is to maintain the incorrect $\overline{\kappa(e)}$ caused by the removal of edges. As $\phi(e)$ is fixed in the whole process, we only need to find the edges whose $\lambda(e)$ changes after the edge removal. To achieve this goal, $\mathsf{TriPeakDecom}$ uses a set S to store the nodes u that $h(u)$ has changed in the iteration since this change may influence $\lambda(e)$ of any edges incident to u (line 11). Regarding a removed edge (u, v), for u, v and each common neighbor w of u and v, it decreases the support of (u, w) and (v, w) (line 18), updates $h(u)$, $h(v)$ and $h(w)$ (line 14, 19) and $\lambda((u, w))$ and $\lambda((v, w))$ (line 19). If a node u whose $h(u)$ is changed, adds u into S (line 15, 20). At the end of iteration, for each node $w \in S$, it updates $\overline{\kappa(e)}$ for all edges incident to w, since the change of $h(w)$ may change $\overline{\kappa(e)}$ (line 21–22). When an iteration finishes, k is decreased and the remaining edges are be taken as the input graph for the next iteration (line 23).

Efficient Maintenance of $\lambda(e, G_k)$. In Algorithm 2, we maintain $\lambda(e, G_k)$ dynamically as the decomposition processes (line 19, 22). Based on the definition of $\lambda(e, G_k)$, for an edge $e = (u, v)$, the key to obtain $\lambda(e, G_k)$ is to compute $h(u, G_k)$ and $h(v, G_k)$. However, the time complexity to compute $h(u, G_k)$ and $h(v, G_k)$ on the fly based on the edge support maintained in Algorithm 2 is at least $O(\max\{\mathsf{deg}(u, G_k), \mathsf{deg}(v, G_k)\})$. Since $h(u, G_k)$ is recomputed frequently in Algorithm 2 (line 14, 19), this approach is inefficient. To improve the efficiency to maintain $h(u)$, for each node u, besides $h(u)$, we maintain the number of edges incident to u with different support values, respectively, i.e., cnt^u_i, which represents the number of edges incident to u with support equals i. Moreover, we also maintain $cnt^u_{\geq h(u)}$ that stores the number of edges incident to u with support not less than $h(u)$. During the decomposition, when the support of an edge $e = (u, v)$ decrease from i to j, we just decrease cnt^u_i and cnt^v_i by 1 and increase cnt^u_j and cnt^v_j by 1. And for the node u (the same as v), if $i \geq h(u)$, $j < h(u)$ and if $cnt^u_{\geq h(u)} > h(u)$, we decrease $cnt^u_{\geq h(u)}$ by 1; and if $cnt^u_{\geq h(u)} = h(u)$, we decrease $h(u)$ by 1 and update $cnt^u_{\geq h(u)}$ as $cnt^u_{\geq h(u)} + cnt^u_{h(u)} - 1$. Otherwise, we just keep $h(u)$ and $cnt^u_{\geq h(u)}$ unchanged. In this way, for the operation updating $h(u)$ regarding a node u in line 14 and 19 in Algorithm 2, we can finish it in $O(1)$ time. As a result, for each edge e, the $\lambda(e, G_k)$ are maintained in $O(1)$ in the decomposition procedure.

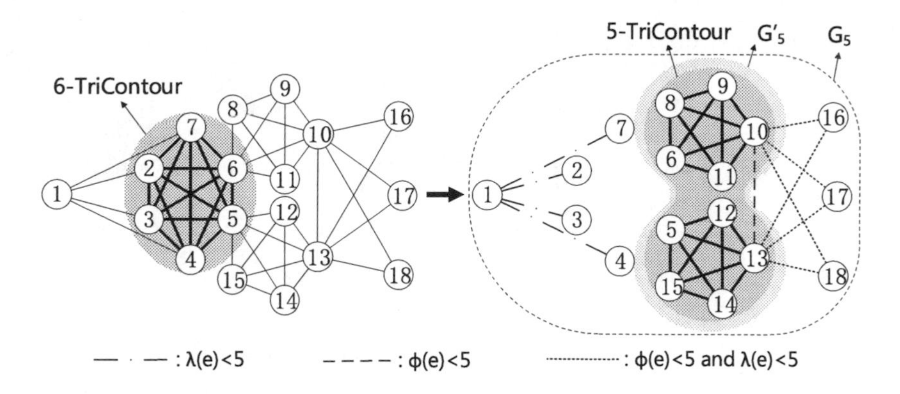

Fig. 5. A running example of Algorithm 2

Example 4. Figure 5 shows a running example of Algorithm 2 on the graph G in Fig. 2. Similar to Algorithm 1, it first performs truss decomposition and finds 6-TriPeak. Then, it removes edges in the 6-TriPeak and performs truss decomposition to find the 5-TriContour. Different from Algorithm 1, when performing the k-truss decomposition to find 5-TriContour, it first prunes the edges with $\kappa(e) < 5$. For example, the edge (v_{10}, v_{13}) is pruned by static upper bound as $\phi((v_{10}, v_{13})) < 5$ though $\lambda((v_{10}, v_{13})) \geq 5$; edge (v_1, v_7) is pruned by dynamic upper bound as $\lambda((v_1, v_7)) < 5$ though $\phi((v_1, v_7)) \geq 5$. As illustrated in Fig. 5, Algorithm 2 significantly reduces the number of unnecessary edges involving the procedure of truss decomposition compared with Algorithm 1.

Theorem 1. *Given a graph G, the running time of Algorithm 2 can be bounded by $O(\kappa_{max} \cdot m^{1.5})$.*

Proof. The whole algorithm can be divided into two stages, the initialization stage (line 1–6) and the main iteration stage (line 7–23). Line 1 invokes maxTruss procedure using $O(m^{1.5})$ time. Line 2–5 can be done in $O(m)$ time. In the main iteration stage, line 9 takes $O(m^{1.5})$ time. Line 13–15 and line 18–20 can be done in constant time, and line 21–22 requires $O(m)$ time. Now the only question left is what is the number of loops in line 12 and 16. For a certain node u, line 16 is bounded by $\mathsf{deg}(u)$, and line 12 is bounded by $|\mathsf{nbr}_{\geq u}|$, which is the number of neighbors of u whose degree is not smaller than u. Thus line 12–20 can be done in $O(\sum_{u \in G_k} (\mathsf{deg}(u) \cdot |\mathsf{nbr}_{\geq u}|))$ time, which is bounded by $O(m^{1.5})$. This is because if $\mathsf{deg}(u) \leq \sqrt{m}$, $|\mathsf{nbr}_{\geq u}| \leq \mathsf{deg}(u) \leq \sqrt{m}$ and $\sum_{u \in G_k} (\mathsf{deg}(u) \cdot |\mathsf{nbr}_{\geq u}|) \leq m^{1.5}$. If $\mathsf{deg}(u) > \sqrt{m}$, $|\mathsf{nbr}_{\geq u}| \leq \sqrt{m}$ as well for $\mathsf{deg}(u) \cdot |\mathsf{nbr}_{\geq u}| \leq \sum_{v \in |\mathsf{nbr}_{\geq u}|} \mathsf{deg}(v) < 2m$, and $\sum_{u \in G_k} (\mathsf{deg}(u) \cdot |\mathsf{nbr}_{\geq u}|) \leq m^{1.5}$. The number of iterations is bounded by κ_{max}. Thus, the the running time of Algorithm 2 can be bounded by $O(\kappa_{max} \cdot m^{1.5})$.

5 Performance Studies

In this section, we evaluate the effectiveness of our model and the efficiency and scalability of our proposed algorithm. The experiments are conducted on a machine with an Intel Xeon 2.20 GHz CPU and 128 GB memory running Red Hat Linux 4.8.5, 64 bit.

Table 1. Datasets used in experiments

Datasets	Type	Number of nodes	Number of edges	Average degree	κ_{max}
DBLP	Citation	317,080	1,049,866	6.62	114
Livemocha	Social	104,103	2,193,083	42.13	27
Flickr	Misc	105,938	2,316,948	43.74	574
Flixster	Social	2,523,386	7,918,801	6.28	47
Skitter	Computer	1,696,415	11,095,298	13.08	68
LiveJournal	Social	3,997,962	34,681,189	17.35	352

Datasets. In our experiments, we evaluate the algorithms on six publicly available real-world datasets as listed in Table 1. Of these, DBLP and LiveJournal are downloaded from *SNAP*[1], and the others are downloaded from *KONECT*[2].

Algorithms. We implement and compare the following four algorithms:

- Baseline: Algorithm 1
- $\mathsf{TriPeakDecom_s}$: TriPeak decomposition algorithm with static upper bound only.
- $\mathsf{TriPeakDecom_d}$: TriPeak decomposition algorithm with dynamic upper bound only.
- TriPeakDecom: Algorithm 2

All algorithms are implemented in C++ and compiled with GNU GCC 4.8.5 using optimization level 2. The time cost of the algorithm is measured as the amount of elapsed wall-clock time during the program execution.

Exp-1: Effectiveness. We evaluate the effectiveness of k-TriPeak and k-Peak by examining the quality of detected subgraph via the clustering coefficient [18] metric. Clustering coefficient (CC) indicates the tendency of nodes in a subgraph to cluster together. Thus, high clustering coefficient means high probability that the connections inside the detected subgraph are dense. In this experiment, we find all k-TriPeaks and k-Peaks with different k values and compute the clustering coefficient of them. Since the distributions of k in findings of k-TriPeak and k-Peak are quite different, here we compare the clustering coefficient of k_1-TriPeak and k_2-Peak of similar size even if $k_1 \neq k_2$. The results are shown in

[1] http://snap.stanford.edu/.
[2] http://konect.uni-koblenz.de/.

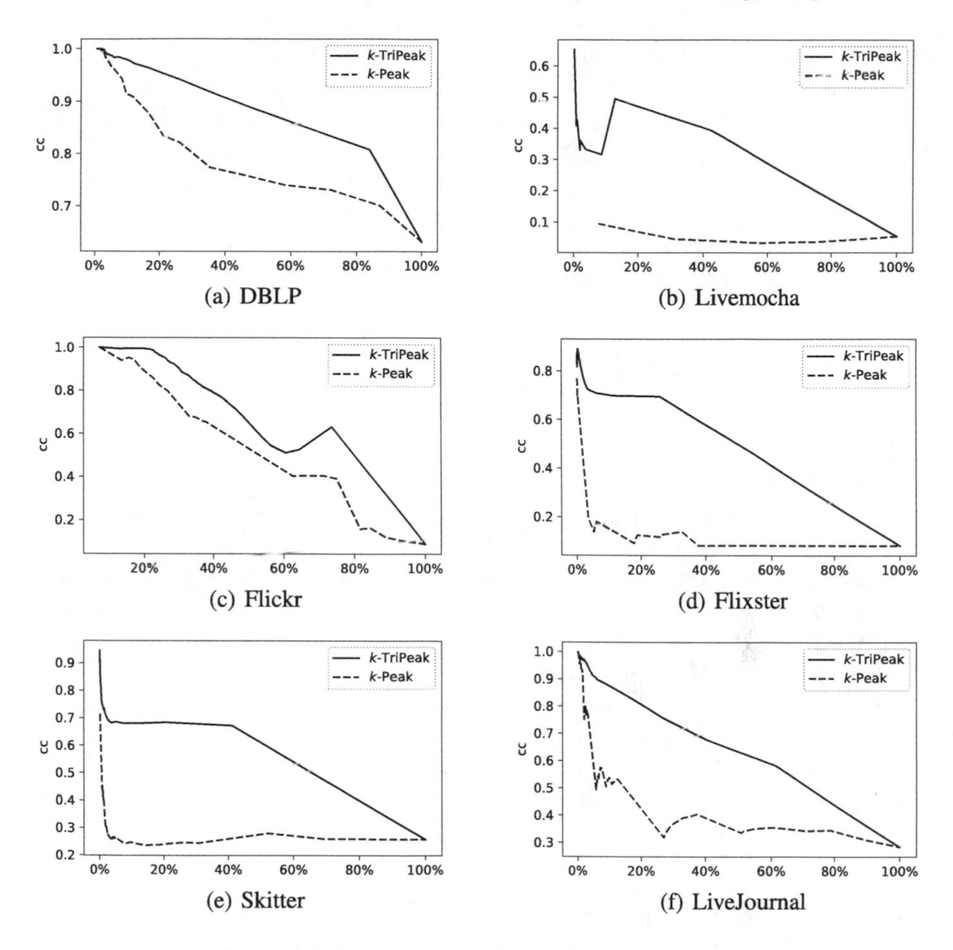

(a) DBLP (b) Livemocha (c) Flickr (d) Flixster (e) Skitter (f) LiveJournal

Fig. 6. Effectiveness of k-TriPeak and k-Peak

Fig. 6. In Fig. 6, the horizontal coordinate denotes the size of k-TriPeak (k-Peak) as a persentage of total graph.

As shown in Fig. 6, as the size of k-TriPeak (k-Peak) increases, the value of clustering coefficient for both of them generally decreases. However, it can be observed that, of similar size, the k-TriPeak is much denser than k-Peak. For example, in *Flixster* dataset, the k-TriPeak of which the size is around 30% of the total graph has clustering coefficient of 0.7, while the k-Peak of similar size only have 0.1. This is because k-TriPeak takes a high-order connectivity structure as the building blocks while k-Peak only considers degree. These results indicate that compared with the k-Peak model, the returned result of k-TriPeak are more cohesive and k-TriPeak is better cohesive subgraph model compared with k-Peak.

Table 2. Running time on real-world datasets

Alg	Dataset					
	DBLP	Livemocha	Flickr	Flixster	Skitter	LiveJournal
Baseline	81.85s	178.5 s	3538.5 s	1017.25 s	1993.32 s	29768.37 s
TriPeakDecom$_s$	19.69 s	62.1 s	2020.35 s	194.41 s	545.78 s	5530.65 s
TriPeakDecom$_d$	15.56 s	98.67 s	1228.3 s	285.02 s	615.49 s	5038.35 s
TriPeakDecom	**14.75 s**	**58.29 s**	**834.79 s**	**150.9 s**	**471.26 s**	**2653.01 s**

Exp-2: Efficiency. In this experiment, we compare the total processing time of those four algorithms on six real-world datasets. The results are reported in Table 2.

Generally, the processing time increases as the size of the graph increases. Baseline takes the most time on all six datasets. It spends more than 8 hours to perform the TriPeak decomposition on LiveJournal dataset. The reason for Baseline's long running time is that edges with small TriPeak number, which make up a large portion of the whole graph, take participate in the k-truss decomposition for the big TriPeak number many times. The algorithms solely adopting static or dynamic upper bounding technique run much faster than Baseline. TriPeakDecom$_s$ is faster than TriPeakDecom$_d$ on Livemocha, Flixster and Skitter and the opposite on DBLP, Flickr and LiveJournal, for they play to their strength on different stages in a decomposition. TriPeakDecom algorithm, which adopts both upper bounding techniques, achieves the best performance on all six datasets. For example, on LiveJournal, it achieves an order of magnitude faster than Baseline.

Exp-3: Scalability. We study the scalability of the four algorithms in this experiment. To test the scalability, we randomly sample the nodes and edges respectively of two largest datasets *Skitter* and *LiveJournal* from 20% to 100% and take the induced subgraph as the input graph. The results are shown in Fig. 7.

As shown in Fig. 7, as the size of the graph increases, the processing times of four algorithms increase due to the increasing of the number of iterations and the number of involved edges in each iteration. Moreover, as the size of the graph increases, the gap in processing times between Baseline and other three algorithms increases. This is because the unnecessary computation on edges, which are reduced by other three algorithms but remained in Baseline, make up larger portion of computation when the size of the graph grows. The gap in processing times between TriPeakDecom$_s$ and TriPeakDecom$_d$ remains small on both datasets. The TriPeakDecom algorithm consumes the least time and its processing time grows the most stably on all datasets. The results show the good scalability of our proposed algorithm.

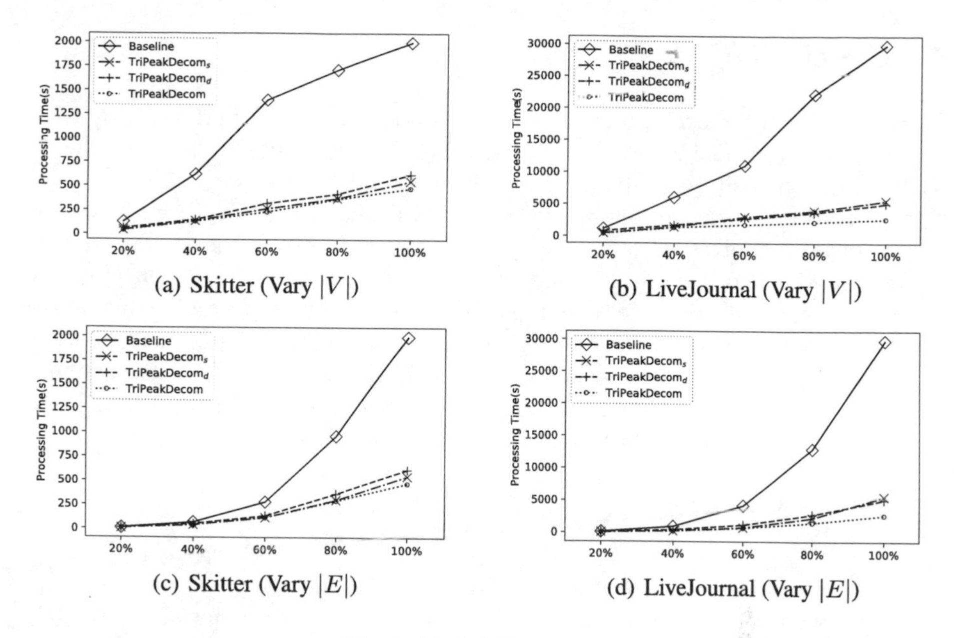

(a) Skitter (Vary $|V|$)

(b) LiveJournal (Vary $|V|$)

(c) Skitter (Vary $|E|$)

(d) LiveJournal (Vary $|E|$)

Fig. 7. Scalability testing

6 Related Work

The most related works to k-TriPeak are k-Peak [7] and k-truss [5], which have been introduced in Sect. 1. Since k-Peak is defined based on degree, the returned results of k-Peak are often not that cohesive [5,14,28] compared with the k-TriPeak [14]. The difference between k-TriPeak and k-truss is that in a k-truss, triangles containing edges from the higher k-truss are taken into consideration while in a k-TriPeak, all these edges are filtered out. This difference leads to it that k-TriPeak is able to find the centers of distinct regions in a graph as k-Peak [7].

Besides k-Peak and k-truss, there are many different models proposed in the literature. One of the most intuitive cohesive subgraph models is the clique model in which each node is adjacent to every other node [4]. More complex models based on the clique model [20,21,25] are also studied in the literature. However, clique is often too restrictive for many applications, thus, more clique relaxation models have been proposed, such as the k-plex [13], n-clan and n-club [11]. Nevertheless, these models always face the problem of computational intractability. To address this problem, more polynomial time solvable cohesive subgraph models are proposed recently, such as k-core [2], triangle k-core [27], (k, s)-core [26], DN-Graph [16], k-edge connected component [22,24] and k-mutual-friend subgraph model [28].

7 Conclusion

Motivated by the recent proposed k-Peak model, in this paper, we propose a new k-TriPeak model based on the triangles in the graph and study the k-TriPeak decomposition problem. To perform the k-TriPeak decomposition, we first present an approach following the idea of k-Peak decomposition. However, this approach involves lots of unnecessary computation. Therefore, we propose a new top-down paradigm to conduct the decomposition. Based on the new paradigm, we devise two effective upper bounds to prune the unnecessary edges involving computation in the baseline approach. Moreover, we explore efficient techniques to maintain the upper bounds during the decomposition. We conduct experiments on large real-world datasets and the experimental results demonstrate the efficiency and effectiveness of our proposed algorithm.

References

1. Alvarez-Hamelin, J.I., Dall'Asta, L., Barrat, A., Vespignani, A.: K-core decomposition of internet graphs: hierarchies, self-similarity and measurement biases. NHM **3**(2), 371–393 (2008)
2. Batagelj, V., Zaveršnik, M.: An o(m) algorithm for cores decomposition of networks. Comput. Sci. **1**(6), 34–37 (2003)
3. Benson, A.R., Gleich, D.F., Leskovec, J.: Higher-order organization of complex networks. Science **353**(6295), 163–166 (2016)
4. Bron, C., Kerbosch, J.: Algorithm 457: finding all cliques of an undirected graph. Commun. ACM **16**(9), 575–576 (1973)
5. Cohen, J.: Trusses: cohesive subgraphs for social network analysis. National Security Agency Technical Report (2008)
6. Feng, X., Chang, L., Lin, X., Qin, L., Zhang, W., Yuan, L.: Distributed computing connected components with linear communication cost. Distrib. Parallel Databases **36**(3), 555–592 (2018)
7. Govindan, P., Wang, C., Xu, C., Duan, H., Soundarajan, S.: The k-peak decomposition: mapping the global structure of graphs. In: Proceedings of WWW, pp. 1441–1450 (2017)
8. Huang, X., Cheng, H., Qin, L., Tian, W., Yu, J.X.: Querying k-truss community in large and dynamic graphs. In: Proceedinhs of SIGMOD, pp. 1311–1322 (2014)
9. Latapy, M.: Main-memory triangle computations for very large (sparse (power-law)) graphs. Theor. Comput. Sci. **407**(1), 458–473 (2008)
10. Luce, R.D., Perry, A.D.: A method of matrix analysis of group structure. Psychometrika **14**(2), 95–116 (1949)
11. Mokken, R.J.: Cliques, clubs and clans. Qual. Quant. **13**(2), 161–173 (1979)
12. Newman, M.E.J., Watts, D.J., Strogatz, S.H.: Random graph models of social networks. PNAS **99**, 2566–2572 (2002)
13. Seidman, S.B., Foster, B.L.: A graph-theoretic generalization of the clique concept. J. Math. Sociol. **6**(1), 139–154 (1978)
14. Shao, Y., Chen, L., Cui, B.: Efficient cohesive subgraphs detection in parallel. In: Proceedings of SIGMOD, pp. 613–624 (2014)
15. Wang, J., Cheng, J.: Truss decomposition in massive networks. PVLDB **5**(9), 812–823 (2012)

16. Wang, N., Zhang, J., Tan, K., Tung, A.K.H.: On triangulation-based dense neighborhood graphs discovery. PVLDB **4**(2), 58–68 (2010)
17. Wasserman, S., Faust, K.: Social Network Analysis: Methods and Applications. Cambridge University Press, Cambridge (1994)
18. Watts, D.J., Strogatz, S.H.: Collective dynamics of 'small-world' networks. Nature **393**(6684), 440 (1998)
19. Yin, H., Benson, A.R., Leskovec, J., Gleich, D.F.: Local higher-order graph clustering. In: Proceedings of SIGKDD, pp. 555–564 (2017)
20. Yuan, L., Qin, L., Lin, X., Chang, L., Zhang, W.: Diversified top-k clique search. In: Proceedings of ICDE, pp. 387–398 (2015)
21. Yuan, L., Qin, L., Lin, X., Chang, L., Zhang, W.: Diversified top-k clique search. VLDB J. **25**(2), 171–196 (2016)
22. Yuan, L., Qin, L., Lin, X., Chang, L., Zhang, W.: I/O efficient ECC graph decomposition via graph reduction. PVLDB **9**(7), 516–527 (2016)
23. Yuan, L., Qin, L., Lin, X., Chang, L., Zhang, W.: Effective and efficient dynamic graph coloring. PVLDB **11**(3), 338–351 (2017)
24. Yuan, L., Qin, L., Lin, X., Chang, L., Zhang, W.: I/O efficient ECC graph decomposition via graph reduction. VLDB J. **26**(2), 275–300 (2017)
25. Yuan, L., Qin, L., Zhang, W., Chang, L., Yang, J.: Index-based densest clique percolation community search in networks. IEEE TKDE **30**(5), 922–935 (2018)
26. Zhang, F., Yuan, L., Zhang, Y., Qin, L., Lin, X., Zhou, A.: Discovering strong communities with user engagement and tie strength. In: Pei, J., Manolopoulos, Y., Sadiq, S., Li, J. (eds.) DASFAA 2018. LNCS, vol. 10827, pp. 425–441. Springer, Cham (2018). https://doi.org/10.1007/978-3-319-91452-7_28
27. Zhang, Y., Parthasarathy, S.: Extracting analyzing and visualizing triangle k-core motifs within networks. In: Proceedings of ICDE (2012)
28. Zhao, F., Tung, A.K.: Large scale cohesive subgraphs discovery for social network visual analysis. PVLDB **6**(2), 85–96 (2012)

Knowledge Graph

Evaluating the Choice of Tags in CQA Sites

Rohan Banerjee(✉), Sailaja Rajanala, and Manish Singh

Indian Institute of Technology Hyderabad,
Sangareddy, India
{cs17mtech11008,cs15resch11009,msingh}@iith.ac.in

Abstract. Tags play a crucial role in CQA sites by facilitating organization, indexing and categorization of the entire post in a few words. The choice of tags determines the audience that is elicited upon to seek a response for any particular post. This could either lead to receiving an accurate response for the question or result in receiving no answers. The choice of tags, thus, directly determines the quality of the post as well as to a large extent the success of the CQA site itself. In this paper, we a present a novel approach to evaluate the choice of tags in any post. We perform tag network analysis to find relationship between tags. We then find the anomalous combination of tags by performing anomaly detection. We demonstrate the robustness of our approach by showing high AUC, in the range of 0.95 to 0.98, on four datasets from Stack Exchange, namely Ask Ubuntu, Server Fault, Super User and Software Engineering.

Keywords: Tag network · Anomaly detection · CQA sites

1 Introduction

Community Questions Answering (CQA) sites have been evolving rapidly in the past few years. In CQA sites questions are posed to a large community of users with the intent of finding an accurate response in the minimum possible time. Although many users in these sites have high expertise in wide range of topics, the knowledge is distributed among a wide range of users. Hence, identifying an appropriate expert for answering a question is a challenging task. In CQA sites, such as Stack Overflow, tags are the predominant means to elect the experts who are notified to answer a question.

Users subscribe to tags that pertain to their topics of interest and get notified when a question is posted with their subscribed tags. Choice of wrong tags spawns numerous problems. For example, an incorrect tagging not only directs the questions to an erroneous group of users, but also attracts down-votes that strongly demotivates subsequent users to pursue the question. Another alternate means to receive responses is by search engines that index the website content. However, the most immediate responses to a question are received from subscribers rather than those from external search engines [3]. Tags, thus play a direct role in influencing the response time of a post in CQA sites.

G. Li et al. (Eds.): DASFAA 2019, LNCS 11446, pp. 625–640, 2019.
https://doi.org/10.1007/978-3-030-18576-3_37

Let's consider the following two example questions that illustrate the importance of choosing appropriate tags while posting a question:

Example 1 – TITLE: *What does the 'yield' keyword do?* TAGS: {`python, iterator, generator, yield, coroutine`} BODY: *What is the use of the yield keyword in Python? What does it do? For example, I'm trying to understand... What happens when the method _get_child_candidates is called? Is a list returned? A single element? Is it called again? When will subsequent calls stop?*

Example 2 – TITLE: *How do genetic algorithms work exactly?* TAGS: {`encoding, score, crossover, mutation, solution`} BODY: *I was looking at the basic genetic algorithm here... You have a problem you want a good enough solution to. You create N random solutions. You evaluate a fitness or score for each for how good a solution it is. Based on each score, they have a higher chance of being picked. Then for those 2 picked, there is a cross over step. There is a chance associated with a crossover... We created a new solution. Do we remove one of the existing ones? If so, which one?*

Then there is a mutation step. This is the part that confuses me most... What do we mutate? A or B or C?...Finally re-evaluate the score for all and repeat all of this until you find a good enough solution. Does anyone know?

The first question is a top voted post from Stack Overflow. It has good choice of tags that spans the entire range of the underlying domain. It has the right mix of generic (`python`, `iterator`) and specific tags (`yield`,`coroutine`). Here the `python` tag is generic because it points to the broad topic Python. On the other hand, the tags like `yield` and `coroutine` are specific tags as they pinpoint to a narrow sub-domain within the entire domain. Thus, experts who are interested in both `python` and `coroutine` would be notified about this question.

The second question demonstrates a bad tagging practice. It is expected that a question on genetic algorithms should include tags on artificial intelligence, evolutionary learning, genetics and algorithms. On the other hand, trivial tags that do not pinpoint to the question have been chosen. Tags like `mutation`, `score` and `solution` do not precisely convey the domain of the question. This would result in the question missing out responses from expert users of the community and instead redirect to a broad range of users who might have no involvement with genetics.

Since the number of posts in CQA sites are increasing drastically, it is difficult to judge high quality incoming questions that needs more attention. One way to address this problem is to evaluate the tags associated with the particular question. Often a question is framed on the intersection of content from a wide range of topics. When a question is framed by an expert a coherent combination of tags is noticed. Tags are chosen such that they exhibit both the breadth as well as the intent of the question clearly. Thus, it is not too far-fetched to assume that most good questions, raised by experts, would necessarily have appropriate tag choices. We utilize this intuition to model normal tagged questions, and then demonstrate in later sections that this characteristic can be utilized to identify low quality questions in CQA sites.

The task of predicting the correctness of tag choices has four main challenges: (a) No standard guidelines for good or bad tagging practice; (b) Due to constant topic evolution, a set of tags which is very unlikely today may become common after few years; (c) High-dimensionality of the tag data, where each tag is a dimension and sites have between between 3K–60K tags, and (d) No existing labelled dataset of good and bad tag choices.

In this paper, we present an unsupervised solution to the problem of evaluating the correctness of tag choices in CQA sites. We use social interaction information from CQA sites to create a dataset of good and bad tag choices. Assuming most of the good questions have good choice of tags, we find the bad choice of tags using anomaly detection. Since the tag dataset is extremely high-dimensional and is also very sparse, we show the Anomaly detection algorithms do not perform well when we use all the dimensions or do dimensionality reduction using common techniques such as PCA. In this paper, we propose a novel tag network analysis based feature reduction algorithm that gives very high AUC for the anomaly detection problem.

The paper is structured as follows: Sect. 2 discusses the related work in this field, Sect. 3 discusses standard dimensionality reduction based approaches, Sect. 4 discusses the proposed tag network based approach, Sect. 5 discusses the results of the implementation and Sect. 6 offers a conclusion to the problem.

2 Related Work

To the best of our knowledge there is no prior work that evaluates the combination of tags in CQA sites. However, lots of work has been done on tag recommendation systems. We use community detection on the tag network to group semantically tags. We compare various community detection algorithms and justify the choice of Infomap for our work. Finally, we substantiate on the need to pose tag evaluation as an anomaly detection problem.

Tag Recommendation – The problem of evaluating tag choices is closely related with tag recommendation. This is because a tag evaluation system can be effectively used to judge the correctness of the set of tags that have been suggested by a tag recommendation algorithm. A tag evaluation system could thus be viewed as a complement to a recommendation system.

There are two broad approaches for tag recommendation, namely graph based and content based approaches [7]. Graph based systems create a folksonomy graph to model mutual tag relations and then leverage this information to perform tag recommendations. For example, [4] proposed a graph based recommendation system built on top of FolkRank, which is a modified version of PageRank for folksonomy graphs. [11, 16] proposed tag recommendations systems based on tensor factorization. The method proposed by [16] represents the data in the form of a 3-order tensor of users, items and tags and uses Higher Order Singular Value Decomposition (HOSVD) to model the relation between the triplets. The Ranking with Tensor Factorization (RTF) model proposed in [11] outperformed both FolkRank and HOSVD. [6] proposed a method based that applies Latent

Dirichlet Allocation (LDA) to a dense core folksonomy graph and extracts topics that are later used to perform tag recommendations.

Content based systems solve the problem of sparsity in the tag graphs by adding the post content information to the folksonomy network. This additional information allows them to perform superior recommendations. [17] created a tag recommendation system that uses both user profiling and NLP techniques to do tag recommendation. [5] proposed a tag recommendation algorithm for social bookmarking systems that uses resource descriptions, previous annotations on the resource and annotations by the same person to perform recommendations. [9] proposed a system that does recommendation by leveraging tag information of the author from previous posts and exploiting the contextual similarity between posts.

Community Detection – Community detection algorithms are used to find relationships between nodes in networks. [21] is nice survey paper that compares popular community detection algorithms. In this paper, we use a hierarchical community detection algorithm known as Infomap, which was proposed by Rosvall et al. [12,13]. Infomap analyzes the information flow of the network and optimizes the map equation by use of random walks. The time complexity of the Infomap algorithm is $O(E)$, where E is the number of edges in the graph.

Anomaly Detection – In CQA sites, if a post has high upvotes, then it is highly likely that the post has appropriate combination of tags. However, a post with lot of downvotes or no votes does not imply that the tag combinations are incorrect. Moreover, the number of good quality tag combinations may be much less. Thus in this paper we model our problem as a one class classification problem. Bellinger et al. [21] demonstrated that the more the class imbalance the more superior the performance of a one-class classifier over a binary classifier.

Diversity in CQA Sites – Since CQA sites are not mediated, there is high diversity in both users and content. Good quality users are expected to post good quality questions with the right set of tags, whereas uninformed users cannot post their question properly. The same holds true for content. Our work attempts to model the diversity in tagging practices of users and evaluate the set of tags that would make an appropriate combination. We view this is as an effective addition in resolving the challenge of diversity in CQA sites.

There are many existing papers that have studied the user diversity aspect in CQA sites. [2] have explored the personality traits of Stack Overflow users by using reputation and attempted to study its effect on upvotes and downvotes. [20] studied the expert behaviour of Stack Overflow users and demonstrated the difference between highly active low quality users and less frequent high quality users. [15] discussed why about half of the users contribute only once in Stack Overflow and how they could be enabled to become more active.

A substantial amount of work has been done that discusses the diversity of posts in CQA sites. [10] proposed a method to detect low quality posts by using both textual features as well as community related aspects to create an automated identification system. [1] addressed why and how the long questions could remained unanswered and factors that lead to the same. [14] evaluated the attributes that contribute to a good code example in Stack Overflow.

3 Tag Anomaly Detection

In this section, we present baseline approaches to find uncanny combinations of tags. In the next section, we present our solution based on tag network analysis.

3.1 Learning Model

There is no simple heuristic to find such uncanny combinations. Since there is no standard metric to find valid tag choices, we use the social interaction information in CQA sites to find questions with good tags. We assume that if a question has got many upvotes, most likely it is a good question and should also have a good combination of tags. However, the vice versa is not true. A post having no votes or negative votes may not necessarily be due to unfavourable tag combinations. Hence, due to the lack of negative training samples, we find it appropriate to model the problem of detecting anomalous tag choices as an outlier detection problem rather than a binary classification problem.

We use one-class SVM to model the valid tag combinations, where the training data consists of tags of the top-k questions. In Stack Exchange, each question has a score based on social interactions, and we use the tags of the top 10K questions. We assume each tag as a dimension and encode our data using one-hot vector representations. This results in a very high-dimensional dataset, where the number of tags could vary from 2K–50K. Since each question can at most have five tags, this data is also very sparse. For the evaluation in Sect. 5, we use this as the baseline method.

3.2 Dimensionality Reduction

To deal with high-dimensionality, we considered standard feature selection and feature reduction techniques. Since the data is already very sparse and we cannot afford to ignore any of the tags given in the question, we do not consider feature selection methods. We next present three different approaches for feature reduction using combination of Principal Component Analysis (PCA) and t-Distributed Stochastic Neighbor Embedding (t-SNE).

PCA is one of the standard feature reduction method for high-dimensional data. However, we found that the prediction accuracy with PCA reduction is even lower compared to the one-hot encoded tag matrix. This is because the tag matrix is extremely sparse and there is no lower dimensional PCA transformation that can capture significant variance in the data without losing too much information.

The problem that arises with PCA is while reducing the dimensions focus is being given on retaining large pairwise distances instead of short pairwise distances. This is results in a crowding problem. In such a case, the datapoints that are close to each other in the high dimensional space get squashed in the low dimensional latent representation [18]. This prevents a good separation of class labels.

This problem can be overcome by use of a manifold learning technique like t-SNE. In contrast to PCA, t-SNE is non-linear approach to dimensionality reduction [19]. t-SNE attempts to model high dimensional points in a low dimensional space such that similar points are mapped close and dissimilar points are mapped far apart with a high probability. This is done by minimizing the Kullback-Leibler divergence between the high dimensional and the low dimensional distributions of the points. t-SNE does not suffer from the crowding problem unlike PCA because a heavy tailed distribution is used to compute the pairwise distances in the low dimensional space [8].

t-SNE yields better results than that obtained without feature reduction or the one obtained using PCA on the tag matrix. However, a drawback of this approach is its quadratic computational complexity in the number of datapoints [19]. Thus, there exists a trade-off between computation time and effective feature representations.

To obtain a good feature representation in unison with a discounted computation time, a third approach is employed. To being with, a significantly higher number of principal components are chosen in order to capture sufficient variance from the matrix. This PCA reduced feature matrix is then further condensed to lower dimensions by use of t-SNE. This approach gives much better results compared to PCA and slightly better than those obtained without any feature reduction. This improvement is due to the superior manifold detection capabilities of t-SNE on the reduced PCA feature matrix. The ROC curves for all the above discussed approaches can be found in Fig. 2a.

4 Anomaly Detection Using Tag Network Analysis

As discussed above, standard feature reduction methods fail to provide any significant improvement in the outlier prediction over a standard one-hot vector tag model. In this section, we present a novel dimensionality reduction method that is based on tag communities. We use the social interactions in CQA sites to group tags into communities and then use anomaly detection to detect tag combinations that are unlikely. This method gives a very high AUC of 0.982, which is very high compared to the best baseline method, namely feature reduction with t-SNE, with AUC of 0.795.

This is primarily due to the following two reasons: Firstly, they do not account for the synonymy of tags. The tags `windows-7` and `windows-10` are treated as two separate topics. For example, if the tag `adobe-acrobat` has occurred with the tag `windows-7` in many previous posts, it is quite probable that the same tag might also occur with `windows-10` tag later. Secondly, the association among similar topics is missing. Consider two topics 'adobe-acrobat' and 'windows'. There are multiple tags belonging to these two topics. It is highly likely that any tag from the 'adobe-acrobat' topic may occur with any tag from the 'windows' topic. They do not capture association among the tags across topics.

To address these challenges, we propose a folksonomy based feature reduction technique, where we use tag communities to do feature reduction. To do community detection we need to create a tag graph, where the edge between tags is derived from the questions posted in the CQA site. Finally, we use the community features to detect anomalous tags using one-class classifier. These steps are explained below in detail.

4.1 Tag Network

We use the connectivity among tags in the CQA site to create the proposed tag graph, which is both weighted and directed each tag forms a unique node in this graph. For each pair of tags, bidirectional edges are created with the edge weight given by the Nearness score, which is defined as:

$$NScore(t_i, t_j) = \frac{\#(t_i, t_j)}{|Q|} \tag{1}$$

where $\#(t_i, t_j)$ is the number of times the tag t_i has co-occurred with the tag t_j and $|Q|$ is the total number of questions. $NScore(t_i, t_j)$ is a symmetric score.

The graph created so far is a directed graph with the number of nodes equal to the number of tags in the entire dataset. The edges in this graph are then pruned using the Edge Direction score, which is defined as:

$$EDScore(t_i \rightarrow t_j) = \frac{\#(t_i, t_j)}{\#t_i} \tag{2}$$

where $\#(t_i, t_j)$ is the number of times the tag t_i has co-occurred with the tag t_j and $\#t_i$ is the number of times the tag t_i. The $EDScore(t_i \rightarrow t_j)$ measures the likelihood that tag t_j will occur given tag t_i. This is an asymmetric score. From the graph we prune edges that have EDScore below some given threshold. This pruning greatly improves the quality of tag communities we get from applying community detection algorithm on the tag graph.

4.2 Finding Tag Communities

To find tag communities we use the Infomap algorithm, which works by optimizing the map equation for the network. The map equation uses a flow based approach that captures the modular structure of the network and is derived from information theory. The Infomap is a two-level algorithm that works similar to the Louvain method used for community detection. Initially, each node is assigned its own module. Each node is moved to the neighbouring module if it decreases the map equation, else it is not moved at all. This process is repeated until no move decreases the map equation. Using, this algorithm, a fairly good graph clustering is obtained in a very short time.

After community detection, the tags that are strongly associated with each other end up in the community. For example, tags like `C#`, `Java` and `.NET` end up in one community, while tags like `html`, `css` and `javascript` end up in another community. Since the flow between the nodes control how clusters are formed, both the edge direction measure and the nearness measure play a vital role in the generation of the communities.

Table 1. Sample tag communities obtained using Infomap algorithm

Communities	NScore	EDScore	NScore + EDScore
Community 1	java, code-smell, graphics, philosophy, ebook, rdbms, friends, ssa, blockchain, google-drive	java, solid, aesthetics, nosql, wcf, linq, vb.net, design-principles, membership-provider	object-oriented, java, interfaces, c++, .net, c, asp.net, mvc, inheritance, c#, architecture
Community 2	agile, wiki, graphical-code, startup, sublime-text, cowboy-coding	budget, xna, peopleware, uml, offshore, visio, brookslaw, srs, agile, prince2	node.js, javascript, web-development, web-applications, html, css, jquery, php, html5, ajax
Community 3	unit-testing, webgl, phpunit, specflow, behat, tdd, selenium-webdriver, xunit, rspec	database, c++11, php, aesthetics, reflector, default-method, liskov-substitution	database, database-design, mysql, sql, nosql, sql-server, orm, mongodb, rdbms
Community 4	easter-eggs, freeware, apache2, advertisement, time-stamp, warranty, fonts	linux, windows, computer-architecture, c, hacking, pronunciation	agile, scrum, project, user-story, waterfall, estimation, teamwork, sprint
Community 5	google-maps, stereotypes, use-case, generalization, activity	calculus, np-complete, dijkstra, geometry, compression, byte, text-processing	licensing, gpl, open-source, mit-license, legal, copyright, lgpl, bsd-license

In Table 1, we show the communities obtained by using only NScore, only EDScore and combination of both NScore and EDScore. In the first approach, we only use nearness as edge weights between the edges and without any edge pruning. The final graph is an undirected graph with NScore as edge weight. For the second approach, we use only EDScore to determine edge direction. If the EDScore between two tags is less than a threshold, then that edge is pruned as there is very low likelihood that one tag implies the other tag. For example, if we consider the tags `Java` and `JButton`, then if a question has the tag `JButton`, then it is very likely that the question will also have the tag `Java`. However, the vice-versa relationship may not hold true. Since NScore is not used, the weights between the edges is set to one for the edges that are not pruned. For the third approach, we use both NScore and EDScore. It uses EDScore as a threshold for pruning while NScore is used as the edge weight between pairs of nodes.

In Sect. 5, we present detailed evaluation on how the two measures, namely NScore and EDScore, affects the community detection and anomaly detection algorithm. From the sample communities shown in Table 1, we can make some interesting observations. For example, in the first community detected by Infomap when only NScore is used, tags like `java`, `code-smell`, `blockchain` and `google-drive` end up in a single community, although they are not so related with each other. When EDScore is used, the results are somewhat better. A clear distinction within communities can be noticed. However, the best results are obtained when NScore and EDScore is combinedly used. Different domains are easily identifiable and therefore make for better representations.

At the conclusion of this step, each tag is assigned to a community. A community is a collection of one or more tags. Further on, each tag is identified by the community it belongs to. This information is used in the next step to model the tag data.

4.3 Detecting Anomalous Tag Choices

In situations where there is abundance of data of one particular class label, a one-class classifier makes for an appropriate choice for modelling rather than a binary classifier. The entire dataset consists of a huge number of questions that have both good and bad choices of tag combinations. Since, it is not feasible to annotate the entire dataset, we make a convenient assumption to extract out those questions that would have good tag combinations. We sort the entire dataset by the upvotes a question has received and choose the top 25% of the posts for modelling the data. The assumption here is that there is a very high probability that the top questions of the dataset must have valid and appropriate tag choices. We model only the top questions and hence, the task of detecting incorrect tag choices is formulated as an outlier detection problem.

Each question has one or more tags assigned to it. To leverage the community data that was obtained in the previous step, we identify each tag by the community it belongs. Each question in the dataset is represented by a one-hot vector of the community of the underlying tags rather than a one-hot vector of the tags. This gives two benefits: (a) Reduction in the number of dimensions, and (b) Handles topic drift by reducing over-fitting.

The number of tags in a website is in the order of thousands. Due to curse of dimensionality, the number of training samples needed to train a model increases exponentially with increasing number of dimensions. We need to ensure that there are sufficient number of samples with each combination of feature values. Since the tag data is very sparse, we cannot get sufficient training data for such high-dimensional data. Since the tag data is represented by the community data, the number of dimensions reduce from the order of thousands to less than a hundred. This naturally improves performance of the classifier.

Which topics stay popular in CQA sites change constantly from time to time and so do their corresponding tags. New domains become popular over time and a combinations of tags which was previously rare may become popular over time. For example, tags like `deep-learning` have become popular in the past few years and were not used a decade before. So, for example, a combination like {`deep-learning, python, resnet`} would be considered valid at the present moment, but would be deemed invalid a decade earlier. The opposite is also true. Tags that are not used frequently are eliminated from the database by the administrators. This poses a challenge on how the tag data must be modelled so that this topic drift characteristic is accommodated.

Communities are a high level projection of the tag data. By representing questions by the community data, we build a model that is more robust to over-fitting. If the tags are used directly in the model, even a slight drift in topic would be identified as an outlier due to a complete change of the associated tags. Further, it is possible to have more than a single correct tag combination for a question. By projecting the high dimensional tag vector to a lower, more concise community vector, a more abstract and meaningful model is created.

This one-hot vector dataset is then used to train a one-class SVM with a RBF kernel. To evaluate tag choices in the test set, the above process is repeated and the community vector data is then forward passed through the model. The invalid tag combinations are detected as outliers by the model. The one-class SVM also outputs the distance from the separating hyperplane, which provides a measure of severity of the outlier.

5 Evaluation

In this section, we present the evaluation of our proposed algorithms for detecting aberrant tag combinations. We validate our results using four big datasets, shown in Table 2, from the Stack Exchange network, namely Software Engineering, Ask Ubuntu, Server Fault and Super User.

Table 2. The four datasets used for evaluation

Dataset	Posts	Questions	Tags	AUC
Software Engineering	1,96,986	48,604	1,640	0.982
Ask Ubuntu	6,79,669	2,94,010	3,045	0.957
Server Fault	6,96,606	2,60,696	3,608	0.96
Super User	9,44,029	3,77,484	5,267	0.972

Due to space constraint, we present detailed analysis only for the Software Engineering dataset. For other three datasets, we only present the result for a single choice of parameter. Software Engineering website is a member of the larger Stack Exchange community, which comprises of several CQA sites of various genres. The dataset had 48,604 questions that were posted in the period of September 2010 to March 2018. Each question had at least one tag and a maximum of five tags. There were a total of 1,639 distinct tags in the dataset.

5.1 Experimental Setup

As discussed before, we do not have any existing annotated dataset of well tagged questions. We therefore rely on the heuristic that questions with high upvotes should have good tag combinations. However, questions having high (or no) downvotes may not necessarily have bad tag combinations. For evaluation, we consider two datasets: *Well Tagged Questions* and *Randomly Tagged Questions*.

Well Tagged Questions – We sort the questions in order of upvotes and select the top 25% questions as the Well Tagged Questions. We split these questions in 9:1 ratio to create the training and the test datasets. We refer to these two datasets as Well Tagged Questions I and Well Tagged Questions II respectively. We consider the latter dataset as the test inliers and any question marked as outlier is considered a false positive.

Randomly Tagged Questions – To generate the bad tag combinations we randomly selected combinations of 3, 4 and 5 tags from the set of all possible tags. We consider these as outliers. There is very low likelihood that a random set of selected tags would collide with a valid tag combination in the training sample. This assumption is valid since tags often co-occur only in particular selected combinations. Since, this is an outlier dataset, any tag combination marked as a non-outlier by any algorithm is considered to be a false negative.

Figure 1 presents a visual representation of the well tagged and the randomly tagged questions. It is observed that the Well Tagged II samples almost entirely coincides with the Well Tagged I samples and the Randomly Tagged questions are significantly isolated from them. This suggests that the heuristics used to generate these datasets are valid.

5.2 Evaluation Metric

To evaluate the effectiveness of the proposed approaches in isolating the anomalous tag combinations from the good tag combinations, a third test dataset is constructed. Around 2% of datapoints from the Randomly Tagged dataset are alloyed with the Well Tagged II dataset. The new dataset thus comprises of 1,020 datapoints, of which 1,000 are inliers and 20 are outliers.

To evaluate the algorithm's performance, a ranking based external validity measure is used. To begin with, outliers are ranked according to their extremity. The extremity can be quantified as the distance of the datapoint from the SVM's hyperplane. Now given this ranking, the top-k outliers are referred to as the declared outliers, denoted by $\mathcal{S}(k)$. The ground truth outliers belonging to the Randomly Tagged dataset, are denoted by $\mathcal{G}$. The ground truth inliers belonging

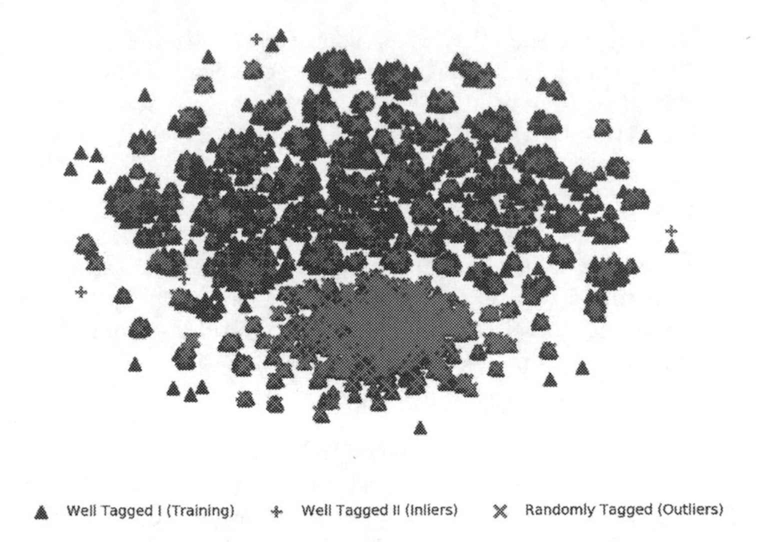

Fig. 1. Visualizing the well tagged and randomly tagged questions using PCA + t-SNE

to the Well Tagged II dataset are denoted by $\mathcal{I}$. Then sensitivity and specificity are defined as follows:

$$Sensitivity = \frac{|\mathcal{S}(k) \cap \mathcal{G}|}{|\mathcal{G}|}$$

$$Specificity = \frac{|\mathcal{S}(k) - \mathcal{G}|}{|\mathcal{I}|}$$

The parameter k is uniformly varied in the range of 0–1020 in steps of 20, and the corresponding ROC curve is produced.

5.3 Results

Figure 2 a shows the ROC curve obtained for the baseline methods described in Sect. 3. Using the one-hot vector of all tags, we get an AUC of 0.776. For the Software Engineering dataset, there are 1640 tags and the encoded matrix is extremely sparse with only 3.57% of non-zero values. To reduce the dimensionality we applied PCA to the tag matrix. The number of principal components are chosen such that atleast 75% of the variance is captured. Since the matrix is extremely sparse, the PCA reduced features gave an AUC of 0.365, which much low compared to using all the features.

To address the problem of linearity with PCA, we tried out a non-linear manifold detection algorithm t-SNE. With t-SNE we get an AUC of 0.795, which higher than both the above cases. However, as discussed in Sect. 3, the drawback is the quadratic time complexity of t-SNE. If we combine PCA and t-SNE, we get an AUC of 0.699, which is lower than using only t-SNE, but takes significantly less time compared to using only t-SNE.

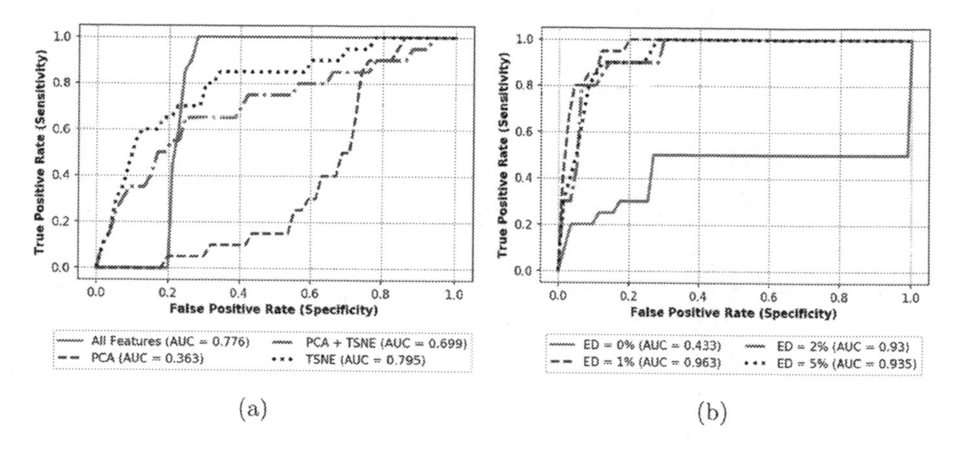

Fig. 2. (a) shows the ROC curves for baseline methods and (b) shows the ROC curves for the proposed method for various EDScores.

Figure 2b depicts the ROC curves obtained when feature reduction is done using tag community detection. An AUC in the range of 0.433 to 0.963 is obtained depending upon the choice of EDScore used to prune the tag network. Without any edge pruning an AUC of 0.433 is obtained. However, a sharp increase in the AUC is obtained when the EDScore is increased. The highest AUC of 0.963 is obtained for an EDScore of 1%.

All community detection algorithms rely a lot on the quality of input graph. EDScore plays a very important role in finding good quality communities. As shown in Fig. 3a, the number of communities obtained from the community detection algorithm is least when we do not use EDScore for pruning edges, and then the number of communities goes on increasing as we increase the threshold for EDScore. When we do not use EDScore, then we get very big communities where distinct communities gets merged. As we increased the threshold, it creates smaller communities by splitting the bigger community. If this parameter is not chosen correctly it will create many small communities, and thereby leading to over-fitting. In other words, under-fitting and over-fitting will occur when the number of communities is very low or very high respectively.

Figure 3b also shows the effect of EDScore in detecting anomalous tag combinations. With the increase in the EDScore, the number of outliers detected in randomly tagged questions increases, peaks, and then begins to fall. There exists an optimum set of communities that best represents the combination of tags in the dataset. Thus, correspondingly there also exists an optimum EDScore. It turns out to be around 1.5% for this dataset. For the other three datasets also we found that EDScore between 1.5% to 2% gives the best result.

In the one-class SVM, the hyperparameter ν determines the effectiveness of detecting the anomalous tag combinations. The value of ν sets an upper bound on the fraction of training errors and a lower bound on the fraction of support vectors. Its value is in the interval (0, 1]. Figure 4a shows the ROC curves for ν

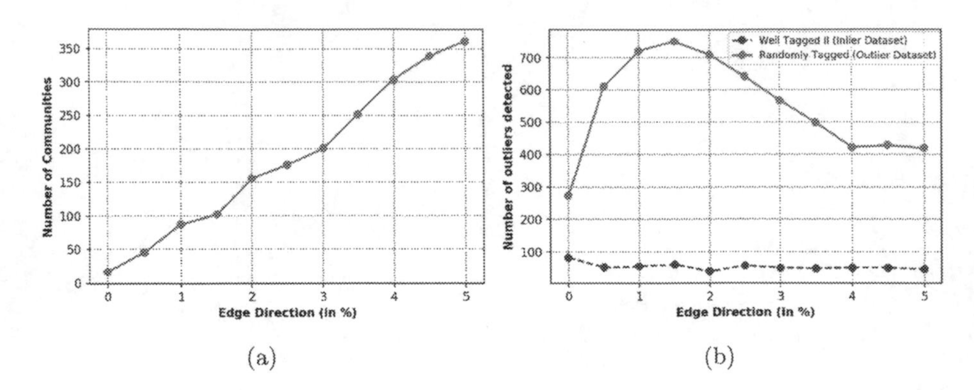

Fig. 3. (a) shows the relation of the number of communities detected by Infomap with the EDScore and (b) shows the relation of the number of outliers detected by the algorithm with the EDScore.

values in the range of 0.01 to 0.5. Decreasing ν causes over-fitting while increasing ν to a very high value causes underfitting. The highest AUC of 0.982 is observed for $\nu = 0.1$.

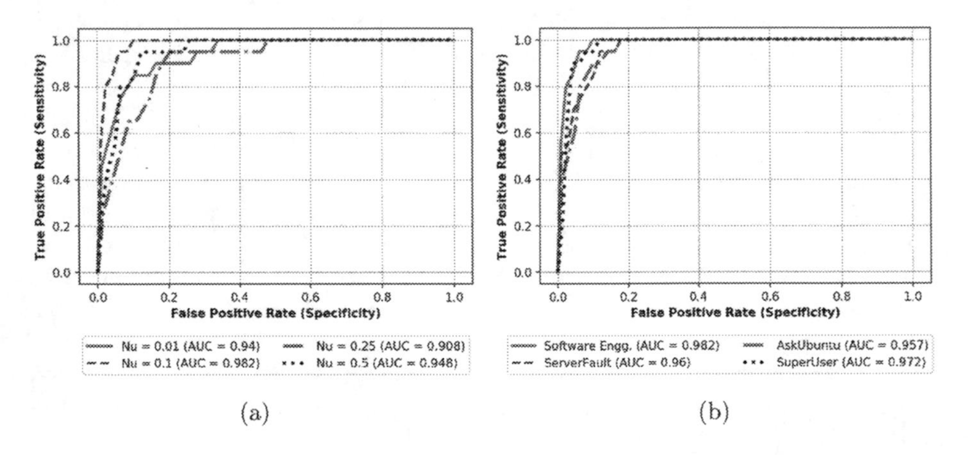

Fig. 4. (a) shows the ROC curves for different values of ν and (b) shows the ROC curves for different datasets present in the Stack Exchange network.

To evaluate the robustness of the algorithm across multiple CQA sites, we consider three other datasets from the Stack Exchange network, namely Ask Ubuntu, Server Fault and Super User. Retaining the optimum parameters for EDScore = 1.5% and $\nu = 0.1$, we repeat the above process for the remaining three datasets. It is observed a high AUC in the range of 0.957 to 0.972 is obtained with all of them.

A slight shift in the AUC scores can be explained as follows: Primarily, the tag network changes from across datasets. This is due to the change in the tags and the associations among them. Table 2 provides a detailed information of these datasets. The number of tags is highest for Super User at 5,267 and lowest for Software Engineering at 1,640. Thus, with a change in the tag network, the optimum value of EDScore suffers a mild shift. This EDScore can be viewed as a hyperparameter in our algorithm that fine tunes the community vector used to represent the tagging, However, this change is only moderate and results indicate that our algorithm still outperforms a naive approach or feature reduced approaches like PCA or t-SNE by wide margins. It is thus safe to assume that this algorithm performs aptly in a wide variety of data.

6 Conclusion

In this paper, we present a novel collaborative approach to detect anomalous tag combinations in CQA sites. We demonstrate that by creating an appropriate folksonomy graph, it is possible to predict whether a certain set of tags is likely to co-occur with a high accuracy. To validate the effectiveness of the algorithm, we compare its results with standard dimensionality reduction techniques, such as PCA and manifold learning techniques like t-SNE. These standard methods manage to produce an AUC in the range of 0.363 to 0.795. We showed that due to sparsity of the tag data and lack of utilization of any tag association information, such methods perform poorly in generating an effective low dimensional representation of the tag network. However, when the underlying tag network is leveraged better feature representations can be obtained. By doing so, we manage to achieve an AUC of 0.982. To be certain of the effectiveness of this technique, we apply the algorithm on a total of four datasets and achieve a consistently high AUC in the range of 0.957 to 0.982. This ensures that the algorithm can effectively deduce anomalous tag combinations in a wide variety of CQA sites.

References

1. Asaduzzaman, M., Mashiyat, A.S., Roy, C.K., Schneider, K.A.: Answering questions about unanswered questions of stack overflow. In: Proceedings of the 10th Working Conference on Mining Software Repositories, pp. 97–100 (2013)
2. Bazelli, B., Hindle, A., Stroulia, E.: On the personality traits of stackoverflow users. In: Proceedings of the 2013 IEEE International Conference on Software Maintenance, pp. 460–463 (2013)
3. Bhat, V., Gokhale, A., Jadhav, R., Pudipeddi, J., Akoglu, L.: Min(e)d your tags: analysis of question response time in stackoverflow. In: Proceedings of the 2014 IEEE/ACM International Conference on Advances in Social Networks Analysis and Mining, pp. 328–335 (2014)
4. Jäschke, R., Marinho, L., Hotho, A., Schmidt-Thieme, L., Stumme, G.: Tag recommendations in folksonomies. In: Kok, J.N., Koronacki, J., Lopez de Mantaras, R., Matwin, S., Mladenič, D., Skowron, A. (eds.) PKDD 2007. LNCS (LNAI), vol. 4702, pp. 506–514. Springer, Heidelberg (2007). https://doi.org/10.1007/978-3-540-74976-9_52

5. Ju, S., Hwang, K.B.: A weighting scheme for tag recommendation in social bookmarking systems. In: Proceedings of the 2009th International Conference on ECML PKDD Discovery Challenge, vol. 497, pp. 109–118 (2009)
6. Krestel, R., Fankhauser, P., Nejdl, W.: Latent Dirichlet allocation for tag recommendation. In: Proceedings of the Third ACM Conference on Recommender Systems, pp. 61–68 (2009)
7. Lipczak, M., Milios, E.E.: Learning in efficient tag recommendation. In: RecSys (2010)
8. der Maaten, L.V.: Learning a parametric embedding by preserving local structure. In: AISTATS (2009)
9. Musto, C., Narducci, F., De Gemmis, M., Lops, P., Semeraro, G.: Star: a social tag recommender system. In: Proceedings of the 2009th International Conference on ECML PKDD Discovery Challenge, vol. 497, pp. 215–227 (2009)
10. Ponzanelli, L., Mocci, A., Bacchelli, A., Lanza, M., Fullerton, D.: Improving low quality stack overflow post detection. In: Proceedings of the 2014 IEEE International Conference on Software Maintenance and Evolution, pp. 541–544 (2014)
11. Rendle, S., Balby Marinho, L., Nanopoulos, A., Schmidt-Thieme, L.: Learning optimal ranking with tensor factorization for tag recommendation. In: Proceedings of the 15th ACM SIGKDD International Conference on Knowledge Discovery and Data Mining, pp. 727–736 (2009)
12. Rosvall, M., Axelsson, D., Bergstrom, C.T.: The map equation. Eur. Phys. J. Special Topics **178**(1), 13–23 (2009)
13. Rosvall, M., Bergstrom, C.T.: An information-theoretic framework for resolving community structure in complex networks. Proc. Nat. Acad. Sci. **104**(18), 7327–7331 (2007)
14. Sillito, J., Maurer, F., Nasehi, S.M., Burns, C.: What makes a good code example?: a study of programming q&a in stackoverflow. In: Proceedings of the 2012 IEEE International Conference on Software Maintenance (ICSM), pp. 25–34 (2012)
15. Slag, R., de Waard, M., Bacchelli, A.: One-day flies on stackoverflow: why the vast majority of stackoverflow users only posts once. In: Proceedings of the 12th Working Conference on Mining Software Repositories, pp. 458–461 (2015)
16. Symeonidis, P., Nanopoulos, A., Manolopoulos, Y.: Tag recommendations based on tensor dimensionality reduction. In: Proceedings of the 2008 ACM Conference on Recommender Systems, pp. 43–50 (2008)
17. Tatu, M., Srikanth, M., D'Silva, T.: RSDC 2008 : tag recommendations using bookmark content (2008)
18. Van Der Maaten, L., Postma, E., Van den Herik, J.: Dimensionality reduction: a comparative review. J. Mach. Learn. Res. **10**, 66–71 (2009)
19. Van der Maaten, L., Hinton, G.: Visualizing high-dimensional data using t-sne. J. Mach. Learn. Res. **9**, 2579–2605 (2008)
20. Yang, J., Tao, K., Bozzon, A., Houben, G.-J.: Sparrows and owls: characterisation of expert behaviour in stackoverflow. In: Dimitrova, V., Kuflik, T., Chin, D., Ricci, F., Dolog, P., Houben, G.-J. (eds.) UMAP 2014. LNCS, vol. 8538, pp. 266–277. Springer, Cham (2014). https://doi.org/10.1007/978-3-319-08786-3_23
21. Yang, Z., Algesheimer, R., Tessone, C.J.: A comparative analysis of community detection algorithms on artificial networks. Sci. Rep. **6** (2016)

Fast Maximal Clique Enumeration for Real-World Graphs

Yinuo Li[1(✉)], Zhiyuan Shao[1(✉)], Dongxiao Yu[2], Xiaofei Liao[1], and Hai Jin[1]

[1] Services Computing Technology and System Lab, Cluster and Grid Computing Lab, School of Computer Science and Technology, Huazhong University of Science and Technology, Wuhan, China
{liyinuo,zyshao,xfliao,hjin}@hust.edu.cn

[2] Institute of Intelligent Computing, School of Computer Science and Technology, Shandong University, Qingdao, China
dxyu@sdu.edu.cn

Abstract. *Maximal Clique Enumeration* (MCE) is one of the most fundamental problems in graph theory, and it has extensive applications in graph data analysis. The state-of-art approach (called as $MCE_{degeneracy}$ in this paper) that solves MCE problem in real-world graphs first computes the degeneracy ordering of the vertices in a given graph, and then for each vertex, conducts the BK_{pivot} algorithm in its neighborhood (called as *degeneracy neighborhood* in this paper). In real-world graphs, the process of degeneracy ordering produces a large number of dense degeneracy neighborhoods. But, the BK_{pivot} algorithm, with its down-to-top nature, adds just one vertex into the result set at each level of recursive calls, and cannot efficiently solve the MCE problem in these dense degeneracy neighborhoods.

In this paper, we propose a new MCE algorithm, called as BK_{rcd}, to improve the efficiency of MCE in a dense degeneracy neighborhood by recursively conducting core decomposition in it. Contrary to BK_{pivot}, BK_{rcd} is a top-to-down approach, that repeatedly chooses and "removes" the vertex with the smallest degree until a clique is reached. We further integrate BK_{rcd} into $MCE_{degeneracy}$ to form a hybrid approach named as $MCE^{hybrid}_{degeneracy}$, that chooses BK_{rcd} or BK_{pivot} adaptively according to the structural properties of the degeneracy neighborhoods. Experimental results conducted in real-world graphs show that $MCE^{hybrid}_{degeneracy}$ achieves high overall performance improvements on the graphs. For example, $MCE^{hybrid}_{degeneracy}$ achieves 1.34× to 2.97× speedups over $MCE_{degeneracy}$ in web graphs taken in our experiments.

1 Introduction

Clique is one of the fundamental structures in an undirected graph that consists of vertices and edges connecting the vertices. In a clique, each vertex is connected to every other vertex via an edge, and a clique is *maximal* if it is not contained by a bigger one. As enumerating all *maximal cliques* (MCs) in a graph facilitates

G. Li et al. (Eds.): DASFAA 2019, LNCS 11446, pp. 641–658, 2019.
https://doi.org/10.1007/978-3-030-18576-3_38

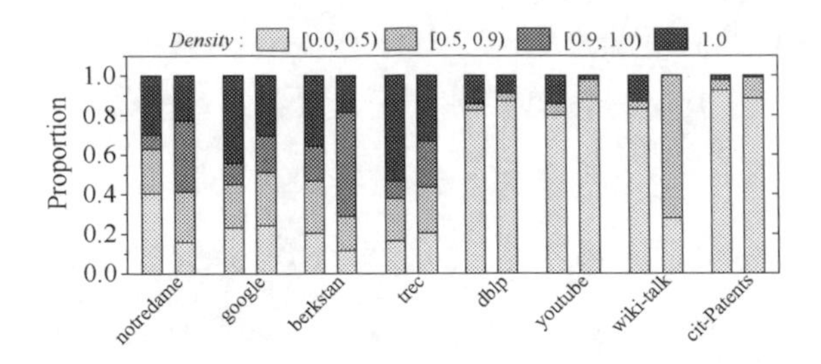

Fig. 1. Density distribution (left columns) of the degeneracy neighborhoods and the corresponding portion of execution times (right columns) of BK_{pivot}

understanding and processing the graph, *Maximal Clique Enumeration* (MCE) problem has been widely studied and used in applications, such as social network analysis [2], community detection [16], computational biology [1]. MCE is also the basis for solving other problems in graph theory, including *Maximal Independent Set* (MIS) [20], graph coloring [3], and many others.

The state-of-art approach [11] (called as $MCE_{degeneracy}$ in this paper) of solving the MCE problem in a given real-world graph first computes the *degeneracy ordering* of the vertices. The degeneracy ordering of a vertex in a graph G is the order that the vertex is removed during a process that repeatedly removes the vertices with the smallest degree from the graph till the graph is empty (i.e., core decomposition [5]). After computing the degeneracy ordering, $MCE_{degeneracy}$ then divide-and-conquers the MCE problem in G into many small MCE problems that are conducted in each vertex's neighborhood by using BK_{pivot} [19]. Figure 1 studies the density distributions of the degeneracy neighborhoods, as well as the corresponding execution times of BK_{pivot} in them when conducting $MCE_{degeneracy}$ in the real-world graphs of Table 2.

From Fig. 1, we can observe that although the original graph is sparse (see the sixth column of Table 2), their degeneracy neighborhoods are rather dense: for the first four web graphs, the densities of about 60% degeneracy neighborhoods are above 0.5, while for social graphs, the densities of about 20% degeneracy neighborhoods are above 0.5. The reason for this phenomenon, as will be discussed in Sect. 6, is that a large portion of these degeneracy neighborhoods are adjacent clique communities that widely exist in real-world graphs [16], and they are dense when these adjacent cliques share many of their vertices in common.

From Fig. 1, we can also observe that the times paid on MCE (using BK_{pivot}) in these degeneracy neighborhoods are approximately proportional to their density distributions, even for the dense degeneracy neighborhoods. This observation is somewhat counter-intuitive: for a dense degeneracy neighborhood (consider a complete graph, or a complete graph with only one edge missing), enumerating its MCs should incur less overhead. Considering a complete degeneracy neighborhood containing k vertices, it only needs a test that consumes $O(k)$ time (by scanning the vertices and comparing their degrees) to discover the MC. Nevertheless, BK_{pivot} needs k recursive calls, each of which chooses the pivot vertex by scanning its adjacency lists and adds the pivot into the result set. This process consumes approximately $O(k^2)$ time in total. Our insight to this inefficiency problem of BK_{pivot} when conducting MCE in dense sub-graphs is that: *BK_{pivot} is, in essence, a down-to-top approach that starts from individual vertices, adds only one vertex into the result set with one independent recursive call, and discovers a MC till all its vertices are added into the result set.*

To improve the efficiency of MCE in the dense degeneracy neighborhoods, in this paper, we propose a new MCE algorithm named as BK_{rcd}, whose basic idea is to conduct core decomposition, that repeatedly chooses and "removes" the vertices with the smallest degrees, in a given (sub-)graph till a clique is reached, and recursively conducts core decomposition in the neighborhoods of the vertices "removed" during the process to prevent from missing MCs. *Different from BK_{pivot}, BK_{rcd} is a top-to-down approach that starts from a (sub-)graph and stops enumeration when the (sub-)graph becomes complete.*

Further, by discussing the efficiencies of BK_{pivot} and BK_{rcd} when solving the MCE problem in a special type graph (named as s,k-graph) that consists of multiple adjacent cliques, and extending the discussion to general graphs by the notion of *divergence*, we give the criteria that a graph should satisfy to benefit from choosing BK_{rcd} than BK_{pivot} to enumerate its MCs. With these criteria, we integrate BK_{rcd} into the framework of $MCE_{degeneracy}$ to form a new hybrid approach named as $MCE^{hybrid}_{degeneracy}$, that chooses BK_{rcd} or BK_{pivot} adaptively by telling if the degeneracy neighborhood to be processed satisfies the criteria.

This paper makes the following contributions:

(1) proposes a new MCE algorithm named as BK_{rcd} that solves the MCE problem in a graph by recursive core decompositions.
(2) gives the criteria that should be satisfied by a (sub-)graph to achieve better performance with BK_{rcd} during MCE than with BK_{pivot}.
(3) integrates BK_{rcd} into the framework of the state-of-art approach, to build a hybrid approach named as $MCE^{hybrid}_{degeneracy}$.
(4) extensively evaluates the performances of $MCE^{hybrid}_{degeneracy}$, and shows that it can improve the efficiency of MCE especially for real-world web graphs.

The rest of this paper is organized as follows: Sect. 2 briefly surveys the related works of this paper. Section 3 lists the notations and gives the background knowledge of this paper. Section 4 presents the algorithm of BK_{rcd}. Section 5 designs the hybrid approach. Section 6 empirically evaluates the result hybrid approach, and Sect. 7 concludes the paper.

2 Related Works

Many algorithms have been proposed to solve and accelerate the solution to the MCE problem. BK algorithm is the most widely-used MCE algorithm proposed in [6], and BK_{pivot} [19] improves BK by avoiding the useless enumerations (Sect. 3 will give more details of these two algorithms). Like BK_{pivot}, BK_{rcd}, which is the MCE algorithm proposed in this paper, can also be considered as a variant of the BK algorithm with its recursive nature. However, different from BK_{pivot}, BK_{rcd} is a top-to-down approach that uses core decomposition to repeatedly remove vertices with the least neighbors in the candidate set till a clique is reached.

With the recent trend of solving the MCE problem in practical settings [7,8,10,11], $MCE_{degeneracy}$ [11] leverages the fact that real-world graphs usually have small degeneracy values, and is the state-of-art approach on solving MCE in real-world graphs (Sect. 3 will also give more details of $MCE_{degeneracy}$). Based on $MCE_{degeneracy}$, our hybrid approach (i.e., $MCE^{hybrid}_{degeneracy}$), proposed in this paper, further improves the solution of MCE in real-world graphs by adaptively choosing BK_{pivot} or BK_{rcd} according to the structural property of the degeneracy neighborhoods that a real-world graph has.

Besides, there are many other approaches proposed to optimize the MCE solution in other special-case graphs. For example, [9] considers the graphs with polynomial amount of MCs (e.g., planar graphs), [18] considers the dynamic graphs, and [15] solves the problem in uncertain graphs.

3 Preliminaries

3.1 Notations

In order to simplify our discussions in this paper, we consider the *simple* undirected graph $G = <V, E>$, where V is the set of vertices and E is the set of edges. That is, we do not consider graphs with self-loops or multi-edges. n is the number of vertices in G, and m is the number of edges.

For a vertex $v \in V$, we use $\Gamma_G(v)$ to denote the set of neighboring vertices of v in G and use $\Gamma_G(S)$ to denote the set of common neighbors for vertices in S in G. $N_G(v)$ denotes the neighborhood of v, which is the induced sub-graph of $\Gamma_G(v)$ with respect to G. With the degeneracy ordering, the set of neighboring vertices of v, i.e., $\Gamma_G(v)$ can be further divided into two parts: $\Gamma_G^+(v)$, which contains vertices with higher (later) degeneracy orders than v, and $\Gamma_G^-(v)$, which contains vertices with lower (earlier) degeneracy orders than v. In this paper, we call the sub-graph, $N_G^+(v)$, which is induced by $\Gamma_G^+(v)$ with respect to $N_G(v)$, as the *degeneracy neighborhood*. In the following discussions, if G is obvious in the context, we will omit G for brevity.

3.2 Background

It is NP-hard to solve the MCE problem in G. Since a graph with n vertices can have up to $3^{n/3}$ MCs [14], and consequently it might be impossible to find a polynomial solution to this problem. BK is a recursive algorithm with three input parameters: R stands for the *result* set that stores the vertices belonging to a MC, P denotes the set for *candidate* vertices that are going to be considered during MCE, and X is the set of vertices that should be *excluded* from enumeration as they have been considered. BK_{pivot}, which is listed in Algorithm 1, improves the efficiency of MCE by avoiding the useless enumerations of the BK algorithm with the following method: choose a *pivot* vertex v from G, the MCs in G can be divided into two types: those containing v, and those not containing v. The MCs containing v will be found by recursively conducting enumeration on $N(v)$, and those not containing v can be discovered by conducting enumeration on the neighborhoods of v's non-neighbors.

Algorithm 1. $BK_{pivot}(R, P, X)$

```
1 if P⋃X is empty then
2 └ output R as a maximal clique
3 choose a pivot v from P⋃X to maximize Γ(v)⋂P
4 foreach u ∈ {v and v's non-neighbors} do
5 │ BK_pivot(R⋃{u}, P⋂Γ(u), X⋂Γ(u))
6 │ P == P − {u}
7 └ X == X⋃{u}
```

$MCE_{degeneracy}$, which is the state-of-art solution on solving the MCE problem in real-world graphs, first computes the degeneracy ordering of the vertices, and then conducts BK_{pivot} in the neighborhood of each vertex. In order to prevent duplicated results (i.e., output the same MC more than once), when conducting BK_{pivot} in the neighborhood of v, $\Gamma^{+}(v)$ is taken as the candidate set, and $\Gamma^{-}(v)$ is taken as the exclusive set, according to the degeneracy orderings. Algorithm 2 lists its pseudo code.

Algorithm 2. $MCE_{degeneracy}(G)$

```
1 Compute the degeneracy orders of all vertices in V
2 foreach v ∈ V do
3 └ BK_pivot(v, Γ⁺(v), Γ⁻(v))
```

4 MCE by Recursive Core Decomposition

The basic idea of BK_{rcd} is to consider a clique as an entity that does not need further enumeration. For an input graph, it repetitively "removes" the vertices that have non-neighbors in the candidate set, until the remaining graph becomes complete (and thus a clique, we call it as the *remaining clique* in this paper). After that, it outputs the remaining clique as a MC if it passes the maximality check. To avoid missing the MCs that contain the vertices "removed" during this process, BK_{rcd} recursively conducts itself in the neighborhoods of the "removed" vertices. Algorithm 3 lists the pseudo-code.

Algorithm 3. $BK_{rcd}(R,P,X)$

1 **if** $P \bigcup X$ *is empty* **then**
2 output R as a maximal clique
3 **while** P *is not clique* **do**
4 choose $v \in P$ which has the most non-neighbors
5 $BK_{rcd}(R \bigcup \{v\},\ P \bigcap \Gamma(v),\ X \bigcap \Gamma(v))$
6 $P = P \setminus \{v\}$
7 $X = X \bigcup \{v\}$
8 **if** $P \neq \emptyset$ *and* $\Gamma(P) \bigcap X == \emptyset$ **then**
9 output $R \bigcup P$ as a maximal clique

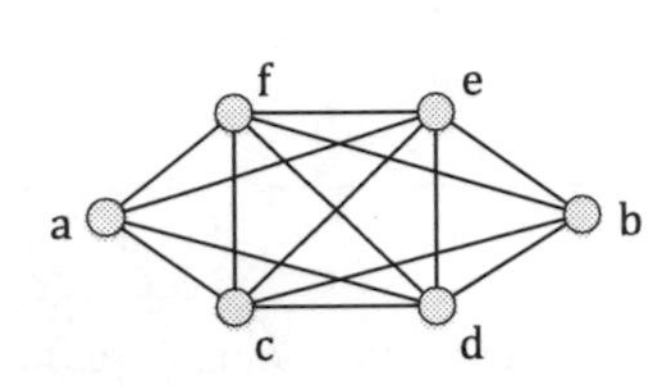

Fig. 2. An example graph

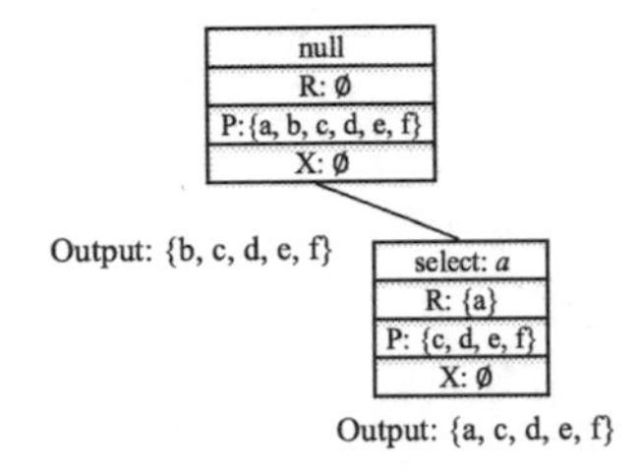

Fig. 3. The search tree of conducting BK_{rcd} in the left example graph

Algorithm 3 first checks if both P and X are empty, and outputs R as a MC if they are (Line 1–2). The algorithm then checks whether the candidate set P is a clique (Line 3): if P is a clique, the algorithm will stop enumeration, and try to output it at Line 8 to 9. If P is not a clique, the algorithm chooses the vertex v that has the largest number of non-neighbors in P, removes v from P (Line 6), and (recursively) conducts the algorithm itself on v's neighborhood (Line 5) before adding it into exclusive set X (Line 7). The algorithm repeats this process until P becomes a clique (i.e., the remaining clique). After that, a check is conducted at Line 8 to tell if the remaining clique is maximal. The criterion of this check is to tell whether X contains any vertex that is a common neighbor to all vertices in P: $R \bigcup P$ is maximal if X does not contain such vertex, and it is not if otherwise. The reason for this check is to prevent the case that some vertices of the MC containing P are "removed" and then added to X (e.g., due to ties) during the process. If such case happens, the MC will be discovered and outputted by conducting Algorithm 3 in the neighborhoods of the removed vertex, not P.

Consider conducting Algorithm 3 in the example graph shown in Fig. 2: in the beginning, the graph is not a clique, and vertex a and b have the same number of non-neighbors. Assume a wins the tie, and is "removed" by the algorithm. After that, P contains 5 vertices: b, c, d, e, and f, and it is maximal, i.e., 5-MC of $\{b, c, d, e, f\}$, as X (containing only a) does not have any common neighbors to the vertices in P. The other 5-MC, i.e., $\{a, c, d, e, f\}$, is discovered by conducting the algorithm in the neighborhood of a. The execution path (search tree) of this process is illustrated in Fig. 3. It is easy to infer that comparing with BK_{pivot} (Algorithm 1), the search tree of BK_{rcd} is much smaller in this example graph, which means that BK_{rcd} spends much less recursive calls, and thus is more efficient, on solving the MCE problem in this example graph.

5 Hybrid Approach

To compare BK_{pivot} and BK_{rcd}, in this section, we consider a special graph structure, the s, k-graph, and analyze the efficiencies of these two algorithms when solving the MCE problem in such special structure. With these analyses, we have the criteria of choosing these algorithms for higher performance, and can further integrate BK_{rcd} into the state-of-art approach of $MCE_{degeneracy}$.

5.1 The s, k-Graph

The definition of s, k-**graph** is given as follow.

Definition 1. (s, k-***graph***) *A s, k-graph, denoted as $G_{s,k}$ ($G_{s,k} = <V_{s,k}, E_{s,k}>$), consists of k adjacent $(s+1)$-cliques, which share s vertices in common.*

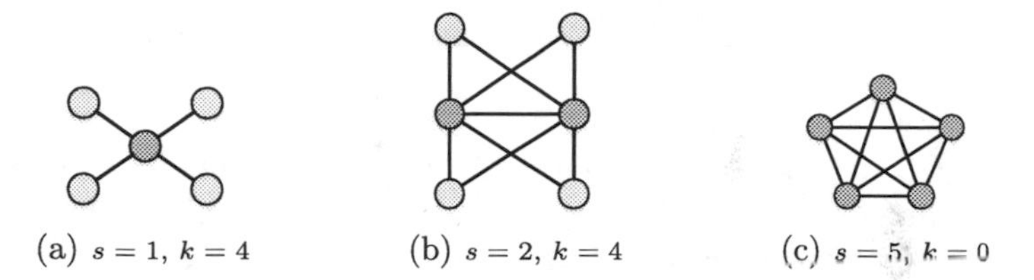

Fig. 4. Three example s, k-graphs (light-yellow vertices are in $\mathcal{K}$ set and dark-grey vertices are in $\mathcal{S}$ set) (Color figure online)

The reason for choosing such graph structure is that as shown in [16], the pattern of adjacent clique communities, that consist of multiple cliques that share a set of vertices in common, widely exists in the real-world graphs. Figure 4 gives three example s, k-graphs with different s and k values.

We further divide the vertex set, i.e., $V_{s,k}$, of a s, k-graph into two sets: the set (denoted as $\mathcal{S}$) that contains the former s vertices which are connecting to all other $(s + k - 1)$ vertices, and the set (denoted as $\mathcal{K}$) containing the latter k vertices which are only connecting to all vertices in $\mathcal{S}$. It is easy to infer that the degree of any vertex taken from $\mathcal{S}$ is $s + k - 1$, and the degree of any vertex taken from $\mathcal{K}$ is s. An independent s, k-graph has k $(s + 1)$-MCs.

5.2 MCE in s, k-Graph

When conducting BK_{pivot} and BK_{rcd} in a s, k-graph, the shapes of the search trees of these two algorithms can be illustrated as in Fig. 5(a) and (b), respectively.

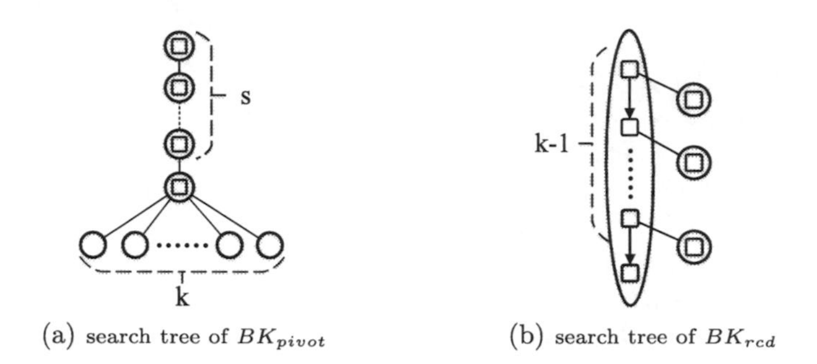

Fig. 5. The shapes of the search trees when conducting BK_{pivot} and BK_{rcd} in a s, k-graph (square denotes vertex choosing, ellipse or circle denotes recursive call)

At the BK_{pivot} side, it first chooses s vertices which have the largest degrees from $\mathcal{S}$ one after another. These recursive calls form first s trunk nodes of the search tree shown in Fig. 5(a), and the last trunk node of the search tree chooses the pivot from $\mathcal{K}$ set. After that, each leaf node of the search tree conducts a recursive call to add a vertex from the $\mathcal{K}$ set to the result set, and outputs a $(s+1)$-MC.

On the other hand, according to Algorithm 3, there will be $k-1$ vertices from the $\mathcal{K}$ set being chosen and "removed" at the beginning, and leaves behind a $(s+1)$-MC as the remaining clique. Each recursive call conducted in the neighborhood of a chosen vertex will output a $(s+1)$-MC after the clique test, whose essence is vertex choosing.

From the above discussion, we find that although these two algorithms have different mechanisms, the overheads incurred by them during MCE are similar, and can broadly classified into two parts: the overhead of the vertex choosing and that of the recursive calls.

In order to choose a vertex with the most number of neighbors (in BK_{pivot}), or a vertex with the most number of non-neighbors (in BK_{rcd}), in a candidate set P, the common practice is to browse the adjacency lists of all vertices in P ($P \subseteq V_{s,k}$ for a s, k-graph), and at each adjacency list, scan each neighboring vertex in the list to judge whether it belongs to P. Assume the time complexity of a judge operation is $O(1)$, the time complexity of vertex choosing can be approximated to the summation of the lengths of the adjacency lists that are scanned during the process. The total cost of vertex choosing in BK_{pivot} (denoted as C^{vc}_{pivot}), when conducting in a s, k-graph, can be approximated by the following equation:

$$C_{pivot}^{vc} = \sum_{i=0}^{s-1}(s-i)\cdot(s+k-1) + k\cdot s^2 + k\cdot s \tag{1}$$

where at the right of the equal sign, the first and second items are the costs of vertex choosing that happens in the vertices in $\mathcal{S}$ and $\mathcal{K}$ respectively (vertices in $\mathcal{K}$ participate each choosing, while fewer and fewer vertices in $\mathcal{S}$ participate the choosing with the algorithm's progress) in the first s trunk nodes, and the third item is the cost of vertex choosing of the last trunk node of the search tree shown in Fig. 5(a).

Using the same method, the total cost of vertex choosing in BK_{rcv} (denoted as C_{rcd}^{vc}), can be approximated by the following equation:

$$C_{rcd}^{vc} = k\cdot s\cdot(s+k-1) + \sum_{i=0}^{k-1}(k-i)\cdot s + s\cdot(s+k-1)\cdot(k-1) \tag{2}$$

Regarding to a recursive call, as its heaviest operation is the set intersection (i.e., $P\bigcap\Gamma(v)$, where v denotes a selected vertex) that happens when preparing the candidate sets, and generally, it is implemented by scanning the adjacency list of $\Gamma(v)$ to identify the neighbors that belong to P, we use the length of $\Gamma(v)$ to approximate the cost of a recursive call. By this way, the total cost of the recursive calls in BK_{pivot} (denoted as C_{pivot}^{rc}), can be approximated as:

$$C_{pivot}^{rc} = (s+k-1)\cdot s + k\cdot s \tag{3}$$

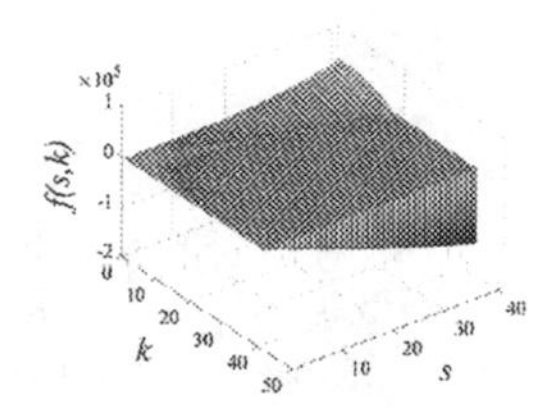

Fig. 6. The benefit function $f(s,k)$ when both s and k vary (the values of $f(s,k)$ are represented as pillars)

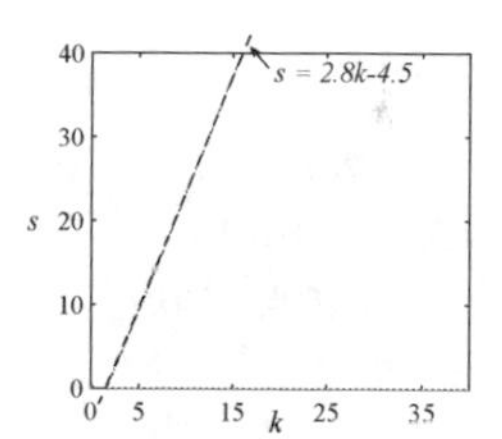

Fig. 7. The intersection of $f(s,k)$ to the sk-plane

Similarly, the total cost of the recursive calls in BK_{rcd} (denoted as C_{rcd}^{rc}), when conducting in a s,k-graph, can be approximated as:

$$C_{rcd}^{rc} = (k-1)\cdot s \tag{4}$$

5.3 Criterion on Choosing BK_{rcd}

In order to pick up the situations that BK_{rcd} harvests higher efficiency, we combine Eqs. 1, 2, 3, and 4 to define a *benefit function* below that approximately depicts the performance gains of choosing BK_{rcd} over BK_{pivot} when solving the MCE problem in a given s, k-graph.

$$\begin{aligned} f(s,k) =& C_{pivot}^{vc} + C_{pivot}^{rc} - C_{rcd}^{vc} - C_{rcd}^{rc} \\ =& \frac{1}{2}s^3 + (-\frac{k}{2} + 2)s^2 + (-\frac{5}{2}k^2 + 5k - \frac{3}{2})s \end{aligned} \tag{5}$$

We plot $f(s, k)$ when both s and k vary in Fig. 6. Figure 7 plots the curve that $f(s, k)$ intersects the sk-plane. According to the shape of $f(s, k)$ illustrated in Fig. 6, we know that when the point of $(s,\ k)$ falls in the left region of the curve in Fig. 7, we will have $f(s, k) > 0$, which means BK_{rcd} achieves higher efficiency than BK_{pivot}. On the contrary, if the point of $(s,\ k)$ falls in the right region of the curve in Fig. 7, we will have $f(s, k) < 0$, which means the opposite. As large portion of the intersection curve in Fig. 7 overlaps with the linear function of $s = 2.8k - 4.5$, we thus have the following approximate conditional formula:

$$f(s,k) \begin{cases} \geq 0, & \text{if } s \geq 2.8k - 4.5 \\ < 0, & \text{if } s < 2.8k - 4.5 \end{cases} \tag{6}$$

With Formula 6, we have the *concise form* of the criterion on choosing BK_{rcd}: for a s, k-graph, when $s \geq 2.8k - 4.5$, BK_{rcd} has higher efficiency than BK_{pivot} on solving the MCE problem in the graph.

5.4 Divergence

Consider an arbitrary undirected graph $G = <V, E>$, suppose there are s vertices that connect to all other vertices in V, but there may be some edges connecting vertices in remaining k vertices. This kind of graph is not a s, k-graph due to the existence of these edges. We further divide V into two subsets like s, k-graph: the $\mathcal{S}$ set of G including s fully connected vertices and $\mathcal{K}$ set including all other k vertices.

Use $d(v)$ to denote the degree of a vertex v, the *divergence* of G, denoted as $div(G)$, is thus defined to express the quantitative difference between G and the s, k-graph which has the same s and k values, by the following formula:

$$div(G) = \max_{v \in \mathcal{K}_G} (d(v) - s) \tag{7}$$

We use the right superscript to annotate the divergence of a graph, and can thus express the arbitrarily taken undirected graph G as $G_{<s,k>}^{div=div(G)}$. We will call the edges that connect vertices in $\mathcal{K}$ as *divergence edges* in the following discussions. Figure 8 gives three examples of s, k-graphs with divergence.

We now discuss the criterion that should be satisfied by $G_{<s,k>}^{div=div(G)}$, to benefit from choosing BK_{rcd} to solve the MCE problem than BK_{pivot}. Consider the

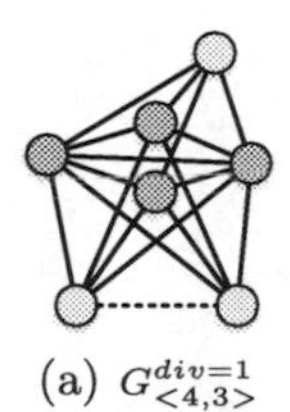
(a) $G^{div=1}_{<4,3>}$

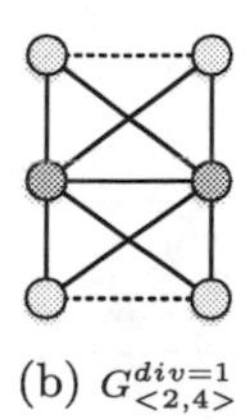
(b) $G^{div=1}_{<2,4>}$

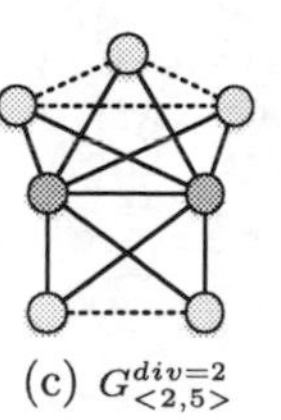
(c) $G^{div=2}_{<2,5>}$

Fig. 8. Example s,k-graphs with divergence (dashed lines denotes the divergence edges)

case of only adding one divergence edge which implies $div(G) = 1$, the introduction of the edge affects the costs of vertex choosing and recursive calls, and following equations lists the variations to these costs (based on Eqs. 1–4):

$$\Delta C^{vc}_{pivot} = 3s + 3 \qquad \Delta C^{rc}_{pivot} = 2$$
$$\Delta C^{vc}_{rcd} = 2(k-1) - [2(k+s-1) \cdot s + s] \qquad \Delta C^{rc}_{rcd} = -s$$

With the introduction of the divergence edge, two of the $(s+1)$-cliques in G merge into one $(s+2)$-clique. The search tree of BK_{pivot} now has $s+1$ trunk nodes, 1 branch node and $k-1$ leaf nodes, as two leaf nodes in Fig. 5(a) become a branch node and a leaf node respectively. On the other hand, the search tree of BK_{rcd} will have only $k-2$ leaf nodes with the introduction of the divergence edge. With these variations in the costs of two MCE algorithms, we will get a new benefit function. By using the same method used in Subsect. 5.3, we have the concise form of the criterion on choosing BK_{rcd} for a s,k-graph with only one divergence edge as the linear function of $s = 2.8k - 8$.

Note that for a $G^{div=1}_{<s,k>}$ graph, it may have multiple divergence edges like Fig. 8(b). Adding more divergence edges into $G^{div=1}_{<s,k>}$ will not change the size of remaining clique in BK_{rcd}. Therefore, the new benefit function always holds for the $G^{div=1}_{<s,k>}$ graph with more divergence edges if we continue above discussions.

By using the above methodology, we have the criterion on benefiting from choosing BK_{rcd} over BK_{pivot} for $G^{div=2}_{<s,k>}$ graphs as linear function $s \geq 2.8k - 11$. For $G^{div\geq 3}_{<s,k>}$ graphs, we continue to use $s \geq 2.8k - 11$ as the criterion on choosing BK_{rcd} over BK_{pivot} for two reasons: first, from the above discussions on the divergences of s,k-graphs, we find that the criterion relaxes with higher divergences introduced to a s,k-graph, at least from $div = 0$ to $div = 2$. Second, in our experiments conducted in Sect. 6, we find that in a real-world graph, there is only a small portion of degeneracy neighborhoods that are the s,k-graphs with much higher divergences (i.e., $div \geq 3$) and satisfy the criterion of the $G^{div=2}_{<s,k>}$ graphs at the same time. We thus use the criterion of the $G^{div=2}_{<s,k>}$ graphs to the graphs with larger divergences, to prevent from complicating our discussions.

We summarize the criteria of choosing BK_{rcd} over BK_{pivot} in Table 1 and thus can combine the merits of both these two algorithms to solve the MCE problem. The idea is to consider the degeneracy neighborhood of each vertex of the input real-world graph as a s,k-graph with divergence, and chooses MCE

Table 1. The criteria that a s, k-graph with divergence should satisfy to benefit from choosing BK_{rcd} over BK_{pivot}

Divergence	Criterion	Notes
$div(G) = 0$	$s \geq 2.8k - 4.5$	s, k-graphs by Definition 1
$div(G) = 1$	$s \geq 2.8k - 8$	$k \geq 3$
$div(G) = 2$	$s \geq 2.8k - 11$	$k \geq 4$
$div(G) \geq 3$	$s \geq 2.8k - 11$	Approximate and empirical criterion

Algorithm 4. $MCE^{hybrid}_{degeneracy}(G(V, E))$

```
1 Compute the degeneracy orders of vertices in V
2 foreach v ∈ V do
3     Consider N+(v) as a s,k-graph with divergence, extract div, s and k.
4     if div, s and k satisfy criteria listed in Table 1 then
5         BK_rcd({v}, Γ+(v), Γ−(v))
6     else
7         BK_pivot({v}, Γ+(v), Γ−(v))
```

algorithms with higher efficiency according to the criteria listed in Table 1. Based on this idea, we build a "hybrid" version (named as $MCE^{hybrid}_{degeneracy}$) of the original $MCE_{degeneracy}$, and list its pseudo-code in Algorithm 4.

Different from $MCE_{degeneracy}$ (Algorithm 2), $MCE^{hybrid}_{degeneracy}$ chooses BK_{rcd} if the structural parameters (i.e., div, s and k) of an input degeneracy neighborhood satisfy the criteria listed in Table 1, and chooses BK_{pivot} if otherwise.

6 Empirical Evaluation

In this section, we evaluate the performance of our proposed hybrid MCE approach on the chosen real-world graphs.

6.1 Experiment Setup

Experiment Environment. We conducted all experiments in this section on a server configured with one Intel Xeon E5-2670 CPU (8 cores, 20 MB shared cache, running at 2.60 GHz), and 220 GB DDR4 memory. The operating system is Ubuntu 16.04 Linux.

Graph Data-Sets. We use the eight graph data-sets listed in Table 2 (left part) to conduct MCE algorithms. notredame, google, berkstan, trec are four real-world web crawler graphs, where vertices represent the web-pages and edges represent the hyper-links connecting the web-pages. Four social networks dblp, youtube,

Table 2. Chosen real-world graph datasets (left part, *deg.* stands for the degeneracy of a graph) and the overall execution times of MCE approaches (right part, digits after the plus sign in the brackets of the "Original approach" column are the execution times paid on computing the degeneracy orderings of the vertices)

Dataset	DataSet properties					Running time (in seconds)		
	Type	n	m	*deg.*	Density	Original approach	Naive approach	Hybrid approach
notredame	Web	325, 729	1, 090, 108	155	2.05e−5	0.54 (+0.32)	0.47	0.34
google		875, 713	4, 322, 051	44	1.13e−5	3.06 (+1.80)	2.77	2.28
berkstan		685, 231	6, 649, 470	201	2.83e−5	8.59 (+1.56)	7.78	2.89
trec		1, 601, 787	6, 679, 248	140	5.20e−6	3.42 (+2.19)	3.04	1.86
dblp	Social	317, 080	1, 049, 866	113	2.09e−5	0.15 (+0.25)	0.15	0.14
youtube		1, 134, 890	2, 987, 624	51	4.64e−6	3.68 (+1.47)	3.77	3.32
wiki-talk		2, 394, 385	4, 659, 565	131	1.63e−6	226.86 (+2.63)	234.15	223.73
cit-Patents		3, 774, 768	16, 518, 947	64	2.32e−6	20.32 (+12.05)	20.41	19.73

wiki-talk, cit-Patents describe the social network or collaboration network. All eight datasets except trec, which could be gotten in [4], are taken from [13].

Evaluated Approaches. We evaluate three approaches listed below:

- *Original approach.* This approach is the state-of-art approach introduced in Subsect. 3.2. It has been introduced in [11] and implemented in [12].
- *Naive approach.* This approach slightly modifies the original approach by adding a complete graph test in BK_{pivot} before it is conducted in a subgraph. With this test, BK_{pivot} quits the enumeration if the input subgraph is complete. The complete test saves the costs of MCE while P is already a clique.
- *Hybrid approach.* This approach uses $MCE^{hybrid}_{degeneracy}$ listed in Algorithm 4 to solve the MCE problem in a given graph. It chooses MCE algorithms (i.e., BK_{rcd} or BK_{pivot}) with higher efficiency according to the structural properties (more specifically, s, k and the divergence) of a degeneracy neighborhood.

All these approaches are implemented based on or by revising the source code of $MCE_{degeneracy}$ that is available at [17]. GNU C++ version 5.4.0 is used to compile and link the source codes with optimization option of -O3.

6.2 Overall Execution Times

Table 2 (right part) reports the overall execution times of MCE when conducting three approaches listed above in the chosen graphs.

From Table 2, we can observe that compared with the original approach, the naive approach improves the performance of MCE in most cases, although the performance improvements are much smaller than our hybrid approach. Occasionally, the naive approach may deteriorate the MCE performance (e.g., in the

Table 3. The statistics of degeneracy neighborhoods ($N^+(v)$) of the chosen graphs (a $N^+(v)$ is "valid" if it has more than one vertex)

Dataset	# of valid $N^+(v)$	# of BK_{rcd} eligible $N^+(v)$			
		$div = 0$	$div = 1$	$div = 2$	$div \geq 3$
notredame	155,250	54,719	4,831	3,943	608
google	686,885	360,944	25,673	19,362	8,397
berkstan	611,599	286,778	13,730	16,358	15,846
trec	1,148,361	721,758	29,461	24,454	7,781
dblp	24,979	3,591	65	17	0
youtube	442,451	65,299	2,588	1,085	6
wiki-talk	621,338	82,099	2,124	726	4
cit-Patents	3,073,519	76,525	9,002	8,435	50

wiki-talk graph). The reason for such phenomenon is that although the added complete graph test in BK_{pivot} reduces the overhead of MCE, when the algorithm finds that the input degeneracy neighborhoods is complete (the reason for the performance improvements), the way of pivot vertex choosing (always chooses the vertex with the most neighbors in the candidate set) in BK_{pivot} prevents it from reaching a "remaining clique" as fast as in BK_{rcd}. In quasi-complete (dense, but not complete) degeneracy neighborhoods, the added complete graph test brings almost no performance benefits while introducing extra overheads (for testing), which counteracts the performance benefits gained from the complete degeneracy neighborhoods, and may even cause worse performance than the original approach.

From Table 2, we can observe that compared with the original approach, our hybrid approach achieves significant performance improvements in the web-graphs. The speedups are 1.59, 1.34, 2.97, and 1.84 respectively for notredame, google, berkstan, and trec. The speedups of our hybrid approach compared to the original approach in the chosen social graphs, however, are small. They are 1.07, 1.11, 1.01, 1.03 respectively for dblp, youtube, wiki-talk, cit-Patents. In order to explain these figures, we need to take deeper looks to the degeneracy neighborhoods of the chosen graphs, since our hybrid approach uses different algorithms to conduct MCE in a degeneracy neighborhood according to its structure.

6.3 The Degeneracy Neighborhoods

We report the statistics of the degeneracy neighborhoods of the chosen graphs in Table 3. From Table 3, we can observe that in web graphs, there are lots of degeneracy neighborhoods that satisfy the criteria listed in Table 1 and are thus eligible to use BK_{rcd} to improve the efficiency during MCE. To facilitate discussion, we call such degeneracy neighborhoods as the BK_{rcd} *eligible degeneracy neighborhoods*. It is easy to calculate that for notredame, google, berkstan,

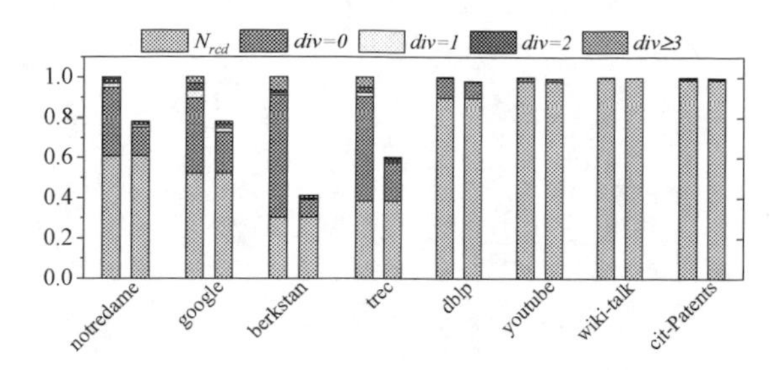

Fig. 9. Breakdown of the execution times paid on MCE in the original approach (left columns) and in our hybrid approach (right columns). (all figures are normalized to the total execution time of the original approach.)

and trec, the portions of the BK_{rcd} eligible degeneracy neighborhoods are about 55.16%, 60.33%, 54.4%, and 68.22% respectively. However, the portions of BK_{rcd} eligible degeneracy neighborhoods are relatively low in our chosen social graphs: they are only about 14.7%, 15.59%, 13.67%, and 3.06% respectively for dblp, youtube, wiki-talk, cit-Patents. Moreover, the majorities of the BK_{rcd} eligible degeneracy neighborhoods in our selected real-world graphs are the s, k-graphs whose divergences equal to 0.

We use the symbol N_{rcd} to denote the set of degeneracy neighborhoods that do *not* satisfy the criteria. Figure 9 reports the execution times of solving the MCE problem in the chosen graphs with the original approach and our hybrid approach. To show the effectiveness of BK_{rcd}, in Fig. 9, the execution times are broken down to illustrate the portion of time spent in various kinds of degeneracy neighborhoods (those of N_{rcd}, and those of BK_{rcd} eligible degeneracy neighborhoods). Note that in the original approach, BK_{pivot} is employed to conduct MCE in the degeneracy neighborhoods, even when they satisfy the criteria listed in Table 1.

From Fig. 9, we can observe that: (1) For the social graphs, their BK_{rcd} eligible degeneracy neighborhoods consume little time even when they are processed by using BK_{pivot} in the original approach (may due to their relatively simple structure). Combined with the fact that they occupy a small portion of all the degeneracy neighborhoods as shown in Table 3, it explains the marginal performance improvements that can be harvested by our hybrid approach. (2) For the web graphs, the majority of performance improvements by choosing BK_{rcd} than BK_{pivot} in the hybrid approach comes from the degeneracy neighborhoods whose divergences equal to zero (i.e., $div = 0$), since they occupy the majority of the BK_{rcd} eligible degeneracy neighborhoods as shown in Table 3.

However, for these BK_{rcd} eligible $div = 0$ degeneracy neighborhoods, the performance improvements vary from one graph to another: the speed-ups are 2.39×, 1.83×, 6.99×, and 2.77× respectively for notredame, google, berkstan, and trec. In order to explain this phenomenon, we choose two graphs (google and berkstan) that lie at two extremes, and plot the size distributions and execution times of their degeneracy neighborhoods in Fig. 10.

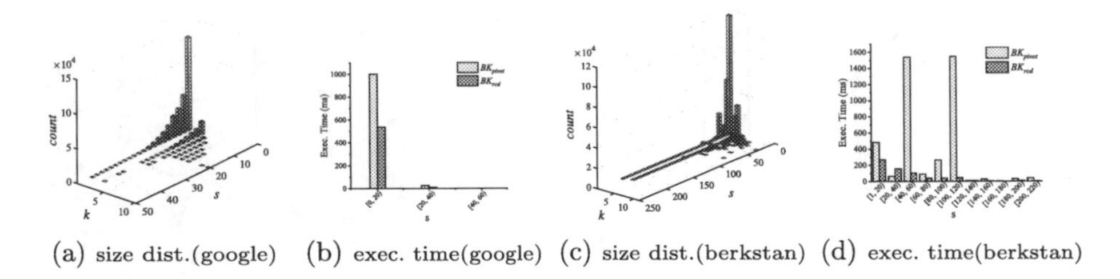

(a) size dist.(google) (b) exec. time(google) (c) size dist.(berkstan) (d) exec. time(berkstan)

Fig. 10. The size distributions and execution times of the BK_{rcd} eligible $div = 0$ degeneracy neighborhoods in google and berkstan

From Fig. 10(a) and (c), we can observe that the BK_{rcd} eligible $div = 0$ degeneracy neighborhoods in google are generally small due to its small degeneracy (i.e., 44, see Table 2), while some of the BK_{rcd} eligible $div = 0$ degeneracy neighborhoods in berkstan are relatively large due to its large degeneracy (i.e., 201). Moreover, most of the $div = 0$ degeneracy neighborhoods that are eligible to use BK_{rcd} are small with small s and k values for both of these two real-world graphs.

From Fig. 10(b), we can observe that the performance improvements by choosing BK_{rcd} than BK_{pivot} in google are limited to the small degeneracy neighborhoods (s in the range of $[1, 20)$), and in these degeneracy neighborhoods, BK_{rcd} spends half of the execution time of BK_{pivot} to solve the MCE problem. This explains the execution times of google in Fig. 9. The situation changes in Fig. 10(d), where large degeneracy neighborhoods (s in the range of $[40, 60)$ and $[100, 120)$) consume the most of the execution times paid on MCE when using BK_{pivot}. For these degeneracy neighborhoods, BK_{rcd} achieves salient performance improvements during enumeration. From Fig. 10(d), we can observe that the performance improvements by choosing BK_{rcd} over BK_{pivot} are higher when the input degeneracy neighborhoods become larger (with larger s values), which validates our approximation method, which uses the lengths of the adjacency lists to approximate the overhead of MCE, and the benefit function illustrated in Fig. 6, as discussed in Sect. 5.

7 Conclusion and Future Works

In this paper, based on the fact that many of the degeneracy neighborhoods in real-world graphs are dense, we propose a new MCE algorithm, named as BK_{rcd}, to improve the efficiency of MCE in these dense sub-graphs. We further integrate BK_{rcd} into the state-of-art approach to form a hybrid approach to improve its overall performance on solving the MCE problem in real-world graphs. By empirical evaluations, we show that our hybrid approach is especially effective in improving the efficiency on solving the problem in real-world web graphs.

Our study conducted in this paper suggests that we can take advantage of the structural properties of an input graph to improve the efficiency of MCE.

We believe that the s, k-graph structure discussed in this paper is *not* the only viable way on improving the efficiency of MCE, and place the analysis on other structures and further discussions to them in our future works.

Acknowledgements. This paper is supported by National Key Research and Development Program of China under grant No. 2018YFB1003500, National Natural Science Foundation of China under grant No. 61825202,61832006, and the "Fundamental Research Funds for the Central Universities of China" under grant No. 2017KFYXJJ066.

References

1. Abu-khzam, F., Baldwin, N., Langston, M., Samatova, N.: On the relative efficiency of maximal clique enumeration algorithms, with application to high-throughput computational biology. In: International Conference on Research Trends in Science and Technology, pp. 1–10 (2005)
2. Alduaiji, N., Datta, A., Li, J.: Influence propagation model for clique-based community detection in social networks. IEEE Trans. Comput. Soc. Syst. **5**(2), 563–575 (2018)
3. Bacsó, G., Gravier, S., Gyárfás, A., Preissmann, M., Sebo, A.: Coloring the maximal cliques of graphs. SIAM J. Discret. Math. **17**(3), 361–376 (2004)
4. Bailey, P., Craswell, N., Hawking, D.: Engineering a multi-purpose test collection for Web retrieval experiments. Inf. Process. Manag. **39**(6), 853–871 (2003)
5. Batagelj, V., Zaversnik, M.: An O(m) algorithm for cores decomposition of networks. Adv. Data Anal. Classif. **5**(2), 129–145 (2011)
6. Bron, C., Kerbosch, J.: Algorithm 457: finding all cliques of an undirected graph. Commun. ACM **16**(9), 575–577 (1973)
7. Chen, Q., Fang, C., Wang, Z., Suo, B., Li, Z., Ives, Z.G.: Parallelizing maximal clique enumeration over graph data. In: DASFAA, pp. 249–264 (2016)
8. Cheng, J., Zhu, L., Ke, Y., Chu, S.: Fast algorithms for maximal clique enumeration with limited memory. In: KDD, pp. 1240–1248 (2012)
9. Chiba, N., Nishizeki, T.: Arboricity and subgraph listing algorithms. SIAM J. Comput. **14**(1), 210–223 (1985)
10. Conte, A., Virgilio, R.D., Maccioni, A., Patrignani, M., Torlone, R.: Finding all maximal cliques in very large social networks. In: EDBT, pp. 173–184 (2016)
11. Eppstein, D., Löffler, M., Strash, D.: Listing all maximal cliques in sparse graphs in near-optimal time. In: ISAAC, pp. 403–414 (2010)
12. Eppstein, D., Strash, D.: Listing all maximal cliques in large sparse real-world graphs. In: SEA, pp. 364–375 (2011)
13. Leskovec, J., Krevl, A.: SNAP datasets: Stanford large network dataset collection (2014). http://snap.stanford.edu/data
14. Moon, J.W., Moser, L.: On cliques in graphs. Isr. J. Math. **3**(1), 23–28 (1965)
15. Mukherjee, A.P., Xu, P., Tirthapura, S.: Enumeration of maximal cliques from an uncertain graph. IEEE Trans. Knowl. Data Eng. **29**(3), 543–555 (2017)
16. Palla, G., Derényi, I., Farkas, I.J., Vicsek, T.: Uncovering the overlapping community structure of complex networks in nature and society. Nature **435**, 814–818 (2005)
17. Strash, D.: Quick cliques: quickly compute all maximal cliques in sparse graphs (2014). https://github.com/darrenstrash/quick-cliques

18. Sun, S., Wang, Y., Liao, W., Wang, W.: Mining maximal cliques on dynamic graphs efficiently by local strategies. In: ICDE, pp. 115–118 (2017)
19. Tomita, E., Tanaka, A., Takahashi, H.: The worst-case time complexity for generating all maximal cliques and computational experiments. Theor. Comput. Sci. **363**(1), 28–42 (2006)
20. Tsukiyama, S., Ide, M., Ariyoshi, H., Shirakawa, I.: A new algorithm for generating all the maximal independent sets. SIAM J. Comput. **6**(3), 505–517 (1977)

Leveraging Knowledge Graph Embeddings for Natural Language Question Answering

Ruijie Wang[1,2,6], Meng Wang[3(✉)], Jun Liu[1,2], Weitong Chen[4], Michael Cochez[5,6,7], and Stefan Decker[5,6]

[1] National Engineering Lab for Big Data Analytics, Xi'an Jiaotong University, Xi'an, China
[2] School of Electronic and Information Engineering, Xi'an Jiaotong University, Xi'an, China
[3] School of Computer Science and Engineering, Southeast University, Nanjing, China
meng.wang@seu.edu.cn
[4] School of Information Technology and Electrical Engineering, The University of Queensland, Brisbane, Australia
[5] Fraunhofer FIT, 53754 Sankt Augustin, Germany
[6] Informatik 5, RWTH University Aachen, Aachen, Germany
[7] Faculty of Information Technology, University of Jyvaskyla, Jyväskylä, Finland

Abstract. A promising pathway for natural language question answering over knowledge graphs (KG-QA) is to translate natural language questions into graph-structured queries. During the translation, a vital process is to map entity/relation phrases of natural language questions to the vertices/edges of underlying knowledge graphs which can be used to construct target graph-structured queries. However, due to linguistic flexibility and ambiguity of natural language, the mapping process is challenging and has been a bottleneck of KG-QA models. In this paper, we propose a novel framework, called KemQA, which stands on recent advances in relation phrase dictionaries and knowledge graph embedding techniques to address the mapping problem and construct graph-structured queries of natural language questions. Extensive experiments were conducted on question answering benchmark datasets. The results demonstrate that our framework outperforms state-of-the-art baseline models in terms of effectiveness and efficiency.

Keywords: Knowledge graph
Natural language question answering · Knowledge graph embedding

1 Introduction

Large-scale knowledge graphs (KGs), such as DBpedia [10] and Wikidata [19], organize web information in the form of KG triples, e.g., (*Renée Zellweger*, *award*, *Academy Award*). Each KG triple contains two vertices (e.g., *Renée Zellweger*,

G. Li et al. (Eds.): DASFAA 2019, LNCS 11446, pp. 659–675, 2019.
https://doi.org/10.1007/978-3-030-18576-3_39

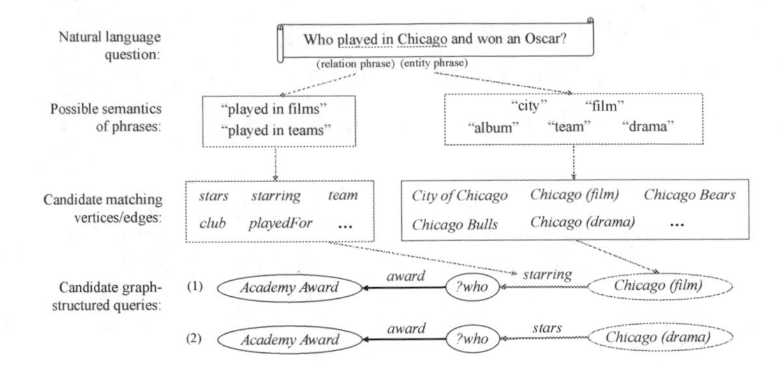

Fig. 1. The general translation process of an example natural language question.

Academy Award) and an edge (e.g., *award*). Consequently, a KG can be represented by an oriented graph, and graph-structured queries, such as SPARQL [7] queries and GraphQL [8] queries, are natural and effective methods for accessing the KG. However, issuing such graph-structured queries requires users to be precisely aware of complicated query syntaxes, and users prefer querying KGs directly with natural language questions (NLQs). Therefore, it is imperative to provide an interface which can translate NLQs into graph-structured queries accurately and efficiently. KG-QA models [1,6,9,23,25] have been proposed in recent years. However, there are still many non-trivial issues due to linguistic flexibility and ambiguity of natural language.

Challenges. Let us consider the translation of the example NLQ "*Who played in Chicago and won an Oscar?*". As illustrated in Fig. 1, a KG-QA model needs to first map entity/relation phrases of the NLQ to vertices/edges of the underlying KG and then assemble the matching vertices/edges into graph-structured queries. There are four major challenges during this process. Firstly, entity/relation phrases of the NLQ may be ambiguous. For example, the relation phrase "played in" may denote "played in films" or "played in teams", and the entity phrase "Chicago" may denote a "city", a "film", or even a "sports team". Secondly, assuming that exact semantics have been determined, searching the matching vertices/edges of an entity/relation phrase in the underlying KG is hard. This is because the entity/relation phrase may be very different from its name in the KG. For example, the candidate matching edges of the phrase "played in films" include *starring* and *stars*, and there is no similarity between them in form. Thirdly, a phrase may have too many candidate matching vertices/edges. For example, if "Chicago" denotes a "sports team", then *Chicago Bulls*, *Chicago Bears*, and all the other Chicago sports teams are candidates. This poses a challenge to efficiency of KG-QA models. Lastly, as illustrated in Fig. 1, we may construct multiple candidate graph-structured queries with the matching vertices/edges, and it is hard to select the optimal one.

Our Solution. In this paper, we focus on the above challenges and propose a novel framework, called KemQA (**K**nowledge Graph **Em**bedding based **Q**uestion **A**nswering Framework), which leverages the relation phrase dictionary [13] and knowledge graph embedding techniques [3,4,11] during the translation of NLQs.

In the off-line stage, KemQA encodes the underlying KG and relation phrase dictionary into a common low-dimensional vector space. Learned embedding vectors of vertices, edges, and relation phrases are essentially latent representations of the KG and relation phrase dictionary, and the embedding vectors can be utilized during the translation of NLQs without any further modification. In the on-line stage, given an NLQ, KemQA first maps entity/relation phrases to their candidate matching vertices/edges. KemQA does not handle the ambiguity issue of phrases during the mapping process, and all candidate matching vertices/edges of each phrase are obtained with the help of learned embedding vectors. Then, KemQA employs a "clustering+translation" strategy to compute the exact semantic of each phrase in the embedding space and generate the matching vertices/edges of each phrase. Finally, KemQA utilizes the matching vertices and edges to construct candidate graph-structured queries and selects the optimal query based on the translation mechanism. In a nutshell, our work makes the following contributions:

- We propose a novel framework, i.e., KemQA, based on KG embedding techniques to answer NLQs by translating NLQs into graph-structured queries.
- We propose a novel embedding method which utilizes the translation mechanism to preserve the structure of the KG while considering the context information to incorporate relation phrases into the common embedding space.
- We propose effective and efficient approaches to map phrases, address the ambiguity issue, and generate target graph-structured queries based on the learned embedding vectors.
- We conduct extensive experiments over the benchmark dataset to evaluate the performance of KemQA. The results prove that KemQA outperforms existing models regarding effectiveness and efficiency.

Organization: The rest of this paper is organized as follows: We introduce our framework in Sect. 2. Experiments are reported in Sect. 3. Related work is discussed in Sect. 4. Finally, conclusions are presented in Sect. 5.

2 Proposed Framework

In this section, we first introduce the notations employed in this paper and then give an overview of our framework.

NLQ and Entity/Relation Phrase. We denote a natural language question (NLQ) as $\mathcal{Q}$. The entity phrase (e.g., "Who", "Chicago", and "Oscar") and the relation phrase (e.g., "played in" and "won") of the NLQ are denoted as ent and rel, respectively.

Knowledge Graph (KG). Let $\mathcal{V}$ be a set of vertices, $\mathcal{E}$ be a set of edges that link vertices. A KG $\mathcal{G} = (\mathcal{V}, \mathcal{E})$ is a finite set of KG triples which are in the form (v_h, e, v_t), where $v_h, v_t \in \mathcal{V}$, and $e \in \mathcal{E}$. A KG triple (v_h, e, v_t) indicates that the head vertex v_h is linked to the tail vertex v_t by the edge e.

Graph-Structured Query. Let $\mathcal{V}_v$ be a set of variables[1], where each variable $v_v \in \mathcal{V}_v$ is distinguished from vertices by a leading question mark symbol, e.g., *?actor*. The triple pattern is similar to the KG triple but allows the usage of variables, e.g., (*?actor*, *award*, *Academy Award*). We define the graph-structured query $\mathcal{G}_Q$ as a finite set of triple patterns.

Relation Phrase Dictionary. We denote a relation phrase dictionary as $\mathcal{D} = (\mathcal{R}, S(\cdot))$, where $\mathcal{R} = \{rel_1, \ldots, rel_m\}$ is a set of relation phrases, and $S(\cdot)$ is a function that returns a set of supporting vertex pairs for each relation phrase in $\mathcal{R}$. Specifically, the supporting vertex pairs of the relation phrase $rel \in \mathcal{R}$ are denoted as $S(rel) = \{(v_h^1, v_t^1), \ldots, (v_h^n, v_t^n)\}$. We employ PATTY [13] as the relation phrase dictionary in this paper, and Table 1 shows two example relation phrases and their supporting vertex pairs.

Table 1. Example relation phrases and their supporting vertex pairs

Relation phrase	Supporting vertex pairs
"played in films"	(*Una Merkel*, *Abraham Lincoln (1930 film)*), (*Brandon Routh*, *Superman Returns*), ...
"played in teams"	(*Al Nesser*, *New York Giants*), (*Sedric Toney*, *Phoenix Suns*), ...

KG Embedding. We use boldface letters to denote learned embedding vectors of $v \in \mathcal{V}$, $e \in \mathcal{E}$, and $rel \in \mathcal{R}$ as $\mathbf{v}$, $\mathbf{e}$, and $\mathbf{rel}$, respectively. We employ the translation mechanism of TransE [3] in KemQA, which refers to that edges of KGs are represented as translation operations from head vertices to tail vertices in the embedding space. Specifically, given a KG triple $(v_h, e, v_t) \in \mathcal{G}$, we learn the embedding vectors $\mathbf{v_h}$, $\mathbf{e}$, and $\mathbf{v_t}$ which hold $\mathbf{v_h} + \mathbf{e} \approx \mathbf{v_t}$ ($\mathbf{v_t}$ should be the closest vertex to $\mathbf{v_h} + \mathbf{e}$ in the embedding space).

2.1 Overview of Our Framework

Our framework KemQA consists of four modules: embedding learning module, phrase mapping module, disambiguation module, and query generation module. We depict an overview of KemQA with a concrete example in Fig. 2.

In the off-line stage, KemQA encodes the underlying KG $\mathcal{G} = (\mathcal{V}, \mathcal{E})$ and the relation phrase dictionary $\mathcal{D} = (\mathcal{R}, S(\cdot))$ into a common low-dimensional vector

[1] In this paper, we focus on the NLQs whose answers are vertices in the underlying KG. Therefore, we only consider vertex variables.

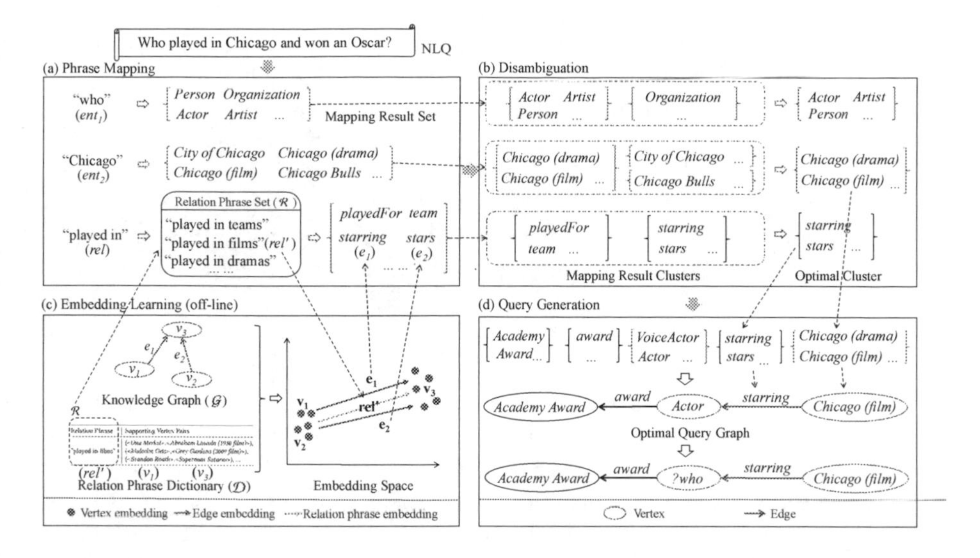

Fig. 2. An overview of our framework.

space. As illustrated in Fig. 2(c), in the embedding space, semantically similar vertices/edges are close to each other, e.g., v_1 is close to v_2, and e_1 is close to e_2, and relation phrases are close to their matching edges, e.g., rel' is close to e_1 and e_2. Meanwhile, the translation mechanism is maintained to capture the inherent structure of the KG and information of the relation phrase dictionary, e.g., $\mathbf{v_1} + \mathbf{e_1} \approx \mathbf{v_3}$ and $\mathbf{v_1} + \mathbf{rel'} \approx \mathbf{v_3}$.

In the on-line stage, given the NLQ "*Who played in Chicago and won an Oscar?*", KemQA first maps each entity/relation phrase of the NLQ to its candidate matching vertices/edges which are referred to as mapping results. For example, KemQA maps "Chicago" to *City of Chicago* and *Chicago (film)*, as shown in Fig. 2(a). Especially for the relation phrase rel, KemQA first searches its similar relation phrases in the relation phrase set $\mathcal{R}$, e.g., rel', then KemQA obtains the edges which are close to rel' in the embedding space as candidate matching edges of rel.

Then, KemQA handles ambiguity of entity/relation phrases by a "clustering+translation" strategy, as illustrated in Fig. 2(b). Firstly, the mapping results of each phrase are clustered into several mapping result clusters according to their semantics. Since we require that semantically similar vertices/edges should be close to each other in the embedding space, the clustering is processed in the embedding space, and we expect that each cluster contains mapping results which have similar semantics. For example, the mapping results of "Chicago" are clustered into three clusters, i.e., {*Chicago (drama)*, *Chicago (film)*, ...}, {*City of Chicago*, ...}, and {*Chicago Bulls*, ...}. Then, KemQA utilizes the mapping result clusters to compute the exact semantic of each phrase based on translation mechanism and select the optimal cluster which contains the exact matching vertices/edges. For example, the exact semantic of "Chicago" is "film", and its optimal mapping result cluster is {*Chicago (drama)*, *Chicago (film)*, ...}.

In the last module, KemQA assembles matching vertices and edges into graph-structured queries which can be evaluated to obtain answers. Since a phrase may correspond to multiple matching vertices/edges, e.g., "Chicago" corresponds to *Chicago (film)* and *Chicago (drama)*, we may construct multiple candidate queries. We represent the candidates with query graphs and select the optimal one based on translation mechanism, as shown in Fig. 2(d).

2.2 Embedding Learning

As introduced above, we have three requirements for the embedding learning module: (1) embedding vectors of semantically similar vertices/edges should be close to each other; (2) relation phrases should be close to their matching edges in the embedding space; (3) translation mechanism should be maintained.

In this module, KemQA encodes the KG $\mathcal{G} = (\mathcal{V}, \mathcal{E})$ and the relation phrase dictionary $\mathcal{D} = (\mathcal{R}, S(\cdot))$ into a common low-dimensional embedding space. Specifically, the embedding vectors of vertices/edges in $\mathcal{G}$ are learned based on their context information, and the embedding vectors of relation phrases in $\mathcal{D}$ are learned based on their supporting vertex pairs. The supporting vertex pairs have been introduced above, and we define the context of vertices/edges as follows:

Definition 1 (Vertex/Edge Context). *Given a KG* $\mathcal{G} = (\mathcal{V}, \mathcal{E})$*, the vertex context of* $v \in \mathcal{V}$ *is* $C_v(v) = \{(e, \hat{v}) | e \in \mathcal{E}, \hat{v} \in \mathcal{V}, (v, e, \hat{v}) \in \mathcal{G} || (\hat{v}, e, v) \in \mathcal{G}\}$. *The edge context of* $e \in \mathcal{E}$ *is* $C_e(e) = \{(v_h, v_t) | v_h, v_t \in \mathcal{V}, (v_h, e, v_t) \in \mathcal{G}\}$.

For the vertex $v \in \mathcal{V}$, the edge $e \in \mathcal{E}$, and the relation phrase $rel \in \mathcal{R}$, we respectively define the conditional probability of v, e, and rel given the context $C_v(v)$, $C_e(e)$, and the supporting vertex pairs $S(rel)$ as follows:

$$P(v|C_v(v)) = \frac{exp(f_1(v, C_v(v)))}{\sum_{v' \in \mathcal{V}} exp(f_1(v', C_v(v)))}, \tag{1}$$

$$P(e|C_e(e)) = \frac{exp(f_2(e, C_e(e)))}{\sum_{e' \in \mathcal{E}} exp(f_2(e', C_e(e)))}, \tag{2}$$

$$P(rel|S(rel)) = \frac{exp(f_3(rel, S(rel)))}{\sum_{rel' \in \mathcal{R}} exp(f_3(rel', S(rel)))}, \tag{3}$$

where $f_1(v', C_v(v))$, $f_2(e', C_e(e))$, and $f_3(rel', S(rel))$ are functions that respectively measure the correlation between an arbitrary vertex $v' \in \mathcal{V}$ and $C_v(v)$, the correlation between an arbitrary edge $e' \in \mathcal{E}$ and $C_e(e)$, and the correlation between an arbitrary relation phrase $rel' \in \mathcal{R}$ and $S(rel)$. Equations 1, 2, and 3 can be considered as the compatibility between the vertex/edge/relation phrase and the vertex context/edge context/supporting vertex pairs. They are formulated as softmax-like representations which have been validated [14,22].

The functions $f_1(v', C_v(v))$, $f_2(e', C_e(e))$, and $f_3(rel', S(rel))$ are formulated as follows:

$$f_1(v', C_v(v)) = -\frac{1}{|C_v(v)|} \sum_{(e,\hat{v}) \in C_v(v)} f_4(v', e, \hat{v}), \tag{4}$$

$$f_2(e', C_e(e)) = -\frac{1}{|C_e(e)|} \sum_{(v_h,v_t) \in C_e(e)} f_5(v_h, e', v_t), \tag{5}$$

$$f_3(rel', S(rel)) = -\frac{1}{|S(rel)|} \sum_{(v_h,v_t) \in S(rel)} f_6(v_h, rel', v_t), \tag{6}$$

where $f_4(v', e, \hat{v})$, $f_5(v_h, e', v_t)$, and $f_6(v_h, rel', v_t)$ are cost functions based on the translation mechanism of TransE, formulated as follows:

$$f_4(v', e, \hat{v}) = \begin{cases} \|\mathbf{v'} + \mathbf{e} - \hat{\mathbf{v}}\|_2^2, & \text{if} \quad (v, e, \hat{v}) \in \mathcal{G}, \\ \|\hat{\mathbf{v}} + \mathbf{e} - \mathbf{v'}\|_2^2, & \text{if} \quad (\hat{v}, e, v) \in \mathcal{G}. \end{cases} \tag{7}$$

$$f_5(v_h, e', v_t) = \|\mathbf{v_h} + \mathbf{e'} - \mathbf{v_t}\|_2^2. \tag{8}$$

$$f_6(v_h, rel', v_t) = \begin{cases} \|\mathbf{v_h} + \mathbf{rel'} - \mathbf{v_t}\|_2^2, & \text{if} \quad \exists e \in \mathcal{E}, (v_h, e, v_t) \in \mathcal{G}, \\ \|\mathbf{v_t} + \mathbf{rel'} - \mathbf{v_h}\|_2^2, & \text{if} \quad \exists e \in \mathcal{E}, (v_t, e, v_h) \in \mathcal{G}. \end{cases} \tag{9}$$

KemQA learns the embedding vectors of vertices, edges, and relation phrases by maximizing the joint probability of all vertices/edges in $\mathcal{G}$ and relation phrases in $\mathcal{D}$, which is formulated as follows:

$$O = \lambda_v \sum_{v \in \mathcal{V}} \log P(v|C_v(v)) + \lambda_e \sum_{e \in \mathcal{E}} \log P(e|C_e(e)) + \lambda_{rel} \sum_{rel \in \mathcal{R}} \log P(rel|S(rel)), \tag{10}$$

where λ_v, λ_e, and λ_{rel} are weighting hyper-parameters.

The intuition of vertex/edge embedding learning is that semantically similar vertices/edges tend to share common context information. For example, *Chicago (drama)* and *Chicago (film)* are semantically similar vertices, and both of them are intensively linked with actors and directors in the KG. *Starring* and *stars* are semantically similar edges, and both of them link a set of "actor-film" vertex pairs. Since we learn the embedding vectors of vertices and edges based on their context, i.e., Eqs. 1 and 2, semantically similar vertices and edges should be close to each other in the embedding space.

The intuition of relation phrase embedding learning is that, if we regard the context of an edge as its supporting vertex pairs, relation phrases and their matching edges tend to share common supporting vertex pairs. For example, given a relation phrase "played in films", its supporting vertex pairs include "actor-film" vertex pairs and its matching edge *starring* also links "actor-film" vertex pairs in the KG. Since we encode relation phrases and edges based on supporting vertex pairs and edge contexts, respectively, i.e., Eqs. 2 and 3, relation phrases should be close to their matching edges in the embedding space.

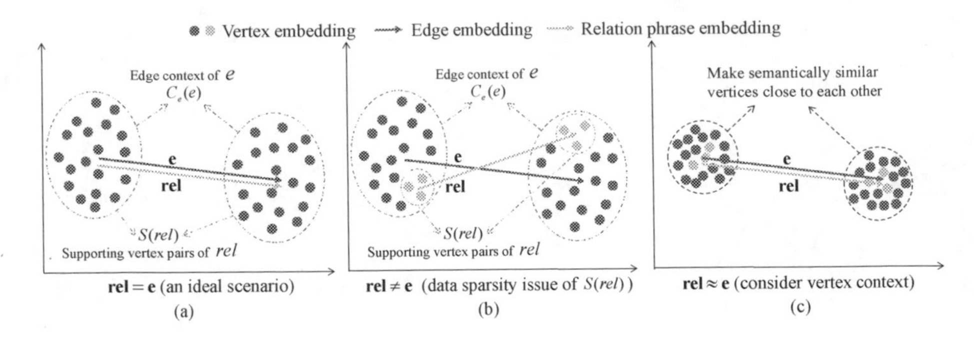

Fig. 3. The learned embedding vectors of the relation phrase rel, the matching edge e, and the vertices in $C_e(e)$ and $S(rel)$.

Note that the translation mechanism is maintained in our embedding module since we adopt it in cost functions, i.e., Eqs. 7, 8, and 9.

In the following, we discuss an important issue during the embedding learning of relation phrases, which is the data sparsity issue of relation phrase dictionaries. We first illustrate an ideal scenario of relation phrase embedding learning in Fig. 3(a), where e is a matching edge of the relation phrase rel, and e links all supporting vertex pairs of rel in the KG, i.e., $C_e(e) = S(rel)$. Then, according to Eqs. 2 and 3, we expect that $\mathbf{e} = \mathbf{rel}$. However, compared to the edge context $C_e(e)$, the size of the supporting vertex pair set $S(rel)$ is usually very limited. For example, in DBpedia, an edge usually links over thousands of vertex pairs[2]. However, in PATTY, a relation phrase only has 11 supporting vertex pairs on average [26]. Therefore, as illustrated in Fig. 3(b), it seems like the learned relation phrase embedding vector **rel** is very possible to be not close to the edge embedding vector **e**. Addressing this issue is another reason why we leverage context information of vertices during embedding learning. Since vertices from the same side of supporting vertex pairs of the relation phrase/the matching edge are usually semantically similar, and considering the context information of vertices helps to make semantically similar vertices be close to each other, the distribution of vertices in $S(rel)$ is close to the distribution of vertices in $C_e(e)$ in the embedding space, as shown in Fig. 3(c). Then, even $S(rel)$ is a small subset of $C_e(e)$, we can still make sure that e is close to rel in the embedding space.

2.3 Phrase Mapping

In this module, we adopt the pre-processing method in [5] and the node-first Super Semantic Query Graph building method in [9] to extract entity/relation phrases of the given NLQ and generate *phrase triples*, each of which consists of one relation phrase and two entity phrases. We denote the *phrase triple* as $t_p = (rel, ent, ent')$ which indicates that there is a relation rel between ent and

[2] http://wiki.dbpedia.org/dbpedia-2016-04-statistics.

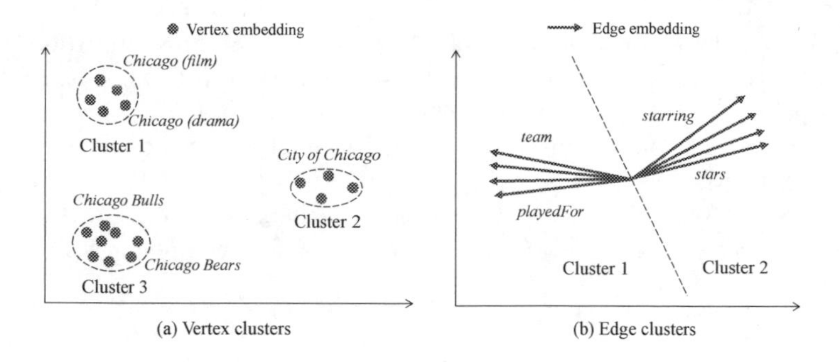

Fig. 4. Mapping result clusters of "Chicago" and "played in".

ent'. For instance, we extract *phrase triples* ("played in", "Who", "Chicago") and ("won", "Who", "Oscar") from the example NLQ in Fig. 2.

The mapping results of entity phrases are obtained during the above process. However, the mapping of relation phrases is generally considered to be a more difficult and problem-specific task [5]. We design a novel method which utilizes the learned embedding vectors to perform the mapping of relation phrases, as shown in Fig. 2. Specifically, given a relation phrase rel of the NLQ, we first employ Levenshtein distance to retrieve similar relation phrases of rel from the relation phrase dictionary $\mathcal{D}$. For example, the similar relation phrases of "played in" include "played in films" and "played in teams". Then we search the edges which are close to the similar relation phrases in the embedding space according to cosine similarity and collect these edges as candidate matching edges of rel. For example, *stars* and *starring* are close to "played in films" in the embedding space, then they are candidate matching edges of "played in".

2.4 Disambiguation

In this module, we employ a "clustering+translation" strategy to compute the exact semantic of each entity/relation phrase and distill the above mapping results of entity/relation phrases in the embedding space.

Firstly, we cluster the mapping results of each entity/relation phrase into several disjoint clusters. Specifically, given an entity phrase ent along with its mapping results $\{v_1, \ldots, v_i, \ldots, v_m\}$, we perform K-means algorithm in the embedding space to cluster them into α clusters $\mathcal{C}_v(ent) = \{Cv_1, \ldots, Cv_j, \ldots, Cv_\alpha\}$, where the value of α is estimated based on the gap statistic [15], and if v_i is clustered into Cv_j, $v_i \in Cv_j$. Analogically, given a relation phrase rel, we cluster its mapping results $\{e_1, e_2, \ldots, e_n\}$ into β clusters $\mathcal{C}_e(rel) = \{Ce_1, Ce_2, \ldots, Ce_\beta\}$. As analyzed in the embedding learning module, we learn embedding vectors of vertices/edges based on their context information, and semantically similar vertices/edges are close to each other in the embedding space. Therefore, vertices/edges in the same cluster share similar semantics, as shown in Fig. 4.

Then, we employ translation mechanism to address the ambiguity issue of entity/relation phrases and determine the optimal mapping result cluster of each phrase. Let us first consider a simple NLQ which contains only one *phrase triple* $t_p = (rel, ent, ent')$. The mapping result clusters of rel, ent, and ent' are respectively denoted as $\mathcal{C}_e(rel)$, $\mathcal{C}_v(ent)$, and $\mathcal{C}_v(ent')$. Then, we represent each entity/relation phrase in t_p by one of its mapping result clusters to construct a *cluster triple* (Cv, Ce, Cv'), where $Cv \in \mathcal{C}_v(ent)$, $Ce \in \mathcal{C}_e(rel)$, and $Cv' \in \mathcal{C}_v(ent')$. According to the translation mechanism of TransE, the cost of the *cluster triple* (Cv, Ce, Cv') is computed by the following equation:

$$cost_c(Cv, Ce, Cv') = \|f_7(Cv) + f_7(Ce) - f_7(Cv')\|_2^2, \tag{11}$$

where $f_7(\cdot)$ is the function that computes the mean embedding vector of a mapping result cluster. $f_7(Cv)$ is formulated as follows:

$$f_7(Cv) = \frac{1}{|Cv|} \sum_{v \in Cv} \mathbf{v}. \tag{12}$$

Since *phrase triple* does not specify the direction of relation phrase, we can construct $2 \cdot |\mathcal{C}_v(ent)| \cdot |\mathcal{C}_e(rel)| \cdot |\mathcal{C}_v(ent')|$ *cluster triples* for the *phrase triple* t_p. If a *cluster triple* $(\hat{Cv}, \hat{Ce}, \hat{Cv'})$ has the minimum cost, we regard $\hat{Cv}$, $\hat{Ce}$, and $\hat{Cv'}$ as optimal mapping result clusters of ent, rel, and ent', respectively.

If the NLQ contains ω *phrase triples* $\{t_p^1, \ldots, t_p^l, \ldots, t_p^\omega\}$, we construct the *cluster triple set* $\mathcal{T}_c = \{(Cv^1, Ce^1, Cv'^1), \ldots, (Cv^l, Ce^l, Cv'^l), \ldots, (Cv^\omega, Ce^\omega, Cv'^\omega)\}$, where (Cv^l, Ce^l, Cv'^l) is the *cluster triple* of t_p^l. The cost is computed as follows:

$$Cost_C(\mathcal{T}_c) = \sum_{l=1}^{\omega} cost_c(Cv^l, Ce^l, Cv'^l). \tag{13}$$

Assuming that the NLQ consists of m entity phrases $\{ent_1, \ldots, ent_i, \ldots, ent_m\}$ and n relation phrases $\{rel_1, \ldots, rel_j, \ldots, rel_n\}$, we can construct $2^n \cdot \prod_{i=1}^{m} |\mathcal{C}_v(ent_i)| \cdot \prod_{j=1}^{n} |\mathcal{C}_e(rel_j)|$ *cluster triple sets*. Analogically, the *cluster triple set* which has the minimum cost contains the optimal mapping results. Considering that real-world NLQs usually consist of less than three phrase triples [17], it is fairly feasible to compute the costs of all possible *cluster triple sets*.

2.5 Graph-Structured Query Generation

In this module, we assemble matching vertices/edges of entity/relation phrases into graph-structured queries. Let us first consider a simple NLQ which contains one *phrase triple* $t_p = (rel, ent, ent')$. $\hat{Cv}$, $\hat{Ce}$, and $\hat{Cv'}$ are optimal mapping result clusters of ent, rel, and ent', respectively. We assemble matching vertices/edges $v \in \hat{Cv}$, $e \in \hat{Ce}$, and $v' \in \hat{Cv'}$ into a candidate *query graph* $\mathcal{G}_q = \{(v, e, v')\}$. There are $|\hat{Cv}| \cdot |\hat{Ce}| \cdot |\hat{Cv'}|$ candidate *query graphs*, and we rank them according to the cost of $\mathcal{G}_q$ which is computed as follows:

$$Cost_q(\mathcal{G}_q) = \sum_{(v,e,v') \in \mathcal{G}_q} \|\mathbf{v} + \mathbf{e} - \mathbf{v}'\|_2^2. \tag{14}$$

For the candidate *query graph* $\hat{\mathcal{G}}_q$ which has the minimum cost, we search class vertices in $\hat{\mathcal{G}}_q$ which represent types of vertices (e.g., *Actor* and *Person*) and replace them with variables to generate the target graph-structured query. Analogically, for the NLQ which consists of multiple *phrase triples*, we also employ the optimal mapping results to generate all candidate *query graphs* and evaluate them by Eq. 14. The target graph-structured query is finally generated by replacing class vertices with variables, as illustrated in Fig. 2(d).

3 Experiments

In this section, to scrutinize the effectiveness and efficiency of our framework, we compare KemQA with several state-of-the-art KG-QA models, including RFF [9], NFF [9], and the models participating QALD-6 [17]. All experiments were conducted on a Linux server with an Intel Core i7 3.40Ghz CPU and 128GB memory running Ubuntu-14.04.1. Note that we employ the Implicit Relation Prediction method in [6] to address implicit relation phrases [6] of NLQs in the following experiments. In the embedding learning, we treat literal values of the underlying KG as vertices and add generalized KG triples [21] to the KG. Considering the large scale of the KG, we also follow [14] to sample vertex/edge context and approximate Eqs. 1, 2, and 3 based on negative sampling. Dimensions of embedding vectors are set to 100, and we set $\lambda_v = 0.5$, $\lambda_e = 0.5$, and $\lambda_{rel} = 1$. In addition, if the graph-structured query generated by the candidate *query graph* which has the minimum cost returns an empty answer due to errors of the embedding learning, we would generate another query based on candidate *query graphs* with higher costs.

DBpedia. DBpedia [10] is a large-scale KG extracted from Wikipedia[3]. DBpedia-2015[4] is the specified benchmark dataset of QALD-6, and it consists of 6.7M vertices, 1.4K edges, and 583M KG triples.

NLQ Dataset. QALD is a series of challenges on KG-QA. We employ QALD-6 [17] as the NLQ dataset in our experiment, which releases 350 training NLQs and 100 test NLQs over DBpedia in its Task-1.

Relation Phrase Dictionary. PATTY is a large resource for text patterns (i.e., relation phrases) that denote binary relationships between vertices in KGs [13]. We employ the Wikipedia version of PATTY[5] which contains 350,569 relation phrases and 3,862,331 supporting vertex pairs.

3.1 Effectiveness Evaluation

In this section, we follow [17] to evaluate the effectiveness of KemQA and report the results in Table 2, where *Processed* states for the number of NLQs processed

[3] https://www.wikipedia.org.
[4] http://wiki.dbpedia.org/develop/datasets.
[5] http://www.mpi-inf.mpg.de/yago-naga/patty/.

Table 2. Results on QALD-6 benchmark (Total number of questions: 100)

	Processed	Recall	Precision	F-1
CANaLI [12]	100	0.89	0.89	0.89
KemQA	**100**	**0.73**	**0.89**	**0.80**
NFF [9]	100	0.70	0.89	0.78
UTQA [18]	100	0.69	0.82	0.75
KWGAnswer [9]	100	0.59	0.85	0.70
RFF [9]	100	0.43	0.77	0.55
NbFramework [17]	63	0.85	0.87	0.54
SemGraphQA [17]	100	0.25	0.70	0.37
UIQA (with manual) [17]	44	0.63	0.54	0.25

by the system, *Recall* indicates the ratio of correct returned answers over all gold answers, *Precision* indicates the ratio of correct returned answers over all returned answers, and *F-1* is a weighted average between the precision and recall [17]. It is worth mentioning that *Recall* and *Precision* are computed with respect to the number of processed NLQs, and *F-1* is computed with respect to the total number of NLQs, i.e., 100.

We make the following observations: (1) Recall, precision, and F-1 of CANaLI are significantly higher than KemQA and other baselines. The reason is that questions answered by CANaLI are posed in Controlled Natural Language [12], and CANaLI is not regarded as a fully developed KG-QA system [9]; (2) Except CANaLI, among the systems processed 100 questions, KemQA achieves the highest recall; (3) KemQA achieves the same precision as CANaLI and NFF. (4) F-1 of KemQA is the highest among all baselines except CANaLI.

In summary, KemQA outperforms the baselines in terms of effectiveness. We conclude the following reasons: (1) KemQA utilizes learned embedding vectors to map relation phrases to their candidate matching edges, which achieves a high mapping precision after addressing the data sparsity issue of relation phrase dictionaries by employing entity context information. (2) KemQA generates a cluster of matching vertices/edges for each phrase during the disambiguation process, which guarantees a high disambiguation recall. (3) The learned embedding vectors are essentially the latent representations of the underlying KG. KemQA computes the semantics of entity/relation phrases and evaluates the generated query graphs based on learned embedding vectors, which makes sure that the generated graph-structured queries are consistent with the underlying KG.

3.2 Efficiency Evaluation

In this section, we evaluate our framework in terms of efficiency. The average time cost of KemQA to answer an NLQ is 473.4 ms, and we report the average time cost of each module in Table 3. The phrase mapping module spends much more

Table 3. Average time cost of each module

Module	Avg. time cost (ms)
Phrase mapping	270.7
Disambiguation	110.2
Query generation and evaluation	92.5

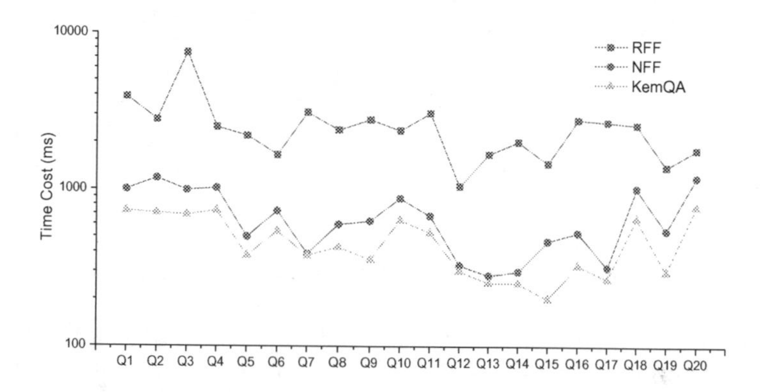

Fig. 5. Time costs of KemQA, RFF, and NFF.

time than other modules because it involves searching in the embedding space. The most time cost of the disambiguation module is spent on the clustering of mapping results. And the evaluation of graph-structured queries spends most time cost of the last module.

Then, we compare KemQA with RFF and NFF which have state-of-the-art efficiency performances. We randomly select 20 NLQs from the QALD-6 dataset and report time costs of answering the 20 NLQs by KemQA, RFF, and NFF in Fig. 5. We can observe that the time cost of KemQA is significantly less than the other two models. We conclude the following reasons: (1) The embedding vectors employed by KemQA are learned in the off-line stage, and we do not need to modify them during the on-line process. (2) Since the phrase mapping and disambiguation of KemQA are performed in the embedding space, KemQA can avoid frequent searches over the large-scale KG. (3) Different from RFF and NFF which obtain question answers by subgraph matching over the underlying KG, KemQA obtains answers by evaluating graph-structured queries which are generated based on numerical calculations according to the translation mechanism. Numerical calculations are time-saving, and the evaluation of graph-structured queries is more efficient than the sub-graph matching of RFF and NFF.

3.3 Failure Analysis

Among the 100 test NLQs of QALD-6, 72 NLQs are answered correctly by KemQA with the F-1 measure of 1. In this section, we analyze the failure reasons of the rest 28 NLQs. As reported in Table 4, for each phase of KemQA, we

Table 4. Failure analysis of KemQA on QALD-6

Failure phase	# (Ratio)	Sample question
Phrase extraction	6 (21.43%)	What is the atmosphere of the Moon composed of?
Phrase mapping	6 (21.43%)	What are the five boroughs of New York?
Disambiguation	5 (17.86%)	Who was on the Apollo 11 mission?
Query generation	2 (7.14%)	Who was Vincent van Gogh inspired by?
Other	9 (32.14%)	Show me all basketball players that are higher than 2 m

count the number of incorrectly answered NLQs caused by this phase. Phrase extraction phase is responsible for 6 failures. Entity/relation phrases of some NLQs are hard to be extracted. For example, the relation phrase of the NLQ "*What is the atmosphere of the Moon composed of?*" is the combination of "atmosphere of" and "composed of" which corresponds to the edge *atmosphere-Composition*. Errors in the phrase mapping phase lead to 6 failures which are mainly due to the limitation of the relation phrase dictionary. For example, given the NLQ "*What are the five boroughs of New York?*", KemQA cannot map the phrase "five boroughs of" based on PATTY. Disambiguation phase is responsible for 5 failures, and the main reason is that some relationships between entity phrases are expressed by ambiguous relation phrases which may have too many semantics. For example, in the NLQ "*Who was on the Apollo 11 mission?*", the relationship between "Who" and "Apollo 11 mission" is expressed by "on". KemQA failed to compute the exact semantic of "on". Query generation phase is responsible for 2 failures, and the reason is that directions of some edges are hard to be determined by the translation mechanism, e.g., *influencedBy*. In addition, KemQA cannot answer 9 questions which require complex operations in target queries.

4 Related Work

A variety of techniques have been leveraged by KG-QA models, including semantic parsing [1,24], templates [16,25], and subgraph matching [9,26]. Given an input NLQ, [1] first constructs candidate queries, each of which is associated with a canonical realization in natural language. Then, [1] scores candidate queries by comparing their associated realizations with the input NLQ using a paraphrase model. The method in [16] produces templates which mirror the internal structures of given NLQs and then instantiates templates by entity identification and predicate detection. Hu et al. [9] propose the semantic query graph to model the query intention of the given NLQ and obtain question answers by matching the semantic query graph over the underlying KG.

With the increasing of scales of underlying KGs, the KG-QA models leveraging conventional graph-based algorithms are compromised by computational inefficiency issues. To address this problem, embedding techniques-based KG-QA models [2,6,20,23] have been proposed in recent years. Bordes et al. [2] learn vector representations of NLQs along with their paired answers and do not translate NLQs into graph-structured queries. Their model requires large-scale question-answer pairs for training and cannot answer NLQs which contain multiple entity phrases. Han et al. [6] propose a more scalable model which leverages the learned embedding representation of the underlying KG to generate query graphs of given NLQs. However, they do not discuss the mapping and disambiguation of phrases which are vital and also the focus of this paper. Wang et al. [20] propose a context-based KG embedding model to answer failing queries by an approximating way, but only focused on the structured SPARQL queries. Yang et al. [23] propose a model which translates NLQs into graph-structured queries using the learned joint relational embeddings. Their model ranks candidate queries by computing the similarity between embeddings of observed features in the NLQ and embeddings of logical features in candidate queries. Different from this model, KemQA generates graph-structured queries based on the translation mechanism which is more interpretable and efficient.

5 Conclusions and Future Work

In this paper, we propose a KG-QA framework, called KemQA, which translates NLQs into graph-structured queries to obtain question answers. In the off-line stage, KemQA encodes the underlying KG and the relation phrase dictionary into a common embedding space. During the embedding learning, we employ the translation mechanism of TransE to capture the inherent structure of the KG and information of the relation phrase dictionary, and the data sparsity issue is well addressed by considering context information. The learned embedding vectors are utilized in all three on-line modules of KemQA which are phrase mapping module, disambiguation module, and query generation module. Results of extensive experiments on the benchmark dataset illustrate that KemQA outperforms the baseline models in both effectiveness and efficiency.

In future work, we intend to employ more complex translation mechanisms, such as the mechanism of TransR [11], in our embedding method to improve the performance of KemQA.

Acknowledgment. This work was supported by National Key Research and Development Program of China (2018YFB1004500), National Natural Science Foundation of China (61532015, 61532004, 61672419, 61672418, and U1736204), Innovative Research Group of the National Natural Science Foundation of China (61721002), Innovation Research Team of Ministry of Education (IRT_17R86), Project of China Knowledge Centre for Engineering Science and Technology, Teaching Reform Project of XJTU (No. 17ZX044), and China Scholarship Council (No. 201806280450).

References

1. Berant, J., Liang, P.: Semantic parsing via paraphrasing. In: ACL, vol. 1, pp. 1415–1425 (2014)
2. Bordes, A., Chopra, S., Weston, J.: Question answering with subgraph embeddings. In: EMNLP, pp. 615–620 (2014)
3. Bordes, A., Usunier, N., Garcia-Duran, A., Weston, J., Yakhnenko, O.: Translating embeddings for modeling multi-relational data. In: Advances in Neural Information Processing Systems, pp. 2787–2795 (2013)
4. Du, Z., Hao, Z., Meng, X., Wang, Q.: CirE: circular embeddings of knowledge graphs. In: Candan, S., Chen, L., Pedersen, T.B., Chang, L., Hua, W. (eds.) DASFAA 2017. LNCS, vol. 10177, pp. 148–162. Springer, Cham (2017). https://doi.org/10.1007/978-3-319-55753-3_10
5. Dubey, M., Banerjee, D., Chaudhuri, D., Lehmann, J.: EARL: joint entity and relation linking for question answering over knowledge graphs. arXiv preprint arXiv:1801.03825 (2018)
6. Han, S., Zou, L., Yu, J.X., Zhao, D.: Keyword search on RDF graphs-a query graph assembly approach. In: CIKM, pp. 227–236. ACM (2017)
7. Harris, S., Seaborne, A., Prud'hommeaux, E.: SPARQL 1.1 query language (2013). http://www.w3.org/TR/sparql11-query/
8. He, H., Singh, A.K.: Graphs-at-a-time: query language and access methods for graph databases. In: SIGMOD, pp. 405–418. ACM (2008)
9. Hu, S., Zou, L., Yu, J.X., Wang, H., Zhao, D.: Answering natural language questions by subgraph matching over knowledge graphs. TKDE **30**(5), 824–837 (2018)
10. Lehmann, J., et al.: DBpedia-a large-scale, multilingual knowledge base extracted from Wikipedia. Semant. Web **6**(2), 167–195 (2015)
11. Lin, Y., Liu, Z., Sun, M., Liu, Y., Zhu, X.: Learning entity and relation embeddings for knowledge graph completion. In: AAAI, pp. 2181–2187 (2015)
12. Mazzeo, G.M., Zaniolo, C.: Answering controlled natural language questions on RDF knowledge bases. In: EDBT, pp. 608–611 (2016)
13. Nakashole, N., Weikum, G., Suchanek, F.: PATTY: a taxonomy of relational patterns with semantic types. In: EMNLP-CoNLL, pp. 1135–1145. Association for Computational Linguistics (2012)
14. Shi, J., Gao, H., Qi, G., Zhou, Z.: Knowledge graph embedding with triple context. In: CIKM, pp. 2299–2302. ACM (2017)
15. Tibshirani, R., Walther, G., Hastie, T.: Estimating the number of clusters in a data set via the gap statistic. J. Roy. Stat. Soc.: Ser. B (Stat. Methodol.) **63**(2), 411–423 (2001)
16. Unger, C., Bühmann, L., Lehmann, J., Ngonga Ngomo, A.C., Gerber, D., Cimiano, P.: Template-based question answering over RDF data. In: WWW, pp. 639–648. ACM (2012)
17. Unger, C., Ngomo, A.-C.N., Cabrio, E.: 6th open challenge on question answering over linked data (QALD-6). In: Sack, H., Dietze, S., Tordai, A., Lange, C. (eds.) SemWebEval 2016. CCIS, vol. 641, pp. 171–177. Springer, Cham (2016). https://doi.org/10.1007/978-3-319-46565-4_13
18. Veyseh, A.P.B.: Cross-lingual question answering using common semantic space. In: Proceedings of TextGraphs-10: the Workshop on Graph-Based Methods for Natural Language Processing, pp. 15–19 (2016)
19. Vrandečić, D., Krötzsch, M.: Wikidata: a free collaborative knowledgebase. Commun. ACM **57**(10), 78–85 (2014)

20. Wang, M., Wang, R., Liu, J., Chen, Y., Zhang, L., Qi, G.: Towards empty answers in SPARQL: approximating querying with RDF embedding. In: Vrandečić, D., et al. (eds.) ISWC 2018. LNCS, vol. 11136, pp. 513–529. Springer, Cham (2018). https://doi.org/10.1007/978-3-030-00671-6_30
21. Wang, R., Wang, M., Liu, J., Yao, S., Zheng, Q.: Graph embedding based query construction over knowledge graphs. In: ICBK, pp. 1–8. IEEE (2018)
22. Wang, S., Chang, X., Li, X., Sheng, Q.Z., Chen, W.: Multi-task support vector machines for feature selection with shared knowledge discovery. Signal Process. **120**, 746–753 (2016)
23. Yang, M.C., Duan, N., Zhou, M., Rim, H.C.: Joint relational embeddings for knowledge-based question answering. In: EMNLP, pp. 645–650 (2014)
24. Yih, W.T., Chang, M.W., He, X., Gao, J.: Semantic parsing via staged query graph generation: question answering with knowledge base. In: ACL-NLP, vol. 1, pp. 1321–1331 (2015)
25. Zheng, W., Zou, L., Lian, X., Yu, J.X., Song, S., Zhao, D.: How to build templates for RDF question/answering: an uncertain graph similarity join approach. In: SIGMOD, pp. 1809–1824. ACM (2015)
26. Zou, L., Huang, R., Wang, H., Yu, J.X., He, W., Zhao, D.: Natural language question answering over RDF: a graph data driven approach. In: SIGMOD, pp. 313–324. ACM (2014)

Measuring Semantic Relatedness with Knowledge Association Network

Jiapeng Li[1,2], Wei Chen[1], Binbin Gu[1], Junhua Fang[1], Zhixu Li[1,3], and Lei Zhao[1(✉)]

[1] School of Computer Science and Technology, Soochow University, Suzhou, China
trajepl@gmail.com, wchzhg@gmail.com, gu.binbin@hotmail.com, {jhfang,zhixuli,zhaol}@suda.edu.cn

[2] Institute of Electronic and Information Engineering of UESTC in Guangdong, Dongguan 523808, Guangdong, China

[3] IFLYTEK Research, Suzhou, China

Abstract. Measuring semantic relatedness between two words is a fundamental task for many applications in both databases and natural language processing domains. Conventional methods mainly utilize the latent semantic information hidden in lexical databases (WordNet) or text corpus (Wikipedia). They have made great achievements based on the distance computation in lexical tree or co-occurrence principle in Wikipedia. However these methods suffer from low coverage and low precision because (1) lexical database contains abundant lexical information but lacks semantic information; (2) in Wikipedia, two related words (e.g. synonyms) may not appear in a window size or a sentence, and unrelated ones may be mentioned together by chance. To compute semantic relatedness more accurately, some other approaches have made great efforts based on free association network and achieved a significant improvement on relatedness measurement. Nevertheless, they need complex preprocessing in Wikipedia. Besides, the fixed score functions they adopt cause the lack of flexibility and expressiveness of model. In this paper, we leverage DBPedia and Wikipedia to construct a **Knowledge Association Network (KAN)** which avoids the information extraction of Wikipedia. We propose a flexible and expressive model to represent entities behind the words, in which attribute and topological structure information of entities are embedded in vector space simultaneously. The experiment results based on standard datasets show the better effectiveness of our model compared to previous models.

Keywords: Semantic relatedness · Knowledge graph · Network embedding

1 Introduction

Computing semantic relatedness between two words is a fundamental task in many databases and natural language processing problems such as lexicon induction [17], Named Entity Disambiguation [7], Keyword Extraction [27], semantic

G. Li et al. (Eds.): DASFAA 2019, LNCS 11446, pp. 676–691, 2019.
https://doi.org/10.1007/978-3-030-18576-3_40

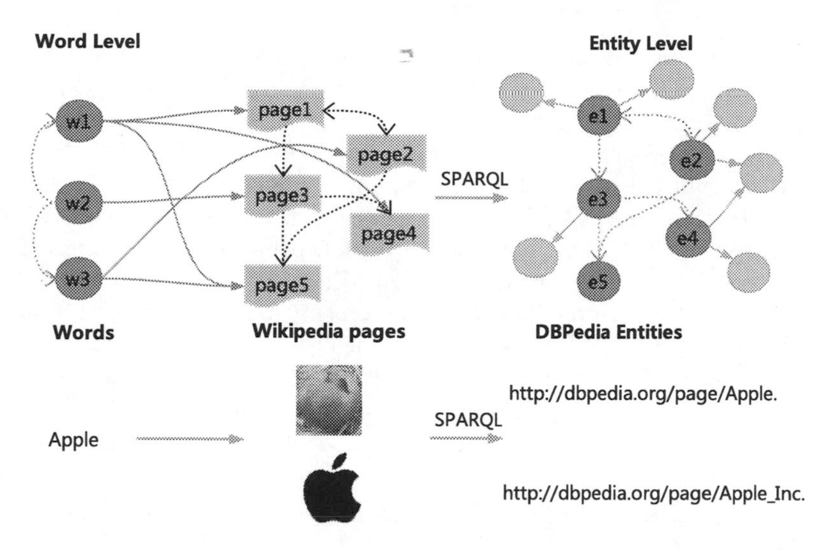

Fig. 1. Knowledge association network (Color figure online)

correspondences discovering [3] and Entity Matching [23]. In the aspect of spam problem [19] and image classification [10], semantic relatedness measurement plays a great role as well.

Due to its importance, plenty of efforts have been made on semantic relatedness measurement. The existing approaches can be roughly divided into three categories as below: (i) The *lexical-based* methods [16,25,28] measure the semantic relatedness between two words based on some lexical databases such as *WordNet* and *Wikitionary*. These methods mainly utilize fixed score functions, such as the path information between two words or the nearest parent common node which two words hold in a lexical tree. Apparently, they only employ pure lexical information but miss semantic information. (ii) The *co-occurrence-based* methods regard two words are related if they appear together in a fixed window size or a sentence. So far plenty of efforts [4,20,25] have applied this co-occurrence principle in the dumps of Wikipedia for semantic relatedness measurement. However, the co-occurrence principle does not always work well. Given that two words are semantically closed, such as synonyms, they do not necessarily appear together. Besides, two words that appear in the same sentence by chance may not necessarily be closely related in semantic space [5]. (iii) The *association network-based* methods propose that for a given word, the first word that comes into human mind is the most related one. To improve the co-occurrence-based methods, a more advanced approach builds an association network based on not only co-occurrences between words, but also the links and shared attributes between entities [5,26]. Based on the association network, some heuristic score functions are adopted to compute the semantic relatedness between entities [5,26]. In this way, they make a great improvement in measuring the semantic relatedness. However, the adopted heuristic score functions are not extensible and cause the lack of flexibility. In addition, to get the structured information of entities, they need significant preprocessing and data transformation efforts in Wikipedia.

To overcome the weaknesses of association network-based approaches mentioned above, we propose a *Knowledge Association Network (KAN)* to better capture the semantic features of words and entities, which consists of word-level and entity-level based on Wikipedia[1] and DBPedia[2]. The word-level leverages the co-occurrence relationship between words to capture the semantic features of words, and the entity-level exploits the semantic features of entities behind words to enhance the word-level relatedness measurement. As shown in Fig. 1, initially, for a word *apple*, we look for related Wikipedia pages *Apple* and *Apple_Inc*, where the semantic information of word *apple* could be captured by text analysis based on co-occurrence principle, and the entities *Apple* and *Apple_Inc* are utilized to reinforce the semantic information of *apple*. Then given that each Wikipedia page has a corresponding entity on DBPedia, we could further capture word-to-entity and entity-to-entity linking information on DBPedia for the input words. In the entity-level, attribute and topological structure space are utilized to represent semantic features of an entity. In Fig. 1, each orange node denotes an attribute of an entity, which constitutes the attribute space. And the relationships among entities form the topological space of entities, where the relationships are mapped from the original topological structure of the entity network on DBPedia.

In our model, we use two different strategies to perform the relatedness measurement in word and entity level respectively. At the word-level, word2vec [11] is carried out to compare the semantic information of words. At the entity-level, we firstly propose a novel entity embedding model by simultaneously considering the attribute space and topological structure space of entities. The attribute space captures the semantic information of attributes around an entity by minimizing a margin ranking loss function inspired by translation embedding on knowledge graph. The topological structure space utilizes random walk to generate sampled sequences and adopts Skip-gram model to get the entities embedding. Compared to existing association network-based methods [5,26], our method could avoid the significant preprocessing on the Wikipedia dump given the natural mapping relations between DBPedia entities and Wikipedia pages. Besides, the entity embedding model also works better than the heuristic score functions used in previous models [5,26].

The contributions made in this paper include:

1. We construct a knowledge association network to compute the relatedness of word-to-word, word-to-entity and entity-to-entity based on Wikipedia and DBPedia for better semantic relatedness measurement.
2. We propose a novel entity embedding model by simultaneously considering the attribute and topological structure space of entities, which works better than heuristic score functions.

[1] https://en.wikipedia.org/wiki/Main_Page.
[2] http://dbpedia.org.

3. Our experiments conducted on standard datasets for semantic relatedness measurement show that our approach outperforms several benchmarking methods.

The rest of the paper is organized as follows: We first cover the related work in semantic relatedness measurement in Sect. 5, and then give the definition and construction process of knowledge association network in Sect. 2. After that, in Sect. 3 we elaborate our approach for computing the semantic relatedness based on the KAN. Next we introduce our experiments in Sect. 4, then finally conclude this paper in Sect. 6.

2 Knowledge Association Network

In this paper, we consider the entities associated with words to enhance the relatedness measurement of word-to-word, and we build a Knowledge Association Network (KAN) to achieve this purpose. The symbols used in this paper are listed in Table 1.

Table 1. Symbols and their meanings

Symbol	Meaning
KAN	Knowledge Association Network
$G = (W, E, R)$	A graph with word set W, entity set E and edge set R
$G_{\{attr,t\}}$	Attribute space G_{attr} and topological structure space G_t
$R_{\{w,we,e\}}$	Three types of edge set
$f_{\{w,we,e\}}$	Three types of relatedness measurement
$W_{\{cnt,tf\_idf\}}$	Two ways of weighting transition probability in G_t
$e(w)$	Entities set related to word w
$\mathbb{R}_{\eth}$	Attribute space in vector space
$\mathbb{R}_{\approx}$	Topological structure space in vector space

Definition 1 Knowledge Association Network (KAN). *Knowledge association network is a graph $G = (W, E, R)$, where W is the word set in vocabulary, E is the entity set associated with the given words, and edge set R denotes the relationships of word-to-word (R_w), entity-to-entity (R_e), and word-to-entity (R_{we}).*

There are many data resources that contain entities which are relevant to words such as Wikipedia, WordNet and DBPedia etc. WordNet provides precise lexical information but lacks adequate semantic information. Wikipedia is a large corpus where entities are described by natural language, that provides abundant unstructured semantic information. Recently, plenty of knowledge graphs are

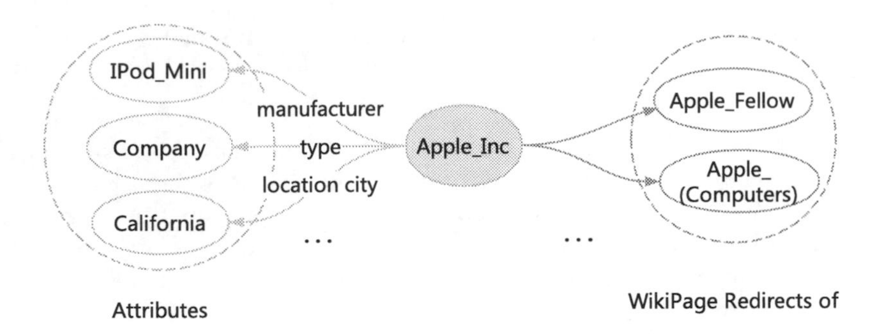

Fig. 2. The semantic features around an entity

established to hold structured knowledge. For example, DBPedia consists of a great number of entities and structured RDF format triples extracted from Wikipedia.

In this paper, we consider the DBPedia as entities database to avoid the significant preprocessing and data transformation efforts in Wikipedia [5,26]. To compute the relatedness of entity-to-entity, we consider two major factors in DBPedia: attributes information and topological structure. The attributes of an entity include the properties, categories, ontology information and some other information which enhance the entity itself. The topological structure reflects the relation between entities on the basis of a special predicate *WikiPageRedirectOf*, that means two entities appear in the same Wikipedia page.

As shown in Fig. 2, for the technology company Apple described as *Apple_Inc* in DBPedia, we get its attributes, that is, "Apple is the manufacturer of IPod Mini (properties)", "Apple is a company (categories)" etc. The relationship descriptions (e.g. "manufacturer", "is-a") are named on the basis of ontology language that contains affluent semantic information. In the aspect of links among other co-occurrence entities in the same Wikipedia page, there are *Apple_Fellow* and *Apple_(Computers)* in accordance with the special relationship *WikipageRedirectOf*.

To distinguish different semantic features of entities conveniently, we denote the attributes of an entity as attributes graph $G_{attr} = \{a_1, a_2, ..., a_j\}$, where a_i denotes an attribute. We define topological structure as $G_t = G(E, R_{redirect})$, where E is a set of entities connected by *WikiPageRedirectOf* (i.e. $R_{redirect}$).

In Wikipedia, a page and it's corresponding DBPedia entity describe the same entity. The predicate called *wikiPageID* reflects this mapping by one unique id, which can be obtained by the *Gensim*[3]. We can get the unique corresponding entity in DBPedia by *wikiPageID* and SPARQL endpoint[4]. For example, the id of Wikipedia page *Apple Inc* is 856, then we can use a simple query to get the corresponding entity name *Apple_Inc*:

[3] https://radimrehurek.com/gensim/wiki.html.
[4] http://dbpedia.org/sparql.

```
PREFIX dbo: <http://dbpedia.org/ontology/>
SELECT ?E WHERE {
    ?E dbo:wikiPageID 856.
}
```

3 Semantic Relatedness Measurement

We give an overview of our model in Fig. 3 where solid lines lead the flow of model and dotted lines demonstrate an additional function from source part to target part, which illustrates the construction of KAN and the relatedness measurement. (1) After an ordinary preprocessing in Wikipedia, for the words in vocabulary, we can get a mapping between words and pages in Wikipedia. (2) Then we query the unique entity by the page id by DBPedia SPARQL endpoint. (3) For the relatedness of entity-to-entity, we divide it into attribute and topological structure and adopt different models within it. Finally we combine three kinds of relatedness measurement word-to-word, word-to-entity and entity-to-entity to form the final semantic relatedness measurement.

3.1 Word-to-Word

The semantic relatedness in word level is mainly measured by (1) distributed vector representation such as word2vec [11] and GloVe [13] etc. (2) word co-occurrence [4,20], which means two words are relevant when they appear in a given window size. Experimental results prove that distributed vector representation works better in computing semantic relatedness [11]. Therefore in this paper, we abandon co-occurrence-based methods and adopt word2vec to train the Wikipedia corpus to product effective vector representation for each word. Formally, let $\overrightarrow{v}_i$ and $\overrightarrow{v}_j$ denote the vector representation of w_i and w_j which can be utilized to calculate the semantic relatedness $f_w(w_i, w_j)$ between w_i and w_j at the word-level based on cosine function, we have:

$$f_w(w_i, w_j) = cos(\overrightarrow{v}_i, \overrightarrow{v}_j) = \frac{\overrightarrow{v}_i \cdot \overrightarrow{v}_j}{\|\overrightarrow{v}_i\| \, \|\overrightarrow{v}_j\|} \quad (1)$$

The word2vec includes skip-gram and CBOW models, using either hierarchical softmax or negative sampling. The combination of skip-gram and negative sampling are used frequently and are effective experimentally. We choose this training program accordingly. The detailed parameters setting can be seen in experiments.

3.2 Word-to-Entity

In KAN, word-level and entity-level hold the one-to-many relationship. For a given word, several relevant entities will rise from KAN due to the word ambiguity. To measure the degree of association between a word (w) and an entity

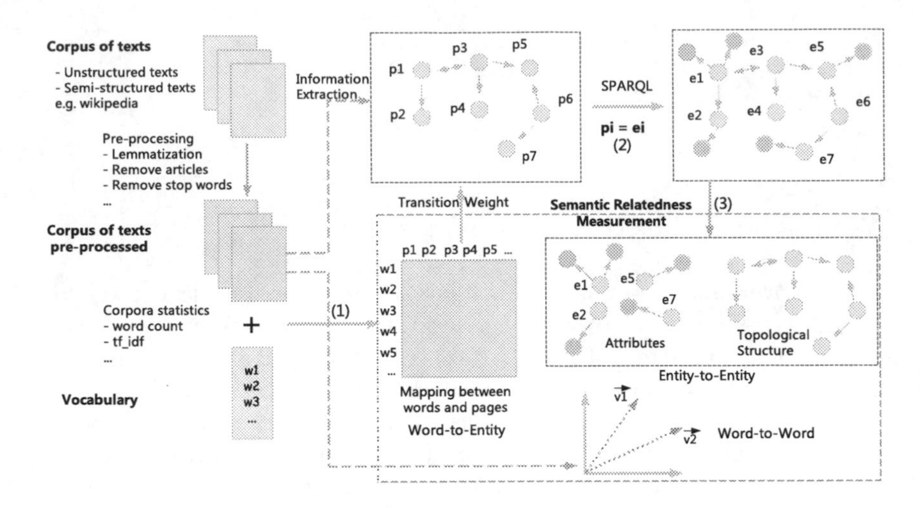

Fig. 3. Semantic relatedness measurement with KAN

(e), (1) some researchers [15] take the co-occurrence times between w and e as the judgement of relatedness, which is insensitive for some common words like *this*, *that* and so on. (2) and some other works [5] consider w and e are closely related if e is the only semantic meaning for word w. They compute the degree of strong connections between only anchor words and their out linked entities based on the *link popularity (LP)* equation,

$$LP(w,e) = \sum_{P} \sum_{w \in S} \frac{\sum_{w' \in S} tf\_idf(w',e)}{\sum_{e' \in e(w)} \sum_{w' \in S} tf\_idf(w',e')} \tag{2}$$

where P indicates a page in Wikipedia, S represents one sentence in P that contains the word w, and w' means every contextual word in S. $e(w)$ is a set of entities which are linked from anchor word w. This method just considers anchor word and out linked entities, but ignores the relevant pages that mention the word. In this paper, we extend the relevant entities $e(w)$ as:

$$e(w) = e_a(w) \cup e_m(w) \tag{3}$$

where $e_a(w)$ is the out linked entities set associated with w, and $e_m(w)$ contains entities that mention the word w but not the out linked entities of w. So we have the *full popularity (FP)* that reflects the degree of connection between w and e:

$$FP(w,e) = \begin{cases} LP(w,e) & e \in e_a(w) \\ \frac{tf\_idf(w,e)}{\sum_{e' \in e_m(w)} tf\_idf(w,e')} & e \in e_m(w) \end{cases} \tag{4}$$

Finally, we have the relatedness of word-to-entity defined as f_{we}:

$$f_{we}(w,e) = \frac{FP(w,e)}{\sum_{e' \in e(w)} FP(w,e')} \tag{5}$$

3.3 Entity-to-Entity

The knowledge association network at the entity-level is fundamentally a multi-relational graph where an entity is described by some discrete attribute and topological structure collectively. It is unreasonable to just consider either of the these information. Two entities may hold totally different attributes but they appear in the similar topological structure and vice verse. The part of attribute holds the detailed semantic information e.g. person A is the friend of B, person B is the member of organization C etc. The topological structure reflects the latent semantic information of co-occurrence relationship of entities. In our model, we adopt two different methods to obtain the vector representation of attribute and topological structure space.

Attribute Space. The straightforward method to embed a set of attributes around an entity is *one-hot*, where when one attribute appears in the attribute space of an entity, the corresponding vector position would be assigned 1, otherwise 0. Nevertheless, a surprisingly large number of attributes in DBPedia bring an insoluble problem for one-hot because of the excessive dimensions. Fortunately, there exists a kind of one-to-many relationship between entities and their attributes, which can be interpreted as a translation operation on the low-dimension entities embedding [2,21]. Suppose that there are N different attributes in our network and the attribute space is denoted as $\mathbb{R}_{\mathfrak{D}}{}^{|N|\times|d|}$, where d is the dimension of vector for one attribute. We combine the relationships and entities to minimize a margin ranking loss over the attribute graph G_{attr}:

$$\mathcal{L} = \sum_{(a,b)\in G^{+}_{attr}} \sum_{b^{-}\in G^{-}_{attr}} [\ell + cos(a,b) - cos(a,b^{-})]_{+} \tag{6}$$

where $[x]_{+} = max(0,x)$, and ℓ is a margin hyperparameter. The G_{attr} contains a set of (h, r, t) triples, that is a head entity h, a relation r and a tail entity t. We select uniformly at random to get positive sample G^{+}_{attr} in two strategies: (i) a consists of the bag of h and r, while b only consists of t; (ii) a consists of h, b consists of r and t. Negative entities b^{-} are sampled from the set of possible triples G^{-}_{attr}. We utilize a k-negative sampling strategy [11] to select k negative pairs for each batch update. The optimization of this method inherits the strategy of stochastic gradient descent (SGD). Each SGD step is one sampling from G^{+}_{attr} in the outer sum.

Topological Structure Space. The topological structure space (G_t) of an entity contains latent semantic information, for example, when somebody browses the Wikipedia page of *Apple_Inc*, there are lots of related entities contained in text description such as *Microsoft_Windows* and *Graphical user interface*, but they are not the attributes of *Apple_Inc*. To consider this latent semantic information, previous works [5,26] make lots of preprocessing in Wikipedia to extract the latent semantic features of entities. To avoid the extraction of link information in Wikipedia, we use DBPedia where a special relation named *WikiPageRedirectOf* connects two entities when one entity's anchor text is mentioned in the Wikipedia page of the other. Then we can get the topological structure space $G_t = G(E, R_{redirect})$, where E is a set of entities and $R_{redirect}$ is the edge set formed by *WikiPageRedirectOf*.

It can be easily seen, G_t is represented as a weighted graph model, where the edges in $R_{redirect}$ hold different transition weights. For instance, somebody is browsing the page of *Apple_Inc* in Wikipedia in which dozens of entities are linked. He wants to know more extended details about *Apple_Inc*, the most several related entities will draw his attention. So he will check the related out linked entities but ignore some other unrelated. It can be seen there are different transition weights from *Apple_Inc* to other entities. Moreover, the transition among different entities is directed, which means G_t is a directed graph as well. Nevertheless, in DBPedia, the raw connections are represented as triples which are unweighted.

To get the weighted graph G_t, suppose that entity e_i and e_j are connected by r_{ij}, the most straightforward way to weight r_{ij} is to consider the occurrence times of the anchor text of e_j in the page of e_i. We regard the anchor text as a single term t_i for e_i. Let $cnt(e_i, e_j)$ denote the co-occurrence times of appearance of t_j in page of e_i. Formally we have the count-based transition weight $W_{cnt}(e_i, e_j)$ from e_i to e_j:

$$W_{cnt}(e_i, e_j) = \frac{cnt(e_i, e_j)}{\sum_{e' \in P_i} cnt(e_i, e')} \tag{7}$$

where P_i denotes the corresponding Wikipedia page of e_i. The $e^{'}$ is one out linked entity in P_i. However, just consider anchor text frequency would give some general frequent terms high degree of relatedness. In order to remedy this weakness, we calculate the *tf_idf*-based (Term Frequency–Inverse Document Frequency) transition weight $W_{tf\_idf}(e_i, e_j)$ from e_i to e_j as follows:

$$W_{tf\_idf}(e_i, e_j) = \frac{tf\_idf(e_i, e_j)}{\sum_{e' \in P_i} tf\_idf(e_i, e')} \tag{8}$$

After getting the weighted G_t, to make the entities are comparable in topological space, we need to embed the entities in expressive vector space. It is easy to understand that the related entities are close to each other in G_t and they hold similar neighborhoods. It requires us to maximize the probability of observing neighborhoods for an entity. Formally, given an entity e_i, we predict its neighborhood entities $(e_0, e_1, ..., e_i, ...e_l)$ with the conditional probability Pr:

$$Pr((e_0, e_1, ..., e_{i-1}, e_{i+1}, ..., e_l)|e_i) \tag{9}$$

How to sample the neighborhood of an entity is widely studied in previous work [6,14]. In this paper, we adopt the randomized walk sampling strategy [6] to get the neighborhoods $N(e_i)$ around the entity e_i.

In order to maximize the probability of Eq. 9 in vector representation, we introduce a mapping function $\Phi : e \in E \rightarrow \mathbb{R}_{\approx}^{|E| \times d}$ where E is the entity set of G_t. Φ is a $|E| \times d$ matrix of parameters which could be obtained by training. For each $e_i \in E$, we can get a d-dimension vector. And our goal is to minimize the following loss function:

$$minimize \ - logPr(N(e_i)|\Phi(e_i)) = -log \prod_{e' \in N(e_i)} Pr(e'|\Phi(e_i)) \quad (10)$$

where $Pr(e'|\Phi(e_i))$ indicates how likely e' appears in neighborhoods of e_i. For each $e' \in N(e_i)$, we adopt the softmax function to normalize the likelihood probability as each e' has a symmetric effect with e_i in feature space [6], so we have conditional probability Pr:

$$Pr(e'|\Phi(e_i)) = \frac{exp(\Phi(e') \cdot \Phi(e_i))}{\sum_{e_k \in N(e_i)} exp(\Phi(e_k) \cdot \Phi(e_i))} \quad (11)$$

Finally, We optimize function 10 using stochastic gradient descent (SGD).

Relatedness of Entity-to-Entity. We can get the embedding for an entity e_i, that consists of attributes embedding ($\overrightarrow{va}_i$) and topological space embedding ($\overrightarrow{vt}_i$). Formally, we formulate the relatedness of entity-to-entity as $f_e(e_i, e_j)$:

$$f_e(e_i, e_j) = \alpha cos(\overrightarrow{va}_i, \overrightarrow{va}_j) + (1 - \alpha) cos(\overrightarrow{vt}_i, \overrightarrow{vt}_j) \quad (12)$$

where $\alpha \in [0, 1]$ is to adjust the weights of two parts.

3.4 Word Semantic Relatedness Measurement *F*

The final semantic relatedness measurement has three parts including word-to-word, word-to-entity and entity-to-entity. We combine the word-to-entity and entity-to-entity as entity-level defined as $F_e(w_i, w_j)$:

$$F_e(w_i, w_j) = \sum_{e_i \in E_i} \sum_{e_j \in E_j} f_{we}(w_i, e_i) f_e(e_i, e_j) f_{we}(w_j, e_j) \quad (13)$$

where E_i is the entities set associated with word w_i. And we denote the word-to-word relatedness as $F_w(w_i, w_j)$ that equals to $f_w(w_i, w_j)$. Finally, we can get the semantic relatedness measurement $F(w_i, w_j)$ in KAN:

$$F(w_i, w_j) = \varphi F_w(w_i, w_j) + (1 - \varphi) F_e(w_i, w_j) \quad (14)$$

where $\varphi \in [0, 1]$ trades off the weight of F_w against F_e.

4 Experiments

In this section, we conduct extensive experiments on different datasets which contain the semantic relatedness measurement by human perceptions. We compute the Pearson correlation coefficient γ, Spearman correlation coefficient ρ and harmonic mean coefficient $\mu = \frac{2\gamma\rho}{\gamma+\rho}$ between results of our experiment and scores of human judgement to evaluate the performance of our model.

4.1 Datasets

The Knowledge Association Network KAN is constructed based on the dump of Wikipedia[5] and DBPedia[6]. The details about the basic datasets are shown in Table 2. The number of entities in DBPedia is larger than that in Wikipedia, since the entities set contain entities extracted from not only Wikipedia but also some other semantic datasets such as ontology language, YAGO and so on. It is necessary to preprocess the Wikipedia before constructing KAN. For each page in Wikipedia, we remove the stop words and punctuations, ignore the shorter pages whose words number less than 50 and some useless namespaces[7] such as *Category, File, Template* without introducing any entity.

Table 2. Wikipedia and DBPedia information

	Entities	Date
Wikipedia	5.5M	2016–10
DBPedia	6.6M	2016–10

4.2 Evaluation

A great number of datasets record the scores of human quantitative judgement for semantic relatedness. We evaluate KAN on three frequently used datasets that are listed in Table 3. Based on the standard datasets, we compare our model with some existing models, containing (1) co-occurrence-based methods: ESA [4], SSA [8], word2vec [11] and SaSA [22]; (2) association network-based methods: AN [26] and HAN [5].

Parameters Tuning. In this paper, it is necessary to determine the following parameters:

- Recall word-to-word, we train word2vec in Wikipedia to get the vector representations for words. And we adopt *100 dimension, 30 window size, Skip-gram model and negative sampling* for word2vec.
- In the section of attributes space embedding, we set *margin* $\ell = 0.05$, *dimension* $d = 100$, *negative sampling number* $k = 50$, and we set the learning rate of SGD as 0.1 to optimize the margin ranking loss.
- In the section of embedding for topological structure space, the Skip-gram model is used for training the sequences of random walk, and we set the *100 dimension, 10 window size* as the basic parameters for training.
- α is proposed for the balance of attributes information and topological structure. φ trades off the weight of word-level against entity-level.

[5] https://dumps.wikimedia.your.org/.
[6] https://wiki.dbpedia.org/downloads-2016-10.
[7] https://en.wikipedia.org/wiki/Wikipedia:Namespace.

Table 3. Word relatedness datasets information

Datasets	Word pairs	Range of score	Reference
MC	30	[0, 4]	Miller and Charles (1991)
RG	65	[0, 4]	Rubenstein and Goodenough (1965)
WS353	353	[0, 10]	Finkelstein et al. (2002)

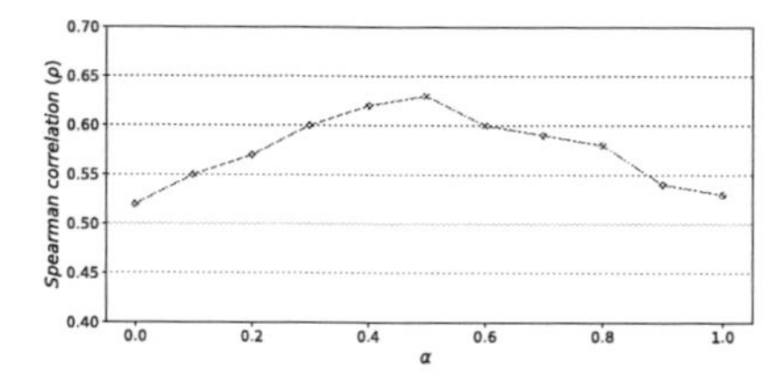

Fig. 4. α tuning on WS-Rel only considering Entity-to-Entity

Fig. 5. Performance with value of λ

In order to get the optimal correlation, we pick *WS-Rel* [1] to tune the parameter α, since there are not many comparison systems in literature report results on this dataset. *WS-Rel* contains 252 pairs of words along with relatedness judgement. We compute word semantic relatedness just on entity-to-entity part (f_e) to tune α, as shown in Fig. 4, Spearman correlation (ρ) increases evidently when the importance of topological structure is raised. And we get the optimal values for α to be 0.5, which means attributes information and topological structure play the same role for semantic relatedness measurement.

Another parameter φ trades off the weight of word-level relatedness F_w against entity-level relatedness F_e. We set tuning rate as 0.1. Figure 5 shows the results w.r.t the multiple value of φ and when $\varphi = 0.2$, we get the largest Spearman correlation (ρ). Obviously, F_w has a leading role and our F_e makes a great supplement for final semantic relatedness measurement.

Comparions Results. Evaluation results of word relatedness on different correlation coefficients are shown in Table 4. Recall embedding for topological structure of our network, there are two strategies to weight the relationship among entities: (1) $W_{cnt}(e_i, e_j)$ denotes the co-occurrence frequency of e_j in page of e_i; (2) $W_{tf\_idf}(e_i, e_j)$ adopts tf_idf to judge how import an entity is to another. Based on these two weight strategies, we construct KAN_{cnt} and KAN_{tf_idf} respectively. We can see that the KAN_{tf_idf} outperforms KAN_{cnt} in different datasets and measurement coefficients, since tf_idf increases proportionally the number of times a term (t) appears in the page of an entity. And the value of tf_idf is offset by the number of pages in Wikipedia that contain the item t, which helps to adjust the weight for the fact that some items appear more frequently in general.

Table 4. Pearson-λ, Spearman-ρ, harmonic mean-μ on the word relatedness datasets

Model	λ			ρ			μ		
	MC	RG	WS353	MC	RG	WS353	MC	RG	WS353
ESA	0.588	- -	0.503	0.727	- -	0.748	0.650	- -	0.602
SSA	0.879	0.861	0.590	0.843	0.833	0.604	0.861	0.847	0.597
word2vec	0.852	0.834	0.633	0.836	0.812	0.645	0.844	0.823	0.639
SaSA	0.886	0.882	0.733	0.855	0.851	0.739	0.870	0.866	0.736
AN_{wiki}	0.865	0.858	0.740	0.848	0.843	0.813	0.856	0.850	0.775
HAN_{wiki}	0.886	0.884	0.772	0.860	0.857	0.826	0.873	0.870	0.798
KAN_{cnt}	0.850	0.826	0.630	0.836	0.805	0.633	0.842	0.816	0.631
KAN_{tf_idf}	**0.892**	**0.887**	**0.783**	**0.866**	**0.861**	**0.835**	**0.879**	**0.874**	**0.808**

When compared with other methods shown in Table 4, our method performs better. AN_{wiki} and HAN_{wiki} get excellent performance on word semantic features relatedness on the idea of *free association network*, which improve the weakness of co-occurrence-based methods. In this paper, we adopt two different model to capture the semantic of attributes (G_{attr}) and topological structure (G_t) in KAN_{tf_idf} and make the model more flexible and expressive.

5 Related Work

Plenty of researchers have studied semantic relatedness between two words and made significant accomplishments, which include:

(i) The *lexical*-based methods measure the semantic relatedness between two words based on some lexical databases such as *WordNet* and *Wikitionary*. WordNet based methods [16] compute semantic relatedness for automatic speech recognition in meetings. However, they do not provide an individual result to reveal the efficiency of semantic relatedness measurements. Wikitionary [25] is introduced as an emerging lexical semantic resource that could be used as a substitute for expert-made resources in AI applications. Other lexical-based methods choose a path based approach [18], which can be utilized with any resource containing concepts connected by lexical semantic relations. Or they adopt a concept vector based approach [4], which is generalized to work on each resource that offers a textual representation of a concept.

(ii) The *co-occurrence*-based methods regard two words are related when they appear in a sentence or a fixed window in corpora texts such as Wikipedia. The initial model WikiRelate! [20] estimates relatedness based on categories in the articles of Wikipedia. Explicit Semantic Analysis (ESA) [4] represents the meaning of articles in a high-dimensional space. WikiRelate! and ESA only leverages texts in Wikipedia but does not consider links among articles. Another model WLM [12] scrutinizes incoming/outgoing links from/to articles instead of exploiting texts in Wikipedia articles. WikiWalk [24] extends the WLM by exploiting not only links that appear in an article but all links, to perform a random walk based on Personalized PageRank. However, those methods are faint to distinguish the different word senses. SensEmbed [9] leverages BabelNet[8] to annotate different word senses in the dump of Wikipedia, and exploits word2vec [11] to train the sense-annotated Wikipedia to get distributed representation of different word senses. Essentially this method is based on the large corpora and needs a significant preprocessing. The REWOrd [15] proposes an approach that exploits the graph nature of RDF and SPARQL query language to access knowledge graph. It not only obtains the comparable result with the state-of-art at that moment, but also avoids the burden of preprocessing and data transformations.

(iii) In order to improve the co-occurrence-based methods, *association network*-based methods is proposed to compute the semantic relatedness between two words utilizing *free association network*, that is, for a given word, the first word that appears in human mind intuitively is the most relevant one. AN [26] is proposed to build an association network based on not only co-occurrences between words, but also the links between Wikipedia pages of entities. Recently, HAN [5] constructs hierarchical association network to capture the association of word-to-word, word-to-entity and entity-to-entity. In this way, they make a great improvement in measuring the semantic relatedness. However, the adopted heuristic score functions are not reliable and cause the lack of flexibility. In addition, to get the semantic information of entities, they need significant preprocessing efforts in Wikipedia.

In this paper, we propose a *Knowledge Association Network* to measure semantic relatedness. Our model avoids the preprocessing of Wikipedia and considers the attribute and topological structure space simultaneously to capture the semantic features of entities. Experimental results show that our model outperforms the benchmarking models.

6 Conclusion

In this work, we focus on computing semantic relatedness to get an approximation to human judgement. We utilize the DBPedia which is derived from Wikipedia as background knowledge to construct a Knowledge Association Network. To measure the word semantic relatedness, we propose a flexible and expressive model to represent entities behind the words, where attribute and topological structure information of entities are embedded in vector space simultaneously. The experiments based on benchmarking datasets show that our model outperforms the state-of-the-art models.

Acknowledgments. This research is partially supported by National Natural Science Foundation of China (Grant No. 61572335, 61632016), the Natural Science Research

[8] http://babelnet.org.

Project of Jiangsu Higher Education Institution (No. 17KJA520003), and the Dongguan Innovative Research Team Program (No.2018607201008).

References

1. Agirre, E., Alfonseca, E., Hall, K.B., Kravalova, J., Pasca, M., Soroa, A.: A study on similarity and relatedness using distributional and WordNet-based approaches. In: Conference of the North American Chapter of the Association of Computational Linguistics, Proceedings, Boulder, Colorado, USA, 31 May–5 June 2009, pp. 19–27 (2009)
2. Bordes, A., Usunier, N., García-Durán, A., Weston, J., Yakhnenko, O.: Translating embeddings for modeling multi-relational data. In: 27th Annual Conference on Neural Information Processing Systems, Lake Tahoe, Nevada, USA, 5–8 December 2013, pp. 2787–2795 (2013)
3. Fan, J., Lu, M., Ooi, B.C., Tan, W., Zhang, M.: A hybrid machine-crowdsourcing system for matching web tables. In: IEEE 30th International Conference on Data Engineering, Chicago, ICDE 2014, pp. 976–987 (2014)
4. Gabrilovich, E., Markovitch, S.: Computing semantic relatedness using Wikipedia-based explicit semantic analysis. In: IJCAI, Hyderabad, India, 6–12 January 2007, pp. 1606–1611 (2007)
5. Gong, X., Xu, H., Huang, L.: HAN: hierarchical association network for computing semantic relatedness. In: AAAI, New Orleans, Louisiana, USA, 2–7 February 2018 (2018)
6. Grover, A., Leskovec, J.: Node2vec: scalable feature learning for networks. In: ACM SIGKDD, San Francisco, CA, USA, 13–17 August 2016, pp. 855–864 (2016)
7. Han, X., Zhao, J.: Structural semantic relatedness: a knowledge-based method to named entity disambiguation. In: ACL, Uppsala, Sweden, 11–16 July 2010, pp. 50–59 (2010)
8. Hassan, S., Mihalcea, R.: Semantic relatedness using salient semantic analysis. In: AAAI, San Francisco, California, USA, 7–11 August 2011 (2011)
9. Iacobacci, I., Pilehvar, M.T., Navigli, R.: SensEmbed: learning sense embeddings for word and relational similarity. In: ACL, Beijing, China, 26–31 July 2015, Volume 1: Long Papers, pp. 95–105 (2015)
10. Leong, C.W., Mihalcea, R.: Measuring the semantic relatedness between words and images. In: IWCS, Oxford, UK, 12–14 January 2011 (2011)
11. Mikolov, T., Chen, K., Corrado, G., Dean, J.: Efficient estimation of word representations in vector space. CoRR (2013)
12. Milne, D., Witten, I.H.: An effective, low-cost measure of semantic relatedness obtained from Wikipedia links. In: AAAI Workshop on Wikipedia and Artificial Intelligence: An Evolving Synergy, pp. 25–30 (2008)
13. Pennington, J., Socher, R., Manning, C.D.: Glove: global vectors for word representation. In: EMNLP, Doha, Qatar, 25–29 October 2014, pp. 1532–1543 (2014)
14. Perozzi, B., Al-Rfou, R., Skiena, S.: DeepWalk: online learning of social representations. In: ACM SIGKDD, KDD 2014, pp. 701–710 (2014)
15. Pirrò, G.: Reword: semantic relatedness in the web of data. In: AAAI, Toronto, Ontario, Canada, 22–26 July 2012 (2012)
16. Pucher, M.: WordNet-based semantic relatedness measures in automatic speech recognition for meetings. In: ACL, Prague, Czech Republic, 23–30 June 2007 (2007)

17. Qadir, A., Mendes, P.N., Gruhl, D., Lewis, N.: Semantic lexicon induction from twitter with pattern relatedness and flexible term length. In: AAAI, Austin, Texas, USA, 25–30 January 2015, pp. 2432–2439 (2015)
18. Rada, R., Mili, H., Bicknell, E., Blettner, M.: Development and application of a metric on semantic nets. IEEE Trans. Syst. Man Cybern. **19**(1), 17–30 (1989)
19. Sandulescu, V., Ester, M.: Detecting singleton review spammers using semantic similarity. In: WWW, Florence, Italy, 18–22 May 2015, Companion Volume, pp. 971–976 (2015)
20. Strube, M., Ponzetto, S.P.: Wikirelate! computing semantic relatedness using Wikipedia. In: AAAI, Boston, Massachusetts, USA, 16–20 July 2006, pp. 1419–1424 (2006)
21. Wu, L.Y., Fisch, A., Chopra, S., Adams, K., Bordes, A., Weston, J.: StarSpace: embed all the things! In: AAAI, New Orleans, Louisiana, USA, 2–7 February 2018, pp. 5569–5577 (2018)
22. Wu, Z., Giles, C.L.: Sense-aware semantic analysis: a multi-prototype word representation model using Wikipedia. In: AAAI, Austin, Texas, USA, 25–30 January 2015, pp. 2188–2194 (2015)
23. Yang, J., Fan, J., Wei, Z., Li, G., Liu, T., Du, X.: Cost-effective data annotation using game-based crowdsourcing. PVLDB **12**(1), 57–70 (2018)
24. Yeh, E., Ramage, D., Manning, C.D., Agirre, E., Soroa, A.: WikiWalk: random walks on Wikipedia for semantic relatedness. In: Proceedings of the Workshop on Graph-based Methods for Natural Language Processing, Singapore, 7 August 2009, pp. 41–49 (2009)
25. Zesch, T., Müller, C., Gurevych, I.: Using wiktionary for computing semantic relatedness. In: AAAI, Chicago, Illinois, USA, 13–17 July 2008, pp. 861–866 (2008)
26. Zhang, K., Zhu, K.Q., Hwang, S.: An association network for computing semantic relatedness. In: AAAI, Austin, Texas, USA, 25–30 January 2015, pp. 593–600 (2015)
27. Zhang, W., Feng, W., Wang, J.: Integrating semantic relatedness and words' intrinsic features for keyword extraction. In: IJCAI, Beijing, China, 3–9 August 2013, pp. 2225–2231 (2013)
28. Zhu, G., Iglesias, C.A.: Computing semantic similarity of concepts in knowledge graphs. IEEE Trans. Knowl. Data Eng. **29**(1), 72–85 (2017)

SINE: Side Information Network Embedding

Zitai Chen[1,2], Tongzhao Cai[1,2], Chuan Chen[1,2](✉), Zibin Zheng[1,2], and Guohui Ling[3]

[1] School of Data and Computer Science, Sun Yat-sen University, Guangzhou, China
chenchuan@mail.sysu.edu.cn
[2] National Engineering Research Center of Digital Life, Sun Yat-sen University, Guangzhou, China
[3] Data Center of Wechat Group, Tencent Technology, Shenzhen, China

Abstract. Network embedding learns low-dimensional features for nodes in a network, which benefits the downstream tasks like link prediction and node classification. Real-world networks are often accompanied with rich side information, such as attributes and labels, while most of the efforts on network embedding are devoted to preserving the pure network structure. Integrating side information is a challenging task since the effects of different attributes vary with nodes and the unlabeled nodes can be influenced by diverse labels from neighbors, not to mention the heterogeneity and incompleteness. To overcome this issue, we propose Side Information Network Embedding (SINE), a novel and flexible framework using multiple side information to learn a node representation. SINE defines a flexible and semantical neighborhood to model the inscape of each node and designs a random walk scheme to explore this neighborhood. It can incorporate different attributes information with particular emphasis depending on the characteristics of each node. And label information can be both explicitly and potentially integrated into the representation. We evaluate our method and existing state-of-the-art methods on the tasks of multi-class classification. The experimental results on 5 real-world datasets demonstrate that our method outperforms other methods on the networks with side information.

Keywords: Network embedding · Random walk · Multilayer network

1 Introduction

Network data are ubiquitous in the real world, ranging from social networks like Wechat and Facebook, marketing networks, airline transportation networks to academic citation networks, to name a few. Abundant useful knowledge is concealed in these networks which can benefit network analysis and applications in reality. For instance, in social networks, link prediction analysis could lower

Z. Chen and T. Cai—Contributed equally to this work.

G. Li et al. (Eds.): DASFAA 2019, LNCS 11446, pp. 692–708, 2019.
https://doi.org/10.1007/978-3-030-18576-3_41

the cost and difficulty for users to seek friends online as well as offer a chance for service providers to improve their user experience. As the size of networks grows, opportunities come with challenges. On the one hand, it enriches the network treasure house and provides ample materials for network researchers. On the other hand, more complex relationships coupled in the networks are increasing the challenges dramatically in the analysis tasks.

Recently, as a novel dimensional reduction technique in analyzing large-scale networks, network embedding is proposed and has attracted a surge of research attention in many researches ranging from data mining, machine learning to mathematics. The main target of network embedding is to preserve as much information as possible from the network with a low dimension representation for each node. To achieve this goal, multiple approaches have been proposed, such as GraRep [2], DeepWalk [14], LINE [17] and SDNE [20]. More importantly, a lot of real-world applications have demonstrated their value in the downstream learning tasks, such as node classification, link prediction and data visualization.

Despite the improvement it gains, current works of network embedding mostly concentrate on preserving the structure of pure networks. In the real world, nodes in a network are usually accompanied with rich side information, such as attributes and labels. The attribute homophily theories [9,10] show the strong connection between node attributes and topological structure. They depend on and influence each other in the network. For instance, articles in Wikipedia might not only cite or be cited by other related articles, but also contain a detailed explanation of the specific object, which helps in link prediction tasks to precisely provide editors with highly related articles. Moreover, labels such as group or community categories also provide useful information to assist in network learning. Even a limited number of labeled nodes can conduct a discriminative embedding. Taking Wechat as an example, users in the same group chat tend to share posts or links of related themes which is informative in precise advertisement targeting. Thus, the importance of side information is self-evident, whilst network embeddings ignoring the side information not only weaken the ability of expression but also blur the representations.

However, it is not easy for the pure network embedding methods like DeepWalk to incorporate additional information during its random walk in the original network since the heterogeneity and incompleteness complicate the situation. Thus, applying the pure network embedding methods directly is problematic. In contrast to the pure network embedding, side information network embedding targets at leveraging the discrepancy of the heterogeneous data sources and distilling the complementary information. What's more, attributes and labels might be sparse, noisy and incomplete. Hence, it is nontrivial to study the problem of fusing labels and attributes into network structure and learning discriminative representations for network nodes. Some recent works have scratched the surface of this topic, yet various problems exist. They either lack careful and specific consideration of side information or are trapped in time-consuming learning models. Exhaustive discussions are given in Sect. 2.

In this paper, we investigate the side information network embedding deeply. Inspired by the groundbreaking work DeepWalk and the constructive follow up work Node2vec on the pure network, we propose an innovative random walk scheme to integrate multiple knowledge on side information network. We aim at answering the following questions: (1) How to incorporate topological structure, attribute information and node labels into a unified representation meanwhile tackling the incomplete, sparse and noisy problem accompanied; (2) How much does this random walk scheme contribute to downstream learning tasks like node classification.

Our main contribution is a flexible framework for learning latent representations for the attributed network with a limited number of labels, called Side Information Network Embedding (SINE). The key ideas of SINE are:

- Measure the node relationships with others on attributes information and then evaluate the importance of attributes and geometric structure for each node individually. In contrast to treating information of each node unanimously, learning on a discriminative data makes the delicate embedding possible.
- Establish *label hubs* and *label hyperlinks* for the labeled nodes to communicate with each other explicitly. And we design a label biased random walk scheme to integrate label information potentially.
- Generate sampled contexts (neighborhoods) for nodes, which contain immediate geometric neighbors, similar nodes in the aspect of attributes and nodes explicitly or latently sharing the same label. Thus, in such all-side neighborhood built with nodes in a heterogeneous relationship, nodes can be modeled with more precise representation. The more frequently two nodes appear in the similar neighborhoods, the more likely they possess similar information.

The rest of this paper is organized as follows: First a brief overview of pure network and side information network embedding is provided, followed by the proposed SINE framework. Then sound experiments are presented. Finally, conclusion and future works are discussed.

2 Related Work

Network embedding can be traced back to the manifold learning, which aims to analyze the structure of manifold and map it into a low dimension Euclidean space to facilitate the machine learning algorithms. However, these methods, such as IsoMap [18], LLE [16], LE [1] and LPP [5], are trapped in the time-consuming eigen-decomposition and not applicable for large scale network embedding.

Recently, inspired by the Skip-Gram [11] learning word representation from its context, [14] propose DeepWalk that generates node neighborhoods with a truncated random walk to simulate the relationship between words and sentences, and bring prosperity to the embedding community. In Node2vec [4], a follow up work of DeepWalk, authors propose a biased random walk which can explore neighborhood under control of extra parameters. To preserve the structure similarity, Struc2vec [15] generates node contexts on the graph which is newly constructed based on structure similarity. On the other line of pure network embedding, a variety of methods [2,15,17,20] are proposed. For example, to preserve first- and second-order proximity, LINE [17] proposes a joint probability and conditional probability model while SDNE [20] adopts an autoencoder model. However, losing sight of labels and attributes may set a limit on the performance of all these topological structure based methods.

Some recent efforts have explored the possibility of integrating side information of the node to learn a better representation. TADW [21] employs an inductive matrix factorization to integrate attributes. SNE [8] proposes a multi-layer perceptron to model the reconstruction error by concatenating attribute record as an input. While they don't model the attribute affinity, which is essential for network analysis. TriDNR [13] learns three kinds of relation node-attribute, inter-node and attribute-label in a coupled deep model. Label information is not used for inter-node relationship modeling, which might weaken its representation power. LANE [7] learns a smooth representation from three individual representations of structure, attribute and label. AANE [6] accelerates the joint learning process of attribute and network structure. However, they equally treat the effect of attribute information on each node, which is too coarse in learning the representations. MMDW [19] only integrates label information by a semi-supervised model, which jointly optimizes the matrix factorization of adjacency matrix and the max-margin classifier of SVM. DANE [3] learns a consistency from the structure and attribute representation which captures the nonlinearity encoded by two autoencoders. However, it suffers from the high computational drawbacks. All in all, the existing methods come across various deficiency. To overcome the problems they meet, we propose a new model with strong pertinence.

3 Framework of SINE

In this section, the problem formulation is firstly given. Then we present the feature learning framework in our method. Next, we introduce attribute embedding module followed by the label embedding module.

3.1 Problem Formulation

We consider the problem of learning node representations in three aspects: structure, attributes and labels. Let $G = \{V, E, W\}$ be a pure network, where V represents the nodes of the network, $E \subseteq (V \times V)$ are the topological connections and W are edge weights (one for unweighted network). With side information, network is further denoted as $G_S = \{V, E, W, X, Y\}$, with multiple attributes $X = \{X^{(i)}\}_{i=1}^{m}$, $X^{(i)} \in \mathbb{R}^{|V| \times s_i}$ where m is the number of attributes and s_i is the size of the ith feature space, and $Y \in \mathbb{R}^{|V| \times |\mathcal{Y}|}$ where $\mathcal{Y}$ is the set of labels. We define a function $\mathcal{L} : V \rightarrow \mathcal{Y}$, and $\mathcal{L}(u) = i$ if node u is labeled with i.

Formally, we aim to learn the low-dimensional representation $H \in \mathbb{R}^{|V| \times d}$ which can incorporate information from three sources of data. As a result, H could achieve better performance in the downstream tasks such as node classification. We denote h_u, a column of H^T, as the representation of node u.

3.2 Feature Learning Framework

We extend the Skip-Gram architecture [11] to the side information network. Formally, in network G_S we maximize the log-probability of observing a network neighborhood $N_S(u)$ for node u conditioned on its representation h_u:

$$\max_H \sum_{u \in V} \log Pr(N_S(u)|h_u). \tag{1}$$

With the assumption of conditional independence and symmetry effects from neighbors, Eq. 1 simplifies to:

$$\max_H \sum_{u \in V} \Big[\log \lambda_u + \sum_{v \in N_S(u)} h_v^T \cdot h_u\Big], \tag{2}$$

where $\lambda_u = \sum_{v \in V} \exp(h_v^T \cdot h_u)$. We can see that nodes in a more similar neighborhood would have similar representations. And a semantically rich neighborhood can more precisely describe the intrinsic correlation on the node. In the following subsections, we will propose our method to integrate side information into the neighborhood. As for the problem of the expensive computational cost on λ_u in Eq. 2, negative sampling [12] is adopted. We optimize Eq. 2 using stochastic gradient ascent over the model parameters defining the features h.

3.3 Measure the Attribute Importance

In contrast to assuming attribute information on different nodes is independent like TADW and SNE, we measure the correlation between nodes with respect to attributes. In detail, a kernel method $\mathcal{K}$ is taken to measure the attribute affinity between any pair of nodes: $\mathcal{K}(u, v) = \phi(X_u) \cdot \phi(X_v)$. We construct the attribute network A that encodes the affinity between two nodes. The edge weight between two nodes u and v is then given by:

$$A_{uv} = \mathcal{K}(u, v), \forall u, v \in V. \tag{3}$$

In G_S, now we have a stack of information networks, 1 topological network G and m weighted networks $\{A^{(i)}\}_{i=1}^m$ built from diverse attributes.

Since the attributes information and topological knowledge are concealed in different networks, we ought to learn the representation from each of the networks. A straightforward way is to build different neighborhoods $N_S^{(i)}(u)$ ($N_S^{(0)}(u)$ denote the neighborhood of node u on G) for each node u on each network of $G \cup \{A^{(i)}\}_{i=1}^m$ and then concatenate their representation learned from respective Skip-Gram models as the final representation. Although from an information preserving view point concatenating different representations could maintain characteristics in diverse networks, it neither alleviates the effect of noise nor distills information hidden across representations from the perspective of integrating information. Furthermore, any incomplete attribute information, which is quite common in real-world datasets, can crash it down for the unobserved node in one of the networks. Another way is to combine neighborhoods $\bigcup_i N_S^{(i)}(u)$ extracted from different networks and then learn the representation from a unique Skip-Gram model. Yet the combination that treats all the side information equally without discrimination for the individual node is careless and unacceptable in the analysis, not to mention the expensive computational cost of building multiple neighborhoods for a node. All in all, how to discriminatingly learn the side information and efficiently sample node neighborhood matters.

Supported by the analysis above, we first propose a measurement on the neighborhood of attribute affinity networks to evaluate the local property. Intuitively, the more similar neighbor is, the more attention should be paid to exploring this neighbor. Exploring the neighborhood shares the same principle. To this end, we define the local cohesion of node u on $A^{(i)}$ as follow:

$$\rho_u^{(i)} = \frac{\bar{a}_u^{(i)}}{\bar{a}^{(i)}} = \frac{avg_t(A_{ut}^{(i)})}{avg_{s,t}(A_{st}^{(i)})}, \tag{4}$$

where $\bar{a}^{(i)}$ is the average edge weight of the ith complete affinity network $A^{(i)}$ and $\bar{a}_u^{(i)}$ is the average weight of edges that associate with node u w.r.t. $A^{(i)}$. Thus, the larger $\rho_u^{(i)}$ is, the more informative immediate neighbors are provided. Then, by comparing the strength of node's local cohesion in different networks, we can distinguish the importance of different attributes for a specific node. In other words, if neighbors are more similar with node u in a certain network than in others, this network should undertake more responsibility for exploring neighbors. For the importance of topology, we can calculate in the same way. We denote $A^{(0)} = G$, $A_{uv}^{(0)} = W_{uv}$ and $\rho_u^{(0)} = 1$ for unweighted network, which is also included in Eq. 4.

Then we propose a multi-network random walk strategy to generate node neighbors $N_S(u)$. Walking across $\{A^{(i)}\}_{i=0}^m$ generates a semantically rich node sequence that incorporates diverse node relationships (or similarities) from different networks. After that, we can construct a neighborhood with multi-relation neighbors for each node. In the proposed random walk scheme, we first decide "which network should be traversed" by $\rho_u^{(i)}$, namely choose the more important data source for node u. The probability is proportional to the importance, in particular:

$$P(u, A^{(i)}) = \frac{\rho_u^{(i)}}{\sum_{i=0}^m \rho_u^{(i)}}. \tag{5}$$

And then we carry out the weighted random walk in the chosen network (e.g. $A^{(i)}$) with the probability as follow:

$$P_A(u,v) = \frac{A^{(i)}_{uv}}{\sum_{x \in V} A^{(i)}_{ux}}. \quad (6)$$

3.4 Fuse the Label Information

Modeling label information is entirely different from attributes, the other kind of side information. Labels are much more refined and scarce in networks. Owning to the sparsity, if we treat labels like attributes to construct an independent network, there would be two problems: Firstly, a great number of nodes without labels will be absent in this network; Secondly, the linkage connecting nodes sharing same the label will build information isolated island, which has no assistance to other nodes. In network analysis, it is always assumed that the node's label is highly correlated to the topological structure and could be affected by its labeled neighbors according to their similarity. We propose two ways to explicitly and potentially fuse labels information in the topological neighborhood as shown in Fig. 1.

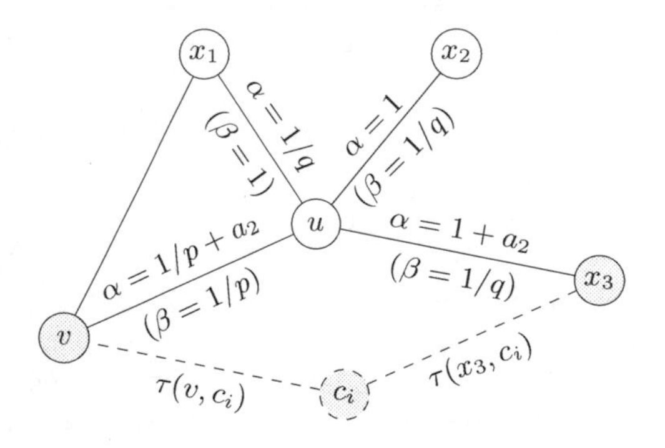

Fig. 1. Illustration of label incorporation way. The nodes colored in cyan are with the same label and the others' label are unknown. The explicit way: The *label hub* c_i is linked with *label hyperlink* (e.g. (v, c_i) and (x_3, c_i)) presented in dashed line. Nodes sharing the same label can walk to each other via their common *label hub*. The potential way: The following example is given to show the influence of node sequence and restrict the influence within 2nd order, which compatible with Node2vec. The walk just transitioned from labeled node v to unlabeled node u and is now evaluating its next step out of node v. Edge notations indicate search bias α for SINE and β for Node2vec. α comprise of bias from topology and label.

To take advantage of label's guidance in gathering node together, we first introduce the notions of *label hyperlink* and *label hub* that help to explicitly learn the label information in the random walk procedure. By building an imaginary *label hub* $c_i, i \in \mathcal{Y}$ for each label on network G, nodes with same label can connect to each other through the *label hyperlink* to the corresponding *label hub*. In particular, the unnormalized probability of walking through *label hyperlink* is defined as follow:

$$\tau(u, c_i) = \begin{cases} \gamma, & \text{if } \mathcal{L}(u) = i, \\ 0, & \text{otherwise}, \end{cases} \tag{7}$$

where γ is a hyperparameter. This explicit method will directly bridge the gap between labeled nodes which are not so close to each other in the topological network. It is also reasonable that nodes with the same label are much closer than those nodes with different labels. In this way, nodes in the neighborhood containing the same label hub are more likely to have similar representations than those who don't.

However, the explicit *label hub* method restricts the influence of label within the labeled nodes, and can not spread the labeled nodes' information to affect the unlabeled neighbors. Thus, we resort to the random walk sequence for a helping hand. Intuitively, a node sequence would be more likely to walk to the related labeled community where it came from. Since nodes in the same community are similar both in topology and label and the node sequence can be regarded as a sampling of the corresponding community, the alternative nodes that are either immediate neighbors of sequence or sharing the same label in the sequence would be more attractive. To measure the attraction of the alternative nodes, we present the biased random walk with two additive parts: topological and labeled parts. Consider a random walk that has a traversed node sequence $T = \{u_i\}_{i=1}^n$ with length n and now resides at node u_n (Fig. 1). The walk now defines the unnormalized transition probability of its neighbor x as follow: $\tau(u_n, x) = \alpha \cdot w_{u_n,x}$, where $\alpha = \alpha_{topo} + \alpha_{label}$,

$$\alpha_{topo} = \begin{cases} 1/p, & \text{if } d(\{u_i\}_{n-m}^{n-1}, x) = 0, \\ 1/q, & \text{if } d(\{u_i\}_{n-m}^{n-1}, x) = 1, \\ 1, & \text{otherwise}, \end{cases} \tag{8}$$

$$\alpha_{label} = \begin{cases} \sum_{i=0}^{m} I(\mathcal{L}(u_{n-i}) = \mathcal{L}(x))a_i, & \text{if } x \text{ is labeled}, \\ 0, & \text{otherwise}, \end{cases} \tag{9}$$

and $d(U, x)$ denotes the shortest distance between node x and nodes in set U, I is the indicator function, m is the range of influence of T and $\{a_i\}_{i=1}^m$ controls the label influence of different distance. To make it clear, α_{topo} controls the sequence to revisit T with bias $1/p$ and walk around T with bias $1/q$. While the α_{label} controls the probability of traversing the neighbors with label that has been visited in T. We perform the label biased random walk with the probability as follow:

$$P_G(u, v) = \frac{\tau(u, v)}{\sum_{x \in V \cup \{c_i\}_{i=1}^{|\mathcal{Y}|}} \tau(u, x)}. \tag{10}$$

It occurs to us that when we restrict the influence of T within the last two nodes (i.e. $m = 1$) when computing α_{topo}, it is similar to Node2vecWalk defined in [4] with exchanging parameters 1 and $1/q$. We denote the bias as β and explain in Fig. 1. The pseudocode for SINE Walk is given in Algorithm 1. The time complexity analysis of this algorithm is given in the experiment section.

To sum up, by generating the node sequences in the newly designed network with the proposed random walk scheme, we can incorporate topology, attributes and labels information into each node's neighborhood $N_S(u)$. Then we can learn the node representation h_u by solving Eq. 2 with stochastic gradient ascent method.

Algorithm 1. The SINE Walk

Input: Start node u, networks $\{A^{(i)}\}_{i=0}^{m}$, walk length l, label hub weight γ, revisit p, look-around q, label influence $\phi(d)$
Output: node sequence T
Initialize T to empty;
Append u to T;
for *iter* = 1 *to* l **do**
  Let *curr* be the last node of T;
  Sample *Graph* from $\{A^{(i)}\}$ with Eq. 5 ;
  V_{curr} = GetNeighbors(*curr* $\in$ *Graph*);
  if *Graph is G* **then**
    $V_C = \{c_i\}_{i=1}^{|\mathcal{Y}|}$;
    Sample *node* from $V_{curr} \cup V_C$ with Eq. 10;
  else
    Sample *node* from V_{curr} with Eq. 6;
  end
  Append s to T;
end
return T;

4 Experiments

In this section, we conduct experiments to evaluate the effectiveness of our proposed framework SINE. In particular, we want to answer the following questions.

(1) What are the impact of attributes information on network embedding and how effective is the multi-network random walk strategy to incorporate attributes?
(2) How effective is the guidance impact of the label in the label biased random walk scheme?
(3) How effective are the node representations learned by SINE compared with other state-of-the-art methods in the downstream tasks?

Table 1. Statistics of the dataset

Dataset	Node	Edge	Attribute	Label
BC	5,196	171,743	8,189	6
Flickr	7,575	239,738	12,047	9
Cora	2,708	5,429	1,433	7
Citeseer	3,312	4,732	3,703	6
Wiki	2,405	17,981	4,973	19

4.1 Datasets

In our experiments, we employ 5 real-world datasets: **BlogCatalog (BC), Flicker, Cora, Citeseer** and **Wiki**. All of them are publicly available, and specially the first two have been used in [7]. **BlogCatalog** and **Flickr** are social media networks. Each node is a user and links are the interaction between them. We take their descriptions as the attributes and the groups or categories they joined as labels. **Cora**, **Citeseer** and **Wiki** are citation networks. Each node is a publication and the links are citation relationships between them. The attribute of each node is the bag-of-words representation of the corresponding paper. Statistics of the datasets are summarized in Table 1. Note that all these datasets provide only one attribute feature.

4.2 Baseline Methods

We compare our method with 7 baseline methods. To evaluate the contribution of the side information, two pure network embedding methods, four attributed network embedding methods and a labeled attributed network embedding method are used for comparison. The first category contains **DeepWalk** [14] and **Node2vec** [4]. The second category includes **AANE** [6], **TADW** [21], **SNE** [8] and **DANE** [3]. The last one contains **LANE** [7].

4.3 Metric and Parameter Settings

We perform the multi-class node classification task to evaluate the quality of node representations learned by different methods. To be more specific, we randomly select some portion of the nodes as training set and the remaining as a test set. We train a one-vs-rest SVM classifier on the training set and evaluate it on the test set. For each training ratio, we repeat the trial for 10 times and report the average results. To measure the classification result, we employ Micro-F1 and Macro-F1 scores as metrics.

In SINE, we compute the attribute affinity network with $\mathcal{K}(\cdot, \cdot)$ defined as cosine similarity of attributes. In experiments, we only preserve the top 20 similar neighbors for each node, randomly sample 20 neighbors when performing label biased random walk, and restrict the attraction of the sequence nodes within two step with $a_1 = r, a_2 = s$, which is the trade-off between the computational cost and the accuracy. The default parameters of SINE are set as follows: window size $k = 5$, walks per node $t = 20$, walk length $l = 20$, label biased random walk parameters $p = 4$, $q = 4$, $r = s = 4$, label hub weight $\gamma = 0.5$. The label ratio used for embedding is 10%. For fairness of comparison, the dimension of embedding vectors d is set to 100 for all the methods. The parameters of DeepWalk and Node2vec are kept the same with SINE. The rest parameters for other algorithms are set following the suggestion in their original papers or source codes.

Table 2. Micro-F1 score of classification

Datasets	Ratio	SINE	LANE	AANE	TADW	SNE	DANE	DW	Node2vec
BC	10%	**0.8459**	0.5696	0.7036	0.7502	0.5714	0.7404	0.3561	0.5750
	20%	**0.8805**	0.6543	0.7756	0.7623	0.6201	0.7907	0.4982	0.6317
	30%	**0.8959**	0.6915	0.8103	0.7972	0.6515	0.8084	0.5295	0.6477
	40%	**0.8991**	0.6987	0.8261	0.8053	0.6744	0.8171	0.5666	0.6524
	50%	**0.9055**	0.7199	0.8353	0.8378	0.6773	0.8348	0.5836	0.6739
Flickr	10%	**0.7897**	0.6212	0.5663	0.2901	0.1164	0.4297	0.1563	0.3089
	20%	**0.8454**	0.7043	0.6301	0.3674	0.1542	0.5655	0.2475	0.3772
	30%	**0.8617**	0.7444	0.6583	0.4210	0.1938	0.6091	0.2760	0.3929
	40%	**0.8678**	0.7664	0.6834	0.4429	0.2171	0.6354	0.2942	0.4171
	50%	**0.8780**	0.7856	0.7034	0.4510	0.2402	0.6530	0.3098	0.4256
Cora	10%	**0.7263**	0.6966	0.3601	0.7166	0.5806	0.5099	0.7301	0.7098
	20%	**0.7987**	0.7666	0.5539	0.7974	0.6631	0.6102	0.7819	0.7694
	30%	0.8176	0.7836	0.6260	**0.8225**	0.7079	0.6529	0.7961	0.7928
	40%	0.8290	0.8057	0.6728	**0.8356**	0.7350	0.6774	0.8191	0.8166
	50%	0.8350	0.8173	0.7029	**0.8471**	0.7555	0.6978	0.8330	0.8291
Citeseer	10%	**0.6651**	0.4977	0.3575	0.5594	0.2138	0.5366	0.4722	0.4095
	20%	**0.7189**	0.5655	0.5101	0.6316	0.3366	0.6210	0.5432	0.5110
	30%	**0.7282**	0.6073	0.5566	0.6595	0.3937	0.6535	0.5846	0.5563
	40%	**0.7390**	0.6281	0.5825	0.6690	0.4354	0.6541	0.6013	0.5925
	50%	**0.7476**	0.6391	0.5915	0.6862	0.4617	0.6734	0.6139	0.5995
Wiki	10%	**0.6601**	0.5684	0.6159	0.4498	0.5624	0.6501	0.4269	0.4427
	20%	**0.7315**	0.6382	0.7066	0.5664	0.6310	0.7087	0.5448	0.5505
	30%	**0.7647**	0.6629	0.7414	0.6168	0.6612	0.7385	0.5780	0.5803
	40%	**0.7765**	0.6832	0.7551	0.6518	0.6871	0.7471	0.6117	0.6190
	50%	**0.7879**	0.6951	0.7698	0.6688	0.7062	0.7609	0.6311	0.6377

4.4 Performance Evaluation

In this section, we will answer the questions proposed in the beginning of Sect. 5 one by one.

Effectiveness of Multi-network Random Walk Strategy. To answer the first question, we evaluate the proposed multi-network random walk strategy which performs random walk cross multiple networks (including topological network and attribute affinity networks) by conducting a series of experiments. We first perform random walk on the attribute affinity network (**Attribute**) and topological network (**Structure**) respectively and feed the node sequences to Skip-Gram model to get the embeddings for each network. Then we mix the node sequences generated on these two networks and use this mixed corpus to produce embeddings in the same way (**Combine**). Finally, we perform the proposed multi-network random walk strategy without labels. The classification results of these four methods on **BlogCatalog** dataset with different training ratios are shown in Table 3.

Table 3. F1-score of classification on BlogCatalog

Training ratio		10%	30%	50%	70%
Micro	Structure	0.3520	0.5317	0.5696	0.5987
	Attribute	0.7677	0.8172	0.8360	0.8446
	Combine	0.7866	0.8575	0.8691	0.8833
	SINE	**0.8183**	**0.8770**	**0.8909**	**0.9015**
Macro	Structure	0.3592	0.5428	0.5810	0.6101
	Attribute	0.7797	0.8241	0.8426	0.8498
	Combine	0.7946	0.8624	0.8735	0.8869
	SINE	**0.8251**	**0.8808**	**0.8946**	**0.9044**

The results in Table 3 illustrate the improvement of our multi-network random walk strategy. Specifically, compared to the first two methods which only utilize either attribute information or network structure, the **Combine** and **SINE** methods always achieve significantly better performance, showing that attribute information is valuable on network embedding. More importantly, our method outperforms other methods in all situations, which proves that our proposed multi-network random walk strategy is effective. In contrast to treating attribute and structure separately, we consider the correlation and interaction between them by a unified random walk sequence to effectively incorporate attribute information, leading to much better node representations.

Effect of Label Information. To answer the second question, we investigate the guidance effect of the label by varing the ratio of labeled nodes from 10% to 90% when performing labeled biased random walk. The training ratios of SVM classifier is fixed to 50%. The result is presented in Fig. 2.

From Fig. 2, we can see that with the increase of label ratio, both metrics are rising, which validates the guidance effect of label on embedding.

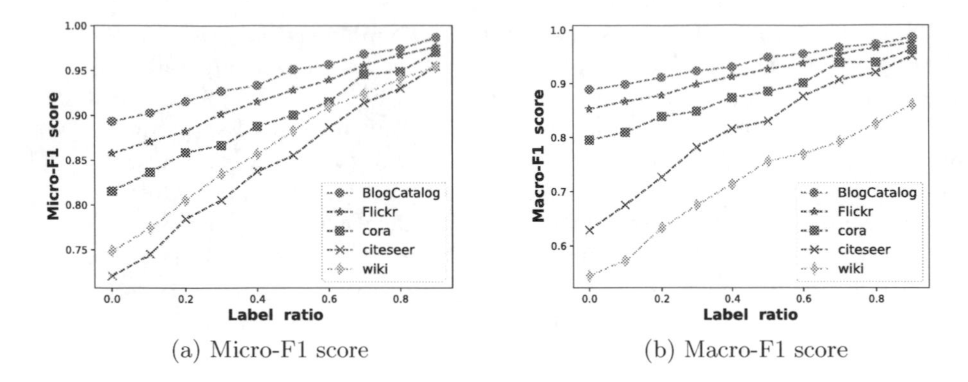

(a) Micro-F1 score

(b) Macro-F1 score

Fig. 2. Classification results of different label ratios

Effectiveness of SINE. To study the effectiveness of our SINE framework which is mentioned in the third question, we compare its performance with all baseline methods with varing the training ratio from 10% to 50%. The classification results of eight methods on five datasets are shown in Table 2. Due to the limitation of space, we only show the result of micro-F1 score and the result of macro-F1 score is similar. From Table 2, we find that our method achieves better performance in most situations with the following observations.

- First, by incorporating the attribute information, most attributed network embedding methods achieve significant improvement compared to the pure network embedding methods.
- Second, with the proposed framework, SINE outperforms other baseline methods in most situations. This is because SINE can effectively integrate the side information and get much more valuable node representations, resulting in better classification results.
- Third, our method performs fairly well when the training ratio is quite small while other baseline methods degrade quickly as the training ratio decreases due to that their representations are noisy and inconsistent in training set and test set. Compared to other algorithms, SINE learns node representations from three data sources, including network structure, attributes information and node labels, which makes the representations more consistent and less noisy.

4.5 Parameter Analysis

In this section, we investigate the effects of parameters, including embedding dimension d, label hub weight γ, and labeled bias r and s. We fix the training ratio to 50% and test the classification F1 scores with different parameters. For dimension d, we vary it from 10 to 100 and conduct experiments on five datasets. Figure 3 shows the variations of classification results with different d. The result suggests that our method is stable when d within a reasonable range. As for label

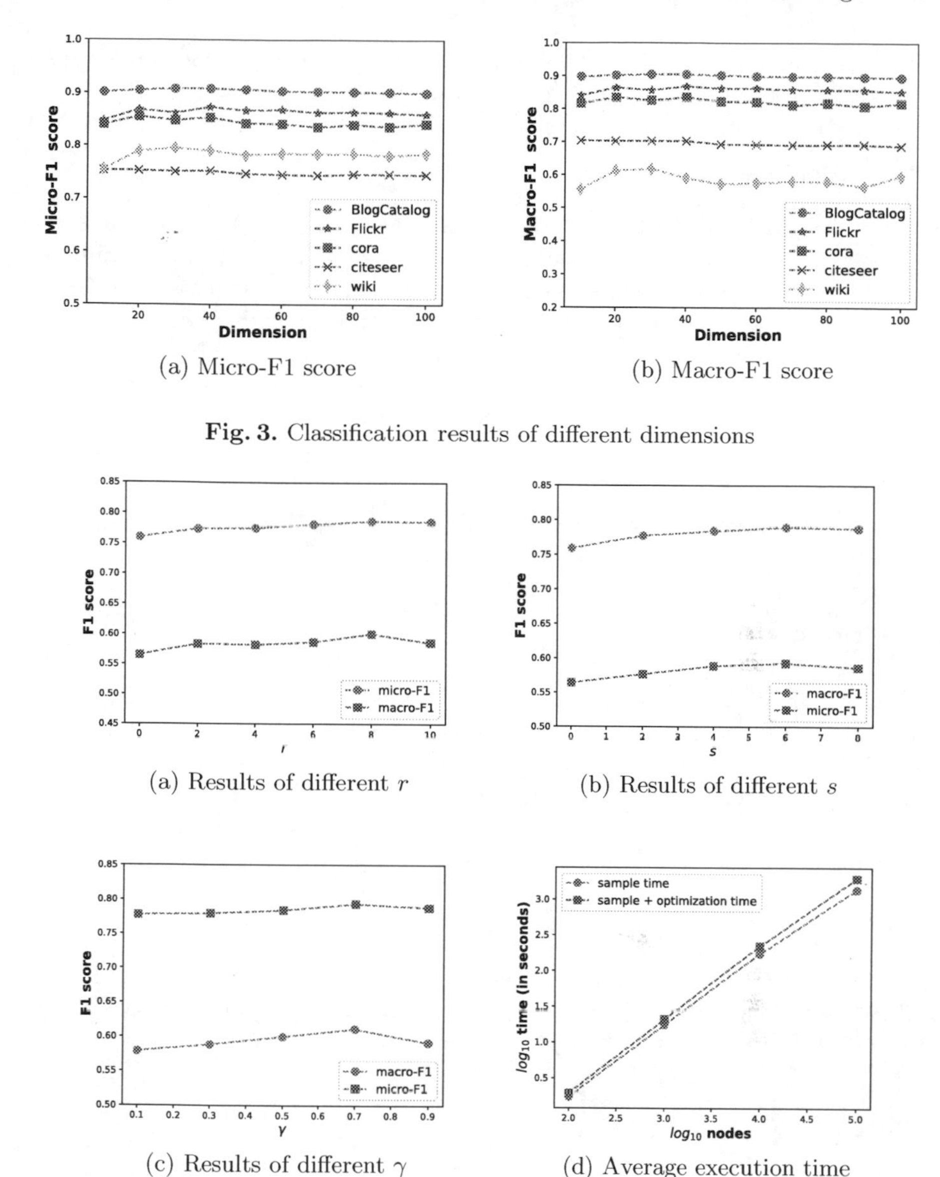

(a) Micro-F1 score

(b) Macro-F1 score

Fig. 3. Classification results of different dimensions

(a) Results of different r

(b) Results of different s

(c) Results of different γ

(d) Average execution time

Fig. 4. Results of parameters analysis and scalability

related parameters γ, r, and s, we set the other two parameters to zeros when analyzing one of them. We vary each of them in different range on the **Citeseer** dataset and the result is presented in Fig. 4. We can find out that the influences of these three parameters are similar. As they increase, the performance becomes better due to the guidance effect of label. However, when they are larger than a threshold, random walk method will always walk to labeled node without walking to its neighbors, losing the information of topological neighborhoods, which reduces the quality of node representations.

4.6 Scalability

The time complexity of our random walk scheme is $O(tle \cdot |V|)$ where e is the average number of edges. In practice, we set $e = 20$ as mentioned in parameter settings so it can be regarded as a constant. The time complexity of Skip-Gram is $O(k(d + d \cdot \log|V|))$ where the window size k and the embedding vector size d are constants so the total complexity is still $O(|V|)$. To test for scalability, we learn node representations using SINE with default parameter values for Erdos-Renyi graphs with node sizes increasing from 10^2 to 10^5. We compute the average running time for 10 independent executions. The result of running time (in log scale) is shown in Fig. 4d. We observe that SINE scales linearly with the size of nodes, which is acceptable in practice. Thus, SINE can be applied to large-scale networks.

5 Conclusion

In this paper, we propose a novel network embedding framework SINE, which can learn high-quality node representations for networks with side information, including attributes and labels. We design a flexible random walk scheme to generate semantically rich neighborhoods for nodes, which contains the proximity in topological structure, node attributes and node labels. The extensive experiments on 5 real-world datasets validate its effectiveness and efficiency.

Acknowlegment. This research was supported by the National Key R&D Program of China (2018YFB1004804), the National Natural Science Foundation of China (11801595, U1811462), the Natural Science Foundation of Guangdong (2018A030 310076), the Program for Guangdong Introducing Innovative and Entrepreneurial Teams (2016ZT06D211) and the CCF Opening Project of Information System.

References

1. Belkin, M., Niyogi, P.: Laplacian eigenmaps for dimensionality reduction anddata representation. Neural Comput. **15**(6), 1373–1396 (2003). https://doi.org/10.1162/089976603321780317
2. Cao, S., Lu, W., Xu, Q.: Grarep: learning graph representations with global structural information. In: Proceedings of the 24th ACM International on Conference on Information and Knowledge Management, pp. 891–900. ACM, New York (2015). https://doi.org/10.1145/2806416.2806512
3. Gao, H., Huang, H.: Deep attributed network embedding. In: Proceedings of the 27th International Joint Conference on Artificial Intelligence, pp. 3364–3370. International Joint Conferences on Artificial Intelligence Organization, July 2018. https://doi.org/10.24963/ijcai.2018/467

4. Grover, A., Leskovec, J.: Node2vec: scalable feature learning for networks. In: Proceedings of the 22nd ACM SIGKDD International Conference on Knowledge Discovery and Data Mining, pp. 855–864. ACM, New York (2016). https://doi.org/10.1145/2939672.2939754
5. He, X., Niyogi, P.: Locality preserving projections. In: Advances in Neural Information Processing Systems, vol. 16, pp. 153–160. MIT Press, Cambridge (2003). http://dl.acm.org/citation.cfm?id=2981345.2981365
6. Huang, X., Li, J., Hu, X.: Accelerated attributed network embedding. In: Proceedings of the 2017 SIAM International Conference on Data Mining, pp. 633–641. SIAM (2017). https://doi.org/10.1137/1.9781611974973.71
7. Huang, X., Li, J., Hu, X.: Label informed attributed network embedding. In: Proceedings of the 10th ACM International Conference on Web Search and Data Mining, pp. 731–739. ACM, New York (2017). https://doi.org/10.1145/3018661.3018667
8. Liao, L., He, X., Zhang, H., Chua, T.: Attributed social network embedding. IEEE Trans. Knowl. Data Eng. 1 (2018). https://doi.org/10.1109/TKDE.2018.2819980
9. Marsden, P.V.: Homogeneity in confiding relations. Soc. Netw. **10**(1), 57–76 (1988). https://doi.org/10.1016/0378-8733(88)90010-X
10. McPherson, M., Smith-Lovin, L., Cook, J.M.: Birds of a feather: homophily insocial networks. Ann. Rev. Sociol. **27**(1), 415–444 (2001). https://doi.org/10.1146/annurev.soc.27.1.415
11. Mikolov, T., Chen, K., Corrado, G., Dean, J.: Efficient estimation of word representations in vector space. CoRR abs/1301.3781 (2013). http://arxiv.org/abs/1301.3781
12. Mikolov, T., Sutskever, I., Chen, K., Corrado, G., Dean, J.: Distributed representations of words and phrases and their compositionality. In: Advances in Neural Information Processing Systems, vol. 26, pp. 3111–3119. Curran Associates, Inc. (2013). http://papers.nips.cc/paper/5021-distributed-representations-of-words-and-phrases-and-their-compositionality.pdf
13. Pan, S., Wu, J., Zhu, X., Zhang, C., Wang, Y.: Tri-party deep network representation. In: Proceedings of the 25th International Joint Conference on Artificial Intelligence, pp. 1895–1901. AAAI Press (2016). http://dl.acm.org/citation.cfm?id=3060832.3060886
14. Perozzi, B., Al-Rfou, R., Skiena, S.: Deepwalk: online learning of social representations. In: Proceedings of the 20th ACM SIGKDD International Conference on Knowledge Discovery and Data Mining, pp. 701 710. ACM, New York (2014). https://doi.org/10.1145/2623330.2623732
15. Ribeiro, L.F., Saverese, P.H., Figueiredo, D.R.: Struc2vec: learning node representations from structural identity. In: Proceedings of the 23rd ACM SIGKDD International Conference on Knowledge Discovery and Data Mining, pp. 385–394. ACM, New York (2017). https://doi.org/10.1145/3097983.3098061
16. Roweis, S.T., Saul, L.K.: Nonlinear dimensionality reduction by locally linear embedding. Science **290**(5500), 2323–2326 (2000). https://doi.org/10.1126/science.290.5500.2323
17. Tang, J., Qu, M., Wang, M., Zhang, M., Yan, J., Mei, Q.: Line: Large-scale information network embedding. In: Proceedings of the 24th International Conference on World Wide Web, pp. 1067–1077. International World Wide Web Conferences Steering Committee, Republic and Canton of Geneva, Switzerland (2015). https://doi.org/10.1145/2736277.2741093

18. Tenenbaum, J.B., de Silva, V., Langford, J.C.: A global geometric framework for nonlinear dimensionality reduction. Science **290**(5500), 2319–2323 (2000). https://doi.org/10.1126/science.290.5500.2319
19. Tu, C., Zhang, W., Liu, Z., Sun, M.: Max-margin deepwalk: discriminative learning of network representation. In: Proceedings of the Twenty-Fifth International Joint Conference on Artificial Intelligence, pp. 3889–3895. AAAI Press (2016). http://dl.acm.org/citation.cfm?id=3061053.3061163
20. Wang, D., Cui, P., Zhu, W.: Structural deep network embedding. In: Proceedings of the 22nd ACM SIGKDD International Conference on Knowledge Discovery and Data Mining, pp. 1225–1234. ACM, New York (2016). https://doi.org/10.1145/2939672.2939753
21. Yang, C., Liu, Z., Zhao, D., Sun, M., Chang, E.Y.: Network representation learning with rich text information. In: Proceedings of the 24th International Conference on Artificial Intelligence, pp. 2111–2117. AAAI Press (2015). http://dl.acm.org/citation.cfm?id=2832415.2832542

A Knowledge Graph Enhanced Topic Modeling Approach for Herb Recommendation

Xinyu Wang[1], Ying Zhang[1], Xiaoling Wang[1](✉), and Jin Chen[2]

[1] Shanghai Key Laboratory of Trustworthy Computing, East China Normal University, Shanghai, China
{xinyuwang,ying.zhang}@stu.ecnu.edu.cn, xlwang@sei.ecnu.edu.cn
[2] Institute for Biomedical Informatics, University of Kentucky, Lexington, KY, USA
chen.jin@uky.edu

Abstract. Traditional Chinese Medicine (TCM) plays an important role in Chinese society and is an increasingly popular therapy around the world. A data-driven herb recommendation method can help TCM doctors make scientific treatment prescriptions more precisely and intelligently in real clinical practice, which can lead the development of TCM diagnosis and treatment. Previous works only analyzing short-text medical case documents ignore rich information of symptoms and herbs as well as their relations. In this paper, we propose a novel model called Knowledge Graph Embedding Enhanced Topic Model (KGETM) for TCM herb recommendation. The modeling strategy we used takes into consideration not only co-occurrence information in TCM medical cases but also comprehensive semantic relatedness of symptoms and herbs in TCM knowledge graph. The knowledge graph embeddings are obtained by TransE, a popular representation learning method of knowledge graph, on our constructed TCM knowledge graph. Then the embeddings are integrated into the topic model by a mixture of Dirichlet multinomial component and latent vector component. In addition, we further propose HC-KGETM incorporating herb compatibility based on TCM theory to characterize the diagnosis and treatment process better. Experimental results on a TCM benchmark dataset demonstrate that the proposed method outperforms state-of-the-art approaches and the promise of TCM knowledge graph embedding on herb recommendation.

Keywords: Traditional Chinese Medicine · Topic model · Knowledge graph embedding · Recommendation

1 Introduction

Traditional Chinese Medicine (TCM) has been assiduously developed over thousands of years. As a comprehensive system which studies disease prevention, diagnosis, treatment, rehabilitation, and healthcare, TCM plays an important

G. Li et al. (Eds.): DASFAA 2019, LNCS 11446, pp. 709–724, 2019.
https://doi.org/10.1007/978-3-030-18576-3_42

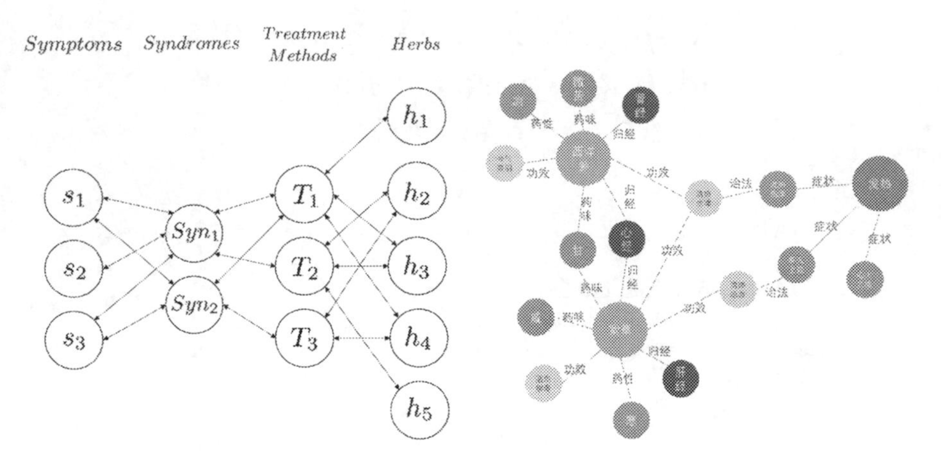

Fig. 1. The relation between symtoms, syndromes, treatments and herbs.

Fig. 2. The schematic diagram of Traditional Chinese Medicine knowledge graph.

role in Chinese society and is increasingly adopted as a complementary therapy around the world. Figure 1 shows the general therapeutic process in TCM. First, the doctor examines the symptoms of the patient, such as "aversion to cold", "headache", "anhidrosis" et al. Then, the syndrome is determined after analyzing symptoms. The syndrome, in this case, is "superficies excess syndrome". Third, the doctor decides the treatment, in this case, is "resolving superficies method". Finally, a prescription which contains a group of herbs is given according to the first three steps and herb compatibility. There are some herb combinations summarized by TCM practicers after long-term development. Some of the herb combinations are often used together to improve treatment effectiveness. While the others are forbidden to be used together since they can have negative effects.

In real clinical practice, the diagnosis and treatment result is given based on TCM theory and the doctors' own experience, which is subjective and lacks standards. The inheriting way of TCM is based on the summarization of previous medical cases literature. Therefore, depicting the diagnosing and prescribing process from medical cases is a vital task, which can be used in objectively giving auxiliary diagnosis and herb recommendation to help TCM doctors making scientific decisions in clinical practice.

We face several challenges on TCM herb recommendation task. First, different from Western Medicine, TCM views the human body as an organic whole. A set of symptoms of the patient are interdependent and interactional. It is inappropriate to treat each symptom of the patient independently. Second, as mentioned above, the therapeutic process in TCM involves four steps and covers a large amount of complex TCM domain knowledge such as herbal compatibility. So there is great difficulty in characterizing the therapeutic process fully and precisely. Third, short texts, such as TCM medical cases, do not contain sufficient statistical information for small amounts of words. Yao et al. [2] proposed a prescription topic model (PTM) for exploring the prescribing patterns in TCM. PTM regarded symptom terms and herb terms as observed variables, and syndrome (the topic of symptoms), treatment (the topic of herbs) as latent variables. However, PTM, as well as the existing method for TCM knowledge mining based on the topic model, have a common drawback: they model symptom terms and herb terms with a bag-of-words model, in other words, they ignore the relatedness between symptoms and relatedness between herbs in fact. These bag-of-word models only consider word co-occurrence in the medical case, without taking full advantage of the prior relevance of symptoms and herbs in the real world. The inference provided by word co-occurrence is single and incomplete, which is not able to explore the relationship of symptoms and herbs sufficiently, especially for short texts like medical cases. While there is TCM literature that records the properties of symptoms and herbs, as well as their relations (such as efficacy, indication, syndrome, or channel tropism). Therefore, how to use this information to improve the existing works is a vital task.

To address the aforementioned challenges, in this paper, we propose a novel method named knowledge graph embedding enhanced topic model (KGETM) based on topic model incorporated with knowledge graph embedding for herb recommendation. We improve the performance of the topic model by incorporating knowledge graph. Knowledge graph is a large scale semantic network consists of triplets. It represents concepts and relations among them adequately and is easy to use. We construct a TCM knowledge graph to express related concepts and relations among them in the form of triplets as auxiliary information for the topic model. TCM theory concepts exist as entities in the knowledge graph, such as herbs, symptoms and their properties. The knowledge graph provides rich semantic relatedness between symptoms and herbs for the topic model, which can improve the abilities of topic model to find the latent connection between symptoms and herbs. For incorporating information of TCM knowledge graph into the topic model, we obtain the embeddings of symptom and herb in the knowledge graph by TransE [9]. In our models, the topic-to-word Dirichlet multinomial component in LDA [3] was replaced with a mixture of a Dirichlet multinomial component and a latent vector component to utilize symptom and herb embeddings.

Unlike LDA [3], there are two topic-to-word distributions in our model correspond to symptom part and herb part of a medical case. For symptom part, as mentioned in the therapeutic process in TCM, doctors determined syndrome by analyzing symptoms. So our models regard symptoms as words and syndrome as the topic. For the herb part, doctors choose herbs according to the treatment for the syndrome. So herbs are viewed as words and treatment is viewed as the topic in our models. The topic syndrome and treatment share the same medical case-topic distribution. In addition, to model the complex TCM therapeutic process as mentioned above, our model incorporated herbal compatibility knowledge. We call this model knowledge graph embedding enhanced topic model with herb compatibility (HC-KGETM), which incorporated herb combinations by encoding herb combinations as sparse constraints.

The contributions of this paper are summarized as follows:

- To the best of our knowledge, we are the first work integrating knowledge graph into TCM clinical data analysis. We propose a Knowledge Graph Embedding Enhanced Topic Model (KGETM) for herb recommendation. The model characterizes the diagnosis and treatment process.
- We further propose HC-KGETM taking herb compatibility into consideration, which is consistent with TCM theory.
- Experimental results demonstrate the effectiveness of the proposed method and the promise of KG embedding for herb recommendation.
- We provide a TCM knowledge graph triplets dataset.

2 Related Work

Topic models are widely used in herb recommendation for TCM researches. Wang et al. [7] propose an asymmetric multinomial probabilistic model for the joint analysis of symptoms, diseases, and herbs, which can be used in herb treatment prediction. Ji et al. [1] proposed a topic model named MCLDA for modeling relationship between the symptoms and herbs. The model considers "pathogenesis" as the latent topic that connects symptoms and herbs. Then a hybrid herb recommendation method is proposed based on the model. Yao et al. [2] proposed a prescription topic model (PTM) which integrates TCM concepts into topic modeling such as "syndrome", "treatment", "herb roles" to better characterize the generative process of prescriptions. However, these topic models find topic patterns in TCM data only by word co-occurrence, without considering the subsistent semantic similarity of symptoms or semantic similarity of herbs.

In recent years, researchers find that incorporating external knowledge into the topic model can extract more semantic coherent topics and learn a better representation of documents. For example, in [11], the authors extend LDA and DMM models by integrating the latent feature vector of words. In [4], the authors replace LDA's categorical distributions of topics with multivariate Gaussian distributions on the embedding space.

Knowledge graph is a semantic network which consists of triple facts (head entity, relation, tail entity). One of the wide usages of knowledge graph is knowledge graph embedding, aiming to embed entities and relations into low-dimensional continuous vector spaces, so as to simplify the manipulation while preserving the properties of the graph. TransE [9] is the most representative model for KG embeddings. It considers a relation $\boldsymbol{r}$ as a translation from head entity $\boldsymbol{h}$ to tail entity $\boldsymbol{t}$, so that $\boldsymbol{h} + \boldsymbol{r} \approx \boldsymbol{t}$ when (h, r, t) holds. Yu et al. [5] designed and constructed a TCM knowledge graph, and applies the knowledge graph in TCM health preservation. Zhang et al. [6] proposed a hybrid recommendation framework which leverages KG embeddings to learn a structural representation of items. Wang et al. [10] construct a medical knowledge graph based on MIMIC-III and DragBank, then they use the KG embeddings to decompose the medicine recommendation into a link prediction process. Yao et al. [8] combine KG embeddings with LDA model by characterizing each topic as a vMF distribution to improve the performance of LDA. However, to the best of our knowledge, there is no existing work incorporating TCM knowledge graph embeddings to analyze TCM clinical data for diagnosis or further herb recommendation.

3 Approach

3.1 Preliminary

The task of herb recommendation is to output a set of herbs given a set of symptoms. By analyzing medical cases, our models can obtain medical case-topic distribution θ, syndrome-symptom distribution ϕ_s and treatment-herb distribution ϕ_h. We calculate the probability for a herb given a set of symptoms using θ, ϕ_s and ϕ_h, and then choose top-ranked herbs as the result according to the probability.

As shown in Fig. 3, M denotes the number of all medical cases. A medical case m is divided into two parts, the symptom part and herb part. The symptom part of m contains a set of symptoms denoted by $m_s = \{m_{s1}, m_{s2}, ..., m_{sN_{ms}}\}$, where N_{ms} is the total number of symptoms in the medical case m. The set of herbs of m denoted by $m^h = \{m_{h1}, m_{h2}, ..., m_{hN_{mh}}\}$, where N_{mh} is the total number of herbs in m. Usually, N_{mh} is bigger than N_{ms}. The vocabulary of symptoms V_s with S distinct symptoms denoted by $V_s = \{V_{s_1}, V_{s_2}, ..., V_{s_S}\}$. The vocabulary of herbs V_h with H distinct herbs denoted by $V_h = \{V_{h_1}, V_{h_2}, ..., V_{h_H}\}$.

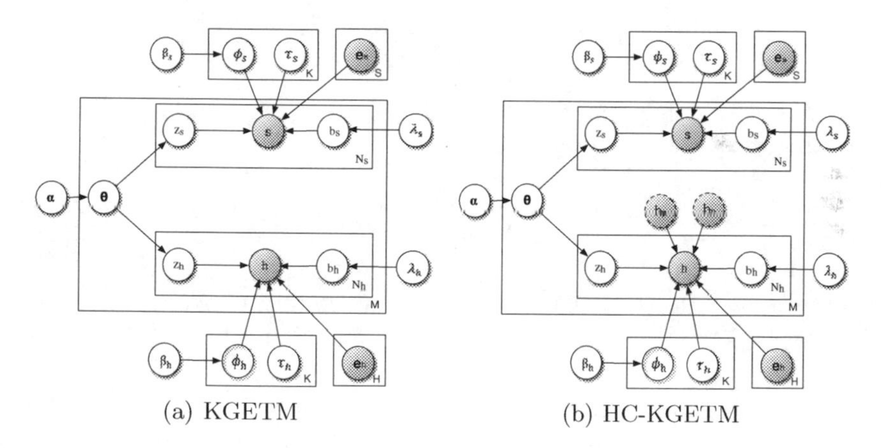

(a) KGETM (b) HC-KGETM

Fig. 3. The graphical representation of KGETM and HC-KGETM.

3.2 KGETM

Here, we introduce the details of Knowledge Graph Embedding Enhanced Topic Model (KGETM). As shown in Fig. 3(a), KGETM has two topic-word distributions correspond to symptom part and herb part in a medical case. In symptom part, the model views symptom s as observed variable, syndrome z_s as latent variable. ϕ_s is S-dimensional syndrome-symptom multinomial for syndrome $k \in 1, ..., K$. In the herb part, herb h is regarded as observed variable, treatment z_h is regarded as latent variable. ϕ_h is H-dimensional treatment-herb multinomial for treatment $k \in 1, ..., K$. To encode symptom and herb embeddings in the topic model, as LFLDA [11], we replace the Dirichlet multinomial component in LDA [3] with a mixture of Dirichlet multinomial component and latent feature component. The latent feature consists of symptom/herb embeddings and topic vectors. As shown in Fig. 3(a), $\boldsymbol{e}_s \in \mathbb{R}^{D_e \times V_s}$, $\boldsymbol{e}_h \in \mathbb{R}^{D_e \times V_h}$ are symptom and herb embedding respectively; $\boldsymbol{\tau}_s \in \mathbb{R}^{D_k \times K}$, $\boldsymbol{\tau}_h \in \mathbb{R}^{D_k \times K}$ are syndrome vector and treatment vector respectively. The latent feature component of KGETM defined as:

$$\text{softmax}(w)_k = \frac{\exp(\boldsymbol{\tau}_{*k} \cdot \boldsymbol{e}_{*w})}{\sum_{w' \in V_*} \exp(\boldsymbol{\tau}_{*k} \cdot \boldsymbol{e}_{*w'})} \tag{1}$$

where w denotes or a symptom or a herb. $\boldsymbol{e}_{*w}$ is the TCM knowledge graph embedding of w, which is obtained by TransE [9] in this work. b_{s_i}/b_{h_j} is a binary indicator variable is sampled from a Bernoulli distribution to determine whether symptom i/herb j is to be generated by the Dirichlet multinomial or latent feature component. V_* denotes the vocabulary of symptoms or herbs. KGETM applies regularized maximum likelihood estimation as [11] to estimate the latent-feature vector of the topic (syndrome/treatment). Log-loss function with ℓ_2 regularization for topic (syndrome/treatment) k is defined as:

$$L_k = -\sum_{w \in V_*} K^{k,w} \left(\boldsymbol{\tau}_{*k} \cdot \boldsymbol{e}_{*w} - \log \left(\sum_{w' \in V_*} \exp\left(\boldsymbol{\tau}_{*k} \cdot \boldsymbol{e}_{*w'} \right) \right) \right) + \mu \|\boldsymbol{\tau}_k\|_2^2 \tag{2}$$

The derivative with respect to the jth element of the vector for syndrome/treatment k is:

$$\frac{\partial L_k}{\partial \boldsymbol{\tau}_{*k,j}} = -\sum_{w \in V_*} K^{k,w} \left(\boldsymbol{e}_{*w,j} - \sum_{w' \in V_*} \boldsymbol{e}_{*w',j} \text{softmax}\left(w'\right)_k \right) + 2\mu \boldsymbol{\tau}_{*k,j} \tag{3}$$

We obtain syndrome/treatment k vector τ_{*k} by minimizing L_k.

The generative process of our model is given as shown in Algorithm 1.

Inference Estimation for KGETM. We use the collapsed Gibbs sampling algorithm to perform inference to calculate the conditional topic assignments z_{mi}^s and z_{mj}^h in KGETM.

The sampling equation for syndrome z_{mi}^s is defined as

$$\begin{aligned} & p(z_{mi}^s = k | s_{mi}, \mathbf{z}_{\neg mi}^s, \boldsymbol{\tau}_s, \mathbf{e}_s) \propto \\ & (N_m^k + \alpha) \times \left((1 - \lambda) \frac{N_{k s_{mi}} + \beta_s}{N_k + S\beta_s} + \lambda \text{softmax}\left(s_{mi}\right)_k \right) \end{aligned} \tag{4}$$

where k is a syndrome, $\mathbf{z}_{\neg mi}^s$ denotes the syndrome assignments of all symptoms ignoring i^{th} symptom of medical case m. N_m^k is the number of times a symptom

Algorithm 1. Generative process of KGETM

1: **for** each each medical case m **do**
2: Draw $\theta_m \sim \text{Dir}(\alpha)$
3: **end for**
4: **for** each syndrome k in 1,...,K **do**
5: Draw $\phi_k^s \sim \text{Dir}(\beta_s)$.
6: **end for**
7: **for** each treatment k in 1,...,K **do**
8: Draw $\phi_k^h \sim \text{Dir}(\beta_h)$.
9: **end for**
10: **for** each ith symptom of the N_{ms} symptoms in medical case m **do**
11: Draw a syndrome $z_{mi}^s \sim \text{Mult}(\theta_m)$.
12: A binary indicator variable $b_{s_{mi}} \sim \text{Ber}(\lambda)$
13: Draw a symptom $s_{mi} \sim (1 - b_{s_{mi}})\text{Mult}(\phi_{z_{mi}^s}^s) + b_{s_{mi}}\text{softmax}(s_{mj})_{z_{mi}^s}$
14: **end for**
15: **for** each jth herb of the N_{mh} herbs in medical case m **do**
16: Draw a treatment $z_{mj}^h \sim \text{Mult}(\theta_m)$.
17: A binary indicator variable $b_{h_{mj}} \sim \text{Ber}(\lambda)$
18: Draw an herb $h_{mj} \sim (1 - b_{h_{mj}})\text{Mult}(\phi_{z_{mj}^h}^h) + b_{h_{mj}}\text{softmax}((h_{mj})_{z_{mj}^h})$
19: **end for**

or a herb in medical case m is generated from topic k (syndrome or treatment) by Dirichlet multinomial component or latent feature component. N_{ms} and N_{mh} is the number of symptoms and herbs in prescription p respectively. $N_{ks_{mi}}$ is the number of times symptom s_{mi} is generated from syndrome k. N_k is the number of times any symptom is generated from syndrome k.

The sampling equation for treatment z_{mj}^h is defined as

$$p(z_{mj}^h = k | h_{mj}, \mathbf{z}_{\neg mj}^h, \boldsymbol{\tau}_h, \mathbf{e}_h) \propto (N_m^k + \alpha) \times \left((1-\lambda) \frac{N_{kh_{mj}} + \beta_h}{N_k' + H\beta_h} + \lambda \cdot \text{softmax}(h_{mj})_{z_{mj}^h} \right) \tag{5}$$

where k is a treatment, $\mathbf{z}_{\neg mj}^h$ denotes the treatment assignments of all herbs ignoring j^{th} herb of medical case m. h_{mj} is i^{th} herb in medical case m. $N_{kh_{mj}}$ is the number of times h_{mj} is generated from k. N_k' is the number of times any herb in all medical cases is generated from k.

With Gibbs sampling, we can make the following parameter estimation:

$$\theta_{mk} = \frac{N_m^k + \alpha}{N_{ms} + N_{mh} + K\alpha} \tag{6}$$

$$\phi_{ks_{mi}}^s = (1-\lambda) \frac{N_{ks_{mi}} + \beta_s}{N_k + S\beta_s} + \lambda \cdot \text{softmax}(s_{mi})_{z_{mi}^s} \tag{7}$$

$$\phi_{kh_{mj}}^h = (1-\lambda) \frac{N_{kh_{mj}} + \beta_h}{N_k' + H\beta_h} + \lambda \cdot \text{softmax}(h_{mj})_{z_{mj}^h} \tag{8}$$

3.3 HC-KGETM

Synergism of set herbs is one important part of Herbal compatibility. There are seven states of herbal compatibility called "seven emotions" in TCM theory. Seven emotions can be simplified into two states: the combination of a set of herbs with a positive influence and the combination of a set of herbs with negative influence. We name the model incorporating herbal compatibility HC-KGETM. Here, we introduce the details of how HC-KGETM incorporates herb combination.

Through the long-term development, TCM theory has summarized some commonly used herb combinations that meet the herbal compatibility principles. As mentioned above, these herb combinations are classified into two kinds: the herb combinations that improve the treatment effectiveness and the herb combinations that have negative effects. The former herb combinations in HC-KGETM called positive-link combinations L^P, the poster herb combinations called negative-link combinations L^N. As shown in Fig. 3(b), h_p and h_n are herbs in L^P and L_N of h respectively. Every L^P and L^N contains at least two herbs. For example, a combination of *Pinellia ternata*, *Baikal Skullcap*, *China Goldthread* and *Dried Ginger* plays a positive role, this combination for *Pinellia ternata* is denoted by $L^N_{\text{Pinellia ternata}} = \{$*Baikal Skullcap*, *China Goldthread*, *Dried Ginger*$\}$. The potential score of sampling topic treatment k for the combination of herb h is

$$f(k,h) = \sum_{i \in L_h^P} \log \max(\gamma, N_{i,k}) + \sum_{j \in L_h^N} \log \frac{1}{\max(\gamma, N_{j,k})} \tag{9}$$

where γ is a hyperparameter. γ has a negative correlation with the influence of herb combinations. The smaller γ is, the stronger correlation between the herbs in the combination is, the stronger influence of the combination is. $N_{i,k}$ and $N_{j,k}$ are the number of times positive-link herb i and negative-link herb j generated from treatment k. Equation 9 increases the probability that herb h will be drawn from the same topics as those of h's positive-link herb set L^P, and decreases its probability of being drawn from the same topics as those of h's negative-link herb set L^N.

The generative process of HC-KGETM is as shown in Algorithm 2.

Algorithm 2. Generative process of HC-KGETM

Process of drawing θ_m, ϕ_k^s, ϕ_k^h and s_{mi} is the same as before.

1: **for** each jth herb of the N_{mh} herbs in medical case m **do**
2: Draw a treatment $\text{z}_{mj}^h \sim \text{Mult}(\theta_m)$.
3: A binary indicator variable $b_{h_{mj}} \sim \text{Ber}(\lambda)$
4: Draw an herb $\text{h}_{mj} \sim ((1 - b_{h_{mj}})\text{Mult}(\phi^h_{z_{mj} x^h_{mj}}) + b_{h_{mj}} \text{softmax}(h_{mj})_{z^h_{mj}}) \times \exp f(z_{mj}^h, h_{mj})$
5: **end for**

Inference Estimation for HC-KGETM. We use the collapsed Gibbs sampling algorithm to perform inference to calculate the conditional topic assignments z^s_{mi} and z^h_{mj}in HC-KGETM.

The inference equation for z^s_{mi} is same as KGETM. The inference equation for z^h_{mj} is defined as

$$p(z^h_{mj} = k|h_{mj}, \mathbf{z}^h_{\neg mj}, \boldsymbol{\tau}'_h, \mathbf{e}_h) \propto (N^k_m + \alpha) \\ \times \left((1-\lambda)\frac{N_{kxh_{mj}} + \beta_h}{N_{kx} + H\beta_h} + \lambda \text{softmax}(h_{mj})\right) \times \exp f(k, h_{mj}) \tag{10}$$

where $f(k, h_{mj})$ denotes potential score of sampling treatment k for combination of herb h_{mj}.

With Gibbs sampling, we can make the following parameter estimation:

$$\phi^h_{kh_{mj}} = \left((1-\lambda)\frac{N_{kxh_{mj}} + \beta_h}{N_{kx} + H\beta_h} + \lambda \text{softmax}(h_{mj})\right) \times \exp f(k, h_{mj}) \tag{11}$$

3.4 Prediction

The aim of the herb recommendation is predicting a set of herbs given a set of symptoms.

The probability of herb h_j given a set of test symptoms s'_m is defined as

$$p(h_j|s'_m) = \frac{1}{N_{s'_m}} \sum_{s_i \in s'_m} p(h_j|s_i) \tag{12}$$

where s'_m is the set of test symptoms, $N_{s'_m}$ is the number of symtoms in s'_m.

The probability of herb h_j given a symptom s_i is

$$p(h_j|s_i) = \sum_{k=0}^{K} p(h_j|k)p(k|s_i) = \sum_{k-0}^{K} p(h_j|k)\frac{p(s_i|k)p(k)}{p(s_i)} \tag{13}$$

where $p(h_i|k)$ is parameter ϕ_{kh_i}, $p(s_j|k)$ is parameter ϕ_{ks_j} in KGETM and HC-KGETM.

The models choose top N herbs as recommendation result for given symptoms according to the probabilities.

4 Experiment

In this section, we conduct experiments for demonstrating the effectiveness of our models on the following research questions:

Question 1: Can our models outperform the state-of-art on herb recommendation task?
Question 2: Can our models outperform other topic models on generalization performance?
Question 3: How do knowledge graph embeddings influence herb recommendation?

4.1 Experimental Settings

TCM Knowledge Graph. We construct a TCM knowledge graph from multiple data sources. Figure 2 shows a part of TCM knowledge graph. The graph views TCM concepts as entities and relatedness between them as relations. The statistical information of the graph is shown in Table 1. The entities contain herbs, herbal properties, efficiencies, symptoms, pathogeny et al. And relations in the graph contains "hasChemical", "treat", "channel tropism", "hasNature" et al. we obtain 100-dimension symptom and herb embeddings by TransE [9].

Table 1. Information about TCM knowledge graph

#entity	55556
#relation	23
#triples	1336421
Data source	TCM-Mesh[a]
	Chinese Traditional Medicine Encyclopedia[b]
	Pharmacopoeia of the People's Republic of China 2005 [19]
	Clinic Terminology of Traditional Chinese Medical Diagnosis [20]
	Clinic Terminology of Traditional Chinese Medical Treatment-Diseases [20]

[a]http://mesh.tcm.microbioinformatics.org.
[b]http://www.a-hospital.com.

Dataset. We experimented with the benchmark TCM dataset [2]. The TCM dataset contains 98,334 raw medical cases and 33,765 processed medical cases (only consist of symptoms and herbs). The total number of symptoms and herbs in 33,765 processed medical cases is 390 and 811 respectively. Among 33,765 processed medical cases, 26,360 of them have both symptoms and herbs appear in our TCM knowledge graph. The 26,360 medical cases are divided into 22,917 for training and 3,443 for testing.

Baseline. We compare our models with eight baselines as follows. The top four methods are topic modeling method, the last three methods are group recommendation method.

- *Author-topic model(ATM)* [13] regards herbs as authors, symptoms as words.
- *LinkLDA* [14] regards herbs as words, symptoms as references.
- *Block-LDA* [15] can model links between herbs.
- *Prescription Topic Model(PTM)* [2] regards symptoms and herbs as words, syndromes and treatments as topics.
- *Bilingual Biterm Topic Model (BiBTM)* [18] regards herbs as the words in a document and symptoms as the words in the translation.
- *COnsensus Model (COM)* [17] regards symptoms as a group of users for herbs recommendation.

- *User-based collaborative filtering with averaging strategy (CF-AVG)* [16] regards symptoms as users and herbs as items. *CF-AVG* uses the average of score of each user gained by collaborative filtering as recommendation score.
- *User-based collaborative filtering with least-misery strategy (CF-LM)* [16] is similar to *CF-AVG*, the difference is *CF-LM* uses smallest score as recommendation score.

Note that there are four models in PTM, we choose the best one as the baseline.

Parameter Setting. The parameter setting of baselines and our models is charted as follows: For *ATM*, $\alpha = 50/K$, $\beta = 0.01$; for *LinkLDA*, $\alpha = 1$, $\beta = 0.1$, $\beta' = 0.01$; for *Block-LDA*, $\alpha_D = \alpha_L = 1$, $\gamma = 0.1$; for *BiBTM*, $\alpha = 1$, $\beta = 0.1$; for *COM*, $\alpha = 50/K$, $\beta = \eta = 0.01$, $\gamma = \gamma_t = 0.5$, $\rho = 0.01$; for *PTM*, $\alpha = 1$, $\beta = \beta' = 0.1$, $\eta = 1$; for *CF-AVG* and *CF-LM*, the similarity measurement is Pearson correlation similarity; for *KGETM*, $\alpha = 0.05$, $\beta_s = \beta_h = 0.01$, $\lambda = 0.6$. for *HC-KGETM*, $\alpha = 0.05$, $\beta_s = \beta_h = 0.01$, $\lambda = 0.6$, $\gamma = 1$.

4.2 Herbs Recommendation (Question 1)

The task of herbs recommendation is outputting a set of herbs given a set of symptoms. We adopt *precision@N* and *recall@N* to measure the results of herb recommendation.

Table 2. precision@N of each model with different K and N

	K = 10			K = 20			K = 30			K = 40		
	p@5	p@10	p@20	p@5	p@10	p@20	p@5	p@10	p@20	p@5	p@10	p@20
ATM	0.0094	0.0099	0.0079	0.0101	0.0086	0.0089	0.0077	0.0077	0.0077	0.0084	0.1069	0.0099
LinkLDA	0.2398	0.1962	0.1418	0.2372	0.1881	0.1343	0.2358	0.1838	0.1309	0.2300	0.1827	0.1296
BlockLDA	0.2342	0.1842	0.1384	0.2323	0.1842	0.1325	0.2289	0.1847	0.1347	0.2143	0.1784	0.1295
BiBTM	0.2243	0.1669	0.1259	0.2243	0.1669	0.1259	0.2243	0.1669	0.1259	0.2243	0.1669	0.1259
CF-AVG	0.2387	0.1993	0.1510	0.2387	0.1993	0.1510	0.2387	0.1993	0.1510	0.2387	0.1993	0.1510
CF-LM	0.2397	0.1995	0.1515	0.2397	0.1995	0.1515	0.2397	0.1995	0.1515	0.2397	0.1995	0.1515
COM	0.2269	0.1784	0.1313	0.2290	0.1800	0.1351	0.2299	0.1830	0.1382	0.2302	0.1819	0.1376
PTM	0.2521	0.1987	0.1472	0.2552	0.2066	0.1547	0.2594	0.2077	0.1558	0.2622	0.2112	0.1572
KGETM	0.2627	0.2078	0.1563	0.2687	0.2138	0.1574	0.2730	0.2162	0.1598	0.2758	0.2162	0.1597
HC-KGETM	**0.2639**	**0.2103**	**0.1561**	**0.2694**	**0.2161**	**0.1597**	**0.2751**	**0.2204**	**0.1622**	**0.2783**	**0.2197**	**0.1626**

Tables 2 and 3 present *Precision@N* and *Recall@N* on herb recommendation of each model with different K and N values respectively. From the result, we can observe that KGETM and HC-KGETM outperform the other methods. HC-KGETM attains 3.2%–6.1% over PTM on *Precision@N*, 4.2%–8.8% over PTM on *Recall@N*. Note the results of group recommendation methods (CF-AVG and CF-LM) are only related to N, because these methods do not have parameter about the number of topics K. *Precision@N* and *Recall@N* of topic model based methods increased with the increase of K. KGETM outperforms

Table 3. recall@N of each model with different K and N

	K = 10			K = 20			K = 30			K = 40		
	r@5	r@10	r@20	r@5	r@10	r@20	r@5	r@10	r@20	r@5	r@10	r@20
ATM	0.0057	0.0148	0.0232	0.0069	0.0129	0.0247	0.0050	0.0110	0.0228	0.0063	0.0156	0.0273
LinkLDA	0.1683	0.2730	0.3919	0.1701	0.2646	0.3476	0.1678	0.2596	0.3636	0.1642	0.2589	0.3591
BlockLDA	0.1603	0.2557	0.3811	0.1649	0.2569	0.3654	0.1631	0.2589	0.3727	0.1513	0.2509	0.3568
BiBTM	0.1537	0.2238	0.3425	0.1537	0.2238	0.3425	1537	0.2238	0.3425	1537	0.2238	0.3425
CF-AVG	0.1724	0.2872	0.4314	0.1724	0.2872	0.4314	0.1724	0.2872	0.4314	0.1724	0.2872	0.4314
CF-LM	0.1728	0.2861	0.4309	0.1728	0.2861	0.4309	0.1728	0.2861	0.4309	0.1728	0.2861	0.4309
COM	0.1557	0.2447	0.3617	0.1568	0.2472	0.3708	0.1570	0.2504	0.3782	0.1569	0.2484	0.3763
PTM	0.1734	0.2728	0.4006	0.1753	0.2848	0.4249	0.1787	0.2863	0.4252	0.1809	0.2937	0.4309
KGETM	0.1822	0.2907	0.4296	**0.1908**	0.2998	0.4384	0.1928	0.3039	0.4441	0.1954	0.3054	0.4426
HC-KGETM	**0.1838**	**0.2912**	**0.4312**	0.1901	**0.3007**	**0.4431**	**0.1948**	**0.3098**	**0.4508**	**0.1959**	**0.3072**	**0.4523**

PTM, that means TCM knowledge graph embeddings in KGETM play an active role. TCM knowledge graph introduces comprehensive semantic correlate about symptoms and herbs, which makes the result better than other methods based on bag-of-words models. HC-KGETM outperforms KGETM, which means herbal compatibility knowledge has positive effects on herb recommendation.

4.3 Evaluation of Topic (Question 2)

Quantitative Results. We adopt perplexity to measure how well the model fits the test data, which is the most common metrics in topic models. A lower perplexity score indicates better overall performance. The perplexity of M medical cases can be defined to:

$$perplexity(h'|s') = \exp\left(-\frac{\sum_{m=1}^{M_{test}} \log p(h'_m|s'_m)}{\sum_{m=1}^{M_{test}} N_{h_m}}\right) \tag{14}$$

where m_{test} are the test medical case, m_{train} are the training medical case. s'_m are symptoms in medical case m of the test set, h'_m are herbs in medical case m of the

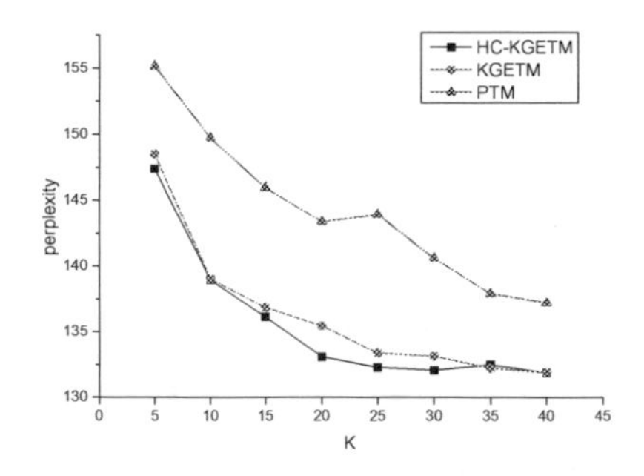

Fig. 4. Herb predictive perplexity of PTM, KGETM and HF-KGETM with different number of topics K.

	K=10	K=20	K=30	K=40
ATM	770.93	793.35	805.53	806.56
Link-LDA	303.07	370.51	357.95	350.69
Block-LDA	386.69	425.33	416.78	397.26
PTM	149.72	143.38	140.59	137.21
KGETM	139.02	135.45	133.14	131.91
HC-KGETM	138.94	133.10	132.07	131.89

Fig. 5. Herb predictive perplexity of six topic models with different number of topics K.

test set. There is a big difference between herb predictive perplexity of these topic models, They cannot be shown in the same figure well. Figure 4 shows herb predictive perplexity of PTM, KGETM, and HC-KGETM with a different number of topics. Figure 5 shows all herb predictive perplexity with a different number of topics. As shown in Fig. 4, KGETM has lower perplexity on herb prediction than PTM, which shows TCM knowledge graph embeddings play a positive role in the herb recommendation task. HC-KGETM improves the performance of KGETM shows the positive effect of herb compatibility. As shown in Fig. 5 Block-LDA outperforms Link-LDA reflects the efficiency of herb pair. KGETM and HC-KGETM significantly outperform ATM, Link-LDA, and Block-LDA.

Qualitative Results. This section evaluates topics learned from all medical cases by our model. Figure 6 shows the top 10 symptoms and herbs of the

PTM		KGETM		HC-KGETM	
symptom	herb	symptom	herb	symptom	herb
癫痫 (Epilepsy)	甘草 (Licorice)	瘰疬 (Crewels)	麝香 (Moschus)	瘰疬 (Crewels)	乳香 (Frankincense)
瘰疬 (Crewels)	防风 (Fangfeng)	臁疮 (Ecthyma)	乳香 (Frankincense)	臁疮 (Ecthyma)	没药 (Myrrh)
杨梅疮 (Syphilis)	当归 (Angelica)	无名肿毒 (Pyogenic infections)	雄黄 (Realgars)	无名肿毒 (Pyogenic infections)	麝香 (Moschus)
皮肤瘙痒 (Itchy skin)	苦参 (Bittergiseng)	壮热 (High fever)	没药 (Myrrh)	瘫痪 (Paralysis)	雄黄 (Realgars)
头痛 (Headache)	丹参 (Dan-shen)	杨梅疮 (Syphilis)	冰片 (Borneol)	耳聋 (Deafness)	轻粉 (Calomelas)
心烦 (Palpitation)	金银花 (Honeysuckle)	牙痛 (Toothache)	朱砂 (Cinnabar)	杨梅疮 (Syphilis)	血竭 (Dragon's Blood)
壮热 (High fever)	川芎 (Chuanxiong Ligusticum)	口臭 (Bad breath)	黄连	半身不遂 (Hemiplegia)	白芷 (Baizhi Angelica)
胸闷 (Chest distress)	白芷 (Baizhi Angelica)	湿疮 (Eczema)	轻粉 (Calomelas)	头痛 (Headache)	草乌 (Kusnezoff Monkshood Mother Root)
鼻衄 (Epistaxis)	人参 (Panax)	疳积 (Malnutrition)	青黛 (Natural Indigo)	湿疮 (Eczema)	冰片 (Borneol)
烦躁 (Dysphoria)	牛黄 (Bezoar)	流泪 (Epiphora)	血竭 (Dragon's Blood)	脱肛 (Prolapse)	川乌 (Common Monkshood Mother Root)

(a) skin-related topic

PTM		KGETM		HC-KGETM	
symptom	herb	symptom	herb	symptom	herb
腹痛 (Stomachache)	桃 (Peach)	腹痛 (Stomachache)	当归 (Angelica)	腹痛 (Stomachache)	当归 (Angelica)
白带 (White vaginal discharge)	桃仁 (Peach seed)	白带 (White vaginal discharge)	白芍 (White Peony Root)	白带 (White vaginal discharge)	白芍 (White Peony Root)
赤白带 (Red and white vaginal discharge)	当归 (Angelica)	胎动不安 (Threatened abortion)	川芎 (Chuanxiong Ligusticum)	赤白带 (Red and white vaginal discharge)	川芎 (Chuanxiong Ligusticum)
浮肿 (Edema)	没药 (Myrrh)	赤白带 (Red and white vaginal discharge)	白术 (Atractylodes macrocephala)	胎动不安 (Threatened abortion)	白术 (Atractylodes macrocephala)
腰痛 (Lumbago)	乳香 (Frankincense)	头晕 (Dizzy)	甘草 (Licorice)	崩漏 (Uterine bleeding)	甘草 (Licorice)
瘰疬 (Crewels)	川芎 (Chuanxiong Ligusticum)	崩漏 (Uterine bleeding)	香附 (Cyperus rotundus)	潮热 Hot flash	人参 (Panax)
瘫痪 (Paralysis)	甘草 (Licorice)	心烦 (Palpitation)	人参 (Panax)	心烦 (Palpitation)	茯苓 (Wolfiporia extensa)
半身不遂 (Hemiplegia)	白芷 (Baizhi angelica)	不孕 (Sterility)	茯苓 (Wolfiporia extensa)	自汗 Spontaneous Sweating	香附 (Cyperus rotundus)
难产 (Difficult labour)	麝香 (Moschus)	浮肿 (Edema)	陈皮 (Tangerine Peel)	头晕 (Dizzy)	陈皮 (Tangerine Peel)
臁疮 (Ecthyma)	杏 (Apricot)	胎漏 (Spotting during pregnancy)	阿胶 (Donkey-hide Glue)	不孕 (Sterility)	芍药 (Peony)

(b) women-related topic

Fig. 6. Top words of topics learned by PTM, KGETM and HC-KGETM (Color figure online)

skin-related topic and female-related topic. We marked the symptoms in red which do not belong to the topic in fact, marked the herbs in red which can not treat the symptoms.

The first topic is about skin-related symptoms and their corresponding herbs. From the result, we can observe that: for PTM, on the one hand, it finds four symptoms are skin-related. On the other hand, PTM finds eight herbs can treat related symptoms. For KGETM, it finds five correct symptoms. Top three symptoms are all skin-related. On the other hand, all of the herbs KGETM finds can treat these symptoms. For HC-KGETM, it finds six symptoms are skin-related and ten correct herbs. In addition, the herbs which HC-KGETM finds accord with herbal compatibility. For example, the combination of realgar, frankincense, moschus, and Myrrh is often used to treat dermatosis.

The second topic is female-related symptoms and their corresponding herbs. We can observe from the result that: for PTM, it finds five correct symptoms, and the top three symptoms are all female-related. On the other hand, PTM finds eight correct herbs. For KGETM, it finds eight female-related symptoms. All of ten herbs KGETM finds can treat female-related symptoms. For HC-KGETM, it finds nine correct symptoms, and ten herbs can treat these symptoms. The combination of Angelica, Chuanxiong Ligusticum and Paeonia lactitol is active against catamenial symptoms.

We can observe that these three models' performance on herbs terms are better than symptoms terms. The performance of KGETM and HC-KGETM are better than PTM's on both herbs and symptoms. Furthermore, the herbs in the result of HC-KGETM accord with herbal compatibility better.

4.4 Comparsion with Different Parameter λ (Question 3)

Hyperparameter λ is the probability of a herb or symptom being generated by latent feature component, that means λ controls the influence of knowledge graph embeddings. The smaller λ is, the weaker influence of knowledge graph embeddings has. To study the influence of knowledge graph embedding, we use

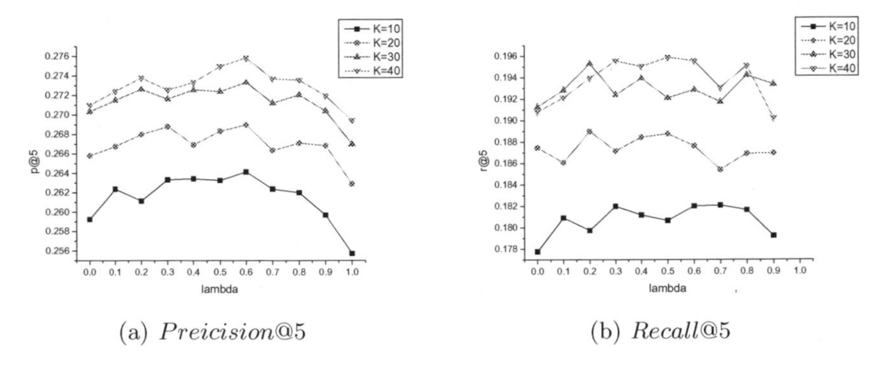

(a) *Preicision*@5 (b) *Recall*@5

Fig. 7. *precision*@5 and *recall*@5 results with different number of topics K, varying the mixture weight λ from 0.0 to 1.0.

KGETM who has no other constraints rather than HC-KGETM. The value of λ is set from 0.0 to 1.0.

Figure 7(a) and (b) shows *precision*@5 and *recall*@5 by KGETM with different number of topics K varied from 10 to 40, and the value of the mixture weight λ varied from 0.0 to 1.0. From the result, we can observe *precision*@5 and *recall*@5 changes with λ changes. KGETM gain the worst result when $\lambda = 0.0$ and $\lambda = 1.0$. $\lambda = 0.0$ means TCM knowledge graph embeddings are not available in models. KGETM changed to bag-of-words based model. Therefore the models gain bad results when $\lambda = 0.0$. $\lambda = 1.0$ means co-occurrence of herb terms and symptom terms are not available in models. KGETM can not mine latent correlations between herb terms and symptom terms well without co-occurrence in medical cases. Therefore the models gain bad results when $\lambda = 1.0$. *precision*@5 and *recall*@5 increase with λ increases when $\lambda \in [0.0, 0.7]$, decrease with λ increases when $\lambda \in [0.8, 1.0]$. We can observe models gain the best result when $\lambda = 0.6$.

5 Conclusion and Future Work

In this paper, we propose a novel method for TCM herb recommendation based on the topic model incorporating TCM knowledge graph embeddings. TCM knowledge graph embeddings introduced comprehensively correlated about symptoms and herbs into the topic model and improved the performance on herb recommendation. In addition, this model incorporated TCM herb compatibility knowledge. Our models outperform the state-of-art methods. The results of herb recommendation can be used to assist TCM doctors to prescribe herbs.

In future work, we will improve the quality of TCM knowledge graph embeddings by using more effective representation learning, increasing the dimension of embeddings. In addition, we will introduce more TCM domain knowledge such as dosage and contraindications of herbs.

Acknowledgements. This work was supported by National Key R&D Program of China (No. 2017YFC0803700), NSFC grants (No. 61532021), Shanghai Knowledge Service Platform Project (No. ZF1213) and SHEITC.

References

1. Ji, W., Zhang, Y., Wang, X., et al.: Latent semantic diagnosis in traditional Chinese medicine. World Wide Web **20**(5), 1071–1087 (2017)
2. Yao, L., Zhang, Y., Wei, B., et al.: A topic modeling approach for traditional Chinese medicine prescriptions. IEEE Trans. Knowl. Data Eng. **30**(6), 1007–1021 (2018)
3. Blei, D.M., Ng, A.Y., Jordan, M.I.: Latent dirichlet allocation. J. Mach. Learn. Res. **3**(Jan), 993–1022 (2003)
4. Das, R., Zaheer, M., Dyer, C.: Gaussian LDA for topic models with word embeddings. In: Proceedings of the 53rd Annual Meeting of the Association for Computational Linguistics and the 7th International Joint Conference on Natural Language Processing (Volume 1: Long Papers), vol. 1, pp. 795–804 (2015)

5. Yu, T., Li, J., Yu, Q., et al.: Knowledge graph for TCM health preservation: design, construction, and applications. Artif. Intell. Med. **77**, 48–52 (2017)
6. Zhang, F., Yuan, N.J., Lian, D., et al.: Collaborative knowledge base embedding for recommender systems. In: Proceedings of the 22nd ACM SIGKDD International Conference on Knowledge Discovery and Data Mining, pp. 353–362. ACM (2016)
7. Wang, S., Huang, E.W., Zhang, R., et al.: A conditional probabilistic model for joint analysis of symptoms, diseases, and herbs in traditional Chinese medicine patient records. In: 2016 IEEE International Conference on Bioinformatics and Biomedicine (BIBM), pp. 411–418. IEEE (2016)
8. Yao, L., Zhang, Y., Wei, B., et al.: Incorporating knowledge graph embeddings into topic modeling. In: AAAI 2017, pp. 3119–3126 (2017)
9. Bordes, A., Usunier, N., Garcia-Duran, A., et al.: Translating embeddings for modeling multi-relational data. In: Advances in Neural Information Processing Systems, pp. 2787–2795 (2013)
10. Wang, M., Liu, M., Liu, J., et al.: Safe medicine recommendation via medical knowledge graph embedding. arXiv preprint arXiv:1710.05980 (2017)
11. Nguyen, D.Q., Billingsley, R., Du, L., et al.: Improving topic models with latent feature word representations. arXiv preprint arXiv:1810.06306 (2018)
12. Yang, Y., Downey, D., Boyd-Graber, J.: Efficient methods for incorporating knowledge into topic models. In: Proceedings of the 2015 Conference on Empirical Methods in Natural Language Processing, pp. 308–317 (2015)
13. Rosen-Zvi, M., Griffiths, T., Steyvers, M., et al.: The author-topic model for authors and documents. In: Proceedings of the 20th Conference on Uncertainty in Artificial Intelligence, pp. 487–494. AUAI Press (2004)
14. Erosheva, E., Fienberg, S., Lafferty, J.: Mixed-membership models of scientific publications. Proc. Nat. Acad. Sci. **101**(Suppl. 1), 5220–5227 (2004)
15. Balasubramanyan, R., Cohen, W.W.: Block-LDA: jointly modeling entity-annotated text and entity-entity links. In: Proceedings of the 2011 SIAM International Conference on Data Mining, pp. 450–461. Society for Industrial and Applied Mathematics (2011)
16. Amer-Yahia, S., Roy, S.B., Chawlat, A., et al.: Group recommendation: semantics and efficiency. Proc. VLDB Endow. **2**(1), 754–765 (2009)
17. Yuan, Q., Cong, G., Lin, C.Y.: COM: a generative model for group recommendation. In: Proceedings of the 20th ACM SIGKDD International Conference on Knowledge Discovery and Data Mining, pp. 163–172. ACM (2014)
18. Wu, T., Qi, G., Wang, H., et al.: Cross-lingual taxonomy alignment with bilingual biterm topic model. In: AAAI 2016, pp. 287–293 (2016)
19. State Pharmacopoeia Commission of the PRC. Pharmacopoeia of the People's Republic of China 2005. BC Decker, Incorporated (2008)
20. State Bureau of Technical Supervision. GB/T 16751.1-1997. Clinic Terminology of Traditional Chinese Medical Diagnosis and Treatment-Diseases. Standards Press of China, Beijing (1997)

Knowledge Base Error Detection with Relation Sensitive Embedding

San Kim[1(✉)], Xiuxing Li[1], Kaiyu Li[1], Jianhua Feng[1], Yan Huang[2], and Songfan Yang[2]

[1] Tsinghua University, Beijing, China
{jins13,lixx16,ky-li18}@mails.tsinghua.edu.cn
[2] TAL Education Group, Beijing, China
{galehuang,yangsongfan}@100tal.com

Abstract. Recently, knowledge bases (KBs) have become more and more essential and helpful data source for various applications and researches. Although modern KBs have included thousands of millions of facts, they still suffer from incompleteness compared with the total amount of facts in real world. Furthermore, a lot of inaccurate and outdated facts may be contained in the KBs. Although there have been many studies dealing with incompleteness of the KBs, very few of works have taken into account detecting the errors in the KBs. Broadly speaking, there are three main challenges in detecting errors in the KBs. (1) Symbolic and logical form of the knowledge representations cannot detect the inconsistencies very well on large scale KBs. (2) It is hard to capture the correlations between relations. (3) There is no golden standard to learn or observe the patterns of inaccurate facts. In this work, we propose a **R**elation **S**ensitive **E**mbedding **A**pproach (RSEA) to detect the inconsistencies from KBs. We first design two correlation functions to measure the relatedness between two relations. Then, a dynamic cluster algorithm is presented to aggregate highly correlated relations into the same clusters. Finally, we encode discrete knowledge facts with effects of correlated relations into continuous vector space, which can effectively detect errors in KBs. We perform extensive experiments on two benchmark datasets, and the results show that our approach achieves high performance in detecting incorrect knowledge facts in these KBs.

Keywords: Knowledge base · Embedding model · Error detection

1 Introduction

Nowadays, *Knowledge Base (KB)* has become the most important cornerstone in many researches and applications such as semantic search, entity linking, question answer system and natural language processing, which can provide data sharing and semantic interpretation [9,11,13,14,19]. Many large scale knowledge bases (Freebase [3], Yago [30], DBpedia [20], WordNet [26], and etc.) have been constructed either automatically or collaboratively in various regions. Although

G. Li et al. (Eds.): DASFAA 2019, LNCS 11446, pp. 725–741, 2019.
https://doi.org/10.1007/978-3-030-18576-3_43

thousands of millions of facts were included in these KBs, they are still far from complete compared with the total number of facts in the real world. On the other hand, due to constantly new emerging facts and outdated information as the world evolves, many inaccurate facts existed in the KB(s) [2,22,31], which were ignored by most of existing applications and works.

In the past few decades, there were many researches leveraged symbolic and logical approaches to augment new facts into the KB(s), but none of them was either tractable or effective. Fortunately, in order to deal with incompleteness of the KBs, many knowledge embedding techniques have been developed recently, which embed all of the entities and relations in the KBs into k-dimensional vector space. Then, given a relation, it predicted new facts by randomly replacing one of the subject or object in the golden triplets, i.e., predicting missing edges in the knowledge graph. The main idea is to define a translation function, in the embedding vector space, which projects the subject through the relation to the corresponding object as similar as possible, if a given triplet is positive, otherwise, as dissimilar as possible. For example, *TransE* [6] is one of the most representative approaches which utilizes a simple summation function to measure whether the subject plus relation approximates to the object in the vector space.

In contrast, detecting errors in the knowledge bases is still a challenging problem, especially, self-detecting incorrect facts without both external data sources and manual labours. There are three main challenges with corresponding limitations. (1) KBs were constructed with discrete representations of the knowledge facts. Thus, by using traditional data cleaning approaches [16,17,21], it is hard to execute a wide range inference over a huge scale knowledge graph to detect global inconsistencies between the triplets. (2) Compared with the entities with a large quantity of descriptive informations, the relations have few contexts to capture their semantic correlations. Thus, traditional embedding techniques only considered single relationships between the entities and ignored the correlations between the multiple relations. (3) Since there isn't any golden standard for reference, previous embedding works assumed all of the triplets in the KB as the golden triplets and automatically generate negative triplets by randomly replacing the entities of either side of the golden triplets. Consequently, they may generate false negative triplets and ignore false positive triplets as well, because the KB is neither complete nor absolutely correct.

In order to address the above challenges, we propose a novel *Relation Sensitive Embedding Approach (RSEA)* which can detect the inconsistencies between the knowledge facts by computing the holistic correlations among the entities and relations in a large scale knowledge graph. Our main idea is that highly correlated relations can be leveraged to detect inconsistencies with each others. For example, in Fig. 1, five relations, *"IsPresidentOf"*, *"IsPoliticianOf"*, *"IsCitizenOf"*, *"BornIn"*, *"LivesIn"*, will be correlated with each other, especially, first three ones. Because the president of a country must be the citizen of this country and a politician of this country, as well as lives in this country and etc. Thus, if there is an error fact ⟨"*MoonJae − in*", "*LivesIn*", "*NorthKorea*"⟩, we can utilize other correlated relations to detect its inconsistency, because most of

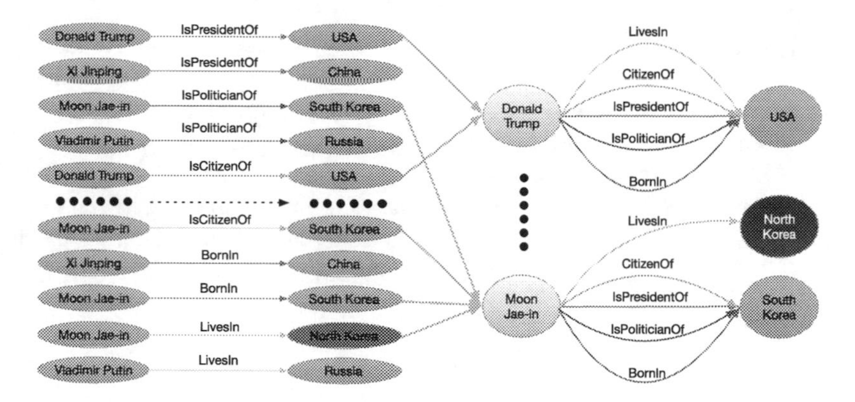

Fig. 1. Example of relation sensitive error detection

relations indicate that "Moon Jae-in" should live in "South Korea". However, since the relations, unlike the entities, have very few descriptive contexts, it is very difficult to represent the relations for computing their correlations. Therefore, we design two correlation functions, based on the probability method and vector representation method, to measure the relatedness between two relations. In order to find dependencies among the relations, we also propose a dynamic cluster algorithm to effectively aggregate highly correlated relations into the same groups. Then, due to the difficulties of long inference on discrete knowledge representations, we encode knowledge facts along with correlation effects of relations into the continuous vector space. Finally, by performing our RSEA, we can learn a model to detect errors in KBs. Also, our RSEA can easily extend previous embedding techniques with enhanced prediction result. According to our knowledge so far, this is the first paper to doubt authority of KBs, which aims to detect the incorrect facts in the KBs, i.e., identifying misplaced edges in the knowledge graph, while KB completion task is to find missing edges.

In this paper, we make the following contributions: (1) We propose a novel relation sensitive embedding approach which involves the effect of correlations between the relations. (2) We define two correlation functions based on probability theory and vector representation methods. (3) We also design a dynamic cluster algorithm to aggregate highly correlated relations effectively. (4) We evaluate our RSEA by conducting extensive experiments on two benchmark datasets. Experimental results demonstrate that our method can not only detect errors in KBs effectively, but also outperform state-of-the-art models significantly in predicting missing facts.

Paper Structure: We review previous embedding techniques in Sect. 2. In Sect. 3, we introduce the basic knowledge base model and formalize our problem, then describe the workflow in details. We present two correlation functions with dynamic cluster algorithm in Sect. 4 and explain our relation sensitive embedding approach in Sect. 5. We exhibit experimental results and summarize the conclusion in Sects. 6 and 7 respectively.

2 Related Works

Recently, many knowledge base embedding techniques were proposed to encode discrete knowledge graph into the continuous vector spaces. We first introduce some frequently used notations. A fact triplet in the KB, i.e., $\langle subject, relation, object\rangle$, were named as head, relation and tail in the embedding works respectively. Hence, a triplet is denoted as $(h,\ r,\ t)$, and their corresponding vector representations are denoted as $(\vec{h},\ \vec{r},\ \vec{t})$. We also denote the matrix as $\vec{M}$ and the tensor as $\vec{T}$. Score function $f_{\vec{r}}(\vec{h},\ \vec{t})$ will obtain higher scores for the positive triplets belonging to the KB and lower scores for the negative triplets generated automatically.

Table 1. Model Comparisons with *score function, the number of parameters* and *time complexity* within the same k dimension embedding space $\mathbb{R}^{k\times k}$. N_e, N_r, N_t, c, h were denoted as the number of entities, relations and triplets in the KB, channels in a tensor and hidden nodes in the neural network, respectively. v is the average number of clusters in a relation.

Models	Score function	# of parameters $O(-)$	Time complexity $O(-)$
TransE	$\Vert\vec{h} + \vec{r} - \vec{t}\Vert_2^2$	$kN_e + kN_r$	N_t
TransH	$\Vert\vec{h}_\perp + d_r - \vec{t}_\perp\Vert_2^2$	$kN_e + 2kN_r$	$2kN_t$
TransR	$\Vert\vec{h}\vec{M}_r + \vec{r} - \vec{t}\vec{M}_r\Vert_2^2$	$kN_e + k(k+1)N_r$	$2k^2N_t$
CTransR	$\Vert\vec{h}\vec{M}_r + \vec{r}_c - \vec{t}\vec{M}_r\Vert_2^2 + \alpha\Vert\vec{r}_c - \vec{r}\Vert_2^2$	$kN_e + k(k+v)N_r$	$2k^2N_t$
TransD	$\Vert\vec{M}_{rh}\vec{h} + \vec{r} - \vec{M}_{rt}\vec{t}\Vert_2^2$	$2kN_e + 2kN_r$	$2kN_t$
SE	$\Vert\vec{M}_{rh}\vec{h} - \vec{M}_{rt}\vec{t}\Vert_1$	$kN_e + 2k^2N_r$	$2k^2N_t$
Unstructured	$\Vert\vec{h} - \vec{t}\Vert_2^2$	kN_e	N_t
SME(linear)	$(\vec{M}_1\vec{h} + \vec{M}_2\vec{r} + \vec{b}_1)^\top(\vec{M}_3\vec{t} + \vec{M}_4\vec{r} + \vec{b}_2)$	$kN_e + kN_r + 4h(k+1)$	$4khN_t$
SME(bilinear)	$((\vec{M}_1\vec{h})\otimes(\vec{M}_2\vec{r}) + \vec{b}_1)^\top((\vec{M}_3\vec{t})\otimes(\vec{M}_4\vec{r}) + \vec{b}_2)$	$kN_e + kN_r + 4h(kc+1)$	$4khcN_t$
SLM	$\vec{u}_r^\top g(\vec{M}_{r1}\vec{h} + \vec{M}_{r2}\vec{t} + \vec{b}_r)$	$kN_e + 2h(k+1)N_r$	$(2k+1)hN_t$
NTN	$\vec{u}_r^\top g(\vec{h}^\top\vec{T}_r\vec{t} + \vec{M}_r\begin{bmatrix}\vec{h}\\ \vec{t}\end{bmatrix} + \vec{b}_r)$	$kN_e + (k^2+2k+2)cN_r$	$(ck^2 + (2h + c)k + h)N_t$
LMF	$\vec{h}^\top\vec{M}_r\vec{t}$	$kN_e + k^2N_r$	$(k^2+k)N_t$

TransE: TransE was proposed by Bordes et al. [6], which defined a simple and efficient score function to translate h through r to t, i.e., $\vec{h}$ plus $\vec{r}$ approximately equals to $\vec{t}$.

$$f_{\vec{r}}(\vec{h},\ \vec{t}) = \Vert\vec{h} + \vec{r} - \vec{t}\Vert_2^2 \qquad (1)$$

Therefore, TransE assumes the score of Eq. 1 will be as low as possible, if $(h,\ r,\ t)$ belongs to the KB as a positive triplet, and high otherwise. Although TransE performed pretty well, when dealing with 1-to-1 relations, it had very poor performance with 1-to-N, N-to-1 and N-to-N relations, as well as reflexive relations. This is because TransE cannot distinguish the entities appeared in the side of N.

TransH: In order to address the above drawbacks, Wang et al. proposed a hyperplane based method, named TransH [32]. Instead of using relation vector $\vec{r}$ to directly translate $\vec{h}$ to $\vec{t}$ in the same embedding space, TransH introduced the relation-specific hyperplane $\vec{w_r}$ (normal vector) on which triplet vector $(\vec{h}, \vec{r}, \vec{t})$ was projected to $(\vec{h_\perp}, \vec{d_r}, \vec{t_\perp})$. The main idea of TransH is that entity vectors of N side can be projected to one point on the hyperplane. Then, similar to TransE, it used the following score function to measure whether a triplet holds, or not.

$$f_{\vec{r}}(\vec{h}, \vec{t}) = \|\vec{h_\perp} + \vec{d_r} - \vec{t_\perp}\|_2^2 \tag{2}$$

where $\vec{h_\perp} = \vec{h} - \vec{w_r}^\top \vec{h} \vec{w_r}$ and $\vec{t_\perp} = \vec{t} - \vec{w_r}^\top \vec{t} \vec{w_r}$ when restricting $\|\vec{w_r}\|_2 = 1$.

TransR/CTransR: Lin et al. introduced TransR [23] which indicates that entities and relations should not embed into the same vector space $\mathbb{R}^k$, because they are completely different objects. Thus, TransR embeds entities and relations into two different embedding spaces, entity space $\mathbb{R}^k$ and relation space $\mathbb{R}^d$, respectively. Then, it uses the following score function to translate h to t in the relation space $\mathbb{R}^d$ by defining a mapping matrix $\vec{M_r}$ w.r.t. each relation r which maps entity vectors from $\mathbb{R}^k$ to $\mathbb{R}^d$.

$$f_{\vec{r}}(\vec{h}, \vec{t}) = \|\vec{h}\vec{M_r} + \vec{r} - \vec{t}\vec{M_r}\|_2^2 \tag{3}$$

where $(\vec{h}, \vec{t}) \in \mathbb{R}^k$, $\vec{r} \in \mathbb{R}^d$ and $\vec{M_r} \in \mathbb{R}^{k \times d}$. They also proposed a cluster-based CTransR as the extension of TransR. Based on the vector offsets of entity pairs of head and tail, i.e., $(\vec{h} - \vec{t})$ obtained from TransE, CTransR aggregates similar entity pairs of a relation into more specific clusters, then assigns a relation vector for each cluster.

TransD: Similar to TransR/CTransR, TranD proposed by Ji et al. [18] also separately embedded entities and relations into different vector spaces, i.e., $(\vec{h}, \vec{t}) \in \mathbb{R}^k$ and $\vec{r} \in \mathbb{R}^d$. However, TransD is more fine-grained model by introducing different projection matrices for h and t respectively, denoted as $\vec{M}_{rh}, \vec{M}_{rt} \in \mathbb{R}^{d \times k}$, due to various types of entities and relations. Hence, for each triplet (h, r, t), it has a triplet of embedding vectors $(\vec{h}, \vec{r}, \vec{t})$ and a triplet of projection vectors $(\vec{h}_p, \vec{r}_p, \vec{t}_p)$, where $\vec{h}, \vec{t}, \vec{h}_p, \vec{t}_p \in \mathbb{R}^k$ and $\vec{r}, \vec{r}_p \in \mathbb{R}^d$. Then, its score function is defined as follows:

$$f_{\vec{r}}(\vec{h}, \vec{t}) = \|\vec{M}_{rh}\vec{h} + \vec{r} - \vec{M}_{rt}\vec{t}\|_2^2 \tag{4}$$

where $\vec{M}_{rh} = \vec{r}_p \vec{h}_p^{\top} + I^{d \times k}$ and $\vec{M}_{rt} = \vec{r}_p \vec{t}_p^{\top} + I^{d \times k}$, and $I^{d \times k}$ is an identity matrix. Thus, TransD has more representative ability to distinguish different types of entities and relations.

Other Methods: Bordes et al. also proposed *Structured Embedding (SE)* model [7], *Unstructured* model [4] and *Semantic Matching Energy (SME)* model [5] in 2011, 2012 and 2014 respectively. SE model employed two relation-specific matrices $\vec{M}_{rh}$ and $\vec{M}_{rt}$ to project head h and tail t, then defined a score function with L_1 distance, $f_{\vec{r}}(\vec{h}, \vec{t}) = \|\vec{M}_{rh}\vec{h} - \vec{M}_{rt}\vec{t}\|_1$. Unstructured model is a simplified version of TransE, which only embedded head h and tail t and ignored all of the relations r, then defined a score function $f_{\vec{r}}(\vec{h}, \vec{t}) = \|\vec{h} - \vec{t}\|_2^2$. Thus, it cannot recognize different relations. SME model utilized the neural network as its score function which is able to consider the correlations between entities and relations by applying matrix products and Hadamard products, then, all of the relations shared parameters of the network. SME defined semantic matching energy functions with linear and bilinear forms, i.e., $f_{\vec{r}}(\vec{h}, \vec{t}) = (\vec{M}_1\vec{h} + \vec{M}_2\vec{r} + \vec{b}_1)^{\top}(\vec{M}_3\vec{t} + \vec{M}_4\vec{r} + \vec{b}_2)$ and $f_{\vec{r}}(\vec{h}, \vec{t}) = ((\vec{M}_1\vec{h}) \otimes (\vec{M}_2\vec{r}) + \vec{b}_1)^{\top}((\vec{M}_3\vec{t}) \otimes (\vec{M}_4\vec{r}) + \vec{b}_2)$, where $\otimes$ is Hadamard product and $\vec{b}_1$, $\vec{b}_2$ are bias vectors. Note that Bordes et al. replaced the matrices in the bilinear form with tensors in after work.

Socher et al. designed two models, *Single Layer Model (SLM)* [29] and *Neural Tensor Network (NTN)* [28]. SLM was a simplified version of NTN model, which leveraged non-linear neural network to define its score function, $f_{\vec{r}}(\vec{h}, \vec{t}) = \vec{u}_r^{\top} g(\vec{M}_{r1}\vec{h} + \vec{M}_{r2}\vec{t} + \vec{b}_r)$, where $\vec{M}_{r1}$, $\vec{M}_{r2}$ are weight matrices w.r.t. relation r, $\vec{b}_r$ is bias vector and $g()$ is $tanh$ operation. NTN model also utilized non-linear neural network to define its score function by combining the second-order correlations. Thus, its score function is $f_{\vec{r}}(\vec{h}, \vec{t}) = \vec{u}_r^{\top} g(\vec{h}^{\top} \vec{T}_r \vec{t} + \vec{M}_r \begin{bmatrix} \vec{h} \\ \vec{t} \end{bmatrix} + \vec{b}_r)$, where $\vec{T}_r$, $\vec{M}_r$, $\vec{b}_r$ denoted 3-way tensor, weighted matrix and bias vector respectively, and $g()$ is $tanh$ operation. Although, NTN had very strong expressive ability, it cannot perform on large scale KB, due to its a large number of parameters.

Latent Factor Model (LMF) was introduced by Jenatton et al. (Sutskever et al.), which used quadratic form to represent second-order correlations between entity vectors. Then, it defined a simple but effective bilinear score function, $f_{\vec{r}}(\vec{h}, \vec{t}) = \vec{h}^{\top} \vec{M}_r \vec{t}$.

Besides those studies, there were many models with strong performances published recently, e.g., *MANIFOLDE* [34], *HOLE* [27], *CONVE* [12], and for more details, please refer to the references. Finally, we summarized score function, the number of parameters and time complexity of the above models in Table 1.

3 Overview

In this section, we first introduce the knowledge base model with notations, then give the formalization of our problem definitions. Finally, we describe the overall workflow in details.

3.1 Knowledge Base

Knowledge Base (KB) is represented in *Resource Description Framework (RDF)* language with a vast amount of resources in the World Wide Web. Now, we define our KB model with a simple version of RDF. A general KB can be regarded as a quintuple $(\mathbb{E}, \mathbb{T}, \mathbb{R}, \mathbb{F}, \mathbb{G})$, where $\mathbb{E}, \mathbb{T}, \mathbb{R}$ denote the set of entities(instances), types(classes) and relations(predicates) respectively. Tuple $\mathbb{F} \subseteq \mathbb{E} \times \mathbb{R} \times \mathbb{E}$ is the set of facts, each of which is represented as a triplet (h, r, t). Tuple $\mathbb{G}$ is a multi-relational knowledge graph in which each of nodes represents an entity and each of edges indicates a relation r from head h to tail t.

3.2 Problem Definitions

As introduced in Sect. 1, KB has many missing facts, as well as a number of incorrect facts, i.e., missing and misplaced edges in the knowledge graph. However, most of previous works only considered filling up missing edges in the knowledge graph and sweep the misplaced edges under the rug. Thus, we aim to detect the inconsistencies between the fact triplets in the KB by utilizing the correlations between the relations. Also, it is able to complete the new facts more accurately. Now, we formalize our problems with three definitions.

Definition 1. CLUSTERING CORRELATED RELATIONS: *Given the set of relations $\mathbb{R}$ in the KB, clustering correlated relations is to aggregate semantic related relations into the same groups by employing our correlation functions.*

Definition 2. DETECTING INACCURATE FACTS: *Given a knowledge graph $\mathbb{G}$ with the set of fact triplets $\mathbb{F}$, inaccurate facts detection problem is to find most likely inaccurate fact triplets in $\mathbb{F}$, i.e., misplaced edges in $\mathbb{G}$.*

Definition 3. PREDICATE MISSING FACTS: *Given a fact triplet,* $(\mathrm{h}, \mathrm{r}, \mathrm{t}) \notin \mathbb{F}$, *missing facts predication problem is to predicate whether this triplet should add to $\mathbb{F}$ with minimum loss, i.e., adding new edge into the $\mathbb{G} \equiv (\mathrm{h}, \mathrm{r}, \mathrm{t}) + \mathbb{G}$, where $\equiv$ means no conflicts.*

3.3 Workflow

In this section, we describe the details of our workflow. As shown in Fig. 2, there are four main parts. Given a KB as an input, we first utilize our correlation functions to measure the correlations between the relations, then leverage our dynamic clustering algorithm to partition highly correlated relations into the

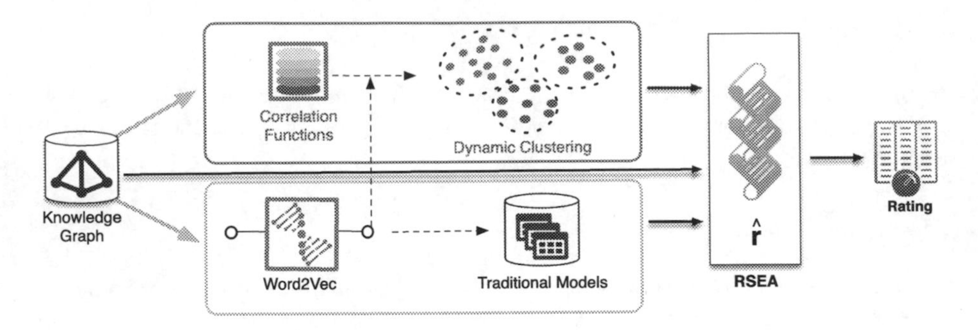

Fig. 2. Illustration of RSEA workflow (Color figure online)

same groups, as shown in red square part. See in green square part, in order to represent the semantics of relations more precisely, we also exploit the traditional models to obtain embeddings to describe the relations semantically by using our relation expression simultaneously. Then we utilize our relation sensitive embedding approach to merge the correlation effects of clusterd relations into the knowledge graph embedding process. Finally, given a fact triplet our RSEA can predicate whether it is true or false with a probability score. Note that our RSEA can also predicate new fact triplets like previous works by rating the triplets in the KB.

4 Correlation Functions with Dynamic Cluster Algorithm

In this section, we propose two correlation functions (probability based function $\mathcal{P}(r_i, r_j)$ and vector based function $\mathcal{V}(r_i, r_j)$) to measure the correlation between two relations r_i and r_j, then design a novel clustering algorithm to group highly correlated relations.

4.1 Probability-Based Correlation Function: $\mathcal{P}(r_i, r_j)$

Intuitively, if two relations have more common elements in the fact triplets in the KB, i.e., entities of subject and object, they will be more likely to correlate with each other. For example, relation *"IsPresidentOf"* is more coherent to relation *"IsCitizenOf"* than relation *"LivesIn"* as they have more common entity pairs of subject and object. Motivated by the aforementioned, we define a probability based correlation function $\mathcal{P}(r_i, r_j)$ by using *Pointwise Mutual Information (PMI)* [10] to measure the correlation between r_i and r_j.

We first introduce some notations. $S(r_i)$ and $O(r_i)$ are the set of subject and object entities of relation r_i in the KB. $\mathcal{N}$ is the total number of facts in the KB. $\mathbf{P_S}(r_i) = \frac{|S(r_i)|}{\mathcal{N}}$ and $\mathbf{P_O}(r_i) = \frac{|O(r_i)|}{\mathcal{N}}$ are the probability of an entity belonging to relation r_i as the subject and object respectively. $\mathbf{P_S}(r_i \cap r_j) = \frac{|P_S(r_i) \cap P_S(r_j)|}{\mathcal{N}}$ is the probability of common subjects of relation r_i and r_j, and $\mathbf{P_O}(r_i \cap r_j) =$

$\frac{|P_O(r_i) \cap P_O(r_j)|}{\mathcal{N}}$ is as same as the above with common objects. Therefore, the subject PMI score of r_i and r_j is:

$$PMI_S(r_i, r_j) = \log \frac{\mathbf{P_S}(r_i \cap r_j)}{\mathbf{P_S}(r_i) \cdot \mathbf{P_S}(r_j)} \tag{5}$$

Definition of the object PMI score, $PMI_O(r_i, r_j)$, is similar. In the following learning tasks, normalization can benefit the model training process [8]. Thus, we normalize the PMI score into $[-1, +1]$ as the follows:

$$NPMI_S(r_i, r_j) = \frac{PMI_S(r_i, r_j)}{-\log \mathbf{P_S}(r_i \cap r_j)} \tag{6}$$

Furthermore, note that there does not exists any negative correlation between relation r_i and r_j because each pair of r_i and r_j have 0 common fact triplets at least. Thus we define our probability based correlation function by constraining score into $[0, 1]$ as the follows:

$$\mathcal{P}(r_i, r_j) = \frac{NPMI_S(r_i, r_j) + NPMI_O(r_i, r_j) + 2}{4} \tag{7}$$

4.2 Vector-Based Correlation Function: $\mathcal{V}(r_i, r_j)$

Note that probability based correlation function is based on the discrete representations of the entity pairs w.r.t. the relation, which may not represent it sufficiently. For example, in the Fig. 1, the entity of "Trump Donald" and "Xi Jinping" indeed have similar meaning of the president of the country but are regarded as different vocabularies in discrete representation. In order to obtain stronger representations of the correlation between r_i and r_j, we measure the correlations within the continuous vector space by defining a vector based correlation function $\mathcal{V}(r_i, r_j)$. However, unlike entities, the relations have few describing context informations. Inspired by the work [33], we utilize all of the pairs of subject and object entity vectors w.r.t. the relation r to represent it.

In this work, we leverage *Word2Vec* to obtain the embedding vector for each of entities. Word2Vec [15,24,25] is a very popular and famous natural language processing tool published by Google in 2013, which takes a huge text corpus as the input to encode words into the vectors by utilizing either continuous bag-of-words (CBOW) model or continuous skip-gram model. Now, we can define the vector correlation function as the follows:

$$\mathcal{V}(r_i, r_j) = cos(E^+(r_i), E^+(r_j)) + \lambda \frac{cos(S^+(r_i), S^+(r_j)) + cos(O^+(r_i), O^+(r_j))}{2} \tag{8}$$

where $S^+(r_i)$ and $O^+(r_i)$ are the summation of the set of subject entity vectors and object entity vectors w.r.t. relation r_i respectively, and the vector representation of the relation r_i is denoted as $E^+(r_i) = S^+(r_i) + O^+(r_i)$.

Equation 8 has two parts, each of which is computed by cosine similarity function. The first part of Eq. 8 only takes into account the overall distance

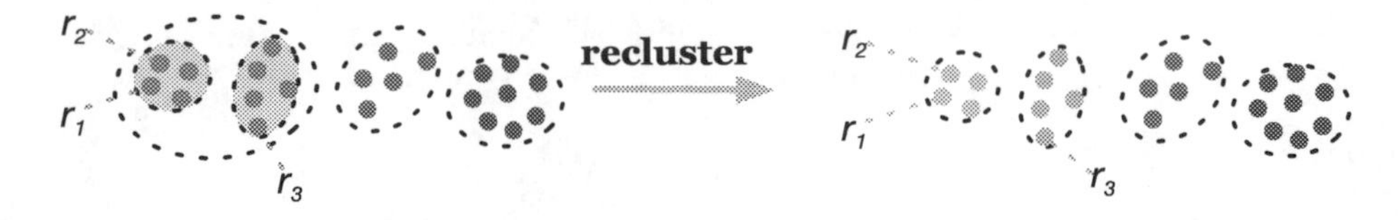

Fig. 3. Dynamic clustering example (Color figure online)

between r_i and r_j in the trained word embedding space, while the second part considered direction of two relations by computing the similarity of the sets of subjects and objects respectively. Finally, we use hyper parameter λ to control its balance which is learned from the experience.

4.3 Dynamic Cluster Algorithm

Our framework first blocks the relations into the clusters and then use highly correlated relations of relation r_i to detect incorrect triplets w.r.t. relation r_i, thus we introduce the clustering algorithm (K-means, DBScan, HAC and etc.) based on proposed correlation functions mentioned in Sects. 4.1 and 4.2.

Intuitively, we could simply use K-means algorithm to divide the relations into the clusters. Although K-means is very fast in practice, such method would have two main drawbacks. First, we do not have prior knowledge about the distributions of the relations w.r.t. the parameter K. Therefore, it is always hard to select a proper K in advance. Second, usually, some relations need to be clustered into more specific groups. For example, in Fig. 3, r_1 and r_2 should be clustered into a more specific cluster, because r_1, r_2 and r_3 should not be in the same cluster. However, K-means algorithm with an inappropriate $K = 3$ will misplace r_1, r_2, r_3 into the same cluster.

Considering the above drawbacks, we propose a *Dynamic Clustering Algorithm*. After clustering, we observe that there exist many clusters which need to be clustered more specifically. For example, in Fig. 3, the green relations should actually fall into two clusters, thus we need to further split them into two clusters. Our insight is that a cluster with uniformly distributed points will have smaller variance than a cluster with skewed distributed points. Based on the above analysis, we first employ K-means algorithm to divide the relations into 2 'big' clusters. For each cluster, we iteratively check if its variance is less than a given threshold, if yes, we believe it is a specific enough cluster, otherwise, we use K-means algorithm again to divide it into two clusters dynamically until achieving convergence. Algorithm 1 illustrates the processing of dynamic clustering with K-means algorithm ($K = 2$).

5 Relation Sensitive Embedding Approach

In this section, we introduce our relation sensitive embedding approach, named *RSEA*, which can take account into the effect of correlations between the relations, so that our model has the ability to detect the errors in the KB.

Algorithm 1. DYNAMIC CLUSTER ALGORITHM(R, δ)

Input: $\mathcal{R}$: set of relations; δ : variance threshold
Output: $\mathcal{C}$: set of relation clusters

```
1  begin
2      L ← null; // List of clustered relations
3      v = 0; //variance of each cluster
4      L ← Kmeans(R); // set k=2
5      while L.length ≠ 0 do
6          for each relation cluster c_i ∈ L do
7              v = CalculateVariance(c_i); // calculate cluster variance of c_i
8              if v < δ then
9                  C.add(c_i);
10                 L.remove(c_i);
11             else
12                 L ← L ∪ Kmeans(c_i); // set k=2
13     return C;
```

5.1 Definition of RSEA

Our RSEA can easily merge with other knowledge base embedding models, e.g., TransE, TransH, TranR and etc. Here, we take TransE model as an example. We combine the relation of TransE and correlations between the relations in a cluster to define an upgrade relation vector $\overrightarrow{\overrightarrow{r}}$ as the followings.

$$\overrightarrow{\overrightarrow{r}} = \overrightarrow{r}_{tr} + \overrightarrow{r}_{co} \tag{9}$$

$$\overrightarrow{r}_{co} = \sum_{i}^{k} w_i \cdot \overrightarrow{r}_i$$

where $\overrightarrow{r}_{tr}$ is the relation vector from traditional embedding model (TransE) and $\overrightarrow{r}_{co}$ is the correlated relation vector, i.e., $w_i = \frac{\mathcal{S}_i}{\sum \mathcal{S}_i}$ and $\overrightarrow{r}_i = E^{+}(r_i)$, where $\mathcal{S}_i$ is the correlation score computed by either $\mathcal{P}$ or $\mathcal{V}$. Thus, upgrade relation vector in RSEA can be revised by its highly correlated relations in its cluster. Then, our score function is defined as the follows:

$$f_{\overrightarrow{\overrightarrow{r}}}(\overrightarrow{h}, \overrightarrow{t}) = \|\overrightarrow{h} + \overrightarrow{\overrightarrow{r}} - \overrightarrow{t}\|_2^2 \tag{10}$$

In practice, we limit the following constrains on the norms of the embeddings h, r, t. $\{||\overrightarrow{h}||_2, ||\overrightarrow{\overrightarrow{r}}||_2, ||\overrightarrow{t}||_2, ||\overrightarrow{r}_{co}||_2\} \leq 1$.

By merging the information contained in the correlated relations, the representations of the relations could be improved, so that the representation result of the relation will be closer to the true semantics. The optimization strategy is based on the fact that, in a knowledge base, the amount of incorrect information is much less than the correct information, as a consequence of which incorporating the auxiliary information obtained form the correlated relations will have a positive impact for the relation representation.

5.2 Training Processing

Here, we define the loss function $\mathcal{L}$ as the following margin based score function:

$$\mathcal{L} = \sum_{(h,r,t)\in\mathbb{F}} \sum_{(h',r,t')\in\mathbb{F}'} [\gamma + f_{\overrightarrow{r}}(\overrightarrow{h}, \overrightarrow{t}) - f_{\overrightarrow{r}}(\overrightarrow{h'}, \overrightarrow{t'})]_+ \tag{11}$$

where $[*]_+ = \max(0, *)$ means the maximum value of two inputs, $\mathbb{F}' \in \{\mathbb{E} \times \mathbb{R} \times \mathbb{E} - \mathbb{F}\}$, i.e., corrupting triplets do not exist in the KB, γ is the margin used to separate positive and negative triplets. To construct $\mathbb{F}'$, we follow the previous works [23,32] by replacing one of the entities in $(h, r, t) \in \mathbb{F}$ with probability of position appearance. During the training processing, we employ *Stochastic Gradient Descent (SGD)* to minimize our loss function.

Table 2. Details of benchmark datasets

Datasets	#Relation	#Entity	#Training	#Valid	#Test	Source
WN18	18	40,943	141,442	5,000	5,000	WordNet
FB15K	1,345	14,951	483,142	50,000	59,071	Freebase

6 Experiments

6.1 Experimental Settings

Datasets. In our experiments, we employ two benchmark datasets, "WN18" generated from WordNet and "FB15K" generated from Freebase [6]. WordNet is a large lexical database in English in which similar entities are corresponding to the same synset. Freebase is one of the largest knowledge bases corresponding to a graph of real world facts constructed by users collaboratively. The detailed statistics of the two datasets are shown in Table 2.

Evaluation. We conduct extensive experiments on two tasks, error detection and link prediction. Knowledge base error detection is to determine whether a given fact triple (h, r, t) is correct. Link prediction is to predict the missing entity h or t in a triple (h, r, t) according to the embedding of another entity and relation. We compared our RSEA with state-of-the-art knowledge base embedding methods (TransE, TransH, TransR and TransD) to justify the effectivity and efficiency of our approach. We merge our RSEA with TransE (TransH, TransR,

TransD) using method in Sect. 5 by running probability (vector) based correlation function in Sect. 4.1 (4.2). We denote these methods as RSEA-(*)-(+). $*\in \{E, H, R, D\}$ where E denotes TransE, H denotes TransH, R denotes TransR and D denotes TransD. $+\in \{P, V\}$ where P denotes probability correlation function in Sect. 4.1 and V denotes vector correlation function in Sect. 4.2.

Metric. For error detection task, "WN18" and "FB15K" have already contained negative triples obtained by corrupting 10% of golden triples in the KB [28]. We compute the accuracy of error detection. For link prediction task, we replace the head/tail of each testing triple (h, r, t) with each entity in the knowledge graph, and give a ranking list of candidate entities in descending order of the scores calculated by score function $f_{\overrightarrow{r}}(\overrightarrow{h}, \overrightarrow{t})$. Based on the entity ranking lists, we adopt two evaluation metrics by aggregating overall testing triples: (1) Hits@10: the proportion of correct entities in Top-10 ranked entities; (2) Mean Rank: the averaged rank of correct entities;

Implementation. All models are trained and tested on a single GTX 1080 TI GPU implemented in *Tensorflow* [1]. We conducted all the experiments for 1000 times and compute the average results.

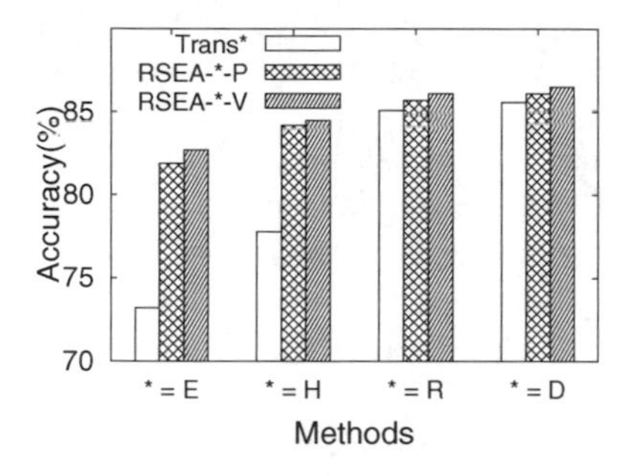

Fig. 4. Error detection of WN18

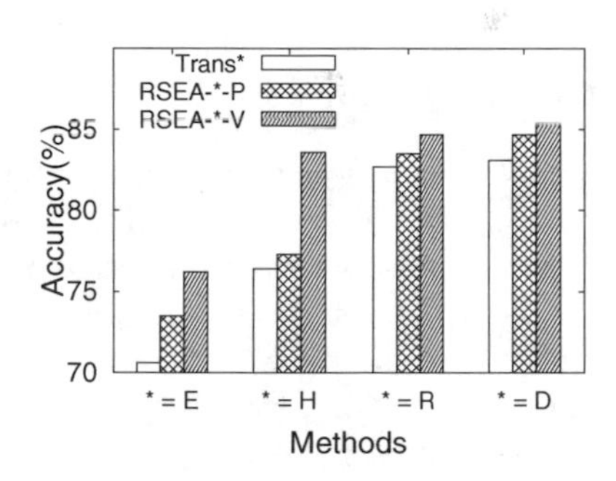

Fig. 5. Error detection of FB15

6.2 Error Detection

For each triple (h, r, t), if the score obtained by score function $f_{\overrightarrow{r}}(\overrightarrow{h}, \overrightarrow{t})$ is below the relation-specific threshold θ, the triple will be classified as an incorrect fact, otherwise as a correct fact. The threshold θ is optimized by maximizing classification accuracy on the validation set. In experiments, we set learning rate α as $\{0.1, 0.01, 0.001\}$, dimension of entities and relations k as $\{50, 100, 150, 200\}$, threshold θ as $\{0.6, 0.7, 0.8, 0.9\}$ and batch size $\mathcal{B}$ as $\{32, 64, 128\}$ in custom. We tune the parameters and finally find that the optimal configurations for RSEA are $\alpha = 0.01$, k $= 200$, $\theta = 0.5$ and $\mathcal{B} = 128$.

The evaluations of error detection on "WN18" and "FB15" are shown in Figs. 4 and 5 respectively. We report the error detection accuracy of our RSEA combining with different methods. As seen in the results, the best error detection accuracy with our RSEA on WN18(FB15) is the RSEA-D-V which achieved 86.5%(85.4%), while the worst one on WN18(FB15) is the RSEA-E-P which achieved 81.9%(73.5%). We observe that: (1) On both WN18 and FB15K, RSEA-(*)-P and RSEA-(*)-V consistently outperform the baselines, especially on FB15K. The reason may be that WordNet itself is a lexical database and different relationships are far apart. (2) RSEA-(*)-V methods outperform the RSEA-(*)-P because the vector correlation function could better mine the semantic information in related relations by taking the characteristics of entities, which could offer more precise on the semantic representation of the relation in the knowledge base. (3) For comparison of accuracy, TranD/RSEA-D-(+) > TranR/RSEA-R-(+) > TranH/RSEA-H-(+) > TranE/RSEA-E-(+). Because the TransD/RSEA-D-(+) distinguished head entities and tail entities with relations in different dimensions of vector space separately, while TransR/RSEA-R-(+) only considered the entities and relations with a same matrix. Thus, it achieved higher accuracy than other methods. TranE/RSEA-E-(+) is only suitable for 1-to-1 relations which leads to the worst performance.

6.3 Link Prediction

In a link prediction task, a corrupted triple may also exist in the knowledge graphs, such a prediction should be considered as correct. The evaluation may rank these corrupted but correct triples with a high score. However, the above evaluations do not deal with the issue and may underestimate the results. To eliminate this factor, we filter out those corrupted triples which appear in either training, validation or testing sets before getting the ranking lists. We name the first evaluation setting as "Raw" and the second one as "Filter". In both settings, a higher Hits@10 is better, while lower Mean Rank is better. The parameter settings are the same as those in the task of error detection task. The best configuration is determined according to the accuracy in validation set.

The experimental results are reported in Table 3. We can observe that: (1) RSEA-(*)-V methods outperform the RSEA-(*)-P because the vector based correlation function could better mine the semantic information in related relations by taking the characteristics of entities, which promises to handle multiple relation semantics. (2) RSEA-(*)-(+) methods outperform other baselines significantly and consistently due to the former introduces more information in the embedding process, which indicates the effectiveness of incorporating semantic information to knowledge graph representation learning. (3) For all the methods, we get a better performance on "Filter" setting than "Raw" setting. Because we eliminate corrupted triples in "filter" setting.

Table 3. Evaluation results on link prediction by mapping properties of relations.

Datasets	WN18				FB15			
Metric	Mean rank		Hits@10		Mean rank		Hits@10	
	Raw	Filter	Raw	Filter	Raw	Filter	Raw	Filter
TransE/RSEA-E-P	261/136	252/127	75.3/80.1	89.6/91.2	241/227	122/81	36.2/44.5	47.5/67.2
TransE/RSEA-E-V	261/132	252/120	75.3/80.4	89.6/91.5	241/221	122/76	36.2/45.3	47.5/68.4
TransH/RSEA-H-P	318/146	312/144	76.4/79.3	90.1/91.7	216/214	118/94	42.5/44.9	65.2/70.1
TransH/RSEA-H-V	318/141	312/140	76.4/80.4	90.1/92.3	216/212	118/91	42.5/52.6	65.2/74.6
TransR/RSEA-R-P	287/204	221/168	78.6/82.9	91.3/91.6	248/237	106/90	48.6/52.1	68.7/70.5
TransR/RSEA-R-V	287/192	221/154	78.6/84.4	91.3/92.7	248/228	106/87	48.6/52.4	68.7/71.4
TransD/RSEA-D-P	275/196	217/161	78.2/82.5	91.5/91.8	239/230	120/92	52.2/54.6	70.1/71.7
TransD/RSEA-D-V	275/190	217/149	78.2/83.9	91.5/92.9	239/221	120/89	52.2/54.9	70.1/72.4

7 Conclusion

In this paper, we propose a relation sensitive embedding approach for error detection in the knowledge bases. In our model, we design two correlation functions for both discrete and continuous representations of relations in knowledge bases. Based on proposed correlation similarities, we present a dynamic clustering algorithm which can give more specific categorization for the relations. Finally, our RSEA encodes the knowledge facts into the vector space which can effectively detect errors in the KB by computing the inconsistent scores. Experimental results show that our model can not only detect errors effectively, but also outperforms existing models in link prediction task on both WordNet and Freebase datasets.

Acknowledgement. This work was supported by the 973 Program of China (2015CB358700), NSF of China (61632016, 61521002, 61661166012), and TAL education.

References

1. Abadi, M., et al.: TensorFlow: a system for large-scale machine learning. In: OSDI, pp. 265–283 (2016)
2. Acosta, M., Zaveri, A., Simperl, E., Kontokostas, D., Auer, S., Lehmann, J.: Crowdsourcing linked data quality assessment. In: Alani, H., et al. (eds.) ISWC 2013. LNCS, vol. 8219, pp. 260–276. Springer, Heidelberg (2013). https://doi.org/10.1007/978-3-642-41338-4_17
3. Bollacker, K.D., Evans, C., Paritosh, P., Sturge, T., Taylor, J.: Freebase: a collaboratively created graph database for structuring human knowledge. In: SIGMOD (2008)
4. Bordes, A., Glorot, X., Weston, J., Bengio, Y.: Joint learning of words and meaning representations for open-text semantic parsing. In: AISTATS, pp. 127–135 (2012)
5. Bordes, A., Glorot, X., Weston, J., Bengio, Y.: A semantic matching energy function for learning with multi-relational data - application to word-sense disambiguation. Mach. Learn. **94**(2), 233–259 (2014)

6. Bordes, A., Usunier, N., García-Durán, A., Weston, J., Yakhnenko, O.: Translating embeddings for modeling multi-relational data. In: NIPS, pp. 2787–2795 (2013)
7. Bordes, A., Weston, J., Collobert, R., Bengio, Y.: Learning structured embeddings of knowledge bases. In: AAAI (2011)
8. Bouma, G.: Normalized (pointwise) mutual information in collocation extraction. In: Proceedings of the Biennial GSCL Conference, pp. 31–40 (2009)
9. Chu, X., et al.: KATARA: a data cleaning system powered by knowledge bases and crowdsourcing. In: SIGMOD (2015)
10. Church, K.W., Hanks, P.: Word association norms, mutual information, and lexicography. Comput. Linguist. **16**(1), 22–29 (1990)
11. Deng, D., Jiang, Y., Li, G., Li, J., Yu, C.: Scalable column concept determination for web tables using large knowledge bases. PVLDB **6**(13), 1606–1617 (2013)
12. Dettmers, T., Minervini, P., Stenetorp, P., Riedel, S.: Convolutional 2D knowledge graph embeddings. In: AAAI, pp. 1811–1818 (2018)
13. Dongo, I., Cardinale, Y., Chbeir, R.: RDF-F: RDF datatype inferring framework: towards better RDF document matching. Data Sci. Eng. **3**(2), 115–135 (2018)
14. Fan, J., Lu, M., Ooi, B.C., Tan, W., Zhang, M.: A hybrid machine-crowdsourcing system for matching web tables. In: ICDE, pp. 976–987 (2014)
15. Goldberg, Y., Levy, O.: Word2vec explained: deriving Mikolov et al'.s negative-sampling word-embedding method. CoRR abs/1402.3722 (2014)
16. Hao, S., Tang, N., Li, G., He, J., Ta, N., Feng, J.: A novel cost-based model for data repairing. In: ICDE, pp. 49–50 (2017)
17. Hao, S., Tang, N., Li, G., Li, J.: Cleaning relations using knowledge bases. In: ICDE, pp. 933–944 (2017)
18. Ji, G., He, S., Xu, L., Liu, K., Zhao, J.: Knowledge graph embedding via dynamic mapping matrix. In: ACL, pp. 687–696 (2015)
19. Kim, S., Li, G., Feng, J., Li, K.: Web table understanding by collective inference. In: CIKM, pp. 217–226 (2018)
20. Lehmann, J., et al.: DBpedia - a large-scale, multilingual knowledge base extracted from Wikipedia. Semantic Web **6**(2), 167–195 (2015)
21. Li, K., Li, G.: Approximate query processing: what is new and where to go? A survey on approximate query processing. Data Sci. Eng. **3**(4), 379–397 (2018)
22. Lin, P., Song, Q., Wu, Y.: Fact checking in knowledge graphs with ontological subgraph patterns. Data Sci. Eng. **3**(4), 341–358 (2018)
23. Lin, Y., Liu, Z., Sun, M., Liu, Y., Zhu, X.: Learning entity and relation embeddings for knowledge graph completion. In: AAAI, pp. 2181–2187 (2015)
24. Mikolov, T., Chen, K., Corrado, G., Dean, J.: Efficient estimation of word representations in vector space. CoRR abs/1301.3781 (2013)
25. Mikolov, T., Sutskever, I., Chen, K., Corrado, G.S., Dean, J.: Distributed representations of words and phrases and their compositionality. In: NIPS, pp. 3111–3119 (2013)
26. Miller, G.A.: WordNet: a lexical database for English. Commun. ACM **38**(11), 39–41 (1995)
27. Nickel, M., Rosasco, L., Poggio, T.A.: Holographic embeddings of knowledge graphs. In: AAAI, pp. 1955–1961 (2016)
28. Socher, R., Chen, D., Manning, C.D., Ng, A.Y.: Reasoning with neural tensor networks for knowledge base completion. In: NIPS, pp. 926–934 (2013)
29. Socher, R., Huval, B., Manning, C.D., Ng, A.Y.: Semantic compositionality through recursive matrix-vector spaces. In: EMNLP-CoNLL, pp. 1201–1211 (2012)
30. Suchanek, F.M., Kasneci, G., Weikum, G.: YAGO: a core of semantic knowledge. In: WWW (2007)

31. Töpper, G., Knuth, M., Sack, H.: DBpedia ontology enrichment for inconsistency detection. In: 8th International Conference on Semantic Systems, I-SEMANTICS 2012, pp. 33–40 (2012)
32. Wang, Z., Zhang, J., Feng, J., Chen, Z.: Knowledge graph embedding by translating on hyperplanes. In: Proceedings of the Twenty-Eighth AAAI Conference on Artificial Intelligence, pp. 1112–1119 (2014)
33. Wang, Z., Li, J.: Text-enhanced representation learning for knowledge graph. In: ICAI, pp. 1293–1299 (2016)
34. Xiao, H., Huang, M., Zhu, X.: From one point to a manifold: knowledge graph embedding for precise link prediction. In: IJCAI, pp. 1315–1321 (2016)

Leon: A Distributed RDF Engine for Multi-query Processing

Xintong Guo, Hong Gao, and Zhaonian Zou(✉)

Harbin Institute of Technology, Harbin, China
{xintong.guo,honggao,znzou}@hit.edu.cn

Abstract. As similar queries keep springing up in real query logs, few RDF systems address this problem. In this paper, we propose Leon, a distributed RDF system, which can also deal with multi-query problem. First, we apply a characteristic-set-based partitioning scheme. This scheme (i) supports the fully parallel processing of join within characteristic sets; (ii) minimizes data communication by applying direct transmission of intermediate results instead of broadcasting. Then, Leon revisits the classical problem of multi-query optimization in the context of RDF/SPARQL. In light of the NP-hardness of the multi-query optimization for SPARQL, we propose a heuristic algorithm that partitions the input batch of queries into groups, and discover the common subquery of multiple SPARQL queries. Our MQO algorithm incorporates with a subtle cost model to generate execution plans.

Our experiments with synthetic and real datasets verify that: (i) Leon's startup overhead is low; (ii) Leon consistently outperforms centralized RDF engines by 1–2 orders of magnitude, and it is competitive with state-of-the-art distributed RDF engines; (iii) Our MQO approach consistently demonstrates 10× speedup over the baseline method.

1 Introduction

The Resource Description Framework (RDF) and SPARQL are W3C recommendations for representing and querying graph data on the Web. As RDF data become larger and wider in range, similar query patterns appear frequently in real query logs. For example, users want to get some information from an academic network, so queries in Fig. 1 may be submitted by different users at the same time. There are overlapped sub-queries between these four queries.

We will revisit the classical problem of multi-query optimization (MQO) in the context of RDF and SPARQL. MQO for SPARQL queries has been proved to be NP-hard. The problem of multi-query optimization has been well studied in relational databases [22,26,31] and semi-structure data [8,12,25]. To apply the MQO techniques developed in these systems to address the MQO problem in SPARQL is a potential solution. However, these off-the-shelf methods are hard to be plugged into RDF query engines seamlessly. The difficulty of using the relational techniques stems mainly from the physical design of RDF data itself. While indexing and storing relational data commonly conform to a carefully

G. Li et al. (Eds.): DASFAA 2019, LNCS 11446, pp. 742–759, 2019.
https://doi.org/10.1007/978-3-030-18576-3_44

calibrated relational schema, the storage scheme for RDF is diversified, e.g., the triple table adopted in RDF-3X [18], the property table in Jena [3], and more recently the use of vertical partitioning in S2RDF [23]. These various storage modules, along with the disparate indexing techniques, make the cost estimation highly error-prone and store dependent.

Some recent works [13,14,16] discuss multi-query optimization algorithms tailored for SPARQL. However, these methods, as well as MQO for graph databases [20], mainly rely on subgraph isomorphism algorithms to detect common sub-queries. This increases the computational complexity, and becomes intolerable when a large amount of queries arrive at the same time.

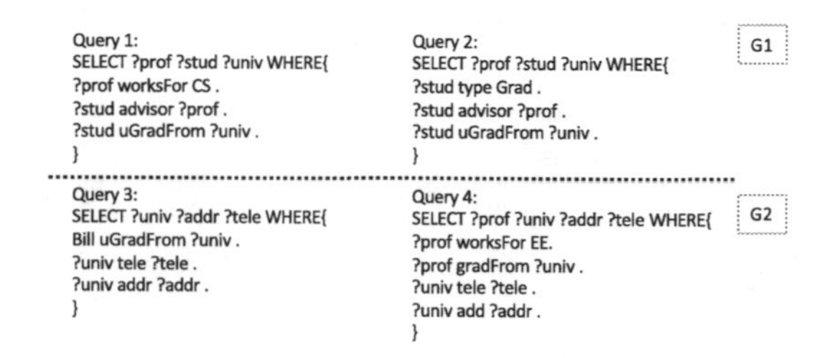

Fig. 1. Multi-query example.

Besides the intrinsic complexity of MQO, the scalability problem becomes more and more severe since the datasets are growing larger and larger. A natural idea is to leverage distributed techniques to handle high latency and highly parallel tasks over large-size datasets. To build a distributed RDF engine for MQO, we face two major challenges: **Partitioning cost:** In systems like [6,7] that use simple hash partitioning heuristics, queries have low chances to be evaluated in parallel without any communication between nodes. Whereas, systems using sophisticated partitioning heuristics [9,28] suffer from high preprocessing cost and sometimes high replication. More importantly, sophisticated methods do not always lead to good query performance. **The complexity of detecting common sub-queries:** There exist an exponential number of ways to partition the input queries. Neither Subgraph-isomorphism-based nor relational-based methods are realistic for a large number of queries. Besides, different query execution plans bring about different query performance, we need a robust cost model to compare candidate strategies.

In this paper, we built a system, called Leon, a distributed in-memory RDF engine specialized for multi-query optimization problem. It mitigates the aforementioned limitations of existing systems. We summarize the novel aspects of our work as follows:

- We present a partitioning scheme based on characteristic sets. It has a low startup cost, and it is also favorable to multi-query optimization.
- We present a heuristic technique for MQO in SPARQL. We leverage characteristic set to detect common sub-structure efficiently and effectively in a batch of SPARQL queries.

- We proposed a robust cost model to reduce communication dramatically.
- We provide an extensive experimental comparison of Leon to the state-of-the-art RDF engines. Leon partitions billion-scale RDF data and starts up in a reasonable time. For single query performance, Leon consistently performs better than centralized RDF engines and MapReduce ones by 1–2 orders of magnitude. It is competitive with the state-of-the-art distributed RDF engines. For large query workloads, MQO approach processes 10x faster over baseline method.

The rest of the paper is organized as follows. Section 2 reviews RDF systems and MQO problem briefly. Section 3 provides essential background on RDF and SPARQL. Section 4 presents the architecture of Leon and an overview of the system components. Section 5 discusses our partitioning technique. Section 6 explains MQO algorithm and locality-aware query planner. Section 7 contains the experimental results. We conclude the whole paper in Sect. 8.

2 Related Work

In this section, we review some representative RDF systems related to Leon, together with multi-query optimization. We refer the readers to [1,5,10] for a comprehensive overview of recent approaches.

2.1 RDF Systems

Triple table is a common practical storage scheme for RDF. It uses a single table with three columns corresponding to subject, predicate and object. An index is created per column for faster join evaluation. This approach scales poorly due to expensive self joins. RDF-3X [18] reduce this cost by using a set of indices that cover all possible permutations of s, p and o. These indices are stored as clustered B+-trees and are compressed using rigorous byte-level techniques. gStore [32] and TripleBit [29] typically employ adjacency lists as a basic building block for storing and processing RDF data. These approaches prune many triples before invoking relational joins to finally generate the row-oriented results of a SPARQL query. However, the performance drop significantly when dealing with complex queries.

It has become infeasible to store all RDF triples in a single machine with the quick proliferation of RDF data. Many distributed RDF engines have sprung up. The quality of partitioning has great impact on the query performance. A popular data partitioning method is hash partitioning, including AdPart-NA [7], H2RDF+[19], and SHARD [21]. This approach distributes RDF triples across different partitions by computing a hash key over the subject or the object of each triple. Hence, hash partitioning can work well for star queries, but for chain or more complex queries, its performance is inefficient.

Several systems employ general graph partitioning techniques, like METIS [11], to partition RDF data in order to improve data locality. Both EAGRE [30] and TriAD [6] use METIS for data partitioning. These algorithms suffer from the skewed data distribution and replication problem.

In a nutshell, systems using simple partitioning methods, have a low startup overhead; however, queries with long paths and complex structures incur high communication costs. Sophisticated partitioning schemes are built specifically for some query shapes, and they do not always guarantee better performance against simple partitioning. MapReduce-based systems offer seamless data distribution and parallelization, while join evaluation suffer from its high overhead.

2.2 Multi-query Optimization

Multi-query optimization is a very classical problem in database history. It has been well studied in relational databases [22,25,26]. The main idea is to identify the common sub-expressions in a batch of queries. Global optimized query plans are constructed by reordering the join sequences and sharing the intermediate results within the same group of queries, therefore minimizing the cost for evaluating the common sub-expressions.

MQO has also been studied on semi-structured data [8,12,25]. Bruno et al. [25] studied navigation and index based path MQO in XML. Unlike the MQO problem in relational and SPARQL cases, path queries can be encoded into a prefix tree where common prefixes share the same branch from the root. This nature provides an important advantage in optimizing concurrent path queries. Nevertheless, the problem of multi-query join optimization was not addressed. However, existing MQO techniques proposed in relational cases cannot be trivially extended to work for SPARQL queries over RDF data, since RDF storage module and indices are highly diversified and prefix-tree notion can not be applied to genetic graph.

There are some works [13,14,16,20] discusses a multi-query optimization algorithm tailored for SPARQL. Nevertheless, these works rely on subgraph isomorphism algorithms to detect common structure of queries. This increases the computational complexity and becomes intolerable when a large amount of queries arrive at the same time.

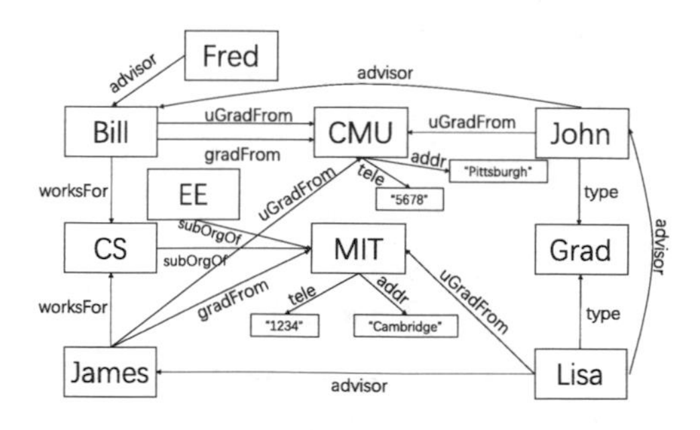

Fig. 2. An RDF graph example.

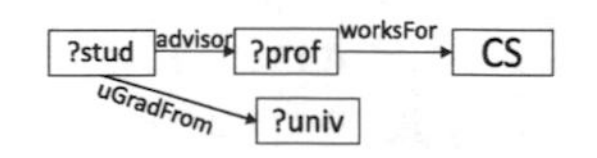

Fig. 3. A query graph.

3 Preliminaries

In this section, we review the terminologies of RDF and SPARQL Query.

RDF models facts about entities in a triple format consisting of a subject s, a predicate p and an object o. A collection of triples D is usually represented as a directed labeled graph with subjects and objects being the nodes, and predicates being the edges of the graph. Formally, let U, B, L be infinite, pairwise disjoint sets of URIs, blank nodes and literals, respectively. Then, an RDF triple t is represented by a triple $(s, p, o) \in (U \cup B) \times (U) \times (U \cup B \cup L)$ and a collection of triples $\{t_1, t_2, \ldots, t_n\}$ is represented by an RDF graph, in which every node $v \in T = (U \cup B \cup L)$ and every edge $e \in U$. Figure 2 shows an example RDF graph of an academic network. An edge and its associated vertices correspond to an RDF triple, e.g., $\langle Bill, worksFor, CS \rangle$.

Following this notation, a SPARQL query Q_i defines a set of triple patterns of the form $(T \cup V) \times (I \cup V) \times (T \cup V)$, where V is the set of variables that can be bound to T. For simplicity, we merely consider Basic Graph Pattern (BGP) in this paper. Figure 3 shows the query graph of Q_1 in Fig. 1, where ?prof, ?stud, ?univ are variables.

As we study a multi-query problem in this work, the input is a set of queries $Q = \{Q_1, Q_2, \ldots, Q_m\}$ that users submit in a given period. We use MQO algorithm to compute a new set Q_{OPT} of queries, evaluate Q_{OPT} over D. There are two requirements for MQO approach: (1) The query results of Q_{OPT} are identical to Q; (2) the overall evaluation time of Q_{OPT}, must be less than the baseline of executing the queries in Q sequentially.

4 System Overview

Leon employs the typical master-slave paradigm and is deployed on a shared-nothing cluster of machines. This architecture is used by many other systems, like [6,24]. Leon uses the standard Message Passing Interface (MPI) [27] for master-worker communication.

The master begins by encoding the RDF triples, and then partitions triples among all workers. Master collect some global statistics during this phase. Each worker loads its triples and collects local statistics. Then, the master aggregates these local statistics and becomes ready for answering queries. Queries are submitted to the master, which decomposes the queries according to partitions. Then, master itself detects the common sub-structures which can be optimized via shared computation. After grouping queries into batch and generating query plans, sub-queries are sent to workers.

Each worker executes a query in two types: distributed and parallel. Queries that require communication are processed in distributed type. The query planner in master devises a global query plan. Each worker gets a copy of this plan and exchange intermediate results. Queries that can be answered without communication are executed in parallel type. In this case, each worker has all the data needed for query evaluation; therefore it generates a local query plan using its local statistics and executes the query without communication.

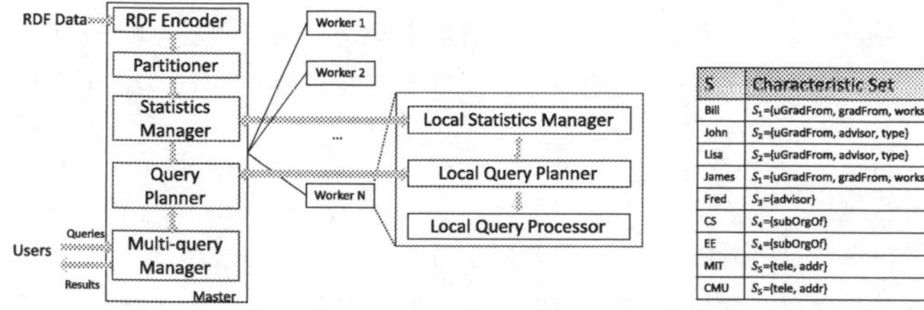

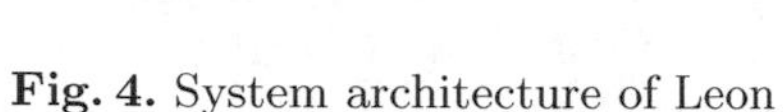

Fig. 4. System architecture of Leon

Fig. 5. Example of five CS(left). The CS_{map} contains the bitmap of CS(right).

5 Partitioning

In this section, we present a heuristic partitioning strategy based on characteristic sets. First, we introduce the definition of characteristic sets, together with its merits in SPARQL query processing. Then, we present an algorithms for CS extraction and a balanced partitioning strategy. Finally, we give four basic data access operations.

5.1 Characteristic Sets (CS)

Every subject in RDF graph has a set of predicates. Subjects of the same category are likely to share the same set of predicates. For example, there are only 615 distinct combinations for more than 845 million triples in UniProt [17]. These observations have been reported for many real-life data sets, such as Yago, LibraryThings, Barton, BTC and UniProt. Inspired by this observation, Neumann and Moerkotte [17] introduced the notion of *characteristic set* as a mean to capture the underlying structure of an RDF dataset. A characteristic set CS identifies node types based on the set of predicates they emit. Formally, given a collection of triples D, the characteristic set $S(s)$ of a subject s is: $S(s) = \{p|\exists o : (s, p, o) \in D\}$. And the set of all CS for a dataset D is: $S(D) = \{S(s)|\exists p, o : (s, p, o) \in D\}$.

Figure 5 shows five CSs derived from Fig. 2. S_1 corresponds to the type of the nodes "Bill" and "James", which comprises the predicates {uGradFrom, gradFrom, worksFor}.

Characteristic sets provide a node-centric partitioning of an RDF dataset, based on the structure of a node. This CS-based partitioning scheme has the following advantages:

- Any join within the same CS can be evaluated without communication, especially join on the subject.
- CS can be directly used to answer certain queries. Many SPARQL queries are specified with constraints on predicates only. For instance, "Select ?x where{?x uGradFrom ?y, ?x gradFrom ?z}", where ?y and ?z are only used

to make sure the existence of the predicates “uGradFrom” and “gradFrom”. In such cases, joins can be eliminate completely by checking the corresponding CSs.

- Cost estimation is accurate. Some query is non-selective in each triple pattern, but final outcome is selective. For instance, “Select ?x where{?x uGradFrom ?y, ?x advisor ?z}”, where the cardinality of “?x uGradFrom ?y” is 4 in Fig. 5, and “?x advisor ?z” is 4. So the estimation of ?x should be 4. If we use CS to estimate instead, we know the combination of “uGradFrom” and “advisor” only belongs to S_2. So the number of subject in S_2 is 2.
- CS can also used to detect similar queries efficiently, since it contains star-shaped structure information. We will elaborate in the next section.

5.2 RDF Encoding and CS Extraction

RDF contains long strings in the form of URIs and literals. To avoid the storage, processing, and communication overheads, it is a common practice to encode RDF strings into numerical IDs and build a bi-directional dictionary. In Leon, triples are stored in memory as four consecutive integers of 4 bytes, one for each triple component, namely cs_id, sub_id, pred_id and obj_id.

Characteristic set $S(D)$ can be easily extracted by linear scan on the triples of a dataset. The algorithm is presented in Algorithm 1. We omit the building progress of bi-directional dictionary due to the space shortage.

First, for every triple, we do string-to-id encoding, and write to new encoded triple table CS_D (Line 2–6). Then, we store every new predicate for each subject in *Sub_preds* (Line 7–8). Multi-value is common in RDF dataset. We only store per p once in each CS. We iterate over the *Sub_preds* and construct a new CS each time a new combination of predicates is found (Line 9–17). Each CS is assigned a unique integer identifier, namely cs_id, and holds a bitmap of the predicates that define it. Every bitmap is stored in CS_{map}. Figure 5 has an example of CS_{map}. Each bit corresponds to the presence of a predicate. This is useful for fast subset checking.

We also store a bi-directional dictionary between CS’s id and subject. A map from CS to sets of subjects is in CS_{sub}. We can use this index to get the triples associated with a specific CS. A map from a subject to CS in Sub_{cs}. This engages locality-awareness of a given subject. When we know the exact location of a subject, we can send intermediate results directly to corresponding machine without broadcasting in the cluster. Finally, we go over CS_D to set the first element of the triple’s vector to cs_id (Line 18–20).

5.3 Balanced CS Partition

After extracting CSs, the next step is to allocate all triples on clusters. In order to achieve simplicity and workload balance, we assure that the triples on each machine is even. Suppose that we want to allocate triples on N machines, so CS_D will be fragmented into a collection of disjoint subsets $P = \{P_1, P_2, \ldots, P_N\}$. The number of triples on each machine $|P_i|$ should be $|CS_D|/N$. As we know the exact number of triples of each CS, we sort it in descending order and put it into machines which will not violate the balancing principle. For example, we sort the CSs in Fig. 5, which is S_2, S_1, S_5, S_4 and S_3. We want to allocate them across two machines, so the ideal number of triples on each machine should be 10. We start from putting S_2 into P_1. When it comes to S_1, it will exceed the size limitation if we put it into P_1. So we put it into P_2. Do the same action for S_5, S_4 and S_3. Finally, we have $P = \{(S_2, S_5), (S_1, S_3, S_4)\}$.

Algorithm 1. Loading Triples and Extracting CSs

```
Input: Triples D
Output: CS_D, CS_map, CS_sub, Sub_cs
1: for every triple t in D do
2:     sub, pred, obj = parse(t)
3:     sub_id, pred_id, obj_id = encode(sub, pred, obj)
4:     CS_D[t][1] = sub_id
5:     CS_D[t][2] = pred_id
6:     CS_D[t][3] = obj_id
7:     if pred_id not in Sub_preds[sub] then
8:         Sub_preds[sub].append(pred_id)
9: for each sub_id in Sub_preds do
10:    if Sub_preds[Sub_id] not in CS then
11:        cs_id=cs_id+1
12:        CS_map[cs_id] = newCS(Sub_preds[Sub_id])
13:        CS_sub[cs_id+1] = {sub_id}
14:        Sub_cs[sub_id] = cs_id
15:    else
16:        CS_sub[cs_id+1].append(sub_id)
17:        Sub_cs[sub_id] = cs_id
18: for every t in CS_D do
19:    sid = CS[t][1]
20:    CS_D[t][0] = Sub_cs[sid]
21: return CS_D, CS_map, CS_sub, Sub_cs
```

5.4 Basic Data Operations

Each machine w_i stores its local set of CSs in an in-memory data structure, where t, s, p, and o are cs, subject, predicate, and object respectively. Assume P is a set of predicates $\{p_1, p_2, \ldots, p_n\}$. We provide the following data operations for the SPARQL query processing module:

- *GetCSs(P)*: given P, return set $\{S_i | P \subseteq S_i\}$;
- *LoadTriples(t)*: given CS t, return set $\{(s,p,o)|(s,p,o) \in CS_D(t)\}$
- *GetPreds(s)*: given s, return set $\{p|(s,p) \in CS_D\}$;
- *GetObjs(s,p)*: given s and p, return set $\{o|(s,p,o) \in CS_D\}$.

Leon has indexed the mapping between predicates and CSs in CS_{map}, the operator *GetCSs(P)* can be executed very fast. *LoadTriples(t)* loads all the triples associated with t according to CS_D's first element. Given s, *GetPreds(s)* is implemented by checking the Sub_{cs} index first to get cs_id associated with s, and then get the p from CS_{map}. With s and p in hand, we can get o by calling *GetObjs(s,p)*. In summary, these operators are able to represent all the SPARQL triple patterns as listed in Table 1.

Table 1. Translating SPARQL triple patterns to basic operations

Patterns	Translated operations
s, p, o	if $o \in GetObjs(s, p)$, true; otherwise, false
s, p, ?o	$GetObjs(s, p)$
s, ?p, o	$\{p \mid p \in GetPreds(s) \cap o \in GetObjs(s, p)\}$
s, ?p, ?o	$\{(p, o) \mid p \in GetPreds(s) \cap o \in GetObjs(s, p)\}$
?s, p, o	$\{s \mid t \in GetCSs(p) \cap (p, o) \in LoadTriples(t)\}$
?s, p, ?o	$\{(s, o) \mid t \in GetCSs(p) \cap (s, p, o) \in LoadTriples(t)\}$
?s, ?p, o	$\{(s, p) \mid s \in S \cap p \in GetPreds(s) \cap o \in GetObjs(s, p)\}$
?s, ?p, ?o	Return all triples

6 Multi-query Optimization

Our MQO algorithm accepts a set $Q = \{Q_1, Q_2, \ldots, Q_m\}$ of queries as input. The algorithm identifies whether there is a cost-effective way to share the evaluation of structurally-overlapped patterns among the queries in Q. Briefly, the algorithm works as follows: (1) It partitions the input queries into groups according to queries' CSs, where queries in the same group are more likely to share common sub-queries that can be optimized. (2) it detects the common parts in the same group, further filter out queries which are unproductive to query in batch, and then generate query plans. We have a robust cost model to decide whether query should be execute in batch or not; (3) it executes the queries according to correspondent cost-efficient query plan.

6.1 Query Decomposing and Coarse-Grained Clustering

Incoming queries are first mapped to CS by the query parser. During this step, the dictionary is used for id resolution of predicates and any other bound nodes in the patterns. Then, each query Q_i is decomposed into a sequence of disjoint forks. A fork is a star-shaped sub-query. The triple patterns in a fork share identical subject join variable. For example, Q_1 is decomposed into two forks, $f_1 = \{?prof\ worksFor\ CS\}$ and $f_2 = \{?stud\ advisor\ ?prof,\ ?stud\ uGradFrom\ ?univ\}$. We use $f = (r, P, L)$ to denote a fork, where r is the identical subject node, $P = \{p_1, p_2, \ldots, p_n\}$ are predicates emitting from r, and $L = \{l_1, \ldots, l_n\}$ is the object nodes. So f_2 can be denoted as ({?stud}, {advisor, uGradFrom}, {?prof, ?univ}). Then, each fork matches to zero or more CSs. In Fig. 1, Q_1 is mapped to $\{(S_1), (S_2)\}$, Q_2 is mapped to $\{(S_2)\}$, Q_3 is mapped to $\{(S_1, S_2), (S_5)\}$, Q_4 is mapped to $\{(S_1), (S_5)\}$.

Finding structural overlaps for a set of queries amounts to finding the isomorphic subgraphs among the corresponding query graphs. This process is computationally expensive, so ideally we would like to find these overlaps only for groups of queries that are promising to be optimized. That is, we want to minimize the computation spent on identifying common subgraphs for query groups that lead to less optimal MQO solutions. Consequently, we implement k-means clustering for an initial partitioning of the input queries into a set G of k query groups, that is $G = \{G^1, \ldots, G^k\}$. k-means algorithm uses Jaccard coefficient of their CS sets as similarity measure. The rational is that, CS could capture not only similarity in predicates, but also the underlying structure of the query. If the CSs' are dissimilar, their structural overlap in original query may also be small; so it is safe to not consider grouping such queries for MQO. We will prove that kmeans by CS is simple but effective in the experiments.

For example, consider the queries in Fig. 1, we partition Q into two groups $G = \{G^1, G^2\}$ according to their similarities in CS, where $G^1 = (Q_1, Q_2)$ and $G_2 = (Q_3, Q_4)$.

6.2 Refining Query Clusters

Starting with the k-means generated groups G, we refine the partitioning of queries further based on their structure similarity and the estimated cost.

We sort CSs according to frequency in each G^i, suppose the most frequent CS is S_j. And then we set the group size $|G^i|$ to $freq(S_j)$. Queries who do not have S_j will be eliminated. In Fig. 1, the most frequent CS in G^1 is S_2, so $|G^1| = 2$; $|G^2| = freq(S_1) = freq(S_5) = 2$, respectively. In this way, each query group shares at least one common CS. We detect the actually common parts within same CSs by scanning all triple patterns in G^i. The predicates and constants associated with nodes are also taken into considerations. So the common part of G^i is $G^1_{com} = \{$(?stud advisor ?prof, ?stud uGradFrom ?univ)$\}$, $G^2_{com} = \{$(?univ addr ?addr, ?univ tele ?tele)$\}$.

The natural routine of multi-query optimization is to precompute the G_{com} part, so the others can use the intermediate results without computation redundancy. However, some queries could not benefit from MQO. Consider Q_3, the selectivity of G^2_{com} may be very low, because all schools have address and telephone in real world, while the selectivity of $Q_3 \backslash G^2_{com} = \{Bill\ underGraduateFrom\ ?univ\}$ is high. If we compute G^2_{com} first, mounting intermediate results will be generated. As the joining variable of $Q_3 \backslash G^2_{com}$ and G^2_{com} is '?univ' and the results of G^2_{com} are not at same machine with $Q_3 \backslash G^2_{com}$, it will incur a broadcasting subject-object join. We could tell from the above example, whether queries should be optimized as a group depends on both selectivity of G_{com} and the join type for each single query.

We design a delicate cost model to refine the query clusters. This model could eliminate the queries who can not benefit from MQO.

6.3 Cost Model

Since data are memory resident, we apply hash joins as it proves to be competitive to more sophisticated join methods. Our dynamic-programming-based query planner devises an ordering of query forks and generates a left-deep join tree, where the right operand of each join is a base fork. We do not use bushy tree plans to avoid building indices for intermediate results.

We first describe the statistics used for cost estimation. Recall that Leon collects and aggregates statistics from workers for global query planning during the query process. We focus on CSs rather than predicates as many other works do; this way the storage complexity of statistics is much smaller and more accurate than statistics based on predicates only.

For each unique CS S_i, we denote the number of subject in each CS as $N(S_i)$. For each predicate $p \in S_i$, we count the distinct object values associated p as: $N(p|S_i) = |\{o|(s,p,o) \in S_i\}|$. For a variable predicate, its cardinality is estimated as the sum of the possible predicates. For a fork query $f = (r, P, L)$ without constraint, the cardinality of its subject r is calculated as: $N(r|f) = \sum_{S_j \in S_q} N(S_j)$, $S_q = GetCSs(P)$. And the cardinality of object l_i whose corresponding predicate is p_i in f is estimated by: $N(l_i|f) = \sum_{S_j \in S_q} N(p_i|S_j)$.

Communication overhead plays a significant role in distributed environment. Consequently, we use communication overhead as our cost function. We set the initial communication cost of DP states to zero. Cardinalities of forks with variable subjects and objects are already captured in the master's global statistics. Hence, we set the cumulative cardinalities of the initial states to the cardinalities of the forks. Furthermore, the master consults the workers to update the cardinalities of fork patterns that are attached to constants or have unbounded predicates. This is done locally at each worker by simple lookups to its indices to update the cardinalities of variables bindings accordingly.

Now, we estimate the cost of expanding a state S with a fork f_j, where c_j is the join variable in f_j. Suppose there are N machines in the cluster. If the join does not incur communication, the cost of the new state $S*$ is zero. Otherwise, the expansion is carried out through hash join and we incur two phases of communication: (i) transmitting intermediate results and (ii) replying with the candidate triples.

Estimating the communication depends on the cardinality of the join variable bindings in S and join type. $C(S)$ denotes the cardinality of state S. There are three cases of expanding S with f_j: **Case 1.** If c_j is a subject, c_j and S are at the same machines, no communication is incurred, then: $cost(S, f_j) = 0$; **Case 2.** If c_j is a subject, c_j and S are not at the same machines. As we know the location of c_j, we send $C(S)$ to the machines of c_j directly, then: $cost(S, f_j) = C(S) + N(c_j|f_j)$; **Case 3.** If c_j is not a subject, we must broadcast to all over the network, and collect the returning results, then: $cost(S, f_j) = (N-1)C(S) + (N-1)N(c_j|f_j)$.

We use the cumulative cardinality when we reach the same state by two different ordering. We finish the cluster refinement here. For each query Q_j in a query group G^i, we estimate the cardinality of $(G^i_{com}, q_j \backslash G^i_{com})$. If precomputing G^i_{com} is beneficial to evaluating Q_j, then Q_j will stay in G^i; otherwise, Q_j will be executed as a single query. The cost model suggests that a good optimization should keep the G_{com} cardinality as small as possible. The result cardinality of $Qi \backslash G_{com}$ is upper bounded by G_{com} since we use left-deep join tree. After the query clusters are finalized, the algorithm precomputes the common parts of each cluster of queries. The overall progress of MQO is presented at Algorithm

2. We use an extra query set G^{k+1} to denote a set of queries which are executed sequentially. First, we decompose each Q_i into forks, and match forks to CSs (Line 1–2). We cluster all queries according to CSs (Line 3). The G_{com} part for each query group is computed by scanning all triple patterns (Line 4–5). Finally, we refine the query cluster by cost estimation (Line 6–12). If precomputing G_{com} is favourable to a query, the query will by optimized in group; otherwise, it will be put into G^{k+1}.

Algorithm 2. MQO

```
Input: Q = {Q_1, ..., Q_m}: a set of queries
Output: Q_OPT = {G^1, ..., G^k, G^{k+1}}: optimized queries
 1: for each Q_i in Q do
 2:     F = CS_decomposition(Q_i)
 3: G = kmeans(F)
 4: for each G^i in G do
 5:     G^i_com = Compute_com(G^i)
 6: for each G^i in G do
 7:     for each Q_j in G_i do
 8:         non-opt = Cost_estimate(Q_j \ G^i_com, G^i_com)
 9:         opt = Cost_estimate(G^i_com, Q_j \ G^i_com)
10:         if opt > non-opt then
11:             G^i = G^i \ Q_j
12:             G^{k+1} = Q_j ∪ G^{k+1}
13: return Q_OPT
```

7 Experiments

Since Leon is designed to serve massive real-life RDF datasets, we want to test it from following aspects: (1) Query performance on single query. (2) Multi-query optimization. This is the highlight of our system.

7.1 Experimental Setup

Datasets: We conducted our experiments using real and synthetic datasets of variable sizes. Table 2 describes these datasets, where S, P, O and CS denote respectively the numbers of unique subjects, predicates, objects and characteristic sets. Their basic information are as follows: (1) We use the synthetic LUBM[1] data generator to generate a dataset called LUBM1k. LUBM is a RDF benchmark widely used for comparing the performance of RDF stores. As we can see from Table 2, the number of CS is only 12, which means entities of the same type have the same set of predicates. (2) WatDiv[2] is a benchmark that enable diversified stress testing of RDF data management systems. In WatDiv, entities of the same type can have the different sets of predicates. For testing our methods,

[1] http://swat.cse.lehigh.edu/projects/lubm/.
[2] http://dsg.uwaterloo.ca/watdiv/.

we use dataset WatDiv1B. (3) We also use YAGO2[3], a real dataset derived from Wikipedia, WordNet and GeoNames. WatDiv and YAGO2 both hold thousands of CS, which will test our system fully.

Hardware Setup: We implement Leon in C++ and use a Message Passing Interface library for synchronization and communication. We set up Leon and its competitors on a cluster of 5 machines each with 198 GB RAM, 480 GB SSD, and two Intel Xeon CPUs (6 cores each). The machines run 64-bit CentOS 7 and are connected by a 10 Gbps Ethernet.

Competitors: We compare Leon against AdPart [7], a recent in-memory RDF system, which is the state-of-the-art. It has two versions, AdPart-NA and AdPart(adapt to workload). We will run AdPart-NA in single query experiments and AdPart in multi-query experiments. We compare with two well-known centralized RDF systems, RDF-3X [18] and TripleBit [29]. We also compare against SHAPE [15], a hadoop-based system relies on semantic hash and RDF-3X as underlying data store. We configure SHAPE with full level semantic hash partitioning and enable the type optimization.

Table 2. Data statistics

Dataset	Triples(M)	$S \cup O$ (M)	P	Size (G)	CS
LUBM1k	137.31	33.52	17	24.7	12
WatDiv1B	1092.16	97.39	86	149	96344
YAGO2	284.42	60.70	98	42	25511

Table 3. Preprocessing time (minute)

System	LUBM1k	WatDiv1B	YAGO2
RDF-3X	26	427	83
TripleBit	10	496	40
SHAPE	26	521	242
AdPart	2	16	4
Leon	3	21	7

7.2 Startup Time

Our first experiment measures the time it takes all systems for preparing the data prior to answering queries. For a fair comparison, we include the overhead of loading data into HDFS for Hadoop.

Table 3 shows the result. AdPart employs hash partitioning on subject, so it starts fastest. SHAPE suffers from the overhead of loading data in HDFS before encoding. TripleBit is faster than RDF-3X, and both of them can finish in a reasonable time. Leon has to extract and partition CSs, so it is slower than AdPart. It takes more time on WatDiv1B and YAGO2, as they have more CSs. But it still outperforms RDF-3X and SHAPE.

7.3 Query Performance

In this section, we focus on single query performance. We demonstrate that Leon is competitive with the state-of-the-art RDF engine.

[3] http://yago-knowledge.org/.

LUBM: In the first experiment, we compare the performance of all systems using LUBM1k dataset and queries Q1–Q7 defined in [2]. Q4 and Q5 are simple selective star queries, whereas Q2 is a simple yet non-selective star query that generates large final results. Q6 is a simple query because it is highly selective. Q1, Q3 and Q7 are complex queries with large intermediate results but very few final results. In Table 4, TripleBit, Leon and AdPart-NA, are close for queries Q4-Q6, due to their high selectivities and star-shapes. They both solves these queries with little communication. SHAPE do not need communication, but it suffers from the non-negligible overhead of using MapReduce to dispatch queries to workers. RDF-3X performs really bad on complex queries.

WatDiv: The WatDiv benchmark defines 20 query templates classified into four categories: linear (L), star (S), snowflake (F) and complex queries (C). We generated 5 queries for each template to get mean time for single query. In Table 6, Leon and AdPart-NA, provide way better performance than other systems. Leon performs better than AdPart-NA for C, S and F queries, as CS reduces the number of joins and prunes search space dramatically. Centralized systems cannot deal with WatDiv datasets gracefully. TripleBit fails to answer C1–C3 queries in all the datasets. Although SHAPE use full level semantic hash partitioning, they do not always outperform the single-machine RDF-3X, and its performance fell significantly as the datasets become large.

YAGO: We use Y1-Y4 defined by AdPart [7], since YAGO has no official benchmark. In Table 5, although object-object joins exist, Leon and TripleBit perform better than others. It is because most of the join data are located on same machine for Leon, and TripleBit is a centralized system which avoids large communication cost. AdPart-NA is slower than Leon, while RDF-3X performs really bad.

Table 4. Query runtimes (ms) for LUBM queries. Bold means the fastest.

System	Q1	Q2	Q3	Q4	Q5	Q6	Q7
RDF-3X	94817	8346	94837	24	57	285	5071
TripleBit	3440	4620	238	42	345	874	2094
SHAPE	17543	759	18709	114	260	275	21356
AdPart-NA	3150	98	**234**	**1**	**1**	27	1756
Leon	**1245**	**64**	263	1	2	**20**	**1306**

Table 5. Query runtimes (ms) for YAGO2. Bold means the fastest.

System	Y1	Y2	Y3	Y4
RDF-3X	60	361550	3700	120
TripleBit	19	**7**	**191**	86
SHAPE	1569	43891	1531	1408
AdPart-NA	**16**	51	683	98
Leon	18	34	310	**56**

7.4 Multi-query Performance

The objective of this section is to evaluates how well MQO algorithm optimizes its alternatives, including the comparison with the baseline approach without any optimization. For this purpose, we define different query workloads on two large scale datasets that have different characteristics, namely, LUBM1k and WatDiv1B. We use 14 queries in LUBM as template, and 20 templates from

Table 6. Query runtimes (ms) for WatDiv queries. '-' means system could not finish the query within 10 min. Bold means the fastest.

System	C1	C2	C3	S1	S2	S3	S4	S5	S6	S7	F1	F2	F3	F4	F5	L1	L2	L3	L4	L5
RDF-3X	1670	18710	258160	23	67	52	43	10	67	27	59	387	878	735	656	213	190	3	132	387
TripleBit	-	-	-	45	789	435	405	603	262	9	1364	659	488	372	73	53	166	136	114	63
SHAPE	5632	14852	-	447	410	604	734	521	935	587	4935	2414	463	685	4375	3018	2405	653	735	145
AdPart-NA	2640	1245	**243**	5	3	**4**	1	**4**	**2**	7	210	54	72	85	**4**	**5**	136	**9**	65	**32**
Leon	**1327**	**583**	341	**4**	**2**	6	2	5	7	**3**	**36**	**23**	**56**	**53**	10	45	**62**	23	**25**	43

WatDiv. We test three algorithms, MQO-non (no optimization), MQO (k-means) and MQO-R (k-means and refinement). MQO-non is the baseline method.

In the experiments, query templates are utilized to generate a seed set, namely Q_{seed}. The generator first choose α seeds from Q_{seed}, and generate remaining parts randomly as W. We change the constants and structures in W, then randomly attach W to seeds to construct the final queries. Each seed group is corresponding to same number of queries in Q, which is $\frac{|Q|}{\alpha}$. The seed in each seed-group is the common sub-query G_{com} what our algorithm is expected to discover.

Impact of kmeans by CS: In this experiment, we want to prove that cluster by CS is meaningful. We adopted the extended Normalized Mutual Information (NMI) [4] to measure the quality of kmeans. NMI is a popular criterion for evaluating the accuracy of clustering result with ground-truth based on information theory. It yields the values between 0 and 1, with 1 corresponding to a perfect matching. We vary α from 10 to 50, and set $k = \alpha$, $|Q| = 1000$. In Fig. 6, kmeans by CS performs consistency better than kmeans by predicate. Because we generate queries who are similar in predicates but dissimilar in structure. kmeans by p relies solely on predicates to determine groups, while CS captures the underlying star-shaped structure. So it is simple but useful to cluster by CS.

Impact of k: We discuss how to set k in kmeans. We set $\alpha = 25$, $|Q| = 1000$, and vary k from 10 to 50. Figure 7 illustrates when $k = \alpha$, the runtime is shortest as it fully utilizes the power of shared computation. When k becomes smaller, the query groups contain fewer similar queries. When k gets larger, G_{com} in each query group is bigger. But there exists some computation redundancy. The principle to choose an appropriate k, is better larger than smaller. In the following experiments, if not said specifically, we set $k = \alpha$.

Impact of $|Q|$: We analyze the evaluating cost spent on query clustering, by fixing $\alpha = 10$. In Fig. 8, we report the clustering time as $|Q|$ getting large, and Q is generated from LUBM templates. MQO-R is slower than MQO obviously, since MQO-R does a selectivity check on G_{com}. Nevertheless, clustering is a small fraction of the total evaluating cost (less than 1%).

Impact of α: We analyze the impact of the number α of seed queries on LUBM1k dataset, by setting $|Q| = 1000$, varying α from 5 to 25. Figure 9 shows that as α increases, less queries can be optimized by MQO and MQO-R. Not

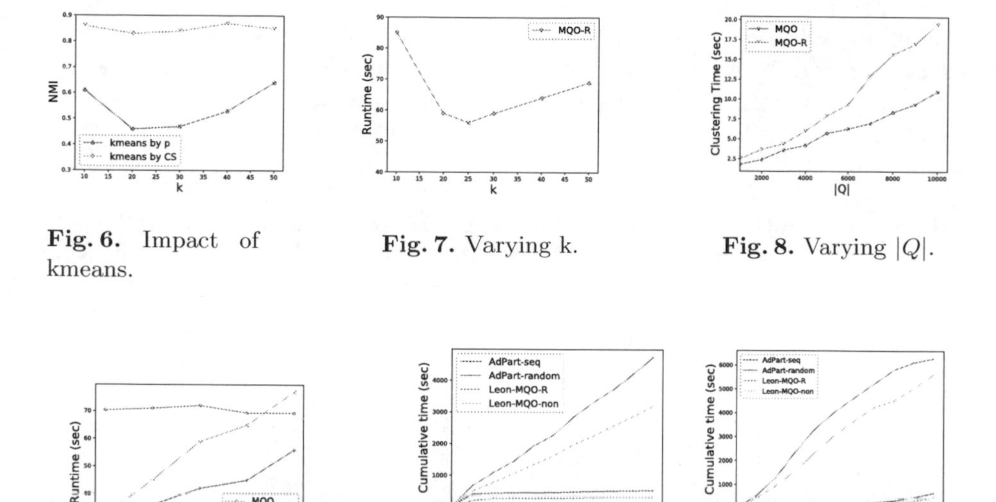

Fig. 6. Impact of kmeans.

Fig. 7. Varying k.

Fig. 8. Varying $|Q|$.

Fig. 9. Varying α.

(a) LUBM workload (b) WatDiv workload

Fig. 10. Workload performance, $\alpha = 10$, $|Q| = 10$k

surprisingly, a larger α increases query diversity and reduces the potential for optimization. This affects evaluation time, but MQO-R is still the best of the three. MQO runtime is larger than MQO-non when $\alpha = 25$. It suggests refinement is essential, and bad optimization strategy will hurt the performance.

Workload Performance. We generate 10k queries for LUBM1k and WatDiv1B workload. We compare four algorithms, namely AdPart-random, AdPart-seq, Leon-MQO-R, Leon-MQO-non. 'seq' suffix means similar queries are putting together, because AdPart is based on hot pattern detection in a period of time and replication among workers.

Figure 10(a) show the cumulative time as the execution progresses in LUBM1k. For AdPart-seq, after every sequence of 1K query executions, the type of queries changes. For AdPart-random, the cumulative time increase sharply although workload adaptivity is available The factor behind this situation lies in that queries from the same seed group do not arrive at the same time, so AdPart could not detect the hot patterns which are accessed a lot. Compare to this, Leon treats a batch of queries as a unit, so the sequence does not influence the query performance. Leon-MQO-R outperforms AdPart-seq as it shares computation as much as possible. The same behavior is observed from the WatDiv workload in Fig. 10(b).

8 Conclusions

In this paper, we presented Leon, a distributed RDF engine which can also deal with multi-query problem. Leon exploits CS-based partitioning to minimize the communication cost during query evaluation, and to detect common structure with a fine-tuned cost model. Our experimental results verify that Leon achieves better query performance than its competitors. More importantly, multi-query optimization algorithm reduces query response time and communication cost dramatically.

Acknowledgment. The work is partially supported by the National Natural Science Foundation of China (No. 61532015, No. 61672189, No. 61832003, No. 61732003 and U1811461).

References

1. Abdelaziz, I., Al-Harbi, R., Khayyat, Z., Kalnis, P.: A survey and experimental comparison of distributed SPARQL engines for very large RDF data. In: PVLDB (2017)
2. Atre, M., Chaoji, V., Zaki, M.J., Hendler, J.A.: Matrix "Bit" loaded: a scalable lightweight join query processor for RDF data. In: WWW (2010)
3. Carroll, J.J., Dickinson, I., Dollin, C., Reynolds, D., Seaborne, A., Wilkinson, K.J.: Jena: implementing the semantic web recommendations. In: WWW (2004)
4. Danon, L., Diaz-Guilera, A., Duch, J., Arenas, A.: Comparing community structure identification. J. Stat. Mech. Theory Exp. (2005)
5. Feng, J., Meng, C., Song, J., Zhang, X., Feng, Z., Zou, L.: SPARQL query parallel processing: a survey. In: 2017 IEEE BigData Congress (2017)
6. Gurajada, S.: TriAD: a distributed shared-nothing RDF engine based on asynchronous message passing. In: SIGMOD Conference (2014)
7. Harbi, R., Abdelaziz, I., Kalnis, P., Mamoulis, N.: Accelerating SPARQL queries by exploiting hash-based locality and adaptive partitioning. VLDB J. **25**(3), 355–380 (2016)
8. Hong, M., Demers, A.J., Gehrke, J., Koch, C., Riedewald, M.: Massively multi-query join processing in publish/subscribe systems. In: SIGMOD Conference (2007)
9. Hose, K., Schenkel, R.: WARP: workload-aware replication and partitioning for RDF. In: 2013 IEEE 29th International Conference on Data Engineering Workshops (ICDEW), pp. 1–6 (2013)
10. Kaoudi, Z., Manolescu, I.: RDF in the clouds: a survey. VLDB J. **24**, 67–91 (2014)
11. Karypis, G., Kumar, V.: A fast and high quality multilevel scheme for partitioning irregular graphs. SIAM J. Sci. Comput. **20**(1), 359–392 (1998)
12. Kementsietsidis, A., Neven, F., de Craen, D.V., Vansummeren, S.: Scalable multi-query optimization for exploratory queries over federated scientific databases. PVLDB **1**, 16–27 (2008)
13. Kim, I., Lee, K.H., Lee, K.C.: SAMUEL: a sharing-based approach to processing multiple SPARQL queries with MapReduce. In: EDBT (2018)
14. Le, W., Kementsietsidis, A., Duan, S., Li, F.: Scalable multi-query optimization for SPARQL. In: 2012 IEEE 28th International Conference on Data Engineering (2012)
15. Lee, K., Liu, L.: Scaling queries over big RDF graphs with semantic hash partitioning. PVLDB **6**, 1894–1905 (2013)
16. Liu, C., Qu, J., Qi, G., Wang, H., Yu, Y.: HadoopSPARQL: a hadoop-based engine for multiple SPARQL query answering. In: ESWC (2012)

17. Neumann, T., Moerkotte, G.: Characteristic sets: accurate cardinality estimation for RDF queries with multiple joins. In: 2011 IEEE 27th International Conference on Data Engineering, pp. 984–994 (2011)
18. Neumann, T., Weikum, G.: RDF-3X: a RISC-style engine for RDF. PVLDB **1**(1), 647–659 (2008)
19. Papailiou, N., Konstantinou, I., Tsoumakos, D.: H_2RDF+: high-performance distributed joins over large-scale RDF graphs. In: BigData Conference (2013)
20. Ren, X., Wang, J.: Multi-query optimization for subgraph isomorphism search. PVLDB **10**, 121–132 (2016)
21. Rohloff, K., Schantz, R.E.: High-performance, massively scalable distributed systems using the MapReduce software framework: the SHARD triple-store. In: PSI EtA (2010)
22. Roy, P., Seshadri, S., Sudarshan, S., Bhobe, S.: Efficient and extensible algorithms for multi query optimization. In: SIGMOD Conference (2000)
23. Schätzle, A., Przyjaciel-Zablocki, M., Skilevic, S., Lausen, G.: S2RDF: RDF querying with SPARQL on spark. PVLDB **9**, 804–815 (2016)
24. Shao, B., Wang, H., Li, Y.: Trinity: a distributed graph engine on a memory cloud. In: SIGMOD Conference (2013)
25. Srivastava, D.: Navigation- vs. index-based XML multi-query processing. In: Proceedings of the ICDE, pp. 139–150 (2003)
26. Trigoni, N., Yao, Y., Demers, A., Gehrke, J., Rajaraman, R.: Multi-query optimization for sensor networks. In: Prasanna, V.K., Iyengar, S.S., Spirakis, P.G., Welsh, M. (eds.) DCOSS 2005. LNCS, vol. 3560, pp. 307–321. Springer, Heidelberg (2005). https://doi.org/10.1007/11502593_24
27. Walker, D.W., Dongarra, J.J.: MPI: a standard message passing interface. Supercomputer **12**, 56–68 (1996)
28. Wu, B., Zhou, Y., Yuan, P., Liu, L., Jin, H.: Scalable SPARQL querying using path partitioning. In: 2015 IEEE 31st International Conference on Data Engineering, pp. 795–806 (2015)
29. Yuan, P., Liu, P., Wu, B., Jin, H., Zhang, W., Liu, L.: TripleBit: a fast and compact system for large scale RDF data. PVLDB **6**, 517–528 (2013)
30. Zhang, X., Chen, L., Tong, Y., Wang, M.: EAGRE: towards scalable I/O efficient SPARQL query evaluation on the cloud. In: 2013 IEEE 29th International Conference on Data Engineering (ICDE), pp. 565–576 (2013)
31. Zhao, Y., Deshpande, P., Naughton, J.F., Shukla, A.: Simultaneous optimization and evaluation of multiple dimensional queries. In: SIGMOD Conference (1998)
32. Zou, L., Mo, J., Chen, L., Özsu, M.T., Zhao, D.: gStore: answering SPARQL queries via subgraph matching. PVLDB **4**(8), 482–493 (2011)

MathGraph: A Knowledge Graph for Automatically Solving Mathematical Exercises

Tianyu Zhao[1(✉)], Yan Huang[2], Songfan Yang[2], Yuyu Luo[1], Jianhua Feng[1], Yong Wang[1], Haitao Yuan[1], Kang Pan[1], Kaiyu Li[1], Haoda Li[1], and Fu Zhu[1]

[1] Tsinghua University, Beijing, China
{zhaoty17,wangy18,yht16,pk16,ky-li18,lhd16,zhuf18}@mails.tsinghua.edu.cn
{luoyuyu,fengjh}@tsinghua.edu.cn
[2] TAL Education Group, Beijing, China
{galehuang,yangsongfan}@100tal.com

Abstract. Knowledge graphs are widely applied in many applications. Automatically solving mathematical exercises is also an interesting task which can be enhanced by knowledge reasoning. In this paper, we design MathGraph, a knowledge graph aiming to solve high school mathematical exercises. Since it requires fine-grained mathematical derivation and calculation of different mathematical objects, the design of MathGraph has major differences from existing knowledge graphs. MathGraph supports massive kinds of mathematical objects, operations, and constraints which may be involved in exercises. Furthermore, we propose an algorithm to align a semantically parsed exercise to MathGraph and figure out the answer automatically. Extensive experiments on real-world datasets verify the effectiveness of MathGraph.

Keywords: Knowledge graph · Mathematical exercise · Knowledge reasoning

1 Introduction

Currently, large scale knowledge graphs are widely used in many real-world applications, such as semantic web search, and question-answer systems, natural language processing, and data analytic. For example, if we ask "What is the highest mountain?" on a web search engine, it may directly show the answer "Everest" with the help of a knowledge graph.

Recently intelligent education is more and more popular and automatically resolving mathematical exercises can help students improve the comprehensive ability. However, it is rather challenging to automatically resolve mathematical exercises without knowledge graphs, because it requires to use complex semantics and extra calculations. In this paper, we propose **MathGraph**, a knowledge graph aiming to solve high school mathematical exercises. MathGraph must be specially designed and differentiated from other knowledge graphs. The reasons are listed as follows:

G. Li et al. (Eds.): DASFAA 2019, LNCS 11446, pp. 760–776, 2019.
https://doi.org/10.1007/978-3-030-18576-3_45

1. ***Knowledge in MathGraph belongs to a very specific domain.*** Building MathGraph requires specific mathematical knowledge. Traditional knowledge graphs are built based on extensive semantic data, e.g., Wikipedia. However, it is very hard to get the semantic data for mathematical problems.
2. ***Knowledge in MathGraph is stored in class-level rather than instance-level.*** Most of the traditional knowledge graphs focus on extracting instances, categories, and relations among instances. For example, a 3-tuple (Beijing, *isCaptialOf*, China) shows a relation between two instances. However, in MathGraph, there is no instance in the origin graph, but only many class-level mathematical objects (such as `Complex Number`, `Ellipse`, etc.). Only if an exercise is given, instances will be created accordingly.
3. ***MathGraph supports mathematical derivation and calculation.*** The reasoning process of mathematical problems is different from other problems, because besides logical relation, mathematical derivation must be included in the knowledge graph to solve mathematical exercises.

Thus, in this paper, we focus on building a knowledge graph MathGraph for resolving mathematical problems. We propose an effective algorithms to align a mathematical problem to MathGraph, and use the aligned sub-graph to resolve a mathematical exercise. The contributions of this paper are as follows.

- We specially design the structure of MathGraph to support mathematical derivation and calculation. We model different mathematical objects, operations and constraints in MathGraph. To the best of our knowledge, this is the first attempt to build a knowledge graph for resolving mathematical problems.
- We propose an algorithm to align a mathematical problem to MathGraph.
- We design a method to resolve mathematical exercises by the help of a semantic parser.
- Experimental study shows great performance of MathGraph and our proposed method.

Figure 1 gives an overview of the exercise-solving process with MathGraph. We detail the structure of MathGraph and the exercise-solving algorithm later.

2 Related Work

2.1 Reasoning with Knowledge Graph

Since knowledge graphs can provide well-structured information and relations of the entities, it is known to be useful to do reasoning in many tasks, such as query answering and relation inference (i.e., to infer missing relations in the knowledge graph [4,11,12]). Guu et al. [7] proposed a technique to answer queries on knowledge graph by "compositionalizing" a broad class of vector space models, which preforms well on query answering and knowledge graph completion. Toutanova et al. [17] proposed a dynamic programming algorithm to incorporate all paths in knowledge graph within a bounded length, and modelled entities

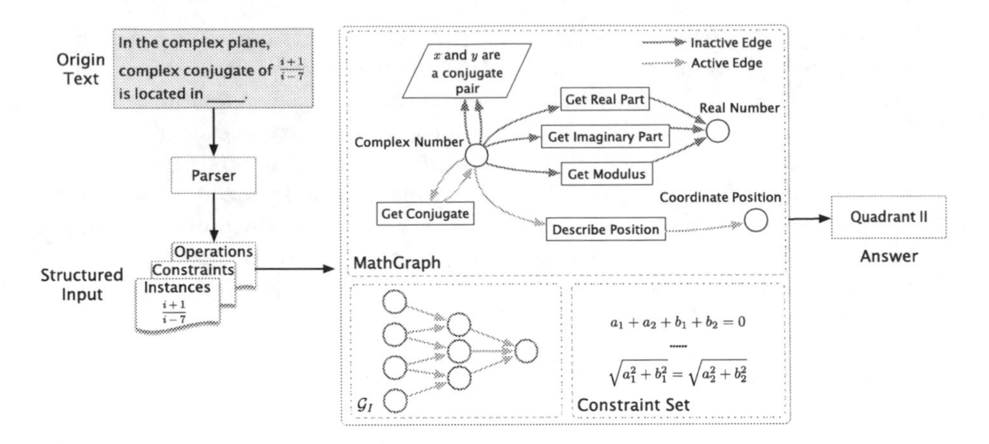

Fig. 1. Overview of using MathGraph to solve a mathematical exercise

and relations in the compositional path representations. Zhang et al. [18] proposed a deep learning architecture and a variational learning algorithm, which can handle noise in the question and do multi-hop reasoning in knowledge graph simultaneously. Zheng et al. [19] used a large number of binary templates rather than semantic parsers to query knowledge graph with natural language. A low-cost technique that can generate a large number of templates automatically is also proposed. Our work is different from above works. Firstly, there are some differences between the structure of MathGraph and existing knowledge graphs (e.g. Freebase and NELL [2]). Secondly, to solve a math exercise usually requires multi-step mathematical derivation, and the derivation procedures need to be output as the problem-solving process. Thirdly, derivation and calculation should be performed simultaneously when solving an exercise to retrieve the answer.

2.2 Automated Solving Mathematical Problems

Automated solving mathematical problems has been studied over years. But they only focused on easy problems, e.g., mathematical problems in primary schools.

Kojiri et al. [10] constructed a mechanism called solution network to automatically generate the answers for mathematical exercises. The solution network is represented as a tree to describe inclusive relations of exercises.

Tomas et al. [16] proposed a framework of Constraint Logic Programming to automatically generate and solve mathematical exercises. This paper proposed to concentrate on the solving procedures rather than many simple exercise templates so that the generation and explanation of these exercises is easy.

Ganesalingam et al. [6] proposed a method that solves elementary mathematical problems using logical derivation and shows solutions which are made difficult to distinguish from human's writing.

However, these works all have their own limits. For example, some can solve problems only involving elementary math (e.g. set theory, basic algebraic operation) and have no deeper theorems; some only support very limited logical derivation. Thus, in this paper, we decide to use a knowledge graph to represent as many mathematical entities and logical relationships as possible.

3 Preliminaries

In this section, we describe the entities that may appear in MathGraph, including mathematical objects and instances, operations, and constraints.

Mathematical Object and Instance. A mathematical object is an abstract object which has a definition, some properties, and can be taken as the target of some operations or derivation. Note that a mathematical object can be defined in terms of other objects. A concrete object that satisfies the definition of the mathematical object is called an instance.

For example, `Complex Number` can be considered as a mathematical object:

- *Definition:* A complex number is a number that can be in the form $a + bi$, where a and b are both real numbers and i is the imaginary unit which satisfies $i^2 = -1$.
- *Property example:* Imaginary part is a property of a complex number. The imaginary part of a complex number $a + bi$ is b.
- *Operation example:* $(a_1 + b_1 i) \cdot (a_2 + b_2 i) = (a_1 a_2 - b_1 b_2) + (a_1 b_2 + a_2 b_1) i$
- *Derivation example:* If $(a_1 + b_1 i)$ and $(a_2 + b_2 i)$ are conjugated to each other, then $a_1 = a_2$ and $b_1 + b_2 = 0$.

And $2 + 3i$ and $(i + 1)(i - 3)$ are instances of `Complex Number`.

Different mathematical objects should be described as different structures in MathGraph. Thus, in MathGraph, a mathematical object is represented with a tuple of **key properties** $(p_1, p_2, \cdots, p_n)$. The key properties of a mathematical object are those properties that together can form and describe all the information of an instance of the object. Table 1 shows examples of key properties of some mathematical objects. Two instances of a mathematical object is equivalent if and only if all the key properties are equivalent.

In a mathematical exercise, instances can be categorized into certain instances and uncertain instances depending on whether it contains some uncertain values as its key properties. An instance is a *certain instance* if all key properties are certain; *uncertain instance* otherwise. For example, a real number 2.3 and a function $f(x) = x + sin(x)$ are certain; a complex number $3 + ai$ (where $a \in \mathbb{R}$) and a random triangle $\triangle ABC$ are uncertain.

Operation. Generally, an operation is an action or procedure which, given one or more mathematical objects as input (known as operands), produces a new object. Simple examples include addition, subtraction, multiplication, division, and exponentiation. In addition, other procedures such as calculating the real part of a complex number, the derivative of a function, and the area of a triangle can also be considered as operations.

Table 1. Examples of key properties of different mathematical objects

Mathematical object	Example instance	Key properties
Complex number	$ai + b$	(a, b)
Elementary function	$f(x) = \langle$an algebraic expression about $x\rangle$	$\langle$the algebraic expression$\rangle$
Triangle	$\triangle ABC$	$(a, b, c, \angle A, \angle B, \angle C)$
Line	$Ax + By + C = 0$	(A, B, C)
Ellipse	$\frac{x^2}{a^2} + \frac{y^2}{b^2} = 1$	(a, b)

Constraint. A constraint is a description or condition about one or more instances, at least one of which is an uncertain instance. There are four types of constraints: *descriptive constraints* (e.g. complex numbers x and y are conjugated), *equality constraints* (e.g. $a + 2 = b$), *inequality constraints* (e.g. $a^2 \leq 5$), and *set constraints* (e.g. $a \in \mathbb{N}$).

Most descriptive constraints cannot be applied directly to solve the exercise, but can be converted into other three types of constraints using some definitions or theorems. For example, if an exercise says "$a + 3i$ and $7 - bi$ are a conjugate pair", by the definition of conjugate complex, we can know that $a = 7$ and $3 + (-b) = 0$ by derivation.

4 The Structure of MathGraph

MathGraph is a directed graph $\mathcal{G} = \langle V, E\rangle$, in which each node $v \in V$ denotes a mathematical object, an operation or a constraint, and each edge $e \in E$ is the relation of two nodes.

4.1 Nodes

In general, nodes are categorized into three different types: **object nodes**, **operation nodes** and **constraint nodes**.

Object Nodes. An object node $v_o = (t, P, C)$ represents a mathematical object, where t denotes an instance template of this mathematical object; $P = (P_1, P_2, \cdots, P_n)$ is a tuple indicating key properties of the mathematical object; and C is a set of constraints that, according to the definition or some theorems, must be satisfied by this mathematical object. Table 2 shows an example of "triangle" as an object node. We can see that properties and theorems of triangles are included in the constraint set.

Operation Nodes. An operation node $v_p = (X_1, X_2, \cdots, X_k, Y, f)$ represents a k-ary operation, where $X_i (i = 1, 2, \cdots, k)$ and Y are object nodes representing the domain of the ith operand x_i and the result of the operation y respectively, and f is a function that implements the operation and can be finished by a series

Table 2. An example of object node: triangle

Mathematical object	Triangle
Instance template	$\triangle ABC$
Key properties	(a, b, c, A, B, C)
Constraint set	$\{a, b, c > 0,$ $0 < A, B, C < \pi,$ $A + B + C = \pi,$ $a + b > c, a + c > b, b + c > a,$ $\frac{a}{\sin A} = \frac{b}{\sin B} = \frac{c}{\sin C},$ $a^2 = b^2 + c^2 - 2bc \sin A,$ $b^2 = a^2 + c^2 - 2ac \sin B,$ $c^2 = a^2 + b^2 - 2ab \sin C\}$

of symbolic execution [1,3,9] process using a symbolic execution library (e.g. SymPy [13], Mathematica [8]) even if some operands are uncertain instances.

For example, *getting the modulus of a complex number* is an unary operation where $X_1 = \langle\texttt{Complex Number}\rangle$, $Y = \langle\texttt{Real Number}\rangle$, and f can be implemented by the following symbolic execution process: (1) Get the real part of x_1; (2) Get the imaginary part of x_1; (3) Return the squared root of the sum of (1) squared and (2) squared.

Constraint Nodes. A constraint node $v_c = (d, X_1, X_2, \cdots, X_k, f)$ represents a descriptive constraints of k instances, where d is the description of the constraint, $X_i (i = 1, 2, \cdots, k)$ are object nodes representing the domain of each involving instance, and f is a function which maps this descriptive constraint into several equality constraints, inequality constraints and set constraints.

For example, a constraint node represents that x_1 and x_2 are a conjugate pair, where $X_1 = X_2 = \langle\texttt{Complex Number}\rangle$, and f can be implemented by the following process: (1) Get the real part of x_1 as a_1; (2) Get the real part of x_2 as a_2; (3) Get the imaginary part of x_1 as b_1; (4) Get the imaginary part of x_2 as b_2; (5) Return two equality constraints: $a_1 = a_2$ and $b_1 + b_2 = 0$.

4.2 Edges

There are two types of edges in MathGraph: the DERIVE edges and the FLOW edges.

The DERIVE Edge. For two object nodes X and Y, there may be a DERIVE edge $e_{\text{DERIVE}} = (X, Y, f)$ to indicate a general-special relationship between them, such as *Triangle* and *Isosceles Triangle*. If $X \xrightarrow{\text{DERIVE}} Y$, an instance of X can be reassigned as an instance of Y if certain conditions are met. These conditions are encapsulated into a function $f : X \to \{\text{False}, \text{True}\}$: if these conditions are met, the function f will return True and reassign the instance from X to Y; otherwise it will simply return False.

For example, there is a DERIVE edge from object node `Triangle` to `Isosceles Triangle`, where the function f can be implemented as: (1) if the values of key properties or a constraint shows that two angles or lengths of two edges of the origin instance are equal, return an instance of *Isosceles Triangle* with the same key properties; (2) return False otherwise.

When solving an exercise, reassigning an instance to a more specific object node will bring more constraints of this object and help find the answer. For example, for a rhombus $ABCD$, if we know that $\angle A = 90°$, we can infer, by the DERIVE edge from object node `Rhombus` to `Square`, that $ABCD$ is a square and has constraints that $\angle A = \angle B = \angle C = \angle D = 90°$.

The FLOW Edge. A FLOW edge $e_{\text{FLOW}} = (X, Y)$ indicates the flow direction of instances during the exercise solving process, which may only exist from an object node to an operation node, from an operation node to an object node, or from an object node to a constraint node.

The FLOW edges between object nodes and operation nodes represent the process of passing instances as parameters before the operation and the process of returning a new instance after it. For example, in Fig. 2, the two FLOW edges pointing to the operation node "addition" indicate that this operation takes two instances of complex number as its input values, and the edge leading from this operation node indicates that it returns a new instance of complex numbers.

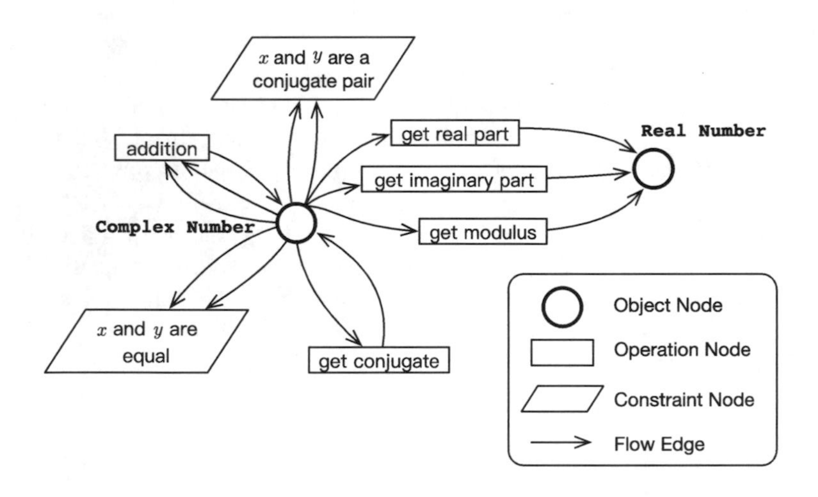

Fig. 2. Example of the FLOW edges

The FLOW edges from object nodes to constraint nodes also represent the process of passing parameters of the constraints. For example, in Fig. 2, the two FLOW edges pointing to the constraint node "x and y are a conjugate pair" indicates that this constraint takes two complex number as its input. Note that constraints nodes only convert descriptive constraints into other types of constraints and generate no instances, so there are no FLOW edges from a constraint node to an object node.

In summary, MathGraph is a well-structured graph supporting different mathematical objects, operations and constraints. Next, we will discuss how to solve mathematical exercises using it.

5 Solving Mathematical Exercises with MathGraph

In this section, we propose a framework to solve a mathematical exercise using MathGraph. First, we use a semantic parser mapping exercise text to the instances, operations and constraints respectively. Then, we solve the constraints and update uncertain instances. Finally, we return the answer of this exercise.

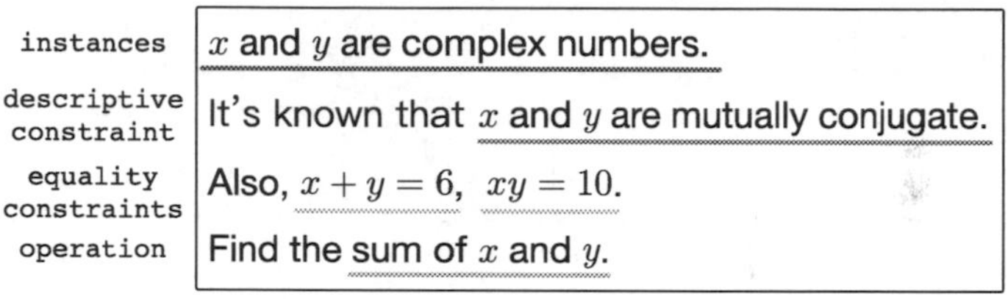

instances	x and y are complex numbers.
descriptive constraint	It's known that x and y are mutually conjugate.
equality constraints	Also, $x + y = 6$, $xy = 10$.
operation	Find the sum of x and y.

Fig. 3. Example of parsing the text into nodes in MathGraph

5.1 Mapping Text in MathGraph

Considering the limited information and expression in the mathematical exercises, we can easily use a rule based semantic parser to parse the exercise text and then map them to corresponding nodes in MathGraph.

The rule-based semantic parser uses a set of rules to parse every sentence of the exercise and recognize the logical relationship in the text. For example, "Let x and y be complex numbers" will be parsed as declaration of two uncertain instances; "Find the coordinates of the conjugate complex of $(i + 1)(i - 1)$" will be parsed as a declaration of a certain instance and two operations.

Mapping Instances. With the semantic parser, every instance generated from the exercise should have already mapped into the corresponding object node. That is, a set of instances $\mathcal{I} = \{(x_1, X_1), \cdots, (x_k, X_k)\}$ is generated by parsing the text of the exercise, where x_i denotes the instance and X_i denotes the corresponding object node.

Instances are classified as certain instances or uncertain instances depending on if the exercise provide certain values or expressions of them. For uncertain instances generated from text, key properties with unknown value should be generated as instances, since they may be used in the operations and constraints of this exercise. For example, for the exercise shown in Fig. 3, x and y are both uncertain instances of object node `Complex Number`. Therefore, we need to generate a_x, b_x, a_y and b_y as four uncertain instances of object node `Real Number`, where a_x and b_x stand for the two key properties of x, and a_y and b_y stand for the key properties of y.

Mapping Operations. The semantic parser can also parse out the a set of operations from the text. Every operation $(o, (x_1, X_1), \cdots, (x_n, X_n))$ in it will be aligned to the corresponding operation node in MathGraph o with its operands, trigger the function in the operation node, and then finally generate a new instance as the output of the operation.

Mapping Constraints. Similar to mapping operations, for every descriptive constraint $(c, (x_1, X_1), \cdots, (x_n, X_n))$ in the exercise, the semantic parser can map it to the corresponding constraint node c with some involving instances, trigger the function in the node, and convert it to several equality/inequality/set constraints.

Also, note that when an uncertain instance is generated, some constraints may also be generated according to the constraint set of the corresponding object node. After that, we gather all the constraints in the exercise as a set for further using.

Algorithm 1 shows the process of mapping text of the exercise.

5.2 Solving Uncertain Instances and Constraints

After parsing all the instances and operations in the exercise, the answer of the exercise should already be generated as an instance (from the text or by an operation). If this instance is a certain instance, we can directly return the value of this instance as the answer; otherwise, we must deal with these uncertain instances and solve the constraints in the exercise to update their values and finally retrieve the answer of the exercise.

Reassign Uncertain Instances. First, we need to check every uncertain instance if it can be reassigned to a more specific object node in MathGraph by a DERIVE edge. For an uncertain instance i that is assigned to an object node v_o, we check every outgoing DERIVE edge of v_o, and if the function of an edge e returns true, then we reassign i to the object node that e points to and add all the constraints in this node to the constraint set. Algorithm 2 shows the pseudo code of this process.

For example, if we have an uncertain instance $\triangle ABC$, and there is a constraint $\angle B = \angle C$ in the constraint set, then the DERIVE edge from `Triangle` to `Isosceles Triangle` should return true. So the instance should be reassigned to `Isosceles Triangle`, and a new constraint $AB = AC$ should be added to the constraint set.

Algorithm 1. MAPPINGTEXT$(t, \mathcal{G})$

Input: t : text of the exercise;
$\mathcal{G}$: MathGraph

Output: $\mathcal{I}_{\text{certain}}$: a set of certain instances;
$\mathcal{I}_{\text{uncertain}}$: a set of uncertain instances;
$\mathcal{C}$: a set of constraints;
$\mathcal{S}_{\text{dependency}}$: a set denoting dependencies of uncertain instances;

1 **begin**
2 Initialize P as a semantic parser;
3 $\mathcal{I}_{\text{certain}}, \mathcal{I}_{\text{uncertain}} \leftarrow P.$MAPPINGINSTANCES$(t, \mathcal{G})$;
4 $\mathcal{O} \leftarrow P.$MAPPINGOPERATIONS$(t, \mathcal{G})$;
5 $\mathcal{C} \leftarrow P.$MAPPINGCONSTRAINTS$(t, G)$;
6 Let $\mathcal{S}_{\text{dependency}}$ be an empty set;
7 **for** each $(x, X) \in \mathcal{I}_{uncertain}$ **do**
8 **for** each key property$(p, X_p) \in x.keyProperties$ **do**
9 **if** p is an uncertain instance **then**
10 $\mathcal{I}_{\text{uncertain}} \leftarrow \mathcal{I}_{\text{uncertain}} \cup \{(p, X_p)\}$;
11 $\mathcal{S}_{\text{dependency}} \leftarrow \mathcal{S}_{\text{dependency}} \cup \{(p, x)\}$;
12 **for** each $(o, (x_1, X_1), \cdots, (x_k, X_k)) \in \mathcal{O}$ **do**
13 $(y, Y) = o.f(x_1, \cdots, x_k)$;
14 **if** y is a certain instance **then**
15 $\mathcal{I}_{\text{certain}} \leftarrow \mathcal{I}_{\text{uncertain}} \cup \{(y, Y)\}$;
16 **else**
17 $\mathcal{I}_{\text{uncertain}} \leftarrow \mathcal{I}_{\text{uncertain}} \cup \{(y, Y)\}$;
18 $C \leftarrow S \cup y.ConstraintSet$;
19 **for** $i = 1$ **to** k **do**
20 **if** x_i is an uncertain instance **then**
21 $\mathcal{S}_{\text{dependency}} \leftarrow \mathcal{S}_{\text{dependency}} \cup \{(x_i, y)\}$;
22 **for** each $(c, (x_1, X_1), \cdots, (x_k, X_k)) \in \mathcal{C}$ **do**
23 **if** c is a descriptive constraint **then**
24 $c \leftarrow c.f(x_1, \cdots, x_k)$;
25 **return** $\mathcal{I}_{certain}, \mathcal{I}_{uncertain}, \mathcal{C}, \mathcal{S}_{dependency}$

Algorithm 2. ReassignUncertainInstances($\mathcal{G}, \mathcal{I}_{\text{uncertain}}, \mathcal{C}$)

Input: $\mathcal{G}$: MathGraph;
$\mathcal{I}_{\text{uncertain}}$: the set of uncertain instances;
$\mathcal{C}$: the constraint set;

```
1 begin
2     for each instance (x, X) ∈ I_uncertain do
3         for each DERIVE edge (X_e, Y_e, f_e) ∈ G do
4             if X_e == X and f_e(x) == True then
5                 C ← C ∪ Y_e.ConstraintSet;
6                 update (x, X) as (x, Y);
```

Algorithm 3. OrganizeUncertainInstances($\mathcal{I}_{\text{uncertain}}, \mathcal{S}_{\text{dependency}}$)

Input: $\mathcal{I}_{\text{uncertain}}$: a set of uncertain instances;
$\mathcal{S}_{\text{dependency}}$: the set denoting dependencies of uncertain instances;
Output: $\mathcal{G}_I$: the graph to organize the uncertain instances;
$\mathcal{S}_I$: a set denoting all instances in $\mathcal{G}_I$ without incoming edges;

```
1 begin
2     Let G_I⟨V_I, E_I⟩ be an empty graph;
3     for (x, y) ∈ S_dependency do
4         V_I ← V_I ∪ {x, y};
5         E_I ← E_I ∪ {(x, y)};
6     S_I ← {v | v ∈ V_I ∧ ∀u ∈ V_I, (u, v) ∉ E_I};
7     return G_I, S_I
```

Organizing Uncertain Instances. Note that for two uncertain instances α and β, there may be a dependency relationship between them, which is caused due to either α is one of the input of an operation node and β is the output or α is one of the key properties of β.

Thus, we use a graph $\mathcal{G}_I = \langle V_I, E_I \rangle$ to describe dependency of all the uncertain instances, where $v \in V_I$ is a node representing an uncertain instance and $e \in E_I$ is a directed edge representing a dependency relationship of two nodes. Note that $\mathcal{G}_I$ is always a DAG, since there will be no dependency loop in it.

Let $\mathcal{S}_I = \{v | v \in V_I \wedge \forall u \in V_I, (u, v) \notin E_I\}$ denote the set containing all node without any incoming edges in $\mathcal{G}_I$. It is obvious that if all nodes in $\mathcal{S}_I$ can turn into certain instances, the instance corresponding to the answer can be derived to a certain instance. Algorithm 3 demonstrates this process.

For example, Fig. 4 shows $\mathcal{G}_I$ of the exercise in Fig. 3, where x and y depend on their respective key properties, and $z = x + y$ depends on its two operands. In this context, $\mathcal{S}_I = \{a_x, b_x, a_y, b_y\}$ and the instance corresponding to the answer is z.

Algorithm 4. PROCESSCONSTRAINTS($\mathcal{C}, \mathcal{G}_I, \mathcal{S}_I$)

Input: $\mathcal{C}$: the constraint set;
$\mathcal{G}_I$: the graph for dependency of uncertain instances;
$\mathcal{S}_I$: the set denoting all instances in $\mathcal{G}_I$ without incoming edges;

1 **begin**
2 **for** each $(c, (x_1, X_1), \cdots, (x_k, X_k)) \in \mathcal{C}$ **do**
3 **for** $i = 1$ **to** k **do**
4 **if** $x_i \notin \mathcal{S}_I$ **then**
5 Replace (x_i, X_i) with its key properties $(p_1, P_1), \cdots, (p_n, P_n)$;
6 SOLVECONSTRAINTS($\mathcal{S}_{\text{constraint}}, \mathcal{S}_I$);

Organizing and Solving Constraints. After the last step, we now have a set of constraints. First, we need to make sure every variable in every constraint is in $\mathcal{S}_I$. If not, this constraint needs to be rewritten by using its key properties as the variable. For example, for the exercise in Fig. 3, the set of the constraint is $\{x + y = 6, xy = 10, a_x = a_y, b_x + b_y = 0\}$. Since $x, y \notin \mathcal{S}_I$, the first two constraints will be rewritten as $a_x+b_xi+a_y+b_yi = 6$ and $(a_x+b_xi)(a_y+b_yi) = 10$.

Now the constraint set includes and formalizes all the constraints in the exercise. So we can apply methods of a symbolic execution library [8,13] or some approximation algorithms [5,15] to solve these equations and/or inequalities. Finally, we will get the value (or range of value) of every instance in $\mathcal{S}_I$. Algorithm 4 shows this process.

Updating Uncertain Instances and Retrieving the Answer. After solving all the constraints in the exercise, we need to update the value of the rest instances in $\mathcal{G}_I$. Since we now know the value of instances in $\mathcal{S}_i$, we can traverse every instance in $\mathcal{G}_I$ in the topological sorting order and update their values in turn. Finally, we return the value of the instance corresponding to the answer. Algorithm 5 shows the complete process of using MathGraph to solve exercise.

Algorithm 5. SOLVINGEXERCISE($t, \mathcal{G}$)

Input: t : text of the exercise;
$\mathcal{G}$: MathGraph
Output: answer of the exercise

1 **begin**
2 $\mathcal{I}_{\text{certain}}, \mathcal{I}_{\text{uncertain}}, \mathcal{C}, \mathcal{S}_{\text{dependency}} \leftarrow$ MAPPINGTEXT($t, \mathcal{G}$);
3 Mark the instance corresponding to the answer as x_{ans};
4 REASSIGNUNCERTAININSTANCES($\mathcal{G}, \mathcal{I}_{\text{uncertain}}, \mathcal{C}$);
5 $\mathcal{G}_I, \mathcal{S}_I \leftarrow$ ORGANIZEUNCERTAININSTANCES($\mathcal{I}_{\text{uncertain}}, \mathcal{S}_{\text{dependency}}$);
6 PROCESSCONSTRAINTS($C, \mathcal{S}_I$);
7 Update the value of every node in $\mathcal{G}_I$ in the topological sorting order;
return value of x_{ans};

6 Experiments

In this section, we conduct extensive experiments on real mathematical datasets to evaluate the performance of our method.

6.1 Datasets and Experiment Setting

We collect four real-world datasets of mathematical exercises of Chinese high schools, namely `Complex`, `Triangle`, `Conic` and `Solid`. The exercises are stored in plain text, and the mathematical expressions are stored in the LaTeX format.

- `Complex`: This dataset contains 1526 mathematical exercises related to calculation and derivation of complex numbers, including basic algebraic operation, the modulus and the conjugate of a complex number, Argand plane, polar representation, etc.
- `Triangle`: This dataset contains 782 mathematical exercises related to solving triangles (using Law of Sines and Law of Cosines), which includes finding missing sides and angles, perimeter, area, radius of the circumscribed circle, etc.
- `Conic`: This dataset contains 1196 exercises related to Conic sections, including calculation and derivation on ellipse, hyperbola and parabola.
- `Solid`: This dataset contains 653 exercises related to solid geometry, which involves a variety of geometries in three-dimension Euclidean space, including pyramids, prisms, etc.

Exercises in the four datasets are categorized into three levels (i.e. easy, medium, and hard) based on the difficulty (which is classified according to the accuracy of many high school students). Table 3 shows the number of exercises with different difficulty levels in the datasets.

Table 3. Summary of exercises in the datasets

	Easy	Medium	Hard	Total
`Complex`	685	634	207	1526
`Triangle`	179	470	133	782
`Conic`	486	602	108	1196
`Solid`	217	336	100	653

In the experiments, we use Neo4j [14] as the graph database platform to build and index MathGraph. For the datasets, we build the knowledge graph manually involving only the instances, operations and constraints that may exist in these exercises. The experiments are implemented in Python 3.7. Sympy is used to do the work of symbolic execution. All the experiments are conducted in a machine with 2.40 GHz Intel Xeon CPU E52630, 48 GB RAM, running Ubuntu 14.04.

6.2 Evaluation and Discussion

We implement a rule-based baseline method as the following procedures:

1. We still use a rule-based semantic parser to parse the text and extract the information.
2. A large quantity of rules are written in advance to match different situations of exercises. Every rule represents an exercise type and has a built-in solving process only for this exercise type.
3. If the exercise matches a rule, then we apply the solving process of the rule and return the answer.

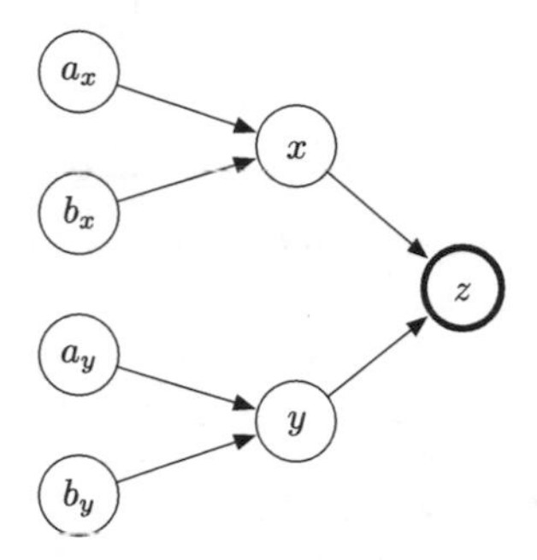

Fig. 4. $\mathcal{G}_I$: a DAG to organize the uncertain instances

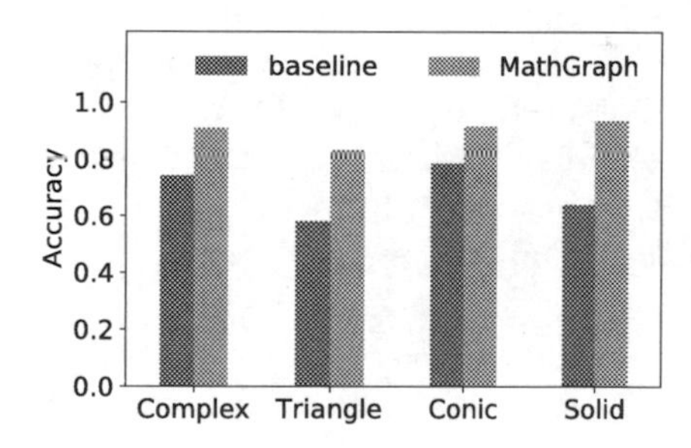

Fig. 5. Overall accuracy on four datasets

Figure 5 shows the exercise-solving accuracy on four datasets. We can see that in every dataset, our method achieves higher accuracy than baseline, e.g., 20% higher accuracy. This result shows the effectiveness of solving problems using MathGraph.

Figure 6 demonstrates the exercise-solving accuracy on different difficulty level. From the experiment result, we have the following observations. Firstly, as

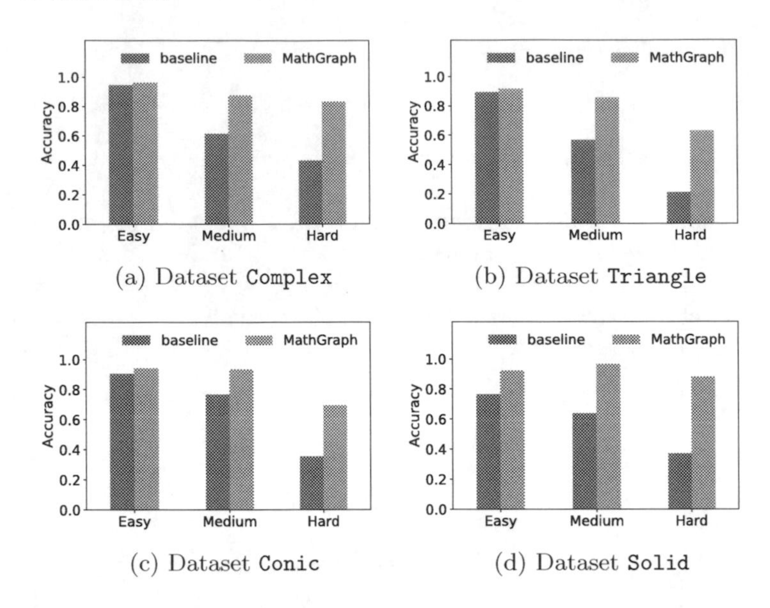

(a) Dataset `Complex` (b) Dataset `Triangle`

(c) Dataset `Conic` (d) Dataset `Solid`

Fig. 6. Accuracy on different difficulty levels

the difficulty of the exercises increases, the accuracy of both methods decreases. Secondly, for easy exercises, the baseline and our method have similar performance; but for medium and hard exercises, MathGraph significantly outperforms the baseline, because our method can use the knowledge graph to do mathematical derivation.

The rule-based baseline considers the exercise as a whole and solving it according to the logic specified by a rule. This means that this method relies on a large amount of rules, and the more complex the exercise is, the more rules and the higher difficult it needs to write. Therefore, this method has a poor performance in hard exercises. However, our method extracts the mathematical objects, calculations, and constraints from these rules and models them into a graph, so it can be used for multi-step calculation and derivation.

7 Conclusion

In this paper, we proposed MathGraph, a knowledge graph for automatically solving mathematical exercises. MathGraph is specially designed to represent different mathematical objects, operations and constraints. Given an exercise, we can use the proposed method to solve it with the help of MathGraph and a pre-built semantic parser. Experimental study on four real-world datasets demonstrates the accuracy of our method.

Acknowledgements. This work was supported by the 973 Program of China (2015CB358700), NSF of China (61632016, 61521002, 61661166012), and TAL education.

References

1. Baldoni, R., Coppa, E., D'Elia, D.C., Demetrescu, C., Finocchi, I.: A survey of symbolic execution techniques. ACM Comput. Surv. **51**(3), 50 (2018)
2. Carlson, A., Betteridge, J., Kisiel, B., Settles, B., Hruschka Jr., E.R., Mitchell, T.M.: Toward an architecture for never-ending language learning. In: Proceedings of the Twenty-Fourth Conference on Artificial Intelligence (AAAI 2010), vol. 5, p. 3. Atlanta (2010)
3. Cousot, P., Cousot, R.: Abstract interpretation: a unified lattice model for static analysis of programs by construction or approximation of fixpoints. In: Proceedings of the 4th ACM SIGACT-SIGPLAN Symposium on Principles of Programming Languages, pp. 238–252. ACM (1977)
4. Dongo, I., Cardinale, Y., Chbeir, R.: RDF-F: RDF datatype inferring framework. Data Sci. Eng. **3**(2), 115–135 (2018)
5. Fletcher, R., Leyffer, S.: Filter-type algorithms for solving systems of algebraic equations and inequalities. In: Di Pillo, G., Murli, A. (eds.) High Performance Algorithms and Software for Nonlinear Optimization, pp. 265–284. Springer, Heidelberg (2003). https://doi.org/10.1007/978-1-4613-0241-4_12
6. Ganesalingam, M., Gowers, W.T.: A fully automatic theorem prover with human-style output. J. Autom. Reason. **58**(2), 253–291 (2017)
7. Guu, K., Miller, J., Liang, P.: Traversing knowledge graphs in vector space. In: Proceedings of the 2015 Conference on Empirical Methods in Natural Language Processing, EMNLP 2015, Lisbon, Portugal, 17–21 September 2015, pp. 318–327 (2015)
8. Mathematica, Version 11.3. Wolfram Research, Inc., Champaign (2018)
9. King, J.C.: Symbolic execution and program testing. Commun. ACM **19**(7), 385–394 (1976)
10. Kojiri, T., Hosono, S., Watanabe, T.: Automatic generation of answers using solution network for mathematical exercises. In: Khosla, R., Howlett, R.J., Jain, L.C. (eds.) KES 2005. LNCS (LNAI), vol. 3683, pp. 1303–1309. Springer, Heidelberg (2005). https://doi.org/10.1007/11553939_181
11. Li, K., Li, G.: Approximate query processing: what is new and where to go? Data Sci. Eng. **3**(4), 379–397 (2018)
12. Lin, P., Song, Q., Wu, Y.: Fact checking in knowledge graphs with ontological subgraph patterns. Data Sci. Eng. **3**(4), 341–358 (2018)
13. Meurer, A., et al.: SymPy: symbolic computing in Python. PeerJ Comput. Sci. **3**, e103 (2017)
14. Neo4j, Inc.: Neo4j, Version 1.1.12. https://neo4j.com/
15. Polyak, B.T.: Gradient methods for solving equations and inequalities. USSR Comput. Math. Math. Phys. **4**(6), 17–32 (1964)
16. Tomás, A.P., Leal, J.P.: A CLP-based tool for computer aided generation and solving of maths exercises. In: Dahl, V., Wadler, P. (eds.) PADL 2003. LNCS, vol. 2562, pp. 223–240. Springer, Heidelberg (2003). https://doi.org/10.1007/3-540-36388-2_16
17. Toutanova, K., Lin, V., Yih, W.t., Poon, H., Quirk, C.: Compositional learning of embeddings for relation paths in knowledge base and text. In: Proceedings of the 54th Annual Meeting of the Association for Computational Linguistics (Volume 1: Long Papers), vol. 1, pp. 1434–1444 (2016)

18. Zhang, Y., Dai, H., Kozareva, Z., Smola, A.J., Song, L.: Variational reasoning for question answering with knowledge graph. In: Proceedings of the Thirty-Second AAAI Conference on Artificial Intelligence (AAAI 2018), pp. 6069–6076 (2018)
19. Zheng, W., Yu, J.X., Zou, L., Cheng, H.: Question answering over knowledge graphs: question understanding via template decomposition. Proc. VLDB Endow. **11**(11), 1373–1386 (2018)

Multi-hop Path Queries over Knowledge Graphs with Neural Memory Networks

Qinyong Wang[1], Hongzhi Yin[1(✉)], Weiqing Wang[2], Zi Huang[1], Guibing Guo[3], and Quoc Viet Hung Nguyen[4]

[1] School of Information Technology and Electrical Engineering, The University of Queensland, Brisbane, Australia
{qinyong.wang,h.yin1}@uq.edu.au, huang@itee.uq.edu.au

[2] Faculty of Information Technology, Monash University, Melbourne, Australia
weiqing405@gmail.com

[3] Software College, Northeastern University, Shenyang, China
guogb@swc.neu.edu.cn

[4] Griffith University, Gold Coast, Australia
quocviethung1@gmail.com

Abstract. There has been increasing research interest in inferring missing information from existing knowledge graphs (KGs) due to the emergence of a wide range of knowledge graph downstream applications such as question answering systems and search engines. Reasoning over knowledge graphs, which queries the correct entities only through a path consisting of multiple consecutive relations/hops from the starting entity, is an effective approach to do this task, but this topic has been rarely studied. As an attempt, the compositional training method equally treats the constructed multi-hop paths and one-hop relations to build training data, and then trains conventional knowledge graph completion models such as TransE in a compositional manner on the training data. However, it does not incorporate additional information along the paths during training, such as the intermediate entities and their types, which can be helpful to guide the reasoning towards the correct destination answers. Moreover, compositional training can only extend some existing models that can be composable, which significantly limits its applicability. Therefore, we design a novel model based on the recently proposed neural memory networks, which have large external memories and flexible writing/reading schemes, to address these problems. Specifically, we first introduce a single network layer, which is then used as the building block for a multi-layer neural network called TravNM, and a flexible memory updating method is developed to facilitate writing intermediate entity information during the multi-hop reasoning into memories. Finally, we conducted extensive experiments on large datasets, and the experimental results show the superiority of our proposed TravNM for reasoning over knowledge graphs with multiple hops.

G. Li et al. (Eds.): DASFAA 2019, LNCS 11446, pp. 777–794, 2019.
https://doi.org/10.1007/978-3-030-18576-3_46

1 Introduction

Automated reasoning refers to the ability for computing systems to make new inferences from the observed evidence, which is an important concept in artificial intelligence, and has attracted extensive attention over the past decades [1,23,29]. For example, one of the recent research trends focuses on reasoning over graphs (networks) [4,12,13,26,35–37]. On the other hand, knowledge graphs (KGs), constructed by relations and entities, have become effective and important underlying representations of knowledge for complex reasoning tasks such as question-answering systems, search engines and recommender systems. Combining both, our work focuses on multi-hop reasoning over knowledge graphs, which aims to infer entities that can be reached by a path only composed of multiple consecutive relations/hops starting from a given entity. For example, *(DonalTrump, PresidentOf, FoundedIn, ?)* is to answer the complex question *Which year was the country that DonalTrump is President of founded in?*. This task could be trivial if the target entity exists in the knowledge graph, but this is usually impossible because most existing knowledge graphs are quite sparse, far from complete and can only capture a fraction of world knowledge [22], e.g., in Freebase [2], 71% of the roughly 3 million people lack place of birth, 94% lack parents, and 99% lack ethnicity. Therefore, how to accurately and effectively perform multi-hop reasoning over those highly incomplete knowledge graphs becomes a challenging task.

Early works mostly focused on logic-based reasoning approaches that employ symbolic representations such as [16,24], which extract rules through mining methods and use these extracted rules to infer new links. While these methods could be rather expressive and interpretable, they have the risk of suffering from the following drawbacks. First, the rules over observed nodes only cover a subset of patterns in knowledge graphs and useful rules could be difficult to capture. Secondly, they explicitly require the rules to be specified by hand before the inference starts running, which could be expensive in the practical applications. Thirdly, when the size of the knowledge graph is large, the inference could easily become intractable.

Another line of work focused on the path ranking algorithm (PRA) and its variants [17,18], which could provide tractable inference over large knowledge bases. However, they require training and maintaining separated models whose parameters are not shared for different relation types, which leads to large number of parameters to learn. Moreover, they operate in a fully discrete space, making the comparison among similar entities and relations in a knowledge graph especially hard. Therefore, these weaknesses make them impractical in real-world downstream KG applications.

Recently, encoding the knowledge graph into a low-dimensional vector space to learn latent representations of entities and relations has become the mainstream, but most existing models in this line of work such as TransE [3] only have one-hop reasoning abilities. Therefore, the most straightforward solution for multi-hop reasoning is to recursively apply the embeddings of those models for every hop in the path. However, this simple solution turns out to be infeasible because it would cause *cascading errors*, that is, minor noise generated in each hop would be added up, causing large deviation from the right answers in the end. Recently, [10] tried to eliminate the cascading errors by compositional training, which equally treats one-hop relations and multi-hop paths to train the traditional knowledge graph completion models such as TransE. One weakness of their method is that during the training phase, it ignores the rich intermediate entity information, e.g., the entity itself or its type, at each hop, which can be helpful to guide the reasoning towards the correct direction. Another weakness is that compositional training can only be applied to models that are *composable* (TransE and Bilinear Model [27]), which limits its applicability. After that, [5] proposed a recurrent neural network (RNN) based model to do multi-hop reasoning, which incorporates entity types during reasoning, but it lacks flexible reading/writing operations and its memory capacities are too small to store sufficient long-term information for complex queries.

To address the problems above, we propose **TravNM** (**Trav**ersing KG with **N**eural **M**emories Networks) which is a multi-layer neural network with carefully designed writing operations. Each network layer as the building block is based on Neural Memory Networks (NemNN) [30] which include two main components: large *augmented memories* that selectively store the information about knowledge graphs and *controller networks* that control the input/output data flow into/from the memories. Moreover, at each hop, TravNM writes information of entities appearing in the path into the memories to better refine them. In particular, TravNM has the following advantages. (1) It encourages local changes instead of global changes in memories. Specifically, it attentively reads the information it needs from the memories, hence, the final output as the predicted entity can be produced after multiple reasoning steps (or hops) are made. (2) It has large external memories instead of using only one hidden state vector in recurrent neural networks (RNN) or long short-term memory (LSTM), and these large memories are critical to memorize longer-term information in order to enhance the reasoning abilities. (3) It explicitly conditions on the outputs stored in memory from previous hops, thus both long- and short-term memories are taken into account.

Finally, we listed four research questions to verify the performance of the proposed model, and extensive experiments on two datasets created from well-known knowledge graphs are conducted to answer them. The experimental results show the superiority of our proposals by comparing with the state-of-the-art models.

2 Related Work

2.1 Reasoning over KG

Over the past decade, there has been extensive research on how to infer missing facts from existing knowledge graphs, and these methods fall into three major categories.

The first category of methods are based on logic rules. Some early work [15, 28] employed Markov Logic Networks (MLN) to reason with first order-logic.

Any rule in first-order logic can be included in an MLN, which makes MLN a powerful tool, but these methods are unable to scale to large knowledge bases. The ProPPR system was proposed to address the scalability problem, which uses random walks through a "proof space" based on a set of logical inference rules to perform the query that is independent of the size of the knowledge base. This allows ProPPR to cope with larger KBs. AMIE [7] is a rule mining system that extracts logical rules based on their support in a knowledge graph, which can handle the open-world assumption of knowledge graphs and has superior efficiency on large knowledge graphs.

The second category of methods are based on the path ranking algorithm (PRA) [17]. The main idea of PRA and its variants is to first find a set of node pairs connected by sequences of edges (or paths) from a graph by random walks and then use those paths as features in a logistic regression model to infer missing edges in the graph.

The third category is the most recent embedding methods, which learn and represent entities and relations as low-dimension embedding vectors in the latent space. These solutions can be easily and naturally applied to one-hop reasoning over knowledge graphs by retrieving the most relevant entities using their defined operations within the latent space. Among these methods, translational models are the most representative ones. TransE [3] was first proposed, which views the relation as the translation from a head entity to a tail entity in the same space. Following TransE, other translational methods were also proposed to address its limitations in dealing with none 1-to-1 relations. TransH [33] and TransR [19] force entities to have different embeddings when involved in different relations. Specifically, TransH introduces relation-specific hyperplanes, where the relation is on its specific hyperplane. TransR extends TransH by modeling entities and relations in separate semantic spaces, and translate entities in the relation space. Despite the huge advancement of one-hop reasoning, the multi-hop reasoning over knowledge graphs has been rarely studied so far. One straightforward solution for this task is based on learned representations of entities and relations in the one-hop reasoning models and recursively apply those models to answer path queries. [10] showed that this solution would cause cascading errors, resulting in wrong answers, and they proposed compositional training that considers two entities connected not only by relations by also relations paths, which can effectively reduce cascading errors. Specifically, based on this idea, the authors proposed to train TransE and Bilinear models [14] in a compositional way, that is, TransE employs the addition composition and Bilinear models employ the multiplication composition. Recently, [5] proposed a chain model based on recurrent neural networks that additionally considers the intermediate entities' types along the path during reasoning. This model achieves the state-of-the-art results.

2.2 Neural Memory Networks

Neural memory network (MemNN) [8], inspired by the physical computer architecture, has attracted increasing attention recently due to its strong abilities in capturing very long ranged dependencies that recurrent neural networks (RNN) and long short-term memory (LSTM) suffer from. Moreover, compared to RNN and LSTM, MemNN can manipulate the memory in a more flexible way. Typically, a MemNN model consists of two parts: a memory to store the information and a controller to interact with data, i.e, to read from and write to the memory. Specifically, to read from the memory for an input vector, similarities (e.g., cosine similarity and inner product) between it and each slot of the memory are first computed, then a softmax function is applied to these similarity values to obtain read weights. To write new information into the memory, an attention mechanism of focusing by content or by location is proposed in [8] to fully utilized all the locations of the memory and make the whole process end-to-end.

MemNN achieved its success in many applications such as recommendation [32], neural language translation [9] and knowledge tracking [38] but it is rarely studied in the field of knowledge graph completion, especially multi-hop reasoning over knowledge graphs. Our model is based on the key-value paired memory networks model [38], a recently proposed MemNN variant, which is a generalization of the way that contexts (e.g., knowledge bases or documents to be read) are stored in memory. The lookup (addressing) stage is based on the key memory while the reading stage (given the returned address) uses the value memory. This provides more flexibility and effectiveness [21].

3 Preliminaries

First, we introduce the notations and concepts used in this paper. Note that in the description below, vectors and matrices are denoted with lowercase bold letters and uppercase bold letters respectively.

$\mathcal{E}$ is a set of entities and $\mathcal{R}$ is a set of relations, and the sizes of $\mathcal{E}$ and $\mathcal{R}$ are N and M, respectively. A knowledge graph $\mathcal{G}$ is defined as a set of triples in the form (h, r, t) representing head entity, relation and tail entity respectively, where $h, t \in \mathcal{E}$ and $r \in \mathcal{R}$, and their r-dimensional embedding representations are $\mathbf{h}$, $\mathbf{r}$ and $\mathbf{t}$. The multi-hop reasoning task can be characterized by path queries that are first introduced in [10]. A path query is represented as a triple $(h, \pi, [\![q]\!]) \in \mathcal{G}^p$, where $\mathcal{G}^p$ is the path query dataset constructed from $\mathcal{G}$ and h is the starting entity, followed by an L-hop path $\pi = (r_1/\ldots/r_L)$ which is a sequence of relations $r_i \in \mathcal{R}$, and $[\![q]\!] \subseteq \mathcal{E}$ is the answer set for this query. We simplify this problem by assuming that $[\![q]\!]$ only contains one entity t, so the path query triple is denoted as (h, π, t). If $[\![q]\!]$ contains multiple entities, we can easily convert $(h, \pi, [\![q]\!]) \in \mathcal{G}^p$ to multiple (h, π, t). π can be augmented with intermediate entity information for training purpose in the form $\pi = r_1/t_1, c_{t_1}/r_2, c_{t_2}\ldots/r_L$ where c_i is the type annotated for the intermediate t_i. Take the toy example mentioned above for illustration, the path query *(DonalTrump, PresidentOf/FoundedIn, ?)* expects the resulting answer to be *1776*.

4 Network Layer

In this section, we describe the single network layer used as a building block for TravNM introduced later. In each layer, $\mathbf{A}$ and $\mathbf{B}$ are entity and relation embedding matrices with the size of $M \times r$ and $N \times r$ respectively, where r is the embedding size. $\mathbf{M^k}$ and $\mathbf{M^v}$ are key memory matrix and value memory matrix, both with the size of $D \times r$, where D and r are the numbers of memory slots and their dimension.

4.1 Memory Addressing

Given a triple (h, r, t), we take h as an index to get its corresponding r-dimensional vector $\mathbf{h}$ from the entity embedding matrix $\mathbf{A}$, and similarly, r is transformed by the relation embedding matrix $\mathbf{B}$ into an r-dimensional vector $\mathbf{r}$. Then their sum $\mathbf{h} + \mathbf{r}$ is fed into the key memory matrix $\mathbf{M^k}$, resulting in attention weights $\mathbf{w}$ by the following steps. First, the similarity between $\mathbf{h} + \mathbf{r}$ and each key memory slot is measured based on their inner product:

$$sim_j = \sum (\mathbf{h} + \mathbf{r})\mathbf{M^k}(j) \tag{1}$$

Then, we apply Softmax, i.e., $Softmax(z_i) = e^{z_i}/\Sigma_j e^{(z_j)}$, to the similarity so that the most similar memory slot produces the largest weight to obtain an attention weight vector $\mathbf{w}$ with size D:

$$w(j) = Softmax(sim_j) \tag{2}$$

The differentiable Softmax function is employed in the soft attention mechanism. This attention process is analogous to physical memory addressing operations in computer architectures [21].

4.2 Memory Reading

In this stage, each value memory slot is multiplied by the corresponding attention weight, resulting in an r-dimensional vector $\bar{\mathbf{t}}$.

$$\bar{\mathbf{t}} = \sum_{j=1}^{D} w(j)\mathbf{M^v}(j) \tag{3}$$

$\bar{\mathbf{t}}$ is a *soft* representation of the translating result of $\mathbf{h} + \mathbf{r}$, which is important for the multi-hop reasoning. Specifically, $\bar{\mathbf{t}}$ relaxes the overstrict requirement of $\mathbf{h} + \mathbf{r} \approx \mathbf{t}$ in TransE, and encourages the intermediate result to only be close to $\bar{\mathbf{t}}$. Therefore, it is a good abstraction of all possible results, which can effectively avoid cascading errors.

5 TravNM

In this section, we introduce how to stack multiple single layers introduced above into TravNM to support multi-hop reasoning over knowledge graphs.

5.1 Structure

In order to adapt to L hops, L network layers could be stacked to process a given path query triple (h, π, t) from training datasets. The input to the ith layer of TravNM, i.e., at the ith hop of reasoning, is $\mathbf{\bar{t}_{i-1}} + \mathbf{r_i}$, where $\mathbf{\bar{t}_{i-1}}$ is the output embedding from the previous $(i-1)^{th}$ layer and $\mathbf{r_i}$ is the embedding of the ith relation in π. Note that in the first hop of the reasoning, the input is $\mathbf{h} + \mathbf{r_i}$, i.e., the model directly utilizes the given head entity. The output of the ith network layer is an intermediate entity embedding $\mathbf{\bar{t}_i}$ which abstracts all possible results reasoned by the first $i-1$ relations in π, i.e., $r_1, \ldots, r_{i-1}$, in sequence.

Algorithm 1: Training TravNM

1 **INPUT:** Training set $\mathcal{G}^p = (h, \pi, t)$, where $\pi = r_1/t_1, c_{t_1}/r_2, c_{t_2} \ldots /r_L$
2 **INITIALIZE:** $\mathbf{A}$, $\mathbf{B}$, $\mathbf{M^k_1}, ldots, \mathbf{M^k_{L_{max}}}$, $\mathbf{M^v_1}, \ldots, \mathbf{M^v_{L_{max}}}$
3 $\mathcal{G}^p_{batch} \leftarrow$ sample($\mathcal{G}^p$, b) ; ▷ `sample a minibatch of size b`
4 $T_{batch} \leftarrow \emptyset$; ▷ T_{batch} `holds positive and negative triples`
5 **for** (h, π, t) *in* $\mathcal{G}^p_{batch}$ **do**
6   **for** *hop* i^{th} **do**
7         $\mathbf{r_i}$ = embed(r_i, $\mathbf{B_i}$);
8   **if** $i = 1$ **then**
9             $\mathbf{h}$ = embed(h,$\mathbf{A_i}$);
10             $\mathbf{w_i} \leftarrow$ attention($\mathbf{h} + \mathbf{r_i}$, $\mathbf{M^k_i}$);
11   **end**
12   **else**
13             $\mathbf{w_i} \leftarrow$ attention($\mathbf{\bar{t}_{i-1}} + \mathbf{r_i}$);
14   **end**
15         $\mathbf{\bar{t}_i} \leftarrow$ read($\mathbf{w_i}$, $\mathbf{M^v_i}$);
16         $\mathbf{t_i}$ = embed(t_i,$\mathbf{A_i}$);
17         $\mathbf{M^v_{i+1}} \leftarrow$ write($\mathbf{w_i}$, $\mathbf{t_i}$, $\mathbf{M^v_i}$);
18         $(h, \pi, t') \leftarrow$ sample($\mathcal{G}^p_{batch_{(h,\pi,t)}}$) ; ▷ `corrupt a triple`
19         $T_{batch} \leftarrow T_{batch} \cup \{((h, \pi, t), (h, \pi, t'))\}$;
20   **end**
21 **end**
22 Update model w.r.t. $\sum_{((h,\pi,t),(h,\pi,t')) \in T_{batch}} \nabla[\gamma + f(h, \pi, t) - f(h, \pi, t')]_+$;

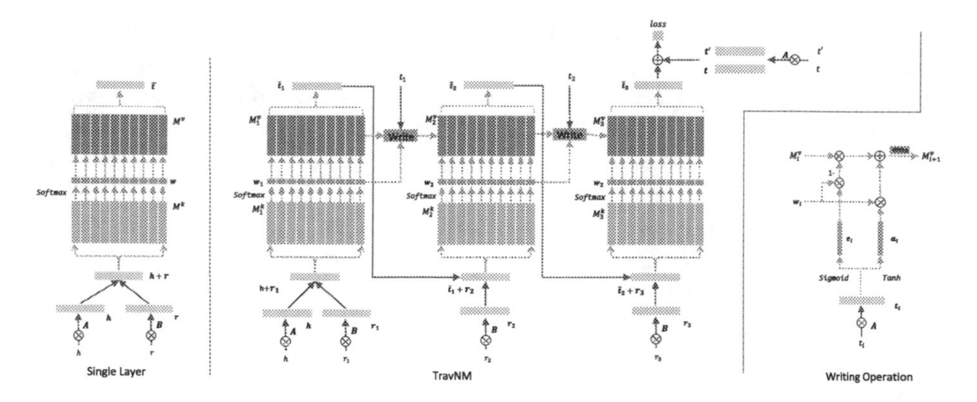

Fig. 1. The architecture of single network layer (Left). The architecture of TravNM (Middle). The multi-hop reasoning with TravNM when $L = 3$; The detailed writing module (Bottom Right).

5.2 Layers Stacking

In order to stack multiple layers, two strategies on parameters tying (i.e., key/value memories) within the unified model are proposed: "sole" and "shared":

1. **"sole":** We proposed this strategy following [30]. In this setting, for the ith network layer that handles its corresponding hop, it has its own independent key memory matrix $\mathbf{M_i^k}$ and value memory matrix $\mathbf{M_i^v}$.
2. **"shared":** Key memory matrices and value memory matrices are shared across different layers, i.e., $\mathbf{M_1^k} = \mathbf{M_2^k} = \cdots = \mathbf{M_L^k}$ and $\mathbf{M_1^v} = \mathbf{M_2^v} = \cdots = \mathbf{M_L^v}$.

5.3 Refining Value Memories

Intermediate entity information (e.g., entities themselves or entity types) appearing in the path could be critical for multi-hop reasoning to eventually reach the correct answer entity, so it should be incorporated into the stacked model in a proper way. Note that this information is only incorporated into the model for training purpose, i.e., this information only exists in training phase in order to refine the value memories, because intermediate entity information is unavailable in testing phase or real-world scenarios. To this end, we propose three methods for writing information into the memories inspired by [8,38], which locally update the value memory matrix with important information. The three methods are developed w.r.t. the information type to be consumed by the stacked model.

"entity". In this method, the value memory matrix at the ith layer is updated with the embedding of the intermediate entity $\mathbf{t_i}$ connected by the ith relation by the writing operation introduced later. For example, regarding *(DonalTrump, PresidentOf/FoundedIn, ?)*, the intermediate entity *the United States* as the answer of the first hop is consumed to update the value memory. We show this model setting in Fig. 1(Middle). Specifically, we take the intermediate entities to be incorporated into the stacked model as an example to explain the writing operation, which is illustrated in Fig. 1(Bottom Right). Say we are at the ith hop, after its memories being read, the embedding of the corresponding intermediate entity $\mathbf{t_i}$ will be written to the value memory matrix $\mathbf{M_i^v}$ using the same attention weights $\mathbf{w}_i$ generated in the memory reading stage. The relevant memory slots are first partially erased and then the information of $\mathbf{t_i}$ is written into the value memory. To update the value memory matrix $\mathbf{M_i^v}$ into $\mathbf{M_{i+1}^v}$ for the next hop, $\mathbf{t_i}$ is first linearly transformed, and the result is fed into the Sigmoid function, then we obtain the *erased vector* $\mathbf{e_i}$:

$$\mathbf{e_i} = Sigmoid(\mathbf{W_e t_i} + \mathbf{b_e}) \tag{4}$$

where $\mathbf{W_e}$ is a linear transformation matrix with size of $r \times r$, and $\mathbf{b_e}$ is an r-dimensional bias vector, and both $\mathbf{W_e}$ and $\mathbf{b_e}$ are shared across all layers. As a result, $\mathbf{e_i}$ is an r-dimensional vector, whose values range from 0 to 1. Then the value memory $\mathbf{M_i^v}$ is partially erased to be $\tilde{\mathbf{M}}_\mathbf{i}^\mathbf{v}$, and each of its memory slots $\tilde{\mathbf{M}}_\mathbf{i}^\mathbf{v}(j)$ is modified by:

$$\tilde{\mathbf{M}}_\mathbf{i}^\mathbf{v}(j) = \tilde{\mathbf{M}}_\mathbf{i}^\mathbf{v}(j) \circ [\mathbf{i} - w_i(j)\mathbf{e_i}] \tag{5}$$

where $\mathbf{i}$ is an r-dimensional unit vector with each element being 1, and $\circ$ is element-wise multiplication. Similar to the reading case, the weight $w_i(j)$ tells the operation where to pay more attention. In this way, the value location is reset to 0 if the corresponding weight and the erased vector element are both 1, otherwise, this location remains unchanged. Following the erasing operation, an r-dimensional *add vector* $\mathbf{a_i}$ is calculated for updating the value memory, i.e., a linear transformation and the *Tanh* function are applied to $\mathbf{v}$ sequentially:

$$\mathbf{a_i} = Tanh(\mathbf{W_a}\mathbf{v} + \mathbf{b_a}) \tag{6}$$

and both of them are linear transformation matrix and bias vector with size $r \times r$ and r, where both $\mathbf{W_a}$ and $\mathbf{b_a}$ are also shared across all layers. Finally, each slot of the value memory matrix is updated to be $\mathbf{M_{i+1}^v}(j)$ as follows:

$$\mathbf{M_{i+1}^v}(j) = \tilde{\mathbf{M}}_\mathbf{i}^\mathbf{v}(j) + w_i(j)\mathbf{a_i} \tag{7}$$

"type". In this method, only the types associated with intermediate entities appearing in the path are exploited to help to reason. Specifically, the entity types are also represented by embeddings through another independent embedding matrix with the same dimension as the entity, then these embeddings are written into the value memory with the same write operation introduced above. Take the same query path mentioned above as an example, the entity type *place* for the intermediate entity *the United States* in the last example is considered. In many knowledge bases, the entity is annotated with types and could be obtained by the provided API. For example, the types of entities in Freebase can be openly accessed by Google Knowledge Graph Search API [1]. There might be more than one types for an entity but we only choose the most popular one.

"relation". This is a baseline method, where only the relations appearing in the path are considered by the model, a common strategy adopted by many existing works [10,25]. We implemented this baseline method by simply removing the writing operation from the model shown in Fig. 1(Middle).

5.4 Training

The score for each training triple in TravNM is defined as:

$$f(h, \pi, t) = \|\bar{\mathbf{t}}_\mathbf{L} - \mathbf{t}\|_2 \tag{8}$$

[1] https://developers.google.com/knowledge-graph/.

where $\bar{\mathbf{t}}_\mathbf{L}$ is the final output embedding from the last (L^{th}) layer, representing the answer for this path query, and $\mathbf{t}$ is the ground truth answer entity's embedding in the path query dataset $\mathcal{G}^p$. We employ the following loss function for each given path query triple (h, π, t):

$$\mathcal{L} = \sum_{(h,\pi,t)\in\mathcal{G}^p,(h,\pi,t')\in\mathcal{G}^{p'}_{(h,\pi,t)}} [m + f(h,\pi,t) - f(h,\pi,t')]_+ + \lambda\sum_i^L(\bar{\mathbf{t}}_\mathbf{i} - \mathbf{t}_i)^2 \quad (9)$$

where $[z]_+ = max(z; 0)$ denotes the standard Hinge-loss with margin m and (h, π, t') is obtained by corrupting (h, π, t), i.e., randomly sampling an entity t' $(t' \neq t)$ from $\mathcal{E}$ to replace the tail entity t:

$$\mathcal{G}^{p'}_{(h,\pi,t)} = \{(h,\pi,t'), t' \in \mathcal{E}\} \quad (10)$$

the second regularization term encourages the generated intermediate entity embeddings to lie near the real intermediate entity embeddings, and λ is a parameter which controls the importance of the regularization term.

If the data points in a high-dimensional space spread too widely, the model would fail due to the curse of dimensionality [6]. To solve it, we bound all entity embeddings within a unit sphere to ensure the model robustness [11], i.e., $\|\mathbf{h}\| \leq 1$ and $\|\mathbf{t}\| \leq 1$. The training algorithm is listed in Algorithm 1, where the "sole" and "entity" settings are adopted.

6 Experiment Setups

In this section, we describe our experimental setups. First, we list four research questions (RQs) to investigate our proposals, then introduce the datasets and the evaluation metrics. Finally, we conduct experiments to answer the RQs.

6.1 Research Questions

We aim to answer the following research questions to evaluate the proposed TravNM model.

RQ1: In terms of layers stacking strategies in TravNM, does "shared" outperforms "sole"?

RQ2: In terms of memory update methods, which of "entity", "type" and "relation" could achieve the best performance?

RQ3: Based on proper layer stacking and memory updates, can TravNM perform better multi-hop reasoning than other comparison models?

RQ4: How do the hyper-parameters affect TravNM in effectiveness and what are the optimal ones?

6.2 Datasets and Metrics

In this section, we introduce the datasets used for evaluating the multi-hop reasoning task.

First, we introduce two large real-world fact knowledge graphs, called Base Datasets.

FB15K-237: The widely used FB15k [3] extracted from Freebase is not adopted in our experiments, since they suffer from test leakage through inverse relations, i.e., many test triples can be obtained by inverting them in the training data [31]. Instead, we use the improved FB15K-237 introduced in [31] that removes inverse relations from FB15k.

YAGO3-10: We also use YAGO3-10 [20] that is a subset of YAGO3 [20], a large knowledge graph extracted from several sources. Each entity in YAGO3-10 co-occurs with at least 10 relations, and most triples deal with descriptive attributes of people like citizenship, gender and profession.

Query path datasets $\mathcal{G}^p$ are then constructed from Base Datasets. We propose a random walk procedure to generate query paths that include intermediate entity information by repeating the following steps, inspired by [10]:

1. Uniformly sample a path length $L \in \{1, \ldots, L_{max}\}$ where L_{max} is the predefined maximum path length, and uniformly sample a head entity $h \in \mathcal{E}$.
2. An L-step random walk is performed starting from h. For the ith hop, uniformly choose a relation r_i from all relations incident on the previous entity t_{i-1} (or h if $i = 1$ for consistence).
3. If $i < L$, uniformly choose the entity t_i from $\mathcal{E}$ that r_i reaches, and look up for its entity type c_{t_i}. If $i = L$, just choose a t_L (denoted as t).
4. Output a query path (h, π, t), where $\pi = r_1/t_1, c_{t_1}/r_2, c_{t_2} \ldots /r_L$ for training set or $\pi = r_1/r_2 \ldots /r_L$ for valid and test set, and add it to $\mathcal{G}^p$.

Statistics of all datasets we use are shown in Table 1.

Following [3,19], we adopt two metrics to evaluate the model.

MRR: Mean Rank Reciprocal Rank calculates the average reciprocal rank of all correctly predicted entities. Compared with Mean Rank (MR), MRR is less sensitive to outliers

Hit@10: This metric calculates the proportion of correct entities in *top*-10 ranked entities. i.e., the proportion of correct entities in *top*-10 ranked entities.

Table 1. Statistics of the experimental datasets

Dataset	$\mid\mathcal{E}\mid$	$\mid\mathcal{R}\mid$	# of triples in train/valid/test		
FB15K-237	14,541	237	272,115	17,535	20,466
FB15K-237-Path			4,455,610	146,611	109, 165
YAGO3-10	123,182	37	1,079,040	5,000	5,000
YAGO3-10-Path			12,174,868	29,684	34,643

6.3 Comparison Methods

Various experiments are designed to answer the four proposed research questions.

Experiment 1. To answer RQ1 and RQ2, we develop different variants of TravNM that employ different layer stacking strategies (i.e., "sole" and "share") and different value memory update methods (i.e., "entity", "type" and "relation"), and they will be evaluated on FB15K-237 w.r.t. the metric MMR and Hit@10 to find out the optimal setting. As for the entity types, there are 56 of them in this dataset.

Experiment 2. To answer RQ3, we compare TravNM, which employs the optimal layer stacking strategy and value memory update method from previous experiments, with Comp. TransE, Comp. Bilinear and Comp. Bilinear Diag that use compositional training proposed in [10], and Rajarshi's Model based on RNN proposed in [5]. The essential idea for compositional training methods including Comp. TransE, Comp. Bilinear and Comp. Bilinear Diag is to train entity pairs connected by both relations and paths:

- ***Comp. TransE:*** Comp. TransE uses the addition composition, and defines the score of (h, π, t) as:

$$f(h, \pi, t) = -\|\mathbf{h} + (\mathbf{r_1} + \cdots + \mathbf{r_L}) - \mathbf{t}\| \tag{11}$$

- ***Comp. Bilinear:*** Comp. Bilinear uses the multiplication composition and defines the score as:

$$f(h, \pi, t) = \mathbf{h}^\intercal(\mathbf{M}_{\mathbf{r_1}} \circ \cdots \circ \mathbf{M}_{\mathbf{r_L}})\mathbf{t} \tag{12}$$

- ***Comp. Bilinear Diag:*** Compared with Comp. Bilinear, this method constraints the relation matrices to be diagonal [34].

Since they have been proved to have better performance than PRA and its variants in [10], we do not compare with them in the experiment.

Rajarshi's Model: This model is based on RNN which allows chains of reasoning across multiple relations and considers entity types as input.

TravNM-Single: We use entity and relation embeddings learned by only one network layer on the fact knowledge graph $\mathcal{G}$ to directly perform multi-hop reasoning. Specifically, a query path is modeled as the sum of the embeddings of all relations in the path, where each embedding is learned from (h, r, t) only by the single layered network.

Furthermore, we show the *top*-3 ranked answers output by TravNM for three different path queries as case studies. We also evaluate the overall performance of our model with respect to different path length.

Experiment 3. To answer RQ4, we run grid search to show how the two important model hyper-parameters (i.e., r, the number of latent factors in $\mathbf{A}, \mathbf{B}$, $\mathbf{M^k}$ and $\mathbf{M^v}$, and D, the number of key/value memory slots) affect its reasoning ability and find out the optimal ones. Specifically, the search for r ranges from 40 to 200 and for D ranges from 24 to 128 on both path datasets, and the hyper-parameters with the best MRR performance are chosen.

Table 2. Results of TravNM variants with different model settings on FB15K-237-Path.

Variant	shared		sole	
	MMR	Hit@10	MMR	Hit@10
relation	0.382	40.7	0.297	35.8
type	0.403	46.6	0.315	37.3
entity	**0.437**	**52.1**	0.380	39.3

7 Experimental Results

In this section, we present the experimental results and discuss them.

7.1 RQ1 & RQ2

To answer RQ1 and RQ2, we run Experiment 1 and its results are shown in Table 2. Similar experimental results are also achieved on the dataset YAGO3-10-Path. Due to the space limitation, we simply omit them.

RQ1. From the column perspective of Table 2, we find that the "shared" strategy achieves better performance than the "sole" strategy. We give several explanations for this. One reason is that "sole" requires a sufficiently large number of training examples for each hop, which is generally not possible, so there are risks that the model would be overfitting during training, but in "shared", the parameters can be trained in every hop. Moreover, for "sole", reasoning with a long path query might not work well because fewer training examples could be provided for the parameters in the higher layers. In real-world applications, the number of relations in each query path (or the path length) is dynamic and even unknown in advance, therefore, the "shared" is more practical and robust.

RQ2. From the row perspective of Table 2, we notice that "entity" that incorporates the intermediate entities has the best performance among "entity", "type" and "relation". This is mainly because the reasoning can be better guided towards the correct final answers by providing the most detailed intermediate routing information among the three. Another reason is that in this way, the value memory could be better trained with more data, i.e., the same value memory can be trained at every hop. Moreover, the "type" setting achieves the second best result, and is better than the "relation" setting as it provides a degree of additional information to the reasoning process compared with "relation", and this conclusion is consistent with that reached in [5]. In conclusion, the optimal model settings for layer stacking and value memory update are "share" and "entity".

7.2 RQ3

We run Experiment 2 to answer RQ3. From Table 3, we can see that TravNM outperforms other models in both datasets in terms of MRR and Hit@10.

Table 3. Experimental results of TravNM and other models in terms of MRR and Hit@10 in the task of multi-hop reasoning. [▼]: Results are obtained by running codes released by their authors (https://github.com/kelvinguu/traversing-knowledge-graphs). [▲]: Results are obtained by running codes released by their authors (https://github.com/rajarshd/ChainsofReasoning).

Method	FB15K-237-Path		YAGO3-10-Path	
	MRR	Hit@10	MRR	Hit@10
Comp. Bilinear [▼]	0.369	44.1	0.467	35.3
Comp. Bilinear Diag [▼]	0.329	42.8	0.463	32.0
Comp. Trans-E [▼]	0.331	45.7	0.604	<u>38.8</u>
Rajarshi's Model [▲]	<u>0.394</u>	<u>48.2</u>	<u>0.618</u>	37.1
TravNM-Single	0.297	38.1	0.446	35.7
TravNM	**0.437**	**52.1**	**0.641**	**42.9**

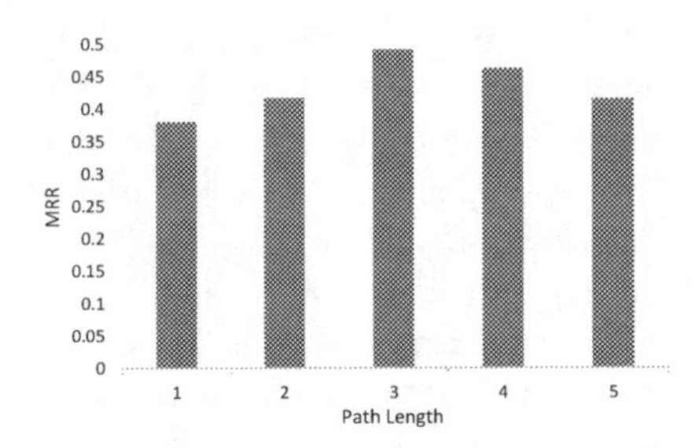

Fig. 2. Performance of TravNM with different path lengths on FB15K-237-Path

Table 4. *Top*-3 ranking answers for 3 example path queries output by TravNM from FB15K-237-Path test dataset, where reverse relations are prefixed with two asterisks (**). The meanings of these complex path queries are: "What country was the movie that Amy Smart was nominated for filmed in?", "What is the citytown that the institute in which the person who wrote the story of 'Man of Steel' studied locates in?" and "What institute does the place where the award the person producing 'The Sopranos' was given to contains?". The ground truth answer is in **bold**.

Path Query	Amy Smart/nominated_for/country
Top-3 Answers	textbfUnited States, England, United Kingdom
Path Query	Man of Steel/story_by/**student/citytown
Top-3 Answers	**East Lansing**, Boston, New York City
Path Query	The Sopranos/**program/award_winner/place_of_birth/contains
Top-3 Answers	Indiana University, **Pratt Institute**, Harvard University

Specifically, TravNM has better performance than the compositional training models that do not consider intermediate entity information. Also, Memory Networks-based TravNM achieves better results than RNN-based Rajarshi's

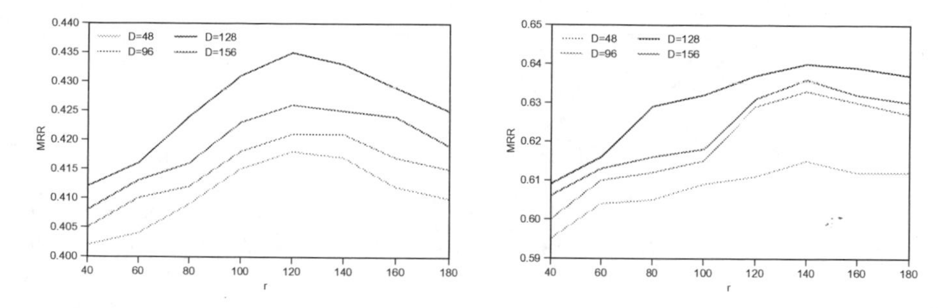

Fig. 3. Model performance w.r.t. to different values of r and D on FB15K-237-Path (left) and YAGO3-10-Path (right)

Model, validating the advantages of Memory Networks in the task of multi-hop reasoning. Another observation is that RNN-based Rajarshi's Model still achieves promising results compared with other methods, showing the significance of memory structures in sequential multi-hop reasoning.

As a case study, we examine some query paths with length 2, 3 and 4 from the test dataset. In Table 4, we show three concrete path queries in the FB15K-237-Path test dataset and their *Top*-3 answers predicted by TravNM, which are retrieved from all entity embeddings based on their similarities with the final output embedding. From this table we can see that the top ranked answers for each query path are similar in concept (*country* for the first query, *city* for the second query and *institute* for the third query) and the correct answer is ranked higher than the others, indicating that given a query path, our model is able to reason out the most probable answer from many entities with the similar concept. Another observation is that although the final relation is 1-to-N in the third query (i.e., the relation *contains*), which is more challenging, our model is still able to rank the correct answer in the top-2.

In addition, Fig. 2 shows the performance of TravNM by varying the path length from 1 to 5, and the experimental results show that our model achieves its best performance on the path queries of length 3, which conforms to the conclusion obtained in [5]. On the other hand, in this table, TravNM-Single achieves poor performance, validating the statement that the existence of cascading errors could really damage performance.

In conclusion, by carefully and smartly stacking multiple single layers, significant improvement is achieved in our model. One reason is that we adopt the strategy of compositional training, which can effectively reduce cascading errors [10].

7.3 RQ4

We run Experiment 3 to find out how different values of two important hype-parameters (r and D) affect the performance of TravNM, and the results are shown in Fig. 3. From those figures, we can observe that the optimal D for both datasets is 128, and the optimal r for FB15K-237-Path and YAGO3-10-Path are 120 and 140 respectively.

8 Conclusion

In this paper, we proposed a Neural Memory Networks-based model called TravNM for the increasingly important but rarely studied multi-hop reasoning task over knowledge graphs. The major weakness of existing methods is that they are unable to incorporate intermediate entity information during training. To address this, we first proposed a single layered network, then stacked multiple network layers in order to adapt to the multi-hop reasoning task, and we also designed proper parameters tying strategy and memory updating method to incorporate entity information. We evaluate TravNM through extensive experiments, and the results validate its superiority compared with other models.

Ackowlegement. This work was supported by ARC Discovery Early Career Researcher Award (DE160100308), ARC Discovery Project (DP170103954; DP19 0101985) and National Natural Science Foundation for Young Scientists of China under Grant No. 61702084.

References

1. Achlioptas, D., Iliopoulos, F.: Random walks that find perfect objects and the lovász local lemma. In: FOCS, pp. 494–503. IEEE (2014)
2. Bollacker, K., Evans, C., Paritosh, P., Sturge, T., Taylor, J.: Freebase: a collaboratively created graph database for structuring human knowledge. In: Proceedings of the 2008 ACM SIGMOD International Conference on Management of Data, pp. 1247–1250 (2008)
3. Bordes, A., Usunier, N., Garcia-Duran, A., Weston, J., Yakhnenko, O.: Translating embeddings for modeling multi-relational data. In: NIPS, pp. 2787–2795 (2013)
4. Chen, H., Yin, H., Wang, W., Wang, H., Nguyen, Q.V.H., Li, X.: PME: projected metric embedding on heterogeneous networks for link prediction. In: SIGKDD, pp. 1177–1186 (2018)
5. Das, R., Neelakantan, A., Belanger, D., McCallum, A.: Chains of reasoning over entities, relations, and text using recurrent neural networks. arXiv preprint arXiv:1607.01426 (2016)
6. Friedman, J., Hastie, T., Tibshirani, R.: The Elements of Statistical Learning, vol. 1 (2001)
7. Galárraga, L., Teflioudi, C., Hose, K., Suchanek, F.M.: Fast rule mining in ontological knowledge bases with AMIE+. VLDB J. Int. J. Very Large Data Bases **24**(6), 707–730 (2015)
8. Graves, A., Wayne, G., Danihelka, I.: Neural turing machines. arXiv (2014)
9. Grefenstette, E., Hermann, K.M., Suleyman, M., Blunsom, P.: Learning to transduce with unbounded memory. In: NIPS, pp. 1828–1836 (2015)
10. Guu, K., Miller, J., Liang, P.: Traversing knowledge graphs in vector space. In: EMNLP, pp. 318–327 (2015)
11. Hsieh, C.K., Yang, L., Cui, Y., Lin, T.Y., Belongie, S., Estrin, D.: Collaborative metric learning. In: WWW, pp. 193–201 (2017)

12. Hung, N.Q., et al.: Answer validation for generic crowdsourcing tasks with minimal efforts. VLDB J. **26**(6), 855–880 (2017)
13. Hung, N.Q.V., Viet, H.H., Tam, N.T., Weidlich, M., Yin, H., Zhou, X.: Computing crowd consensus with partial agreement. TKDE **30**(1), 1–14 (2018)
14. Jenatton, R., Roux, N.L., Bordes, A., Obozinski, G.R.: A latent factor model for highly multi-relational data. In: NIPS, pp. 3167–3175 (2012)
15. Khot, T., Balasubramanian, N., Gribkoff, E., Sabharwal, A., Clark, P., Etzioni, O.: Markov logic networks for natural language question answering. arXiv (2015)
16. Kok, S., Domingos, P.: Statistical predicate invention. In: Proceedings of the 24th International Conference on Machine Learning, pp. 433–440 (2007)
17. Lao, N., Cohen, W.W.: Relational retrieval using a combination of path-constrained random walks. Mach. Learn. **81**(1), 53–67 (2010)
18. Lao, N., Mitchell, T., Cohen, W.W.: Random walk inference and learning in a large scale knowledge base. In: Proceedings of the Conference on Empirical Methods in Natural Language Processing, pp. 529–539. Association for Computational Linguistics (2011)
19. Lin, Y., Liu, Z., Sun, M., Liu, Y., Zhu, X.: Learning entity and relation embeddings for knowledge graph completion. In: AAAI, vol. 15, pp. 2181–2187 (2015)
20. Mahdisoltani, F., Biega, J., Suchanek, F.M.: Yago3: a knowledge base from multilingual wikipedias. In: CIDR (2013)
21. Miller, A., Fisch, A., Dodge, J., Karimi, A.H., Bordes, A., Weston, J.: Key-value memory networks for directly reading documents. arXiv (2016)
22. Min, B., Grishman, R., Wan, L., Wang, C., Gondek, D.: Distant supervision for relation extraction with an incomplete knowledge base. In: PHLT-NAACL, pp. 777–782 (2013)
23. Minsky, M.: The Society of Mind (1986)
24. Muggleton, S.: Inductive logic programming. New Gener. Comput. **8**(4), 295–318 (1991)
25. Neelakantan, A., Roth, B., Mc-Callum, A.: Compositional vector space models for knowledge base inference. In: 2015 AAAI Spring Symposium Series (2015)
26. Nguyen, T.T., Duong, C.T., Weidlich, M., Yin, H., Nguyen, Q.V.H.: Retaining data from streams of social platforms with minimal regret. In: Twenty-Sixth International Joint Conference on Artificial Intelligence. No. EPFL-CONF-227978 (2017)
27. Nickel, M., Tresp, V., Kriegel, H.P.: A three-way model for collective learning on multi-relational data. In: ICML, vol. 11, pp. 809–816 (2011)
28. Niu, F., Ré, C., Doan, A., Shavlik, J.: Tuffy: scaling up statistical inference in markov logic networks using an rdbms. VLDB **4**(6), 373–384 (2011)
29. Poole, D.: First-order probabilistic inference. In: IJCAI, vol. 3, pp. 985–991 (2003)
30. Sukhbaatar, S., Weston, J., Fergus, R., et al.: End-to-end memory networks. In: NIPS, pp. 2440–2448 (2015)
31. Toutanova, K., Chen, D.: Observed versus latent features for knowledge base and text inference. In: Proceedings of the 3rd Workshop on Continuous Vector Space Models and their Compositionality, pp. 57–66 (2015)
32. Wang, Q., Yin, H., Hu, Z., Lian, D., Wang, H., Huang, Z.: Neural memory streaming recommender networks with adversarial training. In: SIGKDD, pp. 2467–2475 (2018)
33. Wang, Z., Zhang, J., Feng, J., Chen, Z.: Knowledge graph embedding by translating on hyperplanes. In: AAAI, vol. 14, pp. 1112–1119 (2014)
34. Yang, B., Yih, W.t., He, X., Gao, J., Deng, L.: Embedding entities and relations for learning and inference in knowledge bases. arXiv (2014)

35. Yin, H., Cui, B., Sun, Y., Hu, Z., Chen, L.: Lcars: a spatial item recommender system. TOIS **32**(3), 11 (2014)
36. Yin, H., Wang, Q., Zheng, K., Li, Z., Yang, J., Zhou, X.: Social influence-based group representation learning for group recommendation. In: ICDE (2019)
37. Yin, H., Zou, L., Nguyen, Q.V.H., Huang, Z., Zhou, X.: Joint event-partner recommendation in event-based social networks. In: ICDE (2018)
38. Zhang, J., Shi, X., King, I., Yeung, D.Y.: Dynamic key-value memory networks for knowledge tracing. In: WWW, pp. 765–774 (2017)

Sentiment Classification by Leveraging the Shared Knowledge from a Sequence of Domains

Guangyi Lv[1], Shuai Wang[2], Bing Liu[2], Enhong Chen[1(✉)], and Kun Zhang[1]

[1] Anhui Province Key Laboratory of Big Data Analysis and Application, School of Computer Science and Technology, University of Science and Technology of China, Hefei, China
{gylv,zhkun}@mail.ustc.edu.cn, cheneh@ustc.edu.cn

[2] University of Illinois at Chicago, Chicago, USA
shuaiwanghk@gmail.com, liub@uic.edu

Abstract. This paper studies sentiment classification in a setting where a sequence of classification tasks is performed over time. The goal is to leverage the knowledge gained from previous tasks to do better on the new task than without using the previous knowledge. This is a lifelong learning setting. This paper proposes a novel deep learning model for lifelong sentiment classification. The key novelty of the proposed model is that it uses two networks: a *knowledge retention network* for retaining domain-specific knowledge learned in the past, and a feature learning network for classification feature learning. The two networks work together to perform the classification task. Our experimental results show that the proposed deep learning model outperforms the state-of-the-art baselines.

1 Introduction

In many sentiment analysis applications, one needs to perform many tasks over time, e.g., to analyze reviews of different products for many different clients, or to analyze reviews of many/all products of a single client (e.g., a retailer). A natural question is whether the system can improve itself over time as it analyzes reviews of more and more categories of products. This is a *lifelong learning* (LL) setting. LL is stated as follows: At any point in time, the learner has performed learning on a sequence of tasks from 1 to $N-1$. When faced with the Nth task, it uses the knowledge gained in the past $N-1$ tasks to help learn the Nth task [7,31,34], which should do better than learning using only the training data of the Nth task alone (without using any past knowledge). In our case, each task is a sentiment classification problem that aims to classify each review into the positive or negative class [19,23].

In [8], a lifelong learning technique is proposed for sentiment classification in the context of naive Bayesian Classification. The method exploited the knowledge of word probabilities under different classes in the past tasks/domains as priors to help optimize the new task learning. In this paper, we propose a novel

G. Li et al. (Eds.): DASFAA 2019, LNCS 11446, pp. 795–811, 2019.
https://doi.org/10.1007/978-3-030-18576-3_47

deep learning model, which outperforms this existing approach. Note that by a domain we mean a category of products. Since each of our task is from a different domain, we use the terms *domain* and *task* interchangeably in this paper.

The proposed deep learning architecture is called SRK (***S****entiment classification by leveraging the* ***R****etained* ***K****nowledge*). It consists of two networks, i.e., the "Feature Learning Network" (FLN) and the "Knowledge Retention Network" (KRN), which jointly perform the supervised learning task. The FLN is like a traditional network for supervised learning. The only difference is that it learns a sequence of tasks. When it comes to learn the next/new task, it has a set of very rich features as the network parameters gained from the previous tasks can help the new task learning. That is because sentiment classification in different domains is quite similar in the sense that sentiment expressions are largely shared and have the same polarities across domains.

As we will see in Sect. 4.2, FLN alone is already able to outperform learning using only the data from each task (without using any previous knowledge or task data). However, we can do better. One weakness with using FLN alone is that if the immediate previous task $N-1$ is very different from the new task N, then the new task may not perform well (see Sect. 4.3) because the short-term memorized/residual parameters from task $N-1$, which can be viewed as a type of parameter initialization, will not be helpful to the new task N. But there may be some other earlier tasks that may be helpful to the new task N. However, it is hard for FLN to use them because of *catastrophic forgetting* [10]. Catastrophic forgetting describes a phenomenon in learning multiple tasks sequentially in a neural network, i.e., after each new task is learned, the knowledge learned from the previous tasks may be forgotten by the network. Because of this problem, the past knowledge that can be leveraged by the new task in FLN is quite limited, which results in weaker performance (see Sect. 4.3). To deal with this problem, we introduce the knowledge retention network KRN with the goal of retaining domain-specific knowledge from the past domains. In order to achieve this goal, KRN needs to deal with catastrophic forgetting.

A novel learning mechanism is proposed for KRN to deal with the forgetting problem. It is thus able to retain domain-specific knowledge learned from individual past domains, which is very helpful in learning similar tasks in the future. The two networks (FLN and KRN) work together through a gate mechanism to provide a robust solution.

Although several models for dealing with catastrophic forgetting exist, their main goal is to preserve the past task learning. That is, after learning the Nth task, the previous $N-1$ tasks will still work. Their main goal is not to leverage the knowledge learned and retained from the previous tasks to help learn the new Nth task better. The focus of the proposed architecture is to leverage the past learned knowledge to do better for the new task. That is why we propose a two-network solution. More related work will be discussed in Sect. 2.

In summary, this paper makes the following contributions:

1. It proposes a deep learning architecture SRK for lifelong sentiment classification, which has not been done before. SRK consists of two sub-networks to

separately capture rich features and also shared knowledge of tasks. A knowledge gate is designed to make the two networks work jointly. As discussed above, this architecture allows the system to flexibly leverages the knowledge learned from previous domains in the new domain.
2. To achieve the goal of retaining domain-specific knowledge to be used in similar tasks in the future, the knowledge retention network (KRN) needs to deal with catastrophic forgetting. A novel partially (gradient) update mechanism is proposed for the purpose based on the observation of activation sparsity in neural networks (see Sect. 3).
3. Experimental results on sentiment classification show that the proposed model SRK outperforms the state-of-the-art baselines from both lifelong learning and continual learning which tries to deal with catastrophic forgetting. It is also more stable/robust as we will see in Sect. 4.3.

2 Related Work

Our work is related to text classification. Traditional text classification mainly uses Bag-of-Words models. In recent years, neural networks have been very popular due to their superior performance [47]. Classic neural networks do not consider the sequential information in text, which is a drawback. To address the issue, various Recurrent Neural Networks (RNNs) have been proposed, e.g., Long Short-Term Memory (LSTM) [13], Gated Recurrent Units (GRU) [9], bidirectional RNN [44], and attention-based RNN [42]. In this work, we adopted GRU as the base for our lifelong RNN model. GRU has been shown to achieve similar performances as LSTM for many NLP tasks.

2.1 Sentiment Classification

Sentiment classification, a key task of sentiment analysis [23], is a special case of text classification. Earlier techniques use hand-crafted features and external resources [21,23,38,43,46], which involve a great deal of human efforts. Recent data-driven neural network methods directly use word features and/or discovered new features by the models themselves. Many neural network based sentiment classification techniques have been reported [4,27,32,41,48]. They don't involve lifelong learning. Many aspect-based sentiment analysis methods are also reported [5,14,17,26,33,36,37,39]. However, these methods also do not accumulate or use knowledge to learn new tasks as we do.

2.2 Lifelong Learning and Catastrophic Forgetting

Our work is most closely related to lifelong learning (LL) [6,7,31,34]. LL has been used for sentiment classification in [8] based on naive Bayesian classifier. However, its method is not applicable to neural networks because LL in [8] exploits word probabilities from past domains, which is not directly usable in neural networks. Furthermore, neural networks suffer from catastrophic forgetting [10] when it learns continually. Similarly, the method in [29] is also hard to apply to neural networks.

Several methods, under the name of *continual learning*, have been proposed to overcome catastrophic forgetting, e.g., system-level consolidation [15,16] and sequential Bayesian [2,11]. However, their main goal is to preserve the previously learned results for past tasks while learning new tasks. Our focus is on how to exploit previously learned knowledge to improve the new task learning. Also, existing works are independent models with specific loss functions. They are hard to be employed in our two-brunch framework.

2.3 Transfer Learning and Multitask Learning

Transfer learning or domain adaptation uses labeled training examples from the source domain to help learning in the target domain that has no or very few labeled training examples [1,12,18,22,40]. One typical use of transfer learning is deep domain adaptation, which usually setups shared weights [20,35] or two-stream architecture [28] to reduce the difference of distributions between the source and target domains. Our feature learning network (FLN) functions like transfer learning as it can transfer some knowledge from past domains to the new domain. However, its transfer is limited as discussed in Sect. 1 because the new domain is mainly affected by its immediate previous domain due to catastrophic forgetting. That is why we use the knowledge-retention network (KRN) to retain past domain-specific knowledge so that the new domain learning can use knowledge from any similar past domains, which transfer learning or domain adaptation cannot do.

LL is also different from multitask learning [3], which tries to optimize the learning of multiple tasks simultaneously [17,30,45]. It does not accumulate past knowledge and does not learn sequentially as LL does.

3 The Proposed Solution

Our lifelong sentiment classification problem can be stated as follows: At a particular point in time, the system has performed a sequence of $N-1$ sentiment classification (SC) tasks using their training data D_1 to D_{N-1}. It now has to learn to perform the Nth new SC task given its training data D_N, where each training example is labeled with a positive or negative polarity (or sentiment). The goal is to leverage the knowledge gained in the past $N-1$ tasks to produce a better classifier than without using the past knowledge. The following subsections present the proposed model.

3.1 Overall Framework

The overall architecture of the proposed SRK model is shown in Fig. 1. It consists of three main components: (1) Feature Learning Network (2) Knowledge Retention Network, and (3) Network Fusion component.

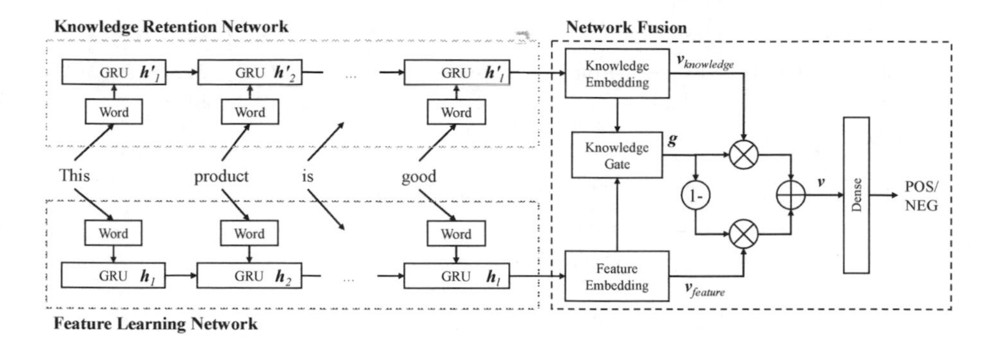

Fig. 1. The overall architecture of the proposed SRK model.

Feature Learning Network. We use the Feature Learning Network (FLN) to learn document representations to perform classification. As shown in the blue box in Fig. 1, the structure of this network is a conventional Gated Recurrent Units (GRU) [9], which is formulated as:

$$\begin{aligned} \boldsymbol{z} &= \sigma(\boldsymbol{U}_z \boldsymbol{x}_t + \boldsymbol{W}_z \boldsymbol{h}_{t-1}), \\ \boldsymbol{r} &= \sigma(\boldsymbol{U}_r \boldsymbol{x}_t + \boldsymbol{W}_r \boldsymbol{h}_{t-1}), \\ \boldsymbol{h}' &= \tanh(\boldsymbol{U}_h \boldsymbol{x}_t + \boldsymbol{W}_h(\boldsymbol{r} \odot \boldsymbol{h}_{t-1})), \\ \boldsymbol{h}_t &= (1 - \boldsymbol{z}) \odot \boldsymbol{h}' + \boldsymbol{z} \odot \boldsymbol{h}_{t-1}, \end{aligned} \tag{1}$$

whose inputs are one-hot representations of words in sentences $\boldsymbol{x} = \{\boldsymbol{w}_1, \boldsymbol{w}_2, \ldots, \boldsymbol{w}_l\}$, where l is the sentence length. These one-hot vectors are first mapped into dense representations by a pre-trained word embedding model $\boldsymbol{E}$. Then, we utilize a GRU cell whose state size is 500 to process the words one by one:

$$\begin{aligned} \boldsymbol{u}_i &= \boldsymbol{E}\boldsymbol{w}_i, \quad i \in 1, 2, \ldots, l, \\ \{\boldsymbol{h}_1, \boldsymbol{h}_2, \ldots, \boldsymbol{h}_l\} &= \mathrm{GRU}(\{\boldsymbol{u}_1, \boldsymbol{u}_2, \ldots, \boldsymbol{u}_l\}), \end{aligned} \tag{2}$$

$\{\boldsymbol{u}_i\}$ and $\{\boldsymbol{h}_i\}$ separately indicate word embedding of word i and GRU's ith state, respectively. We take the final hidden state $\boldsymbol{h}_l$ as the *feature embedding* $\boldsymbol{v}_{feature} = \boldsymbol{h}_l$ or representation of the whole document.

Knowledge Retention Network. To achieve lifelong sentiment classification, we propose a Knowledge Retention Network to learn and to retain domain-specific knowledge from previous tasks. Similar to Feature Learning Network, the structure of this network is also a GRU cell with its state size 500 which has the same input $\boldsymbol{x}$ and gives out states $\{\boldsymbol{h}'_1, \boldsymbol{h}'_2, \ldots, \boldsymbol{h}'_l\}$. The final state $\boldsymbol{h}'_l$ is used as the *knowledge embedding* $\boldsymbol{v}_{knowledge} = \boldsymbol{h}'_l$. To retain knowledge, a different learning method, called *partial update*, is proposed, which will be described in detail in Sect. 3.2.

Network Fusion Component. To make the above networks work together to produce the final result, we designed a *Knowledge Gate* to integrate the two kinds of representations, which is shown in the black box in Fig. 1. By using the knowledge gate, the relevant information for the current task from both two networks will be selectively combined to make the final decision, which is formulated as follows:

$$\begin{aligned} \boldsymbol{g} &= \sigma(\boldsymbol{W}_k \boldsymbol{v}_{feature} + \boldsymbol{U}_k \boldsymbol{v}_{knowledge}), \\ \boldsymbol{v} &= (\mathbf{1} - \boldsymbol{g}) \cdot \boldsymbol{v}_{feature} + \boldsymbol{g} \cdot \boldsymbol{v}_{knowledge}, \\ P(y|\boldsymbol{x}) &= \sigma(\boldsymbol{W}\boldsymbol{v} + \boldsymbol{b}). \end{aligned} \tag{3}$$

Here, $\boldsymbol{g}$ denotes the knowledge gate, $\boldsymbol{W}$ and $\boldsymbol{b}$ are parameters of the last fully connected layer, and σ means the sigmoid activation function.

Finally, we choose Mean-Square Error (MSE) as the training loss function. Note that other loss functions such as cross-entropy can also be applied as alternatives. Furthermore, to ensure both the Feature Learning and the Knowledge Retention Networks learn good representations, we also use additional losses for those two networks:

$$\begin{aligned} L_F &= \text{MSE}(y, \sigma(\boldsymbol{W}\boldsymbol{v}_{feature} + \boldsymbol{b})), \\ L_K &= \text{MSE}(y, \sigma(\boldsymbol{W}\boldsymbol{v}_{knowledge} + \boldsymbol{b})), \end{aligned} \tag{4}$$

where y is the true label. The final objective function is formulated as:

$$L = \text{MSE}(y, P(y|\boldsymbol{x})) + L_F + L_K. \tag{5}$$

3.2 Partial Updating Mechanism

In order to perform sentiment classification tasks sequentially, the *Knowledge Retention Network* needs to learn and retain domain-specific knowledge from every task. However, conventional neural networks have poor performances because they often suffer from the catastrophic forgetting problem as we discussed in the introduction section. Based on our pilot study, we found that even though neural network based models have the ability to learn knowledge from tasks in our case, they have the tendency to remember only knowledge

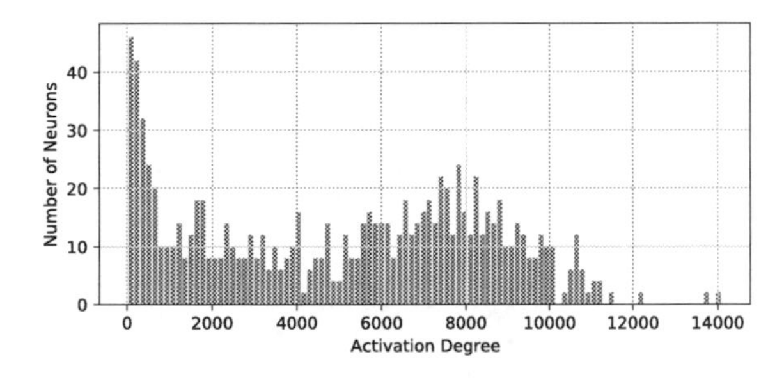

Fig. 2. Activation statistics in the state layer of GRU after convergence. Higher activation degree means more chance to be activated in a task.

that is common across all tasks rather than important specific features in specific domains. They also tend to remember more of the last task learned due to catastrophic forgetting. We propose a *Partial Update Mechanism* to solve the problem.

The idea of partial update is inspired by the observation of activation sparsity in neural networks. Figure 2 shows the histogram of the activation value of state vector $\boldsymbol{h}_t$ in Eq. 1 from a trained GRU network. We can see that only a small number of hidden nodes have very high degrees of activation. Most hidden nodes have relatively tiny activation values. This phenomenon becomes more obvious as the number of model parameters grows.

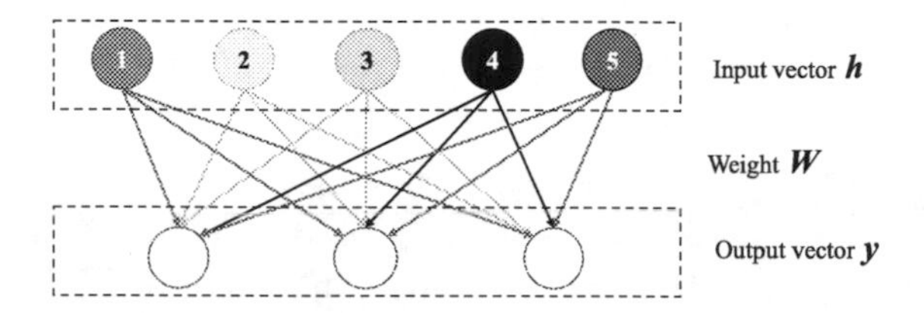

Fig. 3. An example of a dense layer from a network. Colors of the input neurons indicate the activation values. Darker color means a higher activation value or more activated.

This observation indicates that we can exploit the property of activation sparsity for knowledge learning and sharing among tasks. Specifically, for any network converged on task $(\boldsymbol{D}_i, \boldsymbol{T}_i)$, we compute such statistical information for every layer. Shown in Fig. 3, it is clear that the output weights corresponding to those less activated neurons (neurons 2 and 3) can be regarded as less important to task $(\boldsymbol{D}_i, \boldsymbol{T}_i)$. At the same time, we can say those weights corresponding to those frequently activated neurons (neurons 1, 4 and 5) store much of the important knowledge of the task. It is easy to understand that in a matrix multiplication form $\boldsymbol{y} = \boldsymbol{W}\boldsymbol{h}$, when h_2 and h_3 have tiny values, any modification for columns 2 and 3 in matrix $\boldsymbol{W}$ have relatively little impact on the output $\boldsymbol{y}$. Thus, when we train matrix multiplication based networks like RNNs, we can keep some columns in the weight matrix unchanged, and only update those less important ones.

To simplify our description, we name the partially updated weights as *lifelong weights*, their inputs as *control vectors* and the tiny valued components of the control vectors as *free neurons*. Along this line, we can divide the partial update mechanism into two parts: **Free Neuron Detection** and **Gradient Mask**.

Free Neuron Detection. Suppose we are training a model with lifelong weights $\{\boldsymbol{W}^i\}$ and control vectors $\{\boldsymbol{h}^i\}$. To measure the importance of neurons in $\{\boldsymbol{h}^i\}$, we maintain a series of statistical information for $\{\boldsymbol{h}^i\}$ as $\{\boldsymbol{s}^i\}$, where the values of $\boldsymbol{s}^i$ indicate the activation degrees of the corresponding neurons.

For each feed forward pass, values in $\{\boldsymbol{s}^i\}$ are updated as follow:

$$\boldsymbol{s}_t^i = \boldsymbol{s}_{t-1}^i + |\boldsymbol{h}_t^i|, \quad t = 1, 2, \ldots, \quad \boldsymbol{s}_0^i = \boldsymbol{0}, \tag{6}$$

where $|\boldsymbol{h}_t^i|$ indicates the absolute value of the control vector, and $\boldsymbol{s}_t^i$ is actually the accumulation of all $|\boldsymbol{h}_t^i|$ through feed forward propagation during a task training.

As the number of tasks grows, accumulation values $\{\boldsymbol{s}^i\}$ become larger and larger. To avoid overflow, we perform a linear normalization for $\{\boldsymbol{s}^i\}$ before using it directly:

$$\hat{s}_j^i = \frac{s_j^i - \min(\boldsymbol{s}^i)}{\max(\boldsymbol{s}^i) - \min(\boldsymbol{s}^i)}, \quad j = 1, 2, \ldots, d, \tag{7}$$

where d is the dimension of $\boldsymbol{s}^i$, and min() and max() indicate the minimum and maximum values across $\boldsymbol{s}^i$ respectively. Then, in order to find free neurons based on $\hat{\boldsymbol{s}}^i$, we further sort the values of $\hat{\boldsymbol{s}}^i$ in ascending order. The dimensions (neurons) whose values are in the top ϵ percentage will be regarded as free neurons.

Gradient Mask. With the free neurons detected, when we are going to train the model on a new task, *gradient masks* will be used. A gradient mask is a vector which has the same size as the corresponding control vector $\boldsymbol{h}^i$. It tells us which columns in the lifelong weight $\boldsymbol{W}^i$ can be updated. Based on the normalized accumulation $\hat{\boldsymbol{s}}^i$ defined in Eq. 7, we compute the gradient mask $\boldsymbol{m}^i$ for weight $\boldsymbol{W}^i$ as:

$$m_j^i = \begin{cases} 1 - \hat{s}_j^i & \hat{s}_j^i \textit{ is in the top } \epsilon\%, \\ 0 & \textit{otherwise.} \end{cases} \tag{8}$$

Then the update rule for weight $\boldsymbol{W}^i$ is modified as:

$$\boldsymbol{W}^i = \boldsymbol{W}^i - \eta(\boldsymbol{m}^i \odot \frac{\partial L}{\partial \boldsymbol{W}^i}). \tag{9}$$

where η is the learning rate for back-propagation, and $\odot$ represents element-wise multiplication across the rows of the gradient matrix.

We want the ϵ to be self-adaptive in the lifelong learning setting. We propose a self-adjusting strategy for ϵ. It is intuitive that if the model starts the first task with a small ϵ, the knowledge network will suffer from the *cold start* problem, and will have poor performance for a long time. To avoid this, we set ϵ to 100% to allow the parameters to be fully updated in the first task, and then gradually decrease it to a minimum value using the following function:

$$\epsilon = (1 - \tau)e^{-\lambda n} + \tau, \quad n = 0, 1, \ldots, N, \tag{10}$$

where n is the index of the current task, λ and τ separately indicate the scale factor and the minimum value of the threshold, respectively. In our experiments, we found setting both λ and τ to 0.1 usually generate good results.

Specifically, for our *Knowledge Retention Network*, we can find in Eq. 1 that the lifelong weights are $\boldsymbol{U}^z$, $\boldsymbol{W}^z$, $\boldsymbol{U}^r$, $\boldsymbol{W}^r$, $\boldsymbol{U}^h$ and $\boldsymbol{W}^h$. The corresponding control vectors are $\boldsymbol{x}_t$, $\boldsymbol{h}_{t-1}$ and $\boldsymbol{r} \odot \boldsymbol{h}_{t-1}$. In order to increase the value difference between the neurons with high activation degrees and the neurons with low activation degrees, i.e., to enlarge the gap of their activation values in $\boldsymbol{h}_{t-1}$ (in a GRU), we replace the original activation function tanh with the following:

$$f(x) = \begin{cases} 0.9 + kx & x \geq 0.9, \\ x & 0 < x < 0.9, \\ kx & x \leq 0, \end{cases} \tag{11}$$

which is modified from the leaky-ReLU and k is the leak value that is usually set as 0.001. We limit the value of the positive axis to 0.9 in order to make sure that in most cases, the activation value will be no larger than 1.0, which is important to avoid gradient explosion.

4 Experiments

We now evaluate the proposed model SRK and compare it with state-of-the-art baselines. We will see that SRK is not only markedly better than the baselines in general, but also more stable/robust when there is a major domain difference in the successive tasks, a highly imbalanced training data distribution, or a very small training dataset. This is because the accumulated past knowledge in SRK can compensate for these undesirable but realistic training scenarios.

4.1 Experimental Setup

Experiment Data: We use the real-world Amazon review dataset from [8]. This dataset has reviews of 20 categories of diverse products, which we also called 20 domains. Following the existing studies in [1,8,24], for each domain, the reviews with rating score greater than 3 (>3) are regarded as positive reviews and the reviews with rating score less than 3 (<3) are regarded as negative reviews. This gives us 20 sentiment classification tasks (one for each domain). Details about the dataset are given in Table 1. The size in the table is the number of reviews and the neg(%) is the proportion of negative reviews in each domain. For each domain, we further split its dataset into training and testing sets by 80% and 20%. During training, 10% of the training data is used as the development set. Note that our dataset setting here is slightly different from that in [8], which uses only 1000 reviews for each domain before the training and testing split because their method is based on naive Bayes classification. We use the full dataset of each domain for training because deep learning typically needs more training data to learn effective models.

Models for Comparison. We consider the following models for comparison.

- **SRK**. This is our proposed model. Weights of the GRUs are initialized randomly using truncated normal distribution $N(0, 0.001)$. Weights of other dense layers are randomly set following the uniform distribution in the range between $-\sqrt{6/(nin + nout)}$ and $\sqrt{6/(nin + nout)}$, where nin and $nout$ are the number of input and output neurons, respectively. Additionally, we use

Table 1. Data statistics about 20 domains

Domain	Size	Neg (%)	Domain	Size	Neg (%)
Kitchen	5,646	15.18	PC	7,393	21.58
Software	3,633	30.31	Players	5,467	43.23
Sports	4,638	15.17	Camera	4,587	29.28
Music	1,948	2.42	Tools	6,463	29.81
Baby	3,723	13.02	Audio	3,661	19.91
Home	9,355	18.9	Phone	2,703	32.89
Books	2,725	20.9	Laptop	3,686	23.28
Shoes	6,523	12.52	TV	4,579	28.41
Automotive	5,677	10.79	Network	2,815	27.32
Bed	2,750	17.32	Office	4,575	30.01

GloVe [25] as the pre-trained word embedding mentioned in Eq. 2. The model is trained using the RMSProp optimizer[1], where the batch size, learning rate and momentum are set as 32, 0.0001 and 0.9, respectively.

- **LSC** (Lifelong Sentiment Classification). This is the naïve Bayes-based lifelong sentiment classification model in [8], which accumulates knowledge in a Bayesian framework. We use all the parameter settings used in the original work. We obtained the LSC system (and also the Amazon review dataset) from the first author of the paper.
- **I-RNN** (Isolated RNN). A classic RNN model performing each task individually. It does not use knowledge sharing between multiple tasks. Each task is viewed as an isolated one. Its parameters follow the same setting as SRK.
- **FLN**. This model uses only the feature learning network (FLN) of the proposed SRK as described in Sect. 3.1. That is, only the FLN branch of SRK is employed. Its parameters also follow the same setting as SRK.
- **2-FLN**. This model uses two identical feature learning networks (FLNs). Its parameters also follow the same setting as SRK. We use this baseline to see if our SRK results can be achieved by a combination of two FLNs.
- **EWC** (Elastic Weight Consolidation). This is a well-known state-of-the-art algorithm for continual learning that deals with catastrophic forgetting [15]. EWC tries to keep the network parameters close to those of past tasks during the processing of a new task. To achieve that, a constraint is imposed on the network to make the learned parameters staying close to their old values weighted by their importance to the previous tasks. We follow the parameter settings in the original paper.

Evaluation Measure. To have a fair and clear comparison with LSC [8], which is most closely related to our work, we use balanced test sets (by down-sampling the majority class) in our testing (training still uses the original imbalanced

[1] http://www.cs.toronto.edu/~tijmen/csc321/slides/lecture_slides_lec6.pdf.

data as in [8]). Due to the balanced test set, we use accuracy as the evaluation measure. Note that for all our experiments, all compared systems use the same training and test data.

Two Experimental Settings: We use two experimental settings to evaluate SRK: (1) Knowledge accumulation and application. In this setting, we show the importance of knowledge accumulation to the new domain task. Specifically, each domain is treated as the last domain with the rest 19 domains (randomly sequenced) as the previous (or past) domains. (2) Continuous learning and testing. In this case, we evaluate model performance in a continuous learning setting. Specifically, we first set a fixed sequence of all domains, which is randomly chosen. We then let a model process every domain one after another based on the sequence. After each model is built for a domain, it is evaluated based on the test data of that domain. We do this for all candidate models so as to test their performance in this continuous learning setting.

4.2 Results in Experimental Setting (1)

In the first experimental setting, each domain is treated as the last domain in turn. We first compare the performance from all candidate models. The accuracy scores shown in Table 2 are the averaged scores of all 20 domains (averaging the scores of each domain being the last). Based on the overall results, we can see that SRK gives the best accuracy overall.

Table 2. Overall performance of all candidate models.

Model	Accuracy	Model	Accuracy
LSC	86.68	2-FLN	88.44
I-RNN	83.86	EWC	86.29
FLN	87.73	SRK	**89.85**

Comparing SRK with I-RNN, FLN and 2-FLN, we first see that FLN improves I-RNN markedly. This is because I-RNN treats every domain as an independent task, but FLN implicitly retains some common knowledge in its parameters during learning of previous tasks. This also indicates the existence of knowledge sharing in sentiment classification tasks. Comparing 2-FLN and FLN, we see that 2-FLN improves the results obtained by FLN. This is mainly attributed to the capacity of more knowledge learned by the two networks in 2-FLN. Although 2-FLN only learns a single type of features, its two-thread learning mechanism helps capture and keep more useful information from the data. However, SRK outperforms 2-FLN. This is also intuitive because SRK models two types of different information separately, namely, domain-specific (KRN) and cross-domain knowledge (FLN), whereas 2-FLN only considers the same type of information in its two network components. Note that although

Table 3. Performance on 20 domains from all candidate models.

Domain	LSC	I-RNN	FLN	2-FLN	EWC	SRK
Kitchen	74.44	84.79	86.14	87.50	85.99	**89.18**
Software	84.69	85.66	87.33	**88.66**	84.70	88.33
Sports	82.23	79.05	82.90	**86.75**	81.09	84.61
Music	85.52	**87.50**	75.00	75.00	**87.50**	**87.50**
Baby	86.54	76.47	88.82	88.23	88.52	**92.94**
Home	87.37	85.29	88.39	89.21	85.19	**90.52**
Books	84.89	83.65	88.94	85.57	88.68	**89.42**
Shoes	86.48	82.40	87.60	87.20	88.13	**89.20**
Automotive	**90.00**	81.06	83.98	88.34	83.15	87.37
Bed	84.02	80.62	87.50	**90.62**	85.09	**90.62**
PC	87.83	85.53	90.89	**91.25**	85.19	90.71
Players	87.96	86.80	90.95	91.87	88.96	**92.48**
Camera	88.70	85.85	89.64	90.40	88.88	**90.65**
Tools	87.12	85.73	**90.98**	89.67	87.62	90.16
Audio	89.37	82.45	88.15	90.78	87.71	**91.22**
Phone	87.54	82.90	**89.31**	88.46	84.73	88.46
Laptop	89.69	87.69	89.23	**93.07**	90.05	92.69
TV	88.76	85.46	91.13	90.64	83.63	**91.87**
Network	**91.45**	84.95	90.65	86.99	83.93	90.24
Office	**88.81**	83.25	86.92	88.53	86.99	88.76

FLN suffers from catastrophic forgetting to some extend as mentioned in Sect. 1, but since sentiment classification tasks are similar, it retains some cross-domain knowledge in feature learning.

Comparing SRK with LSC and EWC. We can see that SRK achieves a much better result than EWC. The main reason is that, when EWC is enforced to protect its parameters learned in the past domains, its regularization may hinder the model to reach a better parameter fitting for the new domain. Unlike EWC, SRK does not impose restrictions for the past tasks, i.e., no explicit optimization terms for the previous domains, which enable it to fit the current task better, with the knowledge still being retained and used. LSC, a naïve Bayes-based lifelong learning approach, achieves competitive results compared with other neural models, which implies the usefulness of the knowledge accumulation and utilization. However, LSC cannot model the contextual relationship due to its conditional independence assumption on features (words). Its knowledge is also restricted to some extent by only considering the word frequency/count from different (past) domains. SRK, on the other hand, does not have those limitations.

Performance on All 20 Domains: The full results of all models on 20 domains is reported in Table 3. We can make the following observations: (1) SRK achieves the highest scores more than half of the time across all domains (11 out of 20) and outperforms other models by a large margin. (2) In other domains (9 out of 20) where SRK does not perform the best, we can see its performance is competitive, with almost all less than 1% compared to the best model. SRK is thus very stable/robust. (3) Although other models can attain the highest scores in some domains, their performances are quite unstable, i.e., none of them can reach the best scores consistently in those domains where SRK is not the best.

4.3 Results in Experimental Setting (2)

In the second experimental setting, we evaluate the performance gain in a continuous learning fashion. The results from the consecutive learning tasks are presented in Fig. 4, where four models are compared, namely, I-RNN, FLN, EWC and SRK. Since 2-FLN has the closest performance to FLN in this setting and LSC does not learn the same type of parameters like other neural models, their results are excluded here. This simplified visualization and also makes our analysis more focused and clearer.

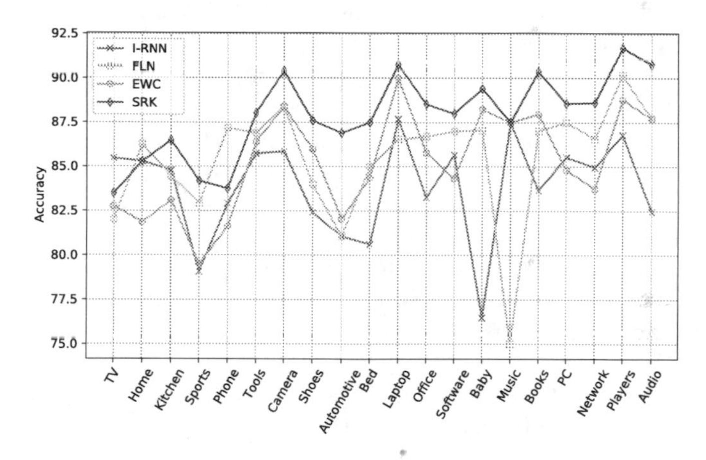

Fig. 4. Accuracy of a specific task sequence.

The conclusion that we can draw from Fig. 4 is three-fold. First, all three continuous learning models (FLN, EWC and SRK) present an overall trend of performance gain over I-RNN from task 1 to task 20, which shows the usefulness of knowledge accumulation during continuous learning. Second, SRK consistently outperforms the other three models. This indicates the superiority of the proposed architecture design in SRK for capturing both domain-specific and cross-domain knowledge. Third, SRK demonstrates its superior stability or robustness to the other models, which we detail next.

Stability/Robustness of SRK. The stability/robustness is a key strength of SRK. Briefly, SRK can address many hard cases where other models can't.

(1) Domain difference. A notable dropping point can be observed from domain *Shoes* to domain *Automotive* in Fig. 4, which can be attributed to their domain differences, i.e., Shoes and Automotive are very different. In this case, the model cannot rely on the parameters learned from its immediate previous (or $(i-1)$ th) domain to improve the current (or i th) domain. In other words, the short-term memorized/residual parameters, which can be viewed as a type of parameter initialization obtained from the last $(i-1)$ task, will not be very helpful for the ith task. This is reflected by both FLN and EWC, where they perform poorly for domain *Automotive*. The proposed SRK is dramatically better in this case because it is able to leverage the knowledge in KRN that is beyond/before the immediate previous domain. However, it is hard for FLN to retain and to exploit such long-term knowledge due to forgetting. EWC mainly aims to deal with catastrophic forgetting by imposing the constraint that its model parameters for learning the i domain are required not to interfere with the old values in the previous domains. It is more likely to ensure the performance of all tasks rather than to learn features or knowledge to help the new task. These explain why SRK reaches a much better result (86.89%) than FLN (81.07%) and EWC (82.04%). A similar situation also occurs from *baby* to *music*.

(2) Imbalanced class distribution in training. In real-world review datasets, the classes are usually imbalanced. Since we used the original class distribution in training, there could be insufficient information to fit the minority class (e.g., the negative class) well and the training/learning might be biased toward the dominating class (e.g., the positive class). This leads to the notably inferior performance for models I-RNN, FLN and EWC for domains *Automotive* and *Shoes*, for models I-RNN and FLN for domains *Baby* and *Music*, respectively. However, this causes no problem for our SRK model as its accumulated knowledge from the past in KRN helps to distinguish positive and negative sentiment signals.

(3) Training data size. When the training data size is small, it causes underfitting. However, since SRK retains the shared knowledge in its KRN (Knowledge Retention Network), this self-learned prior knowledge can provide more supervisory information to help learning to give a better performance. We can see this from domains *Music* and *Bed*, where their dataset sizes are small. In those cases, FLN and I-RNN did not work well but SRK does.

5 Conclusion

In this paper, we proposed a novel deep learning model SRK for lifelong sentiment classification. Specifically, a shared knowledge network is used to retain the knowledge learned from each previous task, and at the same time employed a traditional recurrent neural network to perform feature learning across domains.

Two networks are combined to perform the classification task. For knowledge retention and sharing, a partially update mechanism was designed to preserve the knowledge from individual domains, which has been shown instrumental. Experimental results on real-world datasets shown that the new architecture markedly outperforms the state-of-the-art baselines and is also more stable/robust.

Acknowledgments. This research was partially supported by grants from the National Natural Science Foundation of China (Grants No. 61727809, U1605251) and the program of China Scholarships Council (No. 201706340117). Bing Liu's work was partially supported by National Science Foundation (NSF) under grant nos. IIS1407927 and IIS 1838770, and by Huawei Technologies Co. Ltd with a research gift.

References

1. Blitzer, J., Dredze, M., Pereira, F.: Biographies, bollywood, boom-boxes and blenders: domain adaptation for sentiment classification. In: ACL (2007)
2. Broderick, T., Boyd, N., Wibisono, A., Wilson, A.C., Jordan, M.I.: Streaming variational Bayes. In: NIPS (2013)
3. Caruana, R.: Multitask learning. In: Thrun, S., Pratt, L. (eds.) Learning to Learn. Springer, Boston (1998). https://doi.org/10.1007/978-1-4615-5529-2_5
4. Chen, H., Sun, M., Tu, C., Lin, Y., Liu, Z.: Neural sentiment classification with user and product attention. In: EMNLP (2016)
5. Chen, P., Sun, Z., Bing, L., Yang, W.: Recurrent attention network on memory for aspect sentiment analysis. In: EMNLP (2017)
6. Chen, Z., Liu, B.: Topic modeling using topics from many domains, lifelong learning and big data. In: ICML (2014)
7. Chen, Z., Liu, B.: Lifelong machine learning. Synth. Lect. Artif. Intell. Mach. Learn. **10**, 1–45 (2016)
8. Chen, Z., Ma, N., Liu, B.: Lifelong learning for sentiment classification. In: ACL (2015)
9. Cho, K., Van Merriënboer, B., Bahdanau, D., Bengio, Y.: On the properties of neural machine translation: encoder-decoder approaches. arXiv preprint arXiv:1409.1259 (2014)
10. French, R.M.: Catastrophic forgetting in connectionist networks. Trends Cogn. Sci. **3**, 128–135 (1999)
11. Ghahramani, Z., Attias, H.: Online variational Bayesian learning. In: Slides from Talk Presented at NIPS Workshop on Online Learning (2000)
12. Glorot, X., Bordes, A., Bengio, Y.: Domain adaptation for large-scale sentiment classification: a deep learning approach. In: ICML (2011)
13. Hochreiter, S., Schmidhuber, J.: Long short-term memory. Neural Comput. **9**, 1735–1780 (1997)
14. Katiyar, A., Cardie, C.: Investigating LSTMs for joint extraction of opinion entities and relations. In: ACL (2016)
15. Kirkpatrick, J., et al.: Overcoming catastrophic forgetting in neural networks. PNAS **114**, 3521–3526 (2017)
16. Kumaran, D., Hassabis, D., McClelland, J.L.: What learning systems do intelligent agents need? Complementary learning systems theory updated. Trends Cogn. Sci. **20**, 512–534 (2016)

17. Li, X., Lam, W.: Deep multi-task learning for aspect term extraction with memory interaction. In: EMNLP (2017)
18. Li, Z., Zhang, Y., Wei, Y., Wu, Y., Yang, Q.: End-to-end adversarial memory network for cross-domain sentiment classification. In: IJCAI (2017)
19. Liu, B.: Sentiment analysis and opinion mining. Synth. Lect. Hum. Lang. Technol. **5**, 1–67 (2012)
20. Long, M., Cao, Y., Wang, J., Jordan, M.I.: Learning transferable features with deep adaptation networks. arXiv preprint arXiv:1502.02791 (2015)
21. Nakagawa, T., Inui, K., Kurohashi, S.: Dependency tree-based sentiment classification using CRFs with hidden variables. In: HLT-NAACL (2010)
22. Pan, S.J., Yang, Q.: A survey on transfer learning. TKDE **22**, 1345–1359 (2010)
23. Pang, B., Lee, L.: Opinion mining and sentiment analysis. Found. Trends® Inf. Retr. **2**, 1–135 (2008)
24. Pang, B., Lee, L., Vaithyanathan, S.: Thumbs up? Sentiment classification using machine learning techniques. In: EMNLP (2002)
25. Pennington, J., Socher, R., Manning, C.: GloVe: global vectors for word representation. In: EMNLP (2014)
26. Poria, S., Cambria, E., Gelbukh, A.: Aspect extraction for opinion mining with a deep convolutional neural network. Knowl.-Based Syst. **108**, 42–49 (2016)
27. Qian, Q., Huang, M., Lei, J., Zhu, X.: Linguistically regularized LSTM for sentiment classification. In: ACL (2017)
28. Rozantsev, A., Salzmann, M., Fua, P.: Beyond sharing weights for deep domain adaptation. IEEE Trans. Pattern Anal. Mach. Intell. **41**, 801–814 (2018)
29. Ruvolo, P., Eaton, E.: ELLA: an efficient lifelong learning algorithm. In: ICML (2013)
30. Saha, A., Rai, P., Venkatasubramanian, S., Daume, H.: Online learning of multiple tasks and their relationships. In: AISTATS (2011)
31. Silver, D.L., Yang, Q., Li, L.: Lifelong machine learning systems: beyond learning algorithms. In: AAAI Spring Symposium: Lifelong Machine Learning (2013)
32. Tang, D., Qin, B., Liu, T.: Learning semantic representations of users and products for document level sentiment classification. In: ACL (2015)
33. Tay, Y., Tuan, L.A., Hui, S.C.: Dyadic memory networks for aspect-based sentiment analysis. In: CIKM. ACM (2017)
34. Thrun, S.: Lifelong learning algorithms. In: Thrun, S., Pratt, L. (eds.) Learning to Learn. Springer, Boston (1998). https://doi.org/10.1007/978-1-4615-5529-2_8
35. Tzeng, E., Hoffman, J., Zhang, N., Saenko, K., Darrell, T.: Deep domain confusion: maximizing for domain invariance. arXiv preprint arXiv:1412.3474 (2014)
36. Wang, J., Yu, L.C., Lai, K.R., Zhang, X.: Dimensional sentiment analysis using a regional CNN-LSTM model. In: ACL (2016)
37. Wang, J., et al.: Aspect sentiment classification with both word-level and clause-level attention networks. In: IJCAI (2018)
38. Wang, X., Wei, F., Liu, X., Zhou, M., Zhang, M.: Topic sentiment analysis in Twitter: a graph-based hashtag sentiment classification approach. In: CIKM (2011)
39. Wang, Y., Huang, M., Zhao, L., et al.: Attention-based LSTM for aspect-level sentiment classification. In: EMNLP (2016)
40. Wang, Y., et al.: Dual transfer learning for neural machine translation with marginal distribution regularization. In: AAAI (2018)
41. Xu, J., Chen, D., Qiu, X., Huang, X.: Cached long short-term memory neural networks for document-level sentiment classification. In: EMNLP (2016)
42. Yang, Z., Yang, D., Dyer, C., He, X., Smola, A., Hovy, E.: Hierarchical attention networks for document classification. In: NAACL (2016)

43. Yessenalina, A., Yue, Y., Cardie, C.: Multi-level structured models for document-level sentiment classification. In: EMNLP (2010)
44. Yu, J., Jiang, J.: Learning sentence embeddings with auxiliary tasks for cross-domain sentiment classification. In: EMNLP (2016)
45. Zhang, J., Ghahramani, Z., Yang, Y.: Flexible latent variable models for multi-task learning. Mach. Learn. **73**, 221–242 (2008)
46. Zhang, K., et al.: Image-enhanced multi-level sentence representation net for natural language inference. In: ICDM, pp. 747–756. IEEE (2018)
47. Zhang, X., Zhao, J., LeCun, Y.: Character-level convolutional networks for text classification. In: NIPS (2015)
48. Zhou, X., Wan, X., Xiao, J.: Attention-based LSTM network for cross-lingual sentiment classification. In: EMNLP (2016)

Author Index

Printed in the United States
By Bookmasters